PlantOmics: The Omics of Plant Science

Debmalya Barh
Muhammad Sarwar Khan • Eric Davies
Editors

PlantOmics: The Omics of Plant Science

Springer

Editors
Debmalya Barh
Department of Genomics
Institute of Integrative Omics and Applied Biotechnology (IIOAB)
Nonakuri, West Bengal, India

Muhammad Sarwar Khan
Centre of Biochemistry and Biotechnology
University of Agriculture
Faisalabad, Pakistan

Eric Davies
Department of Plant Biology
North Carolina State University
Raleigh, NC, USA

ISBN 978-81-322-3524-8 ISBN 978-81-322-2172-2 (eBook)
DOI 10.1007/978-81-322-2172-2

Springer New Delhi Heidelberg New York Dordrecht London

Softcover reprint of the hardcover 1st edition 2015

Printed on acid-free paper

Springer (India) Pvt. Ltd. is part of Springer Science+Business Media (www.springer.com)

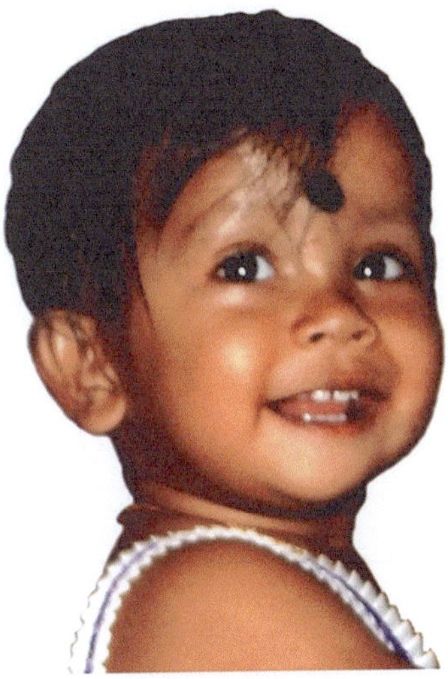

My loving Gul: Shaurya Shree

Foreword

Plants are the primary source of "most of our food, fuel, fibers, fabrics, and pharmaceuticals". Stresses (biotic and abiotic) are the major threats to plants, being the primary cause of crop yield losses worldwide. On the other hand, with the global population expected to reach nine billion by 2050, an increase in crop productivity and quality will be needed to meet the requirements. Each of the 29 chapters of this *PlantOmics: The Omics of Plant Science* book opens a door to exciting cutting-edge omics approaches and their applications to meet the future demands.

The flow of the chapters in the book is highly scientific and strategically organized to be easy to go. It starts with the topic omics approaches in model plants and their applications in improvement of maize and rice like major cereal crops. Chapters 2, 3, and 4 describe very important technologies such as spectroscopy (NIR, MIR, Raman), next generation sequencing (NGS), and functional genomics and their applications in current plant science. Chapters 5 and 6 deal with technical advancements and applications of cyto-mutagenomics and epigenomics in crop improvement. Chapter 7 gives a detailed account on plant miRNA biology, associated technologies, and their tailor-made applications to improve plant stress response.

Each topic dealt in Chaps. 8, 9, 10, 11, 12, 13, 14, 15, 16, 17, 18, 19, 20, 21, 22, 23, 24, and 25 is a unique imprint of this book. These chapters cover well established and several budding omics areas in plant science such as Plant Proteomics, Metabolomics, Glycomics, Lipidomics, Secretomics, Phenomics, Cytomics, Physiomics, Signalomics, Thiolomics, Organelle Omics, Micromorphomics, Microbiomics, Cryobionomics, Nanobiotechnology, and Plant Pharmacogenomics. Each of these chapters describes the latest technologies and applications of the respective omics in a very comprehensive way; therefore, they are up to date, easy to understand, and can be spontaneously adopted to expand the area of our research and development.

Chapters 26, 27, and 28 deal with computational and systems biology approaches in plant science making the book more useful to any kind of plant biology research, whether in a wet lab or *in silico*. The last chapter (Chap. 29) is very brief but interesting where the editors have provided valuable insights on the future directions of omics and plantomics. They have proposed several new areas in omics which we must explore towards development of an integrated meta-omics strategy to ensure the world and earth's health and related issues.

Overall, it is a great effort by Dr. Barh, his editorial team, and 90 expert contributors from 15 countries to make this highly resourceful, up-to-date, thought provoking, and worthwhile unique book for students and researchers in the field of cutting-edge plant omics sciences. I highly recommend the book for keeping you up to date in the field.

Professor
Molecular Biology Division
Head, Unit of Proteomics and Molecular Biomedicine
Instituto Potosino de Investigación
Científica y Tecnológica (IPICyT), Mexico
President, Mexican Proteomics Society, Mexico

Ana Paulina Barba de la Rosa, PhD

Preface

The term "*omics*" depicts completeness. In the last two decades, the term has been suffixed with several biological topics to provide complete information on the subject. With the advent of new technologies, the arena of "*omics*" is increasing rapidly. However, most of the currently available books that deal with omics technologies and their applications are mainly focused on animal system. To fill this gap, we have introduced this *PlantOmics: The Omics of Plant Science* book to provide a complete spectrum of plant related omics to the students and researchers working in the field of cutting-edge plant molecular biotechnology. Equal focus has been given to the technological advancements as well as their specific applications. Therefore, the book provides a comprehensive account of the state-of-the-art latest developments and trends of *omics* approaches in plant science. Several topics have also highlighted the integrative omics strategies enabling the cost-effective development of superior plants for various purposes.

The book consists of *29 chapters* written by *90 experts* from *15 countries* that represent three-fourths of the globe. In the introductory chapter (Chap. 1), Dr. Agrawal and colleagues have described the omics of model plants where genomics, proteomics, transcriptomics, and metabolomics of model plants such as *Arabidopsis*, rice, and maize are dealt in detail. Further, this chapter also provides how these technology derived knowledge can be used for transgenomics, mapping for biotic and abiotic stresses, and marker assisted selection for crop improvement. In Chap. 2, Dr. Cozzolino's group has given a nice overview on the most commonly used spectroscopy techniques such as NIR, MIR, and Raman in plant omic analysis. To make the chapter more resourceful, Dr. Cozzolino has also demonstrated instrumentations and analytic software for these spectroscopy techniques. The hot topic, next generation sequencing (NGS), its technologies, various platforms, algorithms, and *de novo* assembly, annotation, and analysis of plant genome are given by Dr. Tiwary in Chap. 3. Chapter 4, by Dr. Jha and his colleagues, provides a comprehensive account of techniques associated with plant functional genomics and their applications. Drs. Talukdar and Sinjushin in Chap. 5 have described various techniques of cytogenomics and mutagenomics and their cost effective applications in plant breeding and biology. This chapter has also highlighted the mutations that cause alterations in antioxidant defense response to withstand diverse abiotic stresses to reveal intrinsic cellular and metabolic events towards sensitivity of seed plants to salinity, drought, metal toxicity and other stresses, prospecting to formulate effective breeding strategies in different agro-climatic conditions. Epigenomics technologies and their potential applications in crop improvement are summarized in Chap. 6 by Drs. Shafiq and Khan. Especially, this chapter highlights the roles of chromatin remodeling mechanisms in response to environmental stimuli and their role in crop improvement. Chapter 7, by Dr. Boopathi, on Plant miRNomics gives a comprehensive account to explain how the miRNAs fine tune the gene expression and play key roles in developmental timing and patterning of structures in response to external and internal stimuli in plants. This chapter also provides how the miRNAs can be used to improve plant stress responses. Chapter 8, by Dr. Agrawal and his group, describes the recent technological progresses in plant proteomics and highlights the achievements made in understanding the plant proteomes and their applications. In Chap. 9, Dr. Sangwan and colleagues explain various technology platforms in plant metabolomics research and how the metabolomics is used in monitoring and assessing gene functions, stress responses, and to characterize post-genomic processes from a broad perspective along with the challenges the domain is facing. Dr. Khurana's group in Chap. 10 overviews the chemistry and technologies in plant glycomics. This chapter also gives summary of applications of glycomics in biopharming and several biological processes such as plant signaling, stress responses, and immunity. In the next chapter (Chap. 11), Dr. Namasivayam elucidates the chemistry and analytic technologies, lipid signaling in plants, lipidomes in plant defense mechanisms, and several other aspects of plant lipidomics. The comprehensive mechanisms regulating constitutive and induced secretome of diverse plants and their habitat along with technological approaches are discussed by Dr. Yadav and her group in Chap. 12. In Chap. 13, Dr. Rahman and colleagues give a detailed account on integrated-omics approaches in phenomics and its applications in plant and agriculture. Chapter 14, by Drs. Davies and Stankovic, describes how novel methods based on super-fast and super-resolution microscopy can be used in describing proteins, nucleic acids, cytoskeleton, and small molecules of major interest to plants. In Chapter 15, Dr. Karpiński and colleagues educate us on plant physiomics. The chapter provides insights on how the combined molecular-physiological events drive plant growth, development, acclimatization, and defense responses. Dr. Vian et al., in Chap. 16, have introduced the term "Signalomics" and have shown how novel methods can be used to analyze systemic signals including electrical and hydraulic signals in plants. In Chap. 17, Dr. Talukdar and colleagues elucidate the use of latest cutting-edge functional genomics tools to understand the plant thiol metabolism from source (soil) to sink (grains) in diverse arenas of "thiolomics". The next three chapters (Chaps. 18, 19, 20) are dedicated to organelle omics. Chapter 18, by Dr. de Luna Valdez et al., explores how chloroplasts organize their genomes and regulate their transcriptomes, proteomes, and metabolomes, trying to focus on classical knowledge and reviewing new datasets obtained through large-scale research projects and systems approaches that shed light on chloroplast functionality under the chloroplast omics chapter. In Chap. 19, Dr. Khan summarizes the developments from plastid genomics to gene expression and briefly describes how transplastome facilitates expression of vaccines, therapeutics, and plantibodies, in addition to tailoring agronomic traits in plants. Plant mitochondrial omics (Chap. 20), by Dr. Mustafa and his colleagues, describes a detailed account on regulation of mitochondrial genes at transcriptional, post-transcriptional (splicing and RNA editing), translational, and post-translational levels in omics perspective. Chapter 21 describes "Micromorphomics", a term coined by Dr. Tulika Talukdar to explain how plants combat environmental stresses through collective morphological manifestations in their organs architectures. Chapter 21 is dedicated to microbiomics. In this chapter, Dr. Sharma's team has discussed technologies to identify new groups of microorganisms involved in plant diseases from microbiome of rhizosphere and roles of microbiome in plant health and related areas. Drs. Martinez-Montero and Harding in Chap. 22 (Cryobionomics) intend to explore the connections between stability and cryogenic/non-cryogenic stress factors with a view to aiding protocol improvement, optimization, and validation for plant genetic resources conservation with several examples. Chapter 24, by Dr. Kazi and colleagues, focuses on the development and use of "nanotechnology" for formulating agriculturally important chemicals (fertilizers) with more useful properties and their direct delivery as well as their applications in various agricultural sectors. Chapter 25, by the same group, systemically analyzes the recent developments in plant pharmacogenomics and its contributions in the field of molecular and pharmaceutical sciences. Dr. Somvanshi and colleagues in Chap. 26 have attempted to describe several machine learning approaches and their applications in plant biology in a very simple way. Similarly, in Chap. 27, Dr. Sarika's team has emphasized on a number of applications of bioinformatics in agriculture in view of functional genomics, data mining techniques, genome-wide association studies, high-performance computing facilities in agriculture, and various bioinformatics tools/databases important for breeders, biotechnologists, and pathologists. Chapter 28 (Plant systems biology), by Drs. Bhardwaj and Somvanshi, describes recent insights and advancements in systems biology approaches in order to understand how plant systems work. In the brief concluding chapter (Chap. 29), we, the editors, have proposed several omics terms under "*Futuromics*" centraling Plantomics to direct the future perspectives of plant omics in meta-omics era.

We believe that this book will be a valuable resource to all who are working on cutting-edge plant omics. We appreciate your comments and suggestions to improve the next edition.

Nonakuri, India
Faisalabad, Pakistan
Raleigh, NC, USA

Debmalya Barh
Muhammad Sarwar Khan
Eric Davies

Contents

Contributors

Lalit Agrawal, Ph.D. National Botanical Research Institute, Lucknow, India

Pawan Kumar Agrawal, Ph.D. Vivekananda Parvatiya Krishi Anusandhan Sansthan (VPKAS), Almora, India

Mustafeez Mujtaba Babar, Ph.D. Atta-ur-Rahman School of Applied Biosciences (ASAB), National University of Sciences and Technology (NUST), Islamabad, Pakistan

B. Kalyana Babu, Ph.D. Vivekananda Parvatiya Krishi Anusandhan Sansthan (VPKAS), Almora, India

Debmalya Barh, Ph.D. Department of Genomics, Institute of Integrative Omics and Applied Biotechnology (IIOAB), Nonakuri, West Bengal, India

Tulika Bhardwaj, Ph.D. Department of Biotechnology, Teri University, New Delhi, India

Jayant S. Bhat, Ph.D. Regional Research Centre, Indian Agriculture Research Institute, Dharwad, Karnataka, India

N. Manikanda Boopathi, Ph.D. Department of Plant Molecular Biology and Bioinformatics, Centre for Plant Molecular Biology and Biotechnology, Tamil Nadu Agricultural University, Coimbatore, India

Vijaya Brahma, Ph.D. Toronto General Research Institute, University Health Network, Toronto, ON, Canada

Paweł Budiak, Ph.D. Department of Genetics, Breeding and Biotechnology, Warsaw University of Life Sciences, Warszawa, Poland

Sushma Chauhan, Ph.D. Department of Chemical and Biochemical Engineering, Dongguk University, Seoul, South Korea

Daniel Cozzolino, Ph.D. School of Agriculture, Food and Wine, The University of Adelaide, Osmond, SA, Australia

Eric Davies, Ph.D. Department of Plant Biology, North Carolina State University, Raleigh, NC, USA

L.A. de Luna-Valdez, Ph.D. Instituto de Biotecnología, Universidad Nacional Autónoma de México, Cuernavaca, Morelos, Mexico

R.K. Dubey, Ph.D. Department of Floriculture and Landscaping, Punjab Agricultural University, Ludhiana, Punjab, India

S. Encarnación-Guevara, Ph.D. Centro de Ciencias Genómicas, Universidad Nacional Autónoma de México, Cuernavaca, Morelos, Mexico

Alberto Fassio, Ph.D. National Institute for Agricultural Research, INIA La Estanzuela, Colonia, Uruguay

Piotr Gawroński, Ph.D. Department of Genetics, Breeding and Biotechnology, Warsaw University of Life Sciences, Warszawa, Poland

Mustafa Malik Ghulam, Ph.D. Agriculture Biotechnology, National Institute for Biotechnology and Genetic Engineering (NIBGE), Faisalabad, Pakistan

Magdalena Górecka, Ph.D. Department of Genetics, Breeding and Biotechnology, Warsaw University of Life Sciences, Warszawa, Poland

A.A. Guevara-García, Ph.D. Instituto de Biotecnología, Universidad Nacional Autónoma de México, Cuernavaca, Morelos, Mexico

Deepti Bhushan Gupta, Ph.D. TERI University, Vasant Kunj, New Delhi, India

Keith Harding, Ph.D. Damar Research Scientists, Damar, Cuparmuir, Fife, Scotland, UK

Firoz Hossain, Ph.D. Division of Genetics, Indian Agriculture Research Institute, New Delhi, India

M.A. Iquebal, Ph.D. Centre for Agricultural Bioinformatics, Indian Agricultural Statistics Research Institute, New Delhi, India

N. Jagadeeshselvam, Ph.D. Centre for Plant Molecular Biology and Biotechnology, Tamil Nadu Agricultural University, Coimbatore, India

Kaliaperumal Jagatheesh, Ph.D. Department of Biotechnology, Periyar University, Salem, India

Sarika Jaiswal, Ph.D. Centre for Agricultural Bioinformatics, Indian Agricultural Statistics Research Institute, New Delhi, India

Johnravindar D, Ph.D. Department of Biotechnology, Periyar University, Salem, India

Richard L. Jayaraj, Ph.D. Department of Biotechnology, Periyar University, Salem, India

Uday Chand Jha, Ph.D. Indian Institute of Pulses Research (IIPR), Kanpur, India

Stanisław Karpiński, Ph.D. Department of Genetics, Breeding and Biotechnology, Warsaw University of Life Sciences, Warszawa, Poland

Alvina Gul Kazi, Ph.D. Atta-ur-Rahman School of Applied Biosciences (ASAB), National University of Sciences and Technology (NUST), Islamabad, Pakistan

Maria Khalid, Ph.D. Atta-ur-Rahman School of Applied Biosciences (ASAB), National University of Sciences and Technology (NUST), Islamabad, Pakistan

Abdul Rehman Khan, Ph.D. Department of Environmental Sciences, COMSATS I.I.T., Abbottabad, Pakistan

Muhammad Sarwar Khan, Ph.D. Centre for Agricultural Biochemistry and Biotechnology (CABB), University of Agriculture, Faisalabad, Pakistan

S.M. Paul Khurana, Ph.D. Amity Institute of Biotechnology, Amity University Harayana, Gurgaon, Haryana, India

Sumaira Kousar, Ph.D. Industrial Biotechnology, National Institute for Biotechnology and Genetic Engineering (NIBGE), Faisalabad, Pakistan

R. Kowsalya, Ph.D. Department of Biotechnology, Periyar University, Salem, India

Milena Kulasek, Ph.D. Department of Genetics, Breeding and Biotechnology, Warsaw University of Life Sciences, Warszawa, Poland

Dinesh Kumar, Ph.D. Centre for Agricultural Bioinformatics, Indian Agricultural Statistics Research Institute, New Delhi, India

Amit K. Kushwaha, Ph.D. Department of Metabolic and Structural Biology, CSIR-Central Institute of Medicinal and Aromatic Plants, Lucknow, India

P. León-Mejía, Ph.D. Instituto de Biotecnología, Universidad Nacional Autónoma de México, Cuernavaca, Morelos, Mexico

P.K. Malhotra, Ph.D. Schools of Agricultural Biotechnology, Punjab Agricultural University, Ludhiana, Punjab, India

K. Manigandan, Ph.D. Department of Biotechnology, Periyar University, Salem, India

Marcos E. Martinez-Montero, Ph.D. Bioplantas Center, Plant Breeding Laboratory, University of Ciego de Ávila, Ciego de Avila, Cuba

Bhartendu Nath Mishra, Ph.D. Department of Biotechnology, Institute of Engineering and Technology, G.B. Technical University, Lucknow, India

Siddhartha Kumar Mishra, Ph.D. School of Biological Sciences, Dr. Harisingh Gour Central University, Sagar, India

C.S. Mukhopadhyay, Ph.D. School of Animal Biotechnology, Guru Angad Dev Veterinary and Animal Sciences University, Ludhiana, Punjab, India

Raveendran Muthurajan, Ph.D. Centre for Plant Molecular Biology and Biotechnology, Tamil Nadu Agricultural University, Coimbatore, India

Elangovan Namasivayam, Ph.D. Department of Biotechnology, Periyar University, Salem, India

Senthil Natesan, Ph.D. Centre for Plant Molecular Biology and Biotechnology, Tamil Nadu Agricultural University, Coimbatore, India

Khwaja Osama, Ph.D. Department of Biotechnology, Institute of Engineering and Technology, G.B. Technical University, Lucknow, India

Pavan Kumar Padarthi, Ph.D. Department of Biotechnology, Periyar University, Salem, India

Basavanagouda S. Patil, Ph.D. Regional Research Centre, Indian Agriculture Research Institute, Dharwad, Karnataka, India

Hifzur Rahman, Ph.D. Centre for Plant Molecular Biology and Biotechnology, Tamil Nadu Agricultural University, Coimbatore, India

Anil Rai, Ph.D. Centre for Agricultural Bioinformatics, Indian Agricultural Statistics Research Institute, New Delhi, India

Sathishraj Rajendran, Ph.D. Centre for Plant Molecular Biology and Biotechnology, Tamil Nadu Agricultural University, Coimbatore, India

Mahendran Ramachandran, Ph.D. Centre for Plant Molecular Biology and Biotechnology, Tamil Nadu Agricultural University, Coimbatore, India

Valarmathi Ramanathan, Ph.D. Centre for Plant Molecular Biology and Biotechnology, Tamil Nadu Agricultural University, Coimbatore, India

Sasikala Ramasamy, Ph.D. Centre for Plant Molecular Biology and Biotechnology, Tamil Nadu Agricultural University, Coimbatore, India

Saleha Resham, Ph.D. Atta-ur-Rahman School of Applied Biosciences (ASAB), National University of Sciences and Technology (NUST), Islamabad, Pakistan

Ernesto Restaino, Ph.D. National Institute for Agricultural Research, INIA La Estanzuela, Colonia, Uruguay

Navinder Saini, Ph.D. Vivekananda Parvatiya Krishi Anusandhan Sansthan (VPKAS), Almora, India

Neelam S. Sangwan, Ph.D. Department of Metabolic and Structural Biology, CSIR-Central Institute of Medicinal and Aromatic Plants, Lucknow, India

Rajender Singh Sangwan, Ph.D. Department of Metabolic and Structural Biology, CSIR-Central Institute of Medicinal and Aromatic Plants, Lucknow, India

Center for Innovative and Applied Bioprocessing (formerly BioProcessing Unit), (An Autonomous Institute under Department of Biotechnology, Govt. of India), Mohali, Punjab, India

Chiranjib Sarkar, Ph.D. Centre for Agricultural Bioinformatics, Indian Agricultural Statistics Research Institute, New Delhi, India

Sarfraz Shafiq, Ph.D. Department of Environmental Sciences, COMSATS I.I.T., Abbottabad, Pakistan

Anamika Sharma, Ph.D. Department of Dental Microbiology, Theerthankar Mahaveer University, Moradabad, Uttar Pradesh, India

Pankaj Sharma, Ph.D. Department of Plant Breeding and Genetics, Punjab Agricultural University, Ludhiana, Punjab, India

Shubhendu Shekhar, Ph.D. National Institute of Plant Genome Research, New Delhi, India

G.S. Sidhu, Ph.D. Schools of Agricultural Biotechnology, Punjab Agricultural University, Ludhiana, Punjab, India

Andrey Sinjushin, Ph.D. Department of Genetics, M.V. Lomonosov Moscow State University, Moscow, Russia

Pallavi Somvanshi, Ph.D. Department of Biotechnology, Teri University, New Delhi, India

Bratislav Stankovic, Ph.D. University for Information Science and Technology "St. Paul the Apostle", Ohrid, Republic of Macedonia

Pamidimarri D.V.N. Sudheer, Ph.D. Department of Chemical and Biochemical Engineering, Dongguk University, Seoul, South Korea

Magdalena Szechyńska-Hebda, Ph.D. Department of Genetics, Breeding and Biotechnology, Warsaw University of Life Sciences, Warszawa, Poland

The Franciszek Górski Institute of Plant Physiology, Polish Academy of Sciences, Kraków, Poland

Dibyendu Talukdar, Ph.D. Department of Botany, R.P.M. College, University of Calcutta, Uttarpara, Hooghly, West Bengal, India

Tulika Talukdar, Ph.D. Department of Botany, Krishnagar Government College, Nadia, Krishnagar, West Bengal, India

Department of Botany, APC Roy Govt. College, Siliguri, Darjeeling, West Bengal, India

Pragya Tiwari, Ph.D. Department of Metabolic and Structural Biology, CSIR-Central Institute of Medicinal and Aromatic Plants, Lucknow, India

Basant K. Tiwary, Ph.D. Centre for Bioinformatics, Pondicherry University, Pondicherry, India

Swati Tripathi, Ph.D. Vegetable Research Division, National Institute of Horticultural and Herbal Science, Rural Development Administration, Suwon, Republic of Korea

Najam us Sahar Sadaf Zaidi, Ph.D. Atta-ur-Rahman School of Applied Biosciences (ASAB), National University of Sciences and Technology (NUST), Islamabad, Pakistan

Harsh Vardhan, Ph.D. Faculty of Medicine, Department of Immunology/Service of pneumology, University of Sherbrooke, Sherbrooke, QC, Canada

Alain Vian, Ph.D. UMR 1345 IRHS (Université d'Angers, Agrocampus Ouest, INRA), SFR 4207 Quasav, Faculté des Sciences, Angers cedex 01, France

Esteban Vicente, Ph.D. National Institute for Agricultural Research, INIA Salto Grande, Salto, Uruguay

Dinesh K. Yadav, Ph.D. Amity Institute of Biotechnology, Amity University Harayana, Gurgaon, Haryana, India

Neelam Yadav, Ph.D. Amity Institute of Biotechnology, Amity University Harayana, Gurgaon, Haryana, India

Ritesh K. Yadav, Ph.D. Department of Metabolic and Structural Biology, CSIR-Central Institute of Medicinal and Aromatic Plants, Lucknow, India

Sarika Yadav, Ph.D. Amity Institute of Biotechnology, Amity University Harayana, Gurgaon, Haryana, India

About the Editors

Debmalya Barh (M.Sc., M.Tech., M.Phil., Ph.D., PGDM) is a well-known molecular biotechnologist who is an expert in integrative *omics*–based cutting-edge R&D. He is the founder of the Institute of Integrative Omics and Applied Biotechnology (IIOAB) – a first-of-its-kind research organization in India that provides a global platform for multidisciplinary research and advocacy. He works with 400+ well-regarded researchers from 40+ countries and has 150+ high-impact international publications, several books, and book chapters in the field of biomedical and agricultural omics. Dr. Barh is a globally branded editor for editing omics related research reference books from Springer and Taylor & Francis and an editorial and review board member for a number of highly reputed international journals. Due to his significant contributions in the field and in promoting applied biological, agricultural, and bio-medical sciences using unique research strategies, in the year 2010 he has been recognized by *Who's Who in the World* and in 2014 he has been entered into the *Limca Book of Records,* "the Indian equivalent to the Guinness Book of World Records".

Muhammad Sarwar Khan, a Doctorate from the University of Cambridge, is a highly regarded Molecular Biologist from Pakistan who was a Founding Head of Biotech Interdisciplinary Division at NIBGE and is currently serving as the Director of Center of Agricultural Biochemistry and Biotechnology (CABB), University of Agriculture, Faisalabad, Pakistan. He has several awards to his credit, including Civil Award (Presidential Medal for Technology), Gold Medal in Agriculture (Pakistan Academy of Sciences), Performance Gold Medal (Pakistan Atomic Energy Commission), and Biotechnologist Award (National Commission on Biotechnology), and is a Life Fellow of Cambridge Commonwealth Society. He has several high-impact publications in scientific journals including *Nature* and *Nature Biotechnology*, and is an author of a number of books and book chapters. Dr. Khan has made colossal contributions in the field of chloroplast genetic engineering and is pioneer in expressing GFP in plant chloroplasts, developing transplastomic rice and sugarcane. His research focus also includes expression of foreign genes in the chloroplasts to confer agronomic traits such as insect-pest resistance, salinity and herbicide tolerance, and overexpression of antigenic and therapeutic proteins in chloroplasts to develop cost-effective therapeutics and vaccines.

Eric Davies, Ph.D. is currently an Emeritus Professor in the Department of Plant and Microbial Biology, North Carolina State University (NCSU), USA. He is the former Head of the Botany Department at NCSU and the Director of the NASA Specialized Center of Research and Training (NSCORT) in Gravitational Biology. Professor Davies was also a former Visiting Professor at Purdue University, University of Arizona, Ehime University in Japan, University of Brussels in Belgium, Universite Blaise Pascal and Universite d'Angers in France. He is an Honorary Professor at the University of Warmia and Mazuryi in Poland, and has an Honorary Doctorate from Universite Blaise Pascal. He has several refereed publications, book chapters, and books. Professor Davies is an expert in plant wound signaling, especially the rapidly-transmitted physical (turgor and electrical) signals, local changes in ions and hormones, and ultra-rapid, systemic transcript accumulation and degradation. His other research areas are the role of the cytoskeleton in mRNAs attachment to enhance translation in specific sub-cellular regions of the cell; gravity regulated cell signaling and gene expression, especially the response in the maize pulvinus; microwave irradiation; and gene expression in plants.

Omics of Model Plants

Pawan Kumar Agrawal, B. Kalyana Babu, and Navinder Saini

Contents

Abstract

The multiple omics tools and strategies like high-throughput genome-scale genotyping platforms such as whole-genome re-sequencing, proteomics, and metabolomics provide greater opportunities to dissect molecular mechanisms and the discovery of key genes in developing ideal genotypes in the changing climate scenario. The last decade has seen rapid advances in functional genomic research globally. Most of the efforts involve construction of technological and resource platforms for high-throughput DNA sequencing, gene identification, and physical and genetic mapping; functional analysis of genomes for agronomic traits and biological processes; and identification and isolation of functional genes. The functional genomic research aims to understand how the genome functions at the whole-genome level, whereas proteomics looks for the systematic analysis of the protein population in a tissue, cell, or subcellular compartment. Metabolites are the end products of cellular process, and they show the response of biological systems to environmental changes. The current trend in metabolomic

P.K. Agrawal, Ph.D. (✉) • B.K. Babu, Ph.D.
N. Saini, Ph.D.
Vivekananda Parvatiya Krishi Anusandhan Sansthan (VPKAS), Almora, India
e-mail: pawankagrawal@hotmail.com

D. Barh et al. (eds.), *PlantOmics: The Omics of Plant Science*,
DOI 10.1007/978-81-322-2172-2_1, © Springer India 2015

studies is to define the cellular status at a particular time point of development or physiological status. These techniques complement other techniques such as transcriptomics and proteomics and depict precise pictures of the whole cellular process. The growing number of sequenced plant genomes has opened up immense opportunities to study biological processes related to physiology, growth and development, and tolerance to biotic and abiotic stresses at the cellular and whole plant level using a novel systems-level approach. The "omics" approach integrates genome, proteome, transcriptome, and metabolome data into a single data set and can lead to the identification of unknown genes and their regulatory networks involved in metabolic pathways of interest. This will also help in understanding the genotype–phenotype relationship and consequently help to improve the quality and productivity of crop plants for the food and nutritional security of millions of human populations.

Keywords
Arabidopsis • Maize • Rice • *Zea mays* • *Oryza sativa* • Functional genomics • Phenomics • Transcriptomics • Proteomics • Metabolomics • MAS • QTL • Transgenics

Introduction

The last decade has seen rapid advances in functional genomic research globally. Most of the efforts involve construction of technological and resource platforms for high-throughput DNA sequencing, gene identification, and physical and genetic mapping; functional analysis of genomes for agronomic traits and biological processes; and identification and isolation of functional genes. The overall goal of functional genomic research is to understand how the genome functions at the whole-genome level. Similarly, proteomics looks for the systematic analysis of the protein population in a tissue, cell, or subcellular compartment. It enables correlations to be drawn between the range of proteins produced by a cell or tissue and the initiation or progression of a stress or normal metabolism. Metabolites are the end products of cellular process, and they show the response of biological systems to environmental changes. The current trend in metabolomic studies is to define the cellular status at a particular time point of development or physiological status. These techniques complement other techniques such as transcriptomics and proteomics and depict precise pictures of the whole cellular process. The growing number of sequenced plant genomes has opened up immense opportunities to study biological processes related to physiology, growth and development, and tolerance to biotic and abiotic stresses at the cellular and whole plant level using a novel systems-level approach. The "omics" approach integrates genome, proteome, transcriptome, and metabolome data into a single data set and can lead to the identification of unknown genes and their regulatory networks involved in metabolic pathways of interest. This will also help in understanding the genotype–phenotype relationship.

Arabidopsis thaliana

Genomics

Arabidopsis thaliana is an excellent model organism for the analysis of complex biological processes in plants using molecular and biotechnological techniques. The frontiers of plant science, like other branches of the life sciences, have been dominated by genomics over the past 25 years. Many research laboratories are currently putting intensive efforts to isolate *Arabidopsis* genes of biological importance using map-based cloning strategy. Although genetic linkage (Koornneef et al. 1983) and recombinant inbred line (RIL) maps (Lister and Dean 1993) have been reported, the construction of accurate physical maps of the chromosomes will be highly advantageous not only for the genomic sequencing but also for map-based gene cloning (Ward and Jen 1990). Hence, a complete physical map of the *Arabidopsis* genome should be greatly advantageous for cloning the genetic loci of interest as well as sequencing the entire genome. Hence, yeast artificial chromosome (YAC)-based physical maps of chromosomes 2 (Zachgo et al. 1996) and 4 (Schmidt et al. 1995) of *A. thaliana* have been constructed by several workers. Based on the sizes of the YACs and their coverage of the chromosome, the length of chromosome 2 was estimated to be at least 18 Mb. Sato et al. (1998) presented physical map of the entire chromosome 3 which was constructed by ordering the clones from YAC, PI, TAC, and BAC libraries using the information from the sequences of various DNA markers and the terminal sequences of the clones. The sizes of the centromeric regions of *Arabidopsis thaliana* chromosomes 1, 2, and 3 were determined by construction of their physical maps on the basis of restriction analysis (Hosouchi et al. 2002). The sizes of the genetically defined centromeric regions were deduced to be 9 megabases (Mb), 4.2 Mb, and 4.1 Mb, respectively (chromosome 1, from markers T22C23-t7 to T3P8-sp6; chromosome 2, from F5J15-sp6 to T15D9; chromosome 3, from T9G9-sp6 to T15M14) (Copenhaver et al. 1998). Mitochondrial genomes in higher plants are characterized by their high flexibility and variation in size and structure. The mitochondrial genome of *A. thaliana* was physically mapped using cosmid and YAC clones and was found to contain 372 kb size which was relatively large (Klein et al. 1994). The presence of this comparatively large mitochondrial genome in a plant with one of the smallest nuclear genome showed that different size constraints act upon the different genomes in plant cells. *A. thaliana* is known to contain approximately 1,000 copies of 5S rDNA per haploid genome, and they occur in tandem arrays (Campell et al. 1992). The 5S ribosomal RNA genes were mapped to mitotic chromosomes of *Arabidopsis thaliana* by fluorescence in situ hybridization (FISH) by Murata et al. (1997).

Arabidopsis thaliana is widely used as a model for the study of many aspects of plant biology. Because of its small genome size (125 Mb), it was chosen as the subject of the first plant genome sequencing project, an effort that was completed. *Arabidopsis thaliana* was the first plant, and the third multicellular organism after *Caenorhabditis elegans* (The *C. elegans* Sequencing Consortium 1998) and *Drosophila melanogaster* (Adams et al. 2000), to be completely sequenced (The *Arabidopsis* Genome Initiative 2000). Since systematic sequencing was completed in late 2000, the genome sequence has undergone several rounds of reassembly, hole patching, and extension into un-sequenced regions. One of the major features of the *Arabidopsis* genome revealed by the genome sequence was the extent of gene duplication and segmental duplications, which was surprising given the expectation of a functionally compact genome. Approximately 60 % of the genome was thought to be derived from a single duplication event, possibly of the entire genome (The *Arabidopsis* Genome Initiative 2000). The extensive work carried out based on the *Arabidopsis* genome sequence also supports interpretations of the evolution of the vertebrate lineage that propose a central role for genome duplications (Wolfe 2001). Comparison of *Arabidopsis* sequences with genomic sequence from the closely related *Brassica oleracea* (Chinese cabbage) identified regions of high similarity that either identified putative new genes or extended existing gene models. About 30 % of these new genes encoded a transcript. About 25 % of the originally predicted genes had no supporting evidence such as an EST match or reasonable similarity of their putative peptide sequence to any other protein. The decreasing cost along with rapid progress in next-generation sequencing and related bioinformatics computing resources has facilitated large-scale discovery of SNPs in *Arabidopsis* species. Large numbers and genome-wide availability of SNPs make them the marker of choice in partially or completely sequenced genomes. The complete nucleotide sequence of the chloroplast genome of *Arabidopsis thaliana* has been determined (Sato et al. 1999). The genome as a circular DNA composed of 154,478 bp containing a pair of inverted repeats of 26,264 bp, which are separated by small and large single copy regions of 17,780 bp and 84,170 bp, respectively. Cao et al. (2011) presented the first phase of the project, based on population-scale sequencing of 80 strains of *A. thaliana* populations drawn from eight regions throughout the species' native range. They found common small-scale polymorphisms as well as many larger insertions and deletions in the *A. thaliana* pan-genome.

A major goal in evolutionary biology is to identify the genetic basis of adaptive trait variation. In the model plant species *Arabidopsis thaliana*, studies are now being performed exploiting natural variation as a powerful alternative to classical mutant genetics (Koornneef et al. 2004), in particular to identify genes underlying important quantitative trait variation. Benjamin et al. (2010) studied combined analysis of genome-wide association (GWA) study with traditional linkage mapping in order to detect the genetic bases underlying natural variation in flowering time in ecologically realistic conditions in the plant *Arabidopsis thaliana*. It involved phenotyping of nearly 20,000 plants over 2 winters under field conditions in a temperate climate. Simon et al. (2008) studied phenotyping of nearly 20,000 plants over 2 winters under field conditions, including 184 worldwide natural accessions genotyped for 216,509 SNPs and 4,366 RILs derived from 13 independent crosses chosen to maximize genetic and phenotypic diversity. The results showed that combined linkage and association mapping clearly outperforms each method alone when it comes to identifying true associations. Kuittinen et al. (1997) described a quantitative trait locus (QTL) mapping experiment for flowering time in *Arabidopsis*. Five to seven QTLs affecting flowering time were found in a BC_1 population derived from the Finnish Naantali genotype and the German strain Li-5. In a different population, consisting of 165 RILs, Alonso-Blanco et al. (1997) found four QTLs affecting the flowering time. Several loci exhibiting variation in complex traits (quantitative trait loci or QTLs) have been cloned. Examples include using linkage disequilibrium (LD) to fine map the *FRI* and *FLC* loci controlling flowering time (Hagenblad et al. 2004). Natural variation in hypocotyl responses to light was shown to be due to polymorphisms in phytochrome light receptors. Affymetrix expression arrays have also been used for genotyping; total genomic DNA from recombinant inbred lines (RILs) made from a cross of Col and Ler was hybridized to the ATH1 Affymetrix array, and recombination events were identified. Marker and QTL information obtained from a segregating population can be used for the design of efficient breeding strategies. Marker-assisted selection (MAS) has been advocated as a useful tool for rapid genetic advance in the case of quantitative traits (Lande and Thompson 1990; Knapp 1994, 1998). Berloo and Stam (1999) described an experiment using RILs of *A. thaliana* with an objective to compare an MAS breeding strategy, using molecular marker and QTL information, with conventional breeding methods, based on phenotype only. Selection based on marker and QTL information gave approximately the same result as selection based on phenotype. The relative high heritability of flowering time in *Arabidopsis* facilitated successful phenotypic selection. The difference in selection result that was anticipated to be in favor of the marker-assisted approach was therefore not observed.

Functional Genomics

With the availability of complete genome sequences of several organisms, the focus has shifted from structural genomics to functional genomics, specifically in plants where the complete genomic sequences are becoming available (*Arabidopsis* and rice). A variety of approaches are used to clone and gather information about the function(s) of gene(s). Among these, insertional mutagenesis has been extensively used for cloning genes, promoters, enhancers, and other regulatory sequences from *Arabidopsis*. Strategies used for cloning and characterization depend upon the information available about the gene or its product. Expressed sequence tags (ESTs) and microarray-based techniques are some of the powerful approaches in this direction.

A comprehensive molecular-marker-based linkage map exists for *Arabidopsis*, and the map-based cloning of genes conferring specific phenotypes will become even easier with the availability of genomic sequence information. Jun et al. (2011) conducted whole-genome sequencing of *A. thaliana* populations. Here they described the majority of common small-scale polymorphisms as well as many larger insertions and deletions in the *A. thaliana* pan-genome, their effects on gene function, and the patterns of local and global linkage among these variants. The plant hormone auxin, typified by indole-3-acetic acid (IAA), regulates a variety of physiological processes, including apical dominance, tropic responses, lateral root formation, vascular differentiation, embryo patterning, and shoot elongation. Okushima et al. (2005) employed a functional genomic strategy that involves the identification of T-DNA insertion in the ARF gene family members to elucidate some of the biological functions of the ARF transcription factors. Most of the single *ARF* T-DNA insertion mutants fail to show an obvious growth phenotype. However, double mutants, such as arf7 and arf19, showed a strong auxin phenotype that results in the absence of lateral root formation than neither the arf7 nor arf19 single mutant expresses. The completion of the whole-genome sequence of *Arabidopsis thaliana* has made it possible to explore the phytochemical genomics in this species by determining gene-to-metabolite correlation through the comprehensive analysis of metabolite accumulation and gene expression. Takayuki et al. (2007) proposed a strategy that involves the integrated analysis of metabolic profiling, transcriptome coexpression analysis, and the sequence similarity analysis of genes, followed by the reverse genetics and biochemical approaches of narrowed-down candidate genes for a particular reaction or regulation. TILLING (Targeting Induced Local Lesions IN Genomes) is a reverse genetic method that can be employed to generate allelic series of induced mutations in targeted genes for functional analyses. To date, TILLING resources in *Arabidopsis thaliana* are only available in accessions Columbia and *Landsberg erecta*.

A permanent collection of 3,509 independent EMS mutagenized M2 lines was developed in *A. thaliana* accession C24 by Lai et al. (2012). Using the TILLING method identified a total of 73 mutations, comprising 69.6 % missense, 29.0 % sense, and 1.4 % nonsense mutations. Duplication of chromatin following DNA replication requires spatial reorganization of chromatin domains assisted by chromatin assembly factor CAF-1. Nicole et al. (2006) tested the genomic consequences of CAF-1 loss and the function of chromatin assembly factor CAF-1 in heterochromatin formation and suggested that CAF-1 functions in heterochromatin formation. They also suggested the CAF-1 is required only for the complete compaction of heterochromatin but not to maintain transcriptional repression of heterochromatic genes. Extensive work in T-DNA tagging of *Arabidopsis* has become possible because of improvements in techniques for *Agrobacterium*-mediated transformation. Recently, 150,000 transformed plants carrying 225,000 (1.5 insertion per line) independent T-DNA integration events were generated and precise locations determined for >88,000 loci by the Salk Institute for Biological Studies (Alonso et al. 2003). Analysis of the insertion site sequences revealed that insertional mutations had been created in ~74 % of the annotated *Arabidopsis* genes.

Proteomics

Sequencing of complete genomes has advanced our understanding of biological systems, mostly by enabling a broad range of technologies for the analysis of gene functions and by providing information about the theoretical protein-coding capacity of organisms. Proteomics has been defined as "the systematic analysis of the protein population in a tissue, cell, or subcellular compartment" and is often associated with two-dimensional electrophoresis (2-DE). The concept of "proteome" (for PROTEin complement expressed by a genOME, Wilkins et al. 1996) has been emerged recently as a consequence of questions raised from several genome and post-genome projects. The first plant large-scale proteomic work was published in *Arabidopsis*. The *Arabidopsis* proteome map provides a detailed map of 14,867 organ-specific proteotypic peptides, which accounts for the diverse composition of protein samples and confers higher sensitivity to proteotypic peptide selection for targeted and quantitative proteomics.

In terms of proteomics in plants, *Arabidopsis* is currently a unique system. Tom et al. (2006) mapped the *Arabidopsis* organelle proteome. In their study, the density gradient distributions of 689 proteins from *Arabidopsis thaliana* were determined, enabling confident and simultaneous localization of 527 proteins to the endoplasmic reticulum, Golgi apparatus, vacuolar membrane, plasma membrane, mitochondria, and plastids. In plants, progress has been made in determining the proteomes of organelles such as chloroplasts, mitochondria, and peroxisomes. Alison et al. (2005) described the first proteomic analysis of plant (*Arabidopsis thaliana*) nucleoli, in which they identified 217 proteins. This allows a direct comparison of the proteomes of an important nuclear structure between two widely divergent species: human and *Arabidopsis*. The comparison identified many common proteins, plant-specific proteins, proteins of unknown function found in both of the proteomes and proteins that were nucleolar in plants but non-nucleolar in human. In two proteomic studies in *Arabidopsis*, one examining programmed cell death in cell cultures (Swidzinski et al. 2004) and one comparing basal and R gene-mediated defense in leaves (Jones et al. 2004), differentially accumulating proteins were represented by only a small number of relatively abundant proteins, many of which were also transcriptionally regulated during the responses. In 2004, three proteomic studies of the *Arabidopsis* vacuole were published, two using suspension-cultured cells (Shimaoka et al. 2004; Szponarski et al. 2004) and one using mature plants (Carter et al. 2004).

The *A. thaliana*–*P. xylostella* interaction is a model system used to investigate insect resistance in plants, in particular the analysis of inducible defense mechanisms. Richard et al. (2010) investigated the physiological factors affecting feeding behavior by larvae of the insect, *Plutella xylostella*, on herbivore-susceptible and herbivore-resistant *Arabidopsis thaliana*. The leaves of 162 recombinant inbred lines (RILs) were screened to detect genotypes upon which *Plutella* larvae fed least (*P. xylostella* resistant) or most (*P. xylostella* susceptible). The combined results suggest that enhanced production of ROS may be a major pre-existing mechanism of *Plutella* resistance in *Arabidopsis*. To assemble a high-density *Arabidopsis* proteome map, Katja et al. (2008) performed 1354 LTQ (linear trap quadrupole) ion-trap mass spectrometry runs with protein extracts from six different organs. The resulting data files were analyzed with two search algorithms, PeptideProphet (6) and PepSplice (7). They identified 13,029 proteins with 86,456 unique peptides originating from 790,181 tandem mass spectrometry (MS/MS) spectrum assignments at a false-discovery rate below 1 %.

Through tandem mass spectrometry, Kleffmann et al. (2004) identified 690 different proteins from purified *Arabidopsis* chloroplasts. Most proteins could be assigned to known protein complexes and metabolic pathways, but more than 30 % of the proteins have unknown functions, and many are not predicted to localize to the chloroplast. The combined shotgun proteomics and RNA profiling approach is of high potential value to predict metabolic pathway prevalence and to define regulatory levels of gene expression on a pathway scale. Proteomic analysis of glutathione S-transferase of *A. thaliana* identified 20 GSTs at the protein level with a combination of GST antibody detection, LC-MS/MS analysis of 23030 Kda proteins and glutathione affinity chromatography (Pia et al. 2004). Peltier et al. (2004) presented a simple, fast, and scalable off-line procedure based on three-phase partitioning with butanol to fractionate membrane proteomes in combination with both in-gel and in-solution digestions and mass spectrometry. This should help to further accelerate the field of membrane proteomics and revealed new functions of the thylakoid membrane proteome of *Arabidopsis thaliana*.

Metabolomics

The "omics" approach integrates transcriptome and metabolome data into a single data set and can lead to the identification of unknown genes

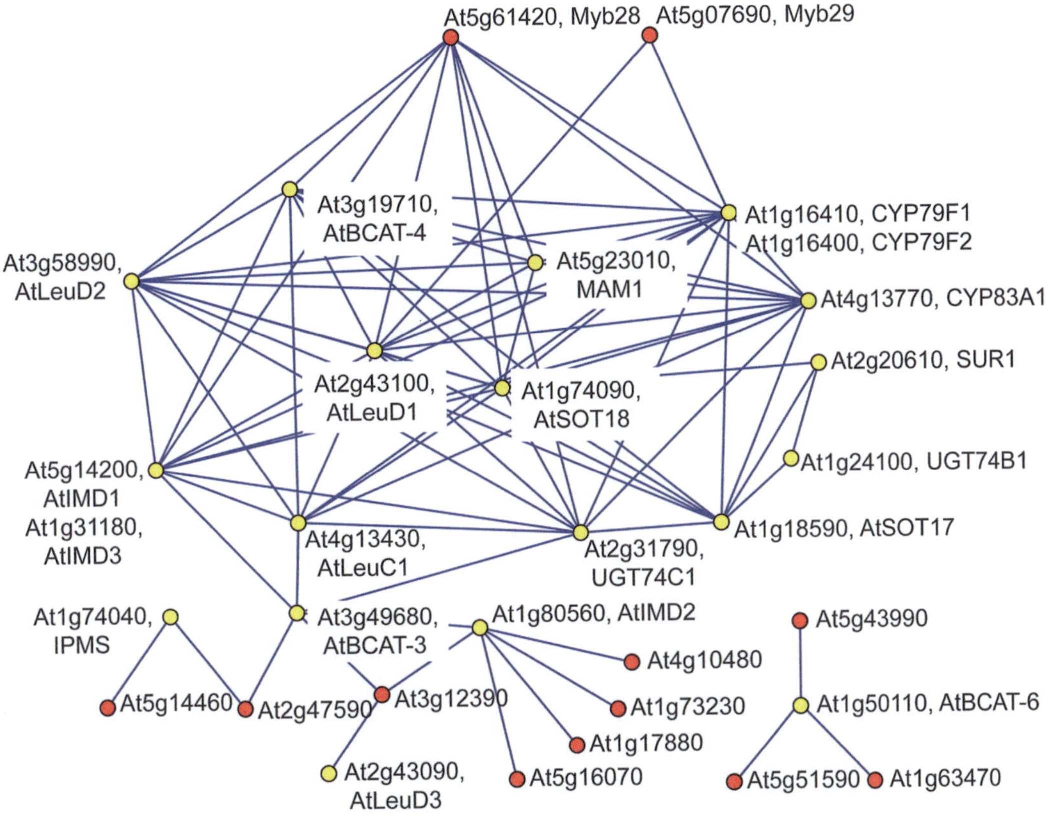

Fig. 1 Coexpression analysis of aliphatic GSL biosynthetic genes and transcription factors. *Yellow* and *red* points indicate genes encoding enzymes and transcription factors, respectively. Transcripts from *AtIMD1* and *AtIMD3* and those from *CYP79F1* and *CYP79F2* were cross-hybridized to the same probe sets on an ATH1 microarray used in AtGenExpress and hence are indistinguishable. Lengths of the lines are valueless in these displays (Masami et al. 2007)

and their regulatory networks involved in metabolic pathways of interest. Masami et al. (2007) reported the discovery of two R2R3-Myb transcription factors that positively control the biosynthesis of GSLs in *Arabidopsis thaliana* by an integrated omics approach. Combined transcriptome coexpression analysis of publicly available, condition-independent data and the condition-specific (i.e., sulfur deficiency) data identified *Myb28* and *Myb29* as candidate transcription factor genes specifically involved in the regulation of aliphatic GSL production (Fig. 1).

The circadian clock is an endogenous timer that anticipates and synchronizes biological processes to the environment. Traditional genetic approaches identified the underlying principles and genetic components, but new discoveries have been greatly impeded by the embedded redundancies that confer necessary robustness to the clock architecture. Brenda and Steve (2013) studied the regulation, and mechanistic connectivity between clock genes and with output processes has substantially broadened through genomic, transcriptomic, proteomic, and metabolomic approaches. Although much evidence supports the regulation of metabolites by the clock (Farre and Weise 2012), few studies have undertaken a global analysis. The metabolomes of the two arrhythmic clock mutants d975 (the triple mutant for PRR9, PRR7, and PRR5) and CCA1-OX (overexpressor of CCA1) have been compared. While both mutants have similar general morphology, their underlying metabolite profiles are quite distinct. Prieurianin (Pri) and prieurianin acetate (Pri-Ac) were identified as two related compounds that caused period shortening, dampening, and reduced expression of GI::LUC and CCA1::LUC. In addition to affecting circadian expression, both Pri and Pri-Ac caused polarity alterations in different cell types that suggested actin cytoskeletal defects.

A metabolite fingerprinting of crude extracts of *A. thaliana* using ^{1}H nuclear magnetic resonance (NMR) spectroscopy and multivariate statistics has been tested for the rapid analysis of unfractionated polar plant extracts, enabling the creation of reproducible metabolite fingerprints (Ward et al. 2003). Salt stress is one of the most important factors limiting plant cultivation. Many investigations of plant response to high salinity have been performed using conventional transcriptomic and/or proteomic approaches. Hence, the functions of the complex stress response pathways are yet to be determined, especially at the metabolic level. Kim et al. (2007) analyzed primary metabolites, especially small polar metabolites such as amino acids, sugars, sugar alcohols, organic acids, and amines, by GC-MS and LC-MS at 0.5, 1, 2, 4, 12, 24, 48, and 72 h after a salt-stress treatment with 100 mM NaCl being the final concentration. The results also suggest the co-induction of glycolysis and sucrose metabolism as well as co-reduction of the methylation cycle as long-term responses to salt stress. Metabolic profiling analyses were performed to determine metabolite temporal dynamics associated with the induction of acquired thermotolerance in response to heat shock and acquired freezing tolerance in response to cold shock (Fatma et al. 2004). This investigation provides new insight into the mechanisms of plant adaptation to thermal stress at the metabolite level, reveals relationships between heat- and cold-shock responses, and highlights the roles of known signaling molecules and protectants. Ralf et al. (2009) studied the putative *A. thaliana* FLS gene family using a combination of genetic and metabolic analysis approaches. They presented evidence that flavonol glycosides found in the fls1-2 mutant are synthesized in plants by the FLS-like side activity of the LDOX enzyme. It is a prerequisite to obtain structural information for the isolated metabolites from the plant of interest. Ryo et al. (2009) isolated metabolites of *Arabidopsis* in a relatively nontargeted way, aiming at the construction of metabolite standards and chemotaxonomic comparison. In addition, 35 known compounds, including six anthocyanins, eight flavonols, one nucleoside, one indole glucosinolate, and four phenylpropanoids and a derivative, together with three indoles, etc., were also isolated and identified from their spectroscopic data. Nuclear magnetic resonance (NMR) has become a key technology in plant metabolomics with the use of stable isotope labeling and advanced heteronuclear NMR methodologies. Jun et al. (2004) performed multidimensional heteronuclear NMR analysis of metabolic movement of carbon and nitrogen nuclei in *Arabidopsis thaliana.* The investigations made by Masami et al. (2004) for gene-to-metabolite networks regulating sulfur and nitrogen nutrition and secondary metabolism in *Arabidopsis*, with integration of metabolomics and transcriptomics, they carried out the transcriptome and metabolome analyses, respectively, with DNA microarray and several chemical analytical methods, including ultrahigh-resolution Fourier transform-ion cyclotron MS.

Rice

Rice is a model crop plant for genomic study because:

1. It has one of the smallest genomes in crop plants, i.e., 390 MB;
2. It has huge economic importance;
3. It is the first genome of crop plants sequenced with high precision;
4. Large amounts of genetic stocks are available to the research community;
5. Rice is rich in germplasm resources including the cultivated forms and its wild relatives which are maintained in germplasm banks;
6. Rice genome shares collinearity with other members of the grass family like corn and wheat;
7. The information available on genetic studies is huge;

8. High-efficiency transformation technology is available and has gained widespread utility;
9. Almost every rice-producing country has very intensive breeding programs; and
10. Large-scale ESTs and insertional mutants are available in rice

Genomics and Synteny

Genetic and Physical Mapping

Brondani et al. (2001) constructed a genetic map with 162 PCR-based markers (155 microsatellite- and 7 sequence-tagged site markers) using a backcross population derived from the cross *O. glumaepatula*, accession RS-16 from the Brazilian Amazon Region x *O. sativa* BG-90-2, an elite rice inbred line. The map included 47 new simple sequence repeat (SSR) markers developed from an *O. glumaepatula* genomic library enriched for AG/TC sequences. All SSR markers were able to amplify the *O. sativa* genome, indicating a high degree of SSR flanking region conservation between *O. glumaepatula* and *O. sativa* species. The map covered 1,500.4 cM, with an average of one marker every 10 cM. Despite some chromosomes being more densely mapped, the overall coverage was similar to the other maps developed for rice. The advantage to construct SSR-based map is to permit the combination of the speed of the PCR reaction and the codominant nature of the SSR marker, facilitating the quantitative trait loci analysis and marker-assisted selection for rice breeding programs. Temnykh et al. (2001) developed a set of 200 Class I SSR markers and integrated into the existing microsatellite map of rice, providing immediate links between the genetic, physical, and sequence-based maps. This contribution brings the number of microsatellite markers that have been rigorously evaluated for amplification, map position, and allelic diversity in *Oryza* spp. to a total of 500. Yamamoto et al. (2001) developed a BC_1F_3 population and constructed an RFLP linkage map. The map was constructed using data from 187 plants and 116 RFLP markers. They compared the chromosomal locations of putative QTLs in their study with those in other studies. They found that several QTLs for these traits, showing a correlation with phenotypic values, were mapped in the same chromosomal regions. Kennard et al. (2000) presented the first genetic map of wild rice (*Zizania palustris*, 2n = 2x = 30), a native aquatic grain of North America. The map is composed principally of previously mapped RFLP genetic markers from rice (*Oryza sativa*, 2n = 2x = 24). The map is important as a foundation for genetic and crop improvement studies, as well as a reference for genome organization comparisons among species in *Gramineae*. A comparative mapping approach with rice is especially useful because wild rice is grouped in the same subfamily, *Oryzoideae*, and no other mapping comparison has yet been made within the subfamily. As rice is the reference point for mapping and gene cloning in cereals, establishing a consensus map within the subfamily identifies conserved and unique regions. The genomes of wild rice and cultivated rice differ in total DNA content (wild rice has twice that of rice) and chromosome pairs (wild rice = 15 versus rice = 12). Rice genome sequence is available and is the first genome of crop plants sequenced with high precision.

BAC Library

A comparison of expressed sequence tag (EST) database from different plants can reveal the diversity in coding sequences between crop species, both closely related and distantly related. Bacterial artificial chromosomes (BACs) are very useful for physical mapping, gene location, cloning, sequencing, and gene discovery in crop plants. Tao et al. (2001) developed a genome-wide BAC-based map of the rice genome. The map consists of 298 BAC contigs and covers 419 Mb of the 430-Mb rice genome. They found that the contigs constituting the map are accurate and reliable. Particularly important to proficiency were (1) a high-resolution, high-throughput DNA sequencing gel-based electrophoretic method for BAC fingerprinting, (2) the use of several complementary large-insert BAC libraries, and (3) computer-aided contig assembly. It has been demonstrated that the fingerprinting method is not significantly influenced by repeated sequences, genome size, and genome complexity. Use of several complementary libraries developed with different restriction enzymes minimized the "gaps" in the physical map. They observed that a clonal coverage of 6.0–8.0 genome equivalents seems to be sufficient for the development of a genome-wide physical map of 95 % genome coverage. Their study indicates that genome-wide BAC-based physical maps can be developed quickly and economically for a variety of plant and animal species by restriction fingerprint analysis *via* DNA sequencing gel-based electrophoresis.

Synteny

The development of genetic and linkage map in many crop species has positional similarity, and information in one crop species will lead to better understanding of crop evolution and functioning of genes in other crop species. This "synteny" allows advances made in one species to have spillover impact in another. Snape et al. (2001) illustrated how genes for vernalization response and cold tolerance on chromosomes 5A and 5D of wheat have been identified and located. They advocated that their relationships to genes in other species, such as barley and rice, could be characterized through comparative mapping approaches, leading to strategies for their isolation using rice genomic tools. Smilde et al. (2001) developed a set of 88 rice expressed sequence tags previously mapped on rice chromosome 1 in the cross "Nipponbare" x "Kasalath." They used these markers for comparative mapping in a cross of the barley. About one-third of the markers were polymorphic between "Igri" and "Franka." These polymorphisms were mapped, and most of these (56 %) confirmed that rice chromosome 1 and barley chromosome 3H are syntenous. The markers that were not fitting in the collinear order were distributed randomly across the barley genome. The comparative maps of barley chromosome 3H and rice chromosome 1 comprise in total 26 common markers covering more than 95 % of the genetic length of both chromosomes. A 30-fold reduction of recombination is seen around the barley centromere, and synteny may be interrupted in this region. In their study, the good overall synteny on a mesoscale (1–10 cM) justifies the use of rice as a platform for map-based cloning in barley.

Dubcovsky et al. (2001) demonstrated collinearity of a large region from barley (*Hordeum vulgare*) chromosome 5H and rice (*Oryza sativa*) chromosome 3 by mapping of several common restriction fragment length polymorphism clones on both regions. One of these clones, WG644, was hybridized to rice and barley bacterial artificial chromosome (BAC) libraries to select homologous clones. A comparison of the rice and barley DNA sequences revealed the presence of four conserved regions, containing four predicted genes. The four genes are in the same orientation in rice, but the second gene is in inverted orientation in barley. The fourth gene is duplicated in tandem in barley but not in rice. Comparison of the homeologous barley and rice sequences assisted the gene identification process and helped determine individual gene structures. General gene structure (exon number, size, and location) was largely conserved between rice and barley and to a lesser extent with homologous genes in *Arabidopsis thaliana*. Similarly, a major rust (*Puccinia melanocephala*) resistance gene identified in a self-progeny of the sugarcane cultivar R570 was known to be linked to a marker revealed by the sugarcane probe CDSR29. Asnaghi et al. (2000) used synteny relationships between sugarcane and three other grasses in an attempt to saturate the region around this rust resistance gene. Comparison of sugarcane, sorghum, corn, and rice genetic maps led to the identification of homeologous chromosome segments at the extremity of sorghum linkage group D, rice linkage group 2, maize linkage group 4, and in the centromeric region of maize linkage group 5.

Gene Expression

The rice blast (*Magnaporthe grisea*) resistance gene *Pi*b is a member of the nucleotide-binding site (NBS) and leucine-rich repeat (LRR) class of plant disease resistance genes and belongs to a small gene family. Wang et al. (2001) isolated and characterized a *Pi*b homologue (PibH8) from rice cultivars Tohoku IL9, Nipponbare, Kasalath, Koshihikari, and IR24 and extensively investigated the expression of the *Pi*b gene family (Pib, PibH8, HPibH8-1, and HPibH8-2) under various environmental and chemical treatments. RNA gel blot analysis revealed that their expression was regulated dramatically by environmental signals, such as temperature, light, and water availability. Their expression was also induced by chemical treatments, such as jasmonic acid, salicylic acid, ethylene, and probenazole. Their findings suggest that expression of the *Pi*b gene family is upregulated by environmental conditions that would favor pathogen infection.

Gene Expression for Salt Tolerance

Kawasaki et al. (2001) investigated transcript regulation in response to high salinity for salt-tolerant rice (var. Pokkali) with microarrays including 1,728 cDNAs from libraries of salt-stressed roots. NaCl at 150 mM reduced photosynthesis to one-tenth of the prestress value within minutes. Hybridizations of RNA to microarray slides probed for changes in transcripts from 15 min to 1 week after salt shock. Beginning 15 min after the shock, Pokkali showed upregulation of transcripts. Approximately 10 % of the transcripts in Pokkali were significantly upregulated or downregulated within 1 h of salt stress. The initial differences between control and stressed plants continued for hours but became less pronounced as the plants adapted over time.

QTL Mapping and Tagging of Useful Genes

Yield and Yield Components

Moncada et al. (2001) used an advanced backcross breeding population to identify quantitative trait loci (QTLs) associated with eight agronomic traits in a BC_2F_2 population derived from an interspecific cross between Caiapo, an upland *Oryza sativa* subsp. *japonica* rice variety from Brazil, and an accession of *Oryza rufipogon* from Malaysia. They concluded that the advanced backcross QTL analysis offers a useful germplasm enhancement strategy for the genetic improvement of cultivars adapted to stress-prone environments. They observed that 56 % of the trait-enhancing QTLs identified in this study were derived from *O. rufipogon*. This figure is similar to the 51 % of favorable QTLs derived from the same parent in crosses with a high-yielding hybrid rice cultivar evaluated under irrigated conditions in a previous study. Similarly, Li et al. (2001) analyzed the quantitative trait loci (QTLs) responsible for the area, perimeter, length, width, and length-width ratio of the flag leaf and the second and third upside-down leaves of a doubled-haploid population of 117 lines derived from a cross between *Oryza sativa* var. *indica* (Zhaiyeqing 8) and var. *japonica* (Jingxi 17) cultivars based on the genetic linkage map of 243 molecular markers. All the traits were continuously distributed with transgressive segregation in the population. For the flag leaf, 12 QTLs were mapped in 8 intervals of 4 chromosomes including 3 QTLs for leaf area, 2 each for leaf perimeter and length, and 5 for ear length-width ratio. The QTLs accounted for 9–27.8 % of phenotypic variation. For the second upside-down leaf, 18 QTLs were mapped for 5 leaf morphological traits in 11 intervals of 5 chromosomes including 3 QTLs for leaf area, 4 each for leaf perimeter and length, 2 for leaf width, and 5 for length-width ratio. The mapped QTLs accounted for 9.1–24.7 % of phenotypic variation. Heading time (HT, days from sowing to heading) is an important agronomic trait in rice. Physiologically, HT can be divided into two stages: vegetative growth time (VGT) and reproductive growth time (RGT). Zhou et al. (2001) in a study mapped HT-related QTLs based on the performance of HT, VGT, and RGT. An *indica* (cv. Gui 630)/*japonica* (cv. Taiwanjing) doubled-haploid population and a corresponding RFLP map were constructed for the study. They used methods of composite interval mapping and multiple-trait composite interval mapping to map QTLs. A total of 19 QTLs were mapped on all 12 rice chromosomes with the exception of chromosomes 1 and 4. Their results showed that (1) more QTLs could be detected by partitioning HT into

VGT and RGT; (2) the genetic variation of HT was largely attributed to VGT; and (3) the two component stages were relatively independent in terms of QTL effects, suggesting that the ratio between VGT and RGT could be genetically adjusted without apparently altering HT.

Mapping for Abiotic Stresses

Abiotic stresses like drought tolerance, cold tolerance, mineral deficiency, mineral toxicity, and submergence tolerance are few major issues in rice.

Drought Tolerance

Drought is one of the main abiotic constraints in rice. A deep root system contributes efficiently in maintaining the water status of the crop through a stress period. Ali et al. (2000) undertook an experiment to map QTLs associated with five root traits using RFLP and AFLP markers in an RIL population developed from two *indica* parents, IR58821-23-B-1-2-1 and IR52561-UBN-1-1-2. A genetic linkage map of 2,022 cM length was constructed comprising 303 AFLP and 96 RFLP markers with an average marker space of 5.0 cM. QTL analysis *via* interval mapping detected 28 QTLs for these five root traits, which were located on chromosomes 1, 2, 3, 4, 6, 7, 10, and 11. Individual QTLs accounted for between 6 and 27 % of the phenotypic variation. Similarly, Zhang et al. (2001) constructed a genetic linkage map consisting of 315 DNA markers and identified a total of 41 QTLs for osmotic adjustment and root traits and individually explained 8–38 % of the phenotypic variance. A region on chromosome 4 harbored major QTLs for several root traits. In another study, Shen et al. (2001) identified QTLs affecting root parameters in a doubled-haploid population of rice derived from the cross, IR64/Azucena.

Cold Tolerance

Norin-PL8 is a cold-tolerant variety of rice (*Oryza sativa*) that was developed by introgressing chromosomal segments from a cold-tolerant *javanica* variety, Silewah. The QTLs for cold tolerance of Norin-PL8 was earlier detected on chromosomes 3 and 4. For fine mapping those genes, Saito et al. 2001 developed a set of near-isogenic lines (NILs) from recombinants in the segregating population. From their study, they concluded that there are at least two QTLs for cold tolerance, tentatively designated as Ctb-1 and Ctb-2, in the introgression on chromosome 4. The map distance between Ctb-1 and Ctb-2 was estimated to be 4.7–17.2 cM. Similarly, Takeuchi et al. (2001) performed QTL analysis with doubled-haploid lines (DHLs) to identify the chromosomal regions controlling cold tolerance (CT) at booting stage of rice (*Oryza sativa* var. *japonica*). Three QTLs controlling CT (qCT-1, qCT-7, and qCT-11) were mapped to chromosomes 1, 7, and 11, respectively. The QTL with the largest effect, qCT-7, was mapped to chromosome 7 and explained 22.1 % of the total phenotypic variation, while qCT-1 and qCT-11 explained approximately 5 % each. For all of the QTLs, alleles from Koshihikari increased the degree of CT.

Flood and Submergence Tolerance

The inheritance and expression of traits associated with submergence stress tolerance at the seedling stage are genetically complex. Using AFLP markers, Sripongpangkul et al. (2000) could identify several genes/QTLs that control plant elongation and submergence tolerance in a recombinant inbred rice population. The most important gene was QIne1 mapped near sd-1 on chromosome 1. The Jalmagna (the deepwater parent) allele at this locus had a very large effect on internode elongation and contributed significantly to submergence tolerance under flooding. The second locus was a major gene, sub1 (t), mapped to chromosome 9, which contributed to submergence tolerance only. The third one was a QTL, QIne4, mapped to chromosome 4. Toojinda et al. (2003) exploited naturally occurring differences between certain rice lines in their tolerance to submergence and used QTL mapping to improve understanding of the genetic and physiological basis of submergence tolerance. Three rice populations, each derived from a single cross between two cultivars differing in their response to submergence, were used to identify QTL associated with plant survival and various linked traits. Several major QTLs determining plant survival, plant height, stimulation of shoot elongation, visual tolerance score, and leaf senescence are each mapped to the same locus on chromosome 9. These QTLs were detected consistently in experiments across all years and in the genetic backgrounds of all three mapping populations. Secondary QTLs influencing tolerance were also identified and located on chromosomes 1, 2, 5, 7, 10, and 11. These QTLs were specific to particular traits, environments, or genetic backgrounds. All identified QTLs contributed to increased submergence tolerance through their effects on decreased underwater shoot elongation or increased maintenance of chlorophyll levels or on both. Septiningsih et al. (2009) developed mega varieties of rice with Sub1 introgression that are submergence tolerant. They observed that all mega varieties with Sub1 introgression had a significantly higher survival rate than the original parents. An intolerant Sub1C allele combined with the tolerant Sub1A-1 allele did not significantly reduce the level of tolerance, and the Sub1C-1 expression appeared to be independent of the Sub1A allele; however, even when Sub1C-1 expression is completely turned off in the presence of Sub1A-2, plants remained intolerant. They concluded that the Sub1 provided a substantial enhancement in the level of tolerance of all the sensitive mega varieties to submergence.

Salt Tolerance

Prasad et al. (2000) mapped QTLs controlling various rice seedling traits conferring salt tolerance by using a doubled-haploid population derived from the cross between IR64 and Azucena. Seven QTLs were identified (threshold LOD _2.00) for seedling traits under salt stress, i.e., two for seed germination (%), one for seedling root length (cm), three for seedling dry matter (mg), and one for seedling vigor. Among the seven QTLs, four were located on chromosome 6. A QTL analysis for root length on chromosome 6 that was flanked by RFLP markers RG162 and RG653 exhibited a very high phenotypic variance of 18.9 % and a peak LOD score of 2.852.

Mapping for Biotic Stresses

Many diseases like bacterial blight, blast, and sheath blight are the major diseases in rice. Genes responsible for resistance against those diseases have been exhaustively worked, mapped, and used in breeding programs. Some of the initial examples are given here. Ammiraju et al. (2000) identified molecular markers linked to a bacterial blight resistance gene from a rice cultivar (Ajaya), which is nonallelic to other known recessive genes conferring resistance to Indian pathotype I of *X. oryzae* pv. *oryzae*. Similarly, Che et al. (2003) in a study used an F_2 rice population from a cross between "4011" and "Xiangzaoxian19" and identified five molecular markers, including three RFLP markers converted from RAPD and AFLP markers and two SSR markers to link with the sheath blight-resistant gene. This dominant resistant gene was named as Rsb 1 and mapped on rice chromosome 5. Ahn et al. (2000) reported the chromosomal localization and molecular mapping of this blast resistance gene and designated it as *Pi*-18, which confers resistance to Korean isolate KI-313 of the blast pathogen. RFLP analysis showed that *Pi*-18 was located near the end of chromosome 11, linked to a single copy clone RZ536 at a distance of 5.4 cM and that this gene was different from *Pi*-1 (t). An allelism test revealed that this gene was also different from *Pi*-k. A study by Fukuoka and Okuno (2001) detected two QTLs for blast resistance on chromosome 4, and one QTL was detected on each of chromosomes 9 and 12. The phenotypic variation explained by each QTL ranged from 7.9 to 45.7 % and the four QTLs explained 66.3 % of the total phenotypic variation. The resistance gene, designated pi21, was mapped on chromosome 4 as a single recessive gene between RFLP marker loci G271 and G317 at a distance of 5.0 cM and 8.5 cM, respectively. Similarly, Fujii et al. (2000) mapped the Pb1 locus for the panicle blast resistance on rice chromosome 11 using RFLP markers. The Pb1 locus was mapped in the middle part of the long arm of chromosome 11. This locus was closely located at 1.2 cM from three RFLP markers: S723, CDO226, and C189. Although based on linkage analysis they determined that S723, CDO226, and C189 were located at the same locus, graphical genotyping analysis, using many progeny cultivars, revealed that S723 was the closest marker to Pb1 among these three.

The major insect pests of rice includes brown planthopper, whitebacked planthopper, stem borer, and stem rot. Some of the initial examples of mapping insect pest resistance are given. Yamasaki et al. (2000) working on brown planthopper phenotyped a set of 71 rice recombinant inbred lines (F_{11}) derived from a cross between a *japonica* cultivar Asominori with ovicidal response and an *indica* cultivar IR24. In composite interval mapping for GWL and EM with 293 RFLP marker loci, two QTLs each on the long arm of chromosome 1 (1 L) and the short arm of chromosome 6 (6S) were detected for both GWL and EM. The 6S QTL explained 72.1 and 85.1 % of the phenotypic variations for GWL and EM, respectively. The QTL on 1 L explained 19.8 and 17.8 % of the phenotypic variations for GWL and EM, respectively. Murai et al. (2001) constructed a high-resolution linkage map as a foundation for map-based cloning of the bph2 locus. An advanced mapping population derived from a cross of "Tsukushibare" (susceptible) with "Norin-PL4" (an authentic bph2-introgression line) was used. Through bulked segregant analysis and linkage analysis, bph2 was located within a 3.2-cM region containing eight AFLP markers. One marker (KAM4) showed complete co-segregation with bph2, and bph2 was mapped within a 1.0-cM region delimited by KAM3 and KAM5, two flanking markers. Huang et al. (2001) conducted a molecular marker-based genetic analysis of the BPH resistance of B5, a highly resistant line that derived its resistant genes from the wild rice, *Oryza officinalis*. Their study revealed two genomic regions on chromosomes 3 and 4, respectively, that contained genes for BPH resistance. These two loci were further assessed by quantitative trait locus analysis resolved to a 14.3 cM interval on chromosome 3 and a 0.4 cM interval on chromosome 4. Similarly, working on WBPH, Ma et al. (2002) studied a rice population consisting of 90 F_3 lines from TN1/Guiyigu to analyze the linkage between DNA markers and a new gene Wbph 6(t) conferring resistance to the whitebacked planthopper. They mapped the Wbph 6(t) gene onto the short arm of rice chromosome 11, with a distance of 21.2 cM to the SSLP marker RM 167.

The Chinese rice cultivar Duokang #1 carries a single dominant gene Gm-6 (t) that confers resistance to the four biotypes of Asian rice gall midge (*Orseolia oryzae*) known in China. Katiyar et al. (2001) performed bulked segregant analysis on progeny of a cross between Duokang #1 and the gall midge-susceptible cultivar Feng Yin Zhan using RAPD method. The RAPD marker OPM06 (1400) amplified a locus linked to Gm-6 (t). The locus was subsequently mapped to rice chromosome 4 in a region flanked by cloned RFLP markers RG214 and RG163. Fine mapping of Gm-6(t) revealed that markers RG214 and RG476 flanked the gene at distances of 1.0 and 2.3 cM, respectively. Another gall midge resistance gene, Gm-2, mapped previously to chromosome 4, is located about 16 cM from Gm-6 (t), to judge by data from a segregating population derived from a cross between Duokang #1 and the Indian cultivar Phalguna that carries Gm-2. Sardesai et al. (2002) identified an AFLP marker SA598 that is linked to Gm7, a gene conferring resistance to biotypes 1, 2, and 4 of the gall midge. A set of PCR primers specific to an RFLP marker, previously identified to be linked to another gall midge resistance gene Gm2, also amplified a 1.5-kb (F8LB) fragment that is linked to Gm7. Gm7 is a dominant gene and nonallelic to Gm2. Hybridization experiments with clones from a YAC library of Nipponbare revealed that Gm7 is tightly linked to Gm2 and is located on chromosome 4 of rice. The germplasm line 87-Y-550 (PI566666) inherited its resistance to stem rot from the wild species *Oryza rufipogon*. Ni et al. (2001) made four crosses of 87-Y-550 (resistant donor) with susceptible lines. One AFLP marker showed significant association with stem rot resistance and accounted for 45.0 % of the phenotypic variation in 59 progenies. This marker was mapped on rice chromosome 2 between the RFLP markers RZ166 and RG139 by using F_2 reference population information. With the strategy of selective genotyping combined with a parental survey, two microsatellite markers, RM232 and RM251, on chromosome 3 were found to be associated with stem rot resistance and accounted for 41.1 % and 37.9 % of the phenotypic variation, respectively.

Meta-analysis of QTLs combines the results of several QTL detection studies and provides narrow confidence intervals for meta-QTLs, permitting easier positional candidate gene identification. It is usually applied to multiple mapping populations but can be applied to one. Khowaja et al. (2009) reported a meta-analysis of drought-related QTLs in the Bala × Azucena mapping population that compiled data from 13 experiments and 25 independent screens providing 1,650 individual QTLs separated into 5 trait categories: drought avoidance, plant height, plant biomass, leaf morphology, and root traits. A heat map of the overlapping 1-LOD confidence intervals provides an overview of the distribution of QTLs. The heat map graphically illustrates the genetic complexity of drought related traits in rice. Formal meta-analysis on chromosome 1, where clusters of QTLs for all trait categories appear close, established that the *sd1* semidwarfing gene coincided with a plant height meta-QTL, that the drought avoidance meta-QTL was not likely to be associated with this gene, and that this meta-QTL was not pleiotropic with close meta-QTLs for leaf morphology and root traits. On chromosome 5, evidence suggests that a drought avoidance meta-QTL was pleiotropic with leaf morphology and plant biomass meta-QTLs but not with meta-QTLs for root traits and plant height 10 cM lower down. A region of dense root QTL activity graphically visible on chromosome 9 was dissected into three meta-QTLs within a space of 35 cM. The confidence intervals for meta-QTLs obtained ranged from 5.1 to 14.5 cM with an average of 9.4 cM, which are approximately 180 genes in rice.

Marker-Assisted Selection

In order to pyramid the useful genes into a single genotype useful for Indian condition, Singh et al. (2001) pyramided three BB resistance genes, xa5, xa13, and Xa21, into cv. PR106, which is widely grown in Punjab, India, using marker-assisted selection. Lines of PR106 with pyramided genes were evaluated after inoculation with 17 isolates of the pathogen from Punjab and six races of Xoo from the Philippines. Genes in combinations were found to provide high levels of resistance to the predominant Xoo isolates from the Punjab and six races from the Philippines. They also evaluated the pyramided lines under natural conditions at 31 sites (including Ludhiana, Jalandhar, Ferozepur, and Sangrur) in commercial fields during 1999. The combination of genes provided a wider spectrum of resistance to the pathogen population prevalent in the region. Similarly, Sanchez et al. (2000) transferred three bacterial blight (BB) resistance genes, xa5, xa13, and Xa21, to IR65598-112 and two other sister lines IR65600-42 and IR65600-96 (new plant type, NPT) rice lines *via* marker-aided backcrossing. The BC_3F_3 NILs having more than one BB resistance gene showed a wider resistance spectrum and manifested increased levels of resistance to the Xoo races, compared with those having a single BB resistance gene. Results for two F_2 populations and the progeny testing of their F_3 lines showed that MAS reached an accuracy of 95 and 96 % of identifying homozygous resistant plants for xa5 and xa13, respectively. Their results demonstrated the usefulness of MAS in gene pyramiding for BB resistance, particularly for recessive genes, such as xa5 and xa13, that are difficult to select through conventional breeding in the presence of a dominant gene such as Xa21. Similarly, Siangliw et al. (2003) crossed three submergence-tolerant cultivars, FR13A, IR67819F2-CA-61, and IR49830-7-1-2-2, with KDML105. Transferring the major QTL for submergence tolerance was facilitated by four backcrossings to the recipient KDML105. Molecular markers tightly linked to the gene(s) involved were developed to facilitate molecular genotyping. They demonstrated that individuals of a BC_4F_3 line that retained a critical region on chromosome 9 transferred from tolerant lines were also tolerant of complete submergence while retaining all the agronomically desirable traits of KDML105. Few successful commercial cultivars released in India using marker-assisted backcrossing are given below.

Commercial Release of MAS-Derived Varieties in India

Improved Samba Mahsuri

Sundaram et al. (2008) introgressed three major bacterial blight resistance genes (*Xa*21, *xa*13, and *xa*5) into Samba Mahsuri (BPT5204), a medium slender grain *indica* rice variety, popular with farmers and consumers of India. Samba Mahsuri is susceptible to several diseases and pests, including bacterial blight (BB). They used PCR-based molecular markers in a backcross breeding program to introgress the genes from a donor line (SS1113) in which all the three genes were present in a homozygous condition. A selected BC_4F_1 plant was selfed to generate homozygous BC_4F_2 plants with different combinations of BB resistance genes. Under conditions of BB infection, the three-gene pyramid lines exhibited a significant yield advantage over Samba Mahsuri which was later on released for commercial cultivation.

Improved Swarna

Swarna is one of the most popular rice varieties in India producing a high yield, good grain eating quality and requires 25 % less nitrogen, as widely claimed by the farmers. Released in Andhra Pradesh in 1982, it spread across the subcontinent and into Bangladesh, where it was never officially released. It is now the number two variety during the wet season in India. The research began in the 1970s when flood-resistant varieties of rice were identified in India. Researchers then crossed and improved the tolerant characteristics to produce higher-yielding rice varieties, which can withstand up to 17 days of complete submergence. Using MAS, this single trait of interest was transferred into commercially valuable rice variety Swarna without losing useful characteristics—such as high yield, good grain quality, or pest and disease resistance. The first variety developed, Swarna-Sub1, showed high survival under submerged conditions compared to the original variety Swarna and gave yield advantages of one to three tons per hectare over Swarna when submerged.

Next-Generation Sequencing

NGS technology is based on massive parallel sequencing as opposed to the Sangers sequencing technology. They rely on the amplification of single isolated DNA molecules and their analysis in a massive parallel way. Hundreds of thousands or even tens of millions of single-stranded DNA molecules are immobilized on a solid surface such as glass slides or on beads, depending on the platform used. The commercially available NGS platforms used for plant genome sequencing include Roche/454 FLX and the Illumina/Solexa Genome Analyzer and the Applied Biosystems SOLiD System.

Next-generation sequencing (NGS) is a powerful tool for the discovery of domestication genes in crop plants and their wild relatives. Re-sequencing of domesticated and commercialized genotypes can identify regions of low diversity associated with domestication and adaptation. The sequence data can be used to design species-specific PCR primers. Novel allelic variations in close or distant relatives can be characterized by NGS. By using NGS populations of large number of individuals can be screened rapidly. NGS also supports in efficient identification and capture of novel genetic variation from related species. NGS allows whole-genome analysis to determine the genetic basis of phenotypic differences. This helps in the facilitation of recombination of traits which are agronomically important both in cultivated and wild relatives. NGS provides very large amounts of DNA sequence data normally with short repeats. As the technology continues to develop, the amount of data and the length of the reads are increasing (Varshney et al. 2009). NGS gives an opportunity to explore genetic variation among plants and their wild relatives in a better way than the earlier technologies. It also allows the most complex plant genome to be studied and understood.

Rice was the first crop to have a reference genome sequence and was a model for the use of NGS. Even if lots of information is available in rice in relation to its origin and evolution, the origin of rice is complex and involves movement of genes between species and groups like *indica* and *japonica* types (Sang and Ge 2007). Whole-genome re-sequencing of rice genotypes has allowed analysis of domestication genes identified in areas of low diversity in the genome as a result of human selection. This approach has shown that both *japonica* and *indica* rice share common areas of low diversity possibly due to introgression from one population to the other following selection. NGS of wild rice population has recently been used to identify the distinctness of Asian and Australian wild rice populations. NGS also provides new tools for the evaluation of grain quality (Henry 2011).

Using existing genetic and genomic resources and tools (mutants, transgenics, cytogenetic stocks, and genomic resources), rice researchers were able to rapidly integrate and apply genome sequence information to understand rice genome structure and evolutions as well as to discover and mine genes, including those underlying complex traits of agricultural importance. Members of the large gene families [viz., transcription factors, peptide transporters, kinases, nucleotide-binding leucine-rich repeats (NB-LRRs), microRNAs, and germins] have been discovered through genome-wide surveys, enabling their cellular functions to be dissected and their roles in plant growth and development to be elucidated. In another example genome sequence-enabled identification and positional cloning of genes responsible for traits selected during domestication, including the seed-shattering trait, led to the identification of molecular changes selected during domestication (Feuillet et al. 2010; Izawa et al. 2009). Rice is now rich in tools for mapping and breeding, including high-density SSRs, comprehensive SNPs, insertion–deletion polymorphism (IDPs), and custom-designed (candidate gene) marker for marker-assisted breeding. One of the outcomes of the NGS in rice for rice genome sequencing has been the development of a comprehensive collection of SNPs. Fukuoka et al. (2009) used sequence-based markers from the blast disease resistance gene *pi*21 region to identify recombinants between *pi*21 and another gene located 37 kb apart that confers poor eating quality. With the information available from NGS, rice varieties with both durable resistance against blast and good eating quality were combined.

Expressed Sequence Tags (ESTs) and Flanking Sequence Tags (FSTs)

Kim et al. (2013a, b) while working with rice observed that transcript levels of the flowering repressor *Ghd7* were decreased while those of its downstream genes, *Ehd1*, *Hd3a*, and *RFT1*, were increased. Sequencing the known flowering-regulator genes revealed mutations in *Ghd7* and *OsPRR37* that cause early translation termination and amino acid substitutions, respectively. Genetic analysis of F_2 progeny from a cross between cv. Kitaake and cv. Dongjin indicated that those mutations additively contribute to the early-flowering phenotype in cv. Kitaake. They generated 10,000 T-DNA tagging lines and deduced 6,758 flanking sequence tags (FSTs), in which 3,122 were genic and 3,636 were intergenic. Among the genic lines, 367 (11.8 %) were inserted into new genes that were not previously tagged. Because the lines were generated by T-DNA that contained the promoterless *GUS* reporter gene, which had an intron with triple splicing donors/acceptors in the right border region, a high efficiency of *GUS* expression was shown in various organs. Sequencing of the GUS-positive lines demonstrated that the third splicing donor and the first splicing acceptor of the vector were extensively used. Similarly, Priya and Jain (2013) constructed a database, RiceSRTFDB, which provided comprehensive expression information for rice TFs during drought and salinity stress conditions and various stages of development. This information will be useful to identify the target TF(s) involved in stress response at a particular stage of development.

Functional Genomics

The last decade has seen rapid advances in rice functional genomic research globally. Most of the efforts can be summarized into three fronts: (1) construction of technological and resource platforms for high-throughput gene identification, (2) functional genomic analysis of agronomic traits and biological processes, and (3) identification and isolation of functional genes (Jiang et al. 2011). The overall goal of rice functional genomic research is to understand how the genome functions by deciphering the information conserved in the sequences, including genes and regulatory elements at the whole-genome level. DNA microarray and expression profiling information of gene expression profile is useful in many ways for functional genomic studies. The temporal and spatial patterns of the gene expression can be studied by the transcript abundance of a gene in multiple tissues. This will give information whether the gene expression is constitutive or tissue specific. The expression pattern of the gene would provide clue to understanding the gene function.

Zhang et al. (2007) proposed Green Super Rice as a new goal for rice breeding. On the premise of continued yield increase and quality improvement, Green Super Rice should possess resistances to multiple insects and diseases, high nutrient efficiency, and drought resistance, promising to greatly reduce the consumption of pesticides, chemical fertilizers, and water. With the further advancement of functional genomics in rice, it is anticipated that more genes for these traits will be available for the development of Green Super Rice. The ideal situation of varietal development is breeding according to designed blueprints to breed for cultivars to meet the diverse needs of global rice production for high yield, superior quality, multiple resistances, and high nutrient use efficiency. However, compared to other crops, especially corn, with major commercial breeding programs, breeding application of genomic technology has been rather limited in rice, despite more advances made in sequencing and functional genomic research in rice than in any other crops. At present, high-throughput and low-cost technologies based on the massive sequence information should be developed for breeding applications, in the forms of multiple sets of oligonucleotide chips to meet the needs of rice breeding programs, such as *indica* vs. *japonica*. With the tremendous efforts and rapid progress in functional genomic research, rice should be expected to be the first crop for practicing breeding by design.

Transgenic for Genomic Studies

Insertion mutants are valuable resources, especially when their insertion positions are annotated into a chromosome. This allows one to identify the knockout mutants in a given gene. Systemic phenotyping of the insertion mutants can be used to determine the phenotypic alteration that co-segregates with the insertion element. Mutants can also be used for the verification of functional analysis for a map-based cloned gene. Transposon tagging and cloning using *Ac/Ds* system and other transposable elements is a novel way to understand gene function and cloning new genes in crop plants. T-DNA insertional mutagenesis is the most successful way of developing population for genomic studies.

T-DNA Insertional Mutagenesis for Functional Genomics in Rice

Efforts have been made to discover the functions of plant genes. Most of the strategies have been based on genetic approaches such as mutant identification and map-based gene isolation. A large number of sequenced rice genes of unknown function have been revealed by large-scale analysis of expressed sequence tags (Yamamoto and Sasaki 1997). Thus, the development of a systematic method for discovering the biological functions of these genes has become extremely important. Insertion of transposon has been employed for functional studies in several species. The insertion of T-DNA is a random process. Insertional mutagenesis is the most suitable method for the systematic functional analysis of a large number of genes in the context of the whole plant. This system allows the production of many mutant lines at one time and the induced mutations can be easily detected by PCR. In *Arabidopsis*, whose entire genomic sequencing has been completed in the past few months, several insertional mutagens like T-DNA, *Ac/Ds*, and *En/Spm* have been employed. The inserted genes are stable over generations (Azpiroz-Leehan and Feldmann 1997). The mutant populations induced by these mutagens are being used for forward and reverse genetics. Rice plants are larger than *Arabidopsis* plants; hence, it is not easy to handle thousands of transgenic rice plants in a greenhouse. For rice, therefore, the use of endogenous insertional mutagens is desirable. Recent comprehensive studies on bacterial artificial chromosome (BAC) end sequences have revealed that the rice genome contains many kinds of transposable elements. Most of them, however, may not be active, because none of the spontaneous waxy mutations were caused by the insertion of transposable elements. Hirochika and his group have shown that a rice retrotransposon, *Tos17*, is highly active during tissue culture and that the activation of *Tos17* is responsible for tissue culture-induced mutations (Hirochika et al. 1996). The unique features of *Tos17* suggest that it can be used for forward and reverse genetic studies. At present many newer methodologies are available for screening T-DNA or transposon insertions within known genes and recovering sequences flanking the insertions (Cooley et al. 1996; Frey et al. 1998; Okushima et al. 2005). Reporter genes as insertional elements have been utilized to aid in the identification of insertions within functional genes. The GUS gene is the most frequently used gene as reporter gene because of the accurate detection of its gene products.

Jeon et al. (2000) produced 11,090 primary transgenic rice plants that carry a T-DNA insertion producing 18,358 fertile lines. They observed that approximately 65 % of the population contains more than one copy of the inserted T-DNA. The transgenic plants contained an average of 1.4 loci of T-DNA inserts. They carried out the histochemical GUS assay in the leaves and roots from 5,353 lines and observed that 1.6–2.1 % of the tested organs were GUS positive in the tested organs. They also observed that the GUS expression patterns were organ or tissue specific or ubiquitous in all parts of the plant. Kohli et al. (2001) developed a population of transgenic indica rice lines (*Oryza sativa* var. *indica* cultivars Bengal and Pusa Basmati) containing the autonomous Activator transposon (Ac) from maize. A transposon excision assay, which uses the reporter green fluorescent protein (GFP) gene driven by the ubiquitin promoter, was employed to monitor Ac excision in various tissues. In their experiment, twenty percent of transformed calluses displayed uniform GFP activity, indicating very early Ac excision, while later excision in another 40 % of calluses was revealed by mosaic GFP activity. They observed that 12 % of all amplified tags display homology to genes or ESTs—five times more than would be expected on the basis of random transpositional insertion. They advocated that the “Tagged Transcriptome Display” (TTD) technique, using a population of Ac insertion-tagged lines, represents a convenient approach for the identification of tagged genes involved in specific processes, as revealed by their expression patterns.

Proteomics

Proteomics is the scientific discipline which studies proteins and searches for proteins that are associated with traits like disease by means of their altered levels of expression and/or posttranslational modification between control and disease states. It enables correlations to be drawn between the range of proteins produced by a cell or tissue and the initiation or progression of a disease state and the effect of therapy (Isabel et al. 2013).

Abiotic Stresses

In recent past, numerous studies in crop proteomics were conducted using different genotypes and tissues, submitted to different stresses, thus complicating comparative analyses. To overcome this problem, several repositories where researchers can find proteomics information for functional analysis were built, such as the International Proteomics Research in *Arabidopsis*, the Maize Proteome DB, the Soybean Proteome DB, or the Organellome among many other databases (reviewed in (Jorrín-Novo et al. 2009). In plant abiotic stress studies, it is common to analyze proteomes by contrasting stressed plants against control ones, attempting to correlate changes in protein accumulation with the plant phenotypic response. Additionally, comparisons between genotypes with different sensitivity toward stress (sensitive *vs*. tolerant) are crucial to understand the putative influence of differentially abundant proteins in tolerant genotypes. As a general tendency, it is observed that tolerant genotypes not only increase oxidative stress metabolism and the scavenging capacity of reactive oxygen species (ROS) but also show enhanced carbohydrate metabolism and more efficient photosynthesis (Hajheidari et al. 2007; Ma et al. 2012; Salekdeh et al. 2002; Witzel et al. 2010). Furthermore, the embryo proteome of six different rice genotypes (with contrasting responses to drought, salt, and cold) showed differences in the posttranslational status of an LEA rice Rab21 (Farinha et al. 2011). This stress protein was found more strongly phosphorylated in the embryos of sensitive genotypes than in tolerant ones. Another proteome comparison between rice (two contrasting genotypes) and *Porteresia coarctata* (a halophytic type of wild rice) allowed identifying in *P. coarctata* 16 proteins involved in osmolyte synthesis, photosystem functioning, RuBisCO activation, cell wall synthesis, and chaperone activity (Sengupta and Majumder 2009). It is interesting to note that two *Porteresia* genes (PcINO1 and PcIMT1, absent in *Oryza sativa*) conferred salt tolerance when overexpressed in tobacco (Das-Chatterjee et al. 2006). However, the products of these genes could not be detected in *Porteresia* proteome in the study conducted by Sengupta and Majumder, using two-dimensional gel electrophoresis (2-DE) followed by MALDI-TOF. Liu et al. (2013b) studied for a comparative proteomic analysis of rice shoots in combination with morphological and biochemical investigations. They observed that arsenate suppressed the growth of rice seedlings, destroyed the cellular ultrastructure, and changed the homeostasis of reactive oxygen species. A total of 38 differentially displayed proteins, which were mainly involved in metabolism, redox, and protein metabolism, were identified. Arsenic was found to inhibit rice growth through negatively affecting chloroplast structure and photosynthesis.

Biotic Stresses

In the case of *Xanthomonas oryzae* pv. *oryzae* (Xoo), infection of transgenic rice suspension-cultured cells overexpressing the disease resistance gene Xa21 with compatible and incompatible races revealed the modulation of 11 proteins in compatible interactions (Chen et al. 2007). This was comparable to modulation of 20 proteins in rice leaf blades inoculated with compatible and incompatible Xoo races (Mahmood et al. 2006). Proteomic studies of the rice–fungus pathosystem have taken advantage of the natural resistance and/or compatibility existing in the rice germplasm and the fungal races. For example, two near-isogenic rice lines CO39 and C101A51 displaying, respectively, compatible and incompatible interactions with *Magnaporthe oryzae* have been used to decipher proteome alteration during fungal infection (Liao et al. 2009). Proteomic study of the apoplastic secretome where pathogenic determinants are potentially accumulating can provide additional insights into the plant–fungus interaction. The majority of identified proteins in the apoplastic secretome using the resistant interaction rice—*Magnaporthe oryzae*— were rice defense-related proteins. On the contrary most of the differentially regulated proteins using 2-D gel approach originated from *Magnaporthe oryzae* in the susceptible interaction (Kim et al. 2013a, b). Investigations of proteome modulation during viral infections in model species have been scarce and mostly restricted to the rice–rice yellow mottle virus (RYMV) pathosystem (Ventelon-Debout et al. 2004; Brizard et al. 2006).

Metabolomics

Metabolites are the end products of cellular process, and they show the response of biological systems to environmental changes. The current trend in metabolomic studies is to define the cellular status at a particular time point of development or physiological status. These techniques complement other techniques such as transcriptomics and proteomics and depict precise pictures of the whole cellular process. The growing number of sequenced plant genomes has opened up immense opportunities to study biological processes related to physiology, growth and development, and tolerance to biotic and abiotic stresses at the cellular and whole plant level using a novel systems-level approach. Rice continues to be the flagship plant model and shares a significant fraction of reported literature on gene functions and phenotypes associated with plant development and metabolism (Hunter et al. 2012). Gene ontology assignments are often enriched by the addition of annotations from *Arabidopsis* and rice based on sequence homology. Depending on the biological question, these annotations are further evaluated to model the metabolic (Dal'Molin et al. 2010; Zhang et al. 2010), regulatory (Yun et al. 2010) and coexpressed networks (Childs et al. 2011) leading to novel discoveries of genes and enzymes regulating important agronomic traits. In order to respond to and survive environmental challenges, plants, as sessile organisms, have developed a multitude of anatomical, morphological, growth habit, and developmental adaptations that are based on underlying genetic variation. Genome sequence, gene structure, and functional annotation provide the basis for understanding a genome. However, in order to understand the physiology, development, and adaptation of a plant and its interaction with the environment, its metabolic network needs to be understood. This network represents a (bio)chemical manifestation of downstream changes in shape, form, and growth and development.

Studies on rice metabolomics have so far focused on the quality of metabolites, such as the types of metabolites that can promote seed germination (Shu et al. 2008), the metabolite variation between mutant and wild-type plants (Wakasa et al. 2006), the profiling of metabolome at different developmental stages (Tarpley et al. 2005), and the observation of natural metabolite variation between rice varieties (Kusano et al. 2007). However, information available on biotic and abiotic stresses is limited. Liu et al. (2013a) for the first time gave a reconstructed and curated genome-scale metabolic model of rice, including gene regulatory network, microRNA target information, and protein–protein interactions. The genome-scale multilevel network provides a detailed reference for rice molecular regulatory analysis and genotype–phenotype mapping. Eventually, a comprehensive molecular regulation database of rice has been developed to systematically store, analyze, and visualize the rice genome-scale multilevel network. Takahara et al. (2010) studied the effects of NADK2 expression in rice by developing transgenic rice plants that constitutively expressed the *Arabidopsis* chloroplastic NADK gene (NK2 lines). NK2 lines showed enhanced activity of NADK and accumulation of the NADP(H) pool, while intermediates of NAD derivatives were unchanged. Comprehensive analysis of the primary metabolites in leaves using capillary electrophoresis mass spectrometry revealed elevated levels of amino acids and several sugar phosphates including ribose-1,5-bisphosphate but no significant change in the levels of the other metabolites. The results suggest that NADP content plays a critical role in determining the photosynthetic electron transport rate in rice and that its enhancement leads to stimulation of photosynthesis metabolism and tolerance of oxidative damages.

Recently, several studies have proposed strategies for the genotype–phenotype mapping. It is clear that approaches of integrating genetics and omics would be a valuable strategy for investigating the regulation of the relationship between plant metabolism and physiology. The potential of metabolomics as a functional genomic tool in addition to transcriptomics and proteomics is well recognized (Carreno-Quintero et al. 2012). Therefore, the integration of different levels of regulatory information (genome, proteome, and metabolome) could probably be a new approach for mapping genotypes to phenotypes. The determination of rice genome sequence (and its annotation), of proteome interactions, and of transcriptome regulatory information has led to the accumulation of sufficient public data to construct systems-level models. These models could increase the understanding of genotype–phenotype relationship and consequently help to improve the quality and productivity of rice.

Maize

Genomics

Maize is the third most important food crop after rice and wheat, both in terms of area and production. In Africa and some of the Asian countries, almost 90 % of maize grown is for human consumption and may account for 80–90 % of the energy intake. The genetic studies in *Zea mays* started with Edward East's 1908 report of inbreeding depression. The earlier genetic studies primarily focused on hybrid vigor in the 1940s and cytogenetic breakthrough studies, e.g., transposable elements by Barbara McClintock. During the last decade, the hegemony of SSRs, declared in the near past as "markers of choice," was broken by single nucleotide polymorphic (SNP) markers (Mammadov et al. 2012). More than 130,000 gene-based SNPs have been identified (Mammadov et al. 2012) and many of these were used for the construction of genetic maps (Shi et al. 2012; Sa et al. 2012). More than 1,800 molecular maps have been developed using different mapping populations in maize were documented in the Maize Genetics and Genomics Database (Maize GDB http://www.maizegdb.org). Molecular marker-facilitated QTL mapping for yield-related traits in maize was first reported by Stuber et al. (1987). Since then, there have been large numbers of studies for identifying associated major genes through gene tagging and quantitative traits locus (QTL) mapping. Many studies have been conducted to identify QTL associated with oil content in maize kernels (Wassom et al. 2008) and provitamin A content (Zhou et al. 2012). Thus, these highly effective and closely linked markers to the QTLs were being used in the marker-assisted selection (MAS) studies for introgressing the targeted genes into locally well-adapted germplasm. Successful MAS applications have been reported for introgression breeding in maize, including introgressions of transgenes (Ragot et al. 1995) and conversions involving simple (Morris et al. 2003) or complex traits (Willcox et al. 2002). Gupta et al. (2009) converted successfully a promising maize hybrid, viz., Vivek Maize Hybrid 9, into QPM version using MAS by introgression of *opaque*2 gene.

MAS for Development of QPM Maize

Quality protein maize (QPM) is the result of *opaque* 2 gene along with associated modifiers. The protein quality of QPM maize is as good as 90 % of the milk protein. Since DNA markers within the exons of this gene were available, it was a viable proposition to use MAS for the introgression of the gene along with the necessary modifiers for the development of QPM maize cultivars. Utilizing this method, many normal maize inbreds and hybrids have been converted into QPM versions and were released for commercial cultivation (Gupta et al. 2009). Vivek QPM 9 was released commercially in 2008 while Vivek QPM 21 was released in 2012.

Vivek QPM 21

Vivek QPM 21 (QPM version of Vivek Maize Hybrid 21) shows >70 % enhancement in tryptophan over the original hybrid, Vivek Maize Hybrid 21. The tryptophan content of Vivek QPM 21 is 0.85, whereas it is 0.49 for Vivek Maize Hybrid 21. Vivek QPM 21 was also tested in the All India Coordinated Trial of *Kharif* 2007, 2008, and 2009, in which it performed equally well in respect of grain yield and other agronomic traits over non-QPM national check, Vivek Maize Hybrid 17, and Vivek Maize Hybrid 21. Vivek Maize Hybrid 21 was released for commercial cultivation in zones I, II, and IV in 2006. The parents of this hybrid have been converted into the QPM version using DNA markers, and this hybrid was reconstituted by crossing VQL 1 and VQL 17. This QPM hybrid shows more than 70 % enhancement in tryptophan over the original hybrid. In the state trials of Uttarakhand under organic condition, this hybrid gave more than 2.4 % higher yield over Vivek Maize Hybrid 21 with an average grain yield of 56.31 quintals per hectare. Vivek QPM 21 was released for the state of Uttarakhand, India, in the year 2012 for commercial cultivation by the State Varietal Release Committee, Uttarakhand, for the hill conditions (Fig. 2).

Fig. 2 Vivek QPM 21- QPM version of Vivek Maize Hybrid 21 developed using MAS. (**a**) Cobs of Vivek QPM 21. (**b**) Grains of Vivek QPM 21

Functional Genomics

Maize is an excellent model plant species for genetics, developmental biology, and physiology especially for those systems for addressing evolutionary dynamics of transposable elements within and between species particularly for large, complex, repetitive genomes such as maize. The maize genomic research is now accelerated by introduction to next-generation sequencing (NGS) technology, and a number of databases are developed for functional genomics, viz., sequencing database (http://www.maizesequence.org), MaizeGDB (http://www.maizegdb.org/), PlantGDB (http://www.plantgdb.org/ZmGDB), TIGR Maize Database (http://maize.jcvi.org), Maize Assembled Genomic Island (http://magi.plantgenomics.iastate.edu/), and Plant Proteome Database (http://ppdb.tc.cornell.edu) and integrated database OPTIMAS-DW(http://www.optimas-bioenergy.org/optimas_dw).

To study the global gene expression, high-density microarrays, ESTs, or full-length cDNAs serve as the central resources (Schadt et al. 2003). There are 364,385 EST sequences of maize available at NCBI database for their further use in the functional analysis of maize genome. Sekhon et al. (2011) studied the gene expression of different developmental stages especially organ- and paralog-specific expression patterns of lignin biosynthetic pathway genes in vegetative organs during the life cycle of a maize plant. Functional annotation of full-length cDNA in gene prediction has sharply improved knowledge about transcriptome of maize (Soderlund et al. 2009). Zinselmeier et al. (2002) characterized the transcriptomes for various traits/organ-specific gene expression like water-stressed maize tissues and female reproductive tissues. While in another transcriptomic study, the gene expression of placenta and endosperm in developing maize kernels, developing immature ear and tassel (Yu and Setter 2003; Andjelkovic and Thompson 2006; Zhuang et al. 2007), leaves and roots (Bassani et al. 2004; Zheng et al. 2004; Jia et al. 2006; Poroyko et al. 2007; Spollen et al. 2008; Li et al. 2010; Lu et al. 2011), salt-stressed tissues (Wang et al. 2003), and aluminum toxicity-stressed tissues such as roots (Maron et al. 2008) was used. Jia et al. (2006) constructed a full-length cDNA library of maize line Han 21 to study the gene regulations in the osmotically stressed maize seedlings. Alexandrov et al. (2009) generated 36,565 full-length cDNAs using different tissues and treatments from diverse hybrids of maize, of which 10,084 were determined to be of high-quality unique clones. Soderlund et al. (2009) generated 27,455 full-length cDNAs from maize inbred B73 (www.maizecdna.org).

RNA interference (RNAi) is a powerful tool for functional genomics, and it causes a reduction in the steady-state RNA levels of not only the target gene but also another closely related gene (McGinnis et al. 2007). Maize is susceptible to *Aspergillus flavus* infection, and using the RNAi technology for pathogenesis-related protein 10 (*PR10*) gene, the expression of *PR10* was reduced by 65 % to more than 99 % in transgenic callus lines (Chen et al. 2010). Virus-induced gene silencing (VIGS) has recently been used for functional genomic studies. Linde et al. (2011) used VIGS to identify maize genes that are functionally involved in the interaction with *Ustilago maydis*, the causative agent of brome mosaic virus (BMV). Shi et al. (2012) used the similar approach VIGS to silence the expression of the gene *ZmTrm2* encoding thioredoxin *m* during sugarcane mosaic virus (SCMV) infection, and it resulted in significant enhancement of systemic SCMV infection.

Proteomics

A significant progress has been made during the last two decades in the analysis of the maize proteome. Protein data sets were generated from different tissues during different physiological and environmental stages. Pechanova et al. (2013) developed the proteomics database from different tissues in maize. Proteomic analysis has been employed to study the important environmental stress factors limiting maize plant growth and productivity. Hu et al. (2010) studied the alterations in protein expression in response to drought and ABA in leaves of maize seedlings. In the ABA-dependent pathway, an anionic peroxidase and two putative uncharacterized proteins were upregulated by drought, while in ABA-independent pathways, glycine-rich RNA-binding protein, pathogenesis-related protein, an enolase, a serine/threonine protein kinase receptor, and a cytosolic ascorbate peroxidase were upregulated by drought. Late embryogenesis abundant (LEA) proteins constitute a complex set of proteins that participate in several plant stress responses. Amara et al. (2012) have identified 20 unfolded maize embryo proteins, 13 of which belong to the LEA family on the basis of the unusual heat stability and acid solubility characteristic of unfolded proteins. Kunpeng et al. (2011) studied extensively the proteome profile of maize leaf tissue at the flowering stage after long-term treatment with rice black-streaked dwarf virus infection. They identified 91 differentially accumulated proteins that belong to multiple metabolic/biochemical pathways. Further analysis of these identified proteins showed that MRDD resulted in dramatic changes in the glycolysis and starch metabolism and eventually the significant differences in morphology and development between virus-infected and normal plants. Mechin et al. (2004) have established a proteome reference map for maize endosperm by 2-D gel electrophoresis and protein identification with LC-MS/MS analysis. Among the 632 protein spots processed, 496 were identified by matching against the NCBInr and ZMtuc-tus databases (using the SEQUEST software).

Metabolomics

Metabolomics is the term coined for essentially comprehensive, nonbiased, high-throughput analyses of complex metabolite mixtures typical of plant extracts. The first International Congress on Plant Metabolomics was held in Wageningen, Netherlands, in April 2002, with the primary goal of bringing together those players who are already active in this field and those who soon plan to be. Metabolomics is driven primarily by recent advances in mass spectrometry (MS) technology and by the goals of functional genomic efforts. The combination of NMR spectroscopy, chemometric methods, and PCA is a useful tool for the discrimination of maize silks in respect to their chemical composition, including rapid authentication of the raw material of current pharmacological interest (Fiehn 2002; Halket et al. 2005; Shulaev 2006). Marcelo et al. (2012) determined metabolic fingerprint and pattern recognition of silk extracts from seven maize landraces cultivated in southern Brazil by NMR spectroscopy and chemometric methods. Metabolomics is also useful for predicting the complex heterotic traits in maize which has been supported by some studies. Christian et al. (2012) crossed 285 diverse Dent inbred lines from worldwide sources with two testers and predicted their combining abilities for seven biomass- and bioenergy-related traits using 130 metabolites. Under drought situations, maize ovule abortion appears to be related to the flux of carbohydrates to the young ear around flowering, and concurrent photosynthesis is required to maintain this above threshold levels (Zinselmeier et al. 1995). The application of an NMR-based metabolic profiling approach to the investigation of saline-induced stress in maize plants was studied by Claire et al. (2011). The maize seedlings were grown in 0, 50, or 150 mM saline solution. Plants were harvested after 2, 4, and 6 days ($n = 5$ per class and time point), and H NMR spectroscopy was performed separately on shoot and root extracts. Metabolomic, transcriptomic, and, to a lesser extent, proteomic studies have been conducted for the high-throughput phenotyping necessary for large-scale physiological, molecular, and quantitative genetic studies, aimed at identifying the function of a particular gene or set of genes involved in the control of complex physiological traits such as NUE (Meyer et al. 2007; Lisec et al. 2008; Kusano et al. 2011). Pavlík et al. (2010) analyzed metabolome of maize plants growing under different nitrogen nutrition conditions and sequential extraction of fresh biomass was used, and isolated fractions were characterized and evaluated using IR spectra. The results showed that an increased induction of oxalic acid in plants after 4 g nitrogen application.

Conclusions

Approaches of integrating genetics and omics would be a valuable strategy for investigating the regulation of the relationship between plant metabolism and physiology. The potential of transcriptomics, proteomics, and metabolomics as tools for functional genomics is well realized. Therefore, the integration of different levels of regulatory information (genome, proteome, and metabolome) could probably be a novel approach for mapping genotypes to phenotypes. The determination of the genome sequence (and its annotation), of proteome interactions, and of transcriptome regulatory information of a crop plant will help in generating sufficient database for developing systems-level models. These models will help in understanding of genotype–phenotype relationship and consequently help to improve the quality and productivity of crop plants for the food and nutritional security of millions of human populations.

References

Abreu IA, Farinha AP, Negrão S, Gonçalves N, Fonseca C, Rodrigues M, Batista R, Saibo NJM, Oliveira MM (2013) Coping with abiotic stress: proteome changes for crop improvement. J Prot http://dx.doi.org/10.1016/j.jprot.2013.07.014

Adams MD, Celniker SE, Holt RA, Evans CA, Gocayne JD, Amanatides PG, Scherer SE, Li PW, Hoskins RA, Galle RF et al (2000) The genome sequence of *Drosophila melanogaster*. Science 287:2185–2195

Ahn SN, Kim YK et al (2000) Molecular mapping of a new gene for resistance to rice blast (*Pyricularia grisea* Sacc.). Euphytica 116(1):17–22

Alexandrov NN, Brover VV, Freidin S, Troukhan ME, Tatarinova TV, Zhang H, Swaller TJ, Lu YP, Bouck J, Flavell RB, Feldmann KA (2009) Insights into corn genes derived from large-scale cDNA sequencing. Plant Mol Biol 69(1–2):179–194

Ali ML, Pathan MS, Zhang J, Bai G, Sarkarung S, Nguyen HT (2000) Mapping QTLs for root traits in a recombinant inbred population from two indica ecotypes in rice. Theor Appl Genet 101:756–766

Alison FP, Gillian PC, Reinier B et al (2005) Proteomic analysis of the *Arabidopsis* nucleolus suggests novel nucleolar functions. Mol Biol Cell 16:260–269

Alonso JM, Stepanova AN, Leisse TJ et al (2003) Genome-wide insertional mutagenesis of *Arabidopsis thaliana*. Science 301:653

Alonso-Blanco C, El-Assal S, Peeters AJM, Koornneef M (1997) Genetic analysis of developmental traits based on the Ler/Cvi natural variation. In: 8th international conference *Arabidopsis* research. Stanford University Press, Madison, p 121

Amara I, Odena A, Oliveira E et al (2012) Insights into maize LEA proteins: from proteomics to functional approaches. Plant Cell Physiol 53(2):312–329

Ammiraju JSS, Jena KK, Reddy APK, Diwaker T, Vaidyanath K (2000) Identification of a molecular marker linked to a gene conferring resistance to Indian pathotype-I of Xanthomonas oryzae pv oryzae in rice. Plant Cell Biotechnol Mol Biol 1:63–66

Andjelkovic V, Thompson R (2006) Changes in gene expression in maize kernel in response to water and salt stress. Plant Cell Rep 25(1):71–79

Asnaghi C, Paulet F, Kaye C, Grivet L, Deu M, Glaszmann JC, D'Hont A (2000) Application of synteny across Poaceae to determine the map location of a sugarcane rust resistance gene. Theor Appl Genet 101:962–969

Azpiroz-Leehan R, Feldmann KA (1997) T-DNA insertion mutagenesis in *Arabidopsis*: going back and forth. Trends Genet 13:152–156

Bassani M, Neumann PM, Gepstein S (2004) Differential expression profiles of growth-related genes in the elongation zone of maize primary roots. Plant Mol Biol 56(3):367–380

Benjamin B, Nathalie F, Matt H et al (2010) Linkage and association mapping of *Arabidopsis thaliana* flowering time in nature. PLoS Genet 6(5):e1000940

Berloo RV, Stam P (1999) Comparison between marker-assisted selection and phenotypical selection in a set of *Arabidopsis thaliana* recombinant inbred lines. Theor Appl Genet 98:113–118

Brenda YC, Steve AK (2013) Global approaches for telling time: Omics and the *Arabidopsis* circadian clock. Semin Cell Dev Biol 24(5):383–392

Brizard JP, Carapito C, Delalande F, Van Dorsselaer A, Brugidou C (2006) Proteome analysis of plant–virus interactome: comprehensive data for virus multiplication inside their hosts. Mol Cell Proteomics 5:2279–2297

Brondani C, Brondani RPV, Rangel PHN, Ferreira ME (2001) Development and mapping of Oryza glumaepatula-derived microsatellite markers in the interspecific cross Oryza glumaepatula x O. sativa. Hereditas 134:59–71

Campell BR, Soung Y, Posch TE, Cullis CA, Town CD (1992) Sequence and organization of 5S ribosomal RNA encoding genes of *Arabidopsis thaliana*. Gene 112:225–228

Cao J, Schneeberger K, Ossowski S et al (2011) Whole-genome sequencing of multiple *Arabidopsis thaliana* populations. Nat Genet 43(10):956–963

Carreno-Quintero N et al (2012) Genetic analysis of metabolome–phenotype interactions: from model to crop species. Trends Genet 29:41–50

Carter C, Pan S, Zouhar J, Avila EL, Girke T, Raikhel NV (2004) The vegetative vacuole proteome of *Arabidopsis thaliana* reveals predicted and unexpected proteins. Plant Cell 16:3285–3303

Che KP, Zhan QC, Xing QH, Wang ZP, Jin DM, He DJ, Wang B (2003) Tagging and mapping of rice sheath blight resistant gene. Theor Appl Genet 106:293–297

Chen F, Yuan Y, Li Q, He Z (2007) Proteomic analysis of rice plasma membrane reveals proteins involved in early defense response to bacterial blight. Proteomics 7:1529–1539

Chen ZY, Brown RL, Damann KE, Cleveland TE (2010) PR10 expression in maize and its effect on host resistance against Aspergillus flavus infection and aflatoxin production. Mol Plant Pathol 11(1):69–81

Childs KL, Davidson RM et al (2011) Gene coexpression network analysis as a source of functional annotation for rice genes. PLoS One 6(7):e22196

Christian R, Eysenberg AC, Grieder C, Lisec J, Technow F, Sulpice R, Altmann T, Stitt M, Willmitzer L, Melchinger AE (2012) Genomic and metabolic prediction of complex heterotic traits in hybrid maize. Nat Genet 44:217–220

Claire LG, Li JV et al (2011) Application of NMR-based metabolomics to the investigation of salt stress in maize (Zea mays). Phytochem Analysis 22(3): 214–224

Cooley MB, Goldsbrough AP, Still DW, Yoder JI (1996) Site-selected insertional mutagenesis of tomato with maize Ac and Ds elements. Mol Gen Genet 252:184–194

Copenhaver GP, Browne WE, Preuss D (1998) Assaying genome-wide recombination and centromere functions with *Arabidopsis* tetrads. Proc Natl Acad Sci U S A 95:247–252

Dal'Molin CG, Quek LE et al (2010) C4GEM, a genome-scale metabolic model to study C4 plant metabolism. Plant Physiol 154(4):1871–1885

Das-Chatterjee A, Goswami L, Maitra S, Dastidar KG, Ray S, Majumder AL (2006) Introgression of a novel salt-tolerant L-myo-inositol 1-phosphate synthase from Porteresia coarctata (Roxb.) Tateoka (PcINO1) confers salt tolerance to evolutionary diverse organisms. FEBS Lett 580:3980–3988

Dubcovsky J, Ramakrishna W, SanMiguel PJ, Busso CS, Yan LL, Shiloff BA, Bennetzen JL, Ramakrishna W, Yan LL (2001) Comparative sequence analysis of colinear barley and rice bacterial artificial chromosomes. Plant Physiol 125(3):1342–1353

Farinha AP, Irar S, de Oliveira E, Oliveira MM, Pages M (2011) Novel clues on abiotic stress tolerance emerge from embryo proteome analyses of rice varieties with contrasting stress adaptation. Proteomics 11: 2389–2405

Farré EM, Weise SE (2012) The interactions between the circadian clock and primary metabolism. Curr Opin Plant Biol 15:293–300

Fatma K, Joachim K, Haskell DW, Wei Z, Cameron SK, Nicole G, Dong YS, Charles LG (2004) Exploring the temperature-stress metabolome of *Arabidopsis*. Plant Physiol 136:4159–4168

Feuillet C, Leach JE, Rogers J, Schnable PS, Eversole K (2010) Crop genome sequencing: lessons and rationales. Trends Plant Sci 16(2):77–88
Fiehn O (2002) Metabolomics – the link between genotypes and phenotypes. Plant Mol Biol 48:155–171
Frey M, Stettner C, Gierl A (1998) A general method for gene isolation in tagging approaches: amplification of insertion mutagenized sites (AIMS). Plant J 13:717–721
Fujii K, Hayano Saito Y, Saito K, Sugiura N, Hayashi N, Tsuji T, Izawa T, Iwasaki M (2000) Identification of a RFLP marker tightly linked to the panicle blast resistance gene, Pb1, in rice. Breed Sci 50:183–188
Fukuoka S, Okuno K (2001) QTL analysis and mapping of pi21, a recessive gene for field resistance to rice blast in Japanese upland rice. Theor Appl Genet 103:185–190
Fukuoka S, Saka N, Koga H et al (2009) Loss of function of a protein containing protein confers durable disease resistance in rice. Science 325:998–1001
Gupta HS, Aggarwal PK, Mahajan V et al (2009) Quality protein maize for nutritional security: rapid development of short duration hybrids through molecular marker assisted breeding. Curr Sci 96:230–237
Hagenblad J, Tang C, Molitor J, Werner J, Zhao K, Zheng H, Marjoram P, Weigel D, Nordborg M (2004) Haplotype structure and phenotypic associations in the chromosomal regions surrounding two Arabidopsis thaliana flowering time loci. Genetics 168:1627–1638
Hajheidari M, Eivazi A, Buchanan BB, Wong JH, Majidi I, Salekdeh GH (2007) Proteomics uncovers a role for redox in drought tolerance in wheat. J Proteome Res 6:1451–1460
Halket JM, Waterman D, Przyborowska AM, Patel RK, Fraser PD, Bramley PM (2005) Chemical derivatization and mass spectral libraries in metabolic profiling by GC/MS and LC/MS/MS. J Exp Bot 56:219–243
Henry RJ (2011) Next generation sequencing for accelerating crop domestication. Brief Funct Genomics 2(1):51–56
Hirochika H, Sugimoto K, Otsuki Y, Tsugawa H, Kanda M (1996) Retrotransposons of rice involved in mutations induced by tissue culture. Proc Natl Acad Sci U S A 93:7783–7788
Hosouchi T, Kumekawa N, Tsuruoka H, Kotani H (2002) Physical map-based sizes of the centromeric regions of Arabidopsis thaliana chromosomes 1, 2, and 3. DNA Res 9:117–121
Hu X, Li Y, Li C, Yang H, Wang W, Lu M (2010) Characterization of small heat shock proteins associated with maize tolerance to combined drought and heat stress. J Plant Growth Regul 29:455–464
Huang Z, He G, Shu L, Li X, Zhang Q (2001) Identification and mapping of two brown planthopper resistance genes in rice. Theor Appl Genet 102:929–934
Hunter S, Jones P et al (2012) InterPro in 2011: new developments in the family and domain prediction database. Nucleic Acids Res 40:D306–D312
Izawa T et al (2009) DNA changes tell us about rice domestication. Curr Opin Plant Biol 12:185–192
Jeon JS, Lee S, Jung K et al (2000) T-DNA insertional mutagenesis for functional genomics in rice. Plant J 22(6):561–570
Jia J, Fu J, Zheng J, Zhou X, Huai J, Wang J, Wang M, Zhang Y et al (2006) Annotation and expression profile analysis of 2073 full-length cDNAs from stress-induced maize (*Zea mays* L.) seedlings. Plant J 48(5):710–727
Jiang Y et al (2011) Rice functional genomics research: progress and implications for crop genetic improvement. Biotechnol Adv. doi:10.1016/j.biotechadv.2011.08.013
Jones AME, Thomas V, Truman B, Lilley K, Mansfield J, Grant M (2004) Specific changes in the *Arabidopsis* proteome in response to bacterial challenge: differentiating basal and R-gene mediated resistance. Phytochemistry 65:1805–1816
Jorrín-Novo JV, Maldonado AM, Echevarría-Zomeño S, Valledor L, Castillejo MA, Curto M et al (2009) Plant proteomics update (2007–2008): second-generation proteomic techniques, an appropriate experimental design, and data analysis to fulfill MIAPE standards, increase plant proteome coverage and expand biological knowledge. J Proteomics 72:285–314
Jun K, Kazuo S, Takashi H (2004) Stable isotope labeling of *Arabidopsis thaliana* for an NMR-based metabolomics approach. Plant Cell Physiol 45(8):1099–1104
Jun C, Korbinian S, Stephan O, Torsten G (2011) Whole-genome sequencing of multiple Arabidopsis thaliana populations. Nat Genet 43(10):956–965
Katiyar SK, Tan Y, Huang B, Chandel G, Xu Y, Zhang Y, Xie Z, Bennett J (2001) Molecular mapping of gene Gm-6(t) which confers resistance against four biotypes of Asian rice gall midge in China. Theor Appl Genet 103:6–7
Katja B et al (2008) Genome-scale proteomics reveals *Arabidopsis thaliana* gene models and proteome dynamics. Science 320:938
Kawasaki S, Borchert C, Deyholos M, Wang H et al (2001) Gene expression profiles during the initial phase of salt stress in rice. Plant Cell 13(4):889–905
Kennard WC, Phillips RL, Porter RA, Grombacher AW (2000) A comparative map of wild rice (Zizania palustris L. 2n = 2x = 30). Theor Appl Genet 101:677–684
Khowaja FS, Norton GJ, Courtois B, PricE AH (2009) Improved resolution in the position of drought-related QTLs in a single mapping population of rice by meta-analysis. BMC Genomics 10:276
Kim JK, Takeshi B, Harada K, Fukusaki E, Kobayashi A (2007) Time-course metabolic profiling in *Arabidopsis thaliana* cell cultures after salt stress treatment. J Exp Bot 58(3):415–424
Kim SG, Wang Y, Lee KH, Park ZY, Park J, Wu J et al (2013a) In-depth insight into in vivo apoplastic secretome of rice–Magnaporthe oryzae interaction. J Proteomics 78:58–71

Kim SL, Choi M, Jung K-H, An G (2013b) Analysis of the early-flowering mechanisms and generation of T-DNA tagging lines in Kitaake, a model rice cultivar. J Exp Bot. doi:10.1093/jxb/ert226

Kleffmann T, Russenberger D, Von Zychlinski A, Christopher W, Sjolander K, Gruissem W, Baginsky S (2004) The *Arabidopsis thaliana* chloroplast proteome reveals pathway abundance and novel protein functions. Curr Biol 14:354–362

Klein M, Ute EO, Ina S, Petra B, Michael U, Axel B, Wolfgang S (1994) Physical mapping of the mitochondrial genome of *Arabidopsis thaliana* by cosmid and YAC clones. Plant J 6(3):447–455

Knapp SJ (1994) Selection using molecular marker indexes. In: Proceedings of the 2nd plant breeding symposium of the Crop Science Society of America, American Society for Horticultural Science, Alexandria, pp 1–11

Knapp SJ (1998) Marker assisted selection as a strategy for increasing the probability of selecting superior genotypes. In: Abstracts plant and animal genome conference VI. Scherago international, San Diego, p PA5

Kohli A, Xiong J, Greco R, Christou P, Pereira A (2001) Tagged Transcriptome Display (TTD) in indica rice using Ac transposition. Mol Genet Genom 266(1):1–11

Koornneef M, Von Elden J, Hanhart CJ et al (1983) Linkage map of *Arabidopsis thaliana*. J Hered 74:265–272

Koornneef M, Alonso-Blanco C, Vreugdenhil D (2004) Naturally occurring genetic variation in *Arabidopsis thaliana*. Annu Rev Plant Biol 55:141–172

Kuittinen H, Sillanpa MJ, Savolainen O et al (1997) Genetic basis of adaptation: flowering time in *Arabidopsis thaliana*. Theor Appl Genet 95:573–583

Kunpeng L, Xu C, Zhang J (2011) Proteome profile of maize (Zea mays L.) leaf tissue at the flowering stage after long-term adjustment to rice black-streaked dwarf virus infection. Gene 485(2):106–113

Kusano M, Fukushima A, Kobayashi M et al (2007) Application of a metabolomic method combining one-dimensional and two-dimensional gas chromatography-time-of-flight/mass spectrometry to metabolic phenotyping of natural variants in rice. J Chromatogr 855:71–79

Kusano M, Fukushima A, Redestig H, Saito K (2011) Metabolomic approaches toward understanding nitrogen metabolism in plants. J Exp Bot 62:1432–1453

Lai KS, Nakayama PK, Iwano M, Takayama S (2012) A TILLING resource for functional genomics in *Arabidopsis thaliana* accession C24. Genes Genet Syst 87:291–297

Lande R, Thompson R (1990) Efficiency of marker-assisted selection in the improvement of quantitative traits. Genetics 124:743–756

Li YangSheng, Zheng XianWu, Li-ShaoQing et al. (2001) Quantitative trait loci analysis of leaf morphology in rice. CRRN, Chin Rice Res Newsl 9(1):2–4

Li L, Li H, Li J, Xu S, Yang X, Yan J (2010) A genome-wide survey of maize lipid-related genes: candidate genes mining, digital gene expression profiling and co-location with QTL for maize kernel oil. Sci China Life Sci 53(6):690–700

Liao M, Li Y, Wang Z (2009) Identification of elicitor-responsive proteins in rice leaves by a proteomic approach. Proteomics 9:2809–2819

Linde K, Kastner C, Kumlehn J, Kahmann R, Doehlemann G (2011) Systemic virus-induced gene silencing allows functional characterization of maize genes during biotrophic interaction with Ustilago maydis. New Phytol 189:471–483

Lisec J, Meyer RC, Steinfath M et al (2008) Identification of metabolic and biomass QTL in *Arabidopsis thaliana* in a parallel analysis of RIL and IL populations. Plant J 53:960–972

Lister C, Dean C (1993) Recombinant inbred lines for mapping RFLP and phenotypic markers in *Arabidopsis thaliana*. Plant J 4:745–750

Liu L, Mei Q, Yu Z, Sun T, Zhang Z, Chen M (2013a) An integrative bioinformatics framework for genome-scale multiple level network reconstruction of rice. J Integr Bioinform 10(2):223

Liu Y, Li M, Han C, Wu F, Tu B, Yang P (2013b) Comparative proteomic analysis of rice shoots exposed to high arsenate. J Integr Plant Biol 55(10):965–978

Lu HF, Dong HT, Sun CB, Qing DJ, Li N, Wu ZK, Wang ZQ, Li YZ (2011) The panorama of physiological responses and gene expression of whole plant of maize inbred line YQ7-96 at the three-leaf stage under water deficit and re-watering. Theor Appl Genet 123(6):943–958

Ma LY, Zhuang JY, Liu GJ, Min SK, Li XM, Ma LY, Zhuang JY, Liu GJ, Min SK, Li XM (2002) Mapping of a new gene Wbph 6(t) resistant to whitebacked planthopper (Sogatella furcifera Horvath) in rice. Chin J Rice Sci 16(1):15–18

Ma HY, Song LR, Shu YJ, Wang S, Niu J, Wang ZK et al (2012) Comparative proteomic analysis of seedling leaves of different salt tolerant soybean genotypes. J Proteomics 75:1529–1546

Mahmood T, Jan A, Kakishima M, Komatsu S (2006) Proteomic analysis of bacterial-blight defense-responsive proteins in rice leaf blades. Proteomics 6:6053–6065

Mammadov JA, Chen W, Ren R, Pai R, Marchione W et al (2012) Development of highly polymorphic SNP markers from the complexity reduced portion of maize [Zea mays L.] genome for use in marker-assisted breeding. Theor Appl Genet 121(3):577–588

Marcelo M, Kuhnen S, Lemos PMM, de Oliveira SK, da Silva DA, Tomazzoli MM, Souza ACV, Pinto RM et al (2012) Metabolomics and chemometrics as tools for chemo(bio)diversity analysis – maize landraces and propolis. In: Dr. Kurt V (ed) Chemometrics in practical applications. InTech, China and Croatia. ISBN 978-953-51-0438-4

Maron LG, Kirst M, Mao C, Milner MJ, Menossi M, Kochian LV (2008) Transcriptional profiling of aluminum toxicity and tolerance responses in maize roots. New Phytol 179:116

Masami YH, Mitsuru Y, Dayan BG et al (2004) Integration of transcriptomics and metabolomics for understanding of global responses to nutritional stresses in *Arabidopsis thaliana*. Proc Natl Acad Sci U S A 101(27):10205–10210

Masami YH, Sugiyama K, Sawada Y (2007) Omics-based identification of *Arabidopsis* Myb transcription factors regulating aliphatic glucosinolate biosynthesis. Proc Natl Acad Sci U S A 104(15):6478–6483

McGinnis K, Murphy N, Carlson AR, Akula A, Akula C, Basinger H et al (2007) Assessing the efficiency of RNA interference for maize functional genomics. Plant Physiol 143(4):1441–1451

Mechin V, Balliau T et al (2004) A two-dimensional proteome map of maize endosperm. Phytochemistry 65:1609–1618

Meyer RC, Steinfath M, Lisec J et al (2007) The metabolic signature related to high plant growth rate in *Arabidopsis thaliana*. Proc Natl Acad Sci U S A 104:4759–4764

Moncada P, Martinez CP, Borrero J, Chatel M, Gauch H Jr, Guimaraes E, Tohme J, McCouch SR (2001) Quantitative trait loci for yield and yield components in an Oryza sativa x Oryza rufipogon BC2F2 population evaluated in an upland environment. Theor Appl Genet 102(1):41–52

Morris M, Dreher K, Ribaut JM, Khairallah M (2003) Money matters (II): costs of maize inbred line conversion schemes at CIMMYT using conventional and marker-assisted selection. Mol Breed 11:235–247

Murai H, Hashimoto Z, Sharma PN, Shimizu T, Murata K, Takumi S, Mori N, Kawasaki S, Nakamura C (2001) Construction of a high-resolution linkage map of a rice brown planthopper (*Nilaparvata lugens* Stal) resistance gene bph2. Theor Appl Genet 103:526–532

Murata M, Heslop-Harrison JS, Fusao M (1997) Physical mapping of the 5S ribosomal RNA genes in *Arabidopsis thaliana* by multi-color fluorescence *in situ* hybridization with cosmid clones. Plant J 12(1):31–37

Ni J, Colowit PM, Oster JJ, Xu K, Mackill DJ (2001) Molecular markers linked to stem rot resistance in rice. Theor Appl Genet 102:511–516

Nicole S, Exner V, Probst A, Gruissem W, Hennig L (2006) Functional genomic analysis of CAF-1 mutants in *Arabidopsis thaliana*. J Biol Chem 281:9560–9568

Okushima Y, Overvoorde PJ, Kazunari A et al (2005) Functional genomic analysis of the AUXIN RESPONSE FACTOR gene family members in *Arabidopsis thaliana*: unique and overlapping functions of ARF7 and ARF19. Plant Cell 17:444–463

Pavlík M, Pavlíková D, Vašíčková S (2010) Infrared spectroscopy-based metabolomic analysis of maize growing under different nitrogen nutrition. Plant Soil Environ 56(11):533–540

Pechanova O, Takáč T, Šamaj J, Pechan T (2013) Maize proteomics: an insight into the biology of an important cereal crop. Proteomics 13(3–4):637–662. doi:10.1002/pmic.201200275

Peltier JB, Ytterberg AJ, Sun Q, Klaas JW (2004) New functions of the thylakoid membrane proteome of *Arabidopsis thaliana* revealed by a simple, fast, and versatile fractionation strategy. J Biol Chem 279:49367–49383

Pia GS, Luis OS, Singh KB, Harvey M (2004) Proteomic analysis of glutathione *S*-transferases of *Arabidopsis thaliana* reveals differential salicylic acid-induced expression of the plant-specific Phi and Tau classes. Plant Mol Biol 54(2):205–219

Poroyko V, Spollen WG, Hejlek LG, Hernandez AG, LeNoble ME, Davis G et al (2007) Comparing regional transcript profiles from maize primary roots under well-watered and low water potential conditions. J Exp Bot 58(2):279–289

Prasad SR, Bagali PG, Shailaja H, Shashidhar HE, Hittalmani S (2000) Molecular mapping of quantitative trait loci associated with seedling tolerance to salt stress in rice (Oryza sativa L.). Curr Sci 78:162–164

Priya P, Jain M (2013) RiceSRTFDB: a database of rice transcription factors containing comprehensive expression, cis-regulatory element and mutant information to facilitate gene function analysis. doi:10.1093/database/bat027

Richard MC, Muhammed A, Deborah AW, Prescott MC, Steven MS, Huw HR, Brian T (2010) Differential proteomic analysis of *Arabidopsis thaliana* genotypes exhibiting resistance or susceptibility to the insect herbivore, *Plutella xylostella*. PLoS One 5(4):e10103, 1–14

Ragot M, Biasiolli M et al. (1995) Marker-assisted backcrossing: a practical example. Techniques et utilisations des marqueurs moléculaires, Les Colloques, INRA, Paris 72: 45–56

Ralf S, De Vos RCH, Bartelniewoehner L, Ishihara H, Sagasser M, Martens S, Weisshaar B (2009) Metabolomic and genetic analyses of flavonol synthesis in *Arabidopsis thaliana* support the in vivo involvement of leucoanthocyanidin dioxygenase. Planta 229(2):427–445

Ryo N, Miyako K, Makoto K et al (2009) Metabolomics-oriented isolation and structure elucidation of 37 compounds including two anthocyanins from *Arabidopsis thaliana*. Phytochemistry 70(8):1017–1029

Sa KJ, Park JY, Park KC, Lee JK (2012) Analysis of genetic mapping in a waxy/dent maize RIL population using SSR and SNP markers. Genes Genomics 34(2): 157–164

Saito K, Miura K, Nagano K, Hayano Saito Y, Araki H, Kato A (2001) Identification of two closely linked quantitative trait loci for cold tolerance on chromosome 4 of rice and their association with anther length. Theor Appl Genet 103:862–868

Salekdeh GH, Siopongco J, Wade LJ, Ghareyazie B, Bennett J (2002) A proteomic approach to analyzing

drought- and salt-responsiveness in rice. Field Crop Res 76:199–219
Sanchez AC, Brar DS, Huang N, Li Z, Khush GS (2000) Sequence tagged site marker-assisted selection for three bacterial blight resistance genes in rice. Crop Sci 40:792–797
Sang T, Ge S (2007) The puzzle of rice domestication. J Integr Plant Biol 49:760–768
Sardesai N, Kumar A, Rajyashri KR, Nair S, Mohan M (2002) Identification and mapping of an AFLP marker linked to Gm7, a gall midge resistance gene and its conversion to a SCAR marker for its utility in marker aided selection in rice. Theor Appl Genet 105:691–698
Sato S, Hirokazu K, Reiko H, Liu Y-G, Daisuke S, Satoshi T (1998) A physical map of *Arabidopsis thaliana* chromosome 3 represented by two contigs of CIC YAC, PI, TAC and BAC clones. DNA Res 5:163–168
Sato S, Nakamura Y, Kaneko T, Asamizu E, Tabata S (1999) Complete structure of the chloroplast genome of *Arabidopsis thaliana*. DNA Res 6:283–290
Schadt EE, Monks SA, Drake TA, Lusis AJ (2003) Genetics of gene expression surveyed in maize, mouse and man. Nature 422(6929):297–302
Schmidt R, West J, Love K et al (1995) Physical map and organization of *Arabidopsis thaliana* chromosome 4. Science 270:480–483
Sekhon RS, Lin H, Childs KL et al (2011) Genome-wide atlas of transcription during maize development. Plant J 66(4):553–563
Sengupta S, Majumder AL (2009) Insight into the salt tolerance factors of a wild halophytic rice, Porteresia coarctata: a physiological and proteomic approach. Planta 229:911–929
Septiningsih EM, Pamplona AM, Sanchez DL, Neeraja CN, Vergara GV, Heuer S, Ismail AM, Mackill DJ (2009) Development of submergence-tolerant rice cultivars: the Sub1 locus and beyond. Ann Bot 103:151–160
Shen L, Courtois B, McNally KL, Robin S, Li Z (2001) Evaluation of near isogenic lines of rice introgressed with QTLs for root depth through marker-aided selection. Theor Appl Genet 103:75–83
Shi LY, Hao ZF, Weng JF, Xie CX et al (2012) Identification of a major quantitative trait locus for resistance to maize rough dwarf virus in a Chinese maize inbred line X178 using a linkage map based on 514 gene-derived single nucleotide polymorphisms. Mol Breed 30:1–11
Shimaoka T, Ohnishi M, Sazuka T, Mitsuhashi N, Hara-Nishimura I, Shimazaki KI, Maeshima M, Yokota A, Tomizawa KI, Mimura T (2004) Isolation of intact vacuoles and proteomic analysis of tonoplast from suspension-cultured cell of *Arabidopsis thaliana*. Plant Cell Physiol 45:672–683
Shu XL, Frank T, Shu QY, Engel KH (2008) Metabolite profiling of germinating rice seeds. J Agric Food Chem 56:11612–11620
Shulaev V (2006) Metabolomics technology and bioinformatics. Brief Bioinform 7:128–139
Siangliw M, Toojinda T, Tragoonrung S, Vanavichit A (2003) Thai jasmine rice carrying QTLch9 (SubQTL) is submergence tolerant, Flooding and plant growth: Selected papers from the 7th conference of the International Society for Plant Anaerobiosis (ISPA), Nijmegen, 12–16 June 2001. Annals Bot 91(Special Issue):255–261
Simon M, Olivier L, Stéphanie D et al (2008) Quantitative trait loci mapping in five new large recombinant inbred line populations of *Arabidopsis thaliana* genotyped with consensus single-nucleotide polymorphism markers. Genetics 178:2253–2264
Singh S, Sidhu JS, Huang N, Vikal Y, Li Z, Brar DS, Dhaliwal HS, Khush GS (2001) Pyramiding three bacterial blight resistance genes (xa5, xa13 and Xa21) using marker-assisted selection into indica rice cultivar PR106. Theor Appl Genet 102:1011–1015
Smilde WD, Haluskova J, Sasaki T, Graner A (2001) New evidence for the synteny of rice chromosome 1 and barley chromosome 3H from rice expressed sequence tags. Genome 44:361–367
Snape JW, Sarma R, Quarrie SA, Fish L, Galiba G, Sutka J (2001) Mapping genes for flowering time and frost tolerance in cereals using precise genetic stocks. Euphytica 120:309–315
Soderlund C, Descour A, Kudrna D (2009) Sequencing, mapping, and analysis of 27,455 maize full-length cDNAs. PLoS Genet 5(11):e1000740
Spollen WG, Tao W, Valliyodan B et al (2008) Spatial distribution of transcript changes in the maize primary root elongation zone at low water potential. BMC Plant Biol 8:32
Sripongpangkul K, Posa GBT, Senadhira DW, Brar D, Huang N, Khush GS, Li ZK (2000) Genes/QTLs affecting flood tolerance in rice. Theor Appl Genet 101:1074–1081
Stuber CW, Edwards MD, Wendel JF (1987) Molecular marker facilitated investigations of quantitative traits loci in maize. II. Factors influencing yield and its component traits. Crop Sci 27:639–648
Sundaram RM, Vishnupriya MR, Biradar SK, Laha GS, Reddy GA, Rani NS, Sarma NP, Sonti RV (2008) Marker assisted introgression of bacterial blight resistance in Samba Mahsuri, an elite indica rice variety. Euphytica 160:411–422
Swidzinski JA, Leaver CJ, Sweetlove LJ (2004) A proteomic analysis of plant programmed cell death. Phytochemistry 65:1829–1838
Szponarski W, Sommerer N, Boyer JC, Rossignol M, Gibrat R (2004) Large-scale characterization of integral proteins from *Arabidopsis* vacuolar membrane by two-dimensional liquid chromatography. Proteomics 4:397–406
Takahara K, Kasajima I, Takahashi H, Hashida S, Itami T, Onodera H, Toki S, Yanagisawa S, Kawai-Yamada M, Uchimiya H (2010) Metabolome and photochemical analysis of rice plants overexpressing Arabidopsis NAD kinase gene. Plant Physiol 152:1863–1873

Takayuki T, Keiko Yonekura S et al (2007) Phytochemical genomics in *Arabidopsis thaliana*: a case study for functional identification of flavonoid biosynthesis genes. Pure Appl Chem 79(4):811–823

Takeuchi Y, Hayasaka H, Chiba B, Tanaka I, Shimano T, Yamagishi M, Nagano K, Sasaki T, Yano M (2001) Mapping quantitative trait loci controlling cool-temperature tolerance at booting stage in temperate Japonica rice. Breed Sci 51:191–197

Tao Q, Chang YL, Wang JZ et al (2001) Bacterial artificial chromosome-based physical map of the rice genome constructed by restriction fingerprint analysis. Genetics 158:1711–1724

Tarpley L, Duran A, Kebrom T, Sumner L (2005) Biomarker metabolites capturing the metabolite variance present in a rice plant developmental period. BMC Plant Biol 5:8

Temnykh S, DeClerck G, Lukashova A, Lipovich L, Cartinhour S, McCouch S (2001) Computational and experimental analysis of microsatellites in rice (Oryza sativa L.): frequency, length variation, transposon associations, and genetic marker potential. Genome Res 11:1441–1452

The *Arabidopsis* Genome Initiative (2000) Analysis of the genome sequence of the flowering plant *Arabidopsis thaliana*. Nature 408:796–815

The *C. elegans* Sequencing Consortium (1998) Genome sequence of the nematode *C. elegans*: a platform for investigating biology. Science 282:2012–2046

Tom PJD, Svenja H, Shadforth IP, John R et al (2006) Mapping the *Arabidopsis* organelle proteome. Proc Natl Acad Sci U S A 103(17):6518–6523

Toojinda T, Siangliw M, Tragoonrung S, Vanavichit A (2003) Molecular genetics of submergence tolerance in rice: QTL analysis of key traits, flooding and plant growth: selected papers from the 7th conference of the International Society for Plant Anaerobiosis (ISPA), Nijmegen, 12–16 June 2001. Annals Bot 91:243–253

Varshney RK, Nayak SN, May GD et al (2009) Next generation sequencing technologies and their implications for crop genetics and breeding. Trends Biotechnol 27:522–530

Ventelon-Debout M, Delalande F, Brizard JP, Diemer H, Van Dorsselaer A, Brugidou C (2004) Proteome analysis of cultivar-specific deregulations of Oryza sativa indica and O. sativa japonica cellular suspensions undergoing rice yellow mottle virus infection. Proteomics 4:216–225

Wakasa K, Hasegawa H, Nemoto H, Matsuda F et al (2006) High-level tryptophan accumulation in seeds of transgenic rice and its limited effects on agronomic traits and seed metabolite profile. J Exp Bot 57:3069–3078

Wang W, Zhai W et al (2001) Chromosome landing at the bacterial blight resistance gene Xa4 locus using a deep coverage rice BAC library. Mol Genet Genomics 265:118–125

Wang H, Miyazaki S, Kawai K, Deyholos M, Galbraith DW, Bohnert HJ (2003) Temporal progression of gene expression responses to salt shock in maize roots. Plant Mol Biol 52(4):873–891

Ward ER, Jen GC (1990) Isolation of single copy sequence clones from a yeast artificial chromosome library of randomly-sheared *Arabidopsis thaliana* DNA. Plant Mol Biol 14:561–568

Ward JL, Harris C, Lewis J, Beale MH (2003) Assessment of 1H NMR spectroscopy and multivariate analysis as a technique for metabolite fingerprinting of *Arabidopsis thaliana*. Phytochemistry 62:949–957

Wassom JJ, Mikkelineni V, Bohn MO, Rocheford TR (2008) QTL for fatty acid composition of maize kernel oil in Illinois high oil 9 B73 backcross-derived lines. Crop Sci 48:69–78

Wilkins MR, Sanchez JC, Gooley AA, Appel RD, Smith IH, Hochstrasser DF et al (1996) Progress with proteome projects: why all proteins expressed by a genome should be identified and how to do it. Biotechnol Genet Eng Rev 13:19–50

Willcox MC, Khairallah MM, Bergvinson D et al (2002) Selection for resistance to southwestern corn borer using marker-assisted and conventional backcrossing. Crop Sci 42:1516–1528

Witzel K, Weidner A, Surabhi GK, Varshney RK, Kunze G, Buck-Sorlin GH et al (2010) Comparative analysis of the grain proteome fraction in barley genotypes with contrasting salinity tolerance during germination. Plant Cell Environ 33:211–222

Wolfe KH (2001) Yesterday's polyploids and the mystery of diploidization. Nat Rev Genet 2:333–341

Yamamoto K, Sasaki T (1997) Large-scale EST sequencing in rice. Plant Mol Biol 35:135–144

Yamamoto T, Taguchi Shiobara F, Ukai Y, Sasaki T, Yano M (2001) Mapping quantitative trait loci for days-to-heading, and culm, panicle and internode lengths in a BC1F3 population using an elite rice variety, Koshihikari, as the recurrent parent. Breed Sci 51:63–71

Yamasaki M, Yoshimura A, Yasui H (2000) Mapping of quantitative trait loci of ovicidal response to brown planthopper (Nilaparvata lugens Stal) in rice (Oryza sativa L.). Breed Sci 50:291–296

Yu LX, Setter TL (2003) Comparative transcriptional profiling of placenta and endosperm in developing maize kernels in response to water deficit. Plant Physiol 131(2):568–582

Yun KY, Park MR et al (2010) Transcriptional regulatory network triggered by oxidative signals configures the early response mechanisms of japonica rice to chilling stress. BMC Plant Biol 10:16

Zachgo EA, Wang ML, Dewdney J et al (1996) A physical map of chromosome 2 of *Arabidopsis thaliana*. Genome Res 6:19–25

Zhang J, Zheng HG et al (2001) Locating genomic regions associated with components of drought resistance in rice: comparative mapping within and across species. Theor Appl Genet 103:19–29

Zhang J, Guo D, Chang YX, You CJ, Li XW, Dai XX et al (2007) Non-random distribution of TDNA insertions at various levels of the genome hierarchy as revealed by analyzing 13804 T-DNA flanking sequences from an enhancer-trap mutant library. Plant J 49:947–959

Zhang P, Dreher K et al (2010) Creation of a genome-wide metabolic pathway database for Populus trichocarpa using a new approach for reconstruction and curation of metabolic pathways for plants. Plant Physiol 153(4):1479–1491

Zheng J, Zhao J, Tao Y, Wang J, Liu Y, Fu J, Jin Y (2004) Isolation and analysis of water stress induced genes in maize seedlings by subtractive PCR and cDNA macroarray. Plant Mol Biol 55(6):807–823

Zhou Y, Li W, Wu W, Chen Q, Mao D, Worland AJ (2001) Genetic dissection of heading time and its components in rice. Theor Appl Genet 102:1236–1242

Zhou ML, Zhang Q, Zhou M, Qi LP et al (2012) Aldehyde dehydrogenase protein superfamily in maize. Funct Integr Genomics 12:683–691

Zhuang Y, Ren G, Yue G, Li Z, Qu X, Hou G, Zhu Y, Zhang J (2007) Effects of water-deficit stress on the transcriptomes of developing immature ear and tassel in maize. Plant Cell Rep 26:2137–2147

Zinselmeier C, Lauer MJ, Boyer JS (1995) Reversing drought-induced losses in grain yield: sucrose maintains embryo growth in maize. Crop Sci 35:1390–1400

Zinselmeier C, Sun Y, Helentjaris T, Beatty M, Yang S, Smith H, Habben J (2002) The use of gene expression profiling to dissect the stress sensitivity of reproductive development in maize. Field Crop Res 75:111–121

Instrumental Techniques and Methods: Their Role in Plant Omics

Daniel Cozzolino, Alberto Fassio, Ernesto Restaino, and Esteban Vicente

Contents

Abstract

Techniques and methods based on vibrational spectroscopy such as near-infrared reflectance (NIR), mid-infrared (MIR) and Raman spectroscopy are known to be non-destructive and low cost. These characteristics are considered as the most important when these methods or techniques are applied in the field of plant omics. This chapter will provide an overview of the most common vibrational spectroscopy techniques used in the field of plant omic analysis (NIR, MIR, Raman). Information about the hardware (instruments) and software (multivariate data methods) will be also presented and discussed.

Keywords

Rapid methods • NIR • MIR • Raman • Spectroscopy • Chemometrics • Plant omics

D. Cozzolino, Ph.D. (✉)
School of Agriculture, Food and Wine,
The University of Adelaide, Waite Campus,
PMB 1 Glen, Osmond, SA 5064, Australia
e-mail: d.cozzolino@adelaide.edu.au

A. Fassio, Ph.D. • E. Restaino, Ph.D.
National Institute for Agricultural Research,
INIA La Estanzuela, Estacion Experimental
Dr Alberto Boerger, Colonia, Uruguay

E. Vicente, Ph.D.
National Institute for Agricultural Research,
INIA Salto Grande, Salto, Uruguay

D. Barh et al. (eds.), *PlantOmics: The Omics of Plant Science*,
DOI 10.1007/978-81-322-2172-2_2, © Springer India 2015

Introduction

The use of new rapid and non-invasive techniques to fingerprint compositional, physiological or biochemical properties that can be used to assist plant scientists, researchers and farmers to target sustainable agriculture is of paramount importance in modern plant science. These techniques can be used also as tools in developing functional databases that can be used in breeding, selection or adaptation of new plant varieties to hostile environments or new production systems. Consequently, the combination of fundamentals in plant science, multivariate data analysis and vibrational spectroscopy methods will enable the development of technologies rapid, non-destructive and low-cost testing in the so-called high-throughput methods or plant omics (Cabrera-Bosquet et al. 2012; White et al. 2012; Walter et al. 2012).

One of the central principles of biology is the concept that a set of genetic instructions, or genotype, interacts with the environment to produce the characteristics, or phenotype, of an organism (Cabrera-Bosquet et al. 2012; White et al. 2012; Walter et al. 2012). Understanding how particular genotypes result in specific phenotypic properties is a core goal of modern biology and enables development of organisms with commercially useful characteristics (Abberton and Marshall 2005; White et al. 2012; Walter et al. 2012). However, prediction of phenotype or plant omics from genotype is generally a difficult task due to the large number of genes and gene products that contribute to most phenotypes in concert with complex and changeable environmental influences (Cabrera-Bosquet et al. 2012; White et al. 2012; Walter et al. 2012).

The last 20 years have created a revolution in our understanding of genotype: while genomes typically are quite large, with millions or billions of nucleotides, the relative chemical simplicity of DNA lends itself to large-scale analysis (Cabrera-Bosquet et al. 2012; Walter et al. 2012). Scientist can now determine genotypes down to the level of individual nucleotides in whole genomes, and the entire genomes are now rapidly sequenced at steadily declining costs and ever-increasing speed (Cabrera-Bosquet et al. 2012; Walter et al. 2012). Next-generation re-sequencing methods provide opportunities to get the complete genotype not only of a single representative of a genus or species but of many representatives of a phylogenetic group or population (Walter et al. 2012). Therefore, genome science is moving beyond the era of reference and a model organism to study in depth any microbe, animal or plant that has characteristics of interest to science and society (Kolukisaoglu and Thurow 2010; Cabrera-Bosquet et al. 2012; Walter et al. 2012).

The term 'omic' is derived from the Latin suffix 'ome' meaning mass or many. The development of metabolomics has depended on advances in a diverse range of instrumental techniques such as liquid chromatography (LC), electrospray ionisation mass spectrometry (ESI-MS), capillary electrophoresis (CE), gas chromatography (GC) and microchip arrays, among other techniques (Sumner et al. 2003; Sweetlove et al. 2004; Dunn and Ellis 2005; Hounsome et al. 2008; Ryan and Robach 2006; Cozzolino 2011, 2012a, b).

Chemical and physiological properties are related to individual or bioactive compounds such as essential oils, terpenoids, flavonoids, volatile compounds and other chemicals which are present in natural products in low concentrations (e.g. ppm or ppb) (McGoverin et al. 2010; Cozzolino 2011; 2012a, b). For many years, the use of classical separation, chromatographic and spectrometric techniques such as high-performance liquid chromatography (HPLC), GC, LC and mass spectrometry (MS) has been used for the elucidation of isolated compounds from different matrices (McGoverin et al. 2010; Cozzolino 2011, 2012a, b). While the use of standard separation, chromatographic and spectrometric methods were found useful in chemical and in both animal and plant physiology studies for fingerprinting as well as to comparing natural and synthetic samples and to identify single

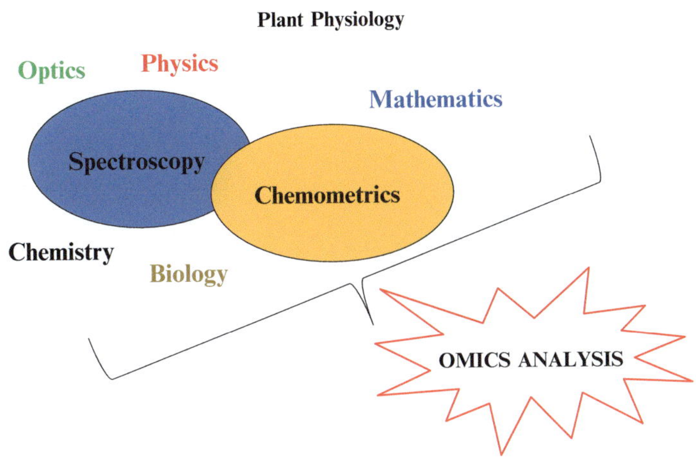

Fig. 1 An integration of different disciplines used in the development of plant omics analysis

active compounds (Cozzolino 2011, 2012a, b). Each of these methods provides unique capabilities to separate different chemical classes of metabolites from several types of samples. At the same time, developments in mathematics, statistics, software and computers have provided algorithms (e.g. principal component analysis, partial least squares, discriminant analysis) and the computational power that enable the analysis and interpretation of complex data sets generated. (Cozzolino 2009, 2011, 2012a, b).

Techniques and methods based on vibrational spectroscopy such as near-infrared reflectance (NIR), mid-infrared (MIR) and Raman spectroscopy are known to be non-destructive and low cost, and they might be considered to be the one of the most appropriate methods or techniques to be used in the field of plant omics (McClure 2004; Montes et al. 2007; Cozzolino 2011, 2012a, b; Smyth and Cozzolino 2011, 2013). Figure 1 shows a simplified scheme of the combination of chemistry, optics, biology and mathematics in the so-called 'omics' methodology applied to plant sciences.

This chapter will provide an overview of vibrational spectroscopy techniques (e.g. NIR, MIR, Raman), instrumentation and multivariate data methods used in plant omic analysis as well as present and discuss applications in the field with some examples.

Vibrational Spectroscopy Methods: NIR, MIR and Raman

Chemical bonds present in the organic matrix of any organic matrix (e.g. plants, vegetables, among others) vibrate at specific frequencies, which are determined by the mass of the constituent atoms, the shape of the molecule, the stiffness of the bonds and the periods of the associated vibrational coupling (Woodcock et al. 2008; Karoui et al. 2010; Rodriguez-Saona and Allendorf 2011). A specific vibrational bond absorb in the infrared (IR) spectral region where diatomic molecules have only one bond that may stretch (e.g. the distance between two atoms may increase or decrease). More complex molecules have many bonds; vibrations can also be conjugated leading to two possible modes of vibration: stretching and bending. Figure 2 shows both the NIR and MIR spectra of starch, amylose and amylopectin from potato (*Solanum tuberosum*). These examples showed how different compounds having specific fingerprint in either wavelength region can be used to analyse plant tissues in the field of omics.

Despite these potential problems, absorption frequencies may be used to identify specific chemical groups, and this capability has traditionally been the main role of Fourier transform (FT)

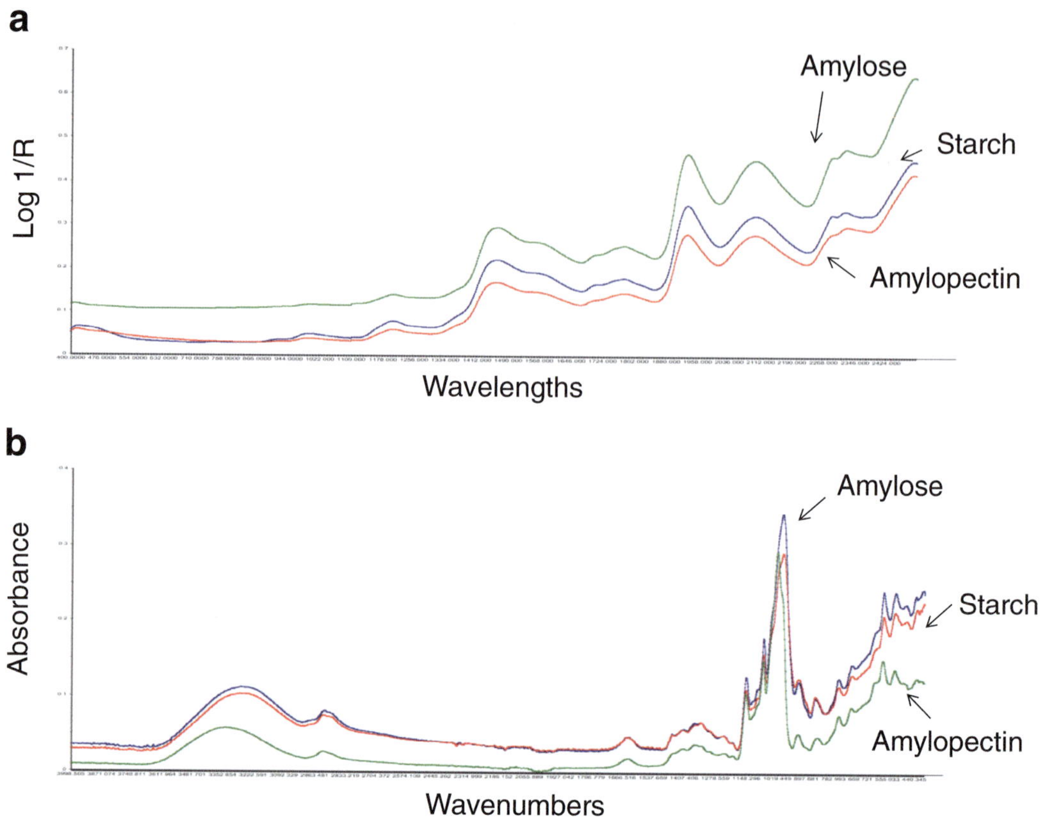

Fig. 2 Near-infrared (**a**) and mid-infrared (**b**) spectra of amylose, amylopectin and starch derived from potato (*Solanum tuberosa*)

mid-infrared (MIR) (FT-MIR) spectroscopy (Woodcock et al. 2008; Karoui et al. 2010; Rodriguez-Saona and Allendorf 2011). Infrared (IR) radiation is the region of the electromagnetic spectrum between the visible (VIS) and the microwave wavelengths (McClure 2004; Blanco and Villaroya 2002; Cozzolino 2009, 2011, 2012a, b).

The nominal range of wavelengths for near-infrared (NIR) is between 750 and 2,500 nm (13,400 to 4,000 cm^{-1}), while for the mid-infrared (MIR) is from 2,500 to 25,000 nm (4,000 to 400 cm^{-1}) (McClure 2004; Blanco and Villaroya 2002; Cozzolino 2009, 2011). NIR spectroscopy is characterised by low molar absorptivities and scattering, which allow nearly effortless evaluation of pure materials.

Although Herschel discovered light in the near-infrared region as early as 1800, spectroscopic scientist of the first half of the last century ignored it, in the belief that it lacked analytical interest (McClure 2004; Cozzolino 2009). The NIR region of the electromagnetic spectrum, once regarded as having little potential for analytical work, has now become one of the most promising techniques for molecular spectroscopy. Broadly speaking, NIR spectroscopy assesses organic chemical structures containing O-H, N-H and C-H bonds through the absorption of energy in the NIR region of the spectrum (Huang et al. 2008).

On the other hand, spectral 'signatures' in the MIR result from the fundamental stretching, bending and rotating vibrations of the sample molecules, while NIR spectra result from complex overtones and high-frequency combinations at shorter wavelengths (McClure 2004; Blanco and Villaroya 2002; Roggo et al. 2007; Cozzolino 2009, 2011). The MIR region of the electromagnetic spectrum lies between 4,000 and 400 cm^{-1} and can be segmented into four broad regions: the X-H stretching region (4,000–2,500 cm^{-1}), the triple-bond region (2,500–2,000 cm^{-1}), the double-bond region (2,000–1,500 cm^{-1}) and the fingerprint region (1,500–400 cm^{-1}) (Woodcock et al. 2008; Karoui et al. 2010; Rodriguez-Saona and Allendorf 2011). Spectral peaks in the MIR frequencies are often sharper and better resolved than in the NIR domain; the higher overtones of the O-H (oxygen-hydrogen), N-H (nitrogen-hydrogen), C-H (carbon-hydrogen) and S-H (sulphur-hydrogen) bands from the MIR wavelengths are still observed in the NIR region, although much weaker than the fundamental frequencies in the MIR. In addition to the NIR spectrum also contains many overlapped bands (McClure 2004; Nicolai et al. 2007; Cozzolino 2009, 2012a, b; Smyth and Cozzolino 2011, 2013).

Although NIR intensities are 10–1,000 times lower than for the MIR, highly sensitive spectrometers can be built through several means including the use of efficient detectors and brighter light sources (McClure 2004; Blanco and Villaroya 2002; Roggo et al. 2007; Cozzolino 2009; Smyth and Cozzolino 2011). In recent years, new instrumentation and computer algorithms (chemometrics) have taken advantage of this complexity and have made the technique much more powerful and simple to use (McClure 2004; Roggo et al. 2007; Nicolai et al. 2007). The arrival of inexpensive and powerful computers has contributed to the surge of wide range of NIR applications in several fields, from medical, pharmaceutical, to the most traditional of food analysis (McClure 2004; Blanco and Villaroya 2002; Roggo et al. 2007; Huang et al. 2008; Cozzolino 2009). The principle of NIR analysis is that the product is irradiated with NIR radiation and the reflected or transmitted radiation is measured. While the radiation penetrates the product, its spectral characteristics change through wavelength-dependent scattering and absorption processes (Nicolai et al. 2007). This change depends on the chemical composition and physical characteristics of the material, as well as on its light scattering properties which are related to the structure.

Reflection is due to three different phenomena: specular reflection, external diffuse reflection and scattering (Nicolai et al. 2007). Most absorption bands in the NIR region are overtone or combination bands of the fundamental absorption bands in the IR region of the electromagnetic spectrum which are due to vibrational and rotational transitions. In large molecules and in complex mixtures, such as foods, the multiple bands and the effect of peak-broadening result in NIR spectra that have a broad envelope with few sharp peaks (Nicolai et al. 2007). However, a major disadvantage of this characteristic overlap, and complexity in the NIR spectra has been the difficulty of quantification and interpretation of data from NIR spectra. On the other hand, the broad overlapping bands can diminish the need for using a large number of wavelengths in calibration and analysis routines. In recent years, new instrumentation and computer algorithms have taken advantage of this complexity and have made the technique much more powerful and simple to use. However, the advent of inexpensive and powerful computers has contributed to the surge of new NIR applications (Nicolai et al. 2007; Cozzolino 2009; Smyth and Cozzolino 2011).

Raman spectroscopy is based on fundamental vibration modes that can be assigned to specific chemical functional groups within a sample molecule and therefore can potentially provide useful information for fingerprinting or qualitative identification (Baranska et al. 2004; Baranska and Schultz 2006; Schulz et al. 2005; Li-Chan 2010). Raman spectroscopy has been also explored as emerging methodology in natural products research and development (Baranska et al. 2004; Baranska and Schultz 2006; Schulz et al. 2005; Li-Chan 2010). Raman spectroscopy, in similar way NIR and MIR, is a constantly developing techniques that also allowed for the rapid, non-destructive, reagent less and high throughput for a diverse range of sample types. Very recently, Raman has also been introduced as a metabolic fingerprinting technique within the plant sciences (Baranska et al. 2004; Baranska and Schultz

Table 1 Characteristics of near-infrared (NIR), mid-infrared (MIR) and Raman spectroscopy

Characteristics	NIR	MIR	Raman
Sample presentation (scanning)	*Easy*	*Needs detailed preparation (except ATR)*	*Easy, similar to NIR*
Path lengths (mm)	*Long path lengths*	*Narrow path length*	*Short path length*
Spectra characteristics	*Highly convoluted (combinations)*	*Fundamental vibrations bands*	
Structure determination (molecular level)	*No*	*Very Good*	*Limited*
Fingerprint identification	*No/yes*	*Yes*	*Limited*
Quantitative analysis (calibration and prediction)	*Yes*	*Yes*	*Yes*
Sample Area	*Large to small*	*Moderate to small*	*Small*

NIR near infrared, *MIR* mid-infrared, *ATR* attenuated total reflectance

2006; Schulz et al. 2005; Li-Chan 2010). Raman spectroscopy coupled with microscopy is recently used with great success, making possible the identification and quantification of photochemical distribution directly from the plant tissues (Baranska et al. 2004; Baranska and Schultz 2006; Schulz et al. 2005; Li-Chan 2010). Despite the advantages of Raman spectroscopy, when compared with MIR or NIR, the development of this technique has been less extensive and slower due to several reasons such as instrumentation, high cost, among other issues (Li-Chan 2010). Table 1 summarises the characteristics of NIR, MIR and Raman spectroscopy for plant omic analysis.

Instrumentation

An overall introduction to the different available platforms or instruments commonly used in plant omics is given below. Spectroscopy combined with chemometric methods has the potential to be a powerful tool for the assessment of chemical composition of any given organic substance (McClure 2004; Blanco and Villaroya 2002; Roggo et al. 2007; Nicolai et al. 2007; Huang et al. 2008; Cozzolino 2009; Smyth and Cozzolino 2011). Recent advances in chemometrics software and computing power have greatly enhanced the development of rapid analytical methods based on spectroscopic data and their subsequent application in a wide range of industries (e.g. agriculture, pharmaceutical, petrochemical) (McClure 2004; Blanco and Villaroya 2002; Roggo et al. 2007; Nicolai et al. 2007; Huang et al. 2008; Cozzolino 2009, 2011), although the instrumentation may require a large capital outlay (up to 150 thousand American dollars) and can be reasonably complex to calibrate and maintain (McClure 2004; Blanco and Villaroya 2002; Roggo et al. 2007; Nicolai et al. 2007; Huang et al. 2008; Cozzolino 2009).

As the technology of spectroscopic instrumentation and chemometrics advances further, the resulting spin-offs may further assist to define and objectively measure chemical composition in different plant tissues (McClure 2004; Blanco and Villaroya 2002; Roggo et al. 2007; Nicolai et al. 2007; Huang et al. 2008; Cozzolino 2009). An NIR spectrophotometer consists of a light source (e.g. a tungsten halogen lamp), sample presentation accessories, monochromator, detector and optical components such as lenses, collimators, beam splitters, integrating spheres and optical fibres (McClure 2004; Blanco and Villaroya 2002; Roggo et al. 2007; Nicolai et al. 2007; Huang et al. 2008; Cozzolino 2009). Basically, two main methods are used to collect the NIR spectra of a given sample, namely, reflectance and transmittance. In reflectance (e.g. generally used to analyse solids or powder samples) mode, the light source and detector are mounted under a specific angle, e.g. 45°, to avoid specular reflection. In transmittance (e.g. generally used to analyse liquid samples) mode, the light source is positioned opposite to the detector, while in interactance mode the light source and detector are positioned parallel to each other in such a way that light due to specular reflection cannot directly enter the detector (McClure 2004; Blanco and Villaroya 2002; Roggo et al. 2007; Nicolai et al. 2007; Huang et al. 2008; Cozzolino 2009).

In both reflectance and transmittance modes, integrated spheres may also be used to collect light and increase the signal-to-noise ratio. Transmission measurements, on the other hand, need very high light intensities which can easily burn the sample surface and alter its spectral properties (McClure 2004; Blanco and Villaroya 2002; Roggo et al. 2007; Nicolai et al. 2007; Huang et al. 2008; Cozzolino 2009). Spectrophotometers are conveniently classified according to the type of monochromator. For example, in a filter instrument, the monochromator is a wheel holding a number of absorption or interference filters, while its spectral resolution is limited (McClure 2004; Blanco and Villaroya 2002; Roggo et al. 2007; Nicolai et al. 2007; Huang et al. 2008; Cozzolino 2009).

In a scanning monochromator instrument, a grating or a prism is used to separate the individual frequencies of the radiation either entering or leaving the sample. The wavelength divider rotates, allowing radiation of the individual wavelengths to subsequently reach the detector (McClure 2004; Blanco and Villaroya 2002; Roggo et al. 2007; Nicolai et al. 2007; Huang et al. 2008; Cozzolino 2009).

In Fourier transform spectrophotometers, interferometers are used to generate modulated light, while the time domain signal of the light reflected or transmitted by the sample onto the sample can be converted into a spectrum via a fast Fourier transform (McClure 2004; Blanco and Villaroya 2002; Roggo et al. 2007; Nicolai et al. 2007; Huang et al. 2008; Cozzolino 2009). Often a Michelson interferometer is used, but also polarisation interferometers can be mounted in some spectrophotometers (McClure 2004; Griffiths and De Haseth 1986; Hashimoto and Kameoka 2008; Subramanian and Rodrigez-Saona 2009).

Recently, photodiode array (PDA) spectrophotometers have been introduced. They consist of a fixed grating that focuses the dispersed radiation onto an array of silicon (Si, 350–1,100 nm) or indium gallium arsenide (InGaAs, 1,100–2,500 nm) photodiode detectors (McClure 2004; Blanco and Villaroya 2002; Roggo et al. 2007; Nicolai et al. 2007; Huang et al. 2008; Cozzolino 2009). There is definitely a shift towards PDA systems because of their high acquisition speed (the integration time is typically 50 ms but can be as low as a few milliseconds) and the absence of moving parts which enables them to be mounted on online fruit grading lines (McClure 2004; Nicolai et al. 2007).

Miniaturised, portable and low-cost versions (less than 20 thousand American dollars) are available from several companies. Laser-based systems do not contain a monochromator but have different laser light sources or a tunable laser. Acousto-optic tunable filter (AOTF) (Stratis et al. 2001) instruments use a diffraction based optical-band-pass filter that can be rapidly tuned to pass various wavelengths of light by varying the frequency of an acoustic wave propagating through an anisotropic crystal medium (McClure 2004; Blanco and Villaroya 2002; Roggo et al. 2007; Nicolai et al. 2007; Huang et al. 2008; Cozzolino 2009).

Liquid crystal tunable filter (LCTF) instruments have been introduced (McClure 2004; Blanco and Villaroya 2002; Roggo et al. 2007; Nicolai et al. 2007; Huang et al. 2008; Cozzolino 2009). They use a birefringent filter to create constructive and destructive interference based on the retardation, in phase between the ordinary and extraordinary light rays passing through a liquid crystal. In this way they act as an interference filter to pass a single wavelength of light (McClure 2004; Blanco and Villaroya 2002; Roggo et al. 2007; Nicolai et al. 2007; Huang et al. 2008; Cozzolino 2009). By combining several electronically tunable stages in series, high spectral resolution can be achieved (McClure 2004; Blanco and Villaroya 2002; Roggo et al. 2007; Nicolai et al. 2007; Huang et al. 2008; Cozzolino 2009).

Microelectromechanical systems (MEMS) combine mechanical parts, sensors, actuators and electronics on a common substrate through the use of microfabrication technology. This technology represents a paradigm shift for industrial spectroscopy and enables a variety of new industrial applications (McClure 2004; Blanco and Villaroya 2002; Roggo et al. 2007; Nicolai et al. 2007; Huang et al. 2008; Cozzolino 2009). In recent years multi- and hyperspectral imaging systems are being incorporated (McClure 2004; Blanco and Villaroya 2002; Roggo et al. 2007; Nicolai et al. 2007; Huang et al. 2008; Cozzolino 2009). In this case multispectral (a few wavelengths) or hyperspectral (a continuous range of wavelengths) imaging systems are required (Montes et al. 2007, 2011; Lu 2007). Such systems produce a spectral data cube—a spectrum at every 2D spatial position (Lu 2007). In early systems a sequence of spatial images was acquired by means of an NIR camera and a set of bandpass filters (McClure 2004; Nicolai et al. 2007; Montes et al. 2007, 2011; Lu 2007). While the acquisition can be fast, the disadvantage of this approach is that only a limited number of wavelengths can be analysed and that they need to be known a priori. In a more recent approach, a line of spatial information with a full spectral range per spatial pixel is captured sequentially to complete a volume of spatial-spectral data (McClure 2004; Nicolai et al. 2007; Montes et al. 2007, 2011; Lu 2007). This is usually achieved by means of a spectrograph which disperses an incoming line of radiation into a spectral and spatial matrix which is captured by the camera (McClure 2004; Nicolai et al. 2007; Montes et al. 2007, 2011). The horizontal and vertical pixels on the camera capture spatial and spectral information, respectively. Such system provides full spectral information at every spatial position. The object must be moved stepwise under the camera by means of an actuator while at each step a line is scanned, but this is not necessarily a disadvantage when the system is mounted on a grading line on which the fruit is physically transported anyway (McClure 2004; Nicolai et al. 2007; Montes et al. 2007, 2011; Lu 2007).

Novel developments include focal plane array cameras in combination with LCTF, AOTF or other monochromator principles which allow for much faster acquisition speeds (McClure 2004; Blanco and Villaroya 2002; Roggo et al. 2007; Nicolai et al. 2007; Huang et al. 2008; Cozzolino 2009). A variety of VIS-NIR spectrophotometers from several manufactures exists today, and these provide a number of different solutions for light dispersion, detectors and sample presentation configurations.

Overall, what instrument to choose is largely dependent on the application, and basically there is a trade-off between price and performance (McClure 2004; Blanco and Villaroya 2002; Roggo et al. 2007; Nicolai et al. 2007; Huang et al. 2008; Cozzolino 2009). However some characteristics must be evaluated or taken in consideration depending on the application such as resolution and noise; for scientific purposes, an instrument with high resolution, 10 nm or better, is favourable. However, there is a direct trade-off between resolution, noise and spectral range; wavelength range covering both VIS (400–780 nm) and the entire NIR region (780–2,500 nm) is recommended for scientific purposes to make sure that as much of the important wavelength bands as possible is included (McClure 2004; Nicolai et al. 2007; Montes et al. 2007, 2011; White et al. 2012). However, if the instrument is to be used for a very specific purpose, the need for full VIS-NIR spectra may not be necessary; other characteristics such as flexibility, possibility for measurements outside the laboratory, as well as other requirements regarding robustness and handling will apply depending on whether the instrument is only intended to be used in the laboratory, or if it will be used for on-line or at-line measurements (McClure 2004; Nicolai et al. 2007; Montes et al. 2007, 2011; White et al. 2012).

Multivariate Data Analysis

The use of vibrational spectroscopy allows the simultaneously analysis of multiple parameters, which provides scientist with a rapid and non-destructive quantitation of major components in many organic substances (Blanco and Villaroya

2002; McClure 2004; Cozzolino 2009; Smyth and Cozzolino 2011). The combination of vibrational spectroscopy methods and chemometrics (multivariate data analysis techniques) is applicable to many foods and agricultural commodities to predict chemical composition with high accuracy (Blanco and Villaroya 2002; McClure 2004; Cozzolino 2009; McGoverin et al. 2010; Smyth and Cozzolino 2011).

The spectrum of an organic material can give a global signature of composition which, with the application of chemometric techniques (e.g. principal component analysis or partial least squares regression), can be used to elucidate particular compositional characteristics in the food matrix not easily detected by targeted chemical analysis (McClure 2004; Cozzolino 2009; Smyth and Cozzolino 2011). Therefore vibrational spectroscopy might be considered as a tool in omics research in order to investigate the functional relationship between several metabolites in the plant. An excellent tutorial describing the steps needed to develop a calibration can be found in Agelet and Hurburgh (2010).

Taken in consideration that many relationships in the omics cannot be expressed in quantitative terms, these relationships are better expressed in terms of similarity and dissimilarity among groups of multivariate data (Lavine 2006; Gottlieb et al. 2004). To develop mathematical models that are suitable for identifying and isolating these groups of classes in multivariate data, pattern recognition techniques must be incorporated as a part of the analysis.

Discriminant or classifications methods allow to assign unclassified objects into predefined categories and are the key objective of such type of tools. Chemometrics, unlike classic statistics, considers multiple variables simultaneously and takes collinearity into account (the variation in one variable, or group of variables, in terms of covariation with other variables). The analysis can mathematically describe the covariation (degree of association) between variables or find a mathematical function (regression model), by which the values of the dependent variables are calculated from values of the measured (independent) variables (Naes et al. 2002; Brereton 2007, 2003). Information lies not only in any individual variable (wavelength) but also in how the variables change with respect to one another (Naes et al. 2002; Brereton 2007, 2003; Smyth and Cozzolino 2011, 2013).

Two main types of pattern recognition methods can be used in omic analysis to interpret the spectra generated by different instruments (NIR, MIR, Raman): supervised and unsupervised. Unsupervised refers to those methods that are based on no presumption of the structure of the data set. Supervised methods assume some structure exists in the data and requires assigning of the samples to prespecified subgroups using ‘artificial’ variables to build classification rules which are later used for allocating new and unknown samples to the most probable subgroup (Cozzolino 2009, 2011, 2012a, b; Smyth and Cozzolino 2011).

Principal component analysis (PCA) is used as a tool for screening, extracting and compressing multivariate data and can be defined as an unsupervised method (Naes et al. 2002; Brereton 2007; Cozzolino 2009, 2011, 2012a, b; Smyth and Cozzolino 2011). PCA employs a mathematical procedure that transforms a set of possibly correlated response variables into a new set of noncorrelated variables, called principal components. PCA can be performed on either a data matrix or a correlation matrix depending on the type of variables being measured (Naes et al. 2002; Brereton 2007; Cozzolino 2009, 2011, 2012a, b; Smyth and Cozzolino 2011).

Discriminant analysis (DA) and partial least squares discriminant analysis (PLS-DA) can be considered a qualitative calibration method. Instead of calibrating for a continuous variable, one calibrates for group membership (categories). The resulting models are evaluated in terms of their predictive ability to predict the new and unknown samples (Naes et al. 2002; Brereton 2008). Discrimination models are usually developed using PLS regression technique as described elsewhere (Naes et al. 2002).

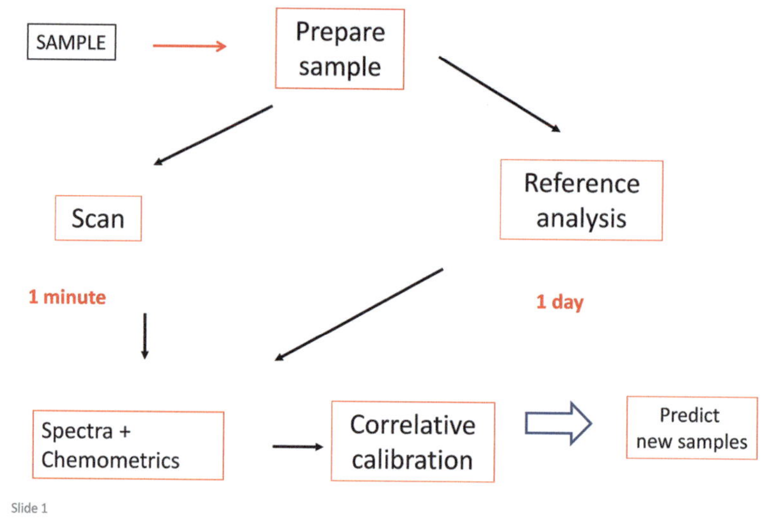

Fig. 3 A simplified scheme of calibration development applied to plant omics

Another discriminant technique extensively used is linear discriminant analysis (LDA) and is defined as a supervised classification technique (Naes et al. 2002; Brereton 2007; Cozzolino 2009, 2011, 2012a, b; Smyth and Cozzolino 2011, 2013). The criterion of LDA for selection of latent variables is maximum differentiation between the categories and minimal variance within categories (Naes et al. 2002; Brereton 2007). This method produces a number of orthogonal linear discriminant functions, equal to the number of categories minus one, that allow the samples to be classified in one or another category. Application of artificial neural networks (ANN) is a more recent technique for data and knowledge processing that is characterised by its analogy with a biological neuron (Naes et al. 2002; Brereton 2007).

Figure 3 provides an example of quantitative analysis or calibration development commonly used when vibrational spectroscopy methods are applied.

Software

There are several commercial software packages specially developed for analysis and calibrations of spectral data providing easy to use functions for spectral analyses and calibrations. Many manufacturers provide their instruments with specific software or have close collaborations with commercial software developers to facilitate exporting and importing of measured spectra and sometimes allowing for real-time predictions. Alternatively, most of the analyses and calibration tools can also be found in commercial, shareware and freeware software and in open-source environments such as the R-project. Some companies supply software packages that are fully integrated and can only calibrate using scans from their instruments.

Others have their software broken into separate components, and the calibration software can be used with scans from several platforms. Several instrument companies have specialised software available to facilitate the development of web-based networks. Examples of third party software are the following: The Unscrambler package (http://www.camo.com) is a menu driven and very easy to use as well as offers a range of pre-processing techniques. The Grams Suite is a general-purpose software package for handling spectroscopic data including chemometrics and is distributed by the Thermo Scientific company (http://www.thermo.com). The Matlab PLS toolbox from the Eigenvector Research (www.eigenvector.com) company offers and combines the flexibility of the Matlab software for applications in which programming is required. Many other statistical packages, such as SAS (http://www.sas.com) and Statistica (http://www.statsoft.com), also provide multivariate calibration but are less convenient to use for processing spectral data.

Examples of Applications of Vibrational Spectroscopy in 'Plant Omics'

Many reports can be found in the literature on several applications of vibrational spectroscopy methods combined with chemometrics on plant omics. In the section below, few examples are given. Note that this is not an exhaustive review rather an overview with some recent examples found in the literature.

Seeds and Grains

Near-infrared reflectance (NIR) spectroscopy is a non-destructive technique, low-cost and high-speed analytical method with broad applications in plant omics (Montes et al. 2007). Cereal grain spectra typically are measured from fine-ground powders or as bulk whole grains (Jespersen and Munck 2009; Williams 2010; Himmelsbasch 2010). Single-seed samples can be used and analysed using NIR to measure phenotypes in segregating populations. Maize kernels can be classified according to characteristics such as starch composition, hardness, avidin or mycotoxin levels (Campbell et al. 2000; Walter et al. 2012). Other applications for measuring wheat attributes included wheat classes, colour, insect infestation, starch composition, among other characteristics (Chambers and Ridgeway 1995; Ridgway and Chambers 1996, 1998; Singh et al. 2010; Jespersen and Munck 2009; Williams 2010). The prediction of constituent concentrations using NIR spectroscopy on intact single seeds has been most successful for plants with small seeds and relatively uniform distribution of seed constituents, such as rapeseed, wheat, sunflower achenes or soybean seeds (Velasco et al. 1999; Fassio and Cozzolino 2004; Agelet et al. 2012; Baranska et al. 2010; Schulz et al. 2005).

The use of hyperspectral imaging combined with NIR data has been reported to analyse different kernel and seed samples (Jespersen and Munck 2009; Williams 2010; Wetzel and Brewer 2010; Singh et al. 2010). For example, Arngren and collaborators (2011) using hyperspectral NIR imaging of single kernels found that wavelengths between 1,400 and 1,550 nm can be associated with water activity related to barley germination as well as with the activity of starch and maltose. In a recent study, Manley and collaborators used hyperspectral NIR imaging to determine water relationships in the wheat grain individual seeds. Although these approaches yield acceptable composition predictions, they are low throughput. Non-destructive screening of single seeds, plants and crops according to specific attributes of interest would make the breeding process much faster, more efficient and cost-effective in order to deliver commercial value of novel genotypes or lines with improved characteristics to be used in the current competitive worldwide market.

The study of the relationship of biological characteristics (in terms of improving malting quality) of barley grain in relation to spectral-structural characteristics may provide useful understanding of how to best utilise and select new genotypes and varieties of barley in order to provide farmers and the industry (e.g. beer and food) with high-quality lines (e.g. malting). In conventional chemical analysis, harsh chemical reagents can unavoidably destroy or even obliterate the histological and physical structure of the biological sample (Yu 2005, 2007; Yu et al. 2008, 2009, 2011; Liu and Yu 2010). The detection of inherent structural and chemical information is beyond the capacity of traditional chemical analysis, owing to the destructive pretreatment and the treatment during chemical analysis. These disadvantages demand new and rapid tools to overcome these limitations (Yu 2005, 2007; Yu et al. 2008, 2009, 2011; Liu and Yu 2010; Zhang and Yu 2012). Recent reports indicated that biological differences in protein availability between barley and bioethanol co-products, which are closely related not only in their total protein content, but also in their inherent protein structures, including protein amide I and amide II and α-helix and β-sheet intensity, can be targeted in terms of spectral peak area and height, as well as their ratios (Yu 2005, 2007; Yu et al. 2008, 2009, 2011; Liu and Yu 2010; Zhang and Yu 2012). Such molecular structural differences can be detected by using synchrotron-based infrared microspectroscopy (SR-IMS), diffuse reflectance infrared Fourier transform (DRIFT) or Fourier transform infrared attenuated total reflectance (ATR) spectroscopy (Yu 2005, 2007; Yu et al. 2008, 2009, 2011; Liu and Yu 2010). Although state-of-the-art research can be found in the literature in relation to the use of SR-IMS, less information is available on the use of molecular spectroscopy techniques such as ATR spectroscopy on single/intact seeds. It is well known that the SR-IMS method allows for the identification and study of the cellular and subcellular structure of the endosperm and other tissues in the specimen under analysis (Yu 2005, 2007; Yu et al. 2008, 2009, 2011; Liu and Yu 2010). However, this technique requires time-consuming and careful preparation of the sample, and is not easily available or accessible in many laboratories around the world. Therefore, this method is limited to few applications. The driving force behind exploring ATR-IR spectroscopy is the availability of simple and state-of-the-art instrumentation that can be used as tools for high-throughput screening or phenotyping in breeding programmes (Yu 2005, 2007; Yu et al. 2008, 2009, 2011; Liu and Yu 2010; Zhang and Yu 2012).

Water Content, Stress and Dry Matter

Infrared wavelength range (above 1,000 nm) is becoming more and more widely used to monitor stomatal conductance of plant canopies in the lab and in the field (for a review, see Zhu et al. 2011; Munns et al. 2010). Recent studies showed the potential of NIR spectroscopy to determine grapevine water potential (leaf and stomata) (De Bei et al. 2011). These authors observed differences in the NIR spectra related to the leaf surface in which the spectra were collected, and this had an effect on the accuracy of the calibration statistics for water potential (De Bei et al. 2011). The global calibrations built using data obtained from glasshouse and field studies on two varieties are indicative that, in the future, a universal calibration, able to predict water potential for all varieties in different environments can be built (De Bei et al. 2011).

It is well known that dry matter yield (DMY) is one of the most important traits as it is directly related to production costs as it influences the concentration of nutrients in the whole plant (Garcia and Cozzolino 2006; Moron et al. 2007; Cozzolino and Labandera 2002; Cabrera-Bosquet et al. 2012; White et al. 2012; Walter et al. 2012). The measurement of DMY in breeding programmes or in plant omics is not straightforward because there is often only poor agreement between yield measured on individual spaced plants and obtained yield in productive swards or crops (Garcia and Cozzolino 2006; Moron et al. 2007; Cozzolino and Labandera 2002; Cabrera-Bosquet et al. 2012; White et al. 2012; Walter et al. 2012). Online measurements for DMY using field-portable near-infrared spectroscopy (NIR) instruments have already facilitated direct selection in the field on family basis (Garcia and Cozzolino 2006; Moron et al. 2007; Cozzolino and Labandera 2002; Cabrera-Bosquet et al. 2012; White et al. 2012; Walter et al. 2012). More complex morphological characters such as tiller density, auxiliary formation, shoot branching and spike/spikelet morphology may be monitored by three-dimensional scanning as recently shown in maize (Montes et al. 2007, 2011). In combination with spectral reflectance, this can be elaborated as a future tool for improving both DMY and morphological characteristics (Montes et al. 2007, 2011). In recent years, 'on the go' or in-field NIR spectroscopy methods have been evaluated to predict nitrogen and water in cereals (Fox et al. 2010; Long et al. 2008).

Nitrogen

With regard to nitrogen economy, forage legumes are primarily selected for improved fixation of atmospheric nitrogen by screening plants for high tissue N concentrations. In forage grasses, nitrogen use efficiency is usually improved by selecting for increased DMY under uniform soil-N conditions (Casler and van Santen 2010). Routine laboratory methods for the determination of N concentrations in grass samples such as Kjeldahl distillation (AOAC 1990) or Dumas are widespread, even though they are time consuming and expensive (Fox et al. 2010; Long et al. 2008; Cabrera-Bosquet et al. 2012; White et al. 2012; Walter et al. 2012). It has been shown that NIR can be implemented to more efficiently determine N concentrations in grass samples (Cabrera-Bosquet et al. 2012; White et al. 2012; Walter et al. 2012). A NIR based approach replacing wet chemistry methods with online field screening constitutes a more direct strategy to select for improved N uptake efficiency and total N concentration (Fox et al. 2010; Long et al. 2008; Cabrera-Bosquet et al. 2012; White et al. 2012; Walter et al. 2012). One of the potential advantages of using NIR spectroscopy in plant omics is the analysis of fresh plant materials (e.g. leaf, whole plant) without the need of drying or grinding (Garcia and Cozzolino 2006; Moron et al. 2007; Cozzolino and Labandera 2002).

Nonstructural and Structural Carbohydrates

Improving forage quality mainly aims at improving dry matter digestibility, increasing the amount of compounds beneficial to livestock such as water-soluble carbohydrates (WSCs) as well as condensed tannins and reducing the amount of unwanted substances such as toxins, oestrogenic compounds or alkaloids (Carbonero et al. 2001; Walter et al. 2012). Due to the moderate to high heritability, genetic gain for forage quality has been substantial in recent decades (Walter et al. 2012). Dry matter digestibility may be increased by breeding for decreased fibre and lignin concentration in the cell wall or by increasing the content of WSCs. These traits are traditionally determined using wet chemistry methods but can be streamlined by NIR spectroscopy. In addition to DMY, N and WSC determination, NIR has proven its value to predict ergovaline (Roberts et al. 1997; Walter et al. 2012) and lignin concentrations in grasses (Andres et al. 2005). A NIR-based lignin prediction would be useful to identify cultivars with beneficial properties for bioenergy production. However, the accuracy of calibration models developed to predict lignin and ergovaline is still limited and needs to be improved for online field applications (Cabrera-Bosquet et al. 2012; White et al. 2012; Walter et al. 2012). Of particular interest to sustain forage quality are fructans, fructose polymers deriving from sucrose and serving as reserve carbohydrates in many plant species (Ritsema and Smeekens 2003). Fructans are key factors in crop plants to respond to abiotic stress, in general, and drought, cold and freezing tolerance in particular (Livingston et al. 2009). A NIR-based approach to quantify fructan concentration in freeze-dried and ground grass samples has recently been reported (Shetty and Gislum 2011). In contrast to WSCs, NIR-based measurements for specific carbohydrates such as fructans are difficult to obtain online in the field. But given the fact that fructans constitute the main part of WSCs in grasses and the high correlation of fructans to total WSCs (Sanada et al. 2007), NIRS-based improvement of WSCs will not only increase digestibility and preference by ruminants but might also provide an innovative approach to develop grasses with improved abiotic stress tolerance (Cabrera-Bosquet et al. 2012; White et al. 2012; Walter et al. 2012) or to gather information about carbohydrates relationships in whole plants (Gergely and Salgo 2005).

Biotic (Diseases) and Abiotic Stress

A combination of the above-mentioned digital and functional phenotyping methods may allow for efficient, simultaneous selection on multiple traits related to persistence (Walter et al. 2012). Durable resistance to major diseases and pests such as crown rust, snow mould, bacterial wilt, fusarium root rot or nematodes is a common objective in any forage and turf breeding programmes. Resistance is usually improved through phenotypic recurrent selection using naturally occurring or artificial infection (Walter et al. 2012). Infrared technologies have been introduced as metabolic fingerprinting techniques within agriculture and food. Such studies have been included into plant-to-plant interactions, where global metabolite changes associated with abiotic and biotic perturbations and interactions (Walter et al. 2012). Other applications included the metabolite fingerprinting of salt-stressed tomatoes, where functional groups of importance related to tomato salinity (Walter et al. 2012). To unravel the genetic basis of complex traits, it is necessary to associate genotypic information with the corresponding phenotypic data. Various spectroscopic and imaging techniques have been studied for the detection of symptomatic and asymptomatic plant diseases. The use of NIR spectroscopy, with a wavelength range between 950 and 1,650 nm, was used to determine the percentage of fungal infection found in rice samples. The total fungal infection and yellow-green Aspergillus infection determined using PLS regression where models were developed from the untreated spectra provided the greatest accuracy in prediction, with a correlation coefficient (r) of 0.67, a standard error of prediction (SEP) of 28 % and a bias of 0.10 %. The result showed that the NIR spectroscopy could be used to detect aflatoxigenic fungal contamination in rice with caution and the technique should be improved to get better prediction model (Dachoupakan Sirisomboon et al. 2013).

Transgenic Organisms, GMO

Combining these novel phenotyping methods with high-throughput genotyping based on different 'omic' technologies and analysing the data with sophisticated mathematical models promises a major breakthrough in the elucidation of the underlying genetic basis of complex plant traits, such as drought or frost tolerance (Rossel et al. 2001; Rui et al. 2005; Alishahi et al. 2010). Munck and collaborators assessed the discriminatory power of NIR spectroscopy combined with PCA to evaluate different lysine mutants in barley. These authors were able to identify different mutants based on the NIR fingerprint of the sample (Munck et al. 2004). It has been shown that it is possible to select improved malting barley genotypes based on its NIR spectrum as a total physical-chemical spectral fingerprint of the samples and interpreted using PCA (Munck et al. 2004, 2001). Near-infrared spectroscopy was also evaluated to discriminate barley flour containing high levels of lysine amino acids (Munck et al. 2004, 2001). It was found that differences between these two barley flours might be observed through the interpretation of the PCA scores (Munck et al. 2004, 2001). Thereby, these two variants of barley flour were classified correctly in two groups according to their different NIR spectra. The use of NIR spectroscopy was also evaluated to assess the effects of different environments on transgenic and non-transgenic plant materials. For example, four genotypes consisting of common genotypes and LYS3 (including four alleles of a, b, c and m) were used (Munck et al. 2004, 2001). The main phenotypic changes caused by manipulating the alleles were recognised by combining NIR spectroscopy with PCA. Several other reports can be found on the use of NIR spectroscopy to discriminate between transgenic and non-transgenic grains (Rossel et al. 2001; Rui et al. 2005; Alishahi et al. 2010). All these authors reported that classification methods such as PCA yield good performances in order to separate completely transgenic from non-transgenic grains, concluding that the utilisation of NIR spectroscopy as a non-destructive method had a suitable efficiency to detect transgenic grains. Munck and collaborators (2001) also applied NIR spectroscopy to discriminate barley flour containing high levels of lysine amino acids. They found that differences between these two barley flours might be observed through the interpretation of the PCA scores. Thereby, these two variants of barley flour were classified correctly in two groups according to their different NIR spectra (Munck et al. 2001, 2004; Alishahi et al. 2010).

Other reports can be found on the use of NIR spectroscopy to discriminate between transgenic and non-transgenic grains (Hurburgh et al. 2000; Campbell et al. 2000; Ahmed 2002; Rossel et al.

2001; Rui et al. 2005). These authors reported that PCA method had a good performance to separate completely transgenic from non-transgenic grains, concluding that the utilisation of NIR spectroscopy as a non-destructive method had a suitable efficiency to detect transgenic grains (Hurburgh et al. 2000; Campbell et al. 2000; Ahmed 2002; Rossel et al. 2001; Rui et al. 2005; Alishahi et al. 2010).

High Throughput in Practice: On Harvest Analysis/On the Go Spectroscopy

In order to unravel the genetic basis of complex traits, it is necessary to associate genotypic information with the corresponding phenotypic data. Recent progress in DNA marker assays and sequencing technologies enable high-throughput genotyping of many individual plants at relatively low cost (Montes et al. 2007, 2011; White et al. 2012). By comparison, the phenotyping of large mapping populations for several traits in field trials is still laborious and expensive. Research and developments in NIR spectroscopy on agricultural harvesters and spectral reflectance of plant canopy present new opportunities to develop novel phenotyping platforms that enable large-scale screenings of genotypes for several traits in several locations within the field trials (Montes et al. 2007, 2011; White et al. 2012). These phenotyping techniques could bring remarkable progress in plant genetic research to unravel, the genetic basis of dynamic traits, such as biomass accumulation and drought or frost stress tolerance. Currently, these traits are treated as static (i.e. are measured only once), but ignoring their dynamic nature entails a tremendous loss of information regarding the analysis of genes and gene networks that are active at different phases of plant development and in reactions to environmental stresses (Montes et al. 2007, 2011; Cabrera-Bosquet et al. 2012; White et al. 2012; Walter et al. 2012). The use of NIR spectroscopy on agricultural harvesters reduces the manpower and expenditure required for the determination of relevant traits. In contrast to conventional sample-based methods, NIR spectroscopy on agricultural harvesters secures a good distribution of measurements within plots and covers substantially larger amounts of plot material (Montes et al. 2007, 2011; Cabrera-Bosquet et al. 2012; White et al. 2012; Walter et al. 2012). Consequently, agricultural harvesters equipped with NIR spectroscopy reduce the sampling error and yield more representative measurements of the plot material and can also be used successfully to determine dry matter, starch and crude protein contents in maize grain (Montes et al. 2007, 2011; Cabrera-Bosquet et al. 2012; White et al. 2012; Walter et al. 2012). In silage maize, the potential of this technology has also been reported for dry matter, starch and soluble sugars (Garcia and Cozzolino 2006; Moron et al. 2007; Cozzolino and Labandera 2002; Montes et al. 2007, 2011; Cabrera-Bosquet et al. 2012; White et al. 2012; Walter et al. 2012). Measurements of crude protein, digestibility, fibre contents and energy-related traits can be used successfully to classify genotypes (Garcia and Cozzolino 2006; Moron et al. 2007; Cozzolino and Labandera 2002), but for the precise quantitative trait evaluation, further technical improvements for an optimal sample presentation are crucial (Montes et al. 2007, 2011; Cabrera-Bosquet et al. 2012; White et al. 2012; Walter et al. 2012). The use of NIR spectroscopy on agricultural harvesters represents a high-throughput phenotyping technique with substantially reduced sampling error, whereas spectral reflectance of plant canopies facilitates the determination of dynamic traits in a non-invasive mode (Montes et al. 2007, 2011; Cabrera-Bosquet et al. 2012; White et al. 2012; Walter et al. 2012).

The use of NIR spectroscopy on agricultural harvesters reduces the manpower and expenditure required for the determination of relevant traits (Garcia and Cozzolino 2006; Moron et al. 2007; Cozzolino and Labandera 2002). In contrast to conventional sample-based methods, NIR spectroscopy on agricultural harvesters secures a good distribution of measurements within plots and covers substantially larger amounts of plot material (Garcia and Cozzolino 2006; Moron et al. 2007; Cozzolino and Labandera 2002;

Montes et al. 2007, 2011). Consequently, agricultural harvesters equipped with NIR spectroscopy reduce the sampling error and yield more representative measurements of the plot material and can also be used successfully to determine dry matter, starch and crude protein contents in maize grain (Montes et al. 2007, 2011; Cabrera-Bosquet et al. 2012; White et al. 2012; Walter et al. 2012). The use of NIR spectroscopy on agricultural harvesters represents a high-throughput phenotyping technique with substantially reduced sampling error, whereas spectral reflectance of plant canopies facilitates the determination of dynamic traits in a non-invasive mode (Garcia and Cozzolino 2006; Moron et al. 2007; Cozzolino and Labandera 2002; Montes et al. 2007, 2011; Cabrera-Bosquet et al. 2012; White et al. 2012; Walter et al. 2012).

Concluding Remarks

Adapting and using advanced omic technologies is a promising way forward to efficiently and reliably improve agronomical important characteristics in the breeding process, to adapt varieties, to improve our understanding about plants (physiology, biochemistry) and to maintain a sustainable agriculture and food security. For example, in combination with optimised breeding schemes and cutting-edge molecular tools, advanced omics has the potential to substantially improve and fasten cultivar development, thereby contributing to a sustainable feed, food and biomass production on both the local and global level.

Infrared and vibrational spectroscopy technologies have been introduced as metabolic fingerprinting techniques within agriculture and food and in recent years have been rediscovered by several research groups in plant sciences. Combining these novel omic methods with high-throughput genotyping based on different IR technologies and analysing the data with sophisticated mathematical models promises a major breakthrough in the elucidation of the underlying genetic basis of complex plant traits, such as drought or frost tolerance (Walter et al. 2012).

Although, vibrational spectroscopy techniques cannot measure molecules with low concentration, the indirect effects of such differences can be observed in the spectrum of a given sample in the so-called fingerprint. These technologies have been successfully applied for natural product composition analysis, product quality assessment and in production control, and it was shown to be a promising technique in plant omics. The spectrum can give a global signature of composition (fingerprint) which, with the application of chemometric techniques (e.g. principal component analysis or discriminant analysis), can be used to elucidate particular compositional characteristics not easily detected by traditional targeted chemical analysis.

The main advantages of this technique over the traditional chemical and chromatographic methods are the rapidity and the ease of use in routine operations. Vibrational spectroscopy-based method is a non-destructive technique which requires minimal or zero sample preparation. Therefore, these techniques can be suggested as the first line of tools to be used in omic studies. The potential savings, reduction of analysis time and cost and the environmentally friendly nature of the technology has positioned vibrational spectroscopy as a very attractive technique with a bright future in the arena of plant omics.

It is clear that the breadth of this type of applications, either in routine use or under developed, is showing no sign of diminishing. The combination of different vibrational spectroscopy analytical techniques with multivariate methods could be used as a tool as a fingerprinting of samples on a large scale. However, the chemical basis of this separation is not addressed using this type of methodology, and other analytical techniques (i.e. GC-MS) need to be used or combined in order to reveal the fundamental causes of the separation/classification. The development of hyperspectral imaging, microspectroscopy and new algorithms will place vibrational spectroscopy as one of the most useful tools in plant omic studies in the near future.

One of the main constraints to further develop and apply these technologies in plant omic studies is the lack of formal education in both vibrational spectroscopy and chemometrics methods. In the last few years, efforts are being made in the International Council for Near-Infrared Spectroscopy (www.icnirs.org) to overcome these issues. Without doubt one of the biggest challenges in the wider use of NIR spectroscopy will be the interpretation of the complex spectra obtained. Although we devoted much time to the interpretation of the models through multivariate analysis or chemometrics, the knowledge of the fundamentals of molecular spectroscopy in omic analysis using infrared spectroscopy is still the main barrier in order to understand the basis and functionality of the models developed.

Acknowledgments This project is supported by Australia's grain growers through their investment body the Grain Research and Development Corporation, with matching funds from the Australian government.

References

Abberton MT, Marshall AH (2005) Progress in breeding perennial clovers for temperate agriculture. J Agric Sci 143:117–135

Agelet LE, Hurburgh CR (2010) A tutorial on near infrared spectroscopy and its calibration. Crit Rev Anal Chem 40:246–260

Agelet LE, Armstrong PR, Romagosa Clariana I, Hurburgh CR (2012) Measurement of single soybean seed attributes by near infrared technologies. A comparative study. J Agric Food Chem 60:8314–8322

Ahmed FE (2002) Detection of genetically modified organisms in foods. Trends Biotechnol 20:215–223

Alishahi H, Farahmand H, Prieto N, Cozzolino D (2010) Identification of transgenic foods using NIR spectroscopy: a review. Spectrochim Acta A Mol Biomol Spectrosc 75:1–7

Andres S, Giraldez FJ, Lopez S, Mantecon AR, Calleja A (2005) Nutritive evaluation of herbage from permanent meadows by near-infrared reflectance spectroscopy: 1. Prediction of chemical composition and in vitro digestibility. J Sci Food Agric 85: 1564–1571

Arngren M, Hansen PW, Eriksen B, Larsen J, Larsen R (2011) Analysis of pregerminated barley using hyperspectral image analysis. J Agric Food Chem 59:11385–11394

AOAC (1990) Official methods of analysis of the Association of Official Analytical Chemists, 15th edn. Association of Official Analytical Chemists, Washington, DC

Baranska M, Schultz H (2006) Application of infrared and Raman spectroscopy for analysis of selected medicinal and spice plants. J Med Spice Plant (Z Arzn Gew Pfl) 2:72–80

Baranska M, Schulz H, Rösch P, Strehle MA, Popp J (2004) Identification of secondary metabolites in medicinal and spice plants by NIR-FT-Raman microspectroscopic mapping. Analyst 129:926–930

Baranska M, Schulz H, Strehle M, Popp J (2010) Applications of vibrational spectroscopy to oilseeds analysis. In: Li-Chan E, Griffiths PR, Chalmers JM (eds) Applications of vibrational spectroscopy in food science. Wiley, Chichester

Blanco M, Villaroya I (2002) NIR spectroscopy: a rapid-response analytical tool. Trends Anal Chem 21: 240–250

Brereton RG (2003) Chemometrics: data analysis for the laboratory and chemical plant. Wiley, Chichester

Brereton RG (2007) Applied chemometrics for scientist. Wiley, Chichester

Brereton G (2008) Applied chemometrics for scientist. Wiley, Chichester

Cabrera-Bosquet L, Crossa J, von Zitzewitz J, Serret MD (2012) High-throughput phenotyping and genomic selection: the frontiers of crop breeding converge. J Integr Plant Biol 54:312–320

Campbell MR, Sykes J, Glover DV (2000) Classification of single and double-mutant corn endosperm genotypes by near-infrared transmittance spectroscopy. Cereal Chem 77:774–778

Carbonero CH, Mueller-Harvey I, Brown TA, Smith L (2001) Sainfoin a beneficial forage legume. Plant Genet Resour Charact Util 9:70–85

Casler M, van Santen E (2010) Breeding objectives in forages. In: Boller B, Posselt U, Veronesi F (eds) Handbook of plant breeding: fodder crops and amenity grasses. Springer, New York, pp 115–160

Chambers J, Ridgeway C (1995) Rapid detection of contaminants in cereals. In: Davies AMC, Williams P (eds) Near infrared spectroscopy: the future waves. NIR Publications, Chichester, pp 484–489

Cozzolino D (2009) Near infrared spectroscopy in natural products analysis. Planta Med 75:746–757

Cozzolino D (2011) Infrared methods for high throughput screening of metabolites: food and medical applications. Comb Chem High Throughput Screen 14:125–131

Cozzolino D (2012a) Recent trends on the use of infrared spectroscopy to trace and authenticate natural and agricultural food products. Appl Spectrosc Rev 47:518–530

Cozzolino D (2012b) Benefits and limitations of infrared technologies in omic research and development of natural drugs and pharmaceutical products. Drug Dev Res 73:504–512

Cozzolino D, Labandera M (2002) Determination of dry matter and crude protein contents of undried forages by near infrared reflectance spectroscopy. J Sci Food Agric 82:380–384

Dachoupakan Sirisomboon C, Putthang R, Sirisomboon P (2013) Application of near infrared spectroscopy to detect aflatoxigenic fungal contamination in rice. Food Control 33:207–214

De Bei R, Sullivan W, Cynkar WU, Cozzolino D, Dambergs RG, Tyerman SD (2011) Non destructive measurement of grapevine water potential using near infrared spectroscopy. Aust J Grape Wine Res 17:62–71

Dunn WB, Ellis DI (2005) Metabolomics: current analytical platforms and methodologies. Trends Anal Chem 24:285–294

Fassio A, Cozzolino D (2004) Non-destructive prediction of chemical composition in sunflower seeds by near infrared spectroscopy. Ind Crop Prod 20:321–329

Fox GP, Bloustein G, Sheppard J (2010) On the go NIT technology to assess protein and moisture during harvest of wheat breeding trials. J Cereal Sci 51:171–173

García J, Cozzolino D (2006) Use of near infrared reflectance spectroscopy to predict chemical composition of forages from breeding programs. Agricultura Tecnica (Chile) 66:41–48

Gergely S, Salgo A (2005) Changes in carbohydrate content during wheat-maturation-what is measured by near infrared spectroscopy? J Near Infrared Spectrosc 13:9–17

Gottlieb DM, Schultz J, Bruun SW, Jacobsen S, Søndergaard I (2004) Multivariate approaches in plant science. Phytochemistry 65:1531–1548

Griffiths PR, De Haseth JA (1986) Fourier transform infrared spectrometry. Wiley, New York

Hashimoto A, Kameoka T (2008) Applications of infrared spectroscopy to biochemical, food, and agricultural processes. Appl Spectros Rev 43:416–451

Himmelsbasch DS (2010) The analysis of rice by vibrational spectroscopy. In: Li-Chan E, Griffiths PR, Chalmers JM (eds) Applications of vibrational spectroscopy in food science. Wiley, Chichester

Hounsome N, Hounsome B, Tomos D, Edward-Jones G (2008) Plant metabolites and nutritional quality of vegetables. J Food Sci 73:R48–R65

Huang H, Yu H, Xu H, Ying Y (2008) Near infrared spectroscopy for on/in-line monitoring of quality in foods and beverages: a review. J Food Eng 87:303–313

Hurburgh CR, Rippke GR, Heithoff C, Roussel SA, Hardy CL (2000) Detection of genetically modified grains by near-infrared spectroscopy. In: Proceedings PITTCON 2000–science for the 21st century, 12–17 March 2000, New Orleans, USA, p 1431

Jespersen BM, Munck L (2009) Cereals and cereal products. In: Sun DW (ed) Infrared spectroscopy for quality analysis and control. Elsevier/Academic, Amsterdam

Karoui R, Downey G, Blecker C (2010) Mid-infrared spectroscopy coupled with chemometrics: a tool for the analysis of intact food systems and the exploration of their molecular structure-quality relationships – a review. Chem Rev 110:6144–6168

Kolukisaoglu U, Thurow K (2010) Future and frontiers of automated screening in plant sciences. Plant Sci 178:476–484

Lavine BK (2006) Clustering and classification of analytical data. In: Meyers RA (ed) Encyclopedia of analytical chemistry. Wiley, Chichester

Li-Chan ECY (2010) Introduction to vibrational spectroscopy. In: Li-Chan E, Griffiths PR, Chalmers JM (eds) Applications of vibrational spectroscopy in food science. Wiley, Chichester, UK

Liu N, Yu P (2010) Characterization of the micro chemical structure of seed endosperm within a cellular dimension among six barley varieties with distinct degradation kinetics, using ultra spatially resolved synchrotron-based infrared micro spectroscopy. J Agric Food Chem 58:7801–7810

Livingston D, Hincha D, Heyer A (2009) Fructan and its relationships to abiotic stress tolerance in plants. Cell Mol Life Sci 66:2007–2023

Long DS, Engel RE, Siemens MC (2008) Measuring grain protein concentration with in line near infrared reflectance spectroscopy. Agron J 100:247–252

Lu R (2007) Quality evaluation of fruit by hyperspectral imaging. In: Sun DW (ed) Computer vision technology for food quality. Elsevier, Amsterdam

McClure FW (2004) 204 years of near infrared technology: 1800–2003. J Near Infrared Spectros 11:487–488

McGoverin CM, Weeranantanaphan J, Downey G, Manley M (2010) The application of near infrared spectroscopy to the measurement of bioactive compounds in food commodities. J Near Infrared Spectros 18:87–111

Montes JM, Melchinger AE, Reif JC (2007) Novel throughput phenotyping platforms in plant genetic studies. Trends Plant Sci 12:433–436

Montes JM, Technow F, Dhillon BS, Mauch F, Melchinger AE (2011) High-throughput non-destructive biomass determination during early plant development in maize under field conditions. Field Crop Res 121:268–273

Morón A, Cozzolino D, García A, Sawchik J (2007) Preliminary study on the use of near infrared reflectance spectroscopy to assess nitrogen content on undried wheat plants. J Sci Food Agric 87:142–157

Munck L, Pram Nielsen J, Møller B, Jacobsen S, Søndergaard I, Engelsen SB, Nørgaard L, Bro R (2001) Exploring the phenotypic expression of a regulatory proteome-altering gene by spectroscopy and chemometrics. Anal Chim Acta 446:171–186

Munck L, Møller B, Jacobsen S, Søndergaard S (2004) Near infrared spectra indicate specific mutant endosperm genes and reveal a new mechanism for substituting starch with (1/3, 1/4)-b-glucan in barley. J Cereal Sci 40:213–222

Munns R, James RA, Sirault XRR, Furbank RT, Jones HG (2010) New phenotyping methods for screening wheat

and barley for beneficial responses to water deficit. J Exp Bot 61:3499–3507
Naes T, Isaksson T, Fearn T, Davies T (2002) A user-friendly guide to multivariate calibration and classification. NIR Publications, Chichester
Nicolai BM, Beullens K, Bobelyn E, Peirs A, Saeys W, Theron KI, Lammertyn J (2007) Non-destructive measurement of fruit and vegetable quality by means of NIR spectroscopy: a review. Post Harv Biol Tech 46:99–118
Ridgway C, Chambers J (1996) Detection of external and internal insect infestation in wheat by near-infrared reflectance spectroscopy. J Sci Food Agric 71(2):251–264
Ridgway C, Chambers J (1998) Detection of insects inside wheat kernels by NIR imaging. J Near Infrared Spectrosc 6(1):115–119
Ritsema T, Smeekens S (2003) Fructans: beneficial for plants and humans. Curr Opin Plant Biol 6:223–230
Roberts CA, Joost RE, Rottinghaus GE (1997) Quantification of ergovalina in tall fescue by near infrared reflectance spectroscopy. Crop Sci 37:281–284
Rodriguez-Saona LE, Allendorf ME (2011) Use of FTIR for rapid authentication and detection of adulteration of food. Annu Rev Food Sci Technol 2:467
Roggo Y, Chalus P, Maurer L, Lema-Martinez C, Edmond A, Jent N (2007) A review of near infrared spectroscopy and chemometrics in pharmaceutical technologies. J Pharm Biomed Anal 44:683–690
Rossel SA, Hardy CL, Hurburgh CR, Rippke GR (2001) Application of near-infrared diffuse reflectance spectroscopy to the detection and identification of transgenic corn. Appl Spectrosc 55:1425–1432
Rui Y, Luo Y, Haung K, Wang W, Zhang L (2005) Discrimination of transgenic corns using NIR diffuse reflectance spectroscopy and back propagation (BP). Spectrosc Spectr Anal 25:1581–1592
Ryan D, Robach K (2006) Metabolomics: the greatest omics of them all? Anal Chem 78:7954–7958
Sanada Y, Takai T, Yamada T (2007) Inheritance of the concentration of water soluble carbohydrates and its relationship with the concentration of fibre and crude protein in herbage of cocksfoot (Dactylis glomerata). Grass Forage Sci 62:322–331
Schulz H, Baranska M, Baranski R (2005) Potential of NIR-FT-Raman spectroscopy in natural carotenoid analysis. Biopolymers 77:212–221
Shetty N, Gislum R (2011) Quantification of fructan concentration in grasses using NIR spectroscopy and PLSR. Field Crop Res 120:31–37
Singh CB, Jayas DS, Paliwal J, White NDG (2010) Identification of insect-damaged wheat kernels using shortwave near-infrared hyperspectral and digital colour imaging. Comput Electron Agric 73(2):118–125
Smyth HE, Cozzolino D (2011) Applications of infrared spectroscopy for quantitative analysis of volatile and secondary metabolites in plant materials. Curr Bioact Compd 7:66–74
Smyth H, Cozzolino D (2013) Instrumental methods (spectroscopy, electronic nose and tongue) as tools to predict taste and aroma in beverages: advantages and limitations. Chem Rev 113:1429–1440
Stratis DN, Eland KL, Carter JC, Tomlinson SJ, Angel SM (2001) Comparison of acoustic-optic and liquid crystal tunable filters for laser induced breakdown spectroscopy. Appl Spectrosc 55:999–1004
Subramanian A, Rodrigez-Saona L (2009) Fourier transform infrared (FTIR) spectroscopy. In: Sun DW (ed) Infrared spectroscopy for food quality analysis and control. Academic, Oxford, pp 146–174
Sumner LW, Mendes P, Dixon RA (2003) Plant metabolomics: large-scale phytochemistry in the functional genomics era. Phytochemistry 62:817–836
Sweetlove LJ, Last RL, Fernie AR (2004) Predictive metabolic engineering: a goal for systems biology. Plant Physiol 132:420–425
Velasco L, Pérez-Vich B, Fernandez-Martinez JM (1999) Non-destructive screening for oleic and linoleic acid in single sunflower achenes by near infrared reflectance spectroscopy. Crop Sci 39:219–222
Walter A, Studer B, Kolliker R (2012) Advanced phenotyping offers opportunities for improved breeding of forages and turfs species. Ann Bot 110:171–1279
Wetzel D, Brewer L (2010) Applications of vibrational spectroscopy to oilseeds analysis. In: Li-Chan E, Griffiths PR, Chalmers JM (eds) Applications of vibrational spectroscopy in food science. Wiley, Chichester, UK
White JW, Andrade-Sanchez P, Gore MA, Bronson KF, Coffelt TA, Conley MM, Feldmann KA, French AN, Heun J, Hunsaker DJ, Jenks MA, Kimball BA, Roth RL, Strand RJ, Thorp KR, Wall GA, Wang G (2012) Field-based phenomics for plant genetics research. Field Crop Res 133:101–112
Williams P (2010) The analysis of wheat by near infrared spectroscopy. Applications of vibrational spectroscopy to oilseeds analysis. In: Li-Chan E, Griffiths PR, Chalmers JM (eds) Applications of vibrational spectroscopy in food science. Wiley, Chichester, UK
Woodcock T, Downey G, O'Donnell CP (2008) Better quality food and beverages: the role of near infrared spectroscopy. J Near Infrared Spectros 16:1–29
Yu P (2005) Applications of hierarchical cluster analysis (CLA) and principal component analysis (PCA) in feed structure and feed molecular chemistry research, using synchrotron-based Fourier transform infrared (FTIR) micro spectroscopy. J Agric Food Chem 53:7115–7127
Yu P (2007) Molecular chemical structure of barley proteins revealed by ultra-spatially resolved synchrotron light source FTIR micro spectroscopy: comparison of barley varieties. Biopolymers 85:308–317
Yu P, Doiron K, Liu D (2008) Shining light on the differences in molecular structural chemical make-up and the cause of distinct degradation behaviour between

malting- and feed-type barley using synchrotron FTIR microspectroscopy: a novel approach. J Agric Food Chem 56:3417–3426

Yu P, Block HC, Doiron K (2009) Understanding the differences in molecular conformation of carbohydrate and protein in endosperm tissues of grains with different biodegradation kinetics using advanced synchrotron technology. Spectrochim Acta A Mol Biomol Spectrosc 71:1837–1844

Yu P, Damiran D, Azarfar A, Niu Z (2011) Detecting molecular features of spectra mainly associated with structural and non-structural carbohydrates in co-products from bioethanol production using DRIFT with uni- and multivariate molecular spectral analyses. Int J Mol Sci 12:1921–1935

Zhang X, Yu P (2012) Using ATR-FT/IR molecular spectroscopy to detect effects of blend DDGS inclusion level on the molecular structure spectral and metabolic characteristics of the proteins in hulless barley. Spectrochim Acta A Mol Biomol Spectrosc 95:53–63

Zhu J, Ingram PA, Benfey PN, Elich T (2011) From lab to field, new approaches to phenotyping root system architecture. Curr Opin Plant Biol 14:310–317

Next-Generation Sequencing and Assembly of Plant Genomes

Basant K. Tiwary

Contents

Abstract

Next-generation sequencing technology produces enormous volume of accurate and inexpensive sequence data in a short span of time. Three available common next-generation sequencing (NGS) platforms for genome sequencing are discussed here. The genome assembly and scaffolding algorithms are described with special emphasis on *de novo* assembly of short-read sequences. The biological applications of next-generation sequencing in plant sciences are covered with examples from plant genomics. An account on future prospects of this technology in plant genome analysis is discussed.

Introduction

The sequence-driven research in molecular biology started with pathbreaking research by two groups led by Sanger and Gilbert (Sanger et al. 1977; Maxam and Gilbert 1977). High-throughput sequencing (HTS) techniques popularly known as next-generation sequencing (NGS) were introduced in 2005 and have revolutionized the biomedical research by substantial increase in scale and resolution of various biological applications. They provide manyfold reads at a markedly reduced cost per sequenced nucleotide than conventional Sanger sequencing. Next-generation sequencing generates a huge amount of data necessitating the development of

B.K. Tiwary, Ph.D. (✉)
Centre for Bioinformatics, Pondicherry University, Pondicherry 605 014, India
e-mail: basant68@email.com; basant@bicpu.edu.in

D. Barh et al. (eds.), *PlantOmics: The Omics of Plant Science*, DOI 10.1007/978-81-322-2172-2_3, powerful computing and efficient algorithms. All commercial platforms have three common phages in their development, namely, preparation of sequencing library by adding adapters (defined sequences), immobilization of DNA fragments of sequencing library to a solid surface, and sequencing (Myllykangas et al. 2011). It can be used for whole genome sequencing, targeted resequencing, and identification of transcription factor binding sites and expression of noncoding RNA. There are several commercial platforms available such as 454 pyrosequencing (Roche Applied Science), the genome analyzer (Illumina), and SOLiD (Applied Biosystems). Next-generation sequencing can be applied to detect molecular variants such as single nucleotide variants, genomic insertions and deletions, and genomic rearrangements. RNA-seq can be used to determine the expression level of known genes and discovery of novel genes. ChIP-seq can be used for screening protein-DNA interaction at genome-wide scale. The whole genome sequencing and assembly of an organism are performed in various phases (Fig. 1). The major focus of this article is to introduce the reader with three common high-throughput sequencing platforms with more emphasis on various computational methods to analyze the next-generation sequencing data obtained from plant genomes.

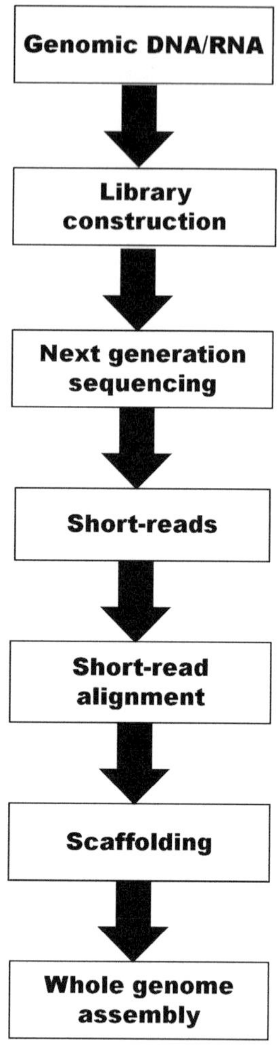

Fig. 1 Flow chart showing various steps in genome sequencing and assembly using next-generation sequencing technology

Next-Generation Sequencing Platforms

The three most popular sequencing platforms widely used to date are Roche 454 pyrosequencing, Illumina (Solexa), and SOLiD (Applied Biosystems).

454 Pyrosequencing

454 is the first next-generation technology introduced by Roche/454 Life Sciences which is based on pyrosequencing. In pyrosequencing, a double-stranded DNA is generated from a single-stranded DNA template by the addition of nucleotides. The addition of nucleotides is detected by the emission of light. It achieves a high throughput (~500 Mbp/run) with 400 bp read lengths. The major demerit associated with this platform is high error rate in homopolymer regions.

Illumina

The Illumina generates a much higher throughput (~1.5 Gbp) with a lower read length (~150 bp) when compared with 454. Although the read length is short, the platform generates high-quality sequences with an error rate less than

1 %. The principle of this method is based on sequencing by synthesis (reversible terminator chemistry). Long homopolymer runs do not affect the quality of sequence due to its chemistry. However, the major challenge associated with this technology is to process a large number of short reads, difficulty in *de novo* assembly, and in resequencing.

SOLiD

The ABI/SOLiD platform generates maximum high throughput achieved to date by any method (~3 Gbp/run) with a read length of 75 bp. This platform requires two kinds of library preparation: fragment or mate paired. Clonal bead populations are prepared in microreactors, and modified microbeads are deposited onto a glass slide. The sequencing was done using multiple cycles of ligation, detection, and cleavage.

Genome Assembly Algorithms

Most of the biological applications of next-generation sequencing such as quantification of transcriptome, assembly of new genomes, and identification of protein binding sites and alignment of sequence reads to a reference sequence as a first step of analysis. The process of aligning short reads into longer sequences is known as assembly. It is like a jigsaw puzzle where each short read is an individual part of the puzzle, and the whole genome sequence is a finished puzzle. There are many alignment tools developed in the last 4 years which are better than classical aligners in terms of speed and accuracy. Alignment algorithms can be based on hash tables, suffix trees, and merge sorting (Miller et al. 2010). The concept of hash table started with basic local alignment search tool (BLAST) which finds significant local alignment comparing exact matches to a k-mer (seeds) in a hash table. Ma et al. (2002) improved this method by creating the spaced seed (i.e., a seed with internal mismatches) which turned out to be the most popular approach for alignment of short reads. Eland, SOAP, SeqMap, MAQ, RMAP, ZOOM, and Novoalign are various popular programs for aligning short reads to a reference genome using spaced seed. Although spaced seed has mismatches within the seed, it never permits any gap in the seed. Eland was the first program developed by Anthony Cox from Illumina that aligns short oligonucleotides against a reference genome. SOAP is an efficient program for gapped and ungapped efficient alignment of short reads onto a reference genome (Li et al. 2008a). SeqMap can map a large amount of short reads to a genome based on index-filtering algorithm (Jiang and Wong 2008). MAQ builds assemblies by mapping short reads to a reference genome using quality score (Li et al. 2008b). RMAP software package has tools to map paired-end reads using a more sophisticated quality score (Smith et al. 2009). ZOOM maps short reads onto a reference genome with improved sensitivity and speed (Lin et al. 2008). Novoalign, a commercial software developed by Novocraft Technologies, is an aligner for short reads from Illumina genome analyzer. Another seeding approach *q*-gram filter builds an index allowing a gap within the seed. Two programs SHRiMP and RazerS are based on *q*-filter algorithm. SHRiMP is highly efficient in mapping short reads to a reference genome with high polymorphism (Rumble et al. 2009). RazerS is a popular read mapper with improved performance for long reads with large numbers of indels (Weese et al. 2009).

The algorithms based on suffix/prefix tries may be represented in a form of suffix tree (McCreight 1976), enhanced suffix array (Abouelhoda et al. 2004), and FM-index (Simpson and Durbin 2010). All algorithms identify exact matches at first and then build inexact alignments based on the exact matches. The suffix trie is a data structure that stores all the suffixes of a string in order to allow fast string matching. A trie needs a huge space and is impractical for even a small genome. Thus, there are various data structures such as suffix tree, suffix array, and FM-index to reduce the space. A suffix tree requires 12–17 bytes per nucleotide and is impractical for holding human genome in memory (Li and Homer 2010). The enhanced suffix array is more space efficient than suffix tree and takes only 6.25 bytes per nucleotide. The FM-index is the most space-efficient data structure using 0.2–2 bytes per nucleotide, and an FM-index of the entire human genome needs 2–8 GB of memory. The most widely used data structure is FM-index due to its small memory footprint. Bowtie, BWA, SOAP2, BWT-SW, and BWA-SW are the most popular programs built upon FM-index. Bowtie is a very fast and memory-efficient aligner for large genomes based on Burrows-Wheeler indexing (Langmead et al. 2009). Burrows-Wheeler alignment (BWA) tool is another efficient short-read aligner for large genomes allowing mismatches and indels based on Burrows-Wheeler transform (Li and Durbin 2009). SOAP2 is a short oligonucleotide alignment program with reduced memory usage and improved alignment speed (Li et al. 2009a). BWT-SW is an efficient tool to find all local alignments (Lam et al. 2008). Burrows-Wheeler Aligner's Smith-Waterman (BWA-SW) alignment is an efficient algorithm to align long reads of up to 1 Mb against a large sequence database (Li and Durbin 2010). However, MUMmer (Kurtz et al. 2004) and OASIS (Meek et al. 2003) are based on suffix tree, whereas Segemehl (Hoffmann et al. 2009) and Vmatch (Abouelhoda et al. 2004) apply enhanced suffix array as data structure. The yet other aligner of biological sequences (YOABS) is a very efficient long-read alignment program having advantages of both hash- and trie-based algorithms (Galinsky 2012).

There are more than 50 short-read alignment software packages available, albeit few of them are popular among users. Table 1 gives a list of popular alignment software packages available for short reads. All programs generate outputs in the form of a Sequence Alignment/Map (SAM; Li et al. 2009b) or Binary Alignment/Map (BAM; Carver et al. 2010) format which can be viewed through alignment viewers (Table 2) such as GBrowse (Stein et al. 2002), LookSeq (Manske and Kwiatkowski 2009), Tablet (Milne et al. 2010, 2013), BamView (Carver et al. 2010; Carver et al. 2013), GenomeView (Abeel et al. 2012), IGV (Thorvaldsdóttir et al. 2013), and MGAviewer (Zhu et al. 2013). The SAM format can be created and manipulated using SAMtools (Li et al. 2009b) which has extensive information regarding a read, its properties, and its alignment to a reference sequence. BAM format is the compressed binary form of SAM format which can be converted to SAM format and *vice versa* using SAMtools.

Table 1 Popular programs for short-read alignment

Program	Algorithm	References
Eland	Spaced seed	Illumina software
SOAP	Spaced seed	Li et al. (2008a)
SeqMap	Spaced seed	Jiang and Wong (2008)
MAQ	Spaced seed	Li et al. (2008b)
RMAP	Spaced seed	Smith et al. (2009)
ZOOM	Spaced seed	Lin et al. (2008)
Novoalign	Spaced seed	Novocraft Tech.
SHRiMP	Q-filter	Rumble et al. (2009)
RazerS	Q-filter	Weese et al. (2009)
BWA	FM-index	Li and Durbin (2009)
Bowtie	FM-index	Langmead et al. (2009)
SOAP2	FM-index	Li et al. (2009b)
BWT-SW	FM-index	Lam et al. (2008)
BWA-SW	FM-index	Li and Durbin (2010)
MUMmer	Suffix tree	Kurtz et al. (2004)
OASIS	Suffix tree	Meek et al. (2003)
Segemehl	Enhanced suffix array	Hoffmann et al. (2009)
Vmatch	Enhanced suffix array	Abouelhoda et al. (2004)
YOABS	Hash and trie based	Galinsky (2012)

Table 2 Alignment viewers of SAM/BAM format

Program	References
GBrowse	Stein et al. (2002)
LookSeq	Manske and Kwiatkowski (2009)
Tablet	Milne et al. (2010, 2013)
BamView	Carver et al. (2010, 2013)
GenomeView	Abeel et al. (2012)
IGV	Thorvaldsdóttir et al. (2013)
MGAviewer	Zhu et al. (2013)

De Novo Assembly of Short Reads

De novo sequence assembly is a method where individual short reads are merged into a long continuous sequence (contig) like the original template. In fact, short reads of 40 nucleotide length can be used to assemble the vast majority of protein encoding genes in most of the prokaryotic genomes albeit having many gaps. Table 3 shows common programs for assembling short reads without any reference genome. All algorithms for assembling capillary-based sequence reads of 400–1,000 nucleotides into long contiguous sequences adopts a common approach known as overlap-layout-consensus (OLC) approach (Li et al. 2012). The OLC algorithm first finds overlaps between sequence reads and then looks for most fitting pairs of reads (layout) and finally derives a consensus sequence from this layout. The overlap step is computationally expensive, and therefore various algorithmic approaches have been adopted to improve the computational efficiency. The OLC approach is adopted by many popular programs such as Arachne (Batzoglou et al. 2002), Celera Assembler (Myers et al. 2000), CAP3 (Huang and Madan 1999), PCAP (Huang et al. 2003), PHRAP (de la Bastide and McCombie 2007), Phusion (Mullikin and Ning 2003), and Newbler (a commercial assembler developed by Roche Diagnostics). Most of the assemblers designed for short-read sequences are based on De Bruijn graph (DBG; Li et al. 2012) and Eulerian path approach (Pevzner et al. 2001). Some of the popular software packages based on DBG and Eulerian paths are Euler (Pevzner et al. 2001), Euler-USR (Chaisson et al. 2009), Velvet (Zerbino and Birney 2008), ABySS (Simpson et al. 2009), ALLPATH-LG (Gnerre et al. 2011), SOAPdenovo (Li et al. 2010), and Gossamer (Conway et al. 2012). The graph-based algorithm assembly creates a model where the string is a node and the relation between strings is represented in a form of edges. In De Bruijn graph (DBG) algorithm, reads are chopped into smaller fragments (k-mers), and k-mers are converted into a DBG for final determination of genome sequence. The optimal solution is obtained through finding a Eulerian path (i.e., a path which covers a node only once) through the graph. However, the string graph assembler (SGA) is a program based on a string graph which keeps all reads intact and generates a graph based on overlaps between reads (Simpson and Durbin 2012).

Table 3 Some programs for *de novo* assembly of short reads

Program	Algorithm	References
Arachne	OLC	Batzoglou et al. (2002)
Celera Assembler	OLC	Myers et al. (2000)
CAP3	OLC	Huang and Madan (1999)
PCAP	OLC	Huang et al. (2003)
PHRAP	OLC	de la Bastide and McCombie (2007)
Phusion	OLC	Mullikin and Ning (2003)
Newbler	OLC	Roche Diagnostics
Euler	DBG and Eulerian path	Pevzner et al. (2001)
Euler-USR	DBG and Eulerian path	Chaisson et al. (2009)
Velvet	DBG and Eulerian path	Zerbino and Birney (2008)
ABySS	DBG and Eulerian path	Simpson et al. (2009)
ALLPATH-LG	DBG and Eulerian path	Gnerre et al. (2011)
SOAPdenovo	DBG and Eulerian path	Li et al. (2010)
Gossamer	DBG and Eulerian path	Conway et al. (2012)
SGA	String graph	Simpson and Durbin (2012)

Scaffolding Algorithms

The large assembled regions of sequence are known as contigs which need to be joined together to get the whole genome sequence. The final process of joining multiple contigs together to form a continuous genome sequence (scaffold or supercontig) is known as scaffolding or finishing. This process is done in four consecutive steps, namely, contig orientation, contig ordering, contig distancing, and gap closing. The orientation of contigs in same direction (5′-3′ direction in prokaryotes) is done using a reverse complementary sequence. The contigs are placed in an appropriate order starting at the origin of replication and extended in 5′-3′ direction of DNA replication. The distance between contigs can be estimated after correct orientation and order. The final step of closing and filling gap can

Table 4 Popular programs for scaffolding

Program	References
SOAPdenovo	Li et al. (2010)
ABySS	Simpson et al. (2009)
Bambus	Pop et al. (2004)
SOPRA	Dayarian et al. (2010)
SSPACE	Boetzer et al. (2011)
OPERA	Gao et al. (2011)
MIP Scaffolder	Salmela et al. (2011)
GRASS	Gritsenko et al. (2012)
RACA	Kim et al. (2013)

result into a finished genome. The paired-end reads provide additional information for grouping two contigs in a genome. Scaffolding process may be based on a graph, where a contig is treated as node and matching pair contigs are connected by edges. The algorithm finds an optimal path through the graph. The scaffolding process may be more accurate using additional information such as reference sequences of related organism, restriction maps, and RNA-seq data. Some of the popular programs (Table 4) for scaffolding are SOAPdenovo (Li et al. 2010), ABySS (Simpson et al. 2009), Bambus (Pop et al. 2004), SOPRA (Dayarian et al. 2010), SSPACE (Boetzer et al. 2011), OPERA (Gao et al. 2011), MIP Scaffolder (Salmela et al. 2011), GRASS (Gritsenko et al. 2012), and RACA (Kim et al. 2013).

Biological Applications of Next-Generation Sequencing

Genome Sequencing

The worldwide effort to understand the genetic basis of common and rare genetic disorder has gained momentum with the advent of next-generation sequencing technology. It will help largely in the identification of single nucleotide polymorphisms (SNPs) and haplotype mapping (International HapMap Consortium 2003) in individual human genomes and lay a foundation for personalized medicine. The 1,000 genome project (http://www.1000genomes.org) turned into reality with the availability of NGS technology. Cancer biology is another area where next-generation sequencing can decipher the novel molecular pathways involved in tumorigenesis (Hahn and Weinberg 2002). Next-generation sequencing will also influence the highly emerging area of synthetic biology where a new enzyme or a novel genetic network may be developed (Khalil and Collins 2010).

Functional Genomics

Functional genomics is focused to apply genomics data for understanding dynamic life processes. RNA-seq is widely used to quantify gene expression levels for different genes like microarray technology (Wang et al. 2009). It has several advantages over microarray analysis such as no prior sequence information is needed; highly expressed and lowly expressed genes are equally detected and allow detailed identification of structure of transcripts including alternative promoters and alternative splicing sites. In RNA-seq technology, the relative abundance of a transcript is estimated by counting the number of times it is hit by the sequence reads. This method accurately estimates relative RNA levels under different experimental conditions or in different cell types.

Epigenetics

Epigenetics deals with heritable regulatory changes in chromosomes without any change in the DNA sequence (Bird 2007). The epigenetic changes such as DNA methylation, histone modification, and ncRNA have an important role in maintaining chromosome structure. The post-translational modifications of histones such as methylation, acetylation, ubiquitination, and phosphorylation generate different "marks" for different functional properties. The DNA-binding proteins, histone modifications, or nucleosomes can be mapped across the genome using *ChIP-seq* approach where a chromatin immunoprecipitation (CHIP) is followed by sequencing (Park 2009). The DNA methylation involves methylation of the cytosine base in DNA and can be identified by a version of NGS known as *Methyl-seq* (Brunner et al. 2009). The active gene regulatory elements can be better understood by using another approach of NGS known as *DNAse-seq* (Song and Crawford 2010). Noncoding RNA (ncRNA) has been implicated in many epigenetic events such as X-chromosome inactivation and gene silencing (Mercer et al. 2009).

Current Status of Next-Generation Sequencing in Plant Genomics

NGS has been used extensively for whole genome sequencing of plants in the last 5 years (Table 5). *Arabidopsis thaliana* (125 Mb) was the first plant completely sequenced in 2000 using Sanger sequencer (*Arabidopsis* genome; Initiative 2000). It was followed by sequencing two major rice varieties, namely, *japonica* (420 Mb; Goff et al. 2002) and *indica* (466 Mb; Yu et al. 2002) in 2002 and first fruit grapevine (487 Mb; Jaillon et al. 2007) in 2007 using the same sequencing method. Subsequently, the draft genomes of papaya (372 Mb; Ming et al. 2008) and legume *Lotus japonicus* (315 Mb; Sato et al. 2008) were developed in 2008. Sorghum genome (730 Mb; Paterson et al. 2009) and maize genome (2.3 Gb; Schnable et al. 2009), and soya bean genome (1.1 Gb; Schmutz et al. 2010) were sequenced in 2009 and 2010, respectively, onto Sanger platform. However, the pace of sequencing plant genomes rapidly increased with the advent of the next-generation sequencing (NGS) technology. Cucumber genome (243.5 Mb; Huang et al. 2009) was sequenced taking advantages of both

Table 5 Overview of plant genomes sequenced applying next-generation sequencing

Plant	Genome size (Mb)	NGS platform	References
Cucumber (*Cucumis sativus*)	244	Illumina	Huang et al. (2009)
Wild grass (*Brachypodium distachyon*)	355	Illumina	International Brachypodium Initiative (2010)
Cocoa (*Theobroma cacao*)	430	Roche 454 and Illumina	Argout et al. (2011)
Apple (*Malus* X *domestica*)	604	Roche 454	Velasco et al. (2010)
Woodland strawberry (*Fragaria vesca*)	210	Roche 454, Illumina, and SOLiD	Shulaev et al. (2011)
Potato (*Solanum tuberosum*)	727	Roche 454 and Illumina	Potato Genome Sequencing Consortium (2011)
Cannabis (*Cannabis sativa*)	534	Roche 454 and Illumina	van Bakel et al. (2011)
Pigeon pea (*Cajanus cajan*)	606	Illumina	Varshney et al. (2012)
Extremophile crucifer (*Thellungiella parvula*)	140	Roche 454 and Illumina	Dassanayake et al. (2011)
Date palm (*Phoenix dactylifera*)	658	Illumina	Al-Dous et al. (2011)
Grape (*Vitis vinifera*)	505	Roche 454	Velasco et al. (2007)
Brassica rapa	288	Illumina	Wang et al. (2011)
Cotton (*Gossypium raimondii*)	775	Illumina	Wang et al. (2012)
Melon (*Cucumis melo*)	375	Roche 454 and Illumina	Garcia-Mas et al. (2012)
Tomato (*Solanum lycopersicom*)	760	Roche 454, Illumina, and SOLiD	Tomato Genome Consortium (2012)
Banana (*Musa acuminata*)	472	Roche 454 and Illumina	D'Hont et al. (2012)
Barley (*Hordeum vulgare*)	4,980	Roche 454 and Illumina	International Barley Genome Sequencing Consortium (2012)
Bread wheat (*Triticum aestivum*)	17,000	Roche 454	Brenchley et al. (2012)
Wheat A (*Triticum urartu*)	4,940	Illumina	Ling et al. (2013)
Sweet orange (*Citrus sinensis*)	367	Illumina	Qiang et al. (2013)
Chickpea (*Cicer arietinum*)	740	Roche 454	Jain et al. (2013)
Sacred lotus (*Nelumbo nucifera*)	929	Roche 454 and Illumina	Ming et al. (2013)

Illumina GA technology (high sequencing depth and low unit cost) and Sanger technology (long read and clone length). In 2010, wild grass, *Brachypodium distachyon* was sequenced using both methods (International Brachypodium Initiative 2010). The cocoa genome (430 Mb; Argout et al. 2011) was sequenced applying two NGS platforms, namely, Roche 454 and Illumina along with Sanger sequencing. The apple genome (604 Mb; Velasco et al. 2010) was sequenced using both Roche 454 technology and Sanger technology. The woodland strawberry (209.8 Mb) was sequenced using three NGS platforms: Roche 454, Illumina Solexa, and Life Technologies SOLiD (Shulaev et al. 2011). In 2011, the potato genome (727 Mbp) was sequenced using two major NGS platforms: Roche 454 and Illumina Genome Analyzer along with conventional Sanger sequencing technology (Potato Genome Sequencing Consortium 2011). The cannabis genome (534 Mb) was sequenced using Roche 454 and Illumina Genome Analyzer IIx or HiSeq platforms (van Bakel et al. 2011). The draft genome of pigeon pea (606 Mb) was sequenced with Illumina technology along with Sanger technology (Varshney et al. 2012). A close relative of *Arabidopsis*, *Thellungiella parvula*, is endemic to saline habitat, and its genome (140 Mb) was investigated using Roche 454 and Illumina GA2 (Dassanayake et al. 2011). The date palm genome (658 Mb) was sequenced *de novo* using Illumina GA2 and Sanger sequencer (Al-Dous et al. 2011). A draft consensus sequence of grape genome (504 Mbp) was developed with 1.7 million SNPs (Velasco et al. 2007). *Brassica rapa* genome was sequenced by *Brassica rapa* genome sequencing project consortium (Wang et al. 2011). The cotton plant draft genome (775 Mb) was sequenced using Illumina HiSeq 2000 platform (Wang et al. 2012). Melon, a close relative of cucumber, was covered for genome (375 Mb) using Roche 454 pyrosequencing, Illumina, and Sanger technologies (Garcia-Mas et al. 2012). The tomato genome (760 Mb) was sequenced using Roche 454 GS FLX, Illumina GA2, and SOLiD along with Sanger sequencing (Tomato Genome Consortium 2012). The banana genome (472 Mb) was sequenced with combined application of Roche 454, Illumina GA2, and Sanger technologies (D'Hont et al. 2012). Recently, both Roche 454 (GS FLX or FLX Titanium) and Illumina (GA2 or HiSeq 2000) have been applied to decipher the genome sequence of barley (4.98 Gbp) (The International Barley Genome Sequencing Consortium 2012). Since the bread wheat has a large genome size (17 Gb) than other cereals and is hexaploid in nature, the successful completion of bread wheat genome sequencing using 454 pyrosequencing and wheat A-genome (4.94 Gb) sequencing on Illumina HiSequation platform is a significant event in the next-generation sequencing of crops (Brenchley et al. 2012; Ling et al. 2013). The completion of wheat genome will not only pave the way for better productivity of wheat crop but decipher the role of polyploidy in plant genome evolution as well. Recently, the whole genome sequencing of sweet orange (*Citrus sinensis*; 367 Mb) was done on Illumina GAII sequencer (Qiang et al. 2013). The draft genome sequence of chickpea (*Cicer arietinum*; 740 Mb) was completed on 454/Roche GS FLX Titanium platform (Jain et al. 2013). The complete genome of sacred lotus (929 Mb) was sequenced with combined application of Illumina and 454 technologies (Ming et al. 2013). Other whole genome sequencing projects are underway in many plant species such as amborella (*Amborella trichopoda*), columbine (*Aquilegia* sp.), sugar beet (*Beta vulgaris*), monkey flower (*Mimulus guttatus*), rose gum tree (*Eucalyptus grandis*), flax (*Linum usitatissimum*), cassava (*Manihot esculenta*), and pear (*Pyrus bretschneideri*). Some species of the lower plant species were sequenced in order to understand the evolution of vascular plants on land. The genome of green alga (Chlamydomonas reinhardtii; 120 Mb) (Merchant et al. 2007), genome of moss (480 Mb; *Physcomitrella patens*) (Rensing et al. 2008), and genome of lycophyte (*Selaginella moellendorffii*; 213 Mb) (Banks et al. 2011) were sequenced using conventional Sanger sequencing and have revealed insights into genomic evolution of land plants.

Future Prospects in Next-Generation Sequencing and Assembly

NGS-based technology has a wide scope for solving many existing problems in genomics. However, the low read length with intrinsic error rate of NGS is a major problem and is a prohibitive factor for *de novo* assembly of large genomes. Therefore, this technology is largely based on the availability of a reference genome. However, this problem will be solved in the future with an increase in size of the read length. Although NGS provides a deep coverage, it has a low throughput in comparison to microarrays. However, this problem may be alleviated by developing large-scale parallel NGS. With an increase in the number of reference genomes, it is expected that whole genome resequencing will become more popular in order to interrogate the diversity of crop genomes. New dedicated algorithms are needed to deal with complex repeats in the plant genome for better quality of assembly. Along with assembly algorithms, the next-generation data quality and quantity should be improved in the near future.

Conclusion

In this work, three common NGS platforms and various computational methods for analysis of NGS-derived sequence data are discussed. The impact of NGS technology on plant genome sequencing especially on crop genomes, till date, is elaborated. It is expected that NGS technology will grow further in sensitivity and speed and will decipher the genomes of other plants to understand the genome evolution and help in revealing key genomic features to agricultural productivity.

References

Abeel T, Van Parys T, Saeys Y, Galagan J, Van de Peer Y (2012) GenomeView: a next-generation genome browser. Nucleic Acids Res 40:e12

Abouelhoda MI, Kurtz S, Ohlebusch E (2004) Replacing suffix trees with enhanced suffix arrays. J Discrete Algorithms 2:53–86

Al-Dous EK, George B, Al-Mahmoud ME, Al-Jaber MY, Wang H, Salameh YM, Al-Azwani EK, Chaluvadi S, Pontaroli AC, DeBarry J et al (2011) De novo genome sequencing and comparative genomics of date palm (*Phoenix dactylifera*). Nat Biotechnol 29:521–527

Arabidopsis Genome Initiative (2000) Analysis of the genome sequence of the flowering plant *Arabidopsis thaliana*. Nature 408:796–815

Argout X, Salse J, Aury J-M, Guiltinan MJ, Droc G, Gouzy J, Allegre M, Chaparro C, Legavre T, Maximova SN et al (2011) The genome of *Theobroma cacao*. Nat Genet 43:101–108

Banks JA, Nishiyama T, Hasebe M, Bowman JL, Gribskov M, dePamphilis C, Albert VA, Aono N, Aoyama T, Ambrose BA, Ashton NW et al (2011) The Selaginella genome identifies genetic changes associated with the evolution of vascular plants. Science 332:960–963

Batzoglou S, Jaffe DB, Stanley K, Butler J, Gnerre S, Mauceli E, Berger B, Mesirov JP, Lander ES (2002) ARACHNE: a whole-genome shotgun assembler. Genome Res 12:177–189

Bird A (2007) Perceptions of epigenetics. Nature 447:396–398

Boetzer M, Henkel CV, Jansen HJ, Butler D, Pirovano W (2011) Scaffolding pre-assembled contigs using SSPACE. Bioinformatics 27:578–579

Brenchley R, Spannagl M, Pfeifer M, Barker GLA, D'Amore R, Allen AM, McKenzie N, Kramer M, Kerhornou A, Bolser D (2012) Analysis of the bread wheat genome using whole-genome shotgun sequencing. Nature 491:705–710

Brunner AL, Johnson DS, Kim SW, Valouev A, Reddy TE, Neff NF, Anton E, Medina C, Nguyen L, Chiao E et al (2009) Distinct DNA methylation patterns characterize differentiated human embryonic stem cells and developing human fetal liver. Genome Res 19:1044–1056

Carver T, Bohme U, Otto T, Parkhill J, Berriman M (2010) BamView: viewing mapped read alignment data in the context of the reference sequence. Bioinformatics 26:676–677

Carver T, Harris SR, Otto TD, Berriman M, Parkhill J, McQuillan JA (2013) BamView: visualizing and interpretation of next-generation sequencing read alignments. Brief Bioinform 14:203–212

Chaisson MJP, Brinja D, Pevzner PA (2009) De novo fragment assembly with short mate-paired reads: does the read length matter? Genome Res 19:336–346

Conway T, Wazny J, Bromage A, Zobel J, Beresford-Smith B (2012) Gossamer—a resource-efficient *de novo* assembler. Bioinformatics 28:1937–1938

D'Hont A, Denoeud F, Aury JM, Baurens FC, Carreel F, Garsmeur O, Noel B, Bocs S, Droc G, Rouard M et al (2012) The banana (*Musa acuminata*) genome and the evolution of monocotyledonous plants. Nature 488:213–217

Dassanayake M, Oh D-H, Haas JS, Hernandez A, Hong H, Ali S, Yun D-J, Bressan RA, Zhu J-K, Bohnert HJ et al (2011) The genome of the extremophile crucifer *Thellungiella parvula*. Nat Genet 43:913–918

Dayarian A, Michael TP, Sengupta AM (2010) SOPRA: scaffolding algorithm for paired reads via statistical optimization. BMC Bioinf 11:345

de la Bastide M, McCombie WR (2007) Assembling genomic DNA sequences with PHRAP. Curr Protoc Bioinform, Chapter 11:Unit 11.4

Galinsky VL (2012) YOABS: yet other aligner of biological sequences—an efficient linearly scaling nucleotide aligner. Bioinformatics 28:1070–1077

Gao S, Sung WK, Nagarajan N (2011) Opera: reconstructing optimal genomic scaffolds with high-throughput paired-end sequences. J Comput Biol 18:1681–1691

Garcia-Mas J, Benjak A, Sanseverino W, Bourgeois M, Mir G, Gonzalez VM, Henaff E, Camara F, Cozzuto L, Lowy E et al (2012) The genome of melon (*Cucumis melo* L.). Proc Natl Acad Sci U S A 109:11872–11877

Gnerre S, Maccallum I, Przybylski D, Ribeiro FJ, Burton JN, Walker BJ, Sharpe T, Hall G, Shea TP, Sykes S et al (2011) High-quality draft assemblies of mammalian genomes from massively parallel sequence data. Proc Natl Acad Sci U S A 108:1513–1518

Goff SA, Ricke D, Lan TH, Presting G, Wang R, Dunn M, Glazebrook J, Sessions A, Oeller P, Varma H et al (2002) A draft sequence of the rice genome (*Oryza sativa* L. *ssp japonica*). Science 296:92–100

Gritsenko AA, Nijkamp JF, Reinders MJT, de Ridder D (2012) GRASS: a generic algorithm for scaffolding next-generation sequencing assemblies. Bioinformatics 28:1429–1437

Hahn WC, Weinberg RA (2002) Mechanisms of disease: rules for making human tumor cells. N Engl J Med 34:1593–1603

Hoffmann S, Otto C, Kurtz S, Sharma CM, Khaitovich P, Vogel J, Stadler PF, Hackermüller J (2009) Fast mapping of short sequences with mismatches, insertions and deletions using index structures. PLoS Comput Biol 5:e1000502

Huang X, Madan A (1999) CAP3: A DNA sequence assembly program. Genome Res 9:868–877

Huang X, Wang J, Aluru S, Yang SP, Hillier L (2003) PCAP: a whole-genome assembly program. Genome Res 13:2164–2170

Huang S, Li R, Zhang Z, Li L, Gu X, Fan W, Lucas WJ, Wang X, Xie B, Ni P, Ren Y et al (2009) The genome of the cucumber, *Cucumis sativus* L. Nat Genet 41:1275–1281

International Brachypodium Initiative (2010) Genome sequencing and analysis of the model grass *Brachypodium distachyon*. Nature 463:763–768

International HapMap Consortium (2003) The International HapMap Project. Nature 426:789–796

Jaillon O, Aury JM, Noel B, Policriti A, Clepet C, Casagrande A, Choisne N, Aubourg S, Vitulo N, Jubin C et al (2007) The grapevine genome sequence suggests ancestral hexaploidization in major angiosperm phyla. Nature 449:463–467

Jain M, Misra G, Patel RK, Priya P, Jhanwar S, Khan AW, Shah N, Singh VK, Garg R, Jeena G et al (2013) A draft genome sequence of the pulse crop chickpea (*Cicer arietinum* L.). Plant J. doi:10.1111/tpj.12173

Jiang H, Wong WH (2008) SeqMap: mapping massive amount of oligonucleotides to the genome. Bioinformatics 24:2395–2396

Khalil AS, Collins JJ (2010) Synthetic biology: applications come of age. Nat Rev Genet 11:367–379

Kim J, Larkin DM, Cai Q, Asan ZY, Ge R-L, Auvil L, Capitanu B, Zhang G, Lewin HA, Ma J (2013) Reference-assisted chromosome assembly. Proc Natl Acad Sci U S A 110:1785–1790

Kurtz S, Phillippy A, Delcher AL, Smoot M, Shumway M, Antonescu C, Salzberg SL (2004) Versatile and open software for comparing large genomes. Genome Biol 5:R12

Lam TW, Sung WK, Tam SL, Wong CK, Yiu SM (2008) Compressed indexing and local alignment of DNA. Bioinformatics 24:791–797

Langmead B, Trapnell C, Pop M, Salzberg SL (2009) Ultrafast and memory-efficient alignment of short DNA sequences to the human genome. Genome Biol 10:R25

Li H, Durbin R (2009) Fast and accurate short read alignment with Burrows-Wheeler transform. Bioinformatics 25:1754–1760

Li H, Durbin R (2010) Fast and accurate long-read alignment with Burrows-Wheeler transform. Bioinformatics 26:589–595

Li H, Homer N (2010) A survey of sequence alignment algorithms for next-generation sequencing. Brief Bioinformatics 11:473–483

Li H, Ruan J, Durbin R (2008a) Mapping short DNA sequencing reads and calling variants using mapping quality scores. Genome Res 18:1851–1858

Li R, Li Y, Kristiansen K, Wang J (2008b) SOAP: short oligonucleotide alignment program. Bioinformatics 24:713–714

Li R, Yu C, Li Y, Lam TW, Yiu SM, Kristiansen K, Wang J (2009a) SOAP2: an improved ultrafast tool for short read alignment. Bioinformatics 25:1966–1967

Li H, Handsaker B, Wysoker A, Fennell T, Ruan J, Homer N, Marth G, Abecasis G, Durbin R, 1000 Genome Project Data Processing Subgroup (2009b) The sequence alignment/map (SAM) format and SAMtools. Bioinformatics 25:2078–2079

Li R, Zhu H, Ruan J, Qian W, Fang X, Shi Z, Li Y, Li S, Shan G, Kristiansen K, Li S, Yang H, Wang J, Wang J (2010) De novo assembly of human genomes with massively parallel short read sequencing. Genome Res 20:265–272

Li Z, Chen Y, Mu D, Yuan J, Shi Y, Zhang H, Gan J, Li N, Hu X, Liu B, Yang B, Fan W (2012) Comparison of the two major classes of assembly algorithms: overlap–layout–consensus and de-bruijn-graph. Brief Funct Genomics 11:25–37

Lin H, Zhang Z, Zhang MQ, Ma B, Li M (2008) ZOOM! Zillions of oligos mapped. Bioinformatics 24:2431–2437

Ling H-Q, Zhao S, Liu D, Wang J, Sun H, Zhang C, Fan H, Li D, Dong L, Tao Y et al (2013) Draft genome of the wheat A-genome progenitor *Triticum urartu*. Nature 496:87–90

Ma B, Tromp J, Li M (2002) PatternHunter: faster and more sensitive homology search. Bioinformatics 18:440–445

Manske HM, Kwiatkowski DP (2009) LookSeq: a browser-based viewer for deep sequencing data. Genome Res 19:2125–2132

Maxam AM, Gilbert W (1977) A new method for sequencing DNA. Proc Natl Acad Sci U S A 74:560–564

McCreight EM (1976) A space-economical suffix tree construction algorithm. J ACM 23:262–272

Meek C, Patel JM, Kasetty S (2003) OASIS: an online and accurate technique for local-alignment searches on biological sequences. In: Proceedings of 29th international conference on Very Large Data Bases (VLDB 2003), Berlin, pp 910–921

Mercer TR, Dinger ME, Mattick JS (2009) Long non-coding RNAs: insights into functions. Nat Rev Genet 10:155–159

Merchant SS, Prochnik SE, Vallon O, Harris EH, Karpowicz SJ, Witman GB, Terry A, Salamov A, Fritz-Laylin LK, Maréchal-Drouard L et al (2007) The *Chlamydomonas* genome reveals the evolution of key animal and plant functions. Science 318:245–250

Miller J, Koren S, Sutton G (2010) Assembly algorithms for next-generation sequencing data. Genomics 14:315–327. doi:10.1016/j.ygeno.2010.03.001

Milne I, Bayer M, Cardle L, Shaw P, Stephen G, Wright F, Marshall D (2010) Tablet–next generation sequence assembly visualization. Bioinformatics 26:401–402

Milne I, Stephen G, Bayer M, Cock PJA, Pritchard L, Cardle L, Shaw PD, Marshall D (2013) Using Tablet for visual exploration of second-generation sequencing data. Brief Bioinform 14:193–202

Ming R, Hou S, Feng Y, Yu Q, Dionne-Laporte A, Saw JH, Senin P, Wang W, Ly BV, Lewis KL et al (2008) The draft genome of the transgenic tropical fruit tree papaya (Carica papaya Linnaeus). Nature 452:991–996

Ming R, VanBuren R, Liu Y, Yang M, Han Y, Li L-T, Zhang Q, Kim M-J, Schatz MC, Campbell M et al (2013) Genome of the long-living sacred lotus (Nelumbo nucifera Gaertn). Genome Biol 14:R41

Mullikin JC, Ning Z (2003) The phusion assembler. Genome Res 13:81–90

Myers EW, Sutton GG, Delcher AL, Dew IM, Fasulo DP, Flanigan MJ, Kravitz SA, Mobarry CM, Reinert KH, Remington KA, Anson EL, Bolanos RA, Chou HH et al (2000) A whole-genome assembly of *Drosophila*. Science 287:2196–2204

Myllykangas S, Buenrostro J, Ji HP (2011) Overview of sequencing technology platforms. In: Rodriguez-Ezpeleta N, Hackenberg M, Aransayet AM (eds) Bioinformatics for high throughput sequencing. Springer, New York

Park PJ (2009) Chip-seq: advantages and challenges of a maturing technology. Nat Rev Genet 10:669–680

Paterson AH, Bowers JE, Bruggmann R, Dubchak I, Grimwood J, Gundlach H, Haberer G, Hellsten U, Mitros T, Poliakov A, Schmutz J et al (2009) The *Sorghum bicolor* genome and the diversification of grasses. Nature 457:551–556

Pevzner PA, Tang H, Waterman MS (2001) An Eulerian path approach to DNA fragment assembly. Proc Natl Acad Sci U S A 98:9748–9753

Pop M, Kosack DS, Salzberg SL (2004) Hierarchical scaffolding with Bambus. Genome Res 14:149–159

Potato Genome Sequencing Consortium (2011) Genome sequence and analysis of the tuber crop potato. Nature 475:189–195

Qiang X, Chen L-L, Ruan X, Chen D, Zhu A, Chen C, Bertrand D, Jiao W-B, Hao B-H, Lyon MP et al (2013) The draft genome of sweet orange (*Citrus sinensis*). Nat Genet 45:59–66

Rensing SA, Lang D, Zimmer AD, Terry A, Salamov A, Shapiro H, Nishiyama T, Perroud P-F, Lindquist EA, Kamisugi YA et al (2008) The *Physcomitrella* genome reveals evolutionary insights into the conquest of land by plants. Science 319:64–69

Rumble SM, Lacroute P, Dalca AV, Fiume M, Sidow A, Brudno M (2009) SHRiMP: accurate mapping of short color-space reads. PLoS Comput Biol 5(5):e1000386. doi:10.1371/journal.pcbi.1000386

Salmela L, Mäkinen V, Välimäki N, Ylinen J, Ukkonen E (2011) Fast scaffolding with small independent mixed integer programs. Bioinformatics 27:3259–3265

Sanger F, Nicklen S, Coulson AR (1977) DNA sequencing with chain-terminating inhibitors. Proc Natl Acad Sci U S A 74:5463–5467

Sato S, Nakamura Y, Kaneko T, Asamizu E, Kato T, Nakao M, Sasamoto S, Watanabe A, Ono A, Kawashima K et al (2008) Genome structure of the legume, *Lotus japonicus*. DNA Res 15:227–239

Schmutz J, Cannon SB, Schlueter J, Ma J, Mitros T, Nelson W, Hyten DL, Song Q, Thelen JJ, Cheng J et al (2010) Genome sequence of the palaeopolyploid soybean. Nature 463:178–183

Schnable PS, Ware D, Fulton RS, Stein JC, Wei F, Pasternak S, Liang C, Zhang J, Fulton L, Graves TA et al (2009) The B73 maize genome: complexity, diversity and dynamics. Science 326:1112–1115

Shulaev V, Sargent DJ, Crowhurst RN, Mockler TC, Folkerts O, Delcher AL, Jaiswal P, Mockaitis K, Liston A, Mane SP et al (2011) The genome of woodland strawberry (*Fragaria vesca*). Nat Genet 43:109–116

Simpson JT, Durbin R (2010) Efficient construction of an assembly string graph using the FM-index. Bioinformatics 26:i367–i373

Simpson JT, Durbin R (2012) Efficient de novo assembly of large genomes using compressed data structures. Genome Res 22:549–556

Simpson JT, Wong K, Jackman SD, Schein JE, Jones SJ, Birol I (2009) ABySS: a parallel assembler for short read sequence data. Genome Res 19:1117–1123

Smith AD, Chung WY, Hodges E, Kendall J, Hannon G, Hicks J, Xuan Z, Zhang MQ (2009) Updates to the RMAP short-read mapping software. Bioinformatics 25:2841–2842

Song L, Crawford GE (2010) DNase-seq: a high-resolution technique for mapping active gene regulatory elements across the genome from mammalian cells. Cold Spring Harb Protoc. doi:10.1101/pdb.prot5384

Stein LD, Mungall C, Shu S, Caudy M, Mangone M, Day A, Nickerson E, Stajich JE, Harris TW, Arva A, Lewis S (2002) The generic genome browser: a building block for a model organism system database. Genome Res 12:1599–1610

The International Barley Genome Sequencing Consortium (2012) A physical, genetic and functional sequence assembly of the barley genome. Nature 491:711–716

Thorvaldsdóttir H, Robinson JT, Mesirov JP (2013) Integrative Genomics Viewer (IGV): high-performance genomics data visualization and exploration. Brief Bioinform 14:178–192

Tomato Genome Consortium (2012) The tomato genome sequence provides insights into fleshy fruit evolution. Nature 485:635–641

van Bakel H, Stout J, Cote A, Tallon C, Sharpe A, Hughes T, Page J (2011) The draft genome and transcriptome of *Cannabis sativa*. Genome Biol 12:R102

Varshney RK, Chen W, Li Y, Bharti AK, Saxena RK, Schlueter JA, Donoghue MTA, Azam S, Fan G, Whaley AM et al (2012) Draft genome sequence of pigeonpea (*Cajanus cajan*), an orphan legume crop of resource-poor farmers. Nat Biotech 30:83–89

Velasco R, Zharkikh A, Troggio M, Cartwright DA, Cestaro A, Pruss D, Pindo M, Fitzgerald LM, Vezzulli S, Reid J et al (2007) A high quality draft consensus sequence of the genome of a heterozygous grapevine variety. PLoS ONE 2(12):e1326. doi:10.1371/journal.pone.0001326

Velasco R, Zharkikh A, Affourtit J, Dhingra A, Cestaro A, Kalyanaraman A, Fontana P, Bhatnagar SK, Troggio M, Pruss D et al (2010) The genome of the domesticated apple (*Malus* X *domestica* Borkh.). Nat Genet 42:833–839

Wang Z, Gerstein M, Snyder M (2009) RNA-Seq: a revolutionary tool for transcriptomics. Nat Rev Genet 10:57–63

Wang X, Wang H, Wang J, Sun R, Wu J, Liu S, Bai Y, Mun J-H, Bancroft I, Cheng F et al (2011) The genome of the mesopolyploid crop species *Brassica rapa*. Nat Genet 43:1035–1039

Wang K, Wang Z, Li F, Ye W, Wang J, Song G, Yue Z, Cong L, Shang H, Zhu S et al (2012) The draft genome of a diploid cotton *Gossypium raimondii*. Nat Genet 44:1098–1103

Weese D, Emde AK, Rausch T, Döring A, Reinert K (2009) RazerS–fast read mapping with sensitivity control. Genome Res 19:1646–1654

Yu J, Hu SN, Wang J, Wong GK, Li S, Liu B, Deng Y, Dai L, Zhou Y, Zhang X et al (2002) A draft sequence of the rice genome (*Oryza sativa* L. *ssp indica*). Science 296:79–92

Zerbino DR, Birney E (2008) Velvet: algorithms for de novo short read assembly using de Bruijn graphs. Genome Res 18:821–829

Zhu Z, Niu B, Chen J, Wu S, Sun S, Li W (2013) MGAviewer: a desktop visualization tool for analysis of metagenomics alignment data. Bioinformatics 29:122–123

Functional Genomics: Applications in Plant Science

Uday Chand Jha, Jayant S. Bhat, Basavanagouda S. Patil, Firoz Hossain, and Debmalya Barh

Contents

Abstract

The completion of the *Arabidopsis* Genome Initiative project in the year 2000 and decoding the whole genome sequence of many plant species thereafter lead to generation of mountains of sequence information. But the function of most of these sequences remains elusive. This grand challenge of defining the function of all the sequence information and genes has been the scope of functional genomics. Functional genomics deals with the function of gene at transcription, at translation and at regulation level. Numerous techniques including microarray, serial analysis of gene expression (SAGE), gene knockout, RNA interference (RNAi), virus-induced gene silencing (VIGS), insertional mutagenesis, Targeting Induced Local Lesions in Genomes (TILLING), EcoTILLING and next-generation sequencing (NGS) technologies have been developed as component of functional genomics to elucidate the function of genes/sequences and can assist in understanding the genetic mechanism of

U.C. Jha, Ph.D. (✉)
Indian Institute of Pulses Research (IIPR), Kanpur 208024, India
e-mail: uday_gene@yahoo.co.in

J.S. Bhat, Ph.D. • B.S. Patil, Ph.D.
Regional Research Centre, Indian Agriculture Research Institute, Dharwad, Karnataka, India

F. Hossain, Ph.D.
Division of Genetics, Indian Agriculture Research Institute, New Delhi, India

D. Barh, Ph.D.
Department of Genomics, Institute of Integrative Omics and Applied Biotechnology (IIOAB), Nonakuri, West Bengal, India
e-mail: dr.barh@gmail.com

D. Barh et al. (eds.), *PlantOmics: The Omics of Plant Science*,
DOI 10.1007/978-81-322-2172-2_4, © Springer India 2015

growth, development and physiological responses of plant. Therefore, functional genomics can aid in discovery of alleles/QTLs, development of high-throughput markers, development of high-density linkage map and ultimately assigning the function of genes/QTLs related with yield and associated traits and biotic and abiotic stress tolerance to develop high yielding genotypes to mitigate the challenges of global food security.

Keywords
Functional genomics • Next-generation sequencing • Molecular markers • Plant science

Introduction

Classical genetics ruled plant breeding until the twentieth century, and its application led to development of numerous high yielding cultivars in various crops. During this period, the concept of "one gene, one enzyme" (Beadle and Tatum 1941) and discovery of double-helical structure of DNA (Watson and Crick 1953) provided the early insights into the functional and structural aspects of the genomes and accelerated the progress of molecular biology. During the last decade of twentieth century, genomes of various organisms (from virus to human) were completely sequenced with the blessing of Sanger's genome sequencing technology (Sanger and Coulson 1975) and later on with next-generation sequencing (NGS) technology (Schuster 2008). Subsequently genomes of many more organisms have been and are being sequenced. This rapid development led to the birth of a new field of molecular biology termed "***genomics***". In fact this term was coined in 1986 itself by mouse geneticist Thomas Roderick (McKusick 1997) to refer "mapping, sequencing and characterizing genomes" based on the term "genome" introduced by Winkler (1920). Hence, the recent developments in the field of molecular biology simply expanded the scope of this term.

Subsequently, whole genome sequencing of the model plant *Arabidopsis* (The *Arabidopsis* Genome Initiative 2000) harbingered a new era of plant genomics. To date genome sequence of more than 50 plants belonging to 49 different plant species have been decoded (Michael and Jackson 2013). However, the function of most of the sequence information generated by genome projects or genes remains elusive. The generation of enormous sequence information and research efforts to understand their role in character expression led to the formation of two new subfields within the main branch of genomics, viz. functional genomics and structural genomics. The application of genomics tools in plants has led to important insights into important biological processes and a wealth of knowledge about development. Now, agriculture can take its share of benefits from genomics.

The genomics-led studies in crop improvement can assist plant breeders in identifying genes/QTLs that could be best utilised to improve crop productivity and quality through genetic engineering and plant breeding. In the past the success of plant breeding largely relied on forward genetics based on screening of natural and induced mutants by phenotypic selection. But introduction of genomics-led high-throughput techniques such as marker-assisted selection (Collard and Mackill 2008; Xu and Crouch 2008), genomic selection (Goddard and Hayes 2007; Jannink et al. 2010) and genotyping by sequencing (GBS) (Huang et al. 2009; Elshire et al. 2011) has offered new opportunities to the plant breeders, enriching the arsenal of classical breeding tools thereby facilitating mapping of desired traits precisely and thus aided in tailoring desirable genotypes. Therefore, genomics in combination with functional genomics-led studies has opened a new avenue to the plant breeders for developing high-throughput markers, developing high-density linkage map and identifying, fine mapping and cloning of gene/QTL followed by analysis of their functions strategically in quicker time. In this chapter we have tried to assemble the functional genomics-driven technologies, their potentials and their limitations followed by the latest relevant applications in plant science.

Functional Genomics

The functional genomics can be defined as "development and application of global (genome-wide or system-wide) experimental approaches to assess gene function by making use of the information and reagents provided by structural genomics" (Hieter and Boguski 1997). Gibson and Muse (2009) defined it as "approaches under development to ascertain the biochemical, cellular, and/or physiological properties of each and every gene product". However, Pevsner (2009) defined it by including non-genic regions as "the genome-wide study of the function of DNA (including genes and non-genic elements), as well as the nucleic acid and protein products encoded by DNA". Functional genomics uses the approaches of both forward genetics (phenotype to gene sequence) and reverse genetics (gene sequence to function). The forward genetics aids in gene discovery, while reverse genetics helps in deciphering the gene function. The main goal of functional genomics is to understand the relationship between the genome and the resultant phenotype of an organism. This could be achieved by knowing the expression, regulation and genome-wide mutagenesis through reverse genetics tools (Alonso and Ecker 2006). The functional genomics contemplates to understand the dynamic properties of an organism at cellular and/or organism levels using genomics and proteomics. This would provide better insights into how the information encoded in an organism's genome could be transformed into biological function. There is a possibility of understanding how a particular mutation leads to a given phenotype. This might have implications in dissecting the genetics of complex traits and understanding of the genetics of traits of economic importance.

Functional Genomics Tools

The field of functional genomics is the result of efforts to understand the significance of the genome sequence information generated. It is necessary to understand the biochemical and physiological function of every gene product and its complexes. The function of gene is manifested at different levels, including at RNA, protein and metabolite levels. Hence, critical analyses at these levels would enable to understand the possible function of a particular gene/gene product and its interaction with other gene/gene products leading to a certain phenotypic expression. Thus, the tools involved have to address the challenges at each level efficiently. However, in this chapter we do not discuss about the functional proteomics (functional genomics of proteins) as there is a separate dedicated chapter on proteomics in this book.

Here, we discuss the tools or techniques involved in defining the functional genomics such as microarray technique; SAGE; transgenic and gene silencing approaches such as gene knockout, RNAi and VIGS; insertional mutagenesis and chemical mutagenesis including TILLING, and EcoTILLING and NGS technique. Each of the techniques and their applications are summarised below.

Expression Profiling

In the early 1990s, scientists took the task of unravelling the gene expression and transcript profile of genome. This led the foundation of functional genomics. After the whole genome sequence information of the various species was available, molecular biologists estimated the number of genes encoded in the genomes of such species. But, this approach provided little information about the gene function. One approach for understanding the function of genes is to establish the identity and abundance of different mRNA transcripts expressed in a tissue or cell. This is termed as "expression profiling". This approach assumes that the genes that have similar expression patterns are functionally related and that changes in gene expression are due to changes in physiological conditions as per the requirement of the cell. The most effective genomics technique first used for gene expression analysis was expressed sequence tags (ESTs). A number of high-throughput technologies used for gene expression or transcriptome analysis are expressed sequence tags (EST) sequencing (Adams et al. 1995), serial analysis of gene expression (Velculescu et al. 1995), massively parallel signature sequencing (Brenner et al. 2000) and microarray technology (Schena et al. 1995). We will discuss here these important technologies widely used for transcript profiling and gene expression analysis across the genome of different plant species.

Microarray Technology

Microarray is a versatile high-throughput tool and is one of the fastest-growing new technologies in the field of genetic research. There are several synonyms of microarray like DNA chips, gene chips, biochips, DNA arrays and gene arrays. This technique was first used in 1995 by Patrick Brown and his group for expression analysis in *Arabidopsis* leaf and root tissues (Schena et al. 1995). DNA microarrays are collection of sequences from thousands of different genes that are immobilised or attached onto a solid support at fixed and known locations/spots in an orderly or fixed way. The supports are usually microscope slides but can also be silicon chips or nylon membranes. The DNA is printed, spotted or actually synthesised directly onto the support. As many as 30,000 cDNAs can be spotted on a microscope slide with each spot corresponding to a unique cDNA. Based on the kind of sequences spotted in microarray, it can be grouped into cDNA microarray (cDNA are immobilised) and oligonucleotide array (synthesised oligonucleotides are attached to the chip). Since the location of each sequence/gene is known, it is used to identify a particular gene sequence. The cDNA microarrays can be used for investigating gene expression in both animal and plant systems. The core principle behind this technique is hybridisation between two DNA strands (target sample DNA and a large set of immobilised DNA sequences), using the complementary nature of DNA strands, specifically paired with each other by forming hydrogen bonds between complementary nucleotide pairs. Based on this, the procedure of microarray analysis is as follows: (1) cDNAs fragments are amplified by PCR. (2) Anchoring of the amplified cDNAs on glass slide having already known positions. (3) Contrasting (positive trait and negative trait) cDNA probes are produced by reverse transcription. (4) The probes hybridised to the DNAs positioned on the glass slide. (5) Confocal microscope driven scanning of array, followed by array image analysis by computer program, for details (see Xiang and Chen 2000; Aharoni and Vorst 2001). (6) Applying statistical methods to infer that differences in gene expression between cell populations or experimental conditions are true or due to chance. (7) Sharing of microarray data. The sharing of microarray data and making it publicly available is important and was highlighted by Pavlidis and Noble (2001). The sharing of data also permits validation of the results.

The databases for microarray data are available in plenty with both public and private (Becker 2001). Two public databases are serving as a repository for data. These are Array Express at the European Bioinformatics Institute (EBI) (http://www.ebi.ac.uk/arrayexpress); Na- 244 J.D. Pollock/Chemistry and Physics of Lipids 121 (2002) 241/256 National Center for Bioinformatics's (NCBI/NIH) Gene Expression Omnibus or GEO (http://www.ncbi.nlm.nih.gov/geo).

Application: Microarray technique is a high-throughput technology for studying the expression profile of genes on whole-genome scale; microarray technique can play a significant role in dissecting the various kinds of simultaneous gene expression in terms of transcription and translation and regulatory network in organism. Its application ranges from model plant *Arabidopsis* to almost all cultivated crop species such as rice, maize, wheat, brassica, potato, cabbage, grape, peanut and soybean given in Table 1. In the early 1990s, it was applied largely in *Arabidopsis* for harnessing the different expression profiles such as organ development (Ruan et al. 1998), phytochrome A-related expression (Spiegelman et al. 2000), systemic acquired resistance (SAR) expression (Maleck et al. 2000), circadian clock regulation response (Harmer et al. 2000) and cold and drought response (Seki et al. 2001). Later, it has been applied to understand the following aspects of plants: (1) Gene expression during plant metabolism: carbon metabolism and starch deposition in tuber have

Table 1 Application of microarray techniques used in functional genomics

Crops	Expression of gene/EST sequences	Functions	References
Arabidopsis	14 candidate genes	Synthesis and modification of outer seed coat	Esfandiari et al. (2013)
Brassica	8 BrDREB1 genes	Associated with abiotic stress responses and development	Lee et al. (2012)
	44 upregulated genes and 33 downregulated genes in pollinated pistil	Expression profile in time of pollination	Jiang et al. (2013)
	B3 family transcription factor genes	Related with plant growth and development	Peng and Weselake (2013)
	4,646 genes	Delayed expression in ogura CMS resulting sterility	Dong et al. (2013)
	XB3 gene	Contributing in plant growth and development	Yuan et al. (2013)
	192 putative polymorphic loci	Associated with development in functional marker	Nishizawa et al. (2012)
	562 differentially expressed genes	Associated with regulation of oleic acid trait	Guan et al. (2012b)
	39 genes	Expression associated with seed oil content	Zhu et al. (2012)
	3 genes, chitinase gene	Associated with biotic stress resistance	Ahmed et al. (2012)
	Expression profile of 33 genes	Associated with salinity stress response	Srivastava et al. (2010)
	674 gene transcript of middle fraction of seed coat, 1,203 gene transcripts found in hypocotyl tissue	Associated with seed coat development	Jiang and Deyholos (2010)
	53 upregulated and 42 downregulated expression transcripts	Improvement of seed oil content	Fu et al. (2009)
Grape	SBP-box genes	Defence against biotic and abiotic stress	Hou et al. (2013)
Maize	74 MAPKKK genes	Involved in signalling pathways and organ development	Kong et al. (2013)
	Water stress gene	Associated with water stress	Yu and Setter (2003)
	ZmALDH9, ZmALDH13 and ZmALDH17	Expressed during drought stress, acid tolerance and pathogen infection	Zhou et al. (2012)
	45 genes	Associated with biotic stress and evolution study	Swanson-Wagner et al. (2012)
	XB3	Involved in plant growth and development	Yuan et al. (2013)
	Differential gene expression between quiescent and germinated maize embryo stages	Germination regulation	Jiménez-López et al. (2011)

(continued)

Table 1 (continued)

Crops	Expression of gene/EST sequences	Functions	References
Rice	Chilling stress gene	Involved in chilling stress	Yamaguchi et al. (2004)
	Drought stress gene	Related with drought response	Hazen et al. (2005) and Lian et al. (2006)
	XB3	Involved in plant growth and development	Yuan et al. (2013)
	NYC4 gene	Degradation of chlorophyll and chlorophyll – protein complexes in time of leaf senescence	Yamatani et al. (2013)
	TaSIP gene	Salinity tolerance	Du et al. (2013a)
Peanut	5,066 EST sequences	Diverse expression during seed, leaf, stem, root and gynophore development	Bi et al. (2010)
Poplar	Calcium-dependent protein kinase (CDPK) gene family	Expressed under various stresses	Zuo et al. (2012)
Potato	Sucrose- and starch-metabolism-associated gene	Carbon metabolism	Watkinson et al. (2008)
Soybean	4,100 unigene ESTs	Involved in organ differentiation in legumes	Maguire et al. (2002)
	27,513 clones of unigene set	Involved in organ development	Vodkin et al. (2004)
	XB3	Associated with plant growth and development	Yuan et al. (2013)
	Genes encoding proteins possibly contributing acid soil tolerance	Related with aluminium tolerance	Mattiello et al. (2012)
Strawberry	Strawberry alcohol acetyltransferase (SAAT) gene	Contributing in fruit ripening	Aharoni et al. (2000)
Wheat	Salt stress gene	Associated with salt stress	Kawaura et al. (2006)
	TaPIMP1	*Bipolaris sorokiniana* and drought stress resistance	Zhang et al. (2012c)
	60 genes of differential expression	Differential response of powdery mildew	Xin et al. (2012)
	Five genes	Formation and development of pistil-like stamens	Yamamoto et al. (2013)

been studied under drought condition in potato (Watkinson et al. 2008). (2) Gene expression during growth and development: microarray expression analysis assisted in analysing multiple expression of seed, leaf, stem, root and gynophore development in peanut (Bi et al. 2010) and analysing expression profile of genes at the time of fertilisation in brassica, revealing role of B3 family transcription factor genes and XB3 gene contributing growth and development (Peng and Weselake 2013; Yuan et al. 2013). (3) Expression pattern under abiotic and biotic stresses: this technique has been used in soybean (Yuan et al. 2013) for abiotic stress expression pattern of eight BrDREB1 genes (Lee et al. 2012). Similarly, salinity tolerance expression profile of 33 genes has been studied (Srivastava et al. 2010). Additionally, the expression of *ZmALDH9, ZmALDH13* and *ZmALDH17* genes of maize during drought stress, acid tolerance and pathogen infection (Zhou et al. 2012) and differential expression of 60 genes taking part in differential response of powdery mildew in wheat have been investigated (Xin et al. 2012). (4) To study expression profile of quality traits: expression analysis of 562 differentially expressed genes accounting for high and low oleic acid trait regulation has been documented (Guan et al. 2012a). In maize 74 MAPKKK genes have been identified by this technique involved in signalling pathways and organ development (Kong et al. 2013).

Advantages: This technique provides advantages over other expression analysis techniques used earlier. (1) Prior information of cDNA of genome is not required in microarray analysis (Aharoni et al. 2000; Harmer and Kay 2000), (2) it can even cover small portion of genomes under investigation (Aharoni et al. 2000) and (3) the *custom* chip used in this technique can easily be fabricated in laboratory (Harmer and Kay 2000).

Disadvantages: It suffers from some drawbacks: (1) There may be a problem in differentiating the closely related gene sequences (Duggan et al. 1999; Ishii et al. 2000) and genome duplication, due to cross hybridisation and misinterpretation of the expression of any single gene (Meyers et al. 2004); (2) it investigates random arbitrary chosen genes (Lee and Lee 2003) and (3) there is no standard set of protocols for doing microarray experiments and hence the comparisons among different experimental datasets are difficult. To overcome this difficulty the MicroArray and Gene Expression group (http://www.mged.org), an organisation consisting of users of microArray data, presented a proposal for the minimum information about microarray experiments (MIAME; Brazma et al. 2001) that would be included in any public database.

Serial Analysis of Gene Expression (SAGE)

Serial analysis of gene expression is an innovative approach for defining the expression profiling of cellular transcripts, gene discovery and analysing metabolic pathway (Gowda et al. 2004). Basically it measures the quantitative profile of expressed genes using the principle of sequencing technology. This technique was first developed by Velculescu et al. (1995). In this method, cDNA is produced from mRNA through reverse transcription from a specific cell, tissue or organ of interest. The cDNA is digested with the enzyme NlaIII, and poly-A ends are ligated with linker fragment with 5′-GGGAC-3′ sequence, which is a recognition site for type IIS restriction enzyme BsmFI. A short tag of 14–15 bp length having well-specified 3′ transcript is isolated by digestion with BsmFI (Velculescu et al. 1995; Zhang et al. 1997). Tags from different expressed sequences are ligated and cloned into a plasmid vector for sequence analysis. After sequencing thousands of tags, the gene corresponding to the tag is identified. The 14–15 bp tag sequence is used as query to search cDNA databases of the organism by BLAST (Altshul et al. 1990). Finally tag annotation is combined to generate a gene expression profile. Many modifications to this technique have been made for increasing its efficiency such as (1) LongSAGE – containing longer-size tags of 21 bp in comparison to conventional SAGE having 14–15 bp tags (Saha et al. 2002; Chen et al. 2002); (2) SuperSAGE – designed by reaping the benefit of high-throughput SAGE and microarray technique simultaneously (Matsumura et al. 2006), it generates 26 bp tags and broadens its use in non-model organisms (Matsumura et al. 2003) and (3) Virtual SAGE – uses the principles of both EST and SAGE analysis for gene expression (Poroyko et al. 2004).

Application: SAGE has been used for discovering new genes, transcript profiling and analysis of cellular metabolic pathways. In plants, it has been applied in analysing the response of expression during host–pathogen interaction in rice (Matsumura et al. 2003) and abiotic stress in chickpea (Molina et al. 2008). SAGE has also been applied to investigate the molecular basis of heterosis and gene regulation pathways in rice by generating 465,679 tags (Bao et al. 2005). By applying SAGE in an elite Chinese super-hybrid rice (LYP9), 10,268 quality tags were generated which helped in assigning candidate genes responsible for heterosis mechanism in rice (Song et al. 2007). Similarly SAGE assisted in searching out 1,183 differentially expressed genes (DGs) in F_1 super-hybrid rice Liangyou-2186, conferring heterosis (Song et al. 2010).

Further, SAGE has also been used for elucidation of plant–pathogen interaction (Gowda et al. 2007; Matsumura et al. 2003). For instance, in rice RL SAGE has been used for understanding the molecular basis of *Rhizoctonia solani* and host resistance response that contributed to fine mapping of the QTLs governing this disease (Venu et al. 2007). LongSAGE disclosed the role of ethylene as induced resistance response against cucumber mosaic virus in tomato (Irian et al. 2007) and gene regulation and expression analysis during grain development in wheat (Poole et al. 2008). Likewise transcript profiling of cyst nematode-infected tomato roots has been performed applying SAGE (Uehara et al. 2007). This powerful tool has been used for investigating the caryopsis development in wheat producing 29,231 unique tags, leading to enhancement of wheat breeding (McIntosh et al. 2007). This technique has been used successfully for comparative analysis in egg plant (Fukuoka et al. 2010).

Advantages: It can be used for evaluation of thousands of genes to obtain quantitative and qualitative information on them (Velculescu et al. 1995; Lee and Lee 2003). Besides, digitisation of generated data is unique feature in this technique (Gowda et al. 2004).

Disadvantage: There are chances of missing analysis of rare transcripts in this technique (Harmer and Kay 2000). This technique also demands large set of expressed sequence tags (Harmer and Kay 2000). Error in expression profile arises due to non-authentic annotation in tags for complex polyploidy genome, viz. wheat (Poole et al. 2008).

Gene Silencing Approaches

Gene Knockout

Gene knockout describes the reciprocal exchange of DNA sequence between two chromosomes harbouring the same genetic loci (Lewin 2004; Tierney and Lamour 2005) starting with a double-strand break and followed by repair of these breaks. It is the most widely accepted model for gene targeting (Iida and Terada 2004). Site-specific recombination gene knockout system (such as Cre-Lox) can be potentially applied in rice and other plants (Iida and Terada 2004). Utility of site-specific recombination-based gene knockout is very limited only used in moss *Physcomitrella patens* (Schaefer and Zryd 1997). So disruption-based gene knockout applying insertion elements has been used for knocking out genes for decoding function of genes (Wisman et al. 1998; Campisi et al. 1999; Tissier et al. 1999). The first report of knocking out *actin* gene in *Arabidopsis* was reported (McKinney et al. 1995). A deletion-based gene knockout system was used in *Arabidopsis* (Li et al. 2002).

Applications

This unique technique has been intensively applied for unravelling the gene function in different plant species. In *Arabidopsis* a list of genes were knocked out for disclosing the function of important morphological and physiological traits (Bouché and Bouchez 2001). This tool has also been used for finding out plant metabolism like lipid, gluconeogenesis, starch, sugar metabolism, etc. (Thorneycroft et al. 2001). In rice the function of gene responsible for plant height has been decoded by knocking out the *gibberellin 2-oxidase (GA2ox)* gene (Hsing et al. 2007), kernel size and yield in maize (Martin et al. 2006); by knocking out *Gln1-3* and *Gln1-4* genes, defence gene and root-knot nematode resistance (Gao et al. 2008) by knocking out *LOX* gene *(ZmLOX3)*. Similarly it has been used for elucidating the function of genes associated with aluminium tolerance (Chen et al. 2012), lutein biosynthesis (Lv et al. 2012), resistance to rice blast (Delteil et al. 2012), root elongation and iron storage (Bashir et al. 2011) in rice, oxidative pentose phosphate pathway (Settles et al. 2007) in maize, drought stress resistance (Malatrasi et al. 2006) in barley and salinity tolerance and ABA signalling (Park et al. 2009) in Arabidopsis. The list of plant genes and their function determined by gene knockout has been given in Table 2.

This unique technique suffers from some drawbacks, it creates pleiotropic effect, and many knockout mutations may be without phenotype (Thorneycroft et al. 2001). Its full potentiality and application in functional genomics are yet to be used fully.

RNAi

RNAi is the process by which expression of a target gene is inhibited by antisense and sense RNAs. RNAi was discovered when injection of double-stranded RNA (dsRNA) into worms led to specific degradation of the corresponding mRNA (Fire et al. 1998). The discovery of RNAi

Table 2 Applications of gene knockout for deciphering gene function in plant species

Crops	Knockout genes	Functions	References
Arabidopsis	*ACT2, ACT4, ACT7*	Reduced fitness	Asmussen et al. (1998) and Gilliland et al. (1998)
	ADC2	Arginine decarboxylase	Soyka and Heyer (1999)
	CCA1	Homeostasis of circadian regulation of genes	Green and Tobin (1999)
	DAG1	Control in seed germination	Papi et al. (2000)
	AlaAT	Conversion of alanine to pyruvate during O_2 deficiency	Miyashita et al. (2007)
	AtSAT32	Salinity tolerance and ABA signalling	Park et al. (2009)
	AtNAR2.1	Inducible high-affinity transport system activity	Feng et al. (2011)
	RPL10	Development under UV stress	Falcone Ferreyra et al. (2010)
	AtJMT promoter	Biotic and abiotic stress response	Seo et al. (2013)
Barley	*HvBCAT-1*	Drought stress	Malatrasi et al. (2006)
Brassica (*Alternaria brassicicola*)	*AbVf19*	Inducing virulence in *A. brassicicola*	Srivastava et al. (2012)
	Gc2	Spike fertility	Friebe et al. (2003)
	sdi1	Sulphur storage	Howarth et al. (2009)
Maize	*knox* gene *lg3/4* genes	–	Bauer et al. (2004)
	Gln1-3 and Gln1-4 genes	Kernel size and yield	Martin et al. (2006)
	6-Phosphogluconate dehydrogenase gene	Oxidative pentose phosphate pathway	Settles et al. (2007)
	LOX gene *(ZmLOX3)*	Defence gene and root-knot nematode resistance	Gao et al. (2008)
	Strigolactones (SL)	Lateral branching	Guan et al. (2012a)
Rice	*Gibberellin 2-oxidase* gene *(GA2ox)*	Plant height	Hsing et al. (2007)
	OsFRDL1	Iron translocation	Yokosho et al. (2009)
	OsARF12	Root elongation and iron storage	Bashir et al. (2011)
	DNG701	DNA hypermethylation and lowers the expression of TOS17	La et al. (2011)
	RAD51C	Reproductive development	Kou et al. (2012)
	OsMGT1	Al tolerance	Chen et al. (2012)
	CYP97A4	Lutein biosynthesis	Lv et al. (2012)
	CslF6	Aids in biosynthesis of mixed-linkage glucan (MLG) – a cell wall polysaccharide	Vega-Sánchez et al. (2012)
	OsIRE1	Endoplasmic reticulum stress response	Lu et al. (2012)
	OsALS1	Al tolerance	Huang et al. (2012a)
	Docs1	Development in root outer cell layer	Huang et al. (2012b)
	OsWRKY28, rTGA2.1 and NH1	Resistance to rice blast	Delteil et al. (2012)

(continued)

Table 2 (continued)

Crops	Knockout genes	Functions	References
Wheat	*GPC-B1 and GPC-B2*	Regulation of senescence and grain protein	Distelfeld et al. (2012)
Wheat (*M. graminicola*)	*MCC1*	Pathogenicity in *M. graminicola*	Choi and Goodwin (2011)

has supported a major paradigm shift from "one gene, one protein" to the concept that noncoding DNA can have profound effects in cells and organisms (Auer and Frederick 2009).

RNAi is a post-transcriptional gene silencing PTGS approach deployed in functional genomics study by switching off the action of a particular gene by breaking down the mRNA and preventing the translational process of the specific gene considered. (1) The first step of RNAi mechanism starts with an enzyme called DICER dsRNA-specific endonuclease (Bernstein et al. 2001; Xie et al. 2004), which cuts dsRNA into pieces called siRNAs (small interference RNAs) (Zamore et al. 2000); (2) these siRNAs attach with a protein and form complexes called RNA-induced silencing complex (RISC) (Hammond et al. 2000); (3) the siRNA with RISC remains apart as single-stranded mRNA and (4) finally the RISC becomes activated and breaks down the target mRNA and inhibits the translational process (Hannon 2002; Schwarz et al. 2003; Kusaba 2004; Fuchs et al. 2004). Recently different types of RNAs based on PTGS and transcriptional gene silencing (TGS) approaches are being used in plants like siRNA (Meister and Tuschl 2004), tasiRNA (Eamens et al. 2008), natsiRNA (Eamens et al. 2008), rasiRNA (Eamens et al. 2008), dsRNA (Fire et al. 1998), miRNA (Vaucheret 2008; Voinnet 2009) and other RNAs (Vazquez 2006; Vazquez et al. 2010).

Application of RNAi in plants: Since its inception, RNAi has been used in transforming the phenotype of cells and whole organisms. Its prolific application has heralded a new era in functional genomics. Biological science is quickly reaping the benefits of this magical technique. Its universal application has been deployed for designing transgenic, recombinant protein production in insect and mammalian system, in metabolic engineering and in plant biotechnology for improving quality and nutritional value (Hebert et al. 2008). The application of RNAi is highlighted in Table 3.

RNAi was described in plants by Waterhouse et al. (1998) for experiments that produced virus-resistant tobacco. Over the last decade, numerous findings have contributed to the view that RNAi is evolutionarily conserved in the plant kingdom and has many diverse functions (Dunoyer and Voinnet 2008; Eamens et al. 2008; Small 2007; Vaucheret 2006; Baulcombe 2004; Kusaba 2004). Our understanding of RNAi has emerged from two areas of plant science, experiments designing transgenic plants and research into virus resistance (Eamens et al. 2008; Vaucheret 2006; Baulcombe 2004; Kusaba 2004; Ghildiyal and Zamore 2009; Hebert et al. 2008). Today, researchers are tailoring varieties of crops to produce small RNAs causing silence of essential genes in insects, nematodes and pathogens, through an approached called host-derived RNAi (HD-RNAi) (Auer and Frederick 2009). The application of RNAi is ever demanding since its discovery and it has been used extensively; it can be used for silencing gene in many organisms other than plant sp. like *Caenorhabditis elegans*, *Drosophila* and animals (Palauqui and Vaucheret 1998; Kamath et al. 2001; Tabara et al. 1998; Matzke et al. 2001; Hannon 2002; Min et al. 2010). In different plants some important RNAi databases are used, for example, for *Medicago truncatula* (www.medicago.org/rnai/) and for *Arabidopsis* (http://2010.cshl.edu/scripts/main2.pl). Therefore, RNAi is one of the gifted reverse genetics tools for functional genomics (Table 4).

Advantages and disadvantages: RNAi technology has many advantages and limitations. The details of advantage and disadvantages have been thoroughly discussed (Gilchrist and Haughn 2010). An advantage of using RNAi is that a specific gene can be silenced if the target sequence is

Table 3 Application of RNAi decoding functions of genes in plant highlighted

Crop	Target gene	Function	Reference
Arabidopsis	*iaaM* and *ipt*	Resistance against crown gall	Escobar et al. (2001)
	KOJAK	Hairless root	Limpens et al. (2004)
	16D10 gene	*Meloidogyne* sp. resistance	Gheysen and Vanholme (2007)
Brassica	*BP1* gene	Increase in photosynthesis	Byzova et al. (2004)
Common bean	*RabA2*	Control of root hair growth during bacterial infection	Blanco et al. (2009)
	AC1 viral gene	Bean golden mosaic virus resistance	Bonfim et al. (2007)
	NADPH oxidase	Colonisation of *Rhizobium*	Montiel et al. (2012)
Cotton	*Cadinene synthase*	Decrease in gossypol	Sunilkumar et al. (2006)
Kiwi fruit (*Actinidia chinensis*)	*CAROTENOID CLEAVAGE DIOXYGENASE* (CCD)	Altered branching	Ledger et al. (2010)
Lotus japonicus	*LjERF1*	Regulation of nodulation	Asamizu et al. (2008)
Maize	*MS45* gene	Male sterility	
	ATPase	Larval mortality resistance	Baum et al. (2007)
Medicago	*DMI2*	Improvement in N_2 fixation	Limpens et al. (2005)
Pea	PsClv2	Increased nodulation	Krusell et al. (2011)
Potato	*HC-Pro*	Crown gall resistance	Waterhouse et al. (1998)
	Polyphenol oxidase gene	Longer shelf life	Wesley et al. (2001)
Rice	*OsGEN-L*	Male sterility	Moritoh et al. (2005)
	NlHT1, *Nlcar* and Nltry	Resistance against brown plant hopper	Zha et al. (2011)
	TPS (trehalose phosphate synthase) gene	Resistance against brown plant hopper	Chen et al. (2011)
	Glycine decarboxylase complex (GDC)	Leaf senescence	Zhou et al. (2013b)
Tobacco	Replication-associated gene	Resistance against viral infection	Zhao et al. (2006) and Tenllado et al. (2003)
Tomato	*DET1* gene	Increase in shelf life	Xiong et al. (2004)
Wheat	*EIN2*	Ethylene insensitivity	Travella et al. (2006)

better chosen (Small 2007). This is also one of the limitations of using RNAi because; unlike insertional mutagenesis, here the exact sequence of the gene is required. This technique can effect on non-/off-target gene (Qiu et al. 2005) and cannot be applied for mutants producing sterile phenotype (Gilchrist and Haughn 2005).

Non-Transgenic Approaches/ Mutational Approach

Virus-Induced Gene Silencing

Virus-induced gene silencing (VIGS) is recently developed as one of the important genomics tools for deciphering gene function. The term (VIGS) was first used by van Kammen (1997). VIGS follows the PTGS mechanism for inducing gene knockdown effect (Baulcombe 1999, 2002; Robertson 2004). The basic principle behind this tool is to insert viral DNA or RNA harbouring the gene sequence of interest into the plants. Specific sequences are inserted in viral genome without disturbing infectivity of virus (Watson et al. 2005). Finally, the introduced virus multiplies and initiates silencing of gene (Lu et al. 2003a; Burch-smith et al. 2004). In this approach plant viruses can be used as vector for induction of sequence-specific VIGS (Baulcombe 1999; Lindbo et al. 2001). This novel approach has been used as an important tool for investigating gene function (Kumagai et al. 1995; Liu et al. 2002a, b; Peart et al. 2002b) described below.

Table 4 Application of VIGS

Crops	Genes	Vectors	Functions	References
Barley	*Rar1, Sgt1 and Hsp90*	Barley stripe mosaic virus (BSMV)	Powdery mildew resistance	Hein et al. (2005)
	P23k	(BSMV)	Synthesis of cell wall polysaccharides	Oikawa et al. (2007)
	necS1	(BSMV)	Increase in resistance against stem rust	Zhang et al. (2009)
	PHYTOENE DESATURASE (BdPDS	(BSMV)	Phosphorus uptake	Pacak et al. (2010)
Brinjal	*PDS, Chl H, Su (sulphur), CLA1*	Tobacco rattle virus (TRV)	Downregulation of the given genes	Liu et al. (2013)
Cotton	*GhNDR1 and GhMKK2*	*Agrobacterium*	*Verticillium* resistance	Gao et al. (2011)
	KATANIN and WRINKLED1	Tobacco rattle virus	Fibre development	Qu et al. (2012)
	GrCla1	TRV	Transiently silence endogenous genes	Gao and Shan (2013)
Oat	*AsPDS*	–	Phosphorus uptake	Pacak et al. (2010)
Pea	*phytoene desaturase* gene	Pea early browning virus (PEBV)	–	Constantin et al. (2004)
Pepper	*CaOvate and CaGA20ox1*	–	Fruit shape determination	Tsaballa et al. (2011)
Potato	*Rx* gene	–	Resistance to potato virus X	Lu et al. (2003b)
	Ten genes	–	Resistance mechanisms to late blight	Du et al. (2013a, b)
Rice	*45 ACE (Avr/Cf-elicited)* genes		Nonhost resistance to *Xanthomonas oryzae* pv. *oryzae*	Li et al. (2012a)
Soybean	*Flavonoid 3'-hydroxylase* gene	Cucumber mosaic virus (CMV)	Flavonoid biosynthesis	Nagamatsu et al. (2007)
	Soybean isoflavone synthase 2	Apple latent spherical virus (ALSV)	Gene functions involved in reproductive stages and early growth stages	Yamagishi and Yoshikawa (2009)
	F3'H		Pigmentation in pubescence	Nagamatsu et al. (2009)
	Rsv1		Soybean mosaic virus resistance	Zhang et al. (2012a)
Strawberry	*FaPYR1*	–	Fruit ripening	Chai et al. (2011)
Tobacco	*N* gene		Resistance to tobacco mosaic virus (TMV)	Liu et al. (2002b) and Peart et al. (2002a)
Tomato	*Pto* gene	–	Resistance response to *Pseudomonas syringae*	Lu et al. (2003b)
	ABA-responsive lea4	–	Functional aspect of moisture-responsive genes	Senthil kumar and Uday Kumar (2006)
	(late embryogenic abundant)	–		
	Sl-EBF1 and Sl-EBF2	–	Fruit ripening	Yang et al. (2010)
	LeRIN, LeACS2, LeACS4 and LeACO1	(TRV)	Ethylene biosynthesis for fruit ripening	Li et al. (2011a)
	GOI	TRV	Fruit development	Fernandez-Moreno et al. (2013)

(continued)

Table 4 (continued)

Crops	Genes	Vectors	Functions	References
Wheat	–		Lr21-mediated leaf rust resistance pathway	Scofield et al. (2005)
	Era1, *Cyp707a* and *Sal1*	BSMV	Drought tolerance	Manmathan et al. (2013)
	Mlo genes		Powdery mildew resistance	Várallyay et al. (2012)
	WRKY53	BSMV	Reduction of aphid resistance	Van Eck et al. (2010)
	Hv-LRR	BSMV	Powdery mildew resistance	Wang et al. (2010)
	Rpg5	–	Produce disease resistance protein domains	Brueggeman et al. (2008)
	(HMW-GS 1Bx14)	BSMV	Glutenin macropolymers	Ma et al. (2012)

Applications of VIGS: This tool has been used in many plants successfully for harnessing the array of gene functions ranging from disease resistance to quality traits. This technique is mostly used in *Nicotiana benthamiana*. The success of VIGS is more responsive in *N. benthamiana* as host in comparison to other plant (Lu et al. 2003a). The different silencing vectors used for VIGS are barley stripe mosaic virus (BSMV) (Holzberg et al. 2002; Van Eck et al. 2010; Wang et al. 2010; Ma et al. 2012; Hein et al. 2005; Zhang et al. 2009), tobacco rattle virus (TRV) (Li et al. 2011a; Qu et al. 2012; Gao and Shan 2013) and pea early browning virus (PEBV) (Constantin et al. 2004). VIGS assisted in disclosing the genes and its function ranging from disease resistance to numerous traits such as fruit ripening, shape, size and many other traits in various plant species is given in Table 5. Additionally, it has been applied in tracking down the role of genes involved in drought tolerance (Manmathan et al. 2013) and other abiotic stress tolerance (Senthil kumar et al. 2008). Recently a modified technique of VIGS called artificial miRNA VIGS (*MIR-VIGS*) has been used in gene silencing in *Nicotiana benthamiana* (Tang et al. 2010). In the near future its application will be indispensible as functional genomics tool for harnessing gene function.

Therefore, this emerging tool has advantages over other reverse genetic approaches: (1) VIGS can be used for initial study of large number of stress genes. This tool can be used without having the complete gene sequence information (Lu et al. 2003a; Burch-Smith et al. 2004). (2) VIGS provides a better option for studying the species that lack mutants and are recalcitrant to genetic transformation (Senthil Kumar et al. 2008), and it can be applied for functional genomics by creating gene knockout phenotype (Scofield and Brandt 2012). Targeted knockdown expression of any gene can be studied by VIGS (Burch-Smith et al. 2006). It can be used as high-throughput functional genomics (Todd et al. 2010; Becker and Lange 2010).

The main drawback of this innovative tool is lacking in adequate VIGS vector and VIGS creates only transient silencing (Padmanabhan and Dinesh-Kumar 2009).

Insertional Mutagenesis

The most popular mutagenesis strategy in functional genomics is insertional mutagenesis, in which a piece of DNA is randomly inserted into the genome causing loss in gene function. The DNA may be transposable elements that can relocate from one genomic location to another (Hayes 2003; Tierney and Lamour 2005). Transposon

Table 5 Some important genes and their functions isolated by T-DNA in different plant species

Plant species	Isolated genes	Functions	References
Arabidopsis	cs (pale mutation)	Associated with novel chloroplast protein synthesis	Koncz et al. (1990)
	MALE STERILITY 2 gene	Involved in male sterility	Aarts et al. (1993)
	CLA1	Involved in chloroplast development	Mandel et al. (1996)
	Cer3	Associated with wax biosynthesis	Hannoufa et al. (1996)
	Pyk20 gene	Acting as transcription factor	Puzio et al. (1999)
	LUMINIDEPENDENS gene	Involved in control of flowering time	Lee et al. (2004)
	hpc1-1D	Developmental and metabolic mutation	Schneider et al. (2005)
	PPR (pentatricopeptide repeat) gene family	Semi-lethal mutation	Kocábek et al. (2006)
Brassica	New promoter	Used as selectable marker gene	Bade et al. (2003)
Rice	OsCP1	Cysteine protease gene	Lee et al. (2004)
Medicago truncatula	Proliferating inflorescence meristem (PIM)/apetela (AP1) squamosa	Floral meristem identity and flower development	Benlloch et al. (2006)

tagging has been used for discovering the function of many important genes in different plant species given in Table 5.

Applications of insertional mutagenesis as reverse genetics tool: Insertional mutagenesis in plant can be achieved using *Agrobacterium* T-DNA (Azpiroz-Leehan and Feldmann 1997; Krysan et al. 1999) and plant transposon (Walbot 2000; Hamer et al. 2001; Ramachandran and Sundersan 2001). T-DNA as insertional mutagen is used in genome-wide mutagenesis programme in *Arabidopsis thaliana* and rice (Walden 2002). Currently over 130,000 T-DNA-tagged *Arabidopsis* lines are made available by the University of Wisconsin *Arabidopsis* Knockout Facility (Krysan et al. 1999). Modified T-DNA vectors have been used in *Arabidopsis* as activation tags (Weigel et al. 2000) as well as gene and promoter traps (Feldmann 1991; Lindesy et al. 1993; Babiyachuk et al. 1997). T-DNA as trap vector can be used in genome-wide screen of rice (Jeon et al. 2000). T-DNA tagging has been used in rice; Zheng and colleagues have created more than 30,000 T-DNA insertional lines (Wu et al. 2003). In rice T-DNA insertion mutagenesis has been used for high-throughput tool for study of in silico reverse genetics. By this approach, 683 T-DNA insertions within genes coding for transcription factors (TF) have been presented; this approach has been successful in unravelling gene function in rice and cereals (Sallaud et al. 2004). Transposon tagging can be applied both as forward and reverse genetic tool for discovering gene function in rice and *Arabidopsis* (Greco et al. 2001; Radhamony et al. 2005). Radhamony et al. (2005) enlisted the gene function of nearly all important genes of *Arabidopsis* deploying insertional mutagenesis as forward and reverse genetics tool. T-DNA-tagged genes in rice have been used in functional genomics analysis using MADS-box genes as test case (Lee et al. 2003).

Some important web based resources for genome-wide random mutagenesis are given such as http://www.biotech.wisc.edu/NewServicesAndResearch/Arabidopsis/default.htm Arabidopsis Knockout Facility at the University of Wisconsin http://www.zmdb.iastate.edu/zmdb/sitemap.html Maize Gene Discovery and *Rescue Mu* Project, http://mtm.cshl.org Maize Targeted Mutagenesis database and http://arabidopsis.org/abrc *Arabidopsis* Biological Resource Center (Primerose and Twyman 2006).

In case of transposon it is widely used for insertional mutagenesis in plants and helped in identifying new genes (Gierl and Saedler 1992). Transposon-based methods have been used in *Arabidopsis*, maize and other plants (Stemple 2004). Many transposons have been used like Ac, suppressor-mutators (spm) and mutator (Mu) from maize, *Tam3* from *Antirrhinum majus* and *Tph1* from petunia. Functional genomics programmes using *Ac* started in *Arabidopsis* (Ito et al. 1999) and in tomato (Meissner et al. 2000). It has also been used in *Brassica napus* (Bade et al. 2003), *Medicago truncatula* (Trieu et al. 2000) and poplar (Groover et al. 2004). In the recent past *Tos17*, endogenous retroposon of rice, *Tnt1* and *Ttol*, retroposon of tobacco, have been used for gene tagging in *Arabidopsis* and in rice (Courtial et al. 2001; Okamoto and Hirochika 2000; Yamazaki et al. 2001; Iantcheva et al. 2009). Recently identified retroelements *MERE* in *M. truncatula* and *LORE* in *Lotus japonicus* can be used in reverse genetic tool (Trichine et al. 2009). T-DNA insertion aided in designing total 372,346 mutant lines contributing in many physiological traits in rice (Chang et al. 2012).

Insertional mutagenesis has certain advantages over traditional form of mutagenesis, the interrupted gene becomes tagged with insertional elements, and the strategy is known as signature tagged mutagenesis (STM). T-DNA insertional mutagenesis also offers the advantage of its heritability and complete loss of function (Gilchrist and Haughn 2010).

In spite of one of the potential reverse genetics tool, insertional mutagenesis/transposon-mediated mutagenesis suffers from some disadvantages like (1) low frequency of mutations causes large number of screening population to find mutations in a given gene (Gilchrist and Haughn 2005), (2) insertion in essential genes usually causes lethality (Till et al. 2003), (3) the precise mechanism of T-DNA integration into the plant genome remains largely unknown (Iida and Terada 2004; Tierney and Lamour 2005), (4) insertional mutagenesis is limited by its host range (McCallum et al. 2000b) and (5) sometimes complex arrangement of T-DNA, multiple insertions, chromosomal duplications and rearrangements and insertion of vector backbone sequence occur (Jorgensen et al. 1987; Radhamony et al. 2005). Therefore, insertional mutagenesis has been used as cost-effective reverse genetics tool for identifying and decoding the gene function of model legumes *Lotus*, soybean, *Medicago, Pisum* (Trichine et al. 2009) and many other plant species.

Chemical Mutagenesis

TILLING and EcoTILLING

One of the groundbreaking discoveries in the history of genetics is the discovery of induced mutations (Muller 1930). Subsequently, it has been widely used for phenotypic screening, mapping of genes and investigating function of genes applying mutagen randomly in genome of a plant species in large populations. A new innovative tool has been designed called TILLING, one of the most dynamic high-throughput reverse genetic tool. This technique includes high density of point mutations and traditional chemical mutagenesis with rapid mutational screening for seeking induced lesions (McCallum et al. 2000b). TILLING accompanies chemical mutagenesis with a sensitive mutation trapping instrument (Koornneef et al. 1982).

The TILLING approach follows the steps: (1) generation of mutagenised population, (2) isolation of DNA and pooling (3), PCR amplification with labelled primer, (4) enzymatic assay and (5) detection of mutation (McCallum et al. 2000b; Till et al. 2007a, b).

Application of TILLING: TILLING is an attractive reverse genetic tool first utilised in *Arabidopsis thaliana* (McCallum et al. 2000a, b) and latter *Arabidopsis* TILLING Project (ATP) was established (Wang et al. 2006). Now this ATP project is known as STP (Seattle TILLING Project) (http://tilling.fhcrc.org:9366/) which enabled the researchers to align more than 100 genes harbouring 1,000 mutations (Till et al. 2003). A computational device has been designed for primer designing and gene modelling called Codons Optimized to Deliver Deleterious Lesion (CODDLe) (http://www.proweb.org/coddle) (McCallum et al. 2000b). Similarly, Sorting

Table 6 Detection of genes/alleles obtained by creating TILLING population in different plant species

Crops	Mutagen used	Detected genes/alleles	Mutation frequency	References
Arabidopsis	EMS	14 genes of TILLer collection	1/89 kb	Martin et al. (2009)
Barley	EMS	*Hin-a, HvFor1* genes	1/1 Mb	Caldwell et al. (2004)
	EMS	–	1/140–800 kb	Gottwald et al. (2009)
	Sodium azide	2 genes (dehydrin genes *Dhn12, Dhn13*)	1/2.5 Mb	Lababidi et al. (2009)
Brassica	EMS	6 genes	1/60 kb	Stephenson et al. (2010)
Common bean	EMS	BAT93 mutant pop	–	Porch et al. (2008)
Lotus japonicus	EMS	12 genes essential for nodule development	1/50 kb	Perry et al. (2009)
Melon	EMS	11 genes involved in ripening	1/573 kb	Dahmani-Mardas et al. (2010)
Musa species	–	80 novel alleles	–	Till et al. (2010)
Oat	–	Beta-glucan biosynthesis	1/20–40 kb	Chawade et al. (2010)
Pea	–	35 genes in Cameor population	1/200 kb	Le Signor et al. (2009)
Rice	–	10 genes	0.5/1 Mb	Wu et al. (2005)
	EMS	10 genes	1/300 kb	Till et al. (2007a)
Sorghum	EMS	*COMT* (caffeic acid O-methyltransferase)	1/526 kb	Xin et al. (2008)
Soybean	NMU (n-nitroso-n-methyl urea) and EMS	–	1/250 kb and 1/550 kb	Cooper et al. (2008)
		–	1/502 kb	Tadege et al. (2009)
		17 gene fragments	6/2.3 Mb	Udvardi et al. (2005)
Tomato	EMS	19 genes related with translation initiation factors eIF4E and eIF4G	1/574 kb	Piron et al. (2010)
	EMS	*Sletr1-1* and *Sletr1-2* genes lowered ethylene response	1/737 kb	Okabe et al. (2011)
Tomato	EMS	7 genes	1/322 kb for 1 % EMS and 1/574 kb for 0.7 % EMS	Minoia et al. (2010)
Wheat	–	52 alleles in hexaploid wheat and 39 alleles in tetraploid wheat	In hexaploid 1/38 kb and in tetraploid 1/51 kb	Uauy et al. (2009)
	EMS	COMT1, HCT2 and 4CL1	In diploid sp. 1/92 kb	Rawat et al. (2012)

Intolerant From Tolerant (SIFT) (http://blocks.fhcrc.org/sift/SIFT.html) web-based program assists in tracking down the neutral and deleterious amino acid changes (Ng and Henikoff 2003), and PARSESNP program (http://www.proweb.org/parsesnp/) predicts the missense mutation and point mutation causing restriction endonuclease sites (Till et al. 2003; Taylor and Greene 2003). This technique has been successfully used in different crops for detection of genes applying various mutagen treatments with varying mutation frequencies given in Table 6. TILLING has been applied successfully in wheat for its polyploidy nature leading to tolerate the high

Table 7 TILLING platform in different plant species

Crops	TILLING platform	Reference
Barley	http://www.scri.sari.ac.uk/programme1/BarleyTILLING.htm	Caldwell et al. (2004)
Brachypodium distachyon	BRACHYTIL	Dalmais et al. (2013)
Glycine max	http://www.soybeantilling.org/index.jsp	Cooper et al. (2008)
Lotus japonicus	RevGenUK (http://revgenuk.jic.ac.uk)	Udvardi et al. (2005)
	http://www.lotusjaponicus.org/tillingpages/homepage.htm	Perry et al. (2003)
Medicago truncatula	http://www.versailles.inra.fr/urgv/tilling.htm	Le Signor et al. (2009) and Perry et al. (2009)
Pisum sativum	http://urgv.evry.inra.fr/UTILLdb	Dalmais et al. (2008)
Tomato	http://zamir.sgn.cornell.edu/mutants/	Menda et al. (2004)
	Micro-Tom TILLING	Okabe et al. (2011, 2013)
Zea mays	http://genome.purdue.edu/maizetilling	Till et al. (2004) and Gilchrist and Haughn (2005)

densities of mutation (Slade et al. 2005), also deployed for investigating extensive allelic series of the *waxy* genes in both hexaploid bread wheat and tetraploid wheat (Uauy et al. 2009). In barley TILLING helped in detecting *Hin-a, HvFor1* genes (Caldwell et al. 2004), dehydrin genes *Dhn12 and Dhn13* (Lababidi et al. 2009), *COMT* gene in sorghum (Xin et al. 2008), 35 genes in Cameor population (Le Signor et al. 2009) and *tendril-less* (Hofer et al. 2009) gene in pea and beta-glucan biosynthesis gene in oat (Chawade et al. 2010). With the advent of this novel tool, it has been extensively applied for functional genomics in different crop plants resulting in development of TILLING platform in various plants given in Table 7.

The main advantages of TILLING in comparison to other reverse genetic approaches are: (1) It is an appropriate tool for genetic modification without introducing a foreign DNA into the genome, irrespective of genome size or reproductive system and ploidy level of organism (Gilchrist and Haughn 2005). (2) The population size for detecting mutation is small in comparison to other reverse genetic approaches (McCallum et al. 2000b). (3) The chances of recovering deletion mutation can be calculated in advance (McCallum et al. 2000b). (4) It is applicable for both small- and large-scale screening. (5) TILLING utilises rapidly advancing technology such as DHPLC utilised in detecting high-throughput polymorphism detection (Gilchrist and Haughn 2005). (6) This strategy can be deployed as high-throughput technique for detecting single base changes within the target gene (Tierney and Lamour 2005).

Although having such sound potentiality, this technique has some drawbacks. In this approach the load of mutations created must be balanced with the recovery of mutants (Till et al. 2003). The fertility must be maintained in the mutagenised organism in the first and in subsequent generations (Perry et al. 2003).

EcoTILLING and its use: Henikoff and Comai (2003) coined the term "EcoTilling" first used for describing *Arabidopsis* ecotypes (hence Ecotilling); it was used to survey variation in five genes in 96 different *Arabidopsis* accessions (Comai et al. 2004). In this approach enzymatic mismatch cleavage and fluorescence detection method are applied similar to TILLING (Colbert et al. 2001; Comai et al. 2004). This technique was first applied in rice crop (Kadaru et al. 2006) followed by wheat (Wang et al. 2006, 2008) and barley (Mejlhede et al. 2006). EcoTILLING can be used for diversity analysis, germplasm screening and functional genomics (Till et al. 2010). In western cottonwood (*Populus trichocarpa*) plant, EcoTILLING aided in tracing 63 new SNPs (Gilchrist et al. 2006). This technique was employed to identify single nucleotide polymorphisms (SNPs) and small insertions/deletions (INDELS) in a collection of *Vigna radiata* accessions (Barkley et al. 2008). Likewise, EcoTILLING has been used to find out orthologous hypoallergenic isoforms of Arah2 in 30 different accessions of *Arachis duranensis* (Ramos et al. 2009; Riascos et al. 2010). It has offered the benefit of SNP haplotype diversity in switchgrass (Weil 2009). Additionally, it can be deployed for identification of gene related with biotic and abiotic stresses (Antollin-Llovere and Parniske 2007). In the recent past EcoTILLING has been deployed in *Musa* species for identification of nucleotide polymorphism; further 800 novel alleles have been discovered from 80 accessions (Till et al. 2010).

De-TILLING (deletion TILLING): In the very recent past another alternative reverse genetic strategy has been discovered called De-TILLING. It includes physically induced genomic deletion and employs fast neutron mutagenesis and PCR-based detection (Rogers et al. 2009). This technique has been used in *Medicago truncatula*. The advantages of this technique are (1) it is independent of plant transformation, of tissue culture and of target size and (2) it recovers knockout mutants (Rogers et al. 2009). Its complete applications in different plant species are yet to be harnessed.

Next-Generation Sequencing

Sanger's DNA sequencing technology in 1975 ruled almost two decades since its development and was considered as one of the most robust techniques for genome sequencing. But its cost and labour becomes a limiting factor for sequencing complete genome. With the progress of cutting-edge technologies, Sanger's sequencing technique has been substituted by next-generation sequencing (Schuster 2008) also known as second-generation sequencing technologies (Pérez-de-Castro et al. 2012), a powerful high-throughput technology, which reduces down the sequencing cost and time and enhances the accuracy of sequencing. NGS technologies include Roche 454 system (the first successfully used NGS in the year 2005) based on sequencing by synthesis principle (Ronaghi 2001), AB SOLiD system, Illumina Golden Gate assay and Compact PGM sequencer different systems. The NGS techniques are classified on the basis of read length, short read (25–75 bp) and long read (400–500 bp) (Shendure and Ji 2008). NGS uses three principles: sequencing by synthesis, sequencing by ligation and single-molecule sequencing (Ansorge 2009; Egan et al. 2012). The principle behind this technique is based on massively parallel sequencing and imaging facilities for generating hundreds of billions of bases per run (Shendure and Ji 2008; Deschamps et al. 2012a). The details of advantages and disadvantages are discussed thoroughly considering throughput, NGS systems used (Metzker 2010; Liu et al. 2012; Pérez-de-Castro et al. 2012).

Applications of NGS in Functional Genomics

Initially, Sanger's sequencing technology was deployed for decoding the genome sequence of model plant *Arabidopsis* (TAGI 2000) and rice (IRGSP 2005). With the arrival of NGS technology, it has heralded a paradigm shift in both plant and animal science by enabling in cracking genome sequences of large number of crops of

Table 8 List of complete genome sequence of different plant species applying next-generation sequencing (NGS)

Crop	NGS technology	References
Beta vulgaris	WGS (NGS)	http://bvseq.molgen.mpg.de/Genome/start.genome.shtml
Brassica rapa	WGS (NGS)	Wang et al. (2011b)
Cicer arietinum	NGS	Varshney et al. (2013) and Jain et al. (2013)
Cajanus cajan	WGS (NGS)	Varshney et al. (2012)
Citrus sinensis	WGS (NGS)	http://www.phytozome.net/citrus.php http://www.jgi.doe.gov/sequencing/why/3128.html
Citrullus lanatus	WGS (NGS)	Ren et al. (2012)
Cucumis melo L.	WGS (NGS)	Garcia-Mas et al. (2012)
Fragaria vesca	WGS (NGS)	Shulaev et al. (2011)
Gossypium raimondii	WGS (NGS)	http://www.phytozome.net/cotton.php
Lagenaria siceraria	WGS (NGS)	Xu et al. (2011a)
Medicago truncatula	WGS (NGS)	Young et al. (2011)
Musa spp.	Roche454 (NGS)	Hribova et al. (2009)
Prunus persica	WGS (NGS)	Ahmad et al. (2011)
Triticum urartu	WGS (NGS)	Ling et al. (2013)
Aegilops tauschii	(NGS)	Jia et al. (2013a)
Triticum aestivum	WGS (NGS)	Brenchley et al. (2012)
Sesamum indicum	WGS (NGS)	Wang et al. (2014)

WGS (whole-genome shotgun strategy)

economic importance in quick time with lower cost. Importantly, NGS has enabled in cracking whole genome sequences of more than one dozen crops are given in Table 8. The application of NGS includes de novo genome sequencing (Velasco et al. 2007; He et al. 2011; Buckler et al. 2010; Lai et al. 2010); transcriptome sequencing including siRNA and miRNA sequencing (Axtell et al. 2006; Jacquier 2009); epigenetic analysis including (1) DNA methylation pattern or methylation profiling (Cokus et al. 2008; Costello et al. 2009), (2) histone modification (Impey et al. 2004; Mikkelsen et al. 2007) and (3) nucleosome pattern analysis (Johnson et al. 2006); genotyping by sequencing (GBS) (Huang et al. 2009); genome-wide association study (GWAS) (Elshire et al. 2011); and single nucleotide polymorphism (SNP) marker development (Davey et al. 2011; Bundock et al. 2009). Some important applications related directly or indirectly with functional genomics are summarised below.

De Novo Whole Genome Sequencing

With the arrival of NGS technology, it has brought a revolution in genome sequencing. It has been applied for de novo whole genome sequencing of plants having no reference genome sequence or with reference genome sequence given in Table 8. For the first time whole 5A chromosome of wheat was sequenced applying NGS (Vitulo et al. 2011). Application of NGS that can facilitate in decoding the complex genome sequence of allopolyploids such as wheat and oilseed rape has been discussed (Edwards et al. 2013). It is also deployed for re-sequencing the plant genome having already reference genome sequence such as rice, maize and *Arabidopsis* (He et al. 2011; Xu et al. 2011b; Yu et al. 2011a; Huang et al. 2013a; Hufford et al. 2012; Cuperus et al. 2010) for marker development, understanding complex traits and QTLs, SNP discovery and allele mining, given in Table 9.

Whole Genome Re-sequencing

Whole genome re-sequencing aims at sequencing of individual member's genome of a species for distinguishing genomic variation in relation to reference genome of that species (Straton 2008). With the blessing of NGS, the sequencing technology has speeded up in re-sequencing the whole genome of populations leading to discovery of markers, development of high-

Table 9 Some important applications of NGS in plant species

Applications	Details	Crops	References
Gene discovery	Discovery of phosphate starvation-responsive genes	Wheat	Oono et al. (2013)
	Cytoplasmic male sterility-associated gene	Rice	Igarashi et al. (2013)
	Fusarium head blight	Wheat	Xiao et al. (2013)
	Xanthomonas TAL-effector activated resistance gene		Deschamps et al. (2012a)
	NSP1 (*nodulation-signalling pathway 1*) and *NSP2*	Brassica	Hayward et al. (2012)
	Phytophthora infestans resistance gene Rpi-dlc2	Potato	Golas et al. (2013)
Candidate gene identification	*Leptosphaeria maculans* (blackleg fungus)	Brassica	Tollenaere et al. (2012)
	Candidate genes controlling 4-methylthio-3-butenyl glucosinolate	Radish	Zou et al. (2013)
MicroRNA discovery	14 novel and 22 conserved miRNAs	Groundnut	Zhao et al. (2010)
	10 novel miRNAs for seed maturation	Brassica	Huang et al. (2013b)
Analysis of miRNA	Effect on plant height and flowering	Wheat	Kantar et al. (2012)
	Identified 292 known miRNAs responsible for pollen development	Rice	Wei et al. (2011)
Mapping by sequencing ("SHORE mapping")	Identification of causal mutations	*Arabidopsis*	Galvão et al. (2012)
Ultradense genetic mapping	Ultradense genetic map containing 10,960 independent loci	*Brassica rapa*	Li et al. (2011b)
Sequence-based linkage map	507 markers including 415 INDELS and 92 SSRs and linkage map of 1234.2 cM	*Brassica rapa*	Wang et al. (2011c)
Saturated linkage mapping	1,063 markers developed 1808.7 cM linkage map	Chickpea	Gaur et al. (2012)
Transcriptome analysis	Facilitated in identification of 728 SSRs, 495 SNPs, 2,088 intron spanning markers and 387 conserved orthologous sequence (COS) markers	Chickpea	Hiremath et al. (2011)
	Bolting control	Sugar beet	Mutasa-Göttgens et al. (2012)
	Mercury-responsive genes	*Medicago truncatula*	Zhou et al. (2013a)
	Response for drought	Sorghum	Dugas et al. (2011)
De novo assembly and transcriptome characterisation	40,100 genes were identified	Groundnut	Zhang et al. (2012b)
SNP discovery	1,359 maize SNPs detected and used for mapping population development	Maize	Liu et al. (2010)
	100,000 putative SNPs	Wheat	Winfield et al. (2012)
		Lentil	Sharpe et al. (2013)

(continued)

Table 9 (continued)

Applications	Details	Crops	References
	4,543 SNPs	Chickpea	Azam et al. (2012)
	575,340 SNPs from three cultivars "Atlantic", "Premier Russet" and "Snowden"	Potato	Hamilton et al. (2011)
	8,784 SNPs	Tomato	Sim et al. (2012)
	20,000 (SNPs)	*Brassica rapa*	Bus et al. (2012)
	11,849 SNPs	*Capsicum annuum*	Nicolai et al. (2012)
	928 candidate SNPs	Wheat	Lai et al. (2012)
	1,022 SNPs	Chickpea	Gaur et al. (2012)
	1,790 SNPs	Soybean	Hyten et al. (2010b)
	1,050 SNPs	Common bean	Hyten et al. (2010a)
	2,659 SNPs	Durum wheat	Trebbi et al. (2011)
	8,207 SNP markers	Lupin	Yang et al. (2012a)
	40,661 candidate SNPs	Alfalfa	Han et al. (2011)
	10,640 SNPs	Sunflower	Bachlava et al. (2012)
	497,118 genome-wide SNPs	*Aegilops tauschii*	You et al. (2011)
SSR marker development	Detection of chromosome-arm-specific microsatellite marker	Wheat	Nie et al. (2012)
	94 reproducible SSRs	Fava bean	Yang et al. (2012b)
	–	Groundnut	Zhang et al. (2012b)
	246 SSRs	*Prunus virginiana*	Wang et al. (2012c)
	SSR marker for distinguishing aphid	Soybean	Jun et al. (2011)

resolution map and QTL mapping (Gao et al. 2012). Re-sequencing of whole genome by NGS has been reported in rice (He et al. 2011; Xu et al. 2011b; Yu et al. 2011a; Huang et al. 2013a), in maize (Buckler et al. 2010) and in soybean (Kim et al. 2010). Seventeen wild species and fourteen cultivated species of soybean have been re-sequenced for substantiating the presence of high allelic diversity in wild species of soybean (Lam et al. 2010). Re-sequencing genome of 446 diverse rice accessions of *Oryza rufipogon* and 1,083 cultivated species of *indica* and *japonica* have given the useful insights of domestication and origin of rice (Huang et al. 2012c). Likewise in maize genome-wide re-sequencing of wild, landrace and improved lines has thrown lights on domestication and evolution of maize (Hufford et al. 2012). In *Arabidopsis* 58 RILs and both parents have been re-sequenced by NGS and 6,159 and 701 SNPs have been identified (Maughan et al. 2010). The mainstay of phenotypic variation has arisen due to evolutionary factors in *Arabidopsis* and its close relatives have been analysed by sequencing 80 strains of *Arabidopsis* (Cao et al. 2011). Re-sequencing of *Arabidopsis* genome led the identification of SNP responsible for causing *ebi-1* phenotype (Ashelford et al. 2011). Similarly re-sequencing served in identification of MIR390a precursor processing–defective mutants in *Arabidopsis* (Cuperus et al. 2010). Additionally, for capturing overall genomic variation, 180 Swedish *Arabidopsis* lines have been sequenced (Long et al. 2013). Likewise, sequencing of 916 foxtail millet varieties endowed in identification of 2.58 million SNPs spearheaded in developing haplotype map of foxtail millet genome (Jia et al. 2013b). Recently whole genome re-sequencing approach has been employed in rice for constructing MutMap (Abe et al. 2012) and MutMap+(Fekih et al. 2013) for capturing mutants contributing quantitative traits; likewise QTL-seq based on whole genome re-sequencing has been applied for quick QTL mapping in rice (Takagi et al. 2013).

Whole Transcriptome Sequencing

Transcriptome sequence enables in translating the functional aspects of genes of an organism (Nagalakshmi et al. 2008; O'Neil and Emrich 2013). NGS technology can facilitate in sequencing the whole transcriptome, which assists in unravelling the functions of the whole mRNA of an organism (Malonae and Oliver 2011; Chu and Corey 2012). Applying NGS, it enabled in enriching the 24,000 ESTs information in switchgrass and 90 % of gene space was covered by transcriptome sequencing (Wang et al. 2012a). Transcriptome sequencing by Illumina paired-end sequencing technology in radish provided 61,554 unigenes and enriched the resources of EST-based SSRs (Wang et al. 2012b); likewise, EST-derived SSRs have been obtained from transcriptome sequencing generating 22,756 unigenes in rubber tree (Li et al. 2012b). In peanut whole transcriptome sequencing enabled to identify 26,048 unigenes and 8,817 unigenes were characterised (Wu et al. 2013); 8,252 unigenes have been annotated (Zhang et al. 2012b). Whole transcriptome sequencing facilitated in discovery of novel SNPs in wheat (Duan et al. 2012), melon (Blanca et al. 2012), black cottonwood (Geraldes et al. 2011), *Brassica napus* (Trick et al. 2009), *Eucalyptus* (Novaes et al. 2008) and lodgepole pine (Parchman et al. 2010). Similarly 192 EST-SSR markers are identified in lentil from transcriptome sequencing (Kaur et al. 2011), EST-derived SSRs in field pea and fava bean (Kaur et al. 2012). In case of chickpea 4,072 SSRs and 36,446 SNPs have been identified from transcriptome sequence of wild chickpea *C. reticulatum* (PI489777) (Jhanwar et al. 2012). Likewise transcriptome sequencing in kabuli chickpea offers repertoire for development of functional markers (Agarwal et al. 2012).

Molecular Marker Discovery

NGS technology has played key role in development of useful high-throughput markers such as SSRs, ESTs and SNPs. Some important molecular markers developed by applying NGS have been discussed. NGS has assisted in the development of 246 SSRs in *Prunus virginiana* (Wang et al. 2012c) and 94 reproducible novel SSRs in fava bean (Yang et al. 2012a) and detection of chromosome arm–specific microsatellite marker in wheat (Nie et al. 2012), SSR marker for distinguishing aphids in soybean (Jun et al. 2011) and SSRs in peanut (Zhang et al. 2012b) given in Table 9. Importantly, the role of NGS in identification of SNP marker is worth mentioning. NGS led resequencing of 8 genotypes of wheat has facilitated in developing exome-based, codominant, SNP marker used for differentiating homozygote and heterozygotes in wheat (Allen et al. 2013) and identification of SNPs in lentil (Sharpe et al. 2013), 20,000 SNPs in *Brassica rapa* (Bus et al. 2012), 8,207 SNP markers and five markers linked with anthracnose disease resistance in lupin (Yang et al. 2012a) and 1,022 SNPs and 4,543 SNPs, respectively, in chickpea (Gaur et al. 2012; Azam et al. 2012). NGS led to identify 575,340 SNPs from three cultivars of potato: "Atlantic", "Premier Russet" and "Snowden" and 96 SNPs were used for allelic diversity measurement (Hamilton et al. 2011). In tomato 8,784 SNPs were detected, derived from transcriptome sequences and utilised for constructing high-density linkage maps for three interspecific F(2) populations (Sim et al. 2012). Zou et al. (2013) searched out 1,953 SNPs associated with QTLs contributing 4-methylthio-3-butenyl glucosinolate contents in roots of radish, *Raphanus sativus* L. By applying NGS technique in durum wheat, 2,659 SNPs have been discovered (Trebbi et al. 2011) and 1,050 SNPs were identified in common bean (Hyten et al. 2010a) further 1,790 SNPs disclosed by NGS in soybean were used for developing high-resolution genetic map (Hyten et al. 2010b). In wheat (*Aegilops tauschii*) without having reference genome sequence, 497,118 genome-wide SNPs have been discovered using NGS (You et al. 2011). By deploying NGS in alfalfa, 40,661 candidate SNPs have been identified which are useful for association mapping and high-resolution mapping (Han et al. 2011). Application of NGS for developing SNP markers and its uses have been thoroughly described by Kumar et al. (2012).

Genotyping by Sequencing (GBS)

Advancement of NGS technology has paved the way for developing high-density linkage map covering all the linkage groups with anchoring thousands of high-throughput markers. These markers constitute especially SNPs, positioned on the linkage map developed earlier, enhancing its resolution, by re-sequencing of genome by NGS platforms in diploid and polyploidy species (Oliver et al. 2011; Eckert et al. 2009). SNP-based high-throughput linkage map has been developed for mapping the recessive mutant loci in maize using RILs developed from B73 x Mo17 (Liu et al. 2010). In tomato 8,784 SNPs developed from transcriptome sequences generated by NGS. These SNPs enabled in developing high-density linkage maps from three interspecific F_2 populations of EXPEN 2000, EXPEN 2012 and EXPIM 2012. The average marker bin intervals were 1.6 cM, 0.9 cM and 0.8 cM, respectively (Sim et al. 2012). NGS-led developed SSR and SNP markers assisted in developing high-density linkage map of 1,227 markers positioned on 9 linkage groups covering 1197.9 cM in brassica (Wang et al. 2012d). Moreover, along with NGS, sequence-related amplified polymorphism (SRAP) markers aided in developing an ever most saturated ultradense genetic map retaining 9,177 SRAP markers, 1,737 integrated unique Solexa paired-end sequences, 46 SSRs and 10,960 independent genetic loci in *B. rapa* (Li et al. 2011b). Similarly a linkage map developed from 114 double haploid lines, harbouring 415 INDELS and 92 SSR markers covering 1234.2 cM length, positioning 152 scaffolds on the chromosomes (Wang et al. 2011a). In chickpea one of the most high-resolution maps has been developed applying NGS comprising 1,063 markers covering map length of 1808.7 cM (Gaur et al. 2012).

Owing to lower cost involved in sequencing by NGS technology, it is utilised for sequencing the entire population contributing in trait mapping along with tracing down markers across the genome, called GBS (Elshire et al. 2011). This innovative approach has been successfully applied in rice by re-sequencing 150 RILs developed from parents of *indica* and *japonica* cultivars, leading to discovery of 1,226,791 SNPs (Huang et al. 2009). A high-quality physical map has been developed harbouring the QTL responsible for green revolution by re-sequencing whole genome of 128 chromosome segment substitution lines (CSSLs) of rice (Xu et al. 2010). Low-density GBS is reported in barley (Chutimanitsakun et al. 2011). In maize and barley GBS approach has been applied (Elshire et al. 2011) and 25,185 bi-allelic SNPs have been detected in maize. Similarly, the approach of GBS has been used in barley and wheat (Poland et al. 2012). GBS has been deployed to sequence the pool of mutants in segregating populations in *Arabidopsis* called SHORE map (Schneeberger et al. 2009), SHORE mapping a GBS strategy used for identification of causal mutation in *Arabidopsis* (Galvão et al. 2012). Likewise, next-generation mapping (NGM) (Austin et al. 2011) technique extension of next-generation genomic sequencing has been employed to map mutations directly from pooled F2 populations in *Arabidopsis*. This technique led to detect three genes associated with cell wall biology in *Arabidopsis* (Austin et al. 2011).

Epigenetic Analysis and Discovery of Small Noncoding/Regulatory RNAs

Epigenetics refers to the gene expression without any alteration in DNA sequences (Liang et al. 2009; Bird 2007). Epigenetic changes are triggered by small RNAs causing changes in DNA methylation and histone modification (acetylation, methylation, phosphorylation and ubiquitinylation) (Simon and Meyers 2011; Rival et al. 2010). Epigenetics mechanisms endow plant to change its gene expression and produce particular phenotype in response to environmental changes (Piferrer 2013). Development of NGS first brought the attention of researchers dealing with epigenetic analysis. Wide application of NGS in epigenetic analysis has been discussed recently (Meaburn and Schulz 2011). Considering the role of methylation of cytosine in DNA contributing in regulation of epigenetics (Laird 2010), NGS has been used for mapping cytosine methylation in *Arabidopsis* (Cokus et al. 2008; Lister et al. 2008). This tool has benefited in rendering the epigenome of plant (Simon and

Meyers 2011), mapping the methylation pattern and regulation of methylation throughout the genome (Lister and Ecker 2009) and generating the epigenetic markers across the genome (Liang et al. 2009). In maize NGS has helped in revealing the relationship between epigenome and transcriptome (Elling and Deng 2009; Eckardt 2009); similarly, it has assisted in development of methylome map in *Arabidopsis* (Zhang et al. 2006). Additionally, this novel technique is used for the discovery of small RNAs including microRNA (miRNA), small interfering RNA (siRNA), transferable RNA (tRNA) and ribosomal RNA (rRNA) playing key role in post-transcriptional gene expression (Xie et al. 2004; Lu et al. 2005; Filipowicz et al. 2008; Morozova and Marra 2008). Considering this, NGS enabled in discovering 14 novel and 22 conserved miRNA families from peanut responsible for growth and development in response to environment stress (Zhao et al. 2010) given in Table 9; similarly, NGS aided in discovering sRNAs accounted for fruit development and ripening in tomato (Mohorianu et al. 2011). Applying NGS, miR156, miR159, miR172, miR167, miR158 and miR166, miRNAs have been identified which are associated with seed development and maturation in *Brassica* sp. (Huang et al. 2013a). In sugarcane 26 conserved families of miRNA have been detected conferring regulation in axillary bud outgrowth sequencing sRNA by NGS (Ortiz-Morea et al. 2013). Similarly applying NGS sequencing, small RNAs that are genome wide in rice assisted in identification of pollen development stage specific 292 miRNAs (Wei et al. 2011). In barley expression profile of miRNA and other noncoding RNAs were analysed in response to phosphorus requirement, and 221 conserved miRNAs as well as 12 novel miRNAs were detected by sequencing sRNAs using Illumina's NGS (Hackenberg et al. 2013). Additionally, NGS analysis led to detection of 66 miRNA genes closely contributing in leaf growth in response to drought in *Brachypodium distachyon* (Bertolini et al. 2013).

RAD Sequencing and Reduced Representation Sequencing

Application of NGS along with restriction enzyme has given birth to some new techniques, restriction-associated digestion (RAD) sequencing, reduced representation sequencing based on reduced representation library (RRL) and complexity reduction of polymorphic sequences (CRoPs) for mostly focusing on sequencing targeted region of genome rather than sequencing whole genome (Davey et al. 2011). RAD sequencing is an innovative method for discovery of SNPs and high-throughput genotyping (Miller et al. 2007; Baird et al. 2008; Davey and Blaxter 2011; Davey et al. 2012). RAD generates two types of markers: codominant markers in targeted region close to restriction site (endonuclease enzyme) and dominant markers within region of the restriction site (Deschamps et al. 2012b). RAD sequencing aided in development of 347 denovo SNPs, which has facilitated in developing linkage map of 1,390 cM in eggplant which developed from cross "305E40" × "67/3" population, and this enabled in tracking down seven QTLs conferring to anthocyanin accumulation (Barchi et al. 2012). A high-density linkage map has been developed from F_1 populations in grape using 1,841 SNPs developed through RAD sequencing (Wang et al. 2012e). RAD sequencing aided in finding out high-throughput 20,000 SNPs and 125 insertions and deletions in *Brassica napus* (Bus et al. 2012); nearly 10,000 SNPs, 1,000 INDELS and 2,000 putative SSRs in eggplant (Barchi et al. 2011) and 34,000 SNPs and nearly 800 INDELS in *C. cardunculus* (Scaglione et al. 2012). RAD sequencing is used for developing markers tagging disease resistance gene against anthracnose disease resistance in lupin (Yang et al. 2012a). In *Arabidopsis* genome-wide genotyping has been done applying 2b RAD (Wang et al. 2012f). RAD sequencing has permitted mapping of QTLs involving in fatty acid synthesis in perennial ryegrass (*Lolium perenne* L.) (Hegarty et al. 2013).

RRL approach has been applied for discovering SNPs in maize (Gore et al. 2009; Deschamps et al. 2012a), soybean (Hyten et al. 2010a) and 4,294 to 14,550 SNPs from four accessions of soybean (Varala et al. 2011); in jointed goat grass (You et al. 2011); in grape (Myles et al. 2010); in common bean (Hyten et al. 2010b) and in rice (Monson-Miller et al. 2012). Rededed representation sequencing based, a new approach called Restriction Enzyme Site Comparative Analysis (RESCAN) has been used for detection of SNP in rice (Kim and Tai 2013). Likewise, reduced representation sequencing helped in identifying SNP residing in major QTL region contributing to pod shattering resistance in rapeseed (Hu et al. 2012), mapping respective QTLs of flowering time and petiole length in *Arabidopsis* (Seymour et al. 2012). Sequencing of RRL contributed in detection of QTLs (*qLpPg1*, *qLpPg2* and *qLpPg3*) responsible for stem rust in *Lolium perenne* (Pfender et al. 2011). RAD based sequencing has facilitated the development of GBS libraries for maize and barley, providing opportunities to the breeder community for practicing genomic selection in the future. Apart from the above applications, some other applications of NGS in association mapping study, evolutionary relationship study, diversity, alien introgression studies and organelle sequencing in plant have been elaborated (Varshney et al. 2009).

Association Mapping for Tagging Genes/Complex QTLs and Exploiting Plant Natural Variation

Traditionally genetic map and QTL mapping were developed by using biparental populations. The details of QTL mapping starting from developing mapping population subjecting contrasting parents to check polymorphism for parental and segregating progenies for the traits and detection of polymorphic markers by genotyping has been reviewed (Collard et al. 2005). An alternative method called association mapping (AM) or linkage disequilibrium has been developed to exploit the natural variation present in germplasm repository and facilitating discovery of QTLs by analysing marker–trait association (Zhu et al. 2008). Association studies provide the opportunity of high-resolution mapping and identification of gene accounting for phenotypic variation, along with high marker coverage in comparison to biparental linkage mapping (Thornsberry et al. 2001; Remington et al. 2001; Wang et al. 2005). This powerful method clarifies complex QTLs using historical and evolutionary recombination episode at population level (Nordborg and Tavare 2002; Risch and Merikangas 1996). AM is practised based on two approaches: (1) candidate gene-based approach and (2) whole-genome based approache genome- wide association study (GWAS) (Zhu et al. 2008).

Initial landmark study of AM was applied in oat (Beer et al. 1997), in rice (Virk et al. 1996) and subsequently in other plants (Thornsberry et al. 2001). The successful application of this technique in different plant species for various traits has been depicted (Zhu et al. 2008). AM offers the benefits of detection of marker–trait relationship ranging from complex qualitative to quantitative traits of interest in various crop plants. AM was implemented for analysing kernel size and milling quality using 36 SSR markers in wheat (Breseghello and Sorrells 2006). The stem rust resistance loci has been identified by association mapping deploying DArT and SSR markers in 276 spring wheat lines (Yu et al. 2011b), and Sr13gene/QTL confirmed in durum wheat earlier using biparental mapping (Letta et al. 2013). Similarly, candidate-based association mapping aided in identification of sclerotium head rot resistance QTLs in sunflower (Fusari et al. 2012), and linkage disequilibrium method enabled to correlate candidate gene marker and resistance to *Verticillium dahliae* QTL in tetraploid potato (Simko et al. 2004). Moreover, candidate gene AM has been applied for analysing drought tolerance in 192 diverse perennial ryegrass (*Lolium perenne* L.) (Yu et al. 2013). Thirteen QTLs accounting for spot blotch in wild barley has been identified using SSR and SNP markers by association analysis (Roy et al. 2010); similarly four QTLs diagnosed for *Septoria* speckled leaf blotch resistance using 3,840 lines of barley (Zhou and Steffenson 2013). In wheat this method has enabled to identify marker–trait association with yield under drought condition (Dodig et al. 2012). Association study led to detect 7QTLs account for preharvest sprouting in wheat using 1,166 DArT and SSR markers in 198 genotypes (Kulwal et al. 2012). In wheat, six candidate genes conferring to flowering time have been unfolded by AM (Rousset et al. 2011). Additionally, AM has aided in the detection of earliness causative QTL region on chromosomes (Le Gouis et al. 2012), main effect QTL for quality traits such as kernel weight, protein content, etc. (Reif et al. 2011), multiple loci offering aluminium resistance (Raman et al. 2010), and russian wheat aphid resistance (RWA2) gene in wheat (Peng et al. 2009). Invertases and starch phosphorylases alleles, contributing in tuber quality (Li et al. 2013) and QTL for tuber sugar content and chip quality (Draffehn et al. 2010), tuber, starch content and yield (Li et al. 2008) in potato have been unravelled by AM. Therefore, AM becomes an alternative approach to the classical breeding approach (biparental mapping) for dissecting complex QTLs concerned with yield, documented in barley (Kraakman et al. 2004). To bridge the benefit of linkage mapping and AM, another robust tool, nested association mapping (NAM), has been applied for the first time in maize for dissection of complex traits. Its advantages and mapping resolution have been discussed thoroughly (Yu et al. 2008). The role of complex QTLs associated with flowering time has been elucidated by studying association studies in 5,000 RILs in NAM population in maize (Buckler et al. 2009) and in barley (Stracke et al. 2009). Backcross nested association (BC-NAM) mapping facilitated in identifying 40 QTLs in *Sorghum* (Mace et al. 2013), the *Dwarf8 (d8)* locus in maize (Larsson et al. 2013) responsible for flowering time. For gaining insight into genetic basis of leaf architecture in maize, NAM has been applied (Tian et al. 2011). Thirty-two QTLs contributing for resistance against southern leaf blight (SLB) disease in maize has been detected applying NAM in 5,000 maize RILs (Kump et al. 2011), and similarly 32 QTLs were investigated causing northern leaf blight in maize deploying NAM (Poland et al. 2011). NAM association study assisted in showing the variation at *bx1* gene conferring DIMBOA content in maize (Butrón et al. 2010). Likewise, deployment of NAM for improvement of kernel composition in maize has been well documented (Cook et al. 2012). However, AM suffers from major drawback of confounding effect of population structure giving rise to false positive (Yu et al. 2006; Zhao et al. 2007; Ingvarsson and Street 2011).

In recent past GWAS has been developed which inspects association of marker and QTLs across the genome (Risch and Merikangas 1996). GWAS is a population-based approach offering the advantage of studying association of SNP and phenotype across the genome from unrelated individuals (Mitchell-Olds 2010). It has been applied to study 107 phenotypes of 200 *Arabidopsis* inbred lines (Atwell et al. 2010). Similarly GWAS is efficient in detecting QTLs from the natural or pre-exiting lines in comparison to biparental mapping strategy. This approach is utilised in finding out for beta-glucan QTL in oat (Newell et al. 2012) and common bacterial blight resistance QTL in *Phaseolus vulgaris* (Shi et al. 2011). GWAS facilitated in marker–trait association for various agronomic and disease resistance traits in barley (Pasam et al. 2012; Berger et al. 2013), saccharification yield in sorghum (Wang et al. 2011a), detecting allelic variation in natural germplasm in rice (Han and Huang 2013) and mapping of 15 traits with 1,536 SNPs in 500 lines of barley (Cockram et al. 2010). Contribution of 61 loci to tocopherol content and composition has been revealed by genome-wide association analysis in *Brassica napus* (Wang et al. 2012g). In recent past GWAS led to identify 512 loci correlated with 47 agronomic traits in foxtail millet (Jia et al. 2013b). From abiotic stress point of view GWAS has been applied in 125 inbred lines of maize and 43 SNPs found to be associated with chilling tolerance (Huang

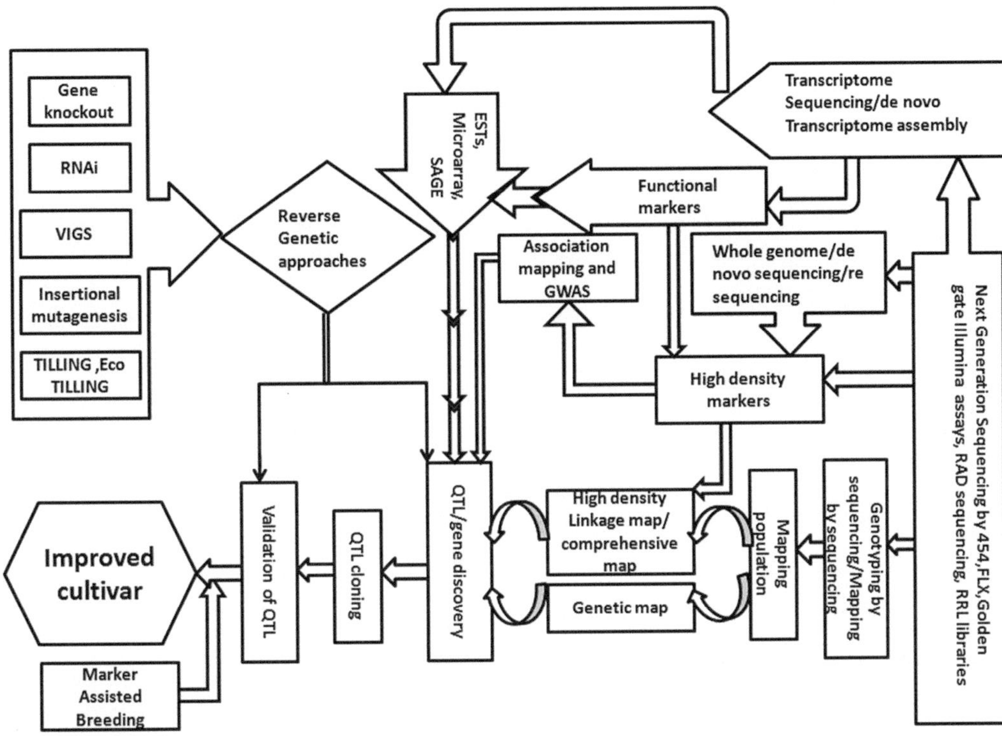

Fig. 1 Integrated approaches of functional genomics for developing improved crop cultivars

et al. 2013c). Therefore, AM can be utilised for genome-wide diversity study, shedding light on marker–trait association and discovery of genes/QTLs. Similarly it has the potentiality for QTL dissection and cloning (Salvio and Tuberosa 2007). To facilitate accurate detection of gene/QTLs coupled with their functions for overall crop improvement, an integrated approach combining all the tools of functional genomics is depicted in Fig. 1.

Perspectives and Future Direction

The advent of fast-evolving DNA sequencing technology has given a new direction in the field of genomics by enabling sequencing of whole genome, extricating precious genomic information of non-model crops and re-sequencing of model crops in quick time and under manageable cost. Reduction of cost for sequencing leads to develop next-next or third-generation sequencing technologies such as single-molecule real-time (SMRT™) sequencing capable in generating longer sequence read (Thudi et al. 2012). Additionally, high-throughput-driven technologies led to development of ultradense linkage map, transcript map, SHORE map and MutMap of plant species mentioned earlier to capture the desired genes/QTLs of interest. Nevertheless, the rapid accumulation of sequence information and decoding of function of these sequence informations are still lagging behind. To date rice and *Arabidopsis* are the only members of plant species for which we have most of the genes with known function. Thus the progress of functional genomic research will benefit the plant science community by unlocking the function of all genes of most of the plant species. Ultimately, this will lead to improvement of crop breeding programme thereby developing cultivar with enhanced tolerance to biotic and abiotic stresses mitigating the challenges of food security in the coming future.

References

Aarts MG, Dirkse WG, Stiekema WJ, Pereira A (1993) Transposon tagging of a male sterility gene in Arabidopsis. Nature 363:715–717

Abe A, Kosugi S, Yoshida K et al (2012) Genome sequencing reveals agronomically-important loci in rice from mutant populations. Nat Biotechnol 30:174–178

Adams MD, Kerlavage AR, Fleischmann RD, Fuldner RA, Bult CJ, Lee NH, Kirkness EF, Weinstock KG, Gocayne JD, White O et al (1995) Initial assessment of human gene discovery and expression patterns based upon 83-million nucleotides of cDNA sequence. Nature 377:3–7

Agarwal G, Jhanwar S, Priya P, Singh VK, Saxena MS, Parida SK, Garg R, Tyagi AK, Jain M (2012) Comparative analysis of kabuli chickpea transcriptome with desi and wild chickpea provides a rich resource for development of functional markers. PLoS One 7:e52443

Aharoni A, Vorst O (2001) DNA microarrays for functional plant genomics. Plant Mol Biol 48:99–118

Aharoni A, Keizer LC, Bouwmeester HJ, Sun Z, Alvarez-Huerta M, Verhoeven HA, Blaas J, van Houwelingen AM, De Vos RC, van der Voet H, Jansen RC, Guis M, Mol J, Davis RW, Schena M, van Tunen AJ, O'Connell AP (2000) Identification of the *SAAT* gene involved in strawberry flavor biogenesis by using DNA microarrays. Plant Cell 12:647–662

Ahmad R, Parfitt DE, Fass J, Ogundiwin E, Dhingra A, Gradziel TM, Lin D, Joshi NA, Martinez-Garcia PJ, Crisosto CH (2011) Whole genome sequencing of peach (Prunus persica L.) for SNP identification and selection. BMC Genomics 12:569

Ahmed NU, Park JI, Seo MS, Kumar TS, Lee IH, Park BS, Nou IS (2012) Identification and expression analysis of chitinase genes related to biotic stress resistance in Brassica. Mol Biol Rep 39:3649–3657

Allen AM, Barker GL, Wilkinson P, Burridge A, Winfield M, Coghill J, Uauy C, Griffiths S, Jack P, Berry S, Werner P, Melichar JP, McDougall J, Gwilliam R, Robinson P, Edwards KJ (2013) Discovery and development of exome-based, co-dominant single nucleotide polymorphism markers in hexaploid wheat (Triticum aestivum L.). Plant Biotechnol J 11:279–295

Alonso JM, Ecker JR (2006) Moving forward in reverse: genetic technologies to enable genome-wide phenomic screens in Arabidopsis. Nat Rev Genet 7:524–536

Altshul SF, Gish W, Miller W, Myers EW, Lipman DJ (1990) Basic local alignment search tool. J Mol Biol 215:403–410

Ansorge WJ (2009) Next- generation DNA sequencing techniques. New Biotechnol 25:195–203

Antollin-Llovere M, Parniske M (2007) TILLING: examples of utilization in plant breeding. Lotus Newsl 37:103–104

Asamizu E, Shimoda Y, Kouchi H, Tabata S, Sato S (2008) A positive regulatory role for LjERF1 in the nodulation process is revealed by systematic analysis of nodule-associated transcription factors of *Lotus japonicas*. Plant Physiol 147:2030–2040

Ashelford K, Eriksson ME, Allen CM, D'Amore R, Johansson M, Gould P, Kay S, Miller AJ, Hall N, Hall A (2011) Full genome resequencing reveals a novel circadian clock mutation in Arabidopsis. Genome Biol 12:R28

Asmussen MA, Gilliland LU, Meagher RB (1998) Detection of deleterious genotypes in multigenerational studies. II. Theoretical and experimental dynamics with selfing and selection. Genetics 149:727–737

Atwell S, Huang YS, Vilhjálmsson BJ, Willems G, Horton M, Li Y, Meng D, Platt A, Tarone AM, Hu TT, Jiang R, Muliyati NW, Zhang X, Amer MA, Baxter I, Brachi B, Chory J, Dean C, Debieu M, de Meaux J, Ecker JR, Faure N, Kniskern JM, Jones JD, Michael T, Nemri A, Roux F, Salt DE, Tang C, Todesco M, Traw MB, Weigel D, Marjoram P, Borevitz JO, Bergelson J, Nordborg M (2010) Genome-wide association study of 107 phenotypes in Arabidopsis thaliana inbred lines. Nature 465:627–631

Auer C, Frederick R (2009) Crop improvement using small RNAs: applications and predictive ecological risk assessments. Trends Biotechnol 27:644–651

Austin RS, Vidaurre D, Stamatiou G, Breit R, Provart NJ, Bonetta D, Zhang J, Fung P, Gong Y, Wang PW, McCourt P, Guttman DS (2011) Next-generation mapping of Arabidopsis genes. Plant J 67:715–725

Axtell MJ, Jan C, Rajagopalan R, Bartel DP (2006) A two hit trigger for siRNA biogenesis in plants. Cell 127:565–577

Azam S, Thakur V, Ruperao P, Shah T, Balaji J, Amindala B, Farmer AD, Studholme DJ, May GD, Edwards D, Jones JD, Varshney RK (2012) Coverage-based consensus calling (CbCC) of short sequence reads and comparison of CbCC results to identify SNPs in chickpea (Cicer arietinum; Fabaceae), a crop species without a reference genome. Am J Bot 99:186–192

Azpiroz-Leehan R, Feldmann KA (1997) T-DNA insertion mutagenesis in *A. thaliana*: going back and forth. Trends Genet 13:146–152

Babiyachuk E, Fuanghthong M, Montagu MV, Inze D, Kushnir S (1997) Efficient gene tagging in *Arabidopsis thaliana* using a gene trap approach. Proc Natl Acad Sci U S A 94:12722–12727

Bachlava E, Taylor CA, Tang S, Bowers JE, Mandel JR, Burke JM, Knapp SJ (2012) SNP discovery and development of a high-density genotyping array for sunflower. PLoS One 7:e29814

Bade J, Van Grinsven E, Custers J, Hoekstra S, Ponstein A (2003) T-DNA tagging in *Brassica napus* as an efficient tool for the isolation of new promoters for selectable marker genes. Plant Mol Biol 52:53–68

Baird NA, Etter PD, Atwood TS, Currey MC, Shiver AL, Lewis ZA, Selker EU, Cresko WA, Johnson EA (2008) Rapid SNP discovery and genetic mapping using sequenced RAD markers. PLoS One 3:e3376

Bao JY, Lee S, Chen C, Zhang XO, Zhang Y, Liu S, Clark T, Wang J, Cao ML, Yang HM, Wang SM, Yu J (2005) Serial analysis of gene expression study of a hybrid rice strain (LYP9) and its parental cultivars. Plant Physiol 138:1216–1231

Barchi L, Lanteri S, Portis E, Acquadro A, Valè G, Toppino L, Rotino GL (2011) Identification of SNP and SSR markers in eggplant using RAD tag sequencing. BMC Genomics 12:304

Barchi L, Lanteri S, Portis E, Valè G, Volante A, Pulcini L, Ciriaci T, Acciarri N, Barbierato V, Toppino L, Rotino GL (2012) A RAD tag derived marker based eggplant linkage map and the location of QTLs determining anthocyanin pigmentation. PLoS One 7:e43740

Barkley NA, Wang ML, Gillaspie AG, Dean RE, Pederson GA, Jenkins TM (2008) Discovering and verifying DNA polymorphisms in a mung bean [*V. radiata* (L.) R. Wilczek] collection by EcoTILLING and sequencing. BMC Res Notes 1:28

Bashir K, Ishimaru Y, Nishizawa NK (2011) Identification and characterization of the major mitochondrial Fe transporter in rice. Plant Signal Behav 6:1591–1593

Bauer P, Lubkowitz M, Tyers R, Nemoto K, Meeley RB, Goff SA, Freeling M (2004) Regulation and a conserved intron sequence of liguleless3/4 knox class-I homeobox genes in grasses. Planta 219:359–368

Baulcombe DC (1999) Fast forward genetics based on virus induced gene silencing. Curr Opin Plant Biol 2:109–113

Baulcombe D (2002) RNA silencing. Curr Biol 12:82–84

Baulcombe D (2004) RNA silencing in plants. Nature 431:356–363

Baum JA, Bogaert T, Clinton W, Heck RG, Feldman P, Ilagan O, Johnson S, PlaetincK G, Munyikwa T, Pleau M, Vaughn T, Robert J (2007) Control of coleopteran insect pests through RNA interference. Nat Biotechnol 25:1322–1326

Beadle GW, Tatum EL (1941) Genetic control of biochemical reactions in Neurospora. Proc Natl Acad Sci U S A 27:499–506

Becker KG (2001) The sharing of cDNA microarray data. Nat Rev Neurosci 2:438–440

Becker A, Lange M (2010) VIGS: genomics goes functional. Trends Plant Sci 15:1–4

Beer SC, Siripoonwiwat W, Donoughue LSO, Souza E, Matthews D, Sorrels ME (1997) Associations between molecular markers and quantitative traits in an oat germplasm pool: can we infer linkages? J Agric Genomics 3:1997

Benlloch R, d'ErfurthI I, Ferrandiz C, Cosson V, Beltron JP, Canas AL, Kondorosi A, Aduen F, Ratet P (2006) Isolation of mtpim proves Tnt1 a useful reverse genetics tool in *Medicago truncatula* and uncovers new aspects of AP1-like functions in legumes. Plant Physiol 142:972–983

Berger GL, Liu S, Hall MD, Brooks WS, Chao S, Muehlbauer GJ, Baik BK, Steffenson B, Griffey CA (2013) Marker-trait associations in Virginia Tech winter barley identified using genome-wide mapping. Theor Appl Genet 126:693–710

Bernstein E, Caudy AA, Hammond SM, Hannon GJ (2001) Role for a bidentate ribonuclease in the initiation step of RNA interference. Nature 409:363–366

Bertolini E, Verelst W, Horner DS, Gianfranceschi L, Piccolo V, Inzé D, Pè ME, Mica E (2013) Addressing the role of microRNAs in reprogramming leaf growth during drought stress in Brachypodium distachyon. Mol Plant 6:423–443

Bi YP, Liu W, Xia H, Su L, Zhao CZ, Wan SB, Wang XJ (2010) EST sequencing and gene expression profiling of cultivated peanut (*Arachis hypogaea* L.). Genome 53:832–839

Bird A (2007) Perceptions of epigenetics. Nature 447:396–398

Blanca J, Esteras C, Ziarsolo P, Pérez D, Fernã Ndez-Pedrosa V, Collado C, Rodrã Guez de Pablos R, Ballester A, Roig C, Cañizares J, Picó B (2012) Transcriptome sequencing for SNP discovery across Cucumis melo. BMC Genomics 13:280

Blanco FA, Meschini EP, Zanetti ME, Aguilar OM (2009) A small GTPase of the Rab family is required for root hair formation and preinfection stages of the common bean-Rhizobium symbiotic association. Plant Cell 21:2797–2810

Bonfim K, Faria JC, Nogueira EO, Mendes EA, Aragão FJ (2007) RNAi-mediated resistance to Bean golden mosaic virus in genetically engineered common bean (Phaseolus vulgaris). Mol Plant Microbe Interact 20:717–726

Bouché N, Bouchez D (2001) Arabidopsis gene knockout: phenotypes wanted. Curr Opin Plant Biol 4:111–117

Brazma A, Hingamp P, Quackenbush J, Sherlock G, Spellman P, Stoeckert C, Aach J, Ansorge W, Ball CA, Causton HC, Gaasterland T, Glenisson P, Holstege FC, Kim IF, Markowitz V, Matese JC, Parkinson H, Robinson A, Sarkans U, Schulze-Kremer S, Stewart J, Taylor R, Vilo J, Vingron M (2001) Minimum information about a microarray experiment (MIAME)-toward standards for microarray data. Nat Genet 29:365–371

Brenchley R, Spannagl M, Pfeifer M, Barker GL, D'Amore R, Allen AM, McKenzie N, Kramer M, Kerhornou A, Bolser D, Kay S, Waite D, Trick M, Bancroft I, Gu Y, Huo N, Luo MC, Sehgal S, Gill B, Kianian S, Anderson O, Kersey P, Dvorak J, McCombie WR, Hall A, Mayer KF, Edwards KJ, Bevan MW, Hall N (2012) Analysis of the bread wheat genome using whole-genome shotgun sequencing. Nature 491:705–710

Brenner S, Johnson M, Bridgham J, Golda G, Lloyd DH (2000) Gene expression analysis by massively parallel signature sequencing (MPSS) on microbead arrays. Nat Biotechnol 18:630–634

Breseghello F, Sorrells ME (2006) Association mapping of kernel size and milling quality in wheat (Triticum aestivum L.) cultivars. Genetics 172:1165–1177

Brueggeman R, Druka A, Nirmala J, Cavileer T, Drader T, Rostoks N, Mirlohi A, Bennypaul H, Gill U, Kudrna D, Whitelaw C, Kilian A, Han F, Sun Y, Gill K, Steffenson B, Kleinhofs A (2008) The stem rust resistance gene Rpg5 encodes a protein with nucleotide-binding-site, leucine-rich, and protein kinase domains. Proc Natl Acad Sci U S A 105:14970–14975

Buckler ES, Holland JB, Bradbury PJ, Acharya CB, Brown PJ, Browne C, Ersoz E, Flint-Garcia S, Garcia A, Glaubitz JC, Goodman MM, Harjes C, Guill K, Kroon DE, Larsson S, Lepak NK, Li H, Mitchell SE, Pressoir G, Peiffer JA, Rosas MO, Rocheford TR, Romay MC, Romero S, Salvo S, Sanchez Villeda H, da Silva HS, Sun Q, Tian F, Upadyayula N, Ware D, Yates H, Yu J, Zhang Z, Kresovich S, McMullen MD (2009) The genetic architecture of maize flowering time. Science 325:714–718

Buckler ES, Warburton ML, Rocheford T (2010) Rare genetic variation at Zea mays crtRB1 increases beta-carotene in maize grain. Nat Genet 42:322–327

Bundock PC, Eliott FG, Ablett G, Benson AD, Casu RE, Aitken KS, Henry RJ (2009) Targeted single nucleotide polymorphism (SNP) discovery in a highly polyploid plant species using 454 sequencing. Plant Biotechnol J 7:347–354

Burch-Smith TM, Anderson JC, Martin GB, Dinesh-Kumar SP (2004) Applications and advantages of virus- induced gene silencing for gene function studies in plants. Plant J 39:734–746

Burch-Smith TM, Schiff M, Liu Y, Dinesh-Kumar SP (2006) Efficient virus-induced gene silencing in Arabidopsis. Plant Physiol 142:21–27

Bus A, Hecht J, Huettel B, Reinhardt R, Stich B (2012) High-throughput polymorphism detection and genotyping in Brassica napus using next-generation RAD sequencing. BMC Genomics 13:281

Butrón A, Chen YC, Rottinghaus GE, McMullen MD (2010) Genetic variation at bx1 controls DIMBOA content in maize. Theor Appl Genet 120:721–734

Byzova M, Verduyn C, De Brouwer D, De Block M (2004) Transforming petals into sepaloid organs in *Arabidopsis* and oilseed rape: implementation of the hairpin RNA-mediated gene silencing technology in an organ-specific manner. Planta 218:379–387

Caldwell DG, McCallum N, Shaw P, Muehlbauer GJ, Marshall DF, Waugh R (2004) A structured mutant population for forward and reverse genetics in Barley (*Hordeum vulgare* L.). Plant J 40:143–150

Campisi L, Yang Y, Yi Y, Heilig E, Herman B, Cassista AJ, Allen DW, Xiang H, Jack T (1999) Generation of enhancer trap lines in *Arabidopsis* and characterization of expression patterns in the inflorescence. Plant J 17:699–707

Cao J, Schneeberger K, Ossowski S, Günther T, Bender S, Fitz J, Koenig D, Lanz C, Stegle O, Lippert C, Wang X, Ott F, Müller J, Alonso-Blanco C, Borgwardt K, Schmid KJ, Weigel D (2011) Whole-genome sequencing of multiple Arabidopsis thaliana populations. Nat Genet 43:956–963

Chai YM, Jia HF, Li CL, Dong QH, Shen YY (2011) FaPYR1 is involved in strawberry fruit ripening. J Exp Bot 62:5079–5089

Chang Y, Long T, Wu C (2012) Effort and contribution of T-DNA Insertion mutant library for rice functional genomics research in China: review and perspective. J Integr Plant Biol 54:953–966

Chawade A, Sikora P, Brautigam M, Larsson M, Vivekanand V, Nakash AM, Chen T, Olsson O (2010) Development and characterization of an oat TILLING-population and identification of mutations in lignin and beta-glucan biosynthesis genes. BMC Plant Biol 10:86

Chen J, Sun M, Lee S, Zhou G, Rowley JD, Wang SM (2002) Identifying novel transcripts and novel genes in the human genome by using novel SAGE tags. Proc Natl Acad Sci U S A 99:12257–12262

Chen J, Zhang D, Yao Q, Zhang J, Dong X, Tian H, Chen J, Zhang W (2011) Feeding-based RNA interference of a trehalose phosphate synthase gene in the brown planthopper, Nilaparvata lugens. Insect Mol Biol 19:777–786

Chen ZC, Yamaji N, Motoyama R, Nagamura Y, Ma JF (2012) Up-regulation of a magnesium transporter gene OsMGT1 is required for conferring aluminum tolerance in rice. Plant Physiol 159:1624–1633

Choi YE, Goodwin SB (2011) Gene encoding a c-type cyclin in Mycosphaerella graminicola is involved in aerial mycelium formation, filamentous growth, hyphal swelling, melanin biosynthesis, stress response, and pathogenicity. Mol Plant Microbe Interact 24:469–477

Chu Y, Corey DR (2012) RNA sequencing: platform selection, experimental design, and data interpretation. Nucleic Acid Ther 22:271–274

Chutimanitsakun Y, Nipper RW, Cuesta-Marcos A, Cistue L, Corey A, Filichkina T, Johnson EA, Hayes PM (2011) Construction and application for QTL analysis of a Restriction Site Associated DNA (RAD) linkage map in barley. BMC Genomics 12:4

Cockram J, White J, Zuluaga DL, Smith D, Comadran J, Macaulay M, Luo Z, Kearsey MJ, Werner P, Harrap D, Tapsell C, Liu H, Hedley PE, Stein N, Schulte D, Steuernagel B, Marshall DF, Thomas WT, Ramsay L, Mackay I, Balding DJ, AGOUEB Consortium, Waugh R, O'Sullivan DM (2010) Genome-wide association mapping to candidate polymorphism resolution in the unsequenced barley genome. Proc Natl Acad Sci U S A 107:21611–21616

Cokus SJ, Feng SH, Zhang XY, Chen ZG, Merriman B, Haudenschild CD, Pradhan S, Nelson SF, Pellegrini M, Jacobson SE (2008) Shotgun bisulphite sequencing of the Arabidopsis genome reveals DNA methylation patterning. Nature 452:215–219

Colbert T, Till BJ, Tompa R, Reynolds S, Steine MN, Yueng AT, McCullum CM, Comai L, Henikoff S (2001) High throughput screening for induced point mutations. Plant Physiol 126:480–484

Collard BCY, Mackill DJ (2008) Marker-assisted selection: an approach for precision plant breeding in the twenty-first century. Philos Trans R Soc B 363:557–572

Collard BCY, Jahufer MZZ, Brouwer JB, Pang ECK (2005) An introduction to markers, quantitative trait loci (QTL) mapping and marker-assisted selection for crop improvement: the basic concepts. Euphytica 142:169–196

Comai L, Young K, Till BJ, Reynolds SH, Greene EA, Codomo CA, Enns LC, Johnson JE, Burtner C, Odden AR, Henikoff S (2004) Efficient discovery of DNA

polymorphisms in natural populations by ecotilling. Plant J 37:778–786
Constantin GD, Krath BN, MacFarlane SA, Nicolaisen M, Johansen IE, Lund OS (2004) Virus-induced gene silencing as a tool for functional genomics in legume species. Plant J 40:622–631
Cook JP, McMullen MD, Holland JB, Tian F, Bradbury P, Ross-Ibarra J, Buckler ES, Flint-Garcia SA (2012) Genetic architecture of maize kernel composition in the nested association mapping and inbred association panels. Plant Physiol 158:824–834
Cooper JL, Till BJ, Laport RG, Darlow MC, Jamai A, Kleffner JM, El- Mellouki T, Liu S, Ritchie R, Nielsen N (2008) TILLING to detect induced mutations in soybean. BMC Plant Biol 8:9
Costello JF, Krzywinski M, Marra MA (2009) A first look at entire human methylomes. Nat Biotechnol 27:1130–1132
Courtial B, Feuerbach F, Eberhard S, Rohmer L, Chiapello H, Camilleri C, Lucas H (2001) *Tnt1* transposition events are induced by in vitro transformation of *Arabidopsis thaliana* and transposed copies integrate into genes. Mol Genet Genomics 265:32–42
Cuperus JT, Montgomery TA, Fahlgren N, Burke RT, Townsend T, Sullivan CM, Carrington JC (2010) Identification of MIR390a precursor processing-defective mutants in Arabidopsis by direct genome sequencing. Proc Natl Acad Sci U S A 107:466–471
Dahmani-Mardas F, Troadec C, Boualem A, Lévêque S, Alsadon AA, Aldoss AA, Dogimont C, Bendahmane A (2010) Engineering melon plants with improved fruit shelf life using the TILLING approach. PLoS One 5:e15776
Dalmais M, Schmidt J, Le Signor C, Moussy F, Burstin J, Savois V, Aubert G, Brunaud V, Oliviera YD, Guichard C, Thompson R, Bendahmane A (2008) UTILLdb, a Pisum sativum in silico forward and reverse genetics tool. Genome Biol 9:R43
Dalmais M, Antelme S, Ho-Yue-Kuang S, Wang Y, Darracq O, d'Yvoire MB, Cézard L, Légée F, Blondet E, Oria N, Troadec C, Brunaud V, Jouanin L, Höfte H, Bendahmane A, Lapierre C, Sibout R (2013) A TILLING platform for functional genomics in Brachypodium distachyon. PLoS One 8:e65503
Davey JW, Blaxter ML (2011) RADSeq: next generation population genetics. Brief Funct Genomics 9:416–423
Davey JW, Hohenlohe PA, Etter PD, Boone JQ, Catchen JM, Blaxter ML (2011) Genome-wide genetic marker discovery and genotyping using next-generation sequencing. Nat Rev Genet 12:499–510
Davey JW, Cezard T, Fuentes-Utrilla P, Eland C, Gharbi K, Blaxter ML (2012) Special features of RAD Sequencing data: implications for genotyping. Mol Ecol. doi:10.1111/mec.12084
Delteil A, Blein M, Faivre-Rampant O, Guellim A, Estevan J, Hirsch J, Bevitori R, Michel C, Morel JB (2012) Building a mutant resource for the study of disease resistance in rice reveals the pivotal role of several genes involved in defence. Mol Plant Pathol 13:72–82
Deschamps S, Llaca V, May GD (2012a) Genotyping-by-sequencing in plants. Biology 1:460–483
Deschamps S, Nannapaneni K, Zhang Y, Hayes K (2012b) Local assemblies of paired-end reduced representation libraries sequenced with the illumina genome analyzer in maize. Int J Plant Genomics 2012:360598
Distelfeld A, Pearce SP, Avni R, Scherer B, Uauy C, Piston F, Slade A, Zhao R, Dubcovsky J (2012) Divergent functions of orthologous NAC transcription factors in wheat and rice. Plant Mol Biol 78:515–524
Dodig D, Zoric M, Kobiljski B, Savic J, Kandic V, Quarrie S, Barnes J (2012) Genetic and association mapping study of wheat agronomic traits under contrasting water regimes. Int J Mol Sci 13:6167–6188
Dong X, Kim WK, Lim YP, Kim YK, Hur Y (2013) Ogura-CMS in Chinese cabbage (Brassica rapa ssp. pekinensis) causes delayed expression of many nuclear genes. Plant Sci 199–200:7–17
Draffehn AM, Meller S, Li L, Gebhardt C (2010) Natural diversity of potato (Solanum tuberosum) invertases. BMC Plant Biol 10:271
Du HY, Shen YZ, Huang ZJ (2013a) Function of the wheat TaSIP gene in enhancing drought and salt tolerance in transgenic Arabidopsis and rice. Plant Mol Biol 81:417–429
Du J, Tian Z, Liu J, Vleeshouwers VG, Shi X, Xie C (2013b) Functional analysis of potato genes involved in quantitative resistance to *Phytophthora infestans*. Mol Biol Rep 40:957–967
Duan J, Xia C, Zhao G, Jia J, Kong X (2012) Optimizing de novo common wheat transcriptome assembly using short-read RNA-Seq data. BMC Genomics 13:392
Dugas DV, Monaco MK, Olsen A, Klein RR, Kumari S, Ware D, Klein PE (2011) Functional annotation of the transcriptome of Sorghum bicolor in response to osmotic stress and abscisic acid. BMC Genomics 12:514
Duggan DJ, Bittner M, Chen Y, Meltzer P, Trent JM (1999) Expression profiling using cDNA microarrays. Nat Genet 21:10–14
Dunoyer P, Voinnet O (2008) Mixing and matching: the essence of plant systemic silencing? Trends Genet 24:151–154
Eamens A, Wang MB, Smith NA, Waterhouse PM (2008) RNA silencing in plants: yesterday, today, and tomorrow. Plant Physiol 147:456–468
Eckardt NA (2009) Deep sequencing maps the maize epigenomic landscape. Plant Cell 21:1024–1026
Eckert AJ, Pande B, Ersoz ES, Wright MH, Rashbrook VK, Nicolet CM, Neale DB (2009) High-throughput genotyping and mapping of single nucleotide polymorphisms in loblolly pine (*Pinus taeda* L.). Tree Genet Genomes 5:225–234
Edwards D, Batley J, Snowdon RJ (2013) Accessing complex crop genomes with next-generation sequencing. Theor Appl Genet 126:1–11
Egan AN, Schlueter J, Spooner DM (2012) Applications of next-generation sequencing in plant biology. Am J Bot 99:175–185
Elling AA, Deng XW (2009) Next-generation sequencing reveals complex relationships between the epigenome

and transcriptome in maize. Plant Signal Behav 4:760–762

Elshire RJ, Glaubitz JC, Sun Q, Poland JA, Kawamoto K, Buckler ES, Mitchell SE (2011) A robust, simple genotyping-by-sequencing (GBS) approach for high diversity species. PLoS One 6:e19379

Escobar MA, Civerolo EL, Summerfelt KR, Dandeker AM (2001) RNAi-mediated oncogene silencing confers resistance to crown gall tumorigenesis. Proc Natl Acad Sci U S A 98:13437–13442

Esfandiari E, Jin Z, Abdeen A, Griffiths JS, Western TL, Haughn GW (2013) Identification and analysis of an outer-seed-coat-specific promoter from Arabidopsis thaliana. Plant Mol Biol 81:93–104

Falcone Ferreyra ML, Pezza A, Biarc J, Burlingame AL, Casati P (2010) Plant L10 ribosomal proteins have different roles during development and translation under ultraviolet-B stress. Plant Physiol 153:1878–1894

Fekih R, Takagi H, Tamiru M, Abe A, Natsume S, Yaegashi H, Sharma S, Sharma S, Kanzaki H, Matsumura H, Saitoh H, Mitsuoka C, Utsushi H, Uemura A, Kanzaki E, Kosugi S, Yoshida K, Cano L, Kamoun S, Terauchi R (2013) MutMap+: genetic mapping and mutant identification without crossing in rice. PLoS One 8(7):e68529

Feldmann KA (1991) T-DNA insertion mutagenesis in *Arabidopsis*: mutational spectrum. Plant J 1:71–82

Feng H, Fan X, Fan X, Liu X, Miller AJ, Xu G (2011) Multiple roles of nitrate transport accessory protein NAR2 in plants. Plant Signal Behav 6:1286–1289

Fernandez-Moreno JP, Orzaez D, Granell A (2013) VIGS: a tool to study fruit development in Solanum lycopersicum. Methods Mol Biol 975:183–196

Filipowicz W, Bhattacharyya SN, Sonenberg N (2008) Mechanisms of posttranscriptional regulation by microRNAs: are the answers in sight? Nat Rev Genet 9:102–114

Fire A, Xu S, Montgomery MK, Kostas SA, Driver SE, Mello CC (1998) Potent and specific genetic interference by double-stranded RNA in *Caenorhabditis elegans*. Nature 391:806–811

Friebe B, Zhang P, Nasuda S, Gill BS (2003) Characterization of a knock-out mutation at the Gc2 locus in wheat. Chromosoma 111:509–517

Fu SX, Cheng H, Qi C (2009) Microarray analysis of gene expression in seeds of Brassica napus planted in Nanjing (altitude: 8.9 m), Xining (altitude: 2261.2 m) and Lhasa (altitude: 3658 m) with different oil content. Mol Biol Rep 36:2375–2386

Fuchs U, Damm-Welk C, Borkhardt A (2004) Silencing of disease-related genes by small interfering RNAs. Curr Mol Med 4:507–517

Fukuoka H, Yamaguchi H, Nunome T, Negoro S, Miyatake K, Ohyama A (2010) Accumulation, functional annotation, and comparative analysis of expressed sequence tags in eggplant (Solanum melongena L.), the third pole of the genus Solanum species after tomato and potato. Gene 450:76–84

Fusari CM, Di Rienzo JA, Troglia C, Nishinakamasu V, Moreno MV, Maringolo C, Quiroz F, Alvarez D, Escande A, Hopp E, Heinz R, Lia VV, Paniego NB (2012) Association mapping in sunflower for sclerotinia head rot resistance. BMC Plant Biol 12:93

Galvão VC, Nordström KJ, Lanz C, Sulz P, Mathieu J, Posé D, Schmid M, Weigel D, Schneeberger K (2012) Synteny-based mapping-by-sequencing enabled by targeted enrichment. Plant J 71:517–526

Gao X, Shan L (2013) Functional genomic analysis of cotton genes with agrobacterium-mediated virus-induced gene silencing. Methods Mol Biol 975:157–165

Gao X, Starr J, Göbel C, Engelberth J, Feussner I, Tumlinson J, Kolomiets M (2008) Maize 9-lipoxygenase ZmLOX3 controls development, root-specific expression of defense genes, and resistance to root-knot nematodes. Mol Plant Microbe Interact 21:98–109

Gao X, Wheeler T, Li Z, Kenerley CM, He P, Shan L (2011) Silencing GhNDR1 and GhMKK2 compromises cotton resistance to Verticillium wilt. Plant J 66:293–305

Gao Q, Yue G, Li W, Wang J, Xu J, Yin Y (2012) Recent progress using high-through put sequencing technologies in plant molecular breeding. J Integr Plant Biol 54:215–227

Garcia-Mas J, Benjak A, Sanseverino W, Bourgeois M, Mir G, González VM, Hénaff E, Câmara F, Cozzuto L, Lowy E, Alioto T, Capella-Gutiérrez S, Blanca J, Cañizares J, Ziarsolo P, Gonzalez-Ibeas D, Rodríguez-Moreno L, Droege M, Du L, Alvarez-Tejado M, Lorente-Galdos B, Melé M, Yang L, Weng Y, Navarro A, Marques-Bonet T, Aranda MA, Nuez F, Picó B, Gabaldón T, Roma G, Guigó R, Casacuberta JM, Arús P, Puigdomènech P (2012) The genome of melon (*Cucumis melo* L.). Proc Natl Acad Sci U S A 109:11872–11877

Gaur R, Azam S, Jeena G, Khan AW, Choudhary S, Jain M, Yadav G, Tyagi AK, Chattopadhyay D, Bhatia S (2012) High-throughput SNP discovery and genotyping for constructing a saturated linkage map of chickpea (Cicer arietinum L.). DNA Res 19: 357–373

Geraldes A, Pang J, Thiessen N, Cezard T, Moore R, Zhao Y, Tam A, Wang S, Friedmann M, Birol I, Jones SJ, Cronk QC, Douglas CJ (2011) SNP discovery in black cottonwood (*Populus trichocarpa*) by population transcriptome resequencing. Mol Ecol Resour 1:81–92

Gheysen G, Vanholme B (2007) RNAi from plants to nematodes. Trends Biotechnol 25:89–92

Ghildiyal M, Zamore PD (2009) Small silencing RNAs: an expanding universe. Nat Rev Genet 10:94–108

Gibson G, Muse SV (2009) A primer of genome science, 3rd edn. Sinauer Associates, Sunderland. ISBN ISBN-13: 978-0878932368

Gierl A, Saedler H (1992) Plant transposable elements and gene tagging. Plant Mol Biol 19:39–49

Gilchrist JE, Haughn WG (2005) TILLING without a plough: a new method with applications for reverse genetics. Curr Opin Plant Biol 8:211–215

Gilchrist E, Haughn G (2010) Reverse genetics techniques: engineering loss and gain of gene function in plants. Brief Funct Genomics 9:103–110

Gilchrist EJ, Haughn GW, Ying CC, Otto SP, Zhuang J, Cheung D, Hamberger B, Aboutorabi F, Kalynyak T, Johnson L, Bohlmann J, Ellis BE, Douglas CJ, Cronk QC (2006) Use of EcoTILLING as an efficient SNP discovery tool to survey genetic variation in wild populations of Populus trichocarpa. Mol Ecol 15:1367–1378

Gilliland LU, McKinney EC, Asmussen MA, Meagher RB (1998) Detection of deleterious genotypes in multigenerational studies. I. Disruptions in individual Arabidopsis actin genes. Genetics 149:717–725

Goddard ME, Hayes BJ (2007) Genomic selection. J Anim Breed Genet 124:323–330

Golas TM, van de Geest H, Gros J, Sikkema A, D'Agostino N, Nap JP, Mariani C, Allefs JJ, Rieu I (2013) Comparative next-generation mapping of the Phytophthora infestans resistance gene Rpi-dlc2 in a European accession of Solanum dulcamara. Theor Appl Genet 126:59–68

Gore MA, Chia J-M, Elshire RJ, Sun Q, Ersoz ES et al (2009) A first-generation haplotype map of maize. Science 326:1115–1117

Gottwald S, Bauer P, Komatsuda T, Lundqvist U, Stein N (2009) TILLING in the two-rowed barley cultivar 'Barke' reveals preferred sites of functional diversity in the gene HvHox1. BMC Res Notes 2:258

Gowda M, Jantasuriyarat C, Dean R, Wang G-L (2004) Robust-LongSAGE (RL-SAGE) for both gene discovery and transcriptome analysis. Plant Physiol 134:890–897

Gowda M, Venu RC, Jia Y, Stahlberg E, Pampanwar V, Soderlund C, Wang GL (2007) Use of robust-long serial analysis of gene expression to identify novel fungal and plant genes involved in host-pathogen interactions. Methods Mol Biol 354:131–144

Greco R, Ouwerkerk PBF, Sallaud C, Kohli A, Colombo L, Puigdomènech P, Guiderdoni E, Christou P, Hoge JHC, Pereira A (2001) Transposon insertional mutagenesis in rice. Plant Physiol 125:1175–1177

Green RM, Tobin EM (1999) Loss of the circadian clock-associated protein 1 in Arabidopsis results in altered clock-regulated gene expression. Proc Natl Acad Sci U S A 96:4176–4179

Groover A, Fontan RJ, Dupper GMAC, Martienssen R, Strauss S, Milan R (2004) Gene and enhancer trap tagging of vascular- expressed genes in poplar trees. Plant Physiol 134:1742–1751

Guan JC, Koch KE, Suzuki M, Wu S, Latshaw S, Petruff T, Goulet C, Klee HJ, McCarty DR (2012a) Diverse roles of strigolactone signaling in maize architecture and the uncoupling of a branching-specific subnetwork. Plant Physiol 160:1303–1317

Guan M, Li X, Guan C (2012b) Microarray analysis of differentially expressed genes between *Brassica napus* strains with high- and low-oleic acid contents. Plant Cell Rep 31:929–943

Hackenberg M, Huang PJ, Huang CY, Shi BJ, Gustafson P, Langridge P (2013) A comprehensive expression profile of MicroRNAs and other classes of non-coding small RNAs in Barley under phosphorus-deficient and -sufficient conditions. DNA Res 20:109–125

Hamer L, DeZwaan TM, Montenegro-Chamorro MV, Frank SA, Hamer JE (2001) Recent advances in large-scale transposon mutagenesis. Curr Opin Chem Biol 5:67–73

Hamilton JP, Hansey CN, Whitty BR, Stoffel K, Massa AN, Van Deynze A, De Jong WS, Douches DS, Buell CR (2011) Single nucleotide polymorphism discovery in elite North American potato germplasm. BMC Genomics 12:302

Hammond SM, Bernstein E, Beach D, Hannon GJ (2000) An RNA-directed nuclease mediates post-transcriptional gene silencing in Drosophila cells. Nature 404:293–296

Han B, Huang X (2013) Sequencing-based genome-wide association study in rice. Curr Opin Plant Biol 16:133–138

Han Y, Kang Y, Torres-Jerez I, Cheung F, Town CD, Zhao PX, Udvardi MK, Monteros MJ (2011) Genome-wide SNP discovery in tetraploid alfalfa using 454 sequencing and high resolution melting analysis. BMC Genomics 12:1–11

Hannon GJ (2002) RNA interference. Nature 418:244–251

Hannoufa A, Negruk V, Eisner G, Lemieux B (1996) The CER3 gene of Arabidopsis thaliana is expressed in leaves, stems, roots, flowers and apical meristems. Plant J 10:459–467

Harmer SL, Kay SA (2000) Microarrays: determining the balance of cellular transcription. Plant Cell 12:613–615

Harmer SL, Hogenesch LB, Straume M, Chang HS, Han B, Zhu T, Wang X, Kreps JA, Kay SA (2000) Orchestrated transcription of key pathways in Arabidopsis by the circadian clock. Science 290:2110–2113

Hayes F (2003) Transposon- based strategies for microbial functional genomics and proteomics. Annu Rev Genet 37:1–7

Hayward A, Vighnesh G, Delay C, Samian MR, Manoli S, Stiller J, McKenzie M, Edwards D, Batley J (2012) Second-generation sequencing for gene discovery in the Brassicaceae. Plant Biotechnol J 10: 750–759

Hazen SP, Pathan MS, Sanchez A, Baxter I, Dunn M, Estes B, Chang HS, Zhu T, Kreps JA, Nguyen HT (2005) Expression profiling of rice segregating for drought tolerance QTLs using rice genome array. Funct Integr Genomics 5:104–116

He Z, Zhai W, Wen H, Tang T, Wang Y, Lu X, Greenberg AJ, Hudson RR, Wu CI, Shi S (2011) Two evolutionary histories in the genome of rice: the roles of domestication genes. PLoS Genet 7:e1002100

Hebert CG, James JV, Bentley WE (2008) Beyond silencing – engineering applications of RNA interference and antisense technology for altering cellular phenotypes. Curr Opin Biotechnol 19:500–505

Hegarty M, Yadav R, Lee M, Armstead I, Sanderson R, Scollan N, Powell W, Skøt L (2013) Genotyping by RAD sequencing enables mapping of fatty acid composition traits in perennial ryegrass (Lolium perenne (L.)). Plant Biotechnol J 11:572–581

Hein I, Barciszewska-Pacak M, Hrubikova K, Williamson S, Dinesen M, Soenderby IE, Sundar S, Jarmolowski A, Shirasu K, Lacomme C (2005) Virus-induced gene silencing-based functional characterization of genes associated with powdery mildew resistance in barley. Plant Physiol 138:2155–2164

Henikoff S, Comai L (2003) Single-nucleotide mutations for plant functional genomics. Annu Rev Plant Biol 54:375–401

Hieter P, Boguski M (1997) Functional genomics: it's all how you read it. Science 278:601–602

Hiremath PJ, Farmer A, Cannon SB, Woodward J, Kudapa H, Tuteja R, Kumar A, Bhanuprakash A, Mulaosmanovic B, Gujaria N, Krishnamurthy L, Gaur PM, Kavikishor PB, Shah T, Srinivasan R, Lohse M, Xiao Y, Town CD, Cook DR, May GD, Varshney RK (2011) Large-scale transcriptome analysis in chickpea (Cicer arietinum L.), an orphan legume crop of the semi-arid tropics of Asia and Africa. Plant Biotechnol J 9:922–931

Hofer J, Turner L, Moreau C, Ambrose M, Isaac P, Butcher S, Weller J, Dupin A, Dalmais M, Le Signor C, Bendahmane A, Ellis N (2009) Tendril-less regulates tendril formation in pea leaves. Plant Cell 21:420–428

Holzberg S, Brosio P, Gross C, Pogue GP (2002) Barley stripe mosaic virus-induced gene silencing in a monocot plant. Plant J 30:315–327

Hou H, Li J, Gao M, Singer SD, Wang H, Mao L, Fei Z, Wang X (2013) Genomic organization, phylogenetic comparison and differential expression of the SBP-box family genes in grape. PLoS One 8:e59358

Howarth JR, Parmar S, Barraclough PB, Hawkesford MJ (2009) A sulphur deficiency-induced gene, sdi1, involved in the utilization of stored sulphate pools under sulphur-limiting conditions has potential as a diagnostic indicator of sulphur nutritional status. Plant Biotechnol J 7:200–209

Hribova E, Neumann P, Macas J, Dolezel J (2009) Analysis of genome structure and organization in banana (Musa acuminata) using 454 sequencing. Plant and Animal Genomes XVII, San Diego, CA

Hsing YI, Chern CG, Fan MJ, Lu PC, Chen KT, Lo SF, Sun PK, Ho SL, Lee KW, Wang YC, Huang WL, Ko SS, Chen S, Chen JL, Chung CI, Lin YC, Hour AL, Wang YW, Chang YC, Tsai MW, Lin YS, Chen YC, Yen HM, Li CP, Wey CK, Tseng CS, Lai MH, Huang SC, Chen LJ, Yu SM (2007) A rice gene activation/knockout mutant resource for high throughput functional genomics. Plant Mol Biol 63:351–364

Hu Z, Hua W, Huang S, Yang H, Zhan G, Wang X, Liu G, Wang H (2012) Discovery of pod shatter-resistant associated SNPs by deep sequencing of a representative library followed by bulk segregant analysis in rapeseed. PLoS One 7:e34253

Huang X, Feng Q, Qian Q, Zhao Q, Wang L, Wang A, Guan J, Fan D, Weng Q, Huang T, Dong G, Sang T, Han B (2009) High-throughput genotyping by whole-genome resequencing. Genome Res 19(6):1068–1076

Huang CF, Yamaji N, Chen Z, Ma JF (2012a) A tonoplast-localized half-size ABC transporter is required for internal detoxification of aluminum in rice. Plant J 69:857–867

Huang CF, Yamaji N, Ono K, Ma JF (2012b) A leucine-rich repeat receptor-like kinase gene is involved in the specification of outer cell layers in rice roots. Plant J 69:565–576

Huang X, Kurata N, Wei X, Wang ZX, Wang A, Zhao Q, Zhao Y, Liu K, Lu H, Li W, Guo Y, Lu Y, Zhou C, Fan D, Weng Q, Zhu C, Huang T, Zhang L, Wang Y, Feng L, Furuumi H, Kubo T, Miyabayashi T, Yuan X, Xu Q, Dong G, Zhan Q, Li C, Fujiyama A, Toyoda A, Lu T, Feng Q, Qian Q, Li J, Han B (2012c) A map of rice genome variation reveals the origin of cultivated rice. Nature 490:497

Huang X, Lu T, Han B (2013a) Resequencing rice genomes: an emerging new era of rice genomics. Trends Genet 29:225–232

Huang D, Koh C, Feurtado JA, Tsang EW, Cutler AJ (2013b) MicroRNAs and their putative targets in Brassica napus seed maturation. BMC Genomics 14:140

Huang J, Zhang J, Li W, Hu W, Duan L, Feng Y, Qiu F, Yue B (2013c) Genome wide analysis of ten chilling tolerance indices at the germination and seedling stages in maize. J Integr Plant Biol. doi:10.1111/jipb.12051

Hufford MB, Xu X, van Heerwaarden J, Pyhäjärvi T, Chia JM, Cartwright RA, Elshire RJ, Glaubitz JC, Guill KE, Kaeppler SM, Lai J, Morrell PL, Shannon LM, Song C, Springer NM, Swanson-Wagner RA, Tiffin P, Wang J, Zhang G, Doebley J, McMullen MD, Ware D, Buckler ES, Yang S, Ross-Ibarra J (2012) Comparative population genomics of maize domestication and improvement. Nat Genet 44:808–811

Hyten DL, Song Q, Fickus EW, Quigley CV, Lim JS, Choi IY, Hwang EY, Pastor-Corrales M, Cregan PB (2010a) High-throughput SNP discovery and assay development in common bean. BMC Genomics 11:475

Hyten DL, Cannon SB, Song Q, Weeks N, Fickus EW, Shoemaker RC, Specht JE, Farmer AD, May GD, Cregan PB (2010b) High-throughput SNP discovery through deep resequencing of a reduced representation library to anchor and orient scaffolds in the soybean whole genome sequence. BMC Genomics 11:38

Iantcheva A, Vassileva V, Ugrinova M, Vlahova M (2009) Development of functional genomic platform for model legumes *Medicago truncatula* in Bulgaria. Biotechnol Biotechnol. Eq 23:1440–1443

Igarashi K, Kazama T, Motomura K, Toriyama K (2013) Whole genomic sequencing of RT98 mitochondria derived from Oryza rufipogon and northern blot analysis to uncover a cytoplasmic male sterility-associated gene. Plant Cell Physiol 54:237–243

Iida S, Terada R (2004) A tale of two integrations, transgene and T-DNA: gene targeting by homologous recombination in rice. Curr Opin Biotechnol 15:132–138

Impey S, McCorkle SR, Cha-Molstad H, Dwyer JM, Yochum GS, Boss JM, McWeeney S, Dunn JJ, Mandel G, Goodman RH (2004) Defining the CREB regulon: a genome-wide analysis of transcription factor regulatory regions. Cell 119:1041–1054

Ingvarsson PK, Street NR (2011) Association genetics of complex traits in plants. New Phytol 189:909–922
International Rice Genome Sequencing Project (2005) The map-based sequence of the rice genome. Nature 436:793–800
Irian S, Xu P, Dai X, Zhao PX, Roossinck MJ (2007) Regulation of a virus-induced lethal disease in tomato revealed by LongSAGE analysis. Mol Plant Microbe Interact 20:1477–1488
Ishii M, Hashimoto S, Tsutsumi S, Wada Y, Matsushima K, Kodama T, Aburatani H (2000) Direct comparison of GeneChip and SAGE on the quantitative accuracy in transcript profiling analysis. Genomics 68:136–143
Ito T, Seki M, Hayashida N, Shibata D, Shinozaki K (1999) Regional insertional mutagenesis of genes on Arabidopsis thaliana chromosome V using the Ac/Ds transposon in combination with a cDNA scanning method. Plant J 17:433–444
Jacquier A (2009) The complex eukaryotic transcriptome: unexpected pervasive transcription and novel small RNAs. Nat Rev Genet 10:833–844
Jain M, Misra G, Patel RK, Priya P, Jhanwar S, Khan AW, Shah N, Singh VK, Garg R, Jeena G, Yadav M, Kant C, Sharma P, Yadav G, Bhatia S, Tyagi AK, Chattopadhyay D (2013) A draft genome sequence of the pulse crop chickpea (Cicer arietinum L.). Plant J. doi:10.1111/tpj.12173
Jannink JL, Lorenz AJ, Iwata H (2010) Genomic selection in plant breeding: from theory to practice. Brief Funct Genomics 9:166–177
Jeon JS, Lee S, Jung KH, Jun SH, Jeong DH, Lee J, Kim C, Jang S, Yang K, Nam J, An J, Han MJ, Sung RJ, Choi HS, Yu JH, Choi JH, Cho SY, Cha SS, Kim SL, An G (2000) T-DNA insertional mutagenesis for functional genomics in rice. Plant J 22:561–570
Jhanwar S, Priya P, Garg R, Parida SK, Tyagi AK, Jain M (2012) Transcriptome sequencing of wild chickpea as a rich resource for marker development. Plant Biotechnol J 10:690–702
Jia J, Zhao S, Kong X, Li Y, Zhao G, He W, Appels R, Pfeifer M, Tao Y, Zhang X, Jing R, Zhang C, Ma Y, Gao L, Gao C, Spannagl M, Mayer KF, Li D, Pan S, Zheng F, Hu Q, Xia X, Li J, Liang Q, Chen J, Wicker T, Gou C, Kuang H, He G, Luo Y, Keller B, Xia Q, Lu P, Wang J, Zou H, Zhang R, Xu J, Gao J, Middleton C, Quan Z, Liu G, Wang J, International Wheat Genome Sequencing Consortium, Yang H, Liu X, He Z, Mao L, Wang J (2013a) Aegilops tauschii draft genome sequence reveals a gene repertoire for wheat adaptation. Nature 496:91–95
Jia G, Huang X, Zhi H, Zhao Y, Zhao Q, Li W, Chai Y, Yang L, Liu K, Lu H, Zhu C, Lu Y, Zhou C, Fan D, Weng Q, Guo Y, Huang T, Zhang L, Lu T, Feng Q, Hao H, Liu H, Lu P, Zhang N, Li Y, Guo E, Wang S, Wang S, Liu J, Zhang W, Chen G, Zhang B, Li W, Wang Y, Li H, Zhao B, Li J, Diao X, Han B (2013b) A haplotype map of genomic variations and genome-wide association studies of agronomic traits in foxtail millet (Setaria italica). Nat Genet. doi:10.1038/ng.2673
Jiang Y, Deyholos MK (2010) Transcriptome analysis of secondary-wall-enriched seed coat tissues of canola (Brassica napus L.). Plant Cell Rep 29:327–342
Jiang J, Jiang J, Qiu L, Miao Y, Yao L, Cao J (2013) Identification of gene expression profile during fertilization in *Brassica campestris* subsp. *chinensis*. Genome 56:39–48
Jiménez-López S, Mancera-Martínez E, Donayre-Torres A, Rangel C, Uribe L, March S, Jiménez-Sánchez G, Sánchez de Jiménez E (2011) Expression profile of maize (Zea mays L.) embryonic axes during germination: translational regulation of ribosomal protein mRNAs. Plant Cell Physiol 52:1719–1733
Johnson SM, Tan FJ, McCullough HL, Riordan DP, Fire AZ (2006) Flexibility and constraint in the nucleosome core landscape of Caenorhabditis elegans chromatin. Genome Res 16:1505–1516
Jorgensen R, Snyder C, Jones JG (1987) T-DNA is organized predominantly in inverted repeat structures in plants transformed with *Agrobacterium tumefaciens* C58 derivatives. Mol Gen Genet 3:1263–1275
Jun TH, Michel AP, Mian MA (2011) Development of soybean aphid genomic SSR markers using next generation sequencing. Genome 54:360–367
Kadaru SB, Yadav AS, Fjellstrom RG, Oard JH (2006) Alternative EcoTILLING protocol for rapid, cost-effective single-nucleotide polymorphism discovery and genotyping in rice (*Oryza sativa* L.). Plant Mol Biol Rep 24:3–22
Kamath RS, Martinez CM, Zipperlen P, Fraser AG, Ahringer J (2001) Effectiveness of specific RNA-mediated interference through ingested double-stranded RNA in *Caenorhabditis elegans*. Genome Biol 21:2.1–2.10
Kantar M, Akpınar BA, Valárik M, Lucas SJ, Doležel J, Hernández P, Budak H (2012) Subgenomic analysis of microRNAs in polyploid wheat. Funct Integr Genomics 12:465–479
Kaur S, Cogan NO, Pembleton LW, Shinozuka M, Savin KW, Materne M, Forster JW (2011) Transcriptome sequencing of lentil based on second-generation technology permits large-scale unigene assembly and SSR marker discovery. BMC Genomics 12:265
Kaur S, Pembleton LW, Cogan NO, Savin KW, Leonforte T, Paull J, Materne M, Forster JW (2012) Transcriptome sequencing of field pea and faba bean for discovery and validation of SSR genetic markers. BMC Genomics 13:104
Kawaura K, Mochida K, Yamazaki Y, Ogihara Y (2006) Transcriptome analysis of salinity stress responses in common wheat using 22k oligo-DNA microarray. Funct Integr Genomics 6:132–142
Kim SI, Tai TH (2013) Identification of SNPs in closely related Temperate Japonica rice cultivars using restriction enzyme-phased sequencing. PLoS One 8:e60176
Kim MY, Lee S, Van K, Kim TH, Jeong SC, Choi IY, Kim DS, Lee YS, Park D, Ma J, Kim WY, Kim BC, Park S, Lee KA, Kim DH, Kim KH, Shin JH, Jang YE, Kim KD, Liu WX, Chaisan T, Kang YJ, Lee YH, Kim KH,

Moon JK, Schmutz J, Jackson SA, Bhak J, Lee SH (2010) Whole-genome sequencing and intensive analysis of the undomesticated soybean (Glycine soja Sieb. And Zucc.) genome. Proc Natl Acad Sci U S A 107:22032–22037

Kocábek T, Repková J, Dudová M, Hoyerová K, Vrba L (2006) Isolation and characterization of a novel semilethal Arabidopsis thaliana mutant of gene for pentatricopeptide (PPR) repeat-containing protein. Genetica 128:395–407

Koncz C, Mayerhofer R, Koncz-Kalman Z, Nawrath C, Reiss B, Redei GP, Schell J (1990) Isolation of a gene encoding a novel chloroplast protein by T-DNA tagging in Arabidopsis thaliana. EMBO J 9: 1337–1346

Kong X, Lv W, Zhang D, Jiang S, Zhang S, Li D (2013) Genome-wide identification and analysis of expression profiles of maize mitogen-activated protein kinase kinase kinase. PLoS One 8:e57714

Koornneef M, Dellbaert LW, van der Veen JH (1982) EMS- and radiation-induced mutation frequencies at individual loci in *Arabidopsis thaliana* (L) Heynh. Mutat Res 93:109–123

Kou Y, Chang Y, Li X, Xiao J, Wang S (2012) The rice RAD51C gene is required for the meiosis of both female and male gametocytes and the DNA repair of somatic cells. J Exp Bot 63:5323–5335

Kraakman AT, Niks RE, Van den Berg PM, Stam P, Van Eeuwijk FA (2004) Linkage disequilibrium mapping of yield and yield stability in modern spring barley cultivars. Genetics 168:435–446

Krusell L, Sato N, Fukuhara I, Koch BE, Grossmann C, Okamoto S, Oka-Kira E, Otsubo Y, Aubert G, Nakagawa T, Sato S, Tabata S, Duc G, Parniske M, Wang TL, Kawaguchi M, Stougaard J (2011) The Clavata2 genes of pea and Lotus japonicus affect autoregulation of nodulation. Plant J 65:861–871

Krysan PJ, Young JC, Sussman MR (1999) T-DNA as an insertional mutagen in *Arabidopsis*. Plant Cell 11:2283–2290

Kulwal P, Ishikawa G, Benscher D, Feng Z, Yu LX, Jadhav A, Mehetre S, Sorrells ME (2012) Association mapping for pre-harvest sprouting resistance in white winter wheat. Theor Appl Genet 125:793–805

Kumagai MH, Donson J, Della-Cioppa G, Harvey D, Hanley K, Grill LK (1995) Cytoplasmic inhibition of carotenoid biosynthesis with virus-derived RNA. Proc Natl Acad Sci U S A 92:1679–1683

Kumar S, Banks TW, Cloutier S (2012) SNP discovery through next-generation sequencing and its applications. Int J Plant Genomics 2012:831460

Kump KL, Bradbury PJ, Wisser RJ, Buckler ES, Belcher AR, Oropeza-Rosas MA, Zwonitzer JC, Kresovich S, McMullen MD, Ware D, Balint-Kurti PJ, Holland JB (2011) Genome-wide association study of quantitative resistance to southern leaf blight in the maize nested association mapping population. Nat Genet 43:163–168

Kusaba M (2004) RNA interference in plants. Curr Opin Biotechnol 15:139–143

La H, Ding B, Mishra GP, Zhou B, Yang H, BellizziMdel R, Chen S, Meyers BC, Peng Z, Zhu JK, Wang GL (2011) A 5-methylcytosine DNA glycosylase/lyase demethylates the retrotransposon Tos17 and promotes its transposition in rice. Proc Natl Acad Sci U S A 108:15498–15503

Lababidi S, Mejlhede N, Rasmussen SK, Backes G, Al-Said W, Bau M, Jahoor A (2009) Identification of barley mutants in the cultivar 'Lux' at the *Dhn* loci through TILLING. Plant Breed 128:332–336

Lai J, Li R, Xu X, Jin W, Xu M, Zhao H, Xiang Z, Song W, Ying K, Zhang M, Jiao Y, Ni P, Zhang J, Li D, Guo X, Ye K, Jian M, Wang B, Zheng H, Liang H, Zhang X, Wang S, Chen S, Li J, Fu Y, Springer NM, Yang H, Wang J, Dai J, Schnable PS, Wang J (2010) Genome-wide patterns of genetic variation among elite maize inbred lines. Nat Genet 42:1027–1030

Lai K, Duran C, Berkman PJ, Lorenc MT, Stiller J, Manoli S, Hayden MJ, Forrest KL, Fleury D, Baumann U, Zander M, Mason AS, Batley J, Edwards D (2012) Single nucleotide polymorphism discovery from wheat next-generation sequence data. Plant Biotechnol J 10:743–749

Laird PW (2010) Principles and challenges of genome-wide DNA methylation analysis. Nat Rev Genet 11:191–203

Lam HM, Xu X, Liu X, Chen W, Yang G, Wong FL, Li MW, He W, Qin N, Wang B, Li J, Jian M, Wang J, Shao G, Wang J, Sun SS, Zhang G (2010) Resequencing of 31 wild and cultivated soybean genomes identifies patterns of genetic diversity and selection. Nat Genet 42:1053–1059

Larsson SJ, Lipka AE, Buckler ES (2013) Lessons from Dwarf8 on the strengths and weaknesses of structured association mapping. PLoS Genet 9:e1003246

Le Gouis J, Bordes J, Ravel C, Heumez E, Faure S, Praud S, Galic N, Remoué C, Balfourier F, Allard V, Rousset M (2012) Genome-wide association analysis to identify chromosomal regions determining components of earliness in wheat. Theor Appl Genet 124:597–611

Le Signor C, Dalamis M, Brunaud V, Thompson R, Bendahmane A (2009) High throughput identification of *Pisum sativum* mutant lines by TILLING: a tool for crop improvement using either forward or reverse genetic approaches. Grain Legumes 52:18–19

Ledger SE, Janssen BJ, Karunairetnam S, Wang T, Snowden KC (2010) Modified CAROTENOID CLEAVAGE DIOXYGENASE8 expression correlates with altered branching in kiwifruit (Actinidia chinensis). New Phytol 188:803–813

Lee JY, Lee DH (2003) Use of serial analysis of gene expression technology to reveal changes in gene expression in Arabidopsis pollen undergoing cold stress. Plant Physiol 132:517–529

Lee S, Kim J, Son SJ, Nam J, Jeong HD, Lee K, Jang S, Yoo J, Lee J, Lee YD, Kang GH, An G (2003) Systemic reverse genetics screening of T- DNA tagged genes in rice for functional genomics analyses: MADS- box genes as a test case. Plant Cell Physiol 12:1403–1411

Lee S, Jung KH, An G, Chung YY (2004) Isolation and characterization of a rice cysteine protease gene, OsCP1, using T-DNA gene-trap system. Plant Mol Biol 54:755–765

Lee SC, Lim MH, Yu JG, Park BS, Yang TJ (2012) Genome-wide characterization of the CBF/DREB1 gene family in Brassica rapa. Plant Physiol Biochem 61:142–152

Letta T, Maccaferri M, Badebo A, Ammar K, Ricci A, Crossa J, Tuberosa R (2013) Searching for novel sources of field resistance to Ug99 and Ethiopian stem rust races in durum wheat via association mapping. Theor Appl Genet 126:1237–1256

Lewin B (2004) Genes VII. Pearson Prentice Hall, Upper Saddle River

Li X, Lassner M, Zhang Y (2002) Deleteagene: a fast neutron deletion mutagenesis-based gene knockout system for plants. Comp Funct Genomics 3:158–160

Li L, Paulo MJ, Strahwald J, Lübeck J, Hofferbert HR, Tacke E, Junghans H, Wunder J, Draffehn A, van Eeuwijk F, Gebhardt C (2008) Natural DNA variation at candidate loci is associated with potato chip color, tuber starch content, yield and starch yield. Theor Appl Genet 116:1167–1181

Li L, Zhu B, Fu D, Luo Y (2011a) RIN transcription factor plays an important role in ethylene biosynthesis of tomato fruit ripening. J Sci Food Agric 91:2308–2314

Li W, Zhang J, Mou Y, Geng J, McVetty PB, Hu S, Li G (2011b) Integration of Solexa sequences on an ultradense genetic map in Brassica rapa L. BMC Genomics 12:249

Li W, Xu YP, Zhang ZX, Cao WY, Li F, Zhou X, Chen GY, Cai XZ (2012a) Identification of genes required for nonhost resistance to *Xanthomonas oryzae pv. oryzae* reveals novel signaling components. PLoS One 7:e42796

Li D, Deng Z, Qin B, Liu X, Men Z (2012b) De novo assembly and characterization of bark transcriptome using Illumina sequencing and development of EST-SSR markers in rubber tree (Hevea brasiliensis Muell. Arg.). BMC Genomics 13:192

Li L, Tacke E, Hofferbert HR, Lübeck J, Strahwald J, Draffehn AM, Walkemeier B, Gebhardt C (2013) Validation of candidate gene markers for marker-assisted selection of potato cultivars with improved tuber quality. Theor Appl Genet 126:1039–1052

Lian X, Wang S, Zhang J, Feng Q, Zhang L, Fan D, Li X, Yuan D, Han B, Zhang Q (2006) Expression profiles of 10,422 genes at early stage of low nitrogen stress in rice assayed using a cDNA microarray. Plant Mol Biol 60:617–631

Liang S, Lu Y, Jelinek J, Estecio M, Li H, Issa JP (2009) Analysis of epigenetic modifications by next generation sequencing. Conf Proc IEEE Eng Med Biol Soc 2009:6730

Limpens E, Ramos J, Franken C, Raz V, Compaan B, Franssen H, Besseling T, Geurts R (2004) RNA interference in *Agrobacterium rhizogenes* transformed roots *of Arabidopsis* and *Medicago truncatula*. J Exp Bot 55:983–992

Limpens E, Mirabella R, Fedorova E, Franken C, Franssen H, Bisseling T, Geurts R (2005) Formation of organelle-like N2-fixing symbiosomes in legume root nodules is controlled by DMI2. Proc Natl Acad Sci U S A 104:10375–10380

Lindbo JA, Fitzmaurice WP, della-Cioppa G (2001) Virus mediated reprogramming of gene expression in plants. Curr Opin Plant Biol 4:181–185

Lindesy K, Wei W, Clarke MC, Mcardle HF, Rooke LM, Topping JF (1993) Tagging genomic sequences that direct transgene expression by activation of a promoter trap in plants. Transgenic Res 2:33–47

Ling HQ, Zhao S, Liu D, Wang J, Sun H, Zhang C, Fan H, Li D, Dong L, Tao Y, Gao C, Wu H, Li Y, Cui Y, Guo X, Zheng S, Wang B, Yu K, Liang Q, Yang W, Lou X, Chen J, Feng M, Jian J, Zhang X, Luo G, Jiang Y, Liu J, Wang Z, Sha Y, Zhang B, Wu H, Tang D, Shen Q, Xue P, Zou S, Wang X, Liu X, Wang F, Yang Y, An X, Dong Z, Zhang K, Zhang X, Luo MC, Dvorak J, Tong Y, Wang J, Yang H, Li Z, Wang D, Zhang A, Wang J (2013) Draft genome of the wheat A-genome progenitor Triticum urartu. Nature 496:87–90

Lister R, Ecker JR (2009) Finding the fifth base: genome-wide sequencing of cytosine methylation. Genome Res 19:959–966

Lister R, O'Malley RC, Tonti-Filippini J, Gregory BD, Berry CC, Millar AH, Ecker JR (2008) Highly integrated single-base resolution maps of the epigenome in Arabidopsis. Cell 133:523–536

Liu Y, Schiff M, Dinesh-Kumar SP (2002a) Virus-induced gene silencing in tomato. Plant J 31:777–786

Liu Y, Schiff M, Marathe R, Dinesh-Kumar SP (2002b) Tobacco Rar1, EDS1 and NPR1/NIM1 like genes are required for N-mediated resistance to tobacco mosaic virus. Plant J 30:415–429

Liu S, Chen HD, Makarevitch I, Shirmer R, Emrich SJ, Dietrich CR, Barbazuk WB, Springer NM, Schnable PS (2010) High-throughput genetic mapping of mutants via quantitative single nucleotide polymorphism typing. Genetics 184:19–26

Liu L, Li Y, Li S, Hu N, He Y, Pong R, Lin D, Lu L, Law M (2012) Comparison of next-generation sequencing systems. J Biomed Biotechnol 2012:1–12

Liu H, Fu D, Zhu B, Yan H, Shen X, Zuo J, Zhu Y, Luo Y (2013) Virus-induced gene silencing in eggplant (Solanum melongena). J Integr Plant Biol 54:422–429

Long Q, Rabanal FA, Meng D, Huber CD, Farlow A, Platzer A, Zhang Q, Vilhjálmsson BJ, Korte A, Nizhynska V, Voronin V, Korte P, Sedman L, Mandáková T, Lysak MA, Seren U, Hellmann I, Nordborg M (2013) Massive genomic variation and strong selection in Arabidopsis thaliana from Sweden. Nat Genet. doi:10.1038/ng.2678

Lu R, Martin-Hernandez AM, Peart JR, Malcuit I, Baulcombe DC (2003a) Virus- Induced gene silencing in plants. Methods 30:296–303

Lu R, Malcuit I, Moffett P, Ruiz MT, Peart J, Wu AJ, Rathjen JP, Bendahmane A, Day L, Baulcombe DC (2003b) High throughput virus-induced gene silencing implicates heat shock protein 90 in plant disease resistance. EMBO J 22:5690–5699

Lu C, Tej SS, Luo S, Haudenschild CD, Meyers BC, Green PJ (2005) Elucidation of the small RNA component of the transcriptome. Science 309:1567–1569

Lu SJ, Yang ZT, Sun L, Sun L, Song ZT, Liu JX (2012) Conservation of IRE1-regulated bZIP74 mRNA unconventional splicing in rice (Oryza sativa L.) involved in ER stress responses. Mol Plant 5:504–514

Lv MZ, Chao DY, Shan JX, Zhu MZ, Shi M, Gao JP, Lin HX (2012) Rice carotenoid β-ring hydroxylase CYP97A4 is involved in lutein biosynthesis. Plant Cell Physiol 53:987–1002

Ma M, Yan Y, Huang L, Chen M, Zhao H (2012) Virus-induced gene-silencing in wheat spikes and grains and its application in functional analysis of HMW-GS-encoding genes. BMC Plant Biol 12:141

Mace ES, Hunt CH, Jordan DR (2013) Supermodels: sorghum and maize provide mutual insight into the genetics of flowering time. Theor Appl Genet 126:1377–1395

Maguire TL, Grimmond S, Forrest A, Iturbe-Ormaetxe I, Meksem K, Gresshoff P (2002) Tissue-specific gene expression in soybean (*Glycine max*) detected by cDNA microarray analysis. J Plant Physiol 159:1361–1374

Malatrasi M, Corradi M, Svensson JT, Close TJ, Gulli M, Marmiroli N (2006) A branched-chain amino acid aminotransferase gene isolated from Hordeum vulgare is differentially regulated by drought stress. Theor Appl Genet 113:965–976

Maleck K, Levine A, Eulgem T, Morgan A, Schmid J, Lawton KA, Dangl JL, Dietrich RA (2000) The transcriptome of *Arabidopsis thaliana* during systemic acquired resistance. Nat Genet 26:403–410

Malonae JH, Oliver B (2011) Microarrays, deep sequencing and the true measure of the transcriptome. BMC Biol 9:34

Mandel MA, Feldmann KA, Herrera-Estrella L, Rocha-Sosa M, León P (1996) CLA1, a novel gene required for chloroplast development, is highly conserved in evolution. Plant J 9:649–658

Manmathan H, Shaner D, Snelling J, Tisserat N, Lapitan N (2013) Virus-induced gene silencing of Arabidopsis thaliana gene homologues in wheat identifies genes conferring improved drought tolerance. J Exp Bot 64:1381–1392

Martin A, Lee J, Kichey T, Gerentes D, Zivy M, Tatout C, Dubois F, Balliau T, Valot B, Davanture M, Tercé-Laforgue T, Quilleré I, Coque M, Gallais A, Gonzalez-Moro MB, Bethencourt L, Habash DZ, Lea PJ, Charcosset A, Perez P, Murigneux A, Sakakibara H, Edwards KJ, Hirel B (2006) Two cytosolic glutamine synthetase isoforms of maize are specifically involved in the control of grain production. Plant Cell 18:3252–3274

Martin B, Ramiro M, Martinez ZJM, Alanso BC (2009) A high- density collection of EMS-induced mutations for TILLING in Landsberg erecta genetic background of Arabidopsis. BMC Plant Biol 9:147

Matsumura H, Nirasawa S, Kiba A, Urasaki N, Saitoh H, Ito M, Kawai-Yamada M, Uchimiya H, Terauchi R (2003) Overexpression of Bax inhibitor suppresses the fungal elicitor-induced cell death in rice (*Oryza sativa* L) cells. Plant J 33:425–434

Matsumura H, Nasir KHB, Yoshida K, Ito A, Kahl G, Kruger DH, Terauchi R (2006) SuperSAGE array: the direct use of 26 base-pair transcript tags in oligonucleotide arrays. Nat Methods 3:469–474

Mattiello L, da Silva FR, Menossi M (2012) Linking microarray data to QTLs highlights new genes related to Al tolerance in maize. Plant Sci 191–192:8–15

Matzke MA, Matzke AJ, Pruss GJ, Vance VB (2001) RNA-based silencing strategy in plants. Curr Opin Genet Dev 11:221–227

Maughan PJ, Yourstone SM, Byers RL, Smith SM, Udall JA (2010) Single-nucleotide polymorphism genotyping in mapping populations via genomic reduction and next-generation sequencing: proof of concept. Plant Genome 3:166–178

McCallum CM, Comai L, Greene EA, Henikoff S (2000a) Target screening for induced mutations. Nat Biotechnol 18:455–457

McCallum CM, Comai L, Greene EA, Henikoff S (2000b) Targeting induced local lesions in genomes (TILLING) for plant functional genomics. Plant Physiol 123:439–442

McIntosh S, Watson L, Bundock P, Crawford A, White J, Cordeiro G, Barbary D, Rooke L, Henry R (2007) SAGE of the developing wheat caryopsis. Plant Biotechnol J 5:69–83

McKinney EC, Ali N, Traut A, Feldmann KA, Belostotsky DA, McDowell JM, Meagher RB (1995) Sequence-based identification of T-DNA insertion mutations in *Arabidopsis*: actin mutants *act2-1* and *act4-1*. Plant J 8:613–622

McKusick VA (1997) Genomics: structural and functional studies of genomes. Genomics 45:244–249

Meaburn E, Schulz R (2011) Next generation sequencing in epigenetics: insights and challenges. Semin Cell Dev Biol. doi:10.1016/j.semcdb.2011.10.010

Meissner R, Chague V, Zhu Q, Emmanuel E, Elkind Y, Levy A (2000) A high throughput system for transposon tagging and promoter trapping in tomato. Plant J 22:265–274

Meister G, Tuschl T (2004) Mechanisms of gene silencing by double-stranded RNA. Nature 431:343–349

Mejlhede N, Kyjovska Z, Backes G, Burhenne K, Rasmussen SK, Jahoor A (2006) EcoTILLING for the identification of allelic variation in the powdery mildew resistance genes *mlo* and *Mla* of barley. Plant Breed 125:461–467

Menda N, Semel Y, Peled D, Eshed Y, Zamir D (2004) *In silico* screening of a saturated mutation library of tomato. Plant J 38:861–872

Metzker ML (2010) Sequencing technologies – the next generation. Nat Rev Genet 11:31–46

Meyers BC, Galbraith DW, Nelson T, Agrawal V (2004) Methods for transcriptional profiling in plants. Be fruitful and replicate. Plant Physiol 135:637–652

Michael T, Jackson M (2013) The first 50 plant genomes. Plant Genome 6:1–7

Mikkelsen TS, Ku MC, Jaffe DB, Issac B, Lieberman E, Giannoukos G, Alvarez P, Brockman W, Kim TK, Koche RP et al (2007) Genome-wide maps of chromatin state in pluripotent and lineage-committed cells. Nature 448:553–560

Miller MR, Dunham JP, Amores A, Cresko WA, Johnson EA (2007) Rapid and cost-effective polymorphism identification and genotyping using restriction site associated DNA (RAD) markers. Genome Res 17:240–248

Min K, Kang J, Lee J (2010) A modified feeding RNAi method for simultaneous knock-down of more than one gene in Caenorhabditis elegans. BioTechniques 48:229–232

Minoia S, Petrozza A, Onofrio DO, Piron F, Mosca G, Giovanni S (2010) A new mutant genetic resource for tomato crop improvement by TILLING technology. BMC Res Notes 3:69

Mitchell-Olds T (2010) Complex-trait analysis in plants. Genome Biol 11:113

Miyashita Y, Dolferus R, Ismond KP, Good AG (2007) Alanine aminotransferase catalyses the breakdown of alanine after hypoxia in Arabidopsis thaliana. Plant J 49:1108–1121

Mohorianu I, Schwach F, Jing R, Lopez-Gomollon S, Moxon S, Szittya G, Sorefan K, Moulton V, Dalmay T (2011) Profiling of short RNAs during fleshy fruit development reveals stage-specific sRNAome expression patterns. Plant J 67:232–246

Molina C, Rotter B, Horres R, Udupa SM, Besser B, Bellarmino L, Baum M, Matsumura H, Terauchi R, Kahl G, Winter P (2008) SuperSAGE: the drought stress-responsive transcriptome of chickpea roots. BMC Genomics 9:553

Monson-Miller J, Sanchez-Mendez DC, Fass J, Henry IM, Tai TH, Comai L (2012) Reference genome-independent assessment of mutation density using restriction enzyme-phased sequencing. BMC Genomics 13:72

Montiel J, Nava N, Cárdenas L, Sánchez-López R, Arthikala MK, Santana O, Sánchez F, Quinto C (2012) A Phaseolus vulgaris NADPH oxidase gene is required for root infection by Rhizobia. Plant Cell Physiol 53:1751–1767

Moritoh S, Miki D, Akiyama M, Kawahara M, Izawa T, Maki H, Shimamoto K (2005) RNAi- mediated silencing of OsGEN-L (OsGEN-like), a new member of the RAD2/XPG nuclease family, causes male sterility by defect of microspore development in rice. Plant Cell Physiol 46:699–715

Morozova O, Marra MA (2008) Applications of next-generation sequencing technologies in functional genomics. Genomics 92:255–264

Muller HJ (1930) Types of visible variations induced by X-rays in Drosophila. J Genet 22:299–334

Mutasa-Göttgens ES, Joshi A, Holmes HF, Hedden P, Göttgens B (2012) A new RNASeq-based reference transcriptome for sugar beet and its application in transcriptome-scale analysis of vernalization and gibberellin responses. BMC Genomics 13:99

Myles S, Chia JM, Hurwitz B, Simon C, Zhong GY, Buckler E, Ware D (2010) Rapid genomic characterization of the genus vitis. PLoS One 5:e8219

Nagalakshmi U, Wang Z, Waern K, Shou C, Raha D, Gerstein M, Snyder M (2008) The transcriptional landscape of the yeast genome defined by RNA sequencing. Science 320:1344–1349

Nagamatsu A, Masuta C, Senda M, Matsuura H, Kasai A, Hong JS, Kitamura K, Abe J, Kanazawa A (2007) Functional analysis of soybean genes involved in flavonoid biosynthesis by virus-induced gene silencing. Plant Biotechnol J 5:778–790

Nagamatsu A, Masuta C, Matsuura H, Kitamura K, Abe J, Kanazawa A (2009) Down-regulation of flavonoid 3′-hydroxylase gene expression by virus-induced gene silencing in soybean reveals the presence of a threshold mRNA level associated with pigmentation in pubescence. J Plant Physiol 166:32–39

Newell MA, Asoro FG, Scott MP, White PJ, Beavis WD, Jannink JL (2012) Genome-wide association study for oat (Avena sativa L.) beta-glucan concentration using germplasm of worldwide origin. Theor Appl Genet 125:1687–1696

Ng PC, Henikoff S (2003) SIFT: predicting amino acid changes that affect protein function. Nucleic Acids Res 31:3812–3814

Nicolaï M, Pisani C, Bouchet JP, Vuylsteke M, Palloix A (2012) Discovery of a large set of SNP and SSR genetic markers by high-throughput sequencing of pepper (Capsicum annuum). Genet Mol Res 11:2295–2300

Nie X, Li B, Wang L, Liu P, Biradar SS, Li T, Dolezel J, Edwards D, Luo M, Weining S (2012) Development of chromosome-arm-specific microsatellite markers in Triticum aestivum (Poaceae) using NGS technology. Am J Bot 99:e369–e371

Nishizawa T, Tamaoki M, Kaneko Y, Aono M, Kubo A, Saji H, Nakajima N (2012) High-throughput capture of nucleotide sequence polymorphisms in three Brassica species (Brassicaceae) using DNA microarrays. Am J Bot 99:e94–e96

Nordborg M, Tavare S (2002) Linkage disequilibrium: what history has to tell us. Trends Genet 18:83–90

Novaes E, Drost DR, Farmerie WG, Pappas GJ Jr, Grattapaglia D, Sederoff RR, Kirst M (2008) High-throughput gene and SNP discovery in Eucalyptus grandis, an uncharacterized genome. BMC Genomics 9:312

Oikawa A, Rahman A, Yamashita T, Taira H, Kidou S (2007) Virus-induced gene silencing of P23k in barley leaf reveals morphological changes involved in secondary wall formation. J Exp Bot 58:2617–2625

Okabe Y, Asamizu E, Saito T, Matsukura C, Ariizumi T, Bres C, Rothan C, Mizoguchi T, Ezura H (2011) Tomato TILLING technology: development of a reverse genetics tool for the efficient isolation of mutants from Micro-Tom mutant libraries. Plant Cell Physiol 52:1994–2005

Okabe Y, Ariizumi T, Ezura H (2013) Updating the Micro-Tom TILLING platform. Breed Sci 63: 42–48

Okamoto H, Hirochika H (2000) Efficient insertion mutagenesis of *Arabidopsis* by tissue culture- induced activation of the tobacco retrotransposon Tto1. Plant J 23:291–304

Oliver RE, Lazo GR, Lutz JD, Rubenfield MJ, Tinker NA, Anderson JM, Wisniewski Morehead NH, Adhikary D, Jellen EN, Maughan PJ, Brown Guedira GL, Chao S, Beattie AD, Carson ML, Rines HW, Obert DE, Bonman JM, Jackson EW (2011) Model SNP development for complex genomes based on hexaploid oat using high-throughput 454 sequencing technology. BMC Genomics 12:77

O'Neil ST, Emrich SJ (2013) Assessing *De Novo* transcriptome assembly metrics for consistency and utility. BMC Genomics 14:465

Oono Y, Kobayashi F, Kawahara Y, Yazawa T, Handa H, Itoh T, Matsumoto T (2013) Characterisation of the wheat (Triticum aestivum L.) transcriptome by de novo assembly for the discovery of phosphate starvation-responsive genes: gene expression in Pi-stressed wheat. BMC Genomics 14:77

Ortiz-Morea FA, Vicentini R, Silva GF, Silva EM, Carrer H, Rodrigues AP, Nogueira FT (2013) Global analysis of the sugarcane microtranscriptome reveals a unique composition of small RNAs associated with axillary bud outgrowth. J Exp Bot 64:2307–2320

Pacak A, Geisler K, Jørgensen B, Barciszewska-Pacak M, Nilsson L, Nielsen TH, Johansen E, Grønlund M, Jakobsen I, Albrechtsen M (2010) Investigations of barley stripe mosaic virus as a gene silencing vector in barley roots and in Brachypodium distachyon and oat. Plant Methods 6:26

Padmanabhan M, Dinesh-Kumar PS (2009) Virus-induced gene silencing as a tool for delivery of dsRNA into plants. Cold Spring Harb Protoc 4:1–4

Palauqui J, Vaucheret H (1998) Transgenes are dispensable for the RNA degradation step of cosuppression. Proc Natl Acad Sci U S A 95:9675–9680

Papi M, Sabatini S, Bouchez D, Camilleri C, Costantino P, Vittorioso P (2000) Identification and disruption of an Arabidopsis zinc finger gene controlling seed germination. Genes Dev 14:28–33

Parchman TL, Geist KS, Grahnen JA, Benkman CW, Buerkle CA (2010) Transcriptome sequencing in an ecologically important tree species: assembly, annotation, and marker discovery. BMC Genomics 11:180

Park MY, Chung MS, Koh HS, Lee DJ, Ahn SJ, Kim CS (2009) Isolation and functional characterization of the Arabidopsis salt-tolerance 32 (AtSAT32) gene associated with salt tolerance and ABA signaling. Physiol Plant 135:426–435

Pasam RK, Sharma R, Malosetti M, van Eeuwijk FA, Haseneyer G, Kilian B, Graner A (2012) Genome-wide association studies for agronomical traits in a world wide spring barley collection. BMC Plant Biol 12:16

Pavlidis P, Noble WS (2001) Analysis of strain and regional variation in gene expression in mouse brain. Genome Biol 2:1–15

Peart JR, Cook G, Feys BJ, Parker JE, Baulcombe DC (2002a) An EDS1 orthologue is required for N-mediated resistance against tobacco mosaic virus. Plant J 29:569–579

Peart JR, Lu R, Sadanandom A, Malcuit I, Moffett P, Brice DC, Schauser L, Jaggard DA, Xiao S, Coleman MJ, Dow M, Jones JD, Shirasu K, Baulcombe DC (2002b) Ubiquitin -associated protein SGT1 is required for host and non –host disease resistance in plants. Proc Natl Acad Sci U S A 99:10865–10869

Peng FY, Weselake RJ (2013) Genome-wide identification and analysis of the B3 superfamily of transcription factors in Brassicaceae and major crop plants. Theor Appl Genet 126:1305–1319

Peng JH, Bai Y, Haley SD, Lapitan NL (2009) Microsatellite-based molecular diversity of bread wheat germplasm and association mapping of wheat resistance to the Russian wheat aphid. Genetica 135:95–122

Pérez-de-Castro AM, Vilanova S, Cañizares J, Pascual L, Blanca JM, Díez MJ, Prohens J, Picó B (2012) Application of genomic tools in plant breeding. Curr Genomics 13:179–195

Perry JA, Wang TL, Welham TJ, Gardner S, Pike JM, Yoshida S, Parniske M (2003) A TILLING reverse genetics tool and a Web-accessible collection of mutants of the legume *Lotus japonicus*. Plant Physiol 131:866–871

Perry J, Brachmann A, Welham T, Binder A, Charpentier M, Groth M, Haage K, Markmann K, Wang TL, Parniske M (2009) TILLING in Lotus japonicus identified large allelic series for symbiosis genes and revealed a bias in functionally defective ethyl methanesulfonate alleles toward glycine replacements. Plant Physiol 151:1281–1291

Pevsner J (2009) Bioinformatics and functional genomics, 2nd edn. Wiley-Blackwell, Hoboken

Pfender WF, Saha MC, Johnson EA, Slabaugh MB (2011) Mapping with RAD (restriction-site associated DNA) markers to rapidly identify QTL for stem rust resistance in Lolium perenne. Theor Appl Genet 122:1467–1480

Piferrer F (2013) Epigenetics of sex determination and gonadogenesis. Dev Dyn 242:360–370

Piron F, Nicolai M, Minoi S, Piednoir E, Moretti A, Salgues A, Zamir D, Caranta C, Bendahmane A (2010) An induced mutation in tomato eIF4E leads to immunity to two potyviruses. PLoS One 5:e11313

Poland JA, Bradbury PJ, Buckler ES, Nelson RJ (2011) Genome-wide nested association mapping of quantitative resistance to northern leaf blight in maize. Proc Natl Acad Sci U S A 108:6893–6898

Poland JA, Brown PJ, Sorrells ME, Jannink JL (2012) Development of high-density genetic maps for barley and wheat using a novel two-enzyme genotyping-by-sequencing approach. PLoS One 7:e32253

Poole RL, Barker GL, Werner K, Biggi GF, Coghill J, Gibbings JG, Berry S, Dunwell JM, Edwards KJ (2008) Analysis of wheat SAGE tags reveals evidence for widespread antisense transcription. BMC Genomics 9:475

Porch TG, Blair MW, Lariguet P, Galeano C, Pankhurst CE, Broughton WJ (2008) Generation of a mutant population for TILLING common bean genotype BAT 93. J Am Soc Hort Sci 134:348–355

Poroyko V, Calugaru V, Fredricksen M, Bohnert HJ (2004) Virtual-SAGE: a new approach to EST data analysis. DNA Res 11:145–152

Primerose SB, Twyman RM (2006) Principles of gene manipulation and genomics, 7th edn. Blackwell Publishing, Oxford

Puzio PS, Lausen J, Almeida-Engler J, Cai D, Gheysen G, Grundler FM (1999) Isolation of a gene from Arabidopsis thaliana related to nematode feeding structures. Gene 239:163–172

Qiu S, Adema CM, Lane T (2005) A computational study of off-target effects of RNA interference. Nucleic Acids Res 33:1834–1847

Qu J, Ye J, Geng YF, Sun YW, Gao SQ, Zhang BP, Chen W, Chua NH (2012) Dissecting functions of KATANIN and WRINKLED1 in cotton fiber development by virus-induced gene silencing. Plant Physiol 160:738–748

Radhamony RN, Prasad AM, Srinivasan R (2005) T-DNA insertional mutagenesis in Arabidopsis: a tool for functional genomics. Electron J Biotechnol 1:82–106

Ramachandran S, Sundersan V (2001) Transposon as tools for functional genomics. Plant Physiol Biochem 39:243–252

Raman H, Stodart B, Ryan PR, Delhaize E, Emebiri L, Raman R, Coombes N, Milgate A (2010) Genome-wide association analyses of common wheat (Triticum aestivum L.) germplasm identifies multiple loci for aluminium resistance. Genome 53:957–966

Ramos ML, Huntley JJ, Maleki SJ, Ozias-Akins P (2009) Identification and characterization of a hypoallergenic ortholog of Ara h 2.01. Plant Mol Biol 69:325–335

Rawat N, Sehgal SK, Joshi A, Rothe N, Wilson DL, McGraw N, Vadlani PV, Li W, Gill BS (2012) BMC Plant Biol 12:205

Reif JC, Gowda M, Maurer HP, Longin CF, Korzun V, Ebmeyer E, Bothe R, Pietsch C, Würschum T (2011) Association mapping for quality traits in soft winter wheat. Theor Appl Genet 122:961–970

Remington DL, Thornsberry JM, Matsuoka Y, Wilson LM, Whitt SR, Doebley J, Kresovich S, Goodman MM, Buckler ES (2001) Structure of linkage disequilibrium and phenotypic associations in the maize genome. Proc Natl Acad Sci U S A 98:11479–11484

Ren Y, Hong Z, Kou Q, Jiang J, Guo S, Zhang H, Hou W, Zou X, Sun H, Gong G, Levi A, Xu Y (2012) A high resolution genetic map anchoring scaffolds of the sequenced watermelon genome. PLoS One 7:e29453

Riascos JJ, Weissinger AK, Weissinger SM, Burks AW (2010) Hypoallergenic legume crops and food allergy: factors affecting feasibility and risk. J Agric Food Chem 58:20–27

Risch N, Merikangas K (1996) The future of genetic studies of complex human diseases. Science 273:1516–1517

Rival A, Beule T, Bertossi FA, Tregear J, Jaligot E (2010) Plant epigenetics: from genomes to epigenomes. Not Bot Hort Agrobot Cluj 38:09–15

Robertson D (2004) VIGS vectors for gene silencing; many targets, many tools. Annu Rev Plant Biol 55:495–519

Rogers C, Wen J, Chen R, Oldroyd G (2009) Deletion-based reverse genetics in *Medicago truncatula*. Plant Physiol 151:1077–1086

Ronaghi M (2001) Pyrosequencing sheds light on DNA sequencing. Genome Res 11:3–11

Rousset M, Bonnin I, Remoué C, Falque M, Rhoné B, Veyrieras JB, Madur D, Murigneux A, Balfourier F, Le Gouis J, Santoni S, Goldringer I (2011) Deciphering the genetics of flowering time by an association study on candidate genes in bread wheat (Triticum aestivum L.). Theor Appl Genet 123:907–926

Roy JK, Smith KP, Muehlbauer GJ, Chao S, Close TJ, Steffenson BJ (2010) Association mapping of spot blotch resistance in wild barley. Mol Breed 26:243–256

Ruan Y, Gilmore J, Conner T (1998) Towards *Arabidopsis* genome analysis: monitoring expression profiles of 1400 genes using cDNA microarrays. Plant J 15:821–833

Saha S, Sparks AB, Rago C, Akmaev V, Wang CJ, Vogelstein B, Kinzler KW, Velculescu VE (2002) Using the transcriptome to annotate the genome. Nat Biotechnol 20:508–512

Sallaud C, Gay C, Larmande P, Be M, Piffanelli P, Piegu B, Droc G, Bourgeois E, Meynard D, Perin C, Sabau X, Ghesquiere A, Glaszmann JC, Delseny M, Guiderdoni E (2004) High throughput T-DNA insertion mutagenesis in rice: a first step towards in silico reverse genetics. Plant J 39:450–464

Salvio S, Tuberosa R (2007) Cloning QTLS in plants. Development and applications. In: Varshney RK, Tuberosa R (eds) Genomic-assisted crop improvement: vol. 1. Genomics approaches and platforms. Springer, Netherlands, pp 207–225

Sanger F, Coulson AR (1975) A rapid method for determining sequences in DNA by primed synthesis with DNA polymerase. J Mol Biol 94:441–448

Scaglione D, Acquadro A, Portis E, Tirone M, Knapp SJ, Lanteri S (2012) RAD tag sequencing as a source of SNP markers in Cynara cardunculus L. BMC Genomics 13:3

Schaefer DG, Zryd JP (1997) Efficient gene targeting in the moss *Physcomitrella patens*. Plant J 11:1195–1206

Schena M, Shalon D, Davis RW, Brown PO (1995) Quantitative monitoring of gene expression patterns with a complementary DNA microarray. Science 270:467–470

Schneeberger K, Ossowski S, Lanz C, Juul T, Petersen AH, Nielsen KL, Jørgensen JE, Weigel D, Andersen

SU (2009) SHOREmap: simultaneous mapping and mutation identification by deep sequencing. Nat Methods 6:550–551
Schneider A, Kirch T, Gigolashvili T, Mock HP, Sonnewald U, Simon R, Flügge UI, Werr W (2005) A transposon-based activation-tagging population in Arabidopsis thaliana (TAMARA) and its application in the identification of dominant developmental and metabolic mutations. FEBS Lett 579:4622–4628
Schuster SC (2008) Next generation DNA sequencing transforms today's biology. Nat Methods 5:16–18
Schwarz DS, Hutvágner G, Du T, Xu Z, Aronin N, Zamore PD (2003) Asymmetry in the assembly of the RNAi enzyme complex. Cell 115:199–208
Scofield SR, Brandt AS (2012) Virus-induced gene silencing in hexaploid wheat using barley stripe mosaic virus vectors. Methods Mol Biol 894:93–112
Scofield SR, Huang L, Brandt AS, Gill BS (2005) Development of a virus-induced gene-silencing system for hexaploid wheat and its use in functional analysis of the *Lr21*-mediated leaf rust resistance pathway. Plant Physiol 138:2165–2173
Seki M, Narusaka M, Abe H, Kasuga M, Yamaguchi-Shinozaki K, Carninci P, Hayashizaki Y, Shinozaki K (2001) Monitoring the expression pattern of 1300 *Arabidopsis* genes under drought and cold stresses by using a full-length cDNA microarray. Plant Cell 13:61–72
Senthil Kumar M, Udayakumar M (2006) High-throughput virus-induced gene-silencing approach to assess the functional relevance of a moisture stress-induced cDNA homologous to lea4. J Exp Bot 57:2291–2302
Senthil Kumar M, RameGowda HV, Hema R, Mysore KS, UdayaKumar M (2008) Virus- induced gene silencing and its application in characterizing gene involved in water-deficit stress tolerance. J Plant Physiol 165:1404–1421
Seo JS, Koo YJ, Jung C, Yeu SY, Song JT, Kim JK, Choi Y, Lee JS, Do Choi Y (2013) Identification of a novel jasmonate-responsive element in the AtJMT promoter and its binding protein for AtJMT repression. PLoS One 8:e55482
Settles AM, Holding DR, Tan BC, Latshaw SP, Liu J, Suzuki M, Li L, O'Brien BA, Fajardo DS, Wroclawska E, Tseung CW, Lai J, Hunter CT, Avigne WT, Baier J, Messing J, Hannah LC, Koch KE, Becraft PW, Larkins BA, McCarty DR (2007) Sequence-indexed mutations in maize using the Uniform Mu transposon-tagging population. BMC Genomics 8:116
Seymour DK, Filiault DL, Henry IM, Monson-Miller J, Ravi M, Pang A, Comai L, Chan SW, Maloof JN (2012) Rapid creation of Arabidopsis doubled haploid lines for quantitative trait locus mapping. Proc Natl Acad Sci U S A 109:4227–4232
Sharpe AG, Ramsay L, Sanderson LA, Fedoruk MJ, Clarke WE, Li R, Kagale S, Vijayan P, Vandenberg A, Bett KE (2013) Ancient orphan crop joins modern era: gene-based SNP discovery and mapping in lentil. BMC Genomics 14:192
Shendure J, Ji H (2008) Next-generation sequencing. Nat Biotechnol 26:1135–1145
Shi C, Navabi A, Yu K (2011) Association mapping of common bacterial blight resistance QTL in Ontario bean breeding populations. BMC Plant Biol 11:52
Shulaev V, Sargent DJ, Crowhurst RN, Mockler TC et al (2011) The genome of woodland strawberry (*Fragaria vesca*). Nat Genet 43:109–116
Sim SC, Durstewitz G, Plieske J, Wieseke R, Ganal MW, Van Deynze A, Hamilton JP, Buell CR, Causse M, Wijeratne S, Francis DM (2012) Development of a large SNP genotyping array and generation of high-density genetic maps in tomato. PLoS One 7:e40563
Simko I, Costanzo S, Haynes KG, Christ BJ, Jones RW (2004) Linkage disequilibrium mapping of a Verticillium dahliae resistance quantitative trait locus in tetraploid potato (Solanum tuberosum) through a candidate gene approach. Theor Appl Genet 108:217–224
Simon SA, Meyers BC (2011) Small RNA-mediated epigenetic modifications in plants. Curr Opin Plant Biol 14:148–155
Slade AJ, Fuerstenberg SI, Loeffler D, Steine MN, Facciotti D (2005) A reverse genetic, nontransgenic approach to wheat crop improvement by TILLING. Nat Biotechnol 23:75–81
Small I (2007) RNAi for revealing and engineering plant gene functions. Curr Opin Biotechnol 18:148–153
Song S, Qu H, Chen C, Hu S, Yu J (2007) Differential gene expression in an elite hybrid rice cultivar (*Oryza sativa*, L) and its parental lines based on SAGE data. BMC Plant Biol 7:49
Song GS, Zhai HL, Peng YG, Zhang L, Wei G, Chen XY, Xiao YG, Wang L, Chen YJ, Wu B, Chen B, Zhang Y, Chen H, Feng XJ, Gong WK, Liu Y, Yin ZJ, Wang F, Liu GZ, Xu HL, Wei XL, Zhao XL, Ouwerkerk PB, Hankemeier T, Reijmers T, van der Heijden R, Lu CM, Wang M, van der Greef J, Zhu Z (2010) Comparative transcriptional profiling and preliminary study on heterosis mechanism of super-hybrid rice. Mol Plant 3:1012–1025
Soyka S, Heyer AG (1999) Arabidopsis knockout mutation of ADC2 gene reveals inducibility by osmotic stress. FEBS Lett 458:219–223
Spiegelman JI, Mindrinos MN, Fankhauser C, Richards D, Lutes J, Chory J, Oefner PJ (2000) Cloning of the arabidopsis RSF1 gene by using a mapping strategy based on high-density DNA arrays and denaturing high-performance liquid chromatography. Plant Cell 12:2485–2498
Srivastava AK, Ramaswamy NK, Suprasanna P, D'Souza SF (2010) Genome-wide analysis of thiourea-modulated salinity stress-responsive transcripts in seeds of Brassica juncea: identification of signalling and effector components of stress tolerance. Ann Bot 106:663–674
Srivastava A, Ohm RA, Oxiles L, Brooks F, Lawrence CB, Grigoriev IV, Cho Y (2012) A zinc-finger-family transcription factor, AbVf19, is required for the induction of a gene subset important for virulence in

Alternaria brassicicola. Mol Plant Microbe Interact 25:443–452

Stemple DL (2004) TILLING- a high-throughput harvest for functional genomics. Nat Rev Genet 5:145–150

Stephenson P, Baker D, Girin T, Perez A, Amoah S, King GJ, Ostergaard L (2010) A rich TILLING resource for studying gene function in *Brassica rapa*. BMC Plant Biol 10:62

Stracke S, Haseneyer G, Veyrieras JB, Geiger HH, Sauer S, Graner A, Piepho HP (2009) Association mapping reveals gene action and interactions in the determination of flowering time in barley. Theor Appl Genet 118:259–273

Straton M (2008) Genome resequencing and genetic variation. Nat Biotechnol 26:65–66

Sunilkumar G, Campbell ML, Puckhaber L, Stipanovic DR, Rathore SK (2006) Engineering cottonseed for use in human nutrition by tissue-specific reduction of toxic gossypol. Proc Natl Acad Sci U S A 103: 18054–18059

Swanson-Wagner R, Briskine R, Schaefer R, Hufford MB, Ross-Ibarra J, Myers CL, Tiffin P, Springer NM (2012) Reshaping of the maize transcriptome by domestication. Proc Natl Acad Sci U S A 109:11878–11883

Tabara H, Grishok A, Mello CC (1998) RNAi in *C. elegans*: soaking in the genome sequence. Science 282:430–431

Tadege M, Wang TL, Wen J, Ratet P, Mysore KS (2009) Mutagenesis and beyond! Tools for understanding legume biology. Plant Physiol 151:978–984

Takagi H, Abe A, Yoshida K, Kosugi S, Natsume S, Mitsuoka C, Uemura A, Utsushi H, Tamiru M, Takuno S, Innan H, Cano LM, Kamoun S, Terauchi R (2013) QTL-seq: rapid mapping of quantitative trait loci in rice by whole genome resequencing of DNA from two bulked populations. Plant J 74:174–183

Tang Y, Wang F, Zhoa J, Xie K, Hong Y, Liu Y (2010) Virus-Based MicroRNA expression for gene functional analysis in plants. Plant Physiol 153:632–641

Taylor NE, Greene EA (2003) PARSESNP: a tool for the analysis of nucleotide polymorphisms. Nucleic Acids Res 31:3808–3811

Tenllado F, Gracia MB, Vargas M, Ruiz RDJ (2003) Crude extracts of bacterially expressed dsRNA can be used to protect plants against virus infections. BMC Biotechnol 3:3–14

The Arabidopsis Genome Initiative (2000) Analysis of the genome sequence of the flowering plant Arabidopsis thaliana. Nature 408:796–815

Thorneycroft D, Sherson SM, Smith SM (2001) Using gene knockouts to investigate plant metabolism. J Exp Bot 52:1593–1601

Thornsberry JM, Goodman MM, Doebley J, Kresovich S, Nielsen D, Buckler ES (2001) Dwarf8 polymorphisms associate with variation in flowering time. Nat Genet 28:286–289

Thudi M, Li Y, Jackson SA, May GD, Varshney RK (2012) Current state-of-art of sequencing technologies for plant genomics research. Brief Funct Genomics 11:3–11

Tian F, Bradbury PJ, Brown PJ et al (2011) Genome-wide association study of leaf architecture in the maize nested association mapping population. Nat Genet 43:159–162

Tierney MB, Lamour KH (2005) An introduction to reverse genetic tools for investigating gene function. The Plant Health Instructor. doi:10.1094/PHI-A-2005-1025-01

Till BJ, Reynolds SH, Greene EA, Codomo CA, Enns LC, Johnson JE, Burtner C, Odden AR, Young K, Taylor NE, Henikoff JG, Comai L, Henikoff S (2003) Large-scale discovery of induced point mutations with high-throughput TILLING. Genome Res 13:524–530

Till BJ, Reynolds SH, Weil C, Springer N, Burtner C, Young K, Bowers E, Codomo CA, Enns LC, Odden AR et al (2004) Discovery of induced point mutations in maize genes by TILLING. BMC Plant Biol 4:12

Till BJ, Cooper J, Tai T, Colowit P, Greene E, Henikoff S, Comai L (2007a) Discovery of chemically induced mutations in rice by TILLING. BMC Plant Biol 7:19

Till BJ, Comai L, Henikoff AS (2007b) TILLING and ECOTILLING for crop improvement. In: Varshney RK, Tuberosa R (eds) Genomics- assisted crop improvement: genomic approaches and platforms. Springer, Dordrecht, pp 333–349

Till BJ, Jankowicz-Cieslak J, Sági L, Huynh OA, Utushi H, Swennen R, Terauchi R, Mba C (2010) Discovery of nucleotide polymorphisms in the Musa gene pool by Ecotilling. Theor Appl Genet 121:1381–1389

Tissier AF, Marillonnet S, Klimyuk V, Patel K, Torres MA, Murphy G, Jones JD (1999) Multiple independent defective suppressor–mutator transposon insertions in *Arabidopsis*: a tool for functional genomics. Plant Cell 11:1841–1852

Todd AT, Liu E, Polvi SL, Pammett RT, Page JE (2010) A functional genomics screen identifies diverse transcription factors that regulate alkaloid biosynthesis in Nicotiana benthamiana. Plant J 62:589–600

Tollenaere R, Hayward A, Dalton-Morgan J, Campbell E, Lee JR, Lorenc MT, Manoli S, Stiller J, Raman R, Raman H, Edwards D, Batley J (2012) Identification and characterization of candidate Rlm4 blackleg resistance genes in Brassica napus using next-generation sequencing. Plant Biotechnol J 10:709–715

Travella S, Klimm TE, Keller B (2006) RNA interference-based gene silencing as an efficient tool for functional genomics in hexaploid bread wheat. Plant Physiol 142:16–20

Trebbi D, Maccaferri M, de Heer P, Sørensen A, Giuliani S, Salvi S, Sanguineti MC, Massi A, van der Vossen EA, Tuberosa R (2011) High-throughput SNP discovery and genotyping in durum wheat (Triticum durum Desf.). Theor Appl Genet 123:555–569

Trichine L, Bourcy M, Ratet P (2009) Insertion mutant collections as genetic tools in model legumes. Grain Legumes 53:14

Trick M, Long Y, Meng J, Bancroft I (2009) Single nucleotide polymorphism (SNP) discovery in the polyploid Brassica napus using Solexa transcriptome sequencing. Plant Biotechnol J 4:334–346

Trieu AT, Burleigh SH, Kardailsky IV, Maldonado-Mendoza IE, Versaw WK, Blaylock LA, Shin H, Chiou TJ, Katagi H, Cewbre GR, Weigel D, Harrison MJ (2000) Transformation of *Medicago truncatula* via infiltration of seedlings or flowering plants with *Agrobacterium*. Plant J 22:531–541

Tsaballa A, Pasentsis K, Darzentas N, Tsaftaris AS (2011) Multiple evidence for the role of an Ovate-like gene in determining fruit shape in pepper. BMC Plant Biol 11:46

Uauy C, Paraiso F, Colasuonno P, Tran RK, Tasi H, Berardi S, Comai L, Dubcovsky J (2009) A modified TILLING approach to detect induced mutations in tetraploid and hexaploid wheat. BMC Plant Biol 9:115

Udvardi MK, Tabata S, Parniske M, Stougaard J (2005) Lotus japonicus: legume research in the fast lane. Trends Plant Sci 10:222–228

Uehara T, Sugiyama S, Masuta C (2007) Comparative serial analysis of gene expression of transcript profiles of tomato roots infected with cyst nematode. Plant Mol Biol 63:185–194

Van Eck L, Schultz T, Leach JE, Scofield SR, Peairs FB, Botha AM, Lapitan NL (2010) Virus-induced gene silencing of WRKY53 and an inducible phenylalanine ammonia-lyase in wheat reduces aphid resistance. Plant Biotechnol J 8:1023–1032

van Kammen A (1997) Virus-induced gene silencing in infected and transgenic plants. Trends Plant Sci 2:409–411

Varala K, Swaminathan K, Li Y, Hudson ME (2011) Rapid genotyping of soybean cultivars using high throughput sequencing. PLoS One 6:e24811

Várallyay E, Giczey G, Burgyán J (2012) Virus-induced gene silencing of Mlo genes induces powdery mildew resistance in *Triticum aestivum*. Arch Virol 157:1345–1350

Varshney RK, Nayak SN, May GD, Jackson SA (2009) Next-generation sequencing technologies and their implications for crop genetics and breeding. Trends Biotechnol 27:522–530

Varshney RK, Chen W, Li Y, Bharti AK, Saxena RK, Schlueter JA, Donoghue MT, Azam S, Fan G, Whaley AM, Farmer AD, Sheridan J, Iwata A, Tuteja R, Penmetsa RV, Wu W, Upadhyaya HD, Yang SP, Shah T, Saxena KB, Michael T, McCombie WR, Yang B, Zhang G, Yang H, Wang J, Spillane C, Cook DR, May GD, Xu X, Jackson SA (2012) Draft genome sequence of pigeonpea (*Cajanus cajan*), an orphan legume crop of resource-poor farmers. Nat Biotechnol 30:83–89

Varshney RK, Song C, Saxena RK, Azam S, Yu S, Sharpe AG, Cannon S, Baek J, Rosen BD, Tar'an B, Millan T, Zhang X, Ramsay LD, Iwata A, Wang Y, Nelson W, Farmer AD, Gaur PM, Soderlund C, Penmetsa RV, Xu C, Bharti AK, He W, Winter P, Zhao S, Hane JK, Carrasquilla-Garcia N, Condie JA, Upadhyaya HD, Luo MC, Thudi M, Gowda CL, Singh NP, Lichtenzveig J, Gali KK, Rubio J, Nadarajan N, Dolezel J, Bansal KC, Xu X, Edwards D, Zhang G, Kahl G, Gil J, Singh KB, Datta SK, Jackson SA, Wang J, Cook DR (2013) Draft genome sequence of chickpea (Cicer arietinum) provides a resource for trait improvement. Nat Biotechnol 31:240–246

Vaucheret H (2006) Post-transcriptional small RNA pathways in plants: mechanisms and regulations. Genes Dev 20:759–771

Vaucheret H (2008) Plant ARGONAUTES. Trends Plant Sci 13:350–358

Vazquez F (2006) Arabidopsis endogenous small RNAs: highways and byways. Trends Plant Sci 11:460–468

Vazquez F, Legrand S, Windels D (2010) The biosynthetic pathways and biological scopes of plant small RNAs. Trends Plant Sci 15:337–345

Vega-Sánchez ME, Verhertbruggen Y, Christensen U, Chen X, Sharma V, Varanasi P, Jobling SA, Talbot M, White RG, Joo M, Singh S, Auer M, Scheller HV, Ronald PC (2012) Loss of Cellulose synthase-like F6 function affects mixed-linkage glucan deposition, cell wall mechanical properties, and defense responses in vegetative tissues of rice. Plant Physiol 159:56–69

Velasco R, Zharkikh A, Troggio M, Cartwright DA, Cestaro A, Pruss D, Pindo M, Fitzgerald LM, Vezzulli S, Reid J (2007) A high quality draft consensus sequence of the genome of a heterozygous grapevine variety. PLoS One 2:e1326

Velculescu VE, Zhang L, Vogelstein B, Kinzler KW (1995) Serial analysis of gene expression. Science 270:484–487

Venu RC, Jia Y, Gowda M, Jia MH, Jantasuriyarat C, Stahlberg E, Li H, Rhineheart A, Boddhireddy P, Singh P, Rutger N, Kudrna D, Wing R, Nelson JC, Wang GL (2007) RL-SAGE and microarray analysis of the rice transcriptome after Rhizoctonia solani infection. Mol Genet Genomics 278:421–431

Virk PS, Ford-Lloyd BV, Jackson MT, Pooni HS, Clemeno TP, Newbury HJ (1996) Predicting quantitative variation within rice germplasm using molecular markers. Heredity 76:296–304

Vitulo N, Albiero A, Forcato C, Campagna D, Dal Pero F, Bagnaresi P, Colaiacovo M, Faccioli P, Lamontanara A, Šimková H, Kubaláková M, Perrotta G, Facella P, Lopez L, Pietrella M, Gianese G, Doležel J, Giuliano G, Cattivelli L, Valle G, Stanca AM (2011) First survey of the wheat chromosome 5A composition through a next generation sequencing approach. PLoS One 6:e26421

Vodkin LO, Khanna A, Shealy R, Clough SJ, Gonzalez DO, Philip R, Zabala G, Thibaud-Nissen F, Sidarous M, Stromvik M et al (2004) Microarrays for global expression constructed with a low redundancy set of 27500 sequenced cDNAs representing an array of developmental stages and physiological conditions of the soybean plant. BMC Genomics 5:73

Voinnet O (2009) Origin, biogenesis, and activity of plant microRNAs. Cell 136:669–687

Walbot V (2000) Saturation mutagenesis using maize transposons. Curr Opin Plant Biol 3:103–107

Walden R (2002) T-DNA tagging in genomics era. Crit Rev Plant Sci 21:143–165

Wang WY, Barratt BJ, Clayton DG, Todd JA (2005) Genome-wide association studies: theoretical and practical concerns. Nat Rev Genet 6:109–118

Wang DK, Sun ZX, Tao YZ (2006) Application of TILLING in plant improvement. Acta Genet Sin 33:957–964

Wang J, Sun JZ, Liu DC, Yang WL, Wang DW, Tong YP, Zhang AM (2008) Analysis of *Pina* and *Pinb* alleles in the microcore collections of Chinese wheat germplasm by EcoTILLING and identification of a novel *Pinb* allele. J Cereal Sci 48:836–842

Wang X, Cao A, Yu C, Wang D, Wang X, Chen P (2010) Establishment of an effective virus induced gene silencing system with BSMV in Haynaldia villosa. Mol Biol Rep 37:967–972

Wang YH, Poudel DD, Hasenstein KH (2011a) Identification of SSR markers associated with saccharification yield using pool-based genome-wide association mapping in sorghum. Genome 54:883–889

Wang X, Wang H, Wang J, Sun R, Wu J, Liu S, Bai Y et al (2011b) The genome of the mesopolyploid crop species *Brassica rapa*. Nat Genet 43:1035–1039

Wang Y, Sun S, Liu B, Wang H, Deng J, Liao Y, Wang Q, Cheng F, Wang X, Wu J (2011c) A sequence-based genetic linkage map as a reference for Brassica rapa pseudochromosome assembly. BMC Genomics 12:239

Wang Y, Zeng X, Iyer NJ, Bryant DW, Mockler TC, Mahalingam R (2012a) Exploring the switchgrass transcriptome using second-generation sequencing technology. PLoS One 7:e34225

Wang S, Wang X, He Q, Liu X, Xu W, Li L, Gao J, Wang F (2012b) Transcriptome analysis of the roots at early and late seedling stages using Illumina paired-end sequencing and development of EST-SSR markers in radish. Plant Cell Rep 31:1437–1447

Wang H, Walla JA, Zhong S, Huang D, Dai W (2012c) Development and cross-species/genera transferability of microsatellite markers discovered using 454 genome sequencing in chokecherry (*Prunus virginiana* L.). Plant Cell Rep 31:2047–2055

Wang W, Huang S, Liu Y, Fang Z, Yang L, Hua W, Yuan S, Liu S, Sun J, Zhuang M, Zhang Y, Zeng A (2012d) Construction and analysis of a high-density genetic linkage map in cabbage (*Brassica oleracea* L. var. *capitata*). BMC Genomics 13:523

Wang N, Fang L, Xin H, Wang L, Li S (2012e) Construction of a high-density genetic map for grape using next generation restriction-site associated DNA sequencing. BMC Plant Biol 12:148

Wang S, Meyer E, McKay JK, Matz MV (2012f) 2b-RAD: a simple and flexible method for genome-wide genotyping. Nat Methods 9:808–810

Wang X, Zhang C, Li L, Fritsche S, Endrigkeit J, Zhang W, Long Y, Jung C, Meng J (2012g) Unraveling the genetic basis of seed tocopherol content and composition in rapeseed (*Brassica napus* L.). PLoS One 7:e50038

Wang L, Yu S, Tong C, Zhao Y, Liu Y, Song C, Zhang Y, Zhang X, Wang Y, Hua W, Li D, Li D, Li F, Yu J, Xu C, Han X, Huang S, Tai S, Wang J, Xu X, Li Y, Liu S, Varshney RK, Wang J, Zhang X (2014) Genome sequencing of the high oil crop sesame provides insight into oil biosynthesis. Genome Biol 15:R39

Waterhouse PM, Graham MW, Wang MB (1998) Virus resistance and gene silencing in plants can be induced by simultaneous expression of sense and antisense RNA. Proc Natl Acad Sci U S A 95:13959–13964

Watkinson JI, Hendricks L, Sioson AA, Heath LS, Bohnert HJ, Grene R (2008) Tuber development phenotypes in adapted and acclimated, drought-stressed Solanum tuberosum ssp. andigena have distinct expression profiles of genes associated with carbon metabolism. Plant Physiol Biochem 46:34–45

Watson JD, Crick FHC (1953) Molecular structure of nucleic acids. Nature 171:737–738

Watson JM, Fusaro AF, Wang M, Waterhouse PM (2005) RNA silencing platforms in plants. FEBS Lett 579:5982–5987

Wei LQ, Yan LF, Wang T (2011) Deep sequencing on genome-wide scale reveals the unique composition and expression patterns of microRNAs in developing pollen of Oryza sativa. Genome Biol 12(6):R53

Weigel D, Ahn JH, Blazquez MA, Borevitz J, Christensen SK, Fankhauser C, Ferrandiz C, Kardailsky I, Malancharuvil EJ, Neff MM, Nguyen JT, Sato S, Wang Z, Xia Y, Dixon RA, Harrison MJ, Lamb CJ, Yanofsky MK, Chory J (2000) Activation tagging in *Arabidopsis*. Plant Physiol 122:1003–1013

Weil CF (2009) TILLING in grass species. Plant Physiol 149:158–164

Wesley SV, Helliwell CA, Smith NA, Wang MB, Rouse DT, Liu Q, Gooding PS, Singh SP, Abbott D, Stoutjesdijk PA (2001) Constructs for efficient, effective and high throughput gene silencing in plants. Plant J 27:581–590

Winfield MO, Wilkinson PA, Allen AM, Barker GL, Coghill JA, Burridge A, Hall A, Brenchley RC, D'Amore R, Hall N, Bevan MW, Richmond T, Gerhardt DJ, Jeddeloh JA, Edwards KJ (2012) Targeted re-sequencing of the allohexaploid wheat exome. Plant Biotechnol J 10:733–742

Winkler HL (1920) Verbreitung und Ursache der Parthenogenesis im Pflanzen-and Tierreiche. Verlag Fischer, Jena

Wisman E, Hartmann U, Sagasser M, Baumann E, Palme K, Hahlbrock K, Saedler H, Weisshaar B (1998) Knockout mutants from an En-1 mutagenized *Arabidopsis thaliana* population generate phenylpropanoid biosynthesis phenotypes. Proc Natl Acad Sci U S A 95:12432–12437

Wu C, Li X, Yuan W, Chen G, Kilian A, Li J, Xu C, Li X, Zhou DX, Wang S, Zhang Q (2003) Developments of enhancer trap lines for functional analysis of the rice genome. Plant J 5:418–427

Wu JL, Wu C, Lei C, Baraoidan M, Bordeos A, Madamba MR, Ramos PM, Mauleon R, Portugal A, Ulat VJ, Bruskiewich R, Wang G, Leach J, Khush G, Leung H (2005) Chemical- and Irradiation-induced mutants of indica rice IR64 for forward and reverse genetics. Plant Mol Biol 59:85–97

Wu N, Matand K, Wu H, Li B, Li Y, Zhang X, He Z, Qian J, Liu X, Conley S, Bailey M, Acquaah G (2013) De novo next-generation sequencing, assembling and annotation of Arachis hypogaea L. Spanish botanical type whole plant transcriptome. Theor Appl Genet 126:1145–1149

Xiang CC, Chen Y (2000) cDNA microarray technology and its applications. Biotechnol Adv 18:35–46

Xiao J, Jin X, Jia X, Wang H, Cao A, Zhao W, Pei H, Xue Z, He L, Chen Q, Wang X (2013) Transcriptome-based discovery of pathways and genes related to resistance against Fusarium head blight in wheat landrace Wangshuibai. BMC Genomics 14:197

Xie Z et al (2004) Genetic and functional diversification of small RNA pathways in plants. PLoS Biol 2:E104

Xin Z, Wang ML, Barkley NA, Burow G, Franks C, Pederson G, Burke J (2008) Applying genotyping (TILLING) and phenotyping analyses to elucidate gene function in a chemically induced sorghum mutant population. BMC Plant Biol 8:103

Xin M, Wang X, Peng H, Yao Y, Xie C, Han Y, Ni Z, Sun Q (2012) Transcriptome comparison of susceptible and resistant wheat in response to powdery mildew infection. Genomics Proteomics Bioinformatics 10:94–106

Xiong AS, Yao QH, Peng RH, Li X, Han PL, Fan HQ (2004) Different effects on *ACC oxidase* gene silencing triggered by RNA interference in transgenic tomato. Plant Cell Rep 23:639–646

Xu Y, Crouch JH (2008) Marker-assisted selection in plant breeding: from publications to practice. Crop Sci 48:391–407

Xu JJ, Zhao QA, Du P, Xu CW, Wang BH, Feng Q, Liu QQ, Tang SZ, Gu MH, Han B, Liang GH (2010) Developing high throughput genotyped chromosome segment substitution lines based on population whole-genome re-sequencing in rice (*Oryza sativa* L.). BMC Genomics 11:656

Xu P, Wu X, Luo J, Wang B, Liu Y, Ehlers JD, Wang S, Lu Z, Li G (2011a) Partial sequencing of the bottle gourd genome reveals markers useful for phylogenetic analysis and breeding. BMC Genomics 12:467

Xu X, Liu X, Ge S, Jensen JD, Hu F, Li X, Dong Y, Gutenkunst RN, Fang L, Huang L, Li J, He W, Zhang G, Zheng X, Zhang F, Li Y, Yu C, Kristiansen K, Zhang X, Wang J, Wright M, McCouch S, Nielsen R, Wang J, Wang W (2011b) Resequencing 50 accessions of cultivated and wild rice yields markers for identifying agronomically important genes. Nat Biotechnol 30:105–111

Yamagishi N, Yoshikawa N (2009) Virus-induced gene silencing in soybean seeds and the emergence stage of soybean plants with Apple latent spherical virus vectors. Plant Mol Biol 71:15–24

Yamaguchi T, Nakayama K, Hayashi T, Yazaki J, Kishimoto N, Kikuchi S, Koike S (2004) cDNA microarray analysis of rice anther genes under chilling stress at the microsporogenesis stage revealed two genes with DNA transposon Castaway in the 5′-flanking region. Biosci Biotechnol Biochem 68:1315–1323

Yamamoto M, Shitsukawa N, Yamada M, Kato K, Takumi S, Kawaura K, Ogihara Y, Murai K (2013) Identification of a novel homolog for a calmodulin-binding protein that is upregulated in alloplasmic wheat showing pistillody. Planta 237:1001–1013

Yamatani H, Sato Y, Masuda Y, Kato Y, Morita R, Fukunaga K, Nagamura Y, Nishimura M, Sakamoto W, Tanaka A, Kusaba M (2013) NYC4, the rice ortholog of Arabidopsis THF1, is involved in the degradation of chlorophyll – protein complexes during leaf senescence. Plant J. doi:10.1111/tpj.12154

Yamazaki M, Tsugawa H, Miyao A, Yano M, Wu J, Yamamoto S, Matsumoto T, Sasaki T, Hirochika H (2001) The rice retrotransposon Tos17 prefers low-copy-number sequences as integration targets. Mol Genet Genomics 265:336–344

Yang Y, Wu Y, Pirrello J, Regad F, Bouzayen M, Deng W, Li Z (2010) Silencing Sl-EBF1 and Sl-EBF2 expression causes constitutive ethylene response phenotype, accelerated plant senescence, and fruit ripening in tomato. J Exp Bot 61:697–708

Yang H, Tao Y, Zheng Z, Li C, Sweetingham MW, Howieson JG (2012a) Application of next-generation sequencing for rapid marker development in molecular plant breeding: a case study on anthracnose disease resistance in Lupinus angustifolius L. BMC Genomics 13:318

Yang T, Bao SY, Ford R, Jia TJ, Guan JP, He YH, Sun XL, Jiang JY, Hao JJ, Zhang XY, Zong XX (2012b) High-throughput novel microsatellite marker of faba bean via next generation sequencing. BMC Genomics 13:602

Yokosho K, Yamaji N, Ueno D, Mitani N, Ma JF (2009) OsFRDL1 is a citrate transporter required for efficient translocation of iron in rice. Plant Physiol 149: 297–305

You FM, Huo N, Deal KR, Gu YQ, Luo MC, McGuire PE, Dvorak J, Anderson OD (2011) Annotation-based genome-wide SNP discovery in the large and complex Aegilops tauschii genome using next-generation sequencing without a reference genome sequence. BMC Genomics 12:59

Young ND, Debelle F, Oldroyd GE, Geurts R et al (2011) The Medicago genome provides insight into the evolution of rhizobial symbioses. Nature 480:520–524

Yu LX, Setter TL (2003) Comparative transcriptional profiling of placenta and endosperm in developing maize kernels in response to water deficit. Plant Physiol 131:568–582

Yu J, Pressoir G, Briggs WH, Vroh BI, Yamasaki M, Doebley JF, McMullen MD, Gaut BS, Nielsen DM, Holland JB et al (2006) A unified mixed-model method for association mapping that accounts for multiple levels of relatedness. Nat Genet 38:203–208

Yu J, Holland JB, McMullen MD, Buckler ES (2008) Genetic design and statistical power of nested association mapping in maize. Genetics 178:539–551

Yu H, Xie W, Wang J, Xing Y, Xu C, Li X, Xiao J, Zhang Q (2011a) Gains in QTL detection using an ultra-high density SNP map based on population sequencing relate to traditional RFLP/SSR markers. PLoS One 6:e17595

Yu LX, Lorenz A, Rutkoski J, Singh RP, Bhavani S, Huerta-Espino J, Sorrells ME (2011b) Association mapping and gene-gene interaction for stem rust resistance in CIMMYT spring wheat germplasm. Theor Appl Genet 123:1257–1268

Yu X, Bai G, Liu S, Luo N, Wang Y, Richmond DS, Pijut PM, Jackson SA, Yu J, Jiang Y (2013) Association of candidate genes with drought tolerance traits in diverse perennial ryegrass accessions. J Exp Bot 64:1537–1551

Yuan X, Zhang S, Liu S, Yu M, Su H, Shu H, Li X (2013) Global analysis of ankyrin repeat domain C3HC4-Type RING finger gene family in plants. PLoS One 8:e58003

Zamore PD, Tuschl T, Sharp PA, Bartel DP (2000) RNAi: double-stranded RNA directs the ATP-dependent cleavage of mRNA at 21 to 23 nucleotide intervals. Cell 101:25–33

Zha W, Peng X, Chen R, Du B, Zhu L, He G (2011) Knockdown of midgut genes by dsRNA-transgenic plant-mediated RNA interference in the hemipteran insect *Nilaparvata lugens*. PLoS One 6(5):e20504

Zhang L, Zhou W, Velculescu VE, Kern SE, Hruban RH, Hamilton SR, Vogelstein B, Kinzler KW (1997) Gene expression profiles in normal and cancer cells. Science 276:1268–1272

Zhang X, Yazaki J, Sundaresan A, Cokus S, Chan SW, Chen H, Henderson IR, Shinn P, Pellegrini M, Jacobsen SE, Ecker JR (2006) Genome-wide high-resolution mapping and functional analysis of DNA methylation in *Arabidopsis*. Cell 126:1189–1201

Zhang L, Lavery L, Gill U, Gill K, Steffenson B, Yan G, Chen X, Kleinhofs A (2009) A cation/proton-exchanging protein is a candidate for the barley NecS1 gene controlling necrosis and enhanced defense response to stem rust. Theor Appl Genet 118: 385–397

Zhang C, Grosic S, Whitham SA, Hill JH (2012a) The requirement of multiple defense genes in soybean Rsv1-mediated extreme resistance to soybean mosaic virus. Mol Plant Microbe Interact 25:1307–1313

Zhang J, Liang S, Duan J, Wang J, Chen S, Cheng Z, Zhang Q, Liang X, Li Y (2012b) De novo assembly and characterisation of the transcriptome during seed development, and generation of genic-SSR markers in peanut (Arachis hypogaea L.). BMC Genomics 13:90

Zhang Z, Liu X, Wang X, Zhou M, Zhou X, Ye X, Wei X (2012c) An R2R3 MYB transcription factor in wheat, TaPIMP1, mediates host resistance to Bipolaris sorokiniana and drought stresses through regulation of defense- and stress-related genes. New Phytol 196:1155–1170

Zhao MM, An RD, Zhao J, Huang GH, He ZH, Chen JY (2006) Transiently expressed short hairpin RNA targeting 126 kDa protein of tobacco mosaic virus interferes with virus infection. Acta Biochim Biophys Sin 38:22–28

Zhao K, Aranzana MJ, Kim S, Lister C, Shindo C, Tang C, Toomajian C, Zheng H, Dean C, Marjoram P et al (2007) An Arabidopsis example of association mapping in structured samples. PLoS Genet 3:e4

Zhao CZ, Xia H, Frazier TP, Yao YY, Bi YP, Li AQ, Li MJ, Li CS, Zhang BH, Wang XJ (2010) Deep sequencing identifies novel and conserved microRNAs in peanuts (*Arachis hypogaea* L.). BMC Plant Biol 10:3

Zhou H, Steffenson BJ (2013) Association mapping of septoria speckled leaf blotch resistance in US barley breeding germplasm. Phytopathology 103:600–609

Zhou ML, Zhang Q, Zhou M, Qi LP, Yang XB, Zhang KX, Pang JF, Zhu XM, Shao JR, Tang YX, Wu YM (2012) Aldehyde dehydrogenase protein superfamily in maize. Funct Integr Genomics 12:683–691

Zhou ZS, Yang SN, Li H, Zhu CC, Liu ZP, Yang ZM (2013a) Molecular dissection of mercury-responsive transcriptome and sense/antisense genes in Medicago truncatula. J Hazard Mater 252–253C:123–131

Zhou Q, Yu Q, Wang Z, Pan Y, Lv W, Zhu L, Chen R, He G (2013b) Knockdown of GDCH gene reveals reactive oxygen species-induced leaf senescence in rice. Plant Cell Environ 36:1476–1489

Zhu C, Gore M, Buckler ES, Yu J (2008) Status and prospects of association mapping in plants. Plant Genome 1:5–20

Zhu Y, Cao Z, Xu F, Huang Y, Chen M, Guo W, Zhou W, Zhu J, Meng J, Zou J, Jiang L (2012) Analysis of gene expression profiles of two near-isogenic lines differing at a QTL region affecting oil content at high temperatures during seed maturation in oilseed rape (Brassica napus L.). Theor Appl Genet 124:515–531

Zuo R, Hu R, Chai G, Xu M, Qi G, Kong Y, Zhou G (2012) Genome-wide identification, classification, and expression analysis of CDPK and its closely related gene families in poplar (*Populus trichocarpa*). Mol Biol Rep 40:2645–2662

Zou Z, Ishida M, Li F, Kakizaki T, Suzuki S, Kitashiba H, Nishio T (2013) QTL analysis using SNP markers developed by next-generation sequencing for identification of candidate genes controlling 4-methylthio-3-butenyl glucosinolate contents in roots of radish, Raphanus sativus L. PLoS One 8:e53541

Cytogenomics and Mutagenomics in Plant Functional Biology and Breeding

Dibyendu Talukdar and Andrey Sinjushin

Contents

D. Talukdar, Ph.D. (✉)
Department of Botany, R.P.M. College, University of Calcutta, Uttarpara, Hooghly 712258, West Bengal, India
e-mail: dibyendutalukdar9@gmail.com

A. Sinjushin, Ph.D.
Department of Genetics, M.V. Lomonosov Moscow State University, Leninskie Gory, 1-12, 119234 Moscow, Russia

Abstract

One of the most important breakthroughs in the history of genetics and plant breeding was the development of plant cytology and experimental mutagenesis, which later brought about plant cytogenetics and mutation breeding and now they have entered in functional biology era with the unprecedented develop-

D. Barh et al. (eds.), *PlantOmics: The Omics of Plant Science*, DOI 10.1007/978-81-322-2172-2_5, © Springer India 2015 ment of plant molecular cytogenetics, genetics, and functional genomics. Application of cell biology particularly chromosome biology in the fields of plant genome structure and function has ushered the development of plant cytogenomics. Development of advanced technology like DNA base-specific fluorescence banding, GISH, and FISH-based chromosome painting has greatly facilitated the identification, localization, and mapping of chromosome-specific markers in plants, which is of high importance in plant molecular systematics, species identification, detection of hybrid nature, alien chromosomes and chromosomal aberrations, analysis of somaclonal variations, and diversity analysis. The dynamism of chromatin architecture and cell cycle, representing chromosome functional biology, is another important part of plant cytogenomics. On the other hand, mutagenomics is defined as applied mutation breeding, in which genomic information and tools are utilized in the designing of breeding strategies, screening, selection and verification/authentication of natural and induced mutants, and the utilization of desirable mutations in the breeding processes. Considerable progress has been made in recent times in breeding of cereals, legumes, oil seeds, vegetables, horticultural crops, spices and condiments, fiber-yielding plants, and medicinal and aromatic plants for diverse types of desirable agronomic and functional traits including disease and abiotic stress resistance/tolerance; herbicide resistance; lowering of anti-nutritional factors; enhancement of proteins, minerals, vitamins, essential amino acids, flavonoids, antioxidants, and dietary fibers; enrichment of soil nutrition; enhancement of ornamental, medicinal, and aromatic values; and development of functional and therapeutic foods and other diverse traits related to nutritional quality and high yield. This can be found in a mutant population which carries variant forms of potentially each and every gene present in a particular genome. The functionality of mutagenomics is greatly enhanced due to integration of classical mutagenesis with modern "omics" technology through the development of desirable diploid mutants, recombinant inbred lines, and aneuploid and polyploid lines as effective cytogenetic tools, utilizable in genome mapping and analysis. Functional sets of aneuploid tools are now available in different edible plants, through which several morphological, biochemical, and molecular traits/markers have been assigned on definite chromosomes to construct linkage maps. Recently, induced mutations showing alterations in antioxidant defense response have been identified and tested against diverse types of abiotic stresses to reveal intrinsic cellular and metabolic events toward sensitivity of seed plants to salinity, drought, metal toxicity, and other stresses. These mutations are giving vital inputs, which can be used in formulating effective breeding strategies in different agroclimatic conditions. Mutagenized population has revealed altered pattern of genome response and can also be exploited in enhancing production of natural plant products like antioxidants and flavonoids. Furthermore, these large mutant populations have the potential in reverse genetics approach by employing various techniques, particularly "Targeting Induced Local Lesions in Genomics (TILLING)" technology to better understand gene functions through high-throughput mutation screening, and have been successfully used in major crop plants along with model plant *Arabidopsis*. The development of mutagenomic approach, thus, provides a cost-effective, clean, and easy-to-use functional tool to increase the genetic diversity and in utilization of this diversity in plant molecular mutation breeding through modern genomic methods.

Keywords

Cytogenomics • Mutagenomics • FISH • Cytogenetic stock • Gene mapping • Molecular mutation breeding • Reverse genetics • TILLING

Introduction

One of the most important breakthroughs in the history of genetics and plant breeding was the development of plant cytology and experimental mutagenesis, which later brought about plant cytogenetics and mutation breeding. Plant cytogenetics has continued to flourish and make essential contributions to advanced genomic projects by delineating gene order, defining contig gaps, and revealing genome rearrangements. With the unprecedented development of advanced technology like DNA base-specific fluorescence banding and GISH as well as FISH-guided chromosome painting along with availability of whole-genome sequencing techniques, classical plant cytogenetics has now entered the functional genomic era, greatly facilitating the identification, localization, and mapping of chromosome-specific markers in plants, which is of high importance in plant breeding, molecular systematics, species identification, detection of hybrid nature, detection of alien chromosomes and chromosomal aberrations, and analysis of somaclonal variations and diversity (Chaudhary et al. 2011). On the other hand, plant mutagenesis is used as mutation breeding, in which genetic information and tools are utilized in the designing of breeding strategies, screening, selection and verification/authentication of induced mutants, and the utilization of desirable mutations in the breeding processes. The integration of classic techniques of plant cytology and mutagenesis with modern "omics" technology has ushered the development of two new concepts, plant cytogenomics and mutagenomics, respectively.

Cytogenomic Techniques, Chromosome Biology, and Genome Analysis

Cytogenomics: Classical Cytogenetics Meets Plant "Omics"

At the dawn of plant cytogenetics, the classic approach of chromosome banding had immense contribution in identification and characterization of plant chromosomes. Early cytogenetic pioneers resolved the structure and dynamic behavior of the chromosomes in rice, sorghum, maize, and tomato and revealed the mystery of allopolyploid nature of wheat, determined the ancestral genome donors, and discovered the nature of chromosome pairing (Gupta 2006). The development of DNA base-specific chromosome banding and in situ hybridization (ISH) techniques revolutionized the classical concepts and flow cytogenetics, opening up opportunities for cytogenetic analysis of essentially any species, regardless of its inherent chromosome morphology and has provided an abundance of knowledge regarding the structural and functional genomics of plants (Gupta 2006). In plants, the use of radioactive tracer or modified nucleotides (attached to biotin, digoxigenin, or fluorescent moieties) to make ISH probes permits microscopic visualization and localization of complementary sequences in cells and nuclei and on individual chromosomes (Lilly et al. 2001; Gupta 2006). All these new advancements facilitated cytology-based molecular analysis of a complete genome or cytogenomics and increased our understanding of the plant genome. The molecular cytogenetics and cytogenomics are often used interchangeably in different literatures, although the latter focus on latest knowledge on molecular cytogenetic techniques (Lilly et al. 2001).

Flow Cytogenetics in Chromosome Functional Biology

Flow karyotyping is a quantitative, statistically accurate, and high-throughput approach for karyotype analysis and the detection of numerical and structural chromosome changes. This technique has been used to detect trisomy of chromosome 6 in barley; estimate the frequency of alien chromosomes in populations of six wheat–rye chromosome addition lines; identify chromosomes carrying interchanges in field bean, garden pea, barley, and wheat and chromosome deletions in wheat; detect accessory B chromosomes in rye and maize; identify alien chromosomes in oat–maize and wheat–rye chromosome addition lines; detect alien chromosome arms in wheat–rye and wheat–barley telosome addition lines; and reveal chromosome polymorphism in barley, maize, rye, and wheat (Kubaláková et al. 2005; Gupta 2006; Šimková et al. 2008). However, as the flow karyotyping is based on chromosome DNA content and/or AT/GC ratio, intrachromosomal rearrangements and reciprocal translocations where equal amounts of DNA are exchanged cannot be identified. Detection of aberrations is also hampered by natural occurrence of chromosome polymorphisms. With these limitations, the labor-intensive character, and a need for expensive equipment, flow karyotyping is now augmented with advanced methods of cytogenetics such as multicolor genomic in situ hybridization (GISH), fluorescence in situ hybridization (FISH), chromosome painting, and DNA arrays. Some of the important uses of flow-sorted chromosomes in plant cytogenomics are

(a) Physical mapping using DNA hybridization and PCR
(b) Physical mapping using FISH
(c) Small-insert DNA libraries
(d) Large-insert DNA libraries
(e) Development of molecular markers
(f) Physical mapping on DNA arrays and array painting
(g) Chromosome sequencing using next-generation technology
(h) Higher-order structure and proteins of mitotic chromosomes

Using microfluidic technology, optical mapping has now been shown to be particularly useful in highly repetitive and duplicated genomes to assemble their sequences and verify finished sequence data, study genome structural polymorphism, and perform genome-wide DNA methylation mapping. A modified approach to construct optical maps employs nanofluidic devices with a series of parallel microchannels through which DNA molecules move and can be analyzed by using nicking enzymes and fluorescent labeling of displaced single strands (Doležel et al. 2012). The use of chromosomal DNA could greatly simplify the assembly of optical maps in organisms with large and polyploid genomes such as bread wheat (Doležel et al. 2012).

Potential DNA markers can be developed from clones of chromosome-specific DNA libraries with large inserts after sequencing their ends, and the markers can be developed directly from only a few nanograms of chromosomal DNA (Doležel et al. 2012). Next-generation sequencing of chromosomal DNA identifies enough sequences from genes and intergenic regions to develop literally an unlimited number of markers, including single nucleotide polymorphisms (SNPs), in crops like wheat, rye, and barley (Mayer et al. 2011; Doležel et al. 2012).

Coupling DNA array technology with flow cytogenetics resulted in development of array painting, which allows high-resolution analysis of the content and breakpoint of aberrant chromosomes. Painting probes are prepared from two derivative chromosomes, each of them is labeled with a different fluorochrome, and both are hybridized to DNA microarray with mapped DNA sequences. Plotting the fluorescence ratio against the clone position along each chromosome provides information on chromosome composition. This approach has been used in mapping of 162 SNP loci on barley chromosome 1H through pilot oligonucleotide pool assay (Šimková et al. 2008) and assignment of 16,804 genes on individual chromosomes using DNA from flow-sorted barley chromosome 1H and arms of chromosomes 2H–7H on barley 44 k Agilent microarray (Mayer et al. 2011). The robust mapping thus constructed is finally compared with barley consensus genetic map, and gene mapping using flow-sorted chromosome arms then permitted the definition of pericentromeric regions in chromosomes 2H–7H (Mayer et al. 2011).

Sequencing single chromosomes reduces costs and simplifies data analysis as compared to whole genomes. In species with sequenced genomes, re-sequencing chromosomes is a rapid means for studying variation at DNA level by aligning short reads to the reference sequence. The low-pass 454 sequencing flow-sorted chromosome was a cost-effective approach to describe gene content, repetitive sequence, assess gene synteny with other species, translocation, and establish comprehensive linear gene-order model, including candidate miRNA (microRNA) precursors, for the chromosome, as successfully applied in wheat (chromosome 4A, 5A, 7DS), barley (chromosome 1H), and rye (chromosome 1R) (Mayer et al. 2011).

In a majority of cases, flow cytogenetics has been employed to aid in analyzing chromosomal DNA. However, there are as yet not fully explored opportunities to analyze the higher-order structure of mitotic chromosomes and their major component—the chromosomal proteins using technique like immunostaining of chromosomal antigens (Gupta 2006).

DNA Base-Specific Fluorescence Banding

Due to phenotypic plasticity, symmetrical karyotype, and occurrence of chromosomal structural rearrangements, instead of conventional banding, DNA base-specific fluorescent chromosome banding has been introduced with the use of two common and effective fluorochromes, chromomycin A3 (CMA) and 4-6′ diamidino 2-phenylindole (DAPI), for the detection of karyotypically visible landmarks, marking of individual chromosome, detection of AT (by DAPI) and GC (by CMA)-rich regions within individual chromosomes, and phylogenetic studies in major crops (Akter and Alam 2005; Talukdar 2010d). This molecular banding technique alone or in combination with ISH has been found very useful in detection of aneuploids, alien chromosome introgression, nature of polyploidy, aneuploids in polyploidy background, and cytogenotoxicity study (Lavania et al. 2010; Talukdar 2010d).

Genomic In Situ Hybridization (GISH)

GISH/multicolor GISH allows the visualization and comparison of chromosomes and genomes of different materials, enabling one to characterize them as polyploids, F_1 hybrids and their progenies, partial allopolyploids, aneuploids, polyhaploids, or recombinant lines. The protocol for GISH is essentially the same as for the FISH, except for the blocking genomic DNA. The GISH protocol is defined to eliminate most of the cross-hybridization between total genomic DNA from the two species. In rice, GISH resolved the distinction between two kinds of genomes in two wild tetraploids, *Oryza minuta* (BBCC) and *O. latifolia* (CCDD), and a highly reproducible complete protocol of chromosome painting by GISH is available. GISH using total genomic DNA of a donor species in combination with an excess amount of unlabeled genomic wheat DNA permits the painting of whole genomes and alien chromosomes in interspecific hybrids (Gupta et al. 2008). GISH is also a very powerful tool for characterizing wheat–alien translocations and permits the determination of translocation breakpoints and sizes of the alien segments. Most of the available wheat–alien translocations conferring resistance to various diseases and pests were recently characterized by GISH and C-banding analyses. Identification of wheat and tritordeum (*Hordeum chilense* × *Aegilops*, hexaploid amphidiploid) chromosomes, introgression of rye chromosomes to wheat and triticale, chromosome pairing in the meiotic metaphase I of wheat–rye hybrids, discrimination of wheat and rye chromosomes, and detection of the individual wheat and rye chromosome arms involved in the chromosome associations are some of the outstanding achievements obtained by GISH (Megyeri et al. 2013). GISH revealed the chromosome constitution of all aneuploids, demonstrating its important role as a tool for genome monitoring in plant breeding. However, the feasibility of chromosome painting by GISH for hybrids consisting of different genomes depends on the diversity of the different contributor genomes. For example, the use of GISH in *Brassica* allopolyploids consisting of a combination of the A, B, and C genomes could not discriminate the three genomes because the repeated sequences are highly homologous among the three genomes. Employing multicolor GISH, somatic hybrids between *Oryza sativa* cv. “Kitaake” (AA, $2n = 24$) and *O. punctata* (BBCC, $2n = 48$) were identified in rice, and the progeny was rescued from embryo culture (Ohmido et al. 2010). In the medicinally important plant, ginseng (*Panax* spp.), GISH analysis using genomic DNA as a probe revealed strong cross-hybridization of genomes between *P. ginseng* and *P. quinquefolius* (Choi et al. 2009).

Fluorescence In Situ Hybridization (FISH)

FISH is a quick and affordable approach to map DNA sequences on specific chromosomal regions. The power of cytogenomic analysis primarily depends on two related aspects of FISH: probe-size detection limit and axial resolution limit. Advances in microscopic sensitivity, signal increase, and noise reduction have all contributed to improved detection limits, whereas advances in cytological resolution of closely linked loci are primarily derived from methods that lengthen the chromosome itself. Although FISH is commonly used to map unique or low-copy-number sequences, it is also used to localize repetitive sequence to produce chromosome recognition cocktails or explore genome relations in polyploid or closely related plant species. FISH permits rapid cytogenetic characterization and chromosome identification by means of a variety of probes such as those from repetitive DNAs, large-fragment clones, or closely related species. The synergy between plant cytogenetics and genomics is strengthened by FISH-guided genome sequencing, FISH-based karyotyping, and mapping of various ribosomal DNA (Lavania et al. 2003, 2005, 2010; Jiang and Gill 2006; Chahota et al. 2011). Mitotic and meiotic FISH continues to be invaluable in genome sequencing efforts. FISH played important roles, for example, in the Solanaceae Genome Project, in the direction of sequencing efforts through BAC-based anchoring of contigs, and in the detection and closure of numerous and significant gaps in euchromatic regions of large genomes (Jiang and Gill 2006). FISH analysis is also unique in differentiation of triticale cultivars through their rye chromosomes, ascertaining the number and chromosomal location of ribosomal 5S and 35S DNA (rDNA) sites along with chromosomal translocation between *Secale cereale*, *Dasypyrum villosum*, and their allotetraploid *S. cereale* × *D. villosum* hybrids; construction of an integrated molecular cytogenetic map of cucumber (*Cucumis sativus* L.) chromosomes 2 through mapping of 11 fosmid clones together with the cucumber centromere-specific type III sequence on meiotic pachytene chromosomes; and identification of 12 metaphase chromosomes and construction of a standardized karyotype of melon (*Cucumis melo* L.) through cross-species fosmid FISH, anchored by SSR markers (Liu et al. 2010; Han et al. 2011; Fradkin et al. 2013). In *Panax* (ginseng), FISH analysis revealed presence of 45S and 5S rRNA genes of *P. notoginseng* ($2n=2x=24$) and *P. ginseng* ($2n=4x=48$) cluster on a single locus on different chromosomes, while *P. quinquefolius* ($2n=4x=48$), *P. japonicus* ($2n=4x=48$), and Korean wild ginseng ($2n=4x=48$) had one locus of the 45S rRNA gene and two loci of the 5S rRNA gene, respectively (Choi et al. 2009; Waminal et al. 2012). The broad applications of FISH in structural, comparative, and functional genomics place plant cytogenetics in a formidable position to complement, accelerate, or guide plant genome research (Table 1). Furthermore, the FISH and the genetic markers generated from the subrepeat variation in the NORs (nucleolus organizer region) in cereals, soybean, and *Arabidopsis* provide anchor points for the construction of cytogenetic, genetic, and physical maps of plant genomes as well as for breeding programs (Yang and Jeong 2008). With the advancement of knowledge, the FISH techniques now have several applied arenas in plant molecular cytogenetics, some of the significant achievements of which are presented in Table 1.

M-FISH and Pachytene-FISH

M-FISH or metaphase-FISH is based on plant meristem tissues, such as root tip, and provides readily available material but can produce variable axial resolution limits (10,000 kb in some cases) depending on whether the probes are in euchromatic or heterochromatic regions. Despite having the poorest axial resolution, M-FISH

Table 1 Important achievements of using FISH techniques in cytogenomics, mapping and genome analysis of some major food crops

Crop	Techniques	Achievements	References
Cereals (rice, wheat, maize, rye, barley, oat, sorghum)	M-FISH, I-FISH, Fiber-FISH, BAC-FISH, YAC-FISH, Pachytene-FISH, ultrasensitive FISH, combined FISH with immunotechniques, extended DNA fiber, superstretched pachytene chromosome	*Rice*, localization of glutelin gene, gall midge resistance gene (*Gm2*) on chromosome 1, bacterial blight resistance locus *Xa21* and blast resistance gene *Pi-b* on chromosome 2, mapping of TrsA repeats, detection of retrotransposons (Ty-1 *copia*, Ty-3 *gypsy*, RIRE 1), mapping of ribosomal DNA; estimating the distance between TrsA and telomere sequences at a chromosomal end, chromosome addition, confirmation of telocentric nature of extra chromosome of telotrisomics; *wheat*: physical arrangement of retrotransposon-related repeats in centromeric regions, copy number, and integration patterns of transgenes in wheat lines obtained by biolistic bombardment, selection, and sorting of chromosomes, analysis of *Gc 2* (gametocidal) knockout mutation and mapping on *Gc2*-carrier chromosome T4B-4Ssh#1; *maize*: painting of all somatic chromosomes, localization of centromere in relation to genetically mapped markers, characterization of centromere-specific histone H3 variant, CENH3 and its association with the kinetochore protein CENPC, identification and distribution of CentC and CRM repetitive sequences and hypomethylation, immune detection of DNA methylation; *barley*: identification and polymorphism of *Hordeum chilense* chromosomes and detection of $BARE_1$ retrotransposon; *oat*: karyotyping of *Avena* species, phylogenomic analysis; new chromosome nomenclature from *Avena* monosomic line; *sorghum*: molecular cytogenetic map, *sh2* gene, liguleless linkage group in sorghum; *rye*: analysis of B chromosomes, telomeric heterochromatin, detection of rare translocation, activity of rRNA genes, identification of chromosome complement in hybrid	Kato et al. (2004), Kubaláková et al. (2005), Jiang and Gill (2006), Marín et al. (2008), Ohmido et al. (2010), Sanz et al. (2010), and Marques et al. (2012)

(continued)

Table 1 (continued)

Crop	Techniques	Achievements	References
Grain legumes (pea, chickpea, grasspea, beans, lentil)	Dual-color FISH, BAC-FISH,	Quantitative karyotyping, identification of NOR (nucleolus organizer region), mapping of 5S and 18S–25S rDNA probes, rDNA evolution, amplification of rDNA, detection of aneuploids (trisomics), analysis of B4 resistance (R) gene cluster in common bean and macrosynteny with *Medicago truncatula* and *Lotus japonicas* in chromosomes *Mt6* and *Lj2*, rDNA localization in chickpea, karyotyping of lentil chromosomes, phylogenetic relationship, location of chromosome telomeres, detection of NOR and satellite DNA sequence families in different *Lathyrus*, *Pisum*, and *Cicer* species	Ali et al. (2000), Balyan et al. (2002), Moscone et al. (2007), David et al. (2009), and Ceccarelli et al. (2010)
Model legumes (*Lotus japonicas*, *Medicago truncatula*)	FISH, BAC-FISH, multicolor FISH, fiber-FISH, RAPD/SSR-aided FISH	Cytogenomic mapping, chromosome characterization, detection of two-linked rDNA loci, comparative genomics with grain legumes like *Phaseolus* and *Arachis*, genome alignment and anchoring with *Arabidopsis* and other nonleguminous crops	Ohmido et al. (2010)
Oil-yielding (Brassica, soybean, peanut, sunflower)	DNA combing-FISH, BAC-BIBAC-FISH, BAC-SSR-FISH	*Brassica*: a 76-kb fragment in a P1-derived artificial chromosome (PAC) clone containing the *SLG* and *SRK* (self-incompatibility locus) genes was used to directly visualize the *S* locus. Using DNA combing and FISH, the positions of the fluorescent signals of *SLG* and *SRK* on the clone are found consistent with their positions on the restriction map; chromosomal rearrangement through homoeologous recombination in *B. napus*, detection of amphidiploids and chromosome triplication across Brassicaceae; *sunflower*: rRNA mapping, mapping of desired agronomic trait, detection of single copy sequence, construction of molecular cytogenetic map; *soybean*: molecular karyotyping, translocation mapping, detection of segmental duplication, validation of mutagenesis and TILLING; *peanut*: confirmation of *Arachis duranensis* and *A. ipaensis* as the wild diploid progenitors of *A. hypogaea*, evidence for origin of autotriploid *A. pintoi*	Seijo et al. (2004), Jiang and Gill (2006), Findley et al. (2011), Lavia et al. (2011), Talia et al. (2011), and Feng et al. (2013)

(continued)

Table 1 (continued)

Crop	Techniques	Achievements	References
Vegetables (potato, tomato, cucumber)	FISH, cross-species, BAC-FISH, multicolor FISH, pooled BAC-FISH, DAPI-FISH, SC-FISH, RAPD-FISH	Karyotyping, DAPI and rDNA mapping, genome sequencing, high-resolution chromosome mapping, detection of chromosomal rearrangement, assignment of genetic linkage map to pachytene chromosome, quantification of euchromatin and heterochromatin, localization of single or low-copy sequences on tomato chromosome, high-resolution karyotyping, mapping of 45SrRNA loci, construction of integrated molecular cytogenetic map in cucumber, synteny between *Cucumis sativus* and *C. melo* and comparative karyotyping with fosmid	Stack et al. (2009), Tang et al. (2008), Liu et al. (2010), and Han et al. (2011)
Sugar-yielding crop (sugarcane, sugar beet)	FISH, multicolor FISH	*Sugarcane*: Determination of basic chromosome number of *Saccharum officinarum*, *S. spontaneum*, origin of modern cultivars, extent of interspecific chromosomal recombination, genome remodeling, origin of new species and introgression; *sugar beet*: High-resolution mapping of YACs and the single-copy gene Hs1(pro-1) on *Beta vulgaris* chromosomes, construction of a reference FISH karyotype for chromosome, identification of chromosome arm, integration of linkage groups and analysis of major repeat family distribution in *B. vulgaris*	D'Hont (2005) and Paesold et al. (2012)
Herbs, fruits and spices	FISH, Fiber-FISH, BAC-FISH, Ultrasensitive FISH with tyramide signal amplification	Localization of single-copy T-DNA insertion in transgenic onion *Allium cepa* L. (Liliaceae), physical localization and measurement of 18S-5.8S-26S and 5S ribosomal RNA in black cumin (Ranunculaceae), *Trigonella foenum-graecum* L. (Fabaceae), sex chromosome differentiation in dioecious *Spinacia oleracea* L. (spinach), distinction between hybrids and non-hybrid accessions of mandarin and mango, molecular karyotyping, genome evolution and cytogenomics of saffron (*Crocus sativus* L.) (Iridaceae), "saffronomics"	Lan et al. (2006), Heslop-Harrison and Schwarzacher (2011), and Chahota et al. (2011)
Medicinal and aromatic plants	FISH and DAPI-banding, fiber-FISH	Karyotyping and chromosome location of rDNA in *Hyoscyamus niger* L. (Solanaceae), *Chlorophytum borivilianum* (Asparagaceae), *Podophyllum hexandrum* Roxb. ex Kunth (Berberidaceae), *Plantago ovata* Forsk., and its wild allies (Plantaginaceae), *Asparagus*, *Achyrocline* (Asteraceae, tribe Gnaphalieae) species, *Coccinia grandis* L. (Cucurbitaceae), *Papaver somniferum* (Papaveraceae) and *Hyoscyamus niger* (Solanaceae), *Artemisia absinthium* (Asteraceae), genomic relationship through rDNA mapping in *Panax* spp. (Araliaceae)	Lavania et al. (2005, 2010), Dhar et al. (2006), Choi et al. (2009), Nag and Rajkumar (2011), and Sousa et al. (2013)

(continued)

Table 1 (continued)

Crop	Techniques	Achievements	References
Horticultural crops	FISH, Fiber-FISH, multicolor FISH, pachytene-FISH	Variation and mapping of rRNA gene in *Lycoris* spp. (Amaryllidaceae), *Paphiopedilum and Dendrobium* (Orchidaceae), mapping of 5S and 45S rRNA and species relationship in *Tagetes*, *Aster*, *Chrysanthemum* (Asteraceae), karyotyping using centromeric repeat in *Antirrhinum majus* (Scrophulariaceae), *Hibiscus* (Malvaceae), *Rosa* spp. (Rosaceae)	Zhang et al. (2005) and Begum et al. (2009)
Fiber crops (cotton, jute)	Meiotic-FSH, BAC-micro satellite-FISH, multicolor FISH	*Cotton*: detection of new rRNA gene location, physical mapping, assignment of six linkage groups to chromosomes 8, 11, 13, 19, 21, and 24 by translocation and BAC-FISH; *Jute*: Chromosome discrimination, detection of major satellite repeats and retrotransposons from *Corchorus olitorius* and *C. capsularis*	Wang et al. (2007) and Begum et al. (2013)

remains a crucial method in plant cytogenetics for rapid assignment of cloned sequences to chromosomes and for ordering loci separated by at least several mega base pairs. Prometaphase chromosomes further improve axial resolving power, whereas interphase-FISH (I-FISH) provides a reported resolution of ~50–100 kb.

Compared to mitotic metaphase chromosomes, pachytene (meiotic) chromosomes may be better substrates for FISH. Instead of two nearby copies of each locus available for FISH on a metaphase chromosome, there are four closely associated copies of each locus available for hybridization on a meiotic bivalent. In spreads of pachytene chromosomes that have been prepared to reveal synaptonemal complex (SC spreads), chromatin extends as a diffuse cloud around each SC, and spreads can be prepared relatively free of overlying debris. The loops of DNA extending from the SC appear to be more accessible to FISH probes than the DNA of condensed metaphase chromosomes. Furthermore, pachytene chromosomes are 5–15 times longer than corresponding metaphase chromosomes, so closely associated loci are more easily resolvable on pachytene chromosomes than that on metaphase chromosomes. The SC-FISH is useful in high-resolution localization of two single-copy sequences and one low-copy sequence on tomato SC *11* (chromosome 11) and suggested that SC-FISH can be used to construct comprehensive maps of single-copy sequences on pachytene chromosomes (Stack et al. 2009).

BAC-FISH

The large genome size and high amount of repetitive DNA have made FISH mapping of single-copy DNA in many crops such as wheat difficult. An alternative approach is the use of large-insert vectors such as bacterial artificial chromosomes (BACs) as FISH probes (BAC-FISH) producing stronger FISH signals. BAC-FISH can facilitate integration of molecular marker-based genetic maps with physical maps, because the technique can readily validate (or dispute) chromosome models, which can be weak in domains high in genomic repeat or in regions with a low density of molecular markers (Findley et al. 2011). BAC-FISH has been successfully used to reveal karyotype change, genome dynamism, phylogenetic relationship, and evolution in major crops (Table 1). In sorghum, BAC clones containing molecular markers mapped to each linkage group were hybridized to *S. bicolor* chromosomes, producing a FISH-based karyotyping and nomenclature system for all ten chromosomes. The strong bimodality of a repetitive sequence and differential FISH signals in pericentromeric regions suggest underlying allotetraploid architecture and occurrence of its remnants in modern sorghum sub-genomes (Jiang and Gill 2006). In sunflower, a set of linkage group-specific bacterial/binary bacterial artificial chromosome (BAC/BIBAC) clones has been used as probes in BAC-FISH to encompass 17 linkage groups, providing valuable tool for identifying sunflower cytogenetic stocks (such as trisomics), tracking alien chromosomes in interspecific crosses and development of molecular cytogenetic resources (Feng et al. 2013; Talia et al. 2011). Using BAC-FISH, the translocation between a (>17.9 Mb) segment of chromosome 13 and a ~4.2 Mb segment of chromosome 11 of *Glycine soja* (a wild relative of cultivated soybean) and six new translocation lines from *G. soja*, *G. max*, and *G. gracilis* are characterized, for which differential chromosome painting with a cocktail of fluorophore-tagged oligonucleotides has been developed (Findley et al. 2011). Besides transgenomic FISH and pooled BAC-PCR methods, introduction of multi-BAC-FISH probe cocktails has markedly facilitated the chromosome- and segment-specific "paints" for analysis of chromosomal structure without the cumbersome difficulties of microdissection, flow-sorting, and DOP-PCR (Feng et al. 2013). Cross-species BAC-FISH painting is useful to reveal undescribed chromosomal rearrangement in potato and tomato. In narrow-leafed lupin (*Lupinus angustifolius* L.), an economically important legume, development of chromosome-specific cytogenomic marker and assignment of the first genetic linkage groups (LGs) to its chromosomal maps are successfully accomplished using the BAC-FISH approach. Based on BAC-end sequences of clones (providing single-locus signals), genetic markers were generated. Eight clones localized on three chromosomes, allowed these chromosomes to be assigned to three linkage groups (LGs), providing a solid foundation for future identification of all chromosomes with specific markers and for complete integration of narrow-leafed lupin LGs (Lesniewska et al. 2011).

BAC Landing-FISH

BAC landing (marker-assisted BACs) is an integral part of comparative genomics and assays of colinearity between *Arabidopsis thaliana* and Brassicaceae species and has been used to "paint" chromosome arms in *Arabidopsis* (Lysak et al. 2003). It is proved to be highly reliable in revealing colinearity between genomes of model grass *Brachypodium distachyon* and that of rice, wheat, barley, and temperate grasses, determining the pattern of divergence of the genomes of related cereals and grasses, the reconstruction of the archetypal grass genome, and the assembly of chromosome "paints" in this species for molecular cytogenetic investigations of chromosome-specific structure and function (Jenkins and Hasterok 2007). The BAC landing is highly effective to develop rapidly tiles of clones syntenic to important regions of much larger Gramineae genomes (Jenkins and Hasterok 2007).

Fiber-FISH

The use of extended DNA fibers as targets for FISH (EDF-FISH) is a powerful cytogenomic tool used to analyze large repetitive regions and has greatly improved the resolution of the FISH technique to about 2.94 kb/μm, which is the range of the Watson–Crick double helix (Lavania et al. 2003; Jiang and Gill 2006). It can be used to gauge the distances between adjacent clones up to ~500 kb and to measure repetitive loci up to ~1.7 mb. Combined with metaphase and interphase nuclei analysis, this tool helps to map loci to specific chromosomes and to determine the distance between loci from a few kb up to several mb. The EDF-FISH has also been used to characterize complex genomic arrangements in plant nuclei or plastids. Besides, Fiber-FISH is applicable with BAC and circular molecules as targets.

Fiber-FISH-based cytogenomic approach was unique in revealing previously undocumented very rare events in organelle genomes of higher plants that cannot be detected by traditional techniques such as DNA gel blot hybridization or polymerase chain reaction. The remarkable flexibility of this approach offers several advantages over earlier techniques employed to decipher chloroplast (cp) DNA, such as (1) reliable analysis of nearly all of the molecules released from a lysed chloroplast, (2) individual intact molecules can be analyzed in a relatively short time, and data from a large number of molecules allow quantification of the percentages of each type of structure in the population of cpDNA molecules, and (3) DNA fragments from different parts of the cpDNA can be labeled and mapped (Lilly et al. 2001). In pea, cucumber, *Arabidopsis*, and tobacco, this technique revealed more structural plasticity of higher plant cpDNA than previously believed and determined the following points (Lilly et al. 2001):

1. About 25–45 % of the cpDNA within developing leaf tissue consists of circular molecules.
2. Pea exhibited fewer circular molecules (25 %) compared with tobacco and *Arabidopsis* but a correspondingly higher percentage (36 %) of linear fibers.
3. The cpDNA from pea showed only one copy of the inverted repeats (IR).
4. Both linear and circular DNA fibers with one to four copies of the chloroplast genome were present, with monomers being the predominant structure.
5. Occurrence of recombination events between the IRs, and random cleavage, resulting in multimeric and aberrantly sized molecules in organelle genomes.
6. Rearranged cpDNA molecules of incomplete genome equivalents.
7. *Arabidopsis* and tobacco chloroplasts contained previously unidentified multimers 900 kb consisting of six to 10 genome equivalents.

Combined FISH

Using combination of molecular methods and chromatin cytology with advanced chromosome preparations and high-resolution imaging apparatus, new insights and models for understanding chromosome organization are being achieved at multiple scales. The power of FISH is strengthened further in combination with techniques like chromatin immunoprecipitation, immunocytochemistry, immunostaining, and pachytene chromosome superstretching, which are used successfully to resolve organization of centromere and DNA methylation pattern in cereals, millets, legumes, and vegetables (Table 1). In rice, pachytene-FISH allows the integration of genetic linkage maps and quantitative chromosome maps. Visualization methods using FISH can reveal the spatial organization of the centromere, heterochromatin/euchromatin, and the terminal structures of rice chromosomes. Furthermore, extended chromatin fiber-FISH and the DNA combing technique can resolve a spatial distance of 1 kb between adjacent DNA sequences, and the detection of even a 300-bp target is now feasible. DNA combing is superior technique for high-resolution measurements of repetitive sequences in plants. Digitally measured distances can also be transformed into kilobases of DNA using the length of a BAC clone of known length along with the length of a standard. The lengths of plant DNA fragments as small as 2 kb have been directly measured on circular BAC molecules using this method.

The combined FISH has also been used successfully in deciphering nuclear architecture and dynamism. Although it is generally believed that chromatin is intertwined and randomly distributed within the space available in the nucleus, recent evidence has demonstrated that the nucleus is a highly compartmentalized structure. The chromatin within the nucleus is organized in the form of chromosome territories (CTs) and interchromatin compartments (IC) (Gupta 2006). While CTs represent individual chromosomes, IC contains macro molecular complexes which are needed for replication, transcription, splicing, and repair. A combination of 3-D FISH and computer-aided deconvolution techniques revealed the following features of chromatin organization and behavior in higher plants (Gupta 2006):

1. Each individual chromosome occupies a discrete space, called the "CT," and that there is little intertwining among chromosomes in an interphase cell,

2. In interphase cells, each chromosome interacts with the nuclear envelope through consistent contact points.
3. Each chromosome interacts with other chromosomes through its heterochromatic regions during interphase.
4. In dividing cells, chromosome movements are nonrandom.

Comparative FISH Mapping

DNA clones from one species can be used as probes for FISH mapping in a related species. Such comparative mapping has several advantages over conventional genetic linkage mapping, i.e., (1) a mapping population is not necessary; (2) it does not require polymorphism, hence any clones from one species can be utilized as a FISH probe provided that they can generate signals in another species; and (3) some evolutionary rearrangements, such as duplication, can be readily detected (Jiang and Gill 2006). High-resolution and ultrasensitive FISH is a powerful tool for comparative genomics, as beautifully demonstrated for members of the Brassicaceae, Solanaceae, and Poaceae. The accessibility of web-based chromosome homology map can facilitate understanding of comparative biology of crop plants (Nagarajan et al. 2008). For example, the liguleless linkage group of *Sorghum bicolor* was physically mapped using rice RFLP-selected sorghum BACs (Kim et al. 2005). A transgenomic sorghum BAC-FISH for maize pachytene chromosome 9 revealed genome hyperexpansion (Kim et al. 2005). In sorghum, comparative cytogenomics are expedited using multi-BAC-FISH to elucidate species relationship between wild and cultivated genotypes. Cross-species BAC-FISH painting of potato and tomato chromosome 6 revealed chromatin structures, resolved physical mapping, and undescribed genomic rearrangement (Tang et al. 2008). Role of genome duplication in expansion of the *Brassica rapa* genome was determined by comparative BAC-FISH of *Arabidopsis thaliana* (Jiang and Gill 2006). The genetic, physical, diversity, and cytomolecular maps of grasses and grains have been integrated through FISH-guided comparative cytogenomics, using the sorghum genome as basis (Kato et al. 2004). Fiber-FISH confirmed that *Arabidopsis thaliana* and *Brassica rapa* divergence was associated with chromosomal duplications. In addition, comparative chromosome painting with pooled BAC probes was used to investigate ancestral relationships among species that diverged within the Brassicaceae (Lysak et al. 2003, 2006). Soybean-based FISH tools, particularly BACs, due to their potential for significant cross-species hybridization, may also be useful in comparative studies of related *Glycine* species (e.g., wild perennial *Glycine* species) (Findley et al. 2011). Furthermore, in the analysis of fast neutron mutagenesis and TILLING (Targeting Induced Local Lesions in Genomes) populations, FISH can validate deletions and trace the fate (integrity and locations) of duplicated sequences.

FISH on Flow-Sorted Chromosomes

Chromosome flow-sorting allows identification and isolation of individual chromosome types. Flow-sorted chromosomes have been used to construct chromosome-specific libraries in plants. However, localization of genes on flow-sorted chromosomes has been accomplished in field bean using chromosome-specific PCR and in maize with FISH mapping genes on maize-sorted chromosomes (Lijia et al. 2006). To overcome the problem of fixation of sorted chromosomes which is a prerequisite for application of in situ hybridization, FISH can be applied directly on sorted chromosomes. The protocol involves flow cytometric sorting of metaphase chromosomes, then fixing them with 4 % paraformaldehyde solution, and re-sorting these chromosomes directly onto a spot on polylysine-coated slides after stained. Sorted chromosomes are advantageous over metaphase chromosomes as targets for FISH mapping studies because a large number of target chromosomes with better chromosome morphology on a small area on the slide are easy to gain by flow-sorting, background is very clear, and hybridization sensitivity is enhanced, as beautifully observed during mapping of 45S and 5S ribosomal DNA in maize, barley, and oat (Lijia et al. 2006), analysis of the intravarietal polymorphism in genomic distribution of GAA clusters in wheat (Kubaláková et al. 2005), and identification of a rare translocation between A and B chromosomes in rye (Kubaláková et al. 2005). A further advantage of using flow-sorted chromosomes for FISH is a possibility to stretch them longitudinally up to a hundredfold compared with untreated chromosomes, making them suitable for high-resolution mapping. This approach is especially attractive for plant species with large genomes as an alternative to FISH on pachytene chromosomes, which are difficult to trace individually (Valárik et al. 2004).

Bar-code FISH

Recently a novel method for high-resolution FISH, using superstretched mitotic chromosomes, was presented, which provided a unique system for controlling stretching degree of mitotic chromosomes and high-resolution bar-code FISH (Valárik et al. 2004).

Primed In Situ (PRINS) DNA Labeling

The primed in situ (PRINS) DNA labeling is an alternative to FISH for the detection of repetitive and low-copy sequences on plant chromosomes. This technique involves labeling of chromosomes by annealing an oligonucleotide DNA primer to the denatured DNA of chromosomes spread on slide glass and extending it enzymatically in situ with incorporation of labeled nucleotides. In plants, the C-PRINS technique has been used to rapid identification and determination of flow-sorted plant chromosomes (Kubaláková et al. 2005), as demonstrated in detection of single-copy sequences in sunflower (Talia et al. 2011).

Chromosome Microdissection and Microcloning

Chromosome microdissection is an advanced technology in cytogenomic research. Unlimited copies of DNA fragments from the isolated chromosome can be obtained, which can be used as probes for chromosome painting and can also be cloned to generate a chromosome-specific DNA library. This would be useful for positional cloning of genes located in the chromosome and for producing mini-DNA libraries of specific single chromosomes or chromosomal segments. Construction of chromosome-specific libraries is a potential strategy for the construction of high-density genetic linkage maps of individual chromosomes and the comprehensive analysis of genomes. Usually, target chromosome must be identified by standard karyotype, followed by its isolation with fine glass microneedles controlled by a micromanipulator. DNA fragments ranging from 0.3 kb to 2 kb are acquired from the isolated single target chromosome via two rounds of PCR mediated by *Sau*3A linker adaptors and then cloned into T-easy vectors to generate a DNA library of that chromosome (Huang et al. 2004). Till date, chromosome-specific libraries have been successfully created in plant species such as wheat, maize, barley, oat, rye, wild beet, and fruit tree *Citrus* (Huang et al. 2004). A few clones from these libraries are utilizable for probing RFLP and tagging important genes. Using a PCR approach based on the DNA of microdissected metaphase chromosomes, STS derivatives of RFLP markers, genetically mapped in oat (*Avena* spp.) linkage maps, have been physically assigned to chromosomes 2, 3, and 7 of diploid oat *Avena strigosa* ($2n=14$). Based on either two or four RFLP-derived STS markers, these three *A. strigosa* chromosomes were found to be homoeologous to the oat linkage groups C, E, and F, respectively (Sanz et al. 2010).

Chromosomal Rearrangement and Genome Synteny

Analysis of the nature of the rearrangements using whole genome sequence comparisons is enabling the history of genome evolution to be reconstructed with unprecedented accuracy. For plant breeders, knowledge of the nature of the changes is important to chalk strategies and candidate accessions for crossing programs. An association between SSR-rich chromosome regions and intergenomic translocation breakpoints is revealed in natural populations of allopolyploid wild wheat (Gupta et al. 2008). In *Cephalanthera* (Orchidaceae), complex rearrangements are involved in chromosome evolution as deduced by analysis of rearranged genomes (Moscone et al. 2007). Similar concept successfully ascertained chromosomal phylogeny and karyotype evolution in Brassicaceae with $x=7$, determined simple and direct macrosyntenic relationship between faba bean and *Medicago truncatula*, revealed occurrence of a common chromosomal rearrangement relative to *M. truncatula* in faba bean and lentil, revealed different levels of conservation in model legume *Medicago truncatula* chromosomes and confirmed phylogenetic relationships, and determined patterns of chromosomal evolution and syntenic relationships among species of Leguminosae, Brassicaceae, and Poaceae (Kato et al. 2004; Mudge et al. 2005; Phan et al. 2006; Lysak et al. 2006). Despite some successes, this technique has been less used in plants, presumably because of the more rapid homogenization of DNA sequences from retrotransposons, so probes from large amounts of DNA become genome specific rather than chromosome- or linkage-group specific. Recent advances in large-insert (BAC or fosmid) hybridization suggest it will be increasingly used to address chromosome evolution (Lysak et al. 2006) and physical linkage mapping of sequences (Han et al. 2011).

Mutagenomics: Linking Classical Mutation Breeding with Modern Plant Genomics

Mutagenesis is a fundamental approach in plant biology to identify gene function, the concept of which is being extensively utilized in modern genomic era (Henikoff et al. 2004; Varshney et al. 2010). Successful development of functional biology tool in crop plants ensures efficient and applicable breeding population/methods. Various agronomic traits with desirable and tractable genetic variations can be developed either through classical or molecular methods or combinations of both.

The Natural Mutations as an Aid to Plant Functional Biology Tool: From Model Plant to Crops

Natural mutants are generated spontaneously during species evolution. A large collection of spontaneous mutants is still available during long evolutionary history and exhibit higher resistance to various abiotic/biotic stresses or have some specific agricultural traits, which are valuable germplasm resources for plant breeding. Natural mutant screens played an important role in the emergence of *Arabidopsis* as a model genetic organism (Koornneef and Meinke 2010). In *Arabidopsis*, cloning of monogenic disease resistance genes, which had a simple inheritance pattern, was successful in the early 1990s. Subsequently, genomic regions of interest for complex traits were identified by association of specific trait values with segregating molecular markers known as quantitative trait loci (QTLs). This eventually led to cloning of the underlying genes (quantitative trait genes, QTGs) through confirmation and validation of QTLs in near-isogenic lines (Nils) followed by fine-mapping and complementation (Alonso-Blanco et al. 2009). Research on natural variation has also led to the identification of functional alleles of genes such as major flowering gene, *FRIGIDA*, and several spontaneous frameshift mutations with biased GC→AT transition in *Arabidopsis* (Johanson et al. 2000). A number of spontaneous mutations affecting plant architecture, leaflet development, and nodulation process have been identified in prominent grain legumes like field pea, chickpea, and ground nut and used to develop erect-growing varieties in different countries. In chickpea and mung bean, several open-flower mutants exhibiting protruded stigma and crumpled petals in large number of flowers were spontaneously appeared and used in hybridization with improved cultivars to produce fertile pods (Sorajjapinum and Srinives 2011; Srinivasan and Gaur 2011). Spontaneous mutation has been discovered and utilized in domestication of narrow-leafed lupine (*Lupinus angustifolius* L.), converting it a suitable grain legume in Western Australia (Lesniewska et al. 2011).

Apart from being a valuable resource for analyzing gene function, natural variation provides an opportunity to study important features of evolutionary ecology and comparative biology at the molecular level. The plant with wide geographical distribution of accessions, coupled with a full toolbox of molecular resources, can be a suitable model for such studies (Weller et al. 2009). The considerable variation that exists among different members of the Brassicaceae is a valuable resource that remains to be exploited through comparative studies with Arabidopsis, as, for example, the analysis of heavy metal accumulation in *Arabidopsis halleri* and the control of flowering in the perennial species *Arabis alpina* (Vernoux et al. 2000). This approach has now been extended beyond the Brassicaceae with the comparisons of flowering time control between *Arabidopsis* and selected cereals (Greenup et al. 2009). Advanced genomic tools in *Arabidopsis*, model legumes, and cereals have also enabled the identification of variation that may underlie speciation events. The natural variation as exist from model plant to wild crop has therefore resulted in major advances of plant functional biology on multiple fronts (Greenup et al. 2009).

Induced Mutagenesis: Development of Mutant Genetic Stocks as Applicable Tool in Plant Functional Biology and Breeding

In the absence of desirable variability for a target trait within the gene pool, induced mutation is the ultimate source of new genetic variations. The goal in mutagenesis is to cause maximal genomic variation with a minimum decrease in viability. Induced mutagenesis in plants usually involves use of chemicals and/or ionizing radiations or biological agents (T-DNA, transposon, etc.). Chemical or physical mutagenesis can introduce random changes throughout the genome, creating a wide variety of mutations in all target genes, and a single plant can contain a large number of different mutations. While fast neutron bombardment and gamma rays result in deletion of DNA fragments of variable length from the genome (deletion mutagenesis), chemical mutagens in general and EMS (ethyl methanesulfonate) in particular can induce very high mutation frequency and trigger point mutation through base pair transitions and have gained popularity since they are easy to use and do not require any specialized equipment (Mba 2013). In model plant *Arabidopsis*, the saturation mutagenesis was an early option as shown by finding multiple mutant alleles of the same gene (Koornneef et al. 2003). Mutations at single nucleotide pairs are always valuable breeding tool. However, the mutagenic event that alters chromosome structure to increase the number of recombination events and breaks undesirable linkages is also significant in plant biology (Parry et al. 2009). The mutagenized populations form huge resources for effective utilization of desirable trait, a list of some of which is in Table 2, and have been subjected to both forward and reverse genetic screening (Parry et al. 2009).

Mutagenesis in Understanding of Plant Stress Tolerance

Mutational approach offers a powerful tool to study the genetic and molecular mechanisms protecting plants against diverse types of biotic and abiotic stresses (Table 2). Induced mutation is effectively utilized to incorporate the resistant gene(s) from the donor parent(s) through the alteration of susceptible alleles. Treatments for inducing mutations to improve yield or morphological traits often lead to improved tolerance to biotic and abiotic stresses, and these are therefore used as donors in the breeding for disease and insect pest resistance (kharkwal and Shu 2009). Pyramiding multiple genes responding to diverse stress factors is successful through classical mutagenesis which affects large parts of genomes, and after proper selection, this can be developed as basic platform to study breeding for stress tolerance. A brief list of successful development of mutagenic stocks in relation to stress tolerance has been given in Table 2.

Table 2 Mutagenesis in generating desirable agroeconomic traits and stress tolerance in model plants to crop species

Plants/crop	Mutagen used	Mutant traits identified	References
Arabidopsis	EMS, γ-rays, carbon ion, cadmium, fast neutron, spontaneous (UV induced) de novo, transposon	Transparent testa tt, glabrous (no trichome) leaf (*gl*), abnormal leaf morphology, flavonoid pathway mutants B, morphogenesis in cytokinin pathway, gibberellin sensitive, male sterility, vitamin C deficiency, stress-related *hos* 1, 2, and 5, freezing tolerant *eskimo 1*, freezing susceptible *sfr*, polygenic, VARICOSE-Related, *SLEEPY1 F-box*, 40S ribosomal protein S3, phosphoglucomutase, and noncoding regions, cuticle biogenesis (*LACERATA*, *FIDDLEHEAD*, *BODYGUARD*), *ebi1* (circadian clock mutant), *MIR390a precursor processing-defective mutants*, conditional meiosis mutant *radially swollen 4* (block chromosome disjunction, loss of separase function, excessive level of cyclin B1;1, disrupt radial microtubule and movement of cohesion complex), cytosine methylation, thymine dimerization, herbicide resistance	Koornneef et al. (2003), Voisin et al. (2009), Koornneef and Meinke (2010), Uchida et al. (2011), and Yang et al. (2011)
Cereals (rice, wheat, maize, barley)/millets (pearl millet, sorghum, oat)	γ-rays, X-rays, NaN_3, EMS, NMU, N-ethyl nitroso urea, colchicine (0.1 %), transposon	Morphological, yield components, grain quality traits, low phytate, adaptation of aromatic rice in different climate, hormone (IAA, GA, ABA), biosynthesis, photoperiod sensitivity, stress responsiveness, tillering *dwarf3* mutant with enhanced leaf longevity, *semi-dwarf* (*sd1*) in rice, *starch branching enzyme (SBE II)* mutant in durum wheat, preharvest sprouting, drought tolerance in sorghum, DELLA domain-related *reduced height1* (*Rht1*) from wheat, *dwarf8* (*d8*) from maize, and *slender1* (*Sln1*) from barley, two semi-dwarf mutants *dwarf & irregular leaf* (*dil1*) in maize, enhanced β-glucan, dietary fiber and fodder value in oat; brown-midrib mutant with reduced lignin content in maize, pearl millet and sorghum for improved forage value	Ahloowalia et al. (2004), Qi et al. (2009), Jiang and Ramachandran (2010), Tomlekova (2010), Chakraborty and Paul (2012), Tiwari et al. (2012), and Mba (2013)
Grain (common bean, pea, chickpea, lentil, grass pea, mung bean, urdbean, faba bean, cow pea, pigeon pea) and underutilized legumes (adzuki bean, senna, lima bean, moth bean, jack bean, lupin, *Clitoria*, horse gram, winged bean)	γ-rays, X-rays, NaN_3, EMS, DES, NMU, NaN_3, ENU, γ-rays + EMS/NMU, fast neutron, hydrazine hydrate, site directed, in vitro, colchicine	Compact, determinate, brachytic, dwarf, erect, tall, multiple leaflets, leaflet shape and size, unifoliate, cock's comb raceme, open-flower (chasmogamous), strong lodging resistance, lobed pod, synchronous pod maturity, attractive testa color (yellow/white), bold seed, higher fertile branches, early maturing, male sterility, high forage values, non-shattering pod, top fruit bearing, high yield and seed protein, super-nodulation, weed competitiveness, disease (powdery mildew, YMV, *Fusarium* wilt, insects, leaf spots, ascochyta blight, mould, rust, pod borer, root-knot nematode, weevil, storage pest, aphids) resistant, herbicide tolerant, and tolerance to abiotic stresses (waterlogging, drought, salinity, nutritional deficiency, acidic soil, sodicity, heavy metal/metalloid toxicity)	Dixit et al. (2000), Barshile et al. (2009), Goyal et al. (2011), Kharkwal and Shu (2009), Kumar et al. (2010, 2012), Pereira and Leitão (2010), Talukdar et al. (2001, 2002), Hussain (2009), Talukdar (2009a, b, 2010c, 2011a, b, c, d, e, 2012a, b, c, e, 2013d, e), Talukdar and Biswas (2002, 2006), Talukdar and Talukdar (2003, 2013), Talukdar (2011g, 2013c), Tsyganov et al. (2013), Talukdar (2014a, b), and Tomlekova (2010)

(continued)

Table 2 (continued)

Plants/crop	Mutagen used	Mutant traits identified	References
Oil-seed legumes (Brassica, peanut, soybean, sunflower)	γ-rays, EMS, MMS, NaN_3, fast neutron, LASER treatment, transposon	Increased tolerance to drought, salinity, biotic stresses, enhanced oil quality (high in unsaturated fatty acid), morphological traits, response and biosynthesis of auxins, high protein accumulation, apetalous flower, dominant petal-closed flower mutation (cleistogamy) in *Bn-clg1A-1D* gene in *Brassica*, dwarf, erect, branching, leaf and floral traits, male sterility, high shelling out-turn, yield and yield components, early/uniform maturity, bold seed, high oil and minerals, enhanced seed quality and pod shattering in soybean: pod shattering, oil quality, disease resistance (mottled virus, nematodes) and tolerance to abiotic (salinity, acidic soil, waterlogging, drought) stress, high yield of sesame	Zou et al. (2003, 2006), Seijo et al. (2004), Pathan and Sleper (2008), Kharkwal and Shu (2009), and Frasch et al. (2011)
Vegetables (potato, tomato, cabbage, cauliflower, radish, lettuce, etc.), grain amaranth	γ-rays, EMS, DES, fast neutron, NEU, NMU, transposon	Salt tolerance, reduced tuber glycoalkaloid content, meiotic mutants, dominant *Ivy* leaf (shape) mutant in potato; morphological traits, hormone response, fertility restoration, male sterility, fruit ripening (non-ripening, never ripe), eIF4E mutant with resistance to potyviruses in tomato, longer shelf life (tomato, melon), improved starch quality (potato), and virus resistance (peppers, tomato); high yield, high starch, carotenoid content of storage roots, disease resistance in sweet potato; HEAD and SINGLE-LEAF, abiotic (freezing, drought, salt) and mosaic virus-resistant mutants, induced β-carotene synthesis in cauliflower; trait improvement in grain amaranth (high dietary fiber, minerals, flavonoids), radish, broccoli, capsicum (quinine reductase induction, secondary alkaloids), carrot (carotene content), lettuce (dwarf, early flowering, male sterility, downy mildew resistance)	Tomlekova (2010), Mou (2011), and Saito et al. (2011)
Sugar-yielding crops	EMS, γ-rays, in vitro mutagenesis, somaclonal variation	Germination percentage, number of tillers per plant, stripped cane height, millable weight per cane, stripped cane yield and sugar recovery, in *Saccharum* spp.	Jung (2004), Wang et al. (2007), and Begum et al. (2013)
Spices, herbs, fruits, medicinal plants	γ-rays, EMS, carbon-ion beam	Male sterility, floral development, season-dependent flower homeotic mutant, increased leaf yields in African *Solanum nigrum-related* species, low oxalate variant in gynoecious *Spinacia oleracea* (spinach), high alkaloid value in *Gloriosa*, fruit quality, waterlogging tolerance, salt tolerance in *Citrus*	Ahloowalia et al. (2004) and Tomlekova (2010)
Horticultural/ornamental/fiber crops	In vitro + radiation, colchicine, EMS, NaN_3	Development of carnation plants, Rosa, Chrysanthemum (salt tolerant), Gerbera (salt tolerant), Gladiolus (flower color, storage capacity of bulb), floral traits, morphogenesis, improved yield and fiber quality, seed protein content, photoperiod conversion in cotton	Ahloowalia et al. (2004) and Hossain et al. (2006)

Mutagenesis in Exploring Developmental Biology and Toward Plant Phenomics

Understandings of morphophysiology and developmental processes are prerequisites to develop effective and functional tools for plant biology and crop breeding. Two classic examples of using mutagenesis in exploring developmental biology of vegetative and reproductive organs of plant have come from model plant *Arabidopsis* in Brassicaceae and *Pisum sativum* (pea) in Fabaceae. While brief life cycle, shorter genome, high and effective self-pollination, and diverse distribution make *Arabidopsis* an ideal type for plant biology, pea has traditionally searched for morphogenetic analysis since the classic work of Mendel on inheritance of seven developmental mutations. To date, the molecular basis has been uncovered only for four Mendelian mutations out of six, and some of them are still of intense agricultural interest (Reid and Ross 2011). A number of valuable mutant stock has been developed in both crops through induced mutagenesis and are now being used to explore intrinsic genetic mechanism of plant architecture, hormonal response, photoperiod sensitivity, and, subsequently, genome mapping of desirable traits, an updated information of which is presented in Table 3, and a diagrammatic view of some pea leaf mutations is given in Fig. 1. These mutants were crossed with each other to get various mutant backgrounds in combinations, facilitating gene mapping and consequent manipulations of different mutations and their interactions in *Arabidopsis* leaf morphogenesis and pea compound leaf development (Mishra et al. 2009; Sinjushin 2011; Kumar et al. 2012). Two classes of photoperiod-responsive induced mutants—(a) early day-neutral mutants that behave under short-day (SD) conditions as if grown under long days (LD) and (b) late day-neutral mutants that behave under LD as if grown under SD—affecting early flowering day-neutral phenotypes in several loci such as *SN*, *DNE*, *PPD* (all recessive), a dominant hypermorphic *phy A* and *COP1* orthologue *LIP1*, and late day-neutral class (*LATE 1, 2,* and *6*) have been genetically characterized along with mutations governing high response (*HR*), rhythmic expression under light/dark cycles, and late flowering (*LF*) during flowering in pea. Besides pea, mutations affecting tendril development (tendril less, simple tendril, anomalous branching, compound), leaf rolling (inward, recurved), jugate arrangement (opposite, alternate, leafletless), floral architecture (open, malformed, extra sepals/keels, long pedicels, distichous pedicels), stem growth, and stipule formations were also isolated in gamma ray-induced progeny of grass pea, another grain, and forage legume and found to be recessive in nature with monogenic and polygenic inheritance (Talukdar 2009b). Several genes governing spikelet development and floral organ formation in cereals are found orthologous to ones of *Arabidopsis* ABC + D and E flower development system. For example, mutations in the *SILKY1* (*SI1*) from maize and *OsMADS16* or *SUPERWOMAN1* (*SPW1*) from rice, orthologues of Arabidopsis *AP3 (APETALA 3)* mutation, induce homeotic changes from stamen into carpels and lodicules into lemma/palea-type organs in cereals (Dwivedi et al. 2008). Similarly, null mutations in the *OsMADS1* or *leafy hull sterile 1* (*lhs1*) genes produce leafy lemma and palea. In addition, the lodicules and stamens are also modified into leafy lemma and palea structures in rice (Dwivedi et al. 2008).

Construction of genomic platforms to high-throughput processes permits the simultaneous generation of very reliable genotypic data from multiple samples and hence abates the problems of generating and evaluating large numbers of putative mutants in quest of invariably low-frequency events (Mba 2013). However, there must be reliable mechanisms for generating the complementary phenotypic data to lead to valid inferences on genotypes of the variants. The adoption of phenomics, as the "the acquisition of high-dimensional phenotypic data on an organism-wide scale" as a component of the detection and deployment of mutation events, holds immense promise in analyzing morphological, physiological, and biochemical data of robust mutant stocks available in model plant *Arabidopsis* and other crops (Houle et al. 2010). In line with *Arabidopsis*, preliminary attempts

Table 3 Functional mutations (non-transgenic) affecting developmental biology in model plant *Arabidopsis thaliana* L. and grain legume *Pisum sativum* L. (pea)

Plant	Type	Mutant traits/genes	Mutant characteristics	References
Arabidopsis thaliana L.	Stem and plant architecture	Stem fasciation, *CLAVATA2, LIKE HETEROCHROMATIN PROTEIN 1 (LHP 1), lfy*	Stem and floral meristems affected in fasciation, plant architecture, leaf development and flowering time (*LHP 1*); meristem activity	Koornneef and Meinke (2010)
	Leaf development	*BLADE-ON-PETIOLE* (*BOP1*, 2); *ASYMMETRIC LEAVES1; fugu1–fugu5; angustifolia3* (*an3*), *erecta* (*er*), *KIP-RELATED PROTEIN2* (*KRP2*); *ROTUNDIFOLIA4, compensating mutants; tonneau (ton), ARGONAUTE1(AGO1)*	Petiole ontogeny; induced enhanced postmitotic cell expansion (*fugu2-1*, *fugu5-1*, *an3-4*, *er 102*); *an* (leaf-cell expansion, the arrangement of cortical microtubules in leaf cells and expression of a gene involved in cell-wall formation); cell proliferation, expansion and leaf shape, short petioles and rounded leaves (*rot*), impaired development by cyclin-dependent kinases, positive regulation of cell proliferation; planes of cell division are altered while the correct position of all the plant organs is maintained (*ton* or *tonneau*); absence of lateral expansion in leaves and floral organs (*ago1*)	Dwivedi et al. (2008) and Koornneef and Meinke (2010)
	Inflorescence	*terminal flower 1(tfl 1)*	Determinate inflorescence, reduced height, more rosette leaves	Johanson et al. (2000)
	Flower structure and development	Mutations in *ABC* model, *APETALA (AP)1, 2, 3, Ap 2-null, AGAMOUS (AG), PISTILLATA (PI), TERMINAL FLOWER 1, SEEDSTICK* (*STK*), *SEPALLATA* (*SEP*)	A-function is triggered by *AP1* and *AP2*, the B-function by *AP3*, *PI* and the C-function by *AG*; formation of reproductive organs in place of petals and sepals (*Ap 2-null*), petal formation instead of stamens, together with additional flower formation in place of carpels (*AG*, loss of function), ovule development (D-function gene, *STK*), triple mutants in *SEP1–3* produce only sepaloid flowers	Johanson et al. (2000) and Koornneef and Meinke (2010)
Pisum sativum L.	Stem form and inflorescence development	*Determinate* (*det*); *determinate habit* (*deh*); *le* and *sln*; *VEGETATIVE* (VEG *1*, *2*), *LATE BLOOMER* (*LATE 5*)	Determinate growth with terminal raceme, less axillary inflorescences (*det*); reduced scale-like stipules (*deh*); dwarf (*le*) and slender and elongated phenotype (*sln*) (gibberellin sensitive); secondary inflorescence development (*veg, late*)	Weller et al. (2009) and Belyakova and Sinjushin (2012)

(continued)

Table 3 (continued)

Plant	Type	Mutant traits/genes	Mutant characteristics	References
		fasciata (fa, fas, fa2)	Ridge-like anomalous enlargement of stem apical meristem, formation of flattened shoot with aberrant phyllotaxis and axillary racemes clustered on the top, flower normal, high yield, lodging prone, axillary raceme becomes terminated with anomalous flower in some fasciated plants	Sinjushin and Gostimskii (2006, 2007)
		Double mutant *det fa*	Weakly fasciated determinate stem ("lupinoid"), high breeding value	Sinjushin and Gostimskii (2007)
	Branching	*Ramosus (rms, 1–5), sax; grafting of rms/sax* mutant	Pattern of branch development, suppression of axillary meristem (*sax*); *rms/sax* revealed startling disconnection between major cytokinin content of xylem sap and shoot tissues of various *rms* mutants, i.e., pea shoots possess powerful homeostatic mechanisms of long-distance signaling for regulation of cytokinin levels during shoot branching	Foo et al. (2007), Mishra et al. (2009) and Kumar et al. (2012)
	Flower number/raceme	*FN* and *FNA*	Multiflowered phenotype similar to distichous pedicel mutant of grass pea	Belyakova and Sinjushin (2012)
	Compound leaf morphology and development	*UNIFOLIATA* (*UNI*), *AFILA; MULTI FOLIATE-PINNA; tendril-less* (*tl*); *tl2* and *insecatus* (*ins*); *COCHLEATA* (*COCH*); *CRISPA* (*CRI*); recombinant *af* uni^{tac}; *tendrilled acacia*-A (tac^{A}); *lld*	Simple leaf (*uni*); leaf with preliminary formation of terminal leaflet instead of distal leaf structures (uni^{tac}); ramified leaf rachis with terminal tendrils with seldom semi-leafless + phenotype (*af*); distal part of leaf in *mfp* plants produces secondary axes; leaflets instead of tendril; partial leaflet-to-tendril transformation (*tl2*, *ins*); pinnate and leaflike stipules, anomalous flower and inflorescence structure together with unusual proliferation of root nodules (*coch*); recombinant "chameleon" phenotype with strongly ramified rachis, leaflets on long petiolules and intermediate tendril-to-leaflet organs (*af* uni^{tac}); ramified rachis (as in *afila*), pinnate lobed leaflets and tendrils with lodging resistant and high breeding value; completely penetrant *leaflet development (lld)*	Mishra et al. (2009), Kumar et al. (2012), and Sharma et al. (2012)

(continued)

Table 3 (continued)

Plant	Type	Mutant traits/genes	Mutant characteristics	References
	Flower structure	Flower zygomorphy- *KEELED WINGS* (*K*), *LOBED STANDARD1* (*LST1*), *ELEPHANT EAR-LIKE LEAF1* and *2* (*ELE1*, *ELE2*)	Homeotic replacement of wings to keel petals (*k*), flag (standard) bears lateral notches in flowers (*lst*), bilaterally symmetrical wings and keel petals together with enlarged stupules (*ele*)	Sharma et al. (2012)
		Floral organ identity- *STAMINA PISTILLOIDA* (*STP*), *PETALOSUS* (*PE*), *PEAM4*, fasciation	Two adaxial stamens of outer whorl converted into carpelloid structures, while other stamens develop normally (multicarpellate) (*stp-1*), organ conversion and development	Sinyushin (2010), Sinjushin (2011), and Kumar et al. (2012)
		Pigmentation, gene *a*, *B*	Absence of anthocyanins in stems, seed coat, leaves, pods and corolla (*a*), flower pigmentation, pink (*b*) by defective flavonoid 3′,5′-hydroxylase	Kumar et al. (2012) and Sharma et al. (2012)
	Pod and seed traits	*Mutant p and v, rugosus* (*r*), *development of funiculus* (*def*), seed testa color,	Unlignified pod (*p, v*), with high forage value, no abscission layer on a boundary between funicle and seed hilum (*def*),	Belyakova and Sinjushin (2012), and Kumar et al. (2012)

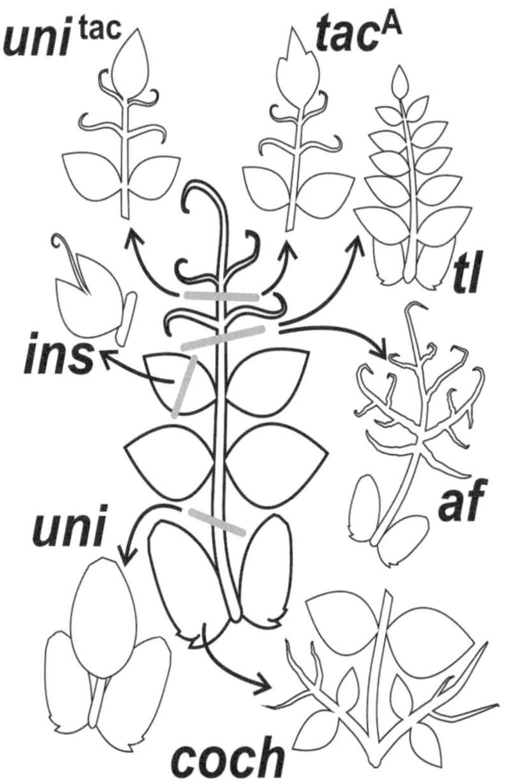

Fig. 1 Diagrammatic view of some prominent pea (*Pisum sativum* L.) leaf mutations

have been made to develop a standardized language of phenomics in model legume *Medicago truncatula* and *Lotus japonicus* using mutant phenotypes such as "late flowering" or "increased internodal distance." Precision in phenotypic descriptions will be critical to genome scale mutant hunts (Mba 2013).

Mutagenesis in Bioresource Development and Functional Food Biology

Instead of costly and unpredictable transgenic-based molecular farming, induced mutagenic technique can be successfully utilized to generate active constituents, antioxidant compounds, carotenoids and flavonoids, insecticide, antifungal and other biocontrol molecules, biomass production, weed-inhibiting allelochemicals, and plant-based industrial raw materials (Mba 2013). Worldwide, several mutant-based bioresource development centers have been established for effective utilization and management of plant resources (Mba 2013).

Of the nearly 3,000 mutant varieties developed globally in different crops, 776 mutants have been induced for nutritional quality (Jain and Suprasanna 2011). Biofortification is a sustainable method of naturally enriching legumes by conventional breeding and modern biotechnology to increase nutritional quality to combat malnutrition in the form of "hidden hunger" (ICARDA-HarvestPlus 2010). In cereals, mutants exhibiting improved protein content and quality with enhanced lysine, easily digestible carbohydrate, and vitreous grains (*floury-2*, *mucronate*, *defective endosperm B30*, *sugary-2 quality protein maize)* have been isolated (Chakraborty and Paul 2012). Chemically induced, nonlethal recessive mutants that decrease seed phytic acid content have been isolated and mapped in maize, rice, wheat, and barley. Low phytic acid crops may improve cooking quality, milling by-product, and nutrition for human population and animal feed that depend upon grains and legumes as staple foods. Several oat (*Avena*) mutants producing heart-healthy high β-glucan and dietary fiber and low glycemic carbohydrate have been isolated (Mba 2013). Among the cultivated crops, grain legumes are nutritionally rich in plant proteins, minerals, fibers, and antioxidant flavonoids but deficient in methionine and cysteine, two important sulfur-containing amino acids (Singh 2003; Talukdar 2012g). Induced mutations for enhancing nutritional quality (high protein and minerals, balanced carbohydrate, low trypsin inhibitor, lectin, high antioxidant capacity, phosphorus, low phytic acid) through genetic biofortification of edible cereals, millets, oil seeds, and crop legumes have generated valuable breeding tools (Piotrowicz-Cieślak et al. 2008; Smulikowska et al. 2008; Talukdar 2009b; Gaikwad and Kothekar 2011). In soybean and peanut, mutations with high oil, protein, methionine, isoflavones, lutein, enhanced oleic (O) acid (*FAD2-1A and FAD2-1B*), and low linolenic (L) (high O/L ratio) acid, without lipoxygenase and low allergenicity (peanut), have been isolated. Mutant lines with a methionine-overproducing phenotype in soybean (Pathan and Sleper 2008) and grass pea (Kumar et al. 2010), iron hyperaccumulation in pea, and high phosphorus in soybean, mung bean, and common beans (Campion et al. 2009; Porch et al. 2009) were isolated and have the potential to be used as parents in hybridization. Metabolic profiling, a useful tool in plant functional genomics, of the wild-type soybeans Taiwan75 and Zhechun No. 3 and the two corresponding *lpa* (*low phytic acid*) mutants *Gm-lpa-TW75-1* and *Gm-lpa-ZC-2* identified significant differences between the wild types and the mutants for the trait (Pathan and Sleper 2008). In grass pea, mutant and segregants developed from mutant×check parent with significantly low (<0.2 %) seed neurotoxin (β-ODAP) and high seed protein, good amount of amino acids L-homoarginine, methionine, and cysteine, and fiber and mineral content have been isolated in gamma ray/EMS irradiated progeny (Smulikowska et al. 2008; Talukdar 2009b). Beneficial oligosaccharide content has been increased, while levels of flatulence-producing raffinose family oligosaccharides (RFO) have been lowered in seeds of grass pea mutants, separately using helium-neon laser light, sodium azide (NaN_3) and N-nitroso-N-methylurea (NMU) as mutagenic agents (Piotrowicz-Cieślak et al. 2008). A recent work by Rao (2010) suggested that the presence of homoarginine in grass pea contributes to a sustained generation of nitric oxide in animal physiology which is highly beneficial in cardiovascular physiology and general well-being. In a major paradigm shift from its usual negative role, the possible therapeutic potentials of multifunctional metabolite β-ODAP (the grass pea neurotoxin, β-N-oxalyl-L-α,β-diaminopropionic acid) in treating Alzheimer's disease, hypoxia, and long-term potentiation of neurons essential for memory through the activation of protein kinase C have been explored (Rao 2010). Mutations producing higher unsaturated fatty acids compared to unhealthy saturated fatty acid have been isolated in *Brassica*, sunflower, safflower, and sesame (Table 2).

Besides food and fodder values, major and underutilized legumes, cereals, spices, and herbs exhibited remarkably high antioxidant activities, flavonoid compositions, and type II diabetes-related enzyme inhibition properties with low glycemic index (slow digestion of carbohydrate) in raw and differentially processed forms (Talukdar 2012f, 2013a, b; Talukdar and Talukdar 2012; Varaprasad et al. 2011). The polyploids have the capacity to generate antioxidant compounds in increased amount and activity (Lavania et al. 2010). Development of the commonly used plants as functional and therapeutic foods needs successful breeding of these value-added traits through utilizable genetic variations which can be achieved through modern mutagenic techniques.

Plant Molecular Mutagenesis

During the past decade, with the unprecedented development in plant molecular genetics and functional genomics, scientific exploration on induced mutation in plants has progressed dramatically from basic research to the development of advanced genomic-based technologies to their unique applications in gene discovery and development of novel crop traits (Kharkwal and Shu 2009; Varshney et al. 2010). Induced mutants are now being used in identifying and ascribing functions to genes through the deductive process of identifying the modified traits and relating the modifications to changes in genomic regions of the induced mutants, in comparison to the wild/normal types. The genomic region(s) responsible for the expression of a trait, i.e., the gene, is detected by analyzing a series of induced mutants vis-à-vis the normal or wild-type variants. These developments are bringing plant mutation breeding into a new dimension—plant molecular mutation breeding (Shu and Lagoda 2007).

Over the last several years, functionally characterized genes, ESTs, and coding genome sequences have been made available to build up molecular markers like SNP (single nucleotide polymorphism), SSR, or COS (conserved orthologous set) (Varshney et al. 2010). These markers are often called perfect or functional markers and are developed from putative coding sequences having known function and consequently have complete association with the QTL or gene. These functional genomic resources are boosting up development of perfect markers in cereals, millets, pulses, herbs, spices, fruit crops, and vegetables (Shu and Lagoda 2007; Varshney et al. 2010). The progress made in using marker-assisted selection (MAS) in tomato, cereals, and pulses has been highlighted in a few recent reviews emphasizing on mapping genes controlling agronomically important traits and molecular breeding of crops in general (Varshney et al. 2010). Several molecular markers such as diversity arrays technology (DArT), amplified fragment length polymorphism (AFLP), random amplified polymorphic DNA (RAPD), and microsatellites and SSR markers have been validated in major cereals, vegetables, and pulses (Varshney et al. 2010). In chickpea, a major grain legume, SSR markers are effectively used to reveal high genetic diversity among the radiation-induced 21 mutant lines and to discriminate the mutants from each other (Khan et al. 2010). DNA markers such as RAPD, SCAR (sequence-cleaved amplified region), and microsatellite linked to induced mutant/*er* (*Erysiphe pisi* Syd.) loci conferring resistance to powdery mildew in pea have been identified as reliable breeding materials (Pereira and Leitão 2010). In soybean, two male-sterile mutants were mapped with molecular markers (Frasch et al. 2011), and marker-assisted backcrossing is attempted to introgress the low phytate traits into cultivars (Kharkwal and Shu 2009). In rice, wheat, and maize, segregating populations are being used to confirm co-segregation between SSR markers and the gene for low phytic acid, identified in mutants. Several mutant traits controlling thermosensitive genic male sterility, fertility restoration, spikelet architecture, and flowering process were mapped in cereal genomes using molecular markers like microsatellites and STS (sequence-tagged sites) (Dwivedi et al. 2008). Both forward and reverse molecular marker (SSR)-assisted selection are effectively utilized in maize for breeding with high protein and good digestibility such as "quality protein maize" associated with opaque-2 (high lysine) and endosperm modification phenotype. In the same crop, two independent, semi-dwarf maize EMS mutants, designated as *dwarf and irregular leaf* (*dil1*), affecting internode length and leaf size were mapped on the long arm of chromosome 6 with SNP, and CAPS (cleaved amplified polymorphic sequence) markers and a candidate gene are identified through positional cloning and validated. In an allopolyploid *Brassica napus*, EMS mutagenesis caused a dominant point mutation in RINGv E3 ubiquitin ligase homoeologous gene by C-to-T transition and induced cleistogamy by the *Bn-clg1A-1D* mutant allele which has been mapped with SNP, STS, AFLP, ACGMs (amplified consensus gene markers), and CAPS markers. Induced mutations facilitated the isolation of genes involved in cascade of nitrogen fixation like controlling autoregulation of nodulation (leucine-rich repeat receptor kinase gene or *GmNARK*), LjHAR1, kinase-associated protein phosphatase (KAPP), several receptor kinases, GmNORK (needed for both nodulation and mycorrhizal symbioses), GmKAPP (encoding kinase-associated protein phosphatase), and GmPOL (poltergeist; another protein phosphatase) in mutants of soybean, pea, and *Lotus japonicas* and revealed complex interplay during nodulation process and hormonal signaling in legumes (Pathan and Sleper 2008; Frasch et al. 2011).

Development of radiation-hybrid (RH) technology is another tool in advancement of molecular mutation breeding, effectively utilizing in cereals, grasses, and some legumes (Riera-Lizarazu et al. 2008). In oat–maize RH where oat plants are carrying radiation-induced subchromosome fragments of a given maize chromosome, the transmission and integrity of maize chromosome 9 rearrangements were evaluated by using a series of DNA-based markers and by GISH, exhibiting much higher transmission of rearrangement (47.6 %) than normal (9.1 %) version of chromosome 9 (Kynast and Riera-Lizarazu 2011). RH mapping of wheat by addition and substitution for chromosome 1D (DWRH-1D), harboring nuclear–cytoplasmic compatibility gene *scs*^{ae}, initially allowed detection of 88 radiation-induced breaks with 39 1D specific markers, and subsequently, it was expanded to a resolution of one break every 199 kb of DNA, utilizing 378 markers. Construction of RH map is being initiated in rye chromosome 3B (Gupta et al. 2008).

Intermingling Cytogenomics with Mutagenomics: Development of Cytogenetic and Molecular Tester Stocks and Genome Mapping

The recent progress in plant cytogenetics has been stimulated mainly by the use of cytogenetic stocks, in which various aneuploids like different trisomics, tetrasomics, double trisomics, monosomics, monosomics alien addition lines, alien addition lines, deletion lines, intergeneric chromosome addition (oat–maize), and chromosome substitutions play pivotal roles in assigning target traits (classical and molecular) on specific chromosomes, defining linkage groups (classical and molecular) and construction of saturated maps (Singh 2003; Kynast and Riera-Lizarazu 2011). A number of programs are exploiting the transfer of important disease resistance genes into rice, wheat, barley, maize, and millet crops through alien addition lines (Heslop-Harrison and Schwarzacher 2011). Induced mutagenesis has been successfully used to develop arrays of diploid and aneuploid mutants, autotetraploids, aneuploids in polyploid background, and reciprocal translocation lines in cereal crops, legumes, millets, grasses, oil seeds, and other plants. A robust stock of aneuploids like different trisomics, tetrasomics, double trisomics, and tertiary trisomics was identified in rice, wheat, maize, barley, potato, cabbage, and legumes (Román et al. 2004; Zou et al. 2003, 2006; Talukdar 2008, 2009c, 2010a, b, c, d, 2011b, 2013f; Talukdar and Biswas 2007, 2008). An updated information of cytogenetic stocks and their uses in assigning classical and molecular linkage groups in prominent crops are presented in Table 4. Several autotetraploids ($2n=4x=28$) exhibiting enhanced growth performances have been isolated through colchicine treatments in grass pea (Talukdar 2010b), a hardy legume crop. These autotetraploids are segregated into progenies of normal tetraploid, tetraploid carrying reciprocal translocation, and aneuploids. Among the aneuploids, pentasomy–trisomy ($2n=4x+1-1=28$), one single pentasomy ($2n=4x+1=29$), and another double trisomy ($2n=4x+1+1=30$) were recovered in polyploid background (Talukdar 2012d). In tomato, a robust mutant in silico databases designated as "TOMATOMA" comprising over 1,000 mutants isolated by EMS, and γ-rays has been developed for mutagenomic analysis, comparative biology, molecular mutation breeding and forward as well as reverse genomic screening (Saito et al. 2011). The cytogenetic stocks are being used to map and linkage analysis of specific morphological, biochemical, and molecular markers on specific chromosomes of various crop plants (Table 4).

Table 4 Cytogenetic and mutation genetic stocks used as functional biology tools in major crops

Crop	Stocks/resources	Trait/gene mapping/linkage analysis/gene expression/genomics	References
Cereals (rice, wheat, maize, barley)	Primary, secondary and tertiary trisomics, set of telotrisomics, isotetrasomics, aneuhaploids, monosomic alien addition lines (MAAL), translocation, B-A translocation, duplicate deficient (maize), polyploids, aneuploids from autotetraploids	*Rice*: localization of morphological markers like photoperiod sensitive, rolled fine strip, slender glume, lax panicle, liguleless, scented rice traits, etc. RFLP mapping, centromere mapping, isozyme alleles (diaphorase 1, esterase, phosphoglucoisomerase), assignment of DNA clones, chromosome microdissection and developing chromosome-specific DNA markers, locating genes to pericentromeric regions and genome-wide synteny between rice and wheat; *wheat*: microsatellite markers, radiation hybrid mapping and map-based cloning by irradiating pollen with 20 Gy gamma rays; *maize*: isozymes of acid phosphatase, β-glucosidase, alcohol dehydrogenase, phosphor hexo isomerase, phosphoglucomutase, 6-phosphogluconate dehydrogenase, malate dehydrogenase linked with increased grain yield, endopeptidase on chromosome 6; transcriptomics of global gene expression; *barley*: analysis of ribosomal RNA cistron multiplicity in chromosomes 6 and 7, acrotrisomic analysis of linkage group, low phytate in all four grains	Subrahmanyam and Azad (1978), Khush et al. (1984), Auger and Birchler (2002), Singh (2003), Qi et al. (2009), and Tiwari et al. (2012)
Millets, rye, buckwheat, pearl millet, oat	Nullisomics, monosomics, trisomics of different types, tetrasomies, translocations, inversions, duplications, deficiencies, polyploids	Meiosis pairing in rye, assignment of S (short style Ss, long style ss) gene of common buckwheat (*Fagopyrum esculentum* Moench) was located on chromosome 4E, chromosomal localization, dosage effect of peroxidase, hybrid seed production from balanced tertiary trisomics in pearl millet, transmission of extra chromosomes in foxtail millet (*Setaria italica* (L.) Beauv.), analysis of B genome of oat (*Avena barbata, A. strigosa*), new chromosome nomenclature by *A. sativa* and *A. byzantina* monosomic lines, mapping of 400 maize sequences including expressed sequence tags and sequence-tagged sites by oat–maize chromosome addition lines, analysis of genome rearrangement by oat–maize radiation hybrids	Jauhar Prem and Hanna (1998), Sybenga (1996), Okagaki et al. (2001), Singh (2003), Chen et al. (2007), and Sanz et al. (2010)

(continued)

Table 4 (continued)

Crop	Stocks/resources	Trait/gene mapping/linkage analysis/ gene expression/genomics	References
Legumes (pea, chickpea, grass pea, pigeon pea, lentil, faba bean, common bean, mung bean, urdbean)	Primary trisomics, tetrasomics, tertiary trisomics, double trisomics, autotetraploids, aneuploids in polyploidy background, reciprocal translocation lines	Morphological (dwarfism, leaflet shape and color, stipule forms, leaf injury, flower color, photoperiod sensitivity, seed coat color, pod indehiscence, bold seed size in lentil, grass pea, determinate growth, solid distribution of pigment on flower, yellow pigment on flower, hilum color, red seed coat in faba bean), stress responsive (flavonoid deficiency, glutathione overproducing, cadmium-sensitive mutations, leaf injury, catalase deficiency), isozymes (esterase and peroxidase, aconitase, acid phosphatase, aspartate aminotransferase, phosphogluconate dehydrogenase, fructokinase, malic enzyme, superoxide dismutase, n-nitrosoglutathione reductase) and molecular (RAPD, RFLP, SSR) markers in beans and peas, dosage effect of superoxide dismutase, ascorbate peroxidase, glutathione reductase, dehydroascorbate reductase and catalases in trisomic and tetrasomic genomes (grass pea)	Satovic et al. (1996), Singh (2003), Román et al. (2004), Talukdar and Biswas (2007, 2008), and Talukdar (2008, 2009a, b, c, 2010a, b, c, 2012d, f, 2013f)
Oil-yielding legume (soybean, peanut)	Primary, tertiary trisomics, MAAL, translocation lines	*Soybean*: chromosome assignment of *v2* (variegated leaf mutant), *p2* (puberulent), *w1* (flower color), *dia 1* (diaphorase), *u 1* (urease), *lx1* (lipoxygenase), *Rps1-k* (resistant to phytophthora root rot), *Rmd* (resistance to *Microsphaera diffusa*), *Rj2* (nodulation response), *y10* (yellow leaf mutant), 11 molecular linkage group using SSR markers, 3 qualitative trait loci (*Pb*, sharp/blunt pubescence, *Y9*, and *Y17*- green/chlorotic foliage) detection of segregation distortion of SSR markers	Palmer and Xu (2008), Cregan et al. (2001), Singh (2003), and Zou et al. (2003, 2006)
Oil-seed crops (sunflower, Brassica, safflower, soybean, peanut)	Primary, secondary trisomic, double trisomics, tetrasomics, monosomics, MAAL, hyper triploids, tetraploids, translocation	*Sunflower*: characterization of chromosomal complement, *Safflower*: detection of translocation homo and heterozygotes, *Soybean*: Mapping of SSR markers in soybean, location of *w1* locus on satellite chromosome, mapping of yellow leaf *y10* mutant (chromosome 3), *ms 1 ms1* locus (male sterile), confirmation of genetic linkage map of the Nucleolus Organizer Region, *peanut*: branching pattern, tetraploid analysis, genome arrangement	Xu et al. (2000), Singh (2003), and Yang and Jeong (2008)

(continued)

Table 4 (continued)

Crop	Stocks/resources	Trait/gene mapping/linkage analysis/ gene expression/genomics	References
Vegetables (cabbage, cucumber, tomato)	Primary trisomic, double trisomic, MAAL, monosomic, telotrisomics	Alien gene introgression, chromosomal assignment of molecular markers or dominant plant traits, molecular mapping of alien genes, the construction of chromosome-specific libraries, and production of disomic addition lines, isozyme of esterase, peroxidase, linkage mapping with tomato telotrisomics	Chen et al. (2004) and Diao et al. (2009)
Sugar crops (sugarcane, sugar beet)	Primary trisomics, MAAL	Assignment of hypocotyls color, early bolting, restorer, RAPD, RFLP, SCAR (sequence characterized amplified region), STS (sequence tagged site), SSR markers in sugarcane; isozymes of leucine aminopeptidase and glutamate oxaloacetate transaminase in sugar beet	Oleo et al. (1993) and Jung (2004)
Forage crops	Primary trisomic	Isozymes of phosphoglucoisomerase and glutamate oxaloacetate transaminase in *Lolium perenne*, narrow-leafed lupin	Singh (2003)
Fiber (cotton)	Primary trisomic, monosomic, monotelodisomic, spontaneous trisomy, hypoaneuploid chromosome substitution	Chromosome location of fertility restorer gene, Rf_I tightly linked with RAPD, TRAP (target region amplified polymorphism marker) and SSR markers, *glandless stem* and *boll (gl1gl6), immature fiber (im), Ligon lintless-2 (Li2), methylation (me), nonpinking (np1np2), Raimondal (Ra1Ra2)* in nine cotton loci through trisomics, assignment of curly cotton (*ac*), LcD_2 light brown lint on chromosome 16, virescent-1 (v_1) in the long arm of chromosome 20, naked-tufted seed (N^1), in the long arm of chromosome 26, mapping of phosphoglucomutase-7 locus on *Gossypium hirsutum* chromosome	Kohel et al. (2002) and Saha et al. (2012)
Medicinal and aromatic plants	Primary trisomics, natural aneuploids	Characterization of additional chromosomes in *Plantago lanceolata* L., natural monosomics, double monosomy, trisomy, tetrasomy, double trisomy in *Betula humilis* (Betulaceae), *Digitalis obscura, Nigella sativa, Asparagus officinalis, Trigonella foenum-graecum,* role of kinetin in dicot embryo formation in *Catharanthus roseus*	Choi et al. (2009) and Jadwiszczak et al. (2011)
Arabidopsis (model plant)	Diploid mutants, meiotic mutants, primary trisomics, telotrisomics	Insensitivity to ethylene (dominant mutation), nutritional mutant, dwarf mutant, seed development (insertional mutagenesis), chlorate-resistant, cadmium-sensitive/rootless mutations, vitamin C-deficient mutant, morphogenesis, assignment of linkage group, morphological markers to chromosome arms and in locating centromeres, detection of chromosome inversion and translocation, deletion within LDOX gene, gene expression, dosage effect analysis	Koornneef et al. (2003), Henry et al. (2010), and Vernoux et al. (2000)

Mutagenomics: Merging with Newer "Omics"

With the rapid advancement of genomics and functional genomics, microarray technology and transcriptomics, metabolite profiling, and spectral models of phenomes, the concept of classical mutagenesis has now merged with more modern "omics" techniques. Microarray analyses reveal that plant mutagenesis may induce more transcriptomic changes than transgene insertion (Varshney et al. 2010). Through comparative genomics, genes and pathways of mutated traits could be identified in a number of crops. The results of metabolite profiling of low phytic acid mutants and their parents were indicative of the genes mutated in rice and soybean, and the deleted genes were identified through comparative genomic analysis in *Citrus*. Recent works suggest that gamma ray and EMS mutagenesis in *Phaseolus* and lentil (catalase-deficient mutants) and aneuploidy (trisomics and tetrasomics) as well as translocation lines of grass pea induce extensive transcriptomic changes of antioxidant defense enzymes (Talukdar D, unpublished). In gamma ray–irradiated progeny of cowpea, cDNA-AFLP showed differential gene expression at different time points of drought stress. The sequenced transcript-derived fragments (TDF) showed high homology to expressed sequence tags of soybean, with a possible function in cell defense/resistance and most importantly signal transduction. In soybean, altered transcriptomic profiling of GmNARK (*glycine max* leucine-rich repeat receptor kinase gene) mutant plants revealed controlling of gene expression involved in the jasmonate pathway by GmNARK-mediated signaling and identity of a second class of GmNARK-controlled genes in a rhizobia-independent manner during nodulation process of soybean (Pathan and Sleper 2008). In grass pea, the candidate genes responsible for condensation and biosynthesis of neurotoxin β-ODAP in varieties differing in content may be identified, cloned, and repressed through functional genomic approach.

Aneuploidy, Dosage Imbalance, and Transcriptomics: Case Study in Maize, *Arabidopsis*, and Grass Pea

For most eukaryotic genomes, the balance in gene dosage is essential for normal function. Aneuploidy leads to severe dosage imbalance of genes on the affected chromosome(s). The alterations in chromosome number that result in aneuploidy are usually associated with phenotypic consequences. However the molecular causes of specific phenotypes and genome-wide expression changes that occur in aneuploids are yet to be fully understandable. The subtle phenotypic differences between different trisomics of a same organism suggest that there might be specific "key" genes on each of the chromosomes that cause these phenotypic effects when their copy number is out of balance with other genes. Stress-response genes, transcription factors, and other potential regulatory genes have been frequently reported to be overrepresented among the genes affected by aneuploidy in plants (Huettel et al. 2008; Makarevitch et al. 2008). Aneuploidy causes greater quantitative changes in gene expression of two maize genes (*sus1* and *sh1*) in 2-week-old plants compared with embryo and endosperm tissues. Maize plants that are trisomic for 90 % of the short arm of chromosome 5 and monosomic for a small distal portion of the short arm of chromosome 6 (segmental aneuploidy) exhibited ectopic expression of knotted-like homeobox gene knox10, which is located on the short arm of chromosome 5, in developing leaves of the aneuploid plants and developed the leaf knotting phenotype (Makarevitch et al. 2008). Expression profiling revealed that approximately 40 % of the expressed genes in the trisomic region manifested the expected 1.5-fold increased transcript levels, while the remaining 60 % of genes did not show altered expression even with increased gene dosage (Makarevitch et al. 2008). Several of such studies on maize suggested that a specific chromosome arm dosage series can affect the expression of multiple genes located throughout the genome through the even slight alteration in the relative expression level of transcription factors, or other regulatory proteins, located in the affected chromosomal region, resulting in both positive and negative correlations of gene expression (upregulation and downregulation) with the dosage of the varied chromosome arm. Furthermore, genes located in the affected region frequently do not exhibit alterations in their expression level, suggesting the occurrence of some level of dosage compensation or, a "buffering" effect, when the level of RNA transcript read from genes present in three copies due to segmental aneuploidy were found to be similar to wild-type levels (Makarevitch et al. 2008). A further investigation on effects of aneuploidy on global gene expression in meristem-enriched and leaf tissues using microarray analysis of over 15,000 genes and on gene expression changes in response to aneuploidy for 30 genes in six different maize tissues at three early developmental stages after germination revealed that at least 23 out of 30 genes analyzed were either ectopically expressed or erroneously silenced in mature aneuploid tissues. Approximately, 50 % of trisomic genes exhibit dosage compensation in each of two tissues. The results also suggested that quantitative changes in gene expression at developmental transition points caused by variation in gene copy number progress through tissue development and result in stable qualitative changes in gene expression patterns (Makarevitch and Harris 2010). In *Arabidopsis thaliana*, a single locus, SENSITIVE TO DOSAGE IMBALANCE (SDI), exhibited segregation distortion in a ploidy-specific manner, and the phenomenon is attributed to increase in the likelihood of retaining genomic rearrangements such as segmental duplications. Additionally, in species where triploids are fertile, aneuploid survival would facilitate gene flow between diploid and tetraploid populations via a triploid bridge and prevent polyploid speciation (Henry et al. 2010). In grass pea, a grain as well as forage legume, dosage-specific response of genomes on antioxidant defense responses has been studied in series of aneuploids such as seven types of trisomics (tr), seven tetrasomics, and 21 different double trisomics (Talukdar 2011f). The switching over of diploid genome to aneuploidy through mutagenesis triggered a massive dosage imbalance, which was manifested in three different directions—extra dosage on activities of superoxide dismutase (tr III), ascorbate peroxidase (tr V), dehydroascorbate reductase (tr II), glutathione reductase (tr IV), inverse dosage on catalase (tr VII), and disomic level of all five enzymes in tr I and tr VI. The dosage effect was magnified in tetrasomics and combined in double trisomics (Talukdar 2011f). Transcriptomic analysis reveals that mRNA gene expression of isozymes of superoxide dismutase, ascorbate peroxidase, dehydroascorbate reductase, and glutathione reductase is upregulated in respective trisomics but is either downregulated or remained in disomic level in case of other trisomics (Talukdar D, unpublished observation). The dosage-specific changes in expression of genes governing metal tolerance are also being investigated in diploid, triploid, and tetraploid genotypes of grass pea (Talukdar D, unpublished). Based on these results of global gene expression profiling of aneuploids, it can be concluded that aneuploidy causes (1) gene dosage effect and predominantly common to multiple tissues, (2) varying degrees of gene dosage compensation for trisomic genes, (3) tissue-specific trans-effects (likely as a result of misregulation due to the slight variation in the presence of a regulatory protein), and (4) tissue-specific fixed qualitative variation in gene expression patterns that is more frequent in mature tissues. However, all of these changes are related with transcriptomic levels, and it is still not clear which of these effects are translated to the protein level and are indeed important for phenotypic abnormalities, considering the occurrence of posttranscriptional and posttranslational regulatory mechanisms in higher plants (Makarevitch and Harris 2010).

Gene-Targeted Mutagenesis

Zinc-finger nucleases (ZFN) can be targeted to specific genes causing a double-stranded break which disables the gene. In the homologous recombination of the targeted gene using ZFN, specific gene can be targeted for mutation in situ leaving the rest of the genome unperturbed. This strategy has several advantages over gene addition procedures, which include the risk of mutations arising from random insertion, because the strategy aims to incorporate exogenous DNA at a predetermined site in the chromosome. Furthermore, the exogenous DNA does not have to include a complete protein coding sequence or separate signals to ensure its expression because it is incorporated at an endogenous locus. Thus, if targeted mutation can be accomplished with high efficiency, other genetic factors affecting the mutant can be ruled out. Employing ZFN technology, the exact roles of three isoforms of isoamylase and pullulanase in starch debranching are being elucidated in pea null mutations (Curtin et al. 2011). An engineered ZFN followed by *Agrobacterium rhizogenes*-mediated hairy root transformation has been used to characterize the mutations related to dicer-like protein and RNA silencing (Curtin et al. 2011). In *Arabidopsis*, ZFN-induced mutagenesis and gene targeting are successful using *Agrobacterium*-mediated floral dip transformation (Koornneef and Meinke 2010). The technique has the potential to be used in both forward and reverse genomics.

Moving Mutagenomics Through Reverse Genomics

With the advent of functional genomics and web-based easily available data, instead of going from phenotype to sequence as in forward genetics, researchers are now opting for reverse genetics in which a gene sequence is known, but its exact function is uncertain. Reverse genetic approaches have permitted the silencing or interruption of individual candidate genes, providing the opportunity to investigate gene function and to relate sequence information to traits. Specific reverse genetic techniques used so far to induce/screen mutations in functional biology of crop legumes includes.

Gene Silencing by RNA Interference (RNAi)

It involves the inhibition of expression of target genes by antisense and sense RNAs. RNAi has recently become a powerful tool to silence the expression of genes and analyze their loss-of-function phenotype, allowing analysis of gene function when mutant alleles are not available. Downregulation of the *starch branching enzyme II* (*SBEII*) gene by RNA interference (RNAi) was previously shown to increase amylose content and resistant starch content in both hexaploid and tetraploid wheat. In polyploidy wheat, dsRNA-expressing constructs containing fragments of genes encoding *phytoene desaturase* (*PDS*) or the signal transducer of ethylene, *ethylene insensitive 2* (*EIN2*), showed stably inherited phenotypes of transformed wheat plants that were similar to mutant phenotypes of the two genes in diploid model (Gupta et al. 2008). In soybean, the RNAi machinery in hairy roots is fully functional in a sequence-specific manner, which allows the rapid analysis of sets of candidate genes for alleles underlying variation. RNAi knockouts have also been used to ascertain homology in floral organ development between cereals, *Petunia*, and *Arabidopsis* (Dwivedi et al. 2008).

Virus-Induced Gene Silencing

This is performed by cloning a 200–1,300 bp cDNA fragment from a plant gene of interest into a DNA copy of the genome of an RNA virus and transfecting the plant with this construct using *Agrobacterium*. Consequently, it leads to knockout or knockdown phenotype for the gene of interest (Gupta et al. 2008). The apple latent spherical virus (ALSV) has been reportedly used with minimal side effects on cucurbits, tomato, tobacco, potato, and different legumes. Recently, *Citrus leaf blotch virus* (CLBV) has been recommended for viral vector in Citrus crops (Tomlekova 2010).

Insertional Mutagenesis/Transposon-Mediated Mutagenesis

T-DNA (the segment of the Ti plasmid of *Agrobacterium tumefaciens* known as T-DNA) or transposon insertion has been exploited to create disruptions in target genes of interest, introduce new genes, or activate endogenous genes in the plant genome. A population of plants each having an insertion(s) at a unique site in the genome is generated either by transformation (T-DNA) or transposon activation. Transposon-tagged non-nodulating mutant blocking the infection thread and nodule primordia formation, designated as *nin* (nodule inception), has been developed in forage legume *Lotus japonicus* (Pathan and Sleper 2008). In soybean, there has been a recent initiative to develop a sizeable number of mutants using the maize Ds element (Dierking and Bilyeu 2009). In the model plant, *Arabidopsis*, a large collection of knockout mutants of T-DNA insertion developed through cloning, transformation, tissue culture methods, and combination with maize transposable elements are valuable tools for reverse genetics (Alonso-Blanco et al. 2009; Koornneef and Meinke 2010). Additional technologies for generating loss-of-function phenotypes such as RNAi and miRNA (Schwab et al. 2006) have also become available. Random insertion libraries have also been generated using activation tagging (Koornneef and Meinke 2010) for dominant mutants and promoterless reporter constructs for selection of insertions at desired intragenic locations coupled with visualization of expression patterns (Koornneef and Meinke 2010). In rice, induced by the insertion of the endogenous retrotransposon *Tos17*, which corresponds to *CesA* (cellulose synthase catalytic subunit) genes, *OsCesA4, OsCesA7*, and *OsCesA9* were expressed in seedlings, culms, premature panicles, and roots but not in mature leaves, revealing their importance in cellulose synthesis in secondary wall. Many mutants appear to result from transposon insertions such as albino rice plants or barley forms differing in susceptibility to powdery mildew. In barley, mutagenesis with EMS, NMU, and NaN3 induces huge genome changes accompanied with morphological variations which are likely driven by activation of various transposons and subsequent deletion as well as insertion, as revealed by sequence-specific amplification polymorphism (SSAP) fingerprints (Polok and Zielinski 2011). SSAP profiles inform about the sites in whole genomes, in which transposons are inserted, coupled with point mutations at target sites. The low copy number and high transposition frequency of *Cs1* in sorghum and its homologous sequences in rice, maize, teosinte, sudan grass, and sugarcane imply that this transposon can be used as an efficient mutagen, indicating its feasibility as a tagging tool (Polok and Zielinski 2011).

TILLING: A High-Throughput Technique for Mutation Discovery

During the last decade, the use of chemically induced mutagenesis has had a renaissance with the development of TILLING (Targeting Induced Local Lesions in Genomes) technology as the most efficient reverse genomic tools in functional biology (Henikoff et al. 2004). However, it was also shown that it could be adopted to use mutant populations developed through physical mutagenesis, such as gamma and fast neutron irradiation. For example, the De-TILLING technique could be effectively used to detect a specific mutant in a pool of 6,000 plants. In TILLING, mutagenesis is complemented by the isolation of chromosomal DNA from every mutated line, and high-throughput screening of induced point mutations at large scale is possible using advanced molecular techniques. A diagrammatic protocol of TILLING is given in Fig. 2. TILLING in legumes has been used either to confirm, by generating additional alleles, a lesion in forward screened mutants, especially those associated with the rhizobium–legume symbiosis, or to generate unique mutants as followed in cereals, millets, vegetable, spices, fruit crops, and of course in the *Arabidopsis* (Perry et al. 2009). TILLING is especially suited to species where there are few genomic resources and where insertion mutagenesis to create knockout mutants is difficult either through a lack of appropriate elements or an inefficient transformation system (Parry et al. 2009). The main advantage of TILLING as a reverse genetics strategy is that it can be applied to any plant species, regardless of its genome size, ploidy level, or method of propagation. Chemical mutagens, which are usually used in TILLING protocols, provide a high frequency of point mutations distributed randomly in the genome. Furthermore, since it is a nongenetically modified technology, it is highly desirable in those crops/countries where application of GM technology is restricted. These advantages have facilitated its swift move from models to crop plants.

Mutant discovery through TILLING process involves: (a) Direct sequencing and next-generation sequencing. (b) Li-Cor—It relies on the specific cleavage of mismatched bases formed as a result of repeated melting and reannealing of a PCR product amplified from a region of interest. (c) Denatured High-Performance Liquid Chromatography (DHPLC). (d) Usual agarose or PAGE-gel analysis. (e) High-Resolution Melt—Intercalating dyes are used that fluoresce only

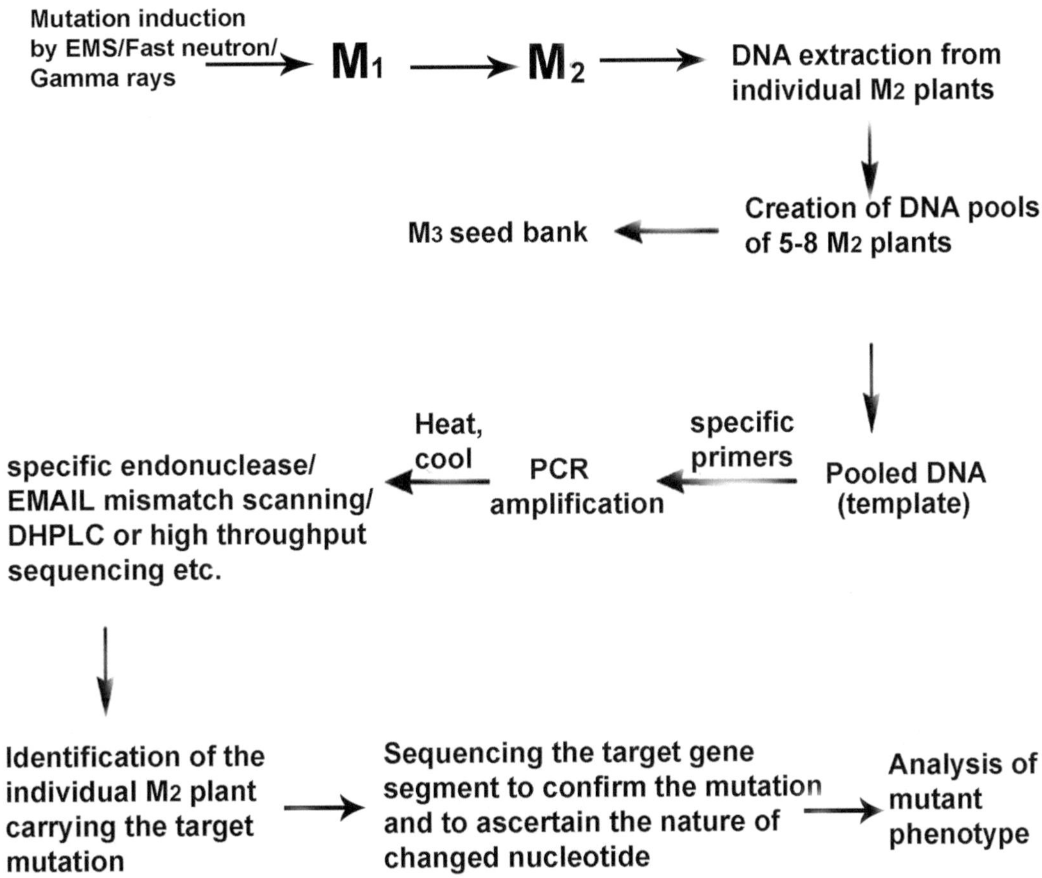

Fig. 2 Diagrammatic representation of generation of a TILLING population in plant

when bound to DNA. In increasing temperature, DNA strands will melt apart causing a release of the dye, and the total fluorescence will decrease in a predictable way. A mutation will cause a shift in the graph as the mismatched base changes the melting temperature. (f) *MALDI-TOF* or matrix-assisted laser desorption ionization time-of-flight spectroscopy. So far, Li-Cor is extensively used in mutant screening. In contrast to conventional protocols of TILLING, which are limited in their ability to detect mismatch cleavage due to nonspecific removal, by the nuclease, of 5′ end-labeled termini, a new highly sensitive and specific mismatch scanning assay being employed in rice called "endonucleolytic mutation analysis by internal labeling" (EMAIL) has been developed using capillary electrophoresis, involving internal amplicon labeling by PCR incorporation of fluorescently labeled deoxynucleotides. Multiple mutations among allelic pools were detected when EMAIL was applied with the mismatch nucleases, greatly enhancing the capacity of mutation detection in specific genes in pooled samples and improving throughput and efficiency and have the potential to be used as reliable and fast-track technology in crop mutation biology and breeding through TILLING (Cross et al. 2008).

Successful genome sequencing in *Arabidopsis* along with model legumes, *Lotus japonicus* and *Medicago truncatula*, and a web-accessible mutant discovery by TILLING approach (Perry et al. 2009) has markedly facilitated the search of genes controlling agronomically desirable traits in *Arabidopsis*, rice, wheat, maize, barley, oat, sunflower, Brassica, sugar beet, potato, tomato, pepper, grapes, Musa, and in the crop as well model legumes, for which several TILLING platforms and various web-based resources have been developed (Table 5). In one of the first mod-

Table 5 Functional mutation discovery through completed and ongoing TILLING projects in major crops

Crop	Platform organizer	Accessible web resources	Target traits/used so far (candidate genes)	References
Arabidopsis thaliana L./EMS	Seattle University, UBC-CAN-TILL	http://tilling.fhcrc.org/	Multiple traits by TILLING/individualized TILLING/Eco-TILLING; identify mutations in the closely linked MAPK/ERK kinase kinase 1 and 3 (i-Tilled)	Henikoff et al. (2004) and Till et al. (2004)
Oryza sativa L. (rice, EMS, NMU, 500 Gy γ-rays), *Triticum* (wheat, EMS), *Zea mays* (maize, pollen mutagenesis by EMS)	University of California, Purdue university, IRRI	http://genome.purdue.edu/maizetilling/; http://www.irri.org	*Alk*, which encodes soluble starch synthase IIa, drought tolerant gene (DREB2a, ERF3, sucrose synthase, actin depolymerizing factor, trehalose-6-phosphate phosphatase), (rice Eco-tilled), Leaf emergence *PLA1* (*plastochron 1*) in rice, *Pina-D1* (puroindoline a), *Pinb-D1* (puroindoline b) for kernel hardness through allelic variations, *Sgp-1* (*starch granule protein 1*), *starch synthase I* (*GBSSI*) in wheat, modified TILLING for polyploid wheat, multiple traits in maize	Henikoff et al. (2004), Gupta et al. (2008), Till et al. (2004), and Mba (2013)
Hordeum vulgare L. (barley), Sorghum bicolor L. (sorghum)	University of Bologna (TILLMore: a TILLING resource in Barley Morex)	http://www.distagenomics.unibo.it/TILLMore/	Row-type morphology (HvHox1 homeodomain-leucine zipper I-class homeobox protein), powdery mildew resistance genes (*mlo, Mla*) by Eco-TILLING, *Rpg1* (*barley stem rust resistance protein gene 1*) in barley; forage digestibility *COMT* (*caffeic acid O-methyltransferase*)	Henikoff et al. (2004) and Parry et al. (2009)
Avena sativa L. (oat)	USDA (GRIN), University of Gothenburg, Denmark (F/R), CropTailor AB	http://www.ars-grin.gov/cgi-bin/npgs/html/taxon.pl?6123; http://www.nordgen.org; http://www.croptailor.com	Increased digestibility (*AsPAL1* phenylalanine ammonia-lyase), food quality (*AsCslF6*-cellulose synthase-like) lignin and β-glucan biosynthesis genes (Swedish oat cv. *SW Belinda*)	Mba (2013)
Arachis hypogaea L. (peanut or groundnut)	UGA, Tifton (F/R)	http://www.caes.uga.edu/commodities/fieldcrops/peanuts/	Improvement of quality traits (genes controlling the oleic to linoleic acid ratio in seed), reduction of allergenicity by conglutin gene (Eco-tilled *Ara h 2.01* in *A. hypogaea* and *Ara d 2.01* in *A. duranensis*), genes for lipoxygenase, phospholipase D, fatty acid desaturase gene (*AhFAD 2*) for biotic/abiotic stresses	Knoll et al. (2011)

Cicer arietinum L. (chickpea) (0.2 % EMS)	WSU, Pullman, USA (F/R), IARI and ICRISAT, India (R)	http://www.intl-pag.org/; http://www.icrisat.org/bt-gene-discovery.htm	Morphological and reproductive traits, drought tolerance	Cooper et al. (2008) and Kumar et al. (2012)
Glycine max L. (soybean) (EMS, NMU)	SIU, Carbondale and University of Queensland (F/R)	http://www.soybeantilling.org/index.jsp; http://www-urgv.versailles.inra.fr/tilling/pepper.htm	Low phytate, higher seed yield and improved seed oil quality, new sources of seed meal and oil composition, improved nodulation, improved resistance to cyst nematode	Cooper et al. (2008) and Dierking and Bilyeu (2009)
Phaseolus vulgaris L. (common bean)	USDA-ARS, Idaho (F) USDA-ARS, Puerto Rico (R) CIAT, CGIAR, Mexico (R)	http://ciat-library.ciat.cigar.org	BAT93 population, low phytate, morphological mutations, Natural polymorphism of traits by Eco (ecotype)-TILLING	Blair et al. (2007) and Porch et al. (2009)
Pisum sativum L. (pea)	Pan-European (R)	http://www.eugrainlegumes.org/	Height, grain quality, stress tolerance, *Tendrilless* gene, multiple traits (le, 20 genes)	Dalmais et al. (2008)
	INRA (F/R)	http://urgv.evry.inra.fr/UTILLdb		
Lotus japonicus	JIC, Norwich (S/R)	http://data.jic.bbsrc.ac.uk/cgi-bin/lotus japonicus/	Starch accumulation, root-nodule symbiosis, nodule development, role of cytosolic invertase in normal plant growth and cellular development without affecting nodule function forage traits	Perry et al. (2009)
Medicago truncatula/EMS/fast neutron	JIC, Norwich	www.jicgenomelab.co.uk	Screening of arbuscule-specific phosphate transporter *MtPT4*, comparative mutagenomic study	Perry et al. (2009)
Solanum lycopersicum L. (tomato)/EMS	INRA (EMS, fast neutron) induced (R), Cornell University, Lyco-TILL, Red Setter, Italy, Canada, India	http://tomatoma.nbrp.jp/; http://zamir.sgn.cornell.edu/mutants/; http://www.agrobios.it/tilling/; https://www.eu-sol.wur.nl/	Identification of six ethylene receptor genes (*SlETR1–SlETR6*) with two allelic mutants of *SlETR1* (*Sletr1-1* and *Sletr1-2*) that resulted in reduced ethylene responses in Micro-Tom, fruit quality (PG, TBG4, EXP 1, RIN, Gr, Lcy-b, e, eIF4E, eIF4G)	Saito et al. (2011) and Okabe et al. (2013)
Cucumis melon L. (melon)/EMS	Boyce Thompson Institute for Plant Research, University of California, Davis, INRA	http://www.icugi.org	Morphological traits, Fruit quality, virus resistance, sex determination, improved shelf life, (eIF4E, ACO 1, 7, PDS, DET, DHS), TILLING/Eco-TILLING	Till et al. (2004) and Mba (2013)
Solanum tuberosum L. (potato)/EMS	INRA (UTILLdb),	www.urgv.versailles.inra.fr/tilling/index.htm	Starch quality (waxy), salt tolerance (SSBN002B23), tuber color (bch, dfr, f3′5′h), TILLING/Eco-TILLING	Mba (2013)

(continued)

Table 5 (continued)

Crop	Platform organizer	Accessible web resources	Target traits/used so far (candidate genes)	References
Helianthus annuus L. (sunflower)/EMS	Sun-TILL, SMART-NUI (F/R),	www.nuigalway.ie	Oil quality (*SAD, fad 2, kasII, kasIII*), multiple traits (12 genes)	Kumar et al. (2013)
Lactuca sativa L. (lettuce)	Warwick University (F/R)	http://www2.warwick.ac.uk/fac/sci/lifesci/research/vegin/lettuce/tilling/	Traits important for sustainability, product quality, increased shelf life (cv. Saladin)	Mou (2011) and Mba (2013)
Brassica oleracea var. *capitata* (cabbage), *B. napus*	UBC (CAN-TILL), JIC, UK (F/R)	www.botany.ubc.ca/can-till/; http://revgenuk.jic.ac.uk	Multiple traits (15 genes); *FAE1* (fatty acid elongase 1) in erucic acid synthesis (Eco-TILLING)	Henikoff et al. (2004) and Mba (2013)
Capsicum annum L. (peppers)/EMS	INRA (R)	www-urgv.versailles.inra.fr/tilling/pepper.htm	Virus resistance [eIF4E, eIF(iso)4E]-Eco-TILLING	Mba (2013)
Vitis vinifera L. (grapevine)	INRA (F/R)	www-urgv.versailles.inra.fr/pub.htm	Fungal resistance (MLO, Pmr6)	Henikoff et al. (2004)
Musa spp. (banana)	FAO/IAEA (F/R)	http://www.genoscope.cns.fr/spip/September-8th-2009-Banana-genome.html	Multiple traits (Tilling/Eco-TILLING)	Mba (2013)
Beta vulgaris L. (sugar beet)	Christian-Albrechts-University of Kiel, Germany (R)	http://www.flowercrop.uni-kiel.de/	Flowering time, bolting behavior, winter hardiness (*BTC1*, *BvFL1* and *BvFT1*) by Eco-TILLING	Frerichmann et al. (2013)

ifications of TILLING technology, mutation detection technology was used to discover polymorphisms in a natural population by Eco-TILLING and individualized TILLING in *A. thaliana*. Besides several yield traits, herbicide resistance is another important target of agriculture. Several classes of herbicides are known to inhibit the *ALS (acetolactate synthase)* gene. The highly selective *ALS*-inhibiting herbicides are very valuable for the weed management for a wide range of crops worldwide. Eco-TILLING was used for the detection of single nucleotide mutations in the *ALS* genes of sulfonylurea (SU) resistant (R) in *Monochoria vaginalis* (Pontederiaceae). Several new virus-resistant alleles from natural population of *Capsicum annuum* (eIF4E and eIF(iso)4E) and *Cucumis* spp. (IF4E) are screened by Eco-TILLING. A modified TILLING system using non-labeled primers and fast capillary gel electrophoresis was applied for high-throughput detection of single nucleotide substitution mutations in rice (Henikoff et al. 2004).

An ideal mutagenesis approach for a highly duplicated paleopolyploid genome like soybean would allow for the simultaneous recovery of plants with single or multiple mutations in each member of a gene family of interest without disruption to the rest of the genetic background. In reverse genetics, a denser mutagenesis compared to diploid has recently been achieved in autotetraploid *Arabidopsis* for a high-efficiency TILLING (Tsai et al. 2013), and this can be applied in other plants where sufficient polyploid stock is available (Talukdar 2012d).

References

Ahloowalia BS, Maluszynski M, Nichterlein K (2004) Global impact of mutation derived varieties. Euphytica 135:187–204

Akter S, Alam SS (2005) Differential fluorescent banding pattern in three varieties of *Cicer arietinum* L. (Fabaceae). Cytologia 70:441–445

Ali HBM, Meister A, Schubert I (2000) DNA content, rDNA loci, and DAPI bands reflect the phylogenetic distance between *Lathyrus* species. Genome 43:1027–1032

Alonso-Blanco C, Aarts MGM, Bentsink L, Keurentjes JJB, Reymond M, Vreugdenhil D, Koornneef M (2009) What has natural variation taught us about plant development, physiology, and adaptation? Plant Cell 21:1877–1896

Auger DL, Birchler JA (2002) Maize tertiary trisomic stocks derived from B-A translocations. J Hered 93(1):42–47

Balyan HS, Houben A, Ahne R (2002) Karyotype analysis and physical mapping of 18S-5.8S-25S and 5S ribosomal RNA loci in species of genus *Lens* Miller (Fabaceae). Caryologia 55:121–128

Barshile JD, Auti SG, Apparao BJ (2009) Genetic enhancement of chickpea through induced mutagenesis. J Food Legumes 22(1):26–29

Begum R, Alam SS, Menzel G, Schmidt T (2009) Comparative molecular cytogenetics of major repetitive sequence families of three *Dendrobium* species (Orchidaceae) from Bangladesh. Ann Bot 104(5):863–872

Begum R, Zakrzewski F, Menzel G, Weber R, Alam SS, Schmidt T (2013) Comparative molecular cytogenetic analyses of a major tandemly repeated DNA family and retrotransposon sequences in cultivated jute *Corchorus* species (Malvaceae). Ann Bot. doi:10.1093/aob/mct103

Belyakova AS, Sinjushin AA (2012) Phenotypic expression and inheritance of *determinate habit* (*deh*) mutation in pea (*Pisum sativum* L.). In: Program and abstract book of VI international conference on legumes genetics and genomics, Hyderabad, India

Blair MW, Porch T, Cichy K, Galeano CH, Lariguet P, Pankhurst C, Broughton W (2007) Induced mutants in common bean (*Phaseolus vulgaris*), and their potential use in nutritional quality breeding and gene discovery. Israel J Plant Sci 55(2):191–200

Campion B, Sparvoli F, Doria E, Tagliabue G, Galasso I, Fileppi M et al (2009) Isolation and characterization of an *lpa* (low phytic acid) mutant in common bean (*Phaseolus vulgaris* L.). Theor Appl Genet 118:1211–1221

Ceccarelli M, Sarri V, Polizzi E, Andreozzi G, Cionini PG (2010) Characterization, evolution and chromosomal distribution of two satellite DNA sequence families in *Lathyrus* species. Cytogenet Genome Res 128:236–244

Chahota RK, Mukai Y, Chaudhary HK, Kishore N, Sharma TR (2011) Karyotyping and *in situ* chromosomal localization of rDNA sites in black cumin *Bunium persicum* (Boiss) B. Fedtsch, 1915 (Apiaceae). Comp Cytogenet 5(4):345–353

Chakraborty NR, Paul A (2012) Role of induced mutations for enhancing nutrition quality and production of food. Int J Bioresour Stress Manag 4(1):14–19

Chaudhary HK, Sood VK, Tayeng T, Kalia V, Sood A (2011) Molecular cytogenetics in physical mapping of genomes and alien introgressions. In: Pratap A, Kumar J (eds) Biology and breeding of food legumes. CAB International, Wallingford

Chen JF, Luo XD, Qian CT, Jahn MM, Staub JE, Zhuang FY, Lou QF, Ren G (2004) *Cucumis* monosomic alien addition lines: morphological, cytological, and genotypic analyses. Theor Appl Genet 108(7):1343–1348

Chen Q-F, Hsam Sai LK, Zeller FJ (2007) Cytogenetic studies on diploid and autotetraploid common buckwheat and their autotriploid and trisomics. Crop Sci 47(6):2340–2345

Choi H-W, Koo D-H, Bang K-H, Paek K-Y, Seong N-S, Bang J-W (2009) FISH and GISH analysis of the genomic relationships among *Panax* species. Genes Genomics 31(1):99–105

Cooper JL, Till BJ, Laport RG, Darlow MC, Kleffner JM, Jamai A, El-Mellouki T et al (2008) TILLING to detect induced mutations in soybean. BMC Plant Biol 8:9

Cregan PB, Kollipara KP, Xu SJ, Singh RJ, Fogarty SE, Hymowitz T (2001) Primary trisomics and SSR markers as tools to associate chromosomes with linkage groups in soybean. Crop Sci 41(4):1262–1267

Cross MJ, Waters DLE, Lee LS, Henry RJ (2008) Endonucleolytic mutation analysis by internal labelling (EMAIL). Electrophoresis 29:1291–1301

Curtin SJ, Zhang F, Sander JF, Haun WJ, Starker C, Baltes NJ, Reyon D, Dahlborg EJ, Goodwin MJ et al (2011) Targeted mutagenesis of duplicated genes in soybean with zinc-finger nucleases. Plant Physiol 156(2):466–473

D'Hont A (2005) Unraveling the genome structure of polyploids using FISH and GISH; examples of sugarcane and banana. Cytogenet Genome Res 109:27–33

Dalmais M, Schmidt J, Le Signor C, Moussy F, Burstin J, Savois V, Aubert G, Brunaud V, de Oliveira Y, Guichard C, Thompson R, Bendahmane A (2008) UTILLdb, a *Pisum sativum* in silico forward and reverse genetics tool. Genome Biol 9:R43

David P, Chen NWG, Pedrosa-Harand A, Thareau V, Sévignac M, Cannon SB, Debouck D, Langin T, Geffroy V (2009) A nomadic subtelomeric disease resistance gene cluster in common bean. Plant Physiol 151(3):1048–1065

Dhar MK, Friebe B, Kaul S, Gill BS (2006) Characterization and physical mapping of ribosomal RNA gene families in *Plantago*. Ann Bot 97(4):541–548

Diao WP, Bao SY, Jiang B, Cui L, Chen JF (2009) Primary trisomics obtained from autotriploid by diploid reciprocal crosses in cucumber. Sex Plant Reprod 22(1):45–51

Dierking EC, Bilyeu KD (2009) New sources of soybean seed meal and oil composition traits identified through TILLING. BMC Plant Biol 9:89

Dixit GP, Tripathi DP, Chandra S, Tewari TN, Tickoo JL (2000) MULLaRP crops: varieties developed during the last fifty years. All India Coordinated Research Project on MULLaRP (ICAR), Indian Institute of Pulses Research, Kanpur

Doležel J, Vrána J, Šafář J, Kubaláková M, Šimková H (2012) Chromosomes in the flow to simplify genome analysis. Funct Integr Genomics 12(3):397–416

Dwivedi S, Perotti E, Ortiz R (2008) Towards molecular breeding of reproductive traits in cereal crops. Plant Biotechnol J 6:529–559

Feng J, Liu Z, Cai X, Jan C-C (2013) Toward a molecular cytogenetic map for cultivated sunflower (*Helianthus annuus* L.) by landed BAC/BIBAC clones. G3 Genes|Genomes|Genetics 3(1):31–40

Findley SD, Psappas AL, Cui Y, Birchler JA, Palmer RG, Stacey G (2011) Fluorescence *in situ* hybridization-based karyotyping of soybean translocation lines. G3 Genes|Genomes|Genetics 1(2):117–129

Foo E, Morris SE, Parmenter K, Young N, Wang H, Jones A, Rameau C, Turmbull CGN, Beveridge CA (2007) Feedback regulation of xylem cytokinin content is conserved in pea and Arabidopsis. Plant Physiol 143(3):1418–1428

Fradkin M, Ferrari MR, Espert SM, Ferreira V, Grassi E, Greizerstein E, Poggio L et al (2013) Differentiation of triticale cultivars through FISH karyotyping of their rye chromosomes. Genome. doi:10.1139/gen-2012-0117

Frasch R, Weigand C, Perez PT, Palmer RG, Sandhu D (2011) Molecular mapping of two environmentally sensitive male-sterile mutants in soybean. J Hered 102:11–16

Frerichmann SLM, Kirchhoff M, Müller AE, Scheidig AJ, Jung C, Kopisch-Obuch FJ (2013) EcoTILLING in *Beta vulgaris* reveals polymorphisms in the *FLC-like* gene *BvFL1* that are associated with annuality and winter hardiness. BMC Plant Biol 13:52

Gaikwad NB, Kothekar VS (2011) Studies on trypsin inhibitor, polyphenol and lectin content in induced mutants of lentil. Indian J Genet 71(2):158–163

Goyal S, Kozgar MI, Fatma S, Khan S (2011) Genetic improvement of blackgram, chickpea and faba bean through conventional and advanced approaches. In: Khan S, Kozgar MI (eds) Breeding of pulse crops. Kalyani Publishers, Ludhiana

Greenup A, Peacock WJ, Dennis ES, Trevaskis B (2009) The molecular biology of seasonal flowering-responses in Arabidopsis and the cereals. Ann Bot 103:1165–1172

Gupta PK (2006) Plant cytogenetics: a re-birth in twenty-first century. Indian J Crop Sci 1(1–2):1–7

Gupta PK, Mir RR, Mohan A, Kumar J (2008) Wheat genomics: present status and future prospects. Int J Plant Genomics 2008:Article ID 896451, 36 pages. doi:10.1155/2008/896451

Han Y, Zhang Z, Huang S, Jin W (2011) An integrated molecular cytogenetic map of *Cucumis sativus* L. chromosome 2. BMC Genet 12:18

Henikoff S, Till BJ, Comai L (2004) TILLING. Traditional mutagenesis meets functional genomics. Plant Physiol 135:630–636

Henry IM, Dilkes BP, Miller ES, Burkart-Waco D, Comai L (2010) Phenotypic consequences of aneuploidy in *Arabidopsis thaliana*. Genetics 186(4):1231–1245

Heslop-Harrison JS (Pat), Schwarzacher T (2011) Organisation of the plant genome in chromosomes. Plant J. doi:10.1111/j.1365-313X.2011.04544.x

Hossain Z, Mandal AKA, Datta SK, Biswas AK (2006) Isolation of a NaCl-tolerant mutant of *Chrysanthemum morifolium* by gamma radiation: *in vitro* mutagenesis and selection by salt stress. Funct Plant Biol 33(1):91–101

Houle D, Govindaraju DR, Omholt S (2010) Phenomics: the next challenge. Nat Rev Genet 11:855–866

Huang D, Wu W, Zhou Y, Hu Z, Lu L (2004) Microdissection and molecular manipulation of single chromosomes in woody fruit trees with small chromosomes using pomelo (*Citrus grandis*) as a model. Theor Appl Genet 108:1366–1370

Huettel B, Kreil DP, Matzke M, Matzke AJ (2008) Effects of aneuploidy on genome structure, expression, and interphase organization in *Arabidopsis thaliana*. PLoS Genet 4(10):e1000226

Hussain H (2009) Mutagenesis of genes for starch debranching enzyme isoforms in pea by zinc-finger endonucleases. In: Shu QY (ed) Induced plant mutations in the genomics era. Food and Agriculture Organization of the United Nations, Rome

ICARDA-HarvestPlus (2010) International Center for Agricultural Research in the dry areas mid-term report: development of high iron and zinc content lentil for nutritional security. HarvestPlus Generation Challenge Program, ICARDA, Aleppo

Jadwiszczak KA, Jabłońska E, Kłosowski S, Banaszek A (2011) Aneuploids in the shrub birch *Betula humilis* populations in Poland. Acta Soc Bot Pol 80(3):233–235

Jain SM, Suprasanna P (2011) Induced mutations for enhancing nutrition and food production. Geneconserve 40:201–215

Jauhar Prem P, Hanna WW (1998) Cytogenetics and genetics of pearl millet. Adv Agron 64:1–26

Jenkins G, Hasterok R (2007) BAC 'landing' on chromosomes of *Brachypodium distachyon* for comparative genome alignment. Nat Protoc 2(1):88–98

Jiang J, Gill BS (2006) Current status and the future of fluorescence in situ hybridization (FISH) in plant genome research. Genome 49:1057–1068

Jiang SY, Ramachandran S (2010) Natural and artificial mutants as valuable resources for functional genomics and molecular breeding. Int J Biol Sci 6(3):228–251

Johanson U, West J, Lister C, Michaels S, Amasino R, Dean C (2000) Molecular analysis of *FRIGIDA*, a major determinant of natural variation in Arabidopsis flowering time. Science 290:344–347

Jung C (2004) Genome analysis: mapping in sugar beet. In: Lörz H, Wenzel G (eds) Biotechnology in agriculture and biotechnology, Molecular marker systems, vol 55. Springer, Berlin/Heidelberg

Kato A, Lamb JC, Birchler JA (2004) Chromosome painting using repetitive DNA sequences as probes for somatic chromosome identification in maize. Proc Natl Acad Sci U S A 101:13554–13559

Khan R, Khan H, Harada F, Harada K (2010) Evaluation of microsatellite markers to discriminate induced mutation lines, hybrid lines and cultigens in chickpea (*Cicer arietinum* L). Aust J Crop Sci 4(5):301–308

Kharkwal MC, Shu QY (2009) The role of induced mutations in world food security. In: Shu QY (ed) Induced plant mutations in the genomics era. Food and Agriculture Organization of the United Nations, Rome

Khush GS, Singh RJ, Sur SC, Librojo AL (1984) Primary trisomics of rice: origin, morphology, cytology and use in linkage mapping. Genetics 107:141–165

Kim JS, Islam-Faridi MN, Klein PE, Stelly DM, Price HJ, Klein RR, Mullet JE (2005) Comprehensive molecular cytogenetic analysis of sorghum genome architecture: distribution of euchromatin, heterochromatin, genes and recombination in comparison to rice. Genetics 171:1963–1976

Knoll JE, Ramos ML, Zeng Y, Holbrook CC, Chow M, Chen S, Maleki S, Bhattacharya A, Ozias-Akins P (2011) TILLING for allergen reduction and improvement of quality traits in peanut (*Arachis hypogaea* L.). BMC Plant Biol 11:81

Kohel RJ, Stelly DM, Yu J (2002) Tests of six cotton (*Gossypium hirsutum* L.) mutants for association with aneuploids. J Hered 93(2):130–132

Koornneef M, Meinke D (2010) The development of Arabidopsis as a model plant. Plant J 61:909–921

Koornneef M, Fransz P, de Jong H (2003) Cytogenetic tools for *Arabidopsis thaliana*. Chromosome Res 11:183–194

Kubaláková M, Kovářová P, Suchánková P, Číhalíková J, Bartoš J, Lucretti S, Watanabe N, Kianian SF, Doležel J (2005) Chromosome sorting in tetraploid wheat and its potential for genome analysis. Genetics 170:823–829

Kumar S, Bejiga G, Ahmed S, Nakkoul H, Sarker A (2010) Genetic improvement of grass pea for low neurotoxin (β-ODAP) content. Food Chem Toxicol 49:589–600

Kumar S, Mishra RK, Kumar A, Chaudhary S, Sharma V, Kumari R (2012) Genetic interaction and mapping studies on the leaflet development (*lld*) mutant in *Pisum sativum*. J Genet 91(3):325–342

Kumar APK, Boualem A, Bhattacharya A, Parikh S, Desai N, Zambelli A, Leon A, Chatterjee M, Bendahmane A (2013) SMART – Sunflower Mutant population And Reverse genetic Tool for crop improvement. BMC Plant Biol 13:38

Kynast RG, Riera-Lizarazu O (2011) Development and use of oat-maize chromosome additions and radiation hybrids. Methods Mol Biol 701:259–284

Lan T, Zhang S, Liu B, Li X, Chen R, Song W (2006) Differentiating sex chromosomes of the dioecious *Spinacia oleracea* L. (spinach) by FISH of 45S rDNA. Cytogenet Genome Res 114:175–177

Lavania UC, Yamamoto M, Mukai Y (2003) Extended chromatin and DNA fibers from active plant nuclei for high-resolution FISH. J Histochem Cytochem 51(10):1249–1253

Lavania UC, Basu S, Srivastava S, Mukai Y, Lavania S (2005) *In situ* chromosomal localization of r-DNA sites in *'Safed Musli'* Chlorophytum Ker. Gawl. and their physical measurement by fiber FISH. J Hered 96:155–160

Lavania UC, Kushwaha JS, Lavania S, Basu S (2010) Chromosomal localization of rDNA and DAPI bands in solanaceous medicinal plant *Hyoscyamus niger* L. J Genet 89:493–496

Lavia GI, Ortiz AM, Robledo G, Fernández A, Seijo G (2011) Origin of triploid *Arachis pintoi* (Leguminosae) by autopolyploidy evidenced by FISH and meiotic behavior. Ann Bot 108(1):103–111

Lesniewska K, Zkiewicz MK, Nelson MN, Mahé F, Aïnouche A, Wolko B, Naganowska B (2011) Assignment of 3 genetic linkage groups to 3 chromosomes of narrow-leafed lupin. J Hered 102(2):228–236

Lijia L, Ma L, Kevin A, Song Y (2006) Flow-sorted chromosomes: a fine material for plant gene physical mapping. Caryologia 59(2):99–103

Lilly JW, Havey MJ, Jackson SA, Jiang J (2001) Cytogenomic analyses reveal the structural plasticity of the chloroplast genome in higher plants. Plant Cell 13:245–254

Liu C, Liu J, Li H, Zhang Z, Han Y, Huang S, Jin W (2010) Karyotyping in melon (*Cucumis melo* L.) by cross-species fosmid fluorescence in situ hybridization. Cytogenet Genome Res 129(1–3):241–249

Lysak MA, Pecinka A, Schubert I (2003) Recent progress in chromosome painting of *Arabidopsis* and related species. Chromosome Res 11:195–204

Lysak MA, Berr A, Pecinka A, Schmidt R, McBreen K, Schubert I (2006) Mechanisms of chromosome number reduction in *Arabidopsis thaliana* and related Brassicaceae species. Proc Natl Acad Sci U S A 103:5224–5229

Makarevitch I, Harris C (2010) Aneuploidy causes tissue-specific qualitative changes in global gene expression patterns in maize. Plant Physiol 152(2):927–938, doi:10.1186/1471-2164-9-7

Makarevitch I, Phillips RL, Springer NM (2008) Profiling expression changes caused by a segmental aneuploid in maize. BMC Genomics 9:7

Marín S, Martín A, Barro F (2008) Comparative FISH mapping of two highly repetitive DNA sequences in *Hordeum chilense* (Roem. et Schult.). Genome 51:580–588

Marques A, Klemme S, Guerra M, Houben A (2012) Cytomolecular characterization of *de novo* formed rye B chromosome variants. Mol Cytogenet 5:34

Mayer KFX, Martis M, Hedley PE, Simkova H, Liu H, Morris JA, Steuernagel B, Taudien S, Roessner S et al (2011) Unlocking the barley genome by chromosomal and comparative genomics. Plant Cell 23:1249–1263

Mba C (2013) Induced mutations unleash the potentials of plant genetic resources for food and agriculture. Agronomy 3:200–231

Megyeri M, Molnár-Láng M, Molnár I (2013) Cytomolecular identification of individual wheat-wheat chromosome arm associations in wheat-rye hybrids. Cytogenet Genome Res 139:128–136

Mishra RK, Chaudhary S, Kumar A, Kumar S (2009) Effects of *MULTIFOLIATE-PINNA, AFILA, TENDRIL-LESS* and *UNIFOLIATA* genes on leaf blade architecture in *Pisum sativum*. Planta 230:177–190

Moscone EA, Samuel R, Schwarzacher T, Schweizer D, Pedrosa- Harand A (2007) Complex rearrangements are involved in *Cephalanthera* (Orchidaceae) chromosome evolution. Chromosome Res 15:931–943

Mou B (2011) Mutations in lettuce improvement. Int J Plant Genomics 2011:Article ID 723518, 7 pages. doi:10.1155/2011/723518

Mudge J, Cannon S, Kalo P, Oldroyd G, Roe B, Town C, Young N (2005) Highly syntenic regions in the genomes of soybean, *Medicago truncatula*, and *Arabidopsis thaliana*. BMC Plant Biol 5(1):15

Nag A, Rajkumar S (2011) Chromosome identification and karyotype analysis of *Podophyllum hexandrum*-Roxb. ex Kunth using FISH. Physiol Mol Biol Plants 17(3):313–316

Nagarajan S, Rens W, Stalker J, Cox T, Smith MF (2008) Chromhome: a rich Internet application for accessing comparative chromosome homology maps. BMC Bioinform 9:168

Ohmido N, Ishimaru A, Kato S, Sato S, Tabata S et al (2010) Integration of cytogenetic and genetic linkage maps of *Lotus japonicus*, a model plant for legumes. Chromosome Res 18:287–299

Okabe Y, Ariizumi T, Ezura H (2013) Updating the Micro-Tom TILLING platform. Breed Sci 63(1):42–48

Okagaki RJ, Kynast RG, Livingston SM, Russell CD, Rines HW, Phillips RL (2001) Mapping maize sequences to chromosomes using oat-maize chromosome addition materials. Plant Physiol 125(3):1228–1235

Oleo M, Lange W, D'Haeseleer M, De Bock TS, Jacobs M (1993) Isozyme analysis of primary trisomies in beet (*Beta vulgaris* L.). Genetical characterization and techniques for chromosomal assignment of two enzyme-coding loci: leucine aminopeptidase and glutamate oxaloacetate transaminase. Theor Appl Genet 86(6):761–768

Paesold S, Borchardt D, Schmidt T, Dechyeva D (2012) A sugar beet (*Beta vulgaris* L.) reference FISH karyotype for chromosome and chromosome-arm identification, integration of genetic linkage groups and analysis of major repeat family distribution. Plant J 72(4):600–611

Palmer RG, Xu M (2008) Positioning 3 qualitative trait loci on soybean molecular linkage group E. J Hered 99(6):674–678

Parry MA, Madgwick J, Bayon PJ, Tearall C, Hernandez-Lopez K, Baudo A, Rakszegi M et al (2009) Mutation discovery for crop improvement. J Exp Bot 60(10):2817–2825

Pathan MS, Sleper DA (2008) Advances in soybean breeding. In: Stacey G (ed) Genetics and genomics of soybean. Springer, New York

Pereira G, Leitão J (2010) Two powdery mildew resistance mutations induced by NEU in *Pisum sativum* L. affect the locus *er1*. Euphytica 171:345–354

Perry J, Brachmann A, Welham T, Binder A, Charpentier M, Groth M, Haage K et al (2009) TILLING in *Lotus japonicus* identified large allelic series for symbiosis genes and revealed a bias in functionally defective ethyl methanesulfonate alleles toward glycine replacements. Plant Physiol 151:1281–1291

Phan HTT, Ellwood SR, Ford R, Thomas S, Oliver R (2006) Differences in syntenic complexity between *Medicago truncatula* with *Lens culinaris and Lupinus albus*. Funct Plant Biol 3(8):775–782

Piotrowicz-Cieślak AI, Rybínski W, Michalczyk DJ (2008) Mutations modulate soluble carbohydrates composition in seeds of *Lathyrus sativus* L. Acta Societatis Botanicorm Poloniae 77(4):281–287

Polok K, Zielinski R (2011) Mutagenic treatment induces high transposon variation in barley (*Hordeum vulgare* L.). Acta Agric Slovenica 97(3):179–188

Porch TG, Blair MW, Lariguet P, Galeano C, Pankhurst CE, Broughton WJ (2009) Generation of a mutant population for TILLING common bean genotype BAT 93. J Am Hortic Soc 134:348–355

Qi L, Friebe B, Zhang P, Gill BS (2009) A molecular-cytogenetic method for locating genes to pericentromeric regions facilitates a genome wide comparison of synteny between the centromeric regions of wheat and rice. Genetics 183(4):1235–1247

Rao SLN (2010) A look at the brighter facets of [beta]-N-oxalyl-l- [alpha], [beta]-diaminopropionic acid, homoarginine and the grasspea. Food Chem Toxicol 49:620–622

Reid JB, Ross JJ (2011) Mendel's genes: toward a full molecular characterization. Genetics 189:3–10

Riera-Lizarazu O, Vales MI, Kianian SF (2008) Radiation hybrid (RH) and HAPPY mapping in plants. Cytogenet Genome Res 120(3–4):233–240

Román B, Satovic Z, Pozarkova D, Macas J, Dolezel J, Cubero JI, Torres AM (2004) Development of a composite map in *Vicia faba*, breeding applications and future prospects. Theor Appl Genet 108(6):1079–1088

Saha S, Wu J, Jenkins JN, McCarty JC, Hayes R, Stelly DM (2012) Interspecific chromosomal effects on agronomic traits in *Gossypium hirsutum* by AD analysis using intermated *G. barbadense* chromosome substitution lines. Theor Appl Genet. doi:10.1007/s00122-012-1965-9

Saito T, Ariizumi T, Okabe Y, Asamizu E, Hiwasa-Tanase K, Fukuda N et al (2011) TOMATOMA: a novel tomato mutant database distributing Micro-Tom mutant collections. Plant Cell Physiol 52(2):283–296

Sanz MJ, Jellen EN, Loarce Y, Irigoyen ML, Ferrer E, Fominaya A (2010) A new chromosome nomenclature system for oat (*Avena sativa* L. and *A. byzantina* C. Koch) based on FISH analysis of monosomic lines. Theor Appl Genet 121(8):1541–1552

Satovic Z, Torres AM, Cubero JI (1996) Genetic mapping of new morphological, isozyme and RAPD markers in *Vicia faba* L. using trisomics. Theor Appl Genet 93(7):1130–1138

Schwab R, Ossowski S, Riester M, Warthmann N, Weigel D (2006) Highly specific gene silencing by artificial microRNAs in Arabidopsis. Plant Cell 18:1121–1133

Seijo JG, Lavia GI, Fernández A, Krapovickas A, Ducasse D, Moscone EA (2004) Physical mapping of the 5S and 18S–25S rRNA genes by FISH as evidence that *Arachis duranensis* and *A. ipaensis* are the wild diploid progenitors of *A. hypogaea* (*Leguminosae*). Am J Bot 91(9):1294–1303

Sharma V, Chaudhary S, Kumar A, Kumar S (2012) *COCHLEATA* controls leaf size and secondary inflorescence architecture via negative regulation of *UNIFOLIATA* (*LEAFY* ortholog) gene in garden pea *Pisum sativum*. J Biosci 37:1041–1059

Shu QY, Lagoda PJL (2007) Mutation techniques for gene discovery and crop improvement. Mol Plant Breed 5:193–195

Šimková H, Šafář J, Suchánková P, Kovářová P, Bartoš J, Kubaláková M, Janda J, Číhalíková J, Mago R, Lelley T, Doležel J (2008) A novel resource for genomics of Triticeae: BAC library specific for the short arm of rye (*Secale cereale* L.) chromosome 1R (1RS). BMC Genomics 9:237

Singh RJ (2003) Plant cytogenetics. CRC Press, Inc., Boca Raton

Sinjushin AA (2011) On the role of genes *DETERMINATE, LATE FLOWERING* and *FASCIATA* in the morphogenesis of pea inflorescence. Ratar Povrt/Field Veg Crop Res 48:313–320

Sinjushin AA, Gostimskii SA (2006) Fasciation in pea: basic principles of morphogenesis. Rus J Dev Biol 37:375–381

Sinjushin AA, Gostimskii SA (2007) Relationship between different fasciated lines of pea. Pisum Genet 39:16–18

Sinyushin AA (2010) Flower fasciation: I. Origin of enlarged meristem. Moscow Univ Biol Sci Bull 65:98–103

Smulikowska S, Rybinski W, Czerwiński J, Taciak M, Mieczkowska A (2008) Evaluation of selected mutants of grasspea (*Lathyrus sativus* L.) var. Krab as an ingredient in broiler chicken diet. J Anim Feed Sci 17:75–87

Sorajjapinum W, Srinives P (2011) Chasmogamous mutant, a novel character enabling commercial hybrid seed production in mungbean. Euphytica 181:217–222

Sousa A, Fuchs J, Renner SS (2013) Molecular cytogenetics (FISH, GISH) of *Coccinia grandis*: a ca. 3 myr-old species of cucurbitaceae with the largest Y/autosome divergence in flowering plants. Cytogenet Genome Res 139:107–118

Srinivasan S, Gaur PM (2011) Genetics and characterization of an open flower mutant in chickpea. J Hered 103(2):297–302

Stack SM, Royer SM, Shearer LA, Chang SB, Giovannoni JJ, Westfall DH, White RA, Anderson LK (2009) Role of fluorescence in situ hybridization in sequencing the tomato genome. Cytogenet Genome Res 124(3–4):339–350

Subrahmanyam NC, Azad AA (1978) Trisomic analysis of ribosomal RNA cistron multiplicity in barley (*Hordeum vulgare* L.). Chromosoma 69(2):255–264

Sybenga J (1996) Aneuploid and other cytological tester sets in rye. Euphytica 89(1):143–151

Talia P, Eduardo J, Greizerstein EJ, Hopp HE, Paniego N et al (2011) Detection of single copy sequences using BAC-FISH and C-PRINS techniques in sunflower chromosomes. Biocell 35:19–28

Talukdar D (2008) Cytogenetic characterization of seven different primary tetrasomics in grass pea (*Lathyrus sativus* L.). Caryologia 61:402–410

Talukdar D (2009a) Dwarf mutations in grass pea (*Lathyrus sativus* L.): origin, morphology, inheritance and linkage studies. J Genet 88(2):165–175

Talukdar D (2009b) Recent progress on genetic analysis of novel mutants and aneuploid research in grass pea (*Lathyrus sativus* L.). Afr J Agric Res 4:1549–1559

Talukdar D (2009c) Development of cytogenetic stocks through induced mutagenesis in grass pea (*Lathyrus sativus*): current status and future prospects in crop improvement. Grain Legume 54:30–31

Talukdar D (2010a) Reciprocal translocations in grass pea (*Lathyrus sativus* L.). Pattern of transmission, detection of multiple interchanges and their independence. J Hered 101:169–176

Talukdar D (2010b) Cytogenetic characterization of induced autotetraploids in grass pea (*Lathyrus sativus* L.). Caryologia 63:62–72

Talukdar D (2010c) Allozyme variations in leaf esterase and root peroxidase isozymes and linkage with dwarfing genes in induced dwarf mutants of grass pea (*Lathyrus sativus* L.). Int J Genet Mol Biol 2(6):112–120

Talukdar D (2010d) Fluorescent-banded karyotype analysis and identification of chromosomes in three improved Indian varieties of grass pea (*Lathyrus sativus* L.). Chromosome Sci 13:3–10

Talukdar D (2011a) Genetics of pod indehiscence in *Lathyrus sativus* L. J Crop Improv 25:1–15

Talukdar D (2011b) Cytogenetic analysis of a novel yellow flower mutant carrying a reciprocal translocation in grass pea (*Lathyrus sativus* L.). J Biol Res-Thessaloniki 15:123–134

Talukdar D (2011c) Bold-seeded and seed coat colour mutations in grass pea (*Lathyrus sativus* L.): origin, morphology, genetic control and linkage analysis. Int J Curr Res 3:104–112

Talukdar D (2011d) Isolation and characterization of NaCl-tolerant mutations in two important legumes, *Clitoria ternatea* L. and *Lathyrus sativus* L.: induced mutagenesis and selection by salt stress. J Med Plants Res 5(16):3619–3628

Talukdar D (2011e) Morpho-physiological responses of grass pea (*Lathyrus sativus* L.) genotypes to salt stress at germination and seedling stages. Legum Res 34(4):232–241

Talukdar D (2011f) The aneuploid switch: extra-chromosomal effect on antioxidant defense through trisomic shift in *Lathyrus sativus* L. Indian J Fundam Appl Life Sci 1(4):263–273

Talukdar D (2011g) Flower and pod production, abortion, leaf injury, yield and seed neurotoxin levels in stable dwarf mutant lines of grass pea (*Lathyrus sativus* L.) differing in salt stress responses. Int J Curr Res 2(1):46–54

Talukdar D (2012a) Ascorbate deficient semi-dwarf *asfL1* mutant of *Lathyrus sativus* exhibits alterations in antioxidant defense. Biol Plant 56(4):675–682

Talukdar D (2012b) Flavonoid-deficient mutants in grass pea (*Lathyrus sativus* L.): genetic control, linkage relationships, and mapping with aconitase and *S* nitrosoglutathione reductase isozyme loci. Sci World J 2012:Article ID 345983, 11 pages, doi:10.1100/2012/345983

Talukdar D (2012c) A glutathione-overproducing mutant in grass pea (*Lathyrus sativus* L.): alterations in glutathione content, modifications in antioxidant defense response to cadmium stress and genetic analysis using primary trisomic. Int J Recent Sci Res 3(4):234–243

Talukdar D (2012d) Meiotic consequences of selfing in grass pea (*Lathyrus sativus* L.) autotetraploids in the advanced generations: cytogenetics of chromosomal rearrangement and detection of aneuploids. Nucleus 55(2):73–82

Talukdar D (2012e) An induced glutathione-deficient mutant in grass pea (*Lathyrus sativus* L.): modifications in plant morphology, alteration in antioxidant activities and increased sensitivity to cadmium. Bioremediat Biodivers Bioavailab 6:75–86

Talukdar D (2012f) Exogenous calcium alleviates the impact of cadmium-induced oxidative stress in *Lens culinaris* Medic. seedlings through modulation of antioxidant enzyme activities. J Crop Sci Biotechnol 15(4):325–334

Talukdar D (2012g) Total flavonoids, phenolics, tannins and antioxidant activity in seeds of lentil and grass pea. Int J Phytomed 4(4):537–542

Talukdar D (2013a) *In Vitro* antioxidant potential and type II diabetes related enzyme inhibition properties of traditionally processed legume-based food and medicinal recipes in Indian Himalayas. J Appl Pharm Sci 3(1):26–32. doi:10.7324/JAPS.2013.30106

Talukdar D (2013b) Antioxidant potential and type II diabetes related enzyme inhibition properties of raw and processed legumes in Indian Himalayas. J Appl Pharm Sci 3(03):013–019

Talukdar D (2013c) Arsenic-induced oxidative stress in the common bean legume, *Phaseolus vulgaris* L. seedlings and its amelioration by exogenous nitric oxide. Physiol Mol Biol Plants 19(1):69–79

Talukdar D (2013d) Selenium priming selectively ameliorates weed– induced phytotoxicity by modulating antioxidant defense components in lentil (*Lens culinaris* Medik.) and grass pea (*Lathyrus sativus* L.). Annu Rev Res Biol 3(3):195–212

Talukdar D (2013e) Floristic compositions and diversity of weed taxa in lentil (*Lens culinaris* Medik.) fields. Bull Environ Pharmacol Life Sci 2(3):33–39

Talukdar D (2013f) Cytogenetics of a reciprocal translocation integrating distichous pedicel and tendril-less leaf mutations in *Lathyrus sativus* L. Caryologia: Int J Cytol Cytosyst Cytogenet 66(1):21–30

Talukdar D, Biswas AK (2002) Characterization of an induced mutant and its inheritance in grass pea (*Lathyrus sativus* L.). Indian J Genet 62:355–356

Talukdar D, Biswas AK (2006) An induced mutant with different flower colour and stipule morphology in *Lathyrus sativus* L. Indian J Genet 66:365–367

Talukdar D, Biswas AK (2007) Seven different primary trisomics in grass pea (*Lathyrus sativus* L.). I Cytogenetic characterization. Cytologia 72:385–396

Talukdar D, Biswas AK (2008) Seven different primary trisomics in grass pea (*Lathyrus sativus* L.). II. Pattern of transmission. Cytologia 73:129–136

Talukdar D, Talukdar T (2003) Inheritance of growth habit and leaf-shape in mungbean [*Vigna radiata* (L.) Wilczek.]. Indian J Genet 63(2):165–166

Talukdar D, Talukdar T (2012) Traditional food legumes in Sikkim Himalayas: preparation of foods, uses and ethnomedicinal perspectives. Int J Curr Res 4(4):64–73

Talukdar D, Talukdar T (2013) Catalase-deficient mutants in lentil (*Lens culinaris* Medik.): perturbations in morpho-physiology, antioxidant redox and cytogenetic parameters. Int J Agric Sci Res 3(2):197–212

Talukdar D (2014a) A common bean (Phaseolus vulgaris) mutant with constitutively low cysteine desulfhydrase activity exhibits growth inhibition but uniquely shows tolerance to arsenate stress. Environ Exp Biol 12:73–81

Talukdar D (2014b) Differential response of cysteine-deficient lentil (Lens culinaris Medik.) mutants impaired in foliar O-acetylserine(thiol)-lyase expression. Plant Gene Trait 5(5):33–39

Talukdar D, Biswas SC, Biswas AK (2001) An induced dwarf mutant of grass pea. Indian J Genet 61(4):383–384

Talukdar D, Biswas SC, Biswas AK (2002) An induced flower colour mutant in grass pea (*Lathyrus sativus* L.). Indian J Genet 62:162

Tang X, Szinay D, Lang C, Ramanna MS, van der Vossen EA, Datema E, Lankhorst RK, de Boer J, Peters SA, Bachem C et al (2008) Cross-species bacterial artificial chromosome-fluorescence in situ hybridization painting of the tomato and potato chromosome 6 reveals undescribed chromosomal rearrangements. Genetics 180(3):1319–1328

Till BJ, Reynolds SH, Weil C, Springer N, Burtner C, Young K, Bowers E et al (2004) Discovery of induced point mutations in maize genes by TILLING. BMC Plant Biol 4:12

Tiwari VK, Riera-Lizarazu O, Gunn HL, Lopez K, Iqbal MJ et al (2012) Endosperm tolerance of paternal aneuploidy allows radiation hybrid mapping of the wheat D-genome and a measure of γ ray-induced chromosome breaks. PLoS One 7(11):e48815

Tomlekova NB (2010) Induced mutagenesis for crop improvement in Bulgaria. Plant Mutat Rep 2(2):4–27

Tsai H, Missirian V, Ngo KJ, Tran RK, Chan SR, Vankatesan S, Comai L (2013) Production of a high efficiency TILLING population through polyploidization. Plant Physiol. doi:10.1104/pp.112.213256

Tsyganov VE, Voroshilova VA, Rozov SM, Borisov AY, Tikhonovich IA (2013) A new series of pea symbiotic mutants induced in the line SGE. Rus J Genet Appl Res 3(2):156–162

Uchida N, Sakamoto T, Kurata T, Tasaka M (2011) Identification of EMS-induced causal mutations in a non-reference *Arabidopsis thaliana* accession by whole genome sequencing. Plant Cell Physiol 52(4):716–722

Valárik M, Bartos J, Kovarova P, Kubalakova M, De Jong JH, Dolezel J (2004) High resolution FISH on super-stretched flow-sorted plant chromosomes. Plant J 37:940–950

Varaprasad KS, Sharma SK, Sivaraj N, Sarker A (2011) Integrated gene resource management of underutilized legumes in India. Euphytica 180:49–56

Varshney RK, Thudi M, May GD, Jackson SA (2010) Legume genomics and breeding. Plant Breed Rev 33:257–304

Vernoux T, Wilson RC, Seeley KA, Reichheld JP, Muroy S, Brown S, Maughan SC et al (2000) The *ROOT MERISTEMLESS/CADMIUM SENSITIVE2* gene defines a glutathione-dependent pathway involved in initiation and maintenance of cell division during post-embryonic root development. Plant Cell 12:97–110

Voisin D, Nawrath C, Kurdyukov S, Franke RB, Reina-Pinto JJ et al (2009) Dissection of the complex phenotype in cuticular mutants of Arabidopsis reveals a role of SERRATE as a mediator. PLoS Genet 5(10):e1000703

Waminal NE, Park HM, Ryu KB, Kim JH, Yang T-J, Kim HH (2012) Karyotype analysis of *Panax ginseng* C.A.Meyer, 1843 (Araliaceae) based on rDNA loci and DAPI band distribution. Comp Cytogenet 6(4):425–441

Wang K, Guo W, Zhang T (2007) Development of one set of chromosome-specific microsatellite-containing BACs and their physical mapping in *Gossypium hirsutum* L. Theor Appl Genet 115:675–682

Weller JL, Hecht V, Liew LC, Sussmilch FC, Bénédicte Wenden B, Knowles CL, Vander Schoor JK (2009) Update on the genetic control of flowering in garden pea. J Exp Bot 60(9):2493–2499

Xu SJ, Singh RJ, Kollipara KP, Hymowitz T (2000) Primary trisomics in soybean: origin, identification, breeding behavior, and use in linkage mapping. Crop Sci 40:1543–1551

Yang K, Jeong S-C (2008) Genetic linkage map of the nucleolus organizer region in the soybean. Genetics 178(1):605–608

Yang X, Boateng KA, Yuan L, Wu S, Baskin TI et al (2011) The *Radially Swollen 4*Separase mutation of *Arabidopsis thaliana* blocks chromosome disjunction and disrupts the radial microtubule system in meiocytes. PLoS One 6(4):e19459

Zhang D, Yang Q, Bao W, Zhang Y, Han B, Xue Y, Cheng Z (2005) Molecular cytogenetic characterization of the *Antirrhinum majus* genome. Genetics 169(1):325–335

Zou JJ, Lee J, Singh R, Xu SS, Cregan PB, Hymowitz T (2003) Assignment of molecular linkage groups to the soybean chromosomes by primary trisomics. Theor Appl Genet 107:745–750

Zou JJ, Singh RJ, Lee J, Xu SJ, Hymowitz T (2006) SSR markers exhibit trisomic segregation distortion in soybean. Crop Sci 46(4):1456–1461

Plant Epigenetics and Crop Improvement

Sarfraz Shafiq and Abdul Rehman Khan

Contents

Abstract

Developmental cues and environmental signals remodel the chromatin structure, thus affecting various processes, including flowering time, imprinting, floral development, and biotic and abiotic stress responses in plants. Chromatin remodeling through histone tail post-translational modifications, DNA methylation, and ATP-dependent nucleosome reorganization represents a ubiquitous mechanism to regulate gene expression. Most of the epigenetic and epigenomic studies for the regulation of gene expression in response to developmental and environmental stimuli have been carried out in *Arabidopsis*. Although genetic modifications have been used for crop improvement, however, the epigenetic modifications are at their beginning. In this chapter, we summarize the roles of chromatin-remodeling mechanisms in response to environmental stimuli and discuss their potential for crop improvement.

Keywords

Histone modifications • Chromatin remodeling • DNA methylation • Biotic and abiotic stresses • Crop improvement

S. Shafiq, Ph.D. (✉) • A.R. Khan, Ph.D.
Department of Environmental Sciences, COMSATS I.I.T., University Road, Abbottabad, Pakistan
e-mail: frazishafiq@gmail.com

D. Barh et al. (eds.), *PlantOmics: The Omics of Plant Science*,
DOI 10.1007/978-81-322-2172-2_6, © Springer India 2015

Introduction

In eukaryotic cell nuclei, genomic DNA is packaged into a highly organized nucleoprotein complex known as chromatin. The fundamental unit of chromatin is the nucleosome, which is composed of ~147 base pairs of DNA wrapped around a core of eight histone molecules (two copies of each of the histones H2A, H2B, H3 and H4). Nucleosomes are not simply static structural units, but are rather dynamic. Nucleosomes can be moved, stabilized/destabilized, and disassembled/reassembled at particular genome locations in response to specific environmental signals or developmental cues. The resulting dynamic of the chromatin structure directly modulates the DNA accessibility, thus regulating all DNA-template processes (i.e., transcription, DNA replication, DNA repair, recombination, transposition, or chromosome segregation) and affecting various processes in plants such as root growth, flowering timing, floral organogenesis, gametophyte or embryo formation, as well as the response to pathogens or environmental changes (Berr et al. 2011). However, not all genes are active at all times. Therefore, cells use several mechanisms along the genome to alter the chromatin structure and the properties of a nucleosome in order to specifically control gene expression. Regulation of gene expression within the chromatin context is controlled by different mechanisms, including nucleosome assembly, ATP-dependent nucleosome reorganization, DNA methylation, and post-translational covalent histone modifications (e.g., acetylation, ubiquitination, methylation, phosphorylation, sumoylation).

Different epigenetic regulators are controlling all the above mechanisms, and the changes in these regulators can influence gene expression of a particular gene or set of genes, while the underlying DNA sequence remains identical (Jablonka and Raz 2009). Most of these changes are reversible developmental effects, and they are part of molecular processes encoding phenotypic plasticity in response to environmental variation (Richards et al. 2010) (Fig. 1). However, inheritable chromatin changes have also been reported (Jablonka and Raz 2009). At this point, it is important to clarify that those modifications which are not inheritable are not included in epigenetics as according to the definition of epigenetics, these modifications must be inheritable (mitotic and/or meiotic). Thus, we can broadly classify these modifications into nonheritable chromatin modifications (chromatin modifications that are the result of processes such as DNA repair or phosphorylation of serine 10 of histone H3, which are observed only at specific times during the cell cycle and are, therefore, unlikely to encode epigenetic information (Springer 2013)) and heritable chromatin modifications. The heritable chromatin modifications can further be classified into mitotically transmissible modifications that are reset in the next generation and meiotically transgenerational chromatin modifications that are inherited/transmitted to the following generations. The mitotically stable epigenetic marks, which accompany development, are mainly histone modification, but there are some examples of involvement of DNA methylation as well (Lauria et al. 2004; Zemach et al. 2010; Khan et al. 2013). However, DNA methylation can exhibit a relatively stable pattern of inheritance even over hundreds of years (Cubas et al. 1999; Manning et al. 2006). Because heritability determines the potential of changes or variations of a trait, it is essential to determine the degree of heritability of epigenetic modifications, their impact on given ecologically important traits (Fisher 1930; Falconer 1996), their role in individual adaptation to changing environment (Visser 2008; Hoffmann and Sgrò 2011) and ultimately in crop improvement.

The heritable epigenetic mutations, i.e., epimutations/epialleles, can be classified into three categories on the basis of relative dependence on the genotype. Pure epialleles constitute the first category, which is solely epigenetic, meaning that they are independent of the genetic variations. The second category is facilitated epialleles, which are not fully dependent on genetic variation, although they are linked and even

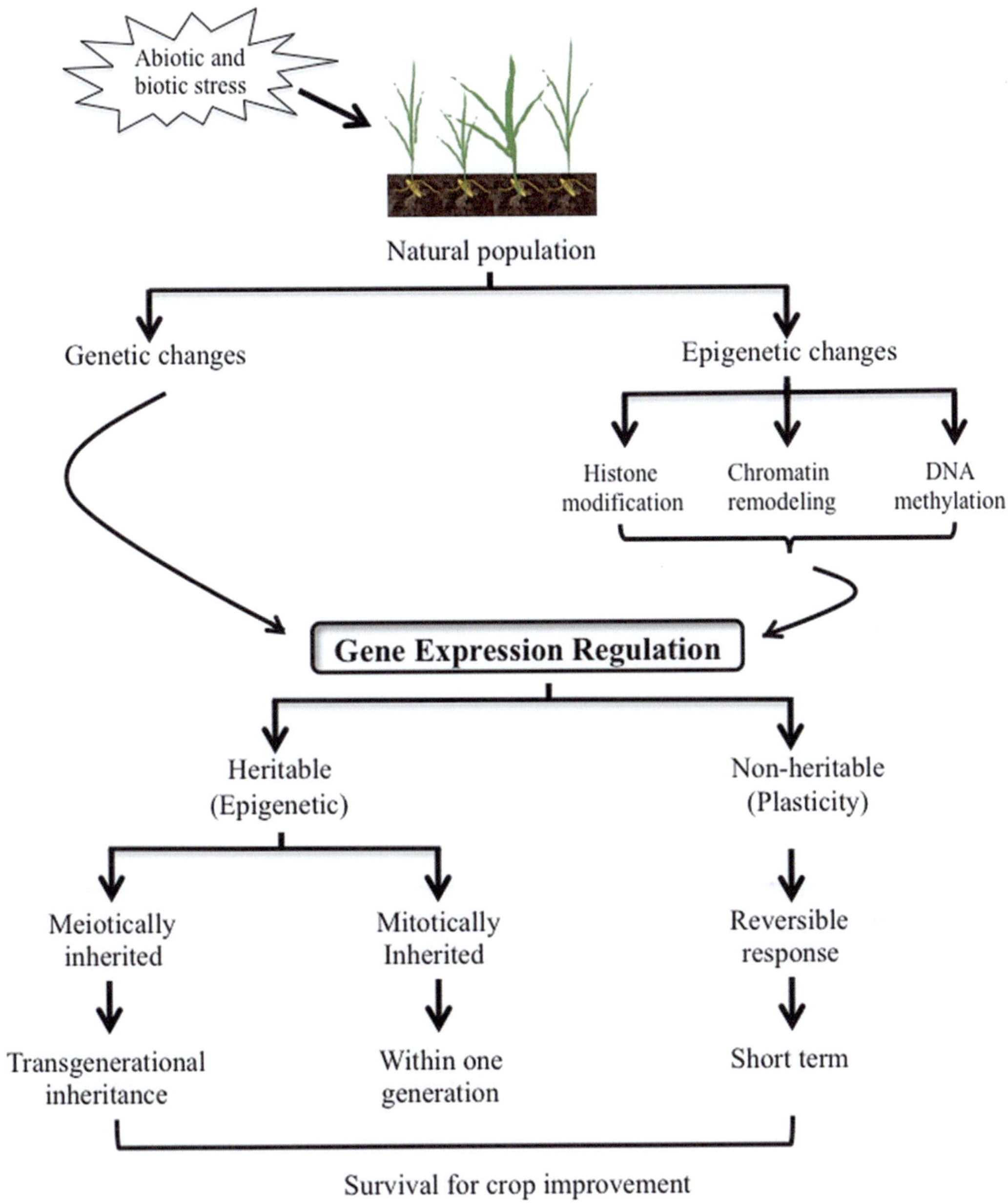

Fig. 1 Gene expression regulation through genetic and epigenetic modifications in natural population in response to environmental stimuli. The genetic and epigenetic changes may act alone or together and regulate the gene expression, which may result in a heritable and nonheritable change and may lead to a survival and/or crop improvement

caused by a genetic variant. The example for this kind of epialleles is DNA methylation spreading into a gene after the insertion of a neighboring transposon where this methylation of the gene is maintained across generations even after the facilitating transposon is excised or segregated away, meaning they could be partly attributable to both genetic and epigenetic differences. The third category is obligate epialleles, which are directly determined by genetic variants and co-segregate with these methylation variants (Woo et al. 2007). For example, methylation of a gene may be dependent on the presence or absence of a nearby transposon. There are many examples of epimutations that provide evidence that the genetic events like transposon insertions, duplications, and other structural rearrangements might trigger the chromatin remodeling which results in epigenetic control for particular haplotypes (Martin et al. 2009; Durand et al. 2012).

Increasing world population and changing climate demand to improve crop species. Although Mendelian-based genetic approaches and DNA sequence variation to select and improve crop varieties capture a substantial portion of heritable variation, dissecting epigenetic mechanisms could lead to more efficient improvement of crops. A crop improvement strategy includes the response to environmental stimuli, and the potential role of chromatin modifications in biotic and abiotic stresses has been recently reported (Chinnusamy and Zhu 2009; Kim et al. 2010; Berr et al. 2012). In view of global climate change, improving our knowledge of epigenetic regulation could have a significant impact on breeding for increased stress tolerance. In this chapter, we summarize the recent advances in epigenetic regulation in response to stress and discuss the potential of epigenetic regulatory mechanisms for crop improvement.

Histone Modifications

The N-terminal tails of histone are subjected to different covalent posttranslational modifications (PTMs) through the addition of acetyl or methyl groups and small peptide such as ubiquitin. Numerous PTMs may occur on one histone or different histones from the same nucleosome. Histone modifications, particularly acetylation/deacetylation and methylation/demethylation, epigenetically regulate the response to various stresses (Table 1). Here below we summarize the current knowledge of the enzymes responsible for histone modifications and involvement to environmental stimuli.

Histone Acetylation

Histone acetylation is linked to transcriptional activation in euchromatin and also related to DNA replication, recombination, and repair (Allfrey et al. 1964; Allis et al. 1985; Unnikrishnan et al. 2010). Acetylation by addition of an acetyl group to histone lysine (K) residues neutralizes the positive charge of lysine and therefore modifies the histone-DNA interaction, relieving DNA from its condensate state and exposing it to the transcriptional machinery. In *Arabidopsis*, lysine residues of histone H3 (K9, K14, K18, K23, and K27) and H4 (K5, K8, K12, K16, and K20) are subjected to acetylation modifications (Earley et al. 2007; Zhang et al. 2007). Histone acetyltransferases (HATs) are divided into four main classes based on the sequence homology with yeast and mammalian HATs and mode of action: GNAT (GCN5-related *N*-terminal acetyltransferases), MYST (MOZ, Ybt2, Sas2, Tip60 like), CBP/p300 (CREB-binding protein), and TAF1/TAF_{II}250 families (Sterner and Berger 2000). AtGCN5 (GENERAL CONTROL NON-REPRESSIBLE 5) was shown to acetylate H3 in vitro (Earley et al. 2007). *Atgcn5* mutant showed reduced levels of global H3 acetylation (Bertrand et al. 2003), particularly on H3K14 and H3K27 at certain gene loci (Benhamed et al. 2006). AtGCN5 was found to be involved in environmental responses (i.e., cold), along with other development pathways (Vlachonasios et al. 2003). AtGCN5 not only interacts with *Arabidopsis* Ada2 homologues AtADA2a and AtADA2b in vitro but also acetylates AtADA2a/b (Stockinger et al. 2001; Mao et al. 2006). *Atada2b* mutants showed a hypersensitive response to salt and abscisic acid (ABA) and altered response to low-temperature stress (Hark et al. 2009). H3 and H4 acetylation was found reduced on *COR6.6* (*COLD-RESPONSIVE 6.6*), *RAB18* (*RESPONSIVE TO ABA 18*), and *RD29b* (*RESPONSIVE TO DESSICATION 29b*) genes under salt stress in *Atada2b* mutants (Kaldis et al. 2011). The cold-induced transcription factor CBF1 (C-repeat/DRE BINDING FACTOR 1) interacts with AtADA2 and AtGCN5 (Mao et al. 2006), and they positively regulate the

Table 1 Histone modifications and chromatin remodeling factors involved in biotic and abiotic stresses

Type	Gene	Plant	Function	References
HATs				
GNAT	*AtGCN5*	Arabidopsis	Cold stress	Vlachonasios et al. (2003)
	AtABO1	Arabidopsis	Drought and oxidative stress tolerance, ABA sensitive	Chen et al. (2006)
	AtELP2	Arabidopsis	Oxidative stress tolerance, ABA sensitive	Zhou et al. (2009)
	AtELP4	Arabidopsis	ABA sensitive	Zhou et al. (2009)
	AtELP6	Arabidopsis	ABA sensitive	Zhou et al. (2009)
	OsHAG702	Rice	Cold and heat stress, ABA sensitive	Liu et al. (2012)
	OsHAG703	Rice	Cold, drought, and salt stress; ABA sensitive	Liu et al. (2012) and Fang et al. (2014)
	OsHAG704	Rice	Heat stress	Liu et al. (2012)
	HvGCN5	Barley	ABA sensitive	Papaefthimiou et al. (2010)
	HvElp3	Barley	ABA sensitive	Papaefthimiou et al. (2010)
MYST	*OsHAM701*	Rice	Drought and salt stress	Liu et al. (2012) and Fang et al. (2014)
	HvMYST	Barley	ABA sensitive	
CBP/p300	*OsHAC701*	Rice	Cold, heat, and salt stress; ABA sensitive	Liu et al. (2012)
	OsHAC703	Rice	Cold, drought, and salt stress; ABA and SA sensitive	Liu et al. (2012) and Fang et al. (2014)
	OsHAC704	Rice	Cold, heat, and salt stress; SA sensitive	Liu et al. (2012)
TAF1	*OsHAF701*	Rice	Cold and drought stress	Liu et al. (2012) and Fang et al. (2014)
HDACs				
RPD3/HDA1	*AtHDA19*	Arabidopsis	Resistance to *A. brassicicola* and *P. syringae*, salt stress tolerance	Zhou et al. (2005), Chen and Wu (2010), and Choi et al. (2012)
	OsHDA705	Rice	SA, JA, and ABA sensitive	Fu et al. (2007)
	OsHDA714	Rice	Cold, salt, and mannitol stress	Fu et al. (2007)
HD2-like	*AtHD6*	Arabidopsis	Freezing tolerance and JA signaling	
	AtHD2C	Arabidopsis	Salt and drought stress tolerance, ABA sensitive	
	OsHDT701	Rice	Resistance to *M. oryzae* and *Xoo*; SA, JA, and ABA sensitive	Fu et al. (2007), Li et al. (2011), and Ding et al. (2012)
	OsHDT702	Rice	SA, JA, and ABA sensitive	Fu et al. (2007)
	HvHDAC2-1	Barley	SA, JA, and ABA sensitive	Demetriou et al. (2009)
	HvHDAC2-2	Barley	SA, JA, and ABA sensitive	Demetriou et al. (2009)
SIR2	*AtSRT2*	Arabidopsis	Resistance to *P. syringae*, SA signaling	Wang et al. (2010)
	OsSIRT701	Rice	Cold, salt, and mannitol stress	Fu et al. (2007)
	OsSIRT702	Rice	Cold, salt, and mannitol stress	Fu et al. (2007)
	OsSRT1	Rice	Oxidative stress tolerance	Huang et al. (2007)

(continued)

Table 1 (continued)

Type	Gene	Plant	Function	References
HMTs				
Lysine	*AtATX1*	Arabidopsis	Drought stress, SA sensitive	Ding et al. (2011)
	AtSDG8	Arabidopsis	Resistance to *A. brassicicola* and *B. cinerea*, JA/ET	Berr et al. (2010) and Palma et al. (2010)
	HvTX1	Barley	Drought stress	Papaefthimiou and Tsaftaris (2012b)
	HvE(Z)	Barley	ABA sensitive	Kapazoglou et al. (2010)
Arginine	*AtPRMT5*	Arabidopsis	Salt stress tolerance	Zhang et al. (2011)
HDMs				
Jumonji (jmj)	*HvPKDM7*	Barley	Drought stress	Papaefthimiou and Tsaftaris (2012a)
Others				
Ubiquitination	*AtHUB1*	Arabidopsis	Resistance to *B. cinerea* and *A. brassicicola*	Dhawan et al. (2009)
PC Complex	*AtMSI1*	Arabidopsis	Drought stress tolerance	Alexandre et al. (2009)
	HvFIE	Barley	ABA sensitive	Kapazoglou et al. (2010)
Remodelers	*AtCHR12*	Arabidopsis	Drought, heat, and salinity stress	Mlynárová et al. (2007)
	AtBRM	Arabidopsis	Drought stress tolerance	Han et al. (2012)
	AtSYD	Arabidopsis	Resistance to *B. cinerea*	Walley et al. (2008)

Abbreviations: *HAT* histone methyltransferases, *HDAC* histone deacetylases, *HMTs* histone methyltransferases, *HDM* histone demethylases, *PC* polycomb

expression of cold-inducible genes during cold stress (Pavangadkar et al. 2010). This suggests that CBF is recruiting GCN5-containing activator complexes to activate the cold-responsive genes. SGF29 (*SAGA-ASSOCIATED FACTOR 29*), another component of GCN5-containing complexes in yeast, has two orthologs in *Arabidopsis* AtSGF29a and AtSGF29b. *Atsgf29a* mutants showed increased tolerance to salt stress (Kaldis et al. 2011), whereas *Atada2b* mutants were hypersensitive. This suggests that different components of GCN5-containing HAT complexes may play a different role in plant stress tolerance. Elongator HAT complex is involved in ABA signaling, drought, and oxidative stress responses in *Arabidopsis* (Chen et al. 2006; Zhou et al. 2009). *AtABO1/ELO2* (*ABA OVERLAY SENSITIVE 1*), an Elp1 homologue of yeast, was identified in a genetic screen of drought-resistant mutant (Chen et al. 2006). *Atabo1/elo2/elp1* mutant showed ABA hypersensitivity in germination and seedling growth and also showed drought- and oxidative-resistant phenotype (Chen et al. 2006). Mutation in the genes coding for the core subcomplex subunits *AtABO1/ELO2/ELP1* and *AtELP2 (ELONGATOR SUBUNIT 2)*, but not in the genes coding for accessory subcomplex subunits *AtELP4 (ELONGATOR SUBUNIT 4)* and *AtELP6 (ELONGATOR SUBUNIT 6)*, caused stomatal closing to be hypersensitive to ABA (Zhou et al. 2009). Furthermore, these single mutants showed resistance to oxidative stress and to CsCl compared to the wild type plant (Zhou et al. 2009). AtELP2 and AtELP3 *(ELONGATOR SUBUNIT 3)* were also required for both basal immunity and effector-triggered immunity (ETI), but not for systemic acquired resistance (SAR) (DeFraia et al. 2010; DeFraia et al. 2013). These results suggest that elongators play crucial roles in ABA signaling pathways and abiotic and biotic stress responses. AtTAF1/HAF2 was shown to be required for light-regulated gene expression (Benhamed et al. 2006). Together, HATs from GNAT family are involved in both biotic and abiotic stresses. However, involvement of HATs from CBP, MYST, and TAF1 classes in biotic and abiotic stresses response is still lacking in *Arabidopsis*.

Until now, the knowledge of HATs in the field crops is very limited. Eight HATs have been identified in rice and divided into four families: GNAT (OsHAG702, OsHAG703, and OsHAG704), MYST (OsHAM701), CBP/p300 (OsHAC701, OsHAC703, OsHAC704), and TAF1/TAF$_{II}$250 (OsHAF701) (Liu et al. 2012). Rice HATs respond to ABA, salicylic acid (SA), and various abiotic stresses, i.e., cold, heat, drought, and salt (Liu et al. 2012; Fang et al. 2014). An increase in transcription of *OsHAG702*, *OsHAG703*, *OsHAC701*, *OsHAC703*, and *OsHAM701* was observed with the exogenous application of ABA, whereas *OsHAC703* and *OsHAC704* transcript levels were reduced with SA application. In addition, *OsHAC701*, *OsHAC703*, *OsHAC704*, and *OsHAG703* transcripts were induced by salt and depressed by cold exposure (Liu et al. 2012). Furthermore, H3 (K9, K18, and K27) and H4 (K5) acetylation and transcripts of *OsHAG703*, *OsHAM701*, *OsHAC703*, and *OsHAF701* were found increased after drought stress in rice seedlings (Fang et al. 2014). Barley HATs belonging to GNAT (HvGCN5 and HvELP3) and MYST (HvMYST) families respond to ABA (Papaefthimiou et al. 2010). The expression of *HvGCN5*, *HvELP3*, and *HvMYST* was induced with exogenous application of ABA (Papaefthimiou et al. 2010). Together, these studies showed that HATs from all the four families are involved in different stresses in field crops. Therefore, the understanding of molecular mechanism may play an important role to cope with various stresses in field crops. It is hoped that this will eventually lead to a long-term improvement of stress tolerance in field crops, which is important for food security.

Histone Deacetylation

The homeostatic balance of histone acetylation is maintained through the antagonistic action between HATs and histone deacetylases (HDACs). In *Arabidopsis*, HDACs are classified into three families: the reduced potassium dependency 3 (RPD3/HDA1) superfamily, the HD2-like family, and the silent information regulator 2 (SIR2) family (Imhof et al. 1997; Sterner and Berger 2000; Strahl and Allis 2000). Functional analysis has demonstrated that HDA1 class of HDACs is involved in both biotic and abiotic stresses response in *Arabidopsis*. Overexpression of *AtHDA19* leads to increased expression of a gene that integrates jasmonic acid (JA) and ethylene (ET) signaling pathway, i.e., *ERF1* (*ETHYLENE RESPONSIVE FACTOR 1*) and *PR* (*PATHOGENESIS RELATED*) genes. This results in increased plant resistance to *Alternaria brassicicola* (Zhou et al. 2005). It is also reported that AtHDA19 (*HISTONE DEACETYLASE 19*) is involved in the repression of SA-mediated defense responses. *Athda19* mutant has increased SA contents and the expression of PR genes, resulting in enhanced resistance to *Pseudomonas syringae* (Choi et al. 2012). AtHDA19 interacts with WRKY38 (*WRKY TRANSCRIPTION FACTOR* 38) and WRKY62 (*WRKY TRANSCRIPTION FACTOR 62*) transcriptional activator to regulate plant basal defense responses (Kim et al. 2008). AtHDA6 (*HISTONE DEACETYLASE 6*), another HDAC, is also involved in JA response, and *Ataxe5/hda6* showed reduced expression of JA-responsive genes *PDF1.2 (PLANT DEFENSIN 1.2)*, *VSP2 (VEGETATIVE STORAGE PROTEIN 2)*, *JIN1 (JASMONATE INSENSITIVE 1)*, and *ERF1* (Wu et al. 2008). *Ataxe5/hda6* mutants also showed reduced freezing tolerance (To et al. 2011), indicating that AtHDA6 has a critical role in freezing tolerance. The expression of ABA and abiotic stress-responsive genes *ABI1 (ABA INSENSITIVE 1)*, *ABI2 (ABA INSENSITIVE 2)*, *KAT1 (POTASSIUM CHANNEL IN ARABIDOPSIS THALIANA 1)*, *KAT2 (POTASSIUM CHANNEL IN ARABIDOPSIS THALIANA 2)*, *DREB2A (DEHYDRATION-RESPONSIVE ELEMENT-BINDING PROTEIN 2A)*, *RD29A (RESPONSIVE TO DESSICATION 29A)*, and *RD29B* was decreased in *Ataxe5/hda6* mutant or *AtHDA6-RNAi* plants (Chen et al. 2010). Similarly, *Athda19* mutant also showed a hypersensitive response to ABA and salt stress (Chen and Wu 2010). This suggests that *AtHDA19* and *AtHDA6* may play a redundant role in modulating ABA and salt stress response. Moreover, AtHDA19 and AtHDA6 play a crucial role in responses to biotic and abiotic stresses. *AtHDA2C*, an HD2-type HDAC, was also shown to be involved in ABA and salt stress response. Overexpression of *AtHD2C* in transgenic plants showed enhanced tolerance to salt and drought stress and ABA-insensitive phenotype (Sridha and Wu 2006). Conversely, *Athd2c* mutant showed a hypersensitive response to ABA and NaCl and decreased tolerance to salt stress (Luo et al. 2012). Furthermore, AtHD2C interacts with AtHDA6 (Luo et al. 2012), suggesting that AtHD2C may functionally associate with AtHDA6 to ABA and salt stress responses and may be a part of HDAC complexes to regulate gene expression through histone modifications. *Arabidopsis SIRTUIN 2* (*AtSRT2*), an SIR2 HDAC expression, is down-regulated upon *Pseudomonas syringae* pv. *tomato* (*Pst*DC 3000) infection. AtSRT2 suppresses the expression of SA biosynthesis genes *PAD4 (PHYTOALEXIN-DEFICIENT 4)*, *EDS5 (ENHANCED DISEASE SUSCEPTIBILITY 5*), and *SID2 (SPAC24B11.11C)*, thereby supressing SA production and expression of defense-regulated genes (Wang et al. 2010).

HDAC and its involvement in biotic and abiotic stresses have also been reported in cereals. Rice has 19 genes coding for HDAC (Hu et al. 2009), which may play an important role in regulating various stress responses. *OsHDA705*, *OsHDT701*, and *OsHDT702* transcripts were found affected by SA, JA, and ABA, whereas *OsHDA714*, *OsSRT701*, and *OsSRT702* expression is modulated by cold, mannitol, and salt (Fu et al. 2007). In *OsSRT1*-RNAi transgenic rice, H3K9 acetylation and H3K9 dimethylation (H3K9me2) levels were decreased and increased, respectively, leading to H2O2 production, DNA fragmentation, cell death, and lesion-mimicking plant hypersensitive responses during incompatible interactions with pathogens. In contrast, *OsSRT1* overexpression showed an enhanced tolerance to oxidative stress (Huang et al. 2007). Overexpression of *OsHDT701*, a plant-specific HD2 HDAC, leads to decreased level of H4 acetylation on flowering and defense-related genes and enhanced susceptibility to the *Magnaporthe oryzae* and *Xanthomonas oryzae* pv. *oryzae* (*Xoo*) pathogens (Li et al. 2011; Ding et al. 2012). In contrast, silencing of *OsHDT701* showed increased levels of H4 acetylation and increased transcription of pattern recognition receptor (PRR) and defense-related genes, elevated generation of reactive oxygen species, as well as enhanced resistance to both *M. oryzae* and *Xoo* (Ding et al. 2012). *HvHDAC2-1* and *HvHDAC2-2* genes, HD2-type HDAC from barley, were found to respond to JA, ABA, and SA treatments, implying an association of these barley genes with plant resistance to biotic and abiotic stresses (Demetriou et al. 2009). *NtHD2a* and *NtHD2b* genes, HD2-type HDAC from tobacco, were found to work as inhibitors of cryptogein-induced cell death (Bourque et al. 2011). Together, HD2-type HDAC from rice and barley carries the same function, suggesting a common function among species for HDAC homologues but also possible species-specific functional diversification, in response to stress. The involvement of HDACs in biotic and abiotic stresses response in agronomically important crops and their underlying molecular mechanism is of utmost importance for sustainable crop improvement.

Histone Methylation

Histone methylation plays an essential role in diverse biological processes ranging from transcriptional regulation to heterochromatin formation. Methylation of histone can occur on lysine (K) or arginine (R) residues leading to either transcriptional activation or repression. Histone methylation not only occurs at different residues (K and R) and distinct sites (e.g., K4, K9, K27, K36, R2, and R17 of H3 and K20 and R3 of H4, etc.) but also differs in the number of methyl groups added (mono-, di-, and tri-methylated). Methylation of lysine residues does not affect their net charge but elevates the hydrophobicity nature of the side chain and may alter intra-

or intermolecular interactions or create new binding surfaces for proteins that bind preferentially to the methylated domains (Liu et al. 2010). Indeed, the arginine residue, after the addition of the methyl group, changes its shape and removes a potential hydrogen bond donor as well (Bedford and Clarke 2009). Mostly the studies have been done on histone modifications only at individual stress-induced plant genes. Very few studies with genome-wide histone methylation analysis have been reported. van Dijk et al. (2010) have studied genome-wide analysis of the histone H3 lysine 4 mono-, di-, and tri-methylation (H3K4me1, H3K4me2, H3K4me3, respectively) patterns in chromatin isolated from *Arabidopsis* rosette leaves before and after dehydration stress. Genome-wide transcript patterns in watered and dehydration-stressed plants were compared in this study. The presence of the H3K4me1, H3K4me2, and H3K4me3 marks is predominantly located on genes, and the distribution of H3K4me1 and H3K4me2 is higher than H3K4me3. Interestingly, H3K4me1, H3K4me2, and H3K4me3 patterns display different dynamics and specific patterns at upregulated, downregulated, and unaffected genes during the response to dehydration stress. A modest change in H3K4me2 and H3K4me1 levels was found at a subset of known stress response genes, but the H3K4me3 abundance over gene bodies changed more dramatically at genes whose transcript levels increased or decreased during dehydration. The different behaviors of each methylation mark during the response process illustrate that each mark plays a distinct role in the transcriptional response of implicated genes. In a recent study, genome-wide profiling of histone H3K4-trimethylation of 25-day-old rice plants under dehydration conditions was done. This analysis uncovered a positive correlation between H3K4me3 accumulation and the expression levels of some drought-responsive genes during dehydration. This correlation could be extended to genes involved in stress-related metabolite and hormone signaling pathways (Zong et al. 2012). These genome-wide histone modification studies help broaden our knowledge on whole genome scale and indicate a need to study histone modifications on a genome-wide level in response to other abiotic stresses as well.

Histone Lysine Methylation

Covalent addition of one, two, or three methyl groups (me1, me2, or me3) mainly occurs on H3K4, H3K9, H3K27, H3K36, and H4K20, and this function is exerted through histone methyltransferases (HMTs). All known plant HMTs have a so-called SET [from the initially identified *Drosophila* HMTs: Suppressor of variegation (Su(var)3-9), Enhancer of Zeste (E(z)), and Trithorax (TRX)] catalytic domain, an evolutionarily conserved sequence of 130–150 amino acids in length. SET Domain Group (SDG) proteins are classified into three subgroups: Su(var)3-9, Enhancer of Zeste (E(z)), and Trithorax (TRX). These subgroups have been shown to establish different chromatin marks, leading to different impacts on transcription. SDGs of the ASH1 and TRX subgroups primarily belong to the Trithorax group (TrxG) and are responsible for methylation on H3K36 and/or H3K4, which are associated with transcriptional activation (Agger et al. 2008; Liu et al. 2010). The E(z) subgroup SDGs catalyze H3K27 methylation associated with transcriptional gene silencing. H3K27 can be mono-, di-, and tri-methylated and seems to be one of the major gene silencing mechanisms in *Arabidopsis* because ~17 % of the coding genes were marked with H3K27me3 (Turck et al. 2007). Classically and conservatively, the Su(var)3-9 subgroup SDGs potentially show an H3K9 methyltransferase activity and are associated with inactive genes located in a euchromatic region and within highly condensed constitutive heterochromatin (Ng et al. 2007). SDG proteins have been involved in diverse biological processes, including flowering time regulation, floral organogenesis, leaf morphogenesis, parental imprinting, and seed development (Liu et al. 2010; Berr et al. 2011; Shafiq et al. 2014). AtATX1/SDG27, a member of the Trithorax group, is a methyltransferase of H3K4me3. AtATX1 was found to be involved in drought and SA pathway responses (Ding et al. 2011; Berr et al. 2012). *Atatx1* mutant displayed larger stomatal apertures, increased transpiration, and decreased tolerance to dehydration stress. AtATX1 is required for the induction of *NECD*, a gene involved in ABA biosynthesis and deposition of H3K4me3 in response to dehydration stress. AtATX1 can influence gene expression by ABA-dependent as well as ABA-independent pathways (Ding et al. 2011). AtATX1 was described as critical for basal resistance against *Pst* DC3000, and it regulates the SA-inducible expression of transcriptional factor *WRKY70* (Alvarez-Venegas et al. 2006; Berr et al. 2012). AtSDG8, another member of the Trithorax group, is the major H3K36me2/me3 methyltransferase (Xu et al. 2008). AtSDG8 was reported in *Pst* DC3000-triggered plant defense through the regulation of particular *R* genes (Palma et al. 2010) and the transcriptional activation of JA/ET signaling-related genes (Berr et al. 2010). *Atsdg8* mutant exhibited reduced resistance to *Alternaria brassicicola* and *Botrytis cinerea* (Berr et al. 2010). H3K36 methylation on defense-related genes is impaired in *Atsdg8* mutant (Berr et al. 2010; Palma et al. 2010), indicating that AtSDG8 mediates the pathogen response by regulating histone methylation of defense-responsive genes. The expression of *AtSUVH2*, *AtSUVH5*, *AtSUVH6*, and *AtSUVH8* genes encoding H3K9 methylation decreased in the progenies of salt-stressed plants (Bilichak et al. 2012). In addition, *Curly leaf* (*CLF*) gene encoding H3K27 methylation was hypermethylated in the progenies of salt-stressed plants (Bilichak et al. 2012). These results suggest that H3K9 and H3K27 methyltransferases are involved in the plant stress adaptation. Until now, HMT involvement in biotic and abiotic stresses response is very limited in crops. HvTX1, barley TRX-like H3K4 methyltransferase, has been shown to be involved in drought stress. The transcripts of *HvTX1* were found increased under drought stress (Shvarts Iu et al. 2010; Papaefthimiou and Tsaftaris 2012b). This suggests that TrxG plays an important role in plant response to environmental stresses. A homologue of polycomb complex subunit from barley HvE(Z) was found to be induced by ABA implying an association with ABA-mediated processes during seed development and stress response (Kapazoglou et al. 2010). Recently, it was shown that 18 genes containing SET domain from maize showed differential expression under salt and drought stress (Qian et al. 2014). Although SET domain proteins are involved in biotic and abiotic stresses in crops, their molecular mechanism is still missing. It is hoped that with emerging new technologies and better understanding of molecular mechanism, the SET domain proteins may have potential for sustainable crop improvement.

Histone Arginine Methylation

Arginine methylation mainly occurs at R2, R8, R17, and R26 of histone H3 and R3 of histone H4 and histone H2A. Arginine methylation can be symmetric or asymmetric and only occurs in mono- and di-methyl states (Aletta et al. 1998; Bedford and Clarke 2009). Arginine methylation is evolutionarily conserved and has been found in fungi, plants, *Caenorhabditis elegans*, *Drosophila melanogaster*, and vertebrates (Krause et al. 2007). Arginine methylation is catalyzed by a small group of protein arginine methyltransferases (PRMTs) that harbor a set of four conserved motifs (i.e., I, post-I, II, III) and a THW loop (Katz et al. 2003). Proteins that are arginine methylated play an essential role in transcriptional regulation, DNA repair, signal transduction, nuclear/cytoplasmic shuttling, RNA processing, and formation of silent chromatin (Bedford and Richard 2005; Bedford and Clarke 2009). In mammals, PRMTs are classified into two classes depending on the nature of the modification introduced. Although both type I and type II catalyze arginine monomethylation, they differ in the final type of arginine modification. The type I enzymes result in asymmetrical dimethylarginine, whereas type II enzymes result in symmetrical dimethylarginine (McBride and Silver 2001; Katz et al. 2003; Jelinic et al. 2006). The involvement of arginine methylation in biotic and abiotic stresses is very poorly understood. AtPRMT5/SKB1, a homologue of the human PRMT5 (*PROTEIN ARGININE METHYLTRANSFERASE 5*), specifically dimethylates symmetrically H4R3 as a type II arginine methyltransferase in *Arabidopsis* (Deng et al. 2010). *Atskb1* mutant displayed salt-hypersensitive phenotype. AtSKB1 suppresses the transcription of stress-responsive genes by increasing the H3R3sm2 (Zhang et al. 2011).

Histone Demethylation

Histone methylation is important for chromatin stability and gene expression and was considered irreversible until the discoveries of demethylases that antagonized or balanced the methylase activities. There are two types of methylases with distinct mechanisms, the lysine-specific demethylases (LSD1) and the Jumonji C (JmjC) domain-containing demethylases. They use different cofactors and act on different substrates to remove methyl groups from methylated lysine residues. LSD1 is catalytically limited to mono- and di-methylated lysine due to the reaction mechanism used to initiate the demethylation (Klose and Zhang 2007). Unlike LSD1, Jmj proteins do not have limitations in their catalytic mechanism and are able to demethylate mono-, di-, and tri-methyl residues (Agger et al. 2008). In *Arabidopsis* and rice, histone demethylases (HDMs) have been found to be involved in many developmental processes and gene silencing (Noh et al. 2004; Sun and Zhou 2008; Chen et al. 2013; Cui et al. 2013; Shafiq et al. 2014). Although the role of HDMs in stress response is not yet clear, evidences suggest that histone demethylation may be involved in stress responses. Increased level of H3K9/K14ac and H3K4me3 and decreased level of H3K9me2 on ABA-responsive genes (*ABI1*, *ABI2*, and *RD29B*) have been found in *Arabidopsis* after ABA treatment (Chen et al. 2010), which suggests that some HDMs are working for the demethylation of H3K9 to activate the ABA-responsive genes. Decreased levels of H3K4me1, H3K4me2, and H3K4me3 and downregulation of stress-responsive genes have been reported upon dehydration stress in *Arabidopsis* (van Dijk et al. 2010). This also suggests that HDMs are modulating the expression of stress-responsive genes. Recent reports describing a putative role of HDMs in stress response are anticipated. Putative plant-specific barley HvPKDM7 histone demethylase was found to be significantly induced by drought stress (Papaefthimiou and Tsaftaris 2012a). Genome-wide analysis of rice showed that a lot of genes were differentially H3K4me3-modified in drought stress (Zong et al. 2012), suggesting that the rice HDMs are involved in stress response.

Other Histone Modifications

Ubiquitination is the covalent attachment of a small (76 amino acids) and highly conserved protein named ubiquitin to the target protein, achieved through the sequential action of the ubiquitin-activating enzyme E1, the ubiquitin-conjugating enzyme E2 (Ubc), and the ubiquitin-protein ligase E3 (Pickart 2001; Smalle and Vierstra 2004). The substrate can remain monoubiquitinated, or the ubiquitin can have several lysine (K) residues that may be substrates themselves for subsequent addition of ubiquitins, resulting in a polyubiquitin chain. H2B monoubiquitination (H2Bub1) in yeast, animals, and *Arabidopsis* is mainly associated with transcriptional activation (Briggs et al. 2002; Dover et al. 2002; Hwang et al. 2003). AtHUB1, catalyzing H2B monoubiquitination, was reported as a regulatory component of plant defense against necrotrophic fungal pathogens (Dhawan et al. 2009). *Athub1* mutant displayed susceptibility to *B. cinerea* and *A. brassicicola*. ET and SA but not JA modulate the resistance of *Athub1* mutants to necrotrophic fungi. *Athub1-6* presents a reduced cell thickness, indicating that HUB1 may regulate resistance by altering plant cell wall-related defense mechanisms (Dhawan et al. 2009). It remains to be explored whether and how H2Bub is involved in plant defense. AtMSI1 (MULTICOPY SUPPRESSOR OF IRAI), a subunit of the Polycomb group (PcG) having H3K27 methylation activity, has been shown to be involved in drought stress (Alexandre et al. 2009). *Atmsi1* mutant displayed increased tolerance to drought stress and increased transcripts of stress and ABA-responsive genes, indicating that AtMSI1 suppresses stress-responsive genes in an ABA-dependent manner. Polycomb complex gene homologue from barley HvFIE

(*FERTILIZATION-INDEPENDENT ENDOSPERM 1*) was found to be induced by ABA implying an association with ABA-mediated processes during seed development and stress response (Kapazoglou et al. 2010).

ATP-Dependent Chromatin Remodeling Factors

ATP-dependent chromatin remodeling complexes use the energy of ATP hydrolysis to alter the structure of chromatin for the regulation of gene expression (Vignali et al. 2000). ATP-dependent chromatin remodeling factors have been found to be involved in biotic and abiotic stresses response. ATP-dependent chromatin complexes can be grouped into three classes: the SWI/SNF ATPases, the imitation switch (ISWI) ATPases, and the chromodomain and helicase-like domain (CHD) ATPases. AtCHR12, an SNF/BRAHMA-type (BRM) chromatin remodeling factor, was shown to be involved in plant growth response to adverse environmental conditions (Mlynárová et al. 2007). Under drought, heat, and salinity stress, *AtCHR12* overexpressing plants exhibited an arrested growth of normally active primary buds as well as reduced growth of primary stem. *Atchr12* mutant plants displayed less growth arrest than the wild type when exposed to stress. However, the molecular mechanism of how the AtCHR12 is involved in growth arrest under adverse environments is not clear. AtBRM, an SWI2/SNF2 chromatin remodeling ATPase, has been demonstrated to be involved in drought stress (Han et al. 2012). *Atbrm* mutant showed increased drought tolerance and regulated the expression of ABA-responsive gene *AB15*. Furthermore, nucleosomes were found destabilized upon the loss of BRM activity, indicating that BRM regulates stress responses through the regulation of nucleosome stability of *AB15*. Using wounding as stimulus, SPLAYED (SYD), a closest homologue of BRM, was shown to be required for the basal as well as stress-induced expression of genes (*PDF1.2*, *VSP2*, and *MYC2*) working downstream of JA/ET signaling pathways (Walley et al. 2008). These results indicate that ATP-dependent chromatin remodeling complexes are playing a crucial role in stress responses, thus influencing plant innate immunity and tolerance. However, the knowledge about the ATP-dependent chromatin remodeling factors in field crops is very limited. After UVB treatment of maize plants, the enrichment of SWI2/SNF2 was found at target genes, implying the involvement of chromatin remodeling factors in abiotic stress responses (Casati et al. 2008). It is expected that by exploiting the rice, maize, and *Brachypodium* genomes, chromatin remodeling complexes and their association with biotic and abiotic stresses will be studied, thus improving crop production.

DNA Methylation

DNA methylation refers to a chemical modification of genomic DNA by the addition/attachment of a methyl (–CH3) group to specific nucleotide bases, which could be cytosine or adenine. It occurs most commonly on cytosine base leading to a 5-methylcytosine. It is conserved in major eukaryotic groups, i.e., plants, animals, and fungi, with few exceptions (Goll and Bestor 2005; Henderson and Jacobsen 2007). Although methylation at cytosine can be explained in a variety of DNA sequence contexts, mechanistically it can be classified broadly into three contexts, CG, CHG, and CHH (where H denotes A, T, or C) (Law and Jacobsen 2010). The pattern of occurrence of DNA methylation varies, i.e., it mainly occurs at CG sites in mammals, but it can occur in CG, CHG, and CHH contexts in plants (Feng et al. 2010). In *Arabidopsis*, the genome-wide DNA methylation level is reported to be 24 %, 6.7 %, and 1.7 % for CG, CHG, and CHH contexts, respectively. Alteration in DNA methylation is associated with gene regulation and transposable element silencing in eukaryotes (Law and Jacobsen 2010). It acts differently in different regions of the genome. In transposable elements (TE), where it appears in all three contexts (CG, CHG, and CHH), it is responsible for transcriptional silencing. In genes, DNA methylation is mainly restricted to CG sites (Law and

Jacobsen 2010; Zhang et al. 2010), although CHG and CHH methylation has also been reported recently (González et al. 2011). The presence of DNA methylation at the gene promoter region is generally negatively correlated with gene expression (Zhang et al. 2006; Li et al. 2012). Furthermore, DNA methylation can also occur within the gene body (i.e., away from the 5′ and 3′ ends of transcription units), in the so-called bell-shaped CG "gene body methylation" pattern. However, the function of gene body methylation is still not clear (Zhang et al. 2010).

DNA METHYLTRANSFERASE 1 (MET1) primarily maintains CG methylation, which is a homologue of the mammalian DNA METHYLTRANSFERASE1 (DNMT1). Moreover, CHROMOMETHYLASE 3 (CMT3) maintains CHG methylation. Maintenance of CHH methylation is complex because it is asymmetrical; it needs to be reacquired *de novo* after each replication, through the action of the plant-specific RNA-dependent DNA methylation (RdDM) pathway (Law and Jacobsen 2010) in which small RNAs (24 nucleotides long) target the *de novo* methyltransferase DOMAINS REARRANGED METHYLTRANSFERASE 2 (DRM2) to homologous genomic loci to establish DNA methylation. The modifications in DNA methylation in response to environmental stress have been reported in both locus-specific and genome-wide studies (Table 2). Some of such examples are explained below: In tomato, altered DNA methylation levels were observed on *Asr1* (*ABSCISIC ACID STRESS RIPENING 1*) gene under drought stress (González et al. 2011). *Asr1* is a non-transposon, protein-coding, water stress-inducible gene of the LEA superfamily in tomato. The expression of *Asr1* increases with the longer duration of drought stress. In addition, the CHH methylation was decreased in drought conditions showing the negative correlation with gene expression (González et al. 2011). Furthermore, the existence of all the three contexts of DNA methylation (CG, CHG, CHH) was reported in the regulatory region of *Asr2* (*ABSCISIC ACID STRESS RIPENING 2*). Interestingly, the gene body methylation was restricted to only one context (CG). The site-specific removal of methyl marks from CNN sites in the regulatory region was observed under drought stress. This response of *Asr2* is heritable through generations and could have evolutionary importance (González et al. 2013). In maize, genome-wide DNA methylation pattern was studied under cold stress. It led to the identification of a fragment named *ZmMI1*, which was transcribed only in cold stress conditions (Steward et al. 2002). About 49 transcription factors showed differential expression in soybean on exposure to salinity stress. Moreover, DNA methylation and expression profiles of one *MYB (MYELOBLASTOSIS)*, one *b-ZIP (BASIC LEUCINE ZIPPER)*, and two *AP2/DREB* transcription factor gene family members were significantly correlated (Song et al. 2012). Choi and Sano (2007) analyzed glycerophosphodiesterase-like protein (*NtGPDL*) gene in tobacco to understand the effect of various stresses, including aluminum, salt, and cold stress. The increased transcription and CG demethylation in the coding regions of *NtGPDL* were observed under stress conditions (Choi and Sano 2007). Change in DNA methylation pattern in response to biotic stress was also reported. In tobacco, a pathogen-responsive gene *NtAlix1 (ALG-2 INTERACTING PROTEIN X 1)* was studied upon the infection of tobacco mosaic virus (TMV). The change of DNA methylation at *NtAlix1* was observed after 24 hours of infection, indicating that DNA methylation pattern undergoes alteration in response to biotic stresses which is closely related to the activation of stress-responsive genes (Wada et al. 2004). One interesting example came from the analysis of *Arabidopsis*, where a putative small RNA target region about 2.6 kb upstream of the ATG start codon of *AtHKT1* (*HIGH-AFFINITY K+TRANSPORTER 1*) gene is normally heavily methylated and its hypomethylation represses the *AtHKT1* gene expression, which is crucial for salt tolerance (Baek et al. 2011). In *Arabidopsis*, the promoter region of *AtRMG1 (RESISTANCE METHYLATED GENE 1)* gene is targeted by RdDM and ROS1-dependent DNA demethylation as a defense response against the *P. syringae* pathogen (Yu et al. 2013). Sharma et al. (2009) character-

Table 2 DNA methylation modifications involved in biotic and abiotic stresses

Sr. No.	Genomic region	Plant species	Stress	Mode of action	References
1	*Asr1*	Tomato	Drought stress	CG hypermethylation and CHH hypomethylation	González et al. (2011)
2	*Asr2*	Tomato	Drought stress	CHH hypomethylation in regulatory region	González et al. (2013)
3	*NtGPDL*	Tobacco	Aluminum, salt, and cold	Hypomethylation	Choi and Sano (2007)
4	*NtAlix1*	Tobacco	Tobacco mosaic virus	Alteration in DNA methylation pattern	Wada et al. (2004)
5	*ZmMI1*	Maize	Cold stress	Root-specific hypomethylation	Steward et al. (2002)
6	*AtHKT1*	Arabidopsis	Salt tolerance	Loss in cytosine methylation in a putative small RNA target region	Baek et al. (2011)
7	*RMG1* promoter	Arabidopsis	*P. syringae*	RMG1 is targeted by RdDM and ROS1-dependent DNA demethylation	Yu et al. (2013)
8	*Glyma11g02400* (Promoter)	Soybean	Salinity stress	−518 to −274 cytosines were demethylated following exposure to salinity stress for 1–24 h	Song et al. (2012)
9	*Glyma16g27950 (Promoter)*	Soybean	Salinity stress	Hypomethylation at transcription start codon (24 to 233)	Song et al. (2012)
10	*Glyma20g30840 (Promoter)*	Soybean	Salinity stress	Hypomethylated cytosines at promoter region 1 (−87 to 163)	Song et al. (2012)
11	Genome-wide	Maize	Cold stress	Global methylation shift. Mainly demethylation of fully methylated fragments	Shan et al. (2013)
12	Genome-wide	Mangrove	Salinity stress	Global hypomethylation	Lira-Medeiros et al. (2010)
13	Genome-wide	Rice	Salinity stress	Differential methylation of salt stress-related genes, retrotransposons, and chromatin modifier genes	Karan et al. (2012)
14	Genome-wide	Rice	Drought stress	Genotype-dependent differential methylation	Wang et al. (2011)
15	Nuclear genome	*Mesembryanthemum crystallinum*	High salinity	CHG hypermethylation	Dyachenko et al. (2006)

ized ten putative DNA methyltransferases in rice. Expression analysis of them was done at different developmental stages and under abiotic stress. High salinity and cold stress induced *OsCMT2*, but drought stress showed no effect. Drought and salinity stress caused *OsCMT3* to exhibit approximately a six- and four fold reduction in mRNA accumulation in rice seedlings subjected to high-salt and dehydration conditions, respectively. In addition to locus-specific stress responses, a good deal of work has been done on the genome-wide level. Genome-wide DNA methylation response to cold stress by MSAP (methylation-sensitive amplification polymorphism) technique in maize revealed global DNA methylation shift. The main part of this shift was attributed to the demethylation of fully methylated fragments (Shan et al. 2013). Lira-Medeiros et al. (2010), in an interesting comparative study of mangrove plants, growing in salt marsh neighborhood and riverside habitat, revealed that riverside plants were much taller and thicker than the plants growing in salt marsh neighborhood. Genome-wide DNA methylation analysis showed considerable hypermethylation in riverside plants in comparison with the plants growing in salt marsh neighborhood suggesting a pivotal role of natural epigenetic variations in a plant population toward environmental adaptation. Similarly, a genome-wide study by MSAP analysis performed in diverse rice genotypes differing in their salt-responsive characteristics highlighted differential methylation and expression of salt stress-related genes, retrotransposons, and chromatin modifier genes (Karan et al. 2012). Another study of genome-wide DNA methylation analysis under drought stress has been reported in rice. In this study, the comparison of two genotypes under drought stress and subsequent recovery revealed the genotype-specific DNA methylation modifications, which were mostly reversed after recovery, but some were maintained even after recovery indicating some sort of stress memory. This study illustrated the importance of these induced epigenetic changes in regulatory mechanisms for adaptation of rice plant to environmental stresses (Wang et al. 2011). Dyachenko et al. (2006) reported hypermethylation of CHG methylation in nuclear genome of *Mesembryanthemum crystallinum* plants during high-salinity stress imposition. These examples provide a glimpse of the importance of epigenetic mechanisms in the plant response to the environmental variation and their potential involvement in the adaptive strategies devised by the plants. In this respect, the reported data sets of various plant methylomes could provide the basis for the selection of differential epigenetic regions as probable targets for the genetic manipulation for crop improvement.

Epigenetic Outlook for Crop Improvement

One very significant part of the success attained in the field of crop yield improvement is attributed to the plant breeding and genetics. The utilization of desirable available variation has been one of the main roles followed by the scientists for the improvement of crop plants. In the last two decades, the researchers across the globe have accumulated the wealth of knowledge that provides the evidence of prevalence of epigenetic variability (natural as well as generated) and its potential to influence the phenotype (agronomic traits) and large crop improvement. Histone modifications are involved in mitotically stable transcriptional activation or repression and exhibit lower level of transgenerational heritability. So, can the mitotically stable epigenetic information be used for crop improvement? The answer is that very similar to transcriptional factors, chromatin changes also control plant morphology and response to the environment, and a greater control over traits may be achieved by understanding these mechanisms, which is highly important for the breeding point of view. Several cases of naturally occurring epialleles (i.e., DNA methylation alleles that are independent of DNA sequence variation causing a visible phenotype) have been described, such as the *Lcyc* locus in *Linaria vulgaris* (Cubas et al. 1999) and an SBP-box gene in tomato (Manning et al. 2006). DNA methylation of natural epialleles has also been described at a larger genomic scale for species such as

Arabidopsis (Cervera et al. 2002; Vaughn et al. 2007), *Spartina anglica* (Salmon et al. 2005), or *Populus trichocarpa* (Raj et al. 2011). These examples indicate the existence of epigenetic variations in natural populations. Therefore, epigenetic variants could be used in breeding programs for the improvement of crops because breeders select for a particular trait rather than a molecular mechanism.

These genome-specific techniques have important implications on the crop improvement by their potential role in the identification of regions, which show epigenetic modifications under various kinds of stress. This identification could lead to the characterization of these regions of interests in the genome. Further studies of these regions could lead to detection of epialleles which could be incorporated into the breeding programs and play their role in crop improvement. Moreover, recent techniques enabled breeders to generate desired allelic variation through the mutagenesis or transgenic modifications to develop a trait not observed in natural population. Epigenetic regulation affects transgene behavior and could be used to establish novel epialleles for breeding purposes. Different approaches have been proposed, speculated, and/or initiated to use the epigenetic diversity in the breeding programs. One big hurdle in producing or developing epigenetic diversity in crop plants is the lack of availability of genome-wide DNA methylation mutants like *met1* and *ddm1*. In such cases, the usage of chemical inhibitors like 5-azadeoxycytidine is a good alternative. Different studies have reported the transgenerational inheritance of the modifications created by its treatment. (Sano et al. 1990; Akimoto et al. 2007). Akimoto et al. (2007) reported that in progenies of 5-azadeoxycytidine-treated rice, some of the altered phenotypes were stably inherited even after 10 years. Some of these phenotypes were of interest from the breeding point of view like resistance of a bacterial pathogen *Xanthomonas oryzae*. Another interesting and exciting approach, which is drawing much attention, is the usage of epigenetic- Recombinant Inbred Lines (epiRILs). These lines are created by artificial crossing of DNA methylation mutants, i.e., *decreased DNA methylation 1* (*ddm1*) or *methyltransferase 1* (*met1*) with their wild types (Johannes et al. 2009; Reinders et al. 2009). Since these mutants are deficient in DNA methylation machinery but genetically similar as that of the wild type, the resulting lines (epiRILs) have almost identical DNA sequences but divergent patterns of DNA methylations. These patterns are reported to be stable across many generations through molecular analysis. Analysis of these lines has shown that they have widespread phenotypic variation for morphological or developmental traits, like flowering time, plant height, as well as biotic and abiotic stresses (Johannes et al. 2009; Reinders et al. 2009; Zhang et al. 2013). Although these methylation variants do not necessarily reveal natural variation, they can be very useful in multiple ways. They can serve as a good material to understand the extent and potential role of the epigenetic variations, which are independent of genetic variations. They can also help to understand the extent of the phenotypic variation caused by the random combination of plant-specific epigenome. The above mentioned hypothesis was confirmed from the various recent publications giving further insight into the basic mechanisms which require DNA methylation, like the effect of DNA methylation on crossing over where the results showed that the distribution of crossing over event is sensitive to DNA methylation, but the rate of crossing over is not affected by it (Colomé-Tatché et al. 2012; Mirouze et al. 2012). Similarly, significant heritable variation in growth rate in response to biotic stresses was reported. All these results further support the opinion that considerable heritable variations in economically important traits could be created by variation in the DNA methylation patterns, and these kinds of approaches could be applicable for crop improvement. Utilization of epialleles generated and/or identified by various researchers in diverse plant/crop species should be exploited in different breeding programs. Such kind of program could start with the identification and understanding of epigenetic pattern in individuals of the selected population. This will lead to the identification of specific phenotypes and, finally, the association studies of that inherited phenotype and epigenetic variation. With the advances in the genome editing technologies, the usage of locus-specific epigenetic modifications could also be used for crop improvement (Chen and Gao 2014).

Techniques Used in Epigenetic/ Epigenomic

There are different techniques in use for the detection of DNA methylation profile and DNA-protein interaction both at locus-specific level and on genome-wide level. Some examples of these techniques are discussed here. On locus-specific level, the DNA methylation profile can be studied through bisulfite treatment technique and through the use of methyl-sensitive restriction enzymes. On a genome-wide level, DNA methylation profile can be studied through MSAP (methyl-sensitive amplification polymorphism) (Yaish et al. 2014), through the use of HPLC (high-performance liquid chromatography) technique (Friso et al. 2002). A short description of these techniques is given below.

DNA Methylation

Bisulfite Treatment

In bisulfite treatment, the genomic DNA is treated with sodium bisulfite, which converts all the non-methylated cytosines into uracil. This conversion is followed by the PCR through specific primers. All the uracil (non-methylated cytosines before bisulfite treatment) and thymine residues (which were always thymines even before bisulfite treatment) are being amplified by PCR as thymine, whereas only 5-methylcytosine residues are amplified as cytosine. After sequencing of PCR product, the analysis of sequences provides the information about the methylated sites in the amplified region as well as the methylation level in a particular genomic region (Frommer et al. 1992). With the advancement in the sequencing technology, this technique can also be used in genome-wide methylation analysis.

Methylation-Sensitive PCR (MSP)

Methylation-sensitive PCR (MSP) is a modification in the above-described bisulfite treatment technique where the amplification is done with primer pair that is specific for methylated DNA and primer pair specific to unmethylated DNA (Herman et al. 1996). This technique can be used in the development of epigenetic markers to be used in marker-assisted selection.

Methyl-Sensitive Restriction Enzyme Technique

In methyl-sensitive restriction enzyme technique, the genomic DNA is digested with the methylation-sensitive restriction enzyme, and this digested DNA is amplified by primers flanking the restriction site. PCR will work only if the restriction site is not cleaved (due to the methylation at that site) (Singer-Sam et al. 1990). This kind of technique along with its various modified forms paves the way for the development of simple and practical epigenetic marker for the epialleles which could have implications in marker-assisted selection and could have important role in the crop improvement strategies.

Methyl-Sensitive Amplification Polymorphism (MSAP)

Methyl-sensitive amplification polymorphism (MSAP) is a technique for the DNA methylation analysis on a genome-wide level, in which digestion of genomic DNA is done with a methylation-sensitive restriction enzyme like *Hpa*II as a first step. This is followed by the ligation of DNA fragments to adaptors, which facilitate the amplification of these fragments. *Msp*I, a methylation-insensitive isoschizomer of HpaII, is used in parallel for digestion, and this digestion serves as a loading control in the experiment. After that, amplification of these fragments through fluorescently labeled primers is done. Comparison of PCR products from different individuals allows the user to identify the interesting fragments. This leads to the isolation and characterization of that fragment (Yaish et al. 2014).

High-Performance Liquid Chromatography Technique (HPLC)

In high-performance liquid chromatography technique (HPLC), the genomic DNA is enzymatically hydrolyzed, and this hydrolyzed DNA is then separated into its four major DNA bases and 5-methyl-2′-deoxycytidine using HPLC. 5-Methyl-2′-deoxycytidine is obtained. The global DNA methylation status is calculated by comparing the amount of 5-methyl-2′-deoxycytidine per microgram of DNA with percent relative standard deviations (%RSD) (Friso et al. 2002).

Methylated DNA Immunoprecipitation

It is another technique to study genome-wide changes in DNA methylation patterns. In this technique, DNA is isolated from cells and sheared through sonication. By the usage of antibodies specifically targeting methylated DNA fragments, isolation of methylated regions occurs, which then can be identified using high-resolution DNA microarrays or next-generation sequencing techniques. The global changes in the methylation patterns across the varied cells can be detected through this technique.

Chromatin Immunoprecipitation (ChIP)

Chromatin immunoprecipitation (ChIP) is a technique used to study the interaction between proteins (e.g., histones) and DNA. The use of highly specific antibodies directed against DNA-binding proteins is required in this technique, and it can be followed by various nucleic acid analysis techniques, including PCR, qPCR, sequencing, and microarray hybridization. It can help to determine whether certain proteins are associated with specific genomic regions and is also useful for identifying regions of the genome associated with specific histone modifications.

Conclusion

Our understanding of epigenetic regulation in plants is rapidly growing. However, up to now, linkage of histone modifications, ATP-dependent chromatin remodeling, and DNA methylation to a specific stress and the origin of this specificity is still unknown. To identify targeted genes and regulatory complexes, genomic binding studies and proteomic analysis will be required, respectively. It is necessary to deepen our investigation of the epigenetic regulators in crops and their underlying molecular mechanism. Understanding the mechanism of epigenetic regulators and their regulatory networks in crops will be a potential tool for further exploitation toward sustainable agriculture. Moreover, it is desirable to design new breeding strategies in which the epigenetic variability should be taken into consideration. This seems even more realistic with the advancement of genomic technologies and cost lowering of next-generation sequencing. Like MAS (marker-assisted selection), epigenetic marker-assisted selection could also be initiated.

References

Agger K, Christensen J, Cloos PA, Helin K (2008) The emerging functions of histone demethylases. Curr Opin Genet Dev 18:159–168

Akimoto K, Katakami H, Kim HJ, Ogawa E, Sano CM, Wada Y et al (2007) Epigenetic inheritance in rice plants. Ann Bot 100:205–217

Aletta JM, Cimato TR, Ettinger MJ (1998) Protein methylation: a signal event in post-translational modification. Trends Biochem Sci 23:89–91

Alexandre C, Moller-Steinbach Y, Schonrock N, Gruissem W, Hennig L (2009) Arabidopsis MSI1 is required for negative regulation of the response to drought stress. Mol Plant 2:675–687

Allfrey V, Faulkner R, Mirsky A (1964) Acetylation and methylation of histones and their possible role in the regulation of RNA synthesis. Proc Natl Acad Sci U S A 51:786–794

Allis CD, Chicoine LG, Richman R, Schulman IG (1985) Deposition-related histone acetylation in micronuclei of conjugating Tetrahymena. Proc Natl Acad Sci U S A 82:8048–8052

Alvarez-Venegas R, Sadder M, Hlavacka A, Baluska F, Xia Y, Lu G et al (2006) The Arabidopsis homolog of trithorax, ATX1, binds phosphatidylinositol 5-phosphate, and the two regulate a common set of target genes. Proc Natl Acad Sci U S A 103:6049–6054

Baek D, Jiang J, Chung J-S, Wang B, Chen J, Xin Z et al (2011) Regulated AtHKT1 gene expression by a distal enhancer element and DNA methylation in the promoter plays an important role in salt tolerance. Plant Cell Physiol 52:149–161

Bedford MT, Clarke SG (2009) Protein arginine methylation in mammals: who, what, and why. Mol Cell 33:1–13

Bedford MT, Richard S (2005) Arginine methylation: an emerging regulator of protein function. Mol Cell 18:263–272

Benhamed M, Bertrand C, Servet C, Zhou D-X (2006) Arabidopsis GCN5, HD1, and TAF1/HAF2 interact to regulate histone acetylation required for light-responsive gene expression. Plant Cell Online 18:2893–2903

Berr A, McCallum EJ, Alioua A, Heintz D, Heitz T, Shen WH (2010) Arabidopsis histone methyltransferase SET DOMAIN GROUP8 mediates induction of the jasmonate/ethylene pathway genes in plant defense response to necrotrophic fungi. Plant Physiol 154:1403–1414

Berr A, Shafiq S, Shen WH (2011) Histone modifications in transcriptional activation during plant development. Biochim Biophys Acta 1809:567–576

Berr A, Ménard R, Heitz T, Shen W-H (2012) Chromatin modification and remodelling: a regulatory landscape for the control of Arabidopsis defence responses upon pathogen attack. Cell Microbiol 14:829–839

Bertrand C, Bergounioux C, Domenichini S, Delarue M, Zhou DX (2003) Arabidopsis histone acetyltransferase AtGCN5 regulates the floral meristem activity through the WUSCHEL/AGAMOUS pathway. J Biol Chem 278:28246–28251

Bilichak A, Ilnystkyy Y, Hollunder J, Kovalchuk I (2012) The progeny of Arabidopsis thaliana plants exposed to salt exhibit changes in DNA methylation, histone modifications and gene expression. PLoS ONE 7:e30515

Bourque S, Dutartre A, Hammoudi V, Blanc S, Dahan J, Jeandroz S, Pichereaux C et al (2011) Type-2 histone deacetylases as new regulators of elicitor-induced cell death in plants. New Phytol 192:127–139

Briggs SD, Xiao T, Sun Z-W, Caldwell JA, Shabanowitz J, Hunt DF et al (2002) Gene silencing: trans-histone regulatory pathway in chromatin. Nature 418:498

Casati P, Campi M, Chu F, Suzuki N, Maltby D, Guan S et al (2008) Histone acetylation and chromatin remodeling are required for UV-B-dependent transcriptional activation of regulated genes in maize. Plant Cell 20:827–842

Cervera MT, Ruiz-García L, Martínez-Zapater JM (2002) Analysis of DNA methylation in Arabidopsis thaliana based on methylation-sensitive AFLP markers. Mol Gen Genet 268:543–552

Chen K, Gao C (2014) Targeted genome modification technologies and their applications in crop improvements. Plant Cell Rep 33:575–583

Chen LT, Wu K (2010) Role of histone deacetylases HDA6 and HDA19 in ABA and abiotic stress response. Plant Signal Behav 5:1318–1320

Chen Z, Zhang H, Jablonowski D, Zhou X, Ren X, Hong X et al (2006) Mutations in ABO1/ELO2, a subunit of holo-Elongator, increase abscisic acid sensitivity and drought tolerance in Arabidopsis thaliana. Mol Cell Biol 26:6902–6912

Chen LT, Luo M, Wang YY, Wu K (2010) Involvement of Arabidopsis histone deacetylase HDA6 in ABA and salt stress response. J Exp Bot 61:3345–3353

Chen Q, Chen X, Wang Q, Zhang F, Lou Z, Zhang Q et al (2013) Structural basis of a histone H3 lysine 4 demethylase required for stem elongation in rice. PLoS Genet 9:e1003239

Chinnusamy V, Zhu JK (2009) Epigenetic regulation of stress responses in plants. Curr Opin Plant Biol 12:133–139

Choi CS, Sano H (2007) Abiotic-stress induces demethylation and transcriptional activation of a gene encoding a glycerophosphodiesterase-like protein in tobacco plants. Mol Genet Genomics 277:589–600

Choi SM, Song HR, Han SK, Han M, Kim CY, Park J et al (2012) HDA19 is required for the repression of salicylic acid biosynthesis and salicylic acid-mediated defense responses in Arabidopsis. Plant J 71:135–146

Colomé-Tatché M, Cortijo S, Wardenaar R, Morgado L, Lahouze B, Sarazin A et al (2012) Features of the Arabidopsis recombination landscape resulting from the combined loss of sequence variation and DNA methylation. Proc Natl Acad Sci U S A 109:16240–16245

Cubas P, Vincent C, Coen E (1999) An epigenetic mutation responsible for natural variation in floral symmetry. Nature 401:157–161

Cui X, Jin P, Gu L, Lu Z, Xue Y, Wei L et al (2013) Control of transposon activity by a histone H3K4 demethylase in rice. Proc Natl Acad Sci U S A 110:1953–1958

DeFraia CT, Zhang X, Mou Z (2010) Elongator subunit 2 is an accelerator of immune responses in Arabidopsis thaliana. Plant J 64:511–523

DeFraia CT, Wang Y, Yao J, Mou Z (2013) Elongator subunit 3 positively regulates plant immunity through its histone acetyltransferase and radical S-adenosylmethionine domains. BMC Plant Biol 13:102

Demetriou K, Kapazoglou A, Tondelli A, Francia E, Stanca MA, Bladenopoulos K et al (2009) Epigenetic chromatin modifiers in barley: I. Cloning, mapping and expression analysis of the plant specific HD2 family of histone deacetylases from barley, during seed development and after hormonal treatment. Physiol Plant 136:358–368

Deng X, Gu L, Liu C, Lu T, Lu F, Lu Z et al (2010) Arginine methylation mediated by the Arabidopsis homolog of PRMT5 is essential for proper pre-mRNA splicing. Proc Natl Acad Sci U S A A107:19114–19119

Dhawan R, Luo H, Foerster AM, Abuqamar S, Du HN, Briggs SD, Scheid OM et al (2009) HISTONE MONOUBIQUITINATION1 interacts with a subunit of the mediator complex and regulates defense against necrotrophic fungal pathogens in Arabidopsis. Plant Cell 21:1000–1019

Ding Y, Avramova Z, Fromm M (2011) The Arabidopsis trithorax-like factor ATX1 functions in dehydration stress responses via ABA-dependent and ABA-independent pathways. Plant J 66:735–744

Ding B, Bellizzi Mdel R, Ning Y, Meyers BC, Wang GL (2012) HDT701, a histone H4 deacetylase, negatively regulates plant innate immunity by modulating histone H4 acetylation of defense-related genes in rice. Plant Cell 24:3783–3794

Dover J, Schneider J, Tawiah-Boateng MA, Wood A, Dean K, Johnston M et al (2002) Methylation of histone H3 by COMPASS requires ubiquitination of histone H2B by Rad6. J Biol Chem 277:28368–28371

Durand S, Bouché N, Perez Strand E, Loudet O, Camilleri C (2012) Rapid establishment of genetic incompatibility through natural epigenetic variation. Curr Biol 22:326–331

Dyachenko OV, Zakharchenko NS, Shevchuk TV, Bohnert HJ, Cushman JC, Buryanov YI (2006) Effect of hypermethylation of CCWGG sequences in DNA of Mesembryanthemum crystallinum plants on their adaptation to salt stress. Biogeosciences 71:461–465

Earley KW, Shook MS, Brower-Toland B, Hicks L, Pikaard CS (2007) In vitro specificities of Arabidopsis co-activator histone acetyltransferases: implications for histone hyperacetylation in gene activation. Plant J 52:615–626

Falconer DS (1996) Introduction to quantitative genetics. Ronald Press Co., New York, pp 365

Fang H, Liu X, Thorn G, Duan J, Tian L (2014) Expression analysis of histone acetyltransferases in rice under drought stress. Biochem Biophys Res Commun 443:400–405

Feng S, Jacobsen SE, Reik W (2010) Epigenetic reprogramming in plant and animal development. Science 330:622–627

Fisher RA (1930) The genetical theory of natural selection. Clarendon Press, Oxford, pp 272

Friso S, Choi S-W, Dolnikowski GG, Selhub J (2002) A method to assess genomic DNA methylation using high-performance liquid chromatography/electrospray ionization mass spectrometry. Anal Chem 74:4526–4531

Frommer M, McDonald LE, Millar DS, Collis CM, Watt F, Grigg GW et al (1992) A genomic sequencing protocol that yields a positive display of 5-methylcytosine residues in individual DNA strands. Proc Natl Acad Sci U S A 89:1827–1831

Fu W, Wu K, Duan J (2007) Sequence and expression analysis of histone deacetylases in rice. Biochem Biophys Res Commun 356:843–850

Goll MG, Bestor TH (2005) Eukaryotic cytosine methyltransferases. Annu Rev Biochem 74:481–514

González RM, Ricardi MM, Iusem ND (2011) Atypical epigenetic mark in an atypical location: cytosine methylation at asymmetric (CNN) sites within the body of a non-repetitive tomato gene. BMC Plant Biol 11:94

González RM, Ricardi MM, Iusem ND (2013) Epigenetic marks in an adaptive water stress-responsive gene in tomato roots under normal and drought conditions. Epigenetics 8:864–872

Han SK, Sang Y, Rodrigues A, Wu MF, Rodriguez PL, Wagner D (2012) The SWI2/SNF2 chromatin remodeling ATPase BRAHMA represses abscisic acid responses in the absence of the stress stimulus in Arabidopsis. Plant Cell 24:4892–4906

Hark AT, Vlachonasios KE, Pavangadkar KA, Rao S, Gordon H, Adamakis ID et al (2009) Two Arabidopsis orthologs of the transcriptional coactivator ADA2 have distinct biological functions. Biochim Biophys Acta 1789:117–124

Henderson IR, Jacobsen SE (2007) Epigenetic inheritance in plants. Nature 447:418–424

Herman JG, Graff JR, Myöhänen S, Nelkin BD, Baylin SB (1996) Methylation-specific PCR: a novel PCR assay for methylation status of CpG islands. Proc Natl Acad Sci U S A 93:9821–9826

Hoffmann AA, Sgrò CM (2011) Climate change and evolutionary adaptation. Nature 470:479–485

Hu Y, Qin F, Huang L, Sun Q, Li C, Zhao Y et al (2009) Rice histone deacetylase genes display specific expression patterns and developmental functions. Biochem Biophys Res Commun 388:266–271

Huang L, Sun Q, Qin F, Li C, Zhao Y, Zhou DX (2007) Down-regulation of a SILENT INFORMATION REGULATOR2-related histone deacetylase gene, OsSRT1, induces DNA fragmentation and cell death in rice. Plant Physiol 144:1508–1519

Hwang WW, Venkatasubrahmanyam S, Ianculescu AG, Tong A, Boone C, Madhani HD (2003) A conserved RING finger protein required for histone H2B monoubiquitination and cell size control. Mol Cell 11:261–266

Imhof A, Yang X-J, Ogryzko VV, Nakatani Y, Wolffe AP, Ge H (1997) Acetylation of general transcription factors by histone acetyltransferases. Curr Biol 7:689–692

Jablonka E, Raz G (2009) Transgenerational epigenetic inheritance: prevalence, mechanisms, and implications for the study of heredity and evolution. Q Rev Biol 84:131–176

Jelinic P, Stehle J-C, Shaw P (2006) The testis-specific factor CTCFL cooperates with the protein methyltransferase PRMT7 in H19 imprinting control region methylation. PLoS Biol 4:e355

Johannes F, Porcher E, Teixeira FK, Saliba-Colombani V, Simon M, Agier N et al (2009) Assessing the impact of transgenerational epigenetic variation on complex traits. PLoS Genet 5:e1000530

Kaldis A, Tsementzi D, Tanriverdi O, Vlachonasios KE (2011) Arabidopsis thaliana transcriptional co-activators ADA2b and SGF29a are implicated in salt stress responses. Planta 233:749–762

Kapazoglou A, Tondelli A, Papaefthimiou D, Ampatzidou H, Francia E, Stanca MA et al (2010) Epigenetic chromatin modifiers in barley: IV. The study of barley polycomb group (PcG) genes during seed development and in response to external ABA. BMC Plant Biol 10:73

Karan R, DeLeon T, Biradar H, Subudhi PK (2012) Salt stress induced variation in DNA methylation pattern and its influence on gene expression in contrasting rice genotypes. PLoS ONE 7:e40203

Katz JE, DlakiÄ‡ M, Clarke S (2003) Automated identification of putative methyltransferases from genomic open reading frames. Mol Cell Proteomics 2:525–540

Khan AR, Enjalbert J, Marsollier A-C, Rousselet A, Goldringer I, Vitte C (2013) Vernalization treatment induces site-specific DNA hypermethylation at the VERNALIZATION-A1 (VRN-A1) locus in hexaploid winter wheat. BMC Plant Biol 13:209

Kim KC, Lai Z, Fan B, Chen Z (2008) Arabidopsis WRKY38 and WRKY62 transcription factors interact with histone deacetylase 19 in basal defense. Plant Cell 20:2357–2371

Kim JM, To TK, Nishioka T, Seki M (2010) Chromatin regulation functions in plant abiotic stress responses. Plant Cell Environ 33:604–611

Klose RJ, Zhang Y (2007) Regulation of histone methylation by demethylimination and demethylation. Nat Rev Mol Cell Biol 8:307–318

Krause CD, Yang Z-H, Kim Y-S, Lee J-H, Cook JR, Pestka S (2007) Protein arginine methyltransferases: evolution and assessment of their pharmacological and therapeutic potential. Pharmacol Ther 113:50–87

Lauria M, Rupe M, Guo M, Kranz E, Pirona R, Viotti A et al (2004) Extensive maternal DNA hypomethylation in the endosperm of Zea mays. Plant Cell 16:510–522

Law JA, Jacobsen SE (2010) Establishing, maintaining and modifying DNA methylation patterns in plants and animals. Nat Rev Genet 11:204–220

Li C, Huang L, Xu C, Zhao Y, Zhou D-X (2011) Altered levels of histone deacetylase OsHDT1 affect differential gene expression patterns in hybrid rice. PLoS ONE 6:e21789

Li X, Zhu J, Hu F, Ge S, Ye M, Xiang H et al (2012) Single-base resolution maps of cultivated and wild rice methylomes and regulatory roles of DNA methylation in plant gene expression. BMC Genomics 13:300

Lira-Medeiros CF, Parisod C, Fernandes RA, Mata CS, Cardoso MA, Ferreira PCG (2010) Epigenetic variation in mangrove plants occurring in contrasting natural environment. PLoS ONE 5:e10326

Liu C, Lu F, Cui X, Cao X (2010) Histone methylation in higher plants. Annu Rev Plant Biol 61:395–420

Liu X, Luo M, Zhang W, Zhao J, Zhang J, Wu K et al (2012) Histone acetyltransferases in rice (Oryza sativa L.): phylogenetic analysis, subcellular localization and expression. BMC Plant Biol 12:145

Luo M, Wang YY, Liu X, Yang S, Lu Q, Cui Y et al (2012) HD2C interacts with HDA6 and is involved in ABA and salt stress response in Arabidopsis. J Exp Bot 63:3297–3306

Manning K, Tör M, Poole M, Hong Y, Thompson AJ, King GJ et al (2006) A naturally occurring epigenetic mutation in a gene encoding an SBP-box transcription factor inhibits tomato fruit ripening. Nat Genet 38:948–952

Mao Y, Pavangadkar KA, Thomashow MF, Triezenberg SJ (2006) Physical and functional interactions of Arabidopsis ADA2 transcriptional coactivator proteins with the acetyltransferase GCN5 and with the cold-induced transcription factor CBF1. Biochim Biophys Acta (BBA) Gene Struct Expr 1759:69–79

Martin A, Troadec C, Boualem A, Rajab M, Fernandez R, Morin H et al (2009) A transposon-induced epigenetic change leads to sex determination in melon. Nature 461:1135–1138

McBride AE, Silver PA (2001) State of the arg: protein methylation at arginine comes of age. Cell 106:5–8

Mirouze M, Lieberman-Lazarovich M, Aversano R, Bucher E, Nicolet J, Reinders J et al (2012) Loss of DNA methylation affects the recombination landscape in Arabidopsis. Proc Natl Acad Sci U S A 109:5880–5885

Mlynárová L, Nap J, Bisseling T (2007) The SWI/SNF chromatin-remodeling gene AtCHR12 mediates temporary growth arrest in Arabidopsis thaliana upon perceiving environmental stress. Plant J 51:874–885

Ng DW, Wang T, Chandrasekharan MB, Aramayo R, Kertbundit S, Hall TC (2007) Plant SET domain-containing proteins: structure, function and regulation. Biochim Biophys Acta 1769:316–329

Noh B, Lee SH, Kim HJ, Yi G, Shin EA, Lee M et al (2004) Divergent roles of a pair of homologous jumonji/zinc-finger-class transcription factor proteins in the regulation of Arabidopsis flowering time. Plant Cell 16:2601–2613

Palma K, Thorgrimsen S, Malinovsky FG, Fiil BK, Nielsen HB, Brodersen P et al (2010) Autoimmunity in Arabidopsis acd11 is mediated by epigenetic regulation of an immune receptor. PLoS Pathog 6:e1001137

Papaefthimiou D, Tsaftaris A (2012a) Significant induction by drought of HvPKDM7-1, a gene encoding a jumonji-like histone demethylase homologue in barley (H. vulgare). Acta Physiol Plant 34:1187–1198

Papaefthimiou D, Tsaftaris AS (2012b) Characterization of a drought inducible trithorax-like H3K4 methyltransferase from barley. Biol Plant 56:683–692

Papaefthimiou D, Likotrafiti E, Kapazoglou A, Bladenopoulos K, Tsaftaris A (2010) Epigenetic chromatin modifiers in barley: III. Isolation and characterization of the barley GNAT-MYST family of histone acetyltransferases and responses to exogenous ABA. Plant Physiol Biochem 48:98–107

Pavangadkar K, Thomashow MF, Triezenberg SJ (2010) Histone dynamics and roles of histone acetyltransferases during cold-induced gene regulation in Arabidopsis. Plant Mol Biol 74:183–200

Pickart CM (2001) Mechanisms underlying ubiquitination. Annu Rev Biochem 70:503–533

Qian Y, Xi Y, Cheng B, Zhu S, Kan X (2014) Identification and characterization of the SET domain gene family in maize. Mol Biol Rep 41:1341–1354

Raj S, Bräutigam K, Hamanishi ET, Wilkins O, Thomas BR, Schroeder W et al (2011) Clone history shapes Populus drought responses. Proc Natl Acad Sci U S A 108:12521–12526

Reinders J, Wulff BBH, Mirouze M, Marí-Ordóñez A, Dapp M, Rozhon W et al (2009) Compromised stability of DNA methylation and transposon immobilization in mosaic Arabidopsis epigenomes. Genes Dev 23:939–950

Richards CL, Bossdorf O, Verhoeven KJF (2010) Understanding natural epigenetic variation. New Phytol 187:562–564

Salmon A, Ainouche ML, Wendel JF (2005) Genetic and epigenetic consequences of recent hybridization and polyploidy in Spartina (Poaceae). Mol Ecol 14:1163–1175

Sano H, Kamada I, Youssefian S, Katsumi M, Wabiko H (1990) A single treatment of rice seedlings with 5-azacytidine induces heritable dwarfism and undermethylation of genomic DNA. Mol Gen Genet MGG 220:441–447

Shafiq S, Berr A, Shen W-H (2014) Combinatorial functions of diverse histone methylations in Arabidopsis thaliana flowering time regulation. New Phytol 201:312–322

Shan X, Wang X, Yang G, Wu Y, Su S, Li S et al (2013) Analysis of the DNA methylation of maize (Zea mays L.) in response to cold stress based on methylation-sensitive amplified polymorphisms. J Plant Biol 56:32–38

Sharma R, Mohan Singh RK, Malik G, Deveshwar P, Tyagi AK, Kapoor S et al (2009) Rice cytosine DNA methyltransferases – gene expression profiling during reproductive development and abiotic stress. FEBSJ 276:6301–6311

Shvarts Iu B, Kahn TG, Pirrotta V (2010) Polycomb and trithorax control genome expression by determining the alternative epigenetic states of chromatin for key developmental regulators. Genetika 46:1413–1416

Singer-Sam J, Grant M, LeBon JM, Okuyama K, Chapman V, Monk M et al (1990) Use of a HpaII-polymerase chain reaction assay to study DNA methylation in the Pgk-1 CpG island of mouse embryos at the time of X-chromosome inactivation. Mol Cell Biol 10:4987–4989

Smalle J, Vierstra RD (2004) The ubiquitin 26S proteasome proteolytic pathway. Annu Rev Plant Biol 55:555–590

Song Y, Ji D, Li S, Wang P, Li Q, Xiang F (2012) The dynamic changes of DNA methylation and histone modifications of salt responsive transcription factor genes in soybean. PLoS ONE 7:e41274

Springer NM (2013) Epigenetics and crop improvement. Trends Genet 29:241–247

Sridha S, Wu K (2006) Identification of AtHD2C as a novel regulator of abscisic acid responses in Arabidopsis. Plant J 46:124–133

Sterner DE, Berger SL (2000) Acetylation of histones and transcription-related factors. Microbiol Mol Biol Rev 64:435–459

Steward N, Ito M, Yamaguchi Y, Koizumi N, Sano H (2002) Periodic DNA methylation in maize nucleosomes and demethylation by environmental stress. J Biol Chem 277:37741–37746

Stockinger EJ, Mao Y, Regier MK, Triezenberg SJ, Thomashow MF (2001) Transcriptional adaptor and histone acetyltransferase proteins in Arabidopsis and their interactions with CBF1, a transcriptional activator involved in cold-regulated gene expression. Nucleic Acids Res 29:1524–1533

Strahl BD, Allis CD (2000) The language of covalent histone modifications. Nature 403:41–45

Sun Q, Zhou DX (2008) Rice jmjC domain-containing gene JMJ706 encodes H3K9 demethylase required for floral organ development. Proc Natl Acad Sci U S A 105:13679

To TK, Nakaminami K, Kim JM, Morosawa T, Ishida J, Tanaka M et al (2011) Arabidopsis HDA6 is required for freezing tolerance. Biochem Biophys Res Commun 406:414–419

Turck F, Roudier F, Farrona S, Martin-Magniette M-L, Guillaume E, Buisine N et al (2007) Arabidopsis TFL2/LHP1 specifically associates with genes marked by trimethylation of histone H3 lysine 27. PLoS Genet 3:e86

Unnikrishnan A, Gafken PR, Tsukiyama T (2010) Dynamic changes in histone acetylation regulate origins of DNA replication. Nat Struct Mol Biol 17:430–437

van Dijk K, Ding Y, Malkaram S, Riethoven JJ, Liu R, Yang J et al (2010) Dynamic changes in genome-wide histone H3 lysine 4 methylation patterns in response to dehydration stress in Arabidopsis thaliana. BMC Plant Biol 10:238

Vaughn MW, Tanurdzić M, Lippman Z, Jiang H, Carrasquillo R, Rabinowicz PD et al (2007) Epigenetic natural variation in Arabidopsis thaliana. PLoS Biol 5:e174

Vignali M, Hassan AH, Neely KE, Workman JL (2000) ATP-dependent chromatin-remodeling complexes. Mol Cell Biol 20:1899–1910

Visser ME (2008) Keeping up with a warming world; assessing the rate of adaptation to climate change. Proc Biol Sci/R Soc 275:649–659

Vlachonasios KE, Thomashow MF, Triezenberg SJ (2003) Disruption mutations of ADA2b and GCN5 transcriptional adaptor genes dramatically affect Arabidopsis growth, development, and gene expression. Plant Cell Online 15:626–638

Wada Y, Miyamoto K, Kusano T, Sano H (2004) Association between up-regulation of stress-responsive genes and hypomethylation of genomic DNA in tobacco plants. Mol Genet Genomics 271:658–666

Walley JW, Rowe HC, Xiao Y, Chehab EW, Kliebenstein DJ, Wagner D, Dehesh K et al (2008) The chromatin remodeler SPLAYED regulates specific stress signaling pathways. PLoS Pathog 4:e1000237

Wang C, Gao F, Wu J, Dai J, Wei C, Li Y (2010) Arabidopsis putative deacetylase AtSRT2 regulates basal defense by suppressing PAD4, EDS5 and SID2 expression. Plant Cell Physiol 51:1291–1299

Wang WS, Pan YJ, Zhao XQ, Dwivedi D, Zhu LH, Ali J et al (2011) Drought-induced site-specific DNA methylation and its association with drought tolerance in rice (Oryza sativa L.). J Exp Bot 62:1951–1960

Woo HR, Pontes O, Pikaard CS, Richards EJ (2007) VIM1, a methylcytosine-binding protein required for centromeric heterochromatinization. Genes Dev 21:267

Wu K, Zhang L, Zhou C, Yu CW, Chaikam V (2008) HDA6 is required for jasmonate response, senescence and flowering in Arabidopsis. J Exp Bot 59:225–234

Xu L, Zhao Z, Dong A, Soubigou-Taconnat L, Renou J-P, Steinmetz A et al (2008) Di- and Tri- but not monomethylation on histone H3 lysine 36 marks active transcription of genes involved in flowering time regulation and other processes in Arabidopsis thaliana. Mol Cell Biol 28:1348–1360

Yaish MW, Peng M, Rothstein SJ (2014) Global DNA methylation analysis using methyl-sensitive amplification polymorphism (MSAP). In: Sanchez-Serrano JJ, Salinas J (eds) Arabidopsis protocols, vol 82. Humana Press, Totowa, pp 285–298

Yu A, Lepère G, Jay F, Wang J, Bapaume L, Wang Y et al (2013) Dynamics and biological relevance of DNA demethylation in Arabidopsis antibacterial defense. Proc Natl Acad Sci U S A 110:2389–2394

Zemach A, Kim MY, Silva P, Rodrigues JA, Dotson B, Brooks MD et al (2010) Local DNA hypomethylation activates genes in rice endosperm. Proc Natl Acad Sci U S A 107:18729–18734

Zhang X, Yazaki J, Sundaresan A, Cokus S, Chan SWL, Chen H et al (2006) Genome-wide high-resolution mapping and functional analysis of DNA methylation in Arabidopsis. Cell 126:1189–1201

Zhang K, Sridhar VV, Zhu J, Kapoor A, Zhu JK (2007) Distinctive core histone post-translational modification patterns in Arabidopsis thaliana. PLoS ONE 2:e1210

Zhang M, Kimatu JN, Xu K, Liu B (2010) DNA cytosine methylation in plant development. J Genet Genomics = Yi chuan xue bao 37:1–12

Zhang Z, Zhang S, Zhang Y, Wang X, Li D, Li Q et al (2011) Arabidopsis floral initiator SKB1 confers high salt tolerance by regulating transcription and pre-mRNA splicing through altering histone H4R3 and small nuclear ribonucleoprotein LSM4 methylation. Plant Cell 23:396–411

Zhang Y-Y, Fischer M, Colot V, Bossdorf O (2013) Epigenetic variation creates potential for evolution of plant phenotypic plasticity. New Phytol 197:314–322

Zhou C, Zhang L, Duan J, Miki B, Wu K (2005) HISTONE DEACETYLASE19 is involved in jasmonic acid and ethylene signaling of pathogen response in Arabidopsis. Plant Cell Online 17:1196–1204

Zhou X, Hua D, Chen Z, Zhou Z, Gong Z (2009) Elongator mediates ABA responses, oxidative stress resistance and anthocyanin biosynthesis in Arabidopsis. Plant J 60:79–90

Zong W, Zhong X, You J, Xiong L (2012) Genome-wide profiling of histone H3K4-tri-methylation and gene expression in rice under drought stress. Plant Mol Biol 81:175–188

Plant miRNomics: Novel Insights in Gene Expression and Regulation

N. Manikanda Boopathi

Contents

Abstract

Advances in transcriptomics have led to the classification of small RNAs (sRNAs) into mainly three categories: miRNAs, siRNAs and piRNAs. However, there are many new types of sRNAs under exploration. Though such sRNAs differ from one another, they collaborate in their mode of action. Among the sRNAs, microRNAs (miRNAs) widely captured the attention of molecular biologists. miRNAs are short, endogenously expressed and non-translated RNAs. Mature plant miRNAs are in general smaller in size (~22 bp) and considered as negative gene regulatory molecules. In general plant miRNAs have the following features: (a) They are coded by miRNA genes with unknown length and are sequentially cleaved from pri-miRNA and pre-miRNA into a short mature miRNA by *Dicer-like 1 (dcl1)* and several other enzymes.

N.M. Boopathi, Ph.D. (✉)
Department of Plant Molecular Biology and Bioinformatics, Centre for Plant Molecular Biology and Biotechnology,
Tamil Nadu Agricultural University,
Coimbatore, India
e-mail: nmboopathi@tnau.ac.in; biotechboopathi@yahoo.com

D. Barh et al. (eds.), *PlantOmics: The Omics of Plant Science*,
DOI 10.1007/978-81-322-2172-2_7, © Springer India 2015

(b) All pre-miRNAs can form a stem-looped hairpin secondary structure with the mature miRNA on one arm and the complementary sequence, termed miRNA*, on the another arm with high negative minimum folding free energy (MFE) and MFE index (MFEI). (c) Typically, miRNAs do negatively regulate target gene expression and the miRNA* sequence is degraded by an unknown mechanism. However, in some cases, the miRNA* sequence also can function to target a specific gene. Over the past few years, microarray technologies, large-scale small RNA and whole genome sequencing projects and data mining have provided a wealth of information about the spectrum of plant miRNAs and their targets. Hitherto identified miRNAs in plant kingdom have shown that they are deeply conserved; nevertheless considerable numbers of species-specific miRNAs also exist. Evidences are gradually mounting to notify that miRNAs have key roles in developmental timing, cell proliferation and cell death, organogenesis, patterning of tissues/organ and more importantly, in response to external environmental stimuli. Thus it is very obvious that plant miRNAs are more numerous and their regulatory impact is more pervasive than was previously suspected.

Keywords
miRNA • Small non-coding RNA • miRNA target • Function of plant miRNA

Introduction

The perfect and complete development of a multicellular organism at normal and abnormal conditions requires specific and coordinated expression of genes. The regulation of gene expression is the most basic level which decides the information encoded in the DNA is decoded into phenotypes. As such it involves complex regulatory networks that direct precise cell division and differentiation patterns. Thus, the key factor that is central to the understanding of the biological systems is how dynamic gene regulatory programmes are generated from the static instructions encoded in the DNA. Though single or several regulators of given gene(s) expression(s) have been described earlier, it is not completely unravelled in several complex trait expressions such as organ formation, behaviour or adaptation to a particular environmental conditions.

However, it is clearly evident that plants and other multicellular organisms need precise spatio-temporal control of gene expression, and this regulatory capacity depends, in part, on small RNAs (sRNAs). Although most genes use RNA in the form of mRNA as a coding intermediate for protein production, there are genes whose final products are RNA and do not code for protein. Such non-coding RNAs range from the transfer and ribosomal RNAs that are involved in protein-synthesising machinery to the more recently discovered regulatory sRNAs. There are several kinds of sRNAs and have specific role in coordinated control of gene expression (see below). This chapter specifically focuses on plant miRNAs, a class of sRNAs and their involvement in negative regulation of plant genes.

miRNAs are small (~20–22 bp in length) RNAs that negatively regulate the expression of genes through specific base pairing with cognate target mRNAs and thereby inducing target mRNA degradation or translational repression or both (Sun 2011). miRNAs were first reported in plants (*Arabidopsis*) during mid-2002 by four different research groups as tiny RNAs with miRNA characteristics (Llave et al. 2002; Mette et al. 2002; Park et al. 2002; Reinhart et al. 2002). Studying the expression profiles of miRNAs, their targets and function in biological system using biotechnological and bioinformatic tools are collectively called as 'miRNomics'. It is expected that integration of miRNomics data with other 'omics' data would help in comprehensive understanding of the precise spatio-temporal control of gene expression and more essentially the specific role of miRNA in gene regulation. Such understanding would also help to develop novel application tools in medicine, agriculture and industries. For example, the miRNA expression profiles provide a powerful support for the characterisation of disease development in plants. The development of micro array technologies, specific for miRNAs, can help to obtain a miRNomics profile for different organs under disease, and high-throughput proteomics analysis can help to identify the miRNAs that control the differential expression of proteins.

Classification and Comparison of Small RNAs

Until a couple of decades before, it has been considered that the most important components of biological systems are DNA (the foundation of heredity) and proteins (the players of the cellular machinery). During those periods, RNA was recognised as an intermediate molecule that bridges the gap between DNA and protein (especially mRNA) or has functional role in splicing (snRNA) or translation machinery (tRNA and rRNA).

On the other hand, advances in molecular biology have led to classify the RNAs into coding and non-coding RNAs. Coding RNAs comprise of messenger RNAs (mRNAs). Non-coding RNAs are subdivided into ribosomal RNAs (rRNAs), transfer RNAs (tRNAs) and small RNAs (sRNAs). Various types of sRNAs have been identified such as microRNAs (miRNAs), small interfering RNAs (siRNAs), piwi-interacting RNAs (piRNAs), small temporal RNAs (stRNAs), tiny non-coding RNAs (tncRNAs) and small modular RNAs (smRNAs). Among them, miRNAs and siRNAs have been well characterised in plant and animal systems (the characteristics of these two sRNAs are described in Table 1) whereas piRNAs have been identified only in animals. Various other types of siRNAs have also been identified, including trans-acting siRNAs (Ta-siRNAs), repeat-associated siRNAs (Ra-siRNAs) and natural-antisense transcript-derived siRNAs (Nat-siRNAs) based on their distinct biogenesis and functions (Kruszka et al. 2012).

Table 1 General features of miRNAs and siRNAs

Property	miRNAs	siRNAs
Description	Regulators of endogenous genes	Defenders of genome integrity when foreign or invasive nucleic acids enter into the cell
Size	20–22 bp	21–24 bp
Biogenesis: precursors	Hairpin-shaped ssRNAs	Long dsRNAs
Biogenesis: nature of precursors	Endogenous precursor such as gene(s) of host's genome	Exogenous precursors such as transposons, transgenes, repeat elements or viruses
Cellular role	mRNA degradation, translational repression	DNA methylation, histone modification and mRNA degradation
Gene regulation mechanism	Post-transcriptionally with partial or full complementarity with target mRNA	Transcriptionally as well as post-transcriptionally with full complementarity with target mRNA
Target genes	miRNAs cannot regulate the genes from which they originate (*cis*) because they are identical, not complementary, to their precursor RNAs	siRNAs can potentially act in both *cis* and *trans* by targeting the elements from which they derive (*cis*) as well as unlinked elements that exhibit substantial complementarity to their sequence (*trans*)
Functions	Cell development, cell differentiation, regulation of developmental process, biotic and abiotic stress response	Defence against transposons and viruses and stress adaptation
First report published in	1993	1999

Thus the latest discovery of an increasing number of small non-coding RNAs with specific regulatory roles has added additional role to the RNA molecules, and it has changed our view on gene expression. It has been shown that small RNAs are known to play essential roles in the multicellular organisms with the surprising exception of the *Saccharomyces cerevisiae* (Sun et al. 2012). sRNAs are involved in a variety of occasions with diverse modes of actions that are essential for genome stability, development and adaptive responses to biotic and abiotic stresses (Jones-Rhoades 2012). For example, they guide DNA elimination during the formation of the macronucleus in protists and heterochromatin assembly in fungi and plants. They target endogenous mRNAs for cleavage and translational repression in plants and animals. sRNAs protect both plant and animal cells against virus infection through an RNA-based immune system. They also control the movement of transposable elements at the transcriptional and post-transcriptional levels in plants and animals.

sRNAs are produced from endogenous genomic loci and repeats and in response to external stimuli such as viruses. sRNAs are invariably bound by argonaute proteins, some of which have endonuclease activity to effect sRNA-guided cleavage of target mRNAs. Argonaute proteins belong to at least two phylogenetic groups: the argonaute subfamily, which binds miRNAs and siRNAs, and the piwi subfamily, which binds piRNAs. Thus the major types of sRNAs are distinguished by their different modes of biogenesis and action (Kruszka et al. 2012). Despite different modes of biogenesis, the sRNAs share similar molecular functions. For example, miRNA, siRNA and piRNAs can direct the cleavage of complementary RNAs. miRNAs and siRNAs can both result in translational inhibition of target mRNAs. siRNAs and piRNAs can both direct chromatin modifications (Chen 2009). Since small RNAs are repressors of gene expression, small RNA-mediated regulation is often referred to as RNA silencing, gene silencing or RNA interference (RNAi). RNA silencing was discovered in plants more than 15 years ago during the course of transgenic experiments that eventually led to silencing of the introduced transgene and, in some cases, of homologous endogenous genes or resident transgenes either by transcription inhibition (transcriptional gene silencing) or RNA degradation (post-transcriptional gene silencing) (Sun et al. 2012).

Among the small RNAs, miRNAs have a number of discrete features as compared to other functional RNA species (Jones-Rhoades 2012). First, most of the known miRNAs are encoded as polycistronic transcripts, proposing that members of the same miRNA family may evolve concurrently and develop in similar ways. Second, it has been known that a significant number of miRNAs are highly conserved in sequences among different related and unrelated organisms. Conserved miRNAs possess a special 'seed' sequence in their 5′ terminus; such conservation suggests that these molecules participate in key cellular processes. Third, miRNAs tend to target and regulate a set of mRNAs instead of a specific mRNA. Fourth, direct experimental evidence supports the notion that the miRNA pathway is an ancient regulatory mechanism evolved before the divergence of multicellular and unicellular organisms. Finally, special cases among viruses are worth mentioning: due to their higher mutation rates and faster evolution processes, most viral miRNAs do not seem to share significant homology with those of their vertebrate counterparts, even within members of the same family, and the lack of homology poses challenges for computational biologists to precisely predict miRNAs based on sequence conservations alone in viruses (Cai et al. 2009).

The past decade has witnessed rapid progress in revealing miRNA diversity, uncovering their mechanisms of action and understanding their biological functions. Here I review our current knowledge of onmiRNAs, with an emphasis on their biogenesis and function in plants. At the same time, I would like to highlight that with the development of the next-generation sequencing technologies and other advances in molecular biology, miRNAs from various species would shoot up in an inconceivable speed in the near future and such discoveries can greatly enrich our knowledge on miRNAs and their function with new dimensions.

Historical Perspectives of miRNA

Although miRNAs were first discovered in the early 1990s (Lee et al. 1993; Wightman et al. 1993), they were not recognised as miRNAs until 2001 (Lee and Ambros 2001; Lau et al. 2001; Lagos-Quintana et al. 2001). Since then, miRNAs have attracted a huge interest from scientists, and large numbers of miRNAs were identified in almost all the biological systems (Bartel 2004). In the year of 1993, the Ambros and Ruvkun labs investigated the first endogenous, non-protein-coding 22 nt RNAs in nematode *Caenorhabditis elegans* as *lin-4* and *let-7*, both of which are key regulatory molecules in the pathway controlling the timing of larval development. *lin-4* is recognised as the founding member of a new class of sRNAs called miRNAs (Lee and Ambros 2001; Lau et al. 2001; Lagos-Quintana et al. 2001). Compared with animal miRNAs, plant miRNAs were identified several years later. In 2002, several researchers identified plant miRNAs by cloning sRNAs and reported the molecular mechanisms of miRNA biogenesis and function in plants (Llave et al. 2002; Mette et al. 2002; Park et al. 2002; Reinhart et al. 2002). In the past couple of years, the total number of miRNAs in economically important crops has dramatically increased as the approaches for efficient identification species-specific miRNAs were developed. The first miRNAs detected in a viral genome were reported in 2004 by Pfeffer and colleagues in Epstein–Barr virus. To date, Release 19 of the miRBase sequence database contains 21264 entries representing hairpin precursor miRNAs, expressing 25141 mature miRNA products, in 193 species. The data are freely available to all through the web interface at http://www.mirbase.org/872 (verified on 08th May, 2013).

Biogenesis of miRNA

miRNA biogenesis is a multi-step enzymatic process which includes transcription, processing, modification and RISC loading (Fig. 1; Jones-Rhoades 2012). First, a miRNA gene is transcribed to a primary miRNA (pri-miRNA), which is usually a long sequence of more than several hundred nucleotides (Fig. 1). This step is controlled by Pol II enzymes (Bartel 2004). After analysing several miRNA-related sequences in *Arabidopsis*, Allen et al. (2004) presented a model for miRNA origin. In their model, plant pre-miRNAs originated from their target genes by formation of inverted duplications which have been transcribed but not modified further. miRNAs may originate anywhere within the plant genome; many genomic regions were found to be sites of miRNAs although these regions were previously considered featureless. These findings suggest that miRNA origin may be more complicated than previously thought by involving many mechanisms such as inversion and duplication (see below).

In the second step, the pri-miRNA is cleaved to a stem-loop intermediate called miRNA precursor or pre-miRNA. This step is controlled by the Drosha RNase III endonuclease in animals (Bartel 2004) or by Dicer-like 1 enzyme (DCL1) in plants (Tang et al. 2003). In animals, pre-miRNAs are then transported by exportin 5 from the nucleus into the cytoplasm (Sun et al. 2010), followed by formation of miRNA–miRNA* duplex and mature miRNAs by another RNase III-like enzyme called Dicer (Bartel 2004). However plant miRNAs differ from animals in this step. Plant miRNAs are cleaved into miRNA–miRNA* duplex possibly by Dicer-like enzyme 1 (DCL1) in the nucleus rather than in the cytoplasm (Bartel 2004); then the duplex is translocated into the cytoplasm by HASTY, the plant orthologue of export in 5 (Jones-Rhoades 2012). DCL1 acts with two partner proteins: HYL1 (a double-stranded RNA binding protein) and SE (a zinc-finger protein). These three proteins co-localise in nuclear Dicing bodies in vivo. SE also localises in numerous nuclear speckles and acts in the splic-

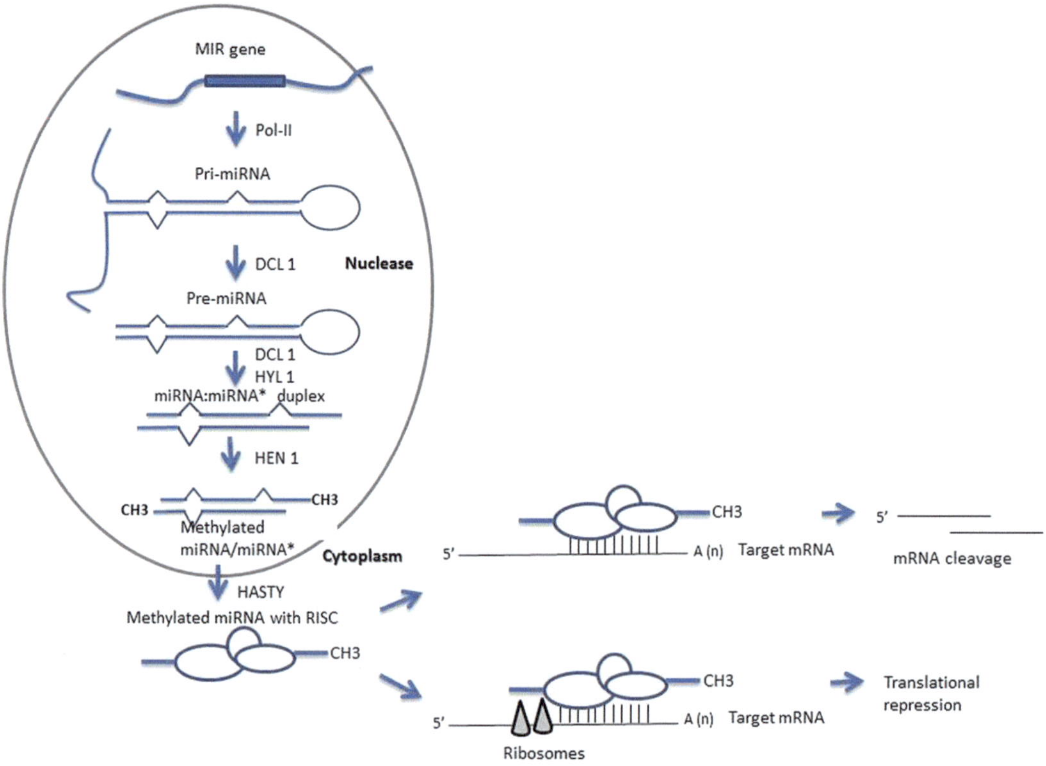

Fig. 1 Successive steps involved in biogenesis of miRNA to target recognition and function

ing of pre-mRNAs. In plants, it is also noticed that the miRNAs are methylated on the ribose of the last nucleotide by the methyltransferase (Chen 2009). The miRNA–miRNA* duplex is methylated on the 2′ OH of the 3′ terminal nucleotides by HEN1. The miRNA is exported from the nucleus to the cytoplasm through export factors including HASTY. In the cytoplasm, both plant and animal miRNAs are unwound into single-strand mature miRNAs by helicase (Bartel 2004).

Finally, the mature miRNAs enter a ribonucleoprotein complex known as the RNA-induced silencing complex (RISC) (Bartel 2004; Chen 2009) where they regulate targeted gene expression. The sRNA degrading nuclease (SDN1) family of exonucleases degrades single-stranded miRNAs to limit their steady-state levels (Chen 2009). Recently it was reported that a large number of 5′ or 3′ truncated miRNAs from various tissues of *Populus trichocarpa* can be isolated. Thus it is suggesting the occurrence of exonucleolytic degradation of miRNAs (Lu et al. 2005a, b). They also showed that a significant portion of the isolated miRNAs contains one or a few post-transcriptionally added adenylic acid residues at the 3′-end. The short adenylate tail of miRNAs is distinct from the longer polyadenylate tail added to other plant RNAs for the exosome-mediated degradation. Results of in vitro miRNA degradation assays revealed that the addition of adenylic acid residues on the 3′-end reduced miRNA degradation rate (Lu et al. 2005a, b).

Current Molecular Understanding of Plant Gene Regulation by miRNAs

In the RISC complex, miRNAs bind to target mRNA and inhibit gene expression through perfect or near-perfect complementarity between the miRNA and the mRNA (Bartel 2004). This causes gene silencing (Almeida and Allshire 2005) and it termed differently in diverse organisms. For example, this process is referred to as RNA interference (RNAi) in animals (Hannon 2002), quelling in fungi (Cogoni et al. 1996) and post-transcriptional gene silencing in plants (Baulcombe 2004). In plants, most target mRNAs contain one single miRNA complementary site, and most corresponding miRNAs typically perfectly complement to these sites and cleave the target mRNAs (Fig. 1; Bartel 2004). Unlike animal miRNA targets, the complementary sites in plants can exist anywhere along the target mRNA rather than at the 3′ UTR. Yet another mechanism was also identified in plant miRNA regulation. Although some miRNAs can perfectly complement to target mRNAs, they regulate gene expression by repressing gene translation possibly through inhibition of ribosome movement (Fig. 1; Chen 2009). This proposes a hypothesis that miRNAs may be involved in more complicated mechanisms to control gene expression in plants than in animals, i.e. plant miRNAs regulate gene expression at the post-transcriptional level not only by repression of mRNA translation but also by direct cleavage of mRNAs.

In general, miRNA*, the complementary strand of mature functional miRNA, is thought to degrade rapidly or accumulate at only very low levels, suggesting that it may not be functional. However, several recent studies have shown that plant miRNA* tends to accumulate at high levels under certain conditions (reviewed in Sunkar et al. 2012) and regulates a different target, i.e. other targets that are regulated by its cognate mature miRNA (Zhang et al. 2011a; Sunkar et al. 2012).

Molecular Structural Features of miRNA Genes and Its Expression

Like that of protein-coding genes, the core structure of a miRNA gene is also composed of a Pol II promoter, a transcribed region and a terminator. These basic structural elements of miRNA genes have been well studied in recent years except for the terminator region. MiRNA promoters contain the basic TATA, CAAT boxes as well as binding sites for specific transcription factors (TFs) (Xie et al. 2005). The promoter activity of an miRNA gene, such as miR390 and miR165/166, can be monitored by fusing it with a reporter gene (e.g. *GUS* or *GFP*) using plant transgenic technology (Montgomery et al. 2008). Like the promoters of protein-coding genes, miRNA promoters also behave in a tissue-specific or a temporally specific manner. Interestingly, the majority of plant miRNAs regulate the expression of TFs (Jones-Rhoades 2012). As a general way of auto-regulation, the promoters of some miRNAs, such as miR156 and miR172, contain the binding sites of their target TFs, further demonstrating a fine-tuning role of miRNAs in their regulatory networks (Wu et al. 2010). The transcribed region of the miRNA, the mature miRNA and the immediate miRNA precursor (pre-miRNA) are small; however, the primary miRNA (pri-miRNA) transcripts tend to be large in size (please see Fig. 1). Thus, the transcribed region of a miRNA gene is usually longer than necessary for miRNA biogenesis. Compared to the promoter and terminator regions, at least part of the miRNA gene transcribed region plays a key role in miRNA biogenesis and is thus an essential part in miRNA gene evolution (Tang et al. 2007). As stated above, unlike miRNA promoters and transcribed regions, miRNA terminators have not been well studied. Their existence is inferred from the fact that pri-miRNAs are polyadenylated. However, it is not known how the terminators or the polyA tail affects miRNA biogenesis or the stability of pri-miRNAs, and thus it necessitates further research (Tang et al. 2007).

Although a large number of miRNA genes disperse over the genome, clustered ones are found co-expressed as polycistronic units that may have functional relationships. In addition, more than half of miRNAs reside in introns of their host genes and co-express with their neighbouring protein-coding sequences, and some may derive from common primary transcripts and even share the same promoters (Jones-Rhoades 2012). Nevertheless, a sizable number of miRNA genes come from regions that are distal from previously annotated protein-coding sequences, and such locations imply that they probably derive from independent transcription units with their own promoters (Cai et al. 2009).

In all the plant species examined, miRNA genes are scattered throughout the genome. A majority (84 %) of miRNA genes are located in intergenic regions. This situation is contrary to that of *Drosophila* species, in which nearly half of the miRNA genes are located in introns (Nozawa et al. 2010). It should be noted that the proportion of intergenic miRNA genes is not correlated with the genome size of the plant species examined. In fact, *Arabidopsis* shows a high proportion of intergenic miRNA genes (86 %) even though its genome size is as small as that of *D. melanogaster* (Nozawa et al. 2010).

A range of techniques are available for miRNA gene expression analysis or quantification, including Northern blotting, dot blotting, RNase protection assay, primer extension analysis, invader assay and quantitative PCR (reviewed in Kim and Nam 2006). Large-scale cDNA cloning can also provide information on the relative expression level of miRNAs in diverse samples. However, most of these techniques involve laborious procedures, making it difficult to determine the level of all known miRNAs. Currently, the most widely used method is based on microarrays. However, the problem of potential cross-hybridisation of related miRNAs still remains unresolved. In addition, systematic bias could be introduced during reverse transcription, PCR amplification, enzymatic labelling or fluorescence tag ligation. These problems were successfully avoided by developing a new procedure called the RNA-primed array-based Klenow enzyme (RAKE) assay (Kim and Nam 2006). The most recent innovation in miRNA detection involves the bead-based flow cytometric method (see below).

Combining all the experimental results, the existence of miRNAs in plant cell is now being verified by the following criteria: (1) a band of about 22-nucleotide mature miRNA could be detected by Northern blotting, usually with precursor miRNA of about 70 nucleotides; (2) the precursor forms a hairpin structure and the mature miRNA is present in one arm of the hairpin; (3) both the mature and the precursor miRNAs are usually phylogenetically conserved; and (4) the precursor miRNAs should be observed when Dicer function is disturbed (Li et al. 2010a, b). Interesting additional mechanisms (such as single nucleotide polymorphisms, RNA editing, methylation and circadian clock) that play important roles in controlling the expression and function of miRNAs were also described by Cai et al. (2009).

In summary, the loci that encode plant miRNAs, the MIR genes and their expression are clearly distinct from previously annotated genes, but their promoters, primary transcripts and responsible RNA polymerase remain to be identified.

Evolution of miRNA

Most plant species contain more than 100 miRNA genes in their genome, and estimation of the numbers of miRNA genes and gene families in ancestral species indicated that the numbers of miRNA genes and gene families increased considerably in the lineage to flowering plants after the divergence from green algae (Jones-Rhoades 2012). With respect to the evolutionary origin of miRNA genes, there are several different mechanisms have been proposed (Jones-Rhoades 2012): (1) miRNA genes may be generated from duplicates of protein-coding genes. This is an attractive hypothesis because amiRNA gene generated from a protein-coding gene would bind to the transcript of the protein-coding gene. In fact, there seem to be a number of miRNA genes generated in this way (Allen et al. 2004; Rajagopalan et al. 2006; Fahlgren et al. 2007). In plants more than 90 % of miRNA gene families which are conserved in nine or more species are multigene families, whereas only 23 % of species-specific miRNA genes from multigene families were reported. These observations suggest that gene duplication has played important roles in increasing the number of miRNA genes in plants (Nozawa et al. 2011). (2) Transposable elements (TEs) may become miRNA genes. In particular, miniature inverted-repeat transposable elements

(MITEs) have a potential to become miRNA genes because they have inverted repeats with a short internal sequence, which can potentially turn into the hairpin structure of amiRNA gene. It has been proposed that dozens of miRNA genes were originated from MITEs or other TEs in *Arabidopsis* and rice (Piriyapongsa and Jordan 2008). (3) New miRNA genes may be generated by duplication of pre-existing miRNA genes with subsequent mutations. This mechanism also seems to be important in plants because each miRNA gene family on average consists of several miRNA genes and there is experimental evidence to support this mechanism (Jones-Rhoades 2012). (4) miRNA genes naturally arise by spontaneous mutations from hairpin structures in the genome. Some miRNA genes have been generated in this way in *Arabidopsis* (De Felippes et al. 2008).

Conservation and Divergence of Plant microRNA Genes

According to published data in miRBase (Griffiths-Jones et al. 2008), in general, miRNAs can be placed into two different categories: the highly abundant and conserved miRNAs and the rare and species-specific miRNAs. In one earlier study, it was found that miRNAs with identification numbers from miR156 to miR408 are exceptionally conserved and their expression levels are generally high. The remaining miRNAs after miR408 in the series are expressed at a low level, species specific or induced by specific conditions (Ma et al. 2010). While the former class contains mostly the ancient miRNAs that are important in plant development and responses to stress, the latter class might be composed of newly evolved miRNAs with their functions being non-essential due to their low activities. For example, in plants, ~20 miRNA families which are well conserved between dicots and monocots are known. Of these, 7 miRNA families, i.e., miR156/157, miR160, miR159, miR319, miR165/166, miR390 and miR408, have been also found in primitive land plants such as *Physcomitrella* and *Selaginella* which suggest that these are deeply conserved (Nozawa et al. 2011). In addition, *Arabidopsis*, rice, *Populus* and *Physcomitrella* possess many non-conserved lineage- or plant species-specific miRNA families (Nozawa et al. 2011).

As stated earlier, experimental and computational analysis has indicated that many plant miRNAs and their targets are conserved between monocots and dicots. Monocots and dicots diverged about 125 million years ago; thus, miRNAs should have existed at that time when they diverged from the same ancestor (Jones-Rhoades 2012). Further, employment of microarray technology and EST database to analyse the expression of several miRNAs in different plant species found that some miRNAs existed not only in dicots and monocots but also in ferns, lycopods and mosses (Zhang et al. 2006a). Both miRNA conservation and miRNA target conservation indicate that plant miRNAs have a very deep origin in plant phylogeny, at least since the last common ancestor of bryophytes and seed plants. Gene regulation by miRNA is an ancient evolutionary mechanism to control gene expression. It is one part of the global gene regulation mechanisms. This suggests that miRNA-mediated gene regulation existed more than 425 million years ago in the plant kingdom (Jones-Rhoades 2012). The age of plant miRNA is comparable to the age of miRNA regulation in metazoans; however, no evidence yet shows that plant miRNAs and animal miRNAs have a common ancestor. However, it is also expected that some non-conserved or lowly conserved miRNAs may be documented in the near future as more miRNAs are discovered in different plants at different developmental stages or under biotic and abiotic stresses. Non-conserved miRNAs may play more specific roles in specific plant species, such as the differentiation and elongation of cotton fibres. Several numbers of novel and differentially expressed miRNAs in contrasting cotton cultivars exposed to water stress were noticed in our laboratory (Boopathi et al. in preparation).

The entire miRNA transcribed region can be divided into two subregions in general: the evolutionarily stable subregion and the variable subregion (Jones-Rhoades 2012). The stable subregion is small in size and includes mainly the mature miRNA and the miRNA, whereas the loop and other sequences beyond the stable subregions, which represent most of the miRNA gene, belong to the variable subregions. The stable and variable subregions are the outcome of evolution over a long period. Except for the stable subregion, the overall sequences of most ancient miRNA genes tend to be variable due to genetic drift, natural selection and fixation in evolution. Sequence variations along the variable subregions of most miRNA genes are helpful in keeping the miRNA genes active (Nozawa et al. 2011). All miRNAs have similar secondary hairpin structures; many of these are evolutionarily conserved. This suggests a powerful approach to predict the existence of new miRNA orthologues or homologues in other species.

Identification of miRNA Genes

There are at least four different methods that are available for identifying miRNAs, and they are based on the major characteristics of miRNAs such as: (a) all miRNAs are small non-coding RNAs, usually consisting of ~20–22 nucleotides for animals and ~ 20–24 nt for plants; (b) all miRNA precursors have a well-predicted stem-loop hairpin structure, and this fold-back hairpin structure has a low free energy; and (c) many miRNAs are evolutionarily conserved, some from worm to human or from ferns to core eudicots or monocots in plants (Zhang et al. 2006a, b). Plant miRNAs are less conserved than animal miRNAs. Usually, only the mature miRNAs are conserved in plants instead of miRNA precursors that are usually conserved in animals. Therefore, when an sRNA (see above) is considered as an miRNA, all of these major characteristics should be included. Further, to evade labelling other sRNAs or fragments of other RNAs as miRNAs, Ambros et al. (2003) developed combined criteria to identify new miRNAs. These combined criteria include both biogenesis and expression criteria, neither of which on its own is sufficient for identifying a candidate gene as a new miRNA (Ambros et al. 2003). So far, all newly predicted or identified miRNAs have conformed to these rules.

The four approaches that are employed for identifying miRNAs are genetic screening or forward genetics (Lee et al. 1993; Wightman et al. 1993), direct cloning and sequencing after isolation of small RNAs (Lu et al. 2005a, b), computational strategy (Brown and Sanseau 2005) and expressed sequence tags (ESTs) analysis (Zhang et al. 2005). Characteristics, advantages and limitations of these methods are given in Table 2, and more details on individual methods are provided below. Earlier studies of miRNA profiling depended on Northern blotting, RT-PCR and cloning, which are labour intensive and time consuming, and they cannot obtain about global miRNA expression patterns. Compared with other methods of protein-coding gene prediction, predicting and annotating miRNA genes still need more work, and hence it is generally suggested that combined use of the above methods will improve the efficiency of miRNA profiling and characterisation.

Genetic Screening or Forward Genetics

This is the first approach by which the first two miRNAs (i.e. the founding members of the miRNAs: *lin4* and *let7*) were identified in *C. elegans* (Lee et al. 1993; Wightman et al. 1993). Since genetic studies are based on clear phenotypes, the in vivo functions of genetically identified miRNAs are well established. This method was similar to methods for identifying other traditional genes. Although this method was useful for identifying some miRNAs, application of this strategy was limited because it is expensive, time consuming and dominated by chance. To overcome some of the shortcomings of genetic screening, another experimental approach was recently described for isolating and identifying new miRNAs. This approach involves direct cloning after isolation of sRNAs.

Table 2 Characteristics, advantages and limitations of different approaches used for miRNA identification

Particulars	Genetic screening/ forward genetics	Direct cloning and sequencing after small RNA isolation	Computational approaches	EST analysis
Requirement of genome sequence	Not required	Not required	Required	Required
Use of specific software	Not required	Not required	Required	Required
Cost of the experiment	High	High but less than genetic screening	Moderate	Low
Efficiency in identification of miRNA	Low	High	Low	High
Experimental confirmation of identified miRNA	Not required	Not required	Required	Required
Efficiency in identification of new miRNA	High	High	High	Low
Can be applied across the species?	Yes	Yes	Yes	No
Can quantitative information on miRNA be obtained?	No	Yes	No	Somewhat
Reference	Lee et al. (1993) and Wightman et al. (1993)	Lu et al. (2005a, b)	Brown and Sanseau (2005)	Zhang et al. (2005)

Direct Cloning and Sequencing

In this approach, sRNA molecules are first isolated by size fractionation. Then these small RNAs are ligated to RNA adapters at their 5′ and 3′ ends (Lu et al. 2005a, b). In the successive steps, they are reverse transcribed into cDNA, which is then amplified and sequenced. Because only sRNAs are isolated and screened by this method, it is a more efficient way to obtain miRNAs than general genetic screening. This method can be further refined by combining it with massively parallel signature sequencing. This method is not only suitable for identifying plant miRNAs but also can quantify miRNA abundance at the same time.

Deep Sequencing

The emergence of next-generation technologies, capable of generating 10^5–10^7 sequences in a single experiment, has revolutionised our ability to meaningfully describe the populations of miRNAs (and other sRNAs) expressed in plant cells (Jones-Rhoades 2012). Deep sequencing has dramatically expanded the number of miRNA families known to exist in plants and has led to revised guidelines of evidence required to justify annotation of plant miRNAs (Meyers et al. 2008).

A simple and apt reason that describes the utility of deep sequencing in miRNA identification is that analysing large numbers of reads increases the chance of recovering rare transcripts. On the other hand, analysing a few hundred reads is often sufficient to obtain clones of highly expressed miRNAs. Since miRNA abundance varies by at least several orders of magnitude, much larger data sets are needed to recover a substantial fraction of lower abundance miRNAs. A second powerful reason for using deep sequencing is that categorising the cloned RNAs is more straightforward when patterns of RNA expression are apparent (Jones-Rhoades 2012). Specifically, the observation of patterns of sRNA accumulation consistent with Dicer processing of a hairpin intermediate (i.e. detection of an miRNA–miRNA* pair, as defined by the predicted secondary structure of a single-stranded precursor) is strong evidence for miRNA-like biogenesis (Meyers et al. 2008). Examples of miRNA–miRNA* pairs are sometimes observed at low depth of sequencing for abundant miRNAs (Reinhart et al. 2002), but deep sequencing makes it feasible for comparatively rare miRNAs (Rajagopalan et al. 2006). At the same time it should be noted that there is considerable evidence that miRNA discovery by deep sequencing has not reached saturation even in well-sampled species like *Arabidopsis* (Fahlgren et al. 2007; Jones-Rhoades 2012).

Computational Methods

The third approach that is used to identify miRNA is the computational approach. This approach is based on a genome sequence and bioinformatic tools (Jones-Rhoades and Bartel 2004). The first miRNA search algorithm was MiRscan, which successfully predicted miRNA genes that display close homology in two nematode worms: *C. elegans* and *C. briggsae*. MiRscan was further improved by defining conserved sequence motifs found in the vicinity of nematode miRNA genes. Since then several tools were developed, and the following are some of the commonly used public resources of machine learning-based approaches for finding miRNA genes: MiRAlign, Mir-abela, Tirplet SVM, MiPred, RNAmicro, Microprocessor SVM, Bayes miRNA find, ProMiR and MiRFinder (Please see Table 3). They have successfully predicted miRNA genes in plants and animals. However, it is generally believed that computational approach is slightly inefficient and certainly not comprehensive. The predicted miRNAs need to be confirmed by experiments such as cloning or Northern blotting.

EST Analysis

The fourth approach is an expressed sequence tag (EST) analysis approach. It is well recognised that several miRNAs are evolutionarily conserved from species to species (Llave et al. 2002; Reinhart et al. 2002; Zhang et al. 2005). This suggests a powerful approach to predict homologies or orthologues of previously known miRNAs (Jones-Rhoades and Bartel 2004). More importantly, this approach is very useful for predicting miRNAs in multiple species, especially in species whose genomes are unknown. It advocates that EST analysis is a good alternative method for identifying miRNAs, especially for species whose genome is poorly understood. However, this method can only identify conserved miRNAs. miRNAs that are more likely non-conserved cannot be identified based on the EST approach.

A novel innovation in miRNA detection involves the bead-based flow cytometric method (Lu et al. 2005a). Each individual bead is marked with fluorescence tags (which can yield up to 100 colours, each representing a single miRNA) and coupled to probes that are complementary to miRNAs of interest. miRNAs are ligated to the 5′ and 3′ adaptors, reverse transcribed, amplified by PCR using a common biotinylated primer, hybridised to the capture beads and stained with streptavidin–phycoerythrin. The beads are then analysed using a flow cytometer capable of measuring bead colour (denoting miRNA identity) and phycoerythrin intensity (denoting miRNA abundance). Since hybridisation takes place in solution, this method offers more specific detection of closely related miRNAs compared with conventional glass-slide microarrays (Lu et al. 2005a).

Documented miRNAs in Plants

Despite notable recent progress in identifying miRNAs, including detailed data on miRNA expression pattern and target genes of miRNAs, they remain widely distributed in published literature. To this end, a cohesive database system is profoundly needed for data deposit and further application. Thus, the availability of a large amount of miRNA information, including the sequence information, structural features and annotated functional roles in crop plants, necessitates in developing databases with specific goals and tools. There are several miRNA and their target databases available in the World Wide Web. Some notable databases are miRBase (the home of microRNA data; the miRBase sequence

Table 3 Publically available resources for miRNA identification using computational methods

Name of the programme/algorithm	Features	Website/reference
Mir-abela	Computed from the entire hairpin structure	http://www.mirz.unibas.ch/cgi/pred_miRNA_genes.cgi
Triplet-SVM	Each hairpin is encoded as a set of 32 triplet elements	http://bioinfo.au.tsinghua.edu.cn/miRNAsvm/
MiPred	32 global and intrinsic hairpin folding attributes based on sequence, structural, statistical thermodynamics and topology	http://web.bii.a-star.edu.sg/_stanley/Publications
RNAmicro	12 features based on structure, sequence composition, conservation, thermodynamic stability and structural conservation	http://www.bioinf.unileipzig.de/~jana/software/
Microprocessor SVM	686 features from structure and sequence; 7 additional Drosha processing sites features	https://demo1.interagon.com/miRNA/
Bayes miRNA find	Secondary structure and sequence features	https://bioinfo.wistar.upenn.edu/miRNA/miRNA/login.php
ProMiR	A hairpin structure is represented as a pairwise sequence; each position of the pairwise sequence has two states: structural and hidden	http://cbit.snu.ac.kr/~ProMiR2/
MiPred	Contiguous structure-sequence composition, minimum of free energy of the secondary structure	http://www.bioinf.seu.edu.cn/miRNA
MiRFinder	18 parameters, including the local secondary structure differences of the stem region of miRNA and the secondary structures of pre-miRNAs	http://www.bioinformatics.org/mirfinder
srnaloop	Use of sequential and structural properties	http://arep.med.harvard.edu/miRNA/pgmlicense.html
miRseeker	Comparative analysis, stem-loop conservation	http://www.ncrna.org/login_form
ERPIN (Easy RNA Profile IdentificatioN)	Sequence or structural alignment	http://tagc.univ-mrs.fr/erpin/
findMiRNA	Seed match, comparative analysis	http://www.molquest.com/help/2.3/programs/Find-miRNA/description.html
MiRAlign	Sequence or structural alignment	http://bioinfo.au.tsinghua.edu.cn/MiRAlign/
PalGrade	Sequential and structural properties	Barad et al. (2004)
miRTour	Automated homology-based discovery of plant miRNA and their targets from sequencing data sets (EST, GSS, SRA, etc.)	http://bio2server.bioinfo.uniplovdiv.bg/miRTour/
miRDeep	Identifies miRNAs using deep-sequencing technique (for worm)	http://www.mdc-berlin.de/en/research/research_teams/systems biology of generegulatory_elements/projects/miRDeep/index.html
miRanalyzer	Identifies miRNAs using deep-sequencing technique (for worm, fly and animals)	http://web.bioinformatics.cicbiogune.es/microRNA/

database is a searchable database of published miRNA sequences and annotation), PMRD (plant microRNA database; this database includes all publicly known plant miRNA sequences – including those in miRBase), MicroCosm Targets (formerly miRBase Targets; a web resource containing computationally predicted targets for microRNAs across many species), miR2Disease-Base (manually curated database documenting known relationships of miRNA dysregulation and human disease), miRecords (manually curated database of experimentally validated miRNA–target interactions), miRvar (database for genomic variations in microRNAs), PASmiR (a literature-curated and web-accessible database, which provides detailed, searchable descriptions of miRNA molecular regulation in different plant abiotic stresses), psRNATarget (a Plant Small RNA Target Analysis Server) and TargetScan (which predicts biological targets of miRNAs by searching for the presence of conserved 8mer and 7mer sites that match the seed region of each miRNA), ASRP (collection of known miRNAs in plants), miRNA map (collection of computationally identified miRNA in metazoan genomes), miRCen (collection and identification of animal miRNA–target interactions using multiple target-prediction programmes), CoGemiR (comparative genomics of miRNAs) and TarBase (collection of experimental miRNA targets).

miRNA Nomenclature

Since hundreds of miRNAs specific to particular crop and thousands across other species are being identified, a system of nomenclature has been adopted and names are designated to specific miRNAs before publication of their discovery (Ambros et al. 2003). Experimentally confirmed microRNAs are given a number that is attached to the prefix mir followed by a dash, e.g. mir-123. The uncapitalised mir- refers to the pre-miRNA and the capitalised miR- refers to the mature form. MiRNAs with similar structures with differences at 1 or 2 nucleotides are noted with added lowercase letter, e.g. miR-1a and miR-1b. It is possible for miRNAs at different loci to produce the same miRNA, and these are shown with additional number, e.g. miR-1-1 and miR-1-2. Strictly speaking, microRNA nomenclature should also be preceded by first letter of the genus and first two letters of the species in which they are observed, e.g. *hsa*-miR-156, i.e. miR156 which is identified in *Homo sapiens*. On the other hand, common miRNA species are also having distinct nomenclature, e.g. viral v-miRNA and drosophila d-miRNA. microRNAs originating from the 3′ end or 5′ end are often denoted with a -3p or 5p suffix, e.g. miR-142-5p and miR-142-3p.

miRNA Targets

As stated earlier, the very first miRNA target was noticed through genetic screens performed by the Ambros laboratory to characterise the heterochronic gene pathway (i.e. the temporal progression of developmental events in *C. elegans*). They uncovered a 22 nt non-coding RNA as the product of the *lin-4* gene (Lee et al. 1993). *lin-4* RNA repressed the protein levels of *lin-14*, a gene that functions in the same developmental pathway. The *lin-4* RNA had the potential to bind, with partial antisense complementarity, to sequences found in the 3′-untranslated region (3′-UTR) of *lin-14* mRNA and repress its translation (Lee et al. 1993). A similar kind of regulation was later noticed when the discovery by the Ruvkun laboratory, of *let-7*, a second 22nt RNA that also functioned in the heterochronic gene pathway, was published (Wightman et al. 1993). As that of *lin-4, let-7* RNA recognised sequences present in the 3′-UTR of its *lin-41* mRNA target and repressed LIN-41 protein levels. Thus it is obvious that each miRNA has its own target gene(s), whose expression(s) is/are negatively regulated by the specific miRNA.

In plants miRNAs were described first in *Arabidopsis* and later in other species. To date, there are 5399 plant miRNAs from 61 species available at the miRBase (www.http://www.mirbase.org/cgi-bin/browse.pl; verified on 09th May, 2013). Most of these miRNAs target transcription factors and thus are implicated in diverse aspects of plant growth and development (Jones-Rhoades 2012). The major challenge in determining miRNA functions is to identify their regulatory targets. By analogy to *lin-4* and *let-7* RNAs, it is reasonable to suppose that miRNAs generally recognise their regulatory targets through base pairing. An indication that target prediction for certain plant miRNAs might be more straightforward came with the identification of miR171, a plant miRNA with perfect antisense complementarity to the mRNAs of three scarecrow-like transcription factors (Llave et al. 2002; Reinhart et al. 2002).

Since thousands of miRNAs are being reported regularly, identification of miRNA targets becomes a crucial phase in appreciating their regulatory function. The methods to do this can be generally classified into two different groups according to their generations (some selective list of computational methods for target prediction are given in Table 4). The first-generation methods such as miRanda, DIANA-microT, RNAhybrid, MicroInspector and TargetScans are based mainly on three characteristic properties: (1) The 5′ seed of the miRNA (nucleotide positions 2–8 of the miRNA) is complementary to the 3′ UTR of the target mRNA. (2) The RNA–RNA duplex has a higher negative folding free energy. (3) Mature miRNAs, binding sites of miRNA to mRNA and miRNA–mRNA duplex all are highly conserved from species to species, particularly within the same kingdom (Li et al. 2010a, b).

The new generation of methods often utilises machine learning-based approaches. PicTar is a typical example. This algorithm scans the alignments of 3′ UTRs for those displaying seed matches to miRNA and then filters the alignments according to their thermodynamic stability. Each predicted target is scored by using a HMM maximum-likelihood fit approach. In the PicTar model, synergistic effects of multiple binding sites of one miRNA or several miRNAs acting together are accounted for along with appropriate scoring of overlapping site and background for binding (Li et al. 2010a, b). The probabilities are assigned according to experimental and computational results. Some experimental studies also suggested that site accessibility was a critical factor for efficient repression. Excellent comparisons of computational methods used to predict miRNAs and their targets are provided in Li et al. (2010a, b).

In plants, the successful targeting reaction requires complementarity of the miRNA at most of the residues (Mallory and Bouché 2008). The consequence of the targeting reaction depends on the nature of the targeted RNA and the extent of complementarity with the miRNA. The target RNA is cleaved and the level of the protein product is reduced if there is near complete complementarity, including positions 9 and 10 of the miRNA. Translational suppression without turnover of the target RNA is mediated by miRNAs with incomplete complementarity to their target (Lanet et al. 2009). In addition, there may be miRNA-mediated targeting of chromatin-associated RNAs that leads directly or indirectly to targeted epigenetic modification (Wu et al. 2010). In some instances, miRNA-mediated gene silencing is a simple negative switch: whenever the miRNA gene is active, the target mRNA is silent. However, these versatile RNA regulators may also participate in feedback loops and carry out more subtle roles in genetic regulation (MacLean et al. 2010). They might dampen fluctuations in target gene expression, for example, influence temporal changes. In some instances, the miRNAs or their precursors may move through plasmodesmata and different stages in the feedback system occur in adjacent cells or in separate roots and shoots. MiRNAs may also initiate regulatory cascades with multiple mRNA targets (MacLean et al. 2010). These cascades involve secondary siRNAs (see above) that associate with argonaute (AGO) proteins, similarly to miRNAs.

It has been shown that an individual miRNA is able to control the expression of more than one target mRNAs and that each mRNA may be regulated by multiple miRNAs (Jones-Rhoades and Bartel 2004). The 5′ region of miRNA usually contributes more to the specificity and activity in binding targets. The interactions between miRNA and mRNA are usually restricted to the ‘seed’

Table 4 Resources used to find miRNA targets using computational strategies

Name of the tool	Characteristic features	Website
PicTar	HMM maximum-likelihood fit approach	http://pictar.mdc-berlin.de/
TargetBoost	Boosted genetic programming algorithm to create weighted sequence motifs	https://demo1.interagon.com/demo
RNA22	Exhaustive pattern discovery based on locally conserved signatures	http://cbcsrv.watson.ibm.com/rna22.html
PITA	Incorporates the role of target-site accessibility within traditional seed finding procedures	http://genie.weizmann.ac.il/pubs/mir07
miTarget	SVM based on structural, thermodynamic and positional features	http://cbit.snu.ac.kr/~miTarget
MicroTar	Complementarity of miRNAs to their target and thermodynamic data	http://tiger.dbs.nus.edu.sg/microtar/
NBmiRTar	Naive Bayes classifier based on sequence and duplex structures feature	http://wotan.wistar.upenn.edu/NBmiRTar
mirWIP	Structural accessibility of target sequences, total free energy of miRNA–target hybridisation and topology of base pairing to the 5′ seed region of the miRNA	http://146.189.76.171/query.php
Sylamer	Nucleotide pattern analysis based on expression profiling data sets	www.ebi.ac.uk/enright/sylamer/
GenMiR++	Bayesian algorithm based on expression data sets	http://www.psi.toronto.edu/genmir
TargetScanS	Seed match (SM), sequence complementarity (SC) and minimal free energy (MFE) of miRNA/target duplex; sequence preferences of target sites	http://www.targetscan.org/
miRanda	SM, SC and MFE	http://www.microrna.org/
RNAhybrid	Measures SC, MFE and statistical significance of miRNA–target interactions	http://bibiserv.techfak.uni-bielefeld.de/rnahybrid/
MirTarget2	Based on machine learning technique; a computational model was trained by a variety of features concerning miRNA–target interactions	http://mirdb.org/miRDB/
DIANA-micro	SM, SC, MFE and sequence preferences of target sites	http://www.diana.pcbi.upenn.edu/cgibin/microt.cgi
miRcheck	Sequence complementarity, allowing gap in miRNA/target duplex (for plants)	http://web.wi.mit.edu/bartel/pub/software.Html
miRU	Sequence complementarity, allowing gap in miRNA/target duplex (for plants)	http://bioinfo3.noble.org/miRNA/miRU.htm
findMiRNA	Sequence complementarity; gap not allowed (for plants)	http://sundarlab.ucdavis.edu/miRNA/

sequence near the 5′ terminus in animals despite the fact that most plant miRNAs regulate their targets based on complete complementarity. The ~ 6 to 8 nt 'seed' sequence is highly conserved among species and even a slight change in sequence may alter its target spectra. It is also suggested that the location of central loop in the miRNA–mRNA duplexes may play a key role in affecting the efficiency of gene regulation mediated by miRNAs (Jones-Rhoades 2012).

miRNAs control the target expression by base pairing to sequence motifs in the 3′UTR of mRNAs with perfect or near-perfect complementarities (Jones-Rhoades and Bartel 2004). Certain AU-rich elements in 3′UTR were uncovered to interact with miRNAs and act both directly and indirectly as potent post-transcriptional regulatory signals. Analysis of the miRNA target sites indicated that genes with longer 3′UTRs usually have higher density of miRNA-binding sites and are mainly involved in developmental modulations, whereas genes with shorter 3′UTRs usually have lower density of miRNA-binding sites and tend to be involved in basic cellular processes (Jones-Rhoades and Bartel 2004). These facts emphasise the importance of 3′UTR in interacting with miRNAs. It is also claimed that a small subset of miRNAs from plants and animals exert repression regulation by specifically targeting the 3′ UTR of some mRNAs. In addition, candidate target sites of miRNAs falling in the protein-coding regions are also identifiable based on computational and experimental approaches. Hence, it is obvious that miRNAs identify their targets by multiple pathways or modes (Cai et al. 2009). Thus the impact of the interaction between miRNAs and their targets is further complicated than it was thought.

Current target-prediction programmes depend on the information from sequence, structure-associated free energy and evolutionary conservation to predict candidate mRNAs. Those bioinformatic methods usually result in the prediction of tens or hundreds of targets for each miRNA with high false positive rates. Therefore, further experiments of gain-of-function and loss-of-function are still needed and will determine how many of these predicted targets are genuinely targeted by miRNAs.

A number of experimental techniques have been used to validate miRNA–target relationships that are predicted by bioinformatics. These include the detection of cleavage activity in vivo or in vitro, the detection of target abundance (at the RNA and/or protein level) following perturbation of miRNA function and the effects of reducing the complementarity between the miRNA and target (Jones-Rhoades et al. 2006). Another approach is to use microarrays or RNA sequencing to globally monitor abundance of RNAs following perturbation of miRNA expression. Another powerful method to empirically identify targets of miRNAs (and other small RNAs) is to sequence libraries of 5′ fragments from uncapped, polyadenylated transcripts (which is also referred to as degredome sequencing). A large number of miRNA–target interactions initially identified via bioinformatics now have confirmed roles in specific aspects of plant biology (Chen 2009). Until now, miRNA target identification has relied mainly on computational approaches since genome-wide experimental approaches have not been developed due to the limitation of technology. This problem has recently been overcome by approaches that combine HITS-CLIP (high-throughput sequencing of RNAs isolated by cross-linking immunoprecipitation) with bioinformatic analysis (Chi et al. 2009).

Thus it can be concluded that the scope of miRNA-mediated regulation in plants, as currently understood, is much more limited in terms of the numbers of genes targeted. Well-characterised plant miRNA–target interactions (as identified by bioinformatic and experimental approaches outlined above) involve extensive complementarity and are relatively few in number. Most plant miRNA families have 0–10 known targets (usually from the same gene family) in a single genome, implying that less than 1 % of plant genes are miRNA targets. It has also been suggested that plant miRNAs, including miRNAs for which verified targets exist, might also have large numbers of yet unidentified targets.

Regulation of the Regulators: Genetic Control of miRNA Biogenesis

Since the discovery of miRNA, bioinformatic and experimental analysis has focused mainly on the identification and predicted target of miRNAs, whereas regulation of miRNA gene expression remained largely unexplored. As more experimental evidence indicated that miRNAs played an important role in the development stage and mechanism of biotic and abiotic responses, it was urgent to understand the regulation process of this petite regulator itself. The bioinformatic studies on the regulation of miRNA could roughly be divided into two parts. The first part focused on prediction of transcription factors (TFs) that governed the expression of miRNAs or location of the promoter region upstream of miRNA sequences. The second part focused on the crosstalk between miRNAs and TFs based on the experimentally proved or computationally predicted TFs and miRNA regulation relationship data sets. miRNA is coordinatively modulated by multifarious effectors such as SNP, miRNA editing, methylation and circadian clock. Cai et al. (2009) and Li et al. (2010a, b) provided a detailed description of such regulations in miRNA biogenesis and function.

Role of miRNA in Plant Growth and Development

miRNAs are playing highly significant roles in regulatory mechanisms operating in plants, including developmental timing, cell differentiation, proliferation and apoptosis. As stated above, miRNAs have several distinct features that impart specific molecular role to them as compared to other regulatory RNA species. Most of the known miRNAs are encoded as polycistronic transcripts and tend to target and regulate a set of mRNAs instead of a specific mRNA substrate. The cross species conserved miRNAs possess a special 'seed' sequence in their 5′ terminus; such conservation suggests that these molecules participate in critical cellular processes (Jones-Rhoades and Bartel 2004; Cai et al. 2009; Jones-Rhoades 2012).

Table 5 illustrates some selected examples of miRNAs and their specific role in plant growth and development. Altogether, these results suggest that miRNAs are implicated in various regulatory processes and signalling pathways that are required for organogenesis in crop plants. Additional investigation of miRNA diversity in tissues or organs of various species may serve to discover new miRNAs, to reveal new targets of conserved miRNAs and to suggest additional roles of miRNAs in these crop species. It has been proposed elsewhere that these new miRNA-modulated patterns may be influential to cellular developmental pattern and that miRNAs may be highly interesting targets to understand the control of cell growth and development in plants (Achard et al. 2004). For example, it was predicted that one of the miR396 targets was the *callose synthase* catalytic subunit (*CFL1, AF085717*), a gene that is differentially expressed during cotton fibre development (Cui et al. 2001). It was also observed that miR396 was expressed in different organs, including fibre and ovules. This suggests that miR396 plays a role in cotton fibre differentiation and development. Further study of the regulatory mechanisms of miRNAs on *CFL1* and other fibre-related genes will allow better understanding of the molecular mechanism of fibre development in cotton. This can lead to efficiently improve cotton fibre yield and quality more precisely and quickly than ever.

Role of miRNA in Biotic and Abiotic Stress Resistance in Plants

Besides the findings of miRNAs acting as master regulators of plant growth and development, other evidences also suggest that miRNAs play a key role in plant stress responses (Kedde et al. 2012). In addition, the expression profiles of most miRNAs that are implicated in plant growth and development are significantly altered during stress. This clearly implies that attenuated plant growth and development under stress may be

Table 5 Representative examples of miRNAs involved in plant growth and development

Name of the miRNA	Crop	Predicted target	Experimentally validated target	Function of miRNA	Reference
miR163	*Arabidopsis*	Members of the plant SABATH methyltransferase family	–	Species divergence and genome hybridisation	Ha et al. (2009)
miR164	*Arabidopsis*	Cup-shaped cotyledon (CUC) genes, namely, CUC1, CUC2 and CUC3	–	Organ initiation from meristematic tissues	Laufs et al. (2004)
miR171 miR396a	Cotton	Regulatory genes such as a predicted histone deacetylase and probable WRKY transcription factor 20 as well as structural genes such as beta-tubulin 7	–	Early boll development	Pang et al. (2011)
miR396	*Arabidopsis*	Growth-regulating factor (GRF) family of transcription factors	The transcription factor bHLH74	Cell proliferation in leaves	Debernardi et al. (2012)
miR165/166	*Arabidopsis*	–	PHABULOSA (PHB) and PHAVOLUTA (PHV) proteins (which are class III homeodomain leucine zipper (HD-ZIP) TFs)	Leaf polarity, shoot apical meristem formation and vascular tissue differentiation	Khan et al. (2009)
miR160	*Arabidopsis*	–	Auxin-responsive factors (ARF10, 16 and 17)	Root development	Khan et al. (2009)
miR396	*Medicagotruncatula*	Growth regulation factors (GRF), which are involved in the promotion of cell proliferation	–	Root tips	Lelandais-Brière et al. (2009)
miR167	Rice	–	Genes of auxin-related pathway	Positive regulator of adventitious root development	Meng et al. (2009)

(continued)

Table 5 (continued)

Name of the miRNA	Crop	Predicted target	Experimentally validated target	Function of miRNA	Reference
miR482	Soybean	–	Resistance (R) gene receptor kinases	Nodule initiation	Li et al. (2010a, b)
miR408	*Populustrichocarpa*	–	Plastocyanin-like, early-responsive to dehydration-related protein	Development of leaves, phloem, xylem, tension xylem and opposite xylem	Lu et al. (2005b)
miR172	*Arabidopsis* and maize	–	APETALA 2 (AP2) and AP2-like mRNAs	Early flowering and disrupting the specification of floral organ identity	Zhang et al. (2007)
miR393	*Citrus trifoliate*	–	Transport inhibitor response-like protein (TIR)	Roots, stems and leaves development	Song et al. (2009)
miR165 and miR166	*Arabidopsis*	–	Class-III homeodomain leucine zipper (HD-ZIP) transcription factor genes: PHABULOSA (PHB), PHAVOLUTA (PHV) and REVOLUTA (REV)	Radialization and adaxialization of leaf and vascular bundles in the stem	Zhang et al. (2007)

under the control of stress-responsive miRNAs. Even subtle and transient changes in miRNA expression during stress can have profound physiological effects (Mendoza-Soto et al. 2012).

Abiotic and biotic stresses are a big issue for plant growth and development and are major constraints to agricultural productivity worldwide. Examples of abiotic stress or adverse environmental factors include drought, submergence, salinity, solar radiation (excess light or high light intensities, UV light), extreme temperatures (heat and low temperature/freezing stress) and pollutants (heavy metals, herbicides). In addition, low concentrations of essential macro- and micronutrients or conditions that result in poor uptake of these nutrients are also perceived as stress by plants. Several field studies showed that environmental stress caused about 20–30 % yield loss and some may completely destroy crop yield (Gepstein and Glick 2013). Similarly, the biotic stresses such pests and diseases also cause huge loss to the farm productivity.

During the evolution, crops have developed different mechanisms to resist different environmental stresses, including salinity, cold, drought, pests and diseases. The molecular basis of plant tolerance to these stresses has been profoundly explored over several decades. These research efforts have identified numerous genes that are induced under such abiotic stress, with the hope that over expression of stress-responsive genes would improve plant stress tolerance. However, these transgenic plants exhibited very small improvements in stress tolerance or no improvement at all (Bartels and Sunkar 2005), largely because the complicated genetic interactions underlying plant stress tolerance are not completely understood, i.e. although several genes have been identified and isolated from plants (http://www.plantstress.com/biotech/index.asp?Flag=1), the principle mechanism of plant resistance still remains unknown. In addition to protein-coding genes, the expression of miRNAs in plants is altered during conditions of stress. Increasing evidences suggest that miRNAs may play an important role in plant response to biotic and abiotic stresses (Tables 6, 7 and 8). The miRNAs that have been identified as stress responsive

Table 6 Representative examples of miRNAs involved in biotic stress response

Name of the miRNA	Crop	Predicted target	Experimentally validated target	Function of miRNA	Reference
miR393	*Arabidopsis*	Genes involved in auxin perception and signalling	–	Protecting plants against pathogenic bacteria	Navarro et al. (2006)
miR482 and miR2118	Tomato	–	Disease resistance proteins with nucleotide binding site (NBS) and leucine-rich repeat (LRR) motifs	Contributes to a novel layer of defence against pathogen attack	Shivaprasad et al. (2012)
miR1885	*Brassica rapa*	–	TIR–NBS–LRR class disease-resistant transcripts	Induced specifically by turnip mosaic virus (TuMV) infection	He et al. (2008)
miR160, miR393 and miR1510	Soybean	–	12 resistance-related genes	In response to soybean mosaic virus infection	Yin et al. (2013)
miR156 variants	*Arabidopsis*	–	At3G15270 (SPL family members)	In response to *Brevicoryne brassicae* attack and *Pseudomonas syringae* infestation	Barah et al. (2013)

Table 7 Representative examples of miRNAs involved in abiotic stress response

Name of the miRNA	Crop	Predicted target	Experimentally validated target	Function of miRNA	Reference
miR169	*Arabidopsis*	–	*NFYA5*	Down-regulation of this miRNA is crucial for adaptation to drought stress	Li et al. (2008)
miR169	*Arabidopsis*	At1g48500 which encodes a ZIM (zinc-finger protein expressed in inflorescence meristem) TF	–	Cold-responsive miRNA	Zhou et al. (2008)
miR827 and miR2111	*Arabidopsis*	–	E3 ligases (At1g02860 and At1g63010 by miR827 and At3g27150 bymiR2111)	Induced during phosphate deprivation	Fujii et al. (2005)
miR399	*Arabidopsis*	–	*PHO2/UBC24* (an E2 ubiquitin-conjugating enzyme)	Mobilisation of internal phosphate from older to younger leaves	Fujii et al. (2005)
miR395	*Arabidopsis*	Low-affinity sulphate transporter (*SULTR2;1/AST68*) and three members of the ATP sulfurylase family (*APS1, APS3, APS4*)	–	Sulphate deprivation	Jones-Rhoades and Bartel (2004)
miR397, miR398, miR408 and miR857	*Arabidopsis*	–	Copper containing proteins such as Cu/Zn SODs (CSDs), plantacyanin and several laccases (laccase-2, laccase-3, laccase-4, laccase-7, laccase-12, laccase-13 and laccase-17)	Save copper for more essential proteins, such as plastocyanin and cytochrome c oxidase, by silencing CSDs, plantacyanin and laccases	Yamasaki et al. (2007)
miR1446	*Populustrichocarpa*	–	GCN5-related N-acetyltransferase (GNAT) family protein, gibberellin response modulator-like protein	In response to cold, drought, hydration, salinity stresses	Lu et al. (2008)
miR319, miR390, miR393, miR398	*Arabidopsis*	TCP (teosinte branched/cycloidea/PCF) TF, TAS3, F-box auxin receptors, TIR1/AFBs and bHLH transcription factors, Cu/Zn superoxide dismutases (CSD) enzymes: cytosolic CSD1 and plastidic CSD2 such as the COX5b.1, the 5b subunit of mitochondrial cytochrome oxidase	–	In response to metal toxicity	Reviewed by Mendoza-Soto et al. (2012)

Table 8 Selective examples of miRNAs that are expressed in response to multiple stresses

Name of the miRNA	Stress under which it was identified	Reference
miR156	Drought, salt, cold, heat, ABA, oxidative, hypoxia, UV B	Reviewed by Sunkar et al. (2012) and Kruszka et al. (2012)
miR159	Biotic, drought, salt, cold, heat, ABA, hypoxia, UV B	
miR160	Biotic, drought, salt, cold, heat, ABA, hypoxia, UV B	
miR162	Drought, salt, cold, hypoxia	
miR165/166	Biotic, drought, cold, heat, hypoxia, UV B	
miR167	Biotic, drought, salt, cold, ABA, hypoxia, UV B	
miR393	Biotic, drought, salt, cold, heat, ABA, UV B	
miR398	Oxidative, Cu and phosphate deficiency, UV, salt, ABA, water deficit, addition of sucrose, paraquat, ozone or plant pathogens	Abdel-Ghany and Pilon (2008), Jagadeeswaran et al. (2009), Jia et al. (2009), Sunkar et al. (2005), and Yamasaki et al. (2007)

in diverse plant species are too numerous to be covered comprehensively here, and hence only selective examples are listed.

On several occasions, miRNAs are shown to be key regulators in plant disease development (Tables 6 and 8). Some of the miRNAs may get involved in virus-induced gene silencing. Helper component-proteinases *(HC-Pro), p19, p21* and *p69*, are unrelated viral suppressors of gene silencing, and they play important roles in the virus response to plant antiviral silencing response (Zhang et al. 2007). Several investigations demonstrated that quite a lot of miRNAs are related to the activity of these viral suppressors. *HC-Pro* inhibited the expression level and activity of miR171 and caused miR171-related developmental deficiency. *P69* enhanced the expression and activity of miRNAs and caused rapid degradation of miRNA-targeted mRNAs and consequently enhance plant resistance to pathogens (Zhang et al. 2007). miRNAs responsive to plant infection by viruses were also identified in many plant species, such as *Brassica rapa* (He et al. 2008), rice (Du et al. 2011), *Arabidopsis* (Hu et al. 2011) and tomato (Lang et al. 2011). The role of individual miRNAs in plant resistance to viruses has not yet been demonstrated. Rather, plants may use the general RNA silencing machinery to degrade viral RNAs or target viral DNAs for methylation (Hohn and Vazquez 2011).

While the role of RNA silencing in defence against viruses was unravelled several years ago, the involvement of miRNA-guided regulations has emerged only recently as one of the many strategies developed by plants to protect against bacterial pathogens (Navarro et al. 2006). Perception of flagellin is crucial for plant resistance to *Pseudomonas syringae* bacterium (Gómez-Gómez and Boller 2002). The miRNA miR393 seems to play an important role in this process (Navarro et al. 2006). Indeed, flagellin-derived peptide induces the expression of miR393a and down-regulates the expression of the auxin receptors AFB1 at the transcriptional level and TIR1, AFB2 and AFB3 at the post-transcriptional level. Although the direct role of miR393 in bacterial resistance has not yet been demonstrated, auxin signalling is important for resistance, and the model predicts that repression of auxin signalling by increased miR393 function would restrict *P. syringae* growth (Navarro et al. 2006). Recently, a group of bacteria-regulated miRNAs that targets genes encoding proteins of the auxin, abscisic acid and jasmonic acid biosynthetic and/or signalling pathways was identified (Zhang et al. 2011b). The expression of 12 target genes, including *ARF8, ARF10, ARF16, ARF17, TIR1, AFB2, AFB3, MYB33* and *MYB65* which are the targets of miR160, miR167, miR393 and miR159, was negatively correlated with the accumulation of their miRNAs upon *Pseudomonas* infection (Zhang et al. 2011b). The results suggested an important role of these miRNAs in plant defence signalling by regulating and fine-tuning multiple plant hormone pathways.

Similarly, high-throughput sequencing has facilitated identification of miRNAs affected in genotype sensitive to Asian soybean rust (ASR), a foliar disease caused by *Phakopsora pachyrhizi* in soybean and powdery mildew strain *Erysiphe graminis* f. sp. *tritici* in wheat (Kulcheski et al. 2011). MIR-Seq11, MIR-Seq13 and MIR-Seq15, which are predicted to target *peroxidases, oxidoreductases* and translational initiation factor transcripts, are down-regulated upon ASR infection. In wheat, 24 miRNAs responsive to powdery mildew infection have also been identified (e.g. miR156, miR164, miR167 and miR393; Kruszka et al. 2012).

Similarly, various miRNAs were identified in *Arabidopsis*, *Oryza*, *Nicotiana*, *Z. mays*, Sorghum, *Populus, Gossypium, Brassica, Vitis, Physcomitrella* and *Chrysanthemum*, and their target genes were found to encode various transcriptional factors or important functional enzymes that play critical roles in plant response to various abiotic stresses (Tables 7 and 8). Specific miRNAs have also been discovered with key roles in protecting the plants against particular abiotic stress.

For example, genome profiling of drought-stressed rice has been carried out at various developmental stages to reveal drought-responsive miRNAs (Zhou et al. 2010). It has led to the identification of 30 miRNA families, which was significantly either up-regulated (such as miR395, miR474, miR845, miR851, miR854, miR901, miR903 and miR1125) or down-regulated (such as miR170, miR172, miR397, miR408, miR529, miR896, miR1030, miR1035, miR1050, miR1088 and miR1126) during drought. Prediction and validation of target genes corresponding to these miRNAs and study of their regulation at the level of transcription factors have evidenced the role of these miRNAs in drought tolerance (Zhou et al. 2010).

In another maize study, 39 miRNAs have been identified with altered expression under submergence stress (Zhang et al. 2008). Among them, expression of 19 miRNAs was up-regulated during the early stages (0–12 h) of submergence, which recovered to normal levels during later stages. However, the expression of 12 miRNAs was down-regulated during the initial stages and up-regulated after 24 h of submergence. Interestingly, seven of these 39 miRNAs were dramatically induced between 24 and 36 h of post-submergence (Zhang, et al. 2008). These miRNAs targeted genes that actively participate in eliminating reactive oxygen species (ROS) and aldehyde groups. Also, target genes possess a *cis-acting* element that is essential to cope with anaerobic conditions. The predicted targets of these miRNAs were classified into three categories (Zhang, et al. 2008). The first category includes various transcription factors involved in plant development and organ formation. For example, *ZAG1*, an agamous-like gene, was detected as a target of miR159. In addition, *HD-ZIP* is a target for miR166 and scarecrow-like family (*SCL*) is a target for miR171. The second category includes several targets of miRNAs that are involved in phytohormone cascade such as GA, Myb and auxin-responsive factors (*ARF12, ARF17* and *ARF25*). The third category includes targets encoding the proteins involved in physiological processes. The predicted targets of submergence stress-responsive miRNAs are involved in carbohydrate and energy metabolism, including *starch synthase, invertase*, malic enzyme and *ATPase*, as well as in elimination of ROS and acetaldehyde (*ALDH*) (Zhang et al. 2008). These findings have highlighted the complexity of adaptive plant responses. These adaptation strategies are helpful for survival of maize seedlings under submergence conditions.

From *Arabidopsis*, sRNA library was constructed to identify sRNAs involved in cold, dehydration and salt stress (Sunkar and Zhu 2004). This study has come out with the identification of two previously known miRNAs (miR171b and miR319c), 24 novel miRNAs constituting 15 new families and 102 novel endogenous siRNAs. From the identified miRNAs, miR393, miR397b and miR402 were up-regulated during cold, dehydration and salt stress whereas miR389a.1 was down-regulated. miR319c was found to be specifically up-regulated during cold stress (Sunkar and Zhu 2004). Microarray-based profiling of cold-responsive miRNAs has also been carried out from rice (Lv et al. 2010). Most of the identified miRNAs were down-regulated during cold stress. miRNAs from miR167 and miR319 families were down-regulated, while miR171 families were reported for variable expression profiles (Lv et al. 2010).

Similarly, salt-responsive miRNAs have also been identified from maize roots (Ding et al. 2009). miRNA microarray hybridisation has led to the identification of 98 salt-responsive miRNAs from 27 plant miRNA families. These miRNAs showed differential expression during salt stress. While 18 miRNAs were expressed in maize salt-tolerant species, 25 miRNAs showed delayed expression in maize salt-sensitive species (Ding et al. 2009). Most of the miRNAs responsive to salt stress directly regulate transcription factors. From *Z. mays*, miR159a/b, miR164a/b/c/d and miR1661m have been cloned that target transcription factors *Myb, NAC1* and homeodomain leucine zipper protein (HD-ZIP) (Ding et al. 2009). Other salt-responsive transcription factors targeted by miRNAs included MADS-box proteins and zinc-finger proteins. Further experimentation has led to the cloning of miRNAs belonging to miR474, miR395 and miR396 families from *Z. mays*. miR474 and miR395 were reported to target negative regulators of salt tolerance. They were up-regulated during salt stress, causing suppression of the respective factors. On the contrary, miR396 was reported to down-regulate in the presence of salt stress (Ding et al. 2009).

In a yet another study, deep-sequencing technology was chosen to determine the small RNA transcriptome of *Saccharum* sp. cultivars grown on saline conditions (Bottino et al. 2013). They constructed four small RNAs libraries prepared from plants grown on hydroponic culture exposed to 170 mMNaCl and harvested after 1 h, 6 h and 24 h. Each library was sequenced individually and together generated more than 50 million short reads, and there were 98 conserved miRNAs and 33 miRNAs*. Several of the microRNA showed considerable differences of expression in the four libraries. The results showed that miRNAs had higher expression in samples treated with severe salt treatment compared to the mild one. Furthermore, the majority of the predicted target genes had an inverse regulation with their correspondent miRNAs. The targets encode a wide range of proteins, including transcription factors, metabolic enzymes and genes involved in hormone signalling, probably assisting the plants to develop tolerance to salinity (Bottino et al. 2013).

Various mechanical stresses involving wind, water or any other entity imposing physical forces upon the plant body have also been found to down- or up-regulate certain miRNAs. For example, a comparative analysis of miRNA expression was performed in *Populus trichocarpa* subjected to mechanical stress via bending the plant stem in an arch for 4 d (Lu et al. 2005b). The expression of miR156, miR162, miR164, miR475, miR480 and miR481 was found to be down-regulated whereas miR408 was up-regulated in the xylem tissue of mechanically stressed plants as compared to the unstressed control (Lu et al. 2005b).

The above findings suggest that miRNA profiles are unique in closely related genotypes with contrasting stress sensitivities. Thus, it is likely that a more comprehensive analysis, including the impact of such regulation on miRNA targets, would provide better insights into miRNA-guided gene regulation that differs in stress-tolerant genotypes. Such molecular mechanisms could then be incorporated into strategies for improving the stress tolerance of crop plants.

miRNAs Responsive to Multiple Stresses

There are some miRNAs that are expressed in response to multiple biotic and abiotic stresses (Table 8). A specific well-studied example is miR398 which is involved in responses to diverse stresses such as oxidative stress, Cu and phosphate deficiency, ultraviolet (UV) stress, salt stress, ABA stress and water deficit and addition of sucrose, paraquat, ozone or plant pathogens (Abdel-Ghany and Pilon 2008; Jagadeeswaran et al. 2009; Jia et al. 2009; Sunkar et al. 2005; Yamasaki et al. 2007). Thus understanding the miR398-mediated stress regulatory network would provide new potential tools for genetic improvement of combined stress tolerance in plants.

Forthcoming Perceptions

The biological functions of the majority of plant miRNAs, including non-conserved and conserved miRNAs in plants, have yet to be uncovered. Up to now, a large body of evidence supports the idea that miRNAs are involved in a broad spectrum of biological progresses involving negative post-transcriptional gene regulation. Based on increasing numbers of specific miRNA functional study, it is indispensable for us to construct a global view about miRNA regulation mechanisms and understand miRNA in different angles (Cai et al. 2009). Thus the regulation of miRNA biogenesis or activity will be a major area of interest. As described by Chen (2009), there are several issues to be resolved including the following: 'Is the processing of specific miRNAs regulated? Are the activities of specific miRNAs regulated? What determines when amiRNA inhibits the translation of its target mRNA rather than cleaving it? How does amiRNA inhibit the translation of its target mRNA?' Similarly, we are yet to know the fates of these miRNAs, i.e. after repressing their targets, what are the molecular mechanisms to get rid of these miRNAs?

In view of the important roles of miRNA in the regulation of gene expression and hence tissue functions and phenotypes, investigations of miRNA offer many opportunities. One application is the use of concepts and techniques for gene targeting to achieve the inhibition of miRNAs in vitro and in vivo (Sun et al. 2010). In the complementary approach, the development of tools for the delivery of miRNAs to suppress the expression of target genes involved in pathogenesis is equally important. These concepts have been adopted for the development of drugs using miRNAs (Sun et al. 2010).

It is generally conceived that miRNAs downregulate gene expression by cleaving mRNA or by repressing mRNA translation. Such understanding has several applications in agriculture. As such, it may be possible to design artificial miRNAs to suppress target gene expression in order to study gene function, similar to the use of antisense mRNA and RNAi which are widely used as tools for studying gene function (see below). Another possibility is the use of miRNA knowledge to improve plant yields, quality or resistance to various environmental stresses including insect and pathogen infection. For example, crop resistance to drought could be improved by down-regulating miR169 (Table 7). Further study of miRNAs could provide us with new tools for increasing crop yield and/or quality.

Antagomirs, a group of modified anti-miRNA oligonucleotides, are currently the most readily available tools for miRNA inhibition. They have been applied successfully to inhibit specific endogenous miRNAs in cell cultures and mice (Krutzfeldt et al. 2005). To enhance the delivery efficiency of antagomirs to target tissues, several techniques were used to conjugate or package the antagomirs. These include methods based on the uses of lipids (e.g. cholesterol or liposomes), peptides (e.g. TAT leading sequences), proteins (e.g. binding proteins or antibodies), viruses (e.g. retroviral and adenoviral vectors), hydrogel and nanoparticles (Sun et al. 2010). Further developments in antagomir oligonucleotide design, packaging and local delivery through novel principles and technologies will serve to enhance the effectiveness of these antagomirs.

In another strategy, synthetic RNAs containing miRNA-targeted sites can serve as a 'decoy' or 'sponge' to compete with miRNA in binding to its target mRNA and thus inhibit miRNA functions. The concept of amiRNA sponge was reported first by Ebert et al. (2007), who engineered the tandem repeats of the putative miRNA-binding sites into the 3′ UTR of green fluorescent protein or luciferase reporter genes. Their results demonstrated that the miRNA sponge effectively suppresses the expression of the reporter gene. Notably, Ebert et al. (2007) have shown that the miRNA sponge outperforms antagomir in most miRNAs tested and that the combination of sponges and antagomir exhibits a synergistic effect. These studies indicate that miRNA sponges can effectively modulate the endogenous miRNA and their target functions. Future directions for bioengineers interested in this area include designing and manufacturing miRNA-sponge expression systems, engineering miRNA sponges to be inducible and tissue specific and developing and improving various delivery vehicles and tools (Sun et al. 2010).

It is possible that the engineering of miRNA constructs may use some of the strategies established for other small RNAs. An example is the tunable RNA interference (RNAi) construct (Deans et al. 2007) with the use of two coupled repressor proteins: one controlling the small hairpin RNA (shRNA) gene expression and another controlling the target gene expression. The shRNA and target gene expressions can thus be controlled by adding inducers specific to their coupled repressors. With such multi-repressor modules, the target gene can be temporally tuned in the presence or absence of shRNA. The various components in the construct are modular in nature, thus allowing regulation of a desired gene in tissue-specific and inducible manners. Although originally designed for shRNA targeting, such a strategy may be applicable for engineering an miRNA-based gene switch (Sun et al. 2010).

As stated in the beginning, artificial miRNAs (amiRNA) that act on specific targets can also be created and they have potential applications in plant functional genomics (Parizotto et al. 2004). Such amiRNAs can reduce the abundance of gene transcripts containing a complementary sequence. Web-based resources have been developed to aid in plant amiRNA design, such as WMD3 (Schwab et al. 2006; http://wmd3.weigelworld.org/). The amiRNA sequence is then integrated into a modified miRNA precursor within a functional miRNA transcript. This amiRNA precursor can be inserted into a transformation vector and introduced into plants for expression. Similar to the native miRNAs, amiRNAs can be processed to their mature forms and direct the RISC to down-regulate the target genes. For plant species where whole genomes have been sequenced, it is possible to avoid off-target suppression by selecting an amiRNA sequence that can distinguish between closely related genes. The amiRNA technique is of great value in functional genomics for crop species because amiRNAs act as dominant suppressors in the first generation of plant transformation. Sun et al. (2012) stated that amiRNA transgenics are more stable and efficient than those created by RNA interference (RNAi).

The greatest gap in our knowledge remains in the unknown functions of the majority of genes in plants. Future miRNA studies will likely be directed to finding functions for genes and dissecting functional redundancy. Thus, miRNAs and amiRNAs are powerful tools for basic research and for genetic modifications. miRNA-based technology will allow the specific down-regulation of a great many genes of unknown function. Using gene-specific suppression with miRNAs or amiRNAs, it is possible to distinguish functions of redundant genes. Genetic manipulation or engineering of new miRNAs could allow the specific regulation of candidate genes for modification of metabolism, growth, development and adaptation of plants. Such modifications would advance breeding programmes in agriculture, horticulture and forestry and improve productivity or response to climate change.

Closing Comments

A tiny spark can burst into mighty flame. Similarly, miRNAs, an evolutionary conserved class of small (~22 nt) non-coding RNAs, have huge roles in cellular processes, and it has recently created much attention among the molecular researchers. During earlier days, several computational approaches have estimated that every biological system may contain ~ 1 % miRNA genes of the total protein-coding genes. However, later bioinformatic studies indicated that the proportion of genes in the genome under miRNA regulation may be much larger than previously thought. For example, it is now considered that about 30 % of all human genes may be regulated by miRNAs (Jones-Rhoades 2012).

However, only few hundred miRNAs have been deposited to date in the miRNA database. This is far from the predicted miRNA number, and it requires more comprehensive strategies to capture all the miRNAs.

Further, there are many questions that remain to be addressed (see above). Although much work is focused on sRNAs, similar mechanisms may also account for the inheritance of other non-coding RNAs, including long non-coding RNAs (lnRNAs). We now have data on miRNAs and miRNA targets in representative species from green algae, non-seed plants and gymnosperms. But we do not have much information about the extent of diversity or conservation of miRNA expression within any of these lineages. Similarly, the analysis of more examples of closely related species and accessions will provide a clearer picture of how miRNAs evolve over shorter time frames. Besides it is unclear whether plants under stress use both modes of target gene regulation, i.e. degradation of transcripts and translational repression, or whether one mode is preferred over the other.

Identification of stress-responsive miRNAs is largely dependent on sequence-based profiling, which is known to have some bias, and thus requires independent validation. Small RNA blot analysis, although lacking sensitivity, is a gold standard for validation. Most published studies have not systematically confirmed profiling data using small RNA blot analysis. Implementation of highly reliable and rigorous assays is essential for firm characterisation of stress-responsive miRNAs in plants. Examining the effect of stress-regulated miRNA on its mRNA target using degradome analysis can provide robust confirmation of the stress responsiveness of miRNA (Sun et al. 2012). In addition to identifying miRNA targets, by analysing degradome libraries from control and stressed samples, it should be possible to quantify the impact of a stress-responsive miRNA on its mRNA target.

Further, studying stress-responsive miRNAs and their target gene expression in individual cell types will provide greater insights into miRNA target networks that operate in a cell- or tissue-specific manner during stress. As our understanding of the roles of miRNAs during stress deepens, the possibilities for using miRNA-mediated gene regulation to enhance plant stress tolerance will become enormous.

Thus, investigations on miRNA offer new and exciting opportunities for scientists. Manipulation of miRNA activities can lead to the integrative understanding of the molecular basis of regulation to systems levels and also help in developing new ways to design the miRNA sequences for enhancement of their agricultural applications and to improve the innovative algorithms and analysis methods for the further advancement of this novel research field. For example, a better understanding of small RNA regulation in hybrids and allopolyploids will help us effectively select the best combinations of parents for producing hybrids and allopolyploid plants and manipulate small RNA expression to overcome species barriers and produce superior hybrids to meet the growing demand in food, feed, cloths and biofuels.

Acknowledgement I wish to thank DBT, India and BMZ-GIZ, Germany, for funding which helped in setting up pilot studies in miRNA. I sincerely apologise to colleagues whose work I could not cite because of space limitations.

References

Abdel-Ghany SE, Pilon M (2008) MicroRNA-mediated systemic down-regulation of copper protein expression in response to low copper availability in Arabidopsis. J Biol Chem 283:15932–15945

Achard P, Herr A, Baulcombe DC, Harberd NP (2004) Modulation of floral development by a gibberellin-regulated microRNA. Development 131:3357–3365

Allen E, Xie Z, Gustafson AM, Sung GH, Spatafora JW, Carrington JC (2004) Evolution of microRNA genes by inverted duplication of target gene sequences in Arabidopsis thaliana. Nat Genet 36:1282–1290

Almeida R, Allshire RC (2005) RNA silencing and genome regulation. Trends Cell Biol 15(5):251–258

Ambros V, Bartel B, Bartel DP, Burge CB, Carrington JC, Chen X, Dreyfuss G, Eddy SR, Griffiths-Jones S, Marshall M, Matzke M, Ruvkun G, Tuschl T (2003) A uniform system for microRNA annotation. RNA 9:277–279

Barad O et al (2004) MicroRNA expression detected by oligonucleotide microarrays: system establishment and expression profiling in human tissues. Genome Res 14:2486–2494

Barah P, Winge P, Kusnierczyk A, Tran DH, Bones AM (2013) Molecular signatures in Arabidopsis thaliana in response to insect attack and bacterial infection. PLoS One 8(3):e58987
Bartel DP (2004) MicroRNAs: genomics, biogenesis, mechanism, and function. Cell 116:281–297
Bartels D, Sunkar R (2005) Drought and salt tolerance in plants. Crit Rev Plant Sci 24:23–58
Baulcombe D (2004) RNA silencing in plants. Nature 431:356–363
Bottino CM et al (2013) High-throughput sequencing of small RNA transcriptome reveals salt stress regulated microRNAs in sugarcane. PLoS One 8(3):e59423
Brown JR, Sanseau P (2005) A computational view of microRNAs and their targets. Drug Discov Today 10:595–601
Cai Y, Yu X, Hu S, Yu J (2009) A brief review on the mechanisms of miRNA regulation. Genomics Proteomics Bioinformatics 7(4):147–154
Chen X (2009) Small RNAs and their roles in plant development. Annu Rev Cell Dev Biol 25:21–44
Chi SW, Zang JB, Mele A, Darnell RB (2009) Argonaute HITS-CLIP decodes microRNA-mRNA interaction maps. Nature 460(7254):479–486
Cogoni C, Irelan JT, Schumacher M, Schmidhauser T, Selker EU, Macino G (1996) Transgene silencing of the al-1 gene in vegetative cells of Neurospora is mediated by a cytoplasmic effector and does not depend on DNA-DNA interactions or DNA methylation. EMBO J 15:3153–3163
Cui XJ, Shin HS, Song C, Laosinchai W, Amano Y, Brown RM (2001) A putative plant homolog of the yeast beta-1, 3-glucan synthase subunit FKS1 from cotton (Gossypium hirsutum L.) fibers. Planta 213:223–230
De Felippes FF, Schneeberger K, Dezulian T, Huson DH, Weigel D (2008) Evolution of Arabidopsis thaliana microRNAs from random sequences. RNA 14:2455–2459
Deans T, Cantor C, Collins J (2007) A tunable genetic switch based on RNAi and repressor proteins for regulating gene expression in mammalian cells. Cell 130:363–372
Debernardi JM, Rodriguez RE, Mecchia MA, Palatnik JF (2012) Functional specialization of the plant miR396 regulatory network through distinct microRNA–target interactions. PLoS Genet 8(1):e1002419
Ding D et al (2009) Differential expression of miRNAs in response to salt stress in maize roots. Ann Bot 103:29–38
Du P et al (2011) Viral infection induces expression of novel phased microRNAs from conserved cellular microRNA precursors. PLoS Pathog 7(8):e1002176
Ebert MS, Neilson JR, Sharp PA (2007) MicroRNA sponges: competitive inhibitors of small RNAs in mammalian cells. Nat Methods 4:721–726
Fahlgren N, Howell MD, Kasschau KD, Chapman EJ, Sullivan CM, Cumbie JS, Givan SA, Law TF, Grant SR, Dangl JL, Carrington JC (2007) High-throughput sequencing of Arabidopsis microRNAs: evidence for frequent birth and death of MIRNA genes. PLoS One 14 2(2):e219
Fujii H, Chiou TJ, Lin SI, Aung K, Zhu JK (2005) A miRNA involved in phosphate-starvation response in Arabidopsis. Curr Biol 15:2038–2043
Gepstein S, Glick BR (2013) Strategies to ameliorate abiotic stress-induced plant senescence. Plant Mol Biol. doi:10.1007/s11103-013-0038-z
Gómez-Gómez L, Boller T (2002) Flagellin perception: a paradigm for innate immunity. Trends Plant Sci 6:251–256
Griffiths-Jones S, Saini HK, van Dongen S, Enright AJ (2008) miRBase: tools for microRNA genomics. Nucleic Acids Res 36:D154–D158
Ha M, Lu J, Tian L, Ramachandran V, Kasschau KD, Chapman EJ, Carrington JC, Chen X, Wang XJ, Chen ZJ (2009) Small RNAs serve as a genetic buffer against genomic shock in Arabidopsis interspecific hybrids and allopolyploids. Proc Natl Acad Sci U S A 106:17835–17840
Hannon GJ (2002) RNA interference. Nature 418:244–251
He XF, Fang YY, Feng L, Guo HS (2008) Characterization of conserved and novel microRNAs and their targets, including a TuMV-induced TIR-NBS-LRR class R gene-derived novel miRNA in Brassica. FEBS Lett 582(16):2445–2452
Hohn T, Vazquez F (2011) RNA silencing pathways of plants: silencing and its suppression by plant DNA viruses. Biochim Biophys Acta 1809(11–12):588–600
Hu Q et al (2011) Specific impact of tobamovirus infection on the Arabidopsis small RNA profile. PLoS One 6(5):e19549
Jagadeeswaran G, Saini A, Sunkar R (2009) Biotic and abiotic stress downregulate miR398 expression in Arabidopsis. Planta 229:1009–1014
Jia X, Wang WX, Ren LG, Chen QG, Mendu V, Willcut B, Dinkins R, Tang XQ, Tang GL (2009) Differential and dynamic regulation of miR398 in response to ABA and salt stress in Populus tremula and Arabidopsis thaliana. Plant Mol Biol 71:51–59
Jones-Rhoades MW (2012) Conservation and divergence in plant miRNAs. Plant Mol Biol 80:3–16
Jones-Rhoades MW, Bartel DP (2004) Computational identification of plant microRNAs and their targets, including a stress-induced miRNA. Mol Cell 14:787–799
Jones-Rhoades MW, Bartel DP, Bartel B (2006) MicroRNAS and their regulatory roles in plants. Annu Rev Plant Biol 57:19–53
Kedde M et al (2012) Role of microRNAs and other sRNAs of plants in their changing environments. J Plant Physiol 169(16):1664–1672
Khan AA, Betel D, Miller ML, Sander C, Leslie CS, Marks DS (2009) Transfection of small RNAs globally perturbs gene regulation by endogenous microRNAs. Nat Biotechnol 27:549–555
Kim VN, Nam JW (2006) Genomics of microRNA. Trends Genet 22(3):165–173

Kruszka K et al (2012) Role of microRNAs and other sRNAs of plants in their changing environments. J Plant Physiol 169(16):1664–1672

Krutzfeldt J et al (2005) Silencing of microRNAs in vivo with 'antagomirs'. Nature 438:685–689

Kulcheski FR et al (2011) Identification of novel soybean microRNAs involved in abiotic and biotic stresses. BMC Genomics 12(1):307

Lagos-Quintana M, Rauhut R, Lendeckel W, Tuschl T (2001) Identification of novel genes coding for small expressed RNAs. Science 294:853–858

Lanet E, Delannoy E, Sormani R, Floris M, Brodersen P, Crété P, Voinnet O, Robaglia C (2009) Biochemical evidence for translational repression by Arabidopsis microRNAs. Plant Cell 21:1762–1768

Lang QL et al (2011) Microarray-based identification of tomato microRNAs and time course analysis of their response to Cucumber mosaic virus infection. J Zhejiang Univ Sci B 12(2):116–125

Lau NC, Lim LP, Weinstein EG, Bartel DP (2001) An abundant class of tiny RNAs with probable regulatory roles in Caenorhabditis elegans. Science 294:858–862

Laufs P et al (2004) MicroRNA regulation of the CUC genes is required for boundary size control in Arabidopsis meristems. Development 131:4311–4322

Lee RC, Ambros V (2001) An extensive class of small RNAs in Caenorhabditis elegans. Science 294:862–864

Lee RC, Feinbaum RL, Ambros V (1993) The C. elegans heterochronic gene lin-4 encodes small RNAs with antisense complementarity to lin-14. Cell 75:843–854

Lelandais-Brière C, Naya L, Sallet E, Calenge F, Frugier F, Hartmann C, Gouzy J, Crespi M (2009) Genome-wide small RNA analysis revealed differentially regulated miRNA isoforms in Medicago truncatula roots and nodules. Plant Cell 21:2780–2796

Li WX et al (2008) The Arabidopsis NFYA5 transcription factor is regulated transcriptionally and post-transcriptionally to promote drought resistance. Plant Cell 20:2238–2251

Li H, Deng Y, Wu T, Subramanian S, Yu O (2010a) Misexpression of miR482, miR1512, and miR1515 increases soybean nodulation. Plant Physiol 153:1759–1770

Li L, Xu J, Yang D, Tan X, Wang H (2010b) Computational approaches for microRNA studies: a review. Mamm Genome 21:1–12

Llave C, Xie Z, Kasschau KD, Carrington JC (2002) Cleavage of scarecrow-like mRNA targets directed by a class of Arabidopsis miRNA. Science 297:2053–2056

Lu J et al (2005a) MicroRNA expression profiles classify human cancers. Nature 435:834–838

Lu SF et al (2005b) Novel and mechanical stress-responsive microRNAs in Populus trichocarpa that are absent from Arabidopsis. Plant Cell 17:2186–2203

Lu S, Sun YH, Chiang VL (2008) Stress-responsive microRNAs in Populus. Plant J 55(1):131–151

Lv DK et al (2010) Profiling of cold-stress-responsive miRNAs in rice by microarrays. Gene 459:39–47

Ma Z, Coruh C, Axtell MJ (2010) Arabidopsis lyrata small RNAs: transient MIRNA and small interfering RNA loci within the Arabidopsis genus. Plant Cell 22:1090–1103

MacLean D, Elina N, Havecker ER, Heimstaedt SB, Studholme DJ, Baulcombe DC (2010) Evidence for large complex networks of plant short silencing RNAs. PLoS One 5:e9901

Mallory AC, Bouché N (2008) MicroRNA-directed regulation: to cleave or not to cleave. Trends Plant Sci 13:359–367

Mendoza-Soto AB, Sánchez F, Hernández G (2012) MicroRNAs as regulators in plant metal toxicity response. Front Plant Sci 3:105

Meng Y, Huang F, Shi Q, Cao J, Chen D, Zhang J, Ni J, Wu P, Chen M (2009) Genome-wide survey of rice microRNAs and microRNA- target pairs in the root of a novel auxin-resistant mutant. Planta 230:883–898

Mette MF, van der Winden J, Matzke M, Matzke AJ (2002) Short RNAs can identify new candidate transposable element families in Arabidopsis. Plant Physiol 130:6–9

Meyers BC, Axtell MJ, Bartel B et al (2008) Criteria for annotation of plant MicroRNAs. Plant Cell 20(12):3186–3190

Montgomery TA, Howell MD, Cuperus JT, Li D, Hansen JE, Alexander AL, Chapman EJ, Fahlgren N, Allen E, Carrington JC (2008) Specificity of ARGONAUTE7-miR390 interaction and dual functionality in TAS3 trans-acting siRNA formation. Cell 133:128–141

Navarro L, Dunoyer P, Jay F, Arnold B, Dharmasiri N, Estelle M, Voinnet O, Jones JDG (2006) A plant miRNA contributes to antibacterial resistance by repressing auxin signaling. Science 312:436

Nozawa M, Miura S, Nei M (2010) Origins and evolution of microRNA genes in Drosophila species. Genome Biol Evol 2:180–189

Nozawa M, Miura S, Nei M (2011) Origins and evolution of microRNA genes in plant species. Genome Biol Evol 564:55–7545

Pang M, Xinga C, Adamsa N, Rodriguez-Uribea L, Hughsc SE, Hansonb SF, Zhang J (2011) Comparative expression of miRNA genes and miRNA-based AFLP marker analysis in cultivated tetraploid cottons. J Plant Physiol 168:824–830

Parizotto EA, Dunoyer P, Rahm N, Himber C, Voinnet O (2004) In vivo investigation of the transcription, processing, endonucleolytic activity, and functional relevance of the spatial distribution of plant miRNA. Genes Dev 18(18):2237–2242

Park W, Li J, Song R, Messing J, Chen X (2002) CARPEL FACTORY, a Dicer homolog, and HEN1, a novel protein, act in microRNA metabolism in Arabidopsis thaliana. Curr Biol 12:1484–1495

Piriyapongsa J, Jordan IK (2008) Dual coding of siRNAs and miRNAs by plant transposable elements. RNA 14:814–821

Rajagopalan R, Vaucheret H, Trejo J, Bartel DP (2006) A diverse and evolutionarily fluid set of microRNAs in Arabidopsis thaliana. Genes Dev 20:3407–3425

Reinhart BJ, Weinstein EG, Rhoades MW, Bartel B, Bartel DP (2002) MicroRNAs in plants. Genes Dev 16:1616–1626

Schwab R, Ossowski S, Riester M, Warthmann N, Weigel D (2006) Highly specific gene silencing by artificial microRNAs in Arabidopsis. Plant Cell 18(5):1121–1133

Shivaprasad PV, Chen HM, Patel K, Bond DM, Santos B, Baulcombe DC (2012) MicroRNA superfamily regulates nucleotide binding site–leucine-rich repeats and other mRNAs. Plant Cell 24:859–874

Song C, Fang J, Li X, Liu H, Thomas Chao C (2009) Identification and characterization of 27 conserved microRNAs in citrus. Planta 230(4):671–685

Sun G (2011) MicroRNAs and their diverse functions in plants. Plant Mol Biol 80(1):17–36

Sun W, Li YSJ, Huang HD, Shyy JYJ, Chien S (2010) microRNA: a master regulator of cellular processes for bioengineering systems. Annu Rev Biomed Eng 12:1–27

Sun YH, Shi R, Zhang XH, Chiang VL, Sederoff RR (2012) MicroRNAs in trees. Plant Mol Biol 80:37–53

Sunkar R, Zhu JK (2004) Novel and stress-regulated microRNAs and other small RNAs from Arabidopsis. Plant Cell 16:2001–2019

Sunkar R, Girke T, Jain PK, Zhu JK (2005) Cloning and characterization of MicroRNAs from rice. Plant Cell 17:1397–1411

Sunkar R, Li YF, Jagadeeswaran G (2012) Functions of microRNAs in plant stress responses. Trends Plant Sci 17(4):196–203

Tang G, Reinhart BJ, Bartel DP, Zamore PD (2003) A biochemical framework for RNA silencing in plants. Genes Dev 17(1):49–63

Tang G, Galili G, Zhuang X (2007) RNAi and microRNA: breakthrough technologies for the improvement of plant nutritional value and metabolic engineering. Metabolomics 3:357–369

Wightman B, Ha I, Ruvkun G (1993) Posttranscriptional regulation of the heterochronic gene lin-14 by lin-4 mediates temporal pattern formation in C. elegans. Cell 75:855–862

Wu L, Zhou H, Zhang Q, Zhang J, Ni F, Liu C, Qi Y (2010) DNA methylation mediated by a microRNA pathway. Mol Cell 38:465–475

Xie Z, Allen E, Fahlgren N, Calamar A, Givan SA, Carrington JC (2005) Expression of Arabidopsis MIRNA genes. Plant Physiol 138:2145–2154

Yamasaki H, Abdel-Ghany SE, Cohu CM, Kobayashi Y, Shikanai T, Pilon M (2007) Regulation of copper homeostasis by microRNA in Arabidopsis. J Biol Chem 282:16369–16378

Yin X, Wang J, Cheng H, Wang X, Yu D (2013) Detection and evolutionary analysis of soybean miRNAs responsive to soybean mosaic virus. Planta 237(5):1213–1225

Zhang BH, Pan XP, Wang QL, Cobb GP, Anderson TA (2005) Identification and characterization of new plant microRNAs using EST analysis. Cell Res 15:336–360

Zhang B, Pan X, Cannon CH, Cobb GP, Anderson TA (2006a) Conservation and divergence of plant microRNA genes. Plant J 46:243–259

Zhang B, Pan X, Cobb GP, Anderson TA (2006b) Plant microRNA: a small regulatory molecule with big impact. Dev Biol 289:3–16

Zhang B, Wang Q, Pan X (2007) MicroRNAs and their regulatory roles in animals and plants. J Cell Physiol 210:279–289

Zhang Z et al (2008) Submergence-responsive microRNAs are potentially involved in the regulation of morphological and metabolic adaptations in Maize Root Cells. Ann Bot 102:509–519

Zhang L, Zheng Y, Jagadeeswaran G, Li Y, Gowdu K et al (2011a) Identification and temporal expression analysis of conserved and novel microRNAs in Sorghum. Genomics 98(6):460–468

Zhang W, Gao S, Zhou X, Chellappan P (2011b) Bacteria-responsive microRNAs regulates plant innate immunity by modulating plant hormone networks. Plant Mol Biol 75:93–105

Zhou X et al (2008) Identification of cold-inducible micro RNAs in plants by transcriptome analysis. Biochim Biophys Acta 1779(11):780–788

Zhou L et al (2010) Genome-wide identification and analysis of drought-responsive microRNAs in Oryza sativa. J Exp Bot 61:4157–4168

Plant Proteomics: Technologies and Applications

Deepti Bhushan Gupta, Shubhendu Shekhar, and Lalit Agrawal

Contents

D.B. Gupta, Ph.D.
TERI University, 10, Institutional Area, Vasant Kunj, New Delhi 110070, India

S. Shekhar, Ph.D.
National Institute of Plant Genome Research, New Delhi 110067, India

L. Agrawal, Ph.D. (✉)
National Botanical Research Institute, Rana Pratap Marg, Lucknow 2260001, India
e-mail: lalit.ncpgr@gmail.com

D. Barh et al. (eds.), *PlantOmics: The Omics of Plant Science*,
DOI 10.1007/978-81-322-2172-2_8,

Abstract

Proteomics is generally defined as the simultaneous and high-throughput study of protein expression profiles in cells, tissues, organs and organisms. It is a relatively new scientific discipline which has developed highly significantly over the last decade and is now recognised as one of the most important tools used in the identification and characterisation of proteins or genes of interest. The researchers have turned to proteomics to study gene products and validate their predicted functions because the availability of the complete genome sequences of a variety of organisms itself is not sufficient to find out biological function. Proteomic data for an experiment includes quantitative expression profiles, profiles of posttranslational modifications (PTMs) and protein interaction networks and linking these towards the understanding of molecular mechanisms associated with endogenous and exogenous cues. The major application of proteomics technologies is to advance our knowledge in crop plant for their development, abiotic and biotic stress tolerance, PTMs and unravelling signal transduction cascades. Further, an in-depth comparative proteome study of subcellular organelles could provide more detailed information about the intrinsic mechanism of developmental or stress response. The success in proteomics research is attributed to advances in various technology platforms associated with MS-based techniques. The accurate quantitation of proteins and peptides in complex biological systems is one of the most challenging areas of proteomics. The discoveries aimed at improving sensitivity, and throughput of both mass analysers and fragmentation technology enabled mass spectrometry (MS)-based proteomics to become the mainstream method for the large-scale analysis of complex proteomes. Along with recent and ongoing improvements in liquid separation technologies and algorithms for protein/peptide identification, MS-based proteomics has become a powerful and valuable analytical tool to study highly complex and dynamic biological systems. In this chapter, we describe the recent progress in plant proteomics and highlight the achievements made in understanding the proteomes of major research area of plant biology.

Keywords

2-DE • Plant proteomics • Mass spectrometry (MS) • Posttranslational modifications (PTMs) • MALDI • ESI • Subcellular proteomics • Secretome • Comparative proteomics • Membrane proteomics • Signalling • Biotic/abiotic stress

Introduction

The establishment of the double helical structure of DNA by Watson and Crick in 1953 followed by the development of DNA sequencing techniques by Sanger et al. in 1977 was a landmark discovery that revolutionised life science research. The genome sequences of several species have been unravelled, from the simplest mycoplasma (Wasinger et al. 1995) to more complex eukaryotes including various plant species. The advent of technology from dideoxy sequencing methods to automated next-generation sequencing platforms has led to the exploration of more and more genomes. However, the exact function of most of the genes remains obscure, and the characterisation of proteins that are encoded by the genome is the present-day challenge. In the post-genomic era, proteomics is one of the fastest-growing areas of biological research, and its objective has moved beyond simple identification and cataloguing of proteins to more comprehensive study of functional and regulatory aspects. Nevertheless, the goal of proteomics remains to obtain a more comprehensive and integrated view of biology by studying all the proteins of a cell rather than an individual protein using the more broad aspects of dynamic protein machinery thereby including many different areas of study under the umbrella of proteomics. The aim of proteomics is not only to catalogue all the proteins in a cell but also to create a complete three-dimensional (3-D) map of the cell indicating its localisation which certainly will require the involvement of a large number of different disciplines such as molecular biology, biochemistry and bioinformatics (Chen and Harmon 2006; Graves and Haystead 2002).

Origins of Proteomics

Proteomics is the study of the proteome, i.e. the whole protein complement of the genome. Proteomics as a technology can be defined as the identification and characterisation of all proteins involved in a particular pathway, organelle, cell, tissue, organ or organism that can be studied in concert to provide an accurate and comprehensive data about that system both qualitatively and quantitatively (Kav et al. 2007; Wright et al. 2012). The terms "proteomics" and "proteome" were coined by Wilkins and colleagues in 1995 as the large-scale characterisation of the entire protein complement of a cell line, tissue or organism that reflects the terms "genomics" and "genome", which describe the entire collection of genes in an organism. The era of proteomics starts with the introduction of the two-dimensional gel by O'Farrell (1975), Klose (1975) and Scheele (1975), who began mapping proteins from *Escherichia coli*, mouse and guinea pig, respectively. However, the development of 2-DE was a major step towards the development of proteomics, but the lack of proper sensitive protein sequencing and identification technologies hampered its progress. Nonetheless, the need of improved sensitive technique was critical for success because protein-loading capacity is one of the major limiting factors for both one-dimensional (1-D) and two-dimensional (2-D) gels. Apart from this, biological samples are also often limiting. The first major breakthrough technology to emerge for the identification of proteins was the inclusion of sequencing of proteins by Edman degradation followed by the development of mass spectrometry (MS) technology. Furthermore, the developments and improvements in microsequencing technology resulted in increased sensitivity of Edman sequencing, and the sensitivity of analysis and accuracy of results for protein identification by MS have increased by several orders of magnitude that even proteins in the femtomolar range can be identified in gels. However, MS is more sensitive, can tolerate protein mixtures and is amenable to high-throughput operations; it has essentially replaced Edman sequencing as a method of choice for the protein identification (Aebersold et al. 1987; Andersen and Mann 2000; Pandey and Mann 2000). The improved techniques in 2-DE allowed comprehensive protein visualisation on 2-D gels. The further development of biological MS and the growth of searchable sequence databases led to the advancement of proteomics. The advancements of the MS techniques that were developed for the ionisation of proteins and peptides include matrix-associated laser desorption ionisation (MALDI) and electrospray ionisation (ESI) combined with the time of flight (TOF), ion trap and triple-quadrupole tandem MS (MS/MS) spectrometers; these offer high sensitivity and mass precision (Karas and Hillenkamp 1988; Fenn et al. 1989; Aebersold and Mann 2003).

Although the separation and visualisation of proteins from crude extracts of tissue samples of an organism or cell culture by two-dimensional gel electrophoresis (2-DGE) followed by identification and characterisation by mass spectrometry (MS) became a common method of choice in proteomic analysis, a well-separated protein mixture within a particular pH range is not only important for obtaining a characteristic MS spectrum for protein identification but also for the quantitative analysis of differentially expressed proteins. However, isolation of proteins and obtaining a reproducible, well-resolved 2-D gel from plant tissue can prove to be particularly challenging due to the high phenolic content and high carbohydrate/protein ratio in most plant tissues. Nevertheless, the resolution of protein spots on a 2-D gel is limited by factors such as abundance, size and other electrophoretic properties; the complete proteome has been fractionated into sub-proteomes such as subcellular compartments, organelles and multiprotein complexes to improve sensitivity and resolution and to reduce the overall intricacy (Sarma et al. 2008; Jung et al. 2000; Park 2004).

Why Proteomics?

The genomic and transcriptomic informations are still fragmentary and can be insufficient to completely understand a complex organism as the DNA–RNA relationship may not be fully correlated to each other as well as not with proteins. Proteins are the functional molecules and, therefore, the most promising candidate to reflect differences in gene expression. Genes may be present and may be mutated, but they are not necessarily transcribed. Some messengers are transcribed but not translated, and thus the number of mRNA copies does not necessarily reflect the number of functional protein molecules, and the ability to qualitatively and quantitatively scrutinise mRNA and protein populations raises the tantalising prospect of deciphering the functional and regulatory networks that represent the bridge between genotype and phenotype. Contrary to the genome and transcriptome, the proteome of a given cell or organism is dynamic. The proteome of a cell reflects the immediate environment in which it is studied. In response to internal or external cues, proteins can be modified by PTMs, undergo translocations within the cell or be synthesised or degraded. Thus, examination of the proteome of a cell is like taking a "snapshot" of the protein environment at any given time. Considering all the possibilities, it is likely that any given genome can potentially give rise to an infinite number of proteomes. Unravelling the proteomes is significantly more challenging and complicated than the genomes for three main reasons. First, in higher eukaryotes, a single gene often produces many different forms of the protein, primarily due to alternative splicing and various posttranslational modifications (PTMs), resulting to its functional diversity. Secondly, genomes are largely stagnant throughout the lifetime of a cell or organism, whereas proteomes are highly versatile. Third, proteomics is currently more challenging than genomics because the technologies required for proteomics are more complex (Hegde et al. 2003; Wright et al. 2012; Graves and Haystead 2002; Rose et al. 2004; Celis et al. 2000). Therefore, the utmost need of proteomics is but natural and can be broadly explained under the following subheads:

Genome Annotation

The primary goal of all the genome sequencing efforts is to ascertain the molecular and the cellular functions of all the gene products. Whilst genome sequencing efforts reveal the basic building blocks of life, a genome sequence alone is insufficient for elucidating biological function. Genome annotation is the means for the identification of genes and assigning its biological function(s) from a particular genome sequence. Current high-throughput genome annotation uses a combination of comparative (sequence-based homology data) and noncomparative (ab initio gene prediction algorithms) methods to identify protein-coding genes in genome sequences. Sequence-based homology to already characterised proteins from other genomes is the basis of annotation of genes of which 30–50 % of predicted gene products either have no known homologs or show too little sequence homology to known proteins making the task of genome annotation difficult. Since approaches used to corroborate the presence of predicted protein-coding genes are typically based on expressed RNA sequences, they cannot independently and unambiguously determine whether a predicted protein-coding gene is translated into a protein. Indeed, the dependency on a sequenced genome or cDNA library may often be restrictive in the scope of studies, particularly for non-model organisms. However, one of the first applications of proteomics is to categorise the total number of genes out of a genome, and this "functional annotation" of a genome is necessary because it is still difficult to predict genes accurately from genomic data. Exact prediction of exon–intron boundaries and structure and alternative splicing of most of the genes, pseudogenes, promoter regulatory regions, untranslated regions and repeats is a difficult problem that cannot be precisely predicted by bioinformatics. Tools for the annotation of genomic and proteomic sequences and their structures have been developed during the last two decades and eventually made accessible to be used with an added advantage to a huge availability of characterised data. The databases having these data often focus in a particular area of annotation and are often most powerful when arranged in such a way in which the data can be probed computationally. For instance, CATH is a database of protein structural domains where an extensive view of a chosen protein family or a narrower view of a particular protein structure can be obtained. Proteomics would not be possible without genomics; however, this does not mean that it is incapable to assist genomics. On the contrary, proteomics provides a fast, relatively despicable and confident method for assembling a large amount of experimental evidence to assist genome annotation. It also has the additional advantage of confirming that transcripts are translated to the proteome stage and can help identify functional details of the mature protein. Over the last few years, there has been a move towards the integration of the wide range of genome and proteome annotation methods and databases in order to provide an overall view of the function of these genes.

Nevertheless, proteogenomics, i.e. an integrated approach of genomic information with that of data obtained from protein studies, is one of the solutions towards this problem to confirm the existence of a particular gene. Proteogenomics allows validation of predicted genes and, more importantly, correction of genome annotation errors such as detection of unannotated genes, reversal of reading frames, identification of translational start sites, stop codon read-throughs or programmed frame shifts and detection of signal peptide processing and other maturation events at the protein level. Several studies dedicated to genome reannotation based on experimental proteomics have paved the way for the proteogenomics approach. A high-throughput tandem mass spectrometry-based proteomics approach can be used to verify coding regions of a genomic sequence due to its ability to directly measure peptides arising from the expressed proteins. Therefore, proteogenomics approaches have the ability to improve the quality of genome annotations (Eisenberg et al. 2000; Yakunin et al. 2004; Liska and Shevchenko 2003; Ansong et al. 2008; Orengo et al. 1997; Wright et al. 2009; Reeves et al. 2009; Baudet et al. 2009/2010).

Protein Expression Studies

Genomic information provides an exceptional platform for cross-correlation between transcriptomic and proteomic characteristics of a particular gene, its expression and biological function(s). However, it is implausible that a simple unidirectional or linear relationship between the transcriptome and the proteome exists, as these two data sets are distinctly different and both have idiosyncratic control and regulation over biological effects. Transcriptome, a dynamic link between the genome, proteome and the cellular phenotype associated with physical characteristics is the subset of genes expressed in a specific cell or tissue type. Recently, a number of techniques have emerged that provide an extremely robust and potent set of tools to study comprehensive and quantitative genome expression. These include differential display PCR, cDNA microarrays and serial analysis of gene expression (SAGE). However, the analysis of mRNA is not a direct reflection of the protein content in the cell consequently having a poor correlation between the mRNA and protein expression levels. Transcription is merely the first step in a long sequence of events resulting in the synthesis of a protein since posttranscriptional control in the form of alternative splicing, polyadenylation and mRNA editing is an important step further. This is a significant step where many different protein isoforms can be generated from a single gene, whereas translational and posttranslational regulation is also an important step further. Proteins, having been formed by translation, are subject to PTMs. It is estimated that up to 300 different types of PTMs exist. Proteins can also be regulated by proteolysis and compartmentalisation. The analysis of protein expression profiles provides an additional information to genomic and mRNA analysis, since a proteome is dynamic and is spatially and temporally expressed. In addition to it, proteins are often functional as interacting molecules that carry out various cellular functions, such as signal transduction and dynamic (e.g. phosphorylation) and/or static modifications (e.g. disulphide linkage) that may not be perceptible from genomic information or from mRNA abundance (Corthals et al. 2000; Revel and Groner 1978; Kwon et al. 2006).

Protein Function

In due course of evolution, a large number of protein families have been produced which share the same three-dimensional architecture and often have detectable sequence and functional similarity. This conservation allows deducing the structural design of all proteins in a family even when only the structure of a single member is known and that eventually allows predicting the biological function(s). Despite the advancements in techniques for determining protein structure, the structures of many proteins are still unknown. With the help of protein prediction programs, computational analysis of genome sequences is producing numerous new *hypothetical* proteins of unknown structure and function. These proteins are called "hypothetical proteins" as they represent the products predicted from the gene sequence; however, no circumstantial evidence for their existence and function is available so far. Several studies revealed that no function can be assigned to about one-third of the sequences in organisms for which the genomes have been sequenced. The complete identification of all proteins in a genome will help the field of structural genomics in which the ultimate goal is to obtain 3-D structures for all proteins in a proteome. This is indispensable since the functions of many proteins can only be inferred by examination of their 3-D structure. Structural genomics or structural proteomics can be defined as the quest to obtain the three-dimensional structures of all proteins. Comparatively, as a recent scientific discipline, proteomics uses a variety of old and new techniques to reveal the structure and conformation, as well as measuring protein concentrations in varying conditions. Structural data can be used to determine the function of various proteins, based on comparison to similar proteins with known functions. The major challenges ahead in structural proteomics include the identification of all the proteins on the genome-wide scale, determining their structure–function relationships and outlining the precise 3-D structures of the proteins. Hitherto, X-ray crystallography or nuclear magnetic resonance (NMR) spectroscopy is the technique typically used to determine the protein structures. Nonetheless, a detailed knowledge of 3-D structure by these techniques is still fragmentary. Therefore, computational methods such as comparative and de novo approaches and molecular dynamic simulations are intensively used as an unconventional tool to predict the 3-D structures and dynamic behaviour of proteins. Computational programs may help to predict the structure of proteins having unknown function. These programs can prove to be a predictive model of the unknown protein's structure by comparing the sequence of the unknown protein to proteins with known 3-D structures and function(s) as templates. This structure-based prediction of functional information is preferred over the sequence-based extrapolations since the similarity in structure is generally more decipherable than the similarity in sequence, and moreover, structure frequently allows a more sensible and informative transfer of functional description than sequence alone. By using the methods that rely only on the structure of the protein to be characterised, such as the matching of 3-D patterns and precise docking of ligands, structural genomics will contribute to functional annotation of proteins in addition to improving homology-based arguments. However, the fecundity of this method depends on the quality of the match between the known template proteins and the unknown target protein. Nevertheless, these prediction programs do not produce structures with the detail or reliability of experimental techniques such as X-ray crystallography or NMR, whereas these prediction methodologies provide a means to critically analyse in a reasonably limited time period resulting into the identification and characterisation of a large number of new proteins identified by the analysis of whole genomes. Therefore, the ultimate aim of structural proteomics is not to obtain the structures or models for all the unknown proteins but to elucidate its functional annotation (Sánchez et al. 2000; Edwards et al. 2000; Liu and Hsu 2005; Christendat et al. 2000; Eisenstein et al. 2000).

Protein Modifications

The cellular integrity and morphology to accomplish numerous biological functions rely on an intricate interplay between the thousands of different biomolecules, whereas the basic biological functions of proteins are encoded by the associated genes. However, the real-time dynamics and regulation of protein structure and function are by and large carried out by specific PTMs of proteins such as phosphorylation, glycosylation and acylation. During recent years, protein PTMs have fascinated the biological and biomedical research especially the plant proteomics to unravel the mechanisms underlying various stress adaptations. Posttranslational modifications (PTMs) are involved in the regulation of a wide range of biological processes such as protein structure, activity and stability. Several hundred PTMs have been known thus so far; nonetheless, relatively very few have been studied using mass spectrometry and proteomics. Initially, different methods for PTM characterisation are developed to study yeast and mammalian biology and later on adopted to explore plants. As a part of a quantitative proteomics strategy, it is helpful to enrich for PTMs on the peptide level to not only identify the PTM but also to establish the functional relevance in the context of regulation, response to different types of biotic and abiotic stresses etc. Using mass spectrometry-based methods, protein phosphorylation is the only PTM that has been studied extensively thus so far at the proteome-wide level in plants. PTMs have been extensively exposed to influence protein–protein interactions, subcellular localisation and an array of both internally and externally generated signal transduction into cellular outcomes often leading to phenotypic variations. A detailed analysis of these modifications presents a formidable challenge; however, their determination generates an indispensable insight into the biological function(s). Methods and techniques developed to characterise individual proteins are now systematically applied to protein populations. A combinatorial approach of function- or structure-based purification of modified "sub-proteomes", such as phosphorylated proteins or modified membrane proteins, with mass spectrometry is particularly successful. Mass spectrometry has become a method of choice for the elucidation of several types of PTMs in both qualitative and quantitative manners. Due to the availability of large data sets on the proteome-wide level, the identification of combinatorial PTM patterns has become feasible. Various reports in this area reveal that many proteins undergo multiple modifications and the sequential or hierarchical patterns exist on many proteins; the biology of these modification patterns is only starting to be unravelled (Jensen 2004, 2006; Ytterberg and Jensen 2010; Zhao and Jensen 2009; Mann and Jensen 2003; Young et al. 2010).

Protein Localisation

Signal hypothesis revealed the existence of zip codes for directing proteins or protein complexes to subcellular compartments such as the nucleus, cytoplasm, mitochondria, endoplasmic reticulum, lysosomes and endosomes, peroxisomes, Golgi and nucleolus so that they interact at defined sites at the correct time. Proteins need to be localised to their proper cellular compartments in order to perform their biological functions. For instance, to promote gene expression, most transcription factors used to localise in the nucleus, whereas some proteins, such as the glucocorticoid receptor, may localise in one compartment (cytoplasm) temporarily and further localise to another compartment (nucleus) in response to a stimulus. Assigning a subcellular location to a protein is very desirable and inevitable to biologists since not only it can help to reveal their role in the cell processes but also it can redefine the knowledge of cellular processes by pinpointing certain activities to specific organelles as proteins are spatially organised according to their function. Protein localisation is one of the most important regulatory mechanisms known as the mislocalisation of proteins is well known to have profound effects on cellular function (e.g. cystic fibrosis). Membrane-bound organelles and discrete cytoskeletal components are the key features of eukaryotic cells that serve to sequester the components into well-defined spaces. Identification of organelle proteins and their macromolecular structure is therefore a key step towards a comprehensive understanding of its biology.

Although recent genomic approaches promise a plethora of information, several fundamental proteomic data sets remain uncatalogued. Protein localisation is assumed to be a strong indicator of gene function and is also useful as a method of evaluating protein information inferred from genetic data, for instance, supporting or refuting putative protein interactions suggested by two-hybrid assay. Furthermore, the subcellular localisation of a protein can often reveal its mechanism of action. Proteomics aims to identify the subcellular localisation of each protein that information can be used to generate a 3-D protein map of the cell, providing novel information about protein regulation. The enrichment of the subcellular compartments followed by the identification of their protein contents by proteomics is a powerful method for rapid protein localisation. To date, very few studies have characterised protein localisation on a large scale, primarily because traditional methods to assign proteins to subcellular locations are mostly targeted to a single protein of interest, and very few high-throughput methods exist by which reporter fusions or epitope-tagged proteins can be generated and subsequently localised. To address this problem, a large data set can be created by integrating the localisation data available thus so far (Davis et al. 2007; Lilley and Dupree 2007; Drumm and Collins 1993; Simpson and Pepperkok 2003; Dunkley et al. 2004; Kumar et al. 2002).

Protein–Protein Interactions

In general, functional proteins interact with each other and very rarely act as single isolated entity. To elucidate the function(s) of an unknown protein, a possible approach is to investigate the function(s) of proteins interacting with it. The systematic study of protein–protein interactions for the purpose of elucidating protein functions is termed "interaction proteomics". One of the elementary significance of proteomics is the understanding of protein–protein interactions since the processes of cell growth, cell cycle, programmed cell death etc. are all regulated by signal transduction through protein complexes. Therefore, revealing the mechanism underlying these cellular processes is important. There are quite a few approaches to analyse protein–protein interactions, i.e. based on (1) the biochemical analysis of multiprotein complexes, for example, pull-down and affinity capture methods; (2) molecular biology approaches that basically include the yeast two-hybrid assay, fluorescence resonance energy transfer and bimolecular fluorescence complementation (*BiFC*); and furthermore (3) the in silico prediction methods. Proteomics aims to develop a complete 3-D map of all protein interactions in the cell to identify the members of functional protein complexes, pathways and protein–ligand binding. Recently, proteome-scale protein physical interaction maps for several organisms have been determined. Further, these physical interactions are complemented by a wealth of information that includes other types of functional relationships between proteins, including genetic interactions, co-expression patterns and mutual evolutionary history. Whole-proteome protein interaction maps can be constructed by taking collectively these pairwise linkages. As protein–protein interactions are fundamental to most biological processes, the systematic and logical identification of all protein–protein interactions is considered a key strategy for revealing the cellular processes. Consequently, several experimental and computational techniques have been developed to methodically determine both the potential and actual protein interactions in selected model organisms. As these interactions are likely to be correlated with the protein's functional properties, protein interaction maps are frequently utilised to reveal in a systematic fashion the potential biological role of proteins of unknown functional classification. Strategies to explore protein–protein interactions, affinity purification and mass spectrometry, yeast two-hybrid, imaging approaches and various diverse databases have been developed. As a result of an increase in the number of identified proteins with the development of MS and large-scale proteome analyses consequently, the false-positive protein identification rate has also increased. Therefore, the universal consensus is to confirm protein–protein interaction data using one or more independent approaches for an accurate evaluation. Furthermore, identification and characterisation of minor protein–protein interactions are fundamental for understanding the functions of transient interactions and low-abundance proteins. The development of new methods and/or improvements in existing methods in addition to the establishment of protein–protein interaction methodologies is highly desirable. These involve detection of minor proteins by MS, multidimensional protein identification technology or OFFGEL electrophoresis analyses, one-shot analysis with a long column or filter-aided sample preparation methods. These sophisticated techniques should permit thousands of proteins to be identified, whereas in-depth proteomic methods should permit the identification and characterisation of transient binding or protein–protein interactions with weaker affinity (Nabieva et al. 2005; Pawson and Nash 2000; Yook et al. 2004; Yanagida 2002; Fukao 2012).

Structural Proteomics

Recently, a large-scale protein structural determination initiated the era of "structural genomics" or "structural proteomics". As protein 3-D structures are more conserved than the sequence, these initiatives also pave the way of biochemical or biophysical functional characterisation through structure. Proteomics studies whose objective is to study a detailed account of the structure of protein complexes or the proteins present in a specific cellular compartment are known as structural proteomics. Structural proteomics emerged as the outcome of synchronised development of high-throughput methodologies and technologies that enabled novel data to be generated with greater efficacy. Structural proteomics attempts to identify all the proteins within a protein complex or organelle, establish their location and characterise all protein–protein interactions. Analysis of the experimental or modelled 3-D structures is one of the key components for the functional understanding of unknown proteins. However, structural proteomics technologies are generating protein structures at an exceptional rate; nevertheless, the current knowledge of 3-D structural detail is still limited. It is usually accepted that quite a few of the structural proteome has a template structure from which reliable conclusions can be drawn; however, 3-D structural coverage of proteins may vary. Nevertheless, these informations will help piece together the overall architecture of cells and explain how expression of certain proteins gives a cell its unique characteristics leading to a unique phenotype. Thus, structural proteomics has a major aim to assemble a map of protein structures that will represent all the proteins included in the "global proteome" (Wild and Saqi 2004; Norin and Sundström 2002; Blackstock and Weir 1999).

Proteome Analysis: Current Technology and Challenges

The era of genomics is well established with an ever-growing number of genomes being sequenced and added to the database everyday. The first plant genome to be sequenced was that of *Arabidopsis thaliana* in 2000, followed by the first crop plant *Oryza sativa*. Thereafter, multiple plant genomes were added to the database, tomato being the most recent (2011). However, the greater job of genome annotation is still in its infancy, and proteomics remains to be the most powerful tool. Unlike animal or microbial genomes, plant genomes have additional complexities in terms of ploidy levels and genome duplication. Besides, the extraction of protein and purification presents a greater challenge. The era of proteomics started with gel-based approaches for the resolution of protein mixtures, which are still considered as the touchstone of proteomics though they suffer from various limitations (O'Farrell et al. 1977). The initial proteome analysis pipelines were based on a 2-DE PAGE separation followed by Edman end sequencing methods. However, the advances in MS/MS technology have led to more powerful tools for the identification of mixture of proteins using gel-free methods. Most gel-free approaches use a bottom-up strategy where proteins are first digested with a proteolytic enzyme, and the obtained complex peptide mixture is then separated via reversed-phase (RP) chromatography coupled to a tandem mass spectrometer. This strategy is currently only successful with partial or simple mixtures. Another major advent of this decade has been the quantitative differential proteomic expression analysis using MS/MS-based ICAT and ITRAQ techniques which has truly revolutionised the field of proteomics. It has led to the automation of the analysis pipeline and created a parallel to the high-throughput platform of genomic studies (Gygi et al. 1999; Gygi et al. 2002; Zhou et al. 2002). However, the challenges with analysis of posttranslational modifications and dynamic resolution of proteins remain to be optimally resolved. In large proteomes, the dynamic resolution is generally limited, and only the most abundant proteins are detected. This has been improved by fractionating a proteome into smaller sub-proteomes. In addition, complex proteomes can be analysed more in depth by a combination of separation techniques combining gel-based and non-gel-based methods.

Mass Spectrometry

Mass spectrometry (MS) is a technique for "weighing" molecules. However, the mass measurement is not done using balance or scale but is based upon the motion of a charged particle, called an ion, in an electric or magnetic field.

A *mass spectrometer* is an instrument that produces ions and separates them in a gas phase according to their mass-to-charge ratio under an electric or magnetic field.

The origin of mass spectrometry lies in the classical experiment by J.J. Thomson more than 100 years ago in the University of Cambridge, England. **J.J. Thomson** discovered that electric discharges in gases produced ions, and these rays of ions would adopt different parabolic trajectories according to their mass when passed through electromagnetic fields. This separation of ions according to their mass (and charge) formed the foundations of modern mass spectrometry experiments.

Francis William Aston, who was a student of J.J. Thomson, designed several mass spectrographs in which ions were dispersed by mass and focused by velocity. This led to improvements in mass resolving power and the subsequent discovery of isotopes for many common naturally occurring elements. Thomson and Aston were honoured for their achievements and received Nobel Prizes in Physics and Chemistry in 1906 and 1922.

Aston FW (1907) Experiments on a new cathode dark space in helium and hydrogen. Proc R Soc Lond Ser A 80(535):45–49

Aston FW (1907) Experiments on the length of the cathode dark space with varying current densities and pressures in different gases. Proc R Soc Lond Ser 79(528):80–95

The power of MS/MS lies in the accuracy of the instrument. Large biomolecules such as peptides can be measured within an accuracy of 0.01 % of the molecular mass. This allows for the identifications of single amino acid differences or a posttranslational modification. Small organic molecules can be measured to an accuracy of 5 ppm which is essential for isotopic detections and deduction of molecular formulae. Structural information can be generated using special types of mass spectrometers with multiple analysers which are known as "tandem mass spectrometers". The sample is fragmented inside the instrument and sequentially analysed. The data generated is useful for the structure elucidation of organic compounds and for peptide or oligonucleotide sequencing.

There is wide range of applications possible for mass spectrometry in academia, research or industry which may be broadly summarised:

- **Accurate molecular mass measurements**: *to confirm sample, to determine the purity of a sample, to verify amino acid substitutions and to detect posttranslational modification*
- **Reaction monitoring**: *to monitor enzyme reactions, chemical modification and protein digestion*
- **Amino acid sequencing**: *sequence confirmation,* de novo *characterisation of peptides and identification of proteins by database searching with a sequence "tag" from a proteolytic fragment*
- **Oligonucleotide sequencing**: *sequencing and quality control of oligonucleotides*
- **Structure determination**: *organic compound structure, protein folding monitored by H/D exchange, protein–ligand complex formation under physiological conditions and macromolecular structure determination*

Mass spectroscopy is now essentially used in any analytical laboratory for applications ranging from analysis of biopolymers, drug discovery, drug testing, environmental monitoring, geological data mining etc.

The Pipeline of Mass Spectroscopy

A mass spectrometer consists of three modules: (1) an *ion source* that converts the gas phase sample molecules into ions (or, in the case of electrospray ionisation, moves ions that exist in solution into the gas phase); (2) a *mass analyser*, which sorts the ions by their masses by applying electromagnetic fields; and (3) a *detector* that measures the value of a signal quality and quantity and generates temporal data for calculating the abundances of each ion present. The inlet introduces the sample into the vacuum of the mass spectrometer. In the source region, neutral sample molecules are ionised and then accelerated into the mass analyser. The mass analyser is the core of any mass spectrometer. Ions are separated in this section, either in space or in time, according to their mass-to-charge ratio. After the ions are separated, they are detected, and the signal is transferred to a data system for analysis. All mass spectrometers also have a vacuum system to maintain the low pressure or high vacuum required for the operation. High vacuum minimises ion–molecule reactions and scattering and neutralisation of the ions (Fig. 1).

A mass spectrometer only recognises "MASS". It produces a spectrum where the *x*-axis is m/z and *y*-axis is the intensity or cps (counts per sec). Each molecule produces characteristic spectra which are considered as the "mass fingerprint" for that molecule. The identification of the unknown is done by comparing with a library of standard mass spectrograms for a wide range of molecules. The sample can be in any form of matter, i.e. liquid, solid or gas, but can only be introduced in a vapour form, which has to be ionised essentially. The instrument reads the charge imparted to the molecule and not the inherent charge of the molecule. Hence, the inlet and the source act as feeders for mass analysers which are actually responsible for mass determination.

Ionisation Methods

The identification in a mass spectrometer necessarily requires the sample to be introduced as gas phase ions. The process of ionisation for nonvolatile, polar or charged molecules can occur by a loss/gain of an electron or of a charged particle (e.g. proton), generating odd or even electron ions, respectively.

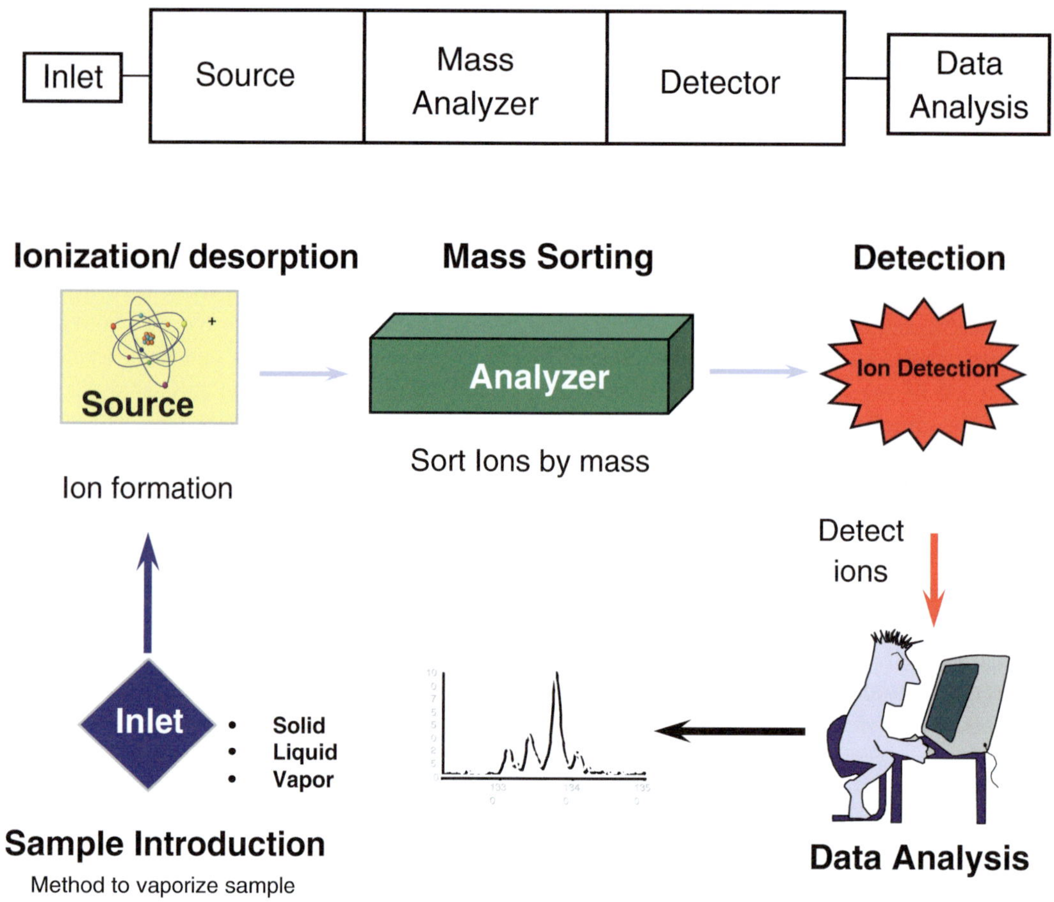

Fig. 1 The pipeline of mass spectrometry

Various ionisation methods can be used, depending on the type of sample under investigation and the mass spectrometer available ("Ionization Methods in Organic Mass Spectrometry", Alison E. Ashcroft, The Royal Society of Chemistry, UK, 1997).

Ionisation Methods Include:

- Electron ionisation (EI)
- Chemical ionisation (CI)
- **Electrospray ionisation (ESI)**
- Fast atom bombardment (FAB)
- Field desorption/field ionisation (FD/FI)
- **Matrix-assisted laser desorption ionisation (MALDI)**
- Thermospray ionisation (TSP)

Most ionisation techniques excite the neutral analyte molecule which then ejects an electron to form a radical cation (M+_)*. Other ionisation techniques involve ion–molecule reactions that produce adduct ions (MH+). The most important considerations are the physical state of the analyte and the ionisation energy. Electron ionisation and chemical ionisation are only suitable for gas phase ionisation and are generally considered as hard techniques used for organic compounds or small molecules. However, soft ionization techniques such electrospray and matrix-assisted laser desorption are used to ionize condensed phase samples and biomolecular investigations. The ionisation energy is important because it controls the amount of fragmentation observed in the mass spectrum. Though fragmentation complicates the mass spectrum, it provides structural information for the identification of unknown compounds. Soft ionisation techniques only produce molecular ions whilst other techniques are more energetic and cause ions to undergo extensive fragmentation. Typically, peptide identification and thermally labile sample applications make use of ESI or MALDI platforms.

ESI: Electrospray Ionisation

ESI (Fenn et al. 1989) is a soft ionisation technique that results in both single and multiply charged ions. The sample is introduced through an ultrafine needle into a strong electric field (typically ±3–5 kV), creating a spray of charged droplets. The charged droplets are desolvated by a counter current drying gas or heat causing the droplet to evaporate. As the solvent evaporates, the droplet shrinks and the charge density at the surface of the droplet increases. The droplet finally reaches a point where the coulombic repulsion from the electric charges exceeds the Rayleigh's stability limit. Electrostatic repulsion greater than the surface tension holding the droplet together causes an explosion in the droplet, creating multiply charged analyte ions. Because electrospray produces multiply charged ions, high-molecular-weight compounds are observed at lower *m/z* value. This increases the mass range of the analyser so that higher-molecular-weight compounds may be analysed with a low-resolution mass spectrometer (Fig. 2).

ESI typically generates singly or doubly charged ions for peptides <2,000 Da whilst higher-molecular-weight peptides yield a series of multiply charged ions. These multiple ion series give an advantage of independent verification of precursor ion mass calculation and deriving various states in the complete reaction. Most commercial mass analysers can now be customised with an ESI source as per the requirement of the type of sample.

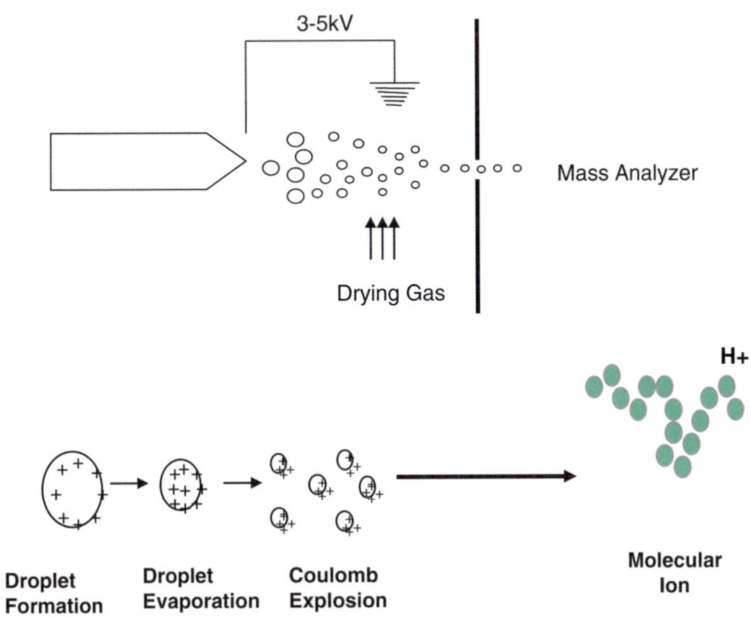

Fig. 2 Schematic representation of ESI process. The electrospray is created by applying a large potential between the metal inlet needle and the electrode plate (3–5 kV) located at a distance of ~0.5–1.0 cm from the needle. The liquid droplets leave the nozzle and the electric field induces a net charge on the small droplets. As the solvent evaporates, the droplet shrinks and the electrostatic repulsions increase. The droplet finally reaches a point where the coulombic repulsion from this electric charge is greater than the surface tension holding it together. This causes the droplet to explode and produce multiply charged analyte ions

MALDI: Matrix-Assisted Laser Desorption/Ionisation

MALDI (Karas and Hillenkamp 1988; Beavis and Chait 1996) is classically used to analyse thermolabile large molecules such as peptides. The sample is mixed with excess of specific wavelength material known as matrix (e.g. ά-cyano-4-hydroxycinnamic acid typically used for peptides. The sample is ionised by exposure to short duration pulses of UV light from nitrogen laser. This leads to ionisation, which is generally protonation, of both the matrix and the analyte. The matrix absorbs the primary energy and transfers it to the sample, and indirect ionisation of the analyte occurs. Application of a high-potential electric field desorbs the ions into the mass analyser. The sample gets converted to gas phase ions directly from a solid state which is suitable for thermolabile molecules such as peptides (Fig. 3).

MALDI is often a preferred method of choice for the analysis of synthetic and natural polymers, proteins and peptides. Analysis of compounds with molecular weights up to 200,000 Da is possible.

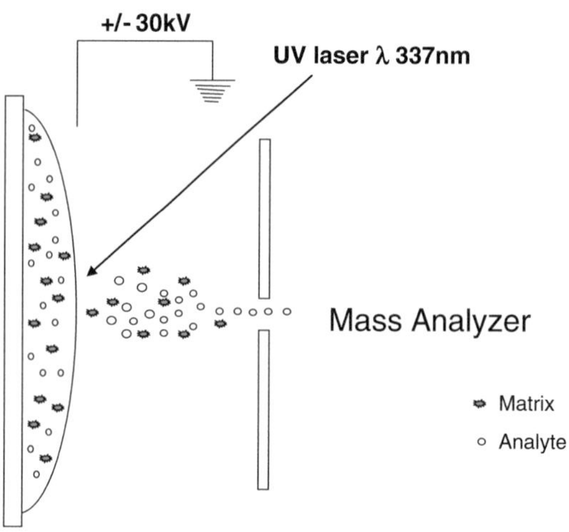

Fig. 3 Schematic representation of MALDI. The sample is prepared by mixing the analyte with a matrix compound which can absorb the UV laser (λ 337 nm). This is placed on a probe tip or a sample plate generally made of inert metal like gold and dried. A laser beam is then focused on this dried mixture, and the energy from a laser pulse is absorbed by the matrix. This energy is transferred to the analyte which both ionises and converts the ions in the gaseous phase. These ions are then desorbed into the mass analyser

Mass Analysers

The ions are created in the source region by any of the above-described methods and are accelerated into the mass analyser by an electric field. The function of the mass analyser is to separate these ions according to their *m/z* value. Each analyser has very different operating characteristics, and the selection of an instrument depends upon the mass range, type of the analyte, time for analysis, resolution etc.

Mass analysers are typically described as:

1. *Continuous analysers*
   - **Quadrupole filters**
   - **Magnetic sectors**
2. *Pulsed analysers*
   - **Time of flight (TOF)**
   - **Quadrupole ion trap mass spectrometers (QUISTOR)**
   - **Fourier transform-ion cyclotron resonance (FT-ICR)**

Continuous analysers allow a single selected *m/z* to the detector, and the mass spectrum is obtained by scanning the mass range so that different mass-to-charge ratio ions are detected. It can be compared to a filter or a monochromator used for optical spectroscopy. At a given mass window set by the instrument, a certain *m/z* is selected whilst other ions are lost. These instruments are useful for single-ion monitoring whilst a complex mixture may not result in good resolution.

Pulsed mass analysers on the other hand scan through the entire mass spectrum from a single pulse of ions. These instruments have a distinct advantage with complex mixtures such as peptides where multiple *m/z* ratios are typically observed. They have a higher signal-to-noise ratio as compared to the continuous analysers. Analysis of peptides generally is done by a TOF or QUISTOR analyser.

Mass Analysis

TOF: Time of Flight

The analysis of mass in a TOF instrument is determined by measuring the time taken by the

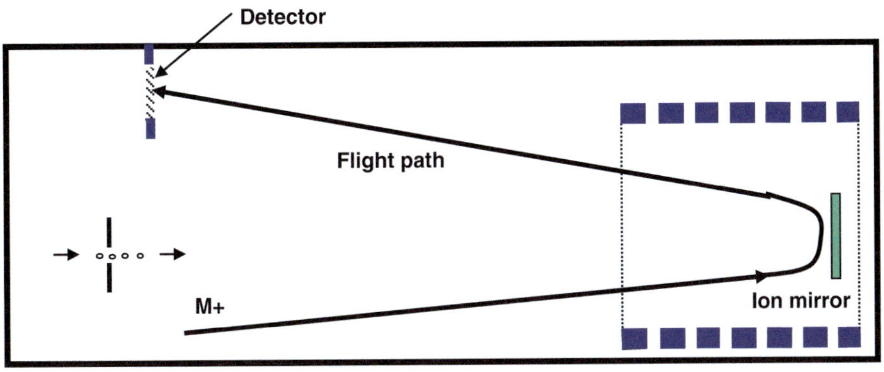

Fig. 4 Time of flight mass analyser

charged ion to cover a known distance in vacuum under a fixed magnetic and electric field. The mass-to-charge ratio (*m*/*z*) can thus be determined as a function of time (Fig. 4):

$$F = Q(E + v?B) \quad \text{Lorentz force law}$$

$$F = ma \quad \text{Newton?s second law}$$

$$\frac{m}{Q} = \frac{(E + v?B)}{a}$$

$$\frac{m}{Q} = \left\{\frac{(E + v?B)}{d}\right\} t^2$$

$$\frac{m}{Q} = Kt^2$$

$$\sqrt{m/z}\,?t$$

F = Force, m = mass, a = acceleration,
E = electric field, $v?B$ = vector cross product of ion velocity and magnetic field,
Q = ion charge, d = distance,
K = instrument constant

The ions are accelerated into the flight tube by an electric field (typically 2–25 kV) under vacuum. Since the force applied is the same, all the ions are accelerated with kinetic energy being directly proportional to mass. The velocity of the ion is thus an inverse function of mass, i.e. the greater the mass of the ion, the longer it takes to reach the detector. The striking of the ions at the detector is recorded with respect to time fractions, and the number of hits gives the abundance of a particular *m*/*z* signal. The plot between time of flight and the signal intensity is converted to a mass spectrograph for calculation of mass of striking fragment ions.

QUISTOR: Quadrupole Ion Trap

The quadrupole ion trap, which was developed in the late 1990s, is a three-dimensional quadrupole, which is capable of first trapping the ion and then analysing the mass of the entire stream of ions from a single source sequentially. The signal-to-noise ratio is high as all the ions are detected. The QUISTOR consists of a hyperbolic ring electrode and two end cap electrodes which form a hollow centre or the ion trap. The space between the two end caps allows the movement of the ions and in and out of the trap. A combination of RF and DC voltages is applied to the electrodes to create a quadrupole electric field. This electric field traps the ions in a potential energy well at the centre of the analyser. To obtain a mass spectrum, electric field is varied such that it sequentially brings ions with increasing *m*/*z* in resonance with the applied frequency. This eventually destabilises the ions and the alteration in the velocity and trajectory of the ions ejects them out of the trap. These sequentially ejected ions are then detected by the detector. The precursor ion and the product ions produced by collision-induced dissociations (CIDs) can be separated in time to produce the entire spectrum of ion products. This is particularly useful in determining structures and building up protein sequences de novo. Both ESI and MALDI sources can be used with an ion trap instrument.

FT-ICR: Fourier Transform-Ion Cyclotron Resonance

FT-ICR (Marshall and Verdun 1990; Amster 1996) instrument captures the ions in a three-dimensional space created by a magnetic field. The mass analyser consists of a reaction cell bound by electrodes known as trapping, excite and detect plates and a magnetic field generated by supermagnets. The *m/z* value of an ion is directly related to the cyclotron frequency. The ICR traps ions in a magnetic field that causes ions to travel in a circular path. The ion's cyclotron frequency (ω) is the angular frequency of an ion's orbit. This frequency is determined by the magnetic field strength (B) and the *m/z* value of the ion. After ions are trapped in this cell, they are detected by measuring the signal at this cyclotron frequency. This type of mass analyser has an extremely high mass resolution and is useful for tandem mass spectrometry experiments.

MS/MS: Tandem Mass Spectrometry

Tandem mass spectrometry involves at least two stages of mass analysis, with some form of fragmentation of the parent ion occurring in between the events. Multiple stages of mass analysis separation can be either achieved in *space* by physically distinct elements linked together in a series. The precursor ion from one unit feeds into another for further fragmentation to form the product ion. These elements could be sectors, transmission quadrupole or time-of-flight instruments. The separation of these linear events could also be achieved in *time* by trapping the precursor ions in three-dimensional space and inducing fragmentation to release the product ion at subsequent time intervals. QUISTOR and FT-ICR are typically used for tandem mass spectrometry experiments.

Protein Mass Spectrometry

One of primary applications of mass spectrometry in proteomics is protein identification or sequencing. Proteins of interest generally are present as a part of a complex proteome. These proteins first need to be fractionated and then subjected to mass analysis. There are various fractionation platforms which are employed before subjecting a protein for identification by MS/MS. The workflow of a typical proteomic analysis is described in Fig. 5. After the isolation of protein mixture, it is subjected to various resolution techniques which may be gel based or non-gel based. The protein is then directionally cleaved using trypsin, which cuts at the *C*-terminal

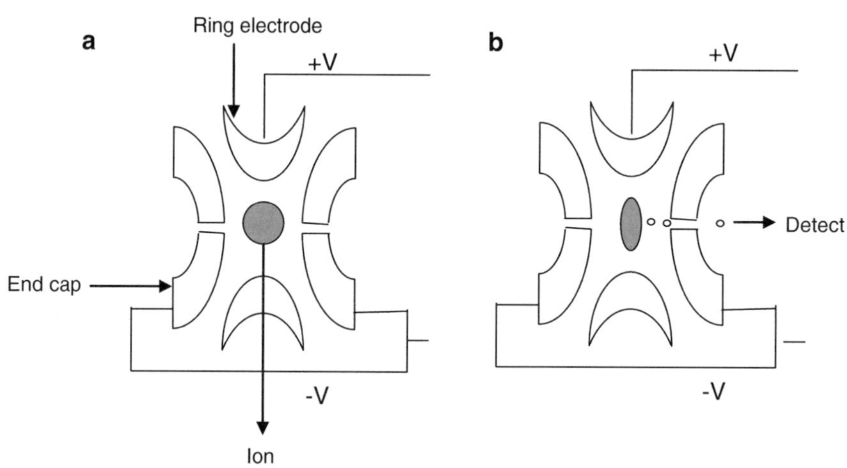

Fig. 5 Quadrupole ion trap. (**a**) Ion trap by the ring and the end cap electrodes. (**b**) Ejection of ions sequentially based on *m/z* to the detector

Fig. 6 A typical workflow for a typical protein mass spectrometry experiment

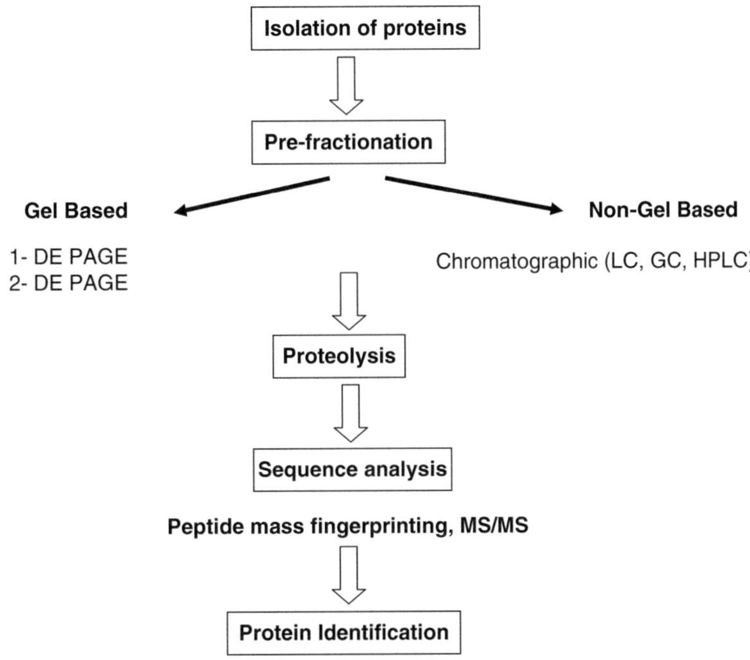

end of arginine or lysine when not followed by proline to generate the peptide fragment. The two major modes of ionisation used are MALDI and ESI which are soft ionisation techniques and generate moderate number of fragments. Peptide fragment masses are determined by MS by either peptide mass fingerprinting or MS/MS platforms (Fig. 6).

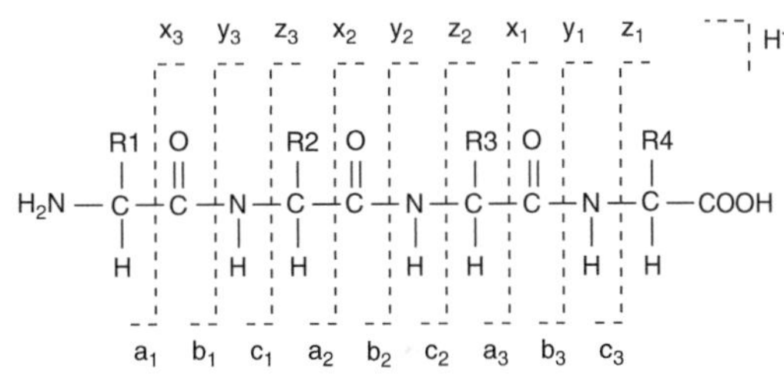

Fig. 7 Roepstorff–Fohlmann–Biemann nomenclature used when the peptide backbone is fragmented by imparting energy onto the molecule

Peptide Fragmentation

The types of fragment ions observed in an MS/MS spectrum depend on the primary sequence, the amount of internal energy, ionisation method, charged state etc. (Fig. 7):

- Fragments can only be detected if they carry at least one charge.
- If this charge is retained on the *N*-terminal fragment, the ion is classed as either *a*, *b* or *c*.
- If the charge is retained on the *C*-terminal, the ion type is either *x*, *y* or *z*.
- A subscript indicates the number of residues in the fragment.

Protein identification process is primarily divided into two parts: (1) assimilation of data generated by MS/MS in silico and (2) interpretation of the data by searching against a database. The mass spectrometer only gives masses of individual fragment ions as a set observed for a single-source ion. Series of *b*-ion and *y*-ion are required to build up the peptide fragment sequence. Multiple fragment sequences are overlapped before arriving at a consensus sequence. The set of fragments generated by MS/MS act as a fingerprint for an individual peptide. Two or more peptides matching to a given protein by searching protein sequence databases give the identity of the protein.

Protein Identification by Peptide Mass Fingerprinting

The protein is subjected to proteolysis by trypsin and peptide fragments are generated. The mass for each peptide fragment is determined by

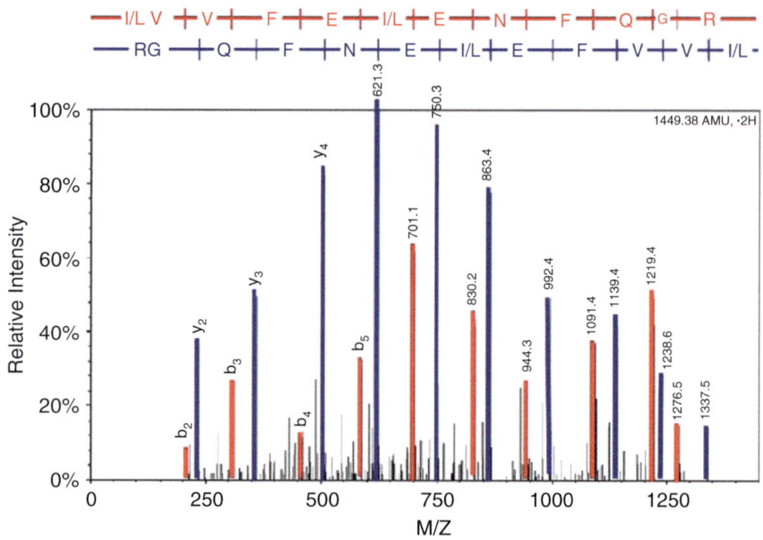

Fig. 8 Example of an annotated MS spectrum. The information about the peptide sequence can be inferred from the mass differences between the peaks (Jonscher 2005)

MS. Each protein results in a set of masses ascribed to fragment ion. The entire set of fragments ion masses generated by a single protein is considered to be unique and serves as a fingerprint to identify the protein. The database for peptide mass fingerprint consists of theoretical digest patterns for all protein sequence entries serving as the directory of mass fingerprint for each protein. The generated data can thus be matched with the library of mass fingerprints in these databases. The success of identification of PMF is however dependent on sequence availability of the required protein in the fingerprint database. Hence, the applications are limited to sequenced genomes.

Protein Identification by Peptide Fragment Ion Searches

This protein identification methodology is a hierarchical process where first a peptide sequence is built up by using the MS/MS data for peptide fragment ions and then all the peptides identified from a single protein are matched against the database for identifying the protein. The individual fragments identified are then assigned to a given protein by searching against protein sequence databases. The fidelity of a match is determined by the total score of the protein identification which is based on number of peptides matching to the same protein, length of the match, percentage coverage of the protein etc. There are various softwares available such as PepIdent and MASCOT being the popular ones (Fig. 8).

Advances in Plant Proteomics

Subcellular Proteomics

Organelle proteomics involves isolating the organelle of interest and producing a catalogue of the proteins present in that organelle by some form of separation of proteins or their proteolytic fragments followed by identification utilising mass spectrometry. The organelle preparation must be free from contamination from other organelle types to determine the specific localisation of a protein with high confidence. Recently, several high-throughput methods have emerged involving quantitative strategies, which have overcome the need to produce a pure organelle for analysis. Each of these methods relies on quantitative proteomics to characterise the distribution pattern of organelles amongst partially enriched fractions generated by various separation technologies and has the potential to discriminate between genuine organelle residents and contaminants without preparation of pure organelles.

Secretome

The plant secretome refers to the set of proteins secreted out of the plant cell into the surrounding extracellular space commonly referred to as the apoplast. Secreted proteins maintain cell structure and act in signalling and are crucial for stress responses where they can interact with pathogen effectors and control the extracellular environment. Secretome studies have firmly established the presence of a substantial level of secreted proteins lacking signal peptides and indicated a large degree of plant species specificity in the composition of secreted proteins. Plant secretomes have been studied in natural conditions (Soares et al. 2007), in different cultivars (Konozy et al. 2013), during nutritional deficiency (Tran and Plaxton 2008), after hormone treatment (Cheng et al. 2009a, b), during temperature change (Gupta and Deswal 2012), during salt stress (Song et al. 2011) and in the presence of pathogens and elicitors (Kim et al. 2009a, b). Martinez-Esteso et al. (2009) studied the grape secretome of SSC in response to methylated cyclodextrins and methyl jasmonate (MeJA) and showed that the expression levels of peroxidases, pathogenesis-related (PR) proteins, SGNH plant lipase-like proteins, xyloglucan endotransglycosylase and subtilisin-like protease were affected. In a similar study, application of elicitors MeJA and cyclodextrins also led to the identification of chitinases and other PR proteins in tomato SSC (Briceno et al. 2012). Gupta et al. (2011) characterised the secretome from SSC of the legume chickpea and identified over 700 proteins by combining 1-D SDS-PAGE and HPLC-MS/MS, and comparing the secretome based on sequence homology to previously published *Arabidopsis, Medicago* and rice data showed a large degree of species specificity in secreted proteins hinting at differences in the apoplast composition between species and monocots and dicots. Several studies have targeted the rhizosphere. Over 100 secreted proteins were identified from rice roots grown in an aseptic hydro-culture (Shinano et al. 2011). These proteins are believed to play an important role in the rhizosphere, and a relatively high number (54 %) had predicted signal peptides.

Nuclear Proteome

Nuclear proteome has recently gained importance as nucleus carries the information necessary for controlled expression of proteins and thus plays an essential role in determining plant response towards any developmental process and biotic or abiotic stress (Ref). Nuclear proteins are predicted to comprise about 10–20 % of the total cellular proteins, suggesting the involvement of the nucleus in a number of diverse functions. Researchers have identified several hundred plant nuclear proteins predominantly from model plants *Arabidopsis* (Bae et al. 2003) and *Medicago* (Repetto et al. 2008, 2012) and crop plants like rice (Choudhary et al. 2009), hot pepper (Lee et al. 2006), soybean (Cooper et al. 2011), xerophyta (Abdalla et al. 2010; Abdalla and Rafudeen 2012) and chickpea (Pandey et al. 2006, 2008). These proteins were presumably associated with a variety of functions, viz. nucleoskeleton structure, development, DNA replication/repair, chromatin assembly/remodelling, signal transduction, mRNA processing, protein folding, transcription regulation, transport, metabolism and cell defence and rescue. The identified proteins revealed the presence of complex regulatory networks that function in this organelle. Many proteins with unknown functions have also been added to the database besides novel proteins which were expected to be present in the nucleus. This clearly displays the power of proteomics where unusual locations of a known protein may result in totally new function.

Mitochondrial Proteome

Plant mitochondrial genomes have features that distinguish them radically from their animal counterparts: a high rate of rearrangement, of uptake and of loss of DNA sequences and an extremely low point mutation rate. Mitochondria act as a cellular powerhouse and also perform numerous other activities like nucleotide and vitamin synthesis, lipid and amino acid metabolisms and involvement in the photorespiratory pathway (Millar et al. 2005). Under stress, the mitochondrial electron transport chain becomes over-reduced, favouring the generation of O_2•, thus affecting plant growth and development (Purvis 1997). Kruft et al. (2001) first used the two-dimensional polyacrylamide gel electrophoresis (2-DE) technique to study the mitochondrial proteins. Thereafter, much attention has been paid to separate proteins for analysing plant mitochondrial proteome under stressful condition. Gel-free method of mitochondrial proteome study using nanoscale 1-D and 2-D liquid chromatography (LC) offers advantages (Kristensen et al. 2004; Brugiere et al. 2004; Heazlewood et al. 2003) over gel-based techniques, as it allows separation of highly acidic or highly basic proteins, very high- and very low-molecular-weight proteins as well as low-abundance proteins. Mitochondria have been a target for subcellular proteomics study, as most of the stresses primarily impair mitochondrial electron transport chain resulting in excess ROS generation. Mitochondrial proteomics were studied from soybean root and hypocotyls under flooding stress (Komatsu et al. 2011), the two contrasting wheat cultivars for salinity tolerance at whole-plant level (Jacoby et al. 2010), *Arabidopsis* for metal homeostasis (Tan et al. 2010), salt stress-induced programmed cell death (PCD) in rice (Chen et al. 2009a, b) and pea under drought, cold and herbicide stresses (Taylor et al. 2005). Mitochondrial proteome has huge importance to evolutionary biologists where it is an important tool to establish the evolutionary linkages owing to its maternal inheritance. Besides, the issues of male sterility are also attributed to the changes in the mitochondrial gene expression.

Extracellular Matrix (Cell Wall) Proteome

Plants, exposed to environmental stress, try to change the composition of cell wall to protect the cellular integrity for prevent mechanical damage. The cell wall or extracellular matrix (ECM) is the first compartment that perceives extracellular signals, transmits them to the cell interior and eventually influences the cell fate decision. Although proteins account for only 10 % of the ECM mass, they comprise several hundreds of different molecules with diverse cellular functions. Increasing evidence suggests that there is continuous cross-talk between the ECM and the cytoskeletal network (Pandey et al. 2010). The combination of ECM extraction and mass spectrometry appears to be a powerful strategy for identification of less abundant and previously unknown protein components involved in different stress responses. Various cell wall proteins have been characterised in *Arabidopsis* (Bayer et al. 2006; Minic et al. 2007; Jamet et al. 2006, 2008; Zhang et al. 2011), *Medicago* (Watson et al. 2004; Soares et al. 2007), chickpea (Bhushan et al. 2006), maize (Zhu et al. 2006), rice (Jung et al. 2008; Chen et al. 2009a, b; Cho et al. 2009) and potato (Lim et al. 2012). In addition, many types of stress-associated cell wall proteins have been identified in crops, including flooding stress-induced proteins in soybean (Komatsu et al. 2010) and wheat (Kong et al. 2009); drought stress-induced proteins in rice (Pandey et al. 2010), maize (Zhu et al. 2007) and chickpea (Bhushan et al. 2007); hydrogen peroxide-induced proteins in rice (Zhou et al. 2011); and/or pathogen-induced proteins in maize or tomato (Chivasa et al. 2005; Dahal et al. 2010). Also, cell wall proteins have been studied in wounded *Medicago* (Soares et al. 2009). Although many proteomics studies of primary cell wall have been conducted in *Arabidopsis* (Chivasa et al. 2002; Boudart et al. 2005; Jamet et al. 2006, 2008), there have been correspondingly fewer proteomics studies devoted to systematic mapping of the proteins of the secondary cell wall (Millar et al. 2009). The utility of plant secondary cell wall biomass for industrial and biofuel purposes depends upon improving cellulose amount, availability and extractability. The possibility of engineering such biomass requires much more knowledge of the genes and proteins involved in the synthesis, modification and assembly of cellulose, lignin and xylans (Millar et al. 2009).

Chloroplast Proteome

The chloroplasts present in plant and algal cells are believed to have descended from an original cyanobacterial endosymbiont. An important link in the development of this highly specialised organelle has been the gene transfer to the nucleus and evolution of protein-sorting machinery which still ensures the targeting of these proteins to the chloroplast. This information can only be unravelled by studying the proteome distribution inside the chloroplast and therefore the special interest in this organelle. Moreover, this is the organelle involved in carbon fixation. The physiology of the chloroplast can be best studied by the corresponding temporal proteome changes. A homology-based comparison of the chloroplast proteome of *Arabidopsis* with the total protein complement of a cyanobacterium (*Synechocystis*) combined with a proteome-wide search for putative chloroplast transit peptides was carried out (Salamini and Leister 2000). The present-day chloroplast was found to be smaller than the cyanobacterial species. The chloroplast proteome has been greatly studied under abiotic stress. Thirty-two differentially expressed chloroplast proteins were found in proteome analysis of soybean chloroplasts responding to ozone stress which revealed downregulation of proteins involved in photosystem I/II and carbon assimilation, and this might be one of the reasons of reduced photosynthetic activity in response to ozone (Ahsan et al. 2010). In contrast, proteins involved in antioxidant defence and carbon metabolism increased under stress. *Arabidopsis* chloroplast proteome using 2D-DIGE technique resulted in minimal change in the plastid proteomes in cold shock, whilst short-term cold acclimation caused major changes in the stroma but few changes in the lumen proteome. In contrast, long-term acclimation resulted in modulation of the proteomes of both compartments, with appearance of new proteins in the lumen and further changes in protein abundance in the stroma (Goulas et al. 2006). In total, 43 differentially displayed proteins were identified which participate in photosynthesis, other plastid metabolic functions, phytohormone biosynthesis and stress sensing and signal transduction, presumably helping the plant in cold sensing and acclimatisation.

Membrane Proteome

The different organelle membranes play important roles in maintaining the homeostasis within organelles, as well as whole-cell level. Approximately 30 % of the cellular proteome is represented by membrane proteins (Schwacke et al. 2004). The membrane-associated proteins perform unique biological roles in development as well as stress adaptation. The composition and dynamics of membrane proteins reflect their diverse function, and their nature and relative amount vary from one organellar membrane to another. These proteins perform some of the most important functions, like regulation of cell signalling, cell–cell interactions and intracellular compartmentalisation (Wu and Yates 2003). The plant membrane proteome is more complex compared to that of animal cells due to the presence of highly specialised organelles such as plastids and vacuoles. Whilst much progress has been made in animal membrane proteomics, far fewer attempts have been made to characterise the plant membrane proteome (Jaiswal et al. 2012; Nouri and Komatsu 2010, Kawamur and Uemura 2003).

Jaiswal et al. (2012) developed a proteome reference map of chickpea to obtain valuable insight into the dynamic repertoire of membrane proteins, using two-dimensional gel electrophoresis, and 91 proteins were identified by MALDI-TOF/TOF and LC-ESI-MS/MS. These proteins were involved in a variety of cellular functions, viz. bioenergy, stress-responsive and signal transduction, metabolism, protein synthesis and degradation, amongst others. Significantly, 70 % of the identified proteins are putative integral membrane proteins, possessing transmembrane domains.

Nouri and Komatsu (2010) investigated the polyethylene glycol-induced osmotic stress impact on plasma membrane proteome of soybean. Using the gel-based proteomics, four and eight protein spots were identified as up- and downregulated, respectively, whereas in the nanoLC-MS/MS approach, 11 and 75 proteins were identified as up- and downregulated, respectively, under polyethylene glycol treatment. Osmotic stress-responsive proteins, for example, transporter proteins and proteins with high number of transmembrane helices as well as low-abundance proteins, were identified by the gel-free proteomics. Mass spectrometric approach was widely used for identification of putative plasma membrane proteins of *Arabidopsis* leaves associated with cold acclimation (Kawamur and Uemura 2003). A significant change in protein profile was observed after cold acclimation.

Comparative Proteomics for GM and Non-GM

Transfer of individual genes that encode specific desirable traits into the host, i.e. genetic modification, has become the fastest adopted technology in the history of modern agriculture which has resulted in improvement in agronomic traits, such as resistance to insects, tolerance to herbicides, improved productivity and quality and other traits not present before genetic modification (Garcia-Canas et al. 2011). However, modifications in a plant genome might result in unintended effects, which may affect human health or the environment (Ioset et al. 2007). With the commercialisation of GM crops, these unintended effects are one of the most controversial issues in debating the biological safety of GM crops. A systematic comparative analysis of molecular features of GM crops and their comparators is needed to clarify unintended effects (Cellini et al. 2004; Garcia-Canas et al. 2011). Profiling techniques allow simultaneous characterisation and comparison of the genome, proteome and metabolome of an organism, thus increasing the chances of detecting the inadvertent effects, and have emerged as useful approaches (Kuiper et al. 2003; Ruebelt et al. 2006). Comparative proteomic strategies combined with 2-DE and MS and with liquid chromatography tandem mass spectrometry (LC-MS/MS) have been extensively used to evaluate the effects of genetic modification on the proteomes of lead GM crops: maize, pea, potato, rice, soybean, tobacco, tomato and wheat. These studies involved safety evaluation of GM crops and functional characterisation of GM crops (for review, see Gong and Wang 2013). Corpillo et al. (2004) first assessed the substantial equivalence of GM tomato, resistance to TSWV, using proteomics approaches and found no qualitative or quantitative differences between the GM tomato and the non-GM control. Similarly, DiCarli et al. (2009) demonstrated that expression of scFv(G4) against the CMV coat protein in tomato did not cause pleiotropic effects. A proteomics study of GM bread wheat overexpressing a low-molecular-weight glutenin subunit (LMW-GS) revealed a series of variations, including overaccumulation of the LMW glutenin and downregulation of all other classes of storage proteins, which constituted a compensatory mechanism (Scossa et al. 2008). Horváth-Szanics et al. (2006) used proteomic methods to identify stress-induced proteins in herbicide-resistant GM wheat lines and found changed level of LMW seed proteins and sensitivity to drought stress in this GM wheat under drought stress. Gong et al. (2012) evaluated proteome differences in seeds from two sets of GM rice (Bar68-1 carrying *bar* and 2036-1a carrying *cry1Ac/sck*) and their controls by 2-DE differential in-gel electrophoresis (2D-DIGE). To obtain relatively objective data, this study included other rice varieties to evaluate proteome variations related to spontaneous genetic variation, genetic breeding and genetic modifications. GM events did not substantially alter protein profiles as compared with conventional genetic breeding and natural genetic variation (Gong et al. 2012). Agrawal et al. (2013) used 2-DE to study comparative proteomics of entire potato tuber life cycle of wild-type and AmA1 transgenic lines and revealed a role for seed storage protein, AmA1, in cellular growth, development and nutrient accumulation.

Comparative Proteomics Under Abiotic Stress

Changes in protein accumulation under stress are closely interrelated to plant phenotypic response to stress determining plant tolerance to stress. Therefore, studies of plant reaction upon stress conditions at protein level can significantly contribute to our understanding of physiological mechanisms underlying plant stress tolerance. Proteomics studies could thus lead to identification of potential protein markers whose changes in abundance can be associated with quantitative changes in some physiological parameters used for a description of genotype's level of stress tolerance. In the field of plant abiotic stress research, the most common case is comparison of proteomes isolated from non-stressed (control) plants and the corresponding proteomes upon stress conditions. Other cases include comparison of proteomes from two different genotypes or plant species with contrasting levels of tolerance to a given stress factor.

The studies aimed at comparison of several proteomes are mostly dominated by 2-DE followed by protein identification via MS analysis, although the sole use of MS techniques not only for protein identification but also for protein quantitation is sometimes applied (e.g. Patterson et al. 2007 (14) used iTRAQ for protein quantitation in two barley cultivars with different sensitivities to elevated concentrations of boron). Differential expression proteomics approach is used for description of sets of proteomes differing both in protein quality and quantity, and it is aimed at protein identification and relative quantitation. However, the differential expression proteomics approach (protein identification and quantitation) itself cannot give any information on protein function since one certain protein can reveal very diverse functions depending on its subcellular localisation, posttranslational modifications or interacting partners.

Low-Temperature Stress

For proteome analysis under cold stress, leaf tissues from *A. thaliana* and *A. thaliana* cold- and salt-tolerant relative *Thellungiella halophila* and poplar (Amme et al. 2006; Gao et al. 2009a, b) as well as root tissue from rice (Hashimoto and Komatsu 2007; Lee et al. 2009) or just trinucleate pollen in anthers of rice (Imin et al. 2004), plant embryos from germinated seeds of soybean (Cheng et al. 2010) or plant seedlings of rice (Cui et al. 2005) were employed. Proteome analysis has been carried out at whole-cell level as well as only at organellar level, e.g. *A. thaliana* nuclear proteome upon cold (Bae et al. 2003) or pea mitochondrial proteome upon cold (Taylor et al. 2005). Most studies have also indicated changes in abundance of enzymes involved in carbohydrate metabolism. Enhanced accumulation of specific dehydration-inducible LEA-II proteins named dehydrins has been repeatedly reported (Kawamur and Uemura 2003; Amme et al. 2006; Degand et al. 2009; Cheng et al. 2010; Vítámvás and Prášil 2008; Vítámvás et al. 2007). Increased levels of RNA-binding protein cp29 have been repeatedly reported (Amme et al. 2006; Gao et al. 2009a, b) as cold significantly affects proteosynthesis. This protein is localised in chloroplast stroma, its activity could be regulated by phosphorylation and it is involved in plastid mRNA processing (Reiland et al. 2009).

Heat

Heat stress is associated with an enhanced risk of improper protein folding and denaturation of several intracellular protein and membrane complexes. Heat-stress response at proteome level has been studied predominantly in rice (Lee et al. 2007), in wheat grain during grain filling period (Skylas et al. 2002; Majoul et al. 2004), in a heat- and drought-tolerant poplar (*Populus euphratica*) (Ferreira et al. 2006) and also in wild plant *Carissa spinarum* inhabiting hot and dry valleys in central China (Zhang et al. 2010). In all cases, a heat-induced increase in several HSPs including proteins from HSP100, HSP70 and sHSP families has been observed. Small HSPs belonging to cytoplasmic-located sHSPs as well as mitochondrial-targeted and chloroplast-targeted sHSPs were detected. In heat-treated grains of two genotypes of common wheat with contrasting tolerance to high temperatures, Skylas et al. (2002) detected seven sHSPs unique to a tolerant genotype which have been proposed biomarkers of heat tolerance and drought strength. Another characteristic feature of heat stress is oxidative damage. Upregulation of several enzymes involved in redox homeostasis such as GST, dehydroascorbate reductase (DHAR), thioredoxin h-type (Trx h) and chloroplast precursors of SOD was reported (Lee et al. 2007). Heat stress also induces profound changes in cytoskeleton composition indicating its reorganisation (Ferreira et al. 2006). In addition, an increased accumulation of some eukaryotic translation initiation factors (eIF4F, eIF5A-3) indicates profound cellular reorganisation leading to programmed cell death (PCD) under a long-term heat treatment (Majoul et al. 2004; Zhang et al. 2010).

Drought

Drought stress is associated with a reduced water availability and cellular dehydration. Therefore, changes in cellular metabolism associated with an osmotic adjustment could be expected. Proteome changes upon drought have been intensively studied in poplar (Bogeat-Triboulot et al. 2007; Bonhomme et al. 2009), maize roots (Zhu et al. 2007), soybean roots (Alam et al. 2010a, b) and sugar beet (Hajheidari et al. 2005), and increased levels of several apoplastic ROS-scavenging enzymes, namely, peroxidases involved in enhance cell wall loosening and proteins involved in pathogenesis and stress defence such as polygalacturonase inhibitor proteins, chitinases and osmotin and nodulin precursors were found (Zhu et al. 2007). Dehydration-induced changes in nuclear proteome of chickpea *Cicer arietinum* and rice *Oryza sativa* (Pandey et al. 2008; Choudhary et al. 2009) and ECM proteome of chickpea (Bhushan et al. 2007) have also been extensively studied. Proteins involved in carbohydrate and nitrogen metabolism, cell wall modification, signal transduction, cell defence and PCD and proteins involved in redox regulation, oxidative stress, chaperone function and photosynthesis (Rubisco) have also been observed.

Waterlogging

Ahsan et al. (2007) and Alam et al. (2010a, b) have analysed proteome changes as well as changes in in vivo hydrogen peroxide (H_2O_2) content and lipid peroxidation in tomato leaves and soybean roots, respectively, affected by waterlogging stress. Interestingly, waterlogging has resulted in enhanced levels of H_2O_2 and lipid peroxidation indicating that this stress factor has an oxidative component. At proteome level, waterlogging induces changes in abundance of proteins involved in several processes, namely photosynthesis, energy metabolism, redox homeostasis, signal transduction, PCD, RNA processing, protein biosynthesis, disease resistance, stress and defence mechanisms.

Salinity

The main crops where effects of salt stress on proteome composition are studied are represented by rice (Abbasi and Komatsu 2004; Kim et al. 2005; Yan et al. 2005; Dooki et al. 2006; Cheng et al. 2009a, b), soybean (Sobhanian et al. 2010) and common and durum wheats (Caruso et al. 2008; Wang et al. 2008) which are all glycophytes. Proteomics experiments carried out on glycophytes also include model plants *A. thaliana* (Ndimba et al. 2005) and tobacco (Dani et al. 2005). In glycophytes (crops), an increased accumulation of enzymes involved in glycolysis and carbohydrate metabolism (fructose-bisphosphate aldolase, ENO) is regularly observed which indicates an enhanced need for energy (Abbasi and Komatsu 2004; Yan et al. 2005; Sobhanian et al. 2010; Ndimba et al. 2005). Other major group of increased proteins are ROS-scavenging enzymes (APX, DHAR, Trx h, peroxiredoxin, SOD) suggesting an oxidative stress (Abbasi and Komatsu 2004; Kim et al. 2005; Dooki et al. 2006; Ndimba et al. 2005). Proteins involved in nucleotide metabolism (nucleoside diphosphate kinase NDPK, guanine nucleotide-binding protein) and fatty acid metabolism (enoyl-ACP reductase) were also upregulated (Dooki et al. 2006).

Comparative Proteomics for Biotic Stress

Plant pathogens are viruses, bacteria, fungi, oomycetes, protozoans and nematodes. Amongst all, the majority, and most destructive, are fungi and oomycetes (Latijnhouwers et al. 2003). However, the lifestyles and strategies of pathogens are diverse, but for their survival and propagation, at least all must colonise the host and overcome its immune system. Conversely, the host must overcome the virulence of the pathogen if it is to remain healthy. In consequence, coevolution of host–pathogen systems has resulted in a complex interplay of pathogen- and host-derived molecules, resulting in systems with a remarkable degree of conservation (Ronald and Beutler 2010).

Crop–Pathogen Interactions

Proteomics is a logical choice for an investigative tool since any plant–pathogen interaction language use proteins. Two-dimensional gel electrophoresis has been initially used for rapidly identifying major proteome differences in healthy versus inoculated plants. The interaction between *Triticum aestivum* and *Fusarium graminearum* (causing *Fusarium* head blight) (Zhou et al. 2006; Wang et al. 2005), wheat and *Puccinia triticina* (leaf rust) (Rampitsch et al. 2006a, b), rice and *Magnaporthe grisea* (rice blast) (Kim et al. 2004a, b), *Brassica napus* (canola) and *Leptosphaeria maculans* (blackleg) (Sharma et al. 2008), *Brassica oleracea* and *X. campestris pv. campestris* (black rot) (Villeth et al. 2009), *Pisum sativum* (pea) and *Peronospora viciae* (downy mildew) (Amey et al. 2008), *rice* and *rice yellow mottle virus* (RYMV) (Ventelon-Debout et al. 2004) and *grapevine* and *Flavescence dore'e* phytoplasma (Margaria and Palmano 2011) were studied through proteomics. 2-DE reveals only gross changes in the proteome in most of the cases, with common results between diverse pathosystems. Metabolic enzymes showed increased presence in all of the pathosystems; in particular, glyceraldehyde-3-phosphate dehydrogenase was reported to increase in abundance in most of pathosystems. The antioxidant enzymes (Zhou et al. 2006; Wang et al. 2005; Rampitsch et al. 2006a, b; Ventelon-Debout et al. 2004; Margaria and Palmano 2011) especially ascorbate peroxidases, thioredoxin (Zhou et al. 2006; Villeth et al. 2009; Amey et al. 2008), fungal cell wall-degrading enzymes (chitinases and b-glucanases) and other pathogenesis-related proteins (Zhou et al. 2006; Wang et al. 2005; Rampitsch et al. 2006a, b; Amey et al. 2008; Margaria and Palmano 2011) were showed increased abundance for combating the pathogens. In grapevine – *Erysiphe necator* (powdery mildew) – study, iTRAQ was used to compare protein expression levels in a susceptible grapevine, Vitis vinifera (Cabernet Sauvignon), compared with mock-inoculated controls. The results support the hypothesis that Cabernet Sauvignon is able to initiate a basal defence response but lacks the necessary R-protein(s) to recognise pathogen Avr gene product(s) and therefore succumbs to disease (Marsh et al. 2010).

Interaction with Bacterial Pathogens and Elicitors

Jones et al. (2006a, b) reported early changes to the defence proteome in three subcellular fractions – total soluble protein, chloroplast enriched and mitochondria enriched – after inoculation with three different strains of Pst DC3000 and provided evidence for the rapid communication between organelles and regulation of primary metabolism through redox-mediated signalling. Jones et al. (2006a, b) identified six differentially phosphorylated proteins robustly changing between a mock-inoculated control, HR and a basal defence response in soluble *A. thaliana* leaf extracts following bacterial challenge, using phosphoprotein affinity enrichment coupled to relative quantification with iTRAQ. This study highlights the reproducibility, utility and problems associated with the quantitative analysis of changes in the complex phosphoproteome from intact green leaf tissue. Casasoli et al. (2008) analysed *A. thaliana* seedling apoplastic proteins elicited by oligogalacturonides (OGs) that accumulate upon challenge by pathogenic microorganisms, using 2-D DIGE and many differentially expressed or posttranslationally modified apoplastic proteins that were identified with either putative defensive roles or with structural features typical of proteins involved in recognition. These findings confirm the role of the cell wall as the first line of defence against pathogens as well as a source of molecules important in plant protection, which help in perception of pathogens. The biotic interactions in the rhizosphere during the communication between the roots of two plants *Medicago sativa* and *A. thaliana* and microbes *P. syringae pv. tomato* DC3000 or *Sinorhizobium meliloti* strain Rm1021 were studied, which revealed a specific, protein-level crosstalk between roots and microbes. It was suggested that secreted proteins may be a critical component in the process of signalling and recognition that occurs between compatible and incompatible interactions (De-la-Pena et al. 2008). The identification of signalling processes and phosphoproteins at the plasma membrane was addressed in large-scale global analyses of protein phosphorylation in model systems with elicitors. Nühse and co-authors used trypsin to digest cytoplasmic face-out vesicles and then enriched phosphopeptides by strong anion exchange (SAX) plus immobilised metal ion affinity chromatography (IMAC) and nanoLC-MS/MS as a strategy for large-scale phosphoproteomics of the plasma membrane from *A. thaliana* suspension cells stimulated with flg22. This identified over 300 phosphorylation sites on approximately 200 putative plasma membrane proteins (Nühse et al. 2003, 2004). In addition, more than 50 sites were mapped on receptor-like kinases revealing an unexpected complexity of the phosphorylation sites' characteristics and regulation. The isotopic quadruplex iTRAQ labelling of peptides was used to achieve quantification of dynamic protein phosphorylation in the same model system of *A. thaliana* cells challenged with flg22 (Nühse et al. 2007).

Plant–Fungus Interaction

Over the past decade, proteomics studies have contributed new knowledge to the *M. grisea*–rice interaction. Indeed, the first descriptive proteomics study of a pathogen-infected host plant focused on this interaction (Konishi et al. 2001). Recent experimental evidence based on differential display analysis of elicitor-responsive proteomes between two rice near-isogenic lines and *M. grisea* glycoprotein elicitor suggested that the incompatible rice line may possess a more sensitive recognition system that can identify and react to specific chemical, biological or physical triggers in a more efficient manner, thus eliciting an early and fast defence response (Liao et al. 2009). The authors also examined extracellular phosphorylation and identified phosphoproteins in both the cell wall (putative lectin receptor-like kinase and endochitinase) and culture filtrate (xyloglucan endo-1,4-b-D glucanases) in chitosan treatment in *A. thaliana* cell suspension cultures, supporting the view that an extracellular kinase activity might be present in plants and an extracellular phosphorylation network could be involved in intercellular communication (Ndimba et al. 2003). Proteomic analysis of chitosan-treated *V. vinifera* cv. Barbera cell suspensions revealed the upregulation of both stilbene and flavonoid pathways, with the resultant production of a wide spectrum of polyphenol antioxidant compounds (Ferri et al. 2009). The proteome changes during the interaction of the model legume *M. truncatula* cells in suspension culture with a pathogen-derived yeast invertase elicitor (YE) and suppressor using *Sinorhizobium meliloti* lipopolysaccharide (LPS) were studied using 2-DE and LC-MS/MS, which revealed upregulated proteins involved in defence only after YE but not LPS treatment (Gokulakannan and Niehaus 2010). The study of an incompatible plant–fungal interaction, the *A. thaliana–A. brassicicola* host–pathogen pair, showed that at least 11 proteins showed reproducible differences in abundance by 2-DE, increasing or decreasing during the progress of the infection. It was demonstrated that the leaf can limit pathogen infection whilst keeping its overall activity largely intact (Kaschani et al. 2009). Differential proteomics study for elicitor-induced sanguinarine biosynthesis in opium poppy cell cultures treated with *B. cinerea* fungal homogenate was done under controlled conditions which provided a platform to characterise the induction of antimicrobial alkaloid biosynthesis and other plant defence pathways (Zulak et al. 2009). The abundance of chaperones, heat shock proteins, protein degradation factors and pathogenesis-related proteins provided a comprehensive proteomics view on the coordination of plant defence responses. The elicitor-induced metabolic enzymes represented the largest category of proteins and included S-adenosylmethionine synthetase, several glycolytic enzymes, a nearly complete set of TCA cycle enzymes, one alkaloid and several other secondary metabolic enzymes.

Comparative Proteomics for Plant Development

Proteomics is an important tool for the analysis of proteins in organisms at the level of organs, cell populations and subcellular compartments under diverse developmental conditions. The number of plant developmental studies using various proteomics approaches is steadily growing. Considerable experimental effort was devoted to the proteomic investigation of hormonal pathways regulating plant development such as brassinosteroid signalling (Tang et al. 2010), auxin signalling (Shi et al. 2008), cytokinin regulation (Xu et al. 2010; Lochmanová et al. 2008), cell proliferation and elongation, cell differentiation and leaf, root, shoot and other plant organ development etc.

Cell Proliferation and Elongation

Proteomics studies in *Medicago truncatula* showed protein expression changes primarily in the cell division-related processes such as metabolism, energy housekeeping or the control of protein synthesis. Further, the stress-related proteins preferentially accumulate dividing tissues, such as root meristem (Holmes et al. 2006) and proliferating protoplasts (De Jong et al. 2007). In both cases, mainly pathogenesis-related proteins, such as PR-10 and heat shock proteins, exhibited higher abundance in dividing tissues. In another proteomics study of a transcription factor NTM (for NAC with transmembrane motif 1) mutant line in *Arabidopsis*, elevations of beta-glucosidase homolog 1 and annexin expression were found altered and exhibiting reduced cell division rate (Lee et al. 2008). Different proteomics studies showed that differential regulation of annexins is also linked to other plant developmental processes including pollen germination (Dai et al. 2007), cotton fibre elongation (Zhao et al. 2010) and somatic embryogenesis (Gómez et al. 2009). These findings also confirm a functional role of some ROS-related proteins such as ascorbate peroxidase (Holmes et al. 2006), dehydroascorbate reductase, glutathione transferase (Lee et al. 2008) and mitochondrial manganese superoxide dismutase (Shi et al. 2008) activity in the cell division regulation. The role of vigorous actin and microtubule cytoskeleton dynamics in cell expansion is also reflected in proteomics studies (Chan et al. 2007). Two independent comparative studies showed downregulation of alpha-tubulin, beta-tubulin and tubulin-folding cofactor A and profiling in mutant cotton fibres with inhibited elongation (Zhao et al. 2010; Pang et al. 2010). Similar results showing upregulation of five actin and two beta-tubulin isoforms were obtained during fibre elongation (Yang et al. 2008). High-throughput proteomics study on *Lilium longiflorum* pollen grain membrane proteins provided valuable contribution to the elucidation of pollen tube polar growth (Pertl et al. 2009). Remarkably, the expression levels of proteins involved in membrane/protein trafficking (Rab 11b GTPase, V-type ATPase and the H+ pyrophosphatase) raised simultaneously with proteins involved in signal transduction, stress response, protein biosynthesis and folding, during the germination of pollen grains. In contrast, proteins involved in cytoskeleton, carbohydrate and energy metabolism and transport of ions were upregulated earlier, when the pollen just started to germinate (Pertl et al. 2009).

Cell Differentiation

In a cell differentiation study, protein profiling of seed-derived calli on different regeneration media with different relative concentrations of cytokinin and auxin showed differences mainly in carbohydrate and energy metabolism and stress-/defence-related proteins (Yin et al. 2008). Interestingly, these protein groups were also activated in *Vanilla planifolia* calli directed for shoot organogenesis (Palama et al. 2010). In addition to cell differentiation, it is possible to reprogram differentiated cells to retain the competency of cell division and organ regeneration by using particular external hormone composition. Kinetin and 2,4-D induced a dedifferentiation of *Arabidopsis* cotyledon cells and was accompanied by protein phosphorylation (Chitteti and Peng 2007a, b). This hormonal treatment induced also protein synthesis, changes in the chromatin structure, cytoskeleton reorganisation and prevalent downregulation of chloroplast proteins (Chitteti et al. 2008). Various proteomics approaches were applied to study somatic embryogenesis of diverse plant species such as cassava (*Manihot esculenta* (Baba et al. 2008; Li et al. 2010)), oak (*Quercus suber* (Gómez et al. 2009)), Valencia sweet orange (*Citrus sinensis* (Pan et al. 2009)), grapevine (*Vitis vinifera*

(Marsoni et al. 2008)) and *Vigna unguiculata* (Nogueira et al. 2007). These reports included studies on protein expression changes during somatic embryogenesis and comparative studies between embryogenic and nonembryogenic calli as well as between gametic and somatic embryogenesis.

Seed Germination

Extensive effort was also dedicated to the proteomic investigation of seed germination. In a study of comparison of the endosperm cap proteome of ABA-inhibited vs. non-inhibited germinating cress (*Lepidium sativum*), seeds showed specific, ABA-responsive, early germination processes, such as lipid mobilisation, energy production, proteolysis and increase in abundance of antioxidant enzymes (Müller et al. 2010) These data suggested that the cress endosperm cap is not a storage tissue similar to cereal endosperm. Instead, it is a metabolically very active tissue regulating the rate of radicle protrusion. The changes in the proteome of rice embryo during germination (Kim et al. 2009a, b) revealed that enzymes detoxifying reactive oxygen species, protein degradation proteins and cytoskeleton-associated proteins play an important role during seed germination. The data of some studies suggest that protein phosphorylation plays an important role in seed germination. One of these studies on protein phosphorylation during maize seed germination revealed 39 protein kinases, 16 phosphatases and 33 phosphoproteins containing 36 phosphorylation sites (Lu et al. 2008). At least one-third of these phosphoproteins represented key components involved in biological processes like DNA repair, gene transcription, RNA splicing and protein translation related to the seed germination.

Seed Development

Seed development studies in Brazilian pine highlighted an active oxidative stress metabolism (ascorbate peroxidase as well as peroxiredoxin) in early seed development along with higher abundance of enzymes involved in cell wall expansion (alpha-xylosidase and type IIIa membrane protein cp-wap13) (Balbuena et al. 2009). Storage proteins (e.g. vicilin-like storage protein) and proteins involved in respiration (triosephosphate isomerase, fructose-bisphosphate aldolase and isocitrate dehydrogenase) were accumulated in the later stages of seed development. The upregulation of glutamine synthase during the early cotyledonary stage indicated active biosynthesis and conversion of glutamine to glutamic acid Balbuena et al. 2009). Similarly protein expression profiles in endosperm and embryo proteomes of dry seed of *Jatropha curcas* indicated some similarities in metabolic pathways between them. However, embryos generally possess proteins mainly involved in anabolic processes and accumulate stress-related proteins, implying increased embryo requirements for protection against stress (Liu et al. 2009).

Plant Organ Development

Nozu et al. (2006) studied developmental changes in root, stem and leaf proteomes in rice during the first 10 weeks after budding and showed that 19 proteins were present in all developmental stages in all tissues which represent metabolic proteins as well as oxidative stress-related proteins such as catalase isozyme A, superoxide dismutase ascorbate peroxidase and peroxiredoxin. Another study in soybean showed that protein transport regulatory proteins, especially those involved in the transport of nuclear-encoded chloroplastic protein into chloroplasts, were presumably involved in leaf development and maturation (Ahsan and Komatsu 2009).

The mechanisms of maize lateral and seminal root formation were extensively studied by comparative proteomics approaches using maize mutant lines. The rum1 (rootless with undetectable meristems 1 (Saleem et al. 2009)) mutant line is altered in both seminal and lateral root formation, whilst the rtcs (rootless concerning crown and seminal roots (Muthreich et al. 2010)) line does not form seminal roots. The comparison of rtcs and wild-type maize embryos showed that changes in disulphide isomerase expression which is involved in protein folding, as well as embryonic protein DC-8, generally seem to have a role in various pathways essential for the formation of different root types (Muthreich et al.

2010). In addition, the proteomics study on rum1 transgenic line revealed that the proteins related to pyridoxine biosynthesis are involved in rum1-dependent pathway of root formation (Saleem et al. 2009). Proteome changes during bud development (Bi et al. 2010) were elucidated in *Pinus sylvestris* L. var. mongolica in order to study mechanisms of bud dormancy induction and release. Stress-induced ascorbate peroxidase, pathogenesis-related proteins and heat shock proteins were involved in bud dormancy induction, and the proteins involved in protein synthesis, cell wall biogenesis and cytoskeleton were upregulated during dormancy release. The comparison of flower and bud proteomes suggested that sucrose generation derived by upregulated phosphoglucomutase and downregulated glycoprotein could serve as an inducer of flavonoid- and anthocyanin-related genes important for petal growth and colour development in mature flower.

The proteomics approach was found to be powerful for the investigation of potato (*Solanum tuberosum*) tuber formation (Agrawal et al. 2008; Lehesranta et al. 2006; Fischer et al. 2008) along with root, leaf and flower development. Changes in the proteome during tuber initiation and growth reflect mainly the processes connected to the accumulation of storage reserves and starch synthesis. Thus, storage proteins, protease inhibitors and proteins involved in secondary metabolism were upregulated during tuber growth. Additionally, some isoforms of patatins, a large family of primary storage proteins, were shown to accumulate in non-swelling stolons, possibly indicating their involvement in tuber initiation (Agrawal et al. 2008; Lehesranta et al. 2006).

Proteomic investigations of corn rachis, which delivers essential nutrients to the developing kernels in maize early during maturation (25 vs. 50 days after silking), revealed significantly increased expression (2.4- to 14.5-fold) of many stress-/defence-related proteins in mature rachis. They included PRm3 (class III chitinase), PR-1, PR-10, beta-1,3-glucanase, endo-1,3-beta-glucanase, germin-like protein subfamily 1 member 17, permatin and Asr protein (Pechanova et al. 2011). Additionally, profilin, an actin-binding protein which regulates actin polymerisation (Staiger et al. 2010), was also upregulated during rachis development and maturation. Previous proteomics study revealed that an inhibition of pollen tube tip growth by latrunculin B (an inhibitor of actin polymerisation) was well correlated with downregulation of profilin (Chen et al. 2006). Recently, profilin2 was identified by proteomics and cell biology approaches as a new cytoskeletal protein modulating vesicular trafficking in *Arabidopsis* roots (Takáč et al. 2011).

Fruit Ripening

Fruit ripening is a developmental complex process which occurs in higher plants and involves a number of stages displayed from immature to mature fruits that depend on the plant species and the environmental conditions. Due to the huge amount of metabolic changes that take place during ripening in fruits from higher plants, the accomplishment of new throughput methods which can provide a global evaluation of this process would be desirable. Differential proteomics of immature and mature fruits would be a useful tool to gain information on the molecular changes which occur during ripening, and also the investigation of fruits at different ripening stages will provide a dynamic picture of the whole transformation of fruits.

The 2-DE of tomatoes in the different ripening stages was analysed for changes in proteome composition. The results showed that an overall intensity increase during ripening was detected in 26 spots, whereas a decrease was seen in 27 spots, and two spots reached their maximum at the breaker or light-red stage (Kok et al. 2008). One important fruit ripening-related protein acid beta-fructofuranosidase was found to be upregulated in the breaker stage, downregulated in the subsequent turning and light-red stages and then once again upregulated in the red stage of ripening (Kok et al. 2008). Parallel studies carried out in three different ripening stages of tomato (unripe, medium ripened and fully ripened) resulted in the identification of pectin esterase and heterotrimeric GTP-binding protein fragment homologous to tobacco (Schuch et al. 1989), which might be implicated in cell wall softening and changes in firmness and are proposed as the ripening specific markers in tomato, since their levels were upregulated during tomato ripening (Schuch et al. 1989). However, the majority of proteins that were characterised corresponded to genes known to be regulated during tomato fruit development. Proteome maps obtained at three stages of ripening were compared to assess the extent to which protein distribution differs in grape skin during ripening. The comparative analysis showed that numerous soluble skin proteins evolved during ripening and revealed specific distributions at different stages. Proteins involved in photosynthesis (Rubisco), carbohydrate metabolism (aconitate hydratase, transketolase, phosphoenolpyruvate carboxylase, oxalyl-CoA decarboxylase and aldehyde dehydrogenase) and stress response (HSP17.7) were identified as being over-expressed at the beginning of colour change (Deytieux et al. 2007). At harvest, the dominant proteins were involved in defence mechanisms. In particular, increases in the abundance of different chitinase and β-1,3-glucanase isoforms were found as the berry ripened. This observation could be correlated with the increase of the activities of both of these enzymes during skin ripening. Thus, the differences observed in proteome maps clearly showed that significant metabolic changes occur in grape skin during this crucial phase of ripening (Deytieux et al. 2007).

Posttranslational Modifications

Often low abundance and/or low concentration including reversible and labile nature of many PTMs create a multifaceted challenge for the analysis of PTMs such as phosphorylation, glycosylation and cysteine oxidation. For improved recognition and site depiction, some novel MS/MS fragmentation strategies such as selective enrichment, electron capture/transfer dissociation (ECD/ETD) and derivatisation/labelling have been used. In addition to it, changes to the analytical setup such as negative ion mode and the use of nonstandard, from time to time basic sample solutions have also been employed. All these are crucial for quantitative plant proteomics where PTMs are playing a substantial role.

Phosphorylation

Protein phosphorylation is indeed an imperative PTM in plants, as well as in animals, involved in various cellular processes. Presence of around 1,000 and 500 protein kinases in *Arabidopsis* and human, respectively, and the recent identification of numerous phosphopeptides in large-scale plant phosphoproteomics further strengthen the importance of phosphorylation as well (Huang et al. 2009; Van Bentem et al. 2008; Reiland et al. 2009). Therefore, the quantitation of phosphoproteomes is of utmost importance to unravel the molecular mechanism behind the cellular processes such as signalling pathways, since phosphorylation and dephosphorylation may be perhaps the initial signalling events, triggering a chain of downstream signalling cascades which ultimately culminates into the differential expression of gene(s).

The tendency of the acidic phosphate group(s) to lower the pI of proteins in 2-DE makes it more suitable approach to study phosphorylation changes since protein isoforms can be resolved. Phosphorylated proteins can be particularly detected using nonradioactive stains such as Pro-Q Diamond (Gerber et al. 2006, 2008; Chitteti and Peng 2007a, b; Boudsocq et al. 2007) and alternatively by incorporated radiolabeled ^{32}P (Rampitsch et al. 2006a, b). The specific use of Pro-Q Diamond is for the quantitation and identification of differentially regulated phosphoprotein isoforms in tobacco elicitation (Gerber et al. 2006) and in the cellular dedifferentiation of *Arabidopsis* (Chitteti and Peng 2007a, b), including the analysis of *Arabidopsis* cells under osmotic stress or ABA-dependent stress (Boudsocq et al. 2007). In addition to it, for the enrichment and quantification of phosphoproteins, Pro-Q Diamond can be used as a purification tool (Ito et al. 2009). Nonetheless, quite often, 2-DE protein spots cannot be used to establish the site of phosphorylation, especially in the case of membrane proteins, which includes several phosphoproteins.

Low stoichiometry and competitive tendency for ionisation due to the presence of non-phosphorylated peptides in vicinity often necessitate the enrichment of phosphopeptides for MS-based phosphoproteomic analysis. Several techniques thus so far are available for selective enrichment (Dunn et al. 2010), particularly in plant proteomics, and several of these techniques have also been used such as immobilised metal affinity chromatography (IMAC) (Grimsrud et al. 2010) and metal oxide affinity chromatography (MOAC) (Hsu et al. 2009) alone or in combination (Sugiyama et al. 2008).

The best employed method prior to any selective enrichment is chemical labelling since phosphoprotein/phosphopeptide enrichment steps can add significantly to the technical bias in quantitative analysis. A recent large-scale phosphoproteomic SILAC study of mouse liver also indicates the preference of metabolic labelling over phosphoprotein/phosphopeptide enrichment steps (Pan et al. 2008). Likewise, the most appropriate quantitation technique for plant phosphoproteomics is the metabolic labelling using ^{15}N salt (Oda et al. 1999), as observed in the case of *Arabidopsis* cells treated with the flagellin bacterial elicitor flg22 and the fungal elicitor xylanase (Benschop et al. 2007). More than 1,000 phosphopeptides from plasma membrane fraction were quantified in this study, and out of that, 76 and 9 phosphopeptides were differentially regulated following flg22 and xylanase elicitation. Nevertheless, in a very similar study of *Arabidopsis*, cells were treated with the flg22 elicitor, and quantitation with iTRAQ was chosen over ^{15}N metabolic labelling due to its multiplexing capabilities. In this investigation, due to the more precise analysis, considering only the ratios with at least a twofold difference, the number of differentially phosphorylated peptides was restricted, i.e. only 12 phosphopeptides were induced (Nuhse et al. 2007). However, the consistency of the data between both the studies implies the identification of the relevant phosphorylation sites.

Redox Proteomics

Environmental stimuli significantly influence the redox status of proteins, predominantly in biotic and abiotic stresses as an oxidative burst associated with the production of reactive oxygen species (ROS) is mainly induced by it (Jaspers and Kangasjarvi 2010; Torres 2010). Usually, cytoplasm is reductive in optimal conditions which favour the reduction of sulfhydryl groups. Nevertheless, cysteines are worst affected amongst the others due to ROS production by the formation of disulphide bonds, unstable sulfonate groups or the irreversible sulfinic or sulfonic acids, even though other residues can also be oxidised. Therefore, to study the redox status of proteins and to quantify reduced cysteines on cysteine-containing peptides, chemical labels that target cysteinyl groups can be used. For the quantification of the reduced cysteines with fluorescent labels, such as monobromobimane (mBBr), cyanide-5-maleimide (cy5m) or CyDyes and DIGE, subsequently protein isoform separation on conventional 2-D gels can be used. Diagonal 2-D native SDS-PAGE is also an alternative (Yano and Kuroda 2008; Stroher and Dietz 2006). A comparative analysis can be possible between the native fluorescent labelled samples with reduced cysteinyl groups and with samples that have been fully reduced by DTT or tris(2-carboxyethyl) phosphine (Fu et al. 2008; Hurd et al. 2009). Labelling of the free SH groups with ICAT reagents can also be a method of choice which can allow a gel-free quantitative study of the redox proteome in plants (Stroher and Dietz 2006; Hagglund et al. 2008, 2010). However, using sequential nonreducing/reducing 2-D SDS-PAGE, redox proteomics has also been undertaken without labels (Cumming et al. 2004). Several of the thioredoxin (Trx) targets or the related glutaredoxin (Grx) targets have been investigated in various redox plant proteome studies (Rinalducci et al. 2008). Both Trx and Grx are involved in the reduction of disulphide bonds in proteins (Montrichard et al. 2009; Rouhier 2010). The redox proteome studied in

Trx-linked reactions during seed germination is commendable one (Yano and Kuroda 2006; Alkhalfioui et al. 2007).

During oxidative stress, one of the most common PTMs is the protein carbonylation by aldehyde or ketone formation on Lys, Arg, Pro or Thr side chains (Rinalducci et al. 2008). Quantification of carbonyl groups is possible by derivatizing them with 2,4-dinitrophenylhydrazine (DNPH) and detecting the DNP adducts with DNP monoclonal antibodies (Tanou et al. 2009). On the other hand, a hydrazide biotin-streptavidin enrichment methodology allows high-throughput findings of carbonylated proteins by MS (Soreghan et al. 2003). A detailed proteomics study of citrus and apple plants subjected to salinity stress and senescence, respectively, revealed a surge of carbonylation events in plant proteomes (Qin et al. 2009). Non-MS-based affinity detection and quantitation techniques coupled with 2-DE are used in these studies.

Nitrosylation

Nitric oxide is a well-established signal molecule involved in plant stress response and development and, to some extent, as ROS as well (Lindermayr and Durner 2009; Qiao and Fan 2008). Stress due to nitric oxide leads to the formation of nitrosylated cysteines or nitrated. Methods such as "biotin switch" and "affinity purification" have also been devised to exclusively target and enrich proteins containing nitrosylated cysteines (Lindermayr and Durner 2009; Torta et al. 2008). For quantitative proteomic analysis of nitrosylated cysteines using straightforward SYPRO Ruby staining intensity values, "biotin switch" enrichment method has also been used coupled with 2-DE and MS (Romero-Puertas et al. 2008). Differentially nitrosylated proteins in HR and subsequent programmed cell death (PCD) in *Arabidopsis* due to the infection with an incompatible bacterial pathogen have been investigated in this study. Similarly, to allow a straightforward quantitation of this PTM, methods have been developed to detect o-nitrotyrosine using iTRAQ reagents (Chiappetta et al. 2009). In sunflower hypocotyls, nitrotyrosine antibodies were used for the detection and quantitation of nitrotyrosine by Chaki et al. (2009). 1-D and 2-D Western blots were used to detect differential nitration following treatment leading to HR in tomato cells by Cecconi et al. (2009). An increase in nitrosylated proteins following salinity stress in citrus plants has been observed by Tanou et al.. However, no nitrated proteins were identified by MS in these studies.

S-Glutathionylation

Glutathionylation, a type of PTM, is an eminent result of cysteine modification. Glutathionylation is a consequence of oxidative or nitrosative stress and is perhaps involved in cellular signalling (Dalle-Donne et al. 2007). Different methods are nowadays available for the detection of this PTM; for instance, 35 S-glutathione labelling, 2-DE separation and biotin-glutathione affinity purification are used to distinguish induced glutathionylation levels of *Arabidopsis* proteins subjected to oxidative stress (Dixon et al. 2005; Gao et al. 2009a, b).

Unravelling Signal Transduction Cascades Using Proteomics Approaches

Signalling processes usually involve direct physical contacts between different components in a pathway, in order to transfer a "signal" from receptors to transcription factors or other intracellular effector proteins. Combinatorial interactions between signalling proteins can be crucial for determining their cell type-specific functions, subcellular localisation and stability. Therefore, the identification of protein complexes and post-translational modifications of signalling proteins are essential to understand signal transduction cascades. The signal is often transmitted from receptors via phosphorylation of intermediate and effector proteins. Protein phosphorylation ensures fast and reversible response to different stimuli. Proteomics approaches are being used to study changes in phosphorylation in response to variation in light or temperature (Bonardi et al. 2005; El-Khatib et al. 2007), invasion of pathogens (for review, see Quirino et al. 2010), hormones (El-Khatib et al. 2007; Li et al. 2009; Chen et al. 2010) and salt stress (Chitteti and Peng 2007a, b). An alternative commonly used mechanism for signal transduction is the targeting of repressor proteins for degradation via ubiquitylation (for review, see Vierstra 2009).

Tandem affinity purification (TAP) approaches, Strep-tags and biotin tags have been successfully used in plants. Alternatively, protein fusions to green fluorescent protein (GFP) are being used, which allow the direct visualisation of the protein expression and subcellular localisation in plants (Karlova et al. 2006). Combination of affinity purification and separation by size exclusion and/or blue native PAGE potentially enables the detection of distinct complexes formed by one protein (Remmerie et al. 2009). Recently, the first systematic proteomics efforts to unravel "interactomes" of specific signalling processes have been accomplished. Proteins of the 14-3-3 family are components of many signalling pathways and bind to a wide variety of client proteins in a phosphorylation-dependent manner. Chang et al. (2009) performed TAP-tag purification of a generic subunit of 14-3-3 protein complexes that was expressed from a constitutive promoter. Complex partners were identified by a quantitative, MudPIT-based strategy. This approach revealed 101 new potential 14-3-3 clients, indicating that 14-3-3s are some of the most connected nodes in the emerging protein–protein interaction network of plants. Another recent proteomics study characterised the core cell cycle interactome in *Arabidopsis* cell cultures; complex partners of 102 cell-cycle associated proteins, constitutively expressed as fusion to an improved version of the TAP tag (GS-tag), were isolated (Van Leene et al. 2010).

Plant Proteomics: Challenges and New Frontiers

Plant proteomics as a discipline has grown multitudes after the release of the model dicot genome sequence of *Arabidopsis* (*Arabidopsis* Genome Initiative 2000) and the monocot genome of rice (Goff et al. 2002). There has been a significant improvement in plant-specific protocols from sample extraction to mass spectrometric analysis. A significant challenge in proteomics when studying plants or any complex biological system is the inability to measure the entire proteome (Ahn et al. 2007). Although limiting, a number of approaches have been used to partially overcome these restrictions. This includes sample fractionation and the enrichment of protein subpopulations or compartments prior to sample analysis by mass spectrometry (Eubel et al. 2008; Huang et al. 2009; Hynek et al. 2009; Ferro et al. 2010).

Techniques involving quantitation by mass spectrometry are now led by the next-generation label-free techniques (Schulze and Usadel 2010). The utilisation of unlabeled targeted approaches (selected reaction monitoring, SRM) has greatly improved sample sensitivity and reproducibility by mass spectrometry (Lange et al. 2008). The ongoing advancement of MS instrumentation and approaches such as sequential window acquisition of all theoretical fragment ion spectra (SWATH; Gillet et al. 2012) has enabled current researchers to employ a wider range of methodological approaches. The label-free technique relies upon the uniqueness of a peptide sequence being monitored which is only a possibility in species with well-characterised genomes such as *Arabidopsis* or rice (Rost et al. 2012). However, in other plants, this remains a challenge, as without complete genome sequences, confidence in a peptide's uniqueness is limited.

Another challenge is to integrate the available proteomic data sources and create community plant resources to create a web of interlinked repositories. Plant research has significantly advanced the field of proteomics by overcoming plant-specific challenges and by contributing to the development of plant-specific-related technologies and analyses. Recently, a coordinated effort was made to create an aggregation portal to summarise the varied *Arabidopsis* proteomic resources in a single interface that was introduced (Joshi et al. 2011). Such integrated approaches are to be fostered for the future of data management and analysis. This resource is significant in that it represents the first example of proteomic data unification by a variety of specialty research groups. Whilst proteomics research in plants will be greatly supported by general advances in the field, there still remains many specific problems that will ultimately require tailored solutions for plant research.

Summary

The primary objectives of plant proteomics in general remain: (1) to get insight into the physiology of different plant species, varieties and their performance towards development parameters, yield indices, pathogen response, abiotic stress management, fruiting etc., (2) to develop improved and safe crops to meet the goals of food security and (3) to develop sustainable agriculture practices and reduce the impact of agriculture on the environment. Proteomics research is the need of the hour and essentially required to integrate the genomic codes to the functional applications. There has been a tremendous development in the technology of proteome analysis from gel-based basic tools to the current quantitative MS/MS-based automated platforms. There has been an exemplary rise in the proteomics studies in the post-genomic era ranging from applications in crop improvement, posttranslational modifications to understanding the natural processes. However, the applications of the proteomic applications need to integrate with the systems biology approach. The genome has limited meaning without a proteome complement which further can only be fully understood by functional characterisation or understanding the metabolome. A broader, interdisciplinary global network should combine multiple strategies simultaneously to integrate the advances in plant biotechnology to reach the larger objective of food security and sustainable development.

References

Abbasi FM, Komatsu S (2004) A proteomic approach to analyze salt-responsive proteins in rice sheath. Proteomics 4:2072–2081

Abdalla KO, Rafudeen MS (2012) Analysis of the nuclear proteome of the resurrection plant Xerophyta viscosa in response to dehydration stress using iTRAQ with 2DLC and tandem mass spectrometry. J Proteomics 18:2361–2374

Abdalla KO, Baker B, Rafudeen MS (2010) Proteomic analysis of nuclear proteins during dehydration of the resurrection plant *Xerophyta viscosa*. Plant Growth Regul 62:279–292

Aebersold R, Mann M (2003) Mass spectrometry-based proteomics. Nature 422:198–207

Aebersold RH, Leavitt J, Saavedra RA, Hood LE, Kent SB (1987) Internal amino acid sequence analysis of proteins separated by one or two-dimensional gel electrophoresis after in situ protease digestion on nitrocellulose. Proc Natl Acad Sci USA 84:6970–6974

Agrawal L, Chakraborty S, Jaiswal DK, Gupta S, Datta A, Chakraborty N (2008) Comparative proteomics of tuber induction, development and maturation reveal the complexity of tuberization process in potato (*Solanum tuberosum* L.). J Proteome Res 7:3803–3817

Agrawal L, Narula K, Basu S, Shekhar S, Ghosh S, Datta A, Chakraborty N, Chakraborty S (2013) Comparative proteomics reveals a role for seed storage protein, AmA1 in cellular growth, development and nutrient accumulation. J Proteome Res. doi:10.1021/pr4007987

Ahn NG, Shabb JB, Old WM, Resing KA (2007) Achieving in-depth proteomics profiling by mass spectrometry. ACS Chem Biol 2:39–52

Ahsan N, Komatsu S (2009) Comparative analyses of the proteomes of leaves and flowers at various stages of development reveal organ-specific functional differentiation of proteins in soybean. Proteomics 9:4889–4907

Ahsan N, Lee DG, Lee SH, Kang KY, Bahk JD, Choi MS et al (2007) A comparative proteomic analysis of tomato leaves in response to water logging stress. Physiol Plant 131:555–570

Ahsan N, Nanjo Y, Sawada H, Kohno Y, Komatsu S (2010) Ozone stress-induced proteomic changes in leaf total soluble and chloroplast proteins of soybean reveal that carbon allocation is involved in adaptation in the early developmental stage. Proteomics 10:2605–2619

Alam I, Lee DG, Kim KH, Park CH, Sharmin SA, Lee H et al (2010a) Proteome analysis of soybean roots under waterlogging stress at an early vegetative stage. J Biosci 35:49–62

Alam I, Sharmin SA, Kim KH, Yang JK, Choi MS, Lee BH (2010b) Proteome analysis of soybean roots subjected to short-term drought stress. Plant Soil 333:491–505

Alkhalfioui F, Renard M, Vensel WH, Wong J et al (2007) Thioredoxin-linked proteins are reduced during germination of *Medicago truncatula* seeds. Plant Physiol 144:1559–1579

Amey RC, Schleicher T, Slinn A, Lewis M et al (2008) Proteomic analysis of a compatible interaction between *Pisum sativum* (pea) and the downy mildew pathogen *Peronospora viciae*. Eur J Plant Pathol 122:41–55

Amme S, Matros A, Schlesier B, Mock HP (2006) Proteome analysis of cold stress response in Arabidopsis thaliana using DIGE-technology. J Exp Bot 57:1537–1546
Amster IJ (1996) Fourier transform mass spectrometry. J Mass Spectrom 31:1325–1337
Andersen JS, Mann M (2000) Functional genomics by mass spectrometry. FEBS Lett 480:25–31
Ansong C, Purvine SO, Adkins JN, Lipton MS, Smith RD (2008) Proteogenomics: needs and roles to be filled by proteomics in genome annotation. Brief Funct Genomic Proteomics 7(1):50–62
Arabidopsis Genome Initiative (2000) Analysis of the genome sequence of the flowering plant *Arabidopsis* thaliana. Nature 408:796–815. doi:10.1038/35048692
Baba A, Nogueira F, Pinheiro C, Brasil J, Jereissati E, Jucá T et al (2008) Proteome analysis of secondary somatic embryogenesis in cassava (*Manihot esculenta*). Plant Sci 175:717–723
Bae MS, Cho EJ, Choi EY, Park OK (2003) Analysis of the *Arabidopsis* nuclear proteome and its response to cold stress. Plant J 36:652–663
Balbuena TS, Silveira V, Junqueira M, Dias LLC, Santa-Catarina C, Shevchenko A et al (2009) Changes in the 2-DE protein profile during zygotic embryogenesis in the Brazilian pine (*Araucaria angustifolia*). J Proteomics 72:337–352
Baudet M, Ortet P, Gaillard JC, Fernandez B et al (2009/2010) Proteomics-based refinement of *Deinococcus deserti* genome annotation reveals an unwanted use of non-canonical translation initiation codons. Mol Cell Proteomics 9(2):415
Bayer EM, Bottrill AR, Walshaw J, Vigouroux M, Naldrett MJ, Thomas CL et al (2006) *Arabidopsis* cell wall proteome defined using multidimensional protein identification technology. Proteomics 6:301–311
Beavis RC, Chait BT (1996) Matrix-assisted laser desorption ionization mass spectrometry of proteins. Method Enzymol 270:519–551, Chap. 22
Benschop JJ, Mohammed S, O'Flaherty M, Heck AJR et al (2007) Quantitative phosphoproteomics of early elicitor signaling in Arabidopsis. Mol Cell Proteomics 6:1198–1214
Bhushan D, Pandey A, Chattopadhyay A, Choudhary MK, Chakraborty S, Datta A, Chakraborty N (2006) Extracellular matrix proteome of chickpea (*Cicer arietinum* L.) illustrates pathway abundance, novel protein functions and evolutionary perspect. J Proteome Res 5:1711–1720
Bhushan D, Pandey A, Choudhary MK, Datta A, Chakraborty S, Chakraborty N (2007) Comparative proteomics analysis of differentially expressed proteins in chickpea extracellular matrix during dehydration stress. Mol Cell Proteomics 6:1868–1884
Bi Y, Wei Z, Shen Z, Lu T, Cheng Y, Wang B et al (2010) Comparative temporal analyses of the *Pinus sylvestris* L. var. mongolica litv. apical bud proteome from dormancy to growth. Mol Biol Rep 38:721–729
Blackstock WP, Weir MP (1999) Proteomics: quantitative and physical mapping of cellular proteins. Trends Biotechnol 17:121–127
Bogeat-Triboulot MB, Brosché M, Renaut J, Jouve L, Le Thiec D, Fayyaz P et al (2007) Gradual soil water depletion results in reversible changes of gene expression, protein profiles, ecophysiology, and growth performance in *Populus euphratica*, a poplar growing in arid regions. Plant Physiol 143:876–892
Bonardi V, Pesaresi P, Becker T, Schleiff E et al (2005) Photosystem II core phosphorylation and photosynthetic acclimation require two different protein kinases. Nature 437:1179–1182
Bonhomme L, Monclus R, Vincent D, Carpin S, Lomenech AM, Plomion C et al (2009) Leaf proteome analysis of eight *Populus × euramericana* genotypes: genetic variation in drought response and in water-use efficiency involves photosynthesis-related proteins. Proteomics 9:4121–4142
Boudart G, Jamet E, Rossignol M, Lafitte C, Borderies G, Jauneau A et al (2005) Cell wall proteins in apoplastic fluids of *Arabidopsis thaliana* rosettes: identification by mass spectrometry and bioinformatics. Proteomics 5:212–221
Boudsocq M, Droillard MJ, Barbier-Brygoo H, Lauriere C (2007) Different phosphorylation mechanisms are involved in the activation of sucrose non-fermenting 1 related protein kinases 2 by osmotic stresses and abscisic acid. Plant Mol Biol 63:491–503
Briceno Z, Almagro L, Sabater-Jara AB, Calderon AA, Pedreno MA, Ferrer MA (2012) Enhancement of phytosterols, taraxasterol and induction of extracellular pathogenesis-related proteins in cell cultures of *Solanum lycopersicum* cv Micro-Tom elicited with cyclodextrins and methyl jasmonate. J Plant Physiol 169:1050–1058
Brugiere S, Kowalski S, Ferro M, Seigneurin-Berny D, Miras S, Salvi D, Ravanel S, d'Herin P, Garin J, Bourguignon J, Joyard J, Rolland N (2004) The hydrophobic proteome of mitochondrial membranes from Arabidopsis cell suspensions. Phytochemistry 65:1693–1707
Caruso G, Cavaliere C, Guarino C, Gubbiotti R, Foglia P, Laganà A (2008) Identification of changes in *Triticum durum* L. leaf proteome in response to salt stress by two-dimensional electrophoresis and MALDI-TOF mass spectrometry. Anal Bioanal Chem 391:381–390
Casasoli M, Spadoni S, Lilley KS, Cervone F et al (2008) Identification by 2-D DIGE of apoplastic proteins regulated by oligogalacturonides in *Arabidopsis thaliana*. Proteomics 8:1042–1054
Cecconi D, Orzetti S, Vandelle E, Rinalducci S et al (2009) Protein nitration during defense response in *Arabidopsis thaliana*. Electrophoresis 30:2460–2468
Celis JE, Kruhoffer M, Gromova I, Frederiksen C, Ostergaard M, Thykjaer T, Gromov P, Yu J, Palsdottir H, Magnusson N, Ornoft TF (2000) Gene expression profiling: monitoring transcription and translation products using DNA microarrays and proteomics. FEBS Lett 480:2–16
Cellini F, Chesson A, Colquhoun I, Constable A, Davies HV, Engel KH et al (2004) Unintended effects and their detection in genetically modified crops. Food Chem Toxicol 42:1089–1125

Chaki M, Valderrama R, Fernandez-Ocana AM, Carreras A et al (2009) Protein targets of tyrosine nitration in sunflower (*Helianthus annuus* L.) hypocotyls. J Exp Bot 60:4221–4234

Chan J, Calder G, Fox S, Lloyd C (2007) Cortical microtubule arrays undergo rotary movements in Arabidopsis hypocotyls epidermal cells. Nat Cell Biol 9:171–175

Chang IF, Curran A, Woolsey R, Quilici D et al (2009) Proteomic profiling of tandem affinity purified 14-3-3 protein complexes in *Arabidopsis thaliana*. Proteomics 9:2967–2985

Chen S, Harmon AC (2006) Advances in plant proteomics. Proteomics 6:5504–5516

Chen Y, Chen T, Shen S, Zheng M, Guo Y, Lin J et al (2006) Differential display proteomic analysis of *Picea meyeri* pollen germination and pollen-tube growth after inhibition of actin polymerization by latrunculin B. Plant J 47:174–195

Chen X, Wang Y, Li J, Jiang A, Cheng Y, Zhang W (2009a) Mitochondrial proteome during salt stress-induced programmed cell death in rice. Plant Physiol Biochem 47(5):407–415

Chen XY, Kim ST, Cho WK, Rim Y, Kim S, Kim SW et al (2009b) Proteomics of weakly bound cell wall proteins in rice calli. J Plant Physiol 166:675–685

Chen Y, Hoehenwarter W, Weckwerth W (2010) Comparative analysis of phytohormone-responsive phosphoproteins in *Arabidopsis thaliana* using TiO2-phosphopeptide enrichment and mass accuracy precursor alignment. Plant J 63:1–17

Cheng FY, Blackburn K, Lin YM, Goshe MB, Williamson JD (2009a) Absolute protein quantification by LC/MS (E) for global analysis of salicylic acid-induced plant protein secretion responses. J Proteome Res 8:82–93

Cheng Y, Qi Y, Zhu Q, Chen X, Wang N, Zhao X et al (2009b) New changes in the plasma-membrane-associated proteome of rice roots under salt stress. Proteomics 9:3100–3114

Cheng L, Gao X, Li S, Shi M, Javeed H, Jing X et al (2010) Proteomic analysis of soybean [*Glycine max* (L.) Meer.] seeds during imbibition at chilling temperature. Mol Breed 26:1–17

Chiappetta G, Corbo C, Palmese A, Marino G, Amoresano A (2009) Quantitative identification of protein nitration sites. Proteomics 9:1524–1537

Chitteti BR, Peng ZH (2007a) Proteome and phosphoproteome dynamic change during cell dedifferentiation in *Arabidopsis*. Proteomics 7:1473–1500

Chitteti BR, Peng ZH (2007b) Proteome and phosphor proteome differential expression under salinity stress in rice (*Oryza sativa*) roots. J Proteome Res 6:1718–1727

Chitteti BR, Tan F, Mujahid H, Magee BG, Bridges SM, Peng Z (2008) Comparative analysis of proteome differential regulation during cell dedifferentiation in *Arabidopsis*. Proteomics 8:4303–4316

Chivasa S, Ndimba BK, Simon WJ, Robertson D, Yu XL, Knox JP et al (2002) Proteomic analysis of the *Arabidopsis thaliana* cell wall. Electrophoresis 23:1754–1765

Chivasa S, Simon WJ, Yu X-L, Yalpani N, Slabas AR (2005) Pathogen elicitor-induced changes in the maize extra cellular matrix proteome. Proteomics 5:4894–4904

Cho WK, Chen XY, Chu H, Rim Y, Kim S, Kim ST et al (2009) Proteomic analysis of the secretome of rice calli. Physiol Plant 135:331–341

Choudhary MK, Basu D, Datta A, Chakraborty N, Chakraborty S (2009) Dehydration-responsive nuclear proteome of rice (*Oryza sativa* L.) illustrates protein network, novel regulators of cellular adaptation, and evolutionary perspective. Mol Cell Proteomics 8:1579–1598

Christendat D, Yee A, Dharamsi A, Kluger Y, Gerstein M, Arrowsmith CH, Edwardsa AM (2000) Structural proteomics: prospects for high throughput sample preparation. Prog Biophys Mol Biol 73:339–345

Cooper B, Campbell KB, Feng J, Garrett WM, Frederick R (2011) Nuclear proteomic changes linked to soybean rust resistance. Mol Biosyst 3:773–783

Corpillo D, Gardini G, Vaira AM, Basso M, Aime S, Accotto GR et al (2004) Proteomics as a tool to improve investigation of substantial equivalence in genetically modified organisms: the case of a virus-resistant tomato. Proteomics 4:193–200

Corthals GL, Wasinger VC, Hochstrasser DF, Sanchez JC (2000) The dynamic range of protein expression: a challenge for proteomic research. Electrophoresis 21:1104–1115

Cui S, Huang F, Wang J, Ma X, Cheng Y, Liu J (2005) A proteomic analysis of cold stress responses in rice seedlings. Proteomics 5:3162–3172

Cumming RC, Andon NL, Haynes PA, Park M et al (2004) Protein disulfide bond formation in the cytoplasm during oxidative stress. J Biol Chem 279:21749–21758

Dahal D, Pich A, Braun HP, Wydra K (2010) Analysis of cell wall proteins regulated in stem of susceptible and resistant tomato species after inoculation with *Ralstonia solanacearum*: a proteomic approach. Plant Mol Biol 73:643–658

Dai S, Wang T, Yan X, Chen S (2007) Proteomics of pollen development and germination. J Proteome Res 6:4556–4563

Dalle-Donne I, Rossi R, Giustarini D, Colombo R, Milzani A (2007) S-glutathionylation in protein redox regulation. Free Radic Biol Med 43:883–898

Dani V, Simon WJ, Duranti M, Croy RRD (2005) Changes in the tobacco leaf apoplast proteome in response to salt stress. Proteomics 5:737–745

Davis JR, Kakar M, Lim CS (2007) Controlling protein compartmentalization to overcome disease. Pharm Res 24(1):17–27

De Jong F, Mathesius U, Imin N, Rolfe BG (2007) A proteome study of the proliferation of cultured *Medicago truncatula* protoplasts. Proteomics 7:722–736

Degand H, Faber AM, Dauchot N, Mingeot D, Watillon B, VanCutsem P et al (2009) Proteomic analysis of chicory root identifies proteins typically involved in cold acclimation. Proteomics 9:2903–2907

De-la-Pena C, Lei Z, Watson BS, Sumner LW, Vivanco J (2008) Root-microbe communication through protein secretion. J Biol Chem 283:25247–25455

Deytieux C, Geny L, Lapaillerie D, Claverol S, Bonneu M, Doneche B (2007) Proteome analysis of grape skins during ripening. J Exp Bot 58:1851–1862

DiCarli M, Villani ME, Renzone G, Nardi L, Pasquo A, Franconi R et al (2009) Leaf proteome analysis of transgenic plants expressing antiviral antibodies. J Proteome Res 8:838–848

Dixon DP, Skipsey M, Grundy NM, Edwards R (2005) Stress-induced protein S-glutathionylation in *Arabidopsis*. Plant Physiol 138:2233–2244

Dooki AD, Mayer-Posner FJ, Askari H, Zaiee A, Salekdeh GH (2006) Proteomic responses of rice young panicles to salinity. Proteomics 6:6498–6507

Drumm ML, Collins FS (1993) Molecular biology of cystic fibrosis. Mol Genet Med 3:33–68

Dunkley TPJ, Watson R, Griffin JL, Dupree P, Lilley KS (2004) Localization of organelle proteins by isotope tagging (LOPIT). Mol Cell Proteomics 3:1128–1134

Dunn JD, Reid GE, Bruening ML (2010) Techniques for phosphopeptide enrichment prior to analysis by mass spectrometry. Mass Spectrom Rev 29:29–54

Edwards AM, Arrowsmith CH, Christendat D, Dharamsi A, Friesen JD, Greenblatt JF, Vedadi M (2000) Protein production: feeding the crystallographers and NMR spectroscopists. Nat Struct Biol 7(Suppl):970–972

Eisenberg D, Marcotte EM, Xenarios I, Yeates TO (2000) Protein function in the post-genomic era. Nature 405:823–826

Eisenstein E, Gilliland GL, Herzberg O, Moult J, Orban J, Poljak RJ, Banerjei L, Richardson D, Howard AJ (2000) Biological function made crystal clear annotation of hypothetical proteins via structural genomics. Curr Opin Biotechnol 11:25–30

El-Khatib RT, Good AG, Muench DG (2007) Analysis of the *Arabidopsis* cell suspension phosphoproteome in response to short-term low temperature and abscisic acid treatment. Physiol Plant 129:687–697

Eubel H, Meyer EH, Taylor NL, Bussell JD, O'Toole N, Heazlewood JL, Castleden I, Small ID, Smith SM, Millar AH (2008) Novel proteins, putative membrane transporters, and an integrated metabolic network are revealed by quantitative proteomic analysis of *Arabidopsis* cell culture peroxisomes. Plant Physiol 148:1809–1829. doi:10.1104/pp. 108.129999

Fenn JB, Mann M, Meng CK, Wong SF, Whitehouse CM (1989) Electrospray ionization for mass spectrometry of large biomolecules. Science 246:64–71

Ferreira S, Hjernø K, Larsen M, Wingsle G, Larsen P, Fey S et al (2006) Proteome profiling of *Populus euphratica* Oliv. upon heat stress. Ann Bot 98:361–377

Ferri M, Tassoni A, Franceschetti M, Righetti L et al (2009) Chitosan treatment induces changes of protein expression profile and stilbene distribution in *Vitis vinifera* cell suspensions. Proteomics 9:610–624

Ferro M, Brugière S, Salvi D, Seigneurin-Berny D, Court M, Moyet L, Ramus C et al (2010) AT_CHLORO, a comprehensive chloroplast proteome database with subplastidial localization and curated information on envelope proteins. Mol Cell Proteomics 9:1063–1084

Fischer L, Lipavska H, Hausman J, Opatrny Z (2008) Morphological and molecular characterization of a spontaneously tuberizing potato mutant: an insight into the regulatory mechanisms of tuber induction. BMC Plant Biol 8:117–130

Fu CX, Hu J, Liu T, Ago T et al (2008) Quantitative analysis of redox-sensitive proteome with DIGE and ICAT. J Proteome Res 7:3789–3802

Fukao Y (2012) Protein–protein interactions in plants. Plant Cell Physiol 53(4):617–625

Gao F, Zhou Y, Zhu W, Li X, Fan L, Zhang G (2009a) Proteomic analysis of cold stress-responsive proteins in *Thellungiella rosette* leaves. Planta 230:1033–1046

Gao XH, Bedhomme M, Veyel D, Zaffagnini M, Lemaire SD (2009b) Methods for analysis of protein glutathionylation and their application to photosynthetic organisms. Mol Plant 2:218–235

Garcia-Canas V, Simo C, Leon C, Ibanez E, Cifuentes A (2011) Ms-based analytical methodologies to characterize genetically modified crops. Mass Spectrom Rev 30:396–416

Gerber IB, Laukens K, Witters E, Dubery IA (2006) Lipopolysaccharide-responsive phosphoproteins in *Nicotiana tabacum* cells. Plant Physiol Biochem 44:369–379

Gerber IB, Laukens K, De Vijlder T, Witters E, Dubery IA (2008) Proteomic profiling of cellular targets of lipopolysaccharide-induced signalling in *Nicotiana tabacum* BY-2 cells. Biochim Biophys Acta 1784:1750–1762

Gillet LC, Navarro P, Tate S, Rost H, Selevsek N, Reiter L et al (2012) Targeted data extraction of the MS/MS spectra generated by data-independent acquisition: a new concept for consistent and accurate proteome analysis. Mol Cell Proteomics 11:O111.016717

Goff SA, Ricke D, Lan TH, Presting G, Wang R, Dunn M, Glazebrook J, Sessions A, Oeller P, Varma H, Hadley D, Hutchison D et al (2002) A draft sequence of the rice genome (*Oryza sativa L. ssp. japonica*). Science 296:92–100. doi:10.1126/science.1068275

Gokulakannan GG, Niehaus K (2010) Characterization of the *Medicago truncatula* cell wall proteome in cell suspension culture upon elicitation and suppression of plant defense. J Plant Physiol 167:1533–1541

Gómez A, López JA, Pintos B, Camafeita E, Bueno MA (2009) Proteomic analysis from haploid and diploid embryos of *Quercus suber* L. identifies qualitative and quantitative differential expression patterns. Proteomics 9:4355–4367

Gong CY, Wang T (2013) Proteomic evaluation of genetically modified crops: current status and challenges. Front Plant Sci 4:41. doi:10.3389/fpls.2013.00041

Gong CY, Li Q, Yu HT, Wang Z, Wang T (2012) Proteomics insight into the biological safety of transgenic modification of rice as compared with conventional genetic breeding and spontaneous genotypic variation. J Proteome Res 11:3019–3029

Goulas E, Schubert M, Kieselbach T, Kleczkowski LA, Gardeström P, Schröder W, Hurry V (2006) The chloroplast lumen and stromal proteomes of Arabidopsis thaliana show differential sensitivity to short- and long-term exposure to low temperature. Plant J 47:720–734

Grave PR, Haystead TAJ (2002) Molecular biologist's guide to proteomics. Microbiol Mol Biol Rev 66:39–63

Grimsrud PA, den Os D, Wenger CD, Swaney DL et al (2010) Large-scale phosphoprotein analysis in *Medicago truncatula* roots provides insight into in vivo kinase activity in legumes. Plant Physiol 152:19–28

Gupta R, Deswal R (2012) Low temperature stress modulated secretome analysis and purification of anti freeze protein from *Hippophae rhamnoides*, a Himalayan wonder plant. J Proteome Res 11:2684–2696

Gupta S, Wardhan V, Verma S, Gayali S, Rajamani U, Datta A et al (2011) Characterization of the secretome of chickpea suspension culture reveals pathway abundance and the expected and unexpected secreted proteins. J Proteome Res 10:5006–5015

Gygi SP, Rist B, Gerber SA et al (1999) Quantitative analysis of complex protein mixtures using isotope coded affinity tags. Nat Biotechnol 17:994–999

Gygi SP, Rist B, Griffin TJ, Eng J, Aebersold R (2002) Proteome analysis of low abundance proteins using multidimensional chromatography and isotope coded affinity tags. J Proteome Res 1:47–54

Hagglund P, Bunkenborg J, Maeda K, Svensson B (2008) Identification of thioredoxin disulfide targets using a quantitative proteomics approach based on isotope-coded affinity tags. J Proteome Res 7:5270–5276

Hagglund P, Bunkenborg J, Yang F, Harder LM et al (2010) Identification of thioredoxin target disulfides in proteins released from barley aleurone layers. J Proteomics 73:1133–1136

Hajheidari M, Abdollahian-Noghabi M, Askari H, Heidari M, Sadeghian SY, Ober ES et al (2005) Proteome analysis of sugar beet leaves under drought stress. Proteomics 5:950–960

Hashimoto M, Komatsu S (2007) Proteomic analysis of rice seedlings during cold stress. Proteomics 7:1293–1302

Heazlewood JL, Howell KA, Whelan J, Millar AH (2003) Towards an analysis of the rice mitochondrial proteome. Plant Physiol 132:230–242

Hegde PS, Whitey IR, Debouckz C (2003) Interplay of transcriptomics and proteomics. Curr Opin Biotechnol 14:647–651

Holmes P, Farquharson R, Hall PJ, Rolfe BG (2006) Proteomic analysis of root meristems and the effects of acetohydroxyacid synthase-inhibiting herbicides in the root of *Medicago truncatula*. J Proteome Res 5:2309–2316

Horváth-Szanics E, Szabó Z, Janáky T, Pauk J, Hajós GJ (2006) Proteomics as an emergent tool for identification of stress-induced proteins in control and genetically modified wheat lines. Chromatographia 63(13 Supplement):S143–S147

Hsu JL, Wang LY, Wang SY, Lin CH et al (2009) Functional phosphoproteomic profiling of phosphorylation sites in membrane fractions of salt-stressed *Arabidopsis thaliana*. Proteome Sci 7:42

Huang S, Taylor NL, Narsai R, Eubel H et al (2009) Experimental analysis of the rice mitochondrial proteome, its biogenesis, and heterogeneity. Plant Physiol 149:719–734

Hurd TR, James AM, Lilley KS, Murphy MR (2009) Measuring redox changes to mitochondrial protein thiols with redox difference gel electrophoresis (Redox-DIGE). Methods Enzymol 456:343–361

Hynek R, Svensson B, Jensen ON, Barkholt V, Finnie C (2009) The plasma membrane proteome of germinating barley embryos. Proteomics 9:3787–3794. doi:10.1002/pmic.200800745

Imin N, Kerim T, Rolfe BG, Weinman JJ (2004) Effect of early cold stress on the maturation of rice anthers. Proteomics 4:1873–1882

Ioset JR, Urbaniak B, Ndjoko-Ioset K, Wirth J, Martin F, Gruissem W et al (2007) Flavonoid profiling among wild type and related GM wheat varieties. Plant Mol Biol 65:645–654

Ito J, Taylor NL, Castleden I, Weckwerth W et al (2009) A survey of the *Arabidopsis thaliana* mitochondrial phosphoproteome. Proteomics 9:4229–4240

Jacoby RP, Millar AH, Taylor NL (2010) Wheat mitochondrial proteomes provide new links between antioxidant defense and plant salinity tolerance. J Proteome Res 9:6595–6604

Jaiswal DK, Ray D, Subba P, Mishra P, Gayali S, Datta A, Chakraborty S, Chakraborty N (2012) Proteomic analysis reveals the diversity and complexity of membrane proteins in chickpea (*Cicer arietinum* L.). Proteome Sci 10:59–71

Jamet E, Canut H, Boudart G, Pont-Lezica RF (2006) Cell wall proteins: a new insight through proteomics. Trends Plant Sci 11:33–39

Jamet E, Boudart G, Borderies G, Charmont S, Lafitte C, Rossignol M et al (2008) Isolation of plant cell wall proteins. Methods Mol Biol 425:187–201

Jaspers P, Kangasjarvi J (2010) Reactive oxygen species in abiotic stress signaling. Physiol Plant 138:405–413

Jensen ON (2004) Modification-specific proteomics: characterization of post-translational modifications by mass spectrometry. Curr Opin Chem Biol 8:33–41

Jensen ON (2006) Interpreting the protein language using proteomics. Nat Rev Mol Cell Biol 7:391–403

Jones AM, Bennett MH, Mansfield JW, Grant M (2006a) Analysis of the defence phosphoproteome of *Arabidopsis thaliana* using differential mass tagging. Proteomics 6:4155–4165

Jones AM, Thomas V, Bennett MH, Mansfield J, Grant M (2006b) Modifications to the *Arabidopsis* defense proteome occur prior to significant transcriptional change in response to inoculation with *Pseudomonas syringae*. Plant Physiol 142:1603–1620

Jonscher KR (2005) Validating sequence assignments for peptide fragmentation patterns: a primer in ms/ms sequence identification. Available at www.ProteomeSoftware.com.

Joshi HJ, Hirsch-Hoffmann M, Baerenfaller K, Gruissem W, Baginsky S, Schmidt R, Schulze WX et al (2011) MASCP gator: an aggregation portal for the visualization of *Arabidopsis* proteomics data. Plant Physiol 155:259–270. doi:10.1104/pp. 110.168195

Jung E, Heller M, Sanchez JC, Hochstrasser DF (2000) Proteomics meets cell biology: the establishment of subcellular proteomes. Electrophoresis 21:3369–3377

Jung YH, Jeong SH, Kim SH, Singh R, Lee JE, Cho YS et al (2008) Systematic secretome analyses of rice leaf and seed callus suspension-cultured cells: workflow development and establishment of high-density two-dimensional gel reference maps. J Proteome Res 7:5187–5210

Karas M, Hillenkamp F (1988) Laser desorption ionization of proteins with molecular masses exceeding 10,000 daltons. Anal Chem 60:2299–2301

Karlova R, Boeren S, Russinova E, Aker J et al (2006) The *Arabidopsis* somatic embryogenesis receptor-like kinase1protein complex includes brassinosteroid-insensitive1. Plant Cell 18:626–638

Kaschani F, Gu C, Niessen S, Hoover H, Cravatt BF, van der Hoorn RA (2009) Diversity of serine hydrolase activities of unchallenged and botrytis-infected Arabidopsis thaliana. Mol Cell Proteomics 8:1082–1093

Kav NV, Srivastava S, Yajima W, Sharma N (2007) Application of proteomics to investigate plant-microbe interactions. Curr Proteomics 4:28–43

Kawamur Y, Uemura M (2003) Mass spectrometric approach for identifying putative plasma membrane proteins of *Arabidopsis* leaves associated with cold acclimation. Plant J 36:141–154

Kim ST, Kim SG, Hwang DH, Kang SY et al (2004a) Proteomic analysis of pathogen-responsive proteins from rice leaves induced by rice blast fungus, *Magnaporthe grisea*. Proteomics 4:3569–3578

Kim ST, Yu S, Kim SG, Kim HJ et al (2004b) Proteome analysis of rice blast fungus (*Magnaporthe grisea*) proteome during appressorium formation. Proteomics 4:3579–3587

Kim DW, Rakwal R, Agrawal GK, Jung YH, Shibato J, Jwa NS et al (2005) A hydroponic rice seedling culture model system for investigating proteome of salt stress in rice leaf. Electrophoresis 26:4521–4539

Kim S, Wang Y, Kang S, Kim S, Rakwal R, Kim Y et al (2009a) Developing rice embryo proteomics reveals essential role for embryonic proteins in regulation of seed germination. J Proteome Res 8:3598–3605

Kim ST, Kang YH, Wang Y, Wu J, Park ZY, Rakwal R et al (2009b) Secretome analysis of differentially induced proteins in rice suspension-cultured cells triggered by rice blast fungus and elicitor. Proteomics 9:1302–1313

Klose J (1975) Protein mapping by combined isoelectric focusing and electrophoresis of mouse tissues. A novel approach to testing for induced point mutations in mammals. Human Genetic 26:231–243

Kok EJ, Lehesranta SJ, van Dijk JP, Helsdingen JR, Dijksma WTP, Van Hoef AMA et al (2008) Changes in gene and protein expression during tomato ripening. Consequences for the safety. Food Sci Technol Int 14:503–518

Komatsu S, Kobayashi Y, Nishizawa K, Nanjo Y, Furukawa K (2010) Comparative proteomics analysis of differentially expressed proteins in soybean cell wall during flooding stress. Amino Acids 39:1435–1449

Komatsu S, Yamamoto A, Nakamura T, Nouri MZ, Nanjo Y, Nishizawa K, Furukawa K (2011) Comprehensive analysis of mitochondria in roots and hypocotyls of soybean under flooding stress using proteomics and metabolomics techniques. J Proteome Res 10(9):3993–4004

Kong FJ, Oyanagi A, Komatsu S (2009) Cell wall proteome of wheat roots under flooding stress using gel-based and LCMS/MS- based proteomics approaches. Biochim Biophys Acta 1804:124–136

Konishi H, Ishiguro K, Komatsu S (2001) A proteomics approach towards understanding blast fungus infection of rice grown under different levels of nitrogen fertilization. Proteomics 1:1162–1171

Konozy EH, Rogniaux H, Causse M, Faurobert M (2013) Proteomic analysis of tomato (*Solanum lycopersicum*) secretome. J Plant Res 126:251–266

Kristensen BK, Askerlund P, Bykova NV, Egsgaard H, Møller IM (2004) Identification of oxidised proteins in the matrix of rice leaf mitochondria by immune precipitation and two-dimensional liquid chromatography-tandem mass spectrometry. Phytochemistry 65:1839–1851

Kruft V, Eubel H, Jeansch L, Werhahn W, Braun HP (2001) Proteomic approach to identify novel mitochondrial proteins in Arabidopsis. Plant Physiol 127(4):1694–1710

Kuiper HA, Kok EJ, Engel KH (2003) Exploitation of molecular profiling techniques for GM food safety assessment. Curr Opin Biotechnol 14:238–243

Kumar A, Agarwal S, Heyman JA, Matson S, Heidtman M, Piccirillo S, Umansky L, Drawid A, Jansen R, Liu Y, Cheung KH, Miller P, Gerstein M, Roeder GS, Snyder M (2002) Subcellular localization of the yeast proteome. Genes Dev 16:707–719

Kwon SJ, Choi EY, Choi YJ, Ahn JH, Park OK (2006) Proteomics studies of post-translational modifications in plants. J Exp Bot 57:1547–1551

Lange V, Picotti P, Domon B, Aebersold R (2008) Selected reaction monitoring for quantitative proteomics: a tutorial. Mol Syst Biol 4:222

Latijnhouwers M, de Wit PJ, Govers F (2003) Oomycetes and fungi: similar weaponry to attack plants. Trends Microbiol 11:462–469

Lee BJ, Kwon SJ, Kim SK, Kim KJ, Park CJ, Kim YJ et al (2006) Functional study of hot pepper 26S proteasome subunit RPN7 induced by tobacco mosaic virus from nuclear proteome analysis. Biochem Biophys Res Commun 351:405–411
Lee DG, Ahsan N, Lee SH, Kang KY, Bahk JD, Lee IJ et al (2007) A proteomic approach in analyzing heat-responsive proteins in rice leaves. Proteomics 7:3369–3383
Lee KH, Kim Y, Park C, Kim H (2008) Proteomic identification of differentially expressed proteins in *Arabidopsis* mutant ntm1-D with disturbed cell division. Mol Cells 25:70–77
Lee DG, Ahsan N, Lee SH, Lee JJ, Bahk JD, Kang KY et al (2009) Chilling stress-induced proteomic changes in rice roots. J Plant Physiol 166:1–11
Lehesranta SJ, Davies HV, Shepherd LVT, Koistinen KM, Massat N, Nunan N et al (2006) Proteomic analysis of the potato tuber life cycle. Proteomics 6:6042–6052
Li H, Wong WS, Zhu L, Guo HW et al (2009) Phosphoproteomic analysis of ethylene-regulated protein phosphorylation in etiolated seedlings of *Arabidopsis* mutant *ein2* using two-dimensional separations coupled with a hybrid quadrupole time-of-flight mass spectrometer. Proteomics 9:1646–1661
Li K, Zhu W, Zeng K, Zhang Z, Ye J, Ou W et al (2010) Proteome characterization of cassava (*Manihot esculenta* Crantz) somatic embryos, plantlets and tuberous roots. Proteome Sci 8:10–21
Liao M, Li Y, Wang Z (2009) Identification of elicitor-responsive proteins in rice leaves by a proteomic approach. Proteomics 9:2809–2819
Lilley KS, Dupree P (2007) Plant organelle proteomics. Curr Opin Plant Biol 10:594–599
Lim S, Chisholm K, Coffin RH, Peters RD, Al-Mughrabi KI, Wang-Pruski G et al (2012) Protein profiling in potato (*Solanum tuberosum* L.) leaf tissues by differential centrifugation. J Proteome Res 11:2594–2601
Lindermayr C, Durner J (2009) S-Nitrosylation in plants: pattern and function. J Proteomics 73:1–9
Liska AJ, Shevchenko A (2003) Expanding the organismal scope of proteomics: cross-species protein identification by mass spectrometry and its implications. Proteomics 3(1):19–28
Liu HL, Hsu JP (2005) Recent developments in structural proteomics for protein structure determination. Proteomics 5:2056–2068
Liu H, Liu Y, Yang M, Shen S (2009) A comparative analysis of embryo and endosperm proteome from seeds of *Jatropha curcas*. J Integr Plant Biol 51:850–857
Lochmanová G, Zdráhal Z, Konečná H, Koukalová S, Malbeck J, Souček P et al (2008) Cytokinin-induced photomorphogenesis in dark-grown *Arabidopsis*: a proteomic analysis. J Exp Bot 59:3705–3719
Lu T, Meng L, Yang C, Liu G, Liu G, Ma W et al (2008) A shotgun phosphoproteomics analysis of embryos in germinated maize seeds. Planta 228:1029–1041
Majoul T, Bancel E, Triboi E, Ben Hamida J, Branlard G (2004) Proteomic analysis of the effect of heat stress on hexaploid wheat grain: characterization of heat-responsive proteins from non-prolamins fraction. Proteomics 4:505–513
Mann M, Jensen ON (2003) Proteomic analysis of post-translational modifications. Nat Biotechnol 21:255–261
Margaria P, Palmano S (2011) Response of the *Vitis vinifera* L. Cv. 'Nebbiolo' proteome to Flavescence dore'e phytoplasma infection. Proteomics 11:212–224
Marsh E, Alvarez S, Hicks LM, Barbazuk WB et al (2010) Changes in protein abundance during powdery mildew infection of leaf tissues of Cabernet Sauvignon grapevine (*Vitis vinifera* L.). Proteomics 10:2057–2064
Marshall AG, Verdun FR (1990) Fourier transforms in NMR, optical, and mass spectrometry: a user's handbook
Marsoni M, Bracale M, Espen L, Prinsi B, Negri AS, Vannini C (2008) Proteomic analysis of somatic embryogenesis in *Vitis vinifera*. Plant Cell Rep 27:347–356
Martinez-Esteso MJ, Selles-Marchart S, Vera-Urbina JC, Pedreno MA, Bru-Martinez R (2009) Changes of defense proteins in the extra cellular proteome of grapevine (*Vitis vinifera* cv. Gamay) cell cultures in response to elicitors. J Proteomics 73:331–341
Millar AH, Heazlewood JL, Kristensen BK, Braun HP, Møller IM (2005) The plant mitochondrial proteome. Trends Plant Sci 10:36–43
Millar DJ, Whitelegge JP, Bindschedler LV, Rayon C, Boudet AM, Rossignol M et al (2009) The cell wall and secretory proteome of a tobacco cell line synthesizing secondary wall. Proteomics 9:2355–2372
Minic Z, Jamet E, Négroni L, Arseneder Garabedian P, Zivy M, Jouanin L (2007) A sub-proteome of *Arabidopsis thaliana* mature stem strapped on Concanavalin A is enriched in cell wall glycoside hydrolases. J Exp Bot 58:2503–2512
Montrichard F, Alkhalfioui F, Yano H, Vensel WH et al (2009) Thioredoxin targets in plants: the first 30 years. J Proteomics 72:452–474
Müller K, Job C, Belghazi M, Job D, Leubner-Metzger G (2010) Proteomics reveal tissue-specific features of the cress (*Lepidium sativum* L.) endosperm cap proteome and its hormone-induced changes during seed germination. Proteomics 10:406–416
Muthreich N, Schützenmeister A, Schütz W, Madlung J, Krug K, Nordheim A et al (2010) Regulation of the maize (*Zea mays* L.) embryo proteome by RTCS which controls seminal root initiation. Eur J Cell Biol 89:242–249
Nabieva E, Jim K, Agarwal A, Chazelle B, Singh M (2005) Whole-proteome prediction of protein function via graph-theoretic analysis of interaction maps. Bioinformatics 21(Suppl 1):i302–i310
Ndimba BK, Chivasa S, Hamilton JM, Simon WJ, Slabas AR (2003) Proteomic analysis of changes in the extracellular matrix of Arabidopsis cell suspension cultures induced by fungal elicitors. Proteomics 3:1047–1059
Ndimba BK, Chivasa S, Simon WJ, Slabas AR (2005) Identification of *Arabidopsis* salt and osmotic stress responsive proteins using two-dimensional difference

gel electrophoresis and mass spectrometry. Proteomics 5:4185–4196

Nogueira F, Goncalves E, Jereissati E, Santos M, Costa J, Oliveira-Neto O et al (2007) Proteome analysis of embryogenic cell suspensions of cowpea (*Vigna unguiculata*). Plant Cell Rep 26:1333–1343

Norin M, Sundström M (2002) Structural proteomics: developments in structure-to-function predictions. Trends Biotechnol 20:79–84

Nouri MZ, Komatsu S (2010) Comparative analysis of soybean plasma membrane proteins under osmotic stress using gel-based and LC MS/MS based proteomics approaches. Proteomics 10:1930–1945

Nozu Y, Tsugita A, Kamijo K (2006) Proteomic analysis of rice leaf, stem and root tissues during growth course. Proteomics 6:3665–3670

Nühse TS, Stensballe A, Jensen ON, Peck SC (2003) Large-scale analysis of *in vivo* phosphorylated membrane proteins by immobilized metal ion affinity chromatography and mass spectrometry. Mol Cell Proteomics 2:1234–1243

Nühse TS, Stensballe A, Jensen ON, Peck SC (2004) Phosphoproteomics of the *Arabidopsis* plasma membrane and a new phosphorylation site database. Plant Cell 16:2394–2405

Nuhse TS, Bottrill AR, Jones AME, Peck SC (2007) Quantitative phosphoproteomic analysis of plasma membrane proteins reveals regulatory mechanisms of plant innate immune responses. Plant J 51:931–940

Oda Y, Huang K, Cross FR, Cowburn D, Chait BT (1999) Accurate quantitation of protein expression and site specific phosphorylation. Proc Natl Acad Sci USA 96:6591–6596

O'Farrell PH (1975) High resolution two-dimensional electrophoresis of proteins. J Biol Chem 250:4007–4021

O'Farrell PZ, Goodman HM, O'Farrell PH (1977) High resolution two-dimensional electrophoresis of basic as well as acidic proteins. Cell 12(4):1133–1141

Orengo CA, Michie AD, Jones S, Jones DT, Swindells MB, Thornton JM (1997) CATH—a hierarchic classification of protein domain structures. Structure 5:1093–1108

Palama T, Menard P, Fock I, Choi Y, Bourdon E, Govinden-Soulange J et al (2010) Shoot differentiation from protocorm callus cultures of *Vanilla planifolia* (Orchidaceae): proteomic and metabolic responses at early stage. BMC Plant Biol 10:82–99

Pan C, Gnad F, Olsen JV, Mann M (2008) Quantitative phosphoproteome analysis of a mouse liver cell line reveals specificity of phosphatase inhibitors. Proteomics 8:4534–4546

Pan Z, Guan R, Zhu S, Deng X (2009) Proteomic analysis of somatic embryogenesis in Valencia sweet orange (*Citrus sinensis* Osbeck). Plant Cell Rep 28:281–289

Pandey A, Mann M (2000) Proteomics to study genes and genomes. Nature 405:837–846

Pandey A, Choudhary MK, Bhushan D, Chattopadhyay A, Chakraborty S, Datta A, Chakraborty N (2006) The nuclear proteome of chickpea (*Cicer arietinum* L.) reveals predicted and unexpected proteins. J Proteome Res 5:3301–3311

Pandey A, Chakraborty S, Datta A, Chakraborty N (2008) Proteomics approach to identify dehydration responsive nuclear proteins from chickpea (*Cicer arietinum* L.). Mol Cell Proteomics 7:88–107

Pandey A, Rajamani U, Verma J, Subba P, Chakraborty N, Datta A et al (2010) Identification of extracellular matrix proteins of rice (*Oryza sativa* L.) involved in dehydration-responsive network: a proteomic approach. J Proteome Res 9:3443–3464

Pang C, Wang H, Pang Y, Xu C, Jiao Y, Qin Y et al (2010) Comparative proteomics indicate that biosynthesis of pectic precursors is important for cotton fiber and *Arabidopsis* root hair elongation. Mol Cell Proteomics 9:2019–2033

Park OK (2004) Proteomic studies in plants. J Biochem Mol Biol 37:133–138

Patterson J, Ford K, Cassin A, Natera S, Bacic A (2007) Increased abundance of proteins involved in phytosiderophore production in boron-tolerant barley. Plant Physiol 144:1612–1631

Pawson T, Nash P (2000) Protein-protein interactions define specificity in signal transduction. Genes Dev 14:1027–1047

Pechanova O, Pechan T, Williams P, Luthe D (2011) Proteomic analysis of the maize rachis: potential roles of constitutive and induced proteins in resistance to Aspergillus flavus infection and aflatoxin accumulation. Proteomics 11:114–127

Pertl H, Schulze WX, Obermeyer G (2009) The pollen organelle membrane proteome reveals highly spatial–temporal dynamics during germination and tube growth of lily pollen. J Proteome Res 8:5142–5152

Purvis AC (1997) Role of the alternative oxidase in limiting superoxide production by plant mitochondria. Physiol Plant 100:165–170

Qiao WH, Fan LM (2008) Nitric oxide signaling in plant responses to abiotic stresses. J Integr Plant Biol 50:1238–1246

Qin GZ, Wang Q, Liu J, Li BQ, Tian SP (2009) Proteomic analysis of changes in mitochondrial protein expression during fruit senescence. Proteomics 9:4241–4253

Quirino BF, Candido ES, Campos PF, Franco OL, Kruger RH (2010) Proteomic approaches to study plant pathogen interactions. Phytochemistry 71:351–362

Rampitsch C, Bykova NV, Mauthe W, Yakandawala N, Jordan M (2006a) Phosphoproteomic profiling of wheat callus labelled in vivo. Plant Sci 171:488–496

Rampitsch C, Bykova NV, McCallum B, Beimcik E, Ens W (2006b) Analysis of the wheat and *Puccinia triticina* (leaf rust) proteomes during a compatible host-pathogen interaction. Proteomics 6:1897–1907

Reeves GA, Talavera D, Thornton JM (2009) Genome and proteome annotation: organization, interpretation and integration. J R Soc Interface 6:129–147

Reiland S, Messerli G, Baerenfaller K, Gerrits B, Endler A, Grossmann J et al (2009) Large-scale Arabidopsis phosphoproteome profiling reveals novel chloroplast

kinase substrates and phosphorylation networks. Plant Physiol 150:889–903
Remmerie N, Roef L, Van De Slijke E, Van Leene J et al (2009) A bioanalytical method for the proteome wide display and analysis of protein complexes from whole plant cell lysates. Proteomics 9:598–609
Repetto O, Rogniaux H, Firnhaber C, Zuber H, Kuster H, Larre C, Thompson R, Gallardo K (2008) Exploring the nuclear proteome of *Medicago truncatula* at the switch towards seed filling. Plant J 56:398–410
Repetto O, Rogniaux H, Larré C, Thompson R, Gallardo K (2012) The seed nuclear proteome. Front Plant Sci 3:289. doi:10.3389/fpls.2012.00289
Revel M, Groner Y (1978) Post-transcriptional and translational controls of gene expression in eukaryotes. Ann Rev Biochem 47:1079–1126
Rinalducci S, Murgiano L, Zolla L (2008) Redox proteomics: basic principles and future perspectives for the detection of protein oxidation in plants. J Exp Bot 59:3781–3801
Romero-Puertas MC, Campostrini N, Matte A, Righetti PG et al (2008) Proteomic analysis of S-nitrosylated proteins in *Arabidopsis thaliana* undergoing hypersensitive response. Proteomics 8:1459–1469
Ronald PC, Beutler B (2010) Plant and animal sensors of conserved microbial signatures. Science 330:1061–1064
Rose JKC, Bashir S, Giovannoni JJ, Jah MM, Saravanan RS (2004) Tackling the plant proteome: practical approaches, hurdles and experimental tools. Plant J 39:715–733
Rost H, Malmstrom L, Aebersold R (2012) A computational tool to detect and avoid redundancy in selected reaction monitoring. Mol Cell Proteomics 11:540–549
Rouhier N (2010) Plant glutaredoxins: pivotal players in redox biology and iron-sulphur centre assembly. New Phytol 186:365–372
Ruebelt MC, Lipp M, Reynolds TL, Schmuke JJ, Astwood JD, Dellapenna D et al (2006) Application of two-dimensional gel electrophoresis to interrogate alterations in the proteome of genetically modified crops. 3. Assessing unintended effects. J Agric Food Chem 54:2169–2177
Salamini F, Leister D (2000) A prediction of the size and evolutionary origin of the proteome of chloroplasts of Arabidopsis. Trends Plant Sci 5:141–142
Saleem M, Lamkemeyer T, Schützenmeister A, Fladerer C, Piepho H, Nordheim A et al (2009) Tissue specific control of the maize (*Zea mays* L.) embryo, cortical parenchyma, and stele proteomes by RUM1 which regulates seminal and lateral root initiation. J Proteome Res 8:2285–2297
Sánchez R, Pieper U, Melo F, Eswar N et al (2000) Protein structure modeling for structural genomics. Nat Struct Biol 7(Suppl):986–999
Sanger F, Nicklen S, Coulson AR (1977) DNA sequencing with chain-terminating inhibitors. Proc Natl Acad Sci USA 74:5463–5467
Sarma AD, Oehrle NW, Emerich DW (2008) Plant protein isolation and stabilization for enhanced resolution of two-dimensional polyacrylamide gel electrophoresis. Anal Biochem 379:192–195
Scheele GA (1975) Two-dimensional gel analysis of soluble proteins. Characterization of guinea pig exocrine pancreatic proteins. J Biol Chem 250:5375–5385
Schuch W, Bird CR, Ray J, Smith CJS (1989) Control and manipulation of gene expression during tomato fruit ripening. Plant Mol Biol 13:303–311
Schulze WX, Usadel B (2010) Quantitation in mass-spectrometry-based proteomics. Annu Rev Plant Biol 61:491–516. doi:10.1146/annurev-arplant-042809-112132
Schwacke R, Flugge UI, Kunze R (2004) Plant membrane proteome database. Plant Physiol Biochem 42:1023–1034
Scossa F, Laudencia-Chingcuanco D, Anderson OD, Vensel WH, Lafiandra D, D'Ovidio R et al (2008) Comparative proteomic and transcriptional profiling of a bread wheat cultivar and its derived transgenic line overexpressing a low molecular weight glutenin subunit gene in the endosperm. Proteomics 8:2948–2966
Sharma N, Hotte N, Rahman MH, Mohammadi M et al (2008) Towards identifying Brassica proteins involved in mediating resistance to *Leptosphaeria maculans*: a proteomics-based approach. Proteomics 8:3516–3535
Shi F, Takasaki H, Komatsu S (2008) Quantitative analysis of auxin-regulated proteins from basal part of leaf sheaths in rice by two-dimensional difference gel electrophoresis. Phytochemistry 69:637–646
Shinano T, Komatsu S, Yoshimur T, Tokutake S, Kong FJ, Watanabe T et al (2011) Proteomic analysis of secreted proteins from aseptically grown rice. Phytochemistry 72:312–320
Simpson JC, Pepperkok R (2003) Localizing the proteome. Genome Biol 4:240
Skylas DJ, Cordwell SJ, Hains PG, Larsen MR, Basseal DJ, Walsh BJ et al (2002) Heat shock of wheat during grain filling: proteins associated with heat-tolerance. J Cereal Sci 35:175–188
Soares NC, Francisco R, Ricardo CP, Jackson PA (2007) Proteomics of ionically bound and soluble extra cellular proteins in *Medicago truncatula* leaves. Proteomics 7:2070–2082
Soares NC, Francisco R, Vielba JM, Ricardo CP, Jackson PA (2009) Associating wound-related changes in the apoplast proteome of Medicago with early steps in the ROS signal-transduction pathway. J Proteome Res 8:2298–2309
Sobhanian H, Razavizadeh R, Nanjo Y, Ehsanpour AA, Jazii FR, Motamed N et al (2010) Proteome analysis of soybean leaves, hypocotyls and roots under salt stress. Proteome Sci 8:19
Song Y, Zhang C, Ge W, Zhang Y, Burlingame AL, Guo Y (2011) Identification of NaCl stress-responsive apoplastic proteins in rice shoot stems by 2D-DIGE. J Proteomics 74:1045–1067

Soreghan BA, Yang F, Thomas SN, Hsu J, Yang AJ (2003) High-throughput proteomic-based identification of oxidatively induced protein carbonylation in mouse brain. Pharm Res 20:1713–1720

Staiger CJ, Poulter NS, Henty JL, Franklin-Tong VE, Blanchoin L (2010) Regulation of actin dynamics by actin-binding proteins in pollen. J Exp Bot 61:1969–1986

Stroher E, Dietz KJ (2006) Concepts and approaches towards understanding the cellular redox proteome. Plant Biol 8:407–418

Sugiyama N, Nakagami H, Mochida K, Daudi A et al (2008) Large-scale phosphorylation mapping reveals the extent of tyrosine phosphorylation in Arabidopsis. Mol Syst Biol 4:193

Takáč T, Pechan T, Richter H, Müller J, Eck C, Böhm N et al (2011) Proteomics on brefeldin A-treated Arabidopsis roots reveals profilin 2 as a new protein involved in the cross-talk between vesicular trafficking and the actin cytoskeleton. J Proteome Res 10:488–501

Tan YF, O'Toole N, Taylor NL, Millar AH (2010) Divalent metal ions in plant mitochondria and their role in interactions with proteins and oxidative stress-induced damage to respiratory function. Plant Physiol 152:747–761

Tang W, Deng Z, Wang Z (2010) Proteomics shed light on the brassinosteroid signaling mechanisms. Curr Opin Plant Biol 13:27–33

Tanou G, Job C, Rajjou L, Arc E et al (2009) Proteomics reveals the overlapping roles of hydrogen peroxide and nitric oxide in the acclimation of citrus plants to salinity. Plant J 60:795–804

Taylor NL, Heazlewood JL, Day DA, Millar AH (2005) Differential impact of environmental stresses on the pea mitochondrial proteome. Mol Cell Proteomics 4:1122–1133

Torres MA (2010) ROS in biotic interactions. Physiol Plant 138:414–429

Torta F, Usuelli V, Malgaroli A, Bachi A (2008) Proteomic analysis of protein S-nitrosylation. Proteomics 8:4484–4494

Tran HT, Plaxton WC (2008) Proteomic analysis of alterations in the secretome of *Arabidopsis thaliana* suspension cells subjected to nutritional phosphate deficiency. Proteomics 8:4317–4326

Van Bentem SD, Anrather D, Dohnal I, Roitinger E et al (2008) Site-specific phosphorylation profiling of *Arabidopsis* proteins by mass spectrometry and peptide chip analysis. J Proteome Res 7:2458–2470

Van Leene J, Hollunder J, Eeckhout D, Persiau G et al (2010) Targeted interactomics reveals a complex core cell cycle machinery in *Arabidopsis thaliana*. Mol Syst Biol 6:397

Ventelon-Debout M, Delalande F, Brizard J-P, Diemar H et al (2004) Proteome analysis of cultivar-specific deregulations of *Oryza sativa indica* and *O. sativa japonica* cellular suspensions undergoing IT Rice yellow mottle virus IT infection. Proteomics 4:216–225

Vierstra RD (2009) The ubiquitin–26S proteasome system at the nexus of plant biology. Nat Rev Mol Cell Biol 10:385–397

Villeth GR, Reis FB, Tonietto A, Huergo L et al (2009) Comparative proteome analysis of *Xanthomonas campestris* pv. *campestris* in the interaction with the susceptible and the resistant cultivars of *Brassica oleracea*. FEMS Microbiol Lett 298:260–266

Vítámvás P, Prášil IT (2008) WCS120 protein family and frost tolerance during cold acclimation, deacclimation and reacclimation of winter wheat. Plant Physiol Biochem 46:970–976

Vítámvás P, Saalbach G, Prášil IT, Čapková V, Opatrná J, Jahoor A (2007) WCS120 protein family and proteins soluble uponboiling in cold-acclimated winter wheat. J Plant Physiol 164:1197–1207

Wang Y, Yang L, Xu H, Li Q et al (2005) Differential proteomic analysis of proteins in wheat spikes induced by *Fusarium graminearum*. Proteomics 5:4496–4503

Wang MC, Peng ZY, Li CL, Li F, Liu C, Xia GM (2008) Proteomic analysis on a high salt tolerance introgression strain of *Triticum aestivum/Thinopyrum ponticum*. Proteomics 8:1470–1489

Wasinger VC, Cordwell SJ, Cerpa-Poljak A, Yan JX, Gooley AA, Wilkins MR, Duncan MW, Harris R, Williams KL, Humphery-Smith I (1995) Progress with gene-product mapping of the Mollicutes: *Mycoplasma genitalium*. Electrophoresis 16(7):1090–1094

Watson JD, Crick FH (1953) Molecular structure of nucleic acids; a structure for deoxyribose nucleic acid. Nature 171:737–738

Watson BS, Lei Z, Dixon RA, Sumner LW (2004) Proteomics of *Medicago sativa* cell walls. Phytochemistry 65:1709–1720

Wild DL, Saqi MAS (2004) Structural proteomics: inferring function from protein structure. Curr Proteomic 1:59–65

Wright JC, Sugden D, Francis-McIntyre S, Riba-Garcia I, Gaskel SJ, Grigoriev IV, Baker SE, Beynon RJ, Hubbard SJ (2009) Exploiting proteomic data for genome annotation and gene model validation in *Aspergillus niger*. BMC Genomics 10:61. doi:10.1186/1471-2164-10-61

Wright PC, Noirel J, Ow SY, Fazeli A (2012) A review of current proteomics technologies with a survey on their widespread use in reproductive biology investigations. Theriogenology 77:738–765.e52

Wu CC, Yates JR (2003) The application of mass spectrometry to membrane proteomics. Nat Biotechnol 21:262–267

Xu Y, Gianfagna T, Huang B (2010) Proteomic changes associated with expression of a gene (ipt) controlling cytokinin synthesis for improving heat tolerance in a perennial grass species. J Exp Bot 6:3273–3289

Yakunin AF, Yee AA, Savchenko A, Edwards AM, Arrowsmith CH (2004) Structural proteomics: a tool for genome annotation. Curr Opin Chem Biol 8:42–48

Yan S, Tang Z, Su W, Sun W (2005) Proteomic analysis of salt stress-responsive proteins in rice root. Proteomics 5:235–244

Yanagida M (2002) Functional proteomics; current achievements. J Chromatogr B 771:89–106

Yang Y, Bian S, Yao Y, Liu J (2008) Comparative proteomic analysis provides new insights into the fiber elongating process in cotton. J Proteome Res 7:4623–4637

Yano H, Kuroda M (2006) Disulfide proteome yields a detailed understanding of redox regulations: a model study of thioredoxin-linked reactions in seed germination. Proteomics 6:294–300

Yano H, Kuroda S (2008) Introduction of the disulfide proteome: application of a technique for the analysis of plant storage proteins as well as allergens. J Proteome Res 7:3071–3079

Yin L, Lan Y, Zhu L (2008) Analysis of the protein expression profiling during rice callus differentiation under different plant hormone conditions. Plant Mol Biol 68:597–617

Yook SH, Oltvai ZN, Barabási AL (2004) Functional and topological characterization of protein interaction networks. Proteomics 4:928–942

Young NL, Plazas-Mayorca MD, Garcia BA (2010) Systems-wide proteomic characterization of combinatorial post-translational modification patterns. Expert Rev Proteomics 7:79–92

Ytterberg AJ, Jensen ON (2010) Modification-specific proteomics in plant biology. J Proteomics 73:2249–2266

Zhang MH, Li GW, Huang W, Bi T, Chen GY, Tang ZC et al (2010) Proteomic study of *Carissa spinarum* in response to combined heat and drought stress. Proteomics 10:3117–3129

Zhang Y, Giboulot A, Zivy M, Valot B, Jamet E, Albenne C (2011) Combining various strategies to increase the coverage of the plant cell wall glycoproteome. Phytochemistry 72:1109–1123

Zhao Y, Jensen ON (2009) Modification-specific proteomics: strategies for characterization of post-translational modifications using enrichment techniques. Proteomics 9:4632–4641

Zhao P, Wang L, Han L, Wang J, Yao Y, Wang H et al (2010) Proteomic identification of differentially expressed proteins in the Ligon lintless mutant of upland cotton (*Gossypium hirsutum* L.). J Proteome Res 9:1076–1087

Zhou W, Eudes F, Laroche A (2006) Identification of differentially regulated proteins in response to a compatible interactions between the pathogen *Fusarium graminearum* and its host *Triticum aestivum*. Proteomic 6:4599–4609

Zhou L, Bokhari SA, Dong CJ, Liu JY (2011) Comparative proteomics analysis of the root apoplasts of rice seedlings in response to hydrogen peroxide. PLoSONE 6:e16723. doi:10.1371/journal.pone.0016723

Zhou Z, Licklider LJ, Gygi SP, Reed R (2002) Comprehensive proteomic analysis of the human spliceosome. Nature 419(6903):182–185

Zhu J, Chen S, Alvarez S, Asirvatham VS, Schachtman DP, Wu Y et al (2006) Cell wall proteome in the maize primary root elongation zone. I. Extraction and identification of water-soluble and lightly ionically bound proteins. Plant Physiol 140:311–325

Zhu J, Alvarez S, Marsh EL, Lenoble ME, Cho IJ, Sivaguru M et al (2007) Cell wall proteome in the maize primary root elongation zone. II. Region-specific changes in water soluble and lightly ionically bound proteins under water deficit. Plant Physiol 145:1533–1548

Zulak KG, Khan MF, Alcantara J, Schriemer DC et al (2009) Plant defense responses in *opium poppy* cell cultures revealed by liquid chromatography-tandem mass spectrometry proteomics. Mol Cell Proteomics 8:86–98

Plant Metabolomics: An Overview of Technology Platforms for Applications in Metabolism

Neelam S. Sangwan, Pragya Tiwari, Siddhartha Kumar Mishra, Ritesh K. Yadav, Swati Tripathi, Amit K. Kushwaha, and Rajender Singh Sangwan

Contents

N.S. Sangwan, Ph.D. (✉) • P. Tiwari, Ph.D.
R.K. Yadav, Ph.D. • A.K. Kushwaha, Ph.D.
Department of Metabolic and Structural Biology, CSIR-Central Institute of Medicinal and Aromatic Plants, P.O. CIMAP, Lucknow 226015, India
e-mail: nsangwan5@gmail.com; nss.cimap@gmail.com

S.K. Mishra, Ph.D.
School of Biological Sciences, Dr. Harisingh Gour Central University, Sagar, India

S. Tripathi, Ph.D.
Vegetable Research Division, National Institute of Horticultural and Herbal Science, Rural Development Administration, Suwon, Republic of Korea

R.S. Sangwan, Ph.D.
Department of Metabolic and Structural Biology, CSIR-Central Institute of Medicinal and Aromatic Plants, P.O. CIMAP, Lucknow 226015, India

Center for Innovative and Applied Bioprocessing (formerly BioProcessing Unit), (An Autonomous Institute under Department of Biotechnology, Govt. of India), Mohali 140306, Punjab, India

D. Barh et al. (eds.), *PlantOmics: The Omics of Plant Science*,
DOI 10.1007/978-81-322-2172-2_9, © Springer India 2015

Abstract

The science of plant metabolomics has revolutionised the underlying platform of research on biological systems. The integration of plant metabolomics with allied branches, namely, system biology, biostatistics and in silico biology, has emerged as significant advancement in comprehensive metabolic profiling of phytomolecules in plant. Further, metabolomics strategies with high level of compositional specificity dealing with both intra- and inter-level organisation of the organism have achieved the predefined standards with overwhelming response. The present chapter entails the achievements and challenges associated with research on plant metabolomics, beginning with a brief introduction on plant metabolomics as interdisciplinary field and recent approaches employed for profiling of plant metabolites. Further, the chapter describes the omics strategies involved in metabolome research and various analytical techniques employed for detection and quantification of plant metabolites. The application ranging from metabolic engineering to abiotic stress response and from in vitro studies to application in functional genomics highlights the rising significance of plant metabolomics as a functional genomics tool. Although studies on plant metabolome have emerged as prospective strategies and contributed tremendously to its growth in recent years, some of the major challenges associated need to be addressed for the science of plant metabolomics to flourish and strengthen in future. Metabolomics is a relatively newer approach aimed at improving understanding of metabolic networks and the subsequent biochemical compositions of the plants and other biological organisms. The important aspects relevant to metabolomics are presented, and perspectives of metabolomics exploitation in the future are outlined. As such, metabolomics is providing new dimensions in the study of systems biology, enabling the in-depth understanding of the intra- and extracellular interactions of plant cells. Metabolomics is also developing into a valuable tool that can be used to monitor and assess gene function and to characterise post-genomic processes from a broad perspective.

Keywords

Bioinformatics databases • Fourier transform ion cyclotron resonance MS (FT-ICR-MS) • Functional genomics • In vitro studies • Plant metabolomics • Metabolic engineering • Statistical analysis

Plant Metabolomics

Integrated biology as a science incorporating 'omics' approaches of analysis and inferences is fast emerging as next level of understanding biological processes collectively as a system of functioning and response. This approach of integrating metabolomics, genomics, proteomics, transcriptomics, ionomics, etc. applies to all biological kingdoms: plants, animals and microbes. However, these systems also differ substantially in their complexity of organisation and micro-level compositions. Therefore, metabolomics of plants, animals and microbes may in essence require the same level of technology platforms but involves specificity. This specificity is most relevant for metabolomics because of compositional specificities of not only organism types but also due to significant differences among families, genera, species, subspecies and at times members. Further, from that perspective, plants are most complex as they differ most in terms of their metabolome particularly secondary metabolome.

Plant metabolomics combines resolution and identification strategies employing sophisticated analytical techniques together with statistical analysis procedures for quantification of cellular metabolic profile and their relationships (Fig. 1). A metabolome truly includes total metabolites produced by an organism during lifetime. Therefore, it is impractical to sample it all times and in all situations. Nevertheless, a reference

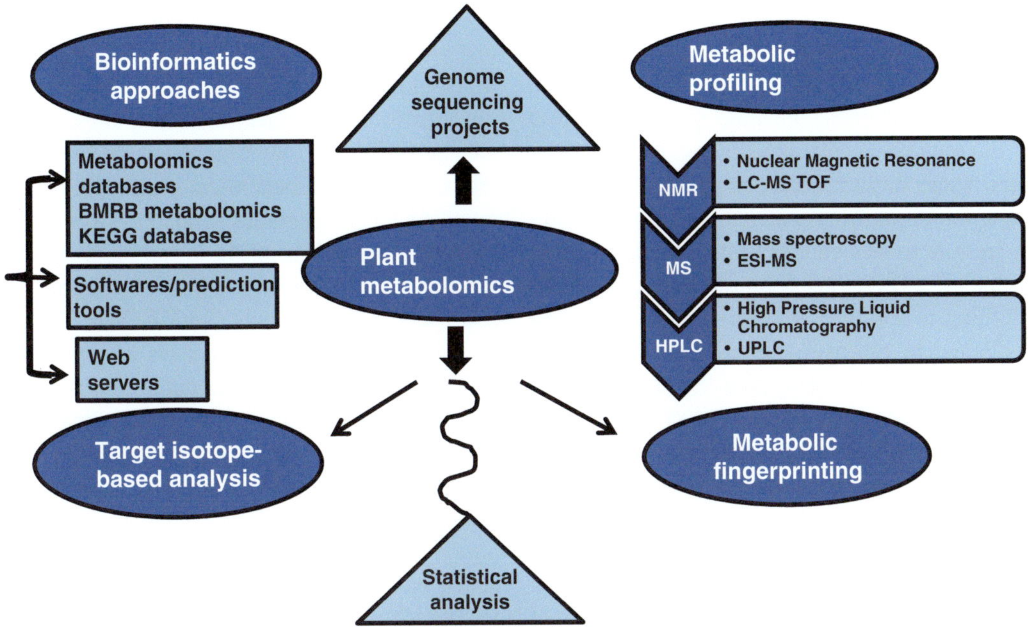

Fig. 1 General schematic representation of techniques utilised in plant metabolomics

metabolome can be practiced to identify variations among known metabolites and expansion of the library by addition of new entities detected and identified. Metabolomics also defines the biochemical networks and the role of a particular metabolite in a biological entity. Further, observations from high-throughput approaches employed for determination of expression patterns of genes (transcriptomics) and identification/quantification of proteins (proteomics) are also considered in manifesting the changes in metabolomics and functioning and response of the biological systems through the statistical applications of correlations and integrations (Schuhmacher et al. 2013). The study of plant metabolomics comprises of the analytical techniques, namely, metabolic profiling, the quantitative estimation of a group of secondary metabolites originating from a particular metabolic pathway (Dunn et al. 2005; Roessner et al. 2001); metabolic fingerprinting, which quantify the complete profiles of a subset of metabolites (Ryan and Robards 2006); and target isotope-based analysis, which aims at the analysis of a specific group of intermediary metabolites of a particular biochemical pathway (Boros et al. 2005). Plant metabolomics is an extensive field encompassing additive inclusion of more than 200,000 secondary metabolites (Fiehn 2002).

The rising popularity of plant metabolomics can be attributed to the fact that it provides an integrated and high-throughput networked linking of metabolites, striking a shift from the conventional phytochemistry or biochemical estimations of only selected core metabolites and thereby restricting to short focus to a big subject.

Primary Plant Metabolomics

Plants produce vast number of chemicals which exhibits chemical diversity as well as complexity among members of a chemical group. They are collectively referred to as plant metabolites. A large number of these chemicals are essential for growth, survival and propagation (reproduction) of the plant under normal environmental conditions as such or by combination/multimerisation within themselves and/or with other molecules (Table 1). They are, therefore, designated as primary metabolites. Carbohydrates, amino acids, peptides, proteins, lipids, nucleotides, hormones, etc. constitute the major categories of primary metabolites (Table 1). The studies focusing on plant metabolism constitute an integral aspect of the science of plant physiology and biochemistry. These studies involve analysing the dynamics of variations in the content of these metabolites in specific tissues under different environmental conditions. Therefore, comprehensive metabolic profiling at a single instance forms the biggest challenge in estimation of diversified metabolites. Because the nature and characteristics of the compounds differ, there is no one method or approach that can facilitate measurement of all the chemicals, different analytical techniques are employed for measuring metabolites based on their principle of identification as well as the sensitivity, accuracy and resolution of the techniques. Nevertheless, comprehensive profiling of metabolites still remains a major challenge.

Carbohydrates

Carbohydrates constitute a major group of compounds in plants. They have been subclassified into groups such as monosaccharides, disaccharides, oligosaccharides and polysaccharides. The mono-, di- and trisaccharides are low-molecular-weight molecules and are referred to as sugars.

Monosaccharides

Monosaccharides are the simplest carbohydrates, structurally defined as aldehydes or ketones with two or more hydroxyl groups. The classification of monosaccharides is based on three parameters: the number of carbon atoms present, location of carbonyl group and chirality. Monosaccharides with three carbon atoms are known as trioses, four are tetroses, five are pentoses, six are hexoses, etc. If the carbonyl group is an aldehyde, it is aldose, and if the carbonyl group is a ketone, then the molecules are referred to as ketose. Except for the first and the last carbon (asymmetric), each possesses a hydroxyl group. These result in chirality with the occurrence of stereoisomers in R or S form. Monosaccharides serve significant function as building blocks for nucleic acid and fuel molecules in cellular mechanisms (glucose). Further, these molecules serve as raw material in biosynthesis of polysaccharides. Metabolomics of carbohydrates in principle does not involve measurement of merely these sugars and their polymers, but it must include intermediates of different biochemical pathways that are involved in the biosynthesis (anabolism) and degradation (catabolism) of these carbohydrates and their conjugates. Therefore, it involves extraction and analysis of metabolic intermediates of several biochemical pathways such as glycolysis, HMP pathway, TCA cycle, starch and cellulose synthesis, photosynthetic carbon reduction cycle, trehalose synthesis, etc. Since the nature of these intermediates vary widely not only as functional groups but also as their conjugates (like phosphorylated metabolites, nucleotide sugars, fatty acids, etc.), measurement approaches and platforms also vary. Therefore, such metabolomics is not only important for understanding physiology of carbohydrates but also of the carbohydrate metabolism-based intermediate that participates in different metabolic pathways of the plant.

Disaccharides and Trisaccharides

Two and three monosaccharide units joined through a glycosidic linkage are termed as a disaccharide and trisaccharides, respectively. These are the simplest polysaccharides, the chemical formula is $C_{12}H_{22}O_{11}$, and examples include lactose, maltose, sucrose, cellobiose and trehalose. The metabolomics of carbohydrates involving these disaccharides and trisaccharides must entail analysing the products of sequence of reactions that are related to the biosynthesis and degradation of these di- and trisaccharides.

Polysaccharides

Polysaccharides are long chains of monosaccharides joined together by glycosidic bonds. They show variation in structures ranging from linear to branched chain structure. Examples include structural polysaccharides such as chitin or cellulose and storage polysaccharides such as starch and glycogen. The general chemical formula is $Cx(H_2O)y$, where x is between 200 and 2,500. These molecules are important class of biopolymers playing significant role in structural architecture or storage of food reserves for the

Table 1 Classification and functions of primary and secondary metabolites in plants

Metabolites	Classification	Biosynthesis	Functional role	Examples	References
Carbohydrates	Monosaccharides	Condensation of amino acids	Building blocks for nucleic acid, fuel molecules in cellular processes	D- and L-glyceraldehydes	Berridge and Irvine (1989) and Coleman and Lee (2004)
				Dihydroxyacetones	
	Disaccharides	Glycosidic linkages of monosaccharides	Sucrose acts as energy reserve in cells; lactose is a nutritive component of milk	Maltose – malt sugar	
				Lactose – milk sugar	
				Sucrose – cane sugar	
	Polysaccharides	2–10 disaccharides joined by glycosidic bonds	Structural components in cells, food deposits	Starch	
				Cellulose	
				Glycogen	
Proteins		Translation	DNA replication, catalyse metabolic reactions, transporter molecules	Amylase	
				Lectins	
				Haemoglobin	
Lipids	Glycerophospholipids	Fatty acid synthesis	Component of lipid bilayer cellular signalling, cellular metabolism	Phosphatidylcholine (lecithin)	Berridge and Irvine (1989)
				Phosphatidylserine	
				Phosphatidylserine	
	Glycerolipids		Energy reserves	Digalactosyldiacylglycerols	Coleman and Lee (2004)
	Sterol lipids		Anti-inflammatory, biomarker for algae, reduces cholesterol levels in blood	Campesterol	
				Brassicasterol	
				Beta-sitosterol	
	Prenol lipids		Transport of oligosaccharides across membranes	Diphosphates	
				Linear alcohol	
	Sphingolipids		Cell signalling in membranes	Ceramides	
	Saccharolipids		Component of membrane bilayer	Acylated glucosamine	
	Polyketides		Antimicrobial, antiparasitic and anticancer	Erythromycins	

(continued)

Table 1 (continued)

Metabolites	Classification	Biosynthesis	Functional role	Examples	References
Flavonoids	Flavonol	Phenylpropanoid pathway	Plant pigment	Kaempferol	Croteau et al. (2000)
				Myricetin	
				Quercetin in *Allium cepa*	
	Anthocyanidins		Plant pigment, fruit and flower colour	Pelargonidin in *Pelargonium graveolens*	
				Cyanidin in *Rosa* spp.	
	Flavan-3-ol (flavanol)		Anticancer	Epicatechin	
				Catechin in *Thea sinensis*	
	Flavanone		Cardiovascular disorder	Naringenin	
				Hesperetin in *Citrus* spp.	
	Flavanonols		Anticancer	Dihydrokaempferol	
				Dihydroquercetin	
	Flavan-3,4-diol		Intermediates in anthocyanidin biosynthesis	Leucoanthocyanidin	
	Isoflavonoids		Antioxidant and anthelmintic	Daidzein	
			Phytoestrogen precursor of pterocarpins	Genistein in *Glycine max*	
	Chalcones		Antituberculosis	Xanthoangelol from *Angelica keiskei*	
Steroids	Sterols	Mevalonate pathway	Constituents of cell membranes	Ergosterol	Tuli et al. (2009)
				Beta-sitosterol	
	Steroids		Anti-inflammatory	Withaferin A	Chaurasiya et al. (2007) and Sangwan et al. (2008)
				Withanolides in *Withania somnifera*	
	Saponins		Antifungal	Gymnemic acids	Tiwari et al. (2013)
			Antidiabetic	Gymnemasaponins	
				Gymnesins in *Gymnema sylvestre*	

Terpenoids	Monoterpenes	MEP and DOXP pathway	Anaesthetic and counterirritant	Menthol	
				Camphor	
				Limonene	
	Diterpenes		Phytohormone	Gibberellins	
	Triterpenes		Phytohormone	Brassinosteroids	
	Sesquiterpenes		Antimalarial	Artemisinin from *Artemisia annua*	Yadav et al. (2014a, b)
	Tetraterpenes		Phytohormone	Abscisic acid	
Cyanogenic glucosides		Biosynthesised from L-valine, L-phenylalanine, L-isoleucine, L-leucine and cyclopentenyl-glycine	Chemical defence system	Cyanogenic glucosides (dhurrin) in *cassava*	
			Plant–insect interaction		
Alkaloids	True alkaloids	Shikimate pathway	Anticholinergic	Nicotine	
			Antibacterial – berberine	Atropine	
			Antihypertension – reserpine	Morphine	
			Spasmolysis – atropine Anticancer –vincristine		
			Antimalarial drug – quinine		
	Polyamine alkaloids			Spermidine	
				Spermine	
				Putrescine	
	Tropane alkaloids			Tropine from *Withania somnifera*	Kushwaha et al. (2013, 2014)
				Pseudotropine	
	Pseudoalkaloids			Caffeine	
				Theobromine	
				Theacrine	
	Peptide and cyclopeptide alkaloids				

cell. Starch is a polymer of glucose, occurs as storage polysaccharide in plants and exists in the form of amylase and branched amylopectin in plant. Other examples of polysaccharides are galactomannan, xylan, arabinoxylan, mannan, fucoidan, laminarin and chrysolaminarin which are found selectively in the plants and often exhibit bioactivities. Metabolomics of polysaccharides includes not only analysis of these end products but also basis of their variation in content and physiological roles. This necessitates understanding the variations in the content of their conjugative molecules, metabolites involved in their storage, structural organisation as well as regulated degradation.

Amino Acids and Proteins

Proteins are large, macromolecules formed by amino acids, the building blocks of proteins. There are 20 amino acids constituting a block of monomers. In addition, there are several nonprotein amino acids that have a specialised role. Metabolomics of amino acids and proteins involves not only separation of the amino acids and proteins but also their identification and estimation of the levels of metabolites that contribute to the level of amino acids. The later involves metabolic intermediate of the pathways of the synthesis and degradation of amino acids and their corresponding carboxylic acids and amines. Amino acids and proteins are resolved and analysed by different techniques; the former involve an LC-MS approach and/or dedicated amino acid analyser, while the latter are analysed by electrophoresis or capillary electrophoresis and identified by MALDI-TOF-TOF and/or LC-MS-MS.

Lipids

Lipids constitute a class of macromolecule playing a pivotal role in cellular processes and are amphiphilic or hydrophobic molecules and include sterols, fats, waxes, phospholipids, fat-soluble vitamins (vitamins A, D, E and K), monoglycerides, diglycerides and triglycerides. Lipids are broadly classified in mainly eight classes: glycerolipids, glycerophospholipids, sphingolipids, fatty acids, saccharolipids, polyketides (derived from ketoacyl subunits condensation), prenol lipids (isoprene unit condensation) and sterol lipids, respectively. Acylglycerols are the lipids composed of fatty acids conjugated with glycerol by an ester bond at one or more of the hydroxyl groups of glycerol and their derivatives. The following is a brief introductory description of the different lipid classes.

Glycerophospholipids

Glycerophospholipids, also referred to as phospholipids, are universal in occurrence and form an integral component of lipid bilayer, cellular signalling and cellular metabolism (Berridge and Irvine 1989). Common examples of glycerophospholipids are phosphatidylcholine (lecithin) and phosphatidylserine.

Glycerolipids

Glycerolipids are composed of three hydroxyl groups of glycerol esterified to different fatty acids, also known as mono-, di- and trisubstituted glycerols (Coleman and Lee 2004). Functionally, they serve as storage deposits in animal tissues and provide cellular energy during pathway reactions. Other subclasses of glycerolipids include glycosylglycerols (one or more sugar residues attached to glycerol through glycosidic linkage), for example, digalactosyldiacylglycerols found in plant membranes (Holzl and Dormann 2007).

Sterol Lipids

Sterol lipids are an essential component of cell membranes together with sphingomyelins and glycerophospholipids, for example, cholesterol and derivatives. Ergosterol is the predominant sterol in fungal cell membranes (Deacon 2005), while plant membranes are composed of campesterol, brassicasterol and beta-sitosterol, respectively. Metabolomics of lipids is a vast field due to chemically diverse nature of different lipids. This entails different pathways like MVA pathway, DOXP pathway, fatty acid synthesis, glycerol production and conjugation of the primary lipid metabolites with phosphate, sulphate, carbohydrates, etc., and therefore, metabolomics demands estimation of the intermediates of all these metabolic pathways and processes.

Plant Secondary Metabolomics

Secondary metabolites can be defined as the group of low-molecular-weight compounds, generally produced in plant besides the primary biosynthetic networks of carbohydrate, lipid and protein metabolism. These are not regarded as essential for plant growth and development but perform significant functions in certain essential biochemical pathways leading to production of bioactive compounds in plant. The biosynthesis of specific secondary metabolites has been favoured during the course of evolution addressing different needs of plant systems from volatiles and pigments responsible for floral scent and thereby pollination or antimicrobial activity against pathogen attack. Different enzymes have evolved to catalyse similar but not identical function catering with the need of specific plant lineages. Furthermore, these confer the plant with several important properties for better adaptability and survival such as defence mechanisms (alkaloid toxins), pollination (pigments), signalling molecules (chemoattractants), drug molecules (antibiotics) and food additives (sweeteners). The secondary metabolites are generally significant natural products and confer pharmacological properties to the plant. Several classes of secondary metabolites serve other important functions, namely, alkaloids as phytoalexins, flavonoids as free radical scavengers and terpenoids as insect pollinators. Others may be involved in cellular signalling or antimicrobial activity. Several classes of secondary metabolites, namely, terpenoids, flavonoids, steroids, alkaloids, phenylpropanoids, cyanogenic glucosides, etc., exist in plant and are classified based on their chemical structure as well as biological properties (Tables 1, 2, and 3). Some of the significant classes are discussed as under.

Flavonoids

Flavonoids are polyphenolic compounds, ubiquitously present in the plant kingdom as secondary metabolites and account for more than 4,500 representatives. Flavonoids are synthesised by the phenylpropanoid pathway which begins with the conversion of phenylalanine to 4-coumaroyl-CoA (Ververidis et al. 2007). It combines with malonyl-CoA to form chalcones (contain two phenyl rings) and forms the backbone of flavonoids. The next step leading to ring closure results in the formation of flavonoids. The enzymatic reactions proceed to form flavanones, dihydroflavanols and anthocyanins. Many products like flavonols, flavan-3-ols, proanthocyanidins (tannins) and other polyphenolics are formed during enzymatic modification. The flavonoids are subdivided into subclasses, namely, flavones, flavonol, flavanone, flavanonol, flavans, anthocyanidins, chalcones, aurones, catechins and isoflavanoids, respectively (Table 1).

The presence of flavonoids in plant promotes plant–animal interactions. Examples include the cyanidins, pelargonidins and delphinidins which attract pollinators and act as seed dispersers (Croteau et al. 2000). These are important plant pigments for flower coloration thereby promoting pollination; they may also act as chemical messengers and cell cycle inhibitors and in floral pigmentation and symbiotic nitrogen fixation. Furthermore, flavonoids have been found to exhibit diverse pharmacological properties. Some flavonoids act as insect feeding attractants (isoquercetin in mulberry), bitterness in plants (proanthocyanidins) and signalling molecules in bacterial interactions (luteolin and apigenin) and as antifungal agents (isoflavonoids), respectively (Croteau et al. 2000).

Steroids

Steroid molecules are found in plants, animals and fungi, and all steroids are formed from cycloartenol (in plants) and from lanosterol (animal and fungi). The structure consists of 17 carbon atoms joined together in the form of four fused

Table 2 Detailed description of flavonoid classification, their chemical structures and examples under different subclasses, respectively

Classes of flavonoids	Description	Chemical structure	Functional groups	Examples	Reference
Flavonol or 3-hydroxyflavone	3-Hydroxy-2-phenylchromen-4-one	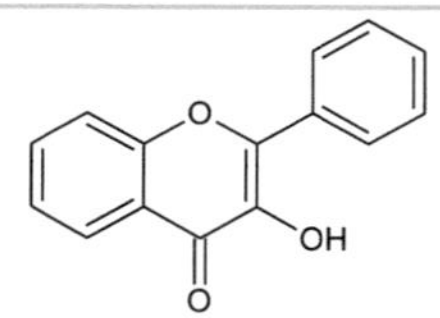	Hydroxyl group	Kaempferol, myricetin, fisetin, quercetin, rhamnazin, pyranoflavonols, furanoflavonols	Croteau et al. (2000)
Anthocyanidins	Aglycones of anthocyanins	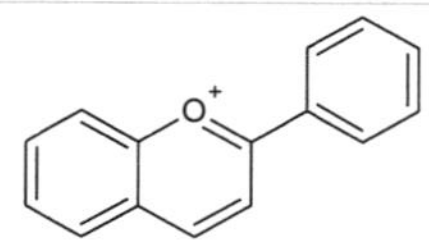	Flavylium (2-phenylchromenylium) ion skeleton	Pelargonidin, peonidin, petunidin, cyanidin, delphinidin, malvidin	
Flavone	2-Phenylchromen-4-one	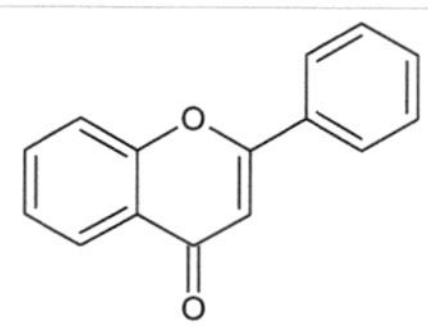	–	Apigenin, luteolin, tangeritin	
Isoflavonoids		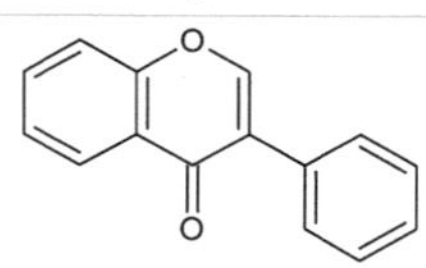	3-Phenylchromen-4-one skeleton	Daidzein, glycitein, genistein	
Flavans					
Flavan-3-ol (flavanol)		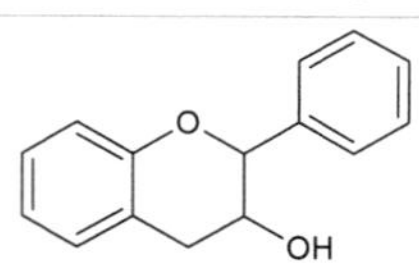	2-Phenyl-3,4-dihydro-2H-chromen-3-ol skeleton	Epicatechins, epigallocatechin, catechin, gallocatechin, catechin-3-gallate	

Flavan-4-ol	2-Phenylchroman-4-ol	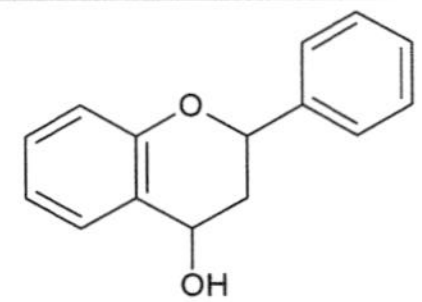	–	Apiforol, luteoforol
Flavan-3,4-diol	2-Phenyl-3,4-dihydro-2H-chromene-3,4-diol	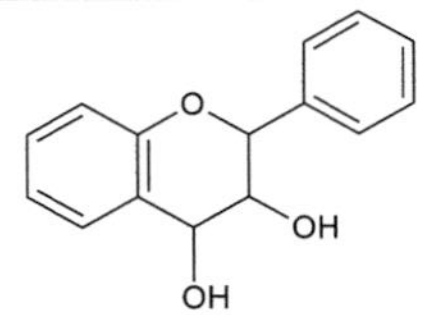	–	Leucoanthocyanidin
Flavanonols or 3-hydroxyflavanone	3-Hydroxy-2,3-dihydro-2-phenylchromen-4-one	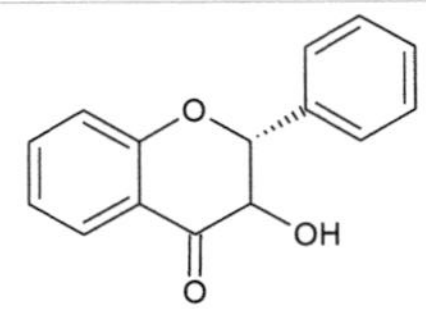	–	Dihydrokaempferol, dihydroquercetin
Flavanone	2,3-Dihydro-2-phenylchromene-4-one	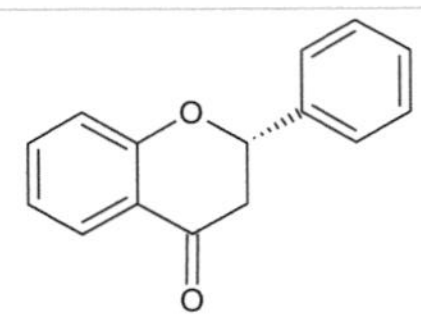	–	Naringenin, eriodictyol, homoeriodictyol, hesperetin

Further, flavonoids comprise of the largest study class of secondary metabolites found ubiquitously in the plant kingdom

Table 3 A summary of sterols, their occurrence in the plant kingdom and their functional properties

Sterols	Chemical structure	Occurrence/functions	References
Cholesterol or (3β)-cholest-5-en-3-ol	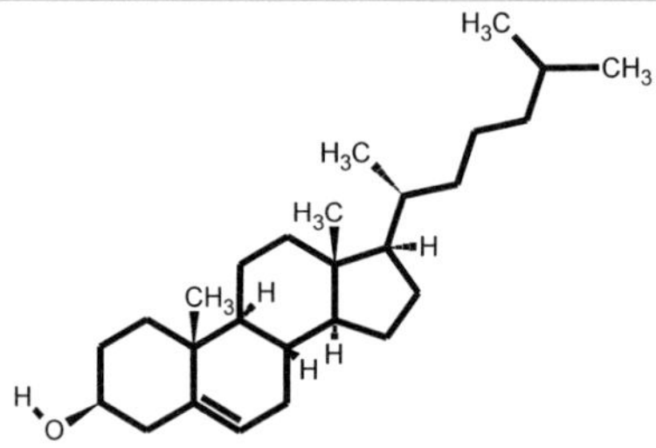	Present in animal cell membranes, essential for membrane permeability and fluidity. Also, precursor for steroid hormones, vitamin D and bile acids	Tuli et al. (2009)
Ergosterol or ergosta-5,7,22-trien-3β-ol	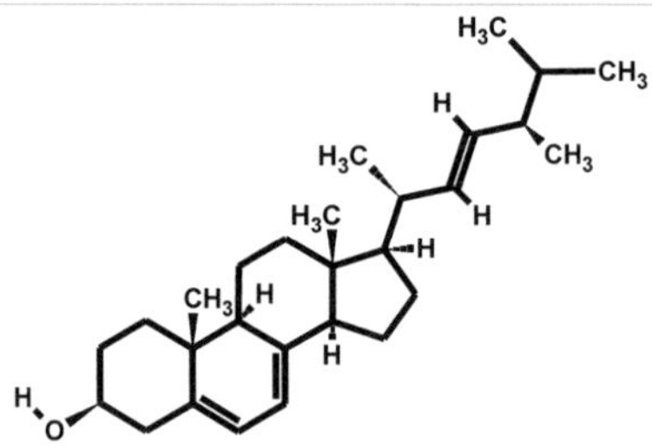	Found in fungi, component of fungal and yeast cell membranes. Precursor of vitamin D2, target for antifungal drugs, for example, *Claviceps* fungus, rye and alfalfa	Deacon (2005)
Campesterol or campestanol; (24*R*)-ergost-5-en-3β-ol	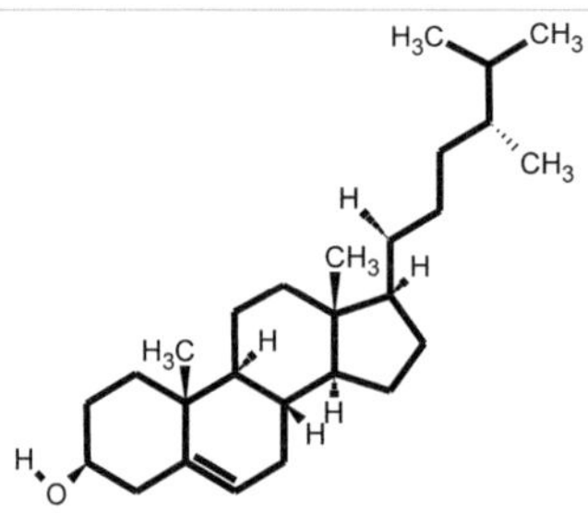	Found in fruits, vegetables and nuts, anti-inflammatory, inhibits mediators causing matrix degradation in osteoarthritis, precursor of anabolic boldenone, for example, lemon grass (citronella)	
Beta-sitosterol or 22,23-dihydrostigmasterol	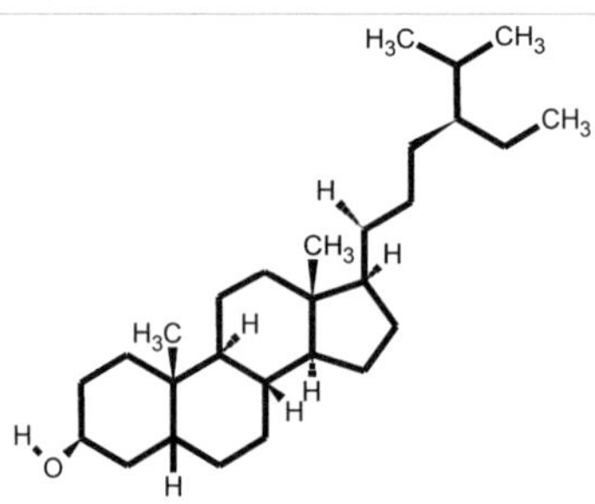	Widely distributed in the plant kingdom, reduces cholesterol levels in blood and precursor of anabolic steroid boldenone, for example, cashew fruits, pumpkin seeds, etc.	
Brassicasterol or ergosta-5,22-dien-3β-ol	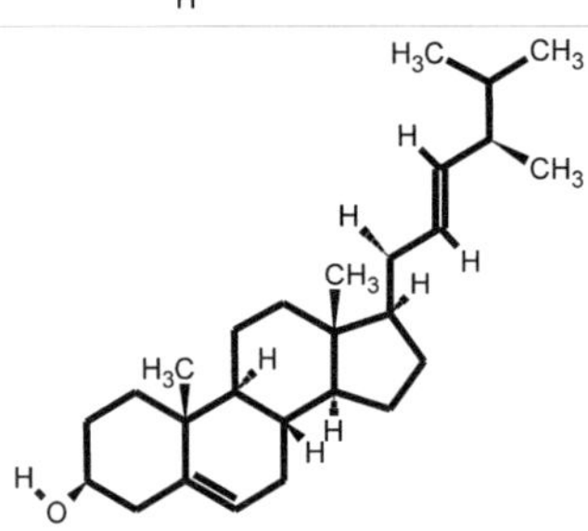	28-Carbon sterol synthesised by unicellular algae and some terrestrial plants, for example, oilseed rape, used as biomarker for the presence of algae, for example, *Mirabilis jalapa*	

(continued)

Table 3 (continued)

Sterols	Chemical structure	Occurrence/functions	References
Stigmasterol	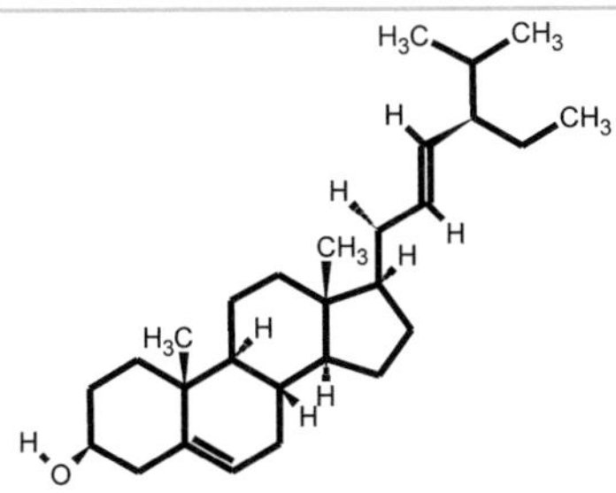	Unsaturated plant sterol found in plants, precursor of semisynthetic progesterone and vitamin D3, used in prostate, ovarian, colon and breast cancer, respectively	
Avenasterol	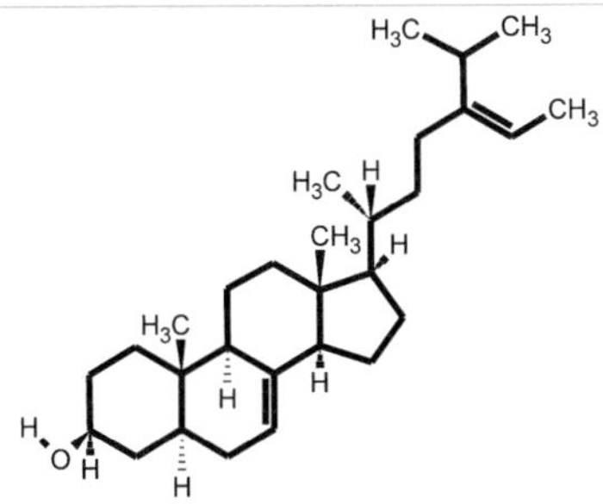	Natural, non-cholesterol-containing sterol	

rings: three cyclohexane rings and one cyclopentane ring. The steroids differ from each other by functional groups present in fused core structure and by the oxidation state of the rings. Functionally, steroids and their metabolites are signalling molecules (steroid hormones) and components of cell membranes (along with phospholipids) and serve as energy reserves for cellular metabolism. The biosynthesis of steroids is an anabolic process occurring through the mevalonate pathway (cytoplasm) in animals employing acetyl-CoA as starting material to form dimethylallyl pyrophosphate (DMAPP) and isopentenyl pyrophosphate (IPP). In plants and bacteria, non-mevalonate/DOXP pathway operates using pyruvate and glyceraldehydes 3-phosphate as substrates.

Sterols

Sterols or steroidal alcohols are an important class of organic molecules and found in fungi, plants and animals (Table 3). Phytosterols are found in plant, for example, stigmasterol, beta-sitosterol and campesterol, while cholesterol is an important zoosterol present in cell membranes. Ergosterol is present in cell membranes of fungi and performs similar function as cholesterol. The biosynthesis of sterols is from acetyl-CoA via the HMG-CoA pathway (Fig. 2). Functionally, sterols play an important role in cell physiology. Further, phytosterols have shown to block cholesterol absorption sites in the human intestine in clinical trials thereby promoting cholesterol-lowering effect. In some plants, sterols occur as glycosides and known as saponins which account for the foaming effect in the plant. The examples include diosgenin (present in *Dioscorea)*, a steroidal sapogenin. Solasodine is a steroidal alkaloid found in family Solanaceae, in potato and tomato (Tuli et al. 2009).

Cyanogenic Glycosides

One of the most extensive classes of secondary metabolites is cyanogenic glycosides. These metabolites can be defined as glycosides of alpha-hydroxynitriles and are stored in plant vacuoles (Vetter 2000). More than 2,500 plant families including ferns, gymnosperms and angiosperms have been found to constitute cyanogenic glucosides (Zagrobelny et al. 2008) Cyanogenic glycosides are mainly found in sorghum, cassava and barley and are biosynthesised

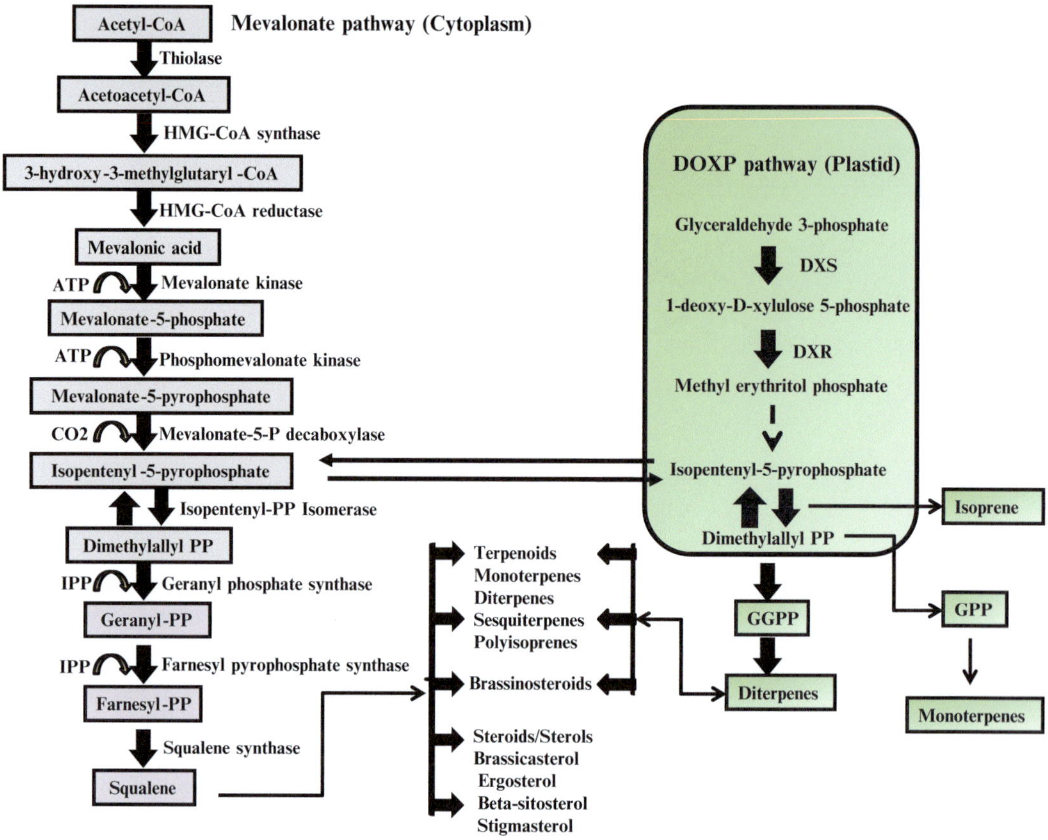

Fig. 2 Schematic representation of MVA and DOXP pathway for the synthesis of steroids and terpenoids, the major secondary metabolome

from six amino acids, namely, L-valine, L-PHENYLALANINE, L-isoleucine, L-leucine and cyclopentenyl-glycine, a nonprotein amino acid (Ganjewala et al. 2010). Studies on sorghum have contributed immensely to the knowledge of biosynthesis and regulation of cyanogenic glycosides in plants. The biosynthetic pathway of CGs includes three essential steps, namely, cytochrome P450–mediated conversion of a precursor amino acid to aldoxime through N-hydroxylation of the amino group of amino acid, conversion of aldoxime to cyanohydrins by P450s and glycosylation of cyanohydrins by UDP-glycosyltransferases. A study has shown the organisation of CYP79A1, CYP71E1 and UDP-glucosyltransferase as metabolon, leading to effective channelling of intermediates in the biosynthesis of dhurrin in sorghum. The presence of cyanogenic glycosides in plants is significant in plant–insect interaction as well as chemical defence system. These also serve as prospective candidates in studies pertaining to chemotaxonomy (Vetter 2000). Furthermore, CGs act as phagostimulant for herbivores specific for CG-containing plants. The catabolism of CGs results in β-cyanoalanine (a neurotoxin), serving as a signalling molecule to restrict pathogens. Further, the research on CGs is essential because certain commercial crops like sorghum, barley and cassava constitutes cyanogenic glycosides as secondary metabolites. Also, deciphering the biosynthetic pathway, identification of pathway enzymes and regulatory mechanisms would serve as an ideal approach in development of cyanogens free crop plants. The science of metabolic engineering of cytochrome P450s has proved successful in creating transgenic cassava with absence of cyanogenic glycosides.

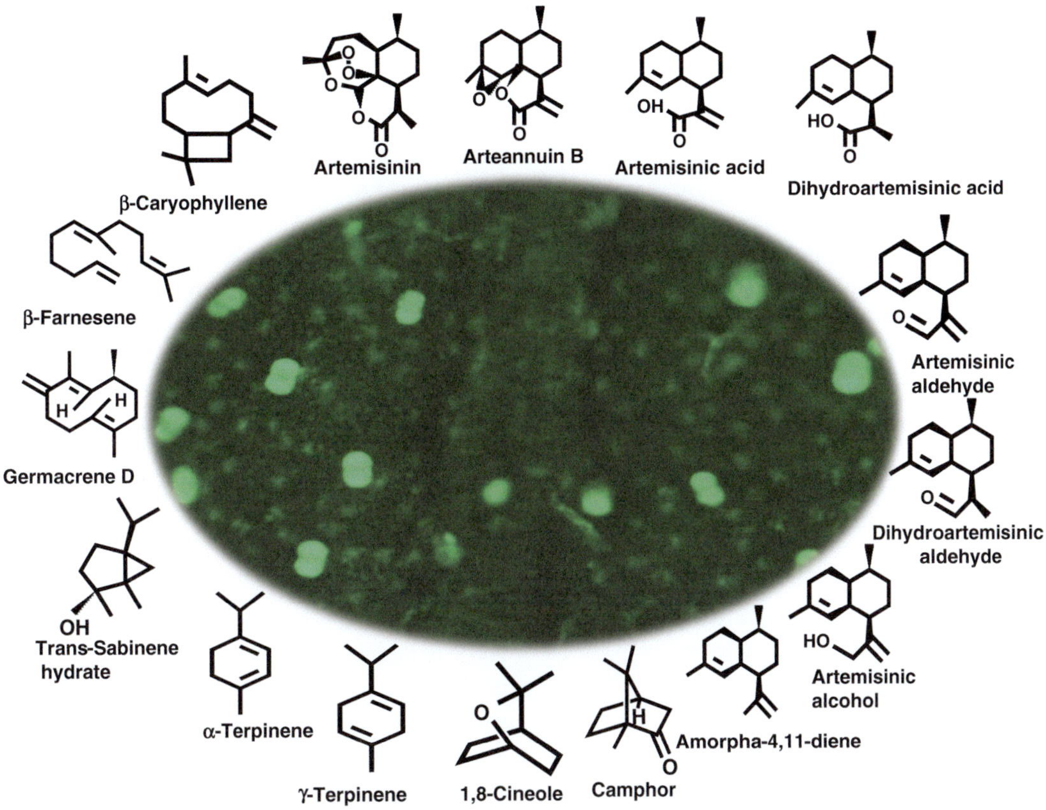

Fig. 3 Microscopic view of glandular trichomes and secondary metabolome of *Artemisia annua* as detected by HPLC and GC-MS

Terpenoids

Terpenoids are one of the largest groups of naturally occurring compounds and are reported to occur from several genera of plants and lower organisms. Often, terpenoids are associated with various pharmacological activities. Terpenoids are composed of basic isoprene units and exist in several diversified structural forms varying in basic isoprene units. These groups of compounds are highly diverse in nature and are synthesised through a complex set of biochemical reactions and crosstalk between cytosol and plastids involving MVA and DOXP pathways (Fig. 2). Various classes of terpenoids are reported to occur in literature and have diverse functional and eco-physiological roles in plants (Sangwan et al. 2001, 2003). Lower terpenoids are basically volatile and attribute specific aromatic characters to the particular aromatic plant and species (Sangwan and Sangwan 2000) and are known to play and modulated under various adaptive conditions (Sangwan et al. 1993, 1994). These volatile compounds are monoterpenoids and sesquiterpenoids and are generally accumulated in specialised structures such as trichomes in plants (Fig. 3). Higher terpenoids found in several medicinal plants are nonvolatile in nature and are reported to have various clinical and pharmacological applications (Shanker et al. 1999; Bose et al. 2013; Sangwan et al. 2011; Tiwari et al. 2013). Using newer omics approaches more and more novel terpenoid compounds are being isolated and reported for their further utilisation and commercial exploitation.

Alkaloids

Alkaloids are also one of the most prominent plant secondary metabolites occurring in plants and other organisms. Alkaloids are nitrogen containing, amino acid-derived low-molecular-weight compounds and have been associated with various clinical and pharmacological activities. Many of the alkaloids are toxic in nature if taken at higher concentration. Of late, there is tremendous omic progress concerning alkaloids. Several studies have been made available in recent past where biosynthesis, accumulation, genomics and metabolomics of alkaloid containing compounds have been taken considerably (Facchini et al. 2004; Kushwaha et al. 2013, 2014).

Recent Approaches in Plant Metabolomics

The study of plant metabolomics has grown significantly from a small hypothetical approach to a promising, scientific methodology during the past few years utilizing several analytical techniques for metabolome analysis (Table 4); mass spectrometry and other chromatographic techniques like liquid or gas chromatography and NMR are being used for simultaneous analysis of metabolites (Roessner and Bowne 2009). These techniques are highly sensitive and sophisticated but fail to resolve certain limitations. The high abundance and complexity of plant metabolites is a major hindrance in complete analysis of metabolome by a single technique (Roessner and Bowne 2009). A range of techniques including metabolic profiling and fingerprinting constitutes the core of metabolomics strategies and detects the presence of all metabolites. An initial screening approach for the presence of metabolites between test and control compound is performed through NMR, Raman spectroscopy, Fourier transform (FT) IR spectroscopy and electrospray ionisation (ESI) MS (Allwood et al. 2008). Additionally, the identification and quantification of an extract coemploying chromatographic techniques, namely, GC or LC coupled to MS or HPLC, is achieved through metabolic profiling. Moreover, MS profiling co-integrated with computational methods allows identification of diverse metabolites and their screening within the system (Goodacre 2005). Spectroscopic techniques offer a rapid and precise estimate of a sample, leading to a high- throughput analysis.

The science of plant metabolomics has emerged as a scientific breakthrough owing to its wide and significant applications ranging from plant metabolic engineering to disease diagnostics and nutritional aspects to pharmacogenomics. The study of metabolites is utilised to delineate the effect of gene deletions and transgenic (Weckwerth et al. 2004), to decode the genetic mechanism of metabolic pathway (Keurentjes et al. 2006), to study the effect of stress such as salinity or temperature (Kim et al. 2007) and to identify biomarker in genetically modified crops to identify desirable or undesirable traits (Catchpole et al. 2005) or mutant screening in plant population. The presence of primary and secondary metabolites which contributes to colour and flavour of tomato were studied through metabolomics approach (Schauer et al. 2005). Other examples include the metabolic sample analysis from conventional and genetically engineered potato tubers to conclude whether the genetically modified crop harbours any undesirable or harmful traits (Catchpole et al. 2005). Subsequently, the science of metabolic profiling was utilized in study constituting degradation of linoleic acid in stored apples (Beuerle and Schwab 1999).

Omics Approaches in the Development of Plant Metabolomics

Bioinformatics forms an integral aspect of metabolomics, and the omics approaches have a significant role in the development of metabolomics, also addressed as 'computational metabolomics'. Certain key areas in bioinformatics which need to be addressed for the development of metabolomics include data management, processing of analysed data, statistical analysis, data organisation and integration and generation of a global mathematical model for defining metabolic network (Shulaev 2006).

Table 4 Technologies related with metabolome analyses

Name of technique	Molecules for separation	References
Thin-layer chromatography (TLC)	Primary and secondary metabolites	Mishra et al. (2012)
Infrared spectroscopy (IR)	Used in quality control, measurement of CO_2 concentrations in greenhouses and measurement of the degree of polymerisation in polymer manufacture	Stuart (2012)
Nuclear magnetic resonance (NMR)	Structural assignment	Chaurasiya et al. (2012), Pauli (2001), Sidhu et al. (2011) and Bharti et al. (2011)
Mass spectrometry (MS)	Determination of mass	Yates (1998)
High-performance liquid chromatography (HPLC) equipped with different kinds of detectors: UV or photodiode array (PDA), fluorescent, electrochemical, etc.	Separation and identification of macromolecules from primary and secondary metabolites	Yadav et al. (2014a, b), Sangwan et al. (2004, 2007, 2008) and Mishra et al. (2008)
Capillary electrophoresis (CE) coupled to different detectors: UV, laser-induced fluorescent (LIF), mass spectrometer (MS or MSMS), etc.	Separation and identification of macromolecules from primary and secondary metabolites	Warren and Adams (2000), Wang et al. (2003) and Markuszewski et al. (2003)
Gas chromatography (GC) coupled to different detectors: MS or MSMS, FID	Separation and identification of mainly volatile macromolecules as well as derivatised phytomolecules of primary and secondary metabolites	Duran et al. (2003), Roessner-Tunali et al. (2003), Colebatch et al. (2004), Fiehn and Weckwerth (2003), Joy et al. (2013) and Chatterjee et al. (2010)
Liquid chromatography tandem mass spectrometry (LC/MS or LC/MS/MS)	Identification of phytomolecules	Tuli et al. (2009)
Fourier transform ion cyclotron mass spectrometry (FTMS)	Identification and mass determination of compounds of primary and secondary metabolites	Marshall et al. (1998)
HPLC coupled to NMR detection (LC/NMR)	Identification and mass determination of compounds of primary and secondary metabolites	Sidhu et al. (2011)
HPLC coupled to NMR and MS detectors (LC/NMR/MS)	Identification and mass determination of compounds of primary and secondary metabolites	Bharti et al. (2011)

Metabolomics databases are the collection of data about diverse metabolites, numerous biochemical reactions and pathways in different organisms and contribute to the better understanding of metabolome coverage and analysis. Such detailed study would decode the mechanisms in plant metabolism together with highlighting the parameters for chemical diversity occurring among plant metabolites. Further, the metabolic pathway databases would serve as an ideal platform in comparative genomics and genome annotation studies (Wishart 2007). However, the information on metabolites is increasing day by day, and the databases need to be constantly upgraded for reference purpose. More information regarding the subcellular location of metabolites, their concentrations, classification, NMR or MS spectra and physical properties should be included to enhance the precision and accuracy of the database. For this, the available data should be experimentally validated, referenced, interpreted and ideally should provide maximum information about the metabolome of an organism (Wishart 2007).

Metabolic correlation network has been designed to incorporate and correlate the enormous metabolic information (Steuer et al. 2003).

Such correlation would facilitate the determination of carbon and nitrogen levels resulting from transcriptional and biochemical regulations of cellular processes and enzymatic conversions (Rhee et al. 2006). The science of plant metabolomics has broadened its perspective to include the data analysis on metabolic fluxes which measures flow between different metabolites. However, the measurement and analysis of fluxes is a difficult task because of limitations in availability of in vivo dat pertaining to location and topology of the metabolic pathways and difficulties associated with modelling intracellular transport of metabolites (Shanks 2005). Stoichiometric analysis, defines the frequently used methodology for the quantification of reactants and products, thereby flux estimation for a single metabolite (Edwards and Palsson 2000). This method has limitation with large metabolic networks, metabolic cycles, parallel metabolic pathways and reversible reactions (Wiechert et al. 2001). A useful in silico tool for metabolic flux analysis is the FluxAnalyzer for MATLAB that co-integrates metabolic pathway and flux analysis (Klamt et al. 2003). The use of radioisotopes such as ^{13}C carbon labelling tends to overcome certain disadvantages of stoichiometric flux analysis (Chaurasiya et al. 2012). However, an extensive and exhaustive approach for analysis of C-constrained flux (stoichiometric model with some flux ratio constraints and stoichiometric balances) is required (Rhee et al. 2006). Finally, all the data corresponding to flux balances, metabolic control, connectivity as well as optimisation of metabolic pathway can be analysed and co-integrated, and simulation can be performed in a cellular modelling environment employing Cell Designer (http://www.systems-biology.org) or E-Cell (http://www.e-cell.org/), respectively.

Data Acquisition

Significant advances over the years have improved manifold our capacity to simultaneously analyse an array of organic components in complex biological mixtures. However, to achieve an entire metabolite complement of a plant in a single step is next to impossible due to dynamic state of metabolism at any given time. The presence of more than a lakh secondary metabolites in plants, including >6,000 flavonoids and >12,000 alkaloids, in the plant kingdom along with large variation of principal components within a tissue makes plant metabolomics analyses more than complex (Hall 2006). This plethora of structurally and functionally related compounds poses a herculean task to separate, detect and document these all together. Therefore, combinations of multiple and parallel extraction and detection techniques using hyphenated systems are employed to cover a comprehensive image of metabolite complement. Initially, a proven protocol might be applied in order to develop a primary insight which might be magnified at certain points following more focused approach towards a set of compounds. The success of a developed procedure may be driven by present metabolites, sample preparation, extraction and detection techniques.

Plant Material and Sample Preparation

One of the most important limitations of analytical techniques is that these capture a metabolic snapshot of the given sample at the time of harvesting and preparation. Although, this information itself is considerably rich as concerned with data acquisition for metabolomics studies, it is nevertheless affected by time and method of tissue harvesting and sampling, as failure in proper handling might yield misleading inferences. In case of plants, it is more cumbersome to maintain uniformity in cultivation compared to microorganisms or animals. Similarly, temporal and spatial metabolic dynamics of plants should also be concerned due to large variations observed across geography and climate. In order to minimise these effects, all the material to be analysed should be grown and harvested together at one place. To minimise such artefactual errors, large-scale cultivation of plant material in a large-volume growth chamber should be preferred (Trethewey 2004). In cases where geographical factor is to be analysed, biological differences might be less clear and would require additional experiments to minimise the gap between observed and actual differences. Artefactual variations among the samples analysed are one of the most important hindrances. Sampling is another important step that needs to be paid careful attention to minimise experimental error. To maintain uniformity, growth stage as well as time of sampling should be carefully monitored. Even after harvesting, a great care should be paid to the treatments of the collected material so that proper metabolic status can be maintained.

Extraction of Metabolites

At present, not a single extraction and detection method could be employed to all known groups of metabolites. Due to chemical complexity, metabolic heterogeneity and dynamic range of metabolites, multiparallel technologies are the only option to develop a broad metabolic snapshot. A vigilant assortment of extraction, separation and detection methods may be deployed to acquire biochemical data of biological samples. In order to develop a refined metabolite profile, analysis of all the extracts, polar, nonpolar, lipophilic and volatile fractions, is desirable. Similarly, the characteristics of the targeted metabolite, the range of metabolites to be examined and their quantaties would form a basis for the development of the optimized protocol. Metabolic profiling by any method is expected to cover a wide range of metabolites. Metabolite extraction methods generally include liquid phase extraction, liquid–liquid extraction and solid–liquid extraction (liquid extraction followed by extraction with solid phase material). In particular, volatile metabolite extraction methods include solid phase extraction (trapping), steam distillation, headspace and solid phase microextraction (SPME) respectively.

Derivatisation of Metabolites

Some of the analytical procedures might require the derivatisation of target metabolites. For example, in case of GC-MS, only volatile compounds are detected, whereas most hydrophilic metabolites should be derivatised by various methods reported in literature such as silylation or other derivatisation methods. If UV or fluorescence detection is employed, HPLC analysis also requires derivatisation of the target sample such as in case of artemisinin detection from plant samples. Artemisinin is converted into Q290 and Q262 which absorb in UV range and make estimation of artemisinin and related compounds in a more precise manner (Yadav et al. 2014a, b). Similarly many secondary metabolites from *W. somnifera* have been detected successfully by using a variety of detectors including ELSD and PDA (Chaurasiya et al. 2007; Sangwan et al. 2008). Specificity as well as efficiency of any protocol should be validated in terms of reproducibility of results in several replicated samples. It is also desirable that the stability of the derivatised metabolites may be retained till the analysis is completed.

Analytical Techniques in Metabolomics

Separation, Detection and Quantification

Separation and detection of metabolites are the most important steps in metabolomics studies. These steps usually employ chromatographic or electrophoretic methods coupled to mass spectrometry. Detection methods other than MS are generally used for quantification purposes. MS is utilised for identification purposes. In this regard, metabolomics data should be viewed in two independent categories, resolution and quantification. The methods chosen for separation as well as detection strongly affect both resolution and quantification. However, as a practical approach, either resolution or quantification should be given priority. Various technologies related with metabolome are listed in Table 1. These techniques are used singly or in combination depending upon the type of metabolites in target. The application of metabolic profiling for the identification of genes involved in the biosynthesis of metabolites, whose production is defined by detailed information encoded by specific genes in plants, specific tissues, or cell types, is a powerful approach in plant metabolomics.

Separation Techniques

Gas chromatography cointegrated with mass spectrometry (GC-MS) is a powerful technique providing high chromatographic resolution. However, the chemical derivatization of biomolecules forms an essential requirement with some exceptions (volatile analysis does not require derivatization). The technique of gas chromatography cannot be used for the analysis of polar metabolites. Comparatively, HPLC is another significant analytical tool, which could be utilized for the quantification of a much wider range of metabolites but with a lower chromatographic resolution. Subsequently, capillary electrophoresis (CE) exhibits better prospects in terms of higher separation efficiency than HPLC and broad analysis of a wider range of metabolites than GC as well as analysis of charged metabolites [52].

Detection Techniques

Various detection methods are utilised for metabolomics analyses depending on the structure and properties of metabolite. Mass spectrometry (MS), the initial technology for metabolome analysis, defines an integrated technology platform incorporating HPLC, GC, or CE, respectively. The technique is utilized for the identification and quantification of metabolites according to its fragmentation pattern. The technique demonstrates specificity and sensitivity (although sensitivity is more important for HPLC since it is affected by charged metabolites and may be subject to ion suppression artifacts) respectively. A resurgence has been witnessed in the past decade with respect to surface-based mass analysis with emerging MS demonstrations on negligible background and reduced sample preparation. The challenges include the analysis of metabolites directly from biofluids and tissues because of the complexity of these samples, which constitutes an enormous range of metabolites. Nanostructure-initiator MS (NIMS) (Northen et al. 2007; Woo et al. 2008), has been developed, an advanced method which facilitates the identification of small molecules. The application of MALDI projects significant background at <1,000 Da, a major hindrance in the analysis of metabolites having low mass ratios as well as limitations consisting of spatial resolution of matrix crystals when compared with tissue imaging. Additionally, another limitation includes the size of resulting matrix crystals, which limits the spatial resolution which could be obtained in tissue imaging. Several other matrix-free desorption/ionization-based approaches have revolutionized the analysis of biofluids and tissues. Secondary ion mass spectrometry (SIMS) was the initial approach applied for the analysis of metabolites from biological samples, the main advantage being high spatial resolution (as small as 50 nm), a powerful feature of tissue imaging with MS. One of the advantages of secondary ion mass spectrometry (SIMS) includes high spatial resolution (<50 nm). However, it's having limited sensitivity (>500 Da). Another matrix-free technique which utilizes a charged solvent spray to desorb ions present on the surface is designated as desorption electrospray ionization (DESI). The benefits of this method are that there is no requirement of a special surface and analysis is performed at ambient pressure with complete monitoring of the samples, while spatial resolution is a limitation in focusing the charged solvent spray. Laser ablation ESI (LAESI), a recently developed technique, solves these disadvantages to a certain extent.

Thin-Layer Chromatography

It is an analytical technique employed for initial screening of metabolites present in a living system. The chromatographic separation relies on separation of substances partitioned between two phases: a mobile phase and a stationary phase. The principle of thin-layer chromatography includes a solid phase (the adsorbent) coated as a thin layer (20 mm thick) onto a solid support. Different kinds of materials like plastics, glass and aluminium are used as solid supports. Further, the mixture of compounds to be separated is dried, dissolved in the respective solvent and spotted on the solid plate. A compound which is adsorbed strongly on the solid phase tends to move less with the mobile phase and vice versa. Another important aspect of TLC estimation includes the solubility percentage of a particular compound in a respective solvent. The fact that a substance show relative solubility in a particular solvent will result in faster elution than the other substance. The compounds in a mixture/extract are separated based on their Rf values. It is defined as the distance travelled by a substance relative to the distance moved by the solvent. It is one of the efficient and robust techniques utilised in identification/formation of respective compounds as well as their separation.

High-Pressure Liquid Chromatography

Plants are good sources of millions of different metabolites with various chemical structures as well as bioactivities. Different analytical techniques are in common use for the analysis, chemical characterization, as well as quantification of these metabolites. High-performance liquid chromatography (HPLC) is one of them and used to separate, identify, and quantify the components in a mixture. In the HPLC method, a liquid sample is passed through a column, which is packed with solid adsorbent material. Different analytes in the sample interact differently with the adsorbent material of the column and inhibit analyte flow. The strength of interaction decides the elution time of the analyte. Weaker interaction results in short elution time, while stronger interaction results in long elution time. The sorbents, granular materials, are the active components of the column and made up of solid particles (2–50 μm) like silica, polymers, etc. The mobile phase passed through the column is a mixture of solvents like water, acetonitrile, and/or methanol. The interactions between analytes and sorbents are very physical in nature (such as hydrophobic, dipole–dipole, and ionic, most often a combination) and greatly influenced by the mobile-phase composition as well as temperature and play important roles in analyte separation. Nowadays, the HPLC instrument is totally controlled by digital microprocessors and software. UV/Vis, photodiode array (PDA), evaporative light scattering detector (ELSD), and mass spectrometry (MS) are various detectors and very common in use for HPLC. A very small volume (in μL) of sample mixture to be analyzed is introduced into the main stream of the mobile phase passing through the column. Different analytes of the sample have different velocities because of specific physical interactions with the sorbent or stationary phase of the column. The time between sample injection and an analyte reaching the detector at the end of the column is the retention time and, under specific conditions, believed to be an identifying characteristic of a given analyte. The reduction of particle size in packing materials results in higher operational pressure as well as improved chromatographic resolution. Two kinds of HPLC, normal-phase chromatography and reverse-phase chromatography, are used very commonly; normal–phase chromatography was developed first by chemists. In this method, separation of analytes is based on their affinity with a polar stationary surface such as silica by polar interaction such as hydrogen bonding or dipole–dipole type of interactions with the sorbent surface. On the contrary, reversed-phase HPLC (RP-HPLC) has a nonpolar stationary phase and an aqueous, moderately polar mobile phase. RP-HPLC operates on the principle of hydrophobic interactions. Nowadays, reversed-phase HPLC is used very commonly (Yadav et al. 2014a, b; Sangwan et al. 2004, 2007, 2008; Chaurasiya et al. 2007; Mishra et al. 2008). Chaurasiya et al. (2007) have analyzed withanolides in the root and leaf of Withania somnifera by HPLC, attached with two different detectors: photodiode array and evaporative light scattering detection. A reversed-phase HPLC method for the simultaneous analysis of nine structurally similar withanolides (17-hydroxy withaferin A, 27-hydroxy withanone, withaferin A, 17-hydroxy-27-deoxy withaferin A, withanolide A, 27-hydroxy withanolide B, withanolide D, 27-deoxywithaferin A and withanone) has been developed. This technique was used by several authors for withanoloid detection in Withania somnifera L. (Chaurasiya et al. 2007; Sangwan et al. 2004, 2007, 2008; Mishra et al. 2008; Sabir et al. 2008). *Artemisia annua*, another important medicinal plant well known for its antimalarial properties, is the only source of artemisinin. RP-HPLC is the favourite technique throughout the world for detection and quantification of artemisinin and its metabolites (Vetter 2000). Van Nieuwerburgh et al. (2006) reported the detection and quantification of artemisinic acid, dihydroartemisinic acid and arteannuin B from *Artemisia* samples by RP-HPLC electrospray quadrupole time-of-flight tandem MS (Van Nieuwerburgh et al. 2006; Mannan et al. 2010). The report of artemisinic acid detection and quantification by HPLC-ELSD was also available in literature (Ferreira and Gonzalez 2008; Kjaer et al. 2013). Recently we have reported the use of HPLC-PDA for detection and quantification of artemisinin, dihydroartemisinic acid and arteannuin B in *Artemisia annua* crude plant extract (Yadav et al. 2014a, b). The solvents used were methanol, water and acetic acid. Yadav et al. (2014a, b) have used two different protocols: one for artemisinin and the other for dihydroartemisinic acid and arteannuin B. Artemisinin was first derivatised to Q260 through Q290 by acid–base reaction and then run through HPLC for detection and quantification, while dihydroartemisinic acid and arteannuin B were detected in intact form without modification (Fig. 3). The application of HPLC is very wide and used in the secondary metabolite profiling of plants, substrate and product identification and quantification in catalytic reaction. For example, glucosyltransferases are enzymes catalysing the conversion of sterol to its glucosidic form. Both sterol as well as sterol glucosides can be detected and quantified by the help of HPLC. The application of HPLC is very wide and applicable to different fields biological sciences.

Flow or Direct Injection/Infusion (FI/DI-MS)

This is essentially mass spectrometry without any separation. MS is one of the most sensitive detection techniques and is a method of choice for plant metabolomics studies. A single mass spectrum thus produced represents a rapid qualitative screening tool appropriate for quality control, mutant screening, biodiversity analysis, etc. Mass spectrometry is often used in parallel to chromatographic separation as a detection technique. Hyphenated techniques usually present high sensitivity and resolution with a reproducible fragmentation pattern of molecules.

Liquid Chromatography–Mass Spectrometry (LC-MS)

LC-MS presents one of the most resourceful techniques for plant metabolomics as it provides an affordable analysis of large group of plant secondary metabolites. Developments of ultra-performance systems and column chemistry have improved the separation and resolution to manifold. However, limited use of the technique with the molecules that could be ionised to be detected in MS put a restriction to its universal adoptability. Nonetheless, an array of techniques are available to ionise the separated molecules in LC systems including electrospray ionisation (ESI), atmospheric pressure chemical ionisation (APCI) and photoionisation (PI). The high-precision analytical separation hyphenated with accurate sensitivity of MS systems has paved the way for well-resolved metabolomics snapshots of complex plant samples in recent times. High-resolution mass spectroscopy coupled with liquid chromatography (LC-HRMS) led to identification of several novel compounds from the metabolite profile of fruits of *W. somnifera*. Several metabolites have been identified in the study, and the mass spectra of compounds aid in structural elucidation on the foundation of accurate molecular masses as well as fragment ions. Plenty of inventive compounds have been elucidated as imitative of withanamides (Tuli et al. 2009).

Liquid chromatography–mass spectrometry (LC-MS) spectral reference libraries of standards are of finite utilization. Multidimensional instrumental techniques like liquid chromatography–mass spectrometry (LC-MS), gas chromatography–mass spectrometry (GC-MS), tandem mass spectrometry (MS/MS) or nuclear magnetic resonance (NMR), either alone or in combination, are in use for initial identification of stranger appearances from biological samples and allow structural elucidation as well as relative profiling in plants. Liquid chromatographic quadrupole tandem time-of-flight mass spectroscopy (LCQTOF-MS/MS) empowers researchers with actual mass and product ion information of metabolites separated by chromatography. These mass data are further used to find an elemental composition of compounds and also used to match with accessible mass information in databases like the NIST and/or KEGG to predict the possible structures. The stepwise fragmentation of isolated compounds by tandem MS does give metabolite ion information to determine the structure. But the final confirmation of the compound identity is necessary and done in two ways: (1) by analysis of an authentic standard substance and (2) by analysis by NMR. NMR is highly chemical sensitive and a method of preference for compound identification. NMR in combination with other chromatographic techniques like LC and MS (LC-MS-NMR) is a supreme technology for both peak identification as well as structure elucidation (Wolfender et al. 2003).

Application of liquid chromatography coupled to liquid chromatography–electrospray ionization–mass spectrometry (LC-ESI-MS) in metabolomics is a supplementary approach to gas chromatography–electron impact–mass spectrometry (GC-EI-MS) and provides twofold advantages: (1) no need of chemical alteration of compounds before analysis and (2) efficiency in separating and quantifying highly polar, thermounstable, and high–molecular weight compounds like oligosaccharides or lipids. For the analysis of plant metabolites, liquid chromatography in combination with different detectors like UV/Vis, photodiode array (PDA), evaporative light scattering detector (ELSD), and mass spectrometry (MS) is already in use and popular in scientific communities since several years. Varieties of columns and different LC methods are reported in scientific literature for separation, detection, and quantification of diversified compounds from plant origin. Coupling of LC with MS is a formidable combination and facilitates further selectivity, sensitivity, and impartial detection along with structural information of detected compounds.

Profiling of Metabolites and Metabolomics Using GC-MS

GC-MS is principally deployed to separate and detect volatile compounds, e.g. low molecular weight alcohols, monoterpenes (C10) and esters, which is a bottleneck as it might miss the thermolabile compounds. The novel reports on the technology of derivatization are extensively exercisable to primary metabolites. Based on the category of metabolomics explication, GC-MS analyses may be carried, and obtained data must be determined by gas chromatography–mass spectrometry (GC-MS) subcomponents, which are part of metabolome estimate components. Comparison of spectra and peak is commonly identified by methods like dot-product approach, probability-based matching and similarity indices (Stein and Scott 1994). The GC-MS profiling of more than 670 metabolites was performed from root and leaf of tobacco, *Nicotiana tabacum* (Birkemeyer et al. 2003). This outcome evaluates generic chemical derivatization by N-methyl-N-(trimethylsilyl)-trifluoroacetamide (MSTFA) reagent and tert-butlydimethylsilylane with the application of N-methyl-N-(tert.-butyldimethylsilyl)- trifluoroacetamide (MTBSTFA) reagent. For a simultaneous study of metabolites from different plant species like Arabidopsis thaliana (Fiehn et al. 2000), Solanum tuberosum (Roessner et al. 2000), Medicago truncatula (Duran et al. 2003), Lycopersicon esculentum (Roessner-Tunali et al. 2003), Saccharum officinarum, Lotus japonicas (Colebatch et al. 2004), and Cucubita maxima (Fiehn and Weckwerth 2003), a gas chromatography–mass spectrometry (GC-MS) instrument is used in the scientific realm. Recent studies have shown that GC-MS-MS analysis is also proven useful in identifying structurally very close substrate and compounds such as tropine, pseudotropine and tropinone which in turn help to identify particular specific reaction in the alkaloid biosynthetic pathway (Kushwaha et al. 2014).

Recently reports have shown unusual existence of elemol rich essential in *Dioscorea* plant species which is traditionally known to be a diosgenin producer. GC-MS studies revealed as much as 70 peaks in *Dioscorea composita* and *D. floribunda* (Joy et al. 2013). Similarly several peaks of lower terpenes and monosesquiterpenoids and sesquiterpenoids were detected and identified (Yadav et al. 2014a, b). Recently, GC-MS approach was followed to generate a metabolomics profile of hexane fraction of *Withania somnifera* extracts (Chatterjee et al. 2010). In this comprehensive analysis of crude extracts of leaf and root of *W. somnifera*, NMR and chromatographic (HPLC and GC-MS) techniques were followed. The added role of gas chromatography belongs to lipidomics that basically instigates lipid metabolites. The chromatographic analysis of polar and complex lipids by conventional chromatography is an arduous and time-consuming method and includes lipid separation into classes, their derivatization, as well as fatty acid chain analysis. Contrary to this, mass spectrometry–dependent analysis of lipids is fast and yields a comprehensive profile. Basically, a lipid profiling process includes solvent extraction of lipids from tissues, amalgamation of phospholipid and/or galactolipid internal standards with suitable solvents, and finally the analysis of the mixture by electrospray ionization tandem mass spectrometry (ESI-MS/MS) (Welti et al. 2003).

Metabolite Profiling and Metabolomics Using NMR

NMR is an abbreviation for nuclear magnetic resonance and a kind of spectroscopic technique with mileage of spin power of nucleus. The nuclear spin is the total angular momentum present in the nucleus of atoms. Only the nuclei that have a nonzero nuclear spin exhibit nuclear magnetic resonance. Among this group of atoms are 1H, 13C, 15N, and 31P, which are elements that are present in bio-organic molecules. NMR has competence to illuminate the comprehensive chemical structure of a molecule and/or compound (Pauli 2001).

NMR is a technique with dynamic capability, which can be applied to various perspectives of metabolomics. The very first perspective is the ability to identify the molecule due to 1H, which is present in practically all bio-organic molecules with elevated natural profusion of 99.9816–99.9974 % and utmost consumed nucleus for NMR dimensions. Basically, the analytical procedure to isolate the compound/molecule of inquisitiveness are used, isolated molecule was solubilised in suitable solvent for accession of 1H two-dimensional (2D)-NMR spectra as 1D-1H NMR spectrum alone is not enough for complete structural elucidation of bio-organic compounds. To elucidate the 3D position of the protons in a molecule, homonuclear 1H-2D spectra such as correlated spectroscopy (COSY), total correlation spectroscopy (TOCSY), and Nuclear Overhauser Effect Spectroscopy (NOESY) are used.

There is a wide diversity of different types of NMR measurements, according to the interest of the user in particular chemical features. Another metabolomics application is in vivo NMR. A recent example of NMR-based metabolomics study is characterisation of 17 metabolites from the fruits of *Withania somnifera* at different developmental stages (Sidhu et al. 2011). The technique is also useful in generating comparative metabolic profiles of plants grown in different geographical regions. Examples include generation of metabolic profiles employing 1H NMR spectroscopy followed by principal component analysis (PCA) and hierarchical clustering analysis (HCA). The 1H NMR spectra showed the presence of diverse metabolites, namely, amino acids, flavonoids, lipids, organic acids, sugars, and withanolides in leaves of *Withania somnifera*. High-resolution magic angle spinning (HR MAS) and nuclear magnetic resonance spectroscopy (MRS) have been utilized to cognize the chemotypic variations of leaves as well as roots for four chemotypes of Withania somnifera (Bharti et al. 2011). A prominent chemotypic variation obtained in 41 diverse primary metabolites has been exposed by statistical analysis (multivariate principal component analysis (PCA)) on HR-MAS 1H NMR spectra of leaves. Another important aspect of NMR spectroscopy is also involved in assigning pathway establishment because the 13C labelling pattern and retrobiosynthetic approaches have revealed the major participation of major pathways in *W. somnifera* (Chaurasiya et al. 2012).

Capillary Electrophoresis–Mass Spectrometry (CE-MS)

Capillary electrophoresis–mass spectrometry (CE-MS) is such a sensitive technology that can cumulatively quantify several metabolites without any chemical derivatization in low amounts of samples. With its highly resolved separation and accurate detection of water-soluble fractions, it is poised to be a strong combinatorial technique for plant metabolomics studies of primary and secondary nature. CE methodology is usually very fast, consumes little sample and reagents and more economical than chromatography and electrophoresis. Capillary electrophoresis (CE) has proven an immense possibility for extensive experiments of biological samples. It can be used for a broad range of substances from small molecules including inorganic ions like carbohydrates, amino acids, peptides, nucleosides, nucleotides, organic acids, vitamins, drugs, and steroids to larger molecules like nucleic acids, proteins, hormones, and alike living cells. The application of capillary electrophoresis (CE) in metabolomics study accounts with pervasive study of innate low–molecular weight molecules and/or compounds. Accordingly, metabolomics methods of capillary electrophoresis (CE) to distinguish, to detect, and to quantify maximum number of metabolites in a single run are usually developed. The concept behind capillary electrophoresis–mass spectrometry (CE-MS) is fine separation of metabolites, ionization of separated metabolites, and identification of ionized metabolites by comparing obtained ions with a range of m/z values. This technique is strong enough for simultaneous assessment of about 1000 charged metabolites in biological samples. Capillary electrophoresis (CE) aided with electrospray ionization–mass spectrometry (CE–ESI-MS) becomes a more specific, reassuring, and disparate technology and can be used for microseparation of metabolites in metabolomics study. Lots of tremendous experiments have been done in plant metabolomics by using capillary electrophoresis–mass spectrometry (CE-MS) instruments and methods. Some evidences are analysis of amino acids and sugars in plant fluids and tissue extracts (Warren and Adams 2000); analysis of organic acids in different plant parts (Wang et al. 2003); analysis of carboxylic acid metabolites, pyridine, and adenine nucleotide metabolites in bacterial cell extracts (Markuszewski et al. 2003); analysis of different secondary metabolites like iridoids, flavonoids, and phenolic compounds in bark and leaves (Cheung et al. 2003) etc. Albeit, the limitation of capillary electrophoresis (CE) is resolution because of less migration time, but competent sensitivity as well as selectivity for metabolomics study can be effectuated with the help of capillary electrophoresis–mass spectrometry (CE-MS).

Fourier Transform Ion Cyclotron Resonance MS (FT-ICR-MS)

FT-ICR-MS is a slowly but strong growing technique for metabolomics studies offering more accurate and highly resolved molecular mass estimations. Nonetheless, high chromatographic resolution becomes less important due to extremely high mass resolution means; however, it creates a bottleneck as isomers, abundant in plant extracts, cannot be distinctly identified. This approach is successfully applied to assess the metabolic status of transgenic tobacco carrying glutamate dehydrogenase gene from *E. coli*. FT-ICR-MS detected 283 ions in roots and 98 ions in leaves that appeared to be changed significantly due to altered GDH activity. 42 % of ions were inferred to known metabolites such as certain amino acids, organic acids, sugars and some fatty acids.

Data Analysis

With the ability to generate never preceding amount of raw data, a whole new world of science has emerged that deals with the handling and analysis in silico of all the metabolites. This has enabled the development of necessary software for collection and pre-processing the data in order to direct comparison of data sets from relative analyses. Data analyses by in silico tools empower searching and processing of data to point out as well as isolate elements of inquisitiveness, aid in demonstration of intricate data in a comprehensible and intellective manner, and facilitate data storage in an effective way.

Data Pre-processing

Data pre-processing is an essential prerequisite for metabolomics studies in plants where it aims to assess broad developmental differences, phenotypic modifications or multifactorial responses of plants to abiotic or biotic stresses. Citing the highly complicated, multidimensional metabolomics data sets, effective manual data handling does not seem feasible. Development of dedicated bioinformatics and statistical tools that convert this data into information has made it rather easy. Moreover, emerging visualisation tools have made these complex data sets relatively and effectively understandable. Unavoidable artefacts or inherent imperfections are natural to any omics approach, and this is true to data generated by any chromatographic procedure representing the first challenge to the effective comparison a large set of chromatograms/mass spectra. With the data generated by HPLC-PDA, warping procedures such as correlative optimised warping (COW), dynamic time warping and permutational time warping appear to be methods of choice. Further, SPECALIGN and METALIGN have been equally applied appropriately for both GC-MS and LC-MS data sets. These packages perform alignment of spectra, baseline correction and noise reduction, effectively reducing size of data sets and comparative analysis time.

Data Mining and Visualisation

In order to convert raw data produced from a machine into biologically relevant information, processed data sets must be subjected to effective analysis followed by standard statistical filtering leading to differential comparisons and reliable conclusions. Unsupervised discriminatory approaches such as principal component analysis (PCA) and hierarchical clustering (HCA) or supervised approaches such as partial least squares (PLS) or SIMCA are generally used methods at present. Effective data visualisation tools enable us to simplify and more easily comprehend the multidimensional complexity generated data sets. Some popular techniques aim for analyses of metabolic data include correlation optimised warping, hierarchical cluster analysis and self-organising mapping.

Correlation Optimised Warping (COW)

Until recently, most of the metabolomics data was analysed based on the selected peaks and their area. This reduced data size also presented a problem of specific peak selection. This limitation could be overcome by taking all the peaks in a chromatogram in consideration. However, it is important to normalise all the peaks as small variation generated by experimental errors may drift the peaks significantly. COW is an approach adopted to align two chromatograms by linear stretching and compression, to the time axis of one profile. The optimal alignment is independent to the compounds in question and decided by the calculated correlation. Any chromatographic data may be subjected to correlation optimised warping.

Data Storage and Database Building

Ability to process data in order to minimise the size while keeping the information intact is a key to effective database building. While aim of a particular experiment is used to be limited, the application of the data generated in the process might be large and might be sought in future provided the efficient and effective storage. Perhaps the biggest challenge of metabolomics stands from the current lack of appropriate databases and data exchange formats. This presents a need for biochemical ontologies to clearly specify each metabolite its relation with others along with suitable databases to store metabolomics data to facilitate relevant queries.

Plant Metabolomics and Statistics

A cluster of small molecules discovered in a cell, organ, or organism is metabolomic. The basis of this metabolomics consideration is dependent on two different perspectives: the chemometric and the targeted or comprehensive profiling. Both perspectives are contrasting in nature; the chemical compounds are not identified first in the chemometric approach, while in the targeted or comprehensive profiling approach, chemical compounds are identified first. In the chemometric approach, a spectral pattern of the chemical compound was used for statistical analysis to differentiate samples, while in the targeted or comprehensive profiling approach, a chemical compound was identified first and then statistically analyzed to identify relevant biomarkers. The chemometric approach attached with an analytical technique is popular as a general statistical approach. Monovariate statistical analysis and multivariate statistical analysis are two different classes of the chemometric approaches.

Monovariate Statistical Analysis

Analysis of chemical samples generates adequate data. These data must be processed to find the relationship among the obtained data. The chemometric approaches are used to discover information from multivariate data. The various aspects and knowledge of statistical and mathematical areas are used in chemometric approaches to rediscover similarity patterns in the data, to address material properties, and to submit and use the multivariate models. Different mathematical as well as statistical laws are in common use to discover experimental data, to endow statistical information about each variable, to find correlations among variables, and to detract data dimensionality. One of the most famous analyses in the scientific world, the analysis of variance, ANOVA (Stuart 2012; Miller and Miller 1993), is practiced to obtain most significant variables in the sample differentiation. One-way and two-way ANOVA analysis was used to differentiate the statistically significant differences between artemisinin accumulation and yield and peltate glandular trichomes in water-stressed samples as well as plant developmental stages (Yadav et al. 2014a, b). Multivariate data analysis (MANOVA) is the extension of ANOVA, and it is applied commonly when two or more related variables are considered and cannot be merged. MANOVA considers only variables which can be distinguished as having noble credibility.

Multivariate Statistical Analysis

Huge amounts of high-dimensional and complex data sets are generated in metabolomics disciplines. It is very arduous to explore and decode by phenomenal inspection and/or by using any traditional univariate statistical method. To rediscover significant knowledge from such huge experimental data sets, mathematical modeling approaches of multivariate data analysis (MVDA) methods are therefore used (Fiehn 2002; Vichi and Saporta 2009; Van den Berg et al. 2009). One variable is considered in monovariate statistical analysis, while correlations among two or more variables are considered in multivariate statistical analysis. Multivariate data analysis is commonly used to track data overview and to classify and/or distinguish among groups of observations and regression modeling between two different sets of data along X- and Y-axes. The huge amount of high-dimensional and complex metabolomics data sets are chemometrically dissected in unsupervised as well as supervised ways (Table 5).

Principal Component Analysis

The objective of principal component analysis (PCA) is to cede optimal sample separation by exercising nodal vectors spanning an n-dimensional space. The reduction in the system dimensionality is also a motive of principal component analysis (PCA). Principal component analysis (PCA) is one type of compression technique (Deacon 2005; Sangwan et al. 2001; Jackson 1991), a classical "unsupervised" technique for the primary analysis of data, and is extensively used. Principal component analysis (PCA) delineates the distinction in a set of multivariate data in terms of orthogonal variables or principal components. As each orthogonal variable or principal component (PC) reckons for a section of the total variance of the data set, principal component analysis (PCA) is a kind of additive model. Principal component analysis (PCA) rediscovers linear relationships of the discrete variables in the original data or principal components (PCs) orthogonally connected and delineates the prime trends of data. The very first step of principal component analysis (PCA) is to discover the minimum number of meaningful orthogonal

Table 5 List of some of unsupervised and supervised chemometric methods generally used in analysis of multivariate data set

Model type	Model	Linear/non-linear	References
Unsupervised model	Canonical correlation analysis (CCA)	Linear	Anderson (1984)
	Clustering and disjoint principal component analysis (CD-PCA)	Linear	Vichi and Saporta (2009)
	Hierarchical clustering analysis (HCA)	Linear	Makretsov et al. (2004)
	Kernel canonical correlation analysis (KCCA)	Non-linear	Larson et al. (2014)
	Kernel principal component analysis (KPCA)	Non-linear	Twining and Taylor (2001)
	Kohonen artificial neural networks (KANN)	Non-linear	Kohonen and Honkela (2011)
	Multilevel simultaneous component analysis (MSCA)	Linear	Ceulemans et al. (2013)
	Principal component analysis (PCA)	Linear	Miller and Miller (1993) and Vichi and Saporta (2009)
	Simultaneous component analysis (SCA)	Linear	Smilde et al. (2005)
	Weighted principal component analysis (WPCA)	Linear	Fan et al. (2011)
Supervised model	Backpropagation artificial neural networks (BANN)	Non-linear	Michalopoulos and Hu (2002)
	ANOVA–simultaneous component analysis (ASCA)	Linear	Smilde et al. (2005)
	Discriminant analysis (DA)	Linear	Guo et al. (2007)
	Kernel-orthogonal partial least squares-DA (K-OPLS-DA)	Non-linear	Boccard and Rutledge (2013)
	Kernel partial least squares-DA (KPLS-DA)	Non-linear	Rosipal and Kramer (2005)
	N-way PLS-DA (N-PLS-DA)	Linear	Perez-Enciso and Tenenhaus (2003)
	Orthogonal partial least squares-DA (OPLS-DA)	Linear	Boccard and Rutledge (2013)
	Bidirectional orthogonal PLS-DA (O2PLS-DA)	Linear	Boccard and Rutledge (2013)
	Orthogonal signal correction (OSC)	Linear	Niazi and Azizi (2008)
	Principal component discriminant analysis (PCDA)	Linear	Jombart et al. (2010)
	Partial least squares (PLS)	Linear	Maitra and Yan (2008)
	Soft independent modelling of class analogy (SIMCA)	Linear	Maesschalck et al. (1999)
	Partial least squares discriminant analysis (PLS-DA)	Linear	Maitra and Yan (2008)

variables or principal components (PC). These orthogonal variables or principal components (PCs) are used to reconstruct the original data matrix. On the basis of the orthogonal variables or principal components (PCs) variable scores, different PCs are clustered together. The descriptor matrix X of principal component analysis (PCA), mathematically projected into a low-dimensional space, is endowing explainable comprehension of the rudimentary complicated

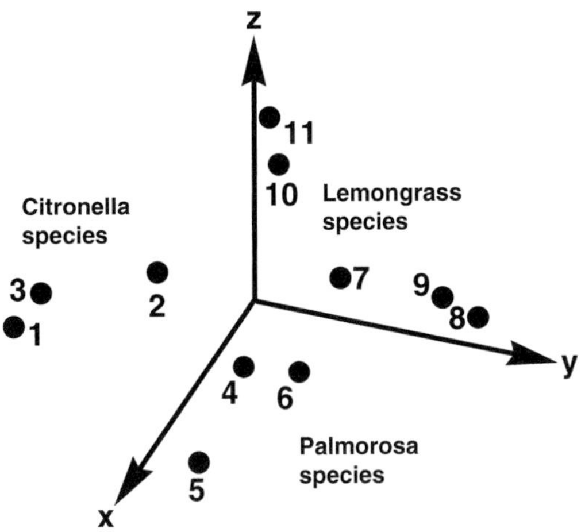

Fig. 4 Representative example of principal component analysis (PCA)-based alignment of *Cymbopogon* species. RAPD profiles were generated for *Cymbopogon* species (Sangwan et al. 2001)

data set by the mean deciphering similarities or differences among PCs (Fig. 4). The obtained score matrix gives intelligence about the kind of relationships between objects (e.g., trends, groupings, and outliers). The principal component analysis can also be used for cultivar colonization on the basis of oil trade type (Fig. 4).

Plant Metabolomics and Abiotic Stress

All organisms grow or develop in a healthy way inside a boundary of environmental circumstances; any divergence in these optimum environmental circumstances is believed to be stress, which is greatly responsible for losses in plant productivity. These stresses are also accountable for spatial or geographical as well as temporal or growing-season restrictions in the cultivation of crops. To comprehend the fundamental network underlying the stress responses, lots of scientific efforts have long been addressed. The search for the identification of specific genes and/or metabolites accountable for tolerance phenotypes in plants is continuing. With the advent of "omics" approaches, high-throughput analysis of changes induced by environmental stresses becomes easy and extends the zone of examination.

A wide range of osmolytes amass in plants upon exposure to osmotic stress to maintain the important cellular functions such as turgor pressure in plant cells. The different solutes amass in plant cells upon exposure to osmotic stress and reckon in different sugars, polyols, betaines, and amino acids (Yadav et al. 2014a; Sabir et al. 2012a; Shulaev et al. 2008). Different compounds have different roles like scrubbing reactive oxygen species (ROS) activity, maintaining structures of enzymes and proteins, membrane integrity, acting as osmoprotectants, acting as low–molecular weight chaperones, acting as chelating agents, redesigning lipids, stabilizing the photosystem II complex, etc. to mitigate the adverse effect caused by osmotic stress. Glycine, betaine, proline, and mannitol are some of the examples (Szabados and Savoure 2010).

Plant responses to different kinds of stress are different. For instance, plant responses to temperature variation represented by the growing season was the induced metabolic changes, plant responses to cold exposure were uniform allocation of metabolic changes, while heat-induced metabolic alterations in plants betide within the first 30 minutes (Gray and Heath 2005; Kaplan et al. 2004). Several compounds like amino acids, β-alanine, GABA, putrescine, fumarate, and malate as well as several carbohydrates were found affected by heat shock, and most of them were discovered to be correlated (Kaplan et al. 2004). Various metabolites like amino acids, polyamines, and several carbohydrates showed a very similar kind of response between the two thermal stresses. Most of these compounds are compatible solutes in the cytosol and/or precursors for secondary metabolite biosynthesis.

Water deficit stress or drought is a significant restriction for plant growth and productivity. The very first responses of the plant upon exposure to drought are retardation of vegetative growth, decline in photosynthetic rate, and stomata closure. A crucial mechanism of adaptation to water deficit stress in plants is osmotic adjustment to maintain cell turgor pressure by accumulating solutes, leading to reduced osmotic potential (Yadav et al. 2014a; Sabir et al. 2012a). The phytohormone abscisic acid (ABA) production in plants gets induced upon exposure to water deficit stress. The scientific literature supports involvement of both ABA-dependent as well as ABA-independent pathways in plant drought responses (Yamaguchi-Shinozaki and Shinozaki 2006). A complicated network of genes and metabolites is involved in the modulation of drought stress–mediated responses in plants, and several drought-inducible genes have been identified and characterized so far in several plant species. Nowadays, researchers are addressing the global metabolic changes induced in plants by water deficit stress (Yadav et al. 2014a; Sangwan et al. 1993, 1994). The levels of various metabolites like proline and phenolics involved in adaptation mechanisms were estimated in plants exposed to stress.

Metabolites involved in adaptation mechanisms and tolerance to different stresses act as antioxidants (Sangwan et al. 1993, 1994; Yadav et al. 2014a; Sabir et al. 2012a), and polyphenols have been highlighted in scientific literature. Dehydration initiates an increase in the content of proline as well as soluble carbohydrates. The localization of different metabolites like carbohydrates, amino acids, phenolics, etc., in plant cells depends on their possible function as cellular protectants against drought stress (Sangwan et al. 1993, 1994; Yadav et al. 2014a, b; Sabir et al. 2008). Osmotic adjustment, ionic balance in cells, and diverse molecular as well as biochemical alterations among different species are different plant adaptation responses addressed in scientific literature (Sangwan et al. 1993, 1994).

Plant Metabolomics-Integrated Functional Genomics

The term *metabolomics* is factually extensive, a nonbiased and high-throughput examination of complex metabolite mixtures of biological samples. This is likely a holistic approach to analyses of metabolomes. Recent advancement and development in different analytical tools as well as functional genomics approaches comprehensively boost the metabolomics study. The overview of metabolic approaches is very broad and very complex and requires establishment of a multifaceted and integrated strategy for best sample extraction, metabolite separation/detection/identification, metabolite quantification, and automated techniques of data gathering/handling/analysis. To achieve this goal, development of both analytical as well as computational techniques is imperative. Microbes as well as plants are the overall richest source of diverse metabolites. Several bacterial genomes are already sequenced, and it is discovered that only few hundred metabolites can be biosynthesized, for example, around 580 different metabolites from Bacillus subtilis and about 800 different metabolites from Escherichia coli. But in case of plants, the scenario is totally different. This value is likely to be in thousands to millions for even individual plants. Merely the number of genes present ranging between 20,000 and 50,000 in plants cannot produce such rich sources of metabolites, but the presence of several multiple-substrate-specific enzymes, the presence of subcellular compartments in a cell, and the occurrence of nonenzymatic reactions as well are also responsible. According to the available scientific literature, about 50,000 different compounds from plants have already been illuminated, and it is anticipated that around 200,000 or even more different compounds from the plant kingdom will approach in future (Pichersky and Gang 2005; Fiehn et al. 2001; Bino et al. 2004, Marja et al. 2005; Tohge et al. 2005). For example, in important medicinal plants, withanolides from Withania somnifera (Sangwan et al. 2008; Chaurasiya et al. 2007; Sangwan et al. 2004; Chaurasiya et al. 2007; Mishra et al. 2008), artemisinic metabolites from Artemisia annua (Yadav et al. 2014a), camphor, different monoterpenoid and sesquiterpenoid volatiles from Artemisia annua essential oil (Yadav et al. 2014a), menthol and related volatile monoterpenoids from Mentha arvensis (Bose et al. 2013), and elemol and volatiles from Dioscorea species (Joy et al. 2013) have been reported. Thus, metabolomics itself presents a ponderable challenge for plant scientists and researchers. With the discovery of high-density microarray techniques, next-generation sequencing technologies, and bioinformatics tools, whole genome and/or transcriptome analysis is possible in a high-throughput manner. Concretely, the description and analysis of the whole transcriptome empower scientists and researchers to identify, characterize, and localize the entire stream of pathway transcripts starting from the transcriptional factors and finally to the effector genes whose expression is regulated by specific transcriptional factors. The transcriptome explication helps researchers to discover coregulatory pathways by exploring up-/down-regulation of huge sets of transcripts associated with common regulatory and/or biosynthetic pathways (Sangwan et al. 2013). Thus, the study of metabolomics-integrated functional genomics is of great relevance to define the genetic elements of plant responses to abiotic stress conditions. The discovery of high-throughput sequencing technologies and bioinformatics tools has significantly increased the genesis as well as analysis of transcriptome/genome data and the competence of acquiring alteration in expression. These advancements in technologies and concept lead to determine transcription start sites at 5' UTR, polyadenylation signals at 3' UTR, alternative splice sites, as well as quantitative data on gene transcript abundance in tissues and/or cells (Sangwan et al. 2013). These next-generation sequencing technologies are deep-sequencing technologies exploring the route for global genomics as well as transcriptomics discoveries and will evidently lead to novel findings about plant abiotic stress responses.

Metabolomics-Embedded Plant Biotechnology

Plant metabolites are to play diverse roles in plants like resistance against pathogens, combat against stress responses, color, taste, and aroma of fruits and flowers (Sangwan et al. 2011). The interactions between the genotype and the environmental stimuli result in the biochemical phenotype of an organism; but these interactions are regulated to maintain the homeostasis by the fluctuations in intracellular physiological conditions of a cell/organ/organism (Weckwerth 2003). Thus, it is very critical to identify and quantify metabolites simultaneously for proper understanding of the dynamics of the metabolome. Fluxes in metabolic pathways under various stimuli decode the positive/negative role of metabolites. The ideal of metabolomics analysis is to discover the alteration in biochemical pathways as well as metabolic networks to find the relation with the physiological and/or developmental phenotype of a cell/tissue/organism (Weckwerth 2003). Scientific literature provides several instances where alteration in metabolic pathways result transformed plants with an augmented nutritional value like the case of genetically modified rice Golden Rice (GR) amass β-carotene in the endosperm (Ye et al. 2000). The use of genetically modified Golden Rice (GR) allowed detracting vitamin A deficiency, which is a major worldwide nutritional problem. GR2, an advanced variety of Golden Rice (GR), was developed by the overexpression of a phytoene synthase gene. GR2 amassed higher amounts of carotenoids (84 % of the total is β-carotene) than GR (Paine et al. 2005). Another example is the genetically modified plant with higher anthocyanin content. Anthocyanins are basically flavonoids, contributing to colors as well as antioxidant properties of plants. These metabolites are to play great roles in human health by caring against human diseases, but low contents in plants are major limitations to optimal benefits. Recently, a genetically modified tomato plant was developed with higher amounts of anthocyanins comparable to the high-anthocyanin-containing plants blackberries and blueberries (Butelli et al. 2008). These new varieties of genetically modified tomatoes have dark purple coloration with threefold higher antioxidant capacity. Nowadays, use of plant metabolomics to understand processes at a cellular level such as the cellular responses to stress conditions is in fashion. The increased production of monoterpenoid and sesquiterpenoid volatiles of Artemisia annua essential oil under prolonged water deficit stress is believed as an environmental perception of drought deficit conditions (Yadav et al. 2014a). Some applications related to metabolomics-embedded plant biotechnology are metabolic engineering of biochemical pathways, discovery of gene function, and engineering biochemical pathways for pharmaceutical production (Table 6).

Table 6 Some applications of metabolomics in plant biotechnology

Organism	Applications
Catharanthus roseus	Improvement of the production of anticancer indole alkaloid by overexpression of ORCA3 and G10H in *C. roseus* plants (Pan et al. 2012; Perez-Enciso and Tenenhaus 2003)
Panicum virgatum	Increased amounts of phenolic acids and a monolignol analog associated with more facile cell wall deconstructions (Niazi and Azizi 2008; Tschaplinski et al. 2012)
Solanum tuberosum (L)	Increased drought tolerance by expression of trehalose-6-phosphate synthase 1 (Jombart et al. 2010; Kondrak et al. 2012)
Oryza sativa	Modulation of salt tolerance by reduction of *OsSUT1* (*O. sativa* sucrose transporter 1) (Maitra and Yan 2008; Siahpoosh et al. 2012)
Arabidopsis thaliana	Distinguish transgenic and non-transgenic plants (Maesschalck et al. 1999; Ren et al. 2009)
Solanum lycopercium	Higher accumulation of flavonoids and thus nutritional value in tomato plants carrying a mutation in HPI/LeDDBI gene (Geladi and Kowalski 1986; Calvenzani et al. 2010)
Withania coagulans	Accumulation of withanoloids in agrobacterium-mediated genetically transformed *Withania coagulans* (Jackson 1991; Mishra et al. 2012)

Metabolomics Resources

Huge amounts of metabolomics data are generated in metabolomics experiments. To properly store and retrieve metabolomics data, need of databases arises. There are various metabolomics resources and metabolite databases available nowadays, as listed in Tables 7 and 8. The Golm Metabolome Database holds knowledge of mass spectra from active metabolites identified and quantified by gas chromatography–mass spectrometry (GC-MS) and gas chromatography (GC). The Madison Metabolomics Database is a resource for metabolomics research based on nuclear magnetic resonance (NMR) and mass spectrometry (MS). Metabolights, another integrative database, is cross species, having information about structures of different metabolites, spectra of different metabolites, as well as biological roles of these metabolites.

Metabolomics at Rothamsted (MeT-RO) is an initiative metabolomics resource holding information about several resources helpful to plant and microbial metabolomics. Another metabolic database, the Metlin Metabolite Database, consists of information about 55,000 metabolites, almost 50,000 higher solution tandem mass spectrometry (MS/MS) spectra and tandem mass spectrometry experiments. PRIMe is a platform for RIKEN metabolomics and is a database concatenating genomic as well as metabolomics data. This database consists of information about metabolites obtained from different instruments like nuclear magnetic resonance (NMR) spectroscopy, gas chromatography–mass spectrometry (GC-MS), liquid chromatography–mass spectrometry (LC-MS), and capilary electrophoresis–mass spectrometry (CEMS). Several databases dedicated to plant species have also been developed.

Plant metabolomics is one example and initially started as a metabolomics and functional genomics tool for explaining the functions of Arabidopsis genes. Databases on particular plants are also available like Metabolome Tomato Database having information on metabolites identified by liquid chromatography–mass spectrometry (LC-MS). A closely related database is Terpmed, specifically for terpenoids, which contains information about plant terpenoids, natural products, and other secondary metabolites of therapeutic uses. The Armec Repository Project was created as a tool to annotate flow injection electrospray MS (FIE-MS) data, but information about HPLCESI-MS can also be found in this database. Besides, the Armec database is expanding and including information about additional species like food crops and human metabolomes for use in nutrition research. There are several database resources that combine metabolite data with metabolic pathways like MetaCyc that hold information about 1,800 pathways from more than 2,000 organisms.

Table 7 Omics approaches constitute an integral aspect of plant metabolomics

S. no.	Name of Web resources	Application of resources
1	Human Metabolome Database (http://www.hmdb.ca/)	Chemical and biological data of human metabolites
2	Golm Metabolome Database (http://gmd.mpimp-golm.mpg.de/)	GC-MS
3	Madison Metabolomics Consortium Database (http://mmcd.nmrfam.wisc.edu/)	NMR and MS
4	MetaboLights (http://www.ebi.ac.uk/metabolights/)	Metabolite structures, spectra, function/cross-species
5	Metabolomics at Rothamsted (http://www.metabolomics.bbsrc.ac.uk/MeT-RO.htm)	Plant and microbial metabolites
6	METLIN Metabolite Database (http://metlin.scripps.edu/)	High-resolution MS/MS spectra and tandem MS experiments
7	PRIMe (http://prime.psc.riken.jp/)	Genomic and metabolomics data/NMR spectroscopy, GC-MS, LC-MS and CE-MS
8	Plantmetabolomics (http://plantmetabolomics.vrac.iastate.edu/ver2/)	*Arabidopsis* and other plants
9	Metabolome Tomato Database (http://appliedbioinformatics.wur.nl)	Metabolites identified by LC-MS
10	TERPMED (http://www.terpmed.eu/databases.html)	Plant terpenoids, natural products, secondary metabolites/therapeutic drugs
11	BiGG Database (http://bigg.ucsd.edu/)	Systems biology simulation and metabolic flux balance modelling, comprehensive information of gene, proteins and metabolites
12	SetupX (http://fiehnlab.ucdavis.edu/projects/binbase_setupx)	Web-based metabolomics, the capture and display of GC-MS metabolomics data
13	BinBase (http://fiehnlab.ucdavis.edu/projects/binbase_setupx)	GC-TOF metabolomics database
14	SYSTOMONAS (SYSTems biology of pseudOMONAS) (http://www.systomonas.de)	Contains extensive transcriptomic, proteomic and metabolomics data as well as metabolic reconstructions of this pathogen
15	KEGG (Kyoto Encyclopedia of Genes and Genomes) (http://www.genome.jp/kegg/)	Information of metabolomes of wide variety of organisms (>700), >15,000 compounds (from animals, plants and bacteria), 7,742 drugs, ~11,000 glycan structures
16	HumanCyc (Encyclopedia of Homo sapiens Genes and Metabolism) (http://humancyc.org/)	A bioinformatics database that describes the human genome and human metabolic pathways
17	BioCyc (Pathway/Genome Databases and Pathway Tools Software) (http://biocyc.org/)	Collection of 371 pathway/genome databases, each database describes the genome organisation and metabolic pathways of a single individual
18	Reactome (http://www.reactome.org/)	Database of biological pathways, protein trafficking and signalling pathways and metabolic pathways
19	Chemical Entities of Biological Interest (ChEBI) (http://www.ebi.ac.uk/chebi/)	Dictionary of small, chemical compounds
20	PubChem (http://pubchem.ncbi.nlm.nih.gov/)	Database consists of chemical structure of small, organic molecules and information of their biological activities
21	ChemSpider (http://www.chemspider.com/)	Database of organic molecules

(continued)

Table 7 (continued)

S. no.	Name of Web resources	Application of resources
22	KEGG GLYCAN (http://www.genome.jp/kegg/glycan/)	Collection of experimentally determined glycan structures
23	In Vivo/In Silico Metabolites Database (IIMDB) (http://metabolomics.pharm.uconn.edu/iimdb/)	Database includes ~23,000 known compounds (mammalian metabolites, drugs, secondary plant metabolites and glycerophospholipids) and >400,000 computationally generated human phase-I and phase-II metabolites of these known compounds
24	DrugBank (http://www.drugbank.ca/)	Comprises of detailed information about drug (i.e. chemical, pharmacological and pharmaceutical) data and comprehensive drug–target information.
25	Therapeutic Target Database (TTD) (http://bidd.nus.edu.sg/group/cjttd/)	Provides information about known therapeutic protein and nucleic acid targets, disease condition, pathway information and drugs/ligands against these targets
26	The Pharmacogenomics Knowledge Base (PharmGKB database) (http://www.pharmgkb.org/)	Repository for genetic, genomic, molecular and cellular phenotype data as well as clinical data of people who participated in pharmacogenomics studies
27	SuperTarget (http://insilico.charite.de/supertarget/)	Contains data set of ~7,300 drug–target relations, tools for 2D drug screening and sequence comparison
28	STITCH (search tool for interactions of chemicals) (http://stitch.embl.de/)	Includes information about metabolic pathways, crystal structure and drug–target interactions
29	BioMagResBank (BMRB) (http://www.bmrb.wisc.edu/)	Repository for experimental NMR spectral data, mainly for macromolecules
30	METLIN Metabolite Database (http://metlin.scripps.edu/)	Repository of metabolic information and tandem mass spectrometry data
31	High-Quality Mass Spectral Database (MassBank) (http://www.massbank.jp/)	Mass spectral database containing high-resolution MS spectra of metabolites
32	Madison Metabolomics Consortium Database (MMCD) (http://mmcd.nmrfam.wisc.edu/)	Database of small molecules of biological significance
33	Birmingham Metabolite Library Nuclear Magnetic Resonance database (BML-NMR) (http://www.bml-nmr.org/)	It contains 3,328 NMR spectra of 208 common metabolite standards (includes both 2-D 1H J-resolved spectra and 1-D 1H spectra, recorded at 500 MHz)
34	Fiehn GC-MS Database (http://fiehnlab.ucdavis.edu/Metabolite-Library-2007)	The collection has data on 713 compounds (name, structure, CAS ID, other links) for which GC/MS data is provided
35	On-Line Metabolic and Molecular Basis of Inherited Disease (OMMBID) (http://www.ommbid.com)	Web-assessable library of number of metabolic disorders including the genetics, metabolism, diagnosis and treatment
36	METAGENE (http://www.metagene.de/)	The database defines the errors in metabolism, (highlighting genetic cause, disease, treatment and the metabolite concentration)
37	COLMAR Metabolomics Web Portal (http://spinportal.magnet.fsu.edu/)	Web server suite which generates a covariance NMR spectrum and NMR data of individual components in metabolomics mixtures

(continued)

Table 7 (continued)

S. no.	Name of Web resources	Application of resources
38	Biological Magnetic Resonance Data Bank (BMRB) (http://www.bmrb.wisc.edu/)	Data from NMR spectroscopy on proteins, peptides, nucleic acids and other biomolecules
39	SUGABASE (http://glycomics.ccrc.uga.edu/GlycomicsPortal/)	A carbohydrate NMR database from the Netherlands
40	ChEMBL (http://www.ebi.ac.uk/chembl/)	Database consists of small drug-like molecules
41	Cambridge Structural Database (CSD)	Repository of 500,000 small molecules and their crystal structures
42	(http://www.ccdc.cam.ac.uk/prods/csd/csd.html)	
43	KEGG LIGAND (http://www.genome.jp/ligand/)	Database includes compound, reactions, Rclass, Rpair and enzymes
44	Klotho (http://www.biocheminfo.org/klotho)	Classification and collection of biological compounds
45	AANT (http://aant.icmb.utexas.edu/)	Nucleotide–amino acid interaction database
46	LIGAND (http://genome.ad.jp/ligand/)	Database has chemical compounds and reactions occurring in biological pathways
47	ZINC (http://zinc.docking.org/)	Consists of commercially available compounds for screening purpose
48	FDA drug (http://www.centerwatch.com/patient/drugs/druglsal.html)	Comprises of drug list, approved by FDA

The table summarises the available bioinformatics resources and databases available for comprehensive analysis of metabolite in plant metabolome

Major Challenges in Plant Metabolomics

The metabolomics analyses require high-throughput technologies aiming to rapid separation, identification and structure elucidation. Despite of its immense significance and valuable contribution to plant biotechnology, the science of plant metabolomics withholds some major challenges which need to be addressed. The science of plant metabolomics has flourished both in the development of new analytical techniques as well as in the expansion of application areas (Mendes 2002). Past studies focused on profiling of metabolites and fingerprinting, but the canvas of metabolomics has expanded its horizon to study of flux pumps. It has become an integral aspect of system biology with the predominance of NMR and mass spectrometry methods. However, computational approaches offer the biggest challenge in comprehensive analysis of metabolome of an organism. The key areas of research includes the identification of metabolic networks which influence metabolite profiles, elucidation of the chemical structures of large number of metabolites which are already detected and to establish standards for data format and analysis (Mendes 2002). A co-integrated approach including proteomics, transcriptomics and metabolomics application would provide the detailed information of the biological system, but the authentication of standards would make this feasible. The next level aims at the integration of these data in global model defining the complete metabolome. Assessment of genetically engineered crops is another key area which requires the application of metabolomics. The genetic alteration in crops might result in undesirable/harmful traits which might prove hazardous for human consumption. Since, metabolomics involves the analysis of metabolites at the basic level, the careful assessment and monitoring of changes occurring at the genome, transcript and metabolic levels in GM crops becomes a necessity. Further, the generated data should be further analysed and integrated in terms of system biology, thereby avoiding the unpredictable effects in GM crops (Oksman-Caldentey and Inze 2004; Rischer and Oksman-Caldentey 2006). Despite the major advances made in the field of plant

Table 8 Bioinformatics resources utilised for analysis of proteins in plant metabolome

S. No.	Web resources	Application on protein analysis
1	CSA (http://www.ebi.ac.uk/thorntonsrv/databases/CSA/)	Contains data on catalytic residues and active site in enzymes with known 3D structures
2	PDBSite (http://srs6.bionet.nsc.ru/srs6/)	3D structure of protein functional regions
3	ANTIMIC (http://research.i2r.a-star.edu.sg/Templar/DB/ANTIMIC)	Database is a collection of antimicrobial peptides
4	Genomic Threading Database GTD (http://bioinf.cs.ucl.ac.uk/GTD)	Structural annotations of the entire proteomes
5	MMDB (http://www.ncbi.nlm.nih.gov/Structure)	Database of 3D structures (part of NCBI, Entrez)
6	E-MSD (http://www.ebi.ac.uk/msd)	Database of macromolecular structure (part of ERI)
7	PDBsum (http://www.ebi.ac.uk/thornton-srv/databases/pdbsum/)	Analysis of total PDB structures
8	GenDiS (http://caps.ncbs.res.in/gendis/home.html)	Classifies the distribution of protein structural superfamilies in genome
9	ISSD (http://www.protein.bio.msu.su/issd)	Includes the integrated data of sequence structure of a protein
10	Protein Folding Database (http://pfd.med.monash.edu.au)	Includes experimental information on protein folding
11	SCOP (http://scop.mrc-lmb.cam.ac.uk/scop)	Provides a detailed account of structural classification of proteins
12	3D-GENOMICS (http://www.sbg.bio.ic.ac.uk/3dgenomics)	Includes structural annotation of complete proteomes
13	SWISS-MODEL Repository (http://swissmodel.expasy.org/repository)	Annotated 3D protein models
14	ASC (http://bioinformatica.isa.cnr.it/ASC)	Collection of biologically active peptides
15	eMOTIF (http://motif.stanford.edu/emotif)	Determination of motifs in protein sequence
16	PASS2 (http://ncbs.res.in/~faculty/mini/campass/pass.html)	Comprises of structural motifs of protein superfamilies
17	GTOP (http://spock.genes.nig.ac.jp/~genome/)	Prediction of protein folds from genome sequences
18	ModBase (http://salilab.org/modbase)	Includes annotated comparative structure models

research, the underlying biological mechanisms that contribute to biosynthesis of metabolites and their enormous diversity remain less understood. The science of plant metabolomics has coevolved with the allied branches of proteomics and transcriptomics; however, certain major task needs to be accomplished for metabolome analysis, for instance, the unavailability of a universal method for isolation and quantification of the total metabolites present in a cell at a particular time. Clustering of the LC-MS and GC-MS data may be a prospective solution to this problem (Jones et al. 2003; Bednarek et al. 2005; Jing et al. 2009; Pu et al. 2009; Banyai et al. 2010; Maes et al. 2010; Mishra and Srivastava 1991; El-Khateeb 1994; Farooqi et al. 1994; Verpoorte et al. 1993; Werrman and Knorr 1993; Fett-Neto et al. 1992; Kyung-Hee et al. 1994; Sabir et al. 2007, 2010, 2011, 2012b; Goossens et al. 2003; Mendes 2002; Rischer and Oksman-Caldentey 2006; Krizevski 2009). Another key approach to decode the complex biosynthetic mechanisms of secondary metabolites includes the combination of data mining metabolic module and science of proteomics and transcriptomics, respectively (Jones et al. 2003; Bednarek et al. 2005; Jing et al. 2009; Pu et al. 2009; Banyai et al. 2010; Maes et al. 2010; Mishra and Srivastava 1991; El-Khateeb 1994; Farooqi et al. 1994; Verpoorte et al. 1993; Werrman and Knorr 1993; Fett-Neto et al. 1992; Kyung-Hee et al. 1994; Sabir et al. 2007, 2010, 2011, 2012b; Goossens et al. 2003; Mendes 2002; Rischer and Oksman-Caldentey 2006; Krizevski 2009). Further, the integration of methodologies with accurate and powerful predictions from different areas of scientific fields is required for complete analysis of metabolic content, the 'metabolome' of a biological system.

Acknowledgements Authors are thankful to DBT, CSIR and DST, New Delhi, for the financial support to various projects run in our laboratory, which have generated data related with metabolism. Authors are thankful to Director CSIR-CIMAP, Lucknow, for constant support and encouragement.

References

Allwood JW, Ellis DI, Goodacre R (2008) Metabolomic technologies and their application to the study of plants and plant-host interactions. Physiol Plant 132:117–135

Anderson TW (1984) An introduction to multivariate statistical analysis, 2nd edn. John Wiley & Sons, New York

Banyai W, Mii M, Supaibulwatana K (2010) Enhancement of artemisinin content and biomass in *Artemisia annua* by exogenous GA3 treatment. Plant Growth Regul 63:45–54

Bednarek P, Schneider B, Svatos A, Oldham NJ, Hahlbrock K (2005) Structural complexity, differential response to infection, and tissue specificity of indolic and phenylpropanoid secondary metabolism in Arabidopsis roots. Plant Physiol 138:1058–1070

Berridge MJ, Irvine RF (1989) Inositol phosphates and cell signalling. Nature 341:197–205

Beuerle T, Schwab W (1999) Metabolic profile of linoleic acid in stored apples: formation of 13\(R)-hydroxy-9(Z), 11(E)-octadecadienoic acid. Lipids 34:375–380

Bharti SK, Bhatia A, Tewari SK, Sidhu OP, Roy R (2011) Application of HR-MAS NMR spectroscopy for studying chemotype variations of *Withania somnifera* (L.) Dunal. Magn Reson Chem 49:659–667

Bino RJ, Hall RD, Fiehn O, Kopka J, Saito K, Draper J, Nikolau BJ, Mendes P, Roessner-Tunali U, Beale MH, Trethewey RN, Lange BM, Wurtele ES, Sumner LW (2004) Potential of metabolomics as a functional genomics tool. Trends Plant Sci 9:418–425

Birkemeyer C, Kolasa A, Kopka J (2003) Comprehensive chemical derivatization for gas chromatography-mass spectrometry-based multi-targeted profiling of the major phytohormones. J Chromatogr A 993:89–102

Boccard J, Rutledge DN (2013) A consensus orthogonal partial least squares discriminant analysis (OPLS-DA) strategy for multiblock omics data fusion. Anal Chim Acta 69:30–39

Boros LG, Lerner MR, Morgan DL, Taylor SL, Smith BJ, Postier RG, Brackett DJ (2005) [1,2-13C2]-D-glucose profiles of the serum, liver, pancreas, and DMBA-induced pancreatic tumors of rats. Pancreas 31:337–343

Bose SK, Yadav RK, Mishra S, Sangwan RS, Singh AK, Mishra B, Srivastava AK, Sangwan NS (2013) Effect of gibberellic acid and calliterpenone on plant growth attributes, trichomes, essential oil biosynthesis and pathway gene expression in differential manner in *Mentha arvensis* L. Plant Physiol Biochem 66C:150–158

Butelli E, Titta L, Giorgioetal M (2008) Enrichment of tomato fruit with health-promoting anthocyanins by expression of select transcription factors. Nat Biotechnol 26:1301–1308

Calvenzani V, Martinelli M, Lazzeri V (2010) Response of wild-type and high pigment-1 tomato fruit to UV-B depletion: flavonoid profiling and gene expression. Planta 231:755–765

Catchpole GS, Beckman M, Enot DP, Mondhe M, Zywicki B, Taylor J, Hardy N, Smith A, King RD, Kell DB, Fiehn O, Draper J (2005) Hierarchical metabolomics demonstrates compositional similarity between genetically modified and conventional potato crops. Proc Natl Acad Sci U S A 102:14458–14462

Ceulemans E, Hubert M, Rousseeuw P (2013) Robust multilevel simultaneous component analysis. Chemom Lab Intell Syst 129:1–20

Chatterjee S, Srivastava S, Khalid A, Singh N, Sangwan RS, Sidhu OP, Roy R, Khetrapal CL, Tuli R (2010) Comprehensive metabolic fingerprinting of *Withania somnifera* leaf and root extracts. Phytochemistry 71:1085–1094

Chaurasiya ND, Uniyal GC, Lal P, Mishra LN, Sangwan NS, Tuli R, Sangwan RS (2007) Analysis of withanolides in root and leaf of *Withania somnifera* by HPLC with photodiode array and evaporating light scattering detection. Phytochem Anal 19:148–154

Chaurasiya ND, Sangwan NS, Sabir F, Misra L, Sangwan RS (2012) Withanolide biosynthesis recruits both mevalonate and DOXP pathways of isoprenogenesis in Ashwagandha, *Withania somnifera* L. (Dunal). Plant Cell Rep 31:1889–1897

Cheung HY, Lai WP, Cheung MS, Leung FM, Hood DJ, Fong WF (2003) Rapid and simultaneous analysis of some bioactive components in *Eucommia ulmoides* by capillary electrophoresis. J Chromatogr A 989:303–310

Colebatch G, Desbrosses G, Ott T, Krusell L, Montanari O, Kloska S, Kopka J, Udvardi MK (2004) Global changes in transcription orchestrate metabolic differentiation during symbiotic nitrogen fixation in Lotus japonicus. Plant J 39:487–512

Coleman RA, Lee DP (2004) Enzymes of triglyceride synthesis and their regulation. Prog Lipid Res 43:134–176

Croteau R, Kutchan TM, Lewis NG (2000) Natural products (secondary metabolites). In: Buchanan BB, Gruissem W, Jones R (eds) Biochemistry and molecular biology of plants. American Society of Plant Biologists, Rockville, pp 1250–1318

Deacon J (2005) Fungal biology. Blackwell Publishers, Cambridge, MA, p 342. ISBN 1-4051-3066

Dunn WB, Bailey NJ, Johnson HE (2005) Measuring the metabolome: current analytical technologies. Analyst 130:606–625

Duran AL, Yang J, Wang L, Sumner LW (2003) Metabolomics spectral formatting, alignment and conversion tools (MSFACTs). Bioinformatics 19:2283–2293

Edwards JS, Palsson BO (2000) The *Escherichia coli* MG1655 *in silico* metabolic genotype: its definition, characteristics, and capabilities. Proc Natl Acad Sci U S A 97:5528–5533

El-Khateeb MA (1994) Effect of some growth regulators on growth, fruit yield and essential oil in dill plant. Bull Fac Sci Cairo Univ 45:187–205

Facchini PJ, Bird DA, St-Pierre B (2004) Can Arabidopsis make complex alkaloids? Trends Plant Sci 9:116–122

Fan Z, Liu E, Xu B (2011) Weighted principal component analysis. Artif Intell Comput Intell 7004:569–574

Farooqi AHA, Shukla YN, Sharma S, Bansal RP (1994) Relationship between gibberellin and cytokinin activity and flowering in *Rosa damascena* Mill. Plant Growth Regul 14:109–113

Ferreira JFS, Gonzalez JM (2008) Analysis of underivatized artemisinin and related sesquiterpene lactones by high-performance liquid chromatography with ultraviolet detection. Phytochem Anal 20:91–97

Fett-Neto AG, DiCosmo F, Reynolds WF, Sakata K (1992) Cell culture of Taxus as a source of the antineoplastic drug taxol and related taxanes. Biotechnology 10:12–15

Fiehn O (2002) Metabolomics-the link between genotypes and phenotypes. Plant Mol Biol 48:155–171

Fiehn O, Weckwerth W (2003) Deciphering metabolic networks. Eur J Biochem 270:579–588

Fiehn O, Kopka J, Dormann P, Altmann T, Trethewey RN, Willmitzer L (2000) Metabolite profiling for plant functional genomics. Nat Biotechnol 18:1157–1161

Fiehn O, Kloska S, Altmann T (2001) Integrated studies on plant biology using multiparallel techniques. Curr Opin Biotechnol 12:82–86

Ganjewala D, Kumar S, Asha Devi S, Ambika K (2010) Advances in cyanogenic glycosides biosynthesis and analyses in plants: a review. Acta Biol Szeged 54:1–14

Geladi P, Kowalski BR (1986) Partial least-squares regression: a tutorial. Anal Chim Acta 185:1–17

Goodacre R (2005) Making sense of the metabolome using evolutionary computation: seeing the wood with the trees. J Exp Bot 56:245–254

Goossens A, Hakkinen ST, Laakso I, Seppanen-Laakso T, Biondi S, De-Sutter V, Lammertyn F, Nuutila AM, Soderlund H, Zabeau M (2003) A functional genomics approach toward the understanding of secondary metabolism in plant cells. Proc Natl Acad Sci U S A 100:8595–8600

Gray GR, Heath D (2005) A global reorganization of the metabolome in Arabidopsis during cold acclimation is revealed by metabolic fingerprinting. Physiol Plant 124:236–248

Guo Y, Hastie T, Tibshirani R (2007) Regularized discriminant analysis and its application in microarray. Biostatistics 8:86–100

Hall RD (2006) Plant metabolomics: from holistic hope, to hype, to hot topic. New Phytol 169:453–468

Holzl G, Dormann P (2007) Structure and function of glycoglycerolipids in plants and bacteria. Prog Lipid Res 46:225–243

Jackson JE (1991) A user's guide to principal components. Wiley, New York. ISBN 1-58025-493-4

Jing F, Zhang L, Li M, Tang Y, Wang Y, Wang Y, Wang Q, Pan Q, Wang G, Tang K (2009) Abscisic acid (ABA) treatment increases artemisinin content in *Artemisia annua* by enhancing the expression of genes in artemisinin biosynthetic pathway. Biologia 64:319–323

Jombart T, Devillard S, Balloux F (2010) Discriminant analysis of principal components: a new method for the analysis of genetically structured populations. BMC Genet 11:94–101

Jones P, Messner B, Nakajima J, Schaffner AR, Saito K (2003) UGT73C6 and UGT78D1, glycosyltransferases involved in flavonol glycoside biosynthesis in *Arabidopsis thaliana*. J Biol Chem 278:43910–43918

Joy IO, Olukemi OA, Yadav RK, Chanotiya CS, Ogbonnia SA, Sangwan NS (2013) A new source of elemol rich essential oil and existence of multicellular oil glands in leaves of the *Dioscorea* species. Sci World J 2013:Article ID 943598, 6 pages. http://dx.doi.org/10.1155/2013/943598

Kaplan F, Kopka J, Haskell DW, Zhao W, Schiller KC, Gatzke N, Sung DY, Guy CL (2004) Exploring the temperature-stress metabolome of Arabidopsis. Plant Physiol 136:4159–4168

Keurentjes JJB, Fu JY, de Vos CHR, Lommen A, Hall RD, Bino RJ, van der Plas LHW, Jansen RC, Vreugdenhil D, Koornneef M (2006) The genetics of plant metabolism. Nat Genet 38:842–849

Kim JK, Bamba T, Harada K, Fukusaki E, Kobayashi A (2007) Time-course metabolic profiling in *Arabidopsis thaliana* cell cultures after salt stress. J Exp Bot 58:415–424

Kjaer A, Verstappen F, Bouwmeester H, Ivarsen E, Frette X, Christensen LP, Grevsen K, Jensen M (2013) Artemisinin production and precursor ratio in full grown *Artemisia annua* L. plants subjected to external stress. Planta 237:955–966

Klamt S, Stelling J, Ginkel M, Gilles ED (2003) Flux analyzer: exploring structure, pathways, and flux distributions in metabolic networks on interactive flux maps. Bioinformatics 19:261–269

Kohonen T, Honkela T (2011) Kohonen network. Scholarpedia, 2:15–68, Retrieved 24 Sept 2012

Kondrak M, Marincs F, Antal F, Juhasz Z, Banfalvi Z (2012) Effects of yeast trehalose-6-phosphate synthase 1 on gene expression and carbohydrate contents of potato leaves under drought stress conditions. BMC Plant Biol 12:74–80

Krizevski RLE (2009) Digitizing the metabolome. J Exp Bot 60:3–4

Kushwaha AK, Sangwan NS, Tripathi S, Sangwan RS (2013) Molecular cloning and catalytic characterization of a recombinant tropine biosynthetic tropinone reductase from *Withania coagulans* leaf. Gene 516:238–247

Kushwaha AK, Sangwan NS, Trivedi PK, Mishra LN, Sangwan RS (2014) Tropine forming tropinine reductase gene from *Withania somnifera*: biochemical characteristics of the recombinant enzyme and novel physiological overtones of tissue wide gene expression patterns. PLoS One 8:744–777

Kyung-Hee J, Sang-Soo K, Cha-Yong C, Jang Ryol L (1994) Development of two stage culture process by optimization of inorganic salts for improving catharanthine production in hairy root cultures of *Catharanthus roseus*. J Ferment Bioeng 77:57–61

Larson NB, Jenkins GD, Larson MC, Vierkant RA, Sellers TA, Phelan CM, Schildkraut JM, Sutphen R, Pharoah PP, Gayther SA, Wentzensen N et al (2014) Kernel canonical correlation analysis for assessing gene-gene interactions and application to ovarian cancer. Eur J Hum Genet 22:126–131. doi:10.1038/ejhg.2013.69

Maes L, Van Nieuwerburgh FCW, Zhang Y, Reed DW, Pollier J, Vande-Casteele SRF, Inze D, Covello PS, Deforce DLD, Gossens A (2010) Dissection of the phytohormonal regulation of trichome formation and biosynthesis of the antimalarial compound artemisinin in *Artemisia annua* plants. New Phytol 189:176–189

Maesschalck RD, Candolfi AD, Massart L, Heuerding S (1999) Decision criteria for soft independent modelling of class analogy applied to near infrared data. Chemom Intell Lab Syst 47:65–77

Maitra S, Yan J (2008) Principle component analysis and partial least squares: two dimension reduction techniques for regression. In: Paper presented at Casualty Actuarial Society, Discussion Paper Program, Seattle, pp 79–90

Makretsov NA, Huntsman DG, Nielsen TO, Yorida E, Peacock M, Cheang MC, Dunn SE, Hayes M, van de Rijn M, Bajdik C, Gilks CB (2004) Hierarchical clustering analysis of tissue microarray immunostaining data identifies prognostically significant groups of breast carcinoma. Clin Cancer Res 15:6143–6151

Mannan A, Liu C, Arsenault PR, Towler MJ, Vail DR, Lorence A, Weathers PJ (2010) DMSO triggers the generation of ROS leading to an increase in artemisinin and dihydroartemisinic acid in *Artemisia annua* shoot cultures. Plant Cell Rep 29:143–152

Marja K, Caldentey O, Saito K (2005) Integrating genomics and metabolomics for engineering plant metabolic pathways. Curr Opin Biotechnol 16:174–179

Markuszewski MJ, Otsuka K, Terabe S, Matsuda K, Nishioka T (2003) Analysis of carboxylic acid metabolites from the tricarboxylic acid cycle in *Bacillus subtilis* cell extract by capillary electrophoresis using an indirect photometric detection method. J Chromatogr A 1010:113–121

Marshall AG, Hendrickson CL, Jackson GS (1998) Fourier transform ion cyclotron resonance mass spectrometry: a primer. Mass Spectrom Rev 17:1–35

Mendes P (2002) Emerging bioinformatics for the metabolome. Brief Bioinform 3:134–145

Michalopoulos D, Hu C-K (2002) An error back-propagation artificial neural networks application in automatic car license plate recognition. Dev Appl Artif Intell 2358:1–8

Miller JC, Miller JN (1993) Statistics for analytical chemistry, 3rd edn. Ellis Horwood, PTR Prentice Hall, New York. ISBN 0130309907

Mishra A, Srivastava NK (1991) Effect of tricontanol formulation 'Miraculan' on photosynthesis, growth, nutrient uptake and essential oil yield of lemongrass (*Cymbopogon flexuosus*). Plant Growth Regul 10:57–63

Mishra LN, Mishra P, Pandey A, Sangwan RS, Sangwan NS, Tuli R (2008) Withanoloids from *Withania somnifera* roots. Phytochemistry 69:1000–1004

Mishra S, Sangwan RS, Bansal S, Sangwan NS (2012) Efficient genetic transformation of *Withania coagulans* (Stocks) Dunal mediated by *Agrobacterium tumefaciens* from leaf explants of *in vitro* multiple shoot culture. Protoplasma 250:451–459

Niazi A, Azizi A (2008) Orthogonal signal correction-partial least squares method for simultaneous spectrophotometric determination of nickel, cobalt, and zinc. Turk J Chem 32:217–228

Northen TR, Yanes O, Northen MT, Marrinucci D, Uritboonthai W, Apon J, Golledge SL, Nordstrom A, Siuzdak G (2007) Clathrate nanostructures for mass spectrometry. Nature 449:1033–1036

Oksman-Caldentey KM, Inze D (2004) Plant cell factories in the post-genomic era: new ways to produce designer secondary metabolites. Trends Plant Sci 9:433–440

Paine JA, Shipton CA, Chaggar S (2005) Improving the nutritional value of Golden Rice through increased pro-vitamin A content. Nat Biotechnol 23:482–487

Pan Q, Wang Q, Yuan F (2012) Overexpression of ORCA3 and G10H in *Catharanthus roseus* plants regulated alkaloid biosynthesis and metabolism revealed by NMR-metabolomics. PLoS One 7:38–45

Pauli GF (2001) qNMR- a versatile concept for the validation of natural product reference compounds. Phytochem Anal 12:28–42

Perez-Enciso M, Tenenhaus M (2003) Prediction of clinical outcome with microarray data: a partial least squares discriminant analysis (PLS-DA) approach. Hum Genet 112:581–592

Pichersky E, Gang DR (2005) Genetics and biochemistry of secondary metabolites in plants: an evolutionary perspective. Trends Plant Sci 5:439–445

Pu GB, Ma DM, Chen JL, Ma LQ, Wang H, Li GF, Ye HC, Liu BY (2009) Salicylic acid activates artemisinin biosynthesis in *Artemisia annua* L. Plant Cell Rep 28:1127–1135

Ren Y, Wang T, Peng Y, Xia B, Qu LJ (2009) Distinguishing transgenic from non-transgenic Arabidopsis plants by 1H NMR-based metabolic fingerprinting. J Genet Genomics 36:621–628

Rhee SY, Dickerson J, Xu D (2006) Bioinformatics and its applications in plant biology. Annu Rev Plant Biol 57:335–360

Rischer H, Oksman-Caldentey KM (2006) Unintended effects in genetically modified crops: revealed by metabolomics? Trends Biotechnol 24:102–104

Roessner U, Bowne J (2009) What is metabolomics all about? Biotechniques 46:363–365

Roessner U, Wagner C, Kopka J, Trethewey RN, Willmitzer L (2000) Technical advance: simultaneous analysis of metabolites in potato tuber by gas chromatography-mass spectrometry. Plant J 23:131–142

Roessner U, Luedemann A, Brust D, Fiehn O, Linke T, Willmitzer L, Fernie AR (2001) Metabolic profiling allows comprehensive phenotyping of genetically or environmentally modified plant systems. Plant Cell 13:11–29

Roessner-Tunali U, Hegemann B, Lytovchenko A, Carrari F, Bruedigam C, Granot D, Fernie AR (2003) Metabolic profiling of transgenic tomato plants over-expressing hexokinase that the influence of hexose phosphorylation diminishes during fruit development. Plant Physiol 133:84–99

Rosipal R, Kramer N (2005) Overview and recent advances in partial least squares. SLSFS 2005. LNCS 3940:34–35

Ryan D, Robards K (2006) Metabolomics: the greatest omics of them all? Anal Chem 78:7954–7958

Sabir F, Sangwan NS, Chaurasiya ND, Misra LN, Tuli R, Sangwan RS (2007) Micro-propagation of *Withania somnifera* L. accessions from axillary meristem for rapid propagation and consistent withanolide productivity. J Herbs Spices Med Plants 13:118–128

Sabir F, Sangwan NS, Chaurasiya ND, Mishra LN, Sangwan RS (2008) *In Vitro* Withanolide production by *Withania somnifera* L. culture. Zeitschrift furNaturforschung 63c:409–412

Sabir F, Kumar A, Tiwari P, Pathak N, Sangwan RS, Bhakuni RS, Sangwan NS (2010) Bioconversion of artemisinin to its nonperoxidic derivative deoxyartemisinin through suspension cultures of *Withania somnifera* Dunal. Zeitschrift furNaturforschung 65:607–612

Sabir F, Sangwan RS, Singh J, Pathak N, Mishra LN, Sangwan NS (2011) Biotransformation of withanolides by cell suspension cultures of *Withania somnifera* (Dunal). Plant Biotechnol Rep 5:112–134

Sabir F, Sangwan RS, Kumar R, Sangwan NS (2012a) Salt stress-induced responses in growth and metabolism in callus cultures and differentiating in vitro shoots of Indian ginseng (*Withania somnifera* Dunal). J Plant Growth Regul 10:344–356

Sabir F, Mishra S, Sangwan RS, Jadaun JS, Sangwan NS (2012b) Qualitative and quantitative variations in withanolides and expression of some pathway genes during different stages of morphogenesis in *Withania somnifera* Dunal. Protoplasma 250:539–549

Sangwan RS, Sangwan NS (2000) Metabolic and molecular analysis of chemotypic diversity in aromatic grasses (Cymbopogon spp.). In: Kumar S, Dwivedi S, Kukreja AK, Sharma JR, Bagchi GD (eds) Aromatic grass monograph. CIMAP, Lucknow, pp 223–247

Sangwan RS, Farooqi AHA, Bansal RP, Sangwan NS (1993) Interspecific variation in physiological and metabolic responses to water stress in five species of *Cymbopogon*. J Plant Physiol 142:618–622

Sangwan NS, Farooqi AHA, Sangwan RS (1994) Effect of drought on growth and essential oil metabolism in lemongrass species. New Phytol 128:173–179

Sangwan NS, Yadav U, Sangwan RS (2001) Molecular analysis of genetic diversity in elite Indian cultivars of essential oil trade types of aromatic grasses (*Cymbopogon* species). Plant Cell Rep 20:437–444

Sangwan NS, Yadav U, Sangwan RS (2003) Genetic diversity among elite varieties of the aromatic grasses, *Cymbopogon martinii*. Euphytica 130:117–130

Sangwan RS, Chaurasiya ND, Mishra LN, Lal P, Uniyal GC, Sharma R, Sangwan NS, Suri AK, Qazi GN, Tuli R (2004) Phytochemical variability in commercial herbal products and preparation of *Withania somnifera* (Ashwagandha). Curr Sci 86:461–465

Sangwan RS, Chaurasiya ND, Lal P, Mishra LN, Uniyal GC, Tuli R, Sangwan NS (2007) Withanoloid A biogeneration, in *in vitro* shoot culture of ashwagandha (*Withania somnifera* Dunal), a main medicinal plant in Ayurveda. Chem Pharm Bull 55:1371–1375

Sangwan RS, Chaurasiya ND, Lal P, Mishra LN, Tuli R, Sangwan NS (2008) Root contained withanolide A is inherently *de novo* synthesized within roots in Ashwagandha (*Withania somnifera*). Physiol Plant 133:278–287

Sangwan NS, Kumar R, Srivastava S, Kumar A, Gupta A, Sangwan RS (2011) Recent developments on secondary metabolite biosynthesis in *Artemisia annua* L. J Plant Biol 37:1–24

Sangwan RS, Tripathi S, Singh J, Narnolia LK, Sangwan NS (2013) *De novo* sequencing and assembly of *Centella asiatica* leaf transcriptome for mapping of structural, functional and regulatory genes with special reference to secondary metabolism. Gene 55:58–76

Schauer N, Steinhauser D, Strelkov S, Schomburg D, Allison G, Moritz T, Lundgren K, Roessner-Tunali U, Forbes MG, Willmitzer L, Fernie AR, Kopka J (2005) GC-MS libraries for the rapid identification of

metabolites in complex biological samples. FEBS Lett 579:1332–1337
Schuhmacher R, Krska R, Weckwerth W, Goodacre R (2013) Metabolomics and metabolite profiling. Anal Bioanal Chem 405:5003–5004
Shanker S, AjayKumar PV, Sangwan NS, Kumar S, Sangwan RS (1999) Oil gland fine structure and metabolic attributes at different phases of leaf development and senescence in *Mentha arvensis*. Biol Plant 42:379–387
Shanks JV (2005) Phytochemical engineering: combining chemical reaction engineering with plant science. AIChE J 51:2–7
Shulaev V (2006) Metabolomics technology and bioinformatics. Brief Bioinform 7:128–139
Shulaev V, Cortes D, Miller G, Mittler R (2008) Metabolomics for plant stress response. Physiol Plant 132:199–208
Siahpoosh MR, Sanchez DH, Schlereth A (2012) Modification of OsSUT1 gene expression modulates the salt response of rice *Oryza sativa* cv. Taipe i309. Plant Sci 182:101–111
Sidhu OP, Annarao S, Chatterjee S, Tuli R, Roy R, Khetrapal CL (2011) Metabolic alterations of *Withania somnifera* (L.) dunal fruits at different developmental stages by NMR spectroscopy. Phytochem Anal 22:492–502
Smilde AK, Jansen JJ, Hoefsloot HC, Lamers R-J, Greef J v d, Timmerman ME (2005) ANOVA-simultaneous component analysis (ASCA): a new tool for analyzing designed metabolomics data. Bioinformatics 21:3043–3048
Stein SE, Scott DR (1994) Optimization and testing of mass spectral library search algorithms for compound identification. J Am Soc Mass Spectrom 5:859–866
Steuer R, Kurths J, Fiehn O, Weckwerth W (2003) Interpreting correlations in metabolomics networks. Biochem Soc Trans 31:1476–1478
Stuart BH (2012) Infrared spectroscopy of biological application: an overview. Encycl Anal Chem. doi:10.1002/97804700
Szabados L, Savoure A (2010) Proline: a multifunctional amino acid. Trends Plant Sci 15:89–97
Tiwari P, Mishra BN, Sangwan NS (2013) Phytochemical and pharmacological properties of *Gymnema Sylvestre*, an important medicinal plant. BioMed Res Int Article ID 830285, 18 pages. http://dx.doi.org/10.1155/2014/830285
Tohge T, Nishiyama Y, Hirai MN, Yano M, Nakajima J, Awazuhara M, Inoue E, Takahashi H, Goodenowe DB, Kitayama M, Noji M, Yamazaki M, Saito K (2005) Functional genomics by integrated analysis of metabolome and transcriptome of Arabidopsis plants overexpressing an MYB transcription factor. Plant J 42:218–235
Trethewey RN (2004) Metabolite profiling as an aid to metabolic engineering in plants. Curr Opin Plant Biol 7:196–201
Tschaplinski TJ, Standaert RF, Engle NL (2012) Downregulation of the caffeic acid O-methyltransferase gene in switch grass reveals a novel monolignol analog. Biotechnol Biofuels 5:71–79
Tuli R, Sangwan RS, Kumar S, Bhattacharya S, Misra LN, Trivedi PK, Tewari SK, Misra P, Chaturvedi P, Sangwan NS, Nair KN, Ojha SK, Mehrotra S, Khajuria A, Suri KA (2009) Ashwagandha (*Withania somnifera*) a model Indian medicinal plant. CSIR Publications, New Delhi, p 294
Twining CJ, Taylor CJ (2001) Kernel principal component analysis and the construction of non-linear active shape models. In: Proceedings of British Machine Vision Conference, Hershey, pp 23–32
Van den Berg RA, Rubingh CM, Westerhuis JA, Van der Werf MJ, Smilde AK (2009) Metabolomics data exploration guided by prior knowledge. Anal Chim Acta 651:173–181
Van Nieuwerburgh FCW, VandeCasteele SRF, Maes L, Goossens A, Inze D, VanBocxlaer J, Deforce DLD (2006) Quantitation of artemisinin and its biosynthetic precursors in *Artemisia annua*. L by high performance liquid chromatography-electrospray quadrupole time-of-flight tandem mass spectrometry. J Chromatogr A 1118:180–187
Verpoorte R, van der Heijden R, Schripsema J, Hoge JHC, Ten Hoopen HJG (1993) Plant cell biotechnology for production of alkaloids: present status and prospects. J Nat Prod 56:186–207
Ververidis FF, Emmanouil T, Carl D, Guenter V, Georg K, Nickolas P (2007) Biotechnology of flavonoids and other phenylpropanoid-derived natural products. Part I: chemical diversity, impacts on plant biology and human health. Biotechnol J 2:12–14
Vetter J (2000) Plant cyanogenic glycosides. Toxicon 38:11–36
Vichi M, Saporta G (2009) Clustering and disjoint principal component analysis. Comput Stat Data Anal 53:3194–3208
Wang M, Qu F, Shan XQ, Lin JM (2003) Development and optimization of a method for the analysis of low-molecular-mass organic acids in plants by capillary electrophoresis with indirect UV detection. J Chromatogr A 989:285–292
Warren CR, Adams MA (2000) Capillary electrophoresis for the determination of major amino acids and sugars in foliage: application to the nitrogen nutrition of Sclerophyllous species. J Exp Bot 51:1147–1157
Weckwerth W (2003) Metabolomics in systems biology. Annu Rev Plant Biol 54:669–689
Weckwerth W, Loureiro ME, Wenzel K, Fiehn O (2004) Differential metabolic networks unravel the effects of silent plant phenotypes. Proc Natl Acad Sci U S A 101:7809–7814
Welti R, Wang X, Williams TD (2003) Electrospray ionization tandem mass spectrometry scan modes for plant chloroplast lipids. Anal Biochem 314:149–152
Werrman U, Knorr D (1993) Conversión of menthyl acétate or neomenthyl acétate into menthol or neomenthol

by cell suspensión cultures of *Mentha canadensis* and *Mentha piperita*. J Agric Food Chem 41:517–520

Wiechert W, Mollney M, Petersen S, de Graaf AA (2001) A universal framework for ^{13}C metabolic flux analysis. Metab Eng 3:265–283

Wishart DS (2007) Current progress in computational metabolomics. Brief Bioinform 8:279–293

Wolfender JL, Ndjoko K, Hostettmann K (2003) Liquid chromatography with ultraviolet absorbance-mass spectrometric detection and with nuclear magnetic resonance spectroscopy: a powerful combination for the on-line structural investigation of plant metabolites. J Chromatogr A 1000:437–455

Woo HK, Northen TR, Yanes O, Siuzdak G (2008) Nanostructure-initiator mass spectrometry: a protocol for preparing and applying NIMS surfaces for high-sensitivity mass analysis. Nat Protoc 3:1341–1349

Yadav RK, Sangwan RS, Sabir F, Srivastava AK, Sangwan NS (2014a) Effect of prolonged water stress on specialized secondary metabolites, peltate glandular trichomes and pathway gene expression in *Artemisia annua* L. Plant Physiol Biochem 74:70–83

Yadav RK, Sangwan RS, Srivastava AK, Maurya S, Sangwan NS (2014b) Comparative profiling and dynamics of artemisinin related metabolites using efficient protocol and expression of biosynthetic pathway genes during developmental span of two elite varieties of *Artemisia annua* L. J Plant Biochem Biotechnol. doi:10.1007/s13562-013-0249-z

Yamaguchi-Shinozaki K, Shinozaki K (2006) Transcriptional regulatory networks in cellular responses and tolerance to dehydration and cold stresses. Annu Rev Plant Biol 57:781–803

Yates JR III (1998) Mass spectrometry and the age of the proteome. J Mass Spectrom 33:1–19

Ye X, Al-Babili S, Kloti A (2000) Engineering the provitamin A (β-carotene) biosynthetic pathway in to (carotenoid-free) rice endosperm. Science 287:303–305

Zagrobelny M, Bak S, Moller BL (2008) Cyanogenesis in plants and arthropods. Phytochemistry 69:1457–1468

Plant Glycomics: Advances and Applications

Sarika Yadav, Dinesh K. Yadav, Neelam Yadav, and S.M. Paul Khurana

Contents

S. Yadav, Ph.D. • D.K. Yadav, Ph.D. • N. Yadav, Ph.D.
S.M.P. Khurana, Ph.D. (✉)
Amity Institute of Biotechnology,
Amity University Harayana,
Gurgaon, Haryana, India
e-mail: smpaulkhurana@gmail.com

Abstract

Glycomics, the study of entire complement of sugars in an organism, helps to analyze the interaction of sugar with other macromolecules like carbohydrates, proteins, and nucleic acid. Greater structural complexity, nonlinear relationship of glycans with genome, and difficulty in isolation, characterization, and synthesis of complex oligosaccharides pose a significant challenge to glycomics. The isolation of plant glycoconjugates from natural sources is a complex process due to the lack of high-throughput user-friendly tools. Recent chemical advances have opened new and exciting possibilities in obtaining pure and chemically defined glycan moieties. Chromatographic techniques, tandem mass spectrometry, MALDI-mass spectrometry, ESI-mass spectrometry, NMR spectroscopy, and carbohydrate/lectin microarray are important tools for glycomics. Glycogene microarrays are useful to identify differentially expressed glycosylation-related genes and to study glycan biosynthesis, structure, and function. Efficient glycoinformatics have considerably enhanced the glycomic research by improving the data quality and reducing

D. Barh et al. (eds.), *PlantOmics: The Omics of Plant Science*,
DOI 10.1007/978-81-322-2172-2_10, © Springer India 2015

experimental costs. Glycans including lectins provide both structural and functional diversity to plants and are useful in transgenic technologies to increase resistance to pathogens and pests. Plant glycomics find their applications in biopharming and biopharmaceutics and provides a novel area of advanced glycome research to understand structure–function relationships of glycans. Unraveling the mysteries of glycomics would indeed be very beneficial as sugars play key role in many biological processes such as signaling, stress responses, and immunity.

Keywords
Biopharming • Glycans • Glycoconjugates • Glycomics • Glycoinformatics • Glycosylation • Lectins • Mass spectrometry • Microarray • Nuclear Magnetic Resonance spectroscopy

Introduction to Glycomics

Glycome refers to the entire complement of sugars, whether free or in the form of glycoconjugate, of an organism, including genetic, physiologic, pathologic, and other aspects. The term "glycomics" is analogous to genomics and proteomics and deals with the comprehensive study of glycome. The term glycomics has been derived from the "glyco" as prefix that refers to sweetness or a sugar and following the "omics" as naming convention established by genomics, which deals with genes and proteomics, which deals with proteins. Glycomics involves the systematic study of all glycan structures of a given cell type or organism and is a subset of glycobiology. Glycomics, since deals with assembly and expression of glycoconjugates in the biological system, helps to analyze, understand, and relate the collection of glycans to biological processes including the interaction of carbohydrates with carbohydrates, proteins, and nucleic acid.

Glycoconjugates are known to be involved in a variety of biological processes in the form of glycopeptides (combination of carbohydrates and proteins), glycolipids (combination of carbohydrates and lipids), glycosaminoglycans, proteoglycans, or other glycoconjugates.

Carbohydrates coat found in variety of cells are intimately involved in various biological processes including viral entry, cell recognition, signal transduction, cell differentiation, cell–cell interactions, bacteria–host interactions, microbial pathogenesis, immunological recognition, fertility, and development. According to the Swiss-Prot database, more than 50 % of the eukaryotic proteins (Apweiler et al. 1999), and about one-third of biopharmaceuticals (Walsh and Jefferis 2006), are glycoproteins.

The major limitation for the advancement in the field of glycomics is the greater structural complexity of these biomolecules as compared to nucleic acids and proteins including branching and linkage diversity. Besides, there exists a nonlinear and indirect relationship of glycans with genome owning to the secondary modifications of the monomers (Fig. 1). Furthermore, the difficulty in isolating, characterizing, and synthesizing complex oligosaccharides has been a significant challenge to progress in the field. Thus, the technologies for rapid assessment of glycan structures (i.e., glycomics) are still in the developmental stages.

Recent chemical advances, such as improved synthetic methods, including the development of an automated solid phase synthesizer, and methods for enzymatic synthesis, have opened new and exciting possibilities in obtaining pure, chemically defined glycan moieties. At the same time, the field has seen growing interest in the development of carbohydrate microarrays and neoglycoconjugates to facilitate otherwise laborious biological studies. By unifying synthetic advances and new biochemical tools, it is now possible to expand the tool chest available to the glycomics researcher.

In the past few years, application of systematic methods has been used to study glycans and their interactions. The increasing number of databases, web sites, and glycan libraries are addressing the needs of glycobiologists and glycochemists. Glycoinformatics, the informatics tools available for assessing primary data

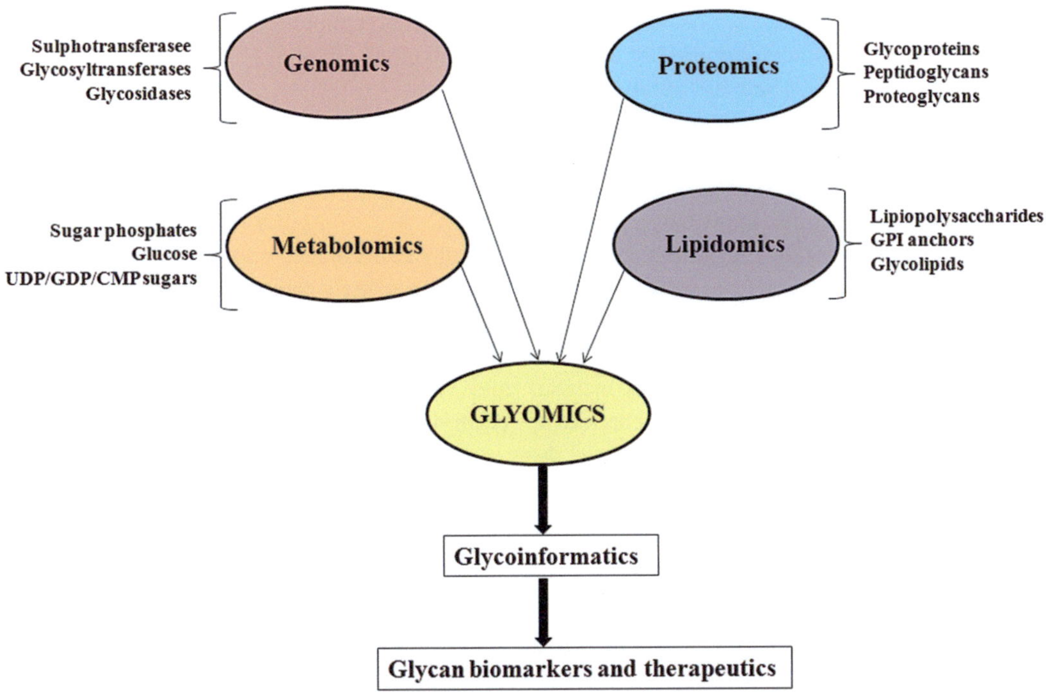

Fig. 1 **Glycomic complexity reflecting cellular complexity**: Given that glycan structures are regulated by metabolism and glyco-enzyme expression and glycans modify both proteins and lipids, functional glycomics also requires the tools of genomics, proteomics, lipidomics, and metabolomics (Adapted from Hart and Copeland 2010)

such as covalent, three-dimensional structures of glycans as well as glycoconjugates and organizing such data into databases for scaling up the generation of primary data, prediction, and characterization of structural or functional relationships. This chapter gives an overview of the status of glycosciences and perspectives on the application of some of the emerging technologies.

Glycans

The term glycan refers to compounds that consist of a large number of monosaccharides linked by a glycosidic bond. D-glucose is the most common constituent of glycans; however, D-fructose, D-galactose, L-galactose, D-mannose, L-arabinose, D-fucose, and D-xylose are also frequent constituents. Some monosaccharide derivatives found in glycans include the amino sugars (D-glucosamine and D-galactosamine) as well as their derivatives (*N*-acetylneuraminic acid and *N*-acetylmuramic acid) and simple sugar acids (glucuronic and iduronic acids). The term glycan refers to the carbohydrate part of a glycoconjugate, viz., glycoproteins, glycolipids, or proteoglycans. Glycans usually consist of an O-glycosidic linkage among the monosaccharides. Cellulose in plants is a glycan composed of β-1,4-linked D-glucose, while chitin in fungi is a glycan composed of β-1,4-linked *N*-acetyl-D-glucosamine.

Types of Glycans on the Basis of Constituent Monomer

On the basis of sugar monomers, the glycans can be classified as homopolymer and heteropolymer which can be further linear or branched.

Homopolysaccharides

Homopolysaccharides are composed of a single type of sugar monomer. These are usually named after the sugar unit they contain, for example, *glucans* (glucose homopolysaccharides) and *mannans* (mannose homopolysaccharides). In general, homopolysaccharides have a well-defined chemical structure with variable molecular weight even from a single source because these are synthesized by an enzyme-catalyzed biological process that lacks any genetic information about the size. Major examples include *cellulose*, an unbranched homopolysaccharide of D-glucose monomers linked together by β-1,4-glycosidic linkage, and *glycogen*, a branched homopolysaccharide of D-glucose monomers joined by α-1,4-glycosidic linkages.

Heteropolysaccharides

Heteropolysaccharides or heteroglycans are the polysaccharides consisting of two or more different monosaccharide or its derivatives. Most of the naturally occurring heteroglycans contain only two different units closely associated with lipid or protein, but few representatives are known to have three or more different monosaccharide units. Thus, heteroglycans have complex structure that are difficult to be studied. The major heteropolysaccharides include glycoproteins (carbohydrates conjugated to proteins) and glycolipids (carbohydrates conjugated to lipids) found in plants and animals.

Glycoconjugates

Glycoconjugates is the classification for carbohydrates covalently linked with other chemical species such as proteins, peptides, lipids, and saccharides. Glycopeptides, peptidoglycans, glycolipids, glycosides, and lipopolysaccharides are some of the major glycoconjugates involved in cell–cell interactions, in cell–cell recognition, in cell–matrix interactions, and in detoxification processes (Lis and Sharon 1993).

Major classes of glycoconjugate include glycoproteins (carbohydrates conjugated to proteins) and glycolipids (carbohydrates conjugated to lipids) found in plants and animals.

Glycoproteins: Protein-Linked Glycans

Carbohydrates attached to protein can be classified into two main categories on the basis of linkage of sugar moities to the amino acid of the protein chain: *N-glycans* and *O-glycans*. The *O*- and *N*-linked glycans are more common in eukaryotes as compared to prokaryotes.

N-Linked Glycans

In plants, like other eukaryotes, the *N*-glycans are covalently linked to the amide group of specific asparagine (Asn) residues constitutive of *N*-glycosylation sites of the protein in endoplasmic reticulum (ER). The *N*-glycosylation sites are made up of tripeptide Asn-X-Ser/Thr where X can be any amino acid except proline and aspartic acid (Kornfeld and Kornfeld 1985). All *N*-glycans share a common minimal structure (Man3GlcNAc2) consisting of an N, N′-diacetyl chitobiose unit, a β-mannose residue linked to the chitobiose, and two α-mannose residues linked to 3rd and 6th hydroxyl of the *β*-mannose residue (Kornfeld and Kornfeld 1985).

The plant *N*-glycans were earlier classified into two major categories, namely, the high-mannose-type and complex-type *N*-glycans. Later, Lerouge et al. (1998), on the basis of nuclear magnetic resonance (NMR)- and mass spectrometry (MS)-based studies, reclassified the plant *N*-glycan into following four classes, viz., high-mannose-type, complex-type, paucimannosidic-type, and hybrid-type *N*-glycans (Fig. 2).

High-Mannose-Type N-Glycans

High-mannose-type *N*-glycans from Man5GlcNAc2 to Man9GlcNAc2 are formed due to the limited trimming of Glc and Man residues from the precursor oligosaccharide Glc3Man9GlcNAc2 (Fig. 3).

In plants, high-mannose-type *N*-glycans were first reported in soybean agglutinin and found to

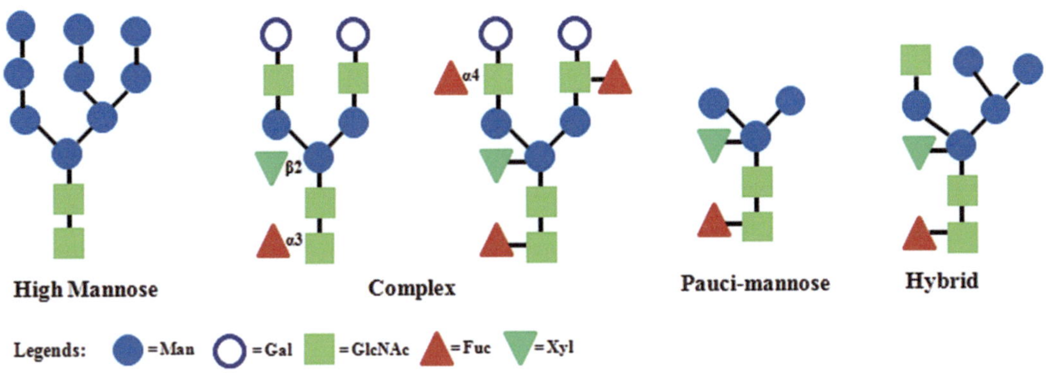

Fig. 2 Major classes of *N*-glycans found in plants

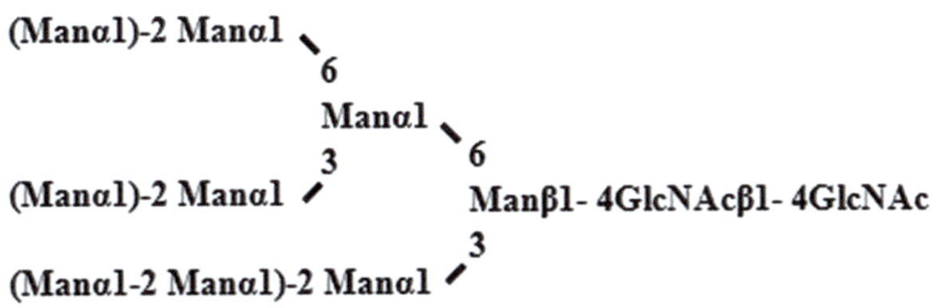

Fig. 3 Structures of high-mannose-type *N*-glycans Man5-GlcNAc2 to Man9GlcNAc2

be *N*-linked to various extracellular and vacuolar glycoproteins (Lis and Sharon 1978). High-mannose-type *N*-glycans of spinach and maize calreticulin with unique *N*-linked oligosaccharides have been identified as glycoprotein-specific chaperone in ER (Navazio et al. 1996; Pagny et al. 2000).

Complex-Type N-Glycans

Like other eukaryotes, plant complex-type *N*-glycans are formed by the glycosidases and glycosyltransferases specific processing of high-mannose-type *N*-glycans in the Golgi apparatus. Complex-type plant *N*-glycans are characterized by α(1,3)-fucose and/or a β(1,2)-xylose residues, respectively, linked to the proximal *N*-acetyl glucosamine and to the β-mannose residues of the core, respectively, and by the presence of β(1,2)-*N*-acetyl glucosamine residues linked to the α-mannose units (Fig. 4).

Larger complex-type plant *N*-glycans identified contain additional α(1,4)-fucose and β(1,3)-galactose residues linked to the terminal *N*-acetyl glucosamine units (Fitchette-Lainé et al. 1997; Melo et al. 1997). Such modifications yield Galβ1-3(Fucα1-4)GlcNAc sequences known as Lewis acid antigens and usually found on cell surface glycoconjugates in mammals (Fig. 4). Similar structures have been isolated from sycamore laccase, miraculin, and a pollen allergen from *Cryptomeria japonica* (Fitchette-Lainé et al. 1997; Melo et al. 1997).

Paucimannosidic-Type N-Glycans

Altmann (1997) proposed this nomenclature for modified form of insect *N*-linked glycans lacking terminal *N*-acetylglucosamine residues linked to the α-mannose residues of the core. The paucimannosidic-type *N*-glycans refer to modified plant oligosaccharides with only an α(1,3)-fucose and/or a β(1,2)-xylose residue linked, to the proximal GlcNAc and the β-mannose residues, respectively, in the core Man3GlcNAc2 (Fig. 5), or to the restricted core Man2GlcNAc2.

Earlier described as phytohemagglutinin-type *N*-glycans, these were identified from various plant glycoproteins (Bollini et al. 1985; Sturm et al. 1992; Gray et al. 1996; Oxley et al. 1996; Yang et al. 1996; Costa et al. 1997). This class of *N*-glycans is formed as a result of the elimination

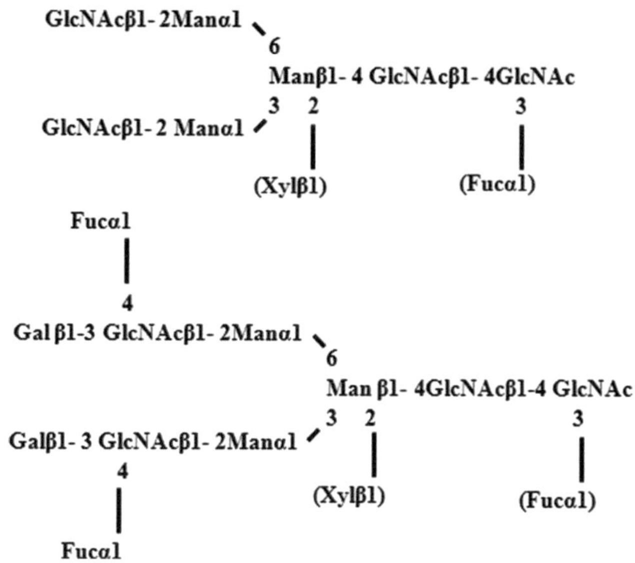

Fig. 4 Structures of complex-type *N*-glycans

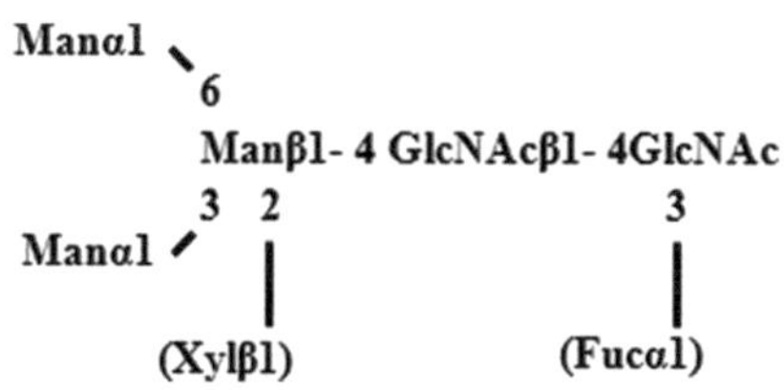

Fig. 5 Structures of paucimannosidic-type *N*-glycans Man3(Xyl)(Fuc)GlcNAc2

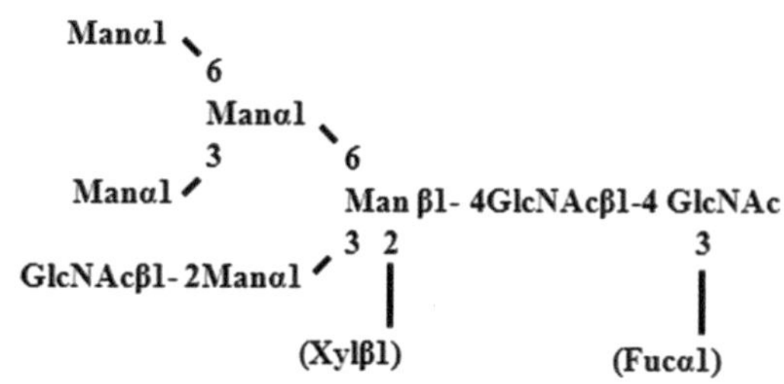

Fig. 6 Structures of hybrid-type *N*-glycans GlcNAcMan5(Xyl)(Fuc)GlcNAc2 from plant glycoproteins

of terminal residues from complex-type *N*-glycans and found typically in vacuole glycoproteins.

Hybrid-Type N-Glycans

Hybrid-type *N*-glycans are formed by the processing of only α(1,3)-mannose branch of the intermediate Man5GlcNAc2 resulting in oligosaccharides having α(1,3)-fucose and/or a β(1,2)-xylose residues linked to GlcNAcMan5GlcNAc2 (Fig. 6) (Oxley et al. 1996).

O-Linked Glycans

The *O*-glycans are linked to the hydroxyl group of serine (Ser), threonine (Thr), hydroxylysine, or hydroxyproline (Hyp) residues in the protein chain (Fig. 7). *O*-linked glycans in eukaryotes are assembled by addition of one sugar moiety at a time on a serine or threonine residue of a peptide chain in the Golgi apparatus (GA). The *O*-linked glycans are formed by the addition of an *N*-acetylgalactosamine (GalNAc) residue on the hydroxyl groups of Ser or Thr. Unlike *N*-linked glycans, there is no known consensus sequence. The presence of a proline residue at either −1 or +3 relative to the serine or threonine is favorable for *O*-linked glycosylation.

Limited information is available about the plant *O*-linked glycans that are usually considered as structurally different from mammalian O-glycans. Humans *O*-glycans are mucin-type *O*-glycans, with xylose residues *O*-linked to Ser or Thr in proteoglycans and glycoproteins. In contrast, plant *O*-linked glycans are formed by the nucleocytoplasmic *O*-glycosylation during the transient addition of a single *N*-acetylglucosamine (GlcNAc) residue on the Ser and Thr residues of the peptide chain (Zachara

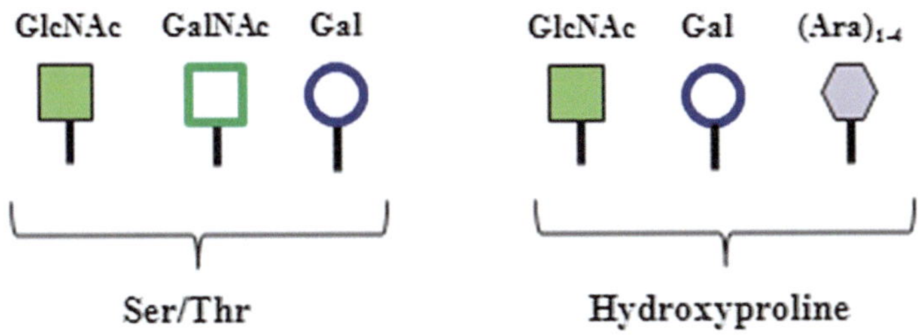

Fig. 7 Types of O-glycans found in plant glycoproteins. Hydroxyproline is unique feature of plants and members of chlorophyceae

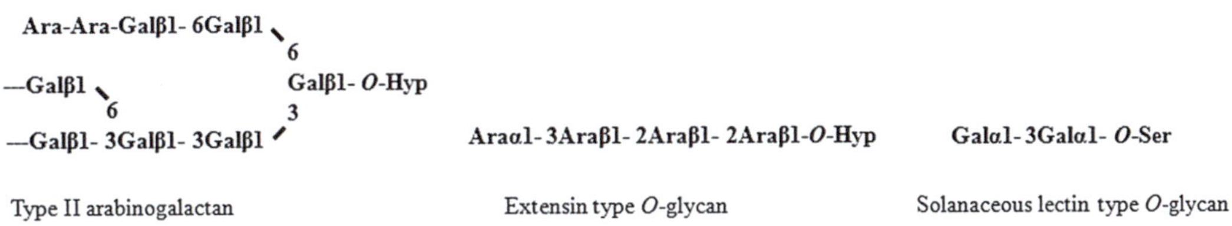

Fig. 8 Structures of different types of *O*-glycans in plants (Wilson 2002)

and Hart 2006). Putative mammalian mucin-type *O*-glycosylation has been reported from some of plant Golgi membrane proteins and rice glutelin (Mitsui et al. 1990; Kishimoto et al. 1999).

The detailed information about the genetic basis for the biosynthesis of *O*-linked glycans (galactose-serine, galactose-hydroxyproline, arabinose-hydroxyproline) is lacking, and it is assumed that some among the many known β(1,3)-glycosyltransferase homologues in plants (20 homologues in *Arabidopsis*) are involved in formation of the β(1,3)-galactose backbone of arabinogalactan side chains (Fig. 8).

Role of arabinogalactans in plant development is established by the effects of arabinogalactans binding reagents on the growth of pollen tube tips, somatic embryogenesis, and cell expansion and/or division (Majewska-Sawka and Nothnagel 2000) which is probably resulting in the plant homologues of animal proteoglycans.

The extensions and arabinogalactan proteins (AGPs) are well-known O-linked glycans containing proteins in plants and belong to Hyp-rich glycoproteins (HRGPs). The O-glycosylation of HRGPs results from two consecutive post-translational modifications involving the hydroxylation of some Pro residues by prolyl 4-hydroxylases (P4Hs) in the ER, and the subsequent O-glycosylation of some, but not all, Hyp residues by glycosyltransferases in the GA. Extensins are extensively O-glycosylated, with one to four arabinosyl residues O-linked onto most Hyp residues, and galactose (Gal) residues bound to many Ser residues (Fig. 9), two types of O-glycosylation that were also described for lectins in Solanaceae (Showalter 2001).

AGPs are the most highly glycosylated HRGPs, with Hyp residues frequently glycosylated by either large arabinogalactan glycomodules (Fig. 9) or short unbranched arabinooligosaccharides (Pope 1977; Qi et al. 1991). HRGPs are involved in many aspects of plant growth and development (Seifert and Roberts 2007), and many effects of O-linked glycosylation on the biological activity of these proteins have been described. Reagents binding or cleaving the O-glycans of AGPs were shown, for instance, to affect growth of pollen tubes, cell expansion, somatic embryogenesis, or xylem differentiation (Motose et al. 2004; van Hengel et al. 2001).

Mechanisms of Glycosylation

Mechanism of N-Glycosylation

N-glycosylation of plant proteins starts in the ER with the transfer of oligosaccharide precursor (Glc3Man9GlcNAc2) by the oligosaccharyl transferase from a dolichol lipid carrier to specific Asn residues on the nascent polypeptide chain. The precursor is subsequently modified by glycosidases and glycosyltransferases during the transport of the glycoprotein to its final localization. Oligosaccharide precursor Glc3Man9GlcNAc2 first undergoes an early trimming of the three terminal glucose units catalyzed by the glucosidases I and II in the ER (Szumilo et al. 1986a, b). A transient reglucosylation by an ER UDPglucose:glycoprotein glucosyltransferase may occur subsequently which is involved in the quality control of glycoproteins in the ER (Hammond et al. 1994). Plant N-glycans can be further modified in the Golgi into complex-type N-glycans during the transport of the glycoprotein from the *cis*, through medial to *trans* cisternae. First, the α-mannosidase I (α-Man I) removes one to four α(1,2)-mannose residues and converts Man9GlcNAc2 to Man5GlcNAc2 (Sturm et al. 1987; Szumilo et al. 1986a, b). Biosynthesis of complex-type N-glycans starts with the addition of a first N-acetylglucosamine residue to the α(1,3)-mannose branch of the Man5GlcNAc2 high-mannose-type glycan. This step is catalyzed by the N-acetylglucosaminyltransferase I (GNT I) to yield GlcNAcMan5GlcNAc2 (Tezuka et al. 1992). Two additional mannose residues are then removed from GlcNAcMan5GlcNAc2 by the α-mannosidase II (α-Man II), and another outer N-acetylglucosamine residue is transferred by the N-acetylglucosaminyltransferase II (GNT II) to the α(1,6)-mannose branch (Tezuka et al. 1992). At this stage, α(1,3)-fucosylation and β(1,2)-xylosylation of the core Man3GlcNAc2 may occur to yield plant-specific N-linked glycans. The sequences of the xylosylation and the fucosylation events are not completely understood. Plant N-linked glycans having only a β(1,2)-xylose or only an α(1,3)-fucose residue have been identified in plant glycoproteins (Costa et al. 1997; Ohsuga et al. 1996). α(1,3)-Fucosylation and β(1,2)-xylosylation are two independent events in plants. Plant hybrid-type N-glycans could result from an uncompleted action of α-Man II (Fig. 10).

Terminal fucose and galactose residues can be further added to complex-type N-glycans to yield one or two Galβ1-3(Fucα1-4)GlcNAc sequences known as Lewis acid antigen involved in cell–cell recognition and cell adhesion processes. After maturation in the ER and the Golgi apparatus, complex-type N-glycans can be further modified during the glycoprotein transport to, or in, the compartment of its final destination. Typical vacuole-type modified N-glycans containing fucose and/or xylose residues but devoid of terminal glucosamine residues are named paucimannosidic-type N-glycans which can only result from post-Golgi modifications occurring on complex-type N-glycans (Lerouge et al. 1998).

Mechanism of O-Glycosylation

In plants, O-glycosylation has been described mainly for the hydroxyl groups of Hyp, Ser, and Thr residues. Plant and mammalian O-glycans usually are considered as structurally different, but with two notable exceptions. A first exception is the nucleocytoplasmic O-glycosylation observed during the transient addition of a single N-acetylglucosamine (GlcNAc) residue onto Ser and Thr residues of the peptide backbone (Zachara and Hart 2006). The second exception is a putative mammalian mucin-type O-glycosylation reported for some plant Golgi membrane proteins and rice glutelin (Mitsui et al. 1990; Kishimoto et al. 1999). The exact mechanism of O-glycosylation is not well established in plants. Unlike *N*-glycosylation, there is no known consensus sequence for *O-glycosylation.* The main O-glycosylated proteins in plants, extensions, and arabinogalactan proteins (AGPs), belong to Hyp-rich glycoproteins (HRGPs). The O-glycosylation of HRGPs results from two consecutive post-translational modifications involving the hydroxylation of some Pro residues by prolyl 4-hydroxylases (P4Hs) in the ER and subsequent O-glycosylation of some, but not all,

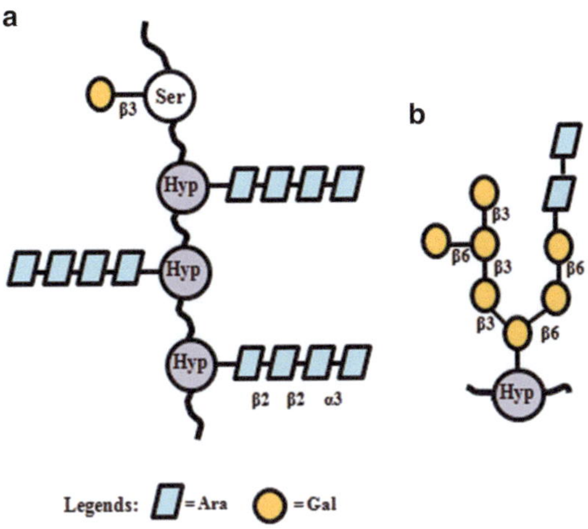

Fig. 9 Plant-specific *O*-glycans: (**a**) extensin-type *O*-glycans and (**b**) arabinogalactan proteins (AGP)-type *O*-glycans

Hyp residues by glycosyltransferases in the GA (Fig. 9). Extensins are extensively glycosylated, with one to four arabinosyl residues O-linked onto most Hyp residues, and galactose (Gal) bound to many Ser residues (Fig. 7), two types of O-glycosylation that were also described for lectins in Solanaceae (Showalter 2001). AGPs are highly glycosylated on Hyp residues by either large arabinogalactan glycomodules or short unbranched arabinooligosaccharides (Qi et al. 1991). HRGPs are involved in many aspects of plant growth and development, and O-linked glycosylation is known to affect the biological activity of these proteins. Removal of O-glycans of AGP affects growth of pollen tubes, cell expansion, somatic embryogenesis, or xylem differentiation (Motose et al. 2004).

Glycolipids: Lipid-Linked Glycans

The term glycolipid is used for amphiphathic molecules containing one or more mono- or oligosaccharide moiety linked by a glycosidic linkage to a hydrophobic ceramide moiety. The asymmetric distribution of hydrophilic head and nonpolar hydrophobic hydrocarbon (acyl) chain in glycolipids leads to aggregation that forms a cluster in an aqueous environment forming the lipid bilayer. Thus, glycolipids play an important role in membrane stabilization. Glycolipids, such as acylglycerol, ceramides (N-acyl sphingoid), prenyl phosphate, glycosylated glycerolipids, sterols, or a sphingolipids, are present in almost all biological membranes. Galactolipids are the most abundant membrane lipids especially in green tissues representing ~75 % of total membrane lipids content (Dörmann and Benning 2002). Glycolipids in plants exist as free sterols, acylated sterols, steryl glycosides, and acylated-steryl glycosides. Naturally occurring steryl glucoside have been identified as 3-β-hydroxyglucosides, where C-1 of the carbohydrate participate in the formation of glycosidic linkage (Grunwald 1978). The most abundant sugar reported in steryl glucoside is glucose, although galactose, mannose, and gentiobiose also have been found in abundance.

Biosynthesis of Glycolipids

Studies on steryl glucoside and esterified steryl glucoside biosynthesis of various plants have suggested that esterified steryl glucoside formation could occur via two pathways (Frasch and Grunwald 1976; Grunwald 1978; Potocka and Zimowski 2008). In the first one, acyl groups are transferred from phospholipids to steryl glucoside by a microsomal enzyme. In the second proposed pathway, a soluble enzyme uses acyl groups originating from galactolipids for steryl glucoside acylation but at a distinctly lower rate. The oligosaccharide of complex glycolipids is synthesized by stepwise addition of sugars catalyzed by specific glycosyltransferases in the Golgi apparatus. Most of the glycosyltransferases that synthesize the glycolipids are typical Golgi-resident type II integral membrane proteins. The components of glycolipid machinery are distributed in the lumen and membrane where membrane-bound sugar transporters and donor sugar nucleotides direct the biosynthesis of basic oligosaccharide structures of different glycolipids (Fig. 11).

Ceramide is the lipid moiety of glycosphingolipids which is transported from the ER to the proximal Golgi by the cytosolic protein CERT. CERT extracts ceramide from ER through its FFAT and START domains and transfers it to

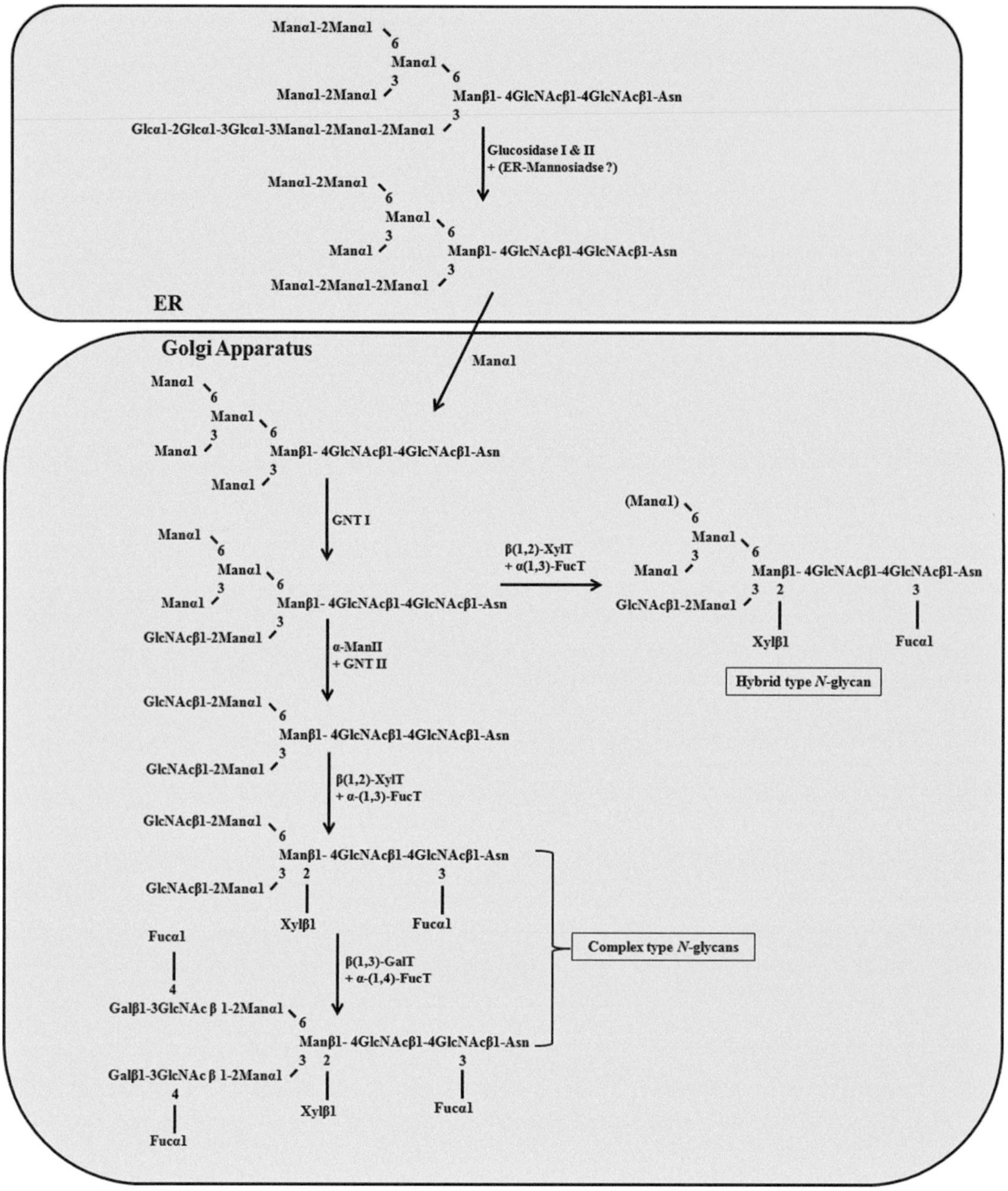

Fig. 10 Processing of plant N-linked glycans in ER and the Golgi apparatus. Abbreviations: α-Man I, α-mannosidase I; α-Man II, α-mannosidase II; GNT I, N-acetylglucosaminyltransferase I; GNT II, N-acetylglucosaminyltransferase II; *β*(1,2)-XylT, β(1,2)-xylosyltransferase; α(1,3)-FucT, α(1,3)-fucosyltransferase; β(1,3)-GalT, β(1,3)-galactosyltransferase; α(1,4)-FucT, α(1,4)-fucosyltransferase (Adapted from Lerouge et al. 1998)

Golgi apparatus in a non-vesicular manner (Hanada et al. 2009). After transport of ceramide in the proximal Golgi, it is converted to glucosylceramide (Cer-Glc) by ceramide glucosyltransferase (GlcT). The catalytic site of GlcT is oriented toward the cytoplasm and has type III topology. Further, Cer-Glc is transported to the distal Golgi by FAPP2, where it is translocated to the luminal leaflet. Cer-Glc is converted into Cer-Lac (Cer-Glc-Gal) and higher glycolipid deriva-

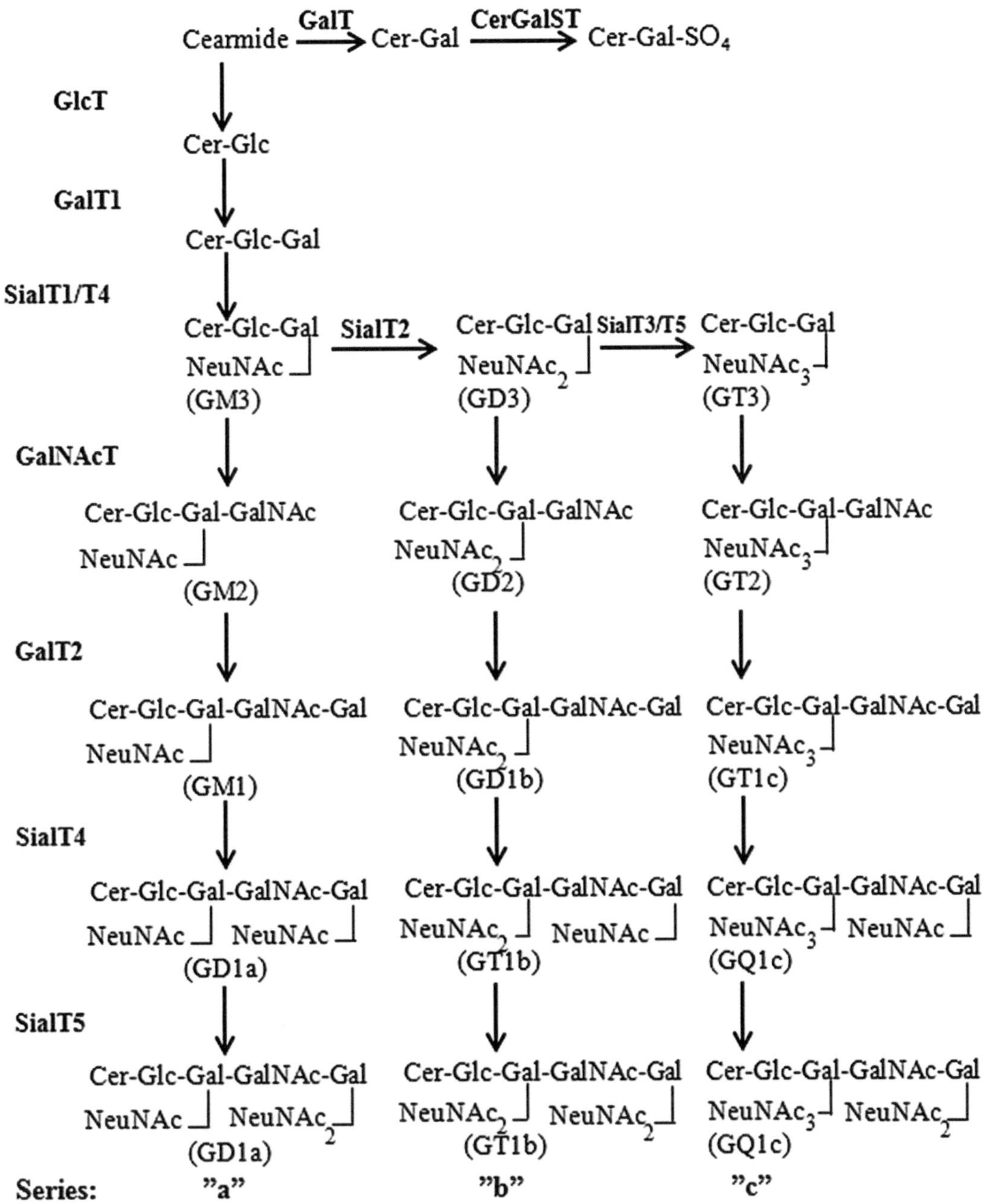

Fig. 11 **Pathway of biosynthesis of glycolipids**: showing only Cer-Glc derived species of the "a", "b," and "c" gangliosides series. GalT, UDPGal:ceramidegalactosyltransferase; CerGalST, PAPS-galactosylceramidesulfotransferase; GlcT, UDP-Glc:ceramideglucosyltransferase; GalT1, UDP-Gal:glucosylceramidegalactosyltransferase; SialT1, CMP-NeuAc:lactosylceramidesialyltransferase; SialT2, CMP-NeuAc:GM3 sialyltransferase; GalNAcT, UDP-GalNAc: LacCer/GM3/GD3/GT3 N-acetylgalactosaminyltransferase; GalT2, UDP-Gal: GA2/GM2/GD2/GT2 galactosyltransferase; SialT3, CMP-NeuAc:GD3 sialyltransferase; SialT4, CMP-NeuAc:GA1/GM1/GD1b/GT1c sialyltransferase; SialT5, CMP-NeuAc:GM1b/GD1a/GT1b/GQ1c sialyltransferase. Nucleotide sugar donors that participate in each transfer step are not shown in Figure (Maccioni et al. 2011)

tives by GT acting in the lumen of the distal Golgi (Fig. 11) (D'Angelo et al. 2007; Maccioni 2007).

Cer-Glc is also transported directly to the plasma membrane but is poorly used for synthesis of more complex glycolipids and turns over rapidly (Warnock et al. 1994). It has been proposed that FAPP2 may also transport Cer-Glc from the *cis* Golgi to the ER, where after translocating to the lumen, it would enter the secretory pathway and is further modified to higher-order derivatives in the Golgi lumen (Halter et al. 2007). Ceramide galactosylation takes place in the ER by a type I ceramide galactosyltransferase (GalT), with a luminal N-terminal domain bearing an ER retrieval signal (KDEL) and the catalytic site (Schulte and Stoffel 1993; Sprong et al. 1998). Cer-Gal is the precursor of sulfatides (Cer-Gal-SO4) by accepting sulfate residues from donor 3′-phosphoadenosine-5′-phosphosulfate in the C-3 of the galactose (Cumar et al. 1968) catalyzed by sulfotransferase.

Importance of Plant Glycomics in Biopharming

Plants can carry out posttranslational protein modifications in a similar manner if not identical as that of mammalian cells. Thus, human like complex glycosylation capability of plants is well known. Moreover, the targeted engineering of plant *N*-glycosylation pathway allows the production of proteins carrying largely homogeneous, human-type oligosaccharides to reduce the risk of allergenicity when those products are aimed for human use.

The agricultural large-scale production of recombinant proteins is known as *biofarming*, whereas it is referred as *biopharming* when deals with medicinal or pharmaceutical proteins. Plants are gaining increasing acceptance as bioreactors since they possess a number of attractive features for the large-scale production of industrial enzymes and pharmaceuticals (Yadav et al. 2013). (Table 1).

Plant cell as well as plant organ culture are of significance for the production of such proteins, like root cell cultures have been reported for the production of mouse immunoglobulins (Doran 2000; Sharp and Doran 2001a, b). Moreover, plant bioreactors are safer as compared to animal expression systems because they act as non-host system for human pathogens, such as viruses and prions (Fischer and Emans 2000; Larrick and Thomas 2001; Daniell et al. 2001). Plant cells are known to have ability to assemble multimeric human proteins such as mammalian immunoglobulins (Larrick and Thomas 2001). Also the genetically engineered plants have been used for the production of such proteins for targeted expression to certain parts of the plant, such as seeds (Giddings et al. 2000; Chadd and Chamow 2001; Russell 1999; Hood and Jilka 1999; Zeitlin et al. 1998; Fiedler and Conrad 1995; Horvath et al. 2000). This enables easy harvesting, storage, and allowing flexibility in subsequent processing of such proteins. Moreover, development of the transgenic plant expression systems such as banana and cassava (Schenk et al. 1999; Zhang et al. 2000) would help to cope with the availability of pharmaceutical and therapeutic products in developing countries where the specialized technology for cell culture is not be widely accessible.

It also includes the production of edible vaccines in genetically modified plants expressing recombinant bacterial or viral proteins in edible parts to stimulate an immune response in the human after consuming them. The mice fed on transgenic alfalfa expressing a foot-and-mouth disease virus protein produced an immune response to the pathogen (Wigdorovitz et al. 1999). Human trials at initial phase using potatoes expressing an antigen, Norwalk virus capsid protein, showed significant titers of specific antigen detectable in the subjects' serum (Tacket et al. 2000).

Plants have emerged as an attractive expression system for the production of recombinant proteins for therapeutic use. However, the plants have differences in their glycosylation state as compared to mammalian system. One of major difference is that plant cells do not synthesize

Table 1 Comparison of different hosts for the production of recombinant therapeutic proteins (Balen and Krsnik-Rasol 2007)

Host	Production cost	Production timescale	Product quality	Glycosylation	Risk of pathogenecity	Storage cost	Ethical concerns
Bacteria	Low	Short	Low	No	Medium	Moderate	Medium
Yeast	Medium	Medium	Medium	Yes (unusual)	Low	Moderate	Medium
Mammalian cells	High	Long	Very high	Yes	High	Expensive	Medium
Insect cells	High	Long	High	Yes (minor differences)	Medium	Expensive	Medium
Transgenic mammals	High	Very long	Very high	Yes	High	Expensive	High
Plant cell cultures	Medium	Medium	High	Yes (minor differences)	Low	Moderate	Medium
Transgenic plants	Low- medium	Long	High	Yes (minor differences)	Low	Low	Medium

sialylated glycans. Presence of terminal sialic acid glycan prevents the rapid opsonization of antigens and, hence, assures the protective response of a vaccine. Also plants synthesize glycan residues that are not found in human glycoconjugates (Lerouge et al. 1998). Plant glycans usually contain xylose, rhamnose, arabinose, and 1–3 fucose residues that are not found in human glycans.

Plant *O*-linked glycans are usually linked to serine, threonine, or hydroxyproline (unlike human). Plant *O*-linked glycans are highly glycosylated therefore called as proteoglycans instead of glycoproteins. These generally include extensins involved in plant cell wall synthesis; lectins and soluble arabinogalactans present in exudates, cell walls, and intercellular spaces. Extensins typically consists of short arabinose chains linked to hydroxyproline and galactose (either monosaccharides or disaccharides of galactose) linked to serine. Soluble arabinogalactans are highly glycosylated with glycans that are rich in rhamnose, arabinose, galactose, glucuronic acid, galacturonic acid, and their methylated derivatives. The *O*-linkage is often between a Gal or Glc residue and a hydroxyproline of the polypeptide backbone.

Plant *N*-linked glycans are slightly different structurally from those of humans, while resemble more to those of insects. These are processed in the ER and Golgi apparatus almost in the same way as the human *N*-glycans. Plants have high-mannose *N*-linked glycans due to incomplete processing of the core intermediate and have a general formula of Man5–9GlcNAc2. Further trimming of core and subsequent chain extension then gives rise to heterogeneous complex-type *N*-glycans having antennae rich in Fuc, xylose, GlcNAc, and Gal (Altmann et al. 2001; Lerouge et al. 1998).

Plant cells synthesize complex glycan structures some of which are not present in human cells and are allergenic to humans. Most importantly $\alpha 1 \rightarrow 3$-linked Fuc and $\beta 1 \rightarrow 2$-linked xylose are known to induce a strong IgE-mediated immune response (Fotisch and Vieths 2001). Thus, it represents a significant barrier to their use for synthesis of glycoproteins for human use. However, genetic humanization of plant-expressed glycoproteins has been successfully shown and suggests the successful use of plants as bioreactor.

One of the most developed areas of research utilizing transgenic plants for the production of human therapeutics and pharmaceuticals includes the production of immunoglobulins or plantibodies (Daniell et al. 2001; Sala et al. 2003). This commonly aims the production of an antibody for topical application to mucosal surfaces and/or for oral administration. Plants exist as only commercially viable expression system for the production of sIgA (Larrick and Thomas 2001). An IgA antibody produced against *Streptococcus mutans* in tobacco has shown promising early results under clinical trials as oral topical administration to prevent dental caries (Larrick et al. 2001). Humanized antibodies against herpes simplex virus glycoprotein expressed in transgenic soybean produced for vaginal administration were shown an effective immunopreventative agent against genital herpes infection in mouse (Hood and Jilka 1999). Tobacco has been used to express antibodies against different lymphomas (McCormick et al. 1999) and carcinoembryonic antigen (CEA) (Vaquero et al. 2002).

Plants *N*-glycosylate the immunoglobulins in the same way as in humans, but the glycoforms synthesized are found to be more variable and highly composed of $\beta 1 \rightarrow 2$-linked xylose and $\alpha 1 \rightarrow 3$-linked Fuc attached to the core Man3GlcNAc2 (Garcia-Casado et al. 1996). The difference in the glycosylation pattern of antibodies produced by plant as compared to human cells is important due to their potential immunogenicity in the human recipients. However, there are reports where a plantibody injected into mice did not elicit an adverse immune reaction (Chargelegue et al. 2000). The glycosylation of both endogenous plant glycoproteins and transgenic glycoproteins expressed in tobacco plant is reported to be altered in the leaves during senescence (Elbers et al. 2001). Thus, suggesting the effect of physiological conditions affecting glycosylation and deliberate modification of plant glycosylation pathways.

Genetically manipulated plant resulting in the premature termination of glycan synthesis, to prevent formation of immunogenic structures, would help to alleviate the problem of potential antigenicity by plant-expressed glycosylated proteins of human use. For example, knockout GlcNAc transferase I would inactivate the pathway by which complex *N*-glycans are extended from the GlcNAc2Man5 core. Another approach would be to prevent the glycoprotein from reaching the Golgi apparatus for final glycan processing that could be achieved by incorporating an ER retention sequence (C-terminal KDEL sequence) with the gene coding for the recombinant glycoprotein (Yadav et al. 2012). Plants with such modification synthesize proteins that carry unmodified high-Man-type glycans. Such glycans are not directly immunogenic to human but poorly suited to human therapeutics as they are being rapidly cleared from circulation, due to recognition by the macrophage Man receptor. Therefore, knocking out plant glycosylation enzymes that build the immunogenic glycan structures would be more logical approach. For example, inactivation of the xylosyltransferase that adds xylose residues to plant glycoproteins (Leiter et al. 1999).

A glycosylated protein can be engineered if the *N*-linked glycosylation sequon Asn-X-Ser/Thr is altered. Addition of Lys-Asp-Glu-Leu (KDEL) at the C-terminus targets proteins to the proximal ER reduces the incidence of core *N*-linked xylose and Fuc in favor of high-Man-type glycosylation patterns (Wandelt et al. 1992; Yadav et al. 2012). Even introduction of an additional human transferase would help to synthesize a more comprehensive array of human glycans. For example, introduction of human β-1,4 galactosyltransferase for stable expression in tobacco cells resulted in conversion of plant-type glycans to more human glycoforms due to synthesis of galactosylated structures (Palacpac et al. 1999). Also, β-1,4 galactosyltransferase transfected tobacco cells expressing mouse immunoglobulins produced partially galactosylated glycoforms of the plantibodies (Bakker et al. 2001).

A major challenge for plant expression system would be to achieve sialylated glycans from transgenic plant cells as plants do not synthesize sialic acids. Thus require to introduce and target not only appropriate and functional sialyltransferases but also enzymes for its synthesis and transport. Misaki et al. (2003) showed that cultured plant cells expressing human β-1,4 galactosyltransferase produced glycoproteins with Gal-extended *N*-linked glycans, appropriate for in vitro sialylation.

Techniques for Detection/ Quantitation of Glycosylation

Both prokaryotes and plants—owing to their diverse, versatile, and easily adaptable biosynthetic machinery—exhibit a large repertoire of monosaccharide building blocks (e.g., l-rhamnose, N-acetyl-d-fucosamine, etc.) as part of their lipopolysaccharides, capsular polysaccharides, antibiotic glycosides, and plant-derived polysaccharides, which, in contrast, are not found in humans. Consequently, the study of glycomics of prokaryotes and the plant kingdom is indeed a daunting task in relation to mammalian genome (Varki et al. 2008). However, there are some common features found with respect to biosynthetic protein *N*-glycosylation pathways between plant and mammalian systems, with the exception of sialic acids which are unique to mammals (Wilson 2002). Both genomics and proteomics are made "relatively" simple due to their template-driven nature arising out of base-pair complementarities and trinucleotide codon-driven transcriptional and translational processes. In comparison to genomics and proteomics—where automated synthesis, amplification, expression, and characterization have become routine—the tools available for glycomics are few. This necessitated the development of unique tools that are currently reaching a reasonable level of sophistication for use by nonspecialists (Bertozzi and Kiessling 2001; Pilobello and Mahal 2007; Prescher and Bertozzi 2006). The

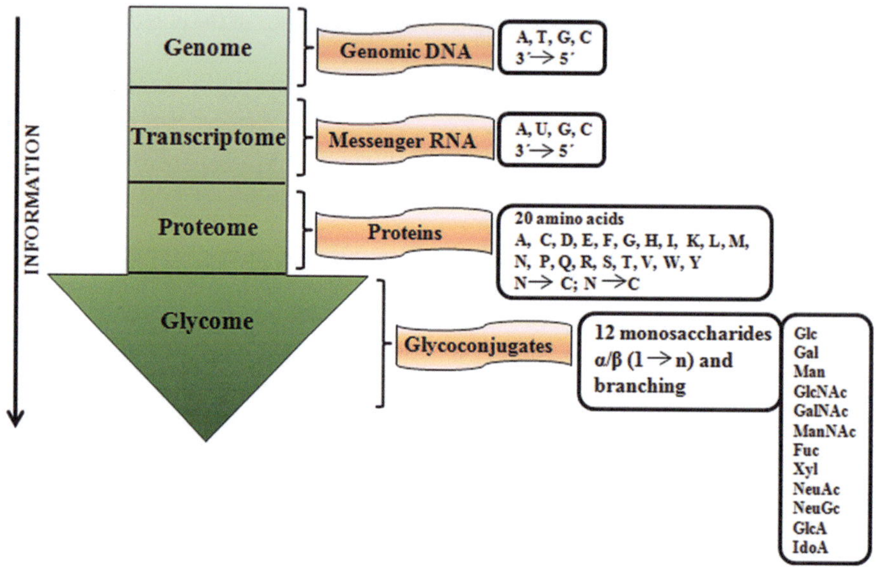

Fig. 12 Complexity and diversity of eukaryotic glycome: the information flow from genome to glycome increases exponentially (diagram not drawn to scale). Glycans provide functional diversity originating from only a single gene. Glycome is generated by complex, non-template-driven biosynthetic glycosylation pathways, governed by availability of carbohydrate substrates, expression, and activity levels of enzymes and sugar–nucleotide transporters

enormous diversity of structures created from limited number of monosaccharides building blocks coupled with its non-template-driven nature has made the task of system-wide glycan profiling a challenging task (Cummings 2009; Gabius et al. 2004) (Fig. 12).

A major obstacle for the advancement of glycomics has been the lack of sensitive and easy-to-use high-throughput analytical tools. The availability of well-characterized enzymes with known substrate specificities, such as PNGase-F (useful for cleaving the intact N-linked glycans out of the asparagine (Asn) side chain), endo-H (useful for cleaving the internal β-(1,4)- GlcNAc residue of N-linked glycans leaving one GlcNAc attached to peptide backbone), and several plant and bacterial glycosidases (useful for distinguishing specific anomeric linkages), and analytical chromatographic methods, based on electrochemical detection and fluorophore derivatization of the reducing end anomeric carbon (Domann et al. 2007), enabled detailed and thorough characterization of the total oligosaccharide structures and linkages (Ghesquiere et al. 2006; Ramachandran et al. 2006). Parallel advances in stereochemically controlled synthesis and purification of oligosaccharides produced homogenous glycans that served as reference standards for biological specimen-derived glycans, haptens for conjugation, production of monoclonal antibodies, and vaccine development (Buskas et al. 2006; Seeberger and Werz 2007; Wang et al. 2007). However, the characterization of a complete structure of a glycan using enzymes and chromatography involves tedious and time-consuming efforts (Ding et al. 2009; Patwa et al. 2009; Zheng et al. 2007).

The advent of proteomics and evolution of powerful user-friendly MS technologies especially methods coupled to chromatography and bioinformatics has enabled rapid analysis of glycans (Bielik and Zaia 2010; von der Lieth et al. 2006; Werz et al. 2007; Zaia 2008). Mass spectrometry, microarray and bioinformatic techniques have the advantage to rapidly scan and analyze samples in a high-throughput manner.

Mass Spectrophotometry

Glycosylation defines the adhesive properties of cell surfaces and the surrounding extracellular environments. Cells respond to stimuli by altering glycan expression; glycan structures vary according to spatial location in tissue and temporal factors. Effective analytical methods are needed for speedy identification of new targets and the development of industrial glycoprotein products. MS is an enabling technology in glycomics. MS has gained widespread use in glycan and glycosylated protein analysis due to its high selectivity, sensitivity, and ability to analyze complex mixtures rapidly. The advent of soft ionization techniques such as ESI and MALDI has revolutionized the glycan research.

MS is capable of providing structural constraints for purified molecules, although complete structural determination typically requires several analytical technologies including MS, linkage analysis, and NMR. MS methods are also used to maximize the structural information produced, given limiting resources in terms of time, labor, and sample quantity. Mass spectrometry has the advantage to rapidly scan and analyze samples in a high-throughput manner, once the glycans or glycopeptides have been isolated, with or without permethylation, or enriched using lectin-based affinity chromatography. In "omics" fields, tasks are usually divided between discovery and targeted analysis. Thus, determination of glycan masses is often an early step in profiling glycan expression. The masses of glycans released from biological source may be determined rapidly and with low sample consumption and high throughput. The resulting information shows the abundances of all glycan compositions present and do not determine the abundances of individual glycan structures, because of the presence of structural isomers. Combinations of separations and tandem mass spectrometry may be used to build information defining the structures of glycans present from product ion patterns. Unlike peptides, which are made of amino acid as building blocks with distinct molecular mass enabling easy identification of unique peptide sequences, the isobaric nature of various monosaccharide components of glycans complicates the assignment. So, although one could easily predict the size of the oligosaccharides and general components such as number of hexosamines, hexoses, sialic acids, etc., using mass spectrometry, the specific monosaccharides and their stereochemical linkages have to be deduced based on known biochemical pathways (North et al. 2009; von der Lieth et al. 2006). High-end techniques such as tandem mass spectrometry (MS)n, which study fragmentation of a given molecular ion peak and generation of fingerprint-like patterns, are being currently developed to identify the component monosaccharides, their linkage positions, and anomeric orientations. The chemical composition of the carbohydrate monosaccharide residues strongly influences the ionization of glycoproteins (Zaia 2008). Glycoconjugate glycosylation tends to increase the acidity of glycoproteins relative to aglycon. Thus, glycosylation increases both the hydrophilicity and surface activity of the modified proteins and peptides. These effects strongly influence the ionization of glycoconjugates, relative to unmodified proteins or peptides. Thus, if analyzing a glycoprotein tryptic digests using positive ion mass spectrometry, the unmodified peptides will be least acidic and more likely to form abundant positive ions, with the result that the ionization of glycosylated proteins will be suppressed. As a general principle, samples to be analyzed using mass spectrometry should be separated into fractions of approximately equal acidities to minimize the extent to which ion suppression occurs.

To simplify the vast number of structures possible for a given oligosaccharide purely based on mathematical permutations and for matching of a given mass spectrometry ion peak to relate to a minimal set of glycans, qualified databases were developed using known mammalian biosynthetic glycosylation pathways as limiting parameters (Cooper et al. 2001; Goldberg et al. 2005; McDonald et al. 2009). Rigorous mathematical models correlating expression levels and activities of enzymes involved in N-glycan biosynthetic pathways to the overall mass spectrometry profile of glycans have been developed to analyze differential glycosylation patterns (Krambeck et al. 2009). Notwithstanding intense efforts by multiple research groups, the use of mass spectrometry as a stand-alone technique for complete characterization of glycans is still far from complete.

Although fast atom bombardment ionization was used to develop many of the principles of modern glycoconjugate mass spectrometry (Zaia 2010), the matrix-assisted laser desorption/ionization (MALDI) (Karas and Hillenkamp 1988; Karas et al. 1987) and electrospray ionization (ESI) (Meng et al. 1988; Whitehouse et al. 1985) are the preferred methods.

Acidic glycan residues dissociate to a substantial degree during the MALDI process. The result is that losses of sialic acid, sulfate, and phosphate residues are likely to be observed for native carbohydrates and glycoconjugates using MALDI. The ESI process confers less vibrational energy on the analyte molecules with the result that fragmentation of glycoconjugates is generally not observed. ESI may be used either by direct infusion of an analyte solution or by direct connection to a liquid chromatography column.

Quantitative measurement is facilitated by chromatographic separation of glycans prior to MS so as to minimize ion suppression effects. Although the structures of glycans separated by chromatographic retention time may be correlated based on retention time to glycan standards (Guile et al. 1996; Knezevic et al. 2009), the use of a mass spectral detector facilitates the development of a retention time library against which unknown glycan compositions may be referenced (Pabst et al. 2007).

Mass spectrometric analysis of glycoconjugates with higher molecular weight is made challenging by both the comparative fragility of glycan chains and their heterogeneity. Thus, although MALDI produces low charge states that are readily interpretable, glycans often undergo fragmentation during the ionization process. Permethylation substantially stabilize glycans as compared to native molecules. Use of ESI is more likely to result in ionization without fragmentation. The ion patterns, however, may be extremely complex due to carbohydrate polymeric complexity and overlapping charge state envelops. These principles are shown in the analysis of a dextran 5,000 polymeric mixture using ESI-MS and MALDI-TOF-MS (Deery et al. 2001). Comparable results on these polysaccharides were obtained using the two techniques for these neutral, relatively stable, polysaccharides. Although MALDI-MS of permethylated glycans was found to be as good as chromatographic techniques, it suffered from frequent loss of terminal sialic acid residues. On the other hand, LC/ESI-MS was found to retain sialic acids intact. The masses have to be mapped with an online MALDI-MS glycomics database. In another inter-laboratory study for structural analysis of N-linked glycans, wide variations were found for levels of sialylation, fucosylation, and antennary branching, highlighting method variability and the need for well-characterized reference standards (Thobhani et al. 2009) (Fig. 13).

More recently, a highly sensitive approach using Fourier transform ion cyclotron mass spectrometry (FT-ICR-MS) was used to characterize different glycans including glycolipids (Park and Labriella 2005; Hakanssan et al. 2001; McFarland et al. 2005). Further, nanoscale liquid delivery using chip-based electrospray interface has been coupled with tandem MS as well as FT-ICR-MS for high sensitivity characterization of glycans (Froesch et al. 2004; Zamfir et al. 2004).

Nuclear Magnetic Resonance (NMR) Spectroscopy

Nuclear magnetic resonance (NMR) spectroscopy is another powerful technique for obtaining important sequence information on glycans. The one-dimensional proton and carbon (anomeric nuclei) spectra of a glycan mixture along with the coupling constants of homonuclear (gradient-selected correlation spectroscopy, gCOSY, and total-correlated spectroscopy, TOCSY) and heteronuclear (heteronuclear multiple-quantum correlation, HMQC, and heteronuclear multiple-bond correlation, HMBC) spectra provide quantitative information about distinct monosaccharide (Guerrini et al. 2001; Manzi et al. 2000; Lopez et al. 1997). The anomeric chemical shifts of

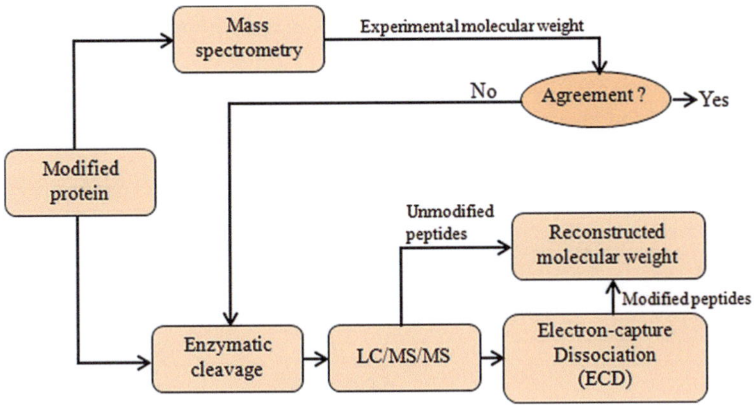

Fig. 13 A workflow diagram showing utilization of traditional MS/MS in combination with liquid chromatography (LC/MS/MS) for automated, high-throughput analysis of glycoproteins

monosaccharides can be classified further on the basis of neighboring monosaccharide (at reducing end) which would provide the abundance of specific linkage between two monosaccharides. When studying biological samples with large mixture of glycans, these techniques have certain limitations and high sample amount is required. Multiple steps of enzymatic and other fragmentation methods are required for larger glycan mixtures which make these steps highly complicated for high-throughput analysis.

Microarray

Carbohydrate Microarray

Carbohydrate microarray technologies are emerging as a novel tools for glycomics that are revolutionizing studies of carbohydrate–protein interactions and the elucidation of carbohydrate ligands involved not only in endogenous receptor systems but also pathogen–host interactions (Paulson et al. 2006; Horlacher and Seeberger 2008; Liang et al. 2008; Liu and Feizi 2008) (Fig. 14). The main advantage of microarray analysis is that a broad range of glycans can be immobilized on solid matrices as minute spots and simultaneously interrogated. The multivalent display of arrayed oligosaccharides can serve to mimic cell surface display and can be used to detect very low affinities of interactions that involve carbohydrates. The miniaturization in microarrays is advantageous and well suited for investigations in glycomics as thousands of binding events can be assessed in parallel on a single chip containing only tiny amounts of sample.

Carbohydrate microarray methods fall into two broad categories: polysaccharide microarrays and oligosaccharide microarrays. Polysaccharide samples derived from natural sources can be readily and randomly immobilized on solid matrices based on hydrophobic physical absorption (Wang et al. 2002; Willats et al. 2002; Moller et al. 2007) or charge-based interaction (Shipp and Hsieh-Wilson 2007) to generate polysaccharide microarrays. Oligosaccharide microarrays, on the other hand, provide detailed information on structure-activity relationships in carbohydrate recognition events. The oligosaccharides, being hydrophilic in nature, are difficult to immobilize. This is usually overcome by conjugating natural or chemically synthesized oligosaccharides to lipids by reductive amination to generate neoglycolipid (NGL) probes with amphipathic properties for arraying (Fukui et al. 2002; Feizi and Chai 2004) (Fig. 15).

Lectin Microarray

Lectins are defined as proteins—that bind carbohydrates present on glycoconjugates as ligands—that are neither enzymes (carbohydrate processing), antibodies (anticarbohydrate antibodies), nor proteins that bind small carbohy-

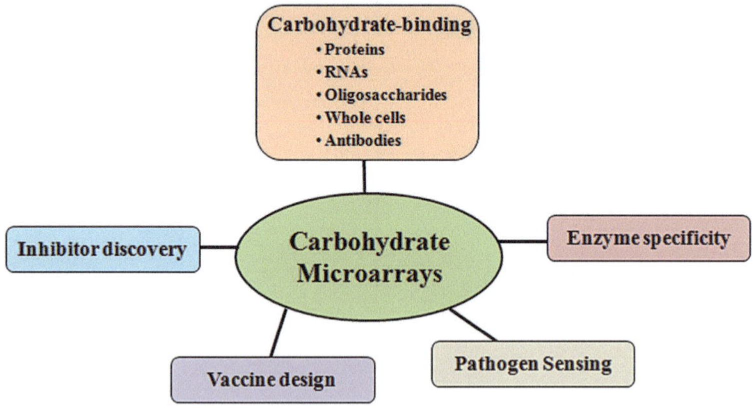

Fig. 14 Application of carbohydrate microarray

drate molecules (energy metabolism) (Goldstein et al. 1980). Lectins, which are available in plenty in plant seeds and could be isolated in enriched forms relatively easily, induced agglutination in blood cells and were instrumental in the study of blood group classification and as affinity matrices for purification of glycoproteins and cell separations (Rudiger and Gabius 2001). Lectins are known to play vital roles in plant physiology, development, and stress response. Most leguminous lectins are metalloproteins with tightly bound Ca^{++} and Mn^{++}, which are essential for carbohydrate-binding activity. Most plant lectins are relatively soluble and can be purified to homogeneity on appropriate immobilized carbohydrate matrices.

Since lectins differ in the types of carbohydrate structures they recognize with high affinity, they are also useful in the characterization of glycoconjugates. Because of the tremendous diversity of carbohydrate-binding specificities among the plant lectins, some researchers classify them according to the small carbohydrate haptens they recognize, e.g., galactose-binding lectins or GlcNAc-binding lectins. On the other hand, lectins act as tools for screening and agglutination, via interaction with bacterial polysaccharides including Nod factors, for attachment of symbiotic bacteria (e.g., rhizobia) and fungi, which help in nitrogen fixation and facilitate mutualism.

Lectin microarray technology has been used to profile the glycosylation of a multitude of biological and clinical samples, leading to new clinical biomarkers and advances in glycobiology. Lectin microarrays, which include >90 plant lectins, recombinant lectins, and selected antibodies, are used to profile N-linked, O-linked, and glycolipid glycans. The specificity of glycan profiling depends upon the carbohydrate-binding proteins arrayed. The current set targets mammalian carbohydrates including fucose, high mannose, branched and complex N-linked, α- and β-galactose and GalNAc, α-2,3- and α-2,6-sialic acid, LacNAc, and Lewis X epitopes (Pilobello et al. 2013).

Several lectins were immobilized to a suitable solid phase (microscopic slide) optimized using multiple chemical methodologies, conditions (humidity, buffer, and temperature), spot size, and spot morphology. Methods for lectin immobilization rely on various methods such as (1) carbine insertion, (2) biotin–avidin bridge, (3) attachment of amine functional group of lysine side chains of protein-backbone of lectins for immobilization to solid surface through epoxy-functionalized (glycidyl derivative) or N-hydroxysuccinimidyl (NHS)-derived esters, (4) exploitation of self-assembled monolayers (SAM) of thiols on gold-coated surfaces, and (5) 3D hydrogel surfaces. In all methods there is a lack of absolute control in guaranteeing the optimal orientation, native multimeric quarternary

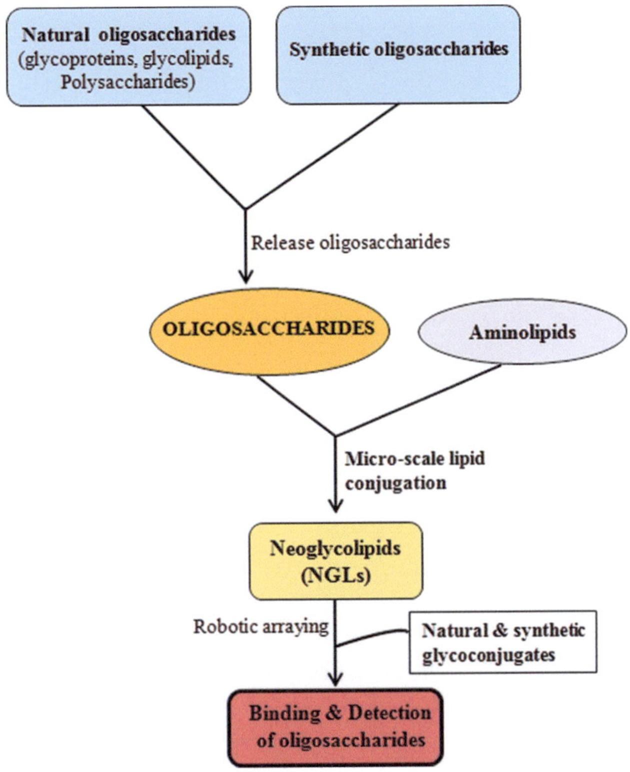

Fig. 15 Schematic representation of NGL-based oligosaccharide microarray

structure, optimal multivalent clustering of carbohydrate recognition domains (CRD) of lectins and their metal ion requirements. A mild periodate-based oxidation of glycan chains of lectin, as employed for glycosylated antibody microarrays (Chen et al. 2007), followed by anchoring to hydroxylamine or hydrazine-containing solid surfaces should be a useful approach (Fig. 16).

This method would leave the CRD of lectins intact, tuck away the crossreacting glycan portions of lectin, and might greatly improve optimal orientation of lectins in their native multimeric forms. Such hydrazide- or hydroxylamine-based methodologies have been used to capture glycans from glycoproteins and mammalian cells (Wollscheid et al. 2009). Lectin microarrays provide a rapid, sensitive, and high-throughput screening tools for glycoprotein profiling. Application of lectin microarrays with complementary glycomics technologies holds potential for decoding the "glycocode" and development of new and effective diagnostics and therapeutics.

Glycogene Microarray

In an attempt to overcome one of the major issues in glycomics, glycan structure characterization through MS, a bioinformatic method to predict glycan structures in a particular cell through the gene expression profiles was developed (Raman et al. 2005). Measurement of simultaneous expression of several thousand genes in different cells to construct genetic networks and pathways has been an important component of a systems approach to molecular and cell biology. Glycogene microarray technologies are used to identify the differentially expressed glycosylation-related genes. In this method, the concept of a "co-occurrence score" was calculated based on the co-occurrence of pairs of links within the same glycan structures. It was expected that by doing so the substrate specificity of glycosyltransferases could be captured in a single numerical matrix. Once this co-occurrence score matrix was developed, it could be used to make predictions from expression data. Development of a gene microarray (glycogene-chip v1) using the "Affymetrix" technology are a step toward

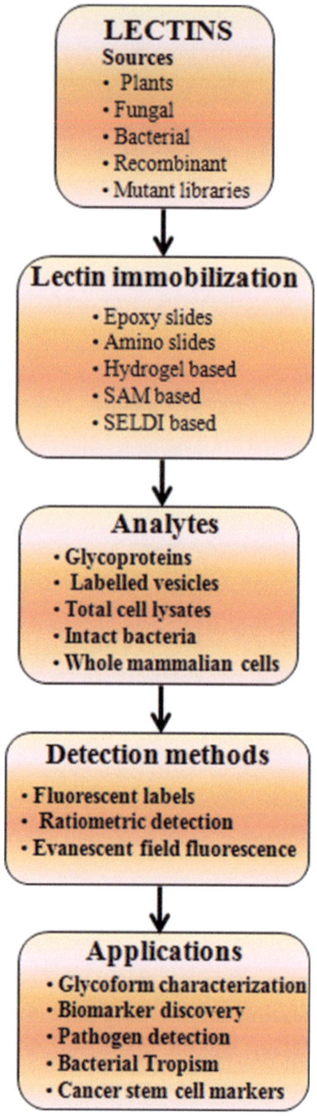

Fig. 16 **Summary of lectin microarray technologies**. Lectin microarray involves various immobilization technologies (1) using lectins derived from multiple sources (2). Glycoconjugates binding from a diverse set of analytes of biological samples (3) have been detected using multiple detection approaches (4). Lectin microarrays provide a rapid, sensitive, and high-throughput screening tools for glyco-profiling (5). SAM, self-assembled monolayers; SELDI, surface-enhanced laser desorption ionization; NHS, N-hydroxysuccinimide esters; Au, gold-coated surfaces

obtaining global information related to glycan biosynthesis, structure, and function (Comelli et al. 2002). The chip contains families of hundreds of genes, termed here "glycogenes," coding for proteins responsible for glycan synthesis and glycan recognition, including glycosyltransferases and other glycan-processing enzymes, enzymes relating to nucleotide synthesis and transport, proteoglycans, and glycan-binding proteins. Thus, these glycogene microarrays provide tremendous information about simultaneous expression of glycan biosynthetic enzymes that can be then correlated with the actual glycan structures, which had been characterized in a given sample.

Glycoinformatics: Bioinformatics for Glycome Analysis

"Glycomics," the scientific attempt to characterize and study carbohydrates, is a rapidly emerging branch of science, for which informatics is just beginning. The structures of glycans as secondary gene products cannot be easily predicted from the DNA sequence. Glycan sequences cannot be described by a simple linear one-letter code as each pair of monosaccharides can be linked in several ways and branched structures can be formed. The availability of comprehensive databases and corresponding bioinformatics tools, to access and analyze the large amounts of experimental data relating to the structure of carbohydrates, will be a prerequisite for the success of the large-scale glycomics projects that aim to decipher new, so far unknown, biological functions of glycans. Efficient bioinformatics descriptions and tools can considerably enhance the efficiency of glycomics research in terms of data quality, analysis, and experimental costs. Glycomics requires sophisticated algorithmic approaches. Several algorithms and models have been developed for glycobiology research in the past several years. Few of the bioinformatics algorithms developed for genomics/proteomics can be directly adapted for glycomics. The development of algorithms, which allow a rapid, automatic interpretation of mass spectra to identify glycans structure, is currently the most active field of research. Bound to proteins, as glycoproteins, glycans are known to affect the functions of proteins. More than half of all protein sequences deposited in the SWISS-PROT databank include potential glycosylation sites and thus may be glycoproteins. Based on an analysis of well-annotated and characterized glycoproteins in SWISS-PROT, it was concluded that more than half of all proteins are glycosylated (Apweiler et al. 1999).

The development and use of informatics tools and databases for glycobiology and glycomics research has increased considerably in recent years. However, the general development in this field can still be considered as being in its infancy when compared to the genomics and proteomics areas. In terms of bioinformatics in glycobiology, there are several paths of research that are currently in progress. The development of algorithms to reliably support the characterization of glycan structures for high-throughput applications is the most immediate demand of the glycomics community. Additionally, several major glyco-related projects (Consortium for Functional Glycomics) (Raman et al. 2006), KEGG Glycan (Hashimoto et al. 2006), GLYCOSCIENCES.de (Lütteke et al. 2006) are maturing and provide well-structured glyco-related data that are awaiting data mining and analysis. With the exciting new developments in carbohydrate arrays and automated MS annotation, the analysis of the glycome has reached a new level of sophistication, which requires broader informatics support. Tables 2 and 3 summarize the Web-based resources for glycomics.

Applications of Plant Glycomics

Carbohydrates conjugated to proteins and lipids are arguably the most abundant and structurally diverse class of molecules in nature. Complex glycans are mainly attached to secreted or cell surface proteins. Unlike the structure of linear DNA and protein sequences, glycans have heterogeneous structures that differ in composition, branching, linkage, and anomericity. Advancement in genomics and proteomics has not helped the glycomics as the sequence information of DNA and proteins have no direct information about biosynthesis of the glycan chains composition, branching, and linkage. These differences have made the identification of the number, structure, and function of glycans in cell physiology a daunting task; hence, the glycomic research lagged far behind genomics and proteomics. However, advancement in technology, rapid growth, and the importance of integral glycans in biology had immensely aided the glycomic research (Varki et al. 2008).

The scope of the glycomics challenge is immense. The covalent addition of glycans to proteins and lipids represents not only the most abundant post-translational modification (PTM) but also by far the most structurally diverse modification. It is estimated that >50 % of polypeptides are covalently modified by glycans (Apweiler et al. 1999); however, this estimate do not include the O-GlcNAc modified myriad of nuclear and cytoplasmic proteins (Hart et al. 2007). Every unique glycan–protein linkage is different and providing both structural and functional diversity. Consideration of the structural diversity of additional branches of glycans and complex terminal saccharides on glycans increases the molecular and functional diversity

Table 2 Major glycomic initiatives and their web resources

Glycomics initiatives	Web resources
Consortium for Functional Glycomics (CFG; USA)	CFG (http://www.functionlaglycomics.org)
Collaborative Glycomics Initiative (Europe)	EuroCarbDB (http://www.eurocarb.org)
Human Disease Glycomics/ Proteomics Initiative (Japan)	HGPI (http://www.hgpi.jp)
Complex Carbohydrate Research (Georgia, USA)	CCRC (http://www.ccrc.uga.edu)

Table 3 Web-based resources for glycomics and their datasets/informations

Web recourse	Datasets or information
Consortium for Functional glycomics (CFG; USA)	
GBP molecule page (http://www.functionalglycomics.org/glycomics/molecule/jsp/gbpMolecule-home.jsp)	Information portal with access to CFG and public databases
Glycan database (http://www.funtionlaglycomics.org/glycomics/molecule/jsp/carbohydrate/carbMoleculeHome.jsp)	Database of glycan structures with search interfaces and links to CFG glycan array and MALDI-MS data
Glycan profiling data (http://www.funtionlaglycomics.org/glycomics/publicdata/glycoprofiling.jsp)	Raw and annotated MALDI-MS profiles of glycans from mouse and human cells and tissues
Glycan array screen data (http://www.funtionlaglycomics.org/glycomics/publicdata/primaryscreen.jsp)	Raw data, bar graph of mean binding signal of GBP to each glycan in the array with links to their structures in glycan database
Gene microarray data (http://www.funtionlaglycomics.org/glycomics/publicdata/microarray.jsp)	Gene expression profiles of glycan biosynthesis enzymes and GBPs in various cells and tissues supplied by the investigators
Transgenic mice phenotyping data (http://www.funtionlaglycomics.org/glycomics/publicdata/phenotypingjsp)	Experimental protocols, data files corresponding to various phenotyping analysis of transgenic mice
Koyoto Encyclopedia of Genes and Genomes (KEGG; Japan)	
KEGG Glycan database (http://glycan.genome.jp)	Database of glycan structures obtained from CarbBank and updated with structures from other labs
KEGG Pathways database (http://www.genome.jp/kegg/pathway.html)	Collection of 15 glycan biosynthesis pathways with links to around 100 glycan biosynthesis enzymes
Other glycomics resources	
Glycosuite database (http://www.glycosuitedb.org)	Commercial database and tools for glycans
Sugarbase (http://boc.chem.uu.nl/sugarbase.html)	Glycan NMR database; chemical shifts of glycan structures
Lectin database (http://imperial.ac.uk/research/animallectins/)	Collection of information on animal lectins
Three-dimensional Lectin database (http://www.cermav.cnrs.fr/lectins)	Three-dimensional structures of lectins in the PDB
Bacterial Glycan database (http://www.glyco.ac.ru/bcsdb)	Database of bacterial glycan structures
CAZy (http://afmb.cnrs-mrs.fr/CAZY/)	Carbohydrate active enzyme database

of varied protein-bound glycans exponentially. Continued research into structural–functional relationships between glycans and proteins also has potential to result in second-generation products engineered to optimize therapeutic efficacy.

Plants represent an unprecedented approach for the production of pharmaceutically important and commercially valuable proteins in plants. Plant lectins with fungicidal and insecticidal activities are used in transgenic technologies to increase plant resistance to pests and phytopathogens (Melnykova et al. 2013). The introduction of lectin-like kinases genes into plant genome was shown to protect plants against environmental stresses and regulate plant growth. Engineering of phytolectins allows obtaining molecules with known carbohydrate specificity that can be applied in various areas. The studies are underway with the aim of design of lectin-based drug delivery systems as well as the pharmaceutical drugs containing plant lectins. Because of the ability of phytolectins to bind to different substances, they can be more widely used in the future.

Plant glycomic research also has significant applications in *biopharming* of proteins and enzymes involving glycosylation. Plants are gaining increasing acceptance as bioreactors and are used for large-scale production of industrial enzymes and pharmaceuticals (Yadav et al. 2013). Plants are potential hosts for the expression of recombinant glycoproteins intended for therapeutic purposes. Many transgenic crops such as tobacco, soybeans, maize, etc. have been used for overexpression of the pharmaceutically important proteins and have significant advantages overexpression in conventional expression systems such as bacterial, yeast, insect, or animal cell lines. One of the major advantages of using transgenic plants, compared with the alternative production systems for biopharmaceuticals, is their ability to perform most of the post-translational modifications (PTMs) required for protein activity, together with their capacity for large-scale production.

However, plants are unable to exactly reproduce human-type glycosylation patterns in biopharmaceuticals (Gomord and Faye 2004; Faye et al. 2005). The *N*-glycans of mammalian glycoproteins produced in transgenic plants differ from their natural counterparts. It has been observed that modified plant *N*-glycans are frequently observed on pollen grains and food allergens of plant origins containing exclusive β(1,2)-xylose linked to the core mannose and α(1,3)-fucose residues substituting proximal *N*-acetylglucosamine and behave as a strong plant glycoallergens. With the prospect of using plants as alternative hosts to mammalian cell lines for the production of therapeutic glycoproteins, significant progress has been made toward the humanization of protein *N*-glycosylation in plant cells. Successful efforts have been made mainly focusing on the compartment targeted expression of therapeutic proteins, the knockout of plant-specific *N*-glycan-processing genes, and/or the introduction of the enzymatic machinery catalyzing the synthesis, transport, and addition of human sugars to optimally utilize the plant glycome.

Nowadays polysaccharide-based biomaterials are an emerging class in several biomedical fields. Most of the polysaccharides used are derived from natural sources such as tamarind seed polysaccharide (TSP), alginate, and chitin. Plant glycomics therefore has huge applications in biomedical and glycotherapeutics and may revolutionize the field of biopharmaceutics.

Glycome profiling, using plant glycomic approaches, also lends itself well to moderate to high-throughput screening of plant cell wall/biomass samples. Glycome profiling may prove to be a broadly applicable experimental approach to a wide variety of studies on plant cell wall/biomass samples (Pattathil et al. 2012). Plant cell wall polysaccharides makes about 75 % of cell wall components. These can be broken down to produce sugar substituent which can be used for biofuel production. Plant glycomic studies using carbohydrate microarray and lectin microarray can also be employed for exploring host–parasite interactions in plants. The field provides huge scope for advance researches.

The emerging field of glycomics relies upon a diverse set of technologies and strategies such as glycoarrays, high-performance liquid chromatography (HPLC), MS, NMR spectroscopy, databases and libraries of natural glycan, and synthetic chemistry. These research tools are supported by bioinformatics and molecular modeling of glycans. The application of these latest advances in biotechnology should result in novel applications of glycomics approaches to diagnostics and glycotherapeutics and in twenty-first-century glycomics may emerge as the new genomics. Figure 17 summarizes some of the applications of plant glycomics.

Summary

Glycomics involves the systematic study of all glycan structures of a given cell type or organism and is a subset of glycobiology. Advance glycomic studies suffer major limitations due to greater structural complexity of glycan biomolecules as compared to nucleic acids and proteins including branching and linkage diversity. The isolation of carbohydrates and glycoconjugates from natural sources is tedious, frequently yields

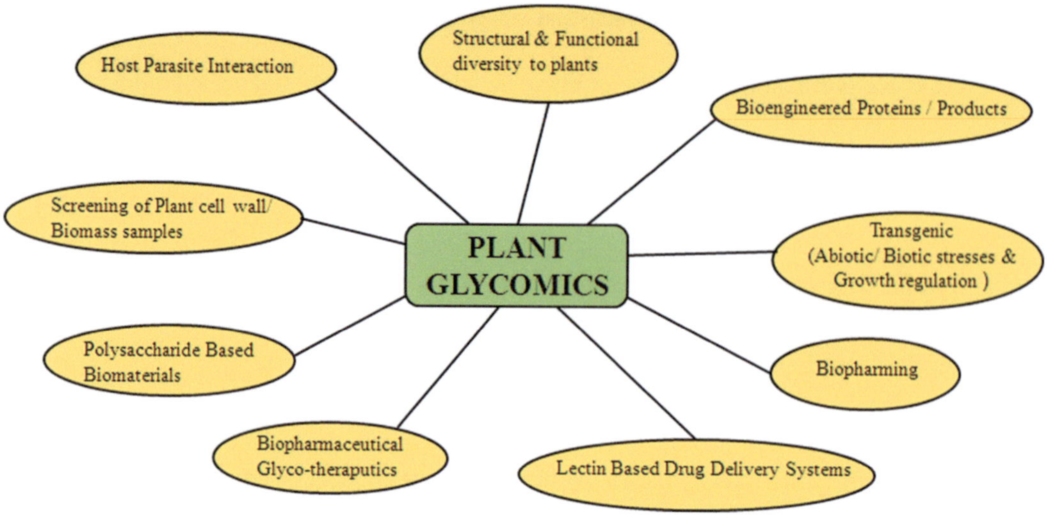

Fig. 17 Applications of plant glycomics. Areas of plant sciences with potential applications of plant glycomics

heterogeneous products, and produces little material. N-linked glycoproteins are the major glycan present in plants and are biosynthesized by a complex mechanism of glycosylation. O-linked glycan are less abundant in plants and unlike *N*-linked glycans, there is no known consensus sequence for glycosylation. In comparison to genomics and proteomics—where automated synthesis, amplification, expression, and characterization have become routine—the tools available for study of glycomics are few. Therefore, there is a need for development of unique tools for glycomic studies. The enormous diversity of structure of monosaccharides has made the task of system-wide glycan profiling a challenging task. Currently used tools for glycomic studies include chromatographic techniques, MALDI-ToF-MS, ESI-MS, NMR spectroscopy, and microarray. Development of high-throughput tandem mass spectrometry and carbohydrate/lectin microarray has revolutionized the field of glycomics. Glycogene microarray technologies are used to identify the differentially expressed glycosylation-related enzyme genes. Development of a gene microarray using the "Affymetrix" technology is a step toward obtaining information about glycan biosynthesis, structure, and function. Efficient bioinformatic tools (glycoinformatics) have considerably enhanced the efficiency of glycomic research, in terms of data quality, analysis, and experimental costs. However, the field can still be considered as being in its infancy. Several glyco-related projects (Consortium for Functional Glycomics), KEGG Glycan, and GLYCOSCIENCES.de are maturing and provide detailed glyco-related data.

Plant glycans including lectins are involved in cell–cell interactions and play significant role in providing both structural and functional diversity to plants. Plant lectins with fungicidal and insecticidal activities are used in transgenic technologies to increase plant resistance to pests and phytopathogens. Plant glycomic studies find their applications in biopharming using transgenic approaches.

References

Altmann F (1997) More than silk and honey-or, can insect cells serve in the production of therapeutic glycoproteins? Glycoconj J 14:643–646

Altmann F, Fabini G, Ahorn H, Wilson IB (2001) Genetic model organisms in the study of Nglycans. Biochimie 83:703–712

Apweiler R, Hermjakob H, Sharon N (1999) On the frequency of protein glycosylation, as deduced from

analysis of the SWISS-PROT database. Biochim Biophys Acta 1473:4–8
Bakker H, Bardor M, Molthoff JW, Gomord V, Elbers I, Stevens LH, Jordi W, Lommen A, Faye L, Lerouge P, Bosch D (2001) Galactose extended glycans of antibodies produced by transgenic plants. Proc Natl Acad Sci U S A 98:2899–2904
Balen B, Krsnik-Rasol M (2007) N-glycosylation of recombinant therapeutic glycoproteins in plant systems. Food Technol Biotechnol 45:1–10
Bertozzi CR, Kiessling LL (2001) Chemical glycobiology. Science 291:2357–2364
Bielik AM, Zaia J (2010) Historical overview of glycoanalysis. Methods Mol Biol 600:9–30
Bollini R, Ceriotti A, Daminati MG, Vitale A (1985) Glycosylation is not needed for the intracellular transport of phytohemagglutinin in developing *Phaseolus vulgaris* cotyledons and for the maintenance of its biological activities. Physiol Plant 65:15–22
Buskas T, Ingale S, Boons GJ (2006) Glycopeptides as versatile tools for glycobiology. Glycobiology 16:113R–136R
Chadd HE, Chamow SM (2001) Therapeutic antibody expression technology. Curr Opin Biotechnol 12(2):188–194
Chargelegue D, Vine ND, van Dolleweerd CJ, Drake PM, Ma JK (2000) A murine monoclonal antibody produced in transgenic plants with plant specific glycans is not immunogenic in mice. Transgenic Res 9:187–194
Chen S, LaRoche T, Hamelinck D, Bergsma D, Brenner D, Simeone D, Brand RE, Haab BB (2007) Multiplexed analysis of glycan variation on native proteins captured by antibody microarrays. Nat Methods 4:437–444
Comelli EM, Amado M, Head SR, Paulson JC (2002) Custom microarray for glycobiologists: consideration for glycosyltransferase gene expression profiling. Biochem Soc Symp 69:135–142
Cooper CA, Gasteiger E, Packer NH (2001) GlycoMod—a software tool for determining glycosylation compositions from mass spectrometric data. Proteomics 1:340–349
Costa J, Ashford DA, Nimtz M, Bento I, Frazao C, Esteves CL, Faro CJ, Kervinen J, Pires E, Verissimo P, Wlodawer A, Carrondo MA (1997) The glycosylation of the aspartic proteinases from barley (*Hordeum vulgare* L.) and cardoon (*Cynara cardunculus* L.). Eur J Biochem 243:695–700
Cumar FA, Barra HS, Maccioni HJ, Caputto R (1968) Sulfation of glycosphingolipids and related carbohydrates by brain preparations from young rats. J Biol Chem 243:3807–3816
Cummings RD (2009) The repertoire of glycan determinants in the human glycome. Mol BioSyst 5:1087–1104
D'Angelo G et al (2007) Glycosphingolipid synthesis requires FAPP2 transfer of glucosylceramide. Nature 449:62–67
Daniell H, Streatfield SJ, Wycoff K (2001) Medical molecular farming: production of antibodies, biopharmaceuticals and edible vaccines. Trends Plant Sci 6:219–226
Deery MJ, Stimson E, Chappell CG (2001) Size exclusion chromatography/mass spectrometry applied to the analysis of polysaccharides. Rapid Commun Mass Spectrom 15:2273–2283
Doran PM (2000) Foreign protein production in tissue cultures. Curr Opin Biotechnol 11:199–204
Ding W, Nothaft H, Szymanski CM, Kelly J (2009) Identification and quantification of glycoproteins using ionpairing normal-phase liquid chromatography and mass spectrometry. Mol Cell Proteomics 8:2170–2185
Domann PJ, Pardos-Pardos AC, Fernandes DL, Spencer DI, Radcliffe CM, Royle L, Dwek RA, Rudd PM (2007) Separation-based glycoprofiling approaches using fluorescent labels. Proteomics 7(Suppl 1):70–76
Dörmann P, Benning C (2002) Galactolipids rule in seed plants. Trends Plant Sci 7:112–118
Elbers IJW, Stoopen GM, Bakker H, Stevens LH, Bardor M, Molthoff JW, Jordi JRM, Bosch D, Lommen A (2001) Influence of growth conditions and developmental stage on N-Glycan heterogeneity of transgenic immunoglobulin G and endogenous proteins in tobacco leaves. Plant Physiol 126(3):1314–1322
Faye L, Boulaflous A, Benchabane M, Gomord V, Michaud D (2005) Protein modifications in the plant secretory pathway: current status and practical implications in molecular pharming. Vaccine 23:1770–1778
Feizi T, Chai W (2004) Oligosaccharide microarrays to decipher the glyco code. Nat Rev Mol Cell Biol 5:582–588
Fiedler U, Conrad U (1995) High-level production and long-term storage of engineered antibodies in transgenic tobacco seeds. Biotechnology 13:1090–1093
Fischer R, Emans N (2000) Molecular farming of pharmaceutical proteins. Transgenic Res 9:279–299
Fitchette-Lainé AC, Gomord V, Cabanes M, Michalski JC, Saint-Macary M, Foucher B, Cavelier B, Hawes C, Lerouge P, Faye L (1997) N-glycans harboring the lewis a epitope are expressed at the surface of plant cells. Plant J 12:1411–1417
Fotisch K, Vieths S (2001) N- and O-linked oligosaccharides of allergenic glycoproteins. Glycoconj J 18:373–390
Frasch W, Grunwald C (1976) Acylatedsteryl glycoside synthesis in seedlings of Nicotianatabacum L. Plant Physiol 58:744–748
Froesch M, Bindila L, Baykut G, Allen M, Peter-Katalinić J, Zamfi AD (2004) Coupling of fully automated chip electrospray to fourier transform ion cyclotron mass spectrometry for high-performance glycoscreening and sequencing. Rapid Commun Mass Spectrom 18:3084–3092
Fukui S, Feizi T, Galustian C, Lawson AM, Chai W (2002) Oligosaccharide microarrays for high-throughput detection and specificity assignments of carbohydrate-protein interactions. Nat Biotechnol 20:1011–1017

Gabius HJ, Siebert HC, Andre S, Jimenez-Barbero J, Rudiger H (2004) Chemical biology of the sugar code. Chembiochem 5:740–764

Garcia-Casado G, Sanchez-Monge R, Chrispeels MJ, Armentia A, Salcedo G, Gomez L (1996) Role of complex asparagine-linked glycans in the allergenicity of plant glycoproteins. Glycobiology 6:471–477

Ghesquiere B, Van Damme J, Martens L, Vandekerckhove J, Gevaert K (2006) Proteome-wide characterization of Nglycosylation events by diagonal chromatography. J Proteome Res 5:2438–2447

Giddings G, Allison G, Brooks D, Carter A (2000) Transgenic plants as factories for biopharmaceuticals. Nat Biotechnol 18:1151–1155

Goldberg D, Sutton-Smith M, Paulson J, Dell A (2005) Automatic annotation of matrix-assisted laser desorption/ ionization N-glycan spectra. Proteomics 5:865–875

Goldstein IJ, Hughes RC, Monsigny M, Osawa T, Sharon N (1980) What should be called a lectin? Nature 285:66

Gomord V, Faye L (2004) Post translational modifications of therapeutic proteins in plants. Curr Opin Plant Biol 7(2):171–181

Gray JSS, Yang BY, Hull SR, Venzke DP, Montgomery R (1996) The glycans of soybean peroxidase. Glycobiology 6:3–32

Grunwald C (1978) Steryl glycoside biosynthesis. Lipids 13:697–703

Guerrini M, Bisio A, Torri G (2001) Combined quantitative ^{1}H and ^{13}C nuclear magnetic resonance spectroscopy for characterization of heparin preparations. Semin Thromb Hemost 27:473–482

Guile GR, Rudd PM, Wing DR, Prime SB, Dwek RA (1996) A rapid high-resolution high-performance liquid chromatographic method for separating glycan mixtures and analyzing oligosaccharide profiles. Anal Biochem 240:210–226

Hakanssan K, Cooper HJ, Emmett MR, Costello CE, Marshall AG, Nilsson CL (2001) Electron capture dissociation and infrared multiphoton dissociation MS/MS of an N-glycosylated tryptic peptic to yield complementary sequence information. Anal Chem 73:4530–4536

Halter D et al (2007) Pre- and post-Golgi translocation of glucosylceramide in glycosphingolipid synthesis. J Cell Biol 179:101–115

Hammond C, Braakman I, Helenius A (1994) Role of N-linked oligosaccharide recognition, glucose trimming, and calnexin in glycoprotein folding and quality control. Proc Natl Acad Sci U S A 91:913–917

Hanada K, Kumagai K, Tomishige N, Yamaji T (2009) CERT-mediated trafficking of ceramide. Biochim Biophys Acta 1791:684–691

Hart GW, Copeland RJ (2010) Glycomics hits the big time. Cell 143:672–676

Hart GW, Housley MP, Slawson C (2007) Cycling of O-linked β-N-acetylglucosamine on nucleocytoplasmic proteins. Nature 446:1017–1022

Hashimoto K, Goto S, Kawano S, Aoki-Kinoshita KF, Ueda N, Hamajima M, Kawasaki T, Kanehisa M (2006) M KEGG: a glycome informatics resource. Glycobiology 16:63R–70R

Horvath PJ, Eagen CK, Fisher NM, Leddy JJ, Pendergast DR (2000) The effects of varying dietary fat on performance and metabolism in trained male and female runners. J Am Coll Nutr 19(1):52–60

Hood EE, Jilka JM (1999) Plant-based production of xenogenic proteins. Curr Opin Biotechnol 10:382–386

Horlacher T, Seeberger PH (2008) Carbohydrate arrays as tools for research and diagnostics. Chem Soc Rev 37:1414–1422

Karas M, Hillenkamp F (1988) Laser desorption ionization of proteins with molecular masses exceeding 10,000 daltons. Anal Chem 60:2299–2301

Karas M, Bachmann D, Bahr U, Hillenkamp F (1987) Matrix-assisted ultraviolet-laser desorption of non-volatile compounds. Int J Mass Spectrom Ion Process 78:53–68

Kishimoto T, Watanabe M, Mitsui T, Mori H (1999) Glutelin basic subunits have a mammalian mucin type O-linked disaccharide side chain. Arch Biochem Biophys 370:271–277

Knezevic A, Polasek O, Gornik O, Rudan I, Campbell H, Hayward C, Wright A, Kolcic I, O'Donoghue N, Bones J, Rudd PM, Lauc G (2009) Variability, heritability and environmental determinants of human plasma N-glycome. J Proteome Res 8:694–701

Kornfeld R, Kornfeld S (1985) Assembly of asparagine-linked oligosaccharides. Annu Rev Biochem 54:631–664

Krambeck FJ, Bennun SV, Narang S, Choi S, Yarema KJ, Betenbaugh MJ (2009) A mathematical model to derive N-glycan structures and cellular enzyme activities from mass spectrometric data. Glycobiology 19:1163–1175

Larrick JW, Thomas DW (2001) Producing proteins in transgenic plants and animals. Curr Opin Biotechnol 12:411–418

Larrick JW, Yu L, Naftzger C, Jaiswal S, Wycoff K (2001) Production of secretory IgA antibodies in plants. Biomol Eng 18:87–94

Leiter H, Mucha J, Staudacher E, Grimm R, Glossl J, Altmann F (1999) Purification, cDNA cloning and expression of GDP-L-Fuc: Asn linked GlcNAcα1,3-fucosyltransferase from mung beans. J Biol Chem 274:830–839

Lerouge P, Cabanes-Macheteau M, Rayon C, Fischette-Lainé AC, Gomord V, Faye L (1998) N-Glycoprotein biosynthesis in plants: recent developments and future trends. Plant Mol Biol 38:31–48

Liang PH, Wu CY, Greenberg WA, Wong CH (2008) Glycan arrays: biological and medical applications. Curr Opin Chem Biol 12:86–92

Lis H, Sharon N (1978) Soybean agglutinin: a plant glycoprotein. J Biol Chem 253:3468–3476

Lis H, Sharon N (1993) Protein glycosylation: structural and functional aspects. Eur J Biochem 218(1):1–27

Liu Y, Feizi T (2008) Microarrays– key technologies and tools for glycobiology. In: Fraser-Reid BO, Tatsuta K, Thiem J (eds) Glycoscience – chemistry and chemical biology, 2nd edn. Springer, Berlin/Heidelberg/New York, pp 2121–2132

Lopez M, Coddeville B, Langridge J, Plancke Y, Sautiere P, Chaabihi H, Chirat F, Harduin-Lepers A, Cerutti M, Ver-bert A, Delannoy P (1997) Microheterogeneity of oligosaccharides carried by the recombinant bovine lactoferrin expressed in Mamestra brassicae cells. Glycobiology 7:635–651

Lütteke T, Bohne-Lang A, Loss A, Goetz T, Frank M, von der Lieth CW (2006) GLYCOSCIENCES.de: an internet portal to support glycomics and glycobiology research. Glycobiology 16:71R–81R

Maccioni HJ (2007) Glycosylation of glycolipids in the Golgi complex. J Neurochem 103(Suppl. 1):81–90

Maccioni HJ, Quiroga R, Spessott W (2011) Organization of the synthesis of glycolipid oligosaccharides in the Golgi complex. FEBS Lett 585:1691–1698

Majewska-Sawka A, Nothnagel EA (2000) The multiple roles of arabinogalactan proteins in plant development. Plant Physiol 122:3–10

Manzi AE, Norgard-Sumnicht K, Argade S, Marth JD, van Halbeek H, Varki A (2000) Exploring the glycan repertoire of genetically modified mice by isolating and profiling of the major glycan classes and nano-NMR analysis of glycan mixtures. Glycobiology 10:669–689

McCormick AA, Kumagai MH, Hanley K, Turpen TH, Hakim I, Grill LK, Tuse D, Levy S, Levy R (1999) Rapid production of specific vaccines for lymphoma by expression of the tumor derived single chain Fv epitopes in tobacco plants. Proc Natl Acad Sci U S A 96:703–708

McDonald CA, Yang JY, Marathe V, Yen TY, Macher BA (2009) Combining results from lectin affinity chromatography and glycocapture approaches substantially improves the coverage of the glycoproteome. Mol Cell Proteomics 8:287–301

McFarland MA, Marshall AG, Hendrickson CL, Nilsson CL, Fredman P, Mansson JE (2005) Structural characterization of GM1 ganglioside by infrared multiphoton dissociation: electron capture dissociation and electron detachment dissociation electrospray ionization FT-ICR-MS/MS. J Am Soc Mass Spectrom 16:752–762

Melnykova NM, Mykhalkiv LM, Mamenko PM, Kots YS (2013) The areas of application for plant lectins. Biopolym Cell 29(5):357–366

Melo NS, Nimtz M, Conradt HS, Fevereiro PS, Costa J (1997) Identification of the human Lewis[a] carbohydrate motif in a secretory peroxidase from a plant cell suspension culture (*Vaccinium myrtillus L.*). FEBS Lett 415:186–191

Meng CK, Mann M, Fenn JB (1988) Of protons or proteins. Z Phys D 10:361–368

Misaki R, Kimura Y, Palacpac NQ, Yoshida S, Fujiyama K, Seki T (2003) Plant cultured cells expressing human beta1,4-galactosyltransferase secrete glycoproteins with galactose-extended N-linked glycans. Glycobiology 13:199–205

Mitsui T, Kimura S, Igaue L (1990) Isolation and characterization of Golgi membranes from suspension-cultured cells of rice. Plant Cell Physiol 31:15–25

Moller I, Sørensen I, Bernal AJ, Blaukopf C, Lee K, Øbro J, Pettolino F, Roberts A, Mikkelsen JD, Knox JP, Bacic A, Willats WGT (2007) High-throughput mapping of cell-wall polymer within and between plants using novel microarrays. Plant J 50:1118–1128

Motose H, Sugiyama M, Fukuda H (2004) A proteoglycan mediates inductive interaction during plant vascular development. Nature 429:873–878

Navazio L, Baldan B, Mariani P, Gerwig GJ, Vliegenthart JFG (1996) Primary structure of the N-linked carbohydrate chains of calreticulin from spinach leaves. Glycoconj J 13:977–983

North SJ, Hitchen PG, Haslam SM, Dell A (2009) Mass spectrometry in the analysis of N-linked and O-linked glycans. Curr Opin Struct Biol 19:498–506

Ohsuga H, Su SN, Takahashi N, Yang SY, Nakagawa H, Shimada I, Arata Y, Lee YC (1996) The carbohydrate moiety of the Bermuda grass antigen BG60. J Biol Chem 271:26653–26658

Oxley D, Munro SLA, Craik DJ, Bacic A (1996) Structure of N glycans on the S3- and S6- allele stylar self-incompatibility ribonucleases of *Nicotiana alata*. Glycobiology 6:611–618

Pabst M, Bondili JS, Stadlmann J, Mach L, Altmann F (2007) Massþretention time¼structure: a strategy for the analysis of N-glycans by carbon LC-ESI-MS and its application to fibrin N-glycans. Anal Chem 79:5051–5057

Pagny S, Cabanes-Macheteau M, Gillikin JW, Leborgne-Castel N, Lerouge P, Boston RS, Faye L, Gomord V (2000) Protein recycling from the Golgi apparatus to the endoplasmic reticulum in plants and its minor contribution to calreticulin retention. Plant Cell 12:739–756

Palacpac NQ, Yoshida S, Sakai H, Kimura Y, Fujiyama K, Yoshida T, Seki T (1999) Stable expression of human beta1,4-galactosyltransferase in plant cells modifies N-linked glycosylation patterns. Proc Natl Acad Sci U S A 96:4692–4697

Park Y, Labriella CB (2005) Applications of fourier transform ion cyclotron mass spectrometry to oligosaccharides. Mass Spectrom Rev 24:232–264

Pattathil S, Avci U, Miller JS, Hahn MG (2012) Immunological approaches to plant cell wall and biomass characterization: Glycome Profiling. Methods Mol Biol 908:61–72

Patwa TH, Qiu Y, Zhao J, Simeone DM, Lubman DM (2009) All-liquid separations, protein microarrays, and mass spectrometry to interrogate serum proteomes: an application to serum glycoproteomics. Methods Mol Biol 520:75–87

Paulson JC, Blixt O, Collins BE (2006) Sweet spots in functional glycomics. Nat Chem Biol 2:238–248

Pilobello KT, Mahal LK (2007) Deciphering the glycocode: the complexity and analytical challenge of glycomics. Curr Opin Chem Biol 11:300–305

Pilobello KT, Agarwal P, Rouse R, Mahal LK (2013) Advances in lectin microarray technology: optimized protocols for piezoelectric print conditions. Curr Protoc Chem Biol 5(1):1–23

Pope DG (1977) Relationships between hydroxyproline containing proteins secreted into the cell wall and medium by suspension-cultured Acer pseudoplatanus cells. Plant Physiol 59:894–900

Potocka A, Zimowski J (2008) Metabolism of conjugated sterols in eggplant. Part 2. Phospholipid: steryl glucoside acyltransferase. Acta Biochim Pol 55:135–140

Prescher JA, Bertozzi CR (2006) Chemical technologies for probing glycans. Cell 126:851–854

Qi W, Fong C, Lamport DTA (1991) Gum arabic glycoprotein is a twisted hairy rope: a new model based on O-galactosylhydroxyproline as the polysaccharide attachment site. Plant Physiol 96:848–855

Ramachandran P, Boontheung P, Xie Y, Sondej M, Wong DT, Loo JA (2006) Identification of N-linked glycoproteins in human saliva by glycoprotein capture and mass spectrometry. J Proteome Res 5:1493–1503

Raman R, Raghuram S, Venkataraman G, Paulson JC, Sasisekhran R (2005) Glycomics: an integrated systems approach to structure-function relationships of glycans. Nat Methods 2(11):817–824

Raman R, Venkataraman M, Ramakrishnan S, Lang W, Raguram S, Sasisekharanet R (2006) Advancing glycomics: implementation strategies at the consortium for functional glycomics. Glycobiology 16:82R–90R

Rudiger H, Gabius HJ (2001) Plant lectins: occurrence, biochemistry, functions and applications. Glycoconj J 18:589–613

Russell DA (1999) Feasibility of antibody production in plants for human therapeutic use. Curr Top Microbiol Immunol 236:119–137

Sala F, Manuela Rigano M, Barbante A, Basso B, Walmsley AM, Castiglione S (2003) Vaccine antigen production in transgenic plants: strategies, gene constructs and perspectives. Vaccine 21:803–808

Schenk D, Barbour R, Dunn W, Grajeda H, Guido T, Huang J, Johnson-Wood K, Khan K, Kholodenko D, Lee M, Liao Z, Lieberburg I, Motter R, Mutter L, Soriano F, Shopp G, Vasquez N, Vandevert C, Walker S, Wogulis M, Yednock T, Games D, Seubert P (1999) Immunization with amyloid-beta attenuates Alzheimer-disease-like pathology in the PDAPP mouse. Nature 400(6740):173–177

Schulte S, Stoffel W (1993) Ceramide UDP galactosyltransferase from myelinating rat brain: purification, cloning and expression. Proc Natl Acad Sci U S A 90:10265–10269

Seeberger PH, Werz DB (2007) Synthesis and medical applications of oligosaccharides. Nature 446:1046–1051

Seifert GJ, Roberts K (2007) The biology of arabinogalactan proteins. Annu Rev Plant Biol 58:137–161

Sharp JM, Doran PM (2001a) Characterization of monoclonal antibody fragments produced by plant cells. Biotechnol Bioeng 73:338–346. doi:10.1002/bit.1067

Sharp JM, Doran PM (2001b) Strategies for enhancing monoclonal antibody accumulation in plant cell and organ cultures. Biotechnol Prog 17:979–992. doi:10.1021/bp010104t

Shipp EL, Hsieh-Wilson LC (2007) Profiling the sulfation specificities of glycosaminoglycan interactions with growth factors and chemotactic proteins using microarrays. Chem Biol 14:195–208

Showalter AM (2001) Arabinogalactan-proteins: structure, expression and function. Cell Mol Life Sci 58:1399–1417

Sprong H, Kruithof B, Leijendekker R, Slot JW, van Meer G, van der Sluijs P (1998) UDP-galactose: ceramide galactosyltransferase is a class I integral membrane protein of the endoplasmic reticulum. J Biol Chem 273:25880–25888

Sturm A, Johnson KD, Szumilo T, Elbein AD, Chrispeels MJ (1987) Subcellular localization of glycosidases and glycosyltransferases involved in the processing of the N-linked oligosaccharides. Plant Physiol 85:741–745

Sturm A, Bergwerff AA, Vliegenthart JFG (1992) ^{1}H-NMR structural determination of the N-linked carbohydrate chains on glycopeptides obtained from the bean lectin phytohemagglutinin. Eur J Biochem 204:313–316

Szumilo T, Kaushal GP, Elbein AD (1986a) Purification and properties of glucosidase I from mung bean seedlings. Arch Biochem Biophys 247:261–271

Szumilo T, Kaushal GP, Hori H, Elbein AD (1986b) Purification and properties of a glycoprotein processing α-mannosidase from mung bean seedling. Plant Physiol 81:383–389

Tacket CO, Mason HS, Losonsky G, Estes MK, Levine MM, Amtzen CJ (2000) Human immune responses to a novel Norwalk virus vaccine delivered in transgenic potatoes. J Infect Dis 182:302–305

Tezuka K, Hayashi M, Ishihara H, Akazawa T, Takahashi N (1992) Studies on synthetic of xylose-containing N-linked oligosaccharides deduced from substrate specificities of the processing enzymes in sycamore cells (*Acer pseudoplatanus* L.). Eur J Biochem 203:401–413

Thobhani S, Yuen CT, Bailey MJ, Jones C (2009) Identification and quantification of N-linked oligosaccharides released from glycoproteins: an inter-laboratory study. Glycobiology 19:210–211

van Hengel AJ, Tadesse Z, Immerzeel P, Schols H, van Kammen A, de Vries SC (2001) N-Acetylglucosamine and glucosamine-containing arabinogalactan proteins control somatic embryogenesis. Plant Physiol 125:1880–1890

Vaquero C, Sack M, Schuster F, Finnern R, Drossard J, Schumann D, Reimann A, Fischer R (2002) A carcino embryonic antigen-specific diabody produced in tobacco. FASEB J 16:408–410

Varki A, Daniel H, Geschwind EEE (2008) Human uniqueness: genome interactions with environment, behaviour and culture. Nat Rev Genet 9:749–763 von der Lieth CW, Lutteke T, Frank M (2006) The role of informatics in glycobiology research with special emphasis on automatic interpretation of MS spectra. Biochim Biophys Acta 1760:568–577

Walsh G, Jefferis R (2006) Post-translational modifications in the context of therapeutic proteins. Nat Biotechnol 24:1241–1252

Wandelt CI, Khan MRI, Craig S, Schroeder HE, Spencer D, Higgins TJV (1992) Vicilin with carboxy-terminal KDEL is retained in the endoplasmic reticulum and accumulates to high levels in leaves of transgenic plants. Plant J 2:181–192

Wang D, Liu S, Trummer BJ, Deng C, Wang A (2002) Carbohydrate microarrays for the recognition of cross-reactive molecular markers of microbes and host cells. Nat Biotechnol 20:275–281

Wang CC, Lee JC, Luo SY, Kulkarni SS, Huang YW, Lee CC, Chang KL, Hung SC (2007) Regioselective one-pot protection of carbohydrates. Nature 446:896–899

Warnock DE, Lutz MS, Blackburn WA, Young WW Jr, Baenziger JU (1994) Transport of newly synthesized glucosylceramide to the plasma membrane by non-golgi pathway. Proc Natl Acad Sci U S A 91(7):2708–2712

Werz DB, Ranzinger R, Herget S, Adibekian A, von der Lieth CW, Seeberger PH (2007) Exploring the structural diversity of mammalian carbohydrates ("glycospace") by statistical databank analysis. ACS Chem Biol 2:685–691

Whitehouse CM, Dreyer RN, Yamashita M, Fenn JB (1985) Electrospray interface for liquid chromatographs and mass spectrometers. Anal Chem 57:675–679

Wigdorovitz A, Carillo C, Dus Santos M, Trono K, Peralta A, Gomez M, Rio R, Franzone SA, Escribano J, Borca M (1999) Induction of a protective antibody response to foot and mouth disease virus in mice following oral or parenteral immunization with alfalfa transgenic plants expressing the viral structural protein VP1. Virology 255:347–353

Willats WG, Rasmussen SE, Kristensen T, Mikkelsen JD, Knox JP (2002) Sugar-coated microarrays: a novel slide surface for the high-throughput analysis of glycans. Proteomics 2(12):1666–1671

Wilson IBH (2002) Glycosylation of proteins in plants and invertebrates. Curr Opin Struct Biol 12:569–577

Wollscheid B, Bausch-Fluck D, Henderson C, O'Brien R, Bibel M, Schiess R, Aebersold R, Watts JD (2009) Mass-spectrometric identification and relative quantification of N-linked cell surface glycoproteins. Nat Biotechnol 27:378–386

Yadav DK, Ashraf S, Singh PK, Tuli R (2012) Localization of rabies virus glycoprotein into the endoplasmic reticulum produces immunoprotective antigen. Protein J 31:447–456

Yadav DK, Yadav N, Khurana SMP (2013) Molecular farming in the decades of Omics. In: OMICS applications in crop sciences. CRC Press-Taylor & Francis Group, USA, pp 563–602

Yang BY, Gray JSS, Montgomery R (1996) The glycans of horseradish peroxidase. Carbohydr Res 287:203–212

Zachara NE, Hart GW (2006) Cell signaling, the essential role of O-GlcNAc. Biochim Biophys Acta 1761:599–617

Zaia J (2008) Mass spectrometry and the emerging field of glycomics. Chem Biol 15:881–892

Zaia J (2010) Mass spectrometry and glycomics. OMICS 14(4):401–418

Zamfir A, Vakhrushev S, Sterling A, Niebel HJ, Allen M, Peter-Katalinic J (2004) Fully automated chip based mass spectrometry for complex carbohydrate system analysis. Anal Chem 76:2046–2054

Zeitlin L, Olmsted SS, Moench TR, Co MS, Martinell BJ, Paradkar M, Russell DR, Queen C, Cone RA, Whaley KJ (1998) A humanized monoclonal antibody produced in transgenic plants for immunoprotection of the vagina against genital herpes. Nat Biotechnol 16:1361–1364. doi:10.1038/4344

Zhang P, Potrykus I, Puonti-Kaerlas J (2000) Efficient production of transgenic cassava using negative and positive selection. Transgenic Res 9:405–415

Zheng T, Yu H, Alexander CM, Beebe DJ, Smith LM (2007) Lectin-modified microchannels for mammalian cell capture and purification. Biomed Microdevices 9:611–617

Plant Lipidomics: Signalling and Analytical Strategies

Elangovan Namasivayam, R. Kowsalya, Pavan Kumar Padarthi, K. Manigandan, Richard L. Jayaraj, Johnravindar D, and Kaliaperumal Jagatheesh

Contents

E. Namasivayam, Ph.D. (✉) • R. Kowsalya, Ph.D.
P.K. Padarthi, Ph.D. • K. Manigandan, Ph.D.
R.L. Jayaraj, Ph.D. • J. D, Ph.D. • K. Jagatheesh, Ph.D.
Department of Biotechnology,
Periyar University, Salem, India
e-mail: elangovannn@gmail.com

D. Barh et al. (eds.), *PlantOmics: The Omics of Plant Science*,
DOI 10.1007/978-81-322-2172-2_11, © Springer India 2015

Abstract

The plant lipidomics is a comprehensive system of all lipids in plants with respect to cell signalling, membrane architecture, transcriptional and translational modulation and cell-cell and cell-protein interactions in response to environmental changes over time. This chapter is mainly focused on the role of plant lipids in signalling pathways during stress conditions, which were described in detail. In order to ameliorate lipid research, various analytical methods developed to characterise and quantify lipids are discussed with the salient points. Various lipid databases are provided which will be useful to access a wide range of information about lipids. Few online resources for the lipids are also described. This can help further research in the field of plant lipidomics. In conclusion, the role of plant lipids over human health and some of biological roles were illustrated.

Keywords

Lipidomics • Plant lipid • Plant defence • Signal transduction

Abbreviations

ABA	Abscisic acid
ATNHX1	Vacuolar Na+/H+ antiporter
ATP	Adenosine triphosphate
ALA	Alpha-linolenic acid
Ca2+	Calcium ion
CAT1	Catalase1 gene
cAMP	Cyclic AMP
CDP-DAG	Cytidine diphosphate diacylglycerol
CID	Collision-induced dissociation
CLA	Conjugated linoleic acid
CMP	Cytidine monophosphate
CoA	Coenzyme A
DAG	Diacylglycerol
dc	Direct current
DGDG	Digalactosyldiacylglycerol
DPG	Diphosphatidylglycerol
DGPP	Diacylglycerol pyrophosphate
DHA	Docosahexaenoic acid
EPA	Eicosapentaenoic acid
ER	Endoplasmic reticulum
ESI	Electrospray ionisation
FT-ICR	Fourier transform ion cyclotron resonance
FTMS	Fourier transform mass spectrometer
GDP	Guanosine diphosphate
GTP	Guanosine triphosphate
H2O2	Hydrogen peroxide
HPLC	High-pressure liquid chromatography
ILCNC	International Lipid Classification and Nomenclature Committee
[Ins(1,4,5)P3]	Inositol (1,4,5) trisphosphate
IP3R	Inositol triphosphate receptors
IR	Infrared lasers
LA	Linoleic acid
LC	Liquid chromatography
LIMSA	Lipid mass spectrum analysis
LIT	Linear ion trap
LTPs	Lipid transfer proteins
LTQ	Linear trap quadrupole
LPC	Lysophosphatidylcholine
MALDI	Matrix-assisted laser desorption/ionisation
MAPK	Mitogen-activated protein kinase
MGDG	Monogalactosyldiacylglycerol
MPIS	Multiple precursor ion scanning
MRM	Multiple reaction monitoring
MS	Mass spectrometry
NMR	Nuclear magnetic resonance
NP	Normal phase
OCN	Oscillating capillary nebuliser
PA	Phosphatidic acid
PAK	Phosphatidic acid kinase
PC	Phosphatidylcholine
PE	Phosphatidylethanolamine
PG	Phosphatidylglycerol
PI	Phosphatidylinositol
PI3K	Phosphoinositide 3-kinase
[PI-PLC]	Phosphoinositide-specific phospholipase C
PIP2	Phosphatidylinositol-4,5-biphosphate
PKC	Phosphokinase C

PK-C	Protein kinase C
PLA2	Phospholipase A2
PLC	Phospholipase C
PLC-DAG	Phospholipase C-Diacylglycerol kinase
PLD	Phospholipase D
[PtdIns(4,5)P_2]	Phosphatidylinositol 4,5-bisphosphate
PtdOH	Substrate phosphatidic acid
PUFAs	Polyunsaturated fatty acids
rf	Radio frequency
ROS	Reactive oxygen species
RP	Reverse phase
SA	Sialic acid
SAR	Systemic acquired resistance
SECD	Spectrum extraction from chromatographic data
SIMS	Secondary ion mass spectrometry
SMILES	Simplified Molecular Line Entry Specification
THAP	2,4,6-Trihydroxyacetophenone
TLC	Thin-layer chromatography
TOF	Time of flight
TP	Tonoplast
TAG	Triacylglycerol
TVP1	Vacuolar-type proton translocating pyrophosphate1
UV	Ultraviolet

Introduction

Lipids are the fundamental components of biological membranes and play an important role in biological systems (Wenk 2005). In plants, lipids and lipid-based derivatives fulfil many key functions like storage of carbon energy, cell compartmentalisation, protection against pathogens and developmental processes. Lipids comprise a wide range of functional and regulatory molecules such as fatty acids, glycerophospholipids, etc., and each cell type exhibits a different lipid composition and distribution. Lipids are essential cellular constituents that have multiple distinct, yet critical roles in cellular function. Membrane lipids act as second messengers through the action of a variety of intracellular enzymes (Roberts 2002).

The origin of lipidomics is intrinsically linked to metabolomics, since lipidomics is a submeadow of metabolomics that aims to detect and quantify the presence of lipids in biological samples. Metabolomics is related to the comprehensive investigation of the metabolome; it represents a vast number of compound classes, including nucleic acids, amino acids, sugars and lipids (Watson 2006). These compounds have diverse physicochemical properties and occur in different concentration ranges that can vary in both space and time (Carrasco-Pancorbo et al. 2009). In the systems biology point of view, plant lipidomics is more than just the complete characterisation of all lipids in a particular cell type, and it could be defined as 'the comprehensive understanding of the influence of all lipids on plant system with respect to cell signalling, membrane architecture, transcriptional and translational modulation, cell-cell and cell-protein interactions, and response to environmental changes over time' (Watson 2006). The ultimate goal of scientists around the world is to connect 'omics' sciences and cross-link available fingerprints of genes, transcripts, proteins and metabolites to get a closer look at the cell interactome and function. The integration of lipidomics with genomics, proteomics and metabolomics will thus provide a powerful tool to decode molecular mechanisms of lipid-associated disorders and to identify biomarkers and novel therapeutic targets.

Plant lipidomics focuses on efficient analysis of the entire spectrum of lipids in plant system, known as a lipidome. This approach provides important insights into the function of a single lipid molecular species.

Plant lipids exhibit immense combinatorial and structural diversity. However, the role of lipids has been complicated by their structural diversity and considerable technical challenges associated with distinguishing pathogenic from non-pathogenic lipid species that contain thousands of isoforms.

Classification of Lipids

To support the emerging field of plant lipidomics, a comprehensive classification system for lipids with a universal platform compatible with bioinformatics requirements was proposed (Fahy et al. 2005) under the leadership of the International Lipid Classification and Nomenclature Committee (ILCNC) which is focused primarily on mammalian lipids. Further, the system has been updated to encompass lipid structures from non-mammalian sources such as plants, bacteria and fungi (Fahy et al. 2009). As a result, lipids were regrouped under the following eight categories that cover eukaryotic and prokaryotic sources: fatty acyls, glycerolipids, glycerophospholipids, sphingolipids, sterol lipids, prenol lipids, saccharolipids and polyketides. Each lipid category contains distinct classes, subclasses, subgroups and subsets (Lipid Maps 2009).

Physical Properties of Lipids

Lipids are hydrophobic or amphipathic molecules that participate in different types of associations (Israelachvili 1992). Neutral or non-polar lipids (e.g. sterol esters, glycerolipids, hydrocarbons and carotenoids) participate in non-covalent interactions through their hydrocarbon chains with other lipids and hydrophobic regions of proteins.

Cellular Functions of Lipids

Plant cellular membranes consist of complex arrangements of lipids and proteins which fulfil diverse functions in the cell (Dowhan 2009) like compartmentalisation of cellular membranes and organelles by defining permeability barriers with respect to their surrounding milieu. The multi-component assemblage of cell membranes exhibits complex phase behaviour, with regions of structural and compositional heterogeneity.

Lipids provide the matrix within which membrane proteins fold that mediates membrane trafficking events, specific lipid molecular species involved in signalling processes by defining membrane domains and by acting as primary and secondary messengers.

Genetics of Plant Lipids

Plants are attractive experimental objects for genetic studies of lipid metabolism for several reasons: (1) There is substantial interest in exploring the degree to which genetic methods can be used to modify the oil composition of seeds. (2) Alterations can be made in the fatty acid composition under certain conditions which helps in the isolation of a wide range of mutants. (3) Producing transgenic plants has opened up entirely new opportunities to apply genetic techniques for the analysis and modification of plant lipid metabolism (Ohlrogge et al. 1991).

Organellar Lipidomics

Plastids

Chloroplast membranes can be divided into two functional regions such as envelope and thylakoid membranes. The envelope is constituted with inner and outer achlorophyllous membranes which spatially separates the cytosol from the plastid compartment. Over 80–90 % of polar lipids are accumulated in plastid envelope and thylakoid membrane (Dorne et al. 1990). The outer leaflet of the thylakoid membrane is enriched with monogalactosyldiacylglycerol (MGDG) and phosphatidylglycerol (PG), whereas the inner leaflet consist of digalactosyldiacylglycerol (DGDG), this variation in the distribution was due to stable asymmetric transmembrane of thylakoid membrane occurring in a variety of monocot and dicot species (Giroud and Siegenthaler 1988; Joyard et al. 1991; Rawyler et al. 1987; Siegenthaler and Giroud 1986).

Mitochondria

The major lipids of plant mitochondrial membranes are phosphatidylcholine (PC), phosphatidylethanolamine (PE) and diphosphatidylglycerol (DPG) and to a lesser extent phosphatidylinositol (PI) and phosphatidylglycerol (PG). Phosphatidic acid is synthesised *via* the stepwise acylation of sn-glycerol-3-phosphate by long-chain fatty acyl-CoA thioesters. The synthesis of the other major phospholipids of plant mitochondria (PC, PE, PG and PI) mainly occurs in the endoplasmic reticulum (ER), and these molecules must be further transported to mitochondria. Hence, plant mitochondria like plastids appear to be semi-autonomous with regard to the formation of their membrane glycerolipids.

Glyoxysomes

Glyoxysomes are specialised peroxisomes found in plants, and it contains enzymes that initiate the breakdown of fatty acids. A plant glyoxysomal membrane contains PC, PG, PI and PE. Further, it is confessed that they are unable to synthesise these lipids, which are delivered from their site of synthesis mainly in the ER and then exported to glyoxysomal membranes.

Membrane Lipidomics

Lipid Content of Plant Membrane

Theoretically, the animal lipidome encompasses 10,000 glycerophospholipid species; 100,000 sphingolipids species; thousands of mono-, di-or/triglycerides and numerous fatty acids with sterol-based structures. The plant lipidome is also believed to display this broad lipid complexity.

The intracellular delivery of soluble and insoluble membrane components to different cellular compartments is required to establish and maintain the latter's identity and specific functions. Protein sorting and transportation to various cellular locations has been extensively investigated in many plant models. These studies have allowed the discovery of multiple targeting sequences, which not only allow a specific targeting to different organelles but also to a sub-compartment of a given organelle (Bar-Peled et al. 1996). All cell membranes have distinct and specific lipid compositions which contribute to their identity.

Lipid Delivery to the Plasma Membrane

The plasma membrane-mediated biosynthesis of most lipids takes place primarily in the ER and to a lesser extent in the Golgi apparatus. Such intracellular region implies that the lipid molecules are more or less specifically transported to the cell surface (Moreau et al. 1998).

Cell-Free Reconstitution of Lipid Transport

The endoplasmic reticulum plays a vital role in vesicular transport and was first studied in vivo. Later, permeabilised cells and cell-free systems were developed to identify membrane and cytosolic proteins involved in vesicular transport both in animal cells and yeast. ATP plays a crucial role in activating downstream chain reaction in both formation and fusion of the vesicles. Coat proteins, GTP binding proteins, cytosolic proteins and membrane proteins of budding, targeting and fusion machineries have also been shown to be involved. However, the vesicular transport is largely accepted in animal and yeast cells, the morphological and biochemical evidence of the existence of vesicular intermediate in plant is restricted to post-Golgi events (Rothman and Wieland 1996).

Several new compounds have been discovered in marine and lacustrine sediments. Instead of terrestrial plants, micro-algae have been considered as better source for lipids (Welti et al. 2007). Plant-derived biomarkers typically include long-chain *n*-alkanes, *n*-fatty acids and *n*-fatty alcohols (Brull et al. 2009). Hence, the majority of the functional plant lipids are phosphatidylinositol and related lipids. Phosphatidylinositol is an important lipid, both as a key membrane constituent and as a participant in essential metabolic processes in all plants, both directly and *via* a number of metabolites. It is an acidic (anionic) phospholipid that in essence consists of a phosphatidic acid backbone, linked *via* the phosphate group to inositol (hexahydroxycyclohexane). In most organisms, the stereochemical form of the last is *myo*-D-inositol (with one axial hydroxyl in position 2 with the remainder equatorial), although other forms (*scyllo*- and *chiro*-) have been found on occasion in plants. As with phosphatidylglycerol, phosphatidylinositol is formed biosynthetically from the precursor cytidine diphosphate diacylglycerol (CDP-DAG) by reaction with inositol and catalysed by the enzyme CDP-DAG inositol phosphatidyltransferase (phosphatidylinositol synthase); the other product of the reaction is cytidine monophosphate (CMP). The enzyme is mainly located in the ER, plasma membrane of many plants and cytosolic side of the bilayer. Phosphatidylinositol is then delivered to other membranes either by vesicular transport or via the agency of specific transfer proteins (McDonald et al. 2007).

The pioneering work of Mable and Lowell Hokin in the 1950s led to the discovery that phosphatidylinositol was converted to polyphosphoinositides with important signalling and other functional activities. This lipid is now known to be phosphorylated by a number of different kinases that place the phosphate moiety on positions 3, 4 and 5 of the inositol ring. Seven different isomers are known, all of which have distinct biological activities. The most significant in quantitative and possibly biological terms were long thought to be phosphatidylinositol 4-phosphate and phosphatidylinositol 4,5-bisphosphate, but it is now recognised that 3-phosphorylated forms are also extremely important (Klose et al. 2012).

Phosphatidylinositol 4-Phosphate

Phosphatidylinositol 4-phosphate is the precursor for 4,5-bisphosphate, and it binds to a protein on the cytoskeleton of the cell with its own characteristic functions. Some of these participate in vesicle formation, while others like the oxysterol-binding protein are involved in lipid transfer and protein enrolment in the Golgi complex. The biological properties of phosphatidylinositol 5-phosphate have taken longer to unravel, because of the difficulties in isomer separation. It is now apparent that it is involved in the osmoregulation process in plants and also has signalling functions (Byrdwell 2001).

Phosphatidylinositol 4,5-Bisphosphate

Phosphatidylinositol 4,5-bisphosphate is especially important as a precursor of further metabolites, because of signalling functions in the plasma membrane. Phosphatidylinositol 4,5-bisphosphate forms complexes with phospholipase C and regulates ion channels for potassium, calcium, sodium and other ions. In most instances, it increases channel activity, while the enzyme phospholipase C hydrolysis reduces.

Lysophosphoinositides

Lysophosphoinositides are water-soluble glycerol phosphoinositides which are fully deacylated into phosphatidylinositol and phosphatidylinositol phosphates. Like other lysophospholipids, lysophosphatidylinositol conjugates with a single fatty acid linked to the glycerol moiety. The formed intermediates are involved in the remodelling of the fatty acid compositions of lipids, whenever arachidonic acid is released for eicosanoid biosynthesis (Fahy et al. 2005).

Lipid Signalling in Plants

Signal transduction is the process in which all cells constantly receive and act in response to new signals from their environment. Many unicellular organisms respond to signalling molecules secreted by adjacent cells for cell-cell communication.

The signalling molecules may be chemical in nature, for instance, hormones, pathogen elicitors, mating receptors, ozone and physical changes (light, temperature and osmotic pressure). Cells must continuously monitor their environmental behaviour and translate this information into an appropriate response *via* receptors. The information carried over the plasma membrane into the cytoplasm will be achieved by initiating specific ion cascade reactions, receptor kinases or via second messenger-mediated effector response. Understanding the molecular mechanism behind these pathways of cell signalling has become the foremost part of the current research.

Receptor-Mediated Signalling Molecules

Many of the molecules transmit the information form one cell to another. Such molecules can act as ligands that bind to receptors expressed by their target cells. There is significant distinction in the structure and function of each type of ligands that serve as signal transmitters. Usually the signalling molecules used by plants vary in complexity from simple gases to proteins.

As mentioned earlier, all signalling molecules act by binding to receptors expressed by their target cells. In several cases, these receptors are expressed on the target cell surface, but some receptors are intracellular proteins located in the cytosol or the nucleus. These intracellular receptors responded to small lipophilic signalling molecules that are able to diffuse across the plasma membrane. Various lipids are involved in intervening plant growth, development and responses to biotic/abiotic cues, and their production is regulated by lipid-signalling enzymes. Lipid-hydrolysing enzymes play a crucial role both in the production of lipid messengers and in other processes, such as cytoskeletal rearrangement, membrane trafficking and degradation. Studies on the downstream targets and modes of action of lipid signals in plants are still in their premature stages, but distinguishing features of plant lipid-based signalling are being recognised. The functions of the major classes of lipid signalling are to trigger cell surface receptors in plant under various environmental conditions.

Lipids signalling in plants are mediated by an ample range of molecules such as glycerolipids (Meijer and Munnik 2003), sphingolipids (Sperling and Heinz 2003), fatty acids (Farmer et al. 2003), oxylipins (Scherer 2002) and sterols (Lindsey et al. 2003). Several advances were made over the past years in understanding lipid signalling and related hydrolytic enzymes. The signalling pathways activated by certain enzymes in plants use a variety of mechanisms that are conserved in animal cells, which are unique to plants.

Phospholipids

Phospholipids are not just structural components of membranes; they can also be essential cofactors for membrane enzymes, signal precursors or signalling molecules of themselves. In every eukaryotic cell, the total phospholipids consist of 1–2 % of phosphatidic acid (PA). Most of it involved in the biosynthesis of structural phospholipids, glycolipids in the ER and plastids by acylating glycerol 3-phosphate and lysophosphatidic acid.

The synthesis of PA is mediated by phospholipase in the plasma membrane activated by a specific membrane receptor binding to an extracellular hormone, pathogen or physical stress. The activated receptor transmits the signal by direct interaction with the phospholipase D or by indirect interaction via DAG kinase-mediated phosphorylation of DAG by G-protein-coupled receptors (Fig. 1).

Regulation of Phospholipid Signalling by G-Protein-Coupled Receptors

As we know from animal systems that the largest family of cell surface receptor transmits signals to intracellular targets by activating heterotrimeric G-proteins. More than thousands of G-protein-coupled receptors have been identified in animals (responsible for smell, sight and taste), and plants induce growth (Scherer 1995), auxin signalling (Millner 1995; Millner and Causier 1996), plant defence response (Beffa et al. 1995)

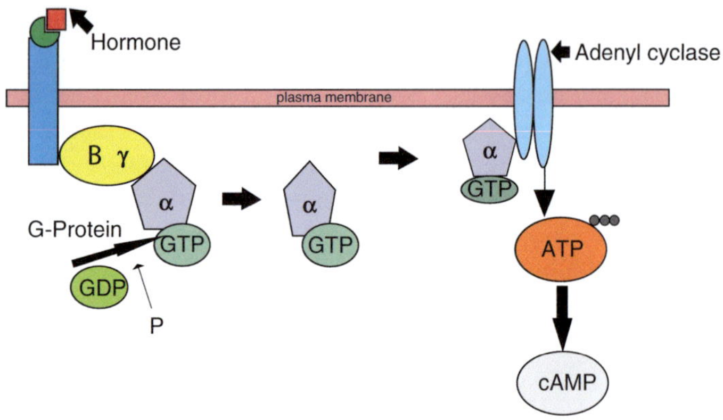

Fig. 1 G-protein-coupled receptor-mediated activation of phospholipids

and mastoparan stimulation (Chandra et al. 1996).

The breakthrough in G-protein came from mammalian hormones (such as epinephrine) that regulate the synthesis of cyclic AMP (cAMP), an important second messenger that mediates cellular response to a variety of hormones. Later, Martin Rodbell in the 1970s discovered that the phosphorylation-dependent GTP is activated followed by activation of cAMP which is mediated by adenyl cyclase.

G-protein consists of three subunits (α, β and γ) commonly called heterotrimeric G-proteins, and the activation involves the binding of α-subunit to guanine nucleotides, which regulates G-protein activity. In the quiescent state, α is bound to GDP (guanine diphosphate) in complex with β and γ. Binding of ligand to receptor, induces conformational changes, such that the cytosolic region of the receptor interacts with G-protein and stimulates the exchange of its GDP for GTP. The activated GTP-α complex then dissociates from β and γ which remain together and function as a βγ complex. Both α and βγ subunits can then modulate different target effectors including phospholipase C, phospholipase D, phospholipase A, PI3K, adenyl cyclase and ion channels (Cho et al. 1995; Kim et al. 1996) dependent on the specificity of the subunits (Fig. 2).

G-protein activators stimulate phospholipase A_2 (PLA_2), phospholipase C (PLC) and phospholipase D in in vivo studies (Legendre et al. 1993; Munnik et al. 1997). Phosphatidic acid kinase (PAK) may also be activated, but the increase in DGPP could just reflect the increase in substrate (PtdOH) concentration from PLC and PLD activity (Munnik et al. 1996). As of now, there is no possible data that PI3K is activated by G-protein signalling.

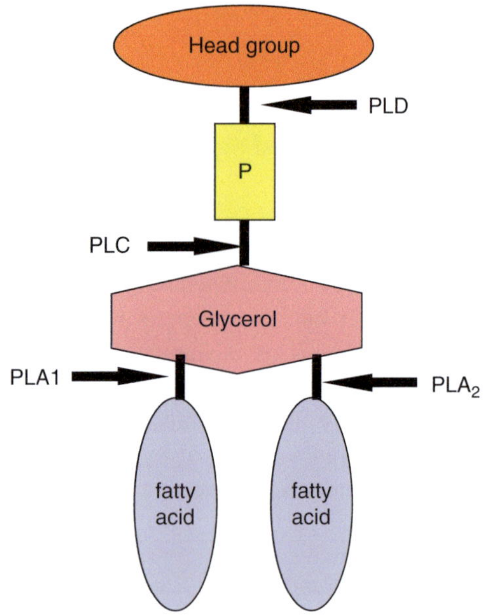

Fig. 2 Structure of phospholipids and its site of hydrolysis

Key Mediators of Phospholipid Signalling

The pathway unique to phospholipid-based signalling is complex sequestered process. The stimulation of surface receptors can trigger not just PLC but a variety of phospholipid-metabolising enzymes such as phospholipase D

(PLD), phosphoinositide 3-kinase (PI3K) and phospholipase A_2 (PLA_2). In *Arabidopsis*, the PLD family consists of five different genes, which have been categorised into four subgroups: α, β, γ and δ, based on their evaluated size, biochemical properties and amino acid sequence. PLDs from many different plant species have been cloned and most of them have been found to be α subfamily (Wang 2000).

Certain agonists, stimulate cells using second messengers which can induce hydrolysis of the minor lipid phosphatidylinositol 4,5-bisphosphate [PtdIns(4,5)P_2]. A variety of hormones stimulate the hydrolysis of PtdIns(4,5)P_2 by a phosphoinositide-specific phospholipase C [PI-PLC] which results in two products, inositol 1,4,5-trisphosphate [Ins(1,4,5)P_3] and DAG, both of which functions as second messengers (Berridge and Irvine 1989; Michell 1975). The aqueous soluble property of Ins(1,4,5)P3 is released into the cytosol where it elicits the downstream of G-protein-coupled receptors and protein tyrosine kinase for the regulation of Ca^{2+} and calmodulin-dependent enzymes which mediated ion-gated channels (Fig. 3). The hydrolysed lipid product of DAG remains in the membrane and activates protein-serine-threonine kinases, a protein kinase C (PKC) family. This family plays an important role in the control of cell growth and differentiation (Munnik 1999; Nobes et al. 1995). Though, the function of DAG remains unclear, certain experimental studies concluded its role in recurrent phosphorylation of phosphatidic acid by DAG-specific kinase (Van der Luit et al. 2000; Den Hartog et al. 2001).

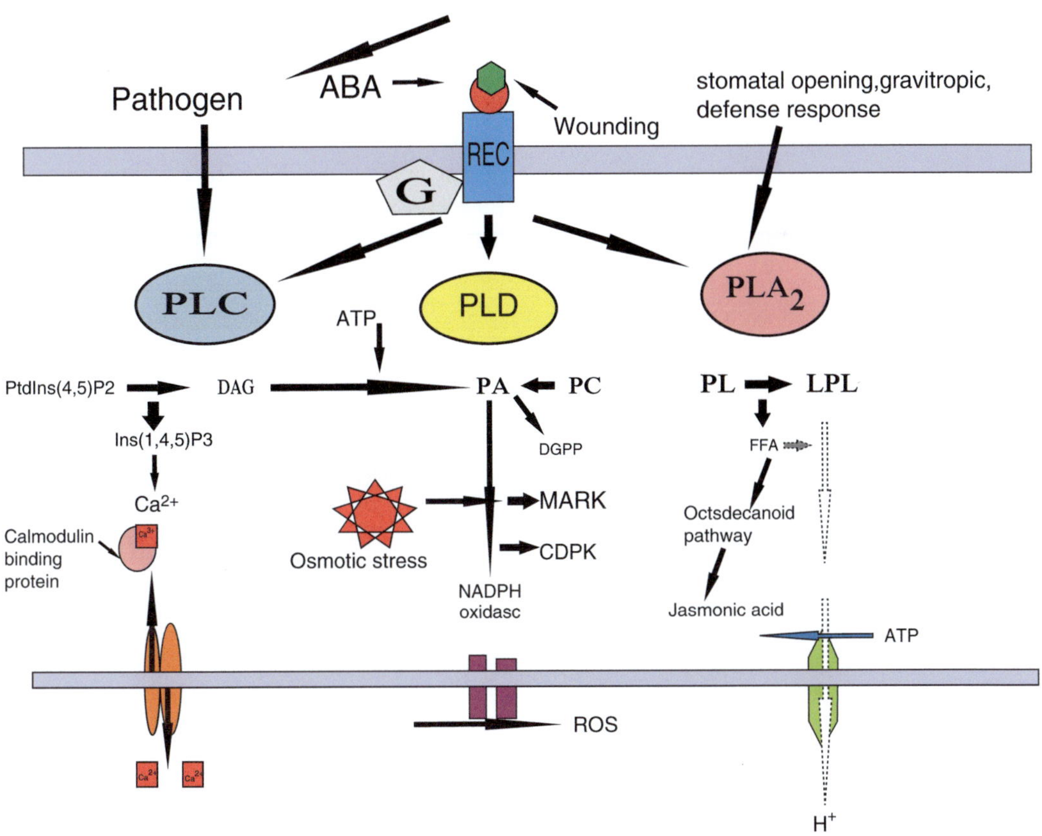

Fig. 3 Phospholipid signalling in plants and its substrates and messengers produced by different phospholipid-hydrolysing enzymes and their downstream cellular and physiological effects under various stress conditions

Lipid Signalling During Plant Stress

Osmotic stress formed during drought, freezing temperatures, salt-contaminated soils and water stress hormone; abscisic acid (ABA) seems to trigger lethal effect on plants (Fig. 4). Both the PLC and the PLD pathways have been extensively implicated in various plants (Wang et al. 1997; Arisz et al. 2003). Addition of salt was found to inhibit PLD activity in tobacco pollen tubes, due to a subclass of tissue-specific PLD response. Finally, it was revealed that PLDδ is the responsible enzyme involved in dehydration under drought condition by stimulating PLD activity-mediated expression of the gene that encodes PLDδ. Silencing this isoform severely reduced the drought-induced PA response (Katagiri et al. 2001).

ROS-Mediated Stress Signalling

External stresses induce accumulation of reactive oxygen species (ROS) such as superoxide, hydrogen peroxide and hydroxyl radicals in plants. These ROS elicit severe damage to cells by contributing to stress injury in plants (Prasad et al. 1994). Because ABA was revealed to induce H_2O_2 production (Pei et al. 2000), hence, ROS may act as intermediate signals for ABA in mediating other response agents like catalase1 gene (*CAT1*) expression (Guan et al. 2000), thermotolerance (Gong et al. 1998), activation of Ca^{2+} channels in guard cells, stomatal closure (Zhang et al. 2001) and ABA biosynthesis (Zhao et al. 2001). In animal cells, declined tyrosine phosphatase activity causes an increase in the stimulation of mitogen-activated protein kinase (MAPK) pathways because tyrosine phosphatase elicits its feedback effect through dephosphorylation in MAPK pathway. Augmented evidences suggested that, transgenic plant showed higher ROS scavenging activity by over expressing free radical scavenger which contributes to stress damage (Kocsy et al. 2001). In hypersensitive responses, sialic acid (SA) is thought to initiate ROS signalling in plants by activating MAPK (Hoyos and Zhang 2000). The transgenic *Arabidopsis* expressing a salicylate hydroxylase (*NahG*) gene, it is concluded that increased osmotic stress tolerance might result from decreased SA-mediated ROS generation in the *NahG*-expressing plants.

In yeast and in animals, mitogen-activated protein kinase (MAPK) pathways are responsible for the production of compatible osmolytes and antioxidants. MAPK pathways are activated by

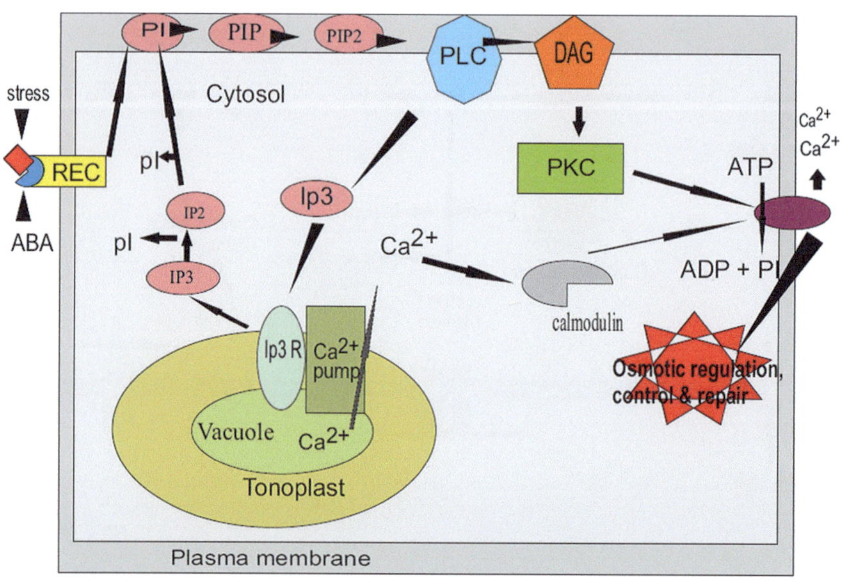

Fig. 4 Lipid signalling pathways in response to cold, ROS, osmotic stress and pathogenic conditions

receptors/sensors such as protein tyrosine kinases, G-protein-coupled receptors and two-component histidine kinases. Among these receptor-type proteins, histidine kinases have been unambiguously identified in plants. An *Arabidopsis* histidine kinase, AtHK1, can complement mutations in the yeast two-component histidine kinase sensor SLN1 and, therefore, may be involved in osmotic stress signal transduction in plants. Understanding the in vivo function of AtHK1 and other putative histidine kinases and their relationship to osmotic stress-activated MAPK pathways will certainly shed light on osmotic stress signal transduction.

Phospholipid Response During Cold Stress

Mechanical strain on the receptor (stress) activates nearby phospholipase C (PLC), which hydrolyses the phosphatidylinositol-4,5-biphosphate (PIP2) in the plasma membrane, and releases IP3 in the cytosol and DAG remains in the membrane lipid. IP3 enters into the tonoplast (TP), where it merges with specific inositol triphosphate receptors (IP3R) that activate Ca^{2+} channels that move Ca^{2+} ions from the vacuole to the cytosol. Calcium ion might also activate other protein kinases, either in free or bound form to calmodulin, and participate in altering the cold stress-stimulated cellular response. Meanwhile, IP3 will be dephosphorylated upon hydrolysis to form IP2 then to IP, which is subsequently converted back to phosphatidylinositol (PI) in the plasma membrane for another cycling. The activation of certain Ca^{2+} channels by cold stress may result from physical alterations in cellular structures (Sangwan et al. 2001; Wang and Nick 2001).

Pathogen-Induced Stress Response

Certain pathogenic elicitors (microbes) activate the phospholipase C-diacylglycerol kinase (PLC-DAG) pathway. Recently, Avr4-specific elicitor, which is derived from *Cladosporium fulvum*, was shown to prompt PA response in transgenic tobacco cells expressing the *Cf-4* resistance gene. The increase in PA is likely to be upstream which corresponds with those for other elicitors and confirms that PA is an early signal in the defence response of plants. PLD is also involved in plant defence, but it appears to play a role in secondary responses. Pseudomonas infection initiates over-expression of PLDs in *Arabidopsis* species. Similarly in rice crop, the gene encoding PLD was increased in response to *Xanthomonas oryzae* bacterial infection (de Torres Zabela et al. 2002). Other experimental studies in tomato cells reported that supplementation of xylanase in cultured cells of tomato showed significant increase in expression of PLD gene (Laxalt et al. 2001).

Enzymes involved in PA cellular responses have been discovered and that many more PA targets have not yet been identified. In earlier studies of PA targets, several mechanisms have emerged by which PA can exert its effects. Most of the PA signalling depend on interaction with other proteins or are activated by the target proteins. Further progress in disentanglement cross-talk mechanisms among the different lipid signalling processes will certainly improve our understanding of lipid-mediated signalling in a vast scope.

The Plant Defence Mechanisms

During evolution, each organism developed a wide range of mechanisms against adverse environmental conditions or infectious organisms. The epidermis of a plant is a very specialised layer, which plays a critical role in the development and survival of the whole organism (Reina-Pinto and Yephremov 2009). There are interesting parallels between animal and plant defence responses as demonstrated by the structural and functional conservation of their signal transduction processes. Lipids are not genetically encoded like other small molecules (proteins). But they are generated and metabolised by enzymes under the influence of biotic and abiotic factors. Certainly, both factors can reduce the average productivity limit by 65–87 %, depending on the crop. Plants are rapidly being attacked by certain fungi, bacteria, nematodes and feeding insects. In order to protect from these lethal effect, they are developing defence mechanisms to keep these predators away from the bay (Laxalt and Munnik 2002). There are numerous stress factors depending on their exposition time and their concentration that can able to reduce the growth rate: in particular, salt stress is certainly one of the most serious abiotic factors limiting the productivity of crop plants. Moreover, improving salt tolerance in plants might have much wider implications, because transgenic salt-tolerant plants often produce tolerance to other stress response chilling, freezing, heat and drought (Bavaro et al. 2007).

Biotic Stress

The stress of living organisms is known as biotic stress. It includes bacterial, viral or fungal attack. Necrotrophy is the process of necrosis in which the plant cells degraded in response to biotic stress. However, plant cell remains alive in biotrophy, whereas the cell remains normal initially and destroyed later in hemibiotrophy (Poltronieri et al. 2011). In all the above cases, the plants respond by activating inositol-3-phosphate for further signalling cascades (Hu et al. 2009).

Abiotic Stress

The abiotic stress can be mechanical wounding, ozone, UV illumination, salinity, drought, etc., have multiplicative effects on abiotic stress consequences and seriously threaten sustainable agricultural production. Therefore, the subject of abiotic stress response in plant metabolism, productivity and sustainability is gaining considerable significance in the contemporary world (Ahmad and Prasad 2004). Drought reduces photosynthesis by decreasing both leaf area and photosynthetic rate per unit leaf area. Withholding of excess water also changes the physical environment for plant growth as well as crop physiology (Kramer 1980). Most studies have reported that mineral uptake declines when water stress intensity increases (Akınc and Losel 2012). Plant response to cold and freezing involves three distinct phases: cold acclimation, freezing and post-freezing recovery. During cold acclimation, the degree of fatty acid unsaturation and the content of phospholipids increase. During freezing, dramatic lipid alterations take place in both extraplastidic and plastidic membranes. Membrane lipids undergo many changes in plants exposed to various stress conditions. Quantitative lipid profiling reveals lipid alterations, and examination of these changes often suggests potential mechanisms for the stress-induced changes.

Role of Secondary Metabolites in Defence Mechanisms of Plants

Plants produce a high diversity of natural products or secondary metabolites with a prominent function in the protection. There are four major groups of secondary metabolites, viz., terpene, phenolics, nitrogen and sulphur containing compounds. Terpenes composed of 5-carbon isopentanoid units are toxins to many herbivores. Phenolics are synthesised primarily from products of the shikimic acid pathway having several important defensive roles in the plants (Boller 1995). Members of N and S containing compounds are synthesised principally from common amino acids. There are 100,000 known secondary metabolites involved in plant defence systems. In the last few years, researchers have isolated numerous plant resistance genes, recognised as R genes which function against fungi, bacteria and nematodes. Most of the R genes are thought to encode receptors that recognise and bind specific molecules originating from pathogens which may be proteins, peptides, lipids, etc. (Mazid et al. 2011).

Stress-Responsive Gene Expression

Lipids have long been recognised as signalling molecules that have the capacity to trigger genes against stress response. Several genes have been characterised for their role on stress protection. ERECTA is a gene regulating transpiration efficiency affecting stomatal closure, while the plant is able to maintain biomass production. Other important regulatory proteins involved in drought stress are proton anti-porters TNHX1 and a proton pyrophosphatase TVP1 are shown to protect against salinity and drought stress in *Arabidopsis*. Plants can also sense and respond to mechanical stimuli, like animals by mechano-transduction, a complex process involving the participation of a multitude of sensors, signalling molecules and genes (Han and Yuan 2009).

Plant Disease Resistance

It is difficult to quantify the damage done by all plant diseases. The microorganism can form symbiotic relationships with the host like *Rhizobium* spp. The microorganism may cause disease in the host plant, the host plant develops resistance to the pathogen and no infection develops. The host plant may show some tolerance to infection; in this case, the pathogen is able to grow and replicate, but symptoms of infection are minimal. The introduction of local defence pathways may lead to the introduction of intercellular signals that produce a systemic response, termed systemic acquired resistance (SAR) (Hong et al. 2008).

Cold Stress

Low temperature is one of the main environmental factors which may influence many physiological processes in plants. Changes in membrane lipid composition play important roles in plant adaptation and survival during cold stress. Plant response to cold and freezing involves three distinct phases: cold acclimation, freezing and post-freezing recovery. During cold acclimation, the degree of fatty acid unsaturation and the content of phospholipids increase. During freezing, dramatic lipid alterations take place in both extra-plastidic and plastidic membranes. Plant stress caused by freezing has been an area of intensive research for many years, but the molecular and cellular mechanisms of freezing injury and tolerance are not well understood. The recovery phase involves tissue thawing, cellular rehydration, restoration of cell structure and resumption of cellular activities. The ability to successfully undergo these processes depends on membranes and is critical for cellular survival after freezing. Plant models such as *Arabidopsis* have led to many important mechanistic insights into plant lipid metabolism in response to freezing (Chen et al. 2013). Expansion of this approach to plants, which harbour unique lipids, such as galactosyl glycerolipids, should greatly facilitate the understanding of lipid functions in plant growth, development and stress responses. One major form of freezing damage is due to the formation of lipid hexagonal II phase in regions where the plasma membrane and the chloroplast envelope are closely placed. Stress induces reduction in nutrient uptake, reduced cell growth and enlargement, leaf expansion, assimilation, translocation and transpiration. More research need to be focus on improvements of crop productivity, the development of high-yielding genotypes, which can survive unexpected environmental changes, particularly in regions dominated by water deficits. Today, advanced tools are required to monitor the alteration in plant lipids and their mechanism of adaptability under stress.

Analytical Strategies in Lipidomics

Analytical strategies in lipidomics are gaining more attention among the researchers due to its importance in various fields. Shifting from basic lipid research, the trend is to understand global lipid metabolite change in a system integrated context to shed light into the pathophysiology (Han and Gross 2003). Various efforts have been taken to promote research in the field of lipidomics. One such effort is 'Lipidomics Expertise Platform', an initiative by the European Union in the year 2005 which offers online resource (http://www.lipidomics.expertise.de) regarding information relating to institutions involved, lipid databases, lipid standards and methods.

To steer this lipid research, advancement in the field of analytical methods to characterise and quantify lipids is increasing. Though lipidomics is a promising field, certain challenge remains unclear. First, the data regarding the lipids in an individual organism is lacking. Second, mapping lipidomes are still out of reach. Third, structural identification of lipids by mass spectrometry (MS) is complicated. Hence, techniques for structural identification should be worked out, and fourth, the diversity of lipids makes it impossible to adapt a common method for extraction, separation and detection of lipids. Advancement in lipidomics from traditional lipid research is focused on two important points: (i) to understand the association between lipid metabolic pathways in biological systems and the metabolic health and (ii) how the changes in these pathways are related to the disease pathology (Wiest and Watkins 2007). Techniques used in the identification and quantification of lipids are broadly classified into

- Mass spectrometry (MS)-based techniques
- Non-MS-based techniques

Mass Spectrometry-Based Techniques

Mass spectrometry is a promising technique in analysing lipid because of its ability to segregate and characterise charged analytes in gaseous phase depending on their mass-to-charge ratio (m/z). Upon characterisation and segregation, information regarding the individual structure can be obtained by fragmenting the lipid ions using collision-induced dissociation (CID). To record these fragmentation reactions, various techniques such as tandem MS or MS/MS or MS are used. Basically, three important sections are involved in mass spectrometry (Fig. 5).

1. An ion source
2. Mass analyser to measure the ionised analytes
3. A detector which records ion signal corresponding to m/z value

Initially, only small and volatile lipids were analysed using MS, and later on, ions of intact biomolecules were analysed using electrospray ionisation (ESI) (Karas and Hillenkamp 1988) and matrix-assisted laser desorption/ionisation (MALDI) (Fenn et al. 1989) without fragmenting high molecular weight and non-volatile lipids. The difference between ESI and MALDI is that ESI probes produces gas-phase ions from molecules in a solution and coupled directly to liquid chromatography. Whereas, in MALDI, intact gas-phase ions are produced from samples embedded in a dry, crystalline matrix *via* laser pulses. Apart from these mass analysers, there are also other analysers such as Paul ion trap, the linear quadrupole, time of flight, Fourier transform ion cyclotron resonance (FT-ICR) and orbitrap. These analysers are often coupled with multistage instruments. Commonly coupled instruments are the quadrupole-linear ion trap and quadrupole-time-of-flight and linear ion trap-orbitrap.

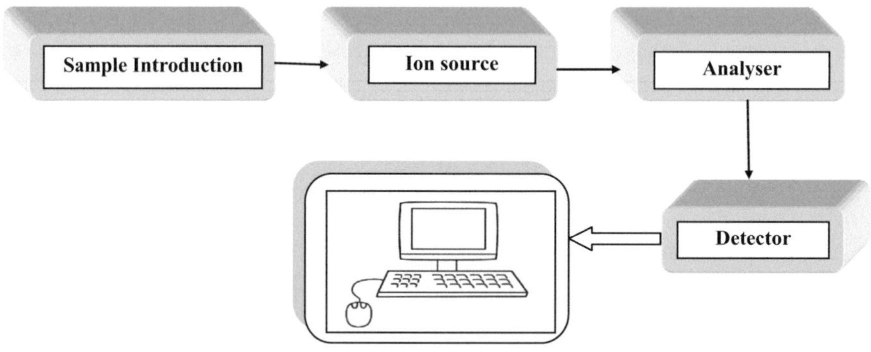

Fig. 5 Schematic representation of the basic components of MS

Ion Trap Mass Spectrometer

Ion trap mass spectrometer is one of the most commonly used MS in laboratories. It is coupled with high-pressure liquid chromatography (HPLC) through the ESI interface. Based upon the user-selected time, ion trap MS can capture ions and trapped ions are subjected to MS, MS/MS and even MS analysis. A couple of advantages are economic viability and high sensitivity. Use of ion trap MS in lipidomics began by Larsen and co-workers (Larsen et al. 2001) to characterise phospholipids up to MS^4. Apart from benefits, few disadvantages are low mass accuracy and low dynamic ranges which often result from the ion trap's limited resolving powers, space charging effects and its low-duty cycle caused from the overhead time required to trap and manipulate ions for MS experiments. The development of a 'linear' or a 'two-dimensional ion trap', linear trap quadrupole (LTQ) or linear ion trap (LIT) can partly expand dynamic range and increase resolution (Schwartz et al. 2002).

Triple Quadrupole

Triple quadrupole has been the analyser of choice for many researchers because of its wide advantages such as its ability to perform precursor ion scans, neutral loss scans and its exquisite sensitivity for identifying specific small molecules for lipid analysis. In brief, quadrupole consists of four accurately matched parallel metal rods. Oscillating high-frequency electric fields are produced when direct current (dc) and radio frequency (rf) are applied to the electrodes, and the mass separation is accomplished by the oscillating motion of ions in this electric field. Ions of specific m/z values pass through quadrupole rods provided with a specific dc and rf potentials. In triple quadrupole instruments, three quadrupoles are arranged sequentially. Q1 and Q3 are operated by both dc and rf potentials, whereas Q2 is operated with only the rf potential (Dass 2007). The Q2 allows all ions to pass through and serves as a total ion containment region and a collision cell. Because ions in the range 0–100 eV can be transmitted through quadrupoles, the MS/MS fragmentations in triple quadrupole instruments are performed via low-energy fragmentation processes.

These instruments can perform tandem MS/MS experiments as follows: precursor ions of a defined m/z are transmitted by first quadrupole (Q1), fragmented *via* CID in second quadrupole (Q2), and the resulting ions are separated in third quadrupole (Q3). Tandem MS runs on three scan modes: (1) a product ion scan done for a selected precursor ion; (2) a precursor ion scan is done for a user-selected product ion; and (3) a neutral loss scan is performed to reveal precursors that fragment by ejecting a neutral fragment of molecular mass selected by the user. Among different scanning modes, multiple reaction monitoring (MRM) where signals are recorded from defined m/z values in both precursor and product ions which help to bring out precise results.

Though triple quadrupole has various advantages, they also have several disadvantages and few of which are (1) low resolving power, (2) medium mass accuracy and (3) very low duty cycle for scanning modes (excluding MRM). This low mass accuracy makes it tougher to explicitly identify the lipids. Alternate to this triple quadrupole, instruments such as QQ-TOF (Ekroos et al. 2002) where third quadrupole (Q3) is replaced by time of flight (TOF) can provide precise mass accuracy but they cannot perform precursor and neutral scans. To overcome these drawbacks, an instrument where Q3 is replaced with LIT named QQ-LIT (or Q trap) is used which can perform precursor and neutral loss scans as well as multistage MS/MS (MS) (Hopfgartner et al. 2004).

Imaging MS by MALDI-TOF

One of the most promising techniques in the field of lipidomics is MALDI-TOF which is used mainly for imaging lipids in tissues. The sample is coated with a solid matrix (any aromatic compound) with a specific absorption spectrum. Within the vacuum chamber, the sample is introduced and a pulsed laser which emits light at a particular wavelength (depending on the solid matrix) is focused on the region of interest. The matrix is then vaporised and the analytes (lipids) are carried along (Fig. 6).

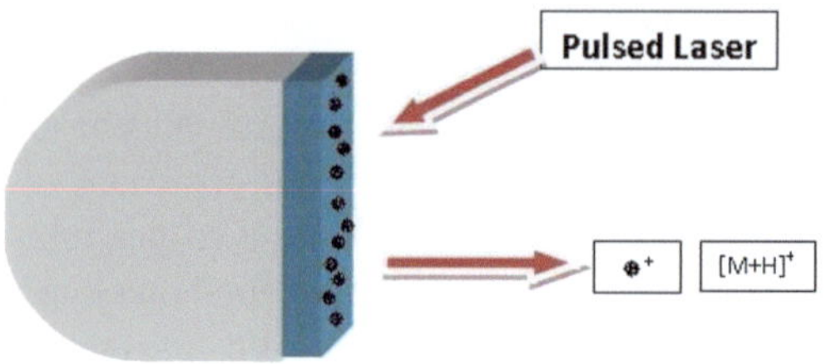

Fig. 6 Diagram representation showing pulsed laser emitting light and analytes are carried with charge

After this process, lipids are charged and these ions are accelerated over a short distance by means of electric field. All the ions, based on their mass, receive identical kinetic energy and travel at different velocities. These ions are then introduced into a TOF mass spectrometer consisting of a long-field free flight-tube maintained under sufficiently high vacuum so that no ion collisions with background gas molecules can occur. Because no external forces (via electric fields) are applied, the lipid ions travel through the flight-tube with the mass-dependent velocities that they acquired during the brief initial acceleration. Therefore, by measuring the time required for the lipid ions to traverse this tube, their m/z values can be deduced. In applications devoted to mapping the lipid profiles across a tissue slide, the process is then repeated by moving the laser beam across the slide.

For tissue imaging using MALDI MS, the tissues are frozen and cut into slices at a range of 5–14 μm and embedded into a matrix usually 2,5-dihydroxybenzoic acid or 2,4,6-trihydroxyacetophenone (THAP) for lipids (Stubiger and Belgacem 2007). MALDI imaging usually uses a N_2 UV wavelength 337 nm for lipids, and for phospholipids, infrared (IR) lasers of wavelength 249 nm are used (McDonnell and Heeren 2007). Apart from these coating techniques, oscillating capillary nebuliser (OCN) system sprays the matrix aerosol as small droplets on the sample which improves matrix homogeneity and solvent control effects (Chen et al. 2008). Deviating from the conventional usage of laser beam, secondary ion MS (SIMS)-TOF uses high-energy particle bombardment with a continuous energetic ion beam such as Bi_3^+ and Ga^+ (Nygren et al. 2004). Though the complexity in analysing the lipids is increasing, elemental identification and sensitivity still need to be addressed.

High-Resolution and High Mass Accuracy Mass Spectrometer

Lipids are identified based upon their mass using Fourier transform mass spectrometer (FTMS) which is one of the most accurate instruments available because it has the ability to trap ions in a strong magnetic field under high vacuum. For analysing lipid mixture, FTMS is coupled with HPLC-ESI. Apart from its accuracy, as like any other complex instrument, their disadvantages are that they are expensive and need skilled labour for operation and maintenance. In FTMS, higher magnetic field is achieved by a superconducting magnet that needs continuous cooling by liquid helium. As an alternative to FTMS, orbitrap has similar resolving capacity and high mass accuracy like FTMS. It is based on oscillating electric field. Orbitrap is less expensive and easier to operate. Linear ion trap-FT and linear ion trap-orbitrap are usually used for lipid quantification and identification (Schwudke et al. 2007). Maxis, a recently developed MS by Bruker, can analyse lipids at sub-ppm mass accuracy, and hence, FTMS will be replaced by Orbitrap and Maxis.

Direct-Infusion ESI-Based MS Technologies

The sensitivity of ESI-MS is more when compared to other fast atom bombardment MS (Han and Gross 1994). Nano-spray MS with a flow rate in nL/min is more sensitive than other conventional methods (Ishida et al. 2004). ESI-MS helps in the detection of lipid species at femtometre amounts (Gross and Han 2006). In spite of certain advantages, ESI is not widely used in lipidomics because during ESI process, phospholipids can either acquire positive or negative charge which can be studied only using positive or negative ESI-MS/MS. At certain time, some phospholipids do not acquire any charge and cannot be analysed. Survey scans help in determining the molecular weight of the lipid; hence, accuracy of molecular weight is directly proportional to the accuracy in lipid prediction. In the second stage, lipids are subjected to collision-induced dissociation (CID) and the fragmentation patterns help in understanding lipid structure (e.g. polar head group and fatty acyl moieties). Lipid subclasses such as lysophosphatidylcholine (LPC), sphingomylein, phosphatidylcholines (PCs), phosphatidylethanolamines and phosphatidylserines can be detected by positive ESI-MS/MS. The fragmentation of protonated PC, LPC and SM ions yield a peak at m/z = 184 which is the diagnostic fragment for the phosphocholine head group. A precursor ion scan of m/z = 184 would therefore highlight the PC-containing lipids out of all the lipids present. Further interest, $[M+H]^+$ ions of PCs appear at even m/z values (the closest integer to accurate mass), while protonated SMs exhibit odd m/z values (Brugger et al. 1997). The fragmentation of protonated PE yields a peak at $[M+H–141]^+$, which corresponds to the neutral loss of the polar head group (PE) and PS ions ($[M+H]^+$) at $[M+H– 185]^+$in a tandem mass spectrum, arising from the loss of the polar head group.

Lipid Separation

Without front-end separation, the extracts of lipids can be analysed directly by electrospray MS (Ivanova et al. 2001). Certain lipid species have low ionisation capabilities, and those lipids cannot be subjected to directed infusion (Wenk et al. 2003). Most of the lipids are separated by vacuum separation techniques like TLC, normal phase (NP) and reversed phase (RP) liquid chromatography (LC) (Houjou et al. 2005) (Fig. 7). Based upon the polarity of their head groups, TLC and normal phase liquid chromatography (NP-LC) separate phospholipids while RP-HPLC separates the phospholipids based on their hydrophobicity and their fatty acyl chains. The principle of separation in RP-HPLC is based on the length of the fatty acyl chain, the longer the chain, the slower the elution. Furthermore, unsaturated fatty acids elute faster than the saturated form. Certain phospholipids have different fatty acyl chains but identical molecular mass which causes problems in direct-infusion ESI-MS/MS. This can be overcome by using RP-HPLC with a C_{30} derivatised silica column. Though there are certain hurdles like sample loss in LC separation (NP and RP) and lipid oxidation in TLC, HPLC with ESI-MS/MS has been a method of choice for lipidomics studies.

Quantification of Lipids

In infusion ESI-MS and in the absence of ionisation suppression effects, the intensity of MS peak is directly proportional to the lipid concentration, but in LC-ESI-MS, the area of the peak determines the concentration of lipids in a mixture. Hence, MS techniques can also be used for lipid quantification (Zacarias et al. 2002). Depending on the phospholipid class, the responses by the instruments are different. Because of the different ionisation efficiency of the polar head groups in lipids, the intensity of the peak cannot blindly determine the abundance of lipids. High intensity peak can sometimes determine lower concentration hence an appropriate internal standard must be added to the sample mixtures (DeLong et al.

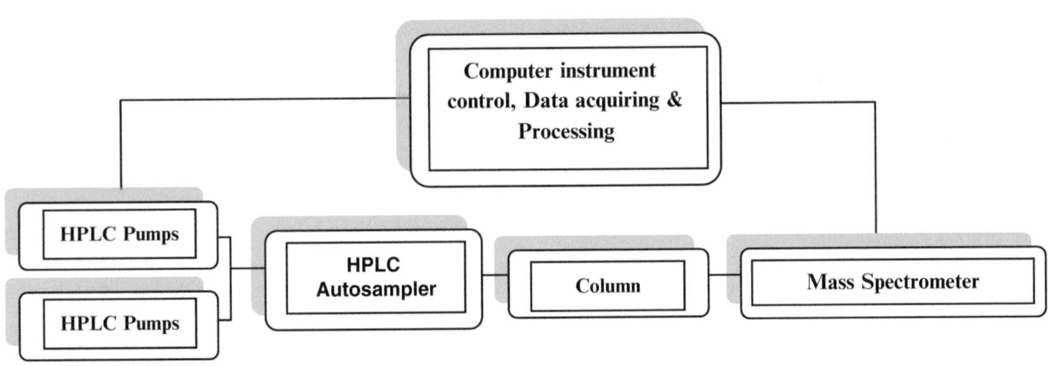

Fig. 7 Schematic diagram of LC-MS system

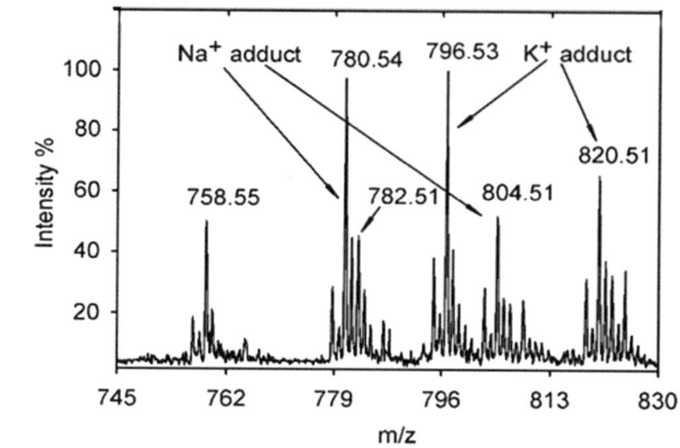

Fig. 8 Positive ion MALDI mass spectra of PCs in potato with 10-mM KCI (Source: Knowles et al. 2001)

1999). Isotopically labelled lipids are also used as an internal standard in the quantification of lipids because isotopically labelled standards and normal sample counterpart will have the same physiochemical properties. An example simple MS spectra used in the quantification of plants phosphatidylcholines (Fig. 8). There are certain difficulties in choosing the right internal standard; hence, quantification is limited to a few targeted lipids.

Non-MS-Based Techniques

Nuclear Magnetic Resonance

Although mass spectrometry techniques are widely used in lipids, certain non-MS techniques like NMR have also been used in analysing abundant lipids (cholesterol and phosphocholine). Adosraku et al. studied lipid profiles in human erythrocytes using proton nuclear magnetic resonance (Adosraku et al. 1994). To characterise phospholipid composition in body tissues and fluids, high-resolution ^{31}P-NMR was used. The latest technique in NMR is *two-dimensionalIH13-C heteronuclear single quantum coherence* NMR (2D HSQC NMR) which was used for lipidomics in mycobacteria. Limited interest in these NMR-based analyses is because of their limited sensitivity, and these advanced techniques are yet to be used in plant lipidomics (Wenk 2005).

High-Throughput Molecular Lipidomics

Various cellular functions such as energy homeostasis and intracellular signalling depend upon regulation of lipid catabolism and anabolism (Woods and Jackson 2006). These metabolisms are altered in various diseases such as cardiovascular diseases, neurodegenerative diseases, cancer and inflammation (Hiukka et al. 2009). Hence, studying the changes in the lipid composition has become a promising field in the biomarker of plant lipidomics. To analyse large number of samples, high-throughput lipidomics uses automated systems and 96-well plates to analyse large number of samples in a short time when compared to manual extraction. Shotgun and LC-MRM platforms are used to analyse extracted lipids to obtain error-free lipidomic data with high sensitivity, stability and reproducibility. Shotgun lipidomics uses a robotic-chip-based device to analyse the sample in a short amount of time at a molecular level. Precursor ion scans and neutral loss scans are used in monitoring and quantifying lipids. The results obtained are analysed using bioinformatics tools.

Bioinformatics for Lipidomics

Unlike genes and proteins data bank, lipid databases are created based on their different scope and organisation due to lack of universal lipid classification scheme. There are three important databases, namely, LipidBank [Yasugi, 2002, 12058481; Watanabe, 2000], LIPIDAT [Caffrey,

Table 1 Important online bioinformatics resources for lipidomics

LMSD	http://www.lipidmaps.org/data/structure/index.html
LIPID MAPS	http://www.lipidmaps.org/
LIPIDAT	http://www.lipidat.ul.ie/
LipidBank	http://lipidbank.jp
LMPD	http://www.lipidmaps.org/data/proteome/index.html
KEGG	http://www.genome.jp/kegg/pathway.html
Cyberlipid Center	http://www.cyberlipid.org/
Lipid Library	http://www.lipidlibrary.co.uk/

1992, 1315624] and LMSD [Sud, 2007, 17098933]. These databases provide the user with a wide range of information about lipids. Few online resources for the lipids and their databases are listed in Table 1. Due to large interest of researchers in lipidomics, a US-based organisation, LIPID MAPS developed and classified lipids into eight categories, namely, (1) fatty acyls, (2) glycerolipids, (3) glycerophospholipids, (4) sphingolipids, (5) sterol lipids, (6) prenol lipids, (7) saccharolipids and (8) polyketides. A total of about 1.68 million lipids can be accessed using this database.

Lipid Analysis Softwares

There are various softwares to draw structures such as ChemDraw and MarvinSketch, but it is difficult to draw lipid structure due to their large and complex structure. David Weininger addressed this issue by initiating the Simplified Molecular Line Entry Specification (SMILES) project. The SMILES format represents the lipid structure as a compact graph with nodes as atoms and edges as bonds. However, the SMILES format does not include 2D coordinates, which makes the visual recognition and comparison difficult. To solve this issue, LIPID MAPS developed various programmes from MS which are coupled with drawing tools (Fahy et al. 2007). LIPID MAPS contain drawing software for fatty acids, sterols, glycerolipids, cardiolipins, glycerophospholipids and sphingolipids. Structures of these lipids can also be visualised in ChemDraw ActiveX/Plugin. There are many different programmes for processing and identifying MS data. One such programme is called Lipid Navigator (http://lipidsearch.jp/LipidNavigator.htm) which is a freely available online resource developed by Mitsui Knowledge Industry in collaboration with Taguchi Laboratory (University of Tokyo). Another free online programme developed by Cracka et al. is TriglyAPCI for analysing APCI-MS for triglycerides. This software identifies each ion in the LC-MS spectrum, and it searches for compounds and gives possible triacylglycerol (TAG) structures. This software is developed using Microsoft Visual Basic 6.0. Other open-source softwares are spectrum extraction from chromatographic data (SECD) and lipid mass spectrum analysis (LIMSA). These softwares can process positive/negative ion mode and identify lipids from MS/MS spectra (Haimi et al. 2006). The output data from SECD can be taken by LIMSA, a dynamic library for batch processing and can perform identification, deconcolution and quantification of lipids. Apart from free softwares, certain commercial softwares such as Lipid Profiler (developed by MDS Sciex) combined with the software called Analyst help in identifying and quantifying lipids detected by multiple precursor ion scanning (MPIS).

Future Prospective of Plant Lipidomics

Lipid profiling is a promising field that desires to be further explored so that routine lipid profiling in plants would cover additional lipids, such as N-acyl phosphatidylethanolamines, sphingolipids, phosphoinositides, free fatty acids and oxylipins, etc. The ESI-MS and ESI-MS/MS are raised areas proven to be most comprehensive for lipid molecular species. But there is a confident need of more than one platform for complete profiling of lipid species. Hence, it may be proper to use gas chromatography/mass spectrometry and/or the coupling of ESI-MS/MS with liquid chromatography for trace compounds and oxylipins. Currently, ESI-MS-/MS-based lipid profiling has been made to order high-throughput analyses and is still a targeted strategy. Recently, the National Institutes of Health funded 35 million US dollars for the 5-year collaborative project entitled 'Lipid Metabolites and Pathway Strategy' to determine the lipidome (LIPIDMAPS) in the mouse macrophage (http://www.lipidmaps.org). Definitely, this initiative will reveal new lipid classes that can be targeted for analysis in plants. Because of significant differences in the known plant and animal lipids, however, a parallel initiative to discover all of the lipids in a plant system would probably give way for priceless information about further lipid targets (Welti and Wang 2004).

Advances in Application of Analytical Methods in Lipid Profiling

This is a dynamic research area in lipid metabolomics (lipidomics), which summarises the different approaches in lipid research, which can be described as targeted lipid analysis, lipid profiling and global lipid profiling. In a targeted lipid analysis approach focus on few lipids which are expected to be important. In a lipid profiling approach, the focus is on a specific group of lipid metabolites, a certain class or pathway. In a global lipid profiling approach, a very wide range of lipids is analysed, as widely as possible. However, the workflow of the different approaches resembles each other to a large extent. Biological samples including cells and tissues spiked with appropriate internal standards are first extracted, and crude lipid extracts are then either pre-fractionated into lipid fractions or kept intact for further separation. MS detections were employed either using direct infusion of a sample or using chromatographic separation (e.g. GC or LC) or substantial lipid fractions or whole lipid extracts to generate lipid data. In this way, lists of lipid metabolites with absolute or relative concentrations are generated from control and diseased plants. Further, the data are normalised and subjected to statistical data analysis for identifying those lipid metabolites which are discriminatory for diseases. MALDI analysis of lipid categories in plant supports the reliability of data generated for lipid profiling.

Till now, many studies have shown that lipidomics appears to be essential in determining novel lipid molecular species that serve as potential biomarkers in many lipid-related diseases. Detailed applications of lipidomics in the discovery of potential lipid biomarkers have been carried out for certain abnormalities in lipid biosynthesis and metabolism of plants.

Significance of Non-phosphorous Membrane Lipids

Phosphate (inorganic phosphate, Pi) is an essential nutrient for all living organisms. However, the amount of Pi available in the soil is often limited, as a major portion of it exist as insoluble salts that cannot be absorbed by roots (Nakamura 2013). Under Pi deprivation, plants start stimulating the uptake of exogenous Pi to maintain endogenous phosphate. However, Pi starvation decreases overall phospholipid content, and phospholipids except for phosphatidyl glycerol (PG) are synthesised extraplastidically, the major research focus has been on the extraplastidic (or endoplasmic reticulum (ER)-derived) supply of DAG.

The DAG metabolism in plants is complex not only because it belongs to phospholipids but also a substrate for the synthesis of phospholipids, galactolipids, sulpholipids and triglycerides. It is important to explore the digalactosyldiacylglycerol (DGDG) transport to other organellar membranes as it is synthesised only at the envelope of plastids; membrane lipid remodelling requires compensating for the loss of phospholipids at different organellar membranes (Zhang et al. 2001).

Importance of Lipid Trafficking/ Transporters

Lipid transfer mechanism plays a vital role in lipid profiling. Lipid biosynthetic enzymes of different subcellular compartments participate in the biogenesis of thylakoid membrane system. This process requires an extensive exchange of lipid precursors between the chloroplast and endoplasmic reticulum (ER). To analyse the underlying lipid trafficking between the ER and chloroplasts, radiolabelled ER membranes were used. The radioactivity associated with the lipids was determined by liquid scintillation counting of lipids isolated from the thin-layer chromatogram. Genetic mutants of the model plant *Arabidopsis thaliana* with disruptions in lipid trafficking between the ER and the chloroplast become available for future work (Xu et al. 2008).

Prospects in Lipid Transfer Protein (LTP) Research

Plant lipid transfer proteins (LTPs) are capable of binding fatty acids and of transferring phospholipids between membranes in vitro, whose biological function is not clearly known. LTPs have been suggested to participate in cutin assembly and in the defence of plants against pathogens and environmental stress. Lipid transfer protein (LTP) genes induced by pathogens are investigated whether these genes are responsive to the infection by the particular micro-organism. The organ specificity of gene expression for the different plant LTP genes was investigated. Further purification, characterisation and biological activity prediction became an emerging area in plant lipidomics (Segura et al. 1993).

Lipid Profiling and High-Throughput Technologies

The combined information from genomics, proteomics and metabolomics will help us to obtain an integrated understanding of a cell or organism. However, these new analytic platforms are high-throughput technologies which substantially increase the dynamic range and number of metabolites and genes that can be detected (Kell 2006). One of the outcomes is the development of informatics tools in the advancement of systems biology. In systems biology, especially metabolomics, data are presently organised with the aim to create computer models simulating biological system.

Algae as a Model for the Study of Lipid Metabolism

Microalgae are receiving more interest from the public and scientific communities due to their vital application in the field of biofuels, commercially important compounds and in bioremediation for agriculture. Biochemical and molecular analysis of microalgae revealed differences in lipid signalling between algal species and in comparison to plants. These differences range from distinct acyl groups present in algal lipids, to a possible more direct role of plastids in the assembly of TAGs during glycerolipid metabolism. Hence, microalgal lipidomics paves a new avenue of genetic engineering in plant lipid metabolism (Liu and Benning 2013).

Significance and Modification of Plant Lipids for Human Health

Polyunsaturated fatty acids (PUFAs) are essential, not synthesised *de novo* in mammals; they must be derived from the diet. Once PUFAs were ingested, they further metabolised and the resulting in long-chain PUFAs, which involved in the synthesis of cellular membranes and serve as precursors for hormone like eicosanoids. Dietary PUFAs are essential in health management and significantly obtained from plant sources (Simopoulos 1999).

The hypolipidemic, antithrombotic, anti-inflammatory and various biological effects of PUFAs have been studied extensively in animal models, tissue cultures and cells. More recently, the effects of fatty acids on gene expression have been investigated, and this focus of interest has led to studies at the molecular level. Fatty acids, either released from membrane phospholipids, act as signalling molecules as well as second messengers in various signal transduction pathways (Graber et al. 1994). They can also act as modulators mediating responses of the cell to extracellular signals. It has been shown that fatty acids rapidly and directly alter the transcription of particular genes involved (Clarke and Jump 1994). Vegetarian diets are relatively low in alpha-linolenic acid (ALA) compared with linoleic acid (LA) with trace amount of eicosapentaenoic acid (EPA) and docosahexaenoic acid (DHA). Clinical studies suggest that tissue levels of long-chain n-3 fatty acids are declined in vegetarians (American Dietetic Association 1997). Nutritional approaches, such as dietary supplementation with ω-3 polyunsaturated fatty acids (ω-3 PUFA), are particularly attractive because they could work additively with established therapies while not exerting negative hemodynamic effects (Harris et al. 2008). Polyunsaturated fatty acids (PUFAs), including conjugated linoleic acid (CLA), are recommended therapeutically to overweight individuals, including young adults and adolescents; there is a need to clarify whether CLA improves or reduces bone mass during a period of bone mineralisation and consolidation (Watson 2006).

In Agriculture Improvement

Lipidomics contributions towards fundamental metabolic pathways that direct to the synthesis of the most important plant glycerolipids are nowadays well documented. A large amount of efforts towards lipid biosynthesis in plants over the recent years has been aimed towards obtaining clones for key enzymes in the pathway. Most of industrial laboratories are contributed towards this attempt because of the probable economic importance that may be obtained from genetic engineering of vegetable oils. TAGs from plant source are the most important agricultural commodity, worth of approximately 25 billion USD/annum. In recent times, a number of victories have been achieved in genetically engineering oil seed fatty acid composition to create new or superior vegetable oils (Ohlrogge et al. 1978; Van Meer 2005; Voelker and Kinney 2001).

Conclusion

Lipidomics, an emerging field in biomedical research, plays an important role in understanding the action of plant lipids on signalling pathways against various biotic and abiotic stresses. Lipidomics in association with genomics, proteomics and metabolomics will contribute in delineating disease mechanism and provide insights into molecular mechanisms of lipid action. Mapping plant lipid content helps in revealing the exact plant defence mechanisms in response to various stresses. Like glycomics and proteomics, lipidomics involves system-level identification and has a vital role in revealing metabolic pathways. Novel analytical techniques have been greatly developed to separate the lipids and to analyse the lipid composition. Though there is few drawbacks, viz., complexity, not economically viable, they provide ample amount of information regarding the difference in regulation of lipid content in different environments. Further improvement in analytical methods and universal bioinformatics databases has to be developed to fuel lipid research in an effective way.

References

Adosraku RK, Choi GT, Constantinou-Kokotos V et al (1994) NMR lipid profiles of cells, tissues, and body fluids: proton NMR analysis of human erythrocyte lipids. J Lipid Res 35:1925–1931

Ahmad P, Prasad MNV (2004) Abiotic stress responses in plants. J Plant Physiol 168:807–815

Akınc Ş, Losel DM (2012) Water stress. In: Ismail M, Mofizur R (eds) Plant water-stress response mechanisms. InTech, Croatia, UK. ISBN 978-953-307-963-9

American Dietetic Association (1997) Position of the American Dietetic Association: vegetarian diets. J Am Diet Assoc 97:1317–1321

Arisz SA et al (2003) Substrate preference of stress-activated phospholipase D in Chlamydomonas and its contribution to PA formation. Plant J 34:595–604

Bar-Peled M, Bassham DC, Raikhel NV (1996) Transport of proteins in eukaryotic cells: more questions ahead. Plant Mol Biol 32:223–249

Bavaro L, Catucci L, Depalo N et al (2007) Lipid content in higher plants under osmotic stress. Bioelectrochemistry 70:12–17

Beffa R, Szell M, Meuwly P et al (1995) Cholera toxin elevates pathogen resistance and induces pathogenesis-related gene expression in tobacco. EMBO J 14:5753–5761

Berridge MJ, Irvine RF (1989) Inositol phosphates and cell signaling. Nature 341:197–205

Boller T (1995) Chemoperception of microbial signals in plant cells. Annu Rev Plant Physiol Plant Mol Biol 46:189–214

Brugger B, Erben G, Sandhoff R et al (1997) Quantitative analysis of biological membrane lipids at the low picomole level by nano-electrospray ionization tandem mass spectrometry. Proc Natl Acad Sci U S A 94:2339–2344

Brull F, Mensink RP, Plat J (2009) Plant sterols: functional lipids in immune function and inflammation. J Lipid Res 4(3):355–365

Byrdwell WC (2001) Atmospheric pressure chemical ionization mass spectrometry for analysis of lipids. Lipids 36(4):327–346

Carrasco-Pancorbo A, Navas-Iglesias N, Cuadros-Rodriguez L (2009) From lipid analysis towards lipidomics, a new challenge for the analytical chemistry of the 21st century. Part I: Modern lipid analysis. Trends Anal Chem 38

Chandra S, Heinstein PF, Low PS (1996) Activation of phospholipase a by plant defense elicitors. Plant Physiol 110:979–986

Chen Y, Allegood J, Liu Y et al (2008) Imaging MALDI mass spectrometry using an oscillating capillary nebulizer matrix coating system and its application to analysis of lipids in brain from a mouse model of Tay-Sachs/Sandhoff disease. Anal Chem 80:2780–2788

Chen D, Yan X, Xu J et al (2013) Lipidomic profiling and discovery of lipid biomarkers in *Stephanodiscus* sp. under cold stress. Metabolomics. doi:10.1007/s11306-013-0515-z

Cho MH, Tan Z, Erneux C et al (1995) The effects of mastoparan on the carrot cell plasma membrane polyphosphoinositide phospholipase C. Plant Physiol 107:845–856

Clarke SD, Jump DB (1994) Dietary polyunsaturated fatty acid regulation of gene transcription. Annu Rev Nutr 14:83–98

Dass C (2007) Fundamentals of contemporary mass spectrometry. Wiley, Hoboken

de Torres Zabela M et al (2002) Differential expression of genes encoding Arabidopsis phospholipases after challenge with virulent or avirulent Pseudomonas isolates. Mol Plant-Microbe Interact 15:808–816

DeLong CJ, Shen YJ, Thomas MJ et al (1999) Molecular distinction of phosphatidylcholine synthesis between the CDP-choline pathway and phosphatidyl ethanolamine methylation pathway. J Biol Chem 274:29683–29688

Den Hartog M et al (2001) Nod factor-induced phosphatidic acid and diacylglycerol pyrophosphate formation: a role for phospholipase C and D in root hair deformation. Plant J 25:55–66

Dorne AJ, Joyard J, Douce R (1990) Do thylakoids really contain phosphatidylcholine? Proc Natl Acad Sci U S A 87:71–74

Dowhan W (2009) Molecular genetic approaches to defining lipid function. J Lipid Res 50:305–310

Ekroos K, Chernushevich IV, Simons K et al (2002) Quantitative profiling of phospholipids by multiple precursor ion scanning on a hybrid quadrupole time-of-flight mass spectrometer. Anal Chem 74:941–949

Fahy E, Subramaniam S, Brown HA et al (2005) A comprehensive classification system for lipids. J Lipid Res 46:839–861

Fahy E, Sud M, Cotter D et al (2007) LIPID MAPS online tools for lipid research. Nucleic Acids Res 35:606–612

Fahy E, Subramaniam S, Murphy RC et al (2009) Update of the LIPID MAPS comprehensive classification system for lipids. J Lipid Res 50:9–14

Farmer EE, Almeras E, Krishnamurthy V (2003) Jasmonates and related oxylipins in plant responses to pathogenesis and herbivory. Curr Opin Plant Biol 6:372–378

Fenn JB, Mann M, Meng CK et al (1989) Electrospray ionization for mass spectrometry of large biomolecules. Science 246:64–71

Giroud C, Siegenthaler PA (1988) Development of oat prothylakoids into thylakoids during greening does not change transmembrane galactolipid asymmetry but preserves the thylakoid bilayer. Plant Physiol 88:412–417

Gong M, Li YJ, Chen SZ (1998) Abscisic acid–induced thermo tolerance in maize seedling is mediated by calcium and associated with antioxidant system. J Plant Physiol 153:488–496

Graber R, Sumida C, Nunez EA (1994) Fatty acids and cell signal transduction. J Lipid Mediat Cell Signal 9:91–116

Gross RW, Han X (2006) Unlocking the complexity of lipids: using lipidomics to identify disease mechanisms, biomarkers, and treatment efficacy. Futur Lipidol 1:539

Guan LM, Zhao J, Scadalios JG (2000) *Cis*-elements and trans-factors that regulate expression of the maize *Cat1* antioxidant gene in response to ABA and osmotic stress: H_2O_2 is the likely intermediary signaling molecule for the response. Plant J 22:87–95

Haimi P, Uphoff A, Hermansson M et al (2006) Software tools for analysis of mass spectrometric lipidome data. Anal Chem 78:8324–8331

Han X, Gross RW (1994) Electrospray ionization mass spectroscopic analysis of human erythrocyte plasma

membrane phospholipids. Proc Natl Acad Sci U S A 91:10635
Han X, Gross RW (2003) Global analyses of cellular lipidomes directly from crude extracts of biological samples by electrospray ionization mass spectrometry: a bridge to lipidomics. J Lipid Res 44:1071
Han PP, Yuan Y-j (2009) Lipidomic analysis reveals activation of phospholipid signaling in mechanotransduction of *Taxus cuspidate* cells in response to shear stress. FASEB J 23:623–630
Harris WS, Miller M, Tighe AP, Davidson MH, Schaefer EJ (2008) Omega-3 fatty acids and coronary heart disease risk: clinical and mechanistic perspectives. Atherosclerosis 197:12–24
Hiukka A, Stahlman M, Pettersson C et al (2009) ApoCIII-enriched LDL in type 2 diabetes displays altered lipid composition, increased susceptibility for sphingomyelinase, and increased binding to biglycan. Diabetes 58:2018–2026
Hong Y, Pan X, Welti R et al (2008) Phospholipase Dα3 is involved in the hyper osmotic response in Arabidopsis. Plant Cell 20:803–816
Hopfgartner G, Varesio E, Tschappat V et al (2004) Triple quadrupole linear ion trap mass spectrometer for the analysis of small molecules and macromolecules. J Mass Spectrom 39:845–855
Houjou T, Yamatani K, Imagawa M et al (2005) A shotgun tandem mass spectrometric analysis of phospholipids with normal-phase and/or reverse-phase liquid chromatography/electrospray ionization mass spectrometry. Rapid Commun Mass Spectrom 19:654–666
Hoyos ME, Zhang S (2000) Calcium-independent activation of salicylic acid-induced protein kinase and a 40-kilodalton protein kinase by hyperosmotic stress. Plant Physiol 122:1355–1363
Hu C, van der Heijden R, Wang M et al (2009) Analytical strategies in lipidomics and applications in disease biomarker discovery. J Chromatogr 877:2836–2846
Ishida M et al (2004) High resolution analysis by nano electrospray ionization Fourier transform ion cyclotron resonance mass spectrometry for the identification of molecular species of phospholipids and their oxidized metabolites. Rapid Commun Mass Spectrom 18(20):2486–2494
Israelachvili J (1992) Relationship between chain protrusions and surfactant critical micelle concentration values. Langmuir 8:1501
Ivanova PT, Cerda BA, Horn DM et al (2001) Electrospray ionization mass spectrometry analysis of changes in phospholipids in RBL-2H3 mastocytoma cells during degranulation. Proc Natl Acad Sci U S A 98: 7152–7157
Joyard J, Block MA, Douce R (1991) Molecular aspects of plastid envelope biochemistry. Eur J Biochem 199:489–509
Karas M, Hillenkamp F (1988) Laser desorption ionization of proteins with molecular masses exceeding 10,000 daltons. Anal Chem 60:2299–2301
Katagiri T et al (2001) Involvement of a novel Arabidopsis phospholipase D, AtPLDd, in dehydration-inducible accumulation of phosphatidic acid in stress signalling. Plant J 26:595–605
Kell DB (2006) Metabolomics, modelling and machine learning in systems biology – towards and understanding of the languages of cells. FEBS J 273:873–894
Kim HY, Cote CG, Crain RC (1996) Inositol 1,4,5-trisphosphate may mediate closure of K+ channels by light and darkness in Samanea saman motor cells. Planta 198:279–287
Klose C, Surma MA, Gerl (2012) Flexibility of eukaryotic lipidome insights from yeast lipidomics. PLoS One 7(4):35063
Kocsy G, Galiba G, Brunold C (2001) Role of glutathione in adaptation and signaling during chilling and cold acclimation in plants. Physiol Plant 113:158–164
Kramer PJ (1980) Drought, stress, and the origin of adaptations. In: Turner NC, Kramer PJ (eds) Adaptations of plants to water and high temperature stress. Wiley, New York, pp 7–20
Larsen A, Uran S, Jacobsen PB et al (2001) Collision-induced dissociation of glycerophospholipids using electrospray ion-trap mass spectrometry. Rapid Commun Mass Spectrom 15:2393–2398
Laxalt AM, Munnik T (2002) Phospholipid signalling in plant defence. Curr Opin Plant Biol 5:1–7
Laxalt AM et al (2001) Characterization of five tomato phospholipase D cDNAs: rapid and specific expression of LePLD*b*1 on elicitation with xylanase. Plant J 26:237–247
Legendre L, Yueh YG, Crain R et al (1993) Phospholipase C activation during elicitation of the oxidative burst in cultured plant cells. J Biol Chem 268:4559–24563
Lindsey K, Pullen ML, Topping JF (2003) Importance of plant sterols in pattern formation and hormone signalling. Trends Plant Sci 8:521–525
Lipid Maps Consortium (2009) Lipid Maps Consortium: lipid metabolites and pathways strategy. www.lipidmaps.org
Liu B, Benning C (2013) Lipid metabolism in microalgae distinguishes itself. Curr Opin Biotechnol 24:300–309
Mazid M, Khan TA, Mohammad (2011) Role of secondary metabolites in defense mechanisms of plants. Biol Med 3(2):232–249
McDonald JG, Thompson BM, McCrum EC et al (2007) Extraction and analysis of sterols in biological matrices by high performance liquid chromatography electrospray ionization mass spectrometry. Methods Enzymol 432:145–170
McDonnell LA, Heeren RM (2007) Imaging mass spectrometry. Mass Spectrom Rev 26:606–643
Meijer HJ, Munnik T (2003) Phospholipid-based signaling in plants. Annu Rev Plant Biol 54:265–306
Michell RH (1975) Inositol phospholipids and cell surface receptor function. Biochim Biophys Acta 415:81–147
Millner PA (1995) The auxin signal. Curr Opin Cell Biol 7:224–231

Millner PA, Causier BE (1996) G-protein coupled receptors in plant cells. J Exp Bot 47:983–992
Moreau P, Bessoule JJ, Mongrand S, Testet T et al (1998) Lipid trafficking in plant cells. Prog Lipid Res 37(6):371–391
Munnik T (1999) Phospholipid metabolism with respect to signal transduction in plant cells. PhD thesis, University of Amsterdam, The Netherlands
Munnik T, de Vrije T, Irvine RF et al (1996) Identification of diacylglycerol pyrophosphate as a novel metabolic product of phosphatidic acid during G-protein activation in plants. J Biol Chem 271:15708–15715
Munnik T, van Himbergen JAJ, Ter Riet T et al (1997) Phospholipid metabolism with respect to signal transduction in plant cells. PhD thesis, University of Amsterdam, The Netherlands
Nakamura Y (2013) Phosphate starvation and membrane lipid remodeling in seed plants. Prog Lipid Res 52:43–50
Nobes CD, Hawkins P, Stephens L, Hall A (1995) Activation of the small GTP-binding proteins rho and rac by growth factor receptors. J Cell Sci 108:225–233
Nygren H, Malmberg P, Kriegeskotte C et al (2004) Bio imaging TOF-SIMS: localization of cholesterol in rat kidney sections. FEBS Lett 566:291–293
Ohlrogge JB, Pollard MR, Stumpf PK (1978) Studies on biosynthesis of waxes by developing jojoba seed tissue. Lipids 13:203–210
Ohlrogge JB, Browse J, Chris R et al (1991) The genetics of plant lipids. Biochim Biophys Acta 1082:1–26
Pei ZM, Murata Y, Benning G, Thomine S et al (2000) Calcium channels activated by hydrogen peroxide mediates abscisic acid signalling in guard cells. Nature 406:731–734
Poltronieri P, Bonsegna S, Domenico SD et al (2011) Molecular mechanisms in plant abiotic stress response. Ratar Povrt Field Veg Crop Res 48:15–24
Prasad TK, Anderson MD, Martin BA et al (1994) Evidence for chilling-induced oxidative stress in maize seedlings and a regulatory role for hydrogen peroxide. Plant Cell 6:65–74
Rawyler A, Unitt MD, Giroud C et al (1987) The transmembrane distribution of galactolipids in chloroplast thylakoids is universal in a wide variety of temperate climate plants. Photosynth Res 11:3–13
Reina-Pinto JJ, Yephremov A (2009) Surface lipids and plant defences. Plant Physiol Biochem 47:540–549
Roberts LJ (2002) Lipids as regulators of cell function. Cell Mol Life Sci 59:727–728
Rothman JE, Wieland FT (1996) Protein sorting by transport vesicles. Science 272(5259):227–234
Sangwan V, Foulds I, Singh J et al (2001) Cold activation of *Brassica napus* BN115 promoter is mediated by structural changes in membranes and cytoskeleton, and requires Ca2 influx. Plant J 27:1–12
Scherer GFE (1995) Activation of phospholipase A2 by auxin and matsporan in hypocotyl segment from zucchini and sunflower. J Plant Physiol 145:483–490
Scherer GF (2002) Secondary messengers and phospholipase A2 in auxin signal transduction. Plant Mol Biol 49:357–372
Schwartz JC, Senko MW, Syka JE (2002) A two-dimensional quadrupole ion trap mass spectrometer. J Am Soc Mass Spectrom 13:659–669
Schwudke D, Hannich JT, Surendranath V et al (2007) Top-down lipidomic screens by multivariate analysis of high-resolution survey mass spectra. Anal Chem 79:4083–4093
Segura A, Moreno M, Garcia-Olmedo F (1993) Purification and anti pathogenic activity of lipid transfer proteins (LTPs) from the leaves of Arabidopsis and spinach. FEBS Lett 332:243–246
Siegenthaler PA, Giroud C (1986) Transversal distribution of phospholipids in prothylakoid and thylakoid membranes from oat. FEBS Lett 201:215–220
Simopoulos AP (1999) Essential fatty acids in health and chronic disease. Am J Clin Nutr 70:560S–569S
Sperling P, Heinz E (2003) Plant sphingolipids: structural diversity, biosynthesis, first genes and functions. Biochim Biophys Acta 632:1–15
Stubiger G, Belgacem O (2007) Analysis of lipids using 2,4,6-trihydroxyacetophenone as a matrix for MALDI mass spectrometry. Anal Chem 79:3206–3213
Van der Luit AH et al (2000) Elicitation of suspension-cultured tomato cells triggers formation of phosphatidic acid and diacylgycerol pyrophosphate. Plant Physiol 123:1507–15154
Van Meer G (2005) Cellular lipidomics. EMBO J 24:3159–3165
Voelker TA, Kinney AJ (2001) Variations in the biosynthesis of seed storage lipids. Annu Rev Plant Physiol Plant Mol Biol 52:261–335
Wang X (2000) Multiple forms of phospholipase D in plants: the gene family, catalytic and regulatory properties, and cellular functions. Prog Lipid Res 39:109–149
Wang QY, Nick P (2001) Cold acclimation can induce microtubular cold stability in a manner distinct from abscisic acid. Plant Cell Physiol 42:999–1005
Wang ZY, Kenigsbuch D, Sun L et al (1997) A Myb-related transcription factor is involved in the phytochrome regulation of an Arabidopsis *Lhcb* gene. Plant Cell 9:491–507
Watson AD (2006) Lipidomics: a global approach to lipid analysis in biological systems. J Lipid Res 47:2101
Welti R, Wang X (2004) Lipid species profiling: High-throughput approaches to identify lipid compositional changes and determine the function of genes involved in lipid metabolism and signaling. Curr Opin Plant Biol 7:337–344
Welti R, Shah J, Li W et al (2007) Plant lipidomics: discerning biological function by profiling plant complex lipids using mass spectrometry. Front Biosci 12:2494–2506
Wenk MR (2005) The emerging field of lipidomics. Nat Rev Drug Discov 4(7):594–610
Wenk MR, Lucast L, Di Paolo G et al (2003) Phosphoinositide profiling in complex lipid mixtures

using electrospray ionization mass spectrometry. Nat Biotechnol 21:813–817

Wiest MM, Watkins SM (2007) Biomarker discovery using high-dimensional lipid analysis. Curr Opin Lipidol 18:181

Woods AS, Jackson SN (2006) Brain tissue lipidomics: direct probing using matrix-assisted laser desorption/ionization mass spectrometry. AAPS J 8:391–395

Xu C, Fan J, Cornish AJ, Benning C (2008) Lipid trafficking between the endoplasmic reticulum and the plastid in *Arabidopsis* requires the extraplastidic TGD4 protein[W]. Plant Cell 20:2190–2204

Zacarias A, Bolanowski D, Bhatnagar A (2002) Comparative measurements of multicomponent phospholipid mixtures by electrospray mass spectroscopy: relating ion intensity to concentration. Anal Biochem 308:152–159

Zhang X, Zhang L, Dong F et al (2001) Hydrogen peroxide is involved in abscisic acid-induced stomatal closure in *Vicia faba*. Plant Physiol 126:1438–1448

Zhao Z, Chen G, Zhang C (2001) Interaction between reactive oxygen species and nitric oxide in drought-induced abscisic acid synthesis in root tips of wheat seedlings. Aust J Plant Physiol 28:1055–1061

Plant Secretomics: Unique Initiatives

Neelam Yadav, S.M. Paul Khurana, and Dinesh K. Yadav

Contents

N. Yadav, Ph.D. (✉) • S.M.P. Khurana, Ph.D.
D.K. Yadav, Ph.D.
Amity Institute of Biotechnology,
Amity University Harayana, Gurgaon, Haryana, India
e-mail: reachtoneelam@gmail.com

Abstract

Plant secretomics is an emerging subfield of proteomics studying proteins globally secreted into the extracellular space (apoplast) by plant cells at defined time under constitutive or induced conditions. Plant secretome has important biological functions in cell wall structure formation, cell-to-cell interaction, extracellular/intracellular signal relay and appropriate cellular response to environmental stimuli. It also regulates the ability or inability of the host to trigger the defence system against the invading pathogen. Defence proteins are secreted via a classical pathway involving N-terminal signal peptide which directs the protein to the ER for routing, modification and subsequent secretion involving the endoplasmic reticulum (ER)–Golgi–trans-Golgi network (TGN)–plasma membrane system. Plant secretome has an increasing number of proteins following unconventional, ER–Golgi-independent or 'leaderless' apoplastic protein secretion mechanisms. Nonconventional mechanisms would be necessary if the presence of a protein in the ER/Golgi disrupts ER functioning or has multiple functions, each occurring in different cellular compartment. A large number of apoplastic leaderless secretome proteins have been identified that play an important role under salinity, low temperature, ion homeostasis and pathogen invasion. Characterisation of secretome is a formidable task, and success can be

D. Barh et al. (eds.), *PlantOmics: The Omics of Plant Science*,
DOI 10.1007/978-81-322-2172-2_12, © Springer India 2015

obliged to the advancement in biochemical, proteomic techniques, mass spectroscopy and bioinformatics. Advanced proteomic technologies established detailed secretome profiles from normal and stressed cell types at a faster pace. Discrimination of the true secretome from those released under environmental stresses is a big challenge. It warrants improved strategies to investigate the secretomes with high sensitivity and reproducibility. The comprehensive mechanisms regulating constitutive and induced secretome of diverse plants and their habitat are future perspective.

Keywords
Apoplast • Leader peptide • Leaderless secretory proteins • Proteomics • Secretome • Secretory pathways

Introduction

Plant secretomics is an emerging field of proteomics studying the secreted proteins of plants called 'secretome'. The term 'secretome' was first used to describe a genome-wide study of the signal peptide-dependent secreted proteins and the protein secretion machineries in *Bacillus subtilis*, a Gram-positive bacterium (Tjalsma et al. 2000). The term is more often limited to include only the secreted proteins (Greenbaum et al. 2001; Hathout 2007; Bouws et al. 2008). Thus, 'secretomics' is defined as the study of proteins globally secreted into the extracellular space (apoplast) of cell, tissue or organ at any given time under specific conditions through various secretory mechanisms under constitutive or induced conditions' (Agrawal et al. 2010).

Plant secretome has important biological functions in the formation of cell wall structure, cell-to-cell interaction, appropriate response to environmental stimuli and defence against pathogens (Isaacson and Rose 2006; Kamoun 2009). The cell wall is a major interface between plant cells and its surrounding environment. Rapid and regulated secretion of specific proteins into this extracellular space (apoplast) is an important defence response (Grant and Lamb 2006).

Apoplastic fluid is a complex mixture of proteins secreted constitutively and proteins secreted in response to environmental stimuli. Secretion of defence proteins or exocytosis in both plants and animals is generally achieved through a conventional pathway involving the endoplasmic reticulum–Golgi–trans-Golgi network–plasma membrane in the plant endomembrane system. It required an N-terminal signal peptide directing the protein to the ER for routing, modification and subsequent secretion via the Golgi apparatus. However, the presence of an increasing number of proteins lacking signal peptide in the apoplastic fluid suggest the existence of unconventional protein secretion mechanism. Numerous ER–Golgi-independent or 'leaderless' eukaryotic secretion mechanisms have been reported. Proteins are secreted by nonconventional mechanisms for a number of reasons. For instance, non-Golgi secretion would be necessary if the presence of a protein in the ER/Golgi would disrupt ER functioning. Non-Golgi secretion could also be desirable if a protein has multiple functions, each occurring in different cellular compartment. The significance of these secretory pathways, particularly in response to stresses, is well studied in animals and yeast (Nickel and Rabouille 2009); however, our information related to the knowledge of the protein population of a plant secretome and related secretory mechanisms remains limited in plants.

Secretomics has increasingly been the focus of biological research, and it has now become an intricate subfield of proteomics. Moreover, the information regarding the number and types of proteins found in the secretome of a specific plant under normal growth and stress conditions is still unknown. Hence, the complete secretome profile has now become a prerequisite rather than an option before we begin to systematically understand the function of secretory processes and proteins. Improvement in sequencing technology has made the genome sequences of more plant species completely known. Currently there are 24 land plants having completed or draft genome sequences available and 72 land plant species with genome sequencing in progress (http://www.ncbi.nlm.nih.gov/genomes/static/gpstat.html). The improvement and automation in proteomic technology is proving increasingly helpful for a systematic identification, qualitative and quantitative profiling and functional characterisation of plant secretome. The parallel development in bioinformatics has multiplied our ability to predict the protein-coding genes and the subcellular topographic locations of the encoded proteins, which is essential for the functional annotation of the genomes. The combined analyses of secretome assisted with genomic and bioinformatic techniques can correlate large-scale plant secretome studies and unravel mechanisms of plant response to various internal and external stimuli. In the present chapter, we have discussed the mechanisms of protein secretion in apoplastic fluid and methods of secretome isolation, separation, identification and annotation of their role in plants successfully completing their life span.

General Pathway of Secretory Proteins

A general characteristic of all prokaryotic or eukaryotic cells is to export the proteins from the cytoplasm to intracellular or extracellular locations. The secreted proteins in the apoplast or the extracellular space mediate major defence responses (Grant and Lamb 2006). The proteins destined to be exported are synthesised with a signal peptide that guides its translocation. Generally, the precursor protein with amino acid sequences of signal peptides is initially recognised by soluble targeting factors for its transport to the target membrane, for its association with translocation machinery. Then polypeptide chain is transported through a proteinaceous channel. The secretion of proteins takes place through secretory pathways involving the endoplasmic reticulum and Golgi apparatus. Reports have shown that the secretion of protein also takes place without classical secretory pathway; in plants and animals, protein secretion is solely mediated by the endoplasmic reticulum and Golgi apparatus. Secreted proteins have a signal peptide at N-terminus to direct them into the ER for sorting, modification and further secretion through the Golgi network. The existence of an alternate secretion mechanism is known which takes place without signal peptide (Auron et al. 1987). Their mechanism of secretion is Golgi-independent or leaderless secretion and is called nonclassical or unconventional secretory pathways (Nickel and Rabouille 2009).

Unconventional protein secretion takes place by two major methods: proteins are either transported in a non-vesicular mode where they pass directly from the cytosol through the plasma membrane or by various vesicular modes with membrane-bounded structures fusing with the plasma membrane before release in the extracellular space (Ding et al. 2012). Recently, a plant-specific compartment named exocyst-positive organelle (EXPO), has been shown to mediate nonclassical protein secretion from cytosol to cellwall without passing proteins via the Golgi apparatus, trans-Golgi network or multivesicular body (Wang et al. 2010a).

Molecular Biology of Secretome

Delivery of proteins through the endomembrane system to plasma membrane or the extracellular space (apoplast) usually starts with cotranslational insertion of proteins into the endoplasmic reticulum and then the cleavage of the signal peptide. The classical secretory pathway is highly conserved in eukaryotes (Burgess and Kelly 1987; Jurgens and Geldner 2007; Marti et al. 2010; Cai et al. 2011, 2012). There are regulated and consecutive secretory pathways that diverge in the trans-Golgi network. The constitutive classical secretory pathway is highly complex and operates in all cells (Fig. 1).

Many soluble proteins are continually secreted from the cell by this pathway and translocate newly synthesised lipids and proteins to the plasma membrane. Specialised secretory cells also have a regulated secretory pathway, by which selected proteins in the trans-Golgi network are diverted into secretory vesicles, where the proteins are stored until an extracellular signal stimulates their secretion.

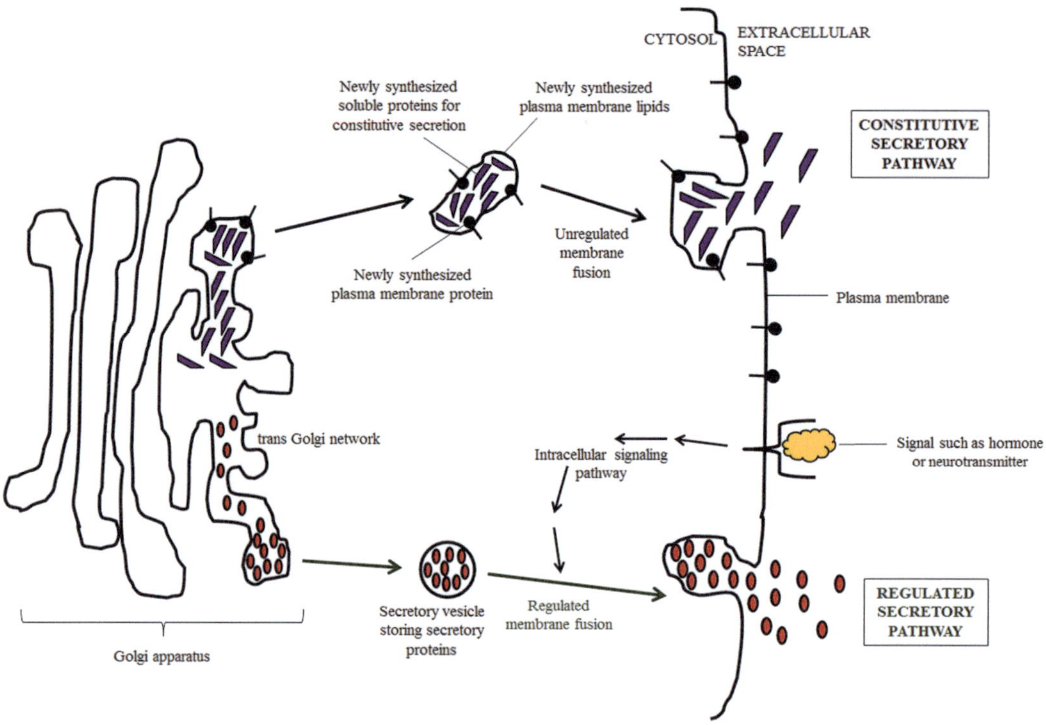

Fig. 1 The constitutive and regulated secretory pathways (Adapted from Alberts et al. 2008)

The secretory pathway transports proteins from one organelle to another within transport vesicles. Vesicular transport mediates a continuous exchange of components between chemically distinct, membrane-enclosed compartments that collectively constitute the biosynthetic–secretory and endocytic pathways. Most transport vesicles form specialised, coated regions of membranes that bud off as coated vesicles, with a distinctive cage of proteins covering their cytosolic surface. Before the vesicles fuse with a target membrane, they discard their coat, as is required for the two cytosolic membrane surfaces to interact directly and fuse. The coat performs two main functions: First is the selection of appropriate molecule transport concentrating specific membrane proteins in a specialised patch, forming vesicle membrane. Second, the coat moulds the vesicle into a curved, basketlike lattice that deforms the membrane patch and thereby shapes the vesicle. Hence, vesicles with the similar type of coat often have relatively the same size and shape. The vesicular transport selectively uses various cytosolic proteins like coat proteins (clathrin, COPI, COPII and retromer), some GTPases (Sar1, Arf1 and Rabs) and the ESCRT complexes (Kirchhausen 2000; Nickel et al. 2002; Gabe Lee et al. 2009; Hurley and Hanson 2010; Gao et al. 2012). There are three well-characterised types of coated vesicles, distinguished by their coat proteins: clathrin-coated, COPl-coated and COPII-coated. Each type is used for different transport steps. Clathrin-coated vesicles mediate transport from the Golgi apparatus and from the plasma membrane, whereas COPI- and COPII-coated vesicles mostly mediate transport from the ER and Golgi cisternae. The correct targeting and fusion of these vesicles to destined organelle depends on organelle-specific tethering factors and SNARE complexes (Cai et al. 2007; Sztul and Lupashin 2009) (Fig. 2).

Many of the secretory proteins from mammals and yeasts are known to follow an unconventional secretory pathway (Auron et al. 1987); such processes are less reported in plants. In plants, more than 50 % of secretory proteins from

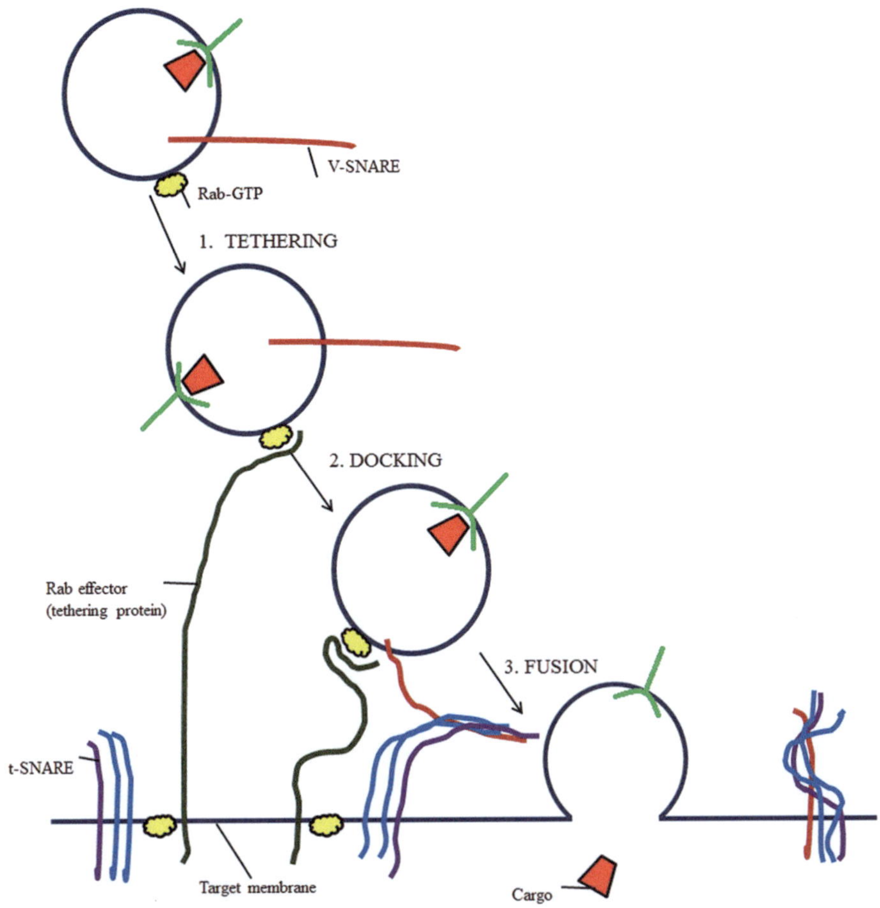

Fig. 2 Vesicle tethering to target membrane: Rab effector proteins interact via active Rab proteins (Rab-GTPs) on the target membrane, vesicle membrane or both to establish the first connection between the two membranes going to fuse. Rab effector is shown here as a filamentous tethering protein; SNARE proteins on the two membranes pair to dock the vesicle to the target membrane and catalyse the fusion of the two apposed lipid bilayer (Adapted from Alberts et al. 2008)

total known plant secretome lack a signal peptide sequence and follow a leaderless secretory pathway (Agrawal et al. 2010). Studies performed using methods that cause least contamination of cytoplasmic proteins during secretome preparation and their analyses using highly sensitive enzymatic, immunoblotting and microarray showed the presence of high percentage of leaderless secretory protein in the plant secretome ruling out the possibility of contaminating nonsecretory proteins (Jung et al. 2007; Tran and Plaxton 2008; Cho et al. 2009).

Classical Secretome with Leader Peptide

The Classical Secretory Pathway for Protein Translocations Across Membrane

Proteins are the workhorses, which are synthesised in the cytoplasm. They ought to transport the entire polypeptide chain across one or two membranes in a unidirectional manner from the site of synthesis to the site of its biological function through the secretory pathway. The classical secretory pathway is a series of steps a cell follows to translocate a protein across a membrane bilayer or out of the cell via the endoplasmic reticulum through a process known as secretion. Translocation of nascent proteins across the membrane of the ER is known to occur in two ways: cotranslational translocation, in which translocation is concurrent with peptide synthesis by the ribosome, or posttranslational translocation, in which the protein is first completely synthesised in the cytosol and released from its polysomal complex and, thereafter, is transported into the ER. Both the methods of translocation are mediated by the same protein channel, known as Sec61 in eukaryotes and SecY in prokaryotes and archaea.

Cotranslational Translocation or Signal Recognition Particle (SRP)-Dependent Pathway

Proteins that follow a secretory pathway are destined for translocation across the ER membrane. The first stretch of the amino acids synthesised, called a signal/leader/transit peptide, allows a series of interactions starting with the recognition and its binding with SRP (Fig. 3).

The amino acid sequences of signal peptides are not conserved. ER targeting is specified by a central stretch of 7–20 hydrophobic amino acids. The extent of hydrophobicity of this region dictates cotranslational import into the ER. Eukaryotic SRP is a complex of six associated polypeptides and an RNA component which target substrates for cotranslational translocation into the ER. The SRP54 binds to the hydrophobic core of signal sequence as it emerges from the ribosome. The SRP complex, when bound to the ribosome and the signal sequence of the nascent peptide, pauses the elongation of the polypeptide by the subcomplex SRP9 and SRP14 by blocking the tRNA (Walter and Johnson 1994; Lutcke 1995). This translational arrest is to ensure proper targeting to the ER membrane before significant portions of the polypeptide emerge from the ribosome and begin to fold.

The ribosome along with its transit peptide–SRP complex is then attached to a docking pro-

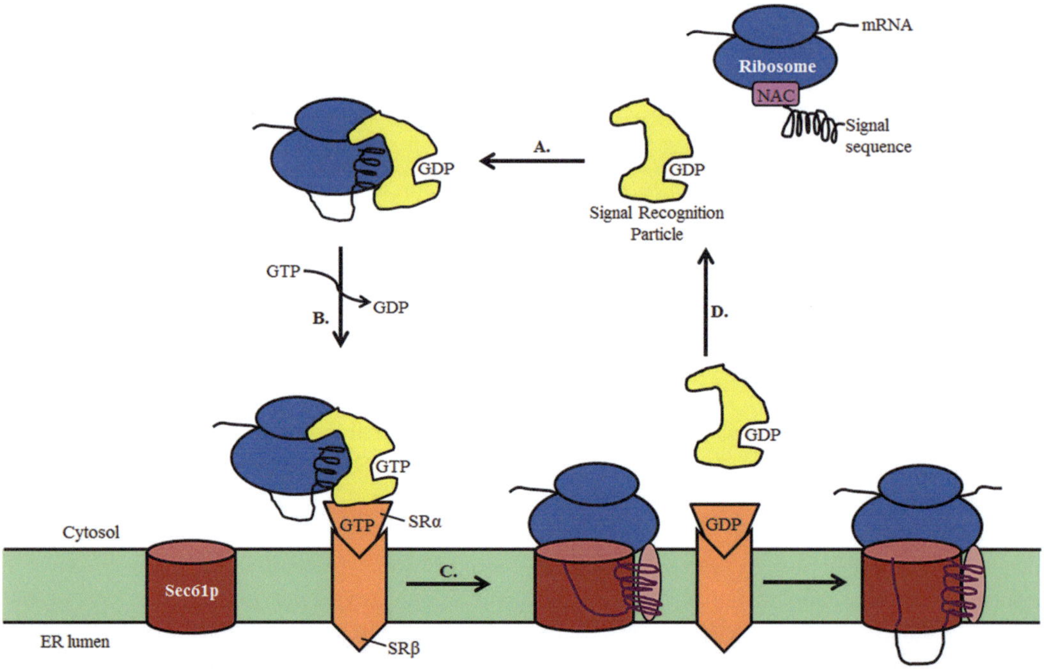

Fig. 3 Mechanism of cotranslational translocation of newly synthesised protein across the membrane (Adapted from Corsi and Schekman 1996)

tein. The docking protein is a heterodimeric SRP receptor (SR) composed of SRα and SRβ subunits. The SRP–nascent chain–ribosome complex binds to the docking protein and transfers the SRP–nascent chain–ribosome complex to the translocon, the Sec61, and then recycles back to the cytosol. The SRP is released from the SRP receptor after receptor-induced GTP hydrolysis by SRP54 component and completes the cycle (Miller et al. 1993). As the SRP and SRP receptor dissociate from the ribosome, the ribosome is able to bind directly with docking protein, Sec61. The Sec61 translocation channel (called SecY in prokaryotes) is a highly conserved heterotrimeric complex composed of α-, β- and γ-subunits. The pore of the channel, formed by the α-subunit, is blocked by a short helical segment which is thought to become unstructured during the beginning of protein translocation, allowing the peptide to pass through the channel. Completion of the synthesis of prepeptide resumes once the nascent signal peptide translocates across the channel into ER lumen. As the synthesis of prepeptide continues, it progressively penetrates into the ER lumen.

During translocation, the signal sequence is cleaved off by a signal peptidase present specifically in the ER lumen, freeing the amino terminus of the growing peptide. Translocated protein undergoes specific posttranslational modifications such as glycosylation or insertion of specific cofactor and is eventually stabilised by attaining a stable functional conformation. The stable functional protein can then be secreted via retrograde/anterograde pathways involving the Golgi apparatus to its final destination. If the secreted protein lacks any secondary signal sequence, they are secreted in the apoplast.

Posttranslational Translocation or SRP-Independent Pathway in Eukaryotes

Unlike cotranslational translocation, posttranslational or SRP-independent translocation of secretory proteins occurs independently of SRP in eukaryotes. The secretory precursor protein is completely translated and released from the ribosomal translation machinery (Fig. 4).

Posttranslational targeting of secretory proteins requires cytosolic components, viz.

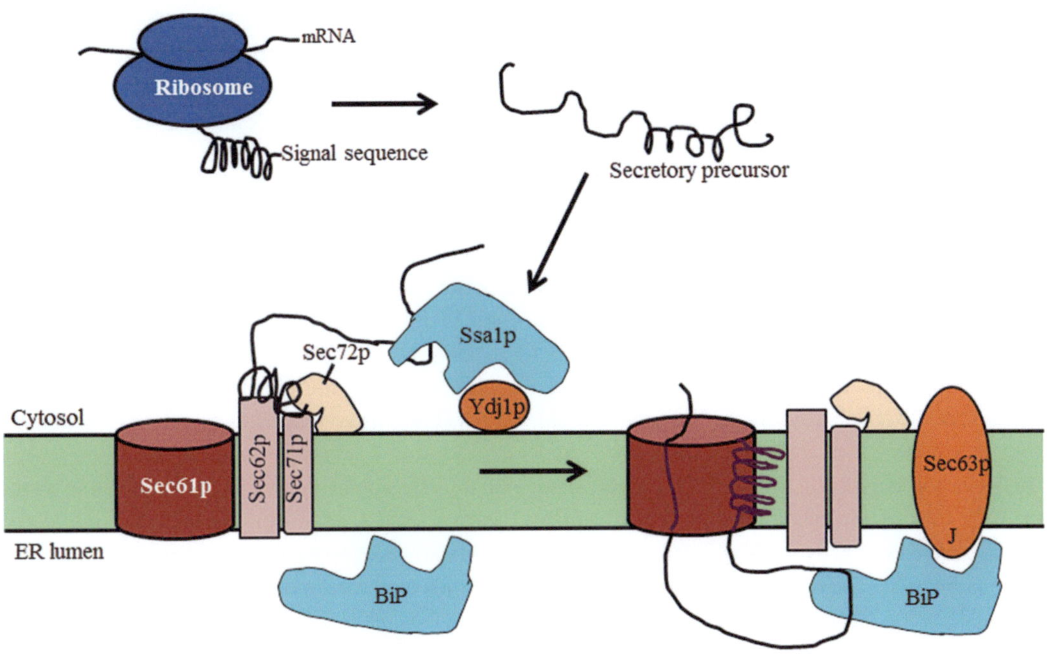

Fig. 4 Mechanism of posttranslational translocation of newly synthesised protein in yeast (Adapted from Corsi and Schekman 1996)

cytosolic heat shock proteins (Hsp70), to maintain the polypeptide in an incompletely folded state. The posttranslational modes of translocation of secretory/membrane proteins prominently involve Sec translocon pathway, and Sec61p is a significant candidate subunit of the translocation channel.

In addition to the Sec61p complex, a second set of proteins called the Sec62p/Sec63p complex is required for posttranslational translocation in yeast. The Sec61 translocon associates with oligomeric membrane protein complex (Rapoport et al. 1999). This oligomeric membrane protein complex includes three integral membrane proteins, Sec62p, Sec63p and Sec71p, as well as Sec72p, which is peripherally associated with the cytosolic face of the ER, probably through association with Sec71p. Sec63p has been shown to form a subcomplex with Sec71p, Sec72p and BiP (Brodsky and Schekman 1993). BiP is a member of the Hsp70 family of ATPases, a group which is characterised as having an N-terminal nucleotide-binding domain and a C-terminal substrate-binding domain, which binds to peptides. Studies have proposed that Sec62p, Sec71p and Sec72p, together, create a surface for secretory precursors to bind before crossing the ER membrane.

Translocation apparatus for posttranslational translocation into a reconstituted proteoliposome consists of Sec61p and Sec62p/Sec63p complexes (Panzner et al. 1995). The Sec62p/Sec63p complex contains a cytoplasmic signal sequence receptor site that binds newly synthesised secretory proteins. The substrates are maintained in an unfolded, translocation-competent conformation with the aid of cytoplasmic chaperones (Chirico et al. 1988). Subsequent to binding, the signal is transferred from Sec62/Sec63 to the signal sequence receptor of the Sec61 translocon, and translocation occurs via the Sec61p channel. The primary role of the membrane protein complex Sec62/Sec63 is to activate the ATPase activity of BiP via Sec63p. The final step in the completion of translocation of precursor secretory proteins is full transfer from the pore into the ER lumen and requires functional Sec63p and BiP. The association of substrate-binding domain of BiP through Sec63p binds nonspecifically to the precursor peptide as it enters the ER lumen and allows the BiP to act as a translocation motor (Glick 1995; Brodsky 1996) and keeps the peptide from sliding backwards in a ratchet-type mechanism.

Unconventional Secretome with Leaderless Peptide or Without Leader Peptide

Secretion of defence proteins in both plants and animals was originally thought to be solely via an endoplasmic reticulum (ER)/Golgi-mediated pathway, with the help of an N-terminal signal peptide directing the protein to the ER.

Leaderless secretory proteins are modified in response to stress, thereafter enabling its interaction with relevant secretory pathways and subsequently resulting in its movement across the membrane (Denny et al. 2000; Backhaus et al. 2004). Multivesicular bodies (MBVs) in plants are prevacuolar compartments (Tse et al. 2004; Miao et al. 2008) and normally considered as endosomes of plants (Lam et al. 2007; Otegui and Spitzer 2008; Wang et al. 2009; Niemes et al. 2010; Robinson et al. 2012). They have been reported in the cytoplasm underlying the invasion papillae surrounding the fungal haustorium. The paramural bodies or lomasome is frequently observed at these sites and considered as the fusion profiles of MVBs with the plasma membrane. Callose is known to accumulate in the papillae and in the multivesicular bodies transported through endocytosis (An et al. 2006; Xu and Mendgen 1994).

Ding et al. (2012) described the three possible pathways for the leaderless secretory proteins (LSP) or nonclassical secretion of proteins.

The first LSP pathway is based on the fusion of multivesicular bodies with the plasma membrane to release the intraluminal vesicles to the apoplast, as exosomes. The release of exosomes depends on the behaviour of cytoplasmic domains of the two plasma membrane-localised SNAREs (syntaxin PEN1 and SNAP33), as their integration into the membrane of early endosome or trans-Golgi network in plants has been shown

(Lam et al. 2007, 2008; Robinson et al. 2008, 2012; Meyer et al. 2009; Bednarek et al. 2010; Wang et al. 2010b). After maturation into the multivesicular bodies, these SNAREs are on the intraluminal vesicles within the multivesicular body (Robinson et al. 2012; Scheuring et al. 2011). These exosomal intraluminal vesicles accumulate SNAREs in the matrix of the papilla after fusion of MBVs to plasma membrane (Fig. 5).

The second LSP pathway is based on the vacuolar fusion to plasma membrane. This was established by the pathogen-induced localised apoptosis at the site pathogen invasion. The localised apoptosis was due to the fusion of vacuole with the plasma membrane and the releasing of hydrolytic vacuolar enzymes with caspase-3-like activity into the apoplast, resulting in the lysis of bacterial and plant cells (Hatsugai and Hara-Nishimura 2010). It suggests that these vacuolar enzymes were originally delivered to the vacuolar lumen through conventional secretory organelles, but their secretion to apoplast is an unconventional secretion (Fig. 5).

The third LSP pathway is mediated by exocyst-positive organelles (EXPOs) discovered from

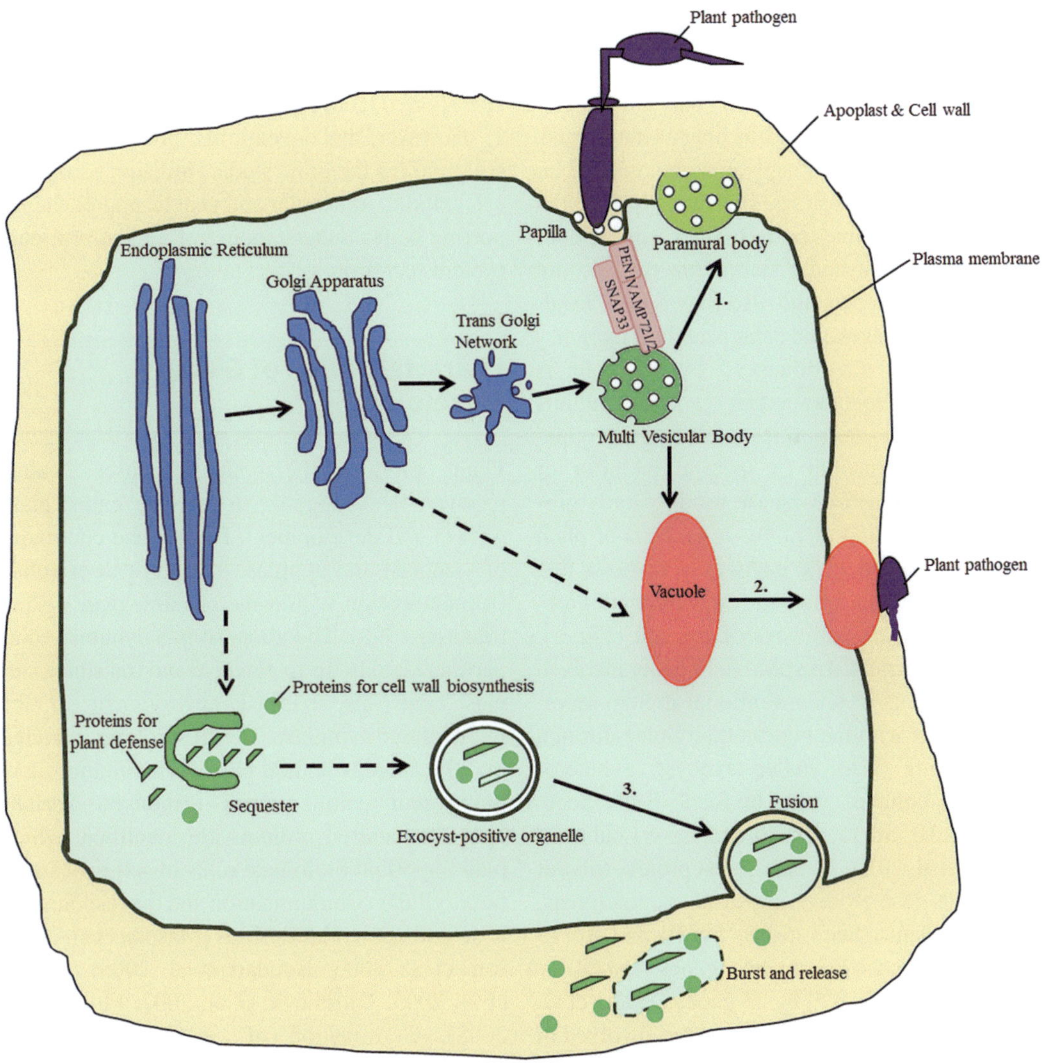

Fig. 5 Unconventional protein secretion pathways in plants (Adapted from Ding et al. 2012)

Table 1 Unconventional protein secretion pathways in plants (Ding et al. 2012)

Markers	Pathways	Regulators	Subcellular localisation of regulators
SAMS-2	Exocyst-positive organelle	AtExo70E2	Exocyst-positive organelle
Mannitol dehydrogenase	Golgi-independent	??	
PMR4	Exosome	PEN1, SNAP33	
GSL5		VAMP721/2	
Aleurain	Central vacuole	PBA1	
CPY			
Aspartyl protease			
Hygromycin phosphotransferase	Golgi-independent	Synaptogamin 2	Golgi

suspension culture of *Arabidopsis* and tobacco BY-2 cells (Wang et al. 2010a). These organelles have also been reported from root tips, root hair cells and pollen grains. EXPOs are double-membrane in the cytoplasm but are single-membrane vesicles outside the plasma membrane. EXPOs are like autophagosomes being double-membrane-bound vesicles. But they are different from autophagosomes because their number does not change much under starvation, they do not fuse with endosomes and also they do not localise with autophagosome using standard marker.

EXPOs are not influenced by Brefeldin A (BFA, a fungal metabolite that reversibly inhibits the anterograde transport from ER to Golgi apparatus) or wortmannin (a specific inhibitor of phosphatidylinositol 3-kinase used to study protein trafficking and identifies organelles of plant secretory and endocytic pathways). It shows that the EXPOs do not follow the conventional secretory or endocytic pathways of the cell (Fig. 5). All vesicle carriers, irrespective of being involved in conventional or unconventional protein secretion, interact with the plasma membrane through the tethering factor called exocyst, a hetero-octameric complex made of Sec3, Sec5, Sec6, Sec8, Sec10, Sec15, Exo70 and Exo84 subunits (Chong et al. 2010). Each exocyst protein subunit is a single-gene product in yeasts and mammals, while in plants, Sec3, Sec5, Sec10 and Sec15 subunits are encoded by two genes, Exo84 by three genes and Exo70 by 23 genes (Zhang et al. 2010). The tethering factor exocyst is involved in conventional secretory processes, self-incompatibility response and pathogen response (Samuel et al. 2009; Zhang et al. 2010; Pecenkova et al. 2011).

The markers, pathways and regulators of unconventional protein secretory pathways which are reported from plants have been listed in Table 1. However, unconventional protein secretion still requires the omic studies utilising biochemical, cellular, molecular and genetic approaches to portray better understanding of nonconventional protein secretion.

Characterisation of Global Secretome

Plants produce metabolic responses against received stress signals to regulate entire plant growth and development. Plants have continuity of symplast and apoplast that helps to establish communication within the physiological system (Sakurai 1998). The apoplast is a dynamic compartment and helps to perceive and transduce signals from the external environment to the intracellular symplast. Hence, proteins secreted into the apoplastic fluid play an important role in biotic and abiotic stress responses. Various apoplast-secreted proteins are identified, which play important biological roles in cell wall structure, cellular communication and the responses to host–pathogen relationships (Masuda et al. 2001; Rep et al. 2003; Boudart et al. 2005; Alvarez et al. 2006; Djordjevic et al. 2007; Floerl et al. 2008). Germination of barley seed showed α-amylase synthesis in the aleurone layer and its secretion into the endosperm to break down starch (Ranki and Sopanen 1984). Apoplastic secretome of apple, peach, pear and plum including xylem sap and leaf apoplast is known to have antioxidative system in response to pox virus (Diaz-Vivancos et al. 2006). In poplar, nearly 300 unique apoplastic proteins have been identified, among which ~144 are from leaf apoplast and ~135 are from stem apoplast (Pechanova et al. 2010), whereas ~97 were root apoplast protein (Dafoe and Constabel 2009). The leaf apoplast proteins have major roles in cell wall metabolism and stress/defence response, whereas root apoplast proteins have the major function of stress/defence with cell wall metabolism as the secondary function (Pechanova et al. 2010).

Detailed studies of secreted proteins under normal, biotic/abiotic stress conditions revealed several types of novel secreted proteins, including the leaderless secretory proteins. These leaderless secretory proteins account for more than 50 % of the total identified secretome from eukaryotes. Presently, about 24 terrestrial plant genomes have been completely sequenced, whereas many are under progress.

The analyses of the different components of cells have revealed that >80 % of the curated secreted proteins are present in the apoplast or exterior to the cell wall. Approximately 1,700 secreted proteins have been manually curated from more than 150 plant species in the UniProtKB/Swiss-Prot database. Their subcellular locations are yet to be verified experimentally. *Arabidopsis thaliana*, being the most extensively studied plant model system, has 941, while *Oryza sativa* (japonica) has 226 curated secreted proteins in the database (Table 2).

Gene ontology analyses based on molecular functions showed that ~40 % of the total plant-secreted proteins and ~50 % of *Arabidopsis-secreted* proteins show hydrolase activity, with almost one third showing binding activity and ~15 % showing the catalytic activity (Lum and Min 2011).

The functional genome annotation requires prediction of protein-coding sequences as well as their subcellular locations. The UniProt Consortium (2010) has a database of plant secretomes allowing better and efficient prediction and analyses of curated and annotated secreted proteins, thus ultimately leading to enhancement of database by accurate prediction of plant secretomes and thus enhancing understanding about the response or action of secreted protein to a variety of internal and external environments.

Table 2 Curated secreted proteins from different plants in UniProt/Swiss-Prot database (Lum and Min 2011)

Plant	No. of proteins
Arabidopsis thaliana	941
Oryza sativa subsp. Japonica	226
Solanum lycopersicum (*Lycopersicon esculentum*)	37
Nicotiana tabacum	28
Hordeum vulgare	27
Triticum aestivum	25
Zea mays	21
Oryza sativa subsp. Indica	16
Capsicum annuum	12
Betula verrucosa (*B. pendula*)	11
Cycas revoluta	10
Phaseolus vulgaris	10
Solanum tuberosum	10
Other species (153 species)	330
Total	**1,704**

Secretome Under Stresses

The plant apoplast research is lagging due to an obsolete concept of apoplast function and difficulties in the extraction and analysis of apoplastic proteins. Apoplast consisting compartments beyond the plasmalemma has a variety of functions during plant growth and development as well as in plant defence responses to stresses (Tian et al. 2009; Pennell 1998). During signal transduction, plant cells transport ligand like ions and other metabolites from the apoplast; therefore, a signal must cross the apoplast and plasmalemma (Sakurai 1998). Stress conditions significantly affect both quantity and quality of apoplastic proteins (Dani et al. 2005). Some stress conditions evidenced to alter the apoplast proteins in response to them include salt (Zhang et al. 2009), low temperature (Marentes et al. 1993), salicylic acid (Cheng et al. 2009a), metal toxicity (Fecht-Christoffers et al. 2003) and pathogen invasion (Oh et al. 2005). The roles of plant apoplastic proteins have been obviously ignored in analysing the plant stress response in comparison to the intracellular signalling pathway components and effectors.

Studies on *ex planta* (suspension cultured cells) and *in planta* systems identified a large number of leaderless secretory proteins in plants (Tran and Plaxton 2008; Cho et al. 2009; Agrawal et al. 2010), accounting for more than 50 % of total secretome, identified under biotic and abiotic stress conditions, exhibiting the existence of signal peptide-independent secretory mechanism.

Secretome Under Abiotic Stresses

Plants have evolved sophisticated systems to cope with adverse environmental conditions such as cold, drought and salinity. Although a number of stress response networks have been proposed, the role of plant apoplast protein stress response is less known.

The monocot model plant rice has salinity as one of the major environmental factors limiting growth and productivity. The rice root apoplastic proteins in response to salt stress have been deciphered (Zhang et al. 2009). The differential expression of rice secretome compared to untreated control revealed its role in response to salt stress and identified approximately 40 proteins, mainly involved in carbohydrate metabolism, oxidoreduction, protein processing and degradation (Song et al. 2011) (Fig. 6).

Low temperature stress decreases the productivity and limits the distribution of crop. Plants have different responses to freezing stress; some are freezing tolerant to withstand extracellular ice formation, and others prevent freezing by supercooling their sap. Apoplast proteome components prevent the lethal cell damage by ice-interacting proteins. The molecular mechanisms and components of the low temperature signalling in the apoplast are little known. The anti-

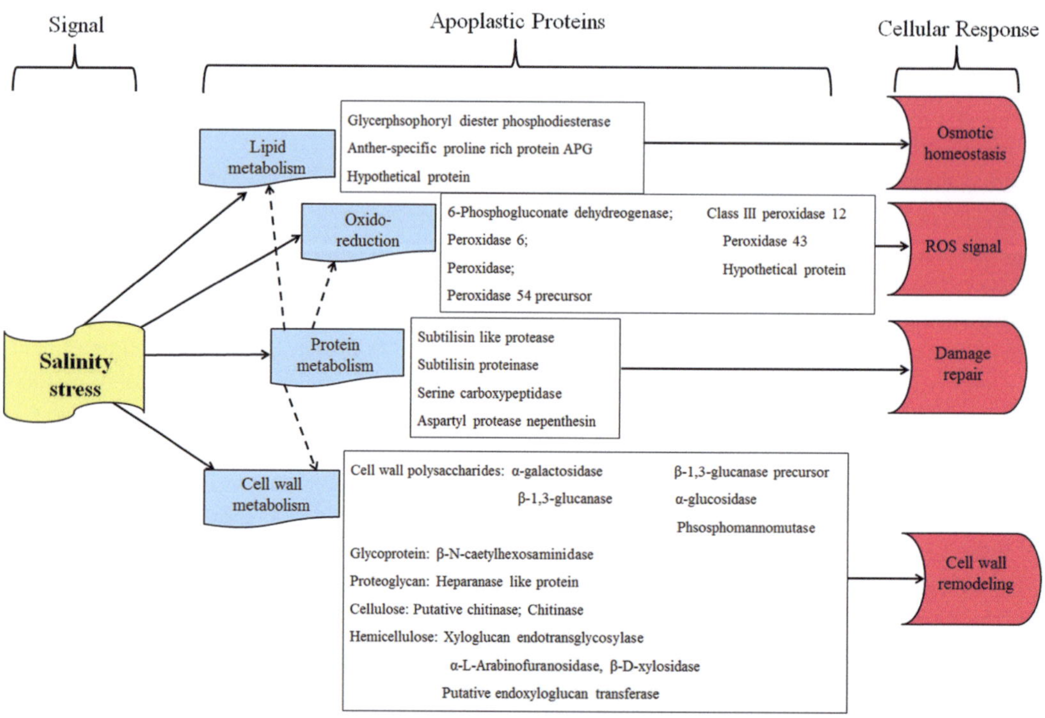

Fig. 6 Salinity stress-responsive apoplastic protein network in rice seedlings (Song et al. 2011)

freezing proteins in the apoplast bind to the ice crystals, thus inhibiting growth of ice crystal rather than ice formation in plants. Some of the plant antifreezing proteins are homologous to pathogenesis-related proteins (chitinase and glucanase) which have hydrolytic activity in addition to antifreezing protein activity (Hon et al. 1995; Yaish et al. 2006).

Apoplast acts as the mediator of cell communication with the environment and is altered by the freezing stress and the analyses of secretome give better understanding to the mechanism of freezing tolerance. Secretome of *Hippophae rhamnoides* (sea buckthorn) identified 60 low-temperature-responsive (LTR) proteins, of which 50 % were upregulated LTR proteins (Gupta and Deswal 2012). SignalP and SecretomeP analysis showed that 76 % of the proteins identified were apoplastic proteins, among which 24 % were following classical and 52 % following the nonclassical secretory pathway (Table 3). Also, the nonsecretory proteins identified were the non-resident apoplast proteins and might be imported in response to any stimulus like low temperature.

Phosphate is a macronutrient important for plant growth and metabolism. The excessive use of phosphate fertilisers results in phosphate deficiency in soil. Plants respond to phosphate deficiency by increased root growth, lateral roots to increase surface area of absorption and reduced shoot growth (Vance et al. 2003). Differentially expressed secretome of *Arabidopsis thaliana* suspension cell cultures under phosphate-sufficient and phosphate-deficient conditions was analysed by proteomic approach which identified 37 unique proteins (Tran and Plaxton 2008). Among them, 24 secreted proteins were phosphorus-deficiency-responsive proteins, while 18 of them were upregulated and six down-regulated secretory proteins (Table 4).

The mannitol dehydrogenase (MDH) is a cytoplasmic enzyme which is secreted in a Golgi-independent manner by tobacco in response to salicylic acid (Cheng et al. 2009b). The mannitol acts as a metabolite as well as an osmoprotectant due to its regulated conversion to mannose by MDH in the cytosol of plants like celery (Stoop et al. 1996). Thus, mannitol and MDH play an important role in plant–pathogen interactions. Golgi-independent secretion of MDH catabolises fungal mannitol in the extracellular space. The apoplastic peroxidases and high levels of leaderless secretory antioxidant protein Cu/Zn superox-

Table 3 Low-temperature stress-responsive secretome of *H. rhamnoides* (Gupta and Deswal 2012)

Functional category	Protein	Predicted secretory pathway
Redox regulation	Lactoylglutathione lyase or glyoxylase 1	Nonclassical
	Superoxide dismutase	Nonclassical
	Thioredoxin	Classical
	Putative lactoylglutathione lyase	Nonclassical
Defence/stress related	Osmotin-like protein	Classical
	Thaumatin-like protein	Classical
	Chitinase	Classical
	GDSL-motif lipase/hydrolase family protein	Nonclassical
	Desiccation-related protein	Classical
	Phenylalanine ammonia lyase	Nonclassical
	Late embryogenesis-like protein	Nonclassical
Signalling	Calmodulin 1	Nonclassical
	Calcium-dependent protein kinase 23	Nonclassical
	GTPase-activating protein	Nonclassical
Metabolism	Putative phosphomannomutase	Nonclassical
Regulation	Cysteine protease	Nonclassical
	Translation-inhibitor protein	Classical

Table 4 Functional characterisation of secretome from *Arabidopsis* cell suspension under phosphate-deficiency condition (Tran and Plaxton 2008)

Protein and putative function	Mode of secretion	Gene expression
Cell wall modifying		
Expansin-like protein	Classical	↓
β-Fructofuranosidase	Classical	**
Galactosyltransferase family protein	Classical	↓
Xyloglucan endo-1,4-β-D-glucanase	Classical	↓
Xyloglucan endotransglycosylase 6	Classical	**
Monocopper oxidase-like protein	Classical	**
Defence/detoxifying		
Glutathione transferase 8	Nonclassical	**
Dehydroascorbate reductase 1	Nonclassical	**
Mn superoxide dismutase 1	Nonclassical	↑
NADPH-dependent thioredoxin reductase 2	Classical	↑
Peroxidase	Classical	↓
Peroxidase 17	Classical	↑
Peroxidase 53	Classical	↑
Glycolysis		
Phosphoglycerate mutase	Nonclassical	**
Enolase	Nonclassical	**
N-Metabolism		
Amidase family protein	Classical	**
Glutamine synthetase	Nonclassical	**
Nucleases		
RNS1	Classical	↑
Proteases		
Leucine aminopeptidase 1	Nonclassical	**
Leucine aminopeptidase 3	Nonclassical	**
Serine carboxypeptidase 50	Classical	**

↑ = Upregulated, ↓ = downregulated, ** = unique proteins

ide dismutase in response to biotic or salicylic acid stress lead to oxidative burst (Bindschedler et al. 2006; Cheng et al. 2009a).

Arabidopsis secretome induced by 1 mM salicylic acid (SA) showed a number of secretory proteins into the apoplast through classical or nonclassical secretory pathway (Cheng et al. 2009a).

Poplar (*Populus spp*) plants growing in riverine ecosystems, characterised by rapid environmental changes, have evolved as multistress response in the apoplast. Apoplast secretome of poplar constitutes a potential adaptive mechanism to inhabit successfully in dynamic riverine ecosystem (Pechanova et al. 2010).

Secretome Under Biotic Stresses

The plants' cell wall acts as a barrier separating plant cells from the external environment. Therefore, proteins that are secreted in the extracellular space or apoplast play an important role in defence response. They reinforce the cell wall and antimicrobial activity via defence proteins such as chitinases, β-1, 3 glucanases, thionins, and defensins and lipid transfer proteins (Grant and Lamb 2006). These secreted proteins include various hydrolytic enzymes that are secreted in the apoplast as self-defence response to pathogen (bacteria, fungi and viruses) attack. Such pathogenesis-related proteins (PRPs) mediate cell-to-cell communication, and many of them follow the nonclassical secretory pathway

(Bowles 1990; Agrawal et al. 2010). Thus, plant secretome might play a key role in the early recognition and defence against pathogen attack.

There are three types of plant responses to pathogen invasion: (1) by sensing the presence of elicitors through the cell surface receptors, (2) by producing localised oxidative burst by release of reactive oxygen species or (3) by releasing different types of antimicrobial compounds. The elicitors arise from the cell wall of pathogens or fragments of the host cell wall released by pathogen activity such as chitin fragments and detection of the fungal attack by release of chitinases like PR3 (Verburg and Huynh 1991; Kaku et al. 2006).

Studies on plant secretome improves our insight of defence mechanism during plant–pathogen interactions. Identification of secreted proteins in *Arabidopsis* suspension-cultured cells (Ndimba et al. 2003; Oh et al. 2005), maize (Chivasa et al. 2005) and tobacco BY2 cell (Okushima et al. 2000) have been reported in response to fungal pathogens. The various pathogen elicitor-responsive proteins identified from plants include lectin receptor-like kinase, endochitinase, xyloglucan endo-1, 4-β-D-glucanases and peroxidise from cell suspension culture filtrates. Also, there are reports from whole-protein extracts that furthers our understanding on plant–pathogen interactions and defence signalling in wheat (Rampitsch et al. 2006), maize (Chivasa et al. 2005), pea (Curto et al. 2006), *Arabidopsis* (Jones et al. 2006) and rice (Ventelon-Debout et al. 2004; Lee et al. 2006).

Kim et al. (2009) identified 21 differentially expressed proteins in the secretome in response to rice blast fungus (*Magnaporthe grisea*) and its elicitor in rice suspension culture. These secreted proteins include chitinases, expansins and germins/oxalate oxidases. The secretory proteins expressed in elicitor-treated suspension cell culture were nearly similar to the *M. grisea*-infected rice leaves. It established the early recognition of pathogens via secreted proteins in resistant rice plants.

Pathogen attack in plants initiates a signalling cascade that leads to the synthesis of salicylic acid, which induces the expression or secretion of various PRPs that play a key role in systemic acquired resistance of plants.

In planta secretome analyses of *Phytophthora capsici*-infected pepper (*Capsicum annuum*) identified 75 secretory proteins (Yeom et al. 2011). Majority of the secreted proteins were defence- and stress-related proteins, proteases, protease inhibitors or cell wall structural proteins.

Water and pathogen stress-mediating PRPs in apoplast are effective in suppressing growth of *Melampsora* causing leaf fungal rust in poplar (Rinaldi et al. 2007). The leaf apoplast secretome showed the presence of proteins like acidic class III chitinase, thaumatin-like protein, blight-associated protein p12, cationic peroxidase 1 and cysteine-rich repeat secretory protein 38 in response to *Melampsora* infection. The acidic class III chitinase expression increased under pathogen challenge with *M. larici-populina* as well as under drought condition. It suggests a broad spectrum of role of apoplastic PRPs under stresses. A group of cysteine-rich peptides, defensins, are conserved in plants, and animals possess antimicrobial activity against a variety of fungus (Thomma et al. 2002). Some of them in plants are antibacterial or even few have a role in anti-insect activity. The defensins have different mechanisms for antifungal activity, such as (1) by interacting fungal cell wall components and causing localised apoptosis (Thevissen et al. 2012), (2) by binding to fungal ion channels to block it (Spelbrink et al. 2004) or (3) by modulating permeability of fungal plasma membrane leading to fungal cell death (Mello et al. 2011). It reveals that apoplast secretome-based defence mechanism works effectively against pathogens by activating pathogenesis-related proteins under biotic stress (Pechanova et al. 2010).

Secretome of Developmental Stages

Multicellular organisms have evolved an efficient system for cellular communication to ensure the ordered development, growth, maintenance and reproduction to successfully complete their life span. It requires cells to coordinate response to environmental stimuli as well as to each other by integrating the wide array of extracellular and intercellular signals. The cellular secretion in the apoplast by plant is an important biological process and serves as an interface between the environment and organism. Cell-to-cell communication during developmental stages predominantly involves secreted peptide ligands and interacts with their receptors present on the plasma membrane on the target cells. Apoplastic secretome contains proteins secreted through the classical ER–Golgi–TGN pathway or secreted by unconventional protein secretion mechanisms. The apoplast is not an empty space bordering a cell but rather participates in functions in plants (Lippmann et al. 2009) including nutrients and growth regulation, water regulation, osmoregulatory homeostasis of solutes, tissue structure, defence against biotic/abiotic stress, transport, osmosis, cell adhesion and gas exchange (Floerl et al. 2012).

The secretome of a tobacco cell suspension culture identified proteins mainly involved in stress defence and cell regeneration processes (Lippmann et al. 2009). Secretome analysis of chickpea (*Cicer arietinum*) callus suspension culture revealed 773 proteins in the extracellular medium (Gupta et al. 2011). Peroxidases, chitinases and other pathogenesis-related proteins were identified in cell cultures of tomato and grapevine after application of elicitors like cyclodextrins and methyl jasmonate (Briceno et al. 2012; Martinez-Esteso et al. 2009). Functional studies have revealed a multitude of secreted peptides involved in diverse biological processes (Table 5). Secreted peptides are categorised into two main groups distinguished on the basis of their biogenesis and overall structure: the CLEs (CLAVATA3/embryo-surrounding region, CLV3/ESR) including related peptide family and the CRPs (cysteine-rich peptides).

In conclusion, secreted peptides play a much more important role in cell–cell communication through relaying signals via ligand–receptor machinery of diverse biological contexts. These recent findings defy the traditional view of plant cells being unable to communicate by ligand–receptor interaction on the surface because of the surrounding cell wall. It is highly likely that better understanding of secretomics of plants growing in diverse environment will reveal more biological processes in which interaction of apoplastic fluid proteins with receptors presents on the plasma membrane of one cell with another cell.

Current Strategies to Study Plant Secretome

Plant cell secretes many proteins through exocytosis to the apoplastic fluid to maintain cell structure and regulate the external environment and as a part of signalling and defence mechanisms.

In recent years, there has been an increased interest in plant and microbe secretomes as the secreted proteins have shown to play an important role in normal growth, stress biology, infection and progression of diseases and subsequent response in plant protection (Agrawal et al. 2010; Stassen et al. 2012). Advancement in the proteomic profiling of plant secretome can be owed to the advancement in biochemical, proteomic techniques, mass spectroscopy and bioinformatics approaches. The complete characterisation of a proteome/secretome is a formidable task, and the degree of success achieved depends on the methods available and their amenability to automation and high-throughput formats. Parameters such as the complexity of the protein mixture, levels of expression and modification and intracellular localisation all impact the choice of proteomics technology to be used. It had established several in-depth secretome profiles from different cell types, apoplastic fluids from normal and diseased conditions at a faster pace. The biggest challenge in secretomic studies is the discrimination between the proteins that are truly secreted from those that are released as the result of non-physiological environmental stresses. Hence, strategies and techniques are being continuously modified to best adapt and to investigate the plant secretomes with high reproducibility. It is therefore important to establish a simple, reproducible

Table 5 Secreted apoplastic peptides and their biological functions in plants (Krause et al. 2013)

Peptide ligand	Peptide ligand	Plant species	Biological function
CLEs			
CLAVATA 3 (CLV3)	CLAVATA 1 (CLV1)	*Arabidopsis thaliana*	Regulation of shoot apical meristem activity
	CLAVATA 2 (CLV2)		
	CORYNE (CRN)		
	Receptor-like protein kinase 2		
	(RPK)/toadstool 2 (TOAD2)		
	Barely any meristem 1/2 (BAM1/2)		
CLAVATA3/ESR-related 1 (CLE 40)	ACR4	*Arabidopsis thaliana*	Regulation of root growth
CLE 14	CLAVATA 2 (CLV2)	*Arabidopsis thaliana*	Regulation of root apical meristem activity
	CORYNE (CRN)		
CLE20	CLAVATA 2 (CLV2)	*Arabidopsis thaliana*	Regulation of root apical meristem activity
	CORYNE (CRN)		
CLE 1		*Arabidopsis thaliana*	Regulation of root apical meristem activity
CLE 7		*Arabidopsis thaliana*	Regulation of shoot apical meristem activity
CLE 41	TDR/phloem intercalated with	*Arabidopsis thaliana*	Regulation of xylem differentiation
	Xylem (PXY)		
CLE 44	TDR/PXY	*Arabidopsis thaliana*	Regulation of xylem differentiation
CLE 42	TDR/PXY	*Arabidopsis thaliana*	Regulation of xylem differentiation
			Regulation of axillary bud formation
TDIF	TDR/PXY	*Zinnia elegans*	Regulation of xylem differentiation
CLE12	Supernumerous nodules (SUNN)	*Medicago truncatula*	Autoregulation during nodulation
CLE13	Supernumerous nodules (SUNN)	*Medicago truncatula*	Autoregulation during nodulation
Cysteine-rich peptide (CRP) (epidermal patterning factors (EPFs))			
(EPFL9)	Too many mouth (TMM)	*Arabidopsis thaliana*	Promotion of stomata development
(EPF2)	Too many mouth (TMM)	*Arabidopsis thaliana*	Inhibition of stomata differentiation
	ERECTA (ER)		
	ERECTA-like 1 (ERL1)		
(EPF1)	ERECTA-like 1 (ERL1)	*Arabidopsis thaliana*	Inhibition of stomata differentiation
(EPF6)/CHALLAH (CHAL)	Too many mouth (TMM)	*Arabidopsis thaliana*	Inhibition of stomata differentiation
	ERECTA (ER)		Regulation of inflorescence architecture
(EPF4)	ERECTA (ER)	*Arabidopsis thaliana*	Regulation of inflorescence architecture

(continued)

Table 5 (continued)

Peptide ligand	Peptide ligand	Plant species	Biological function
Cysteine-rich peptide (CRP) (RALFs)			
RsAFP2		*Raphanus sativus*	Antifungal function
PvD1		*Phaseolus vulgaris*	Antifungal function
ZmES4		*Zea mays*	Pollen tube bursting during fertilisation
Cysteine-rich peptide (CRP)			
Lat52	LePRK2	*Lycopersicon esculentum*	Promotion of pollen hydration and germination and pollen tube growth
LeSTIG1	LePRK2	*Lycopersicon esculentum, Nicotiana tabacum, Petunia hybrida*	Promotion of pollen tube growth in tomato
SLR1-BP	S-locus glycoprotein SLG-like receptor 1 (SLR 1)	*Brassica campestris*	Pollen adhesion on the pistil
LR2-B	S-locus glycoprotein SLG-like receptor 1 (SLR 1)	*Brassica campestris*	Pollen adhesion on the pistil
Cysteine-rich peptide (CRP) (LTP)			
Stigma cysteine-rich adhesion (SCA)		*Lilium longiflorum*	Formation of an adhesive layer in the transmitting tract
LTP5		*Arabidopsis thaliana*	Role in pollen tube growth and pistil function
			Effect on primary shoot growth, elongation of hypocotyls and inflorescence branching
Cysteine-rich peptide (CRP) (DEFL)			
SCR/SP11	SRK	*Brassica*	Determinant of self-incompatibility
LURE1/TfCRP1		*Torenia fournieri*	Pollen tube attraction
LURE2/TfCRP 3		*Torenia fournieri*	Pollen tube attraction
ZmES4		*Zea mays*	Role in pollen tube bursting during Fertilisation
Egg cell 1 (EC1) (DEFL)		*Arabidopsis thaliana*	Role in sperm–egg cell fusion
Other peptides			
Inflorescence deficient in abscission (IDA)	HAESA	*Arabidopsis thaliana*	Role in floral abscission
Phytosulfokine (PSK)	PSKR1, PSKR2	*Arabidopsis thaliana*	Regulation of growth size of roots and shoots
RNase			
S-RNase		*Solanaceae, Rosaceae, Scrophulariaceae*	Inhibition of pollen tube growth
Papaver rhoeas stigma (PrsS)	Papaver rhoeas pollen (PrpS)	*Papaver rhoeas*	Determinant of self-incompatibility

and economic procedure for systematic secretome analysis in plants.

Sample Preparation

The sample preparation of secreted proteins devoid of host-plant proteins is one of the most critical and challenging aspect in the secretome analyses. Most secretome studies to date are performed both on ex vivo suspension cell cultures (SCCs) and *in planta* systems (Fig. 7). *Ex planta* SCCs may not sufficiently complement to the *in planta* environment, thus reducing the correlation to the true physiological secretome.

Ex Planta System

Plant suspension cell cultures are widely used as a convenient tool for secretome analyses of the plant bypassing the structural complexity of the plant. The homogeneity of SCC cell population; the large availability of material; the easiness to maintain, handle and scale up/down; the high rate of cell growth; and the good reproducibility of conditions make it suitable for the analysis of plant secretome. Secreted proteins into the culture medium are used to prepare the secretome.

Different strategies and techniques have been applied to isolate pure secreted proteins suitable for proteomic analysis (Fig. 8).

Particulate free cell culture fluid containing pure secreted proteins can be obtained by filtra-

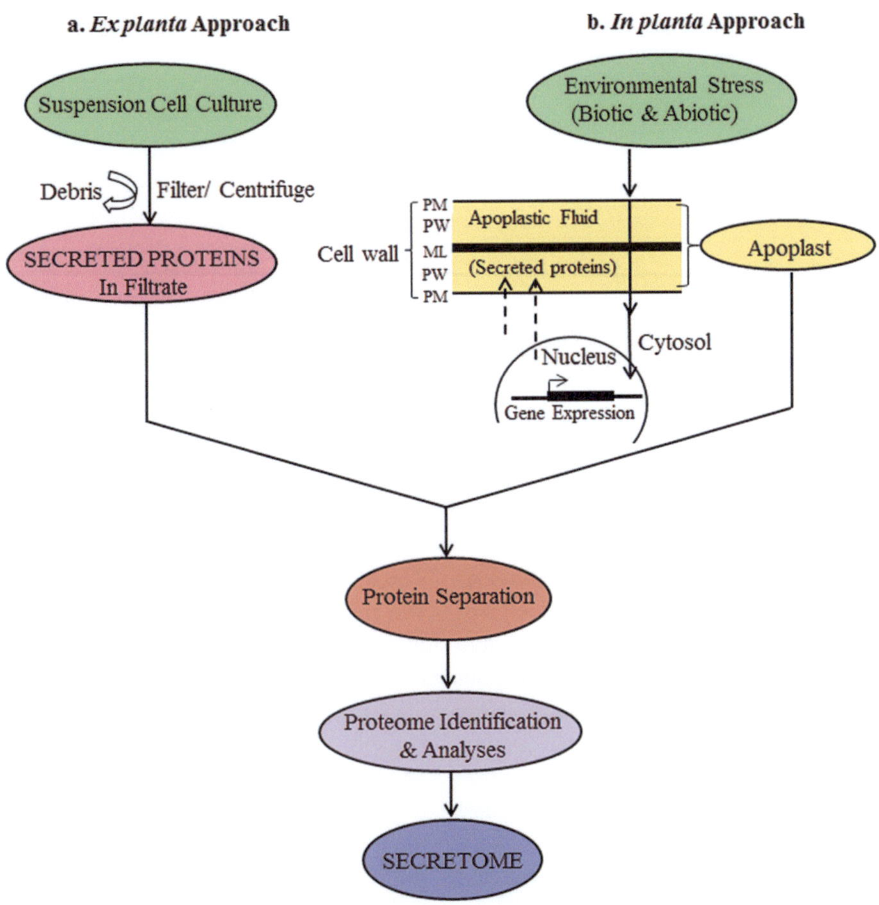

Fig. 7 A general overview of the *ex planta* and *in planta* systems for secretome analysis (Agrawal et al. 2010)

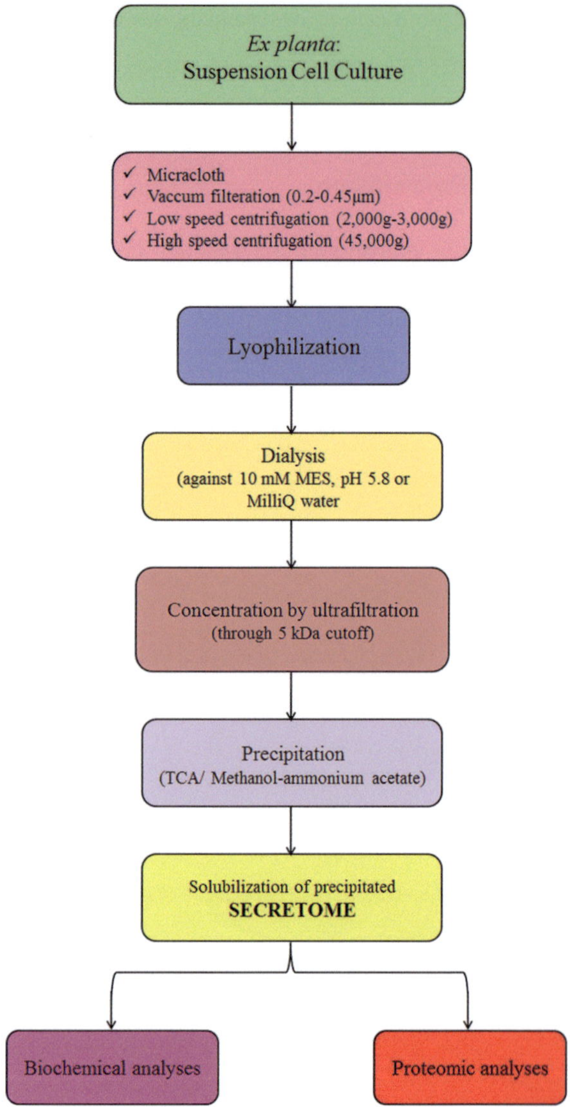

Fig. 8 Workflow showing preparation of secreted proteins using the *ex planta* SCC (Agrawal et al. 2010)

tion and/or high-speed centrifugation. Basically, filtration and centrifugation are a good combination to obtain clear cell culture fluid. Isolated secretome is either snap frozen immediately and stored at −80 °C or freeze-dried in a vacuum lyophiliser followed by dialysis against a suitable buffer. The lyophilised protein sample can be subjected to TCA precipitation to concentrate and effectively remove the salts, small peptides, water-soluble medium components, secondary metabolites and polysaccharides. The secreted protein pellet can be immediately processed for biochemical and proteomic analysis.

In Planta System

The apoplastic fluid between the middle lamella and primary wall is isolated using biochemical methods. The classical vacuum infiltration–centrifugation (VIC) method and the newly introduced gravity extraction method (GEM) are suitable and well-established methods for apoplastic secretome isolation (Fig. 9).

The vacuum infiltration–centrifugation method involves two critical steps: (1) vacuum infiltration with or without appropriate extraction buffer and (2) centrifugation speed and time. The suitability of the classical vacuum infiltration–

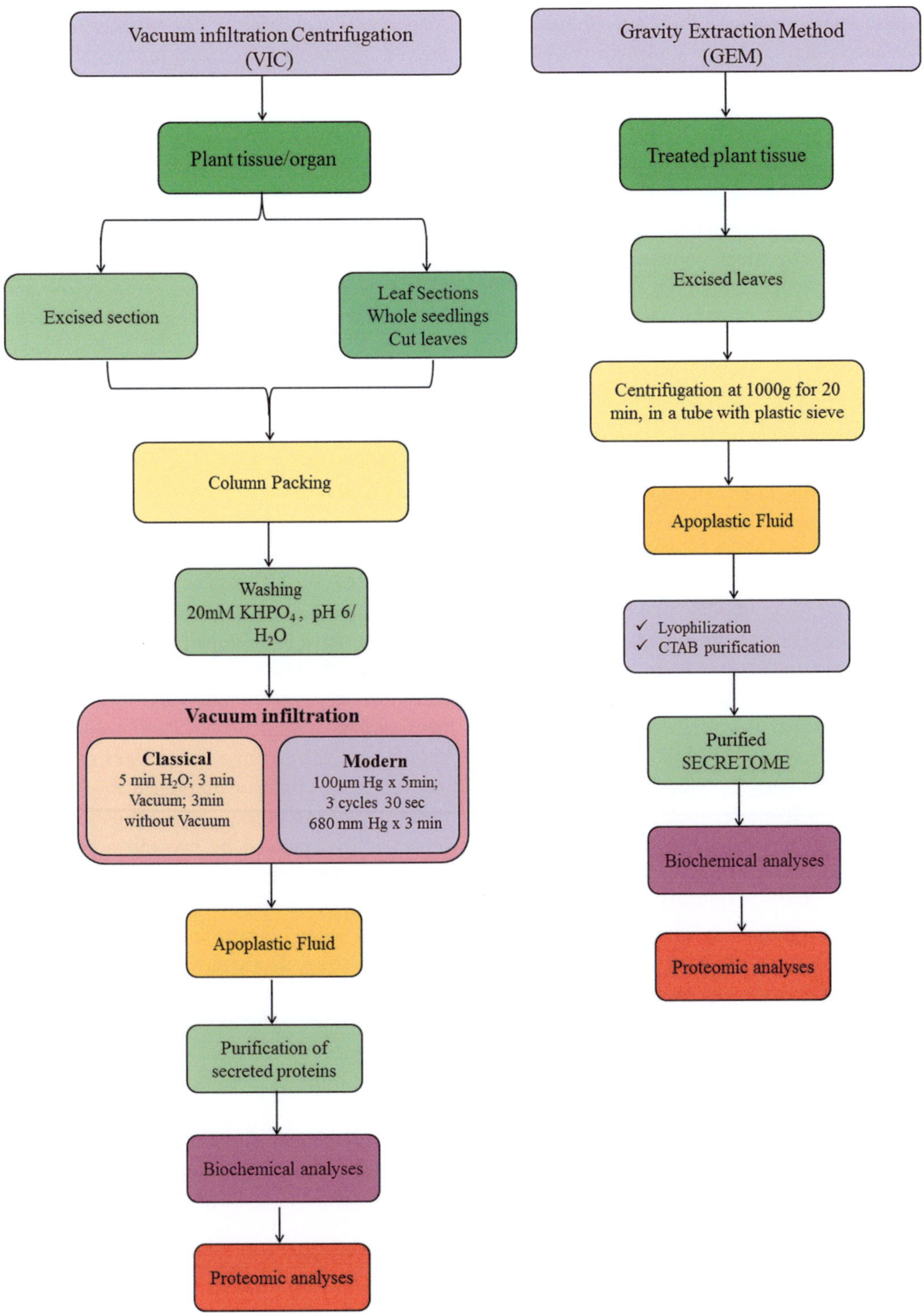

Fig. 9 Workflow for *in planta* preparation of secretome using VIC and GEM (Agrawal et al. 2010)

centrifugation method (Terry and Bonner 1980) for apoplastic secretome collection was further supported by critically evaluating it on intact leaves from different plant species – *Vicia faba* L., *Phaseolus vulgaris* L., *Pisum sativum* L., *Hordeum vulgare* L, *Spinacia oleracea* L., *Beta vulgaris* L. and *Zea mays* L. (Lohaus et al. 2001). Strength of infiltration buffer, incubation time and processing time showed relatively little impact on composition of the apoplastic fluid. In contrast, the pH of infiltrated solution highly influenced the concentration of sucrose and hexoses.

Separation of secretome from the culture medium or extracellular fluid can be easily done by filtration without cell disruption or by low-speed centrifugation. Moreover, the fraction of dead cells can be determined by staining the culture with trypane blue to identify any nonsecreted cytoplasmic proteins as being contaminants. *Ex planta* SCCs in model plants like tobacco, *Arabidopsis*, rice and *Medicago* have been used for secretome studies.

In gravity extraction method, the apoplastic fluid is obtained in a single step. It is a simple, reproducible and novel method for the extraction and preparation of pure secreted proteins (Cottingham 2008).

Two types of biochemical analyses are generally used to assess sample free from contamination of soluble cytoplasmic proteins before starting the proteomic analyses. The enzyme activity and western blotting of soluble cytoplasmic marker proteins, e.g. exclusively cytoplasmic enzymes like glucose 6-phosphate dehydrogenase (GAPDH) (Oh et al. 2005), phosphoenolpyruvate carboxylase (PEP-carboxylase) (Tran and Plaxton 2008) and cytosolic aldolase (Tran and Plaxton 2008), are assessed. The absence of the enzymatic activity and reference band on western blot confirms the purity of apoplastic fluid secretome preparation.

Protein Separation

Proteins are extremely diverse molecules and differ by mass, charge, hydrophobicity, tertiary shape and their affinity for other molecules. Several approaches are being employed for the separation of complex mixtures of protein. It is crucial to obtain a protein sample which contains only the molecule of interest. One-dimensional electrophoresis (SDS-PAGE) and 2D gel electrophoresis are the most popular gel-based protein separation methods. In 1DE, proteins are separated on the basis of molecular mass. Moreover, 1DE is simple to perform, is reproducible and can be used to resolve proteins with molecular masses of 10–300 kDa. However, limited resolving power of a 1DE is a limitation, if a more complex protein mixture, such as a crude cell lysate, is to be separated. Limitation of resolving power can be conquered by the use of 2-dimensional gel electrophoresis.

Two-dimensional (2D) gel electrophoresis is a powerful gel-based method commonly used for 'global' analyses of complex samples, i.e. when we are interested to characterise the entire spectrum of proteins in a sample. One of the greatest advantages of 2DE is the ability to resolve proteins that have undergone some form of post-translational modification. This resolution is possible in 2DE because many types of protein modifications confer a difference in charge as well as a change in mass on the protein. In 2D, proteins are separated by two distinct physical properties of protein. In the first dimension, proteins are resolved according to their net charge and in the second dimension, according to their molecular mass. The combination of these two methods produces resolution far exceeding that obtained in 1DE using a crude protein sample.

In this technique, the proteins are focused first in an immobilised pH gradient strip on the basis of their isoelectric point (acid vs. basic character). The isoelectric focusing gel is then placed over an SDS-PAGE gel and runs in the perpendicular dimension, and proteins are resolved on the basis of mass (Fig. 10). Thus, proteins separate not as bands but as spots, and the position of each protein spot depends on both the size and charge of protein.

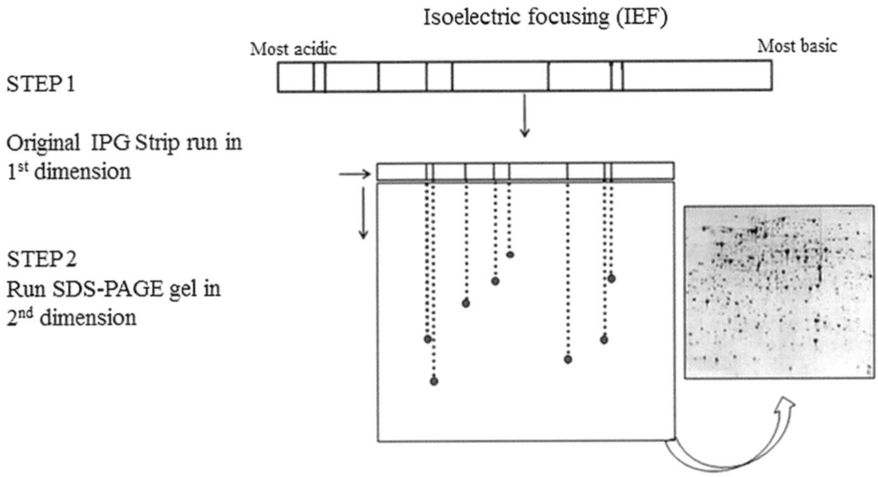

Fig. 10 Two-dimensional gel electrophoresis for secretome separation

Identification of Proteome Make-up of Secretome

After separation of secretome contents on 2D array, its proteome profile is identified by deducing the primary structure of each spot arrayed. Presently, mass spectrometry (MS), coupled with other preparative and analytical methods, is the main method in proteomics for achieving sensitivity and reproducibility. The major advantage of MS analysis is that it can process posttranslationally modified protein sample of any size required in a picomole quantities. Modification of N-terminus of protein cannot be sequenced using Edman's degradation method, and its sensitivity is up to 30 residues. Normally, the combination of 2D gel electrophoresis and MS is the most widely used method to study proteome. The protein spots are in-gel digested with proteolytic enzymes like pepsin, trypsin, chymotrypsin, papain, bromelain and subtilisin. After proteolysis, the peptides are separated on in-line LC-MS instrument. The data obtained from MS analysis of peptides can be taken directly for comparison to protein sequences derived from protein and nucleotide sequence databases.

Recently, a modified version called differential in-gel electrophoresis (DIGE) has improved performance at the gel-based part (Knowles et al. 2003; Marouga et al. 2005; Ye et al. 2010). Multidimensional protein identification technology (MudPIT) (Link et al. 1999) is a gel-free method to analyse the highly complex samples necessary for large-scale proteome analysis by ESI-MS/MS and database searching. As it is most frequently used, MudPIT couples a two-dimensional liquid chromatography (2D-LC) separation of peptides on a microcapillary column with detection in a tandem mass spectrometer. In the MudPIT experiment, a protein or mixture of proteins is reduced, alkylated and digested into a complex mixture of peptides. The digested peptide sample is pressure-loaded directly onto a microcapillary column where they are separated on the basis of their size and hydrophobicity. Once peptides are separated and eluted from the microcapillary column, they are ionised and enter the mass spectrometer, where they are isolated based on their mass-to-charge ratio (m/z). Tandem mass spectra are generated and are searched against a protein database. With the advent of proteomic techniques and mass spectrometry instrumentation, the efficiency of identifying and quantifying proteins in biological samples, including secretomes, has greatly improved.

Bioinformatic Analysis of Secretome Data from 2D and MS Array

High level of precision, sensitivity and resolution of MS have significantly increased high throughput of proteomics. Such high-precision instruments are capable of generating huge volume of high-quality data. Thus, comparison of secretome data obtained from different experiments/laboratories on specific cell/tissue types in defined conditions and specific diseases is warranted. For validation and extraction of significant outcomes has become feasible only by the development of multiple proteomic database and related software. A growing number of prediction tools for the plant secretome enable prediction of SPs, TMDs, GPI anchors or conserved domains in novel secretory proteins. Proteins secreted via the classical ER–Golgi–TGN pathway can be identified by their signal peptide using the SignalP 4.0 server (http://www.cbs.dtu.dk/services/SignalP) (Petersen et al. 2011), Phobius (http://phobius.binf.ku.dk/) (Kall et al. 2004) and TargetP (http://www.cbs.dtu.dk/services/TargetP) (Emanuelsson et al. 2007). The SecretomeP software helps to find LSPs by searching for certain LSP typical protein features apart from the lack of a signal sequence (Bendtsen et al. 2004). The recently developed bioinformatics tool LocTree2 uses a hierarchic, decision tree-like structure imitating the cellular protein sorting cascade to predict the subcellular localisation of proteins and thus also their secretion into the apoplast (Goldberg et al. 2012).

Currently there is an urgent need to deposit both raw mass spectrometry data and the corresponding list of identified proteins in public domains such as PRIDE (www.ebi.ac.uk/pride) and ProteomeXchange (www.proteomexchange.org) for proteomics-related studies.

Acknowledgements We are thankful to Dr. Javed Ahmed, Assistant Professor, King Saud University, and Dr. Abhishek Ojha, Postdoctoral Fellow, International Centre for Genetic Engineering and Biotechnology, New Delhi, for providing us the literatures for consultation and their timely and useful suggestions.

References

Agrawal GK, Jwa NS, Lebrun MH, Job D, Rakwal R (2010) Plant secretome: unlocking secrets of the secreted proteins. Proteomics 10(4):799–827

Alberts B, Johnson A, Lewis J, Raff M, Roberts K, Walter P (2008) Molecular biology of the cell. Garland Science, Taylor & Francis Group, New York

Alvarez S, Goodger JQD, Marsh EL, Chen S, Asirvatham VS, Schachtman DP (2006) Characterization of the maize xylem sap proteome. J Proteome Res 5:963–972

An Q, Hückelhoven R, Kogel KH, Van Bel AJE (2006) Multivesicular bodies participate in a cell wall associated defence response in barley leaves attacked by the pathogenic powdery mildew fungus. Cell Microbiol 8:1009–1019

Auron PE, Warner SJC, Webb AC, Cannon JG, Bernheim HA, McAdam KJPW, Rosenwasser LJ, LoPreste G, Mucci SF, Dinarello CA (1987) Studies on the molecular nature of human interleukin 1. J Immunol 138:1447–1456

Backhaus R, Zehe C, Wegehingel S, Kehlenbach A, Schwappach B, Nickel W (2004) Unconventional protein secretion: membrane translocation of FGF-2 does not require protein unfolding. J Cell Sci 117:1727–1736

Bednarek P, Kwon C, Schulze-Lefert P (2010) Not a peripheral issue: secretion in plant–microbe interactions. Curr Opin Plant Biol 13:378–387

Bendtsen JD, Jensen LJ, Blom N, Von Heijne G, Brunak S (2004) Feature-based prediction of non-classical and leaderless protein secretion. Protein Eng Des Sel 17:349–356

Bindschedler LV, Laurence V, Dewdney J, Blee KA, Stone JM, Asai T, Plotnikov J, Denoux C, Hayes T, Gerrish C, Davies DR, Ausubel FM, Bolwell GP (2006) Peroxidase-dependent apoplastic oxidative burst in Arabidopsis required for pathogen resistance. Plant J 47:851–863

Boudart G, Jamet E, Rossignol M, Lafitte C, Borderies G, Jauneau A, Esquerre-Tugaye M-T, Pont-Lezica R (2005) Cell wall proteins in apoplastic fluids of Arabidopsis thaliana rosettes: identification by mass spectrometry and bioinformatics. Proteomics 5:212–221

Bouws H, Wattenberg A, Zorn H (2008) Fungal secretomes-nature's toolbox for white biotechnology. Appl Microbiol Biotechnol 80:381–388

Bowles DJ (1990) Defence-related proteins in higher plants. Annu Rev Biochem 59:873–907

Briceno Z, Almagro L, Sabater-Jara AB, Calderon AA, Pedreno MA, Ferrer MA (2012) Enhancement of phytosterols, taraxasterol and induction of extracellular pathogenesis-related proteins in cell cultures of Solanum lycopersicum cv Micro-Tom elicited with cyclodextrins and methyl jasmonate. J Plant Physiol 169:1050–1058

Brodsky JL (1996) Post-translational protein translocation: not all hsc70s are created equal. Trends Biochem Sci 21:122–126

Brodsky JL, Schekman R (1993) A Sec63p-BiP complex from yeast is required for protein translocation in a reconstituted proteoliposome. J Cell Biol 123:1355–1363

Burgess TL, Kelly RB (1987) Constitutive and regulated secretion of proteins. Annu Rev Cell Biol 3:243–293

Cai H, Reinisch K, Ferro-Novick S (2007) Coats, tethers, Rabs, and SNAREs work together to mediate the intracellular destination of a transport vesicle. Dev Cell 12:671–682

Cai Y, Jia T, Lam SK, Ding Y, Gao C, San MWY, Pimpl P, Jiang L (2011) Multiple cytosolic and transmembrane determinants are required for the trafficking of SCAMP1 via an ER-Golgi-TGN-PM pathway. Plant J 65:882–896

Cai Y, Zhuang X, Wang J, Wang H, Lam SK, Gao C, Wang X, Jiang L (2012) Vacuolar degradation of two integral plasma membrane proteins, AtLRR84A and OsSCAMP1, is cargo ubiquitination-independent and prevacuolar compartment mediated in plant cells. Traffic 13:1023–1040

Cheng FY, Blackburn K, Lin YM, Goshe MB, Williamson JD (2009a) Absolute protein quantification by LC/MS(E) for global analysis of salicylic acid-induced plant protein secretion responses. J Proteome Res 8:82–93

Cheng FY, Zamski E, Guo W, Pharr DM, Williamson JD (2009b) Salicylic acid stimulates secretion of the normally symplastic enzyme mannitol dehydrogenase (MTD): a possible defense against mannitol secreting fungal pathogens. Planta 230:1093–1103

Chirico WJ, Waters MG, Blobel G (1988) 70 K heat shock related proteins stimulate protein translocation into microsomes. Nature 332:805–810

Chivasa S, Simon WJ, Yu XL, Yalpani N, Slabas AR (2005) Pathogen elicitor–induced changes in the maize extracellular matrix proteome. Proteomics 5:4894–4904

Cho WK, Chen XY, Chu H, Rim Y, Kim S, Kim ST, Kim SW, Park ZY, Kim JY (2009) The proteomic analysis of the secretome of rice calli. Plant Physiol 135:331–341

Chong YT, Gidda SK, Sanford C, Parkinson J, Mullen RT, Goring DR (2010) Characterization of the Arabidopsis thaliana exocyst complex gene families by phylogenetic, expression profiling, and subcellular localization studies. New Phytol 185:401–419

Corsi AK, Schekman R (1996) Mechanism of polypeptide translocation into the endoplasmic reticulum. J Biol Chem 271:30299–30302

Cottingham K (2008) Unlocking the secrets of the rice secretome. J Proteome Res 7:5072

Curto M, Camafeita E, Lopez JA, Maldonado AM, Rubiales D, Jorrin JV (2006) A proteomic approach to study pea (*Pisum sativum*) responses to powdery mildew (*Erysiphe pisi*). Proteomics 6:163–174

Dafoe NJ, Constabel CP (2009) Proteomic analysis of hybrid poplar xylem sap. Phytochemistry 70:856–863

Dani V, Simon WJ, Duranti M, Croy RR (2005) Changes in the tobacco leaf apoplast proteome in response to salt stress. Proteomics 5:737–745

Denny PW, Gokool S, Russell DG, Field MC, Smith DF (2000) Acylation-dependent protein export in Leishmania. J Biol Chem 275:11017–11025

Diaz-Vivancos P, Rubio M, Mesonero V, Periago PM, Ros-Barcelo A, Martinez-Gomez P, Hernandez JA (2006) The apoplastic antioxidant system in Prunus: response to long-term plum pox virus infection. J Exp Bot 57:3813–3824

Ding Y, Wang J, Wang J, Stierhof YD, Robinson DG, Jiang L (2012) Unconventional protein secretion. Trends Plant Sci 17(10):606–615

Djordjevic MA, Oakes M, Li DX, Hwang CH, Hocart CH, Gresshoff PM (2007) The Glycine max xylem sap and apoplast proteome. J Proteome Res 6:3771–3779

Emanuelsson O, Brunak S, von Heijne G, Nielsen H (2007) Locating proteins in the cell using TargetP, SignalP and related tools. Nat Protoc 2(4):953–971

Fecht-Christoffers MM, Braun HP, Lemaitre-Guillier C, VanDorsselaer A, Horst WJ (2003) Effect of manganese toxicity on the proteome of the leaf apoplast in cowpea. Plant Physiol 133:1935–1946

Floerl S, Druebert C, Majcherczyk A, Karlovsky P, Kues U, Polle A (2008) Defence reactions in the apoplastic proteome of oilseed rape (Brassica napus var. napus) attenuate Verticillium longisporum growth but not disease symptoms. BMC Plant Biol 8:129. doi:10.1186/1471-2229-8-129

Floerl S, Majcherczyk A, Possienke M, Feussner K, Tappe H, Gatz C, Feussner I, Kues U, Polle A (2012) Verticillium longisporum infection affects the leaf apoplastic proteome, metabolome, and cell wall properties in Arabidopsis thaliana. PLoS One 7:e31435

Gabe Lee MT, Mishra A, Lambright DG (2009) Structural mechanisms for regulation of membrane traffic by Rab GTPases. Traffic 10:1377–1389

Gao C, Yu CK, Qu S, San MW, Li KY, Lo SW, Jiang L (2012) The Golgi-localized Arabidopsis endomembrane protein12 contains both endoplasmic reticulum export and Golgi retention signals at its C terminus. Plant Cell 24:2086–2104

Glick BS (1995) Can Hsp70 proteins act as force-generating motors? Cell 80:11–14

Goldberg T, Hamp T, Rost B (2012) LocTree2 predicts localization for all domains of life. Bioinformatics 28:i458–i465

Grant M, Lamb C (2006) Systemic immunity. Curr Opin Plant Biol 9:414–420

Greenbaum D, Luscombe NM, Jansen R, Qian J, Gerstein M (2001) Interrelating different types of genomic data, from proteome to secretome: 'oming in on function. Genome Res 11:1463–1468

Gupta R, Deswal R (2012) Low temperature stress modulated secretome analysis and purification of antifreeze protein from Hippophae rhamnoides, a Himalayan

wonder plant. J Proteome Res 11:2684–2696. dx.doi.org/10.1021/pr200944z
Gupta S, Wardhan V, Verma S, Gayali S, Rajamani U, Datta A, Chakraborty S, Chakraborty N (2011) Characterization of the secretome of chickpea suspension culture reveals pathway abundance and the expected and unexpected secreted proteins. J Proteome Res 10:5006–5015
Hathout Y (2007) Approaches to the study of the cell secretome. Expert Rev Proteomics 4:239–248
Hatsugai N, Hara-Nishimura I (2010) Two vacuole-mediated defence strategies in plants. Plant Signal Behav 5:1568–1570
Hon WC, Griffith M, Mlynarz A, Kwok YC, Yang DSC (1995) Antifreeze proteins in winter rye are similar to pathogenesis related proteins. Plant Physiol 109:878–889
Hurley JH, Hanson PI (2010) Membrane budding and scission by the ESCRT machinery: it's all in the neck. Nat Rev Mol Cell Biol 11:556–566
Isaacson T, Rose JKC (2006) The plant cell wall proteome, or secretome. In: Finnie C (ed) Plant proteomics, vol 28, Annual Plant Reviews Series. Blackwell Publishing, Oxford, pp 185–209
Jones AM, Thomas V, Bennett MH, Mansfield J, Grant M (2006) Modifications to the Arabidopsis defense proteome occur prior to significant transcriptional change in response to inoculation with *Pseudomonas syringae*. Plant Physiol 142:1603–1620
Jung C, Lyou SH, Yeu SY, Kim MA, Lee JS, Choi YD, Cheong JJ (2007) Microarray-based screening of jasmonate responsive genes in Arabidopsis thaliana. Plant Cell Rep 26:1053–1063
Jurgens G, Geldner N (2007) The high road and the low road: trafficking choices in plants. Cell 130:977–979
Kaku H, Nishizawa Y, Ishii-Minami N, Akimoto-Tomiyama C, Dohmae N, Takio K, Minami E, Shibuya N (2006) Plant cells recognize chitin fragments for defense signaling through a plasma membrane receptor. Proc Natl Acad Sci U S A 103:11086–11091
Kall L, Krogh A, Sonnhammer EL (2004) A combined transmembrane topology and signal peptide prediction method. J Mol Biol 338(5):1027–1036
Kamoun S (2009) The secretome of plant-associated fungi and oomycetes. In: Deising VH (ed) Plant relationships, 2nd edn, The Mycota. Springer, Berlin/Heidelberg, pp 173–180
Kim ST, Kang YH, Wang Y, Wu J, Park ZY, Rakwal R, Agrawal GK, Lee SY, Kang KY (2009) Secretome analysis of differentially induced proteins in rice suspension-cultured cells triggered by rice blast fungus and elicitor. Proteomics 9:1302–1313
Kirchhausen T (2000) Three ways to make a vesicle. Nat Rev Mol Cell Biol 1:187–198
Knowles MR, Cervino S, Skynner HA, Hunt SP, de Felipe C, Salim K, Meneses-Lorente G, McAllister G, Guest PC (2003) Multiplex proteomic analysis by two-dimensional differential in-gel electrophoresis. Proteomics 3:1162–1171
Krause C, Richter S, Knöll C, Jürgens G (2013) Plant secretome – from cellular process to biological activity. Biochim et Biophys Acta http://dx.doi.org/10.1016/j.bbapap.2013.03.024
Lam SK, Tse YC, Robinson DG, Jiang L (2007) Tracking down the elusive early endosome. Trends Plant Sci 12:497–505
Lam SK, Cai Y, Hillmer S, Robinson DR, Jiang L (2008) SCAMPs highlight the developing cell plate during cytokinesis in tobacco BY-2 cells. Plant Physiol 147:1637–1645
Lee J, Bricker TM, Lefevre M, Pinson SRM, Oard JH (2006) Proteomic and genetic approaches to identifying defence-related proteins in rice challenged with the fungal pathogen *Rhizoctonia solani*. Mol Plant Pathol 7:405–416
Link AJ, Eng J, Schieltz DM, Carmack E, Mize GJ, Morris DR, Garvik BM, Yates JR III (1999) Direct analysis of protein complexes using mass spectrometry. Nat Biotechnol 17:676–682
Lippmann R, Kaspar S, Rutten T, Melzer M, Kumlehn MA, Mock HP (2009) Protein and metabolite analysis reveals permanent induction of stress defense and cell regeneration processes in a tobacco cell suspension culture. Int J Mol Sci 10:3012–3032
Lohaus G, Pennewiss K, Sattelmacher B, Hussmann M, Mühling KH (2001) Is the infiltration-centrifugation technique appropriate for the isolation of apoplastic fluid? A critical evaluation with different plant species. Physiol Plant 111:457–465
Lum G, Min XJ (2011) Plant secretomics: current status and future perspectives. Plant Omics J 4(2):114–119
Lutcke H (1995) Signal recognition particle (SRP), a ubiquitous initiator of protein translocation. Eur J Biochem 228:531–550
Marentes E, Griffith M, Mlynarz A, Brush RA (1993) Proteins accumulate in the apoplast of winter rye leaves during cold acclimation. Physiol Plant 87:499–507
Marouga R, David S, Hawkins E (2005) The development of the DIGE system: 2D fluorescence difference gel analysis technology. Anal Bioanal Chem 382:669–678
Marti L, Fornaciari S, Renna L, Stefano G, Brandizzi F (2010) COPII-mediated traffic in plants. Trends Plant Sci 15:522–528
Martinez-Esteso MJ, Selles-Marchart S, Vera-Urbina JC, Pedreno MA, Bru-Martinez R (2009) Changes of defense proteins in the extracellular proteome of grapevine (Vitis vinifera cv. Gamay) cell cultures in response to elicitors. J Proteomics 73:331–341
Masuda S, Kamada H, Satoh S (2001) Chitinase in cucumber xylem sap. Biosci Biotechnol Biochem 65:1883–1885
Mello EO, Ribeiro SF, Carvalho AO, Santos IS, Da Cunha D, Santa-Catarina C, Gomes VM (2011) Antifungal activity of PvD1 defensin involves plasma membrane permeabilization, inhibition of medium acidification,

and induction of ROS in fungi cells. Curr Microbiol 62:1209–1211

Meyer D, Paionk S, Micali C, O'Connell R, Schulze-Lefert P (2009) Extracellular transport and integration of plant secretory proteins into pathogen-induced cell wall compartments. Plant J 57:986–999

Miao Y, Li KY, Yao X, Jiang L (2008) The vacuolar transport of aleurain-GFP and 2S albumin-GFP fusions is mediated by the same pre-vacuolar compartments in tobacco BY-2 and Arabidopsis suspension cultured cells. Plant J 56:824–839

Miller JD, Wilhelm H, Gierasch L, Gilmore R, Walter P (1993) GTP binding and hydrolysis by the signal recognition particle during initiation of protein translocation. Nature 366:351–354

Ndimba BK, Chivasa S, Hamilton JM, Simon WJ, Slabas AR (2003) Proteomic analysis of changes in the extracellular matrix of *Arabidopsis* cell suspension cultures induced by fungal elicitors. Proteomics 3:1047–1059

Nickel W, Rabouille C (2009) Mechanisms of regulated unconventional protein secretion. Nat Rev Mol Cell Biol 10:148–155

Nickel W, Briigger B, Wieland FT (2002) Vesicular transport: the core machinery of COPI recruitment and budding. J Cell Sci 115:3235–3240

Niemes S, Langhans M, Viotti C, Scheuring D, Yan MSY, Jiang L, Hilmer S, Robinson DG, Pimpl P (2010) Retromer recycles vacuolar sorting receptors from the trans-Golgi network. Plant J 61:107–121

Oh IS, Park AR, Bae MS, Kwon SJ, Kim YS, Lee JE, Kang NY, Lee S, Cheong H, Park OK (2005) Secretome analysis reveals an Arabidopsis lipase involved in defense against *Alternaria brassicicola*. Plant Cell 17:2832–2847

Okushima Y, Koizumi N, Kusano T, Sano H (2000) Secreted proteins of tobacco cultured BY2 cells: identification of a new member of pathogenesis-related proteins. Plant Mol Biol 42:479–488

Otegui MS, Spitzer C (2008) Endosomal functions in plants. Traffic 9:1589–1598

Panzner S, Dreier L, Hartmann E, Kostka S, Rapoport TA (1995) Post-translational protein transport in yeast reconstituted with a purified complex of Sec proteins and Kar2p. Cell 81:561–570

Pecenkova T, Hála M, Kulich I, Kocourková D, Drdová E, Fendrych M, Toupalová H, Žárský V (2011) The role for the exocyst complex subunits Exo70B2 and Exo70H1 in the plant–pathogen interaction. J Exp Bot 62:2107–2116

Pechanova O, Hsu CY, Adams JP, Pechan T, Vandervelde L, Drnevich J, Jawdy S, Adeli A, Suttle JC, Lawrence AM, Tschaplinski TJ, Séguin A, Yuceer C (2010) Apoplast proteome reveals that extracellular matrix contributes to multistress response in poplar. BMC Genomics 11:674. doi:10.1186/1471-2164-11-674

Pennell R (1998) Cell walls: structures and signals. Curr Opin Plant Biol 1:504–510

Petersen TN, Brunak S, von Heijne G, Nielsen H (2011) SignalP 4.0: discriminating signal peptides from transmembrane regions. Nat Methods 8:785–786

Rampitsch C, Bykova NV, McCallum B, Beimcik E, Ens W (2006) Analysis of the wheat and *Puccinia triticina* (leaf rust) proteomes during a susceptible host-pathogen interaction. Proteomics 6:1897–1907

Ranki H, Sopanen T (1984) Secretion of α-amylase by the aleurone layer and the scutellum of germinating barley grain. Plant Physiol 75:710–715

Rapoport TA, Matlack KE, Plath K, Misselwitz B, Staeck O (1999) Post-translational protein translocation across the membrane of the endoplasmic reticulum. Biol Chem 380:1143–1150

Rep M, Dekker HL, Vossen JH, de Boer AD, Houterman PM, de Koster CG, Cornelissen BJC (2003) A tomato xylem sap protein represents a new family of small cysteine-rich proteins with structural similarity to lipid transfer proteins. FEBS Lett 534:82–86

Rinaldi C, Kohler A, Frey P, Duchaussoy F, Ningre N, Couloux A, Wincker P, Thiec DL, Fluch S, Martin F, Duplessis S (2007) Transcript profiling of poplar leaves upon infection with compatible and incompatible strains of the foliar rust Melampsora larici-populina. Plant Physiol 144:347–366

Robinson DG, Langhans M, Saint-Jore-Dupas C, Hawes C (2008) BFA effects are tissue and not just plant specific. Trends Plant Sci 13:405–408

Robinson DG, Pimpl P, Scheuring D, Stierhof YD, Sturm S, Viotti C (2012) Trying to make sense of retromer. Trends Plant Sci 17:431–439

Sakurai N (1998) Dynamic function and regulation of apoplast in the plant body. J Plant Res 111:133–148

Samuel MA, Chong YT, Haasen KE, Aldea-Brydges MG, Stone SL, Goring DR (2009) Cellular pathways regulating responses to compatible and self-Incompatible pollen in Brassica and Arabidopsis stigmas intersect at Exo70A1, a putative component of the exocyst complex. Plant Cell 21:2655–2671

Scheuring D, Viotti C, Krüger F, Künzl F, Sturm S, Bubeck J, Hillmer S, Frigerio L, Robinson DG, Pimpl P, Schumacher K (2011) Multivesicular bodies mature from the trans-Golgi network/early endosome in Arabidopsis. Plant Cell 23:3463–3481

Song Y, Zhang C, Ge W, Zhang Y, Burlingame AL, Guo Y (2011) Identification of NaCl stress-responsive apoplastic proteins in rice shoot stems by 2D-DIGE. J Proteomics 74(7):1045–1067

Spelbrink RG, Dilmac N, Allen A, Smith TJ, Shah DM, Hockerman GH (2004) Differential antifungal and calcium channel-blocking activity among structurally related plant defensins. Plant Physiol 135:2055–2067

Stassen JH, Seidl MF, Vergeer PW, Nijman IJ, Snel B, Cuppen E, Van den Ackervenken G (2012) Effector identification in the lettuce downy mildew Bremia lactucae by massively parallel transcriptome sequencing. Mol Plant Pathol 13(7):719–731

Stoop JMH, Williamson JD, Pharr DM (1996) Mannitol metabolism in plants: a method for coping with stress. Trends Plant Sci 1:139–144

Sztul E, Lupashin V (2009) Role of vesicle tethering factors in the ER–Golgi membrane traffic. FEBS Lett 583:3770–3783

Terry M, Bonner B (1980) An examination of centrifugation as a method of extracting an extracellular solution from peas, and its use for the study of indoleacetic acid-induced growth. Plant Physiol 66:321–325

The UniProt Consortium (2010) The universal protein resource (UniProt) in 2010. Nucleic Acids Res 38:D142–D148. doi:10.1093/nar/gkp846

Thevissen K, de Mello Tavares P, Xu D, Blankenship J, Vandenbosch D, Idkowiak-Baldys J, Govaert G, Bink A, Rozental S, de Groot PW, Davis TR, Kumamoto CA, Vargas G, Nimrichter L, Coenye T, Mitchell A, Roemer T, Hannun YA, Cammue BP (2012) The plant defensin RsAFP2 induces cell wall stress, septin mislocalization and accumulation of ceramides in *Candida albicans*. Mol Microbiol 84:166–180

Thomma BP, Cammue BP, Thevissen K (2002) Plant defensins. Planta 216:193–202

Tian L, Zhang L, Zhang J, Song Y, Guo Y (2009) Differential proteomic analysis of soluble extracellular proteins reveals the cysteine protease and cystatin involved in suspension-cultured cell proliferation in rice. Biochim Biophys Acta 1794:459–467

Tjalsma H, Bolhuis A, Jongbloed JDH, Bron S, van Dijl JM (2000) Signal peptide-dependent protein transport in *Bacillus subtilis*: a genome-based survey of the secretome. Microbiol Mol Biol Rev 64:515–547

Tran HT, Plaxton WC (2008) Proteomic analysis of alterations in the secretome of Arabidopsis thaliana suspension cells subjected to nutritional phosphate deficiency. Proteomics 8:4317–4326. doi:10.1002/pmic.200800292

Tse YC, Mo B, Hilmer S, Zhao M, Lo SW, Robinson DG, Jiang L (2004) Identification of multivesicular bodies as prevacuolar compartments in Nicotiana tabacum BY-2 Cells. Plant Cell 16:672–693

Vance CP, Uhde-Stone C, Allan DL (2003) Phosphorus acquisition and use: critical adaptations by plants for securing a nonrenewable resource. New Phytol 157:427–447

Ventelon-Debout M, Delalande F, Brizard JP, Diemer H, Van Dorselaer A, Brugiduo C (2004) Proteome analysis of cultivar-specific deregulations of *Oryza sativa* indica and *O. sativa* japonica cellular suspensions undergoing rice yellow mottle virus infection. Proteomics 4:216–225

Verburg JG, Huynh QK (1991) Purification and characterization of an antifungal chitinase from Arabidopsis thaliana. Plant Physiol 95:450–455

Walter P, Johnson AE (1994) Signal sequence recognition and protein targeting to the endoplasmic reticulum membrane. Annu Rev Cell Biol 10:87–119

Wang J, Cai Y, Miao Y, Lam SK, Jiang L (2009) Wortmannin induces homotypic fusion of plant prevacuolar compartments. J Exp Bot 60:3075–3083

Wang J, Ding Y, Wang J, Hillmer S, Miao Y, Lo SW, Wang X, Robinson DG, Jiang L (2010a) EXPO, an exocyst-positive organelle distinct from multivesicular endosomes and autophagosomes, mediates cytosol to cell wall exocytosis in *Arabidopsis* and tobacco cells. Plant Cell 22:4009–4030

Wang H, Tse YC, Law AHY, Sun SSM, Sun YB, Xu ZF, Hillmer S, Robinson DG, Jiang L (2010b) Vacuolar sorting receptors (VSRs) and secretory carrier membrane proteins (SCAMPs) are essential for pollen tube growth. Plant J 61:826–838

Xu H, Mendgen K (1994) Endocytosis of 1,3-β-glucans by broad bean cells at the penetration site of the cowpea rust fungus (haploid stage). Planta 195:282–290

Yaish MWF, Doxey AC, McConkey BJ, Moffatt BA, Griffith M (2006) Cold-active winter rye glucanases with ice-binding capacity. Plant Physiol 141:1459–1472

Ye Y, Mar E-C, Tong S, Sammons S, Fang S, Anderson JL, Wang D (2010) Application of proteomics methods for pathogen discovery. J Virol Methods 163:87–95. doi:10.1016/j.jviromet.2009.09.002

Yeom SI, Baek HK, Oh SK, Kang WH, Lee SJ, Lee JM, Seo E, Rose JK, Kim BD, Choi D (2011) Use of a secretion trap screen in pepper following *Phytophthora capsici* infection reveals novel functions of secreted plant proteins in modulating cell death. Mol Plant Microbe Interact 24:671–684

Zhang L, Tian LH, Zhao JF, Song Y, Zhang CJ, Guo Y (2009) Identification of an apoplastic protein involved in the initial phase of salt stress response in rice root by two-dimensional electrophoresis. Plant Physiol 149:916–928

Zhang Y, Liu CM, Emons AMC, Ketelaar T (2010) The plant exocyst. J Integr Plant Biol 52:138–146

Phenomics: Technologies and Applications in Plant and Agriculture

Hifzur Rahman, Valarmathi Ramanathan, N. Jagadeeshselvam, Sasikala Ramasamy, Sathishraj Rajendran, Mahendran Ramachandran, Pamidimarri D.V.N. Sudheer, Sushma Chauhan, Senthil Natesan, and Raveendran Muthurajan

Contents

Abstract

To meet the challenges of global food security in the changing climatic scenario, it would be most imperative to enhance crop productivity under resource competence. It is estimated that approximately 70 % of reduction in crop yield is due to the direct impact of abiotic stresses such as drought, salinity, and extreme temperatures. In the present context, one of the major challenges is large-scale screening of crop performance as a consequence of its genetic makeup. The development of advanced biotechnological and next-generation sequencing tools has lead to the accumulation of enormous data on genomics; nevertheless, data on phenotype and functions is yet to be fully annotated. To exploit the wealth of large-scale genomic data sets, it is necessary to characterize the crop performance quantitatively and link it to the genomic data. High-throughput phenomics studies offer strategies to screen large-scale population (s) for a particular phenotype employing advanced robotics, high-tech sensors, imaging systems, and computing power. Advanced bioinformatics tools further facilitate the analysis of large-scale multi-dimensional,

H. Rahman, Ph.D. (✉) • V. Ramanathan, Ph.D.
N. Jagadeeshselvam, Ph.D. • S. Ramasamy, Ph.D.
S. Rajendran, Ph.D. • M. Ramachandran, Ph.D.
S. Natesan, Ph.D. • R. Muthurajan, Ph.D.
Centre for Plant Molecular Biology and Biotechnology, Tamil Nadu Agricultural University, Coimbatore 641 003, India
e-mail: rahman.biotech@gmail.com

P.D.V.N. Sudheer, Ph.D. • S. Chauhan, Ph.D.
Department of Chemical and Biochemical Engineering, Dongguk University, Seoul 100-715, South Korea

D. Barh et al. (eds.), *PlantOmics: The Omics of Plant Science*,
DOI 10.1007/978-81-322-2172-2_13, © Springer India 2015

high-resolution data collected through phenotyping. With the integrated approach of genotyping and phenotyping, gene functions and environmental responses can be well understood. It will also help in finding more relevant solutions for the major problems that are currently limiting crop production.

Introduction

Adoption of new crop varieties and improved production technologies including modern irrigation practices, pesticides, synthetic nitrogen fertilizers, and other management techniques as a result of Green Revolution during 1960s resulted in dramatic increase in food production. Now as a result of global warming, the pace of environmental change is likely to be unprecedented. Uneven and intense precipitation events, elevated temperatures, drought, and other types of unfavorable weather events are posing pronounced risk on crop yield and its quality, making exceedingly difficult the challenge of feeding an estimated population of nine billion by 2050 (http://www.un.org/en/development/desa/population/). To minimize the impact of climate change on crop production with a view to feed world's ever-growing population, it is necessary to develop new crop varieties with improved resistance against various biotic and abiotic stresses.

Development of crop varieties that can cope up with heat, drought, flood, salinity, and other extremes is one of the most important strategies in improving agricultural crop production under changing climatic scenario. This necessitates use of the biotechnological tools such as marker-assisted breeding and transgenic technology for improving crop varieties for their tolerance against various biotic/abiotic stresses, as the efforts through conventional breeding methods have resulted in slow progress due to complexity of tolerance mechanisms. This approach warrants exploration of unexploited genetic variation for these traits to understand the genetic/molecular basis of tolerance either by QTL mapping or by gene expression profiling (Guarino and Lobell 2011). A meticulous linkage of genotype through phenome level data is essential to understand which genetic architecture affects the phenotype.

During the past two decades, advancements in the field of "genomics" have changed the face of plant biology. Genome sequencing and annotation of the model plant *Arabidopsis* followed by rice has laid the foundation for genomics research in plants (The Arabidopsis Genome Initiative 2000; Yu et al. 2002). As the cost of genome sequencing has come down drastically, researchers were able to sequence multiple genotypes of a crop for the purpose of association mapping and allele mining (Shendure and Ji 2008; Jackson et al. 2011). No matter how fascinating the discoveries in the field of molecular biology are, in the end it is the phenotype that matters. However, the existing whole-genome sequence information has not been adequately annotated to link with plant function and performance. A significant proportion of *Arabidopsis* genome which was sequenced first still remains unannotated or has been annotated based on weak homology clues. Linking of high-throughput physiology and phenomics data to the genome-level sequence data has become a bottleneck in crop improvement (Furbank and Tester 2011). Non-destructive, high-throughput evaluation under controlled environments is lacking in our breeding systems.

Marker-assisted selection of high-yielding crop genotypes adapted to stressful environments is hampered by slow manual phenotyping and laborious destructive sampling. Therefore, precise measurement of different phenotypic/physiological/biochemical traits of a plant and trying to link it with the gene or genotype of that plant has become as important as genotyping. Such relationships are relatively easy to establish in the case of monogenic traits, but it will become complicated if the phenotype is influenced by environment or is controlled by many genes as in case of quantitative traits. Thus, high-throughput screening of genetic variation for yield and component traits controlling tolerance against various biotic and abiotic stresses for QTL mapping/allele mining necessitates a rapid evolution of plant phenomics tools. Bridging the gap between genotype and phenotype will help us to realize the disciplinary goals of both genetics (Bateson 1906) and physiology (Gove 1981).

Efforts towards developing precise high-throughput methods of phenotyping have progressed in a slow manner over past 30 years when compared to that of genotyping (White et al. 2012). In comparison with the advancements in the field of genotyping, phenotyping is still performed manually and most of them are destructive and error prone. This slow progress in the development of phenotyping capability limits our ability to dissect the genetic basis of complex mechanisms especially those related to yield and stress tolerance. To relieve this bottleneck and to complement the extensive genomic information, reliable, automatic, multifunctional, and high-throughput phenotyping platforms integrated with photonics (Kelley 2009), biology (Poorter et al. 2012), computers, and robotics (Furbank and Tester 2011) are essential. Recently the value of phenomic data has been appreciated for several reasons. First, the phenotypic data remains to be the most relevant predictors of any biological function. Second, rapid advances in the field of phenotyping as well as bioinformatics tools has facilitated the analysis of large-scale multi-dimensional phenotypic data (Wang et al. 2010). Therefore, a detailed phenomic data generation has been already initiated for many crop species including rice, *Arabidopsis*, wheat, and barley (Munns et al. 2010; Lu et al. 2008; Yang et al. 2013). A project named RICE2020 has been initiated to functionally characterize every gene in rice genome by the year 2020 using high-throughput plant phenomics (Zhang et al. 2008; Yang et al. 2013).

Phenotype: Interaction Between Genotype and Environment

The terms genotype and phenotype were introduced by Wilhelm Johannsen in 1909. An individual's genotype denotes all of its genetic material, while its phenotype may comprise any observable characteristic or a trait (Gjuvsland et al. 2013). A complex network of interaction between genotype and the environment results in the phenotypic performance of a crop (Pieruschka and Poorter 2012). Plants of the same genotype can have different phenotypes, depending on the environmental conditions they are growing in. So it will be idealistic to collect large number of measures across multiple environments and at different developmental stages. Hence, crop performance as a function of genetic architecture requires a clear understanding on the interaction between genotype and phenotype (Soule 1967; Houle et al. 2010).

Phenomics

David Houle et al. (2010) defined phenomics as the acquisition of multidimensional phenotypic data in an organism as a whole. The word "phenomics" was coined by Steven A. Garan at a guest lecture he gave at the University of Waterloo in 1996. Phenomics, the study of the phenome, is a rapidly emerging area of science which aims at characterizing phenotypes in a rigorous and formal way and links these traits to the associated genes and gene variants (alleles) (Close et al. 2011). Phenomics technology can be used to study plants from the small scale, i.e., individual cell, leaf, or plant to the large scale, i.e., ecosystem. Formally, phenomics is the science of large-scale phenotypic data collection and analysis, whereas the phenome is the actual catalog of measurements. While it shares characteristics with classical mutant screening or quantitative trait analysis, it is distinguished from these traditional approaches in scale and scope (Heffner et al. 2011; Lu et al. 2011). It is further interrelated with other "-omics"-technologies like genomics, transcriptomics, metabolomics, or fluxomics to analyze the plant performance in the field and further link it to the core molecular genetics.

The science of phenomics speeds up phenotyping by using automated high-tech sensors, imaging systems, and computing power. Depending on the trait under observation, phenomics techniques can be used to characterize large number of lines/ individual plants accurately in a fraction of time; it has few advantages over manual phenotyping, viz., reduced time, reduced labor, and cost involved, etc., (Montes et al. 2007; Furbank 2009). The high speed of phenomics-based plant phenotyping accelerate the process of selecting plant varieties/germplasms that perform better in the field under drought, salinity, or high-temperature stress condition or crops with high photosynthetic efficiency or those which can perform better under higher levels of atmospheric carbon dioxide.

Forward Phenomics *vs* Reverse Phenomics

Forward phenomics uses phenotyping tools to "sieve" collections of germplasm for visible and valuable traits. It is used to find out the best germplasm suitable for a particular trait whereas reverse phenomics attempts to find out why the germplasm is behaving better (Furbank and Tester 2011). Forward phenomics speeds up phenotyping of large number of plants (in case of mapping population) or germplasm lines using automated imaging technology which leads to identification of interesting trait/plant suitable to particular situation. Thousands of plants are grown in pots prelabeled with barcodes and screened for interesting traits by automated imaging system (Furbank and Tester 2011). The selected plants with the target traits can then be grown up to produce seed for further analysis and breeding.

In reverse phenomics, the phenotype or desired trait – such as drought tolerance present in a particular germplasm – is already known. Researchers then try to find out the mechanisms that control the trait and the gene(s) underlying the mechanism (Furbank and Tester 2011). This is done by large-scale physiological and biochemical analysis and then linking the data with genes participating in particular biochemical or physiological pathway. Once the candidate gene(s) has been identified by reverse phenomics approach, then expression pattern of the candidate gene(s) will be compared with other genotypes. Thus, reverse phenomics is the detailed dissection of mechanisms underlying specific traits which allows exploitation of this mechanism or the candidate gene(s) associated with the trait which can be introgressed into new varieties or can be transferred to other plant species using genetic transformation technology (Furbank and Tester 2011).

Techniques Used in Plant Phenomics

Plant phenomics is a transdisciplinary science which includes biologists, chemists, physicists, computer scientists, engineers, mathematicians, physiologists, microscopists, geneticists, and plant breeders working together to develop new phenomics tools and methods. It employs screening of large populations with a goal of observing genetic variations present in the population for a particular trait (yield potential, drought, salinity, or high-temperature stress tolerance). Key features of the growth conditions are well defined and closely monitored. Phenotypic data and metadata descriptions of the experimental conditions are captured for detailed data analysis. These analyses would identify relationships between genotype and phenotype as well as reveal correlations between seemingly unrelated phenotypes (Schauer et al. 2006; Lu et al. 2008) and genetic loci (Gerke et al. 2009) (Fig. 1). The following key things, *viz.*, (1) high-resolution imaging systems, (2) automated transport for movement of plants for imaging, (3) powerful algorithms for imaging, and (4) data from other related experiments, need to be kept in mind while setting up the high-throughput phenomics platform. Table 1 describes the applications of various imaging techniques used in phenomics studies for measuring various growth parameters.

Visible Light (Monochromatic or Color) Imaging

Visible light imaging has been widely used in plant science after the invention of first digital camera by Eastman Kodak in 1975 (www.letsgo-digital.org/).With a wavelength (400–700 nm) perception similar to human eye, two-dimensional (2D) images captured by digital cameras can be used to analyze shoot biomass (Tackenberg 2007; Golzarian et al. 2011), yield-related traits (Duan et al. 2011a, b), leaf morphology (Bylesjo et al. 2008), panicle traits (Ikeda et al. 2010), and root architecture (Pascuzzi et al. 2010). Generally rosette plants at the vegetative stage are photographed from the top (Jansen et al. 2009;

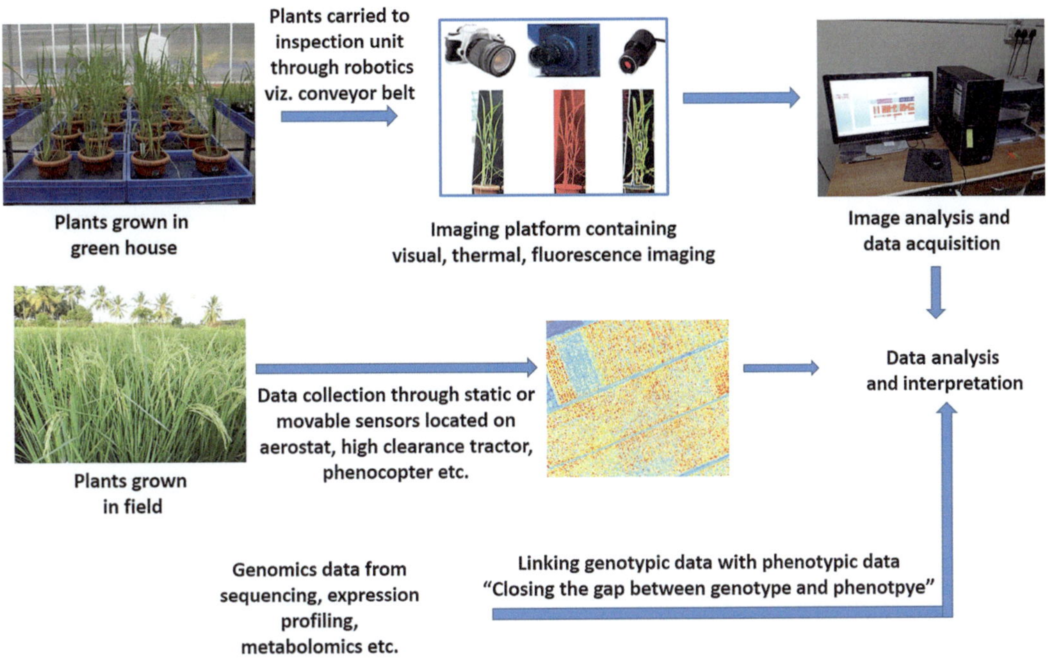

Fig. 1 Steps involved in high-throughput phenomics

Table 1 Applications of various types of imaging techniques in phenotyping of plants for various growth and developmental processes

Imaging technique	Trait studied	Species	Reference
Visible light imaging	Shoot biomass under salinity	Barley	Golzarian et al. (2011)
	Yield traits like number of total spikelets, number of filled spikelets, grain length, grain width, and 1,000-grain weight	Rice	Duan et al. (2011a, b)
	Panicle traits like panicle length and number of various branches and grain number	Rice	Ikeda et al. (2010)
	Root system architecture	Rice	Pascuzzi et al. (2010) and Clark et al. (2011)
	Leaf morphology	*P. tremula*	Bylesjo et al. (2008)
Thermographic imaging	Leaf area index	Rice	Sakamoto et al. (2011) and Shibayama et al. (2011a, b)
	Canopy temperature under water stress	Grapevine	Jones et al. (2009)
	Starch, protein and oil content in kernel for GWAS	Maize	Cook et al. (2012)
	Leaf health for detection of rice blast	Rice	Yang et al. (2012)
	Shoot temperature under water deficit condition	Barley, wheat	[a]Munns et al. (2010)
	Leaf temperature under insect infestation	Wheat	Manickavasagan et al. (2008)
	Osmotic components of salinity tolerance	Wheat and barley	Sirault et al. (2009)

(continued)

Table 1 (continued)

Imaging technique	Trait studied	Species	Reference
Hyperspectral imaging	Leaves and canopy damaged caused by biotic and abiotic stresses	Rice	Huang et al. (2012) and [a]Munns et al. (2010)
	Real-time detection of yellow rust	Wheat	Moshou et al. (2005)
	Leaf growth and nitrogen status	Rice	Nguyen and Lee (2006)
	Panicle health status under biotic stresses	Rice	Liu et al. (2010)
	Kernel damage by various insects	Wheat	Singh et al. (2010)
	Plant health status	Multicrop	[a]Bock et al. (2010)
Chlorophyll fluorescence	Photosynthetic efficiency	Multicrop	[a]Baker (2008)
	Early detection of disease incidence	Wheat	Moshou et al. (2005)
	Leaf growth and photosynthetic efficiency under mutistress conditions	*Arabidopsis*, *Nicotiana tabacum*	Jansen et al. (2009)
	Shoot temperature under water deficit	Barley, wheat	[a]Munns et al. (2010)
	Photosynthetic efficiency under water deficit condition	*Arabidopsis thaliana*	Woo et al. (2008)
X-ray	Root growth parameters under phosphorous fertilization	Wheat	Flavel et al. (2012)
	Root growth and architecture under root/soil interactions	Wheat and rapeseed	Gregory et al. (2003)
MRI	Belowground symptoms caused by cyst bean cyst nematode and *Rhizoctonia solani*	Sugar beet	Hillnhutter et al. (2012)
	Water translocation in tomato truss fruit through xylem	Tomato truss	Windt et al. (2009)
	Water translocation in tomato fruit through xylem	Tomato	Hossain and Nonami (2010)
PET	Shoot to root translocation of photoassimilate	Sugar beet and radish	Jahnke et al. (2009)
	$^{13}NH_4^+$ assimilation and translocation	Rice	Kiyomiya et al. (2011)
	^{52}Fe translocation from roots to young leaves via phloem	Barley	Tsukamoto et al. (2009)

[a]Review articles

Arvidsson et al. 2011; Granier et al. 2006) and cereal shoots from top and two sides at 90° angle (Golzarian et al. 2011). Root tip expansion dynamics at high spatial (mm) and temporal (min) resolutions (Schmundt et al. 1998) or root architecture parameters of entire root systems can be extracted automatically from time series images (Nagel et al. 2009; French et al. 2009). Leinonen and Jones (2004) have used thermal and visible imaging for estimating canopy temperature and identifying plant stress. Similarly Moller et al. (2007) have used thermal and visible imaging for estimating crop water status of irrigated grapevine. Three-dimensional imaging, advancement in visual imaging along with software platform (RootReader3D) has been developed for the rice root system. It converts 2D images captured from different angles into 3D by superimposing them, and this platform can monitor number of roots, its growth and root system architecture with an efficiency of over 100 roots per day (Clark et al. 2011). Paproki et al. (2012) have developed a 3D imaging technique for high-throughput analysis of stem height, leaf width and leaf length in cotton with a mean absolute error of less than 10 %. "3D virtual rice" has been constructed using a 3D digitizer and L-system formalism to demonstrate the differences in the structure and development between cultivars and under different environmental conditions (Watanabe et al. 2005).

Infrared and Hyperspectral Imaging

Nearly, all objects emit infrared radiation at a specific range because of internal molecular movement (Kastberger and Stachl 2003). Infrared imaging is done at two specific range of wavelengths of light one at a range of 0.9–1.55 μm called near-infrared (NIR) and other at a range of 7.5–13.5 μm called far-infrared (Far-IR) wavelength. Near-infrared (NIR) cameras study water content and movement in leaves and soil whereas far-infrared (FIR) cameras are used to study temperature. While using near-infrared (NIR) imaging, plants are grown in clear transparent pots for imaging root and its growth rate (Weirman 2010). The soil NIR measurement is done to calculate rate of water absorption by the roots and find out where and how much water is being used by the plant. Normal digital images of the roots can be overlaid onto the NIR images to understand more about how the roots are working. NIR imaging is also used to study the carbohydrate content of leaves, starch, protein, and oil content in seed kernels (Jones et al. 2009; Cook et al. 2012).

FIR imaging is used to measure and calculate temperature differences either within leaves of one plant or between different plants. So, FIR imaging can be used in the fields to detect plants with cool canopies. The cooler plants are the ones with superior root systems which can absorb more water. FIR cameras can also measure changes in stomatal conductance thus giving an estimate on rate of photosynthesis. FIR imaging is also used in screening germplasms against salinity tolerance in barley (Sirault et al. 2009) as the leaf temperature of a plant increases with increasing salt level. So plant's canopy that can maintain lower temperature when imaged with FIR camera in the presence of salt are more salt tolerant and therefore grow better in saline soils. Using high-speed FIR imaging, it is possible to measure salt tolerance at seedlings stage as well (Sirault et al. 2009).

Healthy green plants reflect a large proportion of NIR light (800–1,400 nm), whereas the soil reflects little NIR light. Soil and unhealthy plants reflect considerably higher red wavelength light as compared to the healthy plants. For these reasons, many studies have combined NIR imaging and visible imaging to detect vegetative indices. Crop Phenology Recording System (CPRS) uses visible light imaging to derive the visible atmospherically resistant index and uses near-infrared imaging (830 nm) to derive the night time relative brightness index and then establishes the relationship between the camera-derived indices and the agronomic traits (Sakamoto et al. 2011). Another approach integrates the visible red image (630–670 nm) and near-IR image (820–900 nm) to assess the rice leaf area index during the before-heading period (Shibayama et al. 2011a) and then uses only near-infrared imaging to predict the Leaf Area Index (LAI) (Shibayama et al. 2011b).

Fluorescence Imaging

Fluorescence occurs when a compound absorbs light at one wavelength and emits light at different wavelength. Fluorescence signals are very weak relative to the excitation light intensity, and for detection it has to be separated from reflected light. Fluorescence imaging systems use cameras having specific spectral cutoff filters or pulsed light to measure fluorescence emission dynamics, making it sensitive to the specific spectral region at which fluorescent signals are emitted (Schreiber et al. 1986; Fiorani et al. 2012). Fluorescence imaging flashes blue light (<500 nm) on the plants, which then emits fluorescence light in the range of 600–750 nm. This fluorescence can be photographed and converted into false-color signals by a computer program, allowing researchers to easily observe differences in fluorescence (Weirman 2010). Chlorophyll fluorescence which is affected by changes in photosynthetic performance is commonly used in phenomics to see the effect of different genes or environmental conditions on the efficiency of photosynthesis at very young stage of the plants (Weirman 2010). Fluorescence imaging is also commonly used to study the effects of pathogens on plants (Balachandran et al. 1997; Osmond et al. 1998), stomatal movements (Cardon et al. 1994), phloem loading and unloading (Siebke and Weis 1995), the correlation between growth and photosynthesis (Walter et al. 2004), and the spatiotemporal variation of photosynthesis under non-limiting and limiting-growth conditions (Osmond et al. 1999; Chaerle and Van Der Straeten 2001; Rascher et al. 2001). Fluorescence imaging has been initially applied at tissue scale such as single leaves, but for analyzing whole plant or canopy, advanced imaging systems such as laser-induced fluorescence transients (LIFT) (Rascher and Pieruschka 2008) or monitoring of sun-induced fluorescence are currently used (Meroni et al. 2009; Malenovsky et al. 2009).

3D (Three-Dimensional) Imaging

Multiple images captured from different angles using several cameras are then combined by computer program to give 3D image of the plant (Weirman 2010). Once a 3D image of the plant is generated, several measurements like leaf number, shape, angle, color, leaf health, tiller number, shoot mass, etc. can be made. Two approaches are most commonly used for 3D imaging and mapping in plants; first, light detection and ranging (LIDAR) techniques that scan a scene (Omasa et al. 2007) and, second, stereo photography that uses two (or more) cameras (Biskup et al. 2007). LIDAR method gives data that are unconnected 3D point cloud which needs individual 3D space points to be assembled to give 3D image, whereas in 3D optical imaging data can be acquired rapidly but the data processing in this system is time consuming (Fiorani et al. 2012). Topp et al. (2013) used a semi-automated 3D in vivo imaging and digital phenotyping for QTL mapping to identify core regions of the rice genome controlling root architecture.

Magnetic Resonance Imaging (MRI)

MRI uses a combination of magnetic field and radio waves to take images and is most commonly used for imaging plant roots. MRI can detect nuclear resonance signals originating from the protons of water and organic molecules present in roots, providing structural information about the internal physiological processes occurring in vivo (Borisjuk et al. 2012). MRI has provided solutions to analyze water distribution and quantification of water content in plant and its different organs nondestructively (Borisjuk et al. 2012). It has also been used to provide the functional information like water diffusion and transportation in plant (through xylem and phloem) in several crops like poplar, castor bean, tomato, and tobacco (Windt et al. 2006). But most commonly, MRI is used to study the root architecture of plants grown in soil-filled pots. Previously the root traits were studied by growing the plant in transparent agar in order to visualize the root systems. This technique does not exactly mimic the soil environment, but with MRI scanning plant roots under analyses can be grown in tubes of soil or sand which has more relevance in validating the findings of roots grown in artificial conditions of pots (Weirman 2010). Besides these applications MRI has been used to study and visualize the symptoms caused by cyst nematodes of sugar beet (Hillnhutter et al. 2012), bean root nodulation, and the root distribution of maize grown with a hetero-specific neighbor (Rascher et al. 2011).

Positron Emission Tomography (PET)

PET is a technique used to image nondestructively the distribution of compounds labeled with positron-emitting radionuclides such as ^{11}C, ^{13}N, or ^{52}Fe (Jahnke et al. 2009; Kiyomiya et al. 2011; Tsukamoto et al. 2009). It visualizes the distribution and transportation of metabolite labeled with positron-emitting radionuclide elements and thus helps in studying metabolism of plants. For dissecting shoot-to-root carbon fluxes, Jahnke et al. (2009) combined PET and MRI and used [^{11}C]-labeled CO_2 and demonstrated sugar beet taproot receiving photo assimilates. Later, Buhler et al. (2011) in a similar experiment demonstrated transport velocity and lateral loss of photoassimilates in plants along a transport path. Thus, MRI and PET when used together, both structural and functional traits (Jahnke et al. 2009) can be analyzed simultaneously and independently as well.

Field-Based Plant Phenomics

Over the last few years, significant developments have been made towards automated phenotyping platforms with the application of robotics, new sensors, and advanced imaging technologies in growth chambers and glasshouses (Granier et al. 2006; Jansen et al. 2009; Furbank and Tester 2011). Several platforms have been designed for complete automated monitoring and analyzing the growth of adult plants, their imaging, data collection, and evaluation (Table 2). These platforms are usually built in to carry plants under screening to a sensing station. The major goal of plant phenomics is to provide quantitative data on dynamic responses of plants to the natural environment. Screening the plants under greenhouse/controlled environment has limitations like changes in solar radiation, wind speed and evaporation rates, limited soil volume, nutrient availability in pots, and interaction with mutualistic, parasitic, or competitor organisms, etc. Phenotyping carried out in such closed environment may fail to characterize the responses relevant to the field conditions; further, the interaction between changing environment and plant phenotype is missing to great extent when plants are screened in controlled environment for a particular trait (Poorter et al. 2012). So, screening and identification of germplasm for a desirable trait is done best under field conditions of soil, climate, and biotic stress agents, similar to where final varieties will be grown.

However, researchers are continuously developing reliable techniques for field-based phenotyping (FBP), where screening at a larger scale and an accurate description of trait expression in cropping system can be predicted. In addition, field measurements represent a significant test for the relevance of the laboratory and greenhouse approaches. For FBP airplane and satellite-based systems are used at field to regional scales, but studies using proximal (close-range) sensing are most commonly the only approach that can provide reliable data at adequate resolution as well as multiple angles and control illumination and at a shorter distance from the target to the sensors (White et al. 2012).

Field-Based Phenotyping Using Static Sensor or Movable Vehicle

For FBP, suitable sensors and automated imaging devices can be fixed at suitable distance, angle, and height for uninterrupted data collection and processing. Phenonet is one of the model data logger developed by CSIRO (Commonwealth Scientific and Industrial Research Organization) linked to sensors like FIR thermometer, chlorophyll fluorescence sensor, soil moisture sensor, camera, and weather station. The data logger can measure canopy temperature and photosynthetic activity of the plant and also monitors environmental variations across the field and continuous crop growth and development. The data recorded is sent to the researcher through web link thus avoiding daily visits to field sites (Weirman 2010).

Phenonet allows the researcher to record the data continuously throughout the life span of plants. The sensors and imaging devices used in phenonet were also mounted on movable vehicles like high-clearance tractor, helicopter, etc. for data collection at a particular time point above the canopy. Phenomobile is a vehicle driven by researcher over crop site, which is equipped with digital cameras to estimate leaf greenness and

Table 2 Examples of automated platforms for whole plant phenome analysis

Name	Plant species	Reference	Web link
PHENOPSIS	*A. thaliana*	Granier et al. (2006)	http://bioweb.supagro.inra.fr/phenopsis/
PHENODYN	Corn, rice	Jung and Bracker (2009)	http://bioweb.supagro.inra.fr/phenodyn/
GROWSCREEN	*A. thaliana*, tobacco	Walter et al. (2007)	http://www.fz-juelich.de/icg/icg-3/jppc/growscreen/
TraitMill	Corn, rice	Reuzeau et al. (2010)	http://www.cropdesign.com/tech traitmill.php
Scanalyzer	*A. thaliana*, duckweed, corn		http://www.lemnatec.de

ground cover, far-infrared thermometer to measure canopy temperature, stereo-imaging system for taking 3D images for determining biomass of plant, spectral reflectance sensors to determine the crop's chemical composition, etc. It travels through the field site at speed of 3–5 km per hour and collects measurements from the plot directly beneath it and on the either side (Weirman 2010; White et al. 2012).

High-clearance tractors and phenomobile work at close proximity to plants and provide better resolution images, but their major limitation is that they can compact the soil, damage leaves and stems, and propagate diseases and pests. Aircraft like manned or unmanned helicopters, aerostats, etc. are also been used for imaging the field covering larger areas rapidly.

The Duncan Technologies MS3100 imager carrying three types of cameras, viz., multispectral, infrared, and conventional digital for visible light, is used by mounting on a helicopter for imaging the canopy. Manned helicopters are capable of carrying a large payload and are able to supply power to an instrument system for longer time. Aircraft can cover large areas rapidly but cannot be used in inclement weather, have high operating costs, and may not permit adequate resolution. However, major limitation in using helicopters is that they cannot be flown close to crop due to fan downwash and regulations on minimum safe altitude (White et al. 2012). Unmanned remote-operated helicopters are better alternative to manned aircraft as they allow flying at lower altitudes and are less expensive to operate. Merz and Chapman (2011) at CSIRO developed remote-controlled gas-powered model helicopter called the "phenocopter." The phenocopter is able to fly itself and make multiple passes over a 1-ha field in just 6 min providing data to measure plant height, canopy cover, lodging, and temperature throughout a day. Berni et al. (2009) have demonstrated the use of unmanned helicopter for thermal and narrowband multispectral remote sensing for vegetation monitoring. Zarco-Tejada et al. (2009) imaged 0.6 ha of citrus orchards to examine fluorescence emission as function of water stress using hyperspectral and infrared cameras.

Helium-filled aerostats which can carry a payload of 2 kg, equipped with digital camera scan, take infrared or color images of a field from 30 to 80 m above the ground (Weirman 2010; White et al. 2012). Jensen et al. (2007) described use of a 1.8 m aerostat to monitor response of wheat to nitrogen using color and near-infrared images acquired with digital cameras. Ritchie et al. (2008) estimated evapotranspiration in a cotton irrigation study using a Normalized Difference Vegetation Index (NDVI) data collected by two-camera system.

Phenotower carrying reflectance and infrared sensors mounted on a trolley can run parallel to the vegetation to collect data on comparative canopy temperature, leaf greenness, and ground-cover between different plant lines at the same time (Weirman 2010; White et al. 2012). Agricultural Irrigation Information System (AgIIS) was used to characterize water and nitrogen stress for plots in a cotton field by obtaining multispectral data for vegetation, nutrient, and water status indices at a spatial resolution of 1 m (Kostrzewski et al. 2003; Colaizzi et al. 2003; Haberland et al. 2010).

Application of Phenomics in Agricultural Research

Screening for Abiotic Stress Tolerance-Related Traits

Abiotic stresses, viz., drought, salinity, submergence and temperature extremes, are the major causes of yield loss in agricultural crops worldwide. Because of complexity regulatory networks behind tolerance against these stresses, phenotyping for tolerance against these abiotic stresses is often a big challenge (Vandenbroucke and Metzlaff 2013). If phenotypic analysis of one plant used to take approximately 15 min, then manual phenotypic analysis of 100 plants would take would require approximately 25 man-hours of work. Screening large mapping population or a collection of germplasms for QTL analysis or association mapping of abiotic stress tolerance will be time consuming and impractical. But with the ability of reliable high-throughput phenotyping techniques, it has now become possible to screen multiple traits for huge population size nondestructively under stress conditions. Here in following sections we have discussed use of high-throughput techniques for biotic, abiotic stress tolerance, yield, and quality improvement.

Measuring Levels of Salinity Tolerance

The response of plants to high external levels of NaCl is phenotypically very dynamic so the methods used for phenotyping should be able to capture the salinity responsive behavior. Responses of a plant to salinity stress occur in two distinct phases (Munns and Tester 2008): First phase is osmotic phase, which starts immediately after the salt concentration around the roots increases to a critical level and inversely affects the growth rate of plants by slowing down expansion of young leaves and inhibiting the initiation of new leaves and the rate of shoot growth falls significantly. Second phase ion-specific response is known as tissue tolerance which starts at a later stage when Na^+ and/or Cl^- has accumulated to toxic levels in the plant tissue leading to premature senescence of older leaves. Both osmotic tolerance and tissue tolerance are very dynamic processes showing reduction in growth rate. Osmotic-tolerant plants show only minor reductions in relative growth rate as compared to osmotic-sensitive plants which will show a severe growth reduction upon exposure to NaCl (Rahman et al. 2014). Changes in growth rate can be measured by daily imaging the plants and reduction in growth rate caused by salinity stress can be measured (Rajendran et al. 2009). An alternative method of measuring osmotic tolerance is by infrared thermography. Because of salinity, stomatal conductance drops down there by increasing the leaf temperature. Based on this principle, Sirault et al. (2009) have developed a screening method for quantifying osmotic tolerance-related components of salinity tolerance for durum wheat and barley germplasms by measuring leaf temperature using infrared thermography. James and Sirault (2012) demonstrated use of infrared thermal imaging (thermography) as a non-invasive, high-throughput technique for the screening of large number of wheat genotypes for salinity tolerance.

In tissue-tolerant phase, tolerant plants are able to compartmentalize high concentrations of shoot Na^+; whereas sensitive plants will not be able to effectively compartmentalize Na^+ away from the cytosol, resulting in early onset of senescence in older leaves and thus leading to change in leaf color. The onset and degree of senescence can be quantified through the color information of visible light imaging thus allowing quantification of the tissue tolerance in plants. Rajendran et al. (2009) have developed a nondestructive method to characterize plant salinity tolerance mechanisms by determining osmotic tolerance and tissue tolerance using visible light imaging techniques with a Scanalyzer 3D. Golzarian et al. (2011) have developed a more accurate high-throughput estimation of biomass for cereal plants under both control and saline conditions using a Scanalyzer 3D designed by LemnaTec. Thus, high through put plant phenomics can be used in QTL mapping and association mapping studies where large population is used and will thus help in linking the phenotype with the genotype.

Measuring Drought Tolerance-Related Traits

Drought tolerance is a complex trait determining crop yield under water-limited conditions which still remain a challenging trait for manipulation by the breeders (Cattivelli et al. 2008). Recent reports indicated the possibility of improving drought tolerance through biotechnological approaches. For this QTLs or underlying candidate genes controlling yield benefit under drought conditions need to be identified using genotypic and phenotypic screens and then incorporated into elite germplasm using modern breeding technologies, such as marker-assisted selection (Tester and Langridge 2010).

Plants use different mechanisms to cope with drought stress, namely, drought escape, drought tolerance, drought recovery, and drought avoidance (Levitt 1972; O'Toole and Chang 1979). Genotypes that have deep, coarse roots with a high ability of branching and penetration, elasticity in leaf rolling, early stomatal closure, and high cuticular resistance are reported as component traits of drought avoidance (Blum et al. 1989; Samson et al. 2002; Wang and Yamauchi 2006).

Among the different mechanisms, drought avoidance is one of the important mechanisms in which roots are associated. Determination of genetic variations for root system architecture is the starting point for developing root system modelling for various crop species. Many new techniques have been developed to characterize the root system under field conditions. Methods such as, non-soil media, optical techniques, soil media, and noninvasive techniques have been used for characterizing the root phenotyping (Gregory et al. 2009). Due to advancements made in the field of phenomics, several automated phenotyping systems are available to study root system architecture using visible, infrared or hyperspectral imaging either in gel-based media or in soil media. Pascuzzi et al. (2010) have used visible light imaging from multiple angles to study root system architecture of rice (number of root and root hairs, root length, and diameter) in gel-based platform. Several techniques are available to study root system architecture in soil media like X-ray computed tomography (Gregory et al. 2003; Perret et al. 2007), NMR (van der Weerd et al. 2001; Jahnke et al. 2009), magnetic resonance imaging (Asseng et al. 2000; van As 2007), and laser scanning (Fang et al. 2009).

In most crops, leaf growth and stem elongation processes are very sensitive to water status and will be the first process affected. One of the early mechanisms of drought avoidance is leaf rolling, reduction in leaf area, reduced stem elongation rate, and reduced transpiration rate (Reddy et al. 2003). High-throughput phenotyping techniques are available to measure leaf area index by using digital infrared imaging (Shibayama et al. 2011a, b). Thus, infrared imaging along with visible imaging can be used to screen the germplasms for drought stress tolerance by measuring reduction in leaf area, stem elongation rate, leaf rolling, and drooping symptoms of drought. Another major mechanism associated with drought resistance is drought escape: plants will try to avoid drought stress by flowering early and completing their life cycle early. Flowering time can be easily recorded using visible imaging and germplasms can be screened for their drought escape behavior.

Canopy temperature is a sensitive and specific indicator of progression of drought stress. Relatively lower canopy temperature in drought-stressed rice indicates the ability of the genotype to extract more soil moisture or indicates the ability to maintain relatively better plant water status. Plant stomatal closure takes place during water stress to avoid loss of water resulting in a decrease of energy dissipation and an increase of canopy temperature (Lafitte et al. 2003). The magnitude of the rise in canopy temperature among water stressed plants is influenced by successive stress imposition. Far-IR (also called IR thermal) imaging is the most general technique used to visualize temperature differences. Zia et al. (2012) have measured crop canopy temperature to assess crop water stress in winter wheat by using infrared thermography. Leinonen and Jones (2004) have used thermal and visible imaging to estimate canopy temperature under both green house and field conditions during water stress conditions. Screening for high leaf temperature has led to the identification of novel mutants in *Arabidopsis* that allowed further insight into the molecular characterization of stomatal regulation (Wang et al. 2004).

It has been proposed that limitation of gas exchange by stomatal closure is the main inhibitory effect on assimilation during drought and increased mesophyll resistance to CO_2 diffusion playing a secondary role (Flexas et al. 2006). Efficient visualization of leaf conductance (thermography) and photosynthetic efficiency (chlorophyll fluorescence) might allow ready resolution of such questions and will further increase throughput and detecting power in screening for drought tolerance (Granier et al. 2006). Chlorophyll fluorescence analysis is useful in investigating photosynthetic performance of plants under stressful environments. Chlorophyll fluorescence has been used to access drought tolerance in coconut varieties (Nainanayake 2007), chickpea (Rahbarian et al.

2011), winter bread wheat (Flagella et al. 1994; Roostaei et al. 2011), transgenic tomato (Mishra et al. 2012), *Arabidopsis* (Woo et al. 2008), etc. Thus, combination of thermal imaging to study the stomatal conductance and fluorescence imaging to study the photosynthetic activity under drought conditions allows us to differentiate high photosynthetic activity lines than those lines which simply close stomata and therefore lose yield potential (Chaerle et al. 2007; Masle et al. 2005).

Biotic Stress Tolerance

As the quality and quantity of crop production reduces due to pest/disease infestation, it has become important to detect and classify the plant infestation at an early stage (Hillocks 1992). Pathogens such as rusts and mildews produce symptoms like yellowing of leaves, necrosis, etc. which can be scored by visible imaging. Pathogens in general influence photosynthetic electron transport and downstream metabolic reactions, resulting in increased chlorophyll fluorescence at early stage of infection. Thus, chlorophyll fluorescence providing quantitative information about photosynthetic function has the potential to address many questions relevant to plant-pathogen interactions (Scholes and Rolfe 2009). Nondestructive digital imaging and imaging via chlorophyll fluorescence have been used to monitor the progress of disease symptoms caused by fungal pathogen (Scholes and Rolfe 2009). High-throughput phenotyping using chlorophyll fluorescence led to early detection of symptoms (before symptoms can be detected by human eyes), quantification of infected tissue area, and potentially the quantification of the susceptible and resistant responses of plant to pathogen attack (Swarbrick et al. 2006; Chaerle et al. 2009; Romer et al. 2011). To identify rice blast disease at the seedling stage, a near-infrared hyperspectral imaging system was used to identify infected and healthy clipped leaves, with 92 % accuracy (Yang et al. 2012). Using visible imaging and color-based corner detection algorithm, plant-hopper infestations on stems of pot-grown rice have been studied (Zhou et al. 2011). Moshou et al. (2005) have developed a ground-based real-time remote-sensing system for detecting yellow rust in wheat at an early stage of disease development (before it can be visibly detected) under field conditions. Fukatsu et al. (2012) have developed a remote pheromone trap monitoring system with image data based on sensor network and image processing and applied to measure the occurrence of the rice bug, *Leptocorisa chinensis*, in a paddy field. Application of phenomics for screening pathogen resistance on foliar symptoms is either in infancy or is developing. Most of the root pathogen causes disruption of xylem tissue resulting in reduced transpiration, stomatal closure and finally leading to hotter canopies, thus offering an opportunity to use thermographic screening of tolerant or resistance germplasms for root pathogen nondestructively. Thus, high-throughput phenotyping is having potential to achieve real-time and dynamic screening of germplasms for biotic stress tolerance high accuracy, surpassing human experts in monitoring plant diseases.

Measuring Yield and Quality-Related Traits

Yield is a complex agronomic trait and in rice it is determined by several parameters like biomass and growth, number of productive tillers, panicle size, and number of grain per panicle and grain weight (influenced by grain size). Traits like biomass, tillering, early vigor, etc. can be measured by taking digital images and then by analyzing the images (Richards et al. 2010). Generally these traits are scored manually but both speed and accuracy for measuring these traits can be improved using high-throughput phenomics tools. Bimodal scanner developed by combining visible light imaging and soft X-ray imaging has been employed to accelerate the counting of spikelet number per panicle and simultaneous counting of filled and unfilled rice spikelet (Duan et al. 2011a). Duan et al. (2011b) have developed an integrated facility to achieve fully automated yield trait scoring with an accuracy of more than 95 % and efficiency of 1,440 plants in 24 working hours.

Integrated use of flatbed scanner and user-coded ImageJ software has made it possible to determine the major orthogonal dimensions of the grains, viz., grain length and width (Igathinathane et al. 2009). Tanabata et al. (2012) have developed SmartGrain software for high-throughput measurement of seed shape parameters, such as seed length, width, area, and perimeter length. They have further validated the efficacy of the software in QTL analysis for rice (*Oryza sativa*) seed shape using a backcross inbred derived from a cross between *japonica* cultivars Koshihikari and Nipponbare.

Yoshioka et al. (2007) used digital image scanner to determine the chalkiness in rice grain with accuracy rate of 90.2 %. Sun et al. (2008) on the principle of digital chroma measurements determined the protein content in rice using red, green, and blue (RGB) chroma and showed that the method is equivalent to Kjeldahl method in estimating protein content in rice grains. Flatbed scanning and image analysis combined with velocity representation method made it possible to identify broken rice kernels (Lin et al. 2012). Digital microscope and appropriate image analysis software, transilluminated imaging were used to assess wheat quality (Venora et al. 2009). X-ray imaging has been used to image single kernels and detect infestation of wheat seed by *Rhyzopertha dominica* (Karunakaran et al. 2004a), red flour beetle (Karunakaran et al. 2004b), *Cryptolestes ferrugineus* (Manickavasagan et al. 2008), sprouting in wheat kernels (Neethirajan et al. 2007b), and vitreousness in durum wheat by scanning seed at different X-ray energies revealing changes in internal density of the wheat grain (Neethirajan et al. 2007a). An automatic inspection system combining a near-infrared instrument and a visible light segregator has been developed to determine rice quality parameters like protein content, moisture content, and sound whole-kernel ratio (Kawamura et al. 2003). Thus, with the appropriate optical imaging, image analysis and robotics tools have the potential to achieve the high-throughput scoring of yield traits and quality, which will be useful in rice (or other cereals) science research.

Role of Bioinformatics in Phenomics

Data Handling and Management

To be successfully applied, plant phenomics should be able to gather high-throughput information covering many traits under study from large numbers of plants in a shorter duration and the ability to transform these raw data into qualitative and quantitative results. High-throughput phenomics will not only increase the number of phenotypes/traits that can be measured but also the sampling size, thus providing large amount of data to be analyzed; for example, phenocopter generates about 3 gigabytes data in one flight whereas high-clearance tractor platform produces a terabyte of data per run (Mogel 2013). Handling and processing such large amounts of data poses a big challenge for high-throughput phenotyping. With the current system of data analysis, it used to take hours or days to process the data. High-throughput phenomics needs high-performance computing and information technology to compute as many parameters as possible. The data needs to be normalized based on control plants. The data flow from the sensors to the lab needs to be highly structured and designed to avoid any wrong assignment of data and samples. Since number of data points is too large to be handled manually, advanced data analysis methods are important from the first step of analyzing the raw data until validating the results (Eberius and Guerra 2009). Plant geneticists and breeders have generated numerous biparental mapping populations that segregate for genetically mapped quantitative trait loci (QTL) and association mapping panel. Information about the map positions of QTL are included in several publications but only a small fraction of raw data is included into existing genomic databases. As opposed to sequence and expression data that need to be deposited in appropriate databases upon publication, the raw data of replicated phenotypic measurements are not deposited in any public repository and are usually lost. A major bioinformatics challenge facing the research community is to develop web-based resources to display the details of complex phenotypes to uncover hidden biological knowledge.

Image-based phenotyping captures not only morphological- and developmental-related phenotypic data but also produce information on the physiological status of the plant through nondestructive close-range or remote-sensing technologies. But major roadblock in performing quantitation and feature extraction is lack of extensible algorithms. For this purpose, a high-throughput image analysis platform such as the Bisque system from the Center for Bioimage Informatics, UC Santa Barbara (http://www.bioimage.ucsb.edu/), has been developed to address the logistical and scalability issues.

There are some key challenges for data management in phenomics research: firstly, the ability to provide a data management service or database that can manage large amount of heterogeneous data in multiple formats (text, image and video) from different measurement platforms and, secondly, it should have the ability to support metadata-related services to provide context and structure for data within the data management service to facilitate effective search, query, and dissemination. And finally it should be able to accommodate evolving and emerging technologies and processes, as phenomics is still in infancy and is a rapidly developing field of research. In phenomics for a meaningful data evaluation and statistical analysis, standard data storage device is required. For this purpose databases need to be developed in which data can be stored on long-term basis and can be further retrieved later without endangering data integrity for metadata analysis. Billiau et al. (2012) have developed "phenotyper" database on data warehouse concept for handling large amount of phenomics data (Billiau et al. 2012). This database is able to manage the data observed simultaneously by 11 separate research groups from four analytical platforms and at 11 different sites. Vankadavath et al. (2009) have developed software "PHENOME" for high-throughput phenotyping, which allows researchers to accumulate, categorize, and manage large volume of phenotypic data. PHENOME software using Personal Digital Assistant (PDA) with built-in barcode scanner aids in collection and analysis of data obtained through large-scale mutagenesis, assessing quantitative trait loci (QTLs), raising mapping population, sampling of several individuals in one or more ecological niches, etc. (Vankadavath et al. 2009). Fabre et al. (2011) have developed PHENOPSIS DB for storing data obtained by *Arabidopsis thaliana* phenotyping. It also allows browsing and sharing of online data generated by the PHENOPSIS platform and offline data collected by experimenters and experimental metadata using an interface for visualization of environmental data of an experiment as well as statistical analysis of phenotypic data and analysis of *Arabidopsis thaliana* plant images. PhenomicDB is another, multispecies (primarily human, mouse, fruit fly, and yeast) resource designed to empower "comparative phenomics" (Kahraman et al. 2005).

The National eResearch Architecture Taskforce (NeAT), Australia, initiated the PODD project which is aimed to design and develop Phenomics Ontology Driven Data (PODD) repository (Li et al. 2010). PODD is created as a repository to record the data; its contextual data (metadata) and data classifiers are in the form of ontological or structured vocabulary terms (Li et al. 2010). The goal of PODD is to capture, manage, annotate, and distribute the data generated by phenotyping platforms (Li et al. 2010). The Phenoscape project (http://kb.phenoscape.org) has been initiated to develop search algorithms capable of linking biological data by relationships between ontological terms and by similarities found between free-text descriptions. Beside these, several other databases have been developed or are under development by different research groups as a repository for plant phenomics studies (Table 3).

Table 3 Databases containing information on phenotypic characteristics of crop plants

Database	Organism	Website	Reference
LycoTILL	Tomato	http://www.agrobios.it/tilling/index.html	Minoia et al. (2010)
OryGenesDB	Rice	http://orygenesdb.cirad.fr/index.html	Droc et al. (2009)
Tos17	Rice	http://pc7080.abr.affrc.go.jp/phenotype	Miyao et al. (2003)
MaizeGDB	Maize	http://www.maizegdb.org/	Schaeffer et al. (2011) and Lawrence et al. (2007)
Panzea	Maize; teosinte	http://www.panzea.org/	Canaran et al. (2008)
PHENOPSIS DB	*Arabidopsis*	http://bioweb.supagro.inra.fr/phenopsis/	Fabre et al. (2011) and Juliette et al. (2011)
Gramene: Genetic Diversity Module	*Arabidopsis*; rice; maize; sorghum; wheat	http://www.gramene.org/db/diversity/diversity_view	Chen et al. (2010)
IonomicsHub	*Arabidopsis*; rice; yeast; soybean; maize	http://www.ionomicshub.org/home/PiiMS	Baxter et al. (2007)
Oryza Tag Line (OTL)	Rice	http://oryzatagline.cirad.fr/	Larmande et al. (2008)
Rice Mutant Database (RMD)	Rice	http://rmd.ncpgr.cn/	Zhang et al. (2006)
T3 Triticeae Toolbox	Wheat; barley	http://triticeaetoolbox.org/	Blake et al. (2012)
Tomato Mutant Database	Tomato	http://zamir.sgn.cornell.edu/mutants/	Menda et al. (2004)
SGN	Solanaceae species	http://solgenomics.net/	Bombarely et al. (2011)
The Phenoscape project	Many species	http://kb.phenoscape.org	Mabee et al. (2012)
GERMINATE	Many plant species	http://germinate.scri.sari.ac.uk/germinate/	Lee et al. (2005)
CEREALAB	Cereals	http://cerelab.com.br/	Milc et al. (2011)
PhenomicDB	Multispecies	http://www.phenomicdb.de/	Kahraman et al. (2005)

Image Analysis Software and Platforms

High-throughput plant phenotyping platforms acquire and record large amounts of image data that must be accurately and robustly calibrated, reconstructed, and analyzed which requires development of sophisticated image understanding and quantification algorithms. High-throughput image analysis for automated phenotyping is used to extract several phenotypic parameters related to growth, yield, and stress tolerance of the plants. Manual analysis of even small-scale data is time consuming and may also lead to inaccuracy. So, robust, flexible, and step-by-step traceable image analysis tools for plant phenotyping are needed. Several image analysis softwares and tools have been developed to analyze either 2D or 3D images of different plant tissues generated by various phenomics platforms (Table 4). One robust software developed by Julich Plant Phenotyping Centre is

Table 4 Softwares and image analysis platforms available for high-throughput phenotyping

Tissue	Software	Parameters measured	Reference
Roots	WinRHIZO Tron	Measures root length, diameter, area, surface area, and root volume	http://www.regent.qc.ca/products/rhizo/RHIZOTron.html
	KineRoot	Measures root growth and curvature	Basu et al. (2007)
	PlaRoM	Measures root extension and growth traits under diurnal or circadian growth rhythms	Yazdanbakhsh and Fisahn (2009)
	EZ-Rhizo	2D analysis of root system architecture	Armengaud et al. (2009)
	GiA Roots	2D analysis of root system architecture	Galkovskyi et al. (2012)
	GROWSCREEN-Rhizo	Root architecture parameters in 2D and shoot biomass evaluation	Nagel et al. (2012)
	RootTrace	Measure of root length and curvature	Naeem et al. (2011) and French et al. (2009)
	DART	2D analysis of root system architecture	Le Bot et al. (2010)
	SmartRoot	**Quantify root growth and architecture** for complex root systems	Lobet et al. (2011)
	RootReader3D	3D analysis of root system architecture	Clark et al. (2011)
	RootReader2D	2D analysis of root system architecture	Clark et al. (2012)
	GROWSCREEN-Root	Predict tree model for root system architecture	http://www.fz-juelich.de/ibg/ibg-2/EN/methods_jppc/GROWSCREEN_root/_node.html
	Growth Explorer	2D analysis of growth patterns of plant roots	Basu and Pal (2012)
	RooTrak	**3D plant root architecture of plant grown in soil**	Mairhofer et al. (2012)
Shoot/leaves	WinFolia	Measures leaf area, morphology, and disease analysis of broad leaves	http://www.regent.qc.ca/products/folia/WinFOLIA.html
	TraitMill	Platform for testing the effect of plant-based transgenes on agronomically valuable traits	Reuzeau et al. (2006)
	PHENOPSIS	Automated measurement of water deficit-related traits like leaf number, leaf area, root growth, and transpiration rate	Granier et al. (2006)

(continued)

Table 4 (continued)

Tissue	Software	Parameters measured	Reference
	LeafAnalyser	Analyzes leaf shape variation	Weight et al. (2007)
	LAMINA	Measures leaf shape (blade dimensions) and size (area)	Bylesjo et al. (2008)
	HYPOTrace	Measures growth rate and hook angle of hypocotyls	Wang et al. (2009)
	HTPheno	Measures plant height, width and projected shoot area	Hartmann et al. (2011)
	LEAFPROCESSOR	Measures different leaf geometries	Backhaus et al. (2010)
	LEAF-GUI	Analyzes macroscopic structure of veins in leaves	Price et al. (2011)
	Canopy Analysis	Extracts forest canopy cover	Korhonen and Heikkinen (2009)
	Assess	Analyzes leaf area, percent disease, root length, lesion count, percent ground cover	Lakhdar Lamari
	LemnaTec 3D Scanalyzer	High-throughput platform for analysis of color, shape, size, and architecture	Golzarian et al. (2011), http://www.lemnatec.com/
	GROWSCREEN FLUORO	Platform for growth analysis and chlorophyll-fluorescence analysis	
Seeds/grain	WinSEEDLE	Measures volume and surface area of seeds and needles	http://www.regent.qc.ca/products/needle/WinSEEDLE.html
	SHAPE	Extracts the contour shape from a full color bitmap image	Iwata and Ukai (2002) and Iwata et al. (2010)
	ImageJ	General image analysis software for area, size, and shape; applied to grain	Herridge et al. (2011), http://rsb.info.nih.gov/ij/
	SmartGrain	High-throughput measurement of seed shape	Tanabata et al. (2012)

GROWSCREEN which is capable of measuring total leaf area, leaf growth, and relative growth rate of large plant populations (Walter et al. 2007). Julich Plant Phenotyping Centre has developed a series of softwares like GROW Map-Root capable of measuring spatial and temporal distribution of root growth (Nagel et al. 2006; Walter et al. 2002). GROWMAP Leaf is capable of spatial and temporal distribution of leaf growth via Digital Image Sequence Processing Setup by digital image sequence processing (Walter and Schurr 2000); GROWSCREEN-Root an image-based software is capable of automatic analysis of root architecture by growing the plants in agar-filled Petri dishes. The software determines parameters like length of primary root, total length and number of lateral roots, and angle of branching and finally draws a false-color image tree model for the root system (Nagel et al. 2009). Similarly GROWSCREEN-Rhizo is a robot-enabled platform for simultaneous measurements of root and shoot growth for plants grown in soil-filled rhizotrons and can be used in plant phenomics studies involving evaluation of root and shoot responses to different water, nutrient (N, P), and soil compaction regimes (Nagel et al. 2012). Several other software packages have been developed to analyze different plant images obtained from phenomics platform (Table 4).

Limitations and Future Perspectives

Increasing population and per capita income necessitates the doubling of food production by 2030 which seems to be exceedingly difficult due to the global climate change. This situation warrants the development of stress-tolerant crop varieties suitable for marginal environments where the occurrence of various biotic/abiotic stresses are more frequent. There is an immense need of identifying new germplasms and landraces who can bear the harsher climate of the future. One of the major challenges for crop researchers is "how to predict crop performance as a function of genetic architecture." While our ability to evaluate phenotypes has grown exponentially in recent years, our ability to translate this wealth of phenotypic data to practical knowledge is still lagging behind.

There is a substantial need for automation of phenotypic evaluation processes (Kolukisaoglu and Thurow 2010), which will not only increase the high throughput but also improve reproducibility of the results and therefore the overall quality of an experiment. Now with the advancements made in the field of high-throughput imaging, robotics, and computational science, it is possible to establish a high-throughput plant phenotyping platform capable of measuring hundreds to thousands of traits in a day. This task of capturing and analyzing maximum amount of data and applying the results in any breeding program requires strong cooperation between agricultural scientists, engineers, and computer scientists.

Most of the abiotic stresses like drought, salinity, and high temperature occur in combination under field conditions. But most of automated phenotyping platforms developed in the public domain have been deployed for use under greenhouse conditions. High-throughput phenotyping techniques meant for use under field conditions has not been developed to a substantial level (White et al. 2012). Mishra et al. (2011) have developed an automated plant screening method for evaluating low-temperature tolerance-related traits in *Arabidopsis* by using an imaging fluorometer capable of estimating chlorophyll fluorescence emission. Automated plant phenotyping methods have been developed for measuring drought and salinity stress tolerance-related traits, but the methods for screening the germplasms for high-temperature stress, cold stress, and submergence tolerance are still lacking. As earth's climate is predicted to warm by an average of 2–4 °C (IPCC 2007) by the end of the twenty-first century, our future focus on phenomics should be based on morphological characteristics, viz., panicle surrounded by many leaves (Wassmann et al. 2009), early flowering and anthesis (Jagadish et al. 2007), genotypes with large anthers (Matsui and Omasa 2002), and anthers with large basal pores (Matsui and Kagata 2003). Efforts are needed to include these observations in high-throughput phenotyping techniques so that more germplasms can be screened for high-temperature stress tolerance both under controlled and open environments. Moreover, the systems developed for large-scale phenotyping is limited to small group of species with similar growth habit and requirements, like in *Arabidopsis* (Arvidsson et al. 2011; Granier et al. 2006; Jansen et al. 2009) and main cereal crops. More generic platforms enabling simultaneous evaluation of multiple species have not been implemented so far.

Automated phenotyping approaches are still expensive because of hardware requirements, viz., robotics, imaging devices, and computing infrastructure. Researchers in developing countries can join hands with phenotyping installations at European Plant Phenotyping Network (http://www.plant-phenotyping-network.eu), Australian Plant Phenomics Facility (http://www.plantphenomics.org.au), Julich Plant Phenotyping Centre-JPPC (http://www.fz-juelich.de/ibg/ibg-2/EN/organisation/JPPC/JPPC_node.html), and International Plant Phenomics Network (http://www.plantphenomics.com) for accessing high-throughput phenomics facilities. Data analysis tools similar to HTPheno (Hartmann et al. 2011) and Integrated Analysis Platform (IAP) (Klukas et al. 2012) need to be developed and popularized. The proliferation of platforms for high-throughput phenomics raises the possibility of bridging the knowledge gap between genotypes and phenotypes and has given the plant scientists new insights into the information encoded in the sequenced plant genomes.

References

Armengaud P, Zambaux K, Hills A, Sulpice R, Pattison RJ, Blatt MR, Amtmann A (2009) EZ-Rhizo: integrated software for the fast and accurate measurement of root system architecture. Plant J 57:945–956

Arvidsson S, Rodriguez PP, Roeber BM (2011) A growth phenotyping pipeline for *Arabidopsis thaliana* integrating image analysis and rosette area modeling for robust quantification of genotype effects. New Phytol 191:895–907

Asseng S, Aylmore LAB, MacFall JS, Hopmans JW, Gregory PJ (2000) Computer-assisted tomography and magnetic resonance imaging. In: Smit AL, Bengough AG, Engels C, van Noordwijk M, Pellerin S, van de Geijn SC (eds) Techniques for studying roots. Springer, Berlin, pp 343–363

Backhaus A, Kuwabara A, Bauch M, Monk N, Sanguinetti G, Fleming A (2010) LEAFPROCESSOR: a new leaf phenotyping tool using contour bending energy and shape cluster analysis. New Phytol 187:251–261

Baker NR (2008) Chlorophyll fluorescence: a probe of photosynthesis *in vivo*. Annu Rev Plant Biol 59:89–113

Balachandran S, Hurry VM, Kelly SE, Osmond CB, Robinson SA, Rohozinski J, Seaton GGR, Sims DA (1997) Concepts of plant biotic stress: some insights into the stress physiology of virus infected plants, from the perspective of photosynthesis. Physiol Plant 100:203–213

Basu P, Pal A (2012) A new tool for analysis of root growth in the spatio-temporal continuum. New Phytol 195(1):264–274

Basu P, Pal A, Lynch JP, Brown KM (2007) A novel image-analysis technique for kinematic study of growth and curvature. Plant Physiol 145:305–316

Bateson W (1906) The progress of genetic research: an inaugural address to the third conference on hybridisation and plant-breeding. In: Punett RS (ed) Scientific papers of William Bateson (1928). Cambridge University Press, Cambridge, pp 142–151

Baxter I, Ouzzani M, Orcun S, Kennedy B, Jandhyala SS, Salt DE (2007) Purdue ionomics information management system: an integrated functional genomics platform. Plant Physiol 143:600–611

Berni JAJ, Tejada PJZ, Suarez L, Fereres E (2009) Thermal and narrowband multispectral remote sensing for vegetation monitoring from an unmanned aerial vehicle. IEEE Trans Geo Sci Remote 47:722–738

Billiau K, Sprenger H, Schudoma C, Walther D, Kohl KI (2012) Data management pipeline for plant phenotyping in a multisite project. Funct Plant Biol 39:948–957

Biskup B, Scharr H, Schurr U, Rascher U (2007) A stereo imaging system for measuring structural parameters of plant canopies. Plant Cell Environ 10:1299–1308

Blake VC, Kling JG, Hayes PM, Jannink JL, Jillella SR, Lee J, Matthews DE, Chao S, Close TJ, Muehlbauer GJ (2012) The hordeum toolbox: the Barley coordinated agricultural project genotype and phenotype resource. Plant Genome 5:81–91

Blum A, Mayer S, Galon G (1989) Agronomic and physiological assessments of genotypic variation for drought resistance in sorghum. Aust J Agric Res 40:49–61

Bock CH, Poole GH, Parker PE, Gottwald TR (2010) Plant disease severity estimated visually, by digital photography and image analysis and by hyperspectral imaging. Crit Rev Plant Sci 29:59–107

Bombarely A, Menda N, Tecle IY, Buels RM, Strickler S, Fischer-York T, Pujar A, Leto J, Gosselin J, Mueller LA (2011) The Sol Genomics Network (http://solgenomics.net): growing tomatoes using Perl. Nucleic Acids Res 39:D1149–D1155

Borisjuk L, Rolletschek H, Neuberger T (2012) Surveying the plant's world by magnetic resonance imaging. Plant J 70:129–146

Buhler J, Huber G, Schmid F, Blumler P (2011) Analytical model for long-distance tracer-transport in plants. J Theor Biol 270:70–79

Bylesjo M, Segura V, Soolanayakanahally RY, Rae AM, Trygg J, Gustafsson P, Jansson S, Street NR (2008) LAMINA: a tool for rapid quantification of leaf size and shape parameters. BMC Plant Biol 8:82. doi:10.1186/1471-2229-8-82

Canaran P, Buckler ES, Glaubitz JC, Stein L, Sun Q, Zhao W, Ware D (2008) Panzea: an update on new content and features. Nucleic Acids Res 36:D1041–D1043

Cardon ZG, Mott KA, Berry JA (1994) Dynamics of patchy stomatal movements, and their contribution to steady-state and oscillating stomatal conductance calculated using gas exchange techniques. Plant Cell Environ 17:995–1007

Cattivelli L, Rizza F, Badeck FW, Mazzucotelli E, Mastrangelo AM, Francia E, Mare C, Tondelli A, Stanca AM (2008) Drought tolerance improvement in crop plants: an integrated view from breeding to genomics. Field Crop Res 105:1–14

Chaerle L, Van Der Straeten D (2001) Seeing is believing: imaging techniques to monitor plant health. Biochim Biophys Acta 1519:153–166

Chaerle L, Leinonen I, Jones HG, Vander Straeten D (2007) Monitoring and screening plant populations with combined thermal and chlorophyll fluorescence imaging. J Exp Bot 58:773–784

Chaerle L, Lenk S, Leinonen I, Jones HG, Van Der Straeten D, Buschmann C (2009) Multi-sensor plant imaging: towards the development of a stress-catalogue. Biotechnol J 4:1152–1167

Chen C, DeClerck G, Casstevens T, Youens-Clark K, Zhang J, Ware D, Jaiswal P, McCouch S, Buckler E (2010) The gramene genetic diversity module: a resource for genotype-phenotype association analysis in grass species. Nat Precedings. doi:http://hdl.handle.net/10101/npre.2010.4645.1

Clark RT, MacCurdy RB, Jung JK, Shaff JE, McCouch SR, Aneshansley DJ, Kochian LV (2011) Three-dimensional root phenotyping with a novel imaging and software platform. Plant Physiol 156:455–465

Clark RT, Famoso AN, Zhao K, Shaff JE, Craft EJ, Bustamante CD, McCouch SR, Aneshansley DJ, Kochian LV (2012) High throughput two dimensional root system phenotyping platform facilitates genetic analysis of root growth and development. Plant Cell Environ. doi:10.1111/j.1365-3040.2012.02587.x

Close T, Riverside UC, Last R et al (2011) National Science Foundation Phenomics: genotype to phenotype, a report of the NIFA-NSF phenomics workshop. (www.nsf.gov/bio/pubs/reports/phenomics_workshop_report.pdf) Assessed on 27 Nov 2014

Colaizzi PD, Barnes EM, Clarke TR, Choi CY, Waller PM, Haberland J, Kostrzewski M (2003) Water stress detection under high frequency sprinkler irrigation with water deficit index. J Irrig Drain Eng-ASCE 129:36–43

Cook JP, McMullen MD, Holland JB, Tian F, Bradbury P, Ibarra JR, Buckler ES, Garcia SAF (2012) Genetic architecture of maize kernel composition in the nested association mapping and inbred association panels. Plant Physiol 158:824–834

Droc G, Perin C, Fromentin S, Larmande P (2009) OryGenes DB 2008 update: database interoperability for functional genomics of rice. Nucleic Acids Res 37:D992–D995

Duan LF, Yang WN, Bi K, Chen SB, Luo QM, Liu Q (2011a) Fast discrimination and counting of filled/unfilled rice spikelets based on bi-modal imaging. Comput Electron Agric 75:196–203

Duan LF, Yang WN, Huang CL, Liu Q (2011b) A novel machine-vision- based facility for the automatic evaluation of yield-related traits in rice. Plant Methods 7:44. doi:10.1186/1746-4811-7-44

Eberius M, Guerra JL (2009) High-throughput plant phenotyping–data acquisition, transformation, and analysis. In: Edwards D et al (eds) Bioinformatics: tools and applications. Springer, New York, pp 259–278

Fabre J, Dauzat M, Negre V, Wuyt N, Tireau A, Gennari E, Neveu P, Tisne S, Massonnet C, Hummel I, Granier C (2011) PHENOPSIS DB: an information system for Arabidopsis thaliana phenotypic data in an environmental context. BMC Plant Biol 11:77

Fang S, Yan X, Liao H (2009) 3D reconstruction and dynamic modeling of root architecture in situ and its application to crop phosphorus research. Plant J 60:1096–1108

Fiorani F, Rascher U, Jahnke S, Schurr U (2012) Imaging plants dynamics in heterogenic environments. Curr Opin Biotechnol 23:227–235

Flagella Z, Pastore D, Campanile RG, Fonzo ND (1994) Photochemical quenching of chlorophyll fluorescence and drought tolerance in different durum wheat (*Triticum durum*) cultivars. J Agric Sci 122:183–192

Flavel RJ, Guppy CN, Tighe M, Watt M, McNeill A, Young IM (2012) Non-destructive quantification of cereal roots in soil using high-resolution X-ray tomography. J Exp Bot 63:2503–2511

Flexas J, Bota J, Galmes J, Medrano H, Carbo MR (2006) Keeping a positive carbon balance under adverse conditions: responses of photosynthesis and respiration to water stress. Physiol Plant 127:343–352

French A, Tomas SU, Holman TJ, Bennett MJ, Pridmore T (2009) High-throughput quantification of root growth using a novel image-analysis tool. Plant Physiol 150:1784–1795

Fukatsu T, Watanabe T, Hu HM, Yoichi H, Hirafuji M (2012) Field monitoring support system for the occurrence of *Leptocorisa chinensis* Dallas (Hemiptera: Alydidae) using synthetic attractants Field Servers, and image analysis. Comput Electron Agric 80:8–16

Furbank RT (2009) Plant phenomics: from gene to form and function. Funct Plant Biol 36:10–11

Furbank RT, Tester M (2011) Phenomics-technologies to relieve the phenotyping bottleneck. Trends Plant Sci 16:635–644

Galkovskyi T, Mileyko Y, Bucksch A, Moore B, Symonova O, Price CA, Topp CN, Iyer-Pascuzzi AS, Zurek PR, Fang S (2012) GiARoots: software for the high throughput analysis of plant root system architecture. BMC Plant Biol 12:116

Gerke J, Lorenz K, Cohen B (2009) Genetic interactions between transcription factors cause natural variation in yeast. Science 323:498–501

Gjuvsland AB, Vik JO, Beard DA, Hunter PJ, Omholt SW (2013) Bridging the genotype–phenotype gap: what does it take? J Physiol 591:2055–2066

Golzarian MR, Frick RA, Rajendran K, Berger B, Roy S, Tester M, Lun DS (2011) Accurate inference of shoot biomass from high-throughput images of cereal plants. Plant Methods 7:2. doi:10.1186/1746-4811-7-2

Gove PB (1981) Webster's third new international dictionary of the English language, unabridged. Merriam, Springfield

Granier C, Aguirrezabal L, Chenu K, Cookson SJ, Dauzat M, Hamard P, Thioux JJ, Rolland G, Bouchier-Combaud S, Lebaudy A (2006) PHENOPSIS, an automated platform for reproducible phenotyping of plant responses to soil water deficit in *Arabidopsis thaliana* permitted the identification of an accession with low sensitivity to soil water deficit. New Phytol 169:623–635

Gregory PJ, Hutchison DJ, Read DB, Jenneson PM, Gilboy WB, Morton EJ (2003) Non-invasive imaging of roots with high resolution X-ray micro-tomography. Plant Soil 255:251–259

Gregory PJA, Bengough G, Grinev D, Schmidt S, Thomas TB, Wojciechowski T, Young IM (2009) Root phenomics of crops: opportunities and challenges. Funct Plant Biol 36:922–929

Guarino L, Lobell D (2011) A walk on the wild side. Nat Clim Chang 8:374–375

Haberland JA, Colaizzi PD, Kostrzewski MA, Waller PM, Choi CY, Eaton FE, Barnes EM, Clarke TR (2010) AgIIS, Agricultural Irrigation Imaging System. Appl Eng Agric 26:247–253

Hartmann A, Czauderna T, Hoffmann R, Stein N, Schreiber F (2011) HTPheno: an image analysis pipeline for high-throughput plant phenotyping. BMC Bioinf 12:148

Heffner EL, Jannink JL, Sorrells ME (2011) Genomic selection accuracy using multifamily prediction models in a wheat breeding program. Plant Genome 4:65–75

Herridge RP, Day RC, Baldwin S, Macknight RC (2011) Rapid analysis of seed size in Arabidopsis for mutant and QTL discovery. Plant Methods 7:3. doi:10.1186/1746-4811-7-3

Hillnhutter C, Sikora RA, Oerke EC, van Dusschoten D (2012) Nuclear magnetic resonance: a tool for imaging belowground damage caused by *Heterodera schachtii* and *Rhizoctonia solani* on sugar beet. J Exp Bot 63:319–327

Hillocks RJ (1992) Cotton diseases. CAB International, Wallingford

Hossain MM, Nonami H (2010) Effects of water flow from the xylem on the growth-induced water potential and the growth-effective turgor associated with enlarging tomato fruit. Environ Control Biol 48:101–116. doi:10.2525/ecb.48.101

Houle D, Govindaraju DR, Omholt S (2010) Phenomics: the next challenge. Nat Rev Genet 11:855–866

Huang JR, Liao HJ, Zhu YB, Sun JY, Sun QH, Liu XD (2012) Hyperspectral detection of rice damaged by rice leaf folder (*Cnaphalocrocis medinalis*). Comput Electron Agric 82:100–107

Igathinathane C, Pordesimo LO, Batchelor WD (2009) Major orthogonal dimensions measurement of food grains by machine vision using image. J Food Res Int 42:76–84

Ikeda M, Hirose Y, Takashi T, Shibata Y, Yamamura T, Komura T, Doi K, Ashikari M, Matsuoka M, Kitano H (2010) Analysis of rice panicle traits and detection of QTLs using an image analyzing method. Breed Sci 60:55–64

IPCC (2007) Working Group II. 4th Assessment Report, 'Impacts, Adaptation and Vulnerability'. http://www.ipcc-wg2.org/

Iwata H, Ukai Y (2002) SHAPE: a computer program package for quantitative evaluation of biological shapes based on elliptic Fourier descriptors. J Hered 93:384–385

Iwata H, Ebana K, Uga Y, Hayashi T, Jannink JL (2010) Genome-wide association study of grain shape variation among Oryza sativa L. germplasms based on elliptic fourier analysis. Mol Breed 25:203–215

Jackson SA, Iwata A, Lee SH, Schmutz J, Shoemaker R (2011) Sequencing crop genomes: approaches and applications. New Phytol 191:915–925

Jagadish SVK, Craufurd PQ, Wheeler TR (2007) High temperature stress and spikelet fertility in rice (*Oryza sativa* L.). J Exp Bot 58:1627–1635

Jahnke S, Menzel MI, van Dusschoten D, Roeb GW, Buhler J, Minwuyelet S, Blumler P, Temperton VM, Hombach T, Streun M (2009) Combined MRI–PET dissects dynamic changes in plant structures and functions. Plant J 59:634–644

James RA, Sirault XR (2012) Infrared thermography in plant phenotyping for salinity tolerance. Methods Mol Biol 913:173–189

Jansen M, Gilmer F, Biskup B, Nagel KA, Rascher U, Fischbach A, Briem S, Dreissen G, Tittmann S, Braun S (2009) Simultaneous phenotyping of leaf growth and chlorophyll fluorescence *via* GROWSCREEN FLUORO allows detection of stress tolerance in *Arabidopsis thaliana* and other rosette plants. Funct Plant Biol 36:902–914

Jensen T, Apan A, Young F, Zeller L (2007) Detecting the attributes of a wheat crop using digital imagery acquired from a low-altitude platform. Comput Electron Agric 59:66–77

Jones HG, Serraj R, Loveys BR, Xiong L, Wheaton A, Price AH (2009) Thermal infrared imaging of crop canopies for the remote diagnosis and quantification of plant responses to water stress in the field. Funct Plant Biol 36:978–989

Juliette F, Myriam D, Vincent N, Nathalie W, Emilie G, Pascal N, Sebastien T, Catherine M, Irene H, Christine G (2011) PHENOPSIS DB: an information system for *Arabidopsis thaliana* phenotypic data in an environmental context. BMC Plant Biol 11:77

Jung C, Bracker G (2009) Automation accelerates crop improvement. Tecan J 1/2009:22–23

Kahraman A, Avramov A, Nashev LG, Popov D, Ternes R, Pohlenz HD, Weiss B (2005) PhenomicDB: a multi-species genotype/phenotype database for comparative phenomics. Bioinformatics 21:418–420

Karunakaran C, Jayas DS, White NDG (2004a) Detection of internal wheat seed infestation by *Rhyzopertha dominica* using X-ray imaging. J Stored Prod Res 40:507–516

Karunakaran C, Jayas DS, White NDG (2004b) Identification of Wheat Kernels damaged by the Red Flour Beetle using X-ray images. Biosyst Eng 87:267–274

Kastberger G, Stachl R (2003) Infrared imaging technology and biological applications. Behav Res Methods Instrum Comput 35:429–439

Kawamura S, Natsuga M, Takekura K, Itoh K (2003) Development of an automatic rice-quality inspection system. Comput Electron Agric 40:115–126

Kelley B (2009) Agri-photonics. SPIE Prof 7:14–17

Kiyomiya S, Nakanishi H, Uchida H, Tsuji A, Nishiyama S, Futatsubashi M, Tsukada H, Ishioka NS, Watanabe S, Ito T (2011) Real time visualization of ^{13}N-translocation in rice under different environmental conditions using positron emitting tracer imaging system. Plant Physiol 125:1743–1753

Klukas C, Pape JM, Entzian A (2012) Analysis of high-throughput plant image data with the information system IAP. J Integr Bioinform 9:191

Kolukisaoglu U, Thurow K (2010) Future and frontiers of automated screening in plant sciences. Plant Sci 178:476–484

Korhonen L, Heikkinen J (2009) Automated analysis of *in situ* canopy images for the estimation of forest canopy cover. For Sci 55(4):323–334

Kostrzewski M, Waller P, Guertin P, Haberland J, Colaizzi P, Barnes E, Thompson T, Clarke T, Riley E, Choi C (2003) Ground-based remote sensing of water and nitrogen stress. Trans ASAE 46:29–38

Lafitte R, Blum A, Atlin G (2003) Using secondary traits to help identify drought-tolerant genotypes. In: Fischer KS, Lafitte R, Fukai S, Atlin G, Hardy B (eds) Breeding rice for drought-prone environments. International Rice Research Institute, Los Banos

Lamari L (2008) ASSESS 2.0: image analysis software for plant disease quantification. American Phytopathological Society, St Paul

Larmande P, Gay C, Lorieux M, Perin C, Bouniol M, Droc G, Sallaud C, Perez P, Barnola I, Biderre-Petit C (2008) *Oryza* tag line, a phenotypic mutant database for the genoplante rice insertion line library. Nucleic Acids Res 36:D1022–D1027

Lawrence CJ, Schaeffer ML, Seigfried TE, Campbell DA, Harper LC (2007) MaizeGDB's new data types, resources and activities. Nucleic Acids Res 35(Database issue):D895–D900

Le Bot J, Serra V, Fabre J, Draye X, Adamowicz S, Pages L (2010) DART: a software to analyse root system architecture and development from captured images. Plant Soil 326:261–273

Lee JM, Davenport GF, Marshall D, Noel Ellis TH, Ambrose MJ, Dicks J, van Hintum TJL, Flavell AJ (2005) GERMINATE. a generic database for integrating genotypic and phenotypic information for plant genetic resource collections. Plant Physiol 139(2):619–631

Leinonen I, Jones HG (2004) Combining thermal and visible imagery for estimating canopy temperature and identifying plant stress. J Exp Bot 55:1423–1431

Levitt J (1972) Responses of plants to environmental stresses. Academic, New York

Li YF (2010) PODD: towards an extensible, domain-agnostic scientific data management system. In: Proceedings of 6th IEEE eScience conference 2010. University of Queensland, Brisbane, Australia, 2010, pp 137–144

Li YF, Kennedy G, Davies F, Hunter J (2010) PODD: an ontology-driven data repository for collaborative phenomics research. In: The role of digital libraries in a time of global change, Lecture notes in computer science, 6102. Springer, Berlin, pp 179–188

Lin P, Chen YM, He Y (2012) Identification of broken rice kernels using image analysis techniques combined with velocity representation method. Food Bioprocess Technol 5:796–802

Liu ZY, Shi JJ, Zhang LW, Huang JF (2010) Discrimination of rice panicles by hyperspectral reflectance data based on principal component analysis and support vector classification. Zhejiang Univ Sci B (Biomed Biotechnol) 11:71–78

Lobet G, Pages L, Draye X (2011) A novel image-analysis toolbox enabling quantitative analysis of root system architecture. Plant Physiol 157:29–39

Lu Y, Savage LJ, Ajjawi I et al (2008) New connections across pathways and cellular processes: industrialized mutant screening reveals novel associations between diverse phenotypes in Arabidopsis. Plant Physiol 146:1482–1500

Lu Y, Savage LJ, Larson MD, Wilkerson CG, Last RL (2011) Chloroplast 2010: a database for large-scale phenotypic screening of Arabidopsis mutants. Plant Physiol 155:1589–1600

Mabee P, Balhoff J, Dahdul W, Lapp H, Midford P, Vision T, Westerfield M (2012) 500,000 fish phenotypes: the new informatics landscape for evolutionary and developmental biology of the vertebrate skeleton. J Appl Ichthyol 28:300–305

Mairhofer S, Zappala S, Tracy SR, Sturrock C, Bennett M, Mooney SJ, Pridmore T (2012) RooTrak: automated recovery of three-dimensional plant root architecture in soil from x-ray microcomputed tomography images using visual tracking. Plant Physiol 158(2):561–569

Malenovsky Z, Mishra KB, Zemek F, Rascher U, Nedbal L (2009) Scientific and technical challenges in remote sensing of plant canopy reflectance and fluorescence. J Exp Bot 60:2987–3004

Manickavasagan A, Jayas DS, White NDG (2008) Thermal imaging to detect infestation by *Cryptolestes ferrugineus* inside wheat kernels. J Stored Prod Res 44:186–192

Masle J, Gilmore SR, Farquhar GD (2005) The ERECTA gene regulates plant transpiration efficiency in Arabidopsis. Nature 436:866–870

Matsui T, Kagata H (2003) Characteristics of floral organs related to reliable self-pollination in rice (*Oryza sativa* L.). Ann Bot 91:473–477

Matsui T, Omasa K (2002) Rice (*Oryza sativa* L.) cultivars tolerant to high temperature at flowering: anther characteristics. Ann Bot 89:683–687

Menda N, Semel Y, Peled D, Eshed Y, Zamir D (2004) In silico screening of a saturated mutation library of tomato. Plant J 38:861–872

Meroni M, Rossini M, Guanter L, Alonso L, Rascher U, Colombo R, Moreno J (2009) Remote sensing of solar induced chlorophyll fluorescence: review of methods and applications. Remote Sens Environ 113:2037–2051

Merz TC, Chapman S (2011) Autonomous unmanned helicopter system for remote sensing missions in unknown environments. Int Arch Photogramm Remote Sens Spat Inf Sci 38-1/C22:1–6

Milc J, Sala A, Bergamaschi S, Pecchioni N (2011) A genotypic and phenotypic information source for marker-assisted selection of cereals: the CEREALAB database. Database Article ID baq038, doi:10.1093/database/baq038

Minoia S, Petrozza A, D'Onofrio O, Piron F, Mosca G, Sozio G, Cellini F, Bendahmane A, Carriero F (2010) A new mutant genetic resource for tomato crop improvement by TILLING technology. BMC Res Notes 3:69

Mishra A, Mishra KB, Hoermiller II, Heyer AJ, Nedbal L (2011) Chlorophyll fluorescence emission as a reporter on cold tolerance in *Arabidopsis thaliana* accessions. Plant Signal Behav 6:301–331

Mishra KB, Iannacone R, Petrozza A, Mishra A, Armentano N, Vecchia GL, Trtílek M, Cellini F,

Nedbal L (2012) Engineered drought tolerance in tomato plants is reflected in chlorophyll fluorescence emission. Plant Sci 182:79–86

Miyao A, Tanaka K, Murata K, Sawaki H, Takeda S, Abe K, Shinozuka Y, Onosato K, Hirochika H (2003) Target site specificity of the *Tos17* retrotransposon shows a preference for insertion within genes and against insertion in retrotransposon-rich regions of the genome. Plant Cell 15:1771–1780

Mogel KH (2013) Taking the phenomics revolution into the field. CSA News Mag 58:4–10

Moller M, Alchanatis V, Cohen Y, Meron M, Tsipris J, Naor A, Ostrovsky V, Sprintsin M, Cohen S (2007) Use of thermal and visible imagery for estimating crop water status of irrigated grapevine. J Exp Bot 58:827–838

Montes JM, Melchinger AE, Reif JC (2007) Novel throughput phenotyping platforms in plant genetic studies. Trends Plant Sci 12:433–436

Moshou D, Bravo C, Oberti R, West J, Bodria L, McCartney A, Ramon H (2005) Plant disease detection based on data fusion of hyper-spectral and multi-spectral fluorescence imaging using Kohonen maps. Real-Time Imag 11:75–83

Munns R, Tester M (2008) Mechanisms of salinity tolerance. Annu Rev Plant Biol 59:651–681

Munns R, James RA, Sirault XRR, Furbank RT, Jones HG (2010) New phenotyping methods for screening wheat and barley for beneficial responses to water deficit. J Exp Bot 61:3499–3507

Naeem A, French AP, Wells DM, Pridmore TP (2011) High throughput feature counting and measurement of roots. Bioinformatics 27:1337–1338

Nagel KA, Schurr U, Walter A (2006) Dynamics of root growth stimulation in *Nicotiana tabacum* in increasing light intensity. Plant Cell Environ 29:1936–1945

Nagel KA, Kastenholz B, Jahnke S, van Dusschoten D, Aach T, Muhlich M, Truhn D, Scharr H, Terjung S, Walter A, Schurr U (2009) Temperature responses of roots: impact on growth, root system architecture and implications for phenotyping. Funct Plant Biol 36:947–959

Nagel KA, Putz A, Gilmer F et al (2012) GROWSCREEN-Rhizo is a novel phenotyping robot enabling simultaneous measurements of root and shoot growth for plants grown in soil-filled rhizotrons. Funct Plant Biol 39:891–904

Nainanayake AD (2007) Use of chlorophyll fluorescence parameters to assess drought tolerance of coconut varieties. COCOS 18:77–105

Neethirajan S, Jayas DS, Karunakaran C (2007a) Dual energy X-ray image analysis for classifying vitreousness in durum wheat. Post Harv Biol Technol 45:381–384

Neethirajan S, Jayas DS, White NDG (2007b) Detection of sprouted wheat kernels using soft X-ray image analysis. J Food Eng 81:509–513

Nguyen HT, Lee BW (2006) Assessment of rice leaf growth and nitrogen status by hyperspectral canopy reflectance and partial least square regression. Eur J Agron 24:349–356

O'Toole JC, Chang TT (1979) Drought resistance in cereals: a case study. In: Mussell H, Staples RC (eds) Stress physiology in crop plants. John Wiley and Sons, New York

Omasa K, Hosoi F, Konishi A (2007) 3D LIDAR imaging for detecting and understanding plant responses and canopy structure. J Exp Bot 58:881–898

Osmond CB, Daley PF, Badger MR, Luttge U (1998) Chlorophyll fluorescence quenching during photosynthetic induction in leaves of *Abutilon striatum* Dicks: infected with Abutilon mosaic virus, observed with a field-portable imaging system. Bot Acta 111:390–397

Osmond CB, Kramer D, Luttge U (1999) Reversible, water stress induced non-uniform chlorophyll fluorescence quenching in wilting leaves of *Potentilla reptans* may not be due to patchy stomatal responses. Plant Biol 1:618–624

Paproki A, Sirault X, Berry S, Furbank R, Fripp J (2012) A novel mesh processing based technique for 3D plant analysis. BMC Plant Biol 12:63

Pascuzzi Iyer AS, Symonova O, Mileyko Y, Hao Y, Belcher H, Harer J, Weitz JS, Benfey PN (2010) Imaging and analysis platform for automatic phenotyping and trait ranking of plant root systems. Plant Physiol 152:1148–1157

Perret JS, Al-Belushi ME, Deadman M (2007) Non-destructive visualization and quantification of roots using computed tomography. Soil Biol Biochem 39:391–399

Pieruschka R, Poorter H (2012) Phenotyping plants: genes, phenes and machines. Funct Plant Biol 39:813–820

Poorter H, Fiorani F, Stitt M, Schurr U, Finck A, Gibon Y, Usadel B, Munns R, Atkin OK, Tardieu F, Pons TL (2012) The art of growing plants for experimental purposes: a practical guide for the plant biologist. Funct Plant Biol 39:821–838

Price CA, Symonova O, Mileyko Y, Hilley T, Weitz JS (2011) Leaf extraction and analysis framework graphical user interface: segmenting and analyzing the structure of leaf veins and areoles. Plant Physiol 155:236–245

Rahbarian R, Ramazanali KN, Ali G, Bagheri A, Farzaneh N (2011) Drought stress effects on photosynthesis, chlorophyll fluorescence and water relations in tolerant and susceptible chickpea (*Cicer arietinum* l.) genotypes. Acta Biol Cracov Ser Bot 53:47–56

Rahman H, Jagadeeshselvam N, Valarmathi R, Sachin B, Sasikala R, Senthil N, Sudhakar D, Robin S, Muthurajan R (2014) Transcriptome analysis of salinity responsiveness in contrasting genotypes of finger millet (Eleusine coracana L.) through RNA-sequencing. Plant Mol Biol 85(4–5):485–503

Rajendran K, Tester M, Roy SJ (2009) Quantifying the three main components of salinity tolerance in cereals. Plant Cell Environ 32:237–249

Rascher U, Pieruschka R (2008) Spatio-temporal variations of photosynthesis: the potential of optical remote sensing to better understand and scale light use efficiency and stresses of plant ecosystems. Precis Agric 9:355–366

Rascher U, Hutt HT, Siebke K, Osmond CB, Beck F, Luttge U (2001) Spatio-temporal variations of metabolism in a plant circadian rhythm: the biological clock as an assembly of coupled individual oscillators. Proc Natl Acad Sci U S A 98:11801–11805

Rascher U, Blossfeld S, Fiorani F, Jahnke S, Jansen M (2011) Non-invasive approaches for phenotyping of enhanced performance traits in bean. Funct Plant Biol 38:968–983

Reddy TY, Reddy VR, Anbumozhi V (2003) Physiological responses of groundnut (Arachis hypogea L.) to drought stress and its amelioration: a critical review. Plant Growth Regul 41:75–88

Reuzeau C, Frankard V, Hatzfeld Y, Sanz A, Van Camp W, Lejeune P, De Wilde C, Lievens K, de Wolf J, Vranken E (2006) TraitmillTM: a functional genomics platform for the phenotypic analysis of cereals. Plant Genet Resour Charact Util 4:20–24

Reuzeau C, Pen J, Frankard V, de Wolf J, Peerbolte R, Broekaert W, van Camp W (2010) TraitMill: a discovery engine for identifying yield-enhancement genes in cereals. Mol Plant Breed 5:753–759

Richards RA, Greg JR, Michelle Watt AG, Spielmeyer W, Dolferus R (2010) Breeding for improved water productivity in temperate cereals: phenotyping, quantitative trait loci, markers and the selection environment. Funct Plant Biol 37:85–97

Ritchie GL, Sullivan DG, Perry CD, Hook JE, Bednarz CW (2008) Preparation of a low-cost digital camera system for remote sensing. Appl Eng Agric 24:885–896

Romer C, Burling K, Hunsche M, Rumpf T, Noga G, Plumer L (2011) Robust fitting of fluorescence spectra for pre-symptomatic wheat leaf rust detection with Support Vector Machines. Comput Electron Agric 79:180–188

Roostaei M, Mohammadi SA, Amri A, Majidi E, Nachit M, Haghparast R (2011) Chlorophyll fluorescence parameters and drought tolerance in a mapping population of winter bread wheat in the highlands of Iran. Russ J Plant Physiol 58:351–358

Sakamoto T, Shibayama M, Kimura A, Takada E (2011) Assessment of digital camera-derived vegetation indices in quantitative monitoring of seasonal rice growth. ISPRS J Photogramm 66:872–882

Samson BK, Hasan H, Wade LJ (2002) Penetration of hardpans by rice lines in the rainfed lowlands. Field Crop Res 76:175–188

Schaeffer ML, Harper LC, Gardiner JM, Andorf CM, Campbell DA, Cannon EK, Sen TZ, Lawrence CJ (2011) MaizeGDB: curation and outreach go hand-in-hand. Database: J Biol Database Curation 2011:bar022. doi:10.1093/database/bar022

Schauer N, Semel Y, Roessner U et al (2006) Comprehensive metabolic profiling and phenotyping of interspecific introgression lines for tomato improvement. Nat Biotechnol 24:447–454

Schmundt D, Stitt M, Jahne B, Schurr U (1998) Quantitative analysis of the local rates of growth of dicot leaves at a high temporal and spatial resolution, using image sequence analysis. Plant J 16:505–514

Scholes JD, Rolfe SA (2009) Chlorophyll fluorescence imaging as a tool for understanding the impact of fungal diseases on plant performance: a phenomics perspective. Funct Plant Biol 36:880–892

Schreiber U, Schliwa U, Bilger W (1986) Continuous recording of photochemical and nonphotochemical chlorophyll fluorescence quenching with a new type of modulation fluorometer. Photosynth Res 10:51–62

Shendure J, Ji H (2008) Next-generation DNA sequencing. Nat Biotechnol 26:1135–1145

Shibayama M, Sakamoto T, Takada E, Inoue A, Morita K, Takahashi W, Kimura A (2011a) Estimating paddy rice leaf area index with fixed point continuous observation of near infrared reflectance using a calibrated digital camera. Plant Prod Sci 14:30–46

Shibayama M, Sakamoto T, Takada E, Inoue A, Morita K, Yamaguchi T, Takahashi W, Kimura A (2011b) Regression-based models to predict rice leaf area index using biennial fixed point continuous observations of near infrared digital images. Plant Prod Sci 14:365–376

Siebke K, Weis E (1995) Assimilation images of leaves of *Glechoma hederacea*: analysis of non-synchronous stomata related oscillations. Planta 196:155–165

Singh CB, Jayas DS, Paliwal J, White NDG (2010) Identification of insect-damaged wheat kernels using short-wave near-infrared hyperspectral and digital colour imaging. Comput Electron Agric 73:118–125

Sirault XRR, James RA, Furbank RT (2009) A new screening method for osmotic component of salinity tolerance in cereals using infrared thermography. Funct Plant Biol 36:970–977

Soule M (1967) Phenetics of natural populations I. Phenetic relationships of insular populations of the side-blotched lizard. Evolution 21:584–591

Sun JP, Hou CY, Feng J, Wang X (2008) Determination of the protein content in rice by the digital chromatic method. J Food Qual 31:250–263

Swarbrick PJ, Schulze-Lefert P, Scholes JD (2006) The metabolic consequences of susceptibility and the activation of race specific or broad spectrum resistance pathways in barley leaves challenged with the powdery mildew fungus. Plant Cell Environ 29:1061–1076

Tackenberg O (2007) A new method for non-destructive measurement of biomass, growth rates, vertical biomass distribution and dry matter content based on digital image analysis. Ann Bot 99:777–783

Tanabata T, Shibaya T, Hori K, Ebana K, Yano M (2012) SmartGrain: high-throughput phenotyping software for measuring seed shape through image analysis. Plant Physiol 160:1871–1880

Tejada ZPJ, Berni JAJ, Subrez L, Sepulcre-Canto G, Morales F, Miller JR (2009) Imaging chlorophyll fluorescence with an airborne narrow-band multispectral camera for vegetation stress detection. Remote Sens Environ 113:1262–1275

Tester M, Langridge P (2010) Breeding technologies to increase crop production in a changing world. Science 327(5967):818–822

The Arabidopsis Genome Initiative (2000) Analysis of the genome sequence of the flowering plant Arabidopsis thaliana. Nature 408(6814):796–815

Topp CN, Anjali SI, Anderson JT et al (2013) 3D phenotyping and quantitative trait locus mapping identify core regions of the rice genome controlling root architecture. PANAS 110:1695–1704

Tsukamoto T, Nakanishi H, Uchida H, Watanabe S, Matsuhashi S, Mori S, Nishizawa NK (2009) ^{52}Fe translocation in barley as monitored by a Positron-Emitting Tracer Imaging System (PETIS): evidence for the direct translocation of Fe from roots to young leaves *via* phloem. Plant Cell Physiol 50:48–57

van As H (2007) Intact plant MRI for the study of cell water relations, membrane permeability, cell-to-cell and long distance water transport. J Exp Bot 58:743–756

van der Weerd L, Claessens MM, Ruttink T, Vergeldt FJ, Schaafsma TJ, Van As H (2001) Quantitative NMR microscopy of osmotic stress responses in maize and pearl millet. J Exp Bot 52:2333–2343

Vandenbroucke K, Metzlaff M (2013) Abiotic stress tolerant crops: genes, pathways and bottlenecks. In: Meyers RA (ed) Encyclopedia of sustainability science and technology. doi: 10.1007/978-1-4419-0851-3

Vankadavath RN, Hussain AJ, Bodanapu R, Kharshiing E, Basha PO, Gupta S, Sreelakshmi Y, Sharma R (2009) Computer aided data acquisition tool for high-throughput phenotyping of plant populations. Plant Methods 5:18. doi:10.1186/1746-4811-5-18

Venora G, Grillo O, Saccone R (2009) Quality assessment of durum wheat storage centres in Sicily: evaluation of vitreous, starchy and shrunken kernels using an image analysis system. J Cereal Sci 49:429–440

Walter A, Schurr U (2000) Spatial variability of leaf development, growth and function. In: Marshall B, Roberts J (eds) Leaf development and canopy growth. Sheffield Academic Press, Sheffield, pp 96–118

Walter A, Spies H, Terjung S, Kuesters R, Kirchgener N, Schurr U (2002) Spatio-temporal dynamics of expansion growth in roots: automatic quantification of diurnal course and temperature response by digital image sequence processing. J Exp Bot 53:689–698

Walter A, Rascher U, Osmond CB (2004) Transition in photosynthetic parameters of midvein and interveinal regions of leaves and their importance during leaf growth and development. Plant Biol 6:184–191

Walter A, Hanno Scharr H, Frank Gilmer F et al (2007) The dynamics of seedling growth acclimation towards altered light conditions can be quantified via GROWSCREEN – a setup designed for rapid optical phenotyping of different plant species. New Phytol 174:447–455

Wang H, Yamauchi A (2006) Growth and function of roots under abiotic stress in soil. In: Huang B (ed) Plant-environment interactions, 3rd edn. CRC Press, New York

Wang Y, Holroyd G, Hetherington AM, Ng CKY (2004) Seeing 'cool' and 'hot'—infrared thermography as a tool for noninvasive, high-throughput screening of Arabidopsis guard cell signalling mutants. J Exp Bot 55:1187–1193

Wang L, Uilecan IV, Assadi AH, Kozmik CA, Spalding EP (2009) HYPOTrace: image analysis software for measuring hypocotyl growth and shape demonstrated on Arabidopsis seedlings undergoing photomorphogenesis. Plant Physiol 149:1632–1637

Wang Z, Liao B, Zhang J (2010) Genomic patterns of pleiotropy and the evolution of complexity. Proc Natl Acad Sci 107:18034–18039

Wassmann R, Jagadish SVK, Heuer S, Ismail A, Redona E, Serraj R, Singh RK, Howell G, Pathak H, Sumfleth K (2009) Climate change affecting rice production: the physiological and agronomic basis for possible adaptation strategies. Adv Agron 101:59–122

Watanabe T, Hanan JS, Room PM, Hasegawa T, Nakagawa H, Takahashi W (2005) Rice morphogenesis and plant architecture: measurement, specification and the reconstruction of structural development by 3D architectural modeling. Ann Bot 2005(95):1131–1143

Weight C, Parnham D, Waites R (2007) TECHNICAL ADVANCE: LeafAnalyser: a computational method for rapid and large-scale analyses of leaf shape variation. Plant J 53:578–586

Weirman A (2010) Plant phenomics teacher resource. http://www.plantphenomics.org.au/files/teacher/Final_Phenomics_for_word_with_images.doc. Accessed on 7 May 2013

White JW, Andrade-Sanchez P, Gore MA, Bronson KF, Coffelt TA, Conley MM, Feldmann KA, French AN, Heun JT, Hunsaker DJ, Jenks MA, Kimball BA, Roth RL, Strand RJ, Thorp KR, Wall GW, Wang G (2012) Field-based phenomics for plant genetics research. Field Crops Res 133:101–112. doi:10.1016/j.for.2012.04.003

Windt CW, Vergeldt FJ, de Jager PA, Van AH (2006) MRI of long-distance water transport: a comparison of the phloem and xylem flow characteristics and dynamics in poplar, castor bean, tomato and tobacco. Plant Cell Environ 29:1715–1729

Windt CW, Gerkema E, Van As H (2009) Most water in the tomato truss is imported through the xylem, not the phloem: a nuclear magnetic resonance flow imaging study. Plant Physiol 151:830–842. doi:10.1104/pp. 109.141044

Woo NS, Badger MR, Pogson BJ (2008) A rapid, non-invasive procedure for quantitative assessment of drought survival using chlorophyll fluorescence. Plant Methods 4:27

Yang Y, Chai RY, He Y (2012) Early detection of rice blast (*Pyricularia*) at seedling stage in Nipponbare rice variety using near-infrared hyper-spectral image. Afr J Biotechnol 11:6809–6817

Yang W, Duan L, Chen G, Xiong L, Liu Q (2013) Plant phenomics and high-throughput phenotyping: accelerating rice functional genomics using multidisciplinary technologies. Curr Opin Plant Biol 16:1–8

Yazdanbakhsh N, Fisahn J (2009) High throughput phenotyping of root growth dynamics, lateral root formation, root architecture and root hair development enabled by PlaRoM. Funct Plant Biol 36:938–946

Yoshioka Y, Iwata H, Tabata M, Ninomiya S, Ohsawa R (2007) Chalkiness in rice: potential for evaluation with image analysis. Crop Sci 47:2113–2120

Yu J, Hu S, Wang J (2002) A draft sequence of the rice genome (*Oryza sativa* L. ssp. *indica*). Science 296:79–92

Zhang J, Li C, Wu C, Xiong L, Chen G, Zhang Q, Wang S (2006) RMD: a rice mutant database for functional analysis of the rice genome. Nucleic Acids Res 34:D745–D748

Zhang QF, Li JY, Xue YB, Han B, Deng XW (2008) Rice 2020: a call for an international coordinated effort in rice functional genomics. Mol Plant 1:715–719

Zhou ZY, Zang Y, Luo XW, Wang P (2011) Color-based corner detection algorithm for rice plant-hopper infestation area on rice stem using the RGB color space. In: ASABE annual international meeting, Louisville, Kentucky, USA paper number 1111374

Zia S, Wenyong D, Spreer W, Spohrer K, Xiongkui H, Muller J (2012) Assessing crop water stress of winter wheat by thermography under different irrigation regimes in North China Plain. Int J Agric Biol Eng 5:3

Plant Cytomics: Novel Methods to View Molecules on the Move

Eric Davies and Bratislav Stankovic

Contents

Abstract

We provide our definition and the brief history of "cytomics" followed by an overview of general methodological approaches of optical imaging, especially fluorescence microscopy. We then go into detail on novel fluor-linking agents (nanobodies, aptamers, and aldehydes) and the array of novel fluors available. We describe many of the new techniques developed for superfast, super-resolution microscopy (photoreactivated localization microscopy, structured illumination microscopy, stimulated emission depletion microscopy, and stochastic optical reconstruction microscopy) followed by quantitative microscopy and image analysis. We then delve into unconventional methods, novel light systems, and alternatives to fluorescence (non-liner optical imaging, single-molecule light absorp-

E. Davies, Ph.D. (✉)
Department of Plant Biology, North Carolina State University, Raleigh, NC, USA
e-mail: edavies.pv@gmail.com

B. Stankovic, Ph.D.
University for Information Science and Technology "St. Paul the Apostle", Partizanska bb, 6000 Ohrid, Republic of Macedonia
e-mail: bratislav.stankovic@fulbrightmail.org

D. Barh et al. (eds.), *PlantOmics: The Omics of Plant Science*,
DOI 10.1007/978-81-322-2172-2_14, © Springer India 2015

tion, luminescent proteins). We then describe how these systems have been employed recently for proteins, nucleic acids, the cytoskeleton, and also small molecules of major interest to plants. We finish with a description of recent findings specific to plant cytomics and furnish several impressive images and other illustrations from the recent plant literature.

Abbreviations

3B	Bayesian analysis of bleaching and blinking
CCD	Charged couple devices
CFP	Cyano fluorescent protein
CMOS	Complementary metal oxide semiconductor
CNOI	Coherent nonlinear optical imaging
DNA	Deoxyribonucleic acid
dSTORM	Direct stochastic optical reconstruction microscopy
FLISM	Fluorescence light sheet microscopy
FOV	Field of view
FRET	Fluorescence (or Förster) resonance energy transfer
FPALM	Fluorescence photoactivation localization microscopy
GFP	Green fluorescent protein
LED	Light-emitting diode
NLDOM	Nonlinear dissipation optical microscopy
OMERO	Open microscopy environment remote objects
PALM	Photoactivation localization microscopy
PGS	Parametric generation spectroscopy
PM	Plasma membrane
PPS	Pump–probe spectroscopy
PY1-ME	Peroxy yellow 1 methyl ester
RNA	Ribonucleic acid
ROS	Reactive oxygen species
RUM	Really unconventional microscopy
SELEX	Systemic evolution of ligands by exponential enrichment
SIM	Structured illumination microscopy
SNAP	Soluble N-ethylmaleimide-sensitive factor-attachment proteins
SPIM	Selective plane illumination microscopy
SR	Super-resolution
SSIM	Saturated structured illumination microscopy
STED	Stimulated emission depletion
STORM	Stochastic optical reconstruction microscopy
tFT	tandem Fluorescent protein timer
TRUE	Time-reversed ultrasound encoded
TULIP	Tunable light-inducible protein tag
YFP	Yellow fluorescent protein

Brief History and Definition(s)

Over a decade ago, we coined the word "cytomics" and defined it as "changes in subcellular location and function of macromolecules (especially proteins) and macromolecular complexes over space and time within intact organisms" (Davies et al. 2001). Our original use of the term was based on ideas put forward earlier (Stankovic et al. 2000) and extended later (Davies and Stankovic 2006; Davies et al. 2012). These ideas arose from our increasing familiarity with the plant cytoskeleton, its complexity of functions, its array of individual proteins forming numerous structures, and its dynamic role in plant metabolism. We were especially interested in proteins with multiple functions and locations, as well as in techniques that would enable us to determine whether individual proteins remained in a single specified location within the cell or whether they could travel from one location to another. We were also interested in whether the same molecule could change its function over time or space. We had become aware of the need for this field of study when we began to work with several multifunctional proteins, which can reside in multiple locations, especially elongation factor 1a (Davies et al. 1998) and apyrase (Shibata et al. 1999). The gene number, subcellular locations, array of isotopes, and multiple functions of these proteins were reviewed recently (Davies et al. 2012).

Despite its recent origins, cytomics has now gained widespread recognition (as of May 2013, there were over 2,000 references in the NCSU library database with cytomics mentioned in the article, but less than 80 involved plant research). The topic has, however, become a major force in medicine (Leif 2009; Támok and Bocsi 2009), pharmaceuticals (Tagore and Gomase 2008; Támok 2010), and basic biology (Murphy 2005; Müller 2008; Robinson 2008; Benndorf et al. 2010). Cytomics is interfaced with proteomics (Bernas et al. 2006) and is considered important for systems biology (Kriete 2005).

General Approach

The structural and functional states of a cell are largely determined by the spatiotemporal organization of its proteome. Understanding the in vivo dynamics of protein localization and their physical interactions is paramount for many problems in biology. With proteins, as with many things, seeing can bring understanding. In particular, in multicellular organisms with cell walls – i.e., plants – it is hard to predict a protein's identity, expression dynamics, or localization solely from its genomic signature. That is why recent advances using specifically localized fluorescent gene markers and image processing have enabled breakthrough automated quantitative analysis of cell growth and genetic activity within living plant tissues. For example, fluorescent protein markers have been used to identify cells and to measure gene expression ratiometrically on a cell level in *Arabidopsis thaliana* (Federici et al. 2012).

Since its identification and isolation from the jellyfish *Aequorea victoria*, the gold standard for imaging in vivo has been the gene encoding green fluorescent protein (GFP). This gene can be inserted into desired model cells or organisms by standard methods, and expression of this gene results in specific fluorescence labeling of live cells without the need for exogenous chemicals. The GFP protein of 238 amino acids can be fused to a protein of interest to visualize its localization and dynamics in a live cell. In addition, GFP can be fused to various protein targeting and localization signals and can thus be subsequently targeted to various cell compartments and sub-compartments. Numerous fluorescent sensors for ions and small molecules, enzymatic activities, signaling cascades, and other important cellular events have all been constructed by fusing GFP with sensitive protein domains (Tsien 1998, 2010; Lukyanov and Belousov 2012).

Since cytomics is based on visualization of specific molecules or macromolecular complexes moving within living tissues in real time, novel methods in this area will be the focus of our chapter. If this chapter had been written more than 3 years ago, we would have stated quite emphatically that the methods would need to involve fluorescence microscopy. However, very recent techniques have been developed that do not use microscopes or fluorescence. Nevertheless the vast bulk of research does, indeed, employ fluorescence microscopy. Basically, fluorescence microscopy involves illuminating a sample from above or below at a specific wavelength and viewing the emitted light at a longer wavelength again from either above or below. If the target (e.g., protein) is not itself fluorescent, it must be made so by tagging it with a fluor. The individual steps involve a linking agent to join the fluor onto the target (macro)molecule, a microscope to view the (moving) fluorescent molecule in vivo, algorithms to convert digital data into pictorial information, and a light source. The light sources of choice now are tunable lasers, since they can be focused on small regions, with varying intensities and very rapid on/off switching. The methods may be useful for a variety of target molecules, complexes, or cell structures, or highly specific to individual targets. The aim of all this research is to make visualization of targets as rapid (superfast), as clear (super-resolution), and as specific as possible.

We will deal with recent publications concerning each of these topics in turn, but as a starting point for beginners in the general field of light microscopy, we suggest a new book (Davidson and Murphy 2012); for fluorescence microscopy we suggest Lichtman and Conchello (2005) and Petty (2007), for super-resolution microscopy

Patterson et al. (2010) and Coltharp and Xiao (2012), and for both fluorescence and super-resolution microscopy Huang et al. (2009), Schermelleh et al. (2010), and Leung and Chou (2011). For more advanced aspects, a series of articles in a *Nature Methods* collection is highly recommended (Nature 2009), available at http://nature.com/nmeth/collections/superresmicroscopy. We will make mention in the appropriate location of any uses that we are aware of, where these very recently developed techniques have been used with plants and then provide a section specifically devoted to recent findings in plants (section "Plant Cytomics").

Novel Fluor-Linking Agents

For molecules that are not innately fluorescent, the most commonly used linking agents are antibodies, specific for the molecule (protein) being investigated, which have the appropriate fluor or set of fluors attached. However the physical size of the antibody and of a fluor such as green fluorescent protein (GFP), or newly developed members of its extended family, limits resolution and may inhibit normal cellular localization, movement, and function, and so efforts to reduce the size (and increase the specificity) of the linking agent are a major area of study.

Nanobodies

The conventional way of linking a fluor to the target protein is by using an antibody specific to that protein. However the bigger the linking agent (and the fluor), the lower the resolution obtained. Further, the antibody (and the fluor) may be larger than the protein of interest and thus can impair the function, localization, and movement of the target protein, thereby generating artifacts. One method to lessen this problem is to use small, high-affinity camelid antibodies of about 13 kDa called nanobodies (Ries et al. 2012). Not only do these lessen the size of the linking agent, they also permit the use of small organic dyes, which are displaced from the target when attached to normal antibodies, which are generally around 150 kDa. These methods have been used recently for high-resolution, single-molecule microscopy of microtubules, living neurons, and intact yeast cells (Ries et al. 2012).

Aptamers

Aptamers are short stretches of single-stranded DNA or RNA that have been subject to SELEX (systemic evolution of ligands by exponential enrichment) which is an in vitro method of generating increasingly effective linking agents to the target-specific proteins (Ulrich et al. 2004). They are generally about the same size as nanobodies, but can be generated not only against immunogenic proteins but also against toxins and non-immunogenic proteins, where antibodies cannot be raised.

Aptamers have been tested recently (Opazo et al. 2012) using primarily the high-resolution technique, STED (stimulated emission depletion) and were found to be as effective as nanobodies. Furthermore, because aptamers (unlike nanobodies) can be conveniently varied in size, these authors (Opazo et al. 2012) were able to show that increasing their size from 15 kDa to 31 kDa and then to 58 kDa greatly reduced the quality of the images obtained when assaying endosomal trafficking.

Aldehydes

A major hurdle for mechanistic studies of many proteins is the lack of a general method for fluorescence labeling with high efficiency, specificity, and speed. This can be overcome with genetic engineering, through incorporation of appropriate motifs into proteins. A novel method for fluorescence labeling for single-molecule imaging employs aldehyde-tagged proteins, and these proteins are made using genetic modification techniques to incorporate the aldehyde motif into the protein (Shi et al. 2012). This method is considered superior to other protein tagging methods in terms of the array of proteins that can be tagged, its approximately 100 % efficiency and specificity of labeling while maintaining the protein's biological function, and resolution of images (Shi et al. 2012). Since the tag (linking agent) is a small part of the protein, it does not add mass, thereby enhancing clarity of images, while the protein is more likely to maintain its normal function and location. Such proteins tagged with aldehyde can be specifically labeled in cell extracts without protein purification and can then be used in single-molecule studies.

Novel Fluors

There has been coevolution in fluors and microscope techniques, with novel fluors begetting novel techniques and novel techniques demanding new fluors; thus the two topics (fluors and microscopes) are not easy to keep separate. Here we will describe the novel fluors first and merely reference by acronym the microscope technique, while details of the techniques will be given later. Fluorescence microscopy in general suffers from the problems stated above, i.e., any modification of the targeted molecule can change its behavior, location, mobility, and function, and the most common technique, immunolabeling, is particularly prone to artifacts. Indeed, a very recent review on immunolabeling artifacts in microscopy (Schnell et al. 2012) discusses pros and cons of immunolabeling in general yet points out the immense value of the cytomics approach (watching protein behavior in living tissue) as being far superior to using fixed tissue and all the artifacts of solubility that this technique can generate.

Super-resolution microscopy (also known as sub-diffraction microscopy – Adam et al. 2011) circumvents the laws of diffraction physics to obtain images with resolution less than the wavelength of light, and these methods generally require special dyes (fluors). This burgeoning field has come about with the development of novel fluors with novel properties. Many of these have been reviewed briefly (Evanko 2012a) and very thoroughly (van de Linde et al. 2012) and include sequential activation of individual fluorophores (Dempsey et al. 2012), photoswitchable proteins (Rego et al. 2012), photoconvertible and biphotochromatic proteins (Adam et al. 2011), tandem fluorescent protein timers (Khmelinskii et al. 2012), and tunable, light-inducible interacting proteins – TULIPs (Strickland et al. 2012).

The properties of 26 organic dyes were tested to determine which combinations were optimal for various forms of super-resolution imaging (PALM, STORM), and the authors (Dempsey et al. 2012) list several dyes, which performed well or excellently in four different spectral ranges thereby yielding super-resolution with 4 colors and low cross talk. Similarly, Adam et al. (2011) report on the rational design of 4-state optical highlighting into a novel fluorescent protein (NijiFP) with reversible on/off switching and biphotochromatic properties that can be used for super-resolution pulse–chase imaging. This supersedes the previous IrisFP with its tetrameric organization making it unsuitable as a fusion tag.

Superfast/Super-Resolution Microscopy Techniques

Super-resolution (SR) microscopy techniques address the shortcomings in both electron and light microscopy, while retaining their strengths (Evanko 2009; Pastrana 2011). These imaging methods provide spatial resolution of up to several tens of nanometers, below the diffraction limit, and approaching virtually molecular resolution (reviewed in van de Linde et al. 2012). The so-called illumination-based SR imaging techniques include stimulated emission depletion (STED) microscopy and saturated structured illumination microscopy (SSIM); the probe-based SR imaging uses photoactivation localization microscopy (PALM) and the related techniques stochastic optical reconstruction microscopy (STORM) and FPALM (fluorescence PALM) which are based on the stochastic activation of fluorescence (Lippincott-Schwarz and Manley 2009).

The point-localization SR method of direct stochastic optical reconstruction microscopy (dSTORM) has been used to image histone H2B proteins in living cells. Histone proteins were labeled with rhodamine and oxazoline dyes by using the genetically encoded chemical trimethoprim or SNAP tags (Wombacher et al. 2010; Klein et al. 2011). In a different example, with the help of a novel high-speed structured illumination microscope, capable of 100-nm resolution at frame rates up to 11 Hz for several hundred time points, it was possible to perform video imaging of tubulin and kinesin dynamics in living *Drosophila melanogaster* S2 cells (Kner et al. 2009). In yet another example, Lee et al. (2012) developed a novel confocal fluorescence microscope with a good optical-sectioning capability (1.0 μm), fast frame rates (<33 fps), and superior fluorescence detection efficiency. Full compatibility of the microscope with conventional cell-imaging techniques allowed single-molecule imaging with great ease at arbitrary depths of living cells, confirmed by monitoring the diffusion motion of fluorescently labeled cAMP receptors of *Dictyostelium discoideum.*

Other methods have been developed that enable the use of any GFP-tagged construct in single-molecule super-resolution microscopy. By further targeting of GFP with small, high-affinity antibodies coupled to organic dyes, it is possible to achieve nanometer spatial resolution and minimal linkage error when analyzing microtubules, living neurons, and yeast cells. In combination with libraries encoding GFP-tagged proteins, virtually any known protein can now be used in SR microscopy (Ries et al. 2012).

Super-resolution microscopy can also be used in plant systems to further our understanding of plant cell structure and function. It has been used successfully to elucidate the lateral organization of the plasma membrane (see Fig. 1 taken from Gutierrez et al. 2010)

Photoreactivated Localization Microscopy (PALM)

Photoreactivated localization microscopy can overcome the diffraction barrier and resolve up to 20 nm in the lateral (x, y) dimensions and about 50 nm in the axial (z) dimension compared with traditional fluorescence microscopy. The strategy of PALM relies on stochastically exciting fluorescent molecules with intermittent laser pulses by employing a photoswitchable fluorescent probe. In an elegant example of the technique, the fluorescent molecules are switched between "on state" (activated state) by the incident light then imaged and "off state" by photobleaching or chemical bleaching. The imaged pixels are analyzed, and the pixel with the highest intensity forms the center of the "located" molecule image. The reconstruction of those imaging cycles and the aggregate position information is then assembled into a super-resolution image (Betzig et al. 2006).

Structured Illumination Microscopy (SIM)

Structured illumination microscopy is a method in which the nonlinearity arises from saturation of the excited state (Rego et al. 2012). SIM enables fluorescence imaging with theoretically unlimited resolution. It takes advantage of the spatially structured illumination light and a nonlinear dependence of the fluorescence emission rate on the illumination intensity. It extends resolution beyond the cutoff by moving information into the observable region, from elsewhere in frequency space, in the form of moiré fringes. The moiré fringes are produced by frequency mixing whenever two signals are multiplied. The multiplication is inherent in fluorescence: the observed emission intensity is the product of the local density of fluorescent dye (i.e., the sample) and the local intensity of excitation light (Gustafsson 2005).

Recent improvements to the technique, such as the use of a photoswitchable fluorescent protein, allow nonlinear structured illumination microscopy of cellular structures at 50-nm resolution (Evanko 2012b). Indeed, SIM was recently used to visualize – at 40-nm resolution – cellular structures including nuclear pores, the actin cytoskeleton, and microtubules labeled with the fluorescent photoswitchable protein Dronpa (Rego et al. 2012).

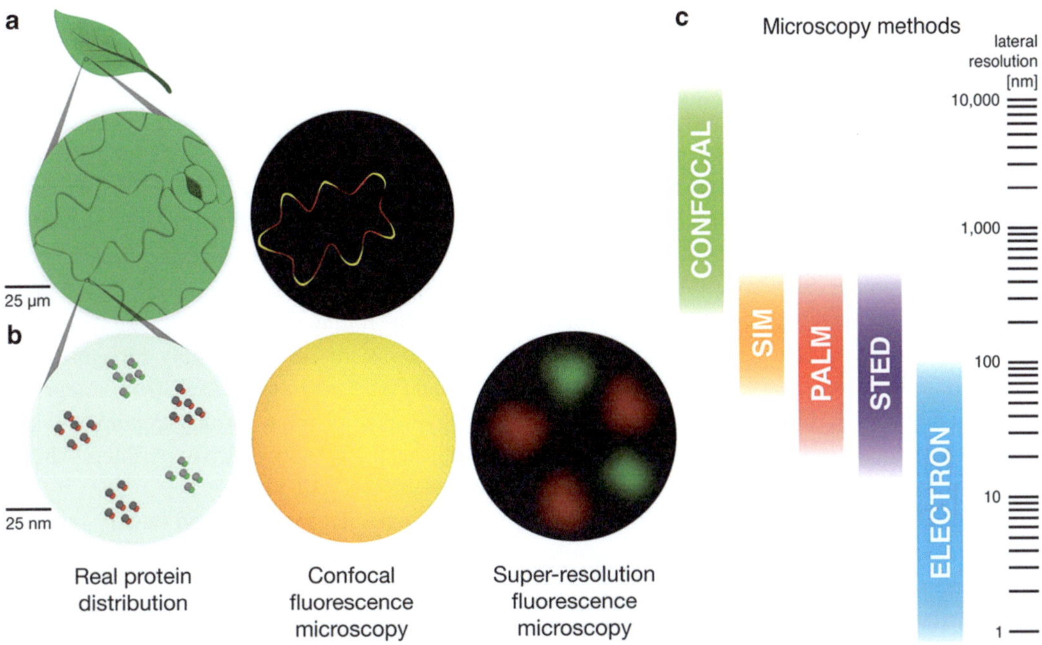

Fig. 1 Super-resolution fluorescence microscopy can reveal plasma membrane (PM) sub-compartmentalization that is unresolvable by conventional techniques. (**a**) Confocal microscopy is sufficient to visualize localization patterns at a scale of hundreds of nanometers. Two hypothetical PM proteins are labeled with different fluorescent tags in a pavement cell. One localizes to the tips of the cell lobes (*green*), much as observed for the small G-protein ROP2, and the other is dispersed more uniformly (*red*). The overlap appears as *yellow*. (**b**) At the nanoscale, these proteins also cluster into disparate microdomains. Super-resolution microscopy is necessary to resolve the two types of microdomains. All images are simulated. (**c**) Comparison of the spatial scales at which different microscopy techniques are useful (Figure 1 from Gutierrez et al. (2010))

Stimulated Emission Depletion (STED) Microscopy

Super-resolution fluorescence microscopy such as stimulated emission depletion (STED), a recently developed laser-scanning technique, can achieve resolution beyond the optical diffraction limit. Stimulated emission depletion microscopy provides SR by selectively deactivating fluorophores to enhance the imaging in the observed area, as initially described by Westphal et al. (2008).

STED immunofluorescence microscopy can be used to elucidate subcellular architecture in vivo. Using STED, it was possible to observe centrioles with a resolution of 60 nm to demonstrate that the centriole distal appendage protein Cep164 localizes in nine clusters spaced around a ring of approximately 300 nm in diameter (Lau et al. 2012). Two-color STED permits the use of popular green-yellow fluorescent labels such as green fluorescent protein, yellow fluorescent protein, Alexa Fluor 488, and calcein green. It is further possible to use a single-excitation/STED laser-beam pair to obtain two-color super-resolution time-lapse imaging, by simultaneously exciting and quenching pairs of these fluorophores, whose signals can then be separated by spectral detection and linear unmixing (Tønnesen et al. 2011).

Stochastic Optical Reconstruction Microscopy (STORM)

Stochastic optical reconstruction microscopy uses photoswitchable fluorescent probes to temporally separate the otherwise spatially overlapping images of individual molecules, allowing the precise localization of individual fluorescent labels in the sample. Similar to PALM, STORM can routinely achieve 25-nm resolution, which corresponds to about 70 base pairs for a linear stretch of double-stranded DNA. STORM thus enables multicolor, three-dimensional fluorescence imaging of molecular complexes, cells, and tissues with near-molecular scale resolution (Zhuang 2009). Whole-cell 3D STORM can reveal interactions between cellular structures with nanometer-scale resolution. This is helpful in particular when trying to understand the spatial relationships of organelles within a cell, e.g., the interactions of mitochondria with microtubules (Huang et al. 2008).

Developments in STORM in combination with the development of small-molecule probes for DNA and RNA labeling now enable visualization of cellular DNA based on direct DNA labeling in cells (Benke and Manley 2012). With improvements that include combining astigmatism imaging with a dual-objective scheme, better than 10-nm lateral resolution and better than 20-nm axial resolution can be obtained when imaging biological specimens (Xu et al. 2012). This level of resolution is sufficient to resolve individual actin filaments in cells and to reveal three-dimensional networks of the actin cytoskeleton (Xu et al. 2012).

Quantitative Microscopy and Image Analysis

Image cytometry provides simple but reliable tools to study molecular processes and structural changes at the level of cells and tissues from microscopic images. Quantitative data resulting from microscopy-based imaging enables the generation of quantitative gene expression data, at the mRNA, protein, and activity level. Preserving the 2D and 3D morphology, volumes, areas, lengths, and numbers of cells and tissues can be calculated and related to these gene expression data (Chieco et al. 2013). Fluorescence fluctuation spectroscopy has enabled quantitative imaging of single mRNAs in living cells (Wu et al. 2012). *In planta* cytometry can now be approached through integrated genetic and computation methods (Federici et al. 2012). Who knows, the future may bring "generative models," systems that do not simply recognize patterns but construct in silico cells from images, as biology transitions into a study system in which scientists readily know which experiments should come next (Baker 2012a).

The molecular imaging field largely relies on advances in high-content fluorescence microscopy, which come with novel requirements for fluorescent probes and labeling techniques. They have also driven the development of data interpretation strategies for extracting meaningful information from rich and complex biological image data. Synthetic fluorophores have a small size, are available in many colors, and can easily be chemically modified and used for stoichiometric labeling of proteins in live cells. A new bottleneck is the analysis of the resulting large and complex data sets. To mitigate this, new easy-to-use mathematical and statistical tools are being developed, which enable bench scientists to rapidly interpret their image data sets. One of those is PhenoRipper (http://www.phenoripper.org/), an open-source software tool designed for rapid exploration of high-content microscopy images. PhenoRipper permits rapid comparison of images obtained under different experimental conditions based on image phenotype similarity (Rajaram et al. 2012a). Another recently developed algorithm for specifying and rendering realistic microscopy images containing diverse cell phenotypes, heterogeneous populations, and imaging artifacts is the open-source SimuCell (http://www.SimuCell.org/) (Cho et al. 2012; Rajaram et al. 2012b).

To localize signals simultaneously emanating from multiple fluorophores, novel hybrid statistical methods are being developed. One of these is the recently developed Bayesian analysis of bleaching and blinking (3B), which can be performed using a common, arc lamp-based widefield fluorescence microscope (Lidke 2012). The 3B analysis does not display the (typically presented) maximum likelihood of each fluorophore's position. Instead the analysis models the entire time series as a set of blinking and bleaching fluorophores. It maps a range of possibilities, thus allowing in vivo resolution of 200-nm intracellular structures (Cox et al. 2012; Baker 2012b).

The techniques of fluorescence microscopy keep growing dramatically, both in terms of technical capabilities and the volume of images generated. Fluorescence microscopy is gradually becoming fully computerized. The sheer volume of data obtained has sparked the creation of online public repositories of microscopic images. This is déjà vu – the amount of data has reintroduced challenges faced years ago, when sequence and structure databases were being established: how to develop fast and effective means of searching for images either by context (e.g., which protein is labeled) or content (e.g., which pattern is displayed) (Cho et al. 2012). One solution to this problem is the creation of content-based image searchers for microscope images. One of the first solutions to this problem is the open microscopy environment remote objects OMERO.searcher software (http://murphylab.web.cmu.edu/software/searcher/), which can be used with any OMERO image database (http://openmicroscopy.org/) (Allan et al. 2012). Exciting times lie ahead.

A major review on quantitative fluorescence imaging in plant samples has appeared very recently (Okumoto et al. 2012). The authors review the current status of generating genetically encoded sensors, sensors suitable for high-resolution imaging, fluorescence measurement in intact plants, and the use of GFP derivatives in discovering novel pathways in plants. They furnish a very useful diagram (see Fig. 2) showing how cyano and yellow fluorescent proteins (CFP/YFP) can be used to target transcription, translation, subcellular location, protein import/export, and metabolism in plant cells (Okumoto et al. 2012). They also show the usefulness of both single fluorescent proteins and FRET-based sensors for examining protein activation, protein translocation, and protein–protein interactions – see Fig. 3 (Okumoto et al. 2012).

Really Unconventional Microscopy (RUM)

Novel non-fluorescence-based approaches allow the observation of individual molecules through their absorption of visible light. These methods open perspectives for single-molecule studies using a much larger group of molecules that absorb light but do not necessarily fluoresce (Hofkens and Roeffaers 2011).

Other unorthodox microscopic optical devices are also being exploited for biological applications and molecular imaging. For example, optical microcavities can be turned into sensitive biosensors, essentially optical tuning forks (Armani et al. 2007). These are made out of silica in the shape of a microtoroid (Evanko 2012c). When light is pumped into the microtoroid, it recirculates with a specific resonance wavelength. The light can interact with molecules binding to the surface of the microtoroid, causing a tiny redshift of the resonance wavelength.

One-dimensional smart probes based on nanowires and nanotubes also show enormous promise for biological applications, as they can safely penetrate the plasma membrane. These smart probes are potentially useful in high-resolution and high-throughput gene and drug delivery, biosensing, and single-cell electrophysiology (Yan et al. 2012). The probes also have a distinct advantage in their capacity to guide light. Optical tweezers have been used to assemble nanowires into optical microcircuits and to create nanowire endoscopes. Attached to the tapered tip of an optical fiber, nanowire endoscopes can guide visible light into intracellular compartments of a living cell and can also detect optical signals from subcellular regions with high spatial resolution. Through light-activated mechanisms these endoscopes can deliver payloads into cells with spatial and temporal specificity (Yan et al. 2012; Evanko 2007).

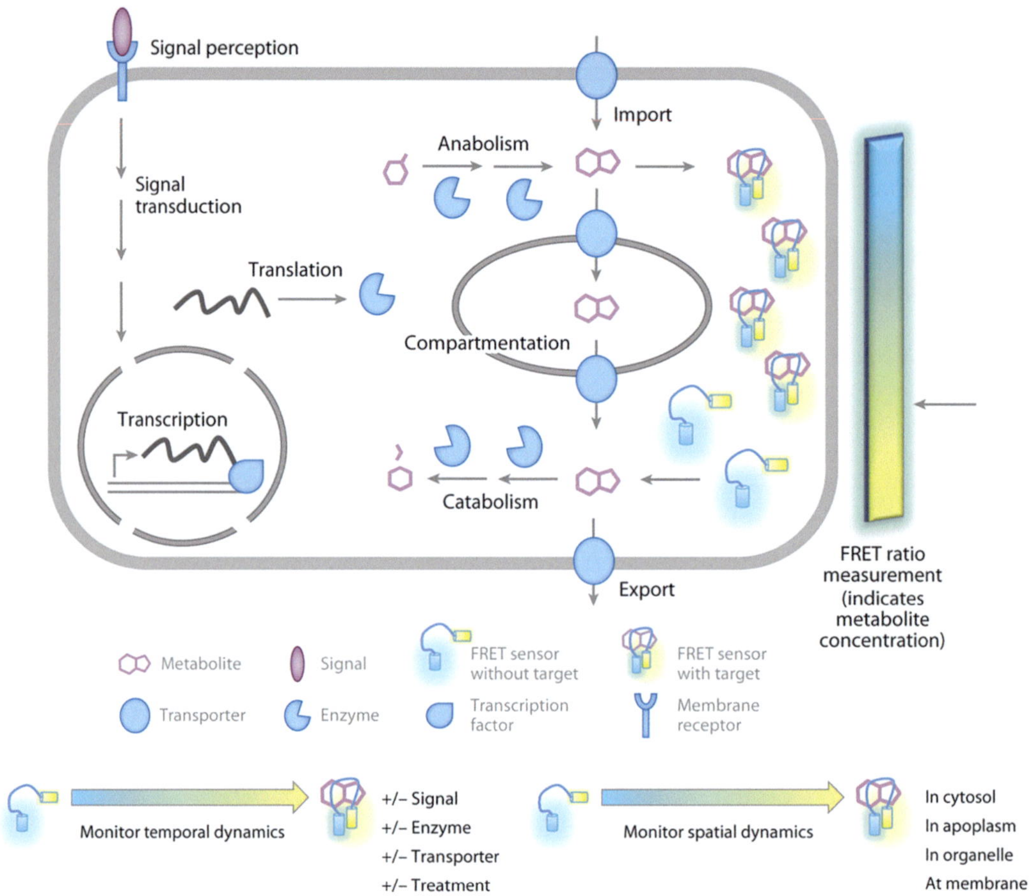

Fig. 2 Generic example of a *cyan* fluorescent protein–*yellow* fluorescent protein (CFP/YFP) Förster resonance energy transfer (FRET) sensor for a metabolite. A single FRET sensor can lead to discoveries of multiple pathways and processes involved in the dynamics of the sensor target. For example, a metabolite sensor reports on the sum of various component fluxes/pathways (e.g., transport, biosynthesis). Perturbations in any of these components, whether by physiological treatment or mutation, can then be detected using the FRET sensor. Once a FRET sensor is properly characterized and expressed in the cellular/subcellular location of interest, measurements can be easily acquired with extremely high temporal and spatial resolution (Figure 1 from Okumoto et al. (2012))

Yet another novel approach capable of achieving super-resolution imaging of biological systems employs lens-free holographic on-chip microscopy device and digital optoelectronic sensor arrays. This computer-based technique is described below (Greenbaum et al. 2012). Other relatively novel and emerging molecular imaging techniques include coherent nonlinear optical imaging, especially the recently developed nonlinear dissipation microscopy (including stimulated Raman scattering and two-photon absorption) and pump–probe microscopy (including excited-state absorption, stimulated emission, and ground-state depletion), providing new image contrasts for nonfluorescent species. Thanks to the high-frequency modulation transfer scheme, these imaging techniques exhibit superb detection sensitivity and offer high molecular specificity (Min et al. 2011).

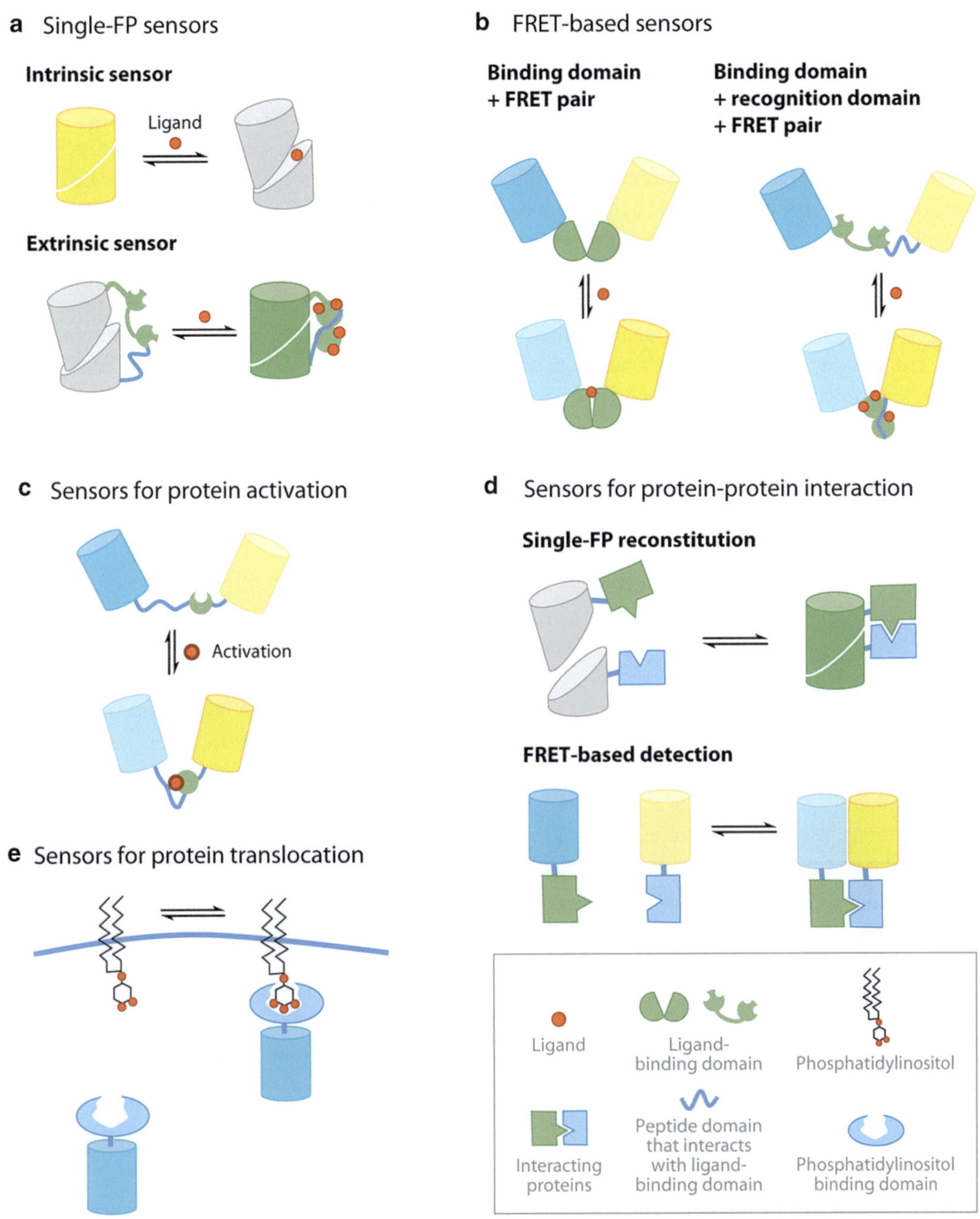

Fig. 3 Schematic representations of various genetically encoded sensors. (**a**) Single fluorescent protein (single-FP) sensors. The emission intensities of FPs are affected either by intrinsic mechanisms (*top*) or by the binding of ligands to extrinsic domains (*bottom*). (**b**) Förster resonance energy transfer (FRET)-based sensors. Ligand binding to the binding module can directly affect FRET efficiency (*left*) or be amplified through a ligand-dependent interaction between two domains (*right*). (**c**) Sensors for protein activation. Modification to the substrate domain (*green*) is recognized by the recognition domain (*blue*), which causes the conformational change to affect FRET efficiency. (**d**) Sensors for protein–protein interaction. Protein–protein interaction is visualized either by complementation of two FP halves (*top*) or by FRET between two fluorophores (*bottom*). (**e**) Sensors for protein translocation. FP fused to the recognition domain is translocated to the cellular domain with an enriched ligand concentration (Figure 2 from Okumoto et al. (2012))

Light Systems/Sources

In conventional fluorescence microscopy, the sample is illuminated from either above or below, but a newly developed technique employs lateral illumination from very thin sheets of light and measures fluorescence from either above or below the sample (Pastrana 2012). This technique is referred to as either fluorescence light sheet microscopy (FLISM) or selective plane illumination microscopy (SPIM) and can be accomplished with single-photon or two-photon excitation (Pastrana 2012). Light sheet microscopy using a scanned Bessel beam in combination with structured illumination or two-photon excitation reduces photobleaching and phototoxicity, improves axial resolution, and allows isotropic three-dimensional imaging (Planchon et al. 2011). The authors demonstrate performance of the method via fast volumetric subcellular imaging of several dynamic processes in single living cells.

A totally different approach termed time-reversed ultrasound-encoded (TRUE) light has been described recently (Wang et al. 2012 and references therein). A light beam is passed through a tissue sample, where it scatters as it propagates through the tissue. A focused pulse of ultrasound is then used to generate a frequency shift of a confined region of the scattered light. The ultrasound pulse acts as a virtual light source within the tissue. The ultrasound frequency-shifted light interferes with a separate reference beam. Both the unshifted and the frequency-shifted lights scatter within the sample and are collected and compared. The resulting interference pattern is imaged onto a camera and is later used in a playback mode to reconstruct an optical focus at the location of ultrasound modulation. The aggregate of generated fluorescence images enables three-dimensional visualization of targeted deep tissue.

Yet another method relies on fluorescence from enhanced GFP, which can – ingeniously – be "turned off" by exposing cells to laser light. The trick is to use a femtosecond laser operating in the near-infrared wavelength range. Such lasers have ultrashort pulse duration, ultrahigh peak power, and low single-photon energy, i.e., low phototoxicity. In vivo, enhanced GFP is turned off by oxidative stress generated by a cascade of events originating from the femtosecond laser irradiation. This allows for a range of cell biology studies to be performed with high spatiotemporal resolution (He et al. 2012; Lukyanov and Belousov 2012).

The combination of light sheet microscopy and localization-based super-resolution imaging allows deep sub-diffraction resolution imaging in thick scattering specimens as demonstrated by three-dimensional super-resolution nanometric imaging of proteins in live 150-μm-diameter cell spheroids (Zanacchi et al. 2011).

Alternatives to Fluorescence Microscopy

Nonlinear Optical Imaging

One alternative to conventional fluorescence microscopy functions in the absence of any lens; it employs digital optoelectronic sensor arrays involving charged couple devices (CCD) or complementary metal oxide semiconductor (CMOS) chips (Greenbaum et al. 2012). These lens-free holographic on-chip microscopy devices need a light source, usually individual or arrays of LEDs, with appropriate sensors (CCD, CMOS), which permit a wide field of view (FOV). As with other computer-based techniques, reconstruction of images is necessary and super-resolution can be attained (Greenbaum et al. 2012).

Single-Molecule Light Absorption

Another alternative to fluorescence microscopy is based on single-molecule light absorption rather than fluorescence (Hofkens and Roeffaers 2011). These authors briefly describe three recent developments in this area and point out how they circumvent former major problems in light absorption techniques. Indeed, the detection of individual molecules in ambient conditions via their absorption signature creates the prospect of single-molecule studies using a much larger group of molecules that absorb light but do not necessarily fluoresce. The most important problems are the tiny attenuation of light that occurs (less than one photon out of a million is absorbed) and the variation in intensity of the light emitted (which can be as much as 1 %) causing the noise-to-signal ratio to be unacceptable. One approach developed by Celebrano et al. (2011) allows for the direct imaging of individual molecules by monitoring the attenuation induced in probe light. The technique splits the laser beam, so that one part focuses on the sample and the other acts as a reference, with the difference nullifying intensity variations of the light source. Instead of using a single wavelength split beam as above, the other methods both employ double beam lasers of different wavelengths. One technique does not directly measure absorption (or attenuation); instead it measures the heat emitted in the immediate vicinity of the absorbing molecule compared with the lack of heat emitted by a transparent second laser (Gaiduk et al. 2010). In the final method, one laser beam irradiates the molecule into its excited state, thereby preventing light absorption by the second beam, but rapid on/off switching of the saturating beam permits absorption measurement by the second beam (Chong et al. 2010). This ground-state depletion microscopy for ultrasensitive detection of absorption contrast is reminiscent of the actinic light used to maintain phytochrome in either of its interconvertible red-absorbing or far-red-absorbing forms, so that its two different absorbance profiles could be determined.

The many advantages of recently developed techniques in fluorescence microscopy, including confocal laser scanning, 2-photon-excited fluorescence, single-molecule microscopy, and super-resolution imaging, are recognized in a major review by Min et al. (2011). However, they focus on the primary disadvantage of fluorescence microscopy – i.e., not all molecules are, or can be made, fluorescent. They espouse coherent nonlinear optical imaging (CNOI), which does not rely on making the target molecule fluoresce with the attendant problems of perturbation of function and location (Min et al. 2011). They list 3 distinct categories of CNOI, parametric generation spectroscopy (PGS), nonlinear dissipation optical microscopy (NLDOM), and pump–probe spectroscopy (PPS) and furnish details of each method.

Luminescent Proteins

By definition fluorescence microscopy demands that the sample be illuminated, but there are times when this is not feasible and chemiluminescence is an alternative (Saito et al. 2012). However most naturally chemiluminescent proteins do not emit enough light to furnish viable images. To remedy this situation, Saito et al. (2012) have developed "nano-lantern" a chimera of luciferase and a fluorescent protein, Venus, which has high-efficiency bioluminescence resonance energy transfer. These methods can not only be used to visualize organelles in intact cells with high resolution but also to detect tumors in freely moving mice and an array of structures in intact plants. Modified nano-lanterns have been developed to visualize Ca^{2+}, cAMP, and ATP dynamics in conditions where fluorescence is impossible – especially light-dark induced changes in chloroplast function in intact leaves (Saito et al. 2012).

Specific Molecules/Macromolecular Complexes

While the section below describes individually classified methods for the analysis of specific molecules, in many instances there is overlap between the topics. For example, in the context of molecular imaging, labeling of mRNA for ß-actin can be discussed under methods for labeling of nucleic acids, under methods for the analysis of the cytoskeleton, or under methods for the analysis of polymers. As well, tagging and imaging of histones can be considered in the context of visualization of both proteins and DNA.

Beyond GFP and its variants (Haseloff 1999; Tsien 1998, 2010), a myriad of newly created fluorescent probes have recently pushed the frontiers of imaging in vivo (Adam et al. 2011). Synthetic fluorophores have a small size, are available in a variety of colors spanning the whole light spectrum, and can easily be chemically modified and used for stoichiometric labeling of proteins in live cells. They are bright and exhibit high signal to noise ratio, they are photostable, and they can be operated as photoswitchable fluorophores even in living cells under physiological conditions. For those reasons, it has been recognized that synthetic fluorophores have the potential to substantially accelerate the broad application of live-cell super-resolution imaging methods (van de Linde et al. 2012).

Proteins

Technologies exist for measuring particular aspects of protein localization and turnover, but comprehensive analysis of protein dynamics is possible only by combining several methods. Fluorescent fusion proteins have revolutionized examination of proteins in living cells. At the same time, studies using these proteins have been scrutinized because fusion proteins are modified and ectopically expressed, in contrast to immunofluorescence (Schnell et al. 2012). The field is wide open and various combinatorial approaches have been tested, with varying success. For example, tandem fluorescent protein timers (tFTs), fusions of two single-color fluorescent proteins that mature with different kinetics, have been used to analyze protein mobility and turnover in living cells (Khmelinskii et al. 2012). By fusing tFTs to other proteins, it is possible to study cellular longevity, inheritance, and segregation. The use of tandem fluorescent protein timers enables (i) investigation of the mobility of proteins between subcellular compartments, (ii) direct measurement of protein degradation kinetics, and (iii) screening for regulators of protein turnover, e.g., regulators of N-end rule-mediated protein degradation (Khmelinskii et al. 2012).

One solution to the problem of determining the spatiotemporal organization of a favorite protein is to use a protein tag (e.g., fluorescent or affinity based). Provided the microscopy is not limiting, such labeling allows one to see exactly where the protein is localized in vivo at a given time. Alternatively, it is possible to use affinity reagents to pull out other molecules the protein associates with (Pastrana 2012). When protein tagging is performed on the large scale of an entire genome, a new TransgeneOmics platform is created (Poser et al. 2008). This type of platform was initially worked out using bacterial artificial chromosomes and eukaryotic cells and then extended to whole worms (Pastrana 2012). The platform is relatively complex, and it combines computer-assisted transgene design, parallel DNA engineering, and next-generation sequencing. The platform also produces outstanding results. In the case of *Caenorhabditis elegans*, the platform "*C. elegans* TransgeneOme" covers 73 % of the proteome. The platform's multipurpose tag allows for localization of proteins in vivo. The platform has been released to the scientific community through a dedicated web application (http://transgeneome.mpi-cbg.de/) (Sarov et al. 2012).

Advanced fluorescence imaging generally relies on fluorophores with controllable emission properties. These are photoactivatable fluorescent proteins capable of reversible on/off photoswitching or irreversible (typically green-to-red) photoconversion. The first engineered fluorescent protein combining these two types of phototransformations is IrisFP (Adam et al. 2011). However, the spectroscopic properties of IrisFP are far from being optimal, and its tetrameric organization complicates its use as a fusion tag. To circumvent this deficiency, four-state optical highlighting can be rationally introduced into photoconvertible fluorescent proteins, to develop a new set of enhanced optical highlighters derived from mEosFP and Dendra2. NijiFP, a promising new fluorescent protein, has photoconvertible and biphotochromatic properties that make it ideal for advanced fluorescence-based imaging applications (Adam et al. 2011).

Another way to directly manipulate the formation of protein complexes in vivo is through naturally photoswitchable proteins. Recent advances in this field include the generation of tunable light-inducible protein tags (TULIPs), which are based on a synthetic interaction between the LOV2 domain of *Avena sativa* phototropin 1 and an engineered PDZ domain (Strickland et al. 2012). TULIPs can recruit proteins to diverse structures in living eukaryotic cells, either globally or with precise spatial control using a steerable laser. This is a highly sophisticated method in which equilibrium binding and kinetic parameters of the interaction are tunable by mutation, making TULIPs readily adaptable to signaling pathways with varying sensitivities and response times (Strickland et al. 2012).

Nucleic Acids

The methods for labeling nucleic acids have markedly improved since the days of our graduate school. Imaging nucleic acids with single-molecule sensitivity in live cells has become an indispensable tool for quantitatively studying nucleic acid biology. Novel approaches have been developed for the visualization of unmodified nucleic acids in vivo. For example, visualization of DNA in living cells was made possible via time-lapse dSTORM imaging, with the help of direct labeling with the commercially available cyanine-based PicoGreen dye (Benke and Manley 2012).

To accomplish imaging of mRNA in living cells, Yamada and coworkers recently labeled transcripts with Pumilio, an RNA binding protein that recognizes the sequence of an RNA transcript (Yamada et al. 2011). Two versions of Pumilio were engineered, which recognize adjacent 8-nucleotide stretches on endogenous, unmodified transcripts of the b-actin gene. Each version of Pumilio also carries different halves of green fluorescent protein. When both versions of Pumilio bind the same transcript, they bring the two GFP halves together, restoring fluorescence. Using this innovative approach, it was possible to resolve singly labeled ß-actin transcripts in vivo. The achieved spatiotemporal resolution even enabled determination of the dynamics of mRNA movement along microtubules (Yamada et al. 2011; Baker 2012c).

Another popular system for fluorescent labeling of mRNA is the MS2 system, which is based on the high-affinity interaction between the MS2 bacteriophage coat protein and its cognate RNA hairpin (Tyagi 2009). It has been extensively used to address a diversity of biological questions in many different model organisms, due to its simplicity and sensitivity (Weil et al. 2010). The MS2 system may show increased background fluorescence. To mitigate the problem of background fluorescence, a single-chain tandem dimer of MCP in the MS3 system was recently used (Wu et al. 2012). This approach significantly increased the uniformity and sensitivity of mRNA labeling, opening the doors to imaging of single mRNAs directly in live cells.

Cytoskeleton

The spatial resolution of optical imaging has been substantially increased through the recent advances in super-resolution fluorescence microscopy (Huang et al. 2010). More recently, combining astigmatism imaging with a dual-objective scheme, Xu and coworkers improved the image resolution of STORM and were thus able to image biological specimens with <10-nm lateral resolution and <20-nm axial resolution (Xu et al. 2012). Using this approach, it was possible to observe two vertically separated layers of actin networks with distinct structural organizations in sheetlike cell protrusions. It was even possible to resolve individual actin filaments in cells and to reveal the three-dimensional ultrastructure of the actin cytoskeleton (Xu et al. 2012).

Advances in three-dimensional (3D) structured illumination microscopy (SIM) – resulting in 3DSIM – have enabled imaging of live samples with 120-nm lateral and 360-nm axial resolution. The technique has been specifically used for imaging of microtubules (Shao et al. 2011).

Nonlinear structured illumination microscopy has recently made progress as a biologically compatible super-resolution imaging method. In an example of cytoskeletal molecular imaging using this technique, reversible photoswitching of the fluorescent protein Dronpa provided 40-nm resolution on purified labeled microtubules, thus enabling detailed visualization of the cytoskeleton (Rego et al. 2012).

Small Molecules

It is often desirable to detect optical signals from small molecules, while targeting subcellular regions with high spatial resolution. In plants, an important and thoroughly investigated small molecule is the hormone auxin, the primary hormone involved in cell enlargement and a key morphogenetic signal. Recent technological advances have enabled researchers to dynamically follow auxin's distribution and signaling during growth and development. The key has been the generation of novel fluorescent protein fusions. This effort resulted in the creation of an in vivo auxin signaling sensor – termed DII-VENUS – in *Arabidopsis thaliana*. The VENUS form of yellow fluorescent protein was fused in-frame to the auxin-interaction domain (termed domain II; DII) and expressed under a constitutive promoter. As DII-VENUS abundance was dependent on auxin, this reporter fusion was used to obtain high-resolution spatiotemporal information about auxin distribution and response during plant growth and development at cellular resolution in different tissues (Brunoud et al. 2012; Nawy 2012).

In plants, other important signaling and regulatory small molecules are the reactive oxygen species (ROS). As the potentially toxic hydrogen peroxide (H_2O_2) emerges as an important signaling molecule in growth and development, molecular imaging with peroxy yellow 1 methyl ester (PY1-ME), a new chemo-selective fluorescent indicator for H_2O_2, can reveal its spatiotemporal distribution in vivo (Miller et al. 2010). Fluorescence techniques for measuring ROS have been reviewed by Swanson et al. (2011) and Choi et al. (2012) describing the same parameters as they did for Ca^{2+} and H^+ (see Table 1).

Perhaps the most important small signal in biology is the ion, Ca^{2+}, which is maintained at very low (10^{-7} M) concentration in the cytoplasm in plant cells, despite being at high (millimolar) concentrations in the cell wall. This huge gradient allows very rapid and large (yet transient) increases in cytoplasmic calcium, which trigger a plethora of downstream events. The most recent comprehensive reviews on the use of fluorescent biosensors to measuring calcium in plants have come from Gilroy's lab (Swanson et al. 2011; Choi et al. 2012). In the former (Swanson et al. 2011), they describe the use of single wavelength probes, ratiometric analysis using Indo-1 and Fura-2, methods for loading Ca^{2+} biosensors into plant cells, the rise of genetically modified FRET-based sensors, and their use for quantitative analysis. In the latter (Choi et al. 2012) among other things, they provide a very useful table (see Table 1) describing the array of Ca^{2+} biosensors used in plants, the cell type and species where they have been used, the subcellular target (compartment) for Ca^{2+} measurement, and targeting methods.

Another important ion in plants is H^+, where the cytoplasmic pH is around 7.7, while the cell wall pH is around 6.5. One of the earliest known signaling functions for H^+ was a result of its auxin-evoked secretion into the cell wall where it could activate cell wall-hydrolyzing enzymes and thus stimulate growth, causing the release of cell wall molecules involved in defense against pathogens (Davies 1987). Again, the most recent reviews of this topic are those of Swanson et al. (2011) and Choi et al. (2012). In the former they stress fluorescence-based techniques for measuring both cytoplasmic and cell wall pH (Swanson et al. 2011), while in the latter, they list the array of biosensors, their cell type/species of use, and their subcellular targeting (Choi et al. 2012).

Plant Cytomics

A number of techniques described above have yet to be used in plants, although many seem entirely feasible. In most aspects of research, plants are inherently far more difficult to work with than animals, and this is especially true in the field of cytomics. First, in mature cells, the "interesting" part, the cytoplasm, is generally a thin smear compressed between the almost inanimate cell wall and the equally inanimate vacuole. Not only does this make isolation of un-degraded biomolecules difficult, it also makes tracking their intra- and intercellular movement more tedious. Second, the cell wall has different refractive properties from the cytoplasm and the cover slip, causing major aberration in the image obtained. Third, and most important, however, is the massive autofluorescence, primarily from chlorophyll and other pigments in photosynthetic tissues and others in the roots. Accordingly, methods successful with plants will have to avoid the use of dyes overlapping the (auto)fluores-

Table 1 Examples of the application of cytosolic and subcellularly targeted GFP-based biosensors for Ca^{2+}, pH, and reactive oxygen species (ROS) in plants

Biosensor	Subcellular locale	Targeting method	Cell types	Species
YC2.1	Cytosol	–	Root hair, root epidermis, guard cells, pollen tubes	*Arabidopsis*, lily, *Medicago*, tobacco
	Nucleus	Nucleoplasmin fusion	Root hairs, pollen tubes	*Medicago*, tobacco
YC3.6	Cytosol	–	Stomatal guard cell, roots, root hairs, cotyledons, pollen tubes	*Arabidopsis*, lotus, tobacco
	Plasma membrane	N-terminal fusion of YC3.6 with the LT16b protein	Roots, cotyledons	*Arabidopsis*
	Nucleus	NLS from SV40 large-T protein	Roots, cotyledons	*Arabidopsis*
YC3.1	Cytosol	–	Pollen, stigmatic papillae	*Arabidopsis*, tobacco
YC4.6	ER	Pumpkin 2S albumin signal peptide/HDEL ER retention signal	Pollen	*Arabidopsis*
D3cpv	Peroxisome	C-terminal KVK–SKL peptide	Roots, cotyledons	*Arabidopsis*
	Tonoplast	N-terminus of CBL2	Roots, cotyledons	*Arabidopsis*
H148D	Cytosol	–	Roots, root hairs	*Arabidopsis*
pHluorin	Cytosol	–	Roots, pollen tubes	*Arabidopsis*, tobacco
	Apoplast	Chitinase signal peptide	Roots	*Arabidopsis*
Pt-GFP	Cytosol	–	Roots, leaves	*Arabidopsis*
Hyper	Cytosol	–	Leaf epidermis, stomatal guard cells, suspension cell culture	*Arabidopsis*
	Peroxisome	C-terminal KSRM peptide	Leaf epidermis, stomatal guard cells	*Arabidopsis*, tobacco
RoGFP1/2	Cytosol	–	Roots, leaves	*Arabidopsis*, tobacco
	Mitochondrion	First 87 amino acids of the tobacco ß-ATPase	Roots, leaves	*Arabidopsis*, tobacco
	ER	Chitinase targeting peptide/HDEL retention signal	Roots, tobacco leaf cells	*Arabidopsis*
	Peroxisome	C-terminal SKL peptide	Leaves	*Arabidopsis*, tobacco
	Plastid	Transketolase target peptide	Leaves	*Arabidopsis*

Table 1 from Choi et al. (2012).

cence range of plant pigments, or avoid the use of fluorescence altogether.

One approach could be single-molecule light absorption (Celebrano et al. 2011; Chong et al. 2010; Gaiduk et al. 2010; Hofkens and Roeffaers 2011) described above (section "Really Unconventional Microscopy (RUM)"), since this involves light absorption and not emission. However, we are unaware of any work done with plant material using these methods. Another approach could be luminescence (Saito et al. 2012). This method has been used successfully with plants, to visualize Ca^{2+}, cAMP, and ATP dynamics after light-dark induced changes in chloroplast function in intact leaves. As stated above (section "Superfast/Super-Resolution

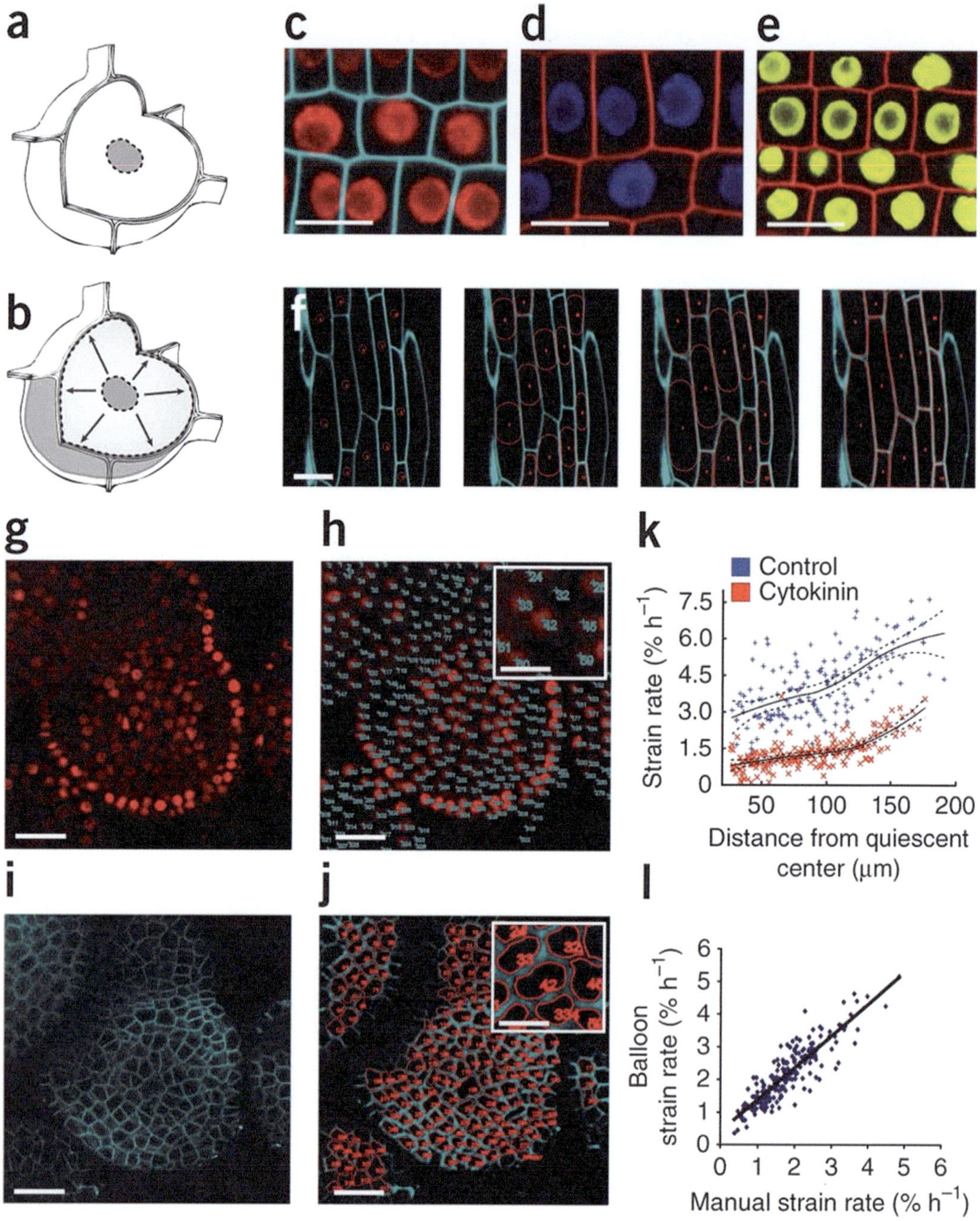

Fig. 4 Automated segmentation of cells in *Arabidopsis*. (**a**, **b**) Schematics of a transgenic plant cell used for *in planta* cytometry: a nuclear fluorescent protein is used for balloon "seeding" and cell indexing (**a**), and a plasma membrane marker is used to obtain cell geometries (**b**). (**c–e**) Confocal microscopy images of *Arabidopsis* root meristem tissues containing *35S::H2B-mRFP1* and *35S::EGFP-LTI6b* (**c**), *35S::H2B-EYFP* and *35S::mCherry-LTI6b* (**d**), and *35S::H2B-ECFP* and *35S::mCherry-LTI6b* (**e**). Histone 2B fusions are localized to nuclei, and LTI6b fusions are localized to the plasma membrane. (**f**) Step-by-step progress of balloon (*red*) inflation during the segmentation of epidermis cells of *Arabidopsis* hypocotyl. (**g–j**) Segmentation of cells in the shoot apical meristem of *Arabidopsis*. Confocal micrographs of a meristem (**g**) showing nuclear localization of H2B-mRFP1 protein, automated identification of labeled nuclei (**h**), plasma membrane-localized distribution of EGFP-LTI6b in the same plant (**i**), and automated segmentation of cells using individual mRFP1-labeled nuclei as seeds for balloon segmentation with EGFP-labeled cell outlines (**j**). Scale bars, 40 μm (**g–j**) and 15 μm (**c–f** and insets in hand **j**). (**k**) Cell-expansion rates at indicated distance from the quiescent center in the root meristem in

Microscopy Techniques"), SR microscopy has been used in plant systems to further our understanding of plant cell structure and function and has successfully been used to elucidate the lateral organization of the plasma membrane (Gutierrez et al. 2010) (see Fig. 1). Furthermore, *in planta* cytometry can now be approached through integrated genetic and computation methods (Federici et al. 2012).

Various GFP-based bioprobes can be combined with different microscopy techniques to study the spatiotemporal aspects of plant biology. In one example, transgenic green fluorescent protein–flotillin1 (GFP-Flot1) in combination with confocal microscopy analysis and transmission electron microscopy immunogold labeling was used to study the spatial and dynamic aspects of GFP-Flot1-positive vesicle formation in *Arabidopsis*. Variable-angle total internal reflection fluorescence microscopy was used to reveal the subcellular localization and the dynamic behavior of GFP-Flot1 (Li et al. 2012). Variable-angle epifluorescence microscopy in combination with GFP-based constructs has already been used as a way to look at cytoskeletal protein dynamics in the plant cell cortex (Konopka and Bednarek 2008). In general, fluorescent protein-based technologies are increasingly used to understand the plant regulatory networks, e.g., the functioning of the plant endomembrane system (Sparkes and Brandizzi 2012). In a recent example, ligand-induced endocytosis of *Arabidopsis thaliana* FLAGELLIN SENSING2 (FLS2) was studied using a GFP-tagged FLS2 that was expressed in *Nicotiana benthamiana* (Choi et al. 2013). In yet another example, cellular signaling was monitored using an array of GFP-based probes for Ca^{2+} (yellow cameleon Ca^{2+} sensors, principally YC2.1 and 3.6), pH (pHluorin and H148D), and reactive oxygen species – ROS (Hyper), all of which are compatible with the standard configurations of confocal microscopes (Choi et al. 2012; Miwa et al. 2006).

Light sheet fluorescence microscopy (LSFM) has been used to for high-resolution live imaging of plant growth in near physiological conditions. LSFM uses a low numerical aperture lens to focus a sheet of light and to illuminate the specimen from one side while collecting the emitted light at a perpendicular axis (Maizel et al. 2011).

One-chromophore fluorescence lifetime microscopy (ocFLM), combined with wavelength-selective fluorescence microscopy, was applied to *Arabidopsis* cells in their tissue environment at high local resolution (Elgass et al. 2009). The ocFLM system consisted of a confocal sample scanning microscope, a spectrally integrating detector for measuring fluorescence intensity and a custom-built time-correlated single-photon counting board for recording fluorescence lifetime decays. This setup, together with a novel method of data analysis, allowed the recording of the spectroscopic and fluorescence lifetime data of GFP fusion proteins in vivo, providing new information about their molecular properties, cell physiological function, and subcellular environment (Elgass et al. 2009).

The level of sophistication of *in planta* cytometry was recently increased using transgenic *Arabidopsis* that expressed spectrally distinct fluorescent proteins in cell nuclei and plasma membranes (Federici et al. 2012). The generated plants allowed the measurement of cellular properties in intact tissue while retaining the cellular context. The technique consists of three automated steps: (1) a genetic marker was first used to specifically label nuclei, index cells, and extract positional information using a particle search algorithm; (2) identified nuclei were used to initiate a contour segmentation algorithm that used a fluorescent signal located at the plasma membrane to extract information regarding cell size, shape, and topology (Fig. 4 – from Federici et al. 2012); and (3) cell-specific gene expression was quantified by ratiometric measurement of the spectrally distinct nuclear fluorescent proteins that were expressed under the control of differ-

Fig. 4 (continued) control and cytokinin-treated roots. Individual values for cortex cell strain rates were obtained from segmented image data. *Solid lines* indicate mean expansion rate and *dashed lines*, ± s.e.m. ($n=20$). (**I**) Comparison of strain rate values obtained from the balloon segmentation technique and from manual measurements of cell size using Image J

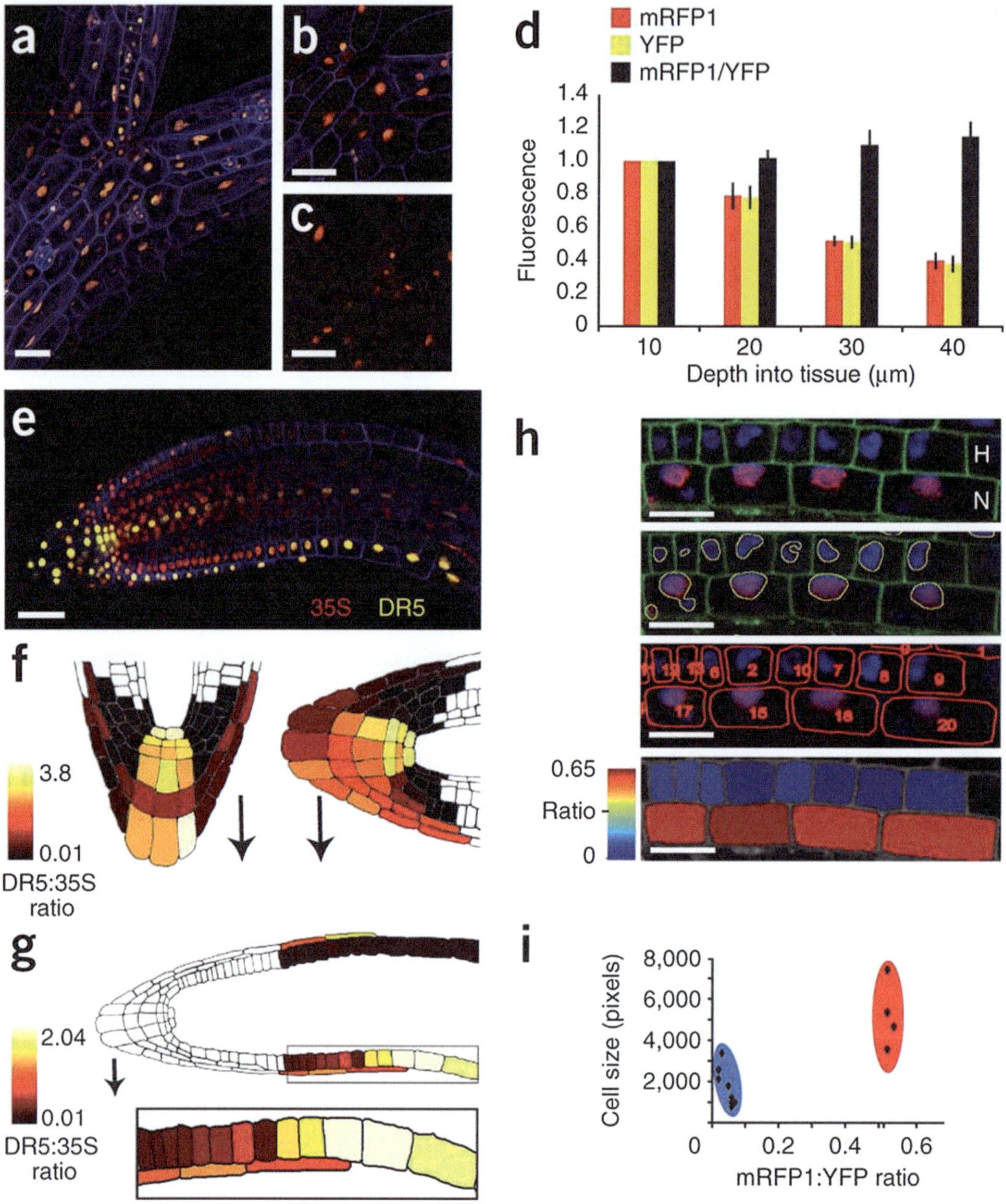

Fig. 5 Nuclear ratiometric measurements for in planta cytometry. (**a**) Confocal projection image of the hypocotyl of a 4-day-old seedling expressing H2B-EYFP (*green channel*), histone H2B-mRFP1 (*red channel*), and EGFP-LTI6b (*blue channel*). (**b**, **c**) Confocal image sections at depths of 10 μm (**b**) and 40 μm (**c**). (**d**) Mean fluorescence intensity (arbitrary units) at indicated depths was normalized to value at 10 μm. Error bars, s.e.m. ($n = 25$ cells); similar data were obtained after the analysis of 8 different seedlings. (**e**) Confocal image of the root of an *Arabidopsis* seedling containing auxin-responsive *DR5rev::3×Venus-N7* (labeled DR5) and a constitutive *35S::H2B-mRFP1* (labeled 35S) and imaged 6 h after shift to a horizontal orientation. (**f**, **g**) Relative levels of expression of the *DR5rev::3×Venus-N7* and *35S::H2B-mRFP1* markers were quantified and plotted on a tracing of the root cell outlines. Ratiometric measurement of auxin-responsive reporter gene expression in the root tip-

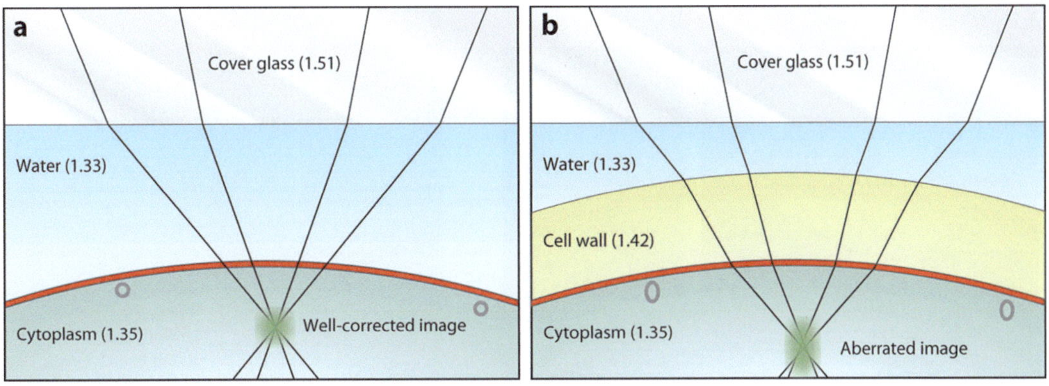

Fig. 6 Image aberrations in plant cells. The plant cell wall acts as an optical element in the imaging light path; numbers in parentheses are refractive indexes. (**a**) For a water immersion objective lens corrected for a 180-μm (#1.5) cover glass, the light rays pass through water and the plasma membrane to the cytoplasm with minimal aberration. (**b**) The plant cell wall is relatively thick and has a higher refractive index than water or cytoplasm, leading to alterations in the imaging light path. The effect of the cell wall is often observed as increased lateral blurring of diffraction-limited objects and axial distortion (Figure 2 from (Shaw and Ehrhardt 2013))

ently regulated promoters (Fig. 5 – from Federici et al. 2012). As mentioned above (section "Quantitative Microscopy and Image Analysis"), a major review on quantitative fluorescence imaging in plant samples has appeared very recently (Okumoto et al. 2012), and some of their work is presented in Figs. 2 and 3.

Finally, just before this chapter was completed, an excellent review on optical imaging in plant systems was published (Shaw and Ehrhardt 2013). They review basic concepts of digital imaging, describe opportunities and challenges for plant biologist, as well as probe development, advances in instrumentation, and computer programs for data analysis. They show schematically (Fig. 6) how the plant cell wall generates aberrations in images as a result of is high refractive index (Shaw and Ehrhardt 2013). They also describe in detail the RootChip microfluidics platform (Fig. 7) with a particularly elegant image of the root tip cytoskeleton of *Arabidopsis* shown in panel F (Shaw and Ehrhardt 2013).

Fig. 5 (continued) (**f**) before (*left*) and 6 h after (*right*) shift to a horizontal orientation and in the non-hair cells of the elongation zone 6 h after shift to a horizontal orientation (**g**). *Arrows* indicate the direction of gravitational force. (**h**) Confocal image of epidermal cells in the elongation zone of an *Arabidopsis* root expressing *IAA2::H2B-mRFP1*, *35S::H2B-EYFP* and *35S::EGFP-LTI6b* (*top*). Root hair (H) and non-root hair (N) cell files are indicated. Nuclear segmentation using the signal from *35S::H2B-EYFP* and cell outline after balloon segmentation are shown in the middle. The ratio of fluorescence signals from *IAA2::H2B-mRFP1* to *35S::H2B-EYFP*, visualized using a color map (*bottom*). (**i**) Ratiometric fluorescence values and size of cells from (**h**). Cells with high mRFP1:EYFP ratios and larger cell sizes are indicated in *red*, cells with low mRFP1:EYFP ratios and smaller cell sizes in *blue*. Scale bars, 20 μm

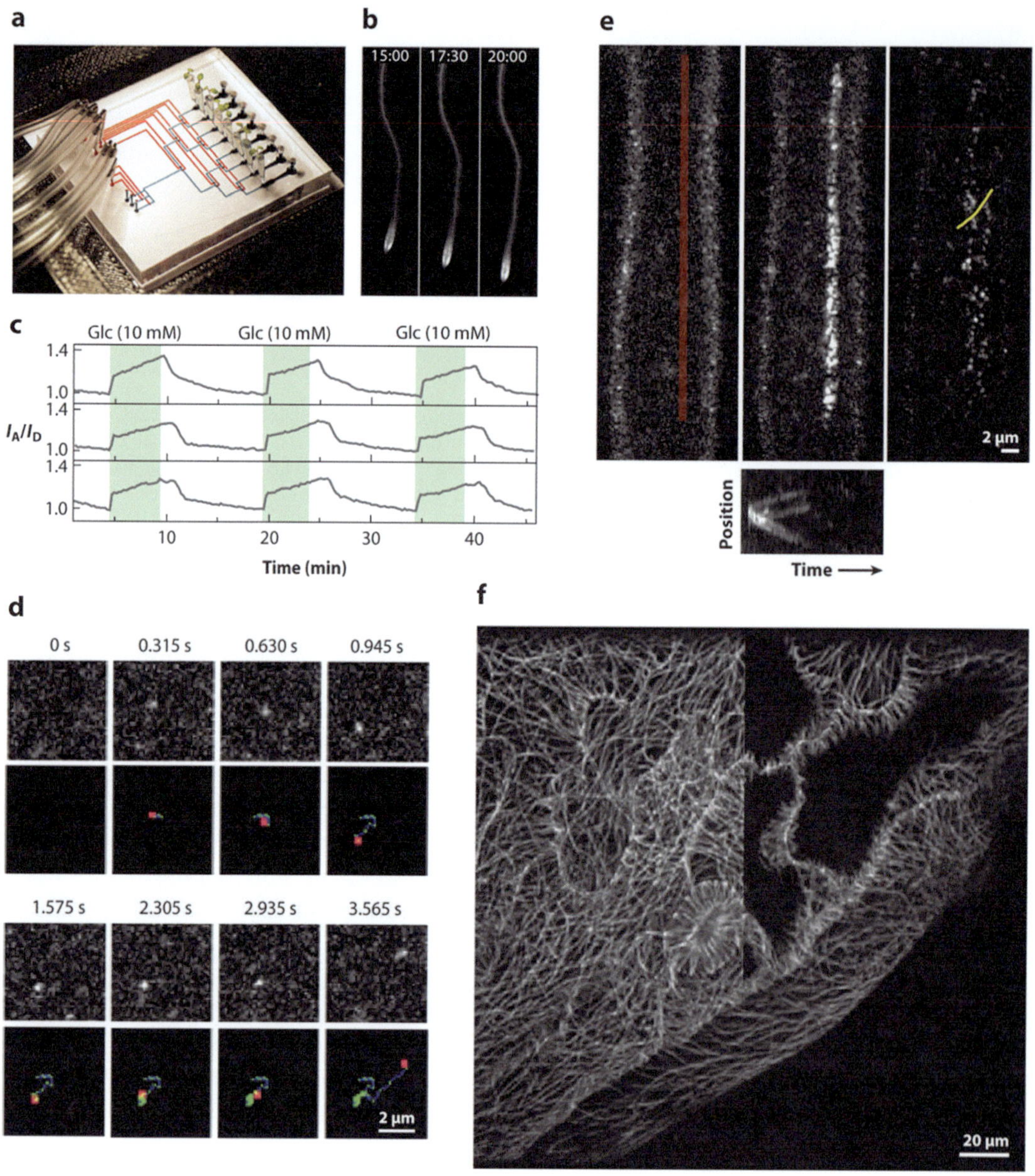

Fig. 7 (**a–c**) The RootChip microfluidics platform. Panel *a* shows a photograph of the RootChip, with contrast dyes revealing the channels used for perfusion of imaging chambers and those used for flow control; this design features eight parallel imaging chambers. Panel *b* shows a time series of roots expressing a glucose nanosensor showing steady growth over 20 h of observation. Panel *c* shows quantitation of the sensor signal in roots growing in three chambers as glucose is repeatedly supplied to and removed from the root environment. Data are plotted as the ratio of acceptor intensity (I_A) over donor intensity (I_D) when the donor is excited at 443 nm. Panels *a* and *c* adapted from Grossmann et al. (2011); panel *b* adapted from Grossmann et al. (2012). (**d**) Image sequence showing how the on-chip amplification of signal over read noise on electron-multiplier charge-coupled device (EMCCD) cameras permits the capture of low signals at high acquisition rates. The sequence shows the motility of a single microtubule nucleation complex labeled with GCP2:GFP in a hypocotyl cell. The images were acquired at 62-ms intervals with a back-thinned EMCCD camera on a spinning-disk confocal microscope. A subset of images from the time sequence are shown. In the diagrams below the images, red dots show the position of the labeled complex in the accompanying image, and *green dots* show the history of complex positions in all previous images, connected in sequence by a *blue line*. Rapid image acquisition permits observation of an episode of confinement for a few seconds before the complex continues more rapid movement. Adapted from

References

Adam V, Moeyaert B, David CC, Mizuno H, Lelimousin M et al (2011) Rational design of photoconvertible and biphotochromic fluorescent proteins for advanced microscopy applications. Chem Biol 18:1241–1251

Allan C, Burel J-M, Moore J, Blackburn C, Linkert M et al (2012) OMERO: flexible, model-driven data management for experimental biology. Nat Methods 9:245–253

Armani AM, Kulkarni RP, Fraser SE, Flagan RC, Vahala KJ (2007) Label-free, single-molecule detection with optical microcavities. Science 317:783–787

Baker M (2012a) Robert E Murphy: creating algorithms to turn images into cell models. Nat Methods 9:629

Baker M (2012b) Susan Cox: using Bayesian statistics to speed super-resolution microscopy. Nat Methods 9:113

Baker M (2012c) RNA imaging *in situ*. Nat Methods 9:787–790

Benke A, Manley S (2012) Live-cell dSTORM of cellular DNA based on direct DNA labeling. ChemBioChem 13:298–301

Benndorf D, von Bergen M, Jehmlich N, Völker U, Schmidt F et al (2010) Advanced tool for characterization of microbial cultures by combining cytomics and proteomics. Appl Microbiol Biotechnol 88:575–584

Bernas T, Gregori G, Asem EK, Robinson JP (2006) Integrating cytomics and proteomics. Mol Cell Proteomics 5:2–13

Betzig E, Patterson GH, Sougrat R, Lindwasser OW, Olenych S et al (2006) Imaging intracellular fluorescent proteins at nanometer resolution. Science 313:1642–1645

Brunoud G, Wells DM, Oliva M, Larrieu A, Mirabet V et al (2012) A novel sensor to map auxin response and distribution at high spatio-temporal resolution. Nature 482:103–106

Celebrano M, Kukura P, Renn A, Sandoghdar V (2011) Single-molecule imaging by optical absorption. Nat Photonics 5:95–98

Chieco P, Jonker A, DeBoer BA, Ruijter JM, Van Noorden CJF (2013) Image cytometry: protocols for 2D and 3D quantification in microscopic images. Prog Histochem Cytochem 47:211–333

Cho BH, Cao-Berg I, Bakal JA, Murphy RF (2012) SimuCell: a flexible framework for creating synthetic microscopy images. Nat Methods 9:634–635

Choi W-G, Swanson SJ, Gilroy S (2012) High-resolution imaging of Ca^{2+}, redox status, ROS and pH using GFP biosensors. Plant J 70:118–128

Choi S, Tamaki T, Ebine K, Uemura T, Ueda T, Nakano A (2013) RABA members act in distinct steps of subcellular trafficking of the FLAGELLIN SENSING2 receptor. Plant Cell 25:1174–1187

Chong S, Min W, Sunney X, Xie XS (2010) Ground-state depletion microscopy: detection sensitivity of single-molecule optical absorption at room temperature. J Phys Chem Lett 1:3316–3322

Coltharp C, Xiao J (2012) Superresolution microscopy for microbiology. Cell Microbiol 14:1808–1818

Cox S, Rosten E, Monypenny J, Jovanovic-Talisman T, Burnette DT et al (2012) Bayesian localization microscopy reveals nanoscale podosome dynamics. Nat Methods 9:195–200

Davidson MW, Murphy DB (2012) Fundamentals of light microscopy and electronic imaging, 2nd edn. Wiley-Blackwell, Hoboken, NJ, USA

Davies E (1987) Wound responses in plants. Biochem Plants 12:243–264

Davies E, Stankovic B (2006) Electrical signals, the cytoskeleton and gene expression: a hypothesis on the coherence of the cellular responses to environmental insult. In: Baluska F, Mancuso S, Volkmann D (eds) Communication in plants – neuronal aspects of plant life. Springer, Berlin/Heidelberg, pp 309–320

Davies E, Abe S, Larkins BA, Clore AM, Quatrano RS, Weidner S (1998) The role of the cytoskeleton in plant protein synthesis. In: Bailey-Serres J, Gallie DR (eds) A look beyond transcription: mechanisms determining mRNA stability and translation in plants. American Society of Plant Physiologists, Rockville, Maryland, USA, pp 115–124

Fig. 7 (continued) Nakamura et al. (2010). (**e**) Images showing how a photoactivatable probe allows labeling and tracking of a select cohort of membrane protein complexes. A hypocotyl cell expressing PA-GFP:CESA6 is excited by a narrow band of 405-nm energy, as indicated by the *red bar* (*left image*). After 15 s, bright punctae are clearly detected in the stimulated region (*center image*). Seven minutes later, the labeled cellulose synthase complexes have moved away (*right image*). Labeled cellulose synthase complexes in the cohort disappear as they are either internalized or bleached (kymograph taken at the position of the *yellow line*; *bottom image*). Images acquired by spinning-disk confocal microscopy (J.J. Lindeboom, R. Gutierrez & D.W. Ehrhardt, unpublished). (**f**) Spinning-disk confocal microscopy to examine image volumes. A living *Arabidopsis* cotyledon expressing a GFP-tubulin-encoding transgene was imaged with a water immersion objective using 120 optical sections at 200-nm intervals. The middle 80 images are projected together on the right side of the image to reveal the cortical nature of the microtubules (Figure 2 from (Shaw and Ehrhardt 2013))

Davies E, Stankovic B, Azuma K, Shibata K, Abe S (2001) Novel components of the plant cytoskeleton: a beginning to plant "cytomics". Plant Sci 160:185–196

Davies E, Stankovic B, Vian A, Woods A (2012) Where has all the message gone? Plant Sci 185:23–32

Dempsey GT, Vaughan JC, Chen KH, Bates M, Zhuang X (2012) Evaluation of fluorophores for optimal performance in localization-based super-resolution imaging. Nat Methods 9:1027–1036

Elgass K, Caesar K, Schleifenbaum F, Stierhof Y-D, Meixner AJ, Harter K (2009) Novel application of fluorescence lifetime and fluorescence microscopy enables quantitative access to subcellular dynamics in plant cells. PLoS One 9(e5716):1–13

Evanko D (2007) Incredible shrinking optics. Nat Methods 4:683

Evanko D (2009) Primer: fluorescence imaging under the diffraction limit. Nat Methods 6:19–20

Evanko D (2012a) More dyes enter the realm of nanoscopy. Nat Methods 9:944

Evanko D (2012b) Better resolution for structured-illumination microscopy. Nat Methods 9:124

Evanko D (2012c) A microscopic endoscope. Nat Methods 9:128

Federici F, Dupuy L, Laplaze L, Heisler M, Haseloff J (2012) Integrated genetic and computation methods for in planta cytometry. Nat Methods 9:483–485

Gaiduk A, Yorulmaz M, Ruijgrok PV, Orrit M (2010) Room-temperature detection of a single molecule's absorption by photothermal contrast. Science 330:353–356

Greenbaum A, Luo W, Su T-W, Gorocs Z, Xue L et al (2012) Imaging without lenses: achievements and remaining challenges of wide-field on-chip microscopy. Nat Methods 9:889–895

Grossmann G, Guo WJ, Ehrhardt DW, Frommer WB, Sit VS, Quake SR (2011) The RootChip: an integrated microfluidic chip for plant science. Plant Cell 23:4234–4240

Grossmann G, Meier M, Cartwright HN, Sosso D, Quake SR et al (2012) Time-lapse fluorescence imaging of Arabidopsis root growth with rapid manipulation of the root environment using the RootChip. J Vis Exp 65:e4290

Gustafsson MGL (2005) Nonlinear structured-illumination microscopy: wide-field fluorescence imaging with theoretically unlimited resolution. Proc Natl Acad Sci U S A 102:13081–13086

Gutierrez R, Grossmann G, Frommer WB, Ehrhardt DW (2010) Opportunities to explore plant membrane organization with super-resolution microscopy. Plant Physiol 154:463–466

Haseloff J (1999) GFP variants for multispectral imaging of living cells. Methods Cell Biol 58:139–151

He H, Li S, Wang S, Hu M, Cao Y, Wang C (2012) Manipulation of light from green fluorescent protein by femtosecond laser. Nat Photonics 6:651–656

Hofkens J, Roeffaers MBJ (2011) Single-cell molecule light absorption. Nat Photonics 5:80–81

Huang B, Jones SA, Brandenburg B, Xhuang X (2008) Whole-cell 3D STORM reveals interactions between cellular structures with nanometer-scale resolution. Nat Methods 5:1047–1054

Huang B, Bates M, Zhuang W (2009) Super-resolution fluorescence microscopy. Annu Rev Biochem 78:993–1016

Huang B, Babcock H, Zhuang XW (2010) Breaking the diffraction barrier: super-resolution imaging of cells. Cell 143:1047–1058

Khmelinskii A, Keller PJ, Bartosik A, Meurer M, Barry JD et al (2012) Tandem fluorescent protein timers for in vivo analysis of protein dynamics. Nat Biotechnol 30:708–714

Klein T, Loschberger A, Proppert S, Wolter S, van de Linde S, Sauer M (2011) Live-cell dSTORM with SNAP-tag fusion proteins. Nat Methods 8:7–9

Kner P, Chhun BB, Griffis R, Winoto L, Gustafsson MGL (2009) Super-resolution video microscopy of live cells by structured illumination. Nat Methods 6:339–342

Konopka CA, Bednarek SY (2008) Variable-angle epifluorescence microscopy: a new way to look at protein dynamics in the plant cell cortex. Plant J 53:186–196

Kriete A (2005) Cytomics in the realm of systems biology. Cytometry 68A:19–20

Lau L, Lee YL, Sahi SJ, Steams T, Moemer WE (2012) STED microscopy with optimized labeling density reveals 9-fold arrangement of centriole protein. Biophys J 102:2925–2935

Lee J, Miyanaga Y, Ueda M, Hohng S (2012) Video-rate confocal microscopy for single-molecule imaging in live cells and superresolution fluorescence imaging. Biophys J 103:1691–1697

Leif RC (2009) Towards the integration of cytomics and medicine. J Biophotonics 2:482–493

Leung BO, Chou KC (2011) Review of super-resolution fluorescence microscopy for biology. Appl Spectrosc 65:967–980

Li R, Liu P, Wan Y, Chen T, Wang Q et al (2012) A membrane microdomain-associated protein, *Arabidopsis* Flot1, is involved in a clathrin-independent endocytic pathway and is required for seedling development. Plant Cell 24:2105–2122

Lichtman JW, Conchello J-A (2005) Fluorescence microscopy. Nat Methods 2:910–919

Lidke KA (2012) Super resolution for common probes and common microscopes. Nat Methods 9:139–141

Lippincott-Schwarz J, Manley S (2009) Putting super-resolution microscopy to work. Nat Methods 6:21–23

Lukyanov KA, Belousov VV (2012) The slow fade of cell fluorescence. Nat Photonics 6:641–643

Maizel A, von Wangenheim D, Federici F, Haseloff J, Stelzer EHK (2011) High-resolution live imaging of plant growth in near physiological bright conditions using light sheet fluorescence microscopy. Plant J 68:377–385

Miller EW, Dickinson BS, Chang CJ (2010) Aquaporin-3 mediates hydrogen peroxide uptake to regulate downstream intracellular signaling. Proc Natl Acad Sci U S A 107:15681–15686

Min W, Freudiger CW, Lu S, Xie XS (2011) Coherent non-linear optical imaging: beyond fluorescence microscopy. Annu Rev Phys Chem 62:507–530

Miwa H, Sun J, Oldroyd GED, Downie JA (2006) Analysis of calcium spiking using a cameleon calcium sensor reveals that nodulation gene expression is regulated by calcium spike number and the developmental status of the cell. Plant J 48:883–894

Müller S (2008) Cytomics reaches microbiology – population heterogeneity on the protein level caused by stress. Cytometry 73A:3–4

Murphy RF (2005) Cytomics and localized proteomics: automated interpretation of the subcellular patterns in fluorescence microscope images. Cytometry 62A:1–3

Nakamura M, Ehrhardt DW, Hashimoto T (2010) Microtubule and katanin-dependent dynamics of microtubule nucleation complexes in the acentrosomal *Arabidopsis* cortical array. Nat Cell Biol 12:1064–1070

Nature (2009) Collections: super-resolution microscopy. Available at http://nature.com/nmeth/collections/superresmicroscopy

Nawy T (2012) Reporting plant hormones levels: a disappearing act. Nat Methods 9:219

Okumoto S, Jones A, Frommer WB (2012) Quantitative imaging with biosensors. Annu Rev Plant Biol 63:663–706

Opazo F, Levy M, Byrom M, Schafer C, Geisler C et al (2012) Aptamers as potential tools for super-resolution microscopy. Nat Methods 9:938–939

Pastrana E (2011) Fast 3D super-resolution fluorescence microscopy. Nat Methods 8:46

Pastrana E (2012) For every protein its tag. Nat Methods 9:941

Patterson G, Davidson M, Manley S, Lippincott-Schwartz J (2010) Super resolution imaging using single-molecule localization. Annu Rev Phys Chem 61:345–367

Petty HR (2007) Fluorescence microscopy: established and emerging methods, experimental strategies, and applications in immunology. Microsc Res Tech 70:687–709

Planchon TA, Gao L, Milkie DE, Davidson MW, Galbraith JA et al (2011) Rapid three-dimensional isotropic imaging of living cells using Bessel beam plane illumination. Nat Methods 5:417–423

Poser I et al (2008) BAC TransgeneOmics: a high throughput method for exploration of protein function in mammals. Nat Methods 5:409–415

Rajaram S, Pavie B, Hac NEF, Altschuler SJ, Wu LF (2012a) PhenoRipper: software for rapidly profiling microscopy images. Nat Methods 9:635–637

Rajaram S, Pavie B, Hac NEF, Altschuler SJ, Wu LF (2012b) SimuCell: a flexible framework for creating synthetic microscope images. Nat Methods 9:634–635

Rego EH, Shao L, Macklin JJ, Winoto L, Johansson GA, Kamps-Hughes N, Davidson MW, Gustafsson MGL (2012) Nonlinear structured-illumination microscopy with a photoswitchable protein reveals cellular structures at 50-nm resolution. Proc Natl Acad Sci U S A 109:135–143

Ries J, Kaplan C, Platonova E, Eghlidi H, Ewers H (2012) A simple, versatile method for GFP-based super-resolution microscopy via nanobodies. Nat Methods 9:582–584

Robinson JP (2008) Cytometry and the dawn of the cytomics generation. Cytometry 73A:51–52

Saito K, Chang Y-F, Horikawa K, Hatsugai N, Higuchi Y et al (2012) Luminescent proteins for high-speed single-cell and whole-body imaging. Nat Commun 3:1262. doi:10.1038/ncomms2248

Sarov M et al (2012) A genome-scale resource for in *vivo* tag-based protein function exploration in *C. elegans*. Cell 150:855–866

Schermelleh L, Heintzmann R, Leonhardt H (2010) A guide to super-resolution fluorescence microscopy. J Cell Biol 190:165–175

Schnell U, Dijk F, Sjollema KA, Giepmans BNG (2012) Immunolabeling artifacts and the need for live-cell imaging. Nat Methods 9:152–158

Shao L, Kner P, Rego EH, Gustafsson MGL (2011) Super-resolution 3D microscopy of live whole cells using structured illumination. Nat Methods 8:1044–1046

Shaw SL, Ehrhardt DW (2013) Smaller, faster, brighter: advances in optical imaging of living plant cells. Annu Rev Plant Biol 64:351–375

Shi X, Jung Y, Lin L-J, Liu C, Wu C, Cann IKO, Ha T (2012) Quantitative fluorescent labeling of aldehyde-tagged proteins for single-molecule imaging. Nat Methods 9:499–503

Shibata K, Morita Y, Abe S, Stankovic B, Davies E (1999) Apyrase from pea stems: isolation, purification, characterization and identification of a NTPase from the cytoskeleton fraction of pea stem tissue. Plant Physiol Biochem 37:1–8

Sparkes I, Brandizzi F (2012) Fluorescent protein-based technologies: shedding new light on the plant endomembrane system. Plant J 70:96–107

Stankovic B, Clore A, Shunnosuke A, Larkins B, Davies E (2000) Actin in protein synthesis and protein body formation. In: Staiger CJ, Baluska F, Volkmann D, Barlow P (eds) Actin: a dynamic framework for multiple cellular functions. Kluwer Acad. Publishers, Dordrecht, pp 129–143

Strickland D, Lin Y, Wagner E, Hope CM, Zayner J et al (2012) TULIPS: tunable, light-controlled interacting protein tags for cell biology. Nat Methods 9:379–384

Swanson SJ, Choi WG, Chanoca A, Gilroy S (2011) In vivo imaging of Ca^{2+}, pH, and reactive oxygen species using fluorescent probes in plants. Annu Rev Plant Biol 62:273–297

Tagore S, Gomase VS (2008) Cytomics. Curr Drug Metab 9:263–266

Támok A (2010) Cytomics for discovering drugs. Cytometry 77A:1–2

Támok A, Bocsi J (2009) Cytomics and regenerative medicine. Cytometry 75A:707–708

Tønnesen J, Nadrigny F, Willig KI, Wedlich-Soldner R, Nagerl UV (2011) Two-color STED microscopy of

living synapses using a single laser-beam pair. Biophys J 101:2545–2552
Tsien RY (1998) The green fluorescent protein. Annu Rev Biochem 67:509–544
Tsien RY (2010) Nobel lecture: constructing and exploiting the fluorescent protein paint box. Integr Biol 2:77–93
Tyagi S (2009) Imaging intracellular RNA distribution and dynamics in living cells. Nat Methods 6:331–338
Ulrich A, Martins AHB, Pesquero JB (2004) RNA and DNA aptamers in cytomics analysis. Cytometry 59A:220–231
van de Linde S, Heilemann M, Sauer M (2012) Live-cell superresolution imaging with synthetic fluorophores. Annu Rev Phys Chem 63:519–540
Wang YM, Judkewitz B, DiMarzio CA, Yang C (2012) Deep-tissue focal fluorescence imaging with digitally time-reversed ultrasound encoded light. Nat Commun 3:928. doi:10.1038/ncomms1925
Weil TT, Parton RM, Davis I (2010) Making the message clear: visualizing mRNA localization. Trends Cell Biol 20:380–390
Westphal V, Rizzoli SO, Lauterbach MA, Kamin D, Jahn R, Hell SW (2008) Video-rate far-field optical nanoscopy dissects synaptic vesicle movement. Science 320:246–249
Wombacher R, Heidbreder M, van de Linde S, Sheetz MP, Heilemann M, Cornish VW, Sauer M (2010) Live-cell super-resolution imaging with trimethoprim conjugates. Nat Methods 7:717–719
Wu B, Chao JA, Singer RH (2012) Fluorescence fluctuation spectroscopy enables quantitative imaging of single mRNAs in living cells. Biophys J 102:2935–2944
Xu K, Babcock HP, Zhuang X (2012) Dual-objective STORM reveals three-dimensional filament organization in the actin cytoskeleton. Nat Methods 9:185–188
Yamada T, Yoshimura H, Inaguma A, Ozawa T (2011) Visualization of non-engineered single mRNAs in living cells using genetically encoded fluorescent probes. Anal Chem 83:5708–5714
Yan R, Park Y-H, Choi Y, Heo C-J, Yang S-M, Lee LP, Yang P (2012) Nanowire-based single-cell endoscopy. Nat Nanotechnol 7:191–196
Zanacchi FC, Lavagnino Z, Donnorso MP, Del Bue A, Furia L, Faretta M, Diaspro A (2011) Live-cell 3D super-resolution imaging in thick biological samples. Nat Methods 8:1047–1049
Zhuang W (2009) Nano-imaging with STORM. Nat Photonics 3:365–367

Plant Physiomics: Photoelectrochemical and Molecular Retrograde Signalling in Plant Acclimatory and Defence Responses

Magdalena Szechyńska-Hebda, Paweł Budiak, Piotr Gawroński, Magdalena Górecka, Milena Kulasek, and Stanisław Karpiński

Contents

Abstract

Plants constantly exposed to fluctuating environmental conditions develop signalling strategies that determine their acclimation, fitness and survival. Coordination between the different cells and tissues requires complex mechanisms of signal communication that lead to overall plant global signalling homeostasis. Numerous pathways controlling signal transduction and gene expression are known. While initial responses of plants to environmental signals rely primarily on electrical signalling, longer-term responses that alter morphology rely on complex physiological networks. Electrochemical signals result from, for example, stimulation of non-photochemical quenching-dependent chloroplast retrograde signalling that with the help of the chloroplast stromules is transmitted to the other organelles and plasma membrane and trigger H^+-ATPase, ion movements and changes in a transmembrane voltage potential. These in turn switch on phytohormones, which are

M. Szechyńska-Hebda, Ph.D.
Department of Genetics, Breeding and Biotechnology, Warsaw University of Life Sciences, ul. Nowoursynowska 159, 02-776 Warszawa, Poland

The Franciszek Górski Institute of Plant Physiology, Polish Academy of Sciences, ul. Niezapominajek 21, 30-239 Kraków, Poland

P. Budiak, Ph.D. • P. Gawroński, Ph.D.
M. Górecka, Ph.D. • M. Kulasek, Ph.D.
S. Karpiński , Ph.D. (✉)
Department of Genetics, Breeding and Biotechnology, Warsaw University of Life Sciences, ul. Nowoursynowska 159, 02-776 Warszawa, Poland
e-mail: stanislaw_karpinski@sggw.pl; stanislaw.karpinski52@gmail.com

D. Barh et al. (eds.), *PlantOmics: The Omics of Plant Science*,
DOI 10.1007/978-81-322-2172-2_15, © Springer India 2015

chemical messengers coordinating cellular activity and anterograde signalling from the nucleus. Their regulatory pathways do not operate independently but rather are linked together in a complex network of interactions as observed, for example, by the antagonistic effect of, for example, cytokinins and abscisic acid. Furthermore, it is accompanied by the constant production of reactive oxygen species (e.g. hydrogen peroxide, singlet oxygen) which are not only harmful agents causing oxidative damage but also have important roles as intra- and intercellular signalling molecules. The simultaneous interplay of electrical signals, multiple hormones and reactive oxygen species influences redox status of the cells that are not directly exposed to primary stress factor and induces retrograde signalling to the nucleus, leading to alterations of gene expression and anterograde signalling. Changes in gene expression profile in response to such direct and indirect environmental stimuli are very complex and largely are depending on specific interactions of *cis*- and *trans*-regulatory elements but also on epigenetic changes, e.g. DNA methylations. All of these finally determine plant growth, development and acclimatory and immune defence responses.

Keywords

Electrochemical retrograde signals • Reactive oxygen species • Hormonal homeostasis • Metaphysiomics • CRE shuffling • DNA methylation • Histone modifications

Introduction

Land plants evolved 510–630 million years ago (Raven and Edwards 2001; Clarke et al. 2011) from a branched, filamentous alga inhabiting shallow fresh water (Kenrick and Crane 1997), probably at the edge of seasonally dry pools (Raven and Edwards 2001). Once plants reached the land, they had to face variable conditions and developed sensory and defence mechanisms that allow them to recognise and precisely respond to various stresses, respectively. In many cases numerous pathways involved in these processes overlap, which complicates the intricate network of plant-environment interactions. Therefore, plants must perform an integrated biological processing in response to external biotic and abiotic stimuli to adjust their metabolism to fluctuating environment. This results in plant tolerance to conditions such as excessive or inadequate light, water, salt and temperature and resistance to pathogens. Despite the importance of these statements, surprisingly little is known about the molecular mechanism that triggers and coordinates the physiological communication network. The knowledge about acclimatory and defence processes in living organisms can be achieved only by systematic study of physiology on the level of cells, tissues, organs and finally whole plant with respect to the whole network of crossing and overlapping signalling pathways (plant physiomics). Physiomics can also employ metaphysiomics to construct networks of physiological features that are associated with genes, proteins and their networks. In the last decade, some new concepts have been suggested in explaining the integrated abiotic- and biotic-induced signalling, and a key role of photoelectrochemical signals, hormonal homeostasis and retrograde signalling was indicated. In this chapter we describe these interactions and mechanisms on various levels of plant organisation. We elucidate intracellular (retrograde signalling) and intercellular (electrochemical signals, hormonal homeostasis) processes and their integration at gene level (metaphysiomics).

Chloroplast-to-Nucleus Signalling

Plastids evolved from cyanobacteria about 1.2–1.5 billion years ago (Dyall et al. 2004; Inaba 2010). During evolution, those organelles exported their genes to the nucleus, at the same time retaining the plastid proteins (Martin and Herrmann 1998). Up to date, plastid genome is reduced in land plants to about 120 genes (Inaba

2010; Martin and Herrmann 1998) that encode 75–80 proteins (Barajas-López et al. 2013; Timmis et al. 2004), while this organelle contains total of the 3,000–4,000 proteins (Barajas-López et al. 2013; Inaba 2010; Soll and Schleiff 2004). By contrast, the genome of the unicellular cyanobacterium *Synechocystis* encodes 3,168 proteins (Martin and Herrmann 1998).

Flow of genes from plastid to nucleus has led to a situation where, for example, subunits of the photosynthetic electron-transporting complexes are encoded in chloroplast (core subunits) and nuclear genome (peripheral subunits) (Barajas-López et al. 2013). The peptide subunits of functional complexes need to be assembled stoichiometrically and, if necessary, reorganised during acclimation. The first process is controlled by anterograde and the latter by retrograde signalling. In retrograde signalling, plastids emit signals that influence nuclear gene expression to match the status of plastids and ensure optimal performance of its processes (operational control) (Barajas-López et al. 2013; Chan et al. 2010; Inaba 2010; Karpiński et al. 2013). Retrograde signalling is also essential during chloroplast development (biogenic control) (Barajas-López et al. 2013; Szechyńska-Hebda and Karpiński 2013).

Environmental changes trigger chloroplast-induced signals and affect many chloroplast-originated biosynthesis pathways and the accumulation of specific metabolites. Flux through the chlorophyll biosynthetic pathway, sugar sensing, reactive oxygen species (ROS) equilibrium and redox state of chloroplast contribute to retrograde signalling (Leister 2005).

Tetrapyrroles play important regulatory roles in retrograde signalling and diverse cellular processes in plants (Barajas-López et al. 2013). Under stress conditions, two intermediates of chlorophyll biosynthesis, Mg-ProtoIX and its methyl ester Mg-ProtoIX-ME, are transported across chloroplast membranes by unknown mechanism (Barajas-López et al. 2013; Zhang et al. 2011), and then they rapidly and transiently accumulate in cytosol. Tetrapyrroles bind to the chaperone HSP90 and inhibit its ATPase activity, which probably leads to the activation of *HY5* transcription factor and *photosynthesis-associated nuclear gene* (*PhANG*) expression (Barajas-López et al. 2013; Kindgren et al. 2012a). Genes encoding components involved in chlorophyll biosynthesis pathway are regulated by Golden 2-like 1 and 2 (GLK1/2). They respond to the plastid signals and thus are sensitive to feedback signalling from the chloroplast (Barajas-López et al. 2013; Waters et al. 2009). Haem is another intermediate of tetrapyrrole biosynthesis pathway. This molecule is produced by plastid ferrochelatase 1 (FC1, haem synthase) and controls *PhANG* expression during chloroplast development (Barajas-López et al. 2013; Woodson et al. 2011).

Recently, a new molecule involved in retrograde signalling has been identified. Phosphonucleotide 3′phosphoadenosine 5′phosphate (PAP) accumulates in the chloroplast under drought or high light stress and is essential for the induction of genes involved in stress response, such as *APX2* and *ELIP2*. Its amount is regulated by SAL1, a phosphatase that dephosphorylates it to AMP. Presumably SAL1 is inhibited during stress and therefore PAP accumulates in the chloroplast. Then it moves to the nucleus and alters RNA catabolism by inhibiting exoribonucleases (XRNs) (Barajas-López et al. 2013; Estavillo et al. 2011).

Reactive oxygen species are known as powerful mediators of different signalling pathways, and among cellular compartments, chloroplasts are able to produce the most massive pools of ROS. This process depends on: photosynthesis, gene expression, chlorophyll (tetrapyrrole) biosynthesis and hormones (Shapiguzov et al. 2012). Changes in photosynthetic electron transport (PET) activity are the source of at least three different ROS: (1) singlet oxygen ($^1O_2$), (2) superoxide radical ($O_2^{\bullet-}$) and (3) hydrogen peroxide (H_2O_2) (Barajas-López et al. 2013). Excess light, drought and nutrient deprivation cause decrease in maximal photosynthetic capacity and overreduction of PET chain. Both processes induce higher leakage of electrons to O_2 and in consequence higher ROS production (Baker 1991; Chan et al. 2010; Li et al. 2009).

Under excess excitation $^1O_2$ is generated by the P680 in PSII reaction centre while PSI produces $O_2^{\bullet-}$ (subsequently metabolised to H_2O_2) in the Mehler reaction (Apel and Hirt 2004; Barajas-López et al. 2013; Chan et al. 2010; Krieger-Liszkay et al. 2008; Mullineaux and Karpinski 2002). $^1O_2$ is the most reactive and its lifetime is very short (200 ns) (Baier and Dietz 2005; Barajas-López et al. 2013). Therefore its action is limited to plastids, where it activates several stress-response pathways. Despite above limitations, its activity is highly specific (op den Camp et al. 2003; Leister 2005). On the other hand, enhanced levels of singlet oxygen $^1O_2$ in chloroplasts trigger programmed cell death. This is not due to the toxicity of singlet oxygen but rather due to EXECUTER-dependent signal. EXECUTER1 and EXECUTER2 are plastid proteins, both required for singlet oxygen signal transmission (Kim et al. 2012; Lee et al. 2007). Some products of oxidation by singlet oxygen play a role in retrograde signalling, for example, volatile β-cyclocitral (Barajas-López et al. 2013; Ramel et al. 2012) and 12-oxophytodienoic acid (OPDA, precursor of jasmonic acid; JA). Functions of both molecules, OPDA and JA, partly overlap (Berger et al. 2007).

Unlike singlet oxygen, hydrogen peroxide is a stable compound (half-life of ~1 ms) and it resembles water molecule. Thus it is transported through membrane with the use of specific aquaporins for H_2O_2 transport (Shapiguzov et al. 2012; Soto et al. 2012; Zardoya 2005), so it can directly influence the cytosolic signalling components. Under normal conditions, its accumulation is correlated with the oxidative/antioxidative capacity of the cell (Szechyńska-Hebda et al. 2007; Miyake 2010), and the stress disrupts this equilibrium (Pnueli et al. 2003; Vanderauwera et al. 2005). By unknown mechanism, a plant is able to recognise the source of H_2O_2 and trigger appropriate response (Barajas-López et al. 2013; Møller and Sweetlove 2010). H_2O_2 signalling regulates the plant growth, development and acclimatory mechanisms.

In ROS-mediated retrograde signalling, the negative correlation is observed between the genes that are modulated by $^1O_2$ and those modulated by H_2O_2 (Alboresi et al. 2011), possibly because H_2O_2 is able to keep Q_A in oxidised state and favours electron flow from PSII (Barajas-López et al. 2013; Krieger-Liszkay et al. 2008).

The triggers of chloroplast-to-nucleus signalling seem to be also the components of the photosynthetic electron transport chain and its redox state (Fey et al. 2005a). Plastoquinone (PQ) is involved in both linear and cyclic electron transports (Allen 2003), which makes it ideal sensor of perturbations in electron flux in thylakoid membranes. Indeed, 54 genes in *Arabidopsis* are specifically activated by signal from redox state of PQ (Fey et al. 2005b). However, probably there is an additional pathway involved in reporting the redox state of the photosynthetic electron transport chain, since reductive redox signal has faster kinetics (30 min) than the oxidative signal (2 h) (Barajas-López et al. 2013; Bräutigam et al. 2009). In green algae the redox state of PQ pool was shown to affect *LHCB* expression (Escoubas et al. 1995; Inaba 2010). In higher plants it is rather correlated with phosphorylation status of LHCB (Inaba 2010; Pursiheimo et al. 2001). Redox state of PQ is monitored through STN7, a kinase that on the one hand optimises photosynthesis (by initiating reallocation of light-harvesting antennas called state transition) and on the other hand leads to retrograde signalling (Barajas-López et al. 2013; Bonardi et al. 2005; Rochaix 2011; Shapiguzov et al. 2012). The latter effect of this kinase activity on *PhANG* expression is not direct (Pesaresi et al. 2009; Pesaresi 2011). Another kinase involved in feedback regulation of PhANGs is chloroplast sensor kinase (CSK) that monitors the flux of electron transport from PSII to PSI and couples PQ redox state to the regulation of chloroplast gene expression (Puthiyaveetil et al. 2012). The CSK protein is synthesised in the cytosol and imported into chloroplasts as a protein precursor. In *Arabidopsis thaliana*, CSK is autophosphorylated and required for control of transcription of chloroplast genes (Puthiyaveetil et al. 2008) and optimising photosynthesis (Puthiyaveetil and Allen 2009; Puthiyaveetil et al. 2008; Shapiguzov et al. 2012).

During redox changes in photosynthetic electron transport, regulation of plastid gene expression encoded in both chloroplast and nucleus is regulated by plastid-encoded RNA polymerase (PEP) (Baginsky et al. 1997; Barajas-López et al. 2013; Kindgren et al. 2012b; Steiner et al. 2009). Signals from at least 4 distinct plastid processes (tetrapyrrole biosynthesis, plastid gene expression, redox, plastid protein import) have been found to be mediated by GUN1 (Inaba 2010; Kakizaki et al. 2009; Koussevitzky et al. 2007). The first three of them are mediated by the following mechanism. Upon stress signals, the chloroplast envelope membrane-bound transcription factor PTM is processed by proteases (process controlled by GUN1) and released to the nucleus. Plant homodomain (PHD) of PTM binds to the promoter region of transcription factor ABI4 and activates its expression by histone modification. ABI4 competitively binds to the G-box *cis-element* of LHCB and inhibits LHCB expression (Barajas-López et al. 2013; Koussevitzky et al. 2007; Sun et al. 2011). Signal from protein import defect uses GLK1, a positive regulator of *LHCB* expression. GUN1 activity is necessary for the repression of GLK1 (Inaba 2010; Kakizaki et al. 2009).

Plastid genes exhibit three distinct patterns of expression. Tetrapyrrole, ROS and sugar signalling are under "master switch" control (Biehl et al. 2005; Leister 2005; Richly et al. 2003). The other genes show "mixed response" (equal numbers are up- and downregulated), or they are specifically co-regulated with nuclear genes and clustered into two regulons (Biehl et al. 2005; Leister 2005; Richly et al. 2003).

Electrical Intra- and Intercellular Signalling

The heterogeneous chloroplast population in a single cell may send contradictory signals to the nucleus, thereby confusing nuclear-controlled developmental and acclimatory processes. However, intra- and intercommunication between individual chloroplasts and coordination of their functioning may resolve this problem. It was recently demonstrated that an overexcited photosystem II (PSII) generates light wavelength-specific electrochemical signalling that simultaneously regulates local and systemic cellular light acclimation and defence responses (Szechyńska-Hebda et al. 2010; Karpiński et al. 2013). These electrical signals can change specific gene expression, and therefore such a mechanism could be crucial in coordination of plant homeostasis under stress conditions.

Electrical signalling in plants was first revealed in the 1870s in insectivorous plants by Burdon-Sanderson (1873) and Darwin (1875). In the twentieth century, evidence for the existence of a variety of electrical phenomena was presented, e.g. action potentials (APs) and variation potentials (VPs) (for review, see Davies 2006) as well as a novel type of electrical signal that propagates systemically and varies with intensity (SP, see Zimmermann et al. 2009). In a broad array of dicot and monocot plant species (Zimmermann et al. 2009), the electrical signals were shown to regulate various processes of plant physiology, including respiration (Dziubinska et al. 1989; Filek and Koscielniak 1997), water uptake (Davies et al. 1991), phloem unloading (Fromm 1991), phloem translocation (Fromm and Bauer 1994), fertilisation (Fromm et al. 1995), plant development (Filek et al. 2002; Dziubinska et al. 2003) and regulation of gene expression (Wildon et al. 1992). This account was extended by a study on the inhibition of photosynthesis (Koziolek et al. 2004), transient changes in chlorophyll fluorescence (PSII electron quantum yield) and leaf gas exchange (Lautner et al. 2005). Recently, photoelectrical signalling was suggested as an alternative way of plastid-to-nucleus retrograde signal transduction and the inter- and intracellular signalling that integrates light acclimation (systemic acquired resistance, SAR), cell death and immune defence responses (systemic acquired acclimation, SAA) (Szechyńska-Hebda et al. 2010; Karpiński et al. 2013). Partial exposure of the *Arabidopsis* rosette to excess light has been shown to cause changes in the membrane polarisation both in locally

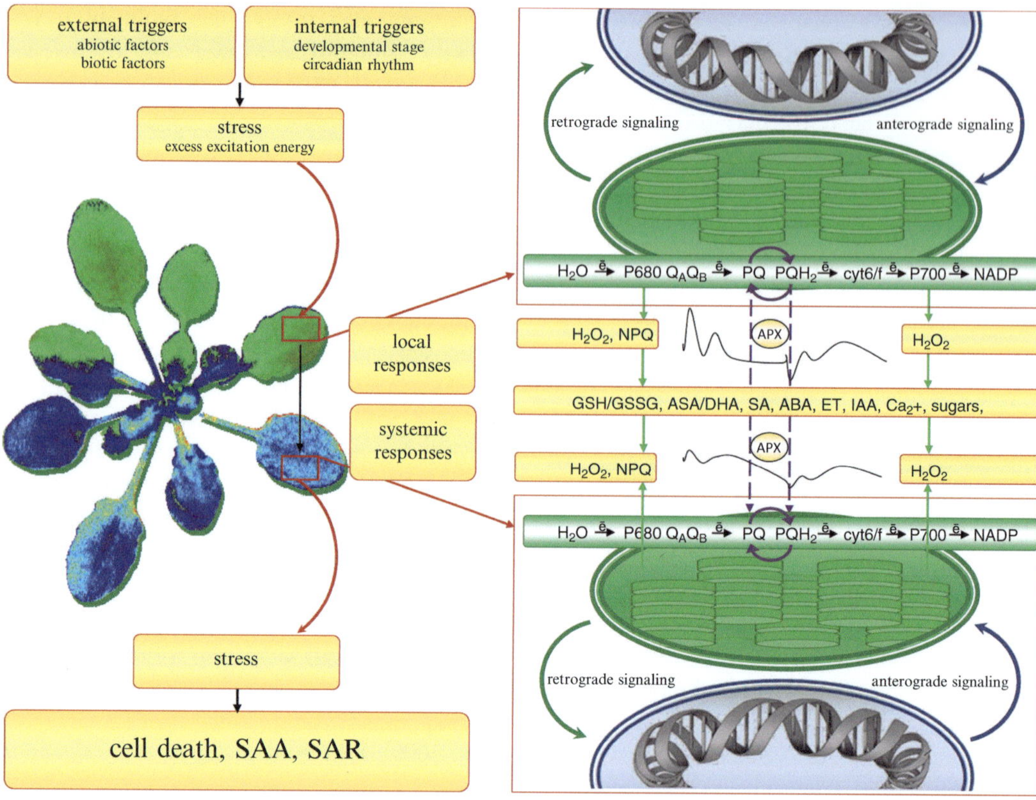

Fig. 1 Image of the non-photochemical changes of the *Arabidopsis thaliana* rosette partially exposed to excess light. Outline of the intra- and intercellular retrograde signalling pathways. Abbreviations: *ABA* abscisic acid, *ASA* ascorbic acid, *ET* ethylene, *GSH* reduced glutathione, *GSSG* oxidised glutathione, *IAA* indole-3-acetic acid, *PQ* plastoquinone, *SA* salicylic acid, *SAA* systemic acquired acclimation, *SAR* systemic acquired resistance

treated leaf and in leaves kept in shadow (undergoing SAA) (Fig. 1). Moreover, various light wavelengths of similar energy, for example, excess white, red (650 nm) and blue (450 nm), are able to induce a specific pattern of changes in the plasma membrane electrical potential and simultaneously induce the plant hypersensitivity reactions and SAR under pathogen attack. Recently it was even proposed that the health of plant may be predicted by recording the electrical signatures of the plants in response to external stimuli (Sharma et al. 2013).

This electric response is thought to be due to the stimulation of the plasma membrane H^+-ATPase and ion movements (Ca^{2+}, K^+, H^+ and Cl^-) (Zimmermann et al. 2009; Szechyńska-Hebda et al. 2010; Sukhov et al. 2013) and changes in a transmembrane voltage potential, usually at the order of $-50 \div -200$ mV. The propagation of such changes within tissues is very rapid (in the order of μs to ms) (Wheeler and Brownlee 2008) and occurs by the plant plasmodesmata (Fromm and Lautner 2007; Burch-Smith et al. 2011) of the bundle sheath cell (BSC) layer (Szechyńska-Hebda et al. 2010; Karpiński et al. 2013). Since, each plasmodesma contains at its axial centre the strands of the endoplasmic reticulum that are continuous between cells (Burch-Smith et al. 2011), electrochemical potential propagation is effective provided the membrane integrity and continuality (chloroplast-stromule-plasma membrane and their extension – the plasmodesmata). Recently, most of the chemistry of the neuromotoric system of animals has been found in plants, for example, neurotransmitters such as acetylcholine and cellular messengers and cellular motors such as calmodulin and actin, respectively (Volkov et al. 2009). Although this nerve-like cellular equipment never develops in plants at the same degree of complexity as in animal nerves, with membrane integrity in the plasmodesmata, a simple neural network is formed, and can be used for the communication over long distances. BSCs were thought to serve as a barrier that protects xylem and phloem, and enable proper distribution of water and inorganic and organic compounds essential for plant growth and development. However, it is documented that BSCs possess highly ion-selective cell membrane; thus they are also involved in the protection of the laminar tissue from toxic ions transported together with water in xylem (Shapira et al. 2009). An exceptional feature of BSC membrane selectivity stays in accordance with the results supporting the hypothesis that BSCs are the main carrier of electrical signals in plants. The metabolism and function of BSCs remain still hardly known but their individual character is unequivocal.

The source of electrochemical signalling seems to be overexcited photosystem II. The generation of the light intensity-specific and wavelength-specific electrochemical signals is dependent on the redox state of the PQ pool. 2,5-Dibromo-3-methyl-6-isopropylbenzoquinone (DBMIB) treatment changed the electrochemical signal amplitude and frequency in leaves exposed to excess light and undergoing SAA, whereas 3-(3,4-dichlorophenyl)-1,1-dimethylurea (DCMU) resulted in the improper transition of photoelectrochemical signals to distinct parts of the plant, similarly as $LaCl_3$ (ion channel blocker). Moreover, if the petiole of a leaf directly exposed to excess light has been treated with DCMU and $LaCl_3$ or damaged mechanically, the expression of the SAA- and SAR-marker genes has not been induced in systemic leaves (Szechyńska-Hebda et al. 2010; Karpiński et al. 2013). Similarly, in poplar tree, both the basipetal, short-distance electrical signalling within the leaf lamina and acropetal, long-distance signalling in the leaf phloem efficiently reduced the quantum yield of electron transport through PSII. Imaging analysis revealed that the signal that can cause yield reduction spreads through the leaf lamina. Cold blocking of the stem proved that the electrical signal transmission via the phloem becomes disrupted, causing the leaf gas exchange to remain unaffected (Lautner et al. 2005).

Propagation of electrical signals is accompanied by wavy-like changes in the NPQ and H_2O_2. The initial burst of light-induced ROS may trigger a cascade of cell-to-cell communication events which results in the formation of a ROS wave that propagates throughout different tissues (leaves and inflorescence stem) and carries the signal for long distances (Miller et al. 2009; Szechyńska-Hebda et al. 2010). H_2O_2 can travel at a rate of 8.4 cm min^{-1}, and it can follow the NPQ wave-like changes and the wave of the membrane electrical potential changes. They are also accompanied by alterations in the leaf water status and temperature and, consequently, by an increased ABA level, decreased stomatal conductance (Fryer et al. 2003; Chang et al. 2004; Grams et al. 2007) and conductance for CO_2 diffusion in the mesophyll (Gallé et al. 2013). Similar to H_2O_2, ABA biosynthesis is initiated in the vascular tissue and activates a signalling network in the neighbouring bundle sheath cells. Together with extracellular H_2O_2, ABA coordinates the redox retrograde signals to activate *APX2* expression and is necessary for the leaf's successful adjustment to repeated episodes of stress, e.g. excess light. The crosstalk between ROS and ABA signalling has also been postulated in the control of pathogen-related cell death and in excess light-induced retrograde signalling (Laloi et al. 2004; Mittler et al. 2011; Karpiński et al. 2013). Moreover, induction of jasmonic acid biosynthesis and changes in salicylic acid content were also linked to electrical signal triggering and propagation (Fisahn et al. 2004; Saeedi et al. 2013). Since the different chemical substances, including hormones, are known to adsorb to the charged surfaces of biological membranes (Filek et al. 2002) and can change membrane structure and permeability (Laggner et al. 2003a, b), the role of SA, JA and ABA in the regulation of the membrane potential via direct cellular membrane alternations can be considered.

The electrochemical signals together with NPQ, H_2O_2 and hormones, which are induced in response to different environmental stimuli (light, cold, heat, pathogen attack, touch) and transmitted within the plant tissue, are components of inter- and intracellular photoelectrochemical retrograde signalling (Karpinski et al. 1999, 2013; Fryer et al. 2003; Karpiński and Szechyńska-Hebda 2010; Szechyńska-Hebda et al. 2010). Indeed, systemic expression of at least *APX1* and *APX2* was specifically regulated by propagating photoelectrical signalling (Szechyńska-Hebda et al. 2010) in leaves that were undergoing SAA as well as in inflorescence stems undergoing SAA. The lack of functional APX2 in *apx2* knock-out mutant causes changes in electrical signal propagation (Szechyńska-Hebda et al. 2010). Moreover, APX2 requires functional photosynthetic electron transport to become induced, and APX induction and H_2O_2 levels were negatively correlated to NPQ changes. The physiological functions of APXs are known quite well. They are one of the key groups of H_2O_2-scavenging enzymes that convert H_2O_2 to water using ascorbate as an electron source in water-water cycle (Asada 1999; Dąbrowska et al. 2007). Thus, they are an important part of an antioxidant system of a plant. H_2O_2 overproduction (i.e. produced upon stress exposure) is damaging to structural components of a cell; however low, non-damaging H_2O_2 amount was proven to play an important signalling role in growth and physiological processes (Foreman et al. 2003; Mittler et al. 2004; Bright et al. 2006). APX1 and APX2 are the most known ascorbate peroxidases located in the cytosol. APX1 steady-state mRNA level in the cell is maintained at a relatively high level but, during stress episode, increases furthermore even several times (Karpiński et al. 1997, 1999; Szechyńska-Hebda et al. 2010). During recovery after stress episode, its induction does not decline to the level observed in non-treated plants. The stress-induced changes in *APX2* expression are different. In non-treated plants, the level of *APX2* mRNA is hardly detectable; however it increases after 7 min of stress (Karpiński et al. 1997, 1999) and reaches the maximum after 30 min of stress. If stress conditions remain, the *APX2* mRNA level begins to decrease after 1 h of stress and several hours after stress reaches the level observed in non-treated plants. The spatiotemporal pattern of expression of these cytosolic APXs suggests that APX1 is the main H_2O_2 scavenger, while APX2 is a "helper enzyme". APX2 may serve as an additional or exclusive protection of the central veins (Koussevitzky et al. 2008; Szechyńska-Hebda et al. 2010; Suzuki et al. 2013) and plays a significant role in signal transmission between plant organs.

APX2 expression has been used as an indicator of stress in plants as was shown to be induced by high light, heat, wounding and heavy metal ion treatment and boosted by simultaneous exposure to a few stress factors (Pnueli et al. 2003; Chang et al. 2004; Rossel et al. 2007; Mühlenbock et al. 2008; Keunen et al. 2013).

The electrical signals dependent on NPQ, APX and H_2O_2 changes arrived in responding tissue much sooner than transcript accumulation began, while transcript accumulation preceded the arrival of the other chemical signals, e.g. hormonal changes (Davies 2004).

Hormonal Homeostasis

The response to stress factor does not only involve changes in the photosynthetic electron flux and corresponding NPQ, ROS and electric changes but is also accompanied by the alterations in the different plant hormones. Although much slower than electrical signals, a propagated intra- and extracellular ROS/hormonal signal was shown to induce acclimatory responses, programmed cell death and immune defences (Karpinski et al. 1999; Fryer et al. 2003; Rossel et al. 2007; Mühlenbock et al. 2008). These processes are known to play a crucial role in the ability of plants to adapt to ever-changing environments by regulating growth and development.

Phytohormones include a few compounds with different chemical structure: abscisic acid (ABA), ethylene (ET), cytokinin (CKs), auxin (IAA) and gibberellin (GA) as well as jasmonate (JA), brassinosteroids (BR) and salicylic acid (SA). These hormones do not operate independently but rather are linked together in a complex network of interactions, and overall homeostasis is driven by the simultaneous interplay of multiple hormones.

One of the most studied topics in the response of plants to abiotic stress is ABA. Its synthesis appears as one of the fastest responses of plants to abiotic stress. It has long been recognised as a plant hormone that is upregulated in response to soil water deficit around the roots. ABA plays a crucial role in the transfer of the signal from stress-exposed root tissues to shoots, causes stomatal closure, leads to the reduction of water loss via transpiration and eventually restricts cellular growth. ABA also modulates aquaporin-related root and shoot hydraulic conductivity for improved nutrients and water distribution (Parent et al. 2009). Moreover, it upregulates processes involved in cell turgor maintenance and desiccation tolerance such as the synthesis of osmotically active solutes and antioxidant enzymes (Chaves et al. 2003). ABA is also known for its role during plant adaptations to cold temperature. Cold stress induces the synthesis of ABA and the exogenous application of ABA improves the cold tolerance of plants (Xue-Xuan et al. 2010). When the winter time is approaching and the temperature goes down, ABA mediates the conversion of the apical meristem into a dormant bud. The newly developing leaves growing above the meristem become converted into stiff bud scales that wrap the meristem closely and will protect it from mechanical damage and drying out during low temperatures. ABA in the bud also acts to enforce dormancy so if an unseasonably warm spell occurs before winter is over, the buds will not sprout prematurely. Only after a prolonged period of cold or the lengthening days of spring will bud dormancy be lifted (Paul and Kumar 2011). A large number of genes associated with de novo ABA biosynthesis and genes encoding ABA receptors have been characterised in *Arabidopsis thaliana*. The catalytic steps of ABA biosynthesis involving the conversion of β-carotene to ABA are mediated by the action of enzymes encoded by *ABA1/LOS6, ABA4, NCED, ABA2* and *ABA3/LOS5*. *ABA3/LOS5* encodes a Mo-cofactor sulphurase (MCSU) that catalyses the final conversion of abscisic aldehyde to ABA. The expression of *ABA3/LOS5* is enhanced when *A. thaliana* plants were exposed to drought or salt. Overexpression of *ABA3/LOS5* under the control of constitutive or drought-inducible promoters resulted in a significant increase in transgenic rice yield under drought conditions in the field (Xiong et al. 2001; Nambara and Marion-Poll 2005).

Cytokinins are considered to be essential regulators, strongly influencing the proliferation and differentiation of plant cells via regulation of ROS homeostasis and control oxidative stress (Szechyńska-Hebda et al. 2007). Increasing evidence has demonstrated that CKs play an important role in the regulation of environmental stress responses as well. Environmental stresses, such as drought and high salinity, decrease the synthesis and transport of CKs from roots. Application of exogenous CKs can increase stomatal apertures and transpiration in many plants and has a positive impact on photosynthetic activity (Tanaka et al. 2006). CKs act as antagonists to ABA in various growth and physiological processes, including environmental stress responses; e.g. during drought stress, ABA content is increased, whereas the active CK content is decreased. A reduction in CK content was also reported to lead to hypersensitivity to ABA and upregulation of stress- and/or ABA-responsive genes, whereas the reverse process was observed with elevation of CK levels (Nishiyama et al. 2011). Hypersensitivity to ABA was suggested as one of the contributors to the increased stress tolerance of CK-deficient mutants. The faster and more profound stress responses of these mutants partially resulted from the activation of the ABA-dependent AREB (ABA-responsive-element (ABRE) binding) pathway (Tran et al. 2010), which governs the majority of ABA-mediated

ABRE-dependent gene expression in response to stress. Constitutive overexpression of genes for CK degradation is associated with an enlarged root system; it also results in stunted shoot growth. This response results in increased drought tolerance and a higher accumulation of nutrients, without a penalty in shoot growth.

Ethylene plays a role in many aspects of plant growth and development and is modulated by different environmental factors e.g. during wounding, pathogen attack, anaerobiosis, heavy metal treatment and oxidative stresses. This effect can be transient (hours) or relatively long-lasting (days) and may vary considerably with species and plant age. ET may control its own production via autoinhibition. During plant development, ET stimulates seed germination, and high ethylene accumulation is traditionally associated with flower and leaf senescence, loss of chlorophyll, stem shortening, abscission of plant parts, fruit ripening and epinasty. Importantly, ET has been shown to be involved in the regulation of PCD (Ludwig et al. 2005). Oxidative stress-induced leaf damage is preceded by a rapid increase in ACC synthase activity, ACC content and ET emission, which are required for ROS accumulation and lesion development (Bouchez et al. 2007). Similarly, external addition of ethylene increased free radical production and caused increased superoxide production and spreading cell death (Hurr et al. 2013). ET is also assigned the role to regulate the pathogen-triggered cell death. In plants challenged with a compatible bacterial pathogen, ethylene signalling was required for symptom development (O'Donnell et al. 2003). ET is known to prevent ABA accumulation and vice versa or to modulate cellular sensitivity to ABA. The interplay of both hormones was observed in *Arabidopsis* mutant *acs7* with reduced ET production, greater ABA concentrations and increased tolerance to salt, heat and drought stresses (Dong et al. 2011).

Salicylic acid is a phenolic compound strictly involved in regulation of the pathogenesis-related (*PR*) gene expression, systemic acquired resistance and the hypersensitive response (Alvarez 2000) as well as in responses to abiotic stresses, such as high light, drought, salt, UV-B and many others. *Arabidopsis* mutants *nahG* (expressing a bacterial salicylate hydroxylase) and *sid2* (SA induction deficient), which have a very low level of SA (below 1 μg/g fresh weight), accumulate a high level of ROS and display severe damages under abiotic stress. On the other hand, when exogenous SA is applied at a dose more than 1 mM or some SA-over-accumulating mutants (*cpr1*, *cpr5* and *cpr6* constitutive inducer of PR proteins) are treated with stress factors, the oxidative burst and cell death are induced (Mateo et al. 2006). Thus, SA and ROS have been proposed to be on a positive feedback loop that amplifies signals leading to defence responses and cell death. It is assumed that SA signalling is mediated by at least two mechanisms, one requiring the *NPR1* (nonexpressor of PR1 genes) gene and the second that is independent of *NPR1* but requires protein kinases (MAPK). The second mechanism is accepted to be the main way of SA-mediated induction of abiotic stress-protective genes (Ludwig et al. 2005; Brodersen et al. 2006). SA is known to inhibit the activity of the last step in the ET biosynthesis pathway, which is catalysed by ACC oxidase. Given the cell-death-promoting role of ET, the antagonism of ET synthesis by SA might be a mechanism by which SA accumulation can contribute to the containment of lesion growth (Mosher et al. 2010).

Jasmonic acid has been shown to play important roles in growth and development, including flower development, tuber formation, tendril coiling, nyctinastic movements, trichome formation and senescence (Wasternack 2007). While JA accumulation generally stimulates the abovementioned processes, it inhibits root and leaf growth and seed germination. In *Arabidopsis* JA biosynthesis is initiated by a wound-mediated release of α-linolenic acid (18:3) from chloroplastic membranes, followed by the activity of several chloroplast-located enzymes, including 13-lipoxygenase (Bonaventure and Baldwin 2010). ET is also involved in wounding-mediated pathway, and therefore ET together with JA is known to rapidly accumulate in herbivore-attacked tissues. Moreover, gene expression analysis with JA-inducible marker genes (*VSP1*, *PR2*) has suggested that JA could be a factor involved in the containment of the ROS-dependent lesion propagation. It was strongly supported by the O_3 sensitivity phenotypes of JA-deficient mutants (*jar1*, *coi1*) (Overmyer et al. 2000). Similarly, the action of the JA and ABA seems to be closely related as the JA-insensitive mutant *jin4* is hypersensitive to ABA during germination (Berger et al. 1996).

Plant Metaphysiomics

In response to stress conditions, hundreds or thousands of genes are induced or repressed to optimise cell metabolism and to persist in nonoptimal conditions. Although particular stress factors induce or suppress a specific set of genes, there is also a set of genes that are similarly regulated across many stress conditions. For example, Ma and Bohnert (2007) using fuzzy k-means clustering method found a set of 197 genes that are induced in most of analysed conditions in *Arabidopsis* including hormonal treatment, different lights and biotic and abiotic stress conditions. These ubiquitous stress responses, similar to those of fungi and animals, employ genes in pathways related to mitogen-activated protein kinases, Snf1-related kinases, vesicle transport, mitochondrial functions and transcription machinery. In this group, genes related to plant-specific hormones ABA and JA were also present. Moreover, this group included genes that previously were characterised as related to biotic stress, but they were also induced by abiotic stress, and vice versa. Interestingly, most of the genes that are downregulated by stress appear to be under developmental regulation. Analysis of promoters clusters from similarly regulated genes revealed that abscisic acid-responsive element (ABRE motif) was over-represented in multiple clusters responding to either light or ABA treatment. Another motif, the W-box, was over-represented in several clusters induced by biotic stresses, and corresponding W-box-binding transcription factors (WRKYs) were themselves induced in these clusters.

Completely different pattern of gene expression is observed when plants are grown in field conditions (Wituszyńska et al. 2013; Richards et al. 2012). Genome-wide gene expression patterns of Bay-0 and Sha *Arabidopsis* accessions in natural environment presented strong differentiation (Richards et al. 2012). Two accessions differed significantly in 3,344 (14 % of the transcripts). Moreover, a gene ontology (GO) enrichment analysis showed that accession differed in the sulphate assimilation and biosynthesis of glucosinolate and glutathione. It was also revealed that in the field a large number of co-expressed gene clusters are enriched in loci responsive to several abiotic and biotic stresses. Authors also confirmed a role of several genes in field environment fluctuations, including APR6, HSP70, ROP10 and AAO3.

Large-scale expression analysis of 18 *Arabidopsis* accessions showed a big variation of transcript levels between analysed genotypes (Gan et al. 2011). 46 % (9,360) of expressed protein-coding genes were differentially expressed between at least one pair of accessions. Of these, 19 % (1,750) had more than tenfold expression changes. For about 60 % of genes, at least five accessions contributed to expression variation. Comparison of DNA sequences revealed that SNPs and single-nucleotide indels were concentrated in the 100-bp promoter region. This corroborates the general findings of extensive *cis* regulation of gene expression in *A. thaliana.* Gene ontology classifications for differentially expressed genes showed over-representation of response to the biotic environment, including pathogen defence and the production of glucosinolates to deter herbivores. These include NB-LRR genes, of which 74 % were differentially expressed at up to 400-fold change. Patterns for housekeeping genes (such as ribosomal proteins, eukaryotic initiation factors or kinesins) were markedly different: although many were differentially expressed, fold changes were generally small, with variation more often being limited to a few accessions.

So far, natural variation among *Arabidopsis* accessions to two hormones SA and auxin was investigated (Delker et al. 2010; van Leeuwen et al. 2007). Significant variation in transcript levels for response to SA was detected among the accessions, with relatively few genes responding similarly across all accessions and time points. The majority of SA-responsive genes showed an SA response in only one or a few accessions, and only 38 genes were SA responsive in the majority of accessions. Interestingly, cluster analysis showed that genes downregulated by SA possessed "chloroplast" and "plastid" GO categories. Light-responsive elements were over-represented in promoters of SA-downregulated genes. Similarly, significant variation in physiological response and gene expression was observed in seven *Arabidopsis* accessions treated with exogenous auxin (Delker et al. 2010). Many of the genes were differentially expressed in three or fewer accessions, whereas only ~100 genes showed twofold or higher expression change in all seven accessions. A relatively large proportion of genes were specifically induced in a single accession. Global auxin-induced expression changes among *Arabidopsis* accessions differ considerably in comparison to each other as well as to the reference accession Col-0, illustrating the large potential for variation in the regulation of diverse auxin-regulated processes. To conclude, it seems that *Arabidopsis* has a highly variable, largely plastic transcriptome.

In many eukaryotes regulation of gene expression by epigenetic changes is necessary for coordinating developmental programmes and specific responses. DNA methylation, the addition of a methyl group to a cytosine, is stable and heritable DNA modification that regulates expression of coding and noncoding genome regions. This DNA modification serves as silencing signal that occurs mainly in transposons and repetitive elements but is also observed in promoters and gene bodies in plants. It is estimated that more than 20 % of genes are methylated in *Arabidopsis* genome (Zilberman et al. 2007). In plants, DNA methylation is deposited at CG, CHG and asymmetric CHH sequences (where H is A, C or T). In eukaryotic organisms DNA methylation is necessary for proper development and is coupled with other epigenetic marks like nucleosome positioning and histone modifications. For example, H3K9me2 histone variant is associated with CHG DNA methylation in *Arabidopsis* (Bernatavichute et al. 2008). Similarly, interaction between DNA methylation, histone modification and gene expression was reported in rice and maize (Li et al. 2008; Wang et al. 2009).

Natural variation in DNA methylation is observed in *Arabidopsis* accessions (Vaughn et al. 2007; Shen et al. 2012). Substantial differences in cytosine methylation were observed between Ler and C24 *Arabidopsis* ecotypes (Shen et al. 2012). Interestingly, the percentage of methylated cytosines was significantly higher in their reciprocal F1 hybrids than in parental lines. Increased methylation of hybrid genomes was predominantly observed in differentially methylated regions in two parents. Moreover, these data suggest that genome-wide remodelling of DNA methylation plays a role in heterosis observed in hybrids. DNA methylation status in particular regions can fluctuate in relatively short timescales. Thirty generations in *Arabidopsis* were enough to accumulate a substantial number of differentially methylated cytosines that can contribute to expression variation (Schmitz et al. 2011; Becker et al. 2011).

DNA methylation is also involved in the regulation of gene expression in response to environmental conditions such as defence responses or salt stress (Bilichak et al. 2012; Dowen et al. 2012). Profiling of DNA methylation of plants exposed to bacterial pathogen *Pseudomonas syringae pv. tomato* DC3000 and SA showed that numerous regions were differentially methylated in response to stress. Differential DNA methylation that occurred mainly in transposons influenced neighbouring genes and changed their expression pattern (Dowen et al. 2012). Moreover, progeny of salt-treated plants exhibited changes in DNA methylation, histone modifications and expression of particular genes (Bilichak et al. 2012). These data suggest that DNA methylation is an important mechanism that allows plants to adapt to changing environment.

Modifications of core histones H2 and H3 affect gene expression in eukaryotes. A few types of histone modifications investigated in plants are described below. Genome-wide location of H2A.Z by chromatin immunoprecipitation coupled with microarray (ChIP-chip) revealed that this histone variant is found in nucleosomes flanking transcription start site (TSS) where it regulates transcription by preventing DNA methylation (Zilberman et al. 2008). H2A.Z is involved in regulation of genes important in regulation of development and environmental responses such as temperature, phosphate deficiency and pathogens. It seems that H2A.Z is a temperature sensor because it was observed that after temperature increase this histone variant was lost from nucleosomes in *Arabidopsis* that allowed appropriate control of gene expression (Kumar and Wigge 2010). Downregulation of expression of phosphate starvation response genes is probably dependent on H2A.Z which is found in histones flanking TSSs of these genes (Smith et al. 2010). Mutant plants with inactive SWR1 complex (required for H2A.Z incorporation into chromatin) showed increased expression of genes involved in systemic acquired resistance implying the role of H2A.Z in defence responses in *Arabidopsis* (March-Díaz et al. 2008).

H3K4me3 is associated with transcriptionally active genes where it is localised mainly in proximity of TSS. This histone modification is found in high frequency in shoots and roots. Interestingly, DNA regions from shoots affected by this modification are more frequent than DNA regions from roots, suggesting tissue-specific epigenetic variation (Wang et al. 2009). H3K4me3 is correlated with H3K9 acetylation (H3K9ac), and their distribution across DNA is associated with a specific gene ontology classification (Ha et al. 2011). Dense distribution near TSS is associated with genes involved in translation, while broad distribution towards coding sequence is observed in genes involved in photosynthesis, chloroplast and thylakoid membrane activities, carbohydrate metabolism and defence responses. These genes are highly expressed in leaves with a high expression variance.

H3K27me3 is often associated with stable silencing of genes involved in development. On average 15–20 % of genes in *Arabidopsis* and 30–40 % of genes in maize and rice are marked with H3K27me3, which is localised mainly in promoters or transcribed regions (He et al. 2010; Zhang et al. 2007; Wang et al. 2009). It was postulated that H3K27me3 is the major silencing mechanisms in plants (Zhang et al. 2007). Small variation of H3K27me3 between *Arabidopsis* accessions (e.g. the comparison of Col and Ler showed 32 Ler-specific and 11 Col-specific targets) cannot explain observed differences in expression patterns (Moghaddam et al. 2011; Dong et al. 2012). Moreover, accession-specific targets are often repressed by other repressive marks like DNA methylation or H3K9me2 (Dong et al. 2012).

Gene expression differences between plant species can be, at least, partially explained by histone modification variation. Locations and levels of H3K4me3, H3K9ac and H3K27me3 histone modifications in proximity of TSS are associated with gene expression variation between *Arabidopsis thaliana* and *Arabidopsis arenosa* (Ha et al. 2011). Similarly, strong positive correlation between H3K4me3 histone mark and differences in gene expression were observed in two rice subspecies and their reciprocal hybrids (He et al. 2010).

Differences in orthologous gene expression between species or ecotypes may result from *cis*- and/or *trans*-regulatory changes. *Trans* effects can result from divergence of transcription factors or chromatin regulators, while *cis* effects can result from changes of *cis*-regulatory elements (CREs). *Cis* and *trans* effects were investigated in *Drosophila*, yeast or *Arabidopsis* (Wittkopp et al. 2002; Tirosh et al. 2009; Shi et al. 2012), and it is supposed that expression differences between species predominantly depend on *cis* effects. Comparison of CREs in promoters of five plant genomes revealed high plasticity of CRE content in analysed genomes that can result in differences of gene expression patterns (Wóycicki et al. 2011). CRE shuffling in promoters of orthologous genes was correlated with three times higher rate of mutations (SNPs and indels) in promoters than in gene bodies. In conclusion, adaptation of plants to new environmental conditions may occur by rapid rewiring of gene regulatory networks driven by CRE shuffling, DNA methylation and histone modifications. We suggest that through evolution, eukaryotic organisms have been equipped with a high degree of freedom that allows for formation of new lines/varieties and species adapted to new ecological niches.

References

Alboresi A, Dall'osto L, Aprile A, Carillo P, Roncaglia E, Cattivelli L, Bassi R (2011) Reactive oxygen species and transcript analysis upon excess light treatment in wild-type Arabidopsis thaliana *vs* a photosensitive mutant lacking zeaxanthin and lutein. BMC Plant Biol 11:62

Allen JF (2003) Cyclic, pseudocyclic and noncyclic photophosphorylation: new links in the chain. Trends Plant Sci 8:15–19

Alvarez ME (2000) Salicylic acid in the machinery of hypersensitive cell death and disease resistance. Plant Mol Biol 44:429–442

Apel K, Hirt H (2004) Reactive oxygen species: metabolism, oxidative stress, and signal transduction. Annu Rev Plant Biol 55:373–399

Asada K (1999) The water-water cycle in chloroplasts: scavenging of active oxygens and dissipation of excess photons. Annu Rev Plant Physiol Plant Mol Biol 50:601–639

Baginsky S, Tiller K, Link G (1997) Transcription factor phosphorylation by a protein kinase associated with chloroplast RNA polymerase from mustard (Sinapis alba). Plant Mol Biol 34:181–189

Baier M, Dietz KJ (2005) Chloroplasts as source and target of cellular redox regulation: a discussion on chloroplast redox signals in the context of plant physiology. J Exp Bot 56:1449–1462

Baker NR (1991) A possible role for photosystem II in environmental perturbations of photosynthesis. Physiol Plant 81:563–570

Barajas-López J d D, Blanco NE, Strand Å (2013) Plastid-to-nucleus communication, signals controlling the running of the plant cell. Biochim Biophys Acta 1833:425–437

Becker C, Hagmann J, Müller J, Koenig D, Stegle O, Borgwardt K, Weigel D (2011) Spontaneous epigenetic variation in the *Arabidopsis thaliana* methylome. Nature 480:245–249

Berger S, Bell E, Mullet JE (1996) Two methyl jasmonate-insensitive mutants show altered expression of AtVsp in response to methyl jasmonate and wounding. Plant Physiol 111:525–531

Berger S, Benediktyová Z, Matouš K, Bonfig K, Mueller MJ, Nedbal L, Roitsch T (2007) Visualization of dynamics of plant–pathogen interaction by novel combination of chlorophyll fluorescence imaging and statistical analysis: differential effects of virulent and avirulent strains of P. syringae and of oxylipins on A. thaliana. J Exp Bot 58:797–806

Bernatavichute YV, Zhang X, Cokus S, Pellegrini M, Jacobsen SE (2008) Genome-wide association of histone H3 lysine nine methylation with CHG DNA methylation in *Arabidopsis* thaliana. PLoS One 3:e3156

Biehl A, Richly E, Noutsos C, Salamini F, Leister D (2005) Analysis of 101 nuclear transcriptomes reveals 23 distinct regulons and their relationship to metabolism, chromosomal gene distribution and co-ordination of nuclear and plastid gene expression. Gene 344:33–41

Bilichak A, Ilnystkyy Y, Hollunder J, Kovalchuk I (2012) The progeny of *Arabidopsis* thaliana plants exposed to salt exhibit changes in DNA methylation, histone modifications and gene expression. PLoS One 7:e30515

Bonardi V, Pesaresi P, Becker T, Schleiff E, Wagner R, Pfannschmidt T, Jahns P, Leister D (2005) Photosystem II core phosphorylation and photosynthetic acclimation require two different protein kinases. Nature 437:1179–1182

Bonaventure G, Baldwin IT (2010) New insights into the early biochemical activation of jasmonic acid biosynthesis in leaves. Plant Signal Behav 5:287–289

Bouchez O, Huard C, Lorrain S, Roby D, Balagué C (2007) Ethylene is one of the key elements for cell death and defense response control in the *Arabidopsis* lesion mimic mutant vad1. Plant Physiol 145:465–477

Bräutigam K, Dietzel L, Kleine T, Ströher E, Wormuth D, Dietz K-J, Radke D, Wirtz M, Hell R, Dörmann P, Nunes-Nesi A, Schauer N, Fernie AR, Oliver SN, Geigenberger P, Leister D, Pfannschmidt T (2009) Dynamic plastid redox signals integrate gene expression and metabolism to induce distinct metabolic states in photosynthetic acclimation in *Arabidopsis*. Plant Cell 21:2715–2732

Bright J, Desikan R, Hancock JT, Weir IS, Neill SJ (2006) ABA-induced NO generation and stomatal closure in *Arabidopsis* are dependent on H2O2 synthesis. Plant J 45:113–122

Brodersen P, Petersen M, Bjørn Nielsen H, Zhu S, Newman MA, Shokat KM, Rietz S, Parker J, Mundy J (2006) *Arabidopsis* MAP kinase 4 regulates salicylic acid- and jasmonic acid/ethylene-dependent responses via EDS1 and PAD4. Plant J 47:532–546

Burch-Smith TM, Brunkard JO, Choi YG, Zambryski PC (2011) Organelle–nucleus cross-talk regulates plant intercellular communication via plasmodesmata. Proc Natl Acad Sci 108:1451–1460

Burdon-Sanderson J (1873) Note on the electrical phenomena which accompany irritation of the leaf of Dionaea muscipula. Proc R Soc (Lond) 21:495–496

Chan KX, Crisp PA, Estavillo GM, Pogson BJ (2010) Chloroplast-to-nucleus communication: current knowledge, experimental strategies and relationship to drought stress signaling. Plant Signal Behav 5:1575–1582

Chang CCC, Ball L, Fryer MJ, Baker NR, Karpinski S, Mullineaux PM (2004) Induction of *ASCORBATE PEROXIDASE 2* expression in wounded *Arabidopsis* leaves does not involve known wound-signaling pathways but is associated with changes in photosynthesis. Plant J 38:499–511

Chaves MM, Maroco JP, Pereira JS (2003) Understanding plant responses to drought: from genes to the whole plant. Funct Plant Biol 30:239–264

Clarke JT, Warnock RCM, Donoghue PCJ (2011) Establishing a time-scale for plant evolution. New Phytol 192:266–301

Dąbrowska G, Kata A, Goc A, Szechyńska-Hebda M, Skrzypek E (2007) Characteristics of the plant ascorbate peroxidase family. Acta Biol Cracov Ser Bot 49:7–17

Darwin C (1875) Insectivorous plants. John Murray, London

Davies E (2004) New functions for electrical signals in plants. New Phytol 16:607–612

Davies E (2006) Electrical signals in plants: facts and hypotheses. In: Volkov AG (ed) Plant electrophysiology. Springer, Berlin

Davies E, Zawadzki T, Witters D (1991) Electrical activity and signal transmission in plants: how do plants know? In: Penel C, Greppin H (eds) Plant signaling, plasma membrane and change of state. University of Geneva, Geneva

Delker C, Pöschl Y, Raschke A, Ullrich K, Ettingshausen S, Hauptmann V, Grosse I, Quint M (2010) Natural variation of transcriptional auxin response networks in *Arabidopsis* thaliana. Plant Cell 22:2184–2200

Dong H, Zhen Z, Peng J, Chang L, Gong Q, Wang NN (2011) Loss of ACS7 confers abiotic stress tolerance by modulating ABA sensitivity and accumulation in *Arabidopsis*. J Exp Bot 62:4875–4887

Dong X, Reimer J, Göbel U, Engelhorn J, He F, Schoof H, Turck F (2012) Natural variation of H3K27me3 distribution between two *Arabidopsis* accessions and its association with flanking transposable elements. Genome Biol 13:R117

Dowen RH, Pelizzola M, Schmitz RJ, Lister R, Dowen JM, Nery JR, Dixon JE, Ecker JR (2012) Widespread dynamic DNA methylation in response to biotic stress. Proc Natl Acad Sci 109:E2183–E2191

Dyall SD, Brown MT, Johnson PJ (2004) Ancient invasions: from endosymbionts to organelles. Science 304:253–257

Dziubinska H, Trebacz K, Zawadzki T (1989) The effect of excitation on the rate of respiration in the liverwort Conocephalum conicum. Physiol Plant 75:417–423

Dziubinska H, Filek M, Szechynska-Hebda M, Trębacz K (2003) Slow vacuolar channels of non-embryogenic and embryogenic cultures of winter wheat. Physiol Plant 25:179–184

Escoubas JM, Lomas M, LaRoche J, Falkowski PG (1995) Light intensity regulation of cab gene transcription is signaled by the redox state of the plastoquinone pool. Proc Natl Acad Sci 92:10237–10241

Estavillo GM, Crisp PA, Pornsiriwong W, Wirtz M, Collinge D, Carrie C, Giraud E, Whelan J, David P, Javot H, Brearley C, Hell R, Marin E, Pogson BJ (2011) Evidence for a SAL1-PAP chloroplast retrograde pathway that functions in drought and high light signaling in *Arabidopsis*. Plant Cell 23:3992–4012

Fey V, Wagner R, Bräutigam K, Pfannschmidt T (2005a) Photosynthetic redox control of nuclear gene expression. J Exp Bot 56:1491–1498

Fey V, Wagner R, Braütigam K, Wirtz M, Hell R, Dietzmann A, Leister D, Oelmüller R, Pfannschmidt T (2005b) Retrograde plastid redox signals in the expression of nuclear genes for chloroplast proteins of *Arabidopsis thaliana*. J Biol Chem 280:5318–5328

Filek M, Koscielniak J (1997) The effect of wounding the roots by high temperature on the respiration rate of the shoot and propagation of electric signal in horse bean seedlings (Vicia faba L. minor). Plant Sci 123:39–46

Filek M, Zembala M, Szechynska-Hebda M (2002) The influence of phytohormones on zeta potential and electrokinetic charges of winter wheat cells. Z Naturforsch C 57:696–704

Fisahn J, Herde O, Willmitzer L, Pena-Cortes H (2004) Analysis of the transient increase in cytosolic Ca2+ during the action potential of higher plants with high temporal resolution: requirement of Ca2+ transients for induction of jasmonic acid biosynthesis and PINII gene expression. Plant Cell Physiol 45:456–459

Foreman J, Demidchik V, Bothwell JHF, Mylona P, Miedema H, Torres MA, Linstead P, Costa S, Brownlee C, Jones JD, Davies JM, Dolan L (2003) Reactive oxygen species produced by NADPH oxidase regulate plant cell growth. Nature 422:442–446

Fromm J (1991) Control of phloem unloading by action potentials in Mimosa. Physiol Plant 83:529–533

Fromm J, Bauer T (1994) Action potentials in maize sieve tubes change phloem translocation. J Exp Bot 45:463–469

Fromm J, Lautner S (2007) Electrical signals and their physiological significance in plants. Plant Cell Environ 30:249–257

Fromm J, Hajirezaei M, Wilke I (1995) The biochemical response of electric signaling in the reproductive system of Hibiscus plants. Plant Physiol 109:375–384

Fryer MJ, Ball L, Oxborough K, Karpinski S, Mullineaux PM, Baker NR (2003) Control of Ascorbate Peroxidase 2 expression by hydrogen peroxide and leaf water status during excess light stress reveals a functional organisation of *Arabidopsis* leaves. Plant J 33:691–705

Gallé A, Lautner S, Flexas J, Ribas-Carbo M, Hanson D, Roesgen J, Fromm J (2013) Photosynthetic responses of soybean (Glycine max L.) to heat-induced electrical signaling are predominantly governed by modifications of mesophyll conductance for CO2. Plant Cell Environ 36:542–552

Gan X, Stegle O, Behr J, Steffen JG, Drewe P, Hildebrand KL, Lyngsoe R, Schultheiss SJ, Osborne EJ, Sreedharan VT, Kahles A, Bohnert R, Jean G, Derwent P, Kersey P, Belfield EJ, Harberd NP, Kemen E, Toomajian C, Kover PX, Clark RM, Rätsch G, Mott R (2011) Multiple reference genomes and transcriptomes for *Arabidopsis thaliana*. Nature 477:419–423

Grams T, Koziolek C, Lautner S, Matyssek R, Fromm J (2007) Distinct roles of electric and hydraulic signals on the reaction of leaf gas exchange upon re-irrigation in Zea mays L. Plant Cell Environ 30:79–84

Ha M, Ng DW-K, Li W-H, Chen ZJ (2011) Coordinated histone modifications are associated with gene expression variation within and between species. Genome Res 21:590–598

He G, Zhu X, Aa E, Chen L, Wang X, Guo L, Liang M, He H, Zhang H, Chen F, Qi Y, Chen R, Deng X-W (2010) Global epigenetic and transcriptional trends among two rice subspecies and their reciprocal hybrids. Plant Cell 22:17–33

Hurr BM, Huber DJ, Vallejos CE, Lee E, Sargent SA (2013) Ethylene-induced overproduction of reactive oxygen species is responsible for the development of watersoaking in immature cucumber fruit. J Plant Physiol 170:56–62

Inaba T (2010) Bilateral communication between plastid and the nucleus: plastid protein import and plastid-to-nucleus retrograde signaling. Biosci Biotechnol Biochem 74:471–476

Kakizaki T, Matsumura H, Nakayama K, Che FS, Terauchi R, Inaba T (2009) Coordination of plastid protein import and nuclear gene expression by plastid-to-nucleus retrograde signaling. Plant Physiol 151:1339–1353

Karpiński S, Szechyńska-Hebda M (2010) Secret life of plants: from memory to intelligence. Plant Signal Behav 5:1391–1394

Karpiński S, Escobar C, Karpińska B, Creissen G, Mullineaux P (1997) Photosynthetic electron transport regulates the expression of cytosolic ascorbate peroxidase genes in *Arabidopsis* during excess light stress. Plant Cell 9:627–640

Karpinski S, Reynolds H, Karpinska B, Wingsle G, Creissen G, Mullineaux P (1999) Systemic signaling and acclimation in response to excess excitation energy in *Arabidopsis*. Science 284:654–657

Karpiński S, Szechyńska-Hebda M, Wituszyńska W, Burdiak P (2013) Light acclimation, retrograde signaling, cell death and immune defenses in plants. Plant Cell Environ 36:736–744

Kenrick P, Crane PR (1997) The origin and early diversification of land plants: a cladistic study. Smithsonian Institution Scholarly Press, Washington, DC

Keunen E, Remans T, Opdenakker K, Jozefczak M, Gielen H, Guisez Y, Vangronsveld J, Cuypers A (2013) A mutant of the *Arabidopsis thaliana* LIPOXYGENASE1 gene shows altered signaling and oxidative stress related responses after cadmium exposure. Plant Physiol Biochem 63:272–280

Kim C, Meskauskiene R, Zhang S, Lee KP, Lakshmanan Ashok M, Blajecka K, Herrfurth C, Feussner I, Apel K (2012) Chloroplasts of *Arabidopsis* are the source and a primary target of a plant-specific programmed cell death signaling pathway. Plant Cell 24:3026–3039

Kindgren P, Norén L, de López J DB, Shaikhali J, Strand A (2012a) Interplay between Heat Shock Protein 90 and HY5 controls PhANG expression in response to the GUN5 plastid signal. Mol Plant 5:901–913

Kindgren P, Kremnev D, Blanco NE, de Dios Barajas López J, Fernández AP, Tellgren-Roth C, Kleine T, Small I, Strand A (2012b) The plastid redox insensitive 2 mutant of *Arabidopsis* is impaired in PEP activity and high light-dependent plastid redox signaling to the nucleus. Plant J Cell Mol Biol 70:279–291

Koussevitzky S, Nott A, Mockler TC, Hong F, Sachetto-Martins G, Surpin M, Lim J, Mittler R, Chory J (2007) Signals from chloroplasts converge to regulate nuclear gene expression. Science 316:715–719

Koussevitzky S, Suzuki N, Huntington S, Armijo L, Sha W, Cortes D, Shulaev V, Mittler R (2008) Ascorbate peroxidase 1 plays a key role in the response of *Arabidopsis thaliana* to stress combination. J Biol Chem 283:34197–34203

Koziolek C, Grams TEE, Schreiber U, Matyssek R, Fromm J (2004) Transient knockout of photosynthesis mediated by electrical signals. New Phytol 161:715–722

Krieger-Liszkay A, Fufezan C, Trebst A (2008) Singlet oxygen production in photosystem II and related protection mechanism. Photosynth Res 98:551–564

Kumar SV, Wigge P (2010) H2A.Z-containing nucleosomes mediate the thermosensory response in *Arabidopsis*. Cell 140:136–147

Laggner P, Filek M, Marcinska I, Szechyńska-Hebda M, Kriechbaum M (2003a) X-ray structure investigations of winter wheat membrane systems. I. Influence of phytohormones on phospholipid orientation in non- and embryogenic cells. Plant Sci 165:265–270

Laggner P, Filek M, Szechynska-Hebda M, Kriechbaum M (2003b) X-ray structure investigations of winter wheat membrane systems. II. Effect of phytohormones on structural properties of mixed phospholipid-sterols membranes. Plant Sci 165:271–275

Laloi C, Mestres-Ortega D, Marco Y, Meyer Y, Reichheld JP (2004) The *Arabidopsis* cytosolic thioredoxin h5 gene induction by oxidative stress and its W-box-mediated response to pathogen elicitor. Plant Physiol 134:1006–1016

Lautner S, Grams TEE, Matyssek R, Fromm J (2005) Characteristics of electrical signals in poplar and responses in photosynthesis. Plant Physiol 138:2200–2209

Lee KP, Kim C, Landgraf F, Apel K (2007) EXECUTER1- and EXECUTER2-dependent transfer of stress-related signals from the plastid to the nucleus of *Arabidopsis thaliana*. Proc Natl Acad Sci 104:10270–10275

Leister D (2005) Genomics-based dissection of the cross-talk of chloroplasts with the nucleus and mitochondria in *Arabidopsis*. Gene 354:110–116

Li X, Wang X, He K, Ma Y, Su N, He H, Stolc V, Tongprasit W, Jin W, Jiang J, Terzaghi W, Li S, Deng XW (2008) High-resolution mapping of epigenetic modifications of the rice genome uncovers interplay between DNA methylation, histone methylation, and gene expression. Plant Cell 20:259–276

Li Z, Wakao S, Fischer BB, Niyogi KK (2009) Sensing and responding to excess light. Annu Rev Plant Biol 60:239–260

Ludwig AA, Saitoh H, Felix G, Freymark G, Miersch O, Wasternack C, Boller T, Jones JD, Romeis T (2005) Ethylene-mediated cross-talk between calcium-dependent protein kinase and MAPK signaling controls stress responses in plants. Proc Natl Acad Sci 102:10736–10741

Ma S, Bohnert HJ (2007) Integration of *Arabidopsis thaliana* stress-related transcript profiles, promoter structures, and cell-specific expression. Genome Biol 8:R49

March-Díaz R, García-Domínguez M, Lozano-Juste J, León J, Florencio FJ, Reyes JC (2008) Histone H2A.Z and homologues of components of the SWR1 complex are required to control immunity in *Arabidopsis*. Plant J 53:475–487

Martin W, Herrmann RG (1998) Gene transfer from organelles to the nucleus: how much, what happens, and why? Plant Physiol 118:9–17

Mateo A, Funck D, Mühlenbock P, Kular B, Mullineaux PM, Karpinski S (2006) Controlled levels of salicylic acid are required for optimal photosynthesis and redox homeostasis. J Exp Bot 57:1795–1807

Miller G, Schlauch K, Tam R, Cortes D, Torres MA, Shulaev V, Dangl JL, Mittler R (2009) The plant NADPH oxidase RBOHD mediates rapid systemic signaling in response to diverse stimuli. Sci Signal 110:8744–8749

Mittler R, Vanderauwera S, Gollery M, Van Breusegem F (2004) Reactive oxygen gene network of plants. Trends Plant Sci 9:490–498

Mittler R, Vanderauwera S, Suzuki N, Miller G, Tognetti VB, Vandepoele K, Gollery M, Shulaev V, Van Breusegem F (2011) ROS signaling: the new wave? Trends Plant Sci 16:300–309

Miyake C (2010) Alternative electron flows (water-water cycle and cyclic electron flow around PSI) in photosynthesis: molecular mechanisms and physiological functions. Plant Cell Physiol 51:1951–1963

Moghaddam AMB, Roudier F, Seifert M, Bérard C, Magniette M-LM, Ashtiyani RK, Houben A, Colot V, Mette MF (2011) Additive inheritance of histone modifications in *Arabidopsis thaliana* intra-specific hybrids. Plant J 67:691–700

Møller IM, Sweetlove LJ (2010) ROS signaling–specificity is required. Trends Plant Sci 15:370–374

Mosher S, Moeder W, Nishimura N, Jikumaru Y, Joo SH, Urquhart W, Klessig DF, Kim SK, Nambara E, Yoshioka K (2010) The lesion-mimic mutant cpr22 shows alterations in abscisic acid signaling and abscisic acid insensitivity in a salicylic acid-dependent manner. Plant Physiol 152:1901–1913

Mühlenbock P, Szechynska-Hebda M, Plaszczyca M, Baudo M, Mateo A, Mullineaux PM, Parker JE, Karpinska B, Karpinski S (2008) Chloroplast signaling and LESION SIMULATING DISEASE1 regulate crosstalk between light acclimation and immunity in *Arabidopsis*. Plant Cell 20:2339–2356

Mullineaux P, Karpinski S (2002) Signal transduction in response to excess light: getting out of the chloroplast. Curr Opin Plant Biol 5:43–48

Nambara E, Marion-Poll A (2005) Abscisic acid biosynthesis and catabolism. Annu Rev Plant Biol 56:165–185

Nishiyama R, Watanabe Y, Fujita Y, Le DT, Kojima M, Werner T, Vankova R, Yamaguchi-Shinozaki K, Shinozaki K, Kakimoto T, Sakakibara H, Schmülling T, Tran LS (2011) Analysis of cytokinin mutants and regulation of cytokinin metabolic genes reveals important regulatory roles of cytokinins in drought, salt and abscisic acid responses, and abscisic acid biosynthesis. Plant Cell 23:2169–2183

O'Donnell PJ, Schmelz E, Block A, Miersch O, Wasternack C, Jones JB, Klee HJ (2003) Multiple hormones act sequentially to mediate a susceptible tomato pathogen defense response. Plant Physiol 133:1181–1189

Op den Camp RGL, Przybyla D, Ochsenbein C, Laloi C, Kim C, Danon A, Wagner D, Hideg E, Göbel C, Feussner I, Natera M, Apel K (2003) Rapid induction of distinct stress responses after the release of singlet oxygen in *Arabidopsis*. Plant Cell 15:2320–2332

Overmyer K, Tuominen H, Kettunen R, Betz C, Langebartels C, Sandermann H Jr, Kangasjärvi J (2000) Ozone-sensitive *Arabidopsis rcd1* mutant reveals opposite roles for ethylene and jasmonate signaling pathways in regulating superoxide-dependent cell death. Plant Cell 12:1849–1862

Parent B, Hachez C, Redondo E, Simonneau T, Chaumont F, Tardieu F (2009) Drought and abscisic acid effects on aquaporin content translate into changes in hydraulic conductivity and leaf growth rate: a trans-scale approach. Plant Physiol 149:2000–2012

Paul A, Kumar S (2011) Responses to winter dormancy, temperature, and plant hormones share gene networks. Funct Integr Genomics 11:659–664

Pesaresi P (2011) Studying translation in *Arabidopsis* chloroplasts. Methods Mol Biol 774:209–224

Pesaresi P, Hertle A, Pribil M, Kleine T, Wagner R, Strissel H, Ihnatowicz A, Bonardi V, Scharfenberg M, Schneider A, Pfannschmidt T, Leister D (2009) *Arabidopsis* STN7 kinase provides a link between short- and long-term photosynthetic acclimation. Plant Cell 21:2402–2423

Pnueli L, Liang H, Rozenberg M, Mittler R (2003) Growth suppression altered stomatal responses and augmented induction of heat shock proteins in cytosolic ascorbate peroxidase (APX 1) – deficient *Arabidopsis* plants. Plant J 34:187–203

Pursiheimo S, Mulo P, Rintamäki E, Aro EM (2001) Coregulation of light-harvesting complex II phosphorylation and lhcb mRNA accumulation in winter rye. Plant J 26:317–327

Puthiyaveetil S, Allen JF (2009) Chloroplast two-component systems: evolution of the link between photosynthesis and gene expression. Proc Biol Sci 276:2133–2145

Puthiyaveetil S, Kavanagh TA, Cain P, Sullivan JA, Newell CA, Gray JC, Robinson C, van der Giezen M, Rogers MB, Allen JF (2008) The ancestral symbiont sensor kinase CSK links photosynthesis with gene expression in chloroplasts. Proc Natl Acad Sci 105:10061–10066

Puthiyaveetil S, Ibrahim IM, Allen JF (2012) Oxidation-reduction signaling components in regulatory pathways of state transitions and photosystem stoichiometry adjustment in chloroplasts. Plant Cell Environ 35:347–359

Ramel F, Birtic S, Ginies C, Soubigou-Taconnat L, Triantaphylidès C, Havaux M (2012) Carotenoid oxidation products are stress signals that mediate gene responses to singlet oxygen in plants. Proc Natl Acad Sci 109:5535–5540

Raven JA, Edwards D (2001) Roots: evolutionary origins and biogeochemical significance. J Exp Bot 52:381–401

Richards CL, Rosas U, Banta J, Bhambhra N, Purugganan MD (2012) Genome-wide patterns of *Arabidopsis* gene expression in nature. PLoS Genet 8:e1002662

Richly E, Dietzmann A, Biehl A, Kurth J, Laloi C, Apel K, Salamini F, Leister D (2003) Covariations in the nuclear chloroplast transcriptome reveal a regulatory master-switch. EMBO Rep 4:491–498

Rochaix JD (2011) Reprint of: regulation of photosynthetic electron transport. Biochim Biophys Acta 1807:878–886

Rossel JB, Wilson PB, Hussain D, Woo NS, Gordon MJ, Mewett OP, Howell KA, Whelan J, Kazan K, Pogson BJ (2007) Systemic and intracellular responses to photooxidative stress in *Arabidopsis*. Plant Cell 19:4091–4110

Saeedi S, Rocher F, Bonmort J, Fleurat-Lessard P, Roblin G (2013) Early membrane events induced by salicylic acid in motor cells of the Mimosa pudica pulvinus. J Exp Bot 64:1829–1836

Schmitz RJ, Schultz MD, Lewsey MG, O'Malley RC, Urich M, Libiger O, Schork NJ, Ecker JR (2011) Transgenerational epigenetic instability is a source of novel methylation variants. Science 334:369–373

Shapiguzov A, Vainonen JP, Wrzaczek M, Kangasjarvi J (2012) ROS-talk – how the apoplast, the chloroplast, and the nucleus get the message through. Front Plant Sci 3:292

Shapira O, Khadka S, Israeli Y, Shani U, Schwartz A (2009) Functional anatomy controls ion distribution in banana leaves: significance of Na+seclusion at the leaf margins. Plant Cell Environ 32:476–485

Sharma S, Singh R, Singh JB, Arya S, Khan S, Lehana P (2013) Effect of electrical impedance due to infliction on Aloe barbadensis Miller (Aloe-vera) leaves. IJCSITY 1:35–43

Shen H, He H, Li J, Chen W, Wang X, Guo L, Peng Z, He G, Zhong S, Qi Y, Terzaghi W, Deng XW (2012) Genome-wide analysis of DNA methylation and gene expression changes in two *Arabidopsis* ecotypes and their reciprocal hybrids. Plant Cell 24:875–892

Shi C, Baldwin IT, Wu J (2012) *Arabidopsis* plants having defects in nonsense-mediated mRNA decay factors UPF1, UPF2, and UPF3 show photoperiod-dependent phenotypes in development and stress responses. J Integr Plant Biol 54:99–114

Smith AP, Jain A, Deal RB, Nagarajan VK, Poling MD, Raghothama KG, Meagher RB (2010) Histone H2A.Z regulates the expression of several classes of phosphate starvation response genes but not as a transcriptional activator. Plant Physiol 152:217–225

Soll J, Schleiff E (2004) Protein import into chloroplasts. Nat Rev Mol Cell Biol 5:198–208

Soto G, Alleva K, Amodeo G, Muschietti J, Ayub ND (2012) New insight into the evolution of aquaporins from flowering plants and vertebrates: orthologous identification and functional transfer is possible. Gene 503:165–176

Steiner S, Dietzel L, Schröter Y, Fey V, Wagner R, Pfannschmidt T (2009) The role of phosphorylation in redox regulation of photosynthesis genes psaA and psbA during photosynthetic acclimation of mustard. Mol Plant 2:416–429

Sukhov V, Akinchits E, Katicheva L, Vodeneev V (2013) Simulation of variation potential in higher plant cells. J Membr Biol 246:287–296

Sun X, Feng P, Xu X, Guo H, Ma J, Chi W, Lin R, Lu C, Zhang L (2011) A chloroplast envelope-bound PHD transcription factor mediates chloroplast signals to the nucleus. Nat Commun 2:477

Suzuki N, Miller G, Sejima H, Harper J, Mittler R (2013) Enhanced seed production under prolonged heat stress conditions in *Arabidopsis thaliana* plants deficient in cytosolic ascorbate peroxidase 2. J Exp Bot 64:253–263

Szechyńska-Hebda M, Karpiński S (2013) Light intensity-dependent retrograde signaling in higher plants. J Plant Physiol. doi:10.1016/j.physletb.2003.10.071

Szechyńska-Hebda M, Skrzypek E, Dąbrowska G, Kościelniak J, Filek M, Wędzony M (2007) The role of oxidative stress induced by growth regulators in the regeneration process of wheat. Acta Physiol Plant 29:327–337

Szechyńska-Hebda M, Kruk J, Górecka M, Karpińska B, Karpiński S (2010) evidence for light wavelength-specific photoelectrophysiological signaling and memory of excess light episodes in *Arabidopsis*. Plant Cell 22:2201–2218

Tanaka Y, Sano T, Tamaoki M, Nakajima N, Kondo N, Hasezawa S (2006) Cytokinin and auxin inhibit abscisic acid-induced stomatal closure by enhancing ethylene production in *Arabidopsis*. J Exp Bot 57:2259–2266

Timmis JN, Ayliffe MA, Huang CY, Martin W (2004) Endosymbiotic gene transfer: organelle genomes forge eukaryotic chromosomes. Nat Rev Genet 5:123–135

Tirosh I, Reikhav S, Levy AA, Barkai N (2009) A yeast hybrid provides insight into the evolution of gene expression regulation. Science 324:659–662

Tran LS, Shinozaki K, Yamaguchi-Shinozaki K (2010) Role of cytokinin responsive two-component system in ABA and osmotic stress signalings. Plant Signal Behav 5:148–150

Van Leeuwen H, Kliebenstein DJ, West ML, Kim K, Van Poecke R, Katagiri F, Michelmore RW, Doerge RW, St Clair D (2007) Natural variation among *Arabidopsis thaliana* accessions for transcriptome response to exogenous salicylic acid. Plant Cell 19:2099–2110

Vanderauwera S, Zimmermann P, Rombauts S, Vandenabeele S, Langebartels C, Gruissem W, Inzé D, Van Breusegem F (2005) Genome-wide analysis of hydrogen peroxide-regulated gene expression in *Arabidopsis* reveals a high light-induced transcriptional cluster involved in anthocyanin biosynthesis. Plant Physiol 139:806–821

Vaughn MW, Tanurdzić M, Lippman Z, Jiang H, Carrasquillo R, Rabinowicz PD, Dedhia N, McCombie WR, Agier N, Bulski A, Colot V, Doerge RW, Martienssen R (2007) Epigenetic natural variation in *Arabidopsis thaliana*. PLoS Biol 5:e174

Volkov AG, Carrell H, Markin VS (2009) Biologically closed electrical circuits in Venus flytrap. Plant Physiol 149:1661–1667

Wang X, Elling AA, Li X, Li N, Peng Z, He G, Sun H, Qi Y, Liu XS, Deng XW (2009) Genome-wide and organ-specific landscapes of epigenetic modifications and their relationships to mRNA and small RNA transcriptomes in maize. Plant Cell 21:1053–1069

Wasternack C (2007) Jasmonates: an update on biosynthesis, signal transduction and action in plant stress response, growth and development. Ann Bot 100:681–697

Waters MT, Wang P, Korkaric M, Capper RG, Saunders NJ, Langdale JA (2009) GLK transcription factors coordinate expression of the photosynthetic apparatus in *Arabidopsis*. Plant Cell 21:1109–1128

Wheeler GL, Brownlee C (2008) Ca2+ signaling in plants and green algae: changing channels. Trends Plant Sci 13:506–514

Wildon DC, Thain JF, Minchin PEH, Gubb IR, Reilly AJ, Skipper YD, Doherty HM, O'Donnell PJ, Bowles DJ (1992) Electrical signaling and systemic proteinase inhibitor induction in the wounded plant. Nature 360:62–65

Wittkopp PJ, True JR, Carroll SB (2002) Reciprocal functions of the *Drosophila* yellow and ebony proteins in the development and evolution of pigment patterns. Development 129:1849–1858

Wituszyńska W, Ślesak I, Vanderauwera S, Szechyńska-Hebda M, Kornaś A, Van Der Kelen K, Karpińska B, Maćkowski S, Van Breusegem F, Karpiński S (2013) LSD1, EDS1 and PAD4 conditionally regulate cellular signaling homeostasis, photosynthesis, water use efficiency and seed yield in *Arabidopsis*. Plant Physiol 161:1795–1805

Woodson JD, Perez-Ruiz JM, Chory J (2011) Heme synthesis by plastid ferrochelatase I regulates nuclear gene expression in plants. Curr Biol 21:897–903

Wóycicki R, Witkowicz J, Gawroński P, Dąbrowska J, Lomsadze A, Pawełkowicz M, Siedlecka E, Yagi K, Pląder W, Seroczyńska A, Śmiech M, Gutman W, Niemirowicz-Szczytt K, Bartoszewski G, Tagashira N, Hoshi Y, Borodovsky M, Karpiński S, Malepszy S, Przybecki Z (2011) The genome sequence of the north-european cucumber (*Cucumis sativus* L.) unravels evolutionary adaptation mechanisms in plants. PLoS One 6:e22728

Xiong L, Ishitani M, Lee H, Zhu JK (2001) The *Arabidopsis* LOS5/ABA3 locus encodes a molybdenum cofactor sulfurase and modulates cold stress- and osmotic stress-responsive gene expression. Plant Cell 13:2063–2083

Xue-Xuan X, Hong-Bo S, Yuan-Yuan M, Gang X, Jun-Na S, Dong-Gang G, Cheng-Jiang R (2010) Biotechnological implications from abscisic acid (ABA) roles in cold stress and leaf senescence as an important signal for improving plant sustainable survival under abiotic-stressed conditions. Crit Rev Biotechnol 30:222–230

Zardoya R (2005) Phylogeny and evolution of the major intrinsic protein family. Biol Cell 97:397–414

Zhang X, Clarenz O, Cokus S, Bernatavichute YV, Pellegrini M, Goodrich J, Jacobsen SE (2007) Whole-genome analysis of histone H3 lysine 27 trimethylation in *Arabidopsis*. PLoS Biol 5:e129

Zhang Z-W, Yuan S, Feng H, Xu F, Cheng J, Shang J, Zhang D-W, Lin H-H (2011) Transient accumulation of Mg-protoporphyrin IX regulates expression of PhANGs – new evidence for the signaling role of tetrapyrroles in mature *Arabidopsis* plants. J Plant Physiol 168:714–721

Zilberman D, Gehring M, Tran RK, Ballinger T, Henikoff S (2007) Genome-wide analysis of *Arabidopsis thaliana* DNA methylation uncovers an interdependence between methylation and transcription. Nat Genet 39:61–69

Zilberman D, Coleman-Derr D, Ballinger T, Henikoff S (2008) Histone H2A.Z and DNA methylation are mutually antagonistic chromatin marks. Nature 456:125–129

Zimmermann MR, Maischak H, Mithofer A, Boland W, Felle HH (2009) System potentials, a novel electrical long-distance apoplastic signal in plants, induced by wounding. Plant Physiol 149:1593–1600

Signalomics: Diversity and Methods of Analysis of Systemic Signals in Plants

Alain Vian, Bratislav Stankovic, and Eric Davies

Contents

Abstract

We provide a brief definition and history of signals, pointing out how differences in body plan between plants and animals require fundamentally different signaling mechanisms, and then list the diversity of chemical and physical signals along with their pathways of transmission, providing details on molecular signals and focusing on the phloem and xylem as being the main conduits for (rapid) systemic signaling. The two major electrical (action potentials and variation potentials) as well as hydraulic signals are then described. The latter part of the chapter deals with methods of analysis of molecular signals, including accessing the phloem and identifying the array of gene products transported therein. A description is provided of the modern methods used in metabolomics and phenotyping to analyze the metabolic consequences of signal action. Conventional techniques for analyzing electrical and hydraulic signals and their ionic components using electrodes are then furnished. Finally we describe novel techniques

A. Vian, Ph.D. (✉)
UMR 1345 IRHS (Université d'Angers, Agrocampus Ouest, INRA), SFR 4207 Quasav, Faculté des Sciences, 2 Bd Lavoisier, 49045 Angers cedex 01, France
e-mail: alain.vian@univ-angers.fr

B. Stankovic, Ph.D.
University for Information Science and Technology "St. Paul the Apostle", Partizanska bb, 6000 Ohrid, Republic of Macedonia
e-mail: bratislav.stankovic@fulbrightmail.org

E. Davies, Ph.D.
Department of Plant Biology, North Carolina State University, Raleigh, NC, USA
e-mail: edavies.pv@gmail.com

D. Barh et al. (eds.), *PlantOmics: The Omics of Plant Science*,
DOI 10.1007/978-81-322-2172-2_16, © Springer India 2015

developed recently in the animal field using fluorescence to monitor real-time changes in membrane potential, which could be adapted for plants to open up new vistas in our understanding of electrical signals in plants.

Abbreviations

ABA(-GE)	Abscisic acid-(glucose-ester conjugate)
CBL	Calcineurin B-like
CIPK	CBL-interacting protein kinase
CML	Calmodulin/calmodulin-like protein
eATP	Extracellular ATP
DAG	Diacylglycerol
FRET	Fluorescence (or Förster) resonance energy transfer
FTIR	Fourier transform infrared spectroscopy
GC-MS	Gas chromatography-mass spectrometry
PLAFP	Phloem lipid-associated family protein
PI	Protease inhibitor
PIIF	Protease inhibitor-inducing factor
RALF	Rapid alkalinization factor
ROS	Reactive oxygen species
SE	Sieve element
SP	System potential
VP	Variation potential
VSFP	Voltage-sensitive fluorescent protein
VOCs	Volatile organic compounds
VT	Voltage transient

Introduction

Signals are packets of biological information generated in one location and transmitted elsewhere, frequently in response to external (environmental) stimuli. After a signal has been transmitted, it acts as an internal stimulus to evoke downstream responses. Hence signals and stimuli are the fundamental units of biological communication (Davies 2004). Signals vary in identity, velocity, informational capacity, and locality (site of generation, transmission, and action).

Over 220 years ago, electrical signals were the only ones known in both plants and animals (Galvani 1791). In the middle of the nineteenth century, chemical signals were discovered in animals, and the field of endocrinology developed (Berthold 1849). Now animals were "superior," insofar as they had both electrical and chemical signals, whereas plants had just electrical signals. It was about 30 years later before Charles Darwin (1881) furnished the first evidence in plants for the existence of a diffusible chemical, which could cause plant cell enlargement. For some reason the discovery of chemical signals (hormones) in plants led to the demise of a role for electrical signals despite the previous 100+ years of research showing their existence in plants (references in Stern 1924). After Darwin (1881), the next attempt to assign signal transmission in plants to a chemical was proposed 35 years later by Ricca (1916) to explain the propagation of the signal that causes pulvinus movements of leaflets in *Mimosa pudica*. While the nature of such molecules remained unknown for a long time, numerous molecules have now proven their ability to carry information a long distance in plants.

Higher animals are heterotrophic and thus need to be motile to locate food and partners for reproduction, while food and gas exchange takes place through interior tubes. This demands that they have a near cylindrical body plan to contain the tubes and appendages for motility, and so they have a very low surface area to volume ratio. In marked contrast, plants are autotrophic, rely on animals or the wind to aid in reproduction, and thus have no need to be motile. Consequently, plants are not able to escape from their abiotic (cold, heat, etc.) or their biotic (herbivores, insects) environment (Maleck and Dietrich 1999). Furthermore their food and gas exchange takes place through external surfaces (leaves for light and CO_2, roots for ions and water), thereby giving them a high surface area to volume ratio with many of their cells in direct contact with the environment (Hallé 1999; Vian et al. 2007). It

Table 1 Signaling distance/pathway in plants

Transmission	Examples	References
Within the cell	Calcium	Dodd et al. (2010) and Stael et al. (2012)
Diffusion, secretion, cytoplasmic streaming	DAG	Jeannette et al. (2010) and Canonne et al. (2011)
	Protons	Kader and Lindberg (2010)
	Kinases	Tena et al. (2011)
	H_2O_2 waves	Vestergaard et al. (2012)
Between cells	IP_3	Gillaspy (2011)
(Apoplasm and/or symplasm)	Auxin, ABA	Blakeslee et al. (2005) and Umezawa et al. (2010)
	ROS	Mittler et al. (2011)
	Oligosaccharides	Shibuya and Nimami (2001)
	Peptides	Murphy et al. (2012)
Between organs	Molecules in xylem	Atkins and Smith (2007) and Krishnan et al. (2011)
Phloem and/or xylem	Molecules in phloem	Dinant and Suárez-López (2012)
	Hydraulic signals	Malone (1993)
	Electrical signals	Davies (2004) and Yan et al. (2009)
Between plants	Phenolic compounds	Treutter (2006) and Cesco et al. (2012)
Airborne transmission, mycorrhiza	Jasmonate and methyl jasmonate	Avanci et al. (2010)
	Ethylene	Wilkinson and Davies (2010)
	Mycorrhiza	Song et al. (2010)
Between generations	DNA methylation	Migicovsky and Kovalchuk (2013)
Gene modification	Histone modification	Bilichak et al. (2012)
	Gene silencing	Qutob et al. (2013)

becomes then tremendously important for plants to sense even minute changes in climatic or mineral resources in the environment and adapt their growth (Läuchli and Grattan 2007) or flower production (Suárez-López 2005) accordingly and also to promote defense strategies. Indeed, environmental factors such as nitrate availability (Forde 2002; Liu et al. 2009), wind (Anten et al. 2010), drought (Hetherington 1998; Jia and Zhang 2008), flooding (Dat et al. 2004), predator attacks, or infection (Heil and Ton 2008; Parker 2009) induce systemic responses in plants that generally result in a reduction in growth and yield.

Plants must, therefore, possess systems to exchange information throughout the entire plant to ensure the coordination of plant development and defense. The systems for transmitting this information are complex and involve multiple components, which are far from being understood. Evidence strongly suggests that information exchange relies on at least two different systems: one involving molecules that are transported within the plant and another that uses electrical and/or hydraulic signals to carry the information throughout the entire plant. Both of these information-transmitting systems involve the primary material-transporting systems (xylem and phloem). Furthermore, the signals themselves vary in identity, diversity, specificity, versatility, rapidity, ubiquity, and locality.

Chemical signals (Tables 1, 2, and 3) and electrical signals are diverse. Chemicals that have been implicated in short- and long-distance and/or organism-to-organism signaling include ions (Ca^{2+}, H^+), volatiles (ethylene, methyl jasmonate), and both small (IAA, GA, NO) and large molecules (proteins, RNA). Electrical signals are less diverse and only three (action potentials, variation potentials, and system potentials) have been described. These signals are more or less specific to the type of stimulation. In many instances, signal production and/or transmission depends upon the strength (injurious or non-injurious) and/or type of the stress (biotic or abiotic). As pointed by Davies (2004), plant signal versatility could result from the initial inability of plants to identify the agent (biotic or abiotic) that

Table 2 Identity/diversity of plant molecular signals

Signal	References
Inorganic molecules/ions	
CO_2	Lake et al. (2002) and Kim et al. (2010)
NO	Molassiotis et al. (2010) and Beaudoin (2011)
ROS, H_2O_2	Mittler et al. (2011) and Suzuki et al. (2012)
Ca^{2+}	Batistič and Kudla (2012) and Kurusu et al. (2013)
H+	Kader and Lindberg (2010)
Others	Liu et al. (2009) and Dinant and Suárez-López (2012)
Small organic molecules	
ABA	Raghavendra et al. (2010) and Nakashima and Yamaguchi-Shinozaki (2013)
IAA	Zhao (2010) and Depuydt and Hardtke (2011)
CK	Choi et al. (2011), Ha et al. (2012), and Hwang et al. (2012)
GA	Lau and Deng (2010) and Hauvermale et al. (2012)
Ethylene	Zhao and Guo (2011) and Wang et al. (2013)
Methyl jasmonate	Staswick (2008) and Wasternack and Hause (2013)
Phosphoinositides, diacylglycerol, lipids	Boss and Im (2012), Dong et al. (2012), Guelette et al. (2012), and Benning et al. (2012)
Phenolic compounds	Mandal et al. (2010)
ATP	Tanaka et al. (2010), Chivasa and Slabas (2012), and Sun et al. (2012)
Salicylic acid, methyl salicylate	Hayata et al. (2010) and Shah and Zeier (2013)
Sugars	Wingler and Roitsch (2008) and Eveland and Jackson (2012)
Volatile organic compounds	Heil and Ton (2008)
Large organic molecules	
Oligosaccharides	Shibuya and Nimami (2001)
RNA/miRNA	Kehr and Buhtz (2008) and Sunkar et al. (2012)
Polypeptides	Matsubayashi and Sakagami (2006), Wang and Fiers (2010), and Sun et al. (2011)

Table 3 Diversity of phloem exudate transcripts and proteins from white lupin

Protein identity	Phloem exudate transcript (%)	Phloem exudate proteins (%)
Metabolism	15.3	23.8
Photosynthesis	3	3.8
Protein modification/turnover	10.5	9.2
Redox regulation	4.7	8.5
Signaling	5.4	2
Stress and defense response	5.5	6.2
Nucleic acid binding/transport	7	2.3
Cell structural components	2.6	6.2
Viral	3	ND
Unknown and unclassified	43	37

Table generated using data from Rodriguez-Medina et al. (2011)

causes the insult. As a consequence, a signal may spread throughout the plant to elicit a general stress response. The ambivalence between specificity and versatility is not yet fully understood and it is quite conceivable that this question has only minor significance from a plant's perspective.

While some signals should be transmitted throughout the plant (i.e., for a global response or to coordinate the different growth sites), other signals are restricted to the cells adjacent to the stimulation site when, for example, they evoke necrosis of tissue. Signals, therefore, differ in the distance they are transmitted, thereby resulting in local vs. systemic responses.

For cell-to-cell transport, chemicals must travel through the plasmodesmata and symplast, while for inter-organ transport they go primarily through the vascular system (phloem or xylem). The rate of transport is not much a problem when the signaling occurs over short distances, but it becomes crucial in the case of long-distance transmission. Indeed, signals move at different speeds, mainly as a function of their nature (i.e., chemical or physical). The transport of solute molecules is often relatively slow, limited to the speed of diffusion and to phloem sieve element (SE) fluxes, although faster transport has been reported in association with a hydraulic surge in the xylem (Malone et al. 1994; Hlaváčková and

Nauš 2007). Indeed, the phloem seems to carry a wide variety of signaling molecules (Dinant and Suárez-López 2012). In this case it is worth noting that the persistence of the signaling molecule might last for a relatively long period of time (up to several days). Thus, molecular signaling is often a slow, diffusion-based, long-term, sustainable response, whereas electrical and hydraulic signaling is rapid but transient.

Another kind of signal termed "memory" exists in plants, which often show a temporal delay in response to specific stimuli. Such a system was first observed with the pioneering work of Thellier et al. (1982) and Desbiez et al. (1984). They studied the long-term (several days) storage of a signal evoked by a minor wound that was able to modify plant morphology (bud outgrowth) when the plant was transferred to the appropriate culture medium. It was also observed with different plants that a primary infection leading to systemic acquired resistance (SAR) in the plant provides a long-term (lifelong) memory of the infection (Spoel and Dong 2012) that enhances the plant's resistance to further infections.

In addition to the signaling mechanisms that stay within the same plant generation, a totally different system, known as epigenetic memory (Saze 2008), transfers the increased resistance potential to the next generation, improving the offspring's capacity to survive various kinds of stress (Mirouze and Paszkowski 2011). This intergenerational memory transmission (Molinier et al. 2006) includes enhanced resistance to infection (Luna et al. 2012; Rasmann et al. 2012; Slaughter et al. 2012) and is based on epigenetic mechanisms (Chinnusamy and Zhu 2009; Pastor et al. 2012) that include chromatin remodeling, DNA methylation, histone methylation and acetylation (Chen and Tiana 2007), and small (mobile) silencing RNA (Molnar et al. 2010). Thellier and Lüttge (2013) reviewed the memory processes in plants and proposed an integrative model.

Signal Identity

Plant signaling systems employ multiple pathways and transmission distances (Table 1). Very-short-distance signaling occurs within the cell (e.g., from the cell membrane to the nucleus), involving several different mediators such as calcium, pH changes, membrane derivatives, and phosphorylation. These events elicit metabolic changes that could be transmitted to the adjacent cells through the symplasm and/or apoplasm (Bloemendal and Kück 2013). Short-distance, cell-to-cell transport is especially important in cell regulation including the maintenance of cell identity in the shoot apical meristem (Clark 2001). It involves mainly small molecules (e.g., ions, inositol phosphate – IP3, auxin, ABA), sugars, and oligosaccharides (Shibuya and Nimami 2001), while peptides are crucial in other aspects of plant development (Murphy et al. 2012). It is still unclear if symplastic pathways could account for long-distance molecule transport. Recently, Sokołowska and Zagórska-Marek (2012) showed that symplastic transport of fluorescent tracers could occur between the cambium and living cells in the secondary xylem in *Acer* and *Populus*. The apoplasm seems to allow transmission of signals, especially the ROS wave (Miller et al. 2009; Mittler et al. 2011). It is also the site of short-distance eATP (external ATP) signaling through the activity of apyrase. As an example, cotton fiber growth is tightly regulated through ectoapyrase activity (Clark et al. 2010). The apoplasm contains numerous protein species that can lead to the generation of local signals, including proteases, cell wall–modifying enzymes, and oxidoreductases (Charmont et al. 2005).

In the present chapter, we will concentrate on long-range, inter-organ signal transmission. The research in this area has been marked with considerable efforts to identify the physical basis of the information. It appears that at least two very different yet complementary systems exist: (1) a chemical system based on the movement of molecules and (2) physical systems based on electrical and hydraulic waves.

These signals move mainly using the conducting system of plants (xylem and phloem). The phloem flow velocities are relatively high, ranging from 1.15 m · h^{-1} in *Phaseolus* and *Cucurbita* plants (Mullendorea et al. 2010) to 3 m · h^{-1} in *Ricinus* (Köckenberger et al. 1997). These speeds are consistent with the systemic transport of solutes throughout most herbaceous plants in less than 30 min. In contrast, Chu et al. (2009) showed in *Pachira macrocarpa* that the xylem flow rate increased under wind stress but did not exceed 36 mm · h^{-1}, 50- to 80-fold slower than in the phloem. Indeed, many stress-related signal molecules (salicylic acid, methyl jasmonate) are found in the xylem (Atkins and Smith 2007; Krishnan et al. 2011) either after initial loading or after translocation from the phloem (Rocher et al. 2006).

Molecular-Based Systems

This part of the signaling system is conceptually the simplest one: the information is contained within a chemical that is produced locally in response to the perception of a stimulus and spread to other parts of the plant, as the source of information and, as a consequence, permitting the systemic responses in distant, unstimulated tissues. According to Heil and Ton (2008), these signal molecules must fulfill four criteria. They must be (1) generated locally at the site of stimulation, (2) transported systemically, (3) accumulated in the distant tissue, and (4) capable of evoking a distant response.

Table 2 summarizes some of the molecules that act as long-distance signals in plants and is based on the chemical structure: inorganic molecules and ions, small organic molecules, and large and complex organic molecules.

Local Ionic and Inorganic Signals

Although the bulk of this chapter deals with systemic signals, we will give brief mention to those that have limited transport – mainly from outside the cytoplasm to inside and vice versa. Calcium is a well-known intracellular messenger (Kudla et al. 2010; Batistič and Kudla 2012), which has a very low (10^{-7} M) steady-state level in the cytoplasm and orders of magnitude higher in the cell wall. Cytoplasmic calcium is capable of undergoing a rapid and transient increase in concentration in response to a wide variety of stimuli. This increase modulates the activity of several calcium-binding proteins (calmodulin, protein kinases) that induce subsequent cellular events (phosphorylation cascades, modification of enzymatic activities, gene expression). The main challenge remaining is to understand how a single molecule (ion) could mediate such different pathways. It seems that spatiotemporal variations in calcium concentration and interactions with a wide variety of calcium-binding proteins (CDPK, CBL/CIPK, CMLs) constitute the basis to encode these diverse responses (Kudla et al. 2010; Batistič and Kudla 2012).

Protons (H+) are highly important since variations in their concentration affect the cell pH homeostasis and could strongly affect the activity of several enzymes (Kader and Lindberg 2010). Acidification of cytosolic pH (ordinarily at around 7.5) is observed after various kinds of environmental perturbations including wounding (Bonnin et al. 1989) and elicitation of *Eschscholzia* protoplasts with a yeast glycoprotein (Roos et al. 2006). Furthermore, auxin-induced increases in proton concentration in the cell wall can activate several enzymes especially wall-weakening enzymes to stimulate cell enlargement (Schopfer 2006).

Carbon dioxide (CO_2) was recognized as an environmental stimulus able to affect stomatal density, and a signal indicating CO_2 deficiency is generated in the mature leaves and, in conjunction with fatty acids, jasmonate, and oxidizing metabolism is capable of moving to young leaves to control stomatal density (Lake et al. 2002). Carbon dioxide also regulates stomatal aperture through ABA metabolism and a CO_2-binding protein (Kim et al. 2010).

Hydrogen peroxide (H_2O_2) concentration is elevated after many environmental stimuli as a result of movement from the cell wall (Cheng and Song 2006) and is implicated in several cellular functions (phosphorylation, gene expression).

Transmitted Ions and Inorganic Molecules

Nitric oxide (NO), a well-known neurotransmitter in mammals, is also produced in plants after mechanical wounding or insect attack and is implicated in many aspects of plant development (Arasimowicz and Floryszak-Wieczorek 2007; Molassiotis et al. 2010; Beaudoin 2011). It is produced in peas by peroxisomes and is primarily located in the vascular bundle (xylem and phloem), which can serve to transport the molecule throughout the plant (Corpas et al. 2004).

Miller et al. (2009) described the genesis of ROS (reactive oxygen species) in response to various stimuli including wounding, heat, high light, and salt stress. This signal moves rapidly ($8\ cm \cdot min^{-1}$) in the apoplasm and can be stopped by the suppression of ROS accumulation. Recently, Mittler et al. (2011) showed the propagation of an ROS wave in response to wounding that spreads from the base to the apex of an *Arabidopsis* plant within 10 min. This auto-propagating ROS signal that moves bidirectionally causes ROS production at a distance from the stimulus site and thus evokes a systematic signaling cascade that follows the production of ROS. It is however not clear if the ROS wave is induced as a generic stress (i.e., nonspecific) signal or if it acts with a specificity that could be encoded through calcium oscillations or protein kinase activity.

Transmitted Simple, Small Organic Molecules

Plant growth regulators (hormones) have long been identified as potential long-distance signals that regulate plant development (Depuydt and Hardtke 2011), although in most cases it remains difficult to establish an unequivocal relationship between the presence of growth regulators in the xylem and/or phloem and morphological and/or metabolic consequences of their translocation (Atkins and Smith 2007).

Abscisic acid or its glucose-ester conjugate (Jiang and Hartung 2008) seems to move from the roots to the stem in water-stressed plants and induces stomatal closure. This molecule appears to be a better candidate for long-distance signal transmission than previously suggested (Peña-Cortés et al. 1991) since a large proportion of ABA could be lost during its transport in the xylem because of the high diffusion of the protonated form, ABA $\cdot$ H^+. In contrast, the membrane permeability to the ABA glucose-ester (ABA-GE) conjugate is very low, keeping its concentration constant during transport in the xylem (Jiang and Hartung 2008). The non-active ABA-GE is rapidly cleaved to ABA in the apoplast by a β-glucosidase (Dietz et al. 2000). The ABA signal could be reinforced by sulfate ions that have been found to be more abundant in the xylem sap after water stress in maize (Goodger and Schachtman 2010). Recent advances in the understanding of ABA interaction with phosphatases to enable SNF1-type kinase action (Raghavendra et al. 2010; Nakashima and Yamaguchi-Shinozaki 2013) open new perspectives in the understanding of hormone action.

The long-distance polar transport of auxin (Cho et al. 2007; Zhao 2010) and its transport in the xylem (Wilkinson and Davies 2002) play key roles in plant morphology under stress conditions. Auxin is also identified as a short-distance mobile messenger in root gravitropism (Swarup et al. 2005) and in the maintenance of root stem cells (Ding and Friml 2010).

Cytokinins (CKs) are implicated in plant responses to the environment (Ha et al. 2012). Cytokinin signaling appears complex since CK produced by bacteria has different effects than the CK originating from plants (Choi et al. 2011). Long-distance transport of CK is implicated in the regulation of stomatal aperture (Jia and Zhang 2008) and in nitrogen starvation signaling (Tamaki and Mercier 2007).

Gibberellins are also present in the phloem (Dinant and Suárez-López 2012) and widely implicated in vegetative plant development (Hauvermale et al. 2012). They are likely to remain intact during transport and modulate plant development.

Flooding increases the synthesis of the ethylene precursor, ACC, in the root and in the xylem sap (Malladi and Burns 2007), causing leaf epinasty. Thus, ACC acts as a soluble mobile signal from roots to shoot (Jackson 2002). This internal long-distance ACC signal could take place in parallel to plant-to-plant airborne transmission of ethylene which is a highly diffusible, volatile compound that elicits and regulates many metabolic processes (fruit ripening, senescence, abscission) and plant responses to stress (reviewed in Bleecker and Kende 2000).

An animal nerve compound, gamma-aminobutyric acid (GABA), a nonprotein amino acid, was proposed as a possible long-distance signaling molecule that upregulates nitrate uptake under conditions of nitrate deprivation (Beuve et al. 2004). However, measuring GABA in the phloem is difficult and the reality of long-distance transport was recently questioned (Shelp 2012).

Systemic responses observed after mechanical wounding rely mainly on the rapid and massive synthesis of jasmonic acid (Hlaváčková and Nauš 2007) and its translocation in the phloem. The synthesis of methyl jasmonate is also observed and it is transported in both the xylem and phloem (Heil and Ton 2008).

Flavonoids are very common, highly diversified phenolic secondary metabolites that are implicated in many metabolic pathways in plants, including resistance to pathogen attacks (Lattanzio et al. 2006). These compounds are transported long distances by using either the symplastic pathway in the upward direction or vascular bundles in the downward direction (Buer et al. 2007). These movements of flavonoids may have considerable significance through their modulation of growth, branching, and auxin transport (reviewed in Buer et al. 2010), although additional work is required to understand these mechanisms.

When plants are attacked by pathogens, they activate a resistance mechanism (systemic acquired resistance, SAR) that operates both locally (at the site of infection) and systemically (throughout the whole plant). Salicylic acid was identified as an inducer of SAR responses both locally and in distant tissue and was therefore proposed as being the mobile chemical transported through the plant to evoke SAR at a distance (Molders et al. 1996). However, it was noticed that SAR could be turned on before salicylic acid increased in the distant tissues and that the genuine signal molecule was in fact methyl salicylate (Park et al. 2007). It was recently demonstrated that light strongly interferes with this signal molecule (Liu et al. 2011) which became unnecessary under high-light conditions (Attaran et al. 2009).

ATP, the main energy molecule of cells, is also a short-distance, signaling molecule when it is exported as eATP (external ATP) outside of the cell (Demidchik et al. 2003) through a leak occurring after cell wall damage (Jeter et al. 2004) or after exocytosis through vesicle fusion (Kim et al. 2006). External ATP, whose presence in the extracellular matrix is tightly controlled through apyrase activity (Sun et al. 2012), could trigger an increase in cytoplasmic [Ca^{2+}] that evokes the accumulation of several transcripts such as MAPK and also in ethylene biosynthesis (Jeter et al. 2004). External ATP also causes the opening of stomata (Hao et al. 2012) and induces the production of nitric oxide in suspension cells (Foresi et al. 2007) that could move in the plant through the vascular bundles. Riewe et al. (2008) demonstrated that apyrase activity in potato is crucial in several aspects of plant development (growth and tuber formation). However, no clear evidence has been given demonstrating long-distance eATP transport in plants.

Sugars are now seen not only as nutrients but also as signaling molecules (Rolland et al. 2006). Sugars are transported mainly as sucrose, whose cleavage takes place in the cytoplasm through the activity of invertase. Hexokinase is the main glucose sensor of the cell, integrating not only sugar but also hormonal signals (Tuteja and Sopory 2008). Indeed, Hammond and White (2011) pointed out a role for sugar and phytohormones (auxin and cytokinin) in changes in long-distance root morphology and phosphate transporter synthesis in root responses to low phosphorus availability. Furthermore, a recent review (Eveland and Jackson 2012) pointed out the potential role for sugars in short-distance signaling (as a key factor in the maintenance of SAM cell identity) and long-distance signaling (in coordination with the plant hormones ethylene, abscisic acid, and auxin) in the regulation of transcription, translation, and protein activity that control plant growth.

This plethora of different systems within the plant is complemented by plant-to-plant communication, i.e., the ability to send a message from one plant to another through the airborne transmission of volatile organic compounds (VOCs) such as methyl jasmonate, methyl salicylate, and several terpenes (Yi et al. 2009; Arimura et al. 2011). Airborne signaling seems to physiologically precede vascular signaling (Heil and Ton 2008). The airborne diffusion of VOCs eliminates the need for an *in planta* transport system (Frost et al. 2007) and could affect plants at a considerable distance (Frost et al. 2008). Several volatile molecules could act as stimuli to elicit responses in distant plants (Heil and Karban 2009; Ueda et al. 2012). For instance, ethylene, methyl salicylate, methyl jasmonate, cis-jasmone, and several terpenoids alone or in combination are able to enhance resistance of plants to herbivore attacks. Interestingly, the methanol vapor surge produced in response to wounding in "emitter" plants enhances the resistance to bacterial infections in distant "receiver" plants yet increases tobacco mosaic virus reproduction and spreading by increasing cell-to-cell communication (Dorokhov et al. 2012).

Transmitted Complex, Large Organic Molecules

In addition to small molecules, larger organic molecules (oligosaccharides, lipids, peptides, nucleic acids, and proteins) are also transmitted, mainly through the phloem, in response to various stimuli (Wu et al. 2002) and contribute to systemic signaling (Fukuda and Higashiyama 2011).

Oligosaccharides were recognized a long time ago as potential elicitors for stress responses (reviewed by Shibuya and Nimami 2001). They could originate from the cell wall of pathogens (β-glucan oligosaccharides, chitin- and chitosan-derived oligosaccharides) or from the damaged cell wall of plants (oligogalacturonides) and are mainly active in dicots. Large oligosaccharides (up to 20 residues) are the most active ones, evoking the accumulation of phytoalexins, defense proteins such as protease inhibitors, the lignification of the attacked regions (but not the hypersensitive response-induced tissue necrosis), and promotion of the antioxidant system in alfalfa roots (Camejo et al. 2012). It is not clear if these compounds contribute to long-distance signaling. Baydoun and Fry (1985), using radiolabeled oligosaccharides, demonstrated that only the small (up to 6 residues), moderately active oligosaccharides were mobile while the larger, more biologically active ones stay close to the site of stimulation. However, Iwai et al. (2003) demonstrated that the xylem sap actually contains oligosaccharides, oligoglycans, and oligogalacturonides that may be involved in long-distance signaling. Simpler molecules such as glycans and sugars also appear to have a signaling function (Lalonde et al. 1999; Etzler and Esko 2009) because of their rapid transport in the phloem.

Recent reports have pointed out the presence of lipids in the phloem (Benning et al. 2012), including phytosterols, glycolipids, and fatty acids, which are not the result of membrane degradation. Some of them are known to be involved in transduction pathways after different environmental stimuli such as wounding, pathogen attacks, and flaming (Guelette et al. 2012). These include phosphatidic acid, phosphatidylglycerol, and phosphatidylinositol that could originate from membranes after lipase activity. They could, in addition to their signal transduction function, themselves be important long-distance signaling molecules to evoke distant responses if their activity is kept intact after being transported. Others, such as jasmonic acid or oxylipin, are specifically produced in response to pathogen infection and are transported in the phloem (Schilmiller and Howe 2005). Lipids are associated with specific proteins (phloem lipid-associated family protein, PLAFP; annexin) that allow the efficient transport of these hydrophobic molecules in the phloem (Benning et al. 2012).

Peptides such as systemin may act as long- or short-distance signaling molecules (Butenko et al. 2009). Systemin, an 18 amino acid peptide, derived from a 200 amino acid precursor prosystemin, isolated from wounded tomato leaves, is able to trigger the systemic expression of protease inhibitor (PI) genes that protect the plant from insect attack (Pearce et al. 1991). Synthetic systemin could be transported in the phloem and was proposed to be the mobile signal responsible for the systemic response (McGurl et al. 1992; Narváez-Vásquez et al. 1994). Jasmonic acid (JA) evokes the expression of protease inhibitor genes (Farmer and Ryan 1990) and is required to initiate the systemic response since the JA-deficient *spr-1* mutant is impaired in the genesis of the systemic signal while maintaining the local response to wounding. Jasmonic acid (whose synthesis is induced by systemin) appears to be the transmitted wound signal (Lee and Howe 2003; Sun et al. 2011). A similarly small (23 amino acid) peptide that triggers systemic wound responses was also identified in *Arabidopsis* as AtPep1 (Huffaker et al. 2006). This peptide binds to membrane receptors (Yamaguchi et al. 2006, 2010) identified as LRR receptor kinases involved in plant responses to aggression. Similar receptors are involved in the binding of phytosulfokine (Matsubayashi et al. 2006), another plant peptide that promotes cellular differentiation and could act as a signaling molecule to reduce (or induce) stress responses (Motose et al. 2009). RALF (rapid alkalinization factor), a 5-kDa peptide (Pearce et al. 2001) that inhibits root growth, may be transported in the xylem sap (Neumann 2007) and constitute a putative long-distance signal molecule. In contrast, the well-characterized CLAVATA3/ESR (CLE) peptide-based signaling pathway only operates for short distances to maintain cell identity in shoot apical meristems (Juna et al. 2008).

RNA can move from cell to cell through the plasmodesmata (Wu et al. 2002) and diffuse short distances. A wide variety of mRNAs (implicated in housekeeping, pathogen resistance, and stress responses) were also found in the phloem (Kehr and Buhtz 2008; Lee and Cui 2009). The phloem SE do not contain translation machinery, and so the different nucleic acids types (mRNA and small RNA) are being transported to distant tissues where they can be translated or influence metabolic activity (Kim et al. 2001; Kehr and Buhtz 2008). Untranslated transcripts are likely to bind to transport proteins (which are consistently detected in the phloem sap). A model for RNA trafficking was recently proposed by Lee and Cui (2009), who suggested that chaperons and mRNA-binding proteins are implicated in mRNA loading into the phloem. The translocation of these mRNAs inducing developmental changes was reported more than 10 years ago (Ruiz-Medrano et al. 1999; Kim et al. 2001), and they have an active role in leaf morphogenesis (Haywood et al. 2005) and tuber formation (Banerjee et al. 2006).

The diversity of phloem mRNA was assayed in *Ricinus* (Doering-Saad et al. 2006) and revealed over 150 unique transcripts. In potato, Hannapel (2010) demonstrated that a StBEL5 mRNA is transported a long distance to the stolon tip where it regulates tuber formation. It seems important to assess the differential phloem transcriptome between control and stimulated plants (both in local and distant tissues) to assess the diversity of mRNAs transported in the phloem that specifically act as signals. Classical molecular methods (i.e., subtractive cloning or global sequencing) could provide valuable information to explore the importance of mRNA trafficking in systemic plant responses.

The phloem sap also contains noncoding small RNAs (Yoo et al. 2004; Buhtz et al. 2008) that could interfere with translation (Zhang et al. 2009). Micro RNAs (miRNA) are small, nontranslated RNAs that regulate gene expression post-transcriptionally (Yoo et al. 2004). They accumulate after various environmental stimuli such as nutrient deficiency (Fujii et al. 2005), wounding (Bozorov et al. 2012), or pathogen attack (Feng et al. 2011). These RNAs are present in the phloem sap and are readily mobile (Buhtz et al. 2010). It is widely admitted that miRNAs play a major role in long-distance information transfer. MicroRNA399 appears to be involved in phosphate starvation (Pant et al. 2008), while Buhtz et al. (2008) identified three miRNAs that accumulate in the phloem when plants were cultivated in conditions deprived of ions. In this context, a challenging task is to identify the targets of these miRNAs since the classical approach by sequence alignment is not sufficiently accurate and leaves a high proportion of nonsignificant putative interactions, mainly because these computerized analyses are performed with little reference to the ongoing metabolism.

The phloem also contains a diverse array of proteins (Kehr 2006) including calcium-related proteins (annexins, calmodulin, kinases), ROS actors (ascorbate peroxidase, glutathione-S-transferase, thioredoxins), protease inhibitors, and lectins, all of which are potential long-distance signal molecules. It is not clear however if the proteins are themselves signals or if they contribute to signal propagation through their metabolic activity. Kehr (2006) compiled data showing that phloem sap contains a large variety of stress-related proteins, including redox metabolism and calcium-related proteins. Giavalisco et al. (2006) characterized the phloem sap of *Brassica napus* and found over 140 protein species, some of them related to redox/stress metabolism and signal transduction, and a large set of metabolic enzymes (e.g., UDP glucose pyrophosphorylase, glyceraldehyde-3-phosphate dehydrogenase, malate dehydrogenase). Recently, Rodriguez-Medina et al. (2011) described the proteome and transcriptome of white lupin phloem sap and found numerous molecules that are similar to those previously described. However, more than one-third of them remain unknown, but the others were categorized into families: basic metabolism, structural components, nucleic acid binding, and stress-related response including redox regulation (Table 3), indicating a wide variety of potential metabolic activities. This latter point is of major interest since the phloem is now understood as a transport system that possesses its own set of proteins and transcripts that are likely to play a central role in systemic responses of plants in response to environmental stresses.

This diversity of protein and enzymatic activity was recently confirmed in *Arabidopsis* by Batailler et al. (2012) where 127 proteins were identified as belonging to pathways involving carbohydrate, lipid, hormone, amino acid, nucleotide, and secondary metabolism, as well as gluconeogenesis, oxidative stress, and the tricarboxylic acid cycle. Aside from these enzymes, the phloem sap also contains all the species (protein transport, cytoskeleton, defense proteins, etc.) that were identified in previous work. The authors strongly suggest that very active metabolic activity takes place in the phloem SE.

Physical Systems

It was pointed out over 30 years ago (Davies and Schuster 1981) that such molecular-based systems fail to explain the rapid, long-distance transmission of signals whose speed is many orders of magnitude greater than chemical transport processes, and so efforts were made to identify the nature of such rapidly generated, bidirectionally transmitted signals (Davies 1987). In the last two decades, it has been shown that, in addition to the chemical signals described above, plants also possess two physically based signals, electrical and hydraulic, and these differ substantially from the chemical-based systems (reviewed in Davies 2006). They are fully systemic, traveling from the root to the stem (and stem to root), exceedingly rapid, and most likely possessing a limited information capacity. Indeed, it seems most likely that the more rapidly a signal is generated and transmitted, the less its information content (Davies and Stankovic 2006).

Electrical Signals

Although electrical signals have been known for over 200 years (Pickard 1973; Davies 2006 and references therein), it is only recently that valid functions have been assigned to them. Indeed, 25 years ago it was still necessary to hypothesize a function (Davies 1987), although rapidly generated, bidirectionally transmitted signals were already known to evoke changes in translation (Davies and Schuster 1981). Electrical signals are now known to affect many aspects of metabolism, including RNA synthesis (Stankovic and Davies 1996, 1997), protein synthesis (Davies and Stankovic 2006), and photosynthesis (Grams et al. 2009).

There are at least four different, yet sometimes intertwined, electrical signals, action potentials (AP), variation potentials (VP), system potentials (SP), and voltage transients (VT), and here we will focus on the first two (reviewed in Davies 2006; Fromm and Lautner 2007). Action potentials are genuine electrical signals, i.e., all-or-none, self-propagating (via the flux of ions through voltage-gated channels), almost constant in velocity and magnitude, and have a refractory period during which a subsequent AP cannot be generated (Davies 2006). In contrast, variation potentials are not genuine electrical signals but local changes in membrane potential evoked by changes in the hydraulic status of the tissue acting on pressure-sensitive ion channels and vary in apparent magnitude and velocity (Davies 2006). Nevertheless, they both seem to involve the same ions, being initiated by an influx of Ca^{2+} and then an efflux of Cl^- and K^+, but they differ in the magnitude and duration of these fluxes (Davies 2006) and in some instances proton fluxes are also involved in long-distance electrical signaling (Grams et al. 2009). Although any cell with voltage-gated channels can generate and transmit signals from cell to cell (via plasmodesmata), long-distance transmission seems to be exclusive through the phloem, primarily the phloem SE (Van Bel et al. 2011; Hafke and Van Bel 2013); thus, methods for "tapping" the phloem are important for electrical as well as chemical signals. APs can be evoked by any external stimulus (electrical, cold, heat, wound) or internal stimulus that can elicit a change in membrane potential sufficiently large to activate voltage-gated channels (Davies 2006).

The rates of transmission of the electrical signals are also variable. The action potential in *Mimosa pudica* propagates rapidly, at a rate of 20–30 mm · s^{-1} (Fromm and Lautner 2007). In sunflower plants, the action potential propagates at a rate of about 7–10 cm · min^{-1}, whereas the variation potential has an initial velocity of almost 30 cm · min^{-1}, rapidly decreasing further from the wounded region to 10–20 % of its initial value (Stankovic et al. 1998).

Hydraulic Signals

According to Malone (1993) hydraulic signals are ubiquitous in plants and can pass through the hydraulic continuum (mainly the xylem) exceedingly rapidly, especially after wounding (disruption of the integrity of the xylem). Indeed, the rate of water movement in the xylem under these conditions can exceed 10 mm/s, while the movement of the pressure front (loss of tension in the xylem) can equal the speed of sound (1,500 m/s). The xylem, therefore, furnishes a multipurpose signaling system. First, the chemicals (e.g., ABA, PIIF) synthesized in or released from the surrounding cut tissue can act as systemic signals with velocities approaching 10 mm/s. Second, changes in pressure will be sensed throughout the xylem and transferred to adjacent living cells as a change in hydrostatic pressure, which, in turn, can affect pressure-sensitive channels in the phloem. Third, the changes in ion flux as a result of these channel openings can be converted into an electrical signal. Fourth, this signal will appear to be a self-propagating signal but is actually the result of local changes in membrane potential accompanying the self-propagating loss of tension. Thus, the loss of tension will diminish with distance from the wounding site and the accompanying electrical changes will diminish, and the resulting signal varies and is thus referred to as a variation potential (VP). In some circumstances the local changes in membrane potential can trigger an action potential (AP), the self-propagating electrical signal. The interplay between the xylem and phloem as long-distance conduits of information and local signaling through the adjacent parenchyma and companion cells is shown very nicely in Fig. 1 (from Fromm and Lautner 2007). It is not surprising that VPs can evoke at least as many downstream consequences as can APs, since both involve influx of Ca^{2+} and efflux of Cl^-, K^+, and perhaps H^+, but these changes in ion concentrations are greater and longer lasting with VPs than with APs. As stated earlier these altered levels of ions in both the symplast and the apoplast can evoke a plethora of changes (Davies 1987; Fromm and Lautner 2007). See Table 4

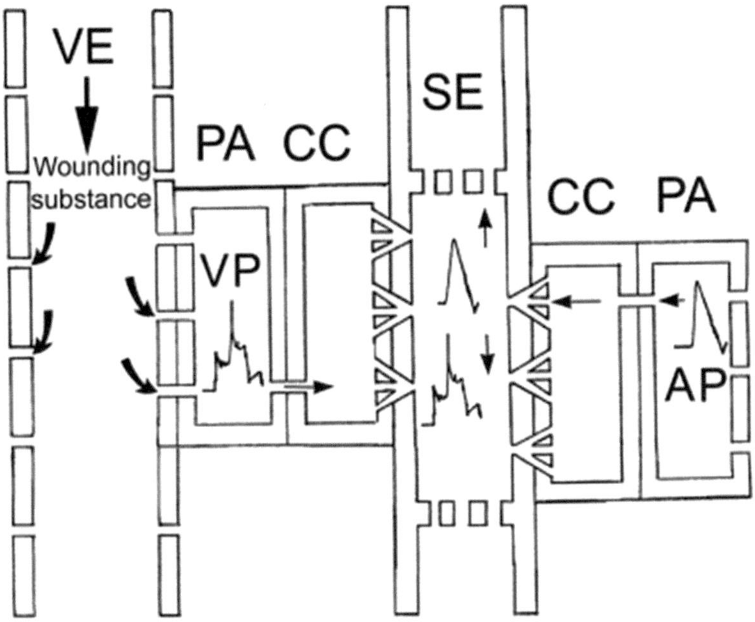

Fig. 1 Electrical communication over long distances. An AP (*right*) can propagate over short distances through plasmodesmata, and after it has reached the sieve element/companion cell (SE/CC) complex, it can travel over long distances along the SE plasma membrane in both directions. In contrast, a VP is generated at the plasma membrane of parenchyma cells (PAs) adjacent to xylem vessels (VEs) by a hydraulic wave or a wounding substance. Because VPs were measured in SEs (Lautner et al. 2005), it is suggested that they also can pass through the plasmodesmal network and can reach the phloem pathway. However, in contrast to APs, their amplitude will be reduced with increasing distance from the site of generation (Figure 3 from Fromm and Lautner (2007). Copyright John Wiley and Sons. Reproduced with permission)

(from Fromm and Lautner 2007 for a list of well-documented responses to electrical signals). Interestingly, it appears that the hydraulic and electrical components can work independently of each of other during the relatively non-damaging treatment of re-irrigation of maize (Grams et al. 2007).

Methods of Analysis and Possible Applications in Agriculture

Molecular-Based Signals

We will concentrate on metabolites present in the phloem since it is the major route for long-distance transport of signaling molecules (Dinant and Lemoine 2010) and thus the most promising and accessible point of investigation to increase overall knowledge and develop practical applications to agriculture. The challenges to fulfill are fourfold: (1) efficient and selective collection of the circulating phloem sap, (2) analysis of its contents, (3) understanding and reconstruction of the metabolic pathways that could take place in the phloem, and (4) testing of individual metabolites or combinations for their effects on plant tolerance to diverse environmental stresses.

Phloem Sampling

First of all, the diversity of molecules that are present before and after the application of a stimulus that elicits a systemic response needs to be assayed as completely as possible, both at the site of stimulus perception and in distant tissues. This step relies on an efficient and selective phloem sampling method. This is a difficult task since the phloem SEs are located within the stem and they possess self-repair mechanisms in case of injury. The phloem is highly reactive to damage (Kehr

Table 4 Well-documented physiological effects of electrical signals in plants (From Fromm and Lautner 2007)

Stimulus	Signal	Plant	Physiological effect	Reference(s)
Mechanical	AP	*Dionaea*	Trap closure Release of digestive enzymes	Sibaoka (1969)
Mechanical	AP	*Drosera*	Tentacle movement to wrap around the insect	Williams and Pickard (1972a, b)
Cold shock, mechanical	AP	*Mimosa*	Regulation of leaf movement	Fromm and Eschrich (1988a, b, c) and Sibaoka (1966, 1969)
Electrical	AP	*Chara*	Cessation of cytoplasmic streaming	Hayama et al. (1979)
Electrical	AP	*Conocephalum*	Increase in respiration	Dziubinska et al. (1989)
Pollination	AP	*Incarvillea, Hibiscus*	Increase in respiration	Sinyukhin and Britikov (1967) and Fromm et al. (1995)
Re-irrigation	AP	*Zea*	Increase in gas exchange	Fromm and Fei (1998)
Cold shock	AP	*Zea*	Reduction in phloem transport	Fromm and Bauer (1994)
Electrical, cooling	AP	*Luffa*	Decrease of elongation growth of the stem	Shiina and Tazawa (1986)
Electrical	AP	*Lycopersicon*	Induction of *pin2* gene expression	Stankovic and Davies (1996)
Heating	VP			
Heating	VP	*Vicia*	Increase in respiration	Filek and Koscielniak (1997)
Heating	VP	*Solanum*	Induction of jasmonic acid biosynthesis and *pin2* gene expression	Fisahn et al. (2004)
Wounding	VP	*Pisum*	Inhibition of protein synthesis, formation of polysomes	Davies et al. (1986) and Davies and Stankovic (2006)
Heating	VP	*Mimosa, Populus*	Transient reduction of photosynthesis	Koziolek et al. (2004) and Lautner et al. (2005)

et al. 2005) and does so by the activation of antioxidant metabolism and the mobilization of P-proteins and callose to repair the damage and plug the intercellular connections. Thus, one needs to keep in mind that the necessary intrusion into the phloem tubes to collect the sap is likely to evoke the accumulation of defense-related molecules. The three methods that were developed to sample the phloem content were recently reviewed from a technical point of view (Dinant and Kehr 2013). All these methods have the same limitation (i.e., they require physical intrusion of the plant tissues that could be perceived as a wound, and therefore modify the steady-state content of undamaged phloem tubes). The spontaneous exudation is the oldest and the simplest one to apply since it consists of the collection of phloem SE contents after localized incision with a razor blade or syringe needle. While this method allows recovery of relatively large amount of phloem SE fluid from plants such as palm and pumpkin, it remains difficult to prepare pure (unpolluted) samples and the rapid accumulation of P-proteins and callose in the conducting tubes prevents the release of SE contents in numerous plants. The use of calcium-chelating agents such as EDTA (EDTA-facilitated exudation) prevents the formation of callose and

P-protein accumulation and thus the closure of phloem tubes (Ernst et al. 2012). In contrast, the use of aphid feeding provides high-quality phloem sampling. The insects that feed on phloem SE contents use their stylet to wend its way between several cells before specifically puncturing the phloem SE and injecting compounds that prevent callose formation, thereby keeping the stylet open. When the stylet is in place, the insect is removed with a laser beam while the stylet is left attached and intact. The phloem SE sap exudate is collected with a glass microcapillary for further analysis. The first drops of sap exudation are usually discarded to prevent pollution (Doering-Saad et al. 2006). While the amount of phloem sap that can be obtained through this method is low, its high quality allows the use of a wide variety of analysis methods and thus should remain the method of choice in herbaceous plants to accurately access and assess the diversity of the phloem content.

Metabolic Activity in the Phloem

Several methods are available to assess the phloem sap content. Fourier transform infrared spectroscopy (FT-IR) is a rapid and versatile method that was successfully used to develop metabolic fingerprints in plants to characterize control vs. salt-stressed tomato (Johnson et al. 2003) or control vs. infected elm (Martína et al. 2005). The method is sufficiently rapid to allow a high number of analyses, and thus the importance of the development stage, diurnal variations, or responses to environmental stimuli can be explored (López-Gresa et al. 2012). These fingerprints could serve as a starting point to develop more sophisticated strategies of metabolome determination. Metabolic profiling aims at the complete description of the different molecular species that are present in a tissue. This technique was successfully used in plants over a decade ago (Fiehn et al. 2000) and is now sensitive enough to assess metabolic diversity on a cellular scale. For instance, micro-sampling of *Arabidopsis* epidermal cells followed by GC-MS-based metabolite profiling allows effective exploration of their metabolome (Ebert et al. 2010). Analytic methods such as high-performance liquid chromatography, gas chromatography, and capillary electrophoresis coupled to mass spectrometry or nuclear magnetic resonance are routinely used to assess metabolome diversity (see Dunn and Ellis 2005; Büscher et al. 2009 for a practical comparison of these methods). Comparative studies of metabolic fingerprinting and/or metabolic profiling of phloem sap from control and stimulated (both local and distant tissues) should provide a very powerful tool to select some metabolites that display differential accumulation. The determination of the phloem protein diversity using classical 1D and 2D gel electrophoresis followed by micro-sequencing or mass spectrometer analysis remains difficult because of the low amount of phloem sap. Strategies using nano-flow liquid chromatography linked to a mass spectrometer (Aki et al. 2008) could constitute a valuable method to assess the phloem proteome.

The phloem sap is an alkaline (pH 8), concentrated solution (osmotic pressure of 1.3–1.5 MPa, Fukumorita and Chino 1982) that may not be optimal for enzyme function. Therefore, the presence of several putative enzymatic activities that are present in the phloem in many unrelated plant species (Kehr et al. 2005) might have two different yet related functions. The first function is the transport of enzymes from their site of synthesis to their site of use via the phloem. This is the simplest explanation since the enzymes and metabolites are considered to be either present or transported "as is" with no interaction between the enzymes and surrounding metabolites. For the second function, the phloem sap is considered as an integrated environment where enzymatic reactions take place and thus modify or transform the molecular signals that originate in the stimulated area. The latter perspective opens new insights for the phloem content to be a metabolic network bearing its own regulation through the general phloem osmotic and ionic environment and the presence of diverse molecules that could interfere with enzymatic reactions. Enzymatic activity in the phloem is demonstrated for oxidative pathways (Walz et al. 2002) and biosynthesis of the plant hormones, ethylene, and jasmonate. It is difficult to decipher which reactions are likely to take place and how they can modify the dynamic of parallel reactions.

Molecular network modeling could constitute a powerful tool to deal with this complexity. Different approaches exist, mainly employing stoichiometric and kinetic models. Stoichiometric models (Llaneras and Picó 2008) are the easiest to set up insofar as they do not require precise information on the steady-state levels of molecular reactions of the molecular network under consideration. In contrast, kinetic models (Schallau and Junker 2010) use different information (enzyme rate, stoichiometry of the reactions, etc.) to feed a precise dynamic mathematical description of the enzymatic processes. In return, the model predicts the time evolution of the different molecular species concentrations (substrates and products of many parallel reactions that could be partially linked). However, this information is not always available, nor easy to measure, making kinetic models more difficult to set up. Both approaches could be used to get information on metabolic fluxes and network-based pathway analysis. This information could in turn be of great interest to better understand the dynamics and distribution of metabolites in diverse tissues and at various time points.

Different kinds of metabolic models have been developed in plants to model the metabolic network (Kruger and Ratcliffe 2012). As an example, the primary metabolism of *Arabidopsis* was modeled (Gomes de Oliveira Dal'Molin et al. 2010). This model (AraGEM) takes into consideration over 1,500 metabolic reactions in several compartments (cytoplasm, vacuole, peroxisome, etc.) and has proven its utility for the prediction of basic metabolism cycle reactions (e.g., photorespiration) and is robust enough to test in silico functional analysis. Recently, Mintz-Orona et al. (2012) published another model (mainly generated from database information) that takes into account the diverse location of enzymes and metabolite fluxes in *Arabidopsis*. Transferring such methods to phloem metabolism networks should be possible since the number of molecular species and enzymes is likely to be much lower than in the cytoplasm. High-purity phloem sampling followed by metabolome analysis and database contribution should furnish metabolic models useful for predicting molecular events on a spatial (local and distant) and temporal basis, thereby identifying the molecular species that differentiate the control vs. stimulated phloem sap. These molecular species could therefore be used as candidate molecules to test their effect on plant tolerance to environmental stimuli.

Effect of Bioactive Molecules on Plants

The effect of these treatments could then be monitored on some of the traits of interest both for agriculture (e.g., growth, yield, resistance to pathogens, etc.) and horticulture (global architecture, leaf morphology, date of flowering, etc.). The general strategy is presented in Fig. 2. The phloem is sampled and analyzed to determine its proteome (2), metabolome (3), and transcriptome (4) both in tissue adjacent to and distant from the stimulus (1). Proteome and metabolomes account for molecules that are immediately active. In contrast, the transcriptome accounts for translatable (mRNA) or regulatory (miRNA) molecules, although the phloem is devoid of translation equipment. The information from these analyses, combined with databases, feeds a metabolic network model (5) that helps to identify putative molecules of interest. These molecules are therefore used (alone or in combination) to treat plants (6); the effects are observed on different items of interest (e.g., increased resistance to pathogens, crop production in stressed environments, plant architecture) using an automated high-throughput phenotyping facility (7).

This strategy requires a considerable number of plants to perform the analyses because of the diversity in metabolites, combinations, and concentrations (several thousand per test is a reasonable estimate). It is therefore an absolute necessity to use an automated, high-throughput system (phenotyping, Furbank and Tester 2011) to evaluate the effect of treatments on traits of interest. Such automated facilities are now functional and some of them were recently used in a very similar context (Tardieu and Tuberosa

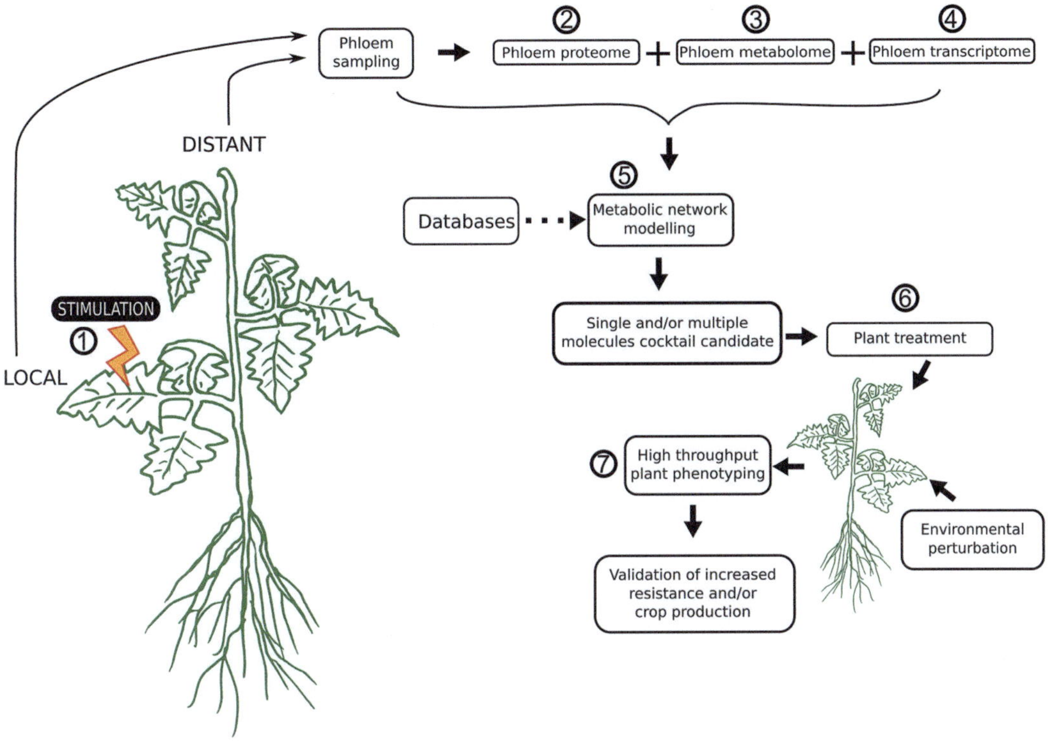

Fig. 2 General strategy to test effect of a molecule (or a combination of molecules) on agricultural or horticultural traits of interest. The plant is stimulated locally to potentially evoke both local and systemic responses (*1*). Phloem sap is sampled and analyzed for proteome (*2*), metabolome (*3*), and transcriptome (*4*). The collected information is used to feed metabolic network modeling (*5*, along with information from databases) to select molecule(s) of interest used to treat plants (*6*) subjected to environmental perturbation. The effect of this treatment (e.g., increased resistance to pathogens, plant architecture, etc.) is evaluated using a high-throughput plant phenotyping facility (*7*)

2010). These authors used a phenotyping facility to identify genetic loci of interest that allow plants to maintain agricultural performances under stress condition (stress tolerance as inheritable traits): the phenotyping facility uses specialized cameras and software (Hartmann et al. 2011) to identify phenotypic variations (e.g., plant architecture) that are related to quantitative trait loci (QTL) of interest. A very similar approach could be used to test the effect of phloem molecules or combinations of molecules (characterized from differential metabolic profiling and/or from metabolic network modeling) on plant characteristics of interest (Fig. 3). These phenotyping methods are applicable to the stem as well as the roots (Iyer-Pascuzzi et al. 2010).

Methods for Electrical-Based Signals

Both of the major electrical signals in plants, action potentials (AP) and variation potentials (VP), must, by definition, involve selective ion movement across membranes leading to changes in membrane potential and thus can be measured by measuring either of these parameters.

Measuring Membrane Potential Using Electrodes

Until recently, the method of (a very limited) choice was to use electrodes, either intracellular or, far more frequently, extracellular, and generally these electrodes measured the difference between the so-called measuring electrode(s) and

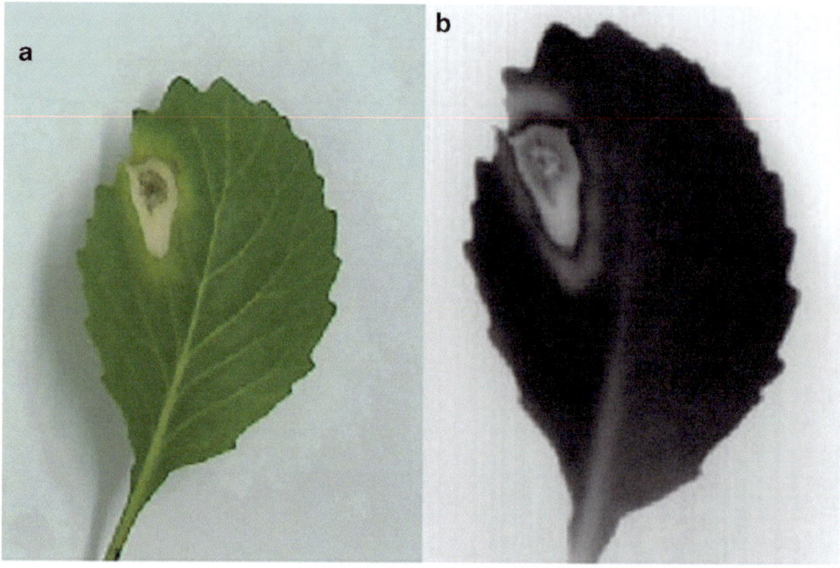

Fig. 3 Example of phenotyping in cabbage after *Alternaria brassicicola* infection. Thermal imaging of a cabbage leaf infected with *Alternaria brassicicola* (5 days after inoculation) to quantify the surface of the leaf infected by the pathogen. (**a**): visible scan image of black spot symptom; (**b**): thermal image. Such images could be use to feed an automated process to identify the potential effect of plant treatment toward an infection (Photographs courtesy of IRHS-FungiSem, obtained with Phenotic technical facilities, SFR 4207 Quasav, Angers, France)

the reference or ground electrode (Davies 2006). The extracellular electrodes measure the global ionic activity within their sphere of influence and can be either surface contact electrodes or inserted electrodes. The former cause little damage to the plant but suffer from drying out and thus a limited duration of use, while the latter damage the tissue but can be used for days or even weeks (Davies 2006). Relatively little skill is needed in conducting these measurements and, except for computer and associated programs for analyzing data, they are comparatively inexpensive. These are the methods we have routinely used, and we eventually chose the extracellular (inserted) electrodes even for experiments on wounding, since we could wait for the plant to recover from the (minor) wound of inserting the electrode before performing the major wound (Stankovic et al. 1998). Since AP and VP have different properties (velocity, magnitude, etc.), it is necessary to have a series of electrodes referenced with the ground electrode in order to differentiate between these signals.

In contrast to extracellular electrodes, intracellular electrodes require far greater skill on the part of the experimenter as well as much more sensitive, delicate, and expensive equipment to make the electrodes, insert them, and conduct the recordings. Work with plants is generally far more difficult than work with animals, and the use of intracellular electrodes is no exception. An animal cell has at most an extracellular matrix surrounding it, which furnishes a minimal physical barrier to an electrode, whereas a plant cell has a tough wall, which can easily break the electrode when attempting to insert it. Furthermore, the animal cell is primarily cytoplasm, whereas the plant cell may be 90 % vacuole and it takes great skill to position the electrode in the cytoplasm and not in the cell wall or the vacuole. Nevertheless, such intracellular electrodes have been use frequently and with great success.

As mentioned earlier, the phloem is a major conduit for chemical and electrical signals but analysis of its contents are made difficult by the wound-induced formation of P-proteins. This has been very nicely circumvented by the use of aphids, which can make their proboscis wend its way between different cells until it encounters a phloem SE. It then punctures the SE and releases an anti-clogging agent. The aphis is severed, leaving the proboscis in the SE. Not only does this allow the release of chemicals from the phloem but it can also act as a recipient for a microelectrode which can then be used to measure action potentials (or specific ions if necessary). This is shown very nicely in Fig. 4 (from Fromm and Lautner 2007).

Measuring Membrane Potential Using Fluorescence

The use of fluorescence as a means to detect macromolecules and small molecules has been around for a few decades, but the use of highly specific fluorescent proteins gained ascendancy with the work from Roger Tsien's lab (Tsien 1998) on green fluorescent protein (GFP) and the tremendous array of genetically modified versions with different absorption and emission spectra that have been developed since (Tsien 2010). The most recent review of the invaluable

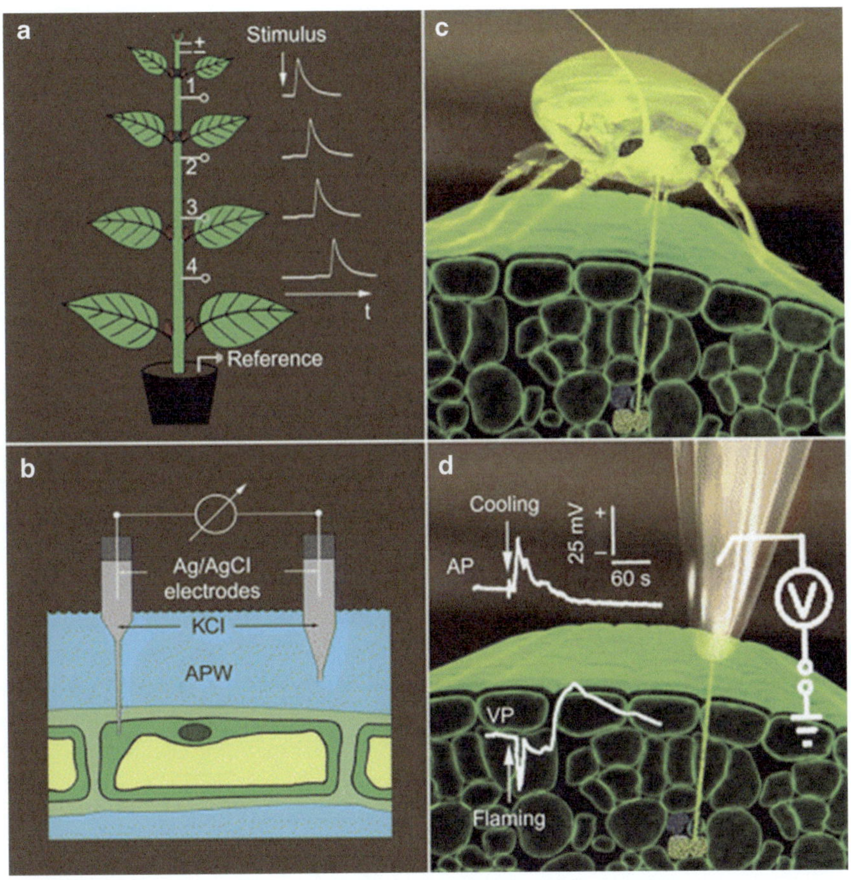

Fig. 4 Techniques for measuring electrical signals in plants. (**a**) Extracellular recording with four channels and a reference electrode inserted in the soil. ±, electrical stimulation. An AP (*right*) generated by electrical stimulation appeared successively at electrodes 1, 2, 3, and 4. (**b**) Intracellular measurement of the membrane potential with a microelectrode inserted into the cytoplasm of an algal cell while the reference electrode is in contact with the artificial pond water (APW) outside the cell. Both electrodes are filled with KCl, clamped in Ag/AgCl pellet holders, and connected to an electrometer. (**c**) Phloem potential measurements; an aphid in feeding position with its stylet inserted into a sieve element on the upper side of a leaf. (**d**) After the aphid is separated from its stylet by a laser pulse, the stylet stump exuded sieve tube sap to which the tip of a microelectrode was attached. Cooling the shoot evoked an AP transmitted acropetally within the phloem, while flaming of a leaf generated a VP with different form and of long duration. *t* time (Figure 1 from Fromm and Lautner (2007). Copyright John Wiley and Sons. Reproduced with permission)

role of fluorescent proteins in plant research (Okumoto et al. 2012) describes their construction, their use in high-resolution imaging *in planta*, and their use in discovery of novel phenomena. More on this topic is included in a companion chapter in this volume (Davies and Stankovic, Plant Cytomics: Novel Methods to View Molecules on the Move).

There has been a recent surge of reports (mainly in *Nature Methods*) describing the use of voltage-sensitive fluorescent proteins to analyze membrane potential and the accompanying action potentials in animal systems. These include the use of fluorescence (or Förster) resonance energy transfer (FRET) of a specially constructed donor and acceptor termed "Mermaid" which was used to record voltage spikes similar to those found with action potentials measured using electrodes (Tsutsui et al. 2008). More recently a bacterial rhodopsin has been shown to "run in reverse" (Looger 2012). This protein is normally activated by light to evoke ion fluxes in vivo, but it can now be used to monitor ion fluxes through voltage-mediated emission of light (Looger 2012). Indeed, these voltage-sensitive fluorescent proteins (VSFP) have now been genetically targeted to specific (animal) cell types (Akemann et al. 2010; Kralj et al. 2011). Pastrana (2012) points out the great value of such techniques in the neurosciences, where immense skill and patience are needed to insert microelectrodes into delicate cells deep within tissues, but far less skill (with perhaps greater accuracy) can be obtained using VSFP. We know of no work with plants using such methods but consider it an excellent path to follow, especially since microelectrode insertion is even trickier with plants. However, it would be difficult to get the appropriate microbial rhodopsin expressed in cells such as phloem SE (lacking translation apparatus) and impossible in xylem vessel elements (dead), which are the major conduits for systemic electrical signals in plants.

Measuring Ion Fluxes Using Electrodes

Both vibrating probes (reviewed in Dorn and Weisenseel 1982) and ion-specific electrodes (reviewed in Ammann 1986 and in Blatt 1991) have been used to determine ion fluxes during action potentials. As with voltage measurements described above, these methods are now being superseded, at least in animal research, by fluorescence techniques.

Measuring Ion Fluxes Using Fluorescence

Fluorescence has been used to measure action potentials (or more specifically the Ca^{2+} influx that takes place during an AP) in plants ever since the development of fluorescent proteins such as the jellyfish protein aequorin (Williamson and Ashley 1982). More recently green fluorescent protein (GFP) and its extended family (Tsien 1998, 2010) have found multiple uses, including in plants (Haseloff 1998), and have come to the fore in calcium measurements, especially in the field of neurophysiology, and again the most relevant articles appear primarily in *Nature Methods*. Genetically modified calcium sensors are being developed (Rochefort and Konnerth 2008) and used to measure very short-lived calcium spikes (Grewe et al. 2010), and ultra-sensitive calcium indictors such as Cameleon-Nano are proving exceedingly useful (Horikawa et al. 2010). Two recent articles from Gilroy's lab have reviewed this topic in plants. One major review lists all the conventional fluorescence techniques for measuring Ca^{2+} (as well as ROS and pH) *in planta* (Swanson et al. 2011) and shows the emission spectra of various constructs including Cameleons. The more recent one (Choi et al. 2012) furnishes a very useful table listing different kinds of biosensors, their subcellular locale of assay, their targeting method, and cell types and plant species (primarily *Arabidopsis*) where they have been used (Table 5).

Hydraulic Signals

As mentioned earlier, when the xylem is massively perturbed by, for instance, flaming a leaf, it loses tension almost immediately, resulting in several signal-like phenomena. First, the loss of tension can be measured by changes in volume in the stem apex by the use of position-sensing transducers (Malone 1993; Stankovic et al. 1998). Second, the loss of tension results in

Table 5 Examples of the application of cytosolic and subcellularly targeted GFP-based biosensors for Ca^{2+}, pH, and reactive oxygen species (ROS) in plants

Biosensor	Subcellular locale	Targeting method	Cell types	Species	References
Ca^{2+}					
YC2.1	Cytosol	–	Root hair, root epidermis, guard cells, pollen tubes	*Arabidopsis*, *Medicago*, lily, tobacco	Miwa et al. (2006), Kosuta et al. (2008), Allen et al. (1999, 2001, 2002), Klusener et al. (2002), Hugouvieux et al. (2001), Capoen et al. (2011), and Watahiki et al. (2004)
	Nucleus	Nucleoplasmin fusion	Root hairs, pollen tubes	*Medicago*, tobacco	Sieberer et al. (2009) and Watahiki et al. (2004)
YC3.6	Cytosol	–	Stomatal guard cell, roots, root hairs, cotyledons, pollen tubes	*Arabidopsis*, lotus, tobacco	Weinl et al. (2008), Krebs et al. (2011), Iwano et al. (2009), and Monshausen et al. (2008, 2009, 2011)
	Plasma membrane	N-terminal fusion of YC3.6 with the LT16b protein	Roots, cotyledons	*Arabidopsis*	Krebs et al. (2011)
	Nucleus	NLS from SV40 large T protein	Roots, cotyledons	*Arabidopsis*	Krebs et al. (2011)
YC3.1	Cytosol	–	Pollen, stigmatic papillae	*Arabidopsis*, tobacco	Iwano et al. (2004), Certal et al. (2008), Michard et al. (2008, 2011), and Watahiki et al. (2004)
YC4.6	ER	Pumpkin 2S albumin signal peptide/HDEL ER retention signal	Pollen	*Arabidopsis*	Iwano et al. (2009)
D3cpv	Peroxisome	C-terminal KVK–SKL peptide	Roots, cotyledons	*Arabidopsis*	Costa et al. (2010)
	Tonoplast	N-terminus of CBL2	Roots, cotyledons	*Arabidopsis*	Krebs et al. (2011)
pH					
H148D	Cytosol	–	Roots, root hairs	*Arabidopsis*	Fasano et al. (2001) and Monshausen et al. (2007, 2009, 2011)
pHluorin	Cytosol	–	Roots, pollen tubes	*Arabidopsis*, tobacco	Gao et al. (2004), Moseyko and Feldman (2001), Certal et al. (2008), and Michard et al. (2008)
	Apoplast	Chitinase signal peptide	Roots	*Arabidopsis*	Gao et al. (2004)
Pt-GFP	Cytosol	–	Roots, leaves	*Arabidopsis*	Schulte et al. (2006)

(continued)

Table 5 (continued)

Biosensor	Subcellular locale	Targeting method	Cell types	Species	References
H_2O_2					
Hyper	Cytosol	–	Leaf epidermis, stomatal guard cells, suspension cell culture	*Arabidopsis*	Costa et al. (2010)
	Peroxisome	C-terminal KSRM peptide	Leaf epidermis, stomatal guard cells	*Arabidopsis*, tobacco	Costa et al. (2010)
Redox					
RoGFP1/2	Cytosol	–	Roots, leaves	*Arabidopsis*, tobacco	Jiang et al. (2006), Meyer et al. (2007), Rosenwasser et al. (2010), and Schwarzlander et al. (2009)
	Mitochondrion	First 87 amino acids of the tobacco β-ATPase	Roots, leaves	*Arabidopsis*, tobacco	Jiang et al. (2006), Rosenwasser et al. (2010), and Schwarzlander et al. (2008, 2009)
	ER	Chitinase targeting peptide/HDEL retention signal	Roots, tobacco leaf cells	*Arabidopsis*	Meyer et al. (2007) and Schwarzlander et al. (2008)
	Peroxisome	C-terminal SKL peptide	Leaves	*Arabidopsis*, tobacco	Rosenwasser et al. (2010) and Schwarzlander et al. (2008)
	Plastid	Transketolase target peptide	Leaves	*Arabidopsis*	Rosenwasser et al. (2010) and Schwarzlander et al. (2008)

Table 1 from Choi et al. (2012). Copyright The Plant Journal. Reproduced with permission

diminished water uptake through the roots, which can be measured by loss of water from a reservoir. Third, the transmitted loss of tension results in the local generation of a variation potential (which gets slower and smaller with distance from the wounded leaf). These three parameters have been measured simultaneously in the same plant (Davies et al. 1991).

References

Akemann W, Mutoh H, Perron A, Rossier J, Knöpfel T (2010) Imaging brain electric signals with genetically targeted voltage-sensitive fluorescent proteins. Nat Methods 7:643–649

Aki T, Shigyo M, Nakano R, Yoneyama T, Yanagisawa S (2008) Nano scale proteomics revealed the presence of regulatory proteins including three FT-like proteins in phloem and xylem saps from rice. Plant Cell Physiol 49(5):767–790

Allen GJ, Chu SP, Harrington CL, Schumacher K, Hoffman T, Tang YY, Grill E, Schroeder JI (2001) A defined range of guard cell calcium oscillation parameters encodes stomatal movements. Nature 411:1053–1057

Allen GJ, Kwak JM, Chu SP, Llopis J, Tsien RY, Harper JF, Schroeder JI (1999) Cameleon calcium indicator reports cytoplasmic calcium dynamics in Arabidopsis guard cells. Plant J 19:735–747

Allen GJ, Murata Y, Chu SP, Nafisi M, Schroeder JI (2002) Hypersensitivity of abscisic acid-induced cytosolic calcium increases in the Arabidopsis farnesyltransferase mutant era1-2. Plant Cell 14:1649–1662

Ammann D (1986) Ion-selective microelectrodes, principles, design and application. Springer, Berlin/Heidelberg

Anten NPR, Alcalá-Herrera R, Schieving F, Onoda Y (2010) Wind and mechanical stimuli differentially

affect leaf traits in Plantago major. New Phytol 188:554–564
Arasimowicz M, Floryszak-Wieczorek J (2007) Nitric oxide as a bioactive signaling molecule in plant stress responses. Plant Sci 172(5):876–887
Arimura GI, Ozawa R, Maffei ME (2011) Recent advances in plant early signaling in response to herbivory. Int J Mol Sci 12:3723–3739
Atkins CA, Smith PMC (2007) Translocation in legumes: assimilates, nutrients, and signaling molecules. Plant Physiol 144(2):550–561
Attaran E, Zeier TE, Griebel T, Zeier J (2009) Methyl salicylate production and jasmonate signaling are not essential for systemic acquired resistance in Arabidopsis. Plant Cell 21:954–971
Avanci NC, Luche DD, Goldman GH, Goldman MHS (2010) Jasmonates are phytohormones with multiple functions, including plant defense and reproduction. Genet Mol Res 9(1):484–505
Banerjee AK, Chatterjee M, Yu Y, Suh SG, Miller WA, Hannapel DJ (2006) Dynamics of a mobile RNA of potato involved in a long distance signaling pathway. Plant Cell 18:3443–3457
Batailler B, Lemaître T, Viulaine F, Sanchez C, Renard D, Cayla T, Beneteau J, Dinant S (2012) Soluble and filamentous proteins in Arabidopsis sieve elements. Plant Cell Environ 35:1258–1273
Batistič O, Kudla J (2012) Analysis of calcium signaling pathways in plants. Biochim Biophys Acta 1820(8):1283–1293
Baydoun EA, Fry SC (1985) The immobility of pectic substances in injured tomato leaves and its bearing on the identity of the wound hormone. Planta 165:269–276
Beaudoin E (2011) The language of nitric oxide signalling. Plant Biol 13(2):233–242
Benning F, Tamot B, Guelette BS, Hoffmann-Benning S (2012) New aspects of phloem-mediated long-distance lipid signaling in plants. Front Plant Sci 3:53
Berthold AA (1849) Transportation der Hoden. Arch Anat Physiol Wiss Med 16:42–46
Beuve N, Rispail N, Laine P, Cliquet JB, Ourry A, Le Deunff E (2004) Putative role of γ-aminobutyric acid (GABA) as a long-distance signal in up-regulation of nitrate uptake in Brassica napus L. Plant Cell Environ 27:1035–1046
Bilichak A, Ilnystkyy Y, Hollunder J, Kovalchuk I (2012) The progeny of Arabidopsis thaliana plants exposed to salt exhibit changes in DNA methylation, histone modifications and gene expression. PLoS One 7(1):e30515. doi:10.1371/journal.pone.0030515
Blakeslee JJ, Peer WA, Murphy AS (2005) Auxin transport. Curr Opin Plant Biol 8(5):494–500
Blatt MR (1991) Ion channel gating in plants: physiological implications and integration for stomatal function. J Membr Biol 124:95–112
Bleecker AB, Kende H (2000) Ethylene: a gaseous signal molecule in plants. Annu Rev Cell Dev Biol 16:1–18
Bloemendal S, Kück U (2013) Cell-to-cell communication in plants, animals, and fungi: a comparative review. Naturwissenschaften 100(1):3–19
Bonnin P, Gendraud M, Desbiez MO (1989) Étude cinetique de l'evolution des pH cytoplasmique et vacuolaire après administration de piqûres sur les cotylédons de Bidens pilosa L. C R Acad Sci Paris 309:459–464
Boss WF, Im YJ (2012) Phosphoinositide signaling. Annu Rev Plant Biol 63:409–429
Bozorov TA, Baldwin TI, Kim SG (2012) Identification and profiling of miRNAs during herbivory reveals jasmonate-dependent and -independent patterns of accumulation in Nicotiana attenuata. BMC Plant Biol 12:209. doi:10.1186/1471-2229-12-209
Buer CS, Muday GK, Djordjevic MA (2007) Flavonoids are differentially taken up and transported long distances in Arabidopsis. Plant Physiol 145:478–490
Buer CS, Imin N, Djordjevic MA (2010) Flavonoids: new roles for old molecules. J Integr Plant Biol 52(1):98–111
Buhtz A, Springer F, Chappell L, Baulcombe DC, Kehr J (2008) Identification and characterization of small RNAs from the phloem of Brassica napus. Plant J 53(5):739–749
Buhtz A, Pieritz J, Springer F, Kehr J (2010) Phloem small RNAs, nutrient stress responses, and systemic mobility. BMC Plant Biol 10:64. doi:10.1186/1471-2229-10-64
Büscher JM, Czernik D, Ewald JC, Sauer U, Zamboni N (2009) Cross-platform comparison of methods for quantitative metabolomics of primary metabolism. Anal Chem 81(6):2135–2143
Butenko MA, Vie AK, Brembu T, Aalen RB, Bones AM (2009) Plant peptides in signaling: looking for new partners. Trends Plant Sci 14(5):255–263
Camejo D, Martí MC, Olmos E, Torres W, Sevilla F, Jiménez A (2012) Oligogalacturonides stimulate antioxidant system in alfalfa roots. Biol Plant 56(3):537–544
Canonne J, Froidure-Nicolas S, Rivas S (2011) Phospholipases in action during plant defense signaling. Plant Signal Behav 6(1):13–18
Capoen W, Sun J, Wysham D, Oteguib MS, Venkateshwaranc M, Hirscha S, Miwaa H, Downiea JA, Morrisa RJ, Anéc JM, Oldroyda GED (2011) Nuclear membranes control symbiotic calcium signaling of legumes. Proc Natl Acad Sci U S A 108:14348–14353
Certal AC, Almeida RB, Carvalho LM, Wong E, Moreno N, Michard E, Carneiro J, Rodriguéz-Léon J, Wu HM, Cheung AY, Feijó JA (2008) Exclusion of a proton ATPase from the apical membrane is associated with cell polarity and tip growth in Nicotiana tabacum pollen tubes. Plant Cell 20:614–634
Cesco S, Mimmo T, Tonon G, Tomasi N, Pinton R, Terzano R, Neumann G, Weisskopf L, Renella G, Landi L, Nannipieri P (2012) Plant-borne flavonoids released into the rhizosphere: impact on soil bioactivities related to plant nutrition. A review. Biol Fertil Soils 48(2):123–149
Charmont S, Jamet E, Pont-Lezica R, Canut H (2005) Proteomic analysis of secreted proteins from Arabidopsis thaliana seedlings: improved recovery

following removal of phenolic compounds. Phytochemistry 66:453–461
Chen ZJ, Tiana L (2007) Roles of dynamic and reversible histone acetylation in plant development and polyploidy. Biochim Biophys Acta 1769:295–307
Cheng Y, Song C (2006) Hydrogen peroxide homeostasis and signaling in plant cells. Sci China C Life Sci 49(1):1–11
Chinnusamy V, Zhu JK (2009) Epigenetic regulation of stress responses in plants. Curr Opin Plant Biol 12:1–7
Chivasa S, Slabas AR (2012) Plant extracellular ATP signaling: new insight from proteomics. Mol Biosyst 8:445–452
Cho M, Lee OR, Ganguly A, Cho HT (2007) Auxin-signaling: short and long. J Plant Biol 50(2):79–89
Choi J, Choi D, Lee S, Ryu CM, Hwang I (2011) Cytokinins and plant immunity: old foes or new friends? Trends Plant Sci 16(7):388–394
Choi WG, Swanson SJ, Gilroy S (2012) High-resolution imaging of Ca2+, redox status, ROS and pH using GFP biosensors. Plant J 70:118–128
Chu CR, Hsieh CI, Wu SY, Phillips NG (2009) Transient response of sap flow to wind speed. J Exp Bot 60(1):249–255
Clark SE (2001) Cell signalling at the shoot meristem. Nat Rev Mol Cell Biol 2:276–284
Clark G, Torres J, Finlayson S, Guan X, Handley C, Lee J, Kays JE, Chen ZJ, Roux SJ (2010) Apyrase (nucleoside triphosphate-diphosphohydrolase) and extracellular nucleotides regulate cotton fiber elongation in cultured ovules. Plant Physiol 152:1073–1083
Corpas FJ, Barroso JB, Carreras A, Quirós M, León AM, Romero-Puertas MC, Esteban FJ, Valderrama R, Palma JM, Sandalio LM, Gómez M, del Rio LA (2004) Cellular and subcellular localization of endogenous nitric oxide in young and senescent pea plants. Plant Physiol 136:2722–2733
Costa A, Drago I, Behera S, Zottini M, Pizzo P, Schroeder JI, Pozzan T, Lo Schiavo F (2010) H2O2 in plant peroxisomes: an in vivo analysis uncovers a Ca(2 +)-dependent scavenging system. Plant J 62:760–772
Darwin C (1881) The power of movement in plants. John Murray, London
Dat JF, Capelli N, Folzer H, Bourgeade P, Badot PM (2004) Sensing and signaling during plant flooding. Plant Physiol Biochem 42:273–282
Davies E (1987) Plant responses to wounding. In: Davies DD (ed) The biochemistry of plants, vol 12. Academic, New York
Davies E (2004) New functions for electrical signals in plants. New Phytol 161(3):607–610
Davies E (2006) Electrical signals in plants: facts and hypotheses. In: Volkov A (ed) Plant electrophysiology – theory and methods. Springer, Berlin/Heidelberg
Davies E, Schuster A (1981) Intercellular communication in plants: evidence for a rapidly generated, bidirectionally transmitted wound signal. Proc Natl Acad Sci USA 78(4):2422–2426
Davies E, Stankovic B (2006) Electrical signals, the cytoskeleton and gene expression: a hypothesis on the coherence of the cellular responses to environmental insult. In: Baluska F, Mancuso S, Volkmann D (eds) Communication in plants, neuronal aspects of plant life. Springer, Berlin/Heidelberg
Davies E, Ramaiah KVA, Abe S (1986) Wounding inhibits protein synthesis yet stimulates polysome formation in aged, excised pea epicotyls. Plant Cell Physiol 27:1377–1386
Davies E, Zawadzki T, Witters D (1991) Electrical activity and signal transmission in plants: how do plants know? In: Penel C, Greppin H (eds) Plant signalling, plasma membrane and change of state. Université de Genève, Geneva
Demidchik V, Nichols C, Oliynyk M, Dark A, Glover BV, Davies JM (2003) Is ATP a signaling agent in plants? Plant Physiol 133:456–461
Depuydt S, Hardtke CS (2011) Hormone signalling crosstalk in plant growth regulation. Curr Biol 21(9):R365–R373
Desbiez MO, Kergosien Y, Champagnat P, Thellier M (1984) Memorization and delayed expression of regulatory messages in plants. Planta 160:392–399
Dietz KJ, Sauter A, Wichert K, Messdaghi D, Hartung W (2000) Extracellular beta-glucosidase activity in barley involved in the hydrolysis of ABA glucose conjugate in leaves. J Exp Bot 51:937–944
Dinant S, Kehr JS (2013) Sampling and analysis of phloem sap. In: Maathuis FJM (ed) Plant mineral nutrients: methods and protocols, methods in molecular biology. Humana Press, New York
Dinant S, Lemoine R (2010) The phloem pathway: new issues and old debates. C R Biol 333:307–319
Dinant S, Suárez-López P (2012) Multitude of long-distance signal molecules acting via phloem. In: Witzany G, Baluška F (eds) Biocommunication of plants. Signaling and communication in plants. Springer, Berlin/Heidelberg
Ding Z, Friml J (2010) Auxin regulates distal stem cell differentiation in Arabidopsis roots. Proc Natl Acad Sci USA 107(26):12046–12051
Dodd AN, Kudla J, Sanders D (2010) The language of calcium signaling. Annu Rev Plant Biol 61:593–620
Doering-Saad C, Newbury HJ, Couldridge CE, Bale JS, Pritchard J (2006) A phloem-enriched cDNA library from Ricinus: insights into phloem function. J Exp Bot 57:3183–3193
Dong W, Lv H, Xia G, Wang M (2012) Does diacylglycerol serve as a signaling molecule in plants? Plant Signal Behav 7(4):472–475
Dorn A, Weisenseel MH (1982) Advances in vibrating probe techniques. Protoplasma 113:89–96
Dorokhov YL, Komarova TV, Petrunia IV, Frolova OY, Pozdyshev DV, Gleba YY (2012) Airborne signals from a wounded leaf facilitate viral spreading and induce antibacterial resistance in neighboring plants. PLoS Pathog 8(4):e1002640. doi:10.1371/journal.ppat.1002640

Dunn WB, Ellis DI (2005) Metabolomics: current analytical platforms and methodologies. Trends Anal Chem 24(4):285–294
Dziubinska H, Trebacz K, Zawadzki T (1989) The effect of excitation on the rate of respiration in the liverwort Conocephalum conicum. Physiol Plant 75:417–423
Ebert B, Zöller D, Erban A, Fehrle I, Hartmann J, Nieh A, Kopka J, Fisahn J (2010) Metabolic profiling of Arabidopsis thaliana epidermal cells. J Exp Bot 61(5):1321–1335
Ernst AM, Jekat SB, Zielonka S, Müllera B, Neumann U, Rüping B, Twymand RM, Krzyzaneke V, Prüfer D, Noll GA (2012) Sieve element occlusion (SEO) genes encode structural phloem proteins involved in wound sealing of the phloem. Proc Natl Acad Sci USA 109(28):E1980–E1989
Etzler ME, Esko JD (2009) Free glycans as signaling molecules. In: Varki A, Cummings RD, Esko JD et al (eds) Essentials of glycobiology, 2nd edn. Cold Spring Harbor, New York
Eveland AL, Jackson DP (2012) Sugars, signalling, and plant development. J Exp Bot 63(9):3367–3377
Farmer EE, Ryan CA (1990) Interplant communication: airborne methyl jasmonate induces synthesis of proteinase inhibitors in plant leaves. Proc Natl Acad Sci USA 87:7713–7716
Fasano JM, Swanson SJ, Blancaflor EB, Dowd PE, Kao TH, Gilroy S (2001) Changes in root cap pH are required for the gravity response of the Arabidopsis root. Plant Cell 13:907–921
Feng J, Liu X, Lai L, Chen J (2011) Spatio-temporal expression of miRNAs in tomato tissues upon Cucumber mosaic virus and Tomato aspermy virus infections. Acta Biochim Biophys Sin 43:258–266
Fiehn O, Kopka J, Dörmann P, Altmann T, Trethewey RN, Willmitzer L (2000) Metabolite profiling for plant functional genomics. Nat Biotechnol 18:1157–1161
Filek M, Koscielniak J (1997) The effect of wounding the roots by high temperature on the respiration rate of the shoot and propagation of electric signal in horse bean seedlings (Vicia faba L. minor). Plant Sci 123:39–46
Fisahn J, Herde O, Willmitzer L, Pena-Cortes H (2004) Analysis of the transient increase in cytosolic Ca2+ during the action potential of higher plants with high temporal resolution: requirement of Ca2+ transients for induction of jasmonic acid biosynthesis and PINII gene expression. Plant Cell Physiol 45:456–459
Forde BG (2002) The role of long distance signaling in plant responses to nitrate and other nutrients. J Exp Bot 53:39–43
Foresi NP, Laxalt AM, Tonón CV, Casalongué CA, Lamattina L (2007) Extracellular ATP induces nitric oxide production in tomato cell suspensions. Plant Physiol 145:589–592
Fromm J, Eschrich W (1988a) Transport processes in stimulated and non-stimulated leaves of Mimosa pudica. I. The movement of 14C-labelled photoassimilates. Trees 2:7–17
Fromm J, Eschrich W (1988b) Transport processes in stimulated and non-stimulated leaves of Mimosa pudica. II. Energesis and transmission of seismic stimulations. Trees 2:18–24
Fromm J, Eschrich W (1988c) Transport processes in stimulated and non-stimulated leaves of Mimosa pudica. III. Displacement of ions during seismonastic leaf movements. Trees 2:65–72
Fromm J, Bauer T (1994) Action potentials in maize sieve tubes change phloem translocation. J Exp Bot 45:463–469
Fromm J, Fei H (1998) Electrical signaling and gas exchange in maize plants of drying soil. Plant Sci 132:203–213
Fromm J, Lautner S (2007) Electrical signals and their physiological significance in plants. Plant Cell Environ 30:249–257
Fromm J, Hajirezaei M, Wilke I (1995) The biochemical response of electrical signaling in the reproductive system of Hibiscus plants. Plant Physiol 109:375–384
Frost CJ, Appel HM, Carlson JE, De Moraes CM, Mescher MC, Schultz JC (2007) Within-plant signaling via volatiles overcomes vascular constraints on systemic signaling and primes responses against herbivores. Ecol Lett 10:490–498
Frost CJ, Mescher MC, Carlson JE, De Moraes CM (2008) Why do distance limitations exist on plant-plant signaling via airborne volatiles? Plant Signal Behav 3(7):466–468
Fujii H, Chiou TJ, Lin SI, Aung K, Zhu JK (2005) A miRNA involved in phosphate-starvation response in Arabidopsis. Curr Biol 15:2038–2043
Fukuda H, Higashiyama T (2011) Diverse functions of plant peptides: entering a new phase. Plant Cell Physiol 52(1):1–4
Fukumorita T, Chino M (1982) Sugar, amino acid and inorganic contents in rice phloem sap. Plant Cell Physiol 23(2):273–283
Furbank RT, Tester M (2011) Phenomics – technologies to relieve the phenotyping bottleneck. Trends Plant Sci 16(12):635–644
Galvani L (1791) De Viribus Electricitatis in Motu Musculari Commentarius. Academy of Science, Bologna
Gao D, Knight MR, Trewavas AJ, Sattelmacher B, Plieth C (2004) Self-reporting Arabidopsis expressing pHand[Ca2 +] indicator sunveil ion dynamics in the cytoplasm and in the apoplast under abiotic stress. Plant Physiol 134:898–908
Giavalisco P, Kapitza K, Kolasa A, Buhtz A, Kehr J (2006) Towards the proteome of Brassica napus phloem sap. Proteomics 6:896–909
Gillaspy GE (2011) The cellular language of myo-inositol signaling. New Phytol 192(4):823–839
Gomes de Oliveira Dal'Molin C, Quek LE, Palfreyman RW, Brumbley SM, Nielsen LK (2010) AraGEM, a genome-scale reconstruction of the primary metabolic network in Arabidopsis. Plant Physiol 152:579–589
Goodger JQ, Schachtman DP (2010) Re-examining the role of ABA as the primary long-distance signal produced by water-stressed roots. Plant Signal Behav 5(10):1298–1301

Grams TEE, Koziolek C, Lautner S, Matyssek R, Fromm J (2007) Distinct roles of electric and hydraulic signals on the reaction of leaf gas exchange upon re-irrigation in Zea mays L. Plant Cell Environ 30:79–84

Grams TEE, Lautner S, Felle HH, Matyssek R, Fromm J (2009) Heat-induced electrical signals affect cytoplasmic and apoplastic pH as well as photosynthesis during propagation through the maize leaf. Plant Cell Environ 32:319–326

Grewe BF, Langer D, Kaspar H, Kampa BM, Helmchen F (2010) High-speed in vivo calcium imaging reveals neuronal network activity with near-millisecond precision. Nat Methods 7:399–405

Guelette BS, Benning UF, Hoffmann-Benning S (2012) Identification of lipids and lipid-binding proteins in phloem exudates from Arabidopsis thaliana. J Exp Bot 63(10):3603–3616

Ha S, Vankova R, Yamaguchi-Shinozaki K, Shinozaki K, Tran LS (2012) Cytokinins: metabolism and function in plant adaptation to environmental stresses. Trends Plant Sci 17(3):172–179

Hafke JB, van Bel AJE (2013) Cellular basis of electrical potential waves along the phloem and impact of coincident Ca2+ fluxes. In: Thompson GA, van Bel AJE (eds) Phloem: molecular cell biology, systemic communication, biotic interactions. Wiley-Blackwell, Ames

Hallé F (1999) Voyage au pays de la forme. Éloge de la plante. Pour une nouvelle biologie. Seuil, Paris

Hammond JP, White PJ (2011) Sugar signaling in root responses to low phosphorus availability. Plant Physiol 156:1033–1040

Hannapel DJ (2010) A model system of development regulated by the long-distance transport of mRNA. J Integr Plant Biol 52(1):40–52

Hao LH, Wang WX, Chen C, Wang YF, Liu T, Li X, Shang ZL (2012) Extracellular ATP promotes stomatal opening of Arabidopsis thaliana through heterotrimeric G protein α subunit and reactive oxygen species. Mol Plant 5(4):852–864

Hartmann A, Czauderna T, Hoffmann R, Stein N, Schreiber F (2011) HTPheno: an image analysis pipeline for high-throughput plant phenotyping. BMC Bioinform 12:148. doi:10.1186/1471-2105-12-148

Haseloff J (1998) GFP variants for multispectral imaging of living cells. Methods Cell Biol 58:139–151

Hauvermale AL, Ariizumi T, Steber CM (2012) Gibberellin signaling: a theme and variations on DELLA repression. Plant Physiol 160(1):83–92

Hayata Q, Hayata S, Irfana M, Ahmadb A (2010) Effect of exogenous salicylic acid under changing environment: a review. Environ Exp Bot 68(1):14–25

Hayama T, Shimmen T, Tazawa M (1979) Participation of Ca2+ in cessation of cytoplasmic streaming induced by membrane excitation in Characeae internodal cells. Protoplasma 99:305–321

Haywood V, Yu TS, Huang NC, Lucas WJ (2005) Phloem long-distance trafficking of Gibberellic acid-insensitive RNA regulates leaf development. Plant J 42:49–68

Heil M, Karban R (2009) Explaining evolution of plant communication by airborne signals. Trends Ecol Evol 25(3):137–144

Heil M, Ton J (2008) Long-distance signaling in plant defense. Trends Plant Sci 13:264–272

Hetherington AM (1998) Plant physiology: spreading a drought warning. Curr Biol 8(25):R911–R913

Hlaváčková V, Nauš J (2007) Chemical signal as a rapid long-distance information messenger after local wounding of a plant? Plant Signal Behav 2(2):103–105

Horikawa K, Yamada Y, Matsuda T, Kobayashi K, Hashimoto M, Matsu-ura T, Miyawaki A, Michikawa T, Mikoshiba K, Nagai T (2010) Spontaneous network activity visualized by ultra-sensitive Ca2+ indicators Cameleon-Nano. Nat Methods 7:729–732

Huffaker A, Pearce G, Ryan CA (2006) An endogenous peptide signal in Arabidopsis activates components of the innate immune response. Proc Natl Acad Sci USA 103:10098–10103

Hugouvieux V, Kwak JM, Schroeder JI (2001) An mRNA cap binding protein, ABH1, modulates early abscisic acid signal transduction in Arabidopsis. Cell 106:477–487

Hwang I, Sheen J, Müller B (2012) Cytokinin signaling networks. Annu Rev Plant Biol 63:353–380

Iwai H, Usui M, Hoshino H, Kamada H, Matsunaga T, Kakegawa K, Ishii T, Satoh S (2003) Analysis of sugars in squash xylem sap. Plant Cell Physiol 44(6):582–587

Iwano M, Shiba H, Miwa T, Che FS, Takayama S, Nagai T, Miyawaki A, Isogai A (2004) Ca2+ dynamics in a pollen grain and papilla cell during pollination of Arabidopsis. Plant Physiol 136:3562–3571

Iwano M, Entani T, Shiba H, Kakita M, Nagai T, Mizuno H, Miyawaki A, Shoji T, Kubo K, Isogai A, Takayama S (2009) Fine-tuning of the cytoplasmic Ca2+ concentration is essential for pollen tube growth. Plant Physiol 150:1322–1334

Iyer-Pascuzzi AS, Symonova O, Mileyko Y, Hao Y, Belcher H, Harer J, Weitz JS, Benfey PN (2010) Imaging and analysis platform for automatic phenotyping and trait ranking of plant root systems. Plant Physiol 152:1148–1157

Jackson MB (2002) Long distance signaling from roots to shoots assessed: the flooding story. J Exp Bot 53:175–181

Jeannette E, Paradis S, Zalejski C (2010) Diacylglycerol pyrophosphate, a novel plant signaling lipid. In: Munnik T (ed) Lipid signal in plants. Springer, Berlin/Heidelberg

Jeter C, Tang W, Henaff E, Butterfield T, Roux SJ (2004) Evidence of a novel cell signaling role for extracellular adenosine triphosphates and diphosphates in Arabidopsis. Plant Cell 16:2652–2664

Jia W, Zhang J (2008) Stomatal movements and long-distance signaling in plants. Plant Signal Behav 3(10):772–777

Jiang K, Schwarzer C, Lally E, Zhang S, Ruzin S, Machen T, Remington SJ, Feldman L (2006) Expression and characterization of a redox-sensing green fluorescent protein (reduction-oxidation-sensitive green fluorescent protein) in Arabidopsis. Plant Physiol 141:397–403

Jiang F, Hartung W (2008) Long-distance signaling of abscisic acid (ABA): the factors regulating the intensity of the ABA signal. J Exp Bot 59(1):37–43

Johnson HE, Broadhurst D, Goodacrev R, Smith AR (2003) Metabolic fingerprinting of salt-stressed tomatoes. Phytochemistry 62:919–928

Juna JH, Fiumea E, Fletchera JC (2008) The CLE family of plant polypeptide signaling molecules. Cell Mol Life Sci 65:743–755

Kader MA, Lindberg S (2010) Cytosolic calcium and pH signaling in plants under salinity stress. Plant Signal Behav 5(3):233–238

Kehr J (2006) Phloem sap proteins: their identities and potential roles in the interaction between plants and phloem-feeding insects. J Exp Bot 57:767–774

Kehr J, Buhtz A (2008) Long distance transport and movement of RNA through the phloem. J Exp Bot 59(1):85–92

Kehr J, Buhtz A, Giavalisco P (2005) Analysis of xylem sap proteins from Brassica napus. BMC Plant Biol 5:11. doi:10.1186/1471-2229-5-11

Kim M, Canio W, Kessler S, Sinha N (2001) Developmental changes due to long-distance movement of a homeobox fusion transcript in tomato. Science 293:287–289

Kim SY, Sivaguru M, Stacey G (2006) Extracellular ATP in plants. Visualization, localization, and analysis of physiological significance in growth and signaling. Plant Physiol 142:984–992

Kim TH, Böhmer M, Hu H, Nishimura N, Schroeder JI (2010) Guard cell signal transduction network: advances in understanding abscisic acid, CO2, and Ca2+ signaling. Annu Rev Plant Biol 61:561–591

Klusener B, Young JJ, Murata Y, Allen GJ, Mori IC, Hugouvieux V, Schroeder JI (2002) Convergence of calcium signaling pathways of pathogenic elicitors and abscisic acid in Arabidopsis guard cells. Plant Physiol 130:2152–2163

Kosuta S, Hazledine S, Sun J, Miwa H, Morris RJ, Downie JA, Oldroyd GE (2008) Differential and chaotic calcium signatures in the symbiosis signaling pathway of legumes. Proc Natl Acad Sci U S A 105:9823–9828

Koziolek C, Grams TEE, Schreiber U, Matyssek R, Fromm J (2004) Transient knockout of photosynthesis mediated by electrical signals. New Phytol 161:715–722

Köckenberger W, Pope JM, Xia Y, Jeffrey KR, Komor E, Callaghan PT (1997) A non-invasive measurement of phloem and xylem water flow in castor bean seedlings by nuclear magnetic resonance microimaging. Planta 201:53–63

Kralj JM, Douglass AD, Hochbaum DR, Maclaurin D, Cohen AE (2011) Optical recording of action potentials in mammalian neurons using a microbial rhodopsin. Nat Methods 9:91–95

Krebs M, Held K, Binder A, Hashimoto K, Den Herder G, Parniske M, Kudla J, Schumacher K (2011) FRET-based genetically encoded sensors allow high-resolution live cell imaging of Ca(2+) dynamics. Plant J 69:181–192

Krishnan HB, Natarajan SS, Bennett JO, Sicher RC (2011) Protein and metabolite composition of xylem sap from field-grown soybeans (Glycine max). Planta 233:921–931

Kruger NJ, Ratcliffe RG (2012) Pathways and fluxes: exploring the plant metabolic network. J Exp Bot 63(6):243–2246

Kudla J, Batistič O, Hashimoto K (2010) Calcium signals: the lead currency of plant information processing. Plant Cell 22:541–563

Kurusu T, Kuchitsu K, Nakano M, Nakayama Y, Lida H (2013) Plant mechanosensing and Ca2+ transport. Trends Plant Sci 18(4):227–233

Lake JA, Woodward FI, Quick WP (2002) Long-distance CO_2 signaling in plants. J Exp Bot 53:183–193

Lalonde S, Boles K, Hellmann H, Barker L, Patrick JW, Frommer WB, Warda JM (1999) The dual function of sugar carriers: transport and sugar sensing. Plant Cell 11:707–726

Lattanzio V, Lattanzio VMT, Cardinali A (2006) Role of phenolics in the resistance mechanisms of plants against fungal pathogens and insects. In: Imperato F (ed) Phytochemistry: advances in research. Research Signpost, Scarborough

Lau OS, Deng XW (2010) Plant hormone signaling lightens up: integrators of light and hormones. Curr Opin Plant Biol 13:571–577

Läuchli A, Grattan SR (2007) Plant growth and development under salinity stress. In: Jenks MA et al (eds) Advances in molecular breeding toward drought and salt tolerant crops. Springer, Berlin/Heidelberg

Lautner S, Grams TEE, Matyssek R, Fromm J (2005) Characteristics of electrical signals in poplar and responses in photosynthesis. Plant Physiol 138:2200–2209

Lee JY, Cui W (2009) Non-cell autonomous RNA trafficking and long-distance signaling. J Plant Biol 52:10–18

Lee GI, Howe GA (2003) The tomato mutant sprl is defective in systemin perception and the production of a systemic wound signal for defense gene expression. Plant J 33:567–576

Liu TY, Chang CY, Chiou TJ (2009) The long-distance signaling of mineral macronutrients. Curr Opin Plant Biol 12(3):312–319

Liu PP, von Dahl CC, Klessig DF (2011) The extent to which methyl salicylate is required for signaling systemic acquired resistance is dependent on exposure to light after infection. Plant Physiol 157:2216–2226

Llaneras F, Picó J (2008) Stoichiometric modelling of cell metabolism. J Biosci Bioeng 105(1):1–11

Looger LJ (2012) Running in reverse: rhodopsins sense voltage. Nat Methods 9:143–144

López-Gresa MP, Lisón P, Kim HK, Choi YH, Verpoorte R, Rodrigo I, Conejero V, Bellés JM (2012) Metabolic

fingerprinting of Tomato Mosaic Virus infected Solanum lycopersicum. J Plant Physiol 169:1586–1596
Luna E, Bruce TJ, Roberts MR, Flors V, Ton J (2012) Next-generation systemic acquired resistance. Plant Physiol 158:844–853
Maleck K, Dietrich RA (1999) Defense on multiple fronts: how do plants cope with diverse enemies? Trends Plant Sci 4(6):215–219
Malladi M, Burns JK (2007) Communication by plant growth regulators in roots and shoots of horticultural crops. Hortic Sci 42(5):1113–1117
Malone M (1993) Hydraulic signals. Philos Trans R Soc Lond B 341:33–39
Malone M, Alarcon JJ, Palumbo L (1994) An hydraulic interpretation of rapid, long-distance wound signaling in the tomato. Planta 193:181–185
Mandal SM, Chakraborty D, Dey S (2010) Phenolic acids act as signaling molecules in plant-microbe symbioses. Plant Signal Behav 5(4):359–368
Martína JA, Sollab A, Coimbrac MA, Gila L (2005) Metabolic distinction of Ulmus minor xylem tissues after inoculation with Ophiostoma novo-ulmi. Phytochemistry 66(20):2458–2467
Matsubayashi Y, Sakagami Y (2006) Peptide hormones in plants. Annu Rev Plant Biol 57:649–674
Matsubayashi Y, Shinohara H, Ogawa M (2006) Identification and functional characterization of phytosulfokine receptor using a ligand-based approach. Chem Rec 6(6):356–364
McGurl B, Pearce G, Orozco-Cardenas M, Ryan CA (1992) Structure, expression, and antisense inhibition of the systemin precursor gene. Science 255:1570–1573
Meyer AJ, Brach T, Marty L, Kreye S, Rouhier N, Jacquot JP, Hell R (2007) Redox-sensitive GFP in Arabidopsis thaliana is a quantitative biosensor for the redox potential of the cellular glutathione redox buffer. Plant J 52:973–986
Michard E, Dias P, Feijo JA (2008) A temporal and spatial analysis of extracellular flux and free cytosolic concentration of calcium and protons in growing pollen tubes of tobacco. Sex Plant Reprod 21:169–181
Michard E, Lima PT, Borges F, Silva AC, Portes MT, Carvalho JE, Gilliham M, Liu LH, Obermeyer G, Feijo JA (2011) Glutamate receptor-like genes form Ca2+ channels in pollen tubes and are regulated by pistil D-serine. Science 332:434–437
Migicovsky Z, Kovalchuk I (2013) Changes to DNA methylation and homologous recombination frequency in the progeny of stressed plants. Biochem Cell Biol 91(1):1–5
Miller G, Schlauch K, Tam R, Cortes D, Torres MA, Shulaev V, Dangl JL, Mittler R (2009) The plant NADPH oxidase RBOHD mediates rapid systemic signaling in response to diverse stimuli. Sci Signal 2:ra45. doi:10.1126/scisignal.2000448
Mintz-Orona S, Meira S, Malitskya S, Ruppin E, Aharonia A, Shlomid T (2012) Reconstruction of Arabidopsis metabolic network models accounting for subcellular compartmentalization and tissue-specificity. Proc Natl Acad Sci USA 109(1):339–344
Mirouze M, Paszkowski J (2011) Epigenetic contribution to stress adaptation in plants. Curr Opin Plant Biol 14:1–8
Mittler R, Vanderauwera S, Nobuhiro Suzuki N, Miller G, Tognetti VB, Vandepoele K, Gollery M, Shulaev V, Breusegem FV (2011) ROS signaling: the new wave? Trends Plant Sci 16(6):300–309
Miwa H, Sun J, Oldroyd GE, Downie JA (2006) Analysis of calcium spiking using a cameleon calcium sensor reveals that nodulation gene expression is regulated by calcium spike number and the developmental status of the cell. Plant J 48:883–894
Molassiotis A, Tanou G, Diamantidis G (2010) NO says more than 'YES' to salt tolerance. Salt priming and systemic nitric oxide signaling in plants. Plant Signal Behav 5(3):209–212
Molders W, Buchala A, Metraux JP (1996) Transport of salicylic acid in tobacco necrosis virus-infected cucumber plants. Plant Physiol 112:787–792
Molinier J, Ries G, Zipfel C, Hohn B (2006) Transgeneration memory of stress in plants. Nature 442:1046–1049
Molnar A, Melnyk CW, Bassett A, Hardcastle TJ, Dunn R, Baulcombe DC (2010) Small silencing RNAs in plants are mobile and direct epigenetic modification in recipient cells. Science 328:872–875
Monshausen GB, Bibikova TN, Messerli MA, Shi C, Gilroy S (2007) Oscillations in extracellular pH and reactive oxygen species modulate tip growth of Arabidopsis root hairs. Proc Natl Acad Sci U S A 104:20996–21001
Monshausen GB, Bibikova TN, Weisenseel MH, Gilroy S (2009) Ca2+ regulates reactive oxygen species production and pH during mechano-sensing in Arabidopsis roots. Plant Cell 21:2341–2356
Monshausen GB, Messerli MA, Gilroy S (2008) Imaging of the Yellow Cameleon 3.6 indicator reveals that elevations in cytosolic Ca2+ follow oscillating increases in growth in root hairs of arabidopsis. Plant Physiol 147:1690–1698
Monshausen GB, Miller ND, Murphy AS, Gilroy S (2011) Dynamics of auxin-dependent Ca2+ and pH signaling in root growth revealed by integrating high-resolution imaging with automated computer vision-based analysis. Plant J 65:309–318
Moseyko N, Feldman LJ (2001) Expression of pH-sensitive green fluorescent protein in Arabidopsis thaliana. Plant Cell Environ 24:557–563
Motose H, Iwamoto K, Endo S, Demura T, Sakagami Y, Matsubayashi Y, Moore KL, Fukuda H (2009) Involvement of phytosulfokine in the attenuation of stress response during the transdifferentiation of Zinnia mesophyll cells into tracheary elements. Plant Physiol 150(1):437–447
Mullendorea DL, Windt CW, Van Asc H, Knoblaucha M (2010) Sieve tube geometry in relation to phloem flow. Plant Cell 22:579–593

Murphy E, Smith E, De Smet I (2012) Small signaling peptides in Arabidopsis development: how cells communicate over a short distance. Plant Cell 24(8):3198–3217

Nakashima K, Yamaguchi-Shinozaki K (2013) ABA signaling in stress-response and seed development. Plant Cell Rep. doi:10.1007/s00299-013-1418-1

Narváez-Vásquez J, Orozco-Cárdenas JM, Ryan CA (1994) A sulfhydryl reagent modulates systemic signaling for wound-induced and systemin-induced proteinase inhibitor synthesis. Plant Physiol 105:725–730

Neumann PM (2007) Evidence for long-distance xylem transport of signal peptide activity from tomato roots. J Exp Bot 58(8):2217–2223

Okumoto S, Jones A, Frommer WB (2012) Quantitative imaging with fluorescent biosensors. Annu Rev Plant Biol 63:663–706

Pant BD, Buhtz A, Kehr J, Scheible WR (2008) MicroRNA399 is a long-distance signal for the regulation of plant phosphate homeostasis. Plant J 53(5):731–738

Park SW, Kaimoyo E, Kumar D, Mosher S, Klessig DF (2007) Methyl salicylate is a critical mobile signal for plant systemic acquired resistance. Science 318:113–116

Parker JE (2009) The quest for long-distance signals in plant systemic immunity. Sci Signal 2(70):pe31. doi:10.1126/scisignal.270pe31

Pastor V, Luna E, Mauch-Manic B, Ton J, Flors V (2012) Primed plants do not forget. Environ Exp Bot. doi:10.1016/j.envexpbot.2012.02.013

Pastrana E (2012) Light based electrophysiology: genetically-encoded voltage sensors are finally measuring up. Nat Methods 9:38. doi:10.1038/nmeth.1825

Pearce G, Strydom D, Johnson S, Ryan CA (1991) A polypeptide from tomato leaves induces wound-inducible proteinase inhibitor proteins. Science 253:895–897

Pearce G, Moura DS, Stratmann J, Ryan CA (2001) RALF, a 5-kDa ubiquitous polypeptide in plants, arrests root growth and development. Proc Natl Acad Sci USA 98:12843–12847

Peña-Cortés H, Willmitzer L, Sanchez-Serrano JJ (1991) Abscisic acid mediates wound induction but not developmental-specific expression of the proteinase inhibitor II gene family. Plant Cell 3(9):963–972

Pickard BG (1973) Action potentials in higher plants. Bot Rev 39:172–201

Qutob D, Chapman BP, Gijzen M (2013) Transgenerational gene silencing causes gain of virulence in a plant pathogen. Nat Commun 4:1349. doi:10.1038/ncomms2354

Raghavendra AS, Gonugunta VK, Christmann A, Grill E (2010) ABA perception and signaling. Trends Plant Sci 15(7):395–401

Rasmann S, De Vos M, Casteel CL, Tian D, Halitschke R, Sun JY, Agrawal AA, Felton GW, Jander G (2012) Herbivory in the previous generation primes plants for enhanced insect resistance. Plant Physiol 158:854–863

Ricca U (1916) Soluzione di un problema di fisiologia: la propagazione di stimolo nella "Mimosa". Nuovo G Bot Ital 23:51–170

Riewe D, Grosman L, Fernie AR, Wucke C, Geigenberger P (2008) The potato-specific apyrase is apoplastically localized and has influence on gene expression, growth, and development. Plant Physiol 147:1092–1109

Rochefort NI, Konnerth A (2008) Genetically encoded Ca2+ sensors come of age. Nat Methods 5:761–762

Rocher F, Chollet JF, Jousse C, Bonnemain JL (2006) Salicylic acid, an ambimobile molecule exhibiting a high ability to accumulate in the phloem. Plant Physiol 141:1684–1693

Rodriguez-Medina C, Atkins CA, Mann AJ, Jordan ME, Smith PMC (2011) Macromolecular composition of phloem exudate from white lupin (Lupinus albus L.). PMC Plant Biol 11:36. doi:10.1186/1471-2229-11-36

Rolland F, Baena-Gonzalez E, Sheen J (2006) Sugar sensing and signaling in plants: conserved and novel mechanisms. Annu Rev Plant Biol 57:675–709

Roos W, Viehweger K, Dordschbal B, Schumann B, Evers S, Steighardt J, Schwartze W (2006) Intracellular pH signals in the induction of secondary pathways – the case of Eschscholzia californica. J Plant Physiol 163(3):369–381

Rosenwasser S, Rot I, Meyer AJ, Feldman L, Jiang K, Friedman H (2010) A fluorometer-based method for monitoring oxidation of redox-sensitive GFP (roGFP) during development and extended dark stress. Physiol Plant 138:493–502

Ruiz-Medrano R, Xoconostle-Cázares B, Lucas WJ (1999) Phloem long-distance transport of CmNACP mRNA: implications for supracellular regulation in plants. Development 126:4405–4419

Saze H (2008) Epigenetic memory transmission through mitosis and meiosis in plants. Semin Cell Dev Biol 19(6):527–536

Schallau K, Junker BH (2010) Simulating plant metabolic pathways with enzyme-kinetic models. Plant Physiol 152(4):1763–1771

Schilmiller AL, Howe GA (2005) Systemic signaling in the wound response. Curr Opin Plant Biol 8:369–377

Schopfer P (2006) Biomechanics of plant growth. Am J Bot 93:1415–1425

Schulte A, Lorenzen I, Bottcher M, Plieth C (2006) A novel fluorescent pH probe for expression in plants. Plant Methods 2:7

Schwarzlander M, Fricker MD, Muller C, Marty L, Brach T, Novak J, Sweetlove LJ, Hell R, Meyer AJ (2008) Confocal imaging of glutathione redox potential in living plant cells. J Microsc 231:299–316

Schwarzlander M, Fricker MD, Sweetlove LJ (2009) Monitoring the in vivo redox state of plant mitochondria: effect of respiratory inhibitors, abiotic stress and assessment of recovery from oxidative challenge. Biochim Biophys Acta 1787:468–475

Shiina T, Tazawa M (1986) Action potential in Luffa cylindrica and its effects on elongation growth. Plant Cell Physiol 27:1081–1089
Sibaoka T (1966) Action potentials in plant organs. Symp Soc Exp Biol 20:49–73
Sibaoka T (1969) Physiology of rapid movements in higher plants. Annu Rev Plant Physiol 20:165–184
Sieberer BJ, Chabaud M, Timmers AC, Monin A, Fournier J, Barker DG (2009) A nuclear-targeted cameleon demonstrates intranuclear Ca2+ spiking in Medicago truncatula root hairs in response to rhizobial nodulation factors. Plant Physiol 151:1197–1206
Sinyukhin AM, Britikov EA (1967) Action potentials in the reproductive system of plants. Nature 215:1278–1280
Shah J, Zeier J (2013) Long-distance communication and signal amplification in systemic acquired resistance. Front Plant Sci 4:30. doi:10.3389/fpls.2013.00030
Shelp BJ (2012) Does long-distance GABA signaling via the phloem really occur? Botany 90:897–900
Shibuya N, Nimami E (2001) Oligosaccharide signalling for defence responses in plant. Physiol Mol Plant Pathol 59:223–233
Slaughter A, Daniel X, Flors V, Estrella L, Hohn B, Mauch-Mani B (2012) Descendants of primed Arabidopsis plants exhibit resistance to biotic stress. Plant Physiol 158:835–843
Sokołowska K, Zagórska-Marek B (2012) Symplasmic, long-distance transport in xylem and cambial regions in branches of Acer pseudoplatanus (Aceraceae) and Populus tremula × P. tremuloides (Salicaceae). Am J Bot 99:1745–1755
Song YY, Zeng RS, Xu JF, Li J, Shen X, Yihdego WG (2010) Interplant communication of tomato plants through underground common mycorrhizal networks. PLoS One 5(10):e13324. doi:10.1371/journal.pone.0013324
Spoel SH, Dong X (2012) How do plants achieve immunity? Defence without specialized immune cells. Nat Rev (Immunol) 12:89–100
Stael S, Wurzinger B, Mair A, Mehlmer N, Vothknecht UC, Teige M (2012) Plant organellar calcium signaling: an emerging field. J Exp Bot 63(4):1525–1542
Stankovic B, Davies E (1996) Both action potentials and variation potentials induce proteinase inhibitor gene expression in tomato. FEBS Lett 390:275–279
Stankovic B, Davies E (1997) Intercellular communication in plants: electrical stimulation of proteinase inhibitor gene expression in tomato. Planta 202:402–406
Stankovic B, Witters D, Zawadzki T, Davies E (1998) Action potentials and variation potentials in sunflower: an analysis of their relationships and distinguishing characteristics. Physiol Plant 103:51–58
Staswick PE (2008) JAZing up jasmonate signaling. Trends Plant Sci 13(2):66–71
Stern K (1924) Electrophysiologie der Pflanzen. Julius Springer, Berlin
Suárez-López P (2005) Long-range signaling in plant reproductive development. Int J Dev Biol 49:761–771
Sun JQ, Jiang HL, Li CY (2011) Systemin/Jasmonate-mediated systemic defense signaling in tomato. Mol Plant 4(4):607–615
Sun J, Zhang C, Zhang X, Deng S, Zhao R, Shen X, Chen S (2012) Extracellular ATP signaling and homeostasis in plant cells. Plant Signal Behav 7(5):566–569
Sunkar R, Li YF, Jagadeeswaran G (2012) Functions of microRNAs in plant stress responses. Trends Plant Sci 17(4):196–203
Suzuki N, Koussevitzky S, Mittler R, Miller G (2012) ROS and redox signalling in the response of plants to abiotic stress. Plant Cell Environ 35(2):259–270
Swanson SJ, Choi W-G, Chanoca A, Gilroy S (2011) In vivo imaging of Ca2+, pH, and reactive oxygen species using fluorescent probes in plants. Annu Rev Plant Biol 62:273–297
Swarup R, Kramer EM, Perry P, Knox K, Ottoline Leyser HM, Haseloff J, Beemster GTS, Bhalerao R, Bennett MJ (2005) Root gravitropism requires lateral root cap and epidermal cells for transport and response to a mobile auxin signal. Nat Cell Biol 7(11):1057–1065
Tamaki V, Mercier H (2007) Cytokinins and auxin communicate nitrogen availability as long-distance signal molecules in pineapple (Ananas comosus). J Plant Physiol 164(11):1543–1547
Tanaka K, Gilroy S, Jones AM, Stacey G (2010) Extracellular ATP signaling in plants. Trends Cell Biol 20(10):601–608
Tardieu F, Tuberosa R (2010) Dissection and modelling of abiotic stress tolerance in plants. Curr Opin Plant Biol 13:206–212
Tena G, Boudsocq M, Sheen J (2011) Protein kinase signaling networks in plant innate immunity. Curr Opin Plant Biol 14(5):519–529
Thellier M, Lüttge U (2013) Plant memory: a tentative model. Plant Biol 15:1–12
Thellier M, Desbiez MO, Champagnat P, Kergosien Y (1982) Do memory processes occur also in plants? Physiol Plant 56(3):281–284
Treutter D (2006) Significance of flavonoids in plant resistance: a review. Environ Chem Lett 4(3):147–157
Tsien RY (1998) The green fluorescent protein. Ann Rev Biochem 67:509–544
Tsien RY (2010) Nobel lecture: constructing and exploiting the fluorescent protein paint box. Integr Biol 2:77–93
Tsutsui T, Karasawa S, Okamura Y, Miyawaki A (2008) Improving membrane potential voltage measurements using FRET with new fluorescent proteins. Nat Methods 5:683–685
Tuteja N, Sopory SK (2008) Chemical signaling under abiotic stress environment in plants. Plant Signal Behav 3(8):525–536
Ueda H, Kikuta Y, Matsuda K (2012) Plant communication: mediated by individual or blended VOCs? Plant Signal Behav 7(2):222–226
Umezawa T, Nakashima K, Miyakawa T, Kuromori T, Tanokura M, Shinozaki K, Yamaguchi-Shinozaki K (2010) Molecular basis of the core regulatory network

in ABA responses: sensing, signaling and transport. Plant Cell Physiol 51(11):1821–1839

Van Bel AJE, Knoblauch M, Furch ACU, Hafke JB (2011) (Questions)(n) on phloem biology. 1. Electropotential waves, Ca2+ fluxes and cellular cascades along the propagation pathway. Plant Sci 181:210–218

Vestergaard CL, Flyvbjerg H, Møller IM (2012) Intracellular signaling by diffusion: can waves of hydrogen peroxide transmit intracellular information in plant cells? Front Plant Sci 3:295. doi:10.3389/fpls.2012.00295

Vian A, Faure C, Girard S, Davies E, Hallé F, Bonnet P, Ledoigt G, Paladian F (2007) Plants respond to GSM-like radiation. Plant Signal Behav 2(6):522–524

Walz C, Juenger M, Schad M, Kehr J (2002) Evidence for the presence and activity of a complete antioxidant defence system in mature sieve tubes. Plant J 31(2):189–197

Wang G, Fiers M (2010) CLE peptide signaling during plant development. Protoplasma 240:33–43

Wang F, Cui X, Sun Y, Dong CH (2013) Ethylene signaling and regulation in plant growth and stress responses. Plant Cell Rep. doi:10.1007/s00299-013-1421-6

Wasternack C, Hause B (2013) Jasmonates: biosynthesis, perception, signal transduction and action in plant stress response, growth and development. An update to the 2007 review in Annals of Botany. Ann Bot. doi:10.1093/aob/mct067

Watahiki MK, Trewavas AJ, Parton RM (2004) Fluctuations in the pollen tube tip-focused calcium gradient are not reflected in nuclear calcium level: a comparative analysis using recombinant yellow cameleon calcium reporter. Sex Plant Reprod 17:125–130

Weinl S, Held K, Schlucking K, Steinhorst L, Kuhlgert S, Hippler M, Kudla J (2008) A plastid protein crucial for Ca2+-regulated stomatal responses. New Phytol 179:675–686

Williams SE, Pickard BG (1972a) Properties of action potentials in Drosera tentacles. Planta 103:193–221

Williams SE, Pickard BG (1972b) Receptor potentials and action potentials in Drosera tentacles. Planta 103:222–240

Wilkinson S, Davies WJ (2002) ABA-based chemical signaling. Plant Cell Environ 25:195–210

Wilkinson S, Davies WJ (2010) Drought, ozone, ABA and ethylene: new insights from cell to plant to community. Plant Cell Environ 33:510–525

Williamson RE, Ashley CC (1982) Free Ca2+ and cytoplasmic streaming in the alga Chara. Nature 296:647–651

Wingler A, Roitsch T (2008) Metabolic regulation of leaf senescence: interactions of sugar signaling with biotic and abiotic stress responses. Plant Biol 10:50–62

Wu X, Weigel D, Wigge PA (2002) Signaling in plants by intercellular RNA and protein movement. Genes Dev 16:151–158

Yamaguchi Y, Pearce G, Ryan CA (2006) The cell surface leucine-rich repeat receptor for AtPep1, an endogenous peptide elicitor in Arabidopsis, is functional in transgenic tobacco cells. Proc Natl Acad Sci USA 103(26):10104–10109

Yamaguchi Y, Huffaker A, Bryan AC, Tax FE, Ryan CA (2010) PEPR2 is a second receptor for the Pep1 and Pep2 peptides and contributes to defense responses in Arabidopsis. Plant Cell 22:508–522

Yan X, Wang Z, Huang L, Wang C, Hou R, Xu Z, Qiao X (2009) Research progress on electrical signals in higher plants. Prog Nat Sci 19(5):531–545

Yi HS, Heil M, Adame-Álvarez RM, Ballhorn DJ, Ryu CM (2009) Airborne induction and priming of plant defenses against a bacterial pathogen. Plant Physiol 151:2152–2161

Yoo BC, Kragler F, Varkonyi-Gasic E, Haywood V, Archer-Evans S, Lee YM, Lough TJ, Lucas WJ (2004) A systemic small RNA signaling system in plants. Plant Cell 16:1979–2000

Zhang S, Sun L, Kragler F (2009) The phloem-delivered RNA pool contains small noncoding RNAs and interferes with translation. Plant Physiol 150(1):378–387

Zhao Y (2010) Auxin biosynthesis and its role in plant development. Annu Rev Plant Biol 61:49–64

Zhao Q, Guo HW (2011) Paradigms and paradox in the ethylene signaling pathway and interaction network. Mol Plant 4(4):626–634

Thiolomics: Molecular Mechanisms of Thiol-Cascade in Plant Growth and Nutrition

Dibyendu Talukdar and Tulika Talukdar

Contents

D. Talukdar, Ph.D. (✉)
Department of Botany, R.P.M. College, University of Calcutta, Uttarpara, Hooghly 712258, West Bengal, India
e-mail: dibyendutalukdar9@gmail.com

T. Talukdar, Ph.D.
Department of Botany, Krishnagar Government College, Nadia, Krishnagar 741101, West Bengal, India

Department of Botany, APC Roy Govt. College, Siliguri, Darjeeling, West Bengal, India
e-mail: talukdartulip12@gmail.com

Abstract

Growing plants have a constitutive demand for thiol (sulfur) to synthesize protein, sulfolipid, and other essential sulfur (S)-containing molecules for growth. The uptake and subsequent distribution of sulfate is regulated in response to demand and environmental factors. Sulfate transport consists of both constitutive and sulfur nutrition-dependent regulated transport. The acquisition of sulfur by plants has become an increasingly important concern

D. Barh et al. (eds.), *PlantOmics: The Omics of Plant Science*,
DOI 10.1007/978-81-322-2172-2_17, © Springer India 2015

for the agriculture due to the decreasing trends of S-emissions from industrial sources and the consequent limitation of inputs from deposition. The recognition of the importance of sulfate for plant growth and vigor and hence crop yield, as well as the nutritional importance of sulfur for human and animal diets, has increasingly been recognized. Cysteine synthesis in plants is a fundamental process for protein biosynthesis and all anabolic pathways that require reduced sulfur. Cysteine is the first committed molecule in plant metabolism that contains both sulfur and nitrogen, and, thus, the regulation of its biosynthesis is of utmost importance for the synthesis of a number of essential metabolites in plant pathways. Cysteine is incorporated into proteins and glutathione directly or serves as a sulfur donor for the synthesis of S-containing compounds such as methionine and its derivatives *S*-adenosylmethionine and *S*-methylmethionine and many secondary compounds. Furthermore, cysteine acts as a general catalyst in redox reactions through the nucleophilic properties of its sulfur atom, utilizing dithiol–disulfide interchange, as displayed in the thioredoxin and the glutaredoxin systems. Molecular characterization involving transcriptomics, proteomics, and metabolomics profiling in major crops like rice, barley, wheat, maize, and legumes along with model plant *Arabidopsis thaliana* revealed that sulfate uptake, distribution, and reductive assimilation are regulated in fine-tune depending on sulfur status and demand and that this cascade is integrated with plant photosynthesis, nutrient transports, antioxidant defense system, hormonal signaling, kinase cascades, carbohydrate metabolism, and during plants' experiences with different biotic and abiotic stresses. This cascade can be manipulated in favor of enhanced plant growth and nutritional benefits—as, for example, effort has been initiated in food and feed legumes (chickpeas, narrow-leafed lupin, soybeans) and other plants with enhanced S-containing amino acids, threonine, glutathione, protein quality, protease inhibitors, and trace elements and with lysine, protein content, and compositions in cereal grains. This emerging prospect can be ushered by using latest cutting-edge functional genomics tools and better understanding of plant thiol-metabolism from source (soil) to sink (grains) in diverse arenas of "thiolomics." In this chapter, the comprehensive knowledge generated in this area has been compiled and analyzed.

Keywords
Antioxidant defense • Cross talk • Functional genomics • Glutathione • Hormone signaling • Nutrition • Plant stress • Regulations • Sulfur metabolisms

Introduction

Sulfur (S) is a critical nutrient for metabolism, plant growth, and development. It represents the ninth and least abundant essential macronutrient in plants (Höfgen and Hesse 2008). The importance of S as a plant nutrient has been recognized since time immemorial, but active research started in the second half of the twentieth century when widespread S deficiencies were observed. S plays an inevitable and imperative role in the formation of amino acids, methionine (Met; 21 %) and cysteine (Cys; 27 %), and synthesis of protein, chlorophyll, and oil in the oilseed crops. Numerous other plant metabolites are also formed from S. Plant S nutrition not only affects crop yield but also quality (Tabe et al. 2003; Chiaiese et al. 2004; Taylor et al. 2008). Therefore, increasing the S utilization efficiency (SUE) of plants is becoming an important issue. SUE was described as "improved capture of resources, the accumulation of greater reserves of S, and improved mechanisms for the remobilization of these reserves" (Hawkesford 2000). The importance of SUE has nicely been demonstrated in two gain-of-function *Arabidopsis* mutants, *sue3* and *sue4*, exhibiting low S tolerance equipped with well-developed root systems, tolerance to heavy metals, and tolerance to oxidative stress (Wu et al. 2010). Cysteine (Cys) is the first stable and committed molecule in plant metabolism that contains both S and nitrogen (N). It is the metabolic precursor for vital cellular components containing reduced S, including glutathione (GSH), homoglutathione, iron–sulfur clusters, vitamin cofactors like biotin and thiamin, and multiple secondary metabolites. GSH is the most abundant low-molecular-weight thiol-buffer with a plethora of functions in plant stress defense, hormone signaling, redox regulation, sexual plant reproduction, and S homeostasis (Noctor et al. 2012). Thus, the regulation of S biosynthesis is of utmost importance for the synthesis of a number of essential metabolites in plant pathways.

A fair part of S incorporated into organic molecules in plants is located in thiol (–SH) groups in proteins (Cys residues) or nonprotein thiols (GSH). The thiol group of Cys in proteins maintains protein structure by forming disulfide bonds between two Cys residues via oxidation. The thiol of Cys and GSH is often involved in the redox cycle by two thiol–disulfide conversions. This interchange is versatile for redox control and mitigation against oxidative stress in nearly all aerobic organisms including plants (Leustek et al. 2000). Sulfate uptake and assimilation share approximately equal control over cellular thiol flux (Vauclare et al. 2002).

Studies in the model plant *Arabidopsis thaliana* (thale cress) provide significant insights on the molecular processes and regulation of thiol-metabolism (Leustek et al. 2000; Saito 2000; Kopriva 2006; Meyer and Rausch 2008; Höfgen and Hesse 2008; Kopriva et al. 2009; Yi et al. 2010). However, the depth of understanding these same pathways in various crop plants is somewhat limited. In general, acquisition of S from soil, its transport, reductive assimilation, formation of downstream metabolites via Cys, and highly versatile functions of different thiolic compounds in diverse events of plant growth and development encompass huge interactive and integrated multilevel networks of thiol-cascade which starts from upstream events of S-metabolisms to produce downstream thiol-metabolites GSH and GSH-mediated entire antioxidant defense within ascorbate (AsA)–GSH cycle and outside it. With the miraculous progress of different "omics" technologies, the biology of plant thiol-metabolism has entered into a new era, reflecting successful utilization of functional genomics in dissecting molecular mechanisms of thiol-cascade in plant growth and nutritional quality through understanding of a new concept, "thiolomics." Transcriptome, proteome, and metabolome analysis of any organism reflects the total biological activities at any given time which are responsible for the adaptation of the organism to the surrounding environmental conditions. *Arabidopsis* is the plant model of choice for global analysis of transcriptome, proteome, and metabolome. S deficiency, the major physiological problem connected with S-metabolism, is being investigated by expression profiling (Hirai et al. 2003; Maruyama-Nakashita et al. 2003; Nikiforova et al. 2003) and combined transcriptome and metabolome analysis (Hirai et al. 2005; Nikiforova et al. 2005a). More than 2,700 genes were found to be affected by S starvation. The genes induced by S deficiency included those coding for sulfate transporters, reduction, and assimilation to downstream thiol-metabolites. Metabolome analysis revealed that from approximately 6,000 analyzed metabolites, 11.5 % were significantly affected by 13 days of S starvation (Nikiforova et al. 2005a, b). The power of the global study of metabolite and transcript networks was demonstrated by Hirai et al. (2005), who used a batch-learning self-organizing mapping analysis to reveal clusters of genes and metabolites regulated by the same mechanism. Altogether, transcriptome, proteome, and metabolome analyses revealed the complexity of the interactions between S, N, and C metabolism and created new domains of the molecular mechanisms of thiol-metabolisms, and the discovery of 49 transcription factor genes responding specifically to S deficiency is of greatest importance in this regard (Nikiforova et al. 2003).

Sulfate Uptake and Transport: Dissecting Fundamental Steps with "Omics" Technologies

The major S form available to plants is sulfate. Uptake of sulfate by plants is considered to be the fundamental step of the S cycle in the nature.

Sulfate is taken up to plant cells by sulfate transporters (Sultr). Recently, the first transcription factors responsible for regulation of sulfate uptake and assimilation have been identified (Maruyama-Nakashita et al. 2006; Hirai et al. 2007). Once sulfate is taken up from the soil solution by dedicated root membrane transport systems, it is either transitorily accumulated in the vacuoles of roots or shoots, or it enters the complex sulfate reductive metabolic pathway (Kopriva 2006). In between the step of sulfate uptake into root cells and its reduction in the faraway leaf chloroplasts, several cell-to-cell inter- and intracellular transports through numerous transmembrane and plasmodesmata exist. Coordination of such short- and long-distance sulfate transport requires the timely regulations of gene expressions encoding proteins involved in sulfate uptake and transport. In order to understand how components of the network interact, it is necessary to analyze the temporal and spatial transcription behavior of all (or most) of the genes in the genome (the transcriptome) simultaneously (Kopriva 2006).

Nascent seedlings and generative tissues have a constitutive demand for S. The uptake and subsequent distribution of sulfate is regulated in response to demand and environmental factors. Many studies have shown that the changes in the capacity of plant S transport were paralleled by changes in the steady-state contents of mRNAs and protein of the group 1 sulfate transporters (Smith et al. 1997; Takahashi et al. 2000; Hawkesford and Wray 2000; Shibagaki et al. 2002; Yoshimoto et al. 2002, 2003; Howarth et al. 2003). The derepression/repression of gene expression seems to be a major factor in the regulation of sulfate uptake in plants. Differential expression of transcripts for S-metabolic enzymes was systematically validated by reverse transcription quantitative PCR in different plants. Since the first reported cloning of a plant sulfate transporter in *Stylosanthes hamata* (Smith et al. 1997), it has gradually become clear that sulfate transport in plants is carried out by a complex system of transporters governed by multigene family. In recent years, many genes controlling expressions of sulfate transporters from diverse plant taxa have been isolated and characterized (reviewed Buchner et al. 2004a). Sulfate transport consists of both constitutive and S nutrition-dependent regulated transport. A decreased intracellular content of sulfate, Cys, and GSH is concomitant with increasing transporter activity (Smith et al. 1997). Gene and protein expression studies have confirmed that regulation occurs predominantly at the level of the mRNA (Smith et al. 1997; Takahashi et al. 2000; Hawkesford and Wray 2000; Yoshimoto et al. 2002; Hawkesford 2003; Kopriva 2006; Takahashi et al. 2011; Talukdar and Talukdar 2013d). Transcriptomic analysis of the sulfate transporter gene family reflects a complex pattern of regulation: (1) cell-specific expression of some groups 1, 2, and 4 transporters under adequate S nutrition which is upregulated by inadequate S nutrition (AtSultr1;2, 1;3, 2;1, 4;1); (2) expression of the group 3 transporters with tissue/organ specificity but no regulation by S nutrition; and (3) cell-/tissue-specific S deficiency-related derepression of some group 1 and 2 transporters (AtSultr1;1, 1;2, 2;1, 2;2) (Buchner et al. 2004a). The process of so-called long-distance sulfate translocation may require several types of transporters responsible for cell-to-cell movement of sulfate across the plasma membrane. Loading of sulfate into the vascular tissues in roots and unloading of sulfate into the leaf cells are assumed to be the two important steps in this process. These two events are controlled by the same sulfate transporter gene, *AST68*, in *A. thaliana* (Takahashi et al. 1997). Using an array hybridization/transcript profiling method in *Arabidopsis* plants subjected to 6, 10, and 13 days of constitutive and induced S starvation, Nikiforova et al. (2003) revealed induction of sulfate transporter *AST68* (Sultr2;1) and Sultr4;1. Analysis of the *Arabidopsis* and rice genome sequences (The Arabidopsis Genome Initiative 2000) has till date enabled the identification of 14 putative sulfate transporter genes in each genome (Vidmar et al. 2000; Buchner et al. 2004a). Alignment and phylogenetic analysis of the 14 *Arabidopsis* and rice proteins subdivides the plant sulfate transporter family into four closely related groups, all with 12 membrane-spanning domains and a STAS (sulfate transporter and anti-sigma antagonist) domain at their carboxy-terminus (Aravind and Koonin 2000), and a fifth more diverse, but clearly related, group with two smaller proteins lacking the STAS domain (Hawkesford 2003). However, nothing is known about plastidic sulfate transporters in higher plants. The sulfate transport in plastids is necessary for the synthesis of many sulfur-containing compounds. For example, in *Spinacia oleracea*, the lack of sulfates leads to considerable changes in the expression of Cys synthesis genes (Lyubetsky et al. 2013). Plastomes of vascular plants lack genes of the sulfate transport system except for rare instances of cyst and cysA. However, the green alga *Helicosporidium* sp. retains cysT. Plastomes of the rhodophyte *Cyanidium caldarium* and *Cyanidioschyzon merolae* and the cyanelle genome of *Cyanophora paradoxa* lack cysT homologues but possess distant homologues of cysA presumably involved in the transport of zinc or manganese (Lyubetsky et al. 2013). In some liverworts, the plastid-encoded sulfate transporter gene *cys*A has been lost up to 29 times, yet intact copies of *cys*A are evolving under selective constraints. Gene loss is more frequent in groups with an increased substitution rate in the plastid genome of liverworts (Wickett et al. 2011). None of the 14 *Arabidopsis* genes encoding putative sulfate transporters seems to be involved in such a crucial function. Plastidic sulfate transporters are, however, identified in *Chlamydomonas reinhardtii* and shown to belong to the bacterial ABC type of transporter (Lyubetsky et al. 2013). Thus, identification of plant plastidic sulfate transporter is undoubtedly one of the greatest challenges in S research (Davidian and Kopriva 2010).

Group 1 and 2 Sulfate Transporters

Group 1 and 2 sulfate transporters, which are localized at the plasma membrane, have been the subject of several studies and are the best characterized groups. Members of group 1 represent high-affinity transporters that facilitate uptake of sulfate by the root (Sultr1;1 and Sultr1;2) or translocation of sulfate from source-to-sink organs (Sultr1;3) (Takahashi et al. 2000, 2011; Shibagaki et al. 2002; Hawkesford 2003; Kopriva 2006; Talukdar and Talukdar 2013d). In wheat, the high-affinity sulfate transporter homologue to the AtSultr1;1 was detected in aleurone cells of wheat grains by immunolocalization (Kopriva 2006). Group 2 is composed of low-affinity sulfate transporters whose gene products may rather play a role in vascular tissues, facilitating the translocation of sulfate around the plant. Differences in the kinetic and expression pattern of AtSultr2;1 (Km 0.41 mM) and AtSultr2;2 (Km 1.2 mM) indicate specific functions in the process of vascular movement of sulfate. AtSultr2;1 is expressed in the xylem parenchyma and phloem cells of leaves, but in the root in xylem parenchyma and pericycle cells. By contrast, AtSultr2;2 is localized specifically in the phloem of roots and in vascular bundle sheath cells of leaves (Takahashi et al. 2000; Maruyama-Nakashita et al. 2003). The observed upregulation of AtSultr2;1 in roots during sulfate starvation, and the increase of the mRNA transcripts level of AtSultr2;1 under selenate treatment, may be an indication of this function (Takahashi et al. 2000; Maruyama-Nakashita et al. 2003). The leaf phloem expression suggests a role in phloem loading for sulfate transport to other organs. The leaf xylem parenchyma localization of AtSultr2;1 might indicate absorption of sulfate from the xylem vessels or reabsorption for further xylem transport. In leaves, however, the expression in the bundle sheath cells surrounding the vascular veins suggests the uptake of sulfate released from xylem vessels at millimolar concentrations for transfer to the primary sites of assimilation in leaf palisade and mesophyll cells. The expression pattern of both Group 2 transporters suggests that the two transporters are involved in balancing the vascular movement of sulfate in relation to the sulfate status of the different tissues (Takahashi et al. 2000; Maruyama-Nakashita et al. 2003). In addition to such transcriptional regulation, both AtSultr 1;1 and AtSultr 1;2, when expressed under control of constitutive 35S promoter, accumulated exclusively in the root and responded to sulfate deficiency in a similar manner to native transporters (Yoshimoto et al. 2007), indicating posttranscriptional control of gene expressions. Possibly, STAS domain is involved in such regulation, as mutations in this domain interfere with correct targeting of sulfate transporters to the plasma membrane (Shibagaki and Grossman 2006). Likewise, protein–protein interactions between two transporters such as between AtSultr 2;1 and AtSultr 3;5 in *Arabidopsis* are important for optimal sulfate transport capacity (Kataoka et al. 2004). In addition to these regulations, plant microRNAs (miRNA) have recently been implicated in regulation of sulfate transport. For example, AtSultr 2;1 expression was downregulated in shoots in response to sulfate starvation and upregulated in the absence of S, which can be explained by an increase in the miRNA in both cases revealing differential roles of miRNAs in different root tissue territories (Takahashi et al. 2000; Jones-Rhoades and Bartel 2004; Kawashima et al. 2009; Davidian and Kopriva 2010). In *Arabidopsis*, a new complex regulatory interplay between Sultr1;1 and Sultr1;2 based upon an extensive comparison of a wide set of growth conditions and metabolite signaling pathways has been proposed (Rouached et al. 2009). Differential expression and alternative splicing of rice sulfate transporter family members regulate S status during plant growth, development, and stress conditions (Kumar et al. 2011a, b).

Group 3 and 4 Sulfate Transporters

Group 3 is composed of low-affinity transporters localized at the plasma membrane. Unlike groups 1 to 3, group 4 sulfate transporters have been localized to the vacuolar membrane/tonoplast. The Sultr4;1 gene is shown to be expressed in roots under S-sufficient and S-deficient conditions, where it may play a role in the efflux of sulfate from the vacuolar lumen into the cytoplasm and influence the vacuolar storage capacity for sulfate (Kataoka et al. 2004). Contrastingly, Sultr4;2 gene expression is shown to be highly inducible by S limitation in the same tissue. The Sultr4;1/Sultr4;2 double knockout mutants contained higher amounts of sulfate than did wild-type plants. Comparison of single and Sultr4;1/Sultr4;2 double knockout mutants suggested that Sultr4;1 plays a major role and Sultr4;2 has a supplementary function (Kataoka et al. 2004). In addition to the verification of the subcellular localization of this group 4 sulfate transporters in *Arabidopsis*, analysis of T-DNA mutations showed an increased accumulation of sulfate and decrease of Cys and GSH contents when plants were grown on low sulfate (Takahashi et al. 2003). The drastic reduction of root sulfate concentrations under sulfate deficiency is accompanied by an upregulation of sulfate transporter 4;1 in *Brassica* (Hawkesford 2003). Increased expression of this transporter maximizes the vacuolar efflux of stored sulfate under these conditions. Analysis of qRT-PCR expression profiles reveals that *Arabidopsis* Sultr4;1 gene is strongly expressed (10-fold higher than the Sultr4;2 gene) in developing seeds and that its disruption significantly increases seed sulfate content, suggesting that Sultr 4;1 is involved in the efflux of sulfate from vacuoles within developing seeds. Furthermore, a proteome analysis of Sultr4;1 mutant seeds reveals metabolic modulations suggesting adaptations to altered sulfate compartmentation. This event implicates Sultr4;1-mediated sulfate transport in establishment of defense mechanisms against oxidative stress during seed development (Zuber et al. 2010). This study also pointed out that Sultr4;1 is highly expressed during grain filling stage, whereas Sultr4;2 is expressed at almost constitutively from embryogenesis to the dry mature stage. Also, the qRT-PCR data revealed that the Sultr4;1 gene was more highly expressed than Sultr4;2 in most plant organs. Interestingly, expression studies of both Sultr 4 transporters in oilseed rape (*Brassica napus*) leaves revealed a differential S-dependent expression pattern: BnSultr 4;2 was more highly expressed than BnSultr 4;1 in response to sulfate depletion (Parmar et al. 2007; Dubousset et al. 2009). In the same sulfur-depletion conditions, rapeseed leaf sulfate content decreased significantly in relation to upregulation of BnSultr 4;1, confirming the involvement of both Sultr4 members in vacuolar sulfate remobilization in *Brassica* (Dubousset et al. 2009). A transcriptional regulator, Sulfur LIMitation1 (SLIM1), regulating sulfate uptake and assimilation has been shown to induce AtSultr1;1, AtSultr1;2, and AtSultr 4;2 gene expression in response to sulfate starvation (Maruyama-Nakashita et al. 2006). Clearly, cytosolic and plastidic sulfate homeostasis is important to avoid toxification when excess sulfate is accumulated in the vacuole.

Group 5 Sulfate Transporter

Among the fifth group, the *Arabidopsis Sultr5;2* gene has recently been demonstrated to encode a high-affinity root molybdate (Mo) transporter, MOT1 (Tomatsu et al. 2007), which raises the question of the role of group 5 genes in sulfate transport. However, cotransport of Mo through plant sulfate transporter SHST1 has been reported (Fitzpatrick et al. 2008). Indeed, expression of the sulfate transporter SHST1 from *Stylosanthes hamata* in a *Saccharomyces cerevisiae* mutant defective in sulfate transport, YSD1, increased its capacity to take up Mo when grown in the presence of low Mo concentrations (Fitzpatrick et al. 2008). While sulfate did not inhibit the transport of Mo through this transporter, Mo reduced sulfate transport via SHST1 (Fitzpatrick et al. 2008). However, a complex interaction between S, Mo, and selenium in S-deficient and S-sufficient plant has recently been studied in Indian mustard, *Brassica juncea* (Schiavon et al. 2012).

Induction, Regulations, and Tissue Distributions of Sulfate Transporters

The two distinct group 1 sulfate transporters are different in their inducibilities in relation to the nutritional status of the plant. One transporter (AtSultr1;2, LeSultr1;1) mediates the uptake of sulfate under both S-replete and S-deficient conditions, and expression is relatively insensitive to external sulfate concentrations. The second transporter (AtSultr1;1, LeSultr1;2) is highly inducible under sulfate limitation but almost absent in non-S-stressed plants (Yoshimoto et al. 2002, 2003; Hawkesford 2003; Howarth et al. 2003). The higher inducibility of AtSultr1;1 compared to AtSultr1;2 can be explained by the absence of the SURE (sulfur-responsive element) element in the promoter of AtSultr1;2. This suggests that the upregulation of the root major transporter AtSultr1;2 is probably essentially controlled by SLIM1 when sulfate availability in the external medium is restricted but that an additional regulatory mechanism dependent on the SURE element drives a strong expression of AtSultr1;1 during S-deficient conditions. This dual inducible uptake system was confirmed by the identification of selenate-resistant (sel) mutants of *Arabidopsis* (Shibagaki et al. 2002), exhibiting that a lesion in the AtSultr1;2 sulfate transporter isoform restricted the uptake of both sulfate and its toxic analog, selenate. Analysis of another mutation of AtSultr1;2, (*sel1-10*) indicated that AtSultr1;2 serves as a major facilitator for the acquisition of sulfate, and despite upregulation of AtSultr1;1 expression in the *sel1-10* mutant, growth is reduced (Maruyama-Nakashita et al. 2003). Besides the use of *sel* mutant, the power of genetic approach for dissecting unexplored events in thiol-metabolisms has also been reflected by green fluorescent protein (GFP). GFP expression from the well-characterized bSR promoter fragment from conglycinin (Awazuhara et al. 2002) was used as a tool in the search for mutants with altered S deficiency response. Seeds from plants harvesting the bSR::GFP construct were mutagenized and mutants were selected with increased GFP expression at normal S supply (Ohkama-Ohtsu et al. 2004). In one of these mutants, the level of OAS was increased and, in addition to GFP expression, the mRNA levels of several other genes responsive to sulfate starvation were increased even at normal S concentration. Map-based cloning and sequence analysis identified a thiol reductase to be responsible for the elevated OAS levels (Ohkama-Ohtsu et al. 2004).

Expression pattern of sulfate transporters greatly differs in different plant species and during different growth stages and tissues of the same plant. Spatial expression analysis of AtSultr1;3 in *Arabidopsis*, and of group 1 high-affinity sulfate transporters in other plant species, indicated that sulfate transport in vascular tissues is not restricted to low-affinity transport. In sulfate-deficient barley roots, HvSultr1;1 was expressed within the stele (Rae and Smith 2002). This was not observed for the homologous *Arabidopsis* AtSultr1;1 and 1;2. Presumably, in barley, a single transporter is responsible for functions carried out by more than one transporter in *Arabidopsis* (Rae and Smith 2002). In the rice genome, two group 2 transporters are present, but pattern of their spatial expression is unknown. In tomato, LeSultr1;1 is expressed under sulfate-deprived conditions in the pericycle (Howarth et al. 2003). This indicates that under sulfate stress, plants are able to induce additional high-affinity sulfate transport to maintain vascular movement of sulfate under low-sulfate concentrations. In *Arabidopsis*, AtSultr1;3 seems to mediate the interorgan transport of sulfate as high-affinity transporter by specific expression exclusively in the phloem of all *Arabidopsis* organs analyzed (Yoshimoto et al. 2003). Analysis of a AtSultr1;3 T-DNA insertion mutant provided direct evidence for this function by restricting movement of labeled sulfate from the cotyledon to the other organs, indicating importance of AtSultr1;3 for source-to-sink transport of sulfate (Yoshimoto et al. 2003). The specific expression of lower high-affinity transporters in the root tip, as well as in axillary buds, indicates the importance of an adequate sulfate supply to fast-growing tissues (Takahashi et al. 1997, 2000; Rae and Smith 2002). In addition, in root tips, high levels of expression are likely to be of functional value to facilitate "foraging" (Buchner et al. 2004a). Developing leaves are strong S-sinks. The upregulation of *Arabidopsis* AtSultr2;2 and 1;3 in leaves under sulfate starvation revealed participation of both transporters in the vascular allocation of sulfate from leaves to other tissues (Takahashi et al. 2000; Yoshimoto et al. 2003). In *Arabidopsis*, AtSultr1;1 and AtSultr 1;2 expression is upregulated in leaves subjected to S starvation. AtSultr1;2 promoter activity was found in the guard cells under normal growth conditions (Yoshimoto et al. 2002). The upregulation of the *Arabidopsis* group 4 transporter, AtSultr4;1, under S deficiency in roots as well as in leaves (Takahashi et al. 2000) indicated the importance of vacuolar efflux of sulfate regulated by the S demand. In tomato, high-affinity sulfate transporter, LeSultr1;2, was overexpressed in the stem and in the leaves due to S-stress (Howarth et al. 2003). LeSultr1;2 expression was also observed induced in the vascular tissue of the *Verticillium*-resistant tomato line, GCR 218, after infection by *Verticillium dahlia*. This suggested prominent roles played by the sulfate transporter in the mechanism of *Verticillium* resistance involving elemental sulfur formation.

Among prominent food legumes, transcript levels of two sulfate transporters, Sultr1;2 and Sultr3;3, were elevated in the active phase of storage protein accumulation in common beans, *Phaseolus vulgaris* L. (Liao et al. 2012). Transcriptomic profiling identifies predominant expression of Sultr3 members in *P. vulgaris*, whereas expression of Sultr5 members was predominant in *Vigna mungo* (Liao et al. 2013). Understanding the role of S in food legumes growth is important from the point of view that the deficiency of the S-containing amino acids Cys, cystine, and Met may limit the nutritional value of food and feed (Khan and Mazid 2011; Liao et al. 2012). The increased levels of Sultr1;2 and Sultr 3;3 transcripts in common beans are consistent with *Arabidopsis* studies implicating the high-affinity sulfate transporter Sultr1;3 in phloem transport of sulfate (Yoshimoto et al. 2003) and low-affinity group 3 sulfate transporters in the transport of sulfate from the seed coat to the embryo (Zuber et al. 2010). In chickpeas, expression of the low-affinity AtSultr2;1 homologue was not restricted to vegetative tissues but was also found in pods and in the testa of developing embryos (Tabe et al. 2003). The group 3 sulfate transporter homologue of AtSultr3.1 is also expressed in the pods, in the testa, and, additionally, in developing chickpea embryos. Expression of the sulfate transporter homologue of AtSultr3;3 was detectable in almost all organs with high abundance in the developing chickpea embryo, suggesting participation in sulfate transport in many cell types (Tabe et al. 2003).

Unexpectedly, group 3 sulfate transporter was identified as essential for nitrogen fixation in legume nodules (Krusell et al. 2005). In lentil (*Lens culinaris* Medik.) seedlings, two putative sulfate transporter genes, LcSultr1;1 and LcSultr1;2, were upregulated in roots of L 414 genotype in response to arsenate (Talukdar and Talukdar 2013d). Temporal expression pattern revealed initial downregulation of both LcSultr1;1 and LcSultr1;2 in L 414, but their significant elevations were observed during later stages of metalloid exposures. Transcripts of LcSultr2;1 and LcSultr2;2 initially changed nonsignificantly in L 414, followed by their downregulations. Expression levels of two group 1 transporters were initially low in lentil genotype DPL 59 and further downregulated during prolonged arsenate exposure. Transcripts of both LcSultr2 ;1 and LcSultr2 ;2 in DPL 59 were initially unchanged but increased after 24 h of metalloid exposures (Talukdar and Talukdar 2013d). This observation indicates genotypic differences in regulation of gene expression of sulfate transporters in lentil crop. A detailed discussion of the current status of sulfate transporter regulation in *Arabidopsis* and crop plants can be found in Takahashi et al. (2011).

Sulfate Assimilation and Cys Biosynthesis-Cellular Regulations, Homeostasis, and Functional Interplay

Plant S-assimilation is the effective delivery of sulfate to the plastid, the major site of the assimilatory reductive pathway. For assimilation into Cys, sulfate is activated by adenylation to adenosine 5′ phosphosulfate (APS) in a reaction catalyzed by ATP sulfurylase (ATPS; EC 2.7.7.4). APS is reduced to sulfite by APS reductase (APR; EC 1.8.4.9) with electrons derived from GSH. Sulfite is further reduced by a ferredoxin-dependent sulfite reductase (SiR; EC 1.8.7.1) to sulfide, which is incorporated by O-acetylserine (thiol)lyase (OAS-TL; 2.5.1.47) into the amino acid skeleton of O-acetylserine (OAS) to form Cys. OAS is synthesized by acetylation of serine with acetyl-Coenzyme A catalyzed by serine acetyltransferase (SAT; EC 2.3.1.30) (Leustek et al. 2000; Suter et al. 2000; Kopriva et al. 2001, 2009) (Fig. 1). Transcript levels of APR in soybean decrease significantly in the absence of N and increase under S-deprivation conditions. Likewise, expression of both ATPS and APR undergoes concomitant changes in both gene expressions and enzyme activity across developmental stages of soybean (Phartiyal et al. 2006, 2008). In vast majority of reports, changes in APR activity correlated well with changes in mRNA and protein accumulation, indicating a simple transcriptional regulation of the corresponding genes, with a rare occurrence of additional level of posttranscriptional redox regulation (Koprivova et al. 2008). In contrast to ATPS and APR genes, the SiR gene shows little transcriptional regulation. Nevertheless, the SiR single copy gene At5g04590 is not only essential for survival; the encoded SiR activity can become limiting for the flux through the reduction pathway (Khan et al. 2010). OAS and sulfide are the substrates for the Cys synthesis and thus are fundamental for the homeostasis of reduced S in the plant. OAS has now been considered as the best signal, if not sulfate itself, of sulfate starvation in plants. Feeding of OAS to *Arabidopsis* induced low S-responsive genes such as sulfate transporters and APR (Koprivova et al. 2000; Hesse et al. 1999, 2004; Hirai et al. 2003). Evidence that OAS functions as a putative signaling molecule has come from computational analysis of time-series experiments and on studies of transgenic plants, conditionally displaying increased OAS levels (Hubberten et al. 2012). Transcripts whose levels correlated with the transient and specific increase in OAS levels observed in leaves of *Arabidopsis thaliana* plants 5–10 min after transfer to darkness and with diurnal oscillation of the OAS content, showing a characteristic peak during the night, were identified. Induction of a SAT in transgenic *A. thaliana* plants expressing the genes under the control of an inducible promoter resulted in a specific time-dependent increase in OAS levels (Hubberten et al. 2012). Monitoring the transcriptome response at time points at which no changes in

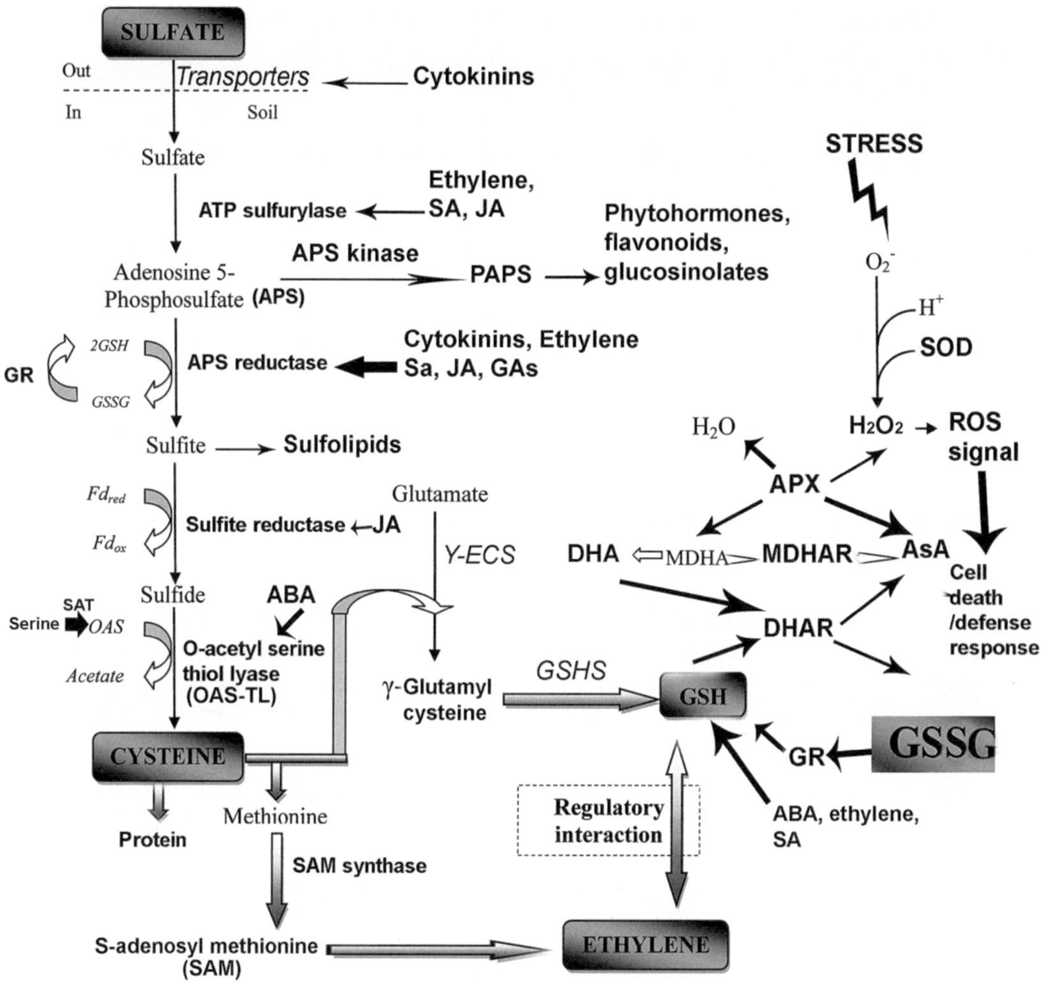

Fig. 1 Overview of thiol-cascade and integration with stress response and hormonal regulations; *JA* jasmonate, *SA* salicylic acid, *ABA* abscisic acid, *SOD* superoxide dismutase, *APX* ascorbate (AsA) peroxidase, *DHA* dehydroascorbate, *MDHAR* monodehydroascorbate reductase, *DHAR* DHA reductase, *GR* GSH reductase, *GSH* reduced glutathione, *OAS* O-acetyl serine, *PAPS* 3′-phosphate-5′-adenosine phosphosulfate, *SAT* serine acetyl transferase, *γ-ECS* γ-glutamyl cysteine synthetase, *GSHS* GSH synthase

S-related metabolites except OAS were observed and correlating this with the light/dark transition and diurnal experiments resulted in identification of six genes (adenosine-5′-phosphosulfate reductase 3, sulfur-deficiency-induced 1, sulfur-deficiency-induced 2, low-sulfur-induced 1, serine hydroxymethyltransferase 7, and ChaC-like protein) whose expression was highly correlated with that of OAS. These data suggest that OAS displays a signaling function leading to changes in transcript levels of a specific gene set irrespective of the S status of the plant (Hubberten et al. 2012). Multiple experimental approaches indicate that SAT catalyzes the limiting step in Cys biosynthesis (Droux 2003; Bonner et al. 2005). OAS-TL and SAT physically interact to form multimeric complex known as cysteine synthase (CS) complex (Hell and Wirtz 2011) or recently called as cysteine regulatory complex (CRC) (Yi et al. 2010). Five SAT and nine OAS-TL genes are found in the *Arabidopsis* genome (Kopriva et al. 2009), while the soybean genome contains

8 putative SAT and 15 putative β-cyanoalanine synthase (BSAS) (OAS-TL plus related enzymes) genes (Yi et al. 2010). Characterization of OAS-TL-encoding genes reveals their distinct but cooperative expression in Cys synthesis of soybean (Zhang et al. 2008). In *Vicia sativa*, a putative OAS-TL gene, designated as *Voas-tl5* (GenBank Accession No. DQ456491), was cloned and characterized (Novero et al. 2008). The mRNA transcription patterns generated from semiquantitative RT-PCR revealed that the *Voas-tl5* gene was highly transcribed in leaf, pod, and seed tissues. Phylogenetic analysis revealed that the gene belonged to the BSAS 5 subgroup (Novero et al. 2008). At the transcript level, BSAS isoforms show dynamic but partially overlapping expression pattern depending on organ types and developmental stages (Chronis and Krishnan 2004; Zhang et al. 2008). Presumably, overall rate of Cys synthesis at a certain time and location in soybean may be determined by interplay among BSAS isoforms expressed: some carry out Cys biosynthesis while the others are more involved for β-cyanoalanine synthesis and desulfuration using Cys as substrate. A discrepancy in OAS-TL activity and mRNA expression level of a BSAS isoform in wild soybean (*Glycine soja*) may be related to differential expression patterns of multiple isoforms and/or difference in preferred biochemical activity among BSAS isoforms (Zhang et al. 2008). Although the mRNA expression level of BSAS isoforms studied so far largely corresponds to the total OAS-TL activity during seed development, it is not yet clear exactly how many BSAS isoforms are expressed during seed development and what is the in vivo function of each enzyme expressed (Chronis and Krishnan 2004; Zhang et al. 2008). The relevance of compartmentation of Cys biosynthesis in phototrophic organisms has been studied in vascular model plant *Arabidopsis*, unicellular green alga *Chlamydomonas reinhardtii*, and the colonizer moss, *Physcomitrella patens* (Birke et al. 2012). In *Arabidopsis thaliana*, synthesis of Cys and its precursors OAS and sulfide is compartmentalized between the cytosol, chloroplasts, and mitochondria, contributing regulation of Cys synthesis. The predominant cytosolic isoforms are SAT2, SAT4, SAT5, and OAS-TL A, whereas SAT1 and OAS-TL B are targeted to the plastids and SAT3 and OAS-TL C localize to mitochondria. Mitochondrial SAT3 contributes to approximately 80 % of total SAT activity in the *Arabidopsis* leaf cell, suggesting prominent role of mitochondria for total OAS production (Birke et al. 2012; Wirtz et al. 2004, 2012). Conversely, contribution of mitochondrial OAS-TL C to total OAS-TL activity in leaves is very low (<5 %). Cytosolic and plastidic isoforms of OAS-TL in *Arabidopsis* account both for more than 45 % of total OAS-TL activity; however only loss of a cytosolic isoform leads to decrease in total Cys production demonstrating its predominant role in the synthesis of Cys (Watanabe et al. 2008a, b; Krueger et al. 2009). In contrast, Cys synthesis is exclusively restricted to chloroplasts in the unicellular green alga *Chlamydomonas reinhardtii*. The moss *Physcomitrella patens* colonizes land but is still characterized by a simple morphology compared to vascular plants. Native OAS-TL proteins demonstrated the presence of five OAS-TL protein species encoded by two genes in *Physcomitrella*. At least one of the gene products is dual targeted to plastids and cytosol, as shown by combination of GFP fusion localization studies, purification of chloroplasts, and identification of N termini from native proteins (Birke et al. 2012). The bulk of OAS-TL protein is targeted to plastids, whereas there is no evidence for a mitochondrial OAS-TL isoform and only a minor part of OAS-TL protein is localized in the cytosol. This demonstrates that subcellular diversification of Cys synthesis is already initialized in *Physcomitrella* but appears to gain relevance later during evolution of vascular plants. Multiple lines of evidence suggest a critical role for formation of the CS complex in different subcellular compartments in plants. The evidence that mitochondrial CS complex regulates OAS biosynthesis in plants has come from biochemical analyses of recombinant plant SAT and OAS-TL. This indicates that the reversible association of the proteins in the CS complex controls cellular S homeostasis (Wirtz et al. 2010, 2012). In this study, NMR spectroscopy of isolated mitochondria from wild type, *serat (SAT)2;2*, and *oastl-C*

mutant plants exhibited SAT-dependent export of OAS. The presence of Cys resulted in reduced OAS export in mitochondria of *oastl-C* mutants but not in wild-type mitochondria. This is in agreement with the stronger in vitro feedback inhibition of free SAT by Cys compared with CS complex-bound SAT and explains the high OAS export rate of wild-type mitochondria in the presence of Cys. The predominant role of mitochondrial OAS synthesis was validated in planta by feeding [^{3}H]serine to the wild-type and loss-of-function mutants for OAS-TLs in the cytosol, plastids, and mitochondria (Wirtz et al. 2012). Computational modeling coupled with transgenomic and transcriptomic analysis provided ample proof of more complex interaction of both enzymes underlying the mechanism of their reciprocal regulation (Wawrzyńska et al. 2013). At the mRNA level, SAT and OAS-TL are constitutively expressed, although expressions of some isoforms increase/decrease under nutritional and environmental stress conditions (Yamaguchi et al. 1999; Domínguez-Solís et al. 2004; Kawashima et al. 2005; Yi et al. 2010; Talukdar and Talukdar 2013d). The fact that SAT and OAS-TL expression does not compensate at either the RNA or protein level in mutants, lacking one or two isoforms, further supports that transcriptional control plays a limited role in regulating expression of Cys synthesis (Haas et al. 2008; Heeg et al. 2008; Watanabe et al. 2008a, b). Alternatively, interaction between SAT and OAS-TL, which is mediated by C-terminal tail of SAT and active site pocket of OAS-TL, appears to provide an effective regulatory mechanism that readily responds to cellular concentration of sulfide and OAS (Bonner et al. 2005; Francois et al. 2006; Kumaran et al. 2009). High concentrations of OAS and/or low levels of thiol compounds in the cell are likely signals that induce high-affinity sulfate transporter and restore the conditions favoring formation of the CS in *Arabidopsis* and potato (Hirai et al. 2003; Hopkins et al. 2005). The importance of OAS in plant growth and development has been elucidated in *Arabidopsis* point mutations impaired in cytosolic OAS-TL expressions, leading to early leaf death (old 3–1). The early leaf death phenotype is temperature dependent and is associated with increased expression of defense-response and oxidative stress marker genes. Independent of the presence of the *odd-ler* gene, *OAS-A1* is involved in maintaining S and thiol levels and is required for resistance against (Cd) stress (Shirzadian-Khorramabad et al. 2010). In addition to increasing SAT activity, CS formation can alleviate the inhibitory effect of Cys on SAT activity (Kumaran et al. 2009). Transient expression of soybean GmSerat (SAT) 2;1 fused with GFP revealed its dual targeting to cytosol and plastid (Liu et al. 2006). ATSAT 2;1 is also found in both cytosol and plastid in the later developmental stage but is exclusively targeted to the plastid in the earlier stage of *Arabidopsis* (Noji et al. 2001). Transcriptomic profiling revealed that expression of the cytosolic SAT1;1 and SAT1;2 was approximately fourfold higher in *Phaseolus vulgaris* while expression of the plastidic SAT2;1 was twofold higher in *Vigna mungo* (Liao et al. 2013). Among *BSAS* family members, BSAS4;1, encoding a cytosolic cysteine desulfhydrase, and BSAS1;1, encoding a cytosolic OAS-TL, were most highly expressed in both species. This was followed by BSAS3;1 encoding a plastidic BSAS which was more highly expressed by 10-fold in *P. vulgaris*. The data identify BSAS3;1 as a candidate enzyme for the biosynthesis of *S*-methylcysteine through the use of methanethiol as substrate instead of cyanide. Expression of GLC1 would provide a complete sequence leading to the biosynthesis of γ-Glutamyl-*S*-methylcysteine in plastids. The detection of *S*-methylhomoglutathione in *P. vulgaris* suggested that homoglutathione synthetase may accept, to some extent, γ-Glutamyl-*S-methylcysteine* as substrate, which might lead to the formation of *S*-methylated phytochelatins (Liao et al. 2013). Using unique quadruple knockout mutants of SAT that retained only one functional isoform in *Arabidopsis*, Watanabe et al. (2010) nicely compared metabolite and transcriptome data from these mutants with N-, P-, K-, and S-depleted plants. The study revealed many similarities with general nutrient-depletion-induced senescence (NuDIS), indicating the recruitment of existing regulatory programs for nutrient-starvation responses. Several candidate genes that could be involved in these processes were identified, including transcription factors and other regulatory proteins, as well as the functional categories of their target genes. These results outline components of the regulatory network controlling plant development under sulfate stress. For example, increased gene expression of Sultr1;2, Sultr1;3, Sultr4, and APR in the group 2 SAT mutant indicates that these genes, which have been assumed to be regulated by OAS or thiols, are also regulated by OAS/thiol-independent factors, as OAS and thiol contents remain unchanged in the mutants (Watanabe et al. 2010).

The transcriptional and metabolic responses of plants to S supply limitation have been investigated by several groups (Nikiforova et al. 2003, 2005a, b, 2006; Maruyama-Nakashita et al. 2003, 2004; Maruyama-Nakashita and Takahashi 2005; Hirai et al. 2003, 2005). Using array hybridization experiments, Nikiforova et al. (2003) examined transcript levels of 16 128 *Arabidopsis* EST clones representing approximately 7,200 individual genes, corresponding to about 30 % of the total *Arabidopsis* genome, and 18 genes linked to S-related pathways were sorted out. Besides Sultr 2;1 and 4;1, the other prominent members are GSH-dependent dehydroascorbate reductase (DHAR), cytosolic isoform of glutathione reductase (GR), glutathione peroxidase (GPX), SAT-1, OAS-TLC, S-adenosylmethionine synthetase 2, and γ-ECS (Nikiforova et al. 2003). Further, transcriptome data were overlaid with >100 nonredundant compounds of known chemical structure (Nikiforova et al. 2005a). Integration of these data sets allowed the first multifactorial correlation network to be created, revealing potential relationships among genes and metabolites under S limitation (Nikiforova et al. 2005b). Transgenic poplars overexpressing γ-ECS offered an opportunity to address the effects of increased GSH synthesis on the sulfate assimilation pathway because Cys availability is most critical for the rate of GSH synthesis (Noctor et al. 2012). Although leaf GSH levels were three- to fourfold higher in transgenic poplars overexpressing γ-ECS in the cytosol, foliar activities of enzymes of sulfate assimilation, ATPS, APR, SiR, SAT, and OAS-TL, and their mRNA levels were not different from those of wild-type poplars (Hartmann et al. 2004). This indicates the fact that sulfate reduction in poplar is sufficient to provide the additional Cys necessary to accommodate the enhanced GSH synthesis. Remarkably enough, the increased GSH level in transgenic poplars did not downregulate ATPS and APR as commonly observed in several herbaceous plant species and crops (Vauclare et al. 2002; Talukdar and Talukdar 2013d). The lack of regulation of APR and ATPS in the transgenic poplar lines must be caused by a second signal that positively influences APR mRNA accumulation and activity and overrides the negative signal of GSH.

Glutathione: The Center of Thiol-Cascade

In plants, the thiol-containing tripeptide GSH (in reduced form) is a major regulator of cellular redox state as well as an essential contributor to processes such as the detoxification of xenobiotics, regeneration of ascorbate in AsA–GSH cycle, thiol–disulfide exchange reactions through GRXs (glutaredoxins), protein *S*-glutathionylation or thiolation by forming stable mixed disulfide bonds with protein Cys residues, the sequestration of heavy metals, the storage of excess S in the form of Cys, weed-induced phytotoxicity, hormone signaling, as an electron donor for APR in sulfur assimilation, and numerous other metabolic and cellular processes involved in plant growth and development (Mullineaux and Rausch 2005; Meyer and Rausch 2008; Rouhier et al. 2008; Noctor et al. 2012; Talukdar 2013c, e, f). The role of GSH in cell cycle progression during promotion of root growth has earlier been demonstrated in *Arabidopsis rml 1* (*root meristem less 1*) mutant (Vernoux et al. 2000) and in an ascorbate-deficient mutant line *asfL-1* of hardy legume grass pea (Talukdar 2012a). Versatility of this thiol peptide as efficient buffer has also been demonstrated in two catalase-deficient mutants of lentil where failure of defense cross talk between thiol-linked enzymes led to onset of oxidative stress and cell-division anomalies (Talukdar and Talukdar 2013a). Besides pea and beans, grass pea and lentil have drawn increasing attention due to their low-input requirement, availability of robust cytogenetic and biochemical mutant stocks, and escalating demand for legume food and forage (Vaz Patto et al. 2006; Kumar et al. 2011a, b; Talukdar et al. 2002; Talukdar and Biswas 2005, 2007a, b; Talukdar 2008, 2009a, b, 2010a, b, c, 2011a, d, 2012b, c, d, 2013g). Very recently, the role of GSH redox in leaf photosynthesis and mitigating oxidative metabolisms in different mating types of a critically endangered legume tree, *Gymnocladus assamicus*, has been elucidated (Talukdar and Talukdar 2014). Legumes are the only plant family with significant amounts of a GSH analog homoglutathione, which contains a β-alanine instead of glycine. Nodules are organs with the highest GSH and/or hGSH content in legumes because of their role in defense of the nitrogenase against reactive oxygen species (ROS) (Matamoros et al. 2003). However, GSH and hGSH are important also for establishing the symbiosis. Our understanding of the synthesis and function of GSH is primarily based on the Brassicaceae, in particular *Arabidopsis* and Indian Mustard (*Brassica juncea*), although additional studies completed in legumes, such as soybean, grass pea, and lentil, expand the role of this peptide to species-specific analogs. GSH synthesis requires the activities of two dedicated ATP-dependent enzymes: γ-glutamyl cysteine synthase (γ-ECS) and glutathione synthetase (GS). The first enzyme, γ-ECS, utilizes L-glutamate and L-Cys to generate γ-glutamyl cysteine and is a rate-limiting enzyme (Hell and Bergmann 1990; Jez et al. 2004). From this dipeptide and glycine, GSHS, the second enzyme in the pathway, then synthesizes GSH (Jez et al. 2004; Noctor et al. 2012). Transcript analysis and activity assays in common bean (*Phaseolus vulgaris*) nodules revealed localization of γ-ECS to only chloroplasts and GSHS to both plastids and the cytoplasm as also occurs in *Arabidopsis* (Wachter et al. 2005) but differs with cowpea (*Vigna unguiculata*) nodules where γ-ECS is localized to both plastids and cytosol (Moran et al. 2000). In Brassicaceae, differential targeting of GSH1 and GSH2 is achieved by multiple transcription initiations in different cellular compartments (Wachter et al. 2005).

GSH-redox pool in cellular environment is maintained by delicate balance between its production and catabolism. GSH supply in cell is generally maintained by its synthesis and regeneration by the FAD-linked action of GR enzymes within AsA–GSH cycle (Noctor et al. 2012; Talukdar 2012a). In a magnificent study, Mhamdi et al. (2010) addressed the role of GR1 isoforms in H_2O_2 responses through a combined genetic, transcriptomic, and redox profiling approach. To identify the potential role of changes in GSH status in H_2O_2 signaling, gr1 mutants, which show a constitutive increase in oxidized glutathione (GSSG), were compared with a catalase-deficient background (cat2), in which GSSG accumulation is conditionally driven by H_2O_2. Parallel transcriptomics analysis of gr1 and cat2 identified overlapping gene expression profiles that in both lines were dependent on growth day length. Overlapping genes included phytohormone-associated genes, in particular implicating GSH oxidation state in the regulation of jasmonic acid (JA) signaling. GSH synthesis is regulated by the supply of the constituent amino acids and by feedback inhibition of γ-ECS by GSH (Noctor et al. 2012). To address this regulation in more detail, a poplar hybrid *Populus tremula* X *P. alba* (INRA clone no. 717-1-B4, Versailles, France) was transformed to express bacterial γ-ECS or GSHS either in the cytosol or in the chloroplast (Noctor et al. 2012). Overexpression of γ-ECS, but not of GSHS, increased foliar and root GSH concentration (Noctor et al. 2012), thus confirming the major role of γ-ECS in the control of GSH synthesis. Experiments with poplar leaf disks revealed that feeding of γ-EC dramatically enhanced GSH synthesis compared with feeding of Cys and Glu, this effect being more profound in the GSHS-overexpressing plants, whereas Cys was more effective in the γ-ECS-overexpressing poplar (Noctor et al. 2012). Interestingly, overexpression of γ-ECS either in the cytosol or in the chloroplast did not decrease Cys and Met concentrations (Herschbach et al. 2010). Presumably, sulfate reduction and Cys formation are adjusted to the higher demand for GSH synthesis in γ-ECS transgenic trees, the detail mechanism of which is still unknown. Changes in the redox state of γ-ECS provide a posttranslational mechanism for regulation of activity (Jez et al. 2004; Hicks et al. 2007; Noctor et al. 2012). The soybean genome contains two full-length copies of the GSHS gene that are 91 % identical. Localization tag analysis suggests that both of these transcripts are targeted multiple locations (Yi et al. 2010). Several studies describe transcriptional regulation of the pathway and report increased expression of the genes encoding γ-ECS and GSHS under different stress conditions (Xiang and Oliver 1998). Interestingly, in *Arabidopsis* suspension cells, transcriptional upregulation of γ-ECS in response to various oxidative stresses was not observed, even though both GSHS activity and cellular GSH levels increased and new concepts on post-transcriptional regulations and diversity in synthesis pathway have emerged (Galant et al. 2011). GSH is efficient in repressing sulfate uptake after a period of sulfate starvation but also during normal S nutrition (Lappartient et al. 1999; Vauclare et al. 2002). S is important for chelating heavy metals through metallothioneins, i.e., Cys-rich proteins and phytochelatins (PCs), small polypeptides with repeating γ-EC units. PCs are synthesized from GSH by phytochelatin synthase (PCS). The sulfhydryl groups of Cys residues bind the heavy metal ions, and the resulting complexes are excreted to the vacuole. GSH is also important for resistance against herbicides. Differences in herbicide toxicity are often based on the capacity of the plants to detoxify the herbicide, e.g., through the glutathione S-transferase (GST) reaction and subsequent excretion of the conjugate into the vacuole (Edwards and Dixon 2005). In plants, GSTs are encoded by a large gene family with approximately 50 members in *Arabidopsis* and rice (Edwards and Dixon 2005), highlighting the importance of GSH conjugate formation for the metabolism of endogenous compounds and the detoxification of noxious compounds such as herbicides. GSH conjugates are predominantly generated in the cytosol, with minor GST activities in the nucleus, chloroplast, and mitochondrion (Dixon and Edwards 2009). Glutathionylation of compounds is an important reaction in the detoxification of electrophilic xenobiotics and in the biosynthesis of endogenous molecules. The GSH conjugates are further processed by peptidic cleavage reactions. In animals and plants, γ-glutamyl transpeptidases initiate the turnover by removal of the glutamate residue from the conjugate. Plants have a second route leading to the formation of γ-glutamylcysteinyl (γ-GluCys) conjugates. PCS is well known to mediate the synthesis of heavy metal-binding PCs. In addition, the enzyme is also able to catabolize GSH conjugates to the γ-GluCys derivative. Cellular compartmentalization of PCS and its role in the plant-specific γ-GluCys conjugate pathway have been studied in *Arabidopsis thaliana* (Blum et al. 2010; Noctor et al. 2012). Localization studies of both *Arabidopsis* PCS revealed a ubiquitous presence of AtPCS1 in *Arabidopsis* seedlings, while AtPCS2 was only detected in the root tip. A functional AtPCS1:eGFP (enhanced green fluorescent protein) fusion protein was localized to the cytosolic compartment (Blum et al. 2010).

Defining Cross Talk with Plant Metabolisms: Hormonal Response, Photosynthesis, Carbohydrate and Lipid Metabolisms, Kinase Cascades, and Other Metabolisms

Thiol-Cascade and Plant Hormone Metabolisms

Recent developments indicate that plant hormones play pivotal roles in regulating S-metabolisms (Ohkama et al. 2002; Maruyama-Nakashita et al. 2004, 2006), a brief outline of which is presented in Table 1 and Fig. 1. Whereas abscisic acid (ABA), indole-3-acetic acid (IAA), 1-aminocyclopropane 1-carboxylic acid (ACC, precursor of ethylene), gibberellic acid (GA_3), and jasmonic acid (JA) were not able to induce expression of GFP derived from the S-responsive element and, thus, mimic the S starvation response, trans-zeatin caused an increase in GFP synthesis both in S-sufficient and in S-deficient

Table 1 Phytohormone-mediated regulations of thiol-cascade gene expressions in *Arabidopsis*

Hormone	Gene(s) expressed	Functions in thiol-cascade	References
Auxins	*SULTR1;2,1;4* *SULTR4;1*	Sulfur (S) uptake, transport through root, and vacuole overexpressed	Nikiforova et al. (2003)
	At2g44460 (thioglucosidase gene), *APR2*	Repressed expressions, impeded S-reductive assimilation	Nikiforova et al. (2003)
Gibberellins	*APR1, APR2*	Differential regulations	Koprivova et al. (2008)
Cytokinins	*APR1, SULTR2;2*	Induced during vascular transport of S and during S-assimilation through an increase in sucrose concentrations	Ohkama et al. (2002)
	SULTR1; 1, SULTR1; 2	Repressed expressions in roots through involvement of CRE/WOL/AHK4-mediated signal	Maruyama-Nakashita et al. (2003)
Abscisic acid (ABA)	*AtGSTUI7, OsMCSU*	Repressed GSH pool in *Arabidopsis*. The molybdenum (Mo) cofactor sulfurase (MCSU) in rice transfers the sulfur ligand to aldehyde oxidase-bound MoCo	Maruyama-Nakashita et al. (2003)
Ethylene	*APR1, APR3*	Overexpressed	Koprivova et al. (2008)
Jasmonate	*APR 1, APR 2, SIR, SAT3, γ-ECS, APS2, APR3, ATPS*	Methyl jasmonate upregulates the expressions	Jost et al. (2000)
Salicylic acid	*APR1, APR2, APR3, γ-ECS, GSHS*	Overexpressed	Kopriva (2006)

conditions. In addition, zeatin treatment resulted in an increased accumulation of mRNA for APR and a low-affinity sulfate transporter (Ohkama et al. 2002). By contrast, cytokinins repress the expression of high-affinity sulfate transporters and sulfate uptake capacity of *Arabidopsis* roots, since feeding with zeatin downregulated transcript levels of AtSultr 1;1 and AtSultr 1;2 and also root sulfate acquisition (Maruyama-Nakashita et al. 2004). Cytokinin acts through the cytokinin response receptor (CRE1) to regulate sulfate uptake and transporter expression. In the *cre1-1* mutant, application of cytokinin only partly reduces sulfate uptake, suggesting redundancy as noted for the case of phosphate deprivation. The effect of zeatin is dependent on the CRE1/WOL/AHK4 cytokinin receptor (Maruyama-Nakashita et al. 2004).

The physiological responses to auxin involve changes in gene expressions. A NIT3 nitrilase, involved in synthesis of IAA, belongs to genes strongly induced by S deficiency (Kutz et al. 2002). In addition, the cis-acting element conferring S starvation response recently identified in *Arabidopsis* Sultr1;2 promoter contains an auxin response factor (ARF) binding sequence (Maruyama-Nakashita et al. 2006). Auxin regulates the cell-specific transcription of target genes via two types of transcription factors, ARFs and Aux/IAA proteins. In *Arabidopsis thaliana*, the Aux/IAA and ARF gene families are represented by 29 and 23 loci, respectively. Several transcription factors of both families, including IAA13, IAA28, and ARF2, are moderately upregulated during S depletion (Nikiforova et al. 2005b, 2006). IAA28 exhibits a high degree of connectivity and so was identified as a hub of the transcript/metabolite co-response network responding to S starvation (Nikiforova et al. 2005b). A gain-of-function *Arabidopsis* mutant of IAA28 has been reported to exhibit suppressed lateral root formation (Rogg et al. 2001) by repressing transcription, perhaps of genes that promote lateral root initiation in response to auxin signals. ARF2 has been identified as a transcription factor binding to AuxRE in promoters of auxin response genes, thus activating flowering, senescence, and abscission. It also functions as a light-independent repressor of cell growth and of differential hypocotyl growth during seedling hook formation (Li et al. 2004a, b; Ellis et al. 2005; Okushima et al. 2005a, b). ARF2, a pleiotropic developmental regulator, knockout mutants develop a phenotype with increased leaf size, enhanced flower formation, and increased seed size (Ellis et al. 2005). Overexpression of ARF2 is likely to be lethal as it was impossible to retrieve ARF2 overexpressing lines, while co-suppression lines could be isolated. Alterations were observed in the contents of thiol-metabolites in response to the manipulation of Aux/IAA and ARF transcription factor expression, especially of the key thiol-metabolites Cys and GSH (Falkenberg et al. 2008). These changes were interpreted as broad spectrum shifts in amino acid metabolism and carbon/nitrogen balance caused by changes in the expression of AUX/IAA and ARF target genes, rather than a direct effect of auxin on sulfate metabolism (Falkenberg et al. 2008). Combined transcriptomic data and metabolic profiling indicated, among other potential links, a relationship between serine metabolism and tryptophan (trp)–glucosinolate–auxin metabolism. Based on these data, it was hypothesized that auxin might be involved in communicating the nutrient status of the shoot to the root, inducing lateral root formation when S is limiting (Nikiforova et al. 2003, 2005b).

JA did not affect the expression of the S-responsive promotor element (Ohkama et al. 2002) but is nevertheless involved in regulation of sulfate assimilation. Transcriptomic analysis revealed fast but transient increase in mRNA levels of many genes involved in sulfate assimilation and GSH synthesis without affecting sulfur metabolite levels in *Arabidopsis* treated with methyl jasmonate (Jost et al. 2000; Harada et al. 2001). The fact that the mRNA for sulfate transporters was unaffected confirmed that JA may not be involved in the regulation by sulfur nutrition although genes of JA biosynthesis are induced by S starvation (Hirai et al. 2003, 2004; Maruyama-Nakashita et al. 2003; Nikiforova et al. 2003). The increase of level of GSH in plants treated with ABA and salicylic acid (SA) indicates a complex interaction of sulfate assimilation, GSH synthesis, and hormone metabolisms. ABA induces mRNA accumulation of cytosolic OAS-TL (Barroso et al. 1999). Thus, it seems that this compound may have a more profound effect on the control of S-metabolism. On the other hand, SA plays a central role in plant defense against pathogens. Treatment with the biologically active SA analog 2,6-dichloroisonicotinic acid increased the GSH level leading to a reduction of NPR1, a regulator of systemic acquired resistance (SAR), and expression of the *PR1* gene for a pathogenesis-related protein (Mou et al. 2003). SA was also implicated in the mechanism of nickel tolerance in hyperaccumulator *Thlaspi* species. Elevated SA levels engineered in *Arabidopsis* led to an increase in SAT activity and GSH content and, subsequently, increased tolerance to Ni (Noctor et al. 2012). Whether SA regulates the expression of γ-ECS and GS, if and how it affects SAT, or if it utilizes another mechanism to increase GSH synthesis remains to be elucidated.

The interaction between ethylene and S has been shown to control the regulation of plant processes and abiotic stress tolerance. The main pathway for ethylene biosynthesis comes from Met. Met is a fundamental metabolite in plant cells because it controls the level of several key metabolites, such as ethylene, polyamines, and biotin, through its first metabolite, s-adenosylmethionine (SAM). It is first converted to SAM, then ACC (1-amino cyclopropane 1-carboxylic acid), and finally ethylene in three consecutive reactions catalyzed by the enzymes of SAM synthetase, ACS, and ACO, respectively (Iqbal et al. 2013). Bürstenbinder et al. (2007) using an *mtk* mutant, that has a disruption of the Yang cycle, reported that the Yang cycle contributes to SAM homeostasis, especially when de novo SAM synthesis is limited, such as at S starvation. S availability and ethylene have been shown to regulate GSH synthesis and stress tolerance to ozone (Yoshida et al. 2009) and Cd stress (Masood et al. 2012). Ethylene plays important roles in selenite resistance in *Arabidopsis*. A comprehensive gene expression analysis showed that transcripts regulating ethylene synthesis (ACS6) and signaling (ERF) were upregulated by selenate treatment, and plants overexpressing ERF1 exhibited an increase in selenium (Se) resistance (Van Hoewyk et al. 2008). These results indicate that Se resistance achieved through ethylene signaling is not mediated by S starvation resulting from the Se treatment but is a Se-specific response. The resistance mechanism may involve ethylene-enhanced S uptake and assimilation, as observed in *Arabidopsis thaliana* accessions, Columbia (Col)-0 (Iqbal et al. 2013). Koprivova et al. (2008) reported that the application of 0.2 mM ACC, which stimulates ethylene production, increased accumulation of APR activity. Recently, it has been shown that ethylene action in mustard is dependent on S availability (Masood et al. 2012). There is an indication that APR activity is increased in salt stress if ethylene signaling is disturbed, but GSH will not accumulate suggesting that components of GSH biosynthesis are under the control of ethylene (Koprivova et al. 2008).

The other level of interaction of nutrient and hormone may be visualized at the level of ROS production. The regulation and interaction between ROS and AsA–GSH cycle impact the synthesis of plant hormones such as SA, GA, ABA, and ethylene, which may signal plant response to nutrient deficiency. The involvement of ROS in S signaling may be more complex than that of K deprivation because the AsA–GSH cycle, i.e., downstream of sulfate assimilation, is involved in the removal of H_2O_2. Direct analysis of H_2O_2–GSH interactions in cat2 gr1 double mutants of *Arabidopsis* established that GR1-dependent GSH status is required for multiple responses to increased H_2O_2 availability, including limitation of lesion formation, accumulation of SA, induction of pathogenesis-related genes, and signaling through JA pathways (Mhamdi et al. 2010).

Thiol-Cascade and Photosynthesis

Photosynthesis is known to be sensitive to sulfate because sulfate is a competitive inhibitor of ribulose-1,5-biphosphate carboxylase and inhibits photophosphorylation (Dietz and Pfannschmidt 2011). Sulfate uptake is well coordinated with the uptake and assimilation of carbon. Redox proteomics is an emerging technology aimed at defining the redox protein inventory of the cells and cell compartments and analyzing the redox state of target proteins on a broad scale. Both gel- and chromatography-based redox protein screening systems have been applied to plant and chloroplast protein fractions and resulted in lists of thylakoid lumenal, stromal, and chloroplast membrane-bound candidate redox proteins that undergo thiol modifications, most commonly dithiol–disulfide transitions (Mullineaux and Rausch 2005; Rouhier et al. 2008). Novel regulators including components of thiol-cascade such as Cys, GSH, and thioredoxins in photosynthetic redox control of plant metabolism (starch biosynthesis, lipid synthesis) and gene expressions have recently been explored (Dietz and Pfannschmidt 2011).

Thiol-Cascade and Anthocyanin Biosynthesis

Like N- and P-nutrient starvation, S starvation is tightly linked to anthocyanin biosynthesis (Lillo et al. 2008). The transcription factors PAP1 (At1g56650) and PAP2 (At1g66390) are positive regulators of the anthocyanin pathway (Borevitz et al. 2000) and are upregulated in the dwarfed quadruple mutants of *Arabidopsis* (Watanabe et al. 2010). Additionally, several known transcription factors downstream of the PAP genes, TT8 (At4g09820), TTG2 (At2g37260), and EGL3 (At1g63650), are induced along with target genes within the anthocyanin biosynthetic pathway, such as anthranilate synthase (ANS; At4g22880), dihydroflavonol reductase (DFR; At5g42800), and various others. Anthocyanin accumulation is usually augmented to various stress responses and is consistent with the upregulation of ROS network genes in the dwarfed quadruple mutants in *Arabidopsis* (Watanabe et al. 2010), as has been shown, for example, in the old1/cpr5 mutant (Jing et al. 2008) and the old5 mutant (Schippers et al. 2008), which are in a state of high-cellular oxidative stress and show early senescence phenotypes.

Thiol-Cascade and Nodulation

The importance of S for nodule function was evidenced by analysis of *Lotus* sym mutants, showing nonfunctional nodules (Krusell et al. 2005). Both the sym13 and sym81 mutations, which display N deficiency syndromes under symbiotic but not nonsymbiotic growth conditions and form smaller nodules with reduced nitrogenase content and N-fixing capacity, exhibit a defective SST1 sulfate transporter. SST1 is a group 3 sulfate transporter expressed in a nodule-specific manner and located in the symbiosome membrane (Krusell et al. 2005). However, the strong sym phenotype of the sst1 mutants is surprising as at least one additional sulfate transporter is expressed specifically in *Lotus* nodules (Krusell et al. 2005). In addition, the reduction of total S content of 20–25 % in the mutants compared with wild-type nodules can hardly explain the severe disruption in N-fixation as plants can reduce S content by up to 70 % without phenotypic changes (Nikiforova et al. 2003). The mechanism by which the loss of SST1 aborts N-fixation thus needs to be addressed in more detail. GSH and hGSH play a critical role in the nodulation process of *Medicago truncatula* (Frendo et al. 2005).

Thiol-Cascade and Plant C_4 Metabolisms

In a significant development, transcriptomic analysis revealed that in plants like *Zea mays* with C_4 metabolisms, the mRNAs for APR, ATPS, and SiR accumulated in bundle sheath cells only, whereas OAS-TL transcript was detected in both mesophyll cells (MCs) and bundle sheath cells (BSCs) (Kopriva et al. 2001, 2009). A coordinate increase in mRNA levels for sulfate transporters ATPS and APR was observed in maize roots and leaves upon sulfate starvation (Bolchi et al. 1999; Hopkins et al. 2005), and the ATPS mRNA level was repressed in the presence of reduced S compounds (Bolchi et al. 1999). Not only sulfate assimilation but also the synthesis and reduction of GSH seem to be differently localized in C_4 plants. However, GSH is not equally distributed between MCs and BSCs in maize. GSHS activity is greater in MCs than in BSCs, resulting in GSH synthesis predominantly in the MCs and higher GSH levels in this cell type (Kopriva et al. 2001). Cys is transported from BSC protoplast as reduced S to the MCs. The enzymes of GSH synthesis and corresponding mRNAs were, however, found to be localized in both MCs and BSCs (Gómez et al. 2004). Both enzymes were detected in chloroplasts and in the cytosol (Gómez et al. 2004) which is in sharp contrasts with the Brassicaceae, where γ-ECS is localized in plastids and GSHS is prevalently cytosolic (Wachter et al. 2005). Interestingly, unlike the usual observations that GSH exerts the feedback repression of sulfate assimilation as found in members of Brassicaceae (Vauclare et al. 2002) and in lentil of Fabaceae (Talukdar and Talukdar 2013d), in maize Cys acts directly without conversion to GSH probably due to BSC localization of sulfate assimilation in maize (Kopriva 2006). Presumably due to the low capacity for NADPH formation in BSCs, GR is found exclusively in MCs of maize (Kopriva et al. 2001). Thus, it seems likely that there are species-specific differences in the intercellular localization of GSH biosynthetic enzymes, which are dependent on increased capacity for transport of various thiol compounds and possibly result in different regulatory mechanisms for S-assimilation in C_4 plants.

Thiol-Cascade and Plant Kinase Cascade

A screen for soybean expressed sequence tags (EST) showing sequence similarity to known SATs identified SSAT1 (GLYMA16G03080), and an interaction screen isolated GmSerat (SAT)2;1 (GLYMA18G08910) as a substrate for a calcium-dependent protein kinase (CDPK) (Chronis and Krishnan 2004; Liu et al. 2006).

Both characterized soybean SAT isoforms are sensitive to feedback inhibition by Cys but to varying degrees (Chronis and Krishnan 2004; Liu et al. 2006). Nonetheless, GmSAT2;1 lacking the N-terminal localization sequence becomes insensitive to Cys when it is phosphorylated by CDPK at a site close to the C-terminus by CDPK (Liu et al. 2006). The finding that full-length GmSAT2;1 does not display similar phosphorylation-dependent sensitivity to Cys suggests that a combination of subcellular localization and phosphorylation determines the effect of feedback inhibition. In the case of SSAT1, which does not have a putative CDPK-phosphorylation site at its C-terminus, CS formation with OAS-TL provides a similar protection to SSAT1 against Cys (Kumaran et al. 2009). Whereas none of five *Arabidopsis* SAT isoforms contain a putative CDPK-phosphorylation site, GmSAT2;1 and four other SAT isoforms in soybean carry potential CDPK-dependent phosphorylation sites (B-X-X-S/T: where B is a basic residue lysine or arginine, X is any residue, and S/T is serine or threonine) near the C-terminus (Liu et al. 2006). Considering the fact that putative CDPK-phosphorylation sites are also found near the C-terminus in the SAT from other plants, including tobacco, sunflower, and poplar (Liu et al. 2006), it needs to be determined whether C-terminal phosphorylation of SAT affects CRC formation and whether this posttranslational modification is more widely used to modulate feedback inhibition by Cys. Integration of different signaling factors such as mitogen-activated protein kinase (MAPK), MAPK kinase kinase, and CDPK was upregulated in rice under As stress (Huang et al. 2012). Moreover, As(V) markedly increased the activity of MAPKs and CDPK-like kinases, and CDPK and NADPH oxidases were involved in As-induced MAPK activation (Huang et al. 2012).

Thiol-Cascade and Other Plant Metabolisms

Sulfate uptake is well coordinated with the uptake and assimilation of N. During N limitation, sulfate uptake is strongly reduced (Koprivova et al. 2000). This reduction of uptake corresponds to significantly lower accumulation of transcripts for AtSultr1;1 and AtSultr1;2 (Maruyama-Nakashita et al. 2004). In addition, the genes for HASulT are induced by sucrose (Maruyama-Nakashita et al. 2004). Proteomic analysis in *Arabidopsis* Sultr 4;1 mutant seed revealed strong cross talk between plant thiol-metabolisms and lipid and sterol metabolism, as well as sugar and polysaccharide metabolism. Lipid synthesis that occurs in the plastids is a strong sink for electrons. Plastid redox state affects lipid metabolism. Acetyl-CoA carboxylase (ACCase) catalyzes the committed step of malonyl-CoA production in plastid lipid synthesis. Isolated ACCase in vitro is inactive without reductant and activated after addition of DTT or reduced thioredoxins (TRXs). The chloroplast ACCase consists of four polypeptides, in which the Cys residue constitutes the important part. Biotin carboxyl carrier subunit of ACCase in *Chlamydomonas reinhardtii* is subjected to *S*-thiolation with GSH (Dietz and Pfannschmidt 2011). Biotin carboxylase is a target of glutathionylation in *Arabidopsis* cell culture. Thus, each of the subunits of ACCase is potentially controlled by redox regulation using diverse mechanisms. This fact underlines the link between thiol-linked redox state and lipid metabolism. Furthermore, it is now known that envelope-bound monogalactosyldiacylglycerol synthase (MGD) is a major lipid component of chloroplasts. Plant MGD possesses nine conserved Cys residues. Its regulation by thiol redox state is suggested to enable galactolipid synthesis along with photosynthetic activity and to foster replacement of eventually oxidized lipids under conditions that cause oxidative stress (Dietz and Pfannschmidt 2011). Several spots corresponding to enzymes involved in polysaccharide catabolism, a β-glucosidase and a β-galactosidase, increased in mutant seeds, probably as a way to sustain glycolysis and fatty-acid biosynthesis. Two proteins of amino acid metabolism were also overaccumulated in the Sultr4;1 mutant seeds that correspond to glutamine synthase and Met synthase, although level of storage proteins was not affected (Zuber et al. 2010). OAS-TL in its unbound form catalyzes the synthesis of Cys from OAS and sulfide and becomes inactivated when bound to the complex. The CS thus constitutes a branch point where reduced S gets incorporated into a carbon backbone. At least for N and P starvation, an induction of S uptake and assimilation has been documented (Kopriva et al. 2009, 2012). Limitation of both nutrients led to an increase in APR1, APR3, Sultr4;1, Sultr1;3, and different SATs. Further confirmation regarding importance of S-metabolism pathway in controlling many other primary and secondary metabolisms in plants has come from two independent *Arabidopsis thaliana* T-DNA insertion lines deficient in SiR transcripts and enzyme activity; sir1-2 seedlings had 14 % SiR transcript levels compared with the wild type and were early seedling lethal. sir1-1 seedlings had 44 % SiR transcript levels and were viable but strongly retarded in growth (Khan et al. 2010). This unique study pointed out that disruption in SiR expression resulted in concomitant decrease in ATPS4, APR2, Sultr 2, and pathogen defense genes and increase in vegetative storage proteins (VSP1 and VSP2), NITRILASE1 (NIT1) and NIT2, and chlorophyll-degrading gene CHLOROPHYLLASE1, and steady-state levels of most of the S-related metabolites, as well as the expression of many primary metabolism genes, were changed in leaves of sir1-1. Hexose and starch contents were decreased, while free amino acids increased. Inorganic carbon, N, and S composition were also severely altered, demonstrating strong perturbations in metabolism that differed markedly from known sulfate deficiency responses (Khan et al. 2010).

Partitioning of Thiol-Cascade

In a competing pathway, APS-kinase (APSK) phosphorylates APS at the 3′-ribose position to yield 3′-phosphate-5′-adenosine phosphosulfate (PAPS) (Lee and Leustek 1998) as APS + ATP ↔ PAPS + ADP. PAPS provides an S-donor for various sulfotransferases involved in the synthesis of plant hormones, sulfolipids, flavonoids, and glucosinolates. Functional analysis reveals multiple APSK isoforms in *Arabidopsis* (Leustek et al. 2000; Mugford et al. 2009). Interestingly, T-DNA insertional knockout mutants of APSK-1 and APSK-2 in *Arabidopsis* resulted in a dwarfed phenotype and a 450 % increase in Cys content but decrease in secondary metabolites (Mugford et al. 2009). This suggests that partitioning of sulfate flux between the reductive assimilatory and APS phosphorylation pathways is important for growth and development.

Thiol-Metabolisms and Plant Stress Response

Plant homoeostasis is a result of complex regulatory processes that keep plant metabolic events and physiological responses in balance, despite regular, temporary, or localized changes in inputs such as light, nutrients, and water, or other environmental conditions as well as biotic stresses (Tripathi et al. 2012a, b). The system is generally thought to be able to buffer imbalances to a certain extent but may then have to shift to new homoeostatic states when severe deficiencies in nutrient supply or limitations in growth conditions occur. S-metabolism into thiol-containing compounds is critical for protecting plants from oxidative and environmental stresses (Zagorchev et al. 2013). CS overexpression in tobacco confers tolerance to S-containing environmental pollutants (Noji et al. 2001). Although stress signaling pathways are ascertained by transcriptomic analysis and promoter–reporter constructs, recently it has become apparent that posttranscriptional and posttranslational control plays an important role as well. This has been demonstrated, as, for example, in the overexpression of SOS1 under control of constitutive promoter which did not lead to accumulation of the mRNA in control plants but only under salt stress. Genes involved in proline biosynthesis are either under control of miRNA or undergo posttranscriptional regulation. Phosphorylation and ubiquitination are well-recognized processes in salt stress signaling (Kopriva et al. 2012). Therefore, the responses of genes to treatments may not always be correlated with the responses of the encoding proteins and enzyme activities. This has been demonstrated in reports on individual genes and pathways, as well as on a much larger scale using omics technologies. Proteomics analysis of *Arabidopsis* roots treated with NaCl detected 215 differently abundant spots which resulted in the identification of 86 proteins. Among these, many known stress-related proteins were present, as well as proteins involved in metabolism, protein synthesis, and signal transduction but poorly correlated with their corresponding mRNA data. Similarly, proteome data generated upon infection with bacterial pathogen and circadian changes of 23 enzyme activities did not match greatly with corresponding variations in transcriptomics data from the same material, confirming a large contribution of posttranscriptional and posttranslational regulation (Srivastava et al. 2009). Nevertheless, transcript analysis and especially microarrays using rich repositories are still the most utilizable methods of choice for dissecting plant stress signaling.

Regulations of Thiol-Metabolisms During Salt and Drought Stresses

The S-containing group called thiol is strongly nucleophilic. This is preferably suitable for biological redox reactions and plays an important role in protection against salt and water stress-induced oxidative damage. Substantial level of study indicates that salt tolerance in plants has been associated with their capability to synthesize GSH and indicates a potential role of the S nutrition (Anjum et al. 2012; Talukdar 2011b, c, e, 2012g, 2013j, l, m; Astolfi and Zuchi 2013). Salt stress increases the activities of SAT and OAS-TL leading to higher rate of Cys biosynthesis, which results in increased accumulation of GSH for defense responses to salt stress (Astolfi and Zuchi 2013). Ruiz and Blumwald (2002) reported that S-assimilation rate and biosynthesis of Cys and GSH were greatly increased in *Brassica napus* plant exposed to saline conditions. Furthermore, changes in S-assimilation enzymes have been reported due to salt stress in *Arabidopsis* and broccoli (Lopez-Berenguera et al. 2007), respectively. Overexpression of sulfate transporters, ATP sulfurylase, Cys, OAS, and GSH resulted in increased resistance to oxidative stress (Astolfi and Zuchi 2013). The control of S partitioning and manipulating the synthesis of S-containing compounds in plants using genetic engineering may be a potential option for increasing salt tolerance. In roots of wild-type (WT) *Arabidopsis*, APR activity, protein accumulation, and mRNA levels were increased threefold after 5 h of exposure to 150 mM NaCl. Analysis of various mutants in hormone signaling revealed that the regulation was ABA insensitive; however, the response of APR activity was uncoupled from the mRNA response. However, treatment with EGTA to disrupt Ca^{2+} signaling prevented the increase in both mRNA and enzyme activity upon the salt treatment. In most of the mutants, the APR activity was not increased upon the salt treatment or even decreased, despite the increased mRNA accumulation. Assimilatory sulfate reduction may induce salt tolerance by coordinating various physiological processes and molecular mechanisms which are likely to be induced by phytohormones (Fatma et al. 2013). The involvement of phytohormones in S signaling and salt stress is a complex phenomenon as phytohormones may affect S availability and control gene expression related to S-metabolism (reviewed by Fatma et al. 2013). In a unique study, drought and salt stress tolerance of an *Arabidopsis* glutathione *S*-transferase U17 knockout mutant are attributed to the combined effect of GSH and ABA. GSH contents increased in response to NaCl stress in leaves but not in roots, the primary site of salt exposure, in gray poplar hybrid (Herschbach et al. 2010). The increasing leaf GSH concentrations correlated with stress-induced decreases in transpiration and net CO_2 assimilation rates at light saturation. Enhanced rates of photorespiration could also be involved in preventing ROS formation in chloroplasts and, thus, in protecting PS II from damage. Accumulation of glycine and serine in leaves suggested increasing rates of photorespiration. Since serine and glycine are both immediate precursors of GSH that can limit GSH synthesis, and

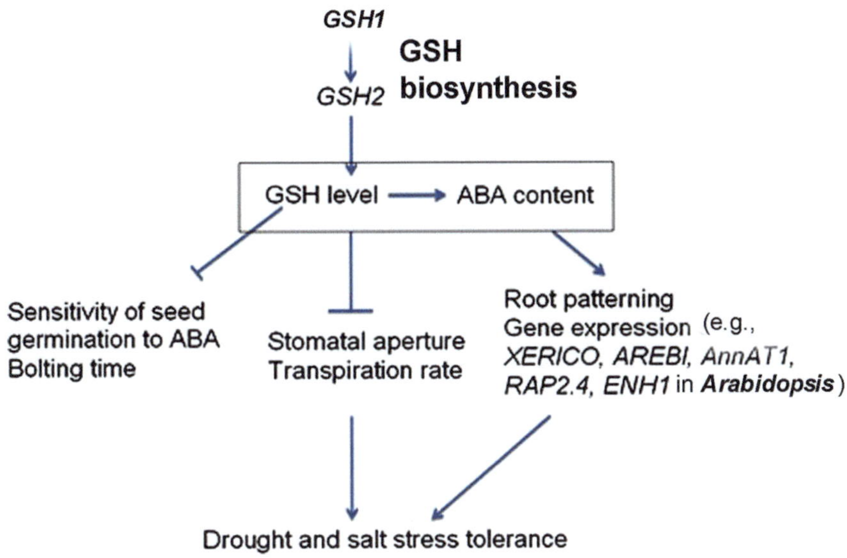

Fig. 2 A simple model for the involvement of GSH, GST, hormonal regulations, and plant phenotypes during modulation of drought and salt stress tolerance in plants

serine being involved in Cys biosynthesis, it is concluded that the salt-induced accumulation of leaf thiol-metabolites such as GSH results from enhanced photorespiration and is thus probably restricted to the cytosol (Herschbach et al. 2010). Similar roles of GSH in modulating plant growth and development have been explored in grass pea, a hardy legume, and in lentil genotypes under water stress (Talukdar 2013l). A model of involvement of GSH in drought and salt stress tolerant has been presented in Fig. 2.

Thiol-Metabolisms and Heavy Metal Stress

The pivotal roles played by thiol-cascade in conferring plants to heavy metal stress were evidenced much earlier by works of Harada et al. (2001) who engineered tobacco plants expressing a rice CS gene to make it tolerant to toxic levels of Cd, a well-known toxic heavy metal. In a significant work, Hossain and Komatsu (2012) used high-throughput comparative proteomic approaches to dissect heavy metal stress responses in soybeans and observed alteration of thiol redox homeostasis. Coordinated expression of GSH metabolic genes has been observed in *Arabidopsis* in response to heavy metal and JA (Xiang and Oliver 1998). Prominent roles of GSH-redox state, its regeneration by GR in AsA–GSH cycle, and modulation of GSH-mediated antioxidant defense have been demonstrated in grass pea mutants overproducing GSH as well as in GSH-deficient condition under Cd stress (Talukdar 2012c, d). In lentil genotypes differing in Cd tolerance, exogenous Ca significantly ameliorated Cd-induced oxidative stress by modulating antioxidant defense in which GSH, GSSG, and GSH redox played essential roles (Talukdar 2012e). These studies clearly pointed out cascading roles of GSH and GSH-derived PCs in mitigating ROS-mediated oxidative stress through induction of the entire AsA–GSH cycle components and their fine-tuned integration with defense components outside this cycle (Talukdar 2012c, d, e). In a unique study, van de Mortel et al. (2008) found expression differences for genes involved in lignin, GSH, and sulfate metabolism in response to Cd in *Arabidopsis thaliana* and the related Zn/Cd-hyperaccumulator *Thlaspi caerulescens*. Addition of 50 mM $CdCl_2$ to the irrigation medium of mature *Arabidopsis* plants induces a rapid accumulation of the mRNA for cytosolic OAS-TL isoforms coupled with high increase in Cys content to meet growing demand for GSH and PCs throughout the leaf lamina, the root and stem cortex, and stem vascular tissues, and this suggests that increased Cys availability is responsible for Cd tolerance (Domínguez-Solís et al. 2004). Remarkably, expression of the gene encoding SAT4 responds to sulfate deprivation and to Cd exposure despite presumably contributing little SAT activity in vivo, pointing to a function in thiol-mediated stress response (Kawashima et al. 2005). Transcripts encoding SAT1 also increase 10-fold when a catalase2-deficient mutant of *Arabidopsis* is transferred from high CO_2 concentrations to ambient air, presumably to provide more Cys for GSH synthesis to be used in the detoxification of H_2O_2 by the AsA–GSH cycle (Queval et al. 2009). In the latter case the increase in SAT1 mRNA is accompanied by an increase in total levels of Cys and GSH. It seems therefore that the expression of gene encoding plastidic SAT1 is triggered by oxidative stress even though plastidic SAT activity contributes only about 10 % to the total SAT activity in nonstressed leaves and that metabolic regulation of the CS complexes in the three compartments is responsible for increased production and contents of Cys in response to environmental challenges (Hell and Wirtz 2011). Molecular responses to Cd exposure have also been identified in plants, such as *Brassica juncea*, *Arabidopsis* (Xiang and Oliver 1998), and the other plant species in which Cd exposure induced a coordinated transcriptional regulation of genes encoding γ-ECS, GSHS, and PCS (Sun et al. 2005). Combined transcript, enzymatic, and metabolic profiling in the moss *Physcomitrella patens* revealed vital involvement of sulfate assimilatory genes under Cd stress (Rother et al. 2006). Cd-induced GSH and PC synthesis have a well-established relationship with sulfate uptake, transport, and assimilation in plants in agreement with the concept of demand-driven regulation (Nocito et al. 2006). Besides GSH and hGSH in some legumes, there are other thiol-containing tripeptides in plants. Grasses and rice produce hydroxymethylglutathione, in which a serine replaces the terminal glycine, following exposure to heavy metals (Klapheck et al. 1994). Corn synthesizes an analog with a terminal glutamate, and horseradish generates a tripeptide with a glutamine in place of the glycine following Cd exposure. RNA blot analysis of transcript amounts obtained from roots and leaves of heavy metal accumulator Indian mustard, *Brassica juncea*, indicated enhanced gene expression of ATPS and APR and high level of Cys but decreased expressions of low-affinity sulfate transporters and low level of GSH (Heiss et al. 1999). Examination of steady-state mRNA levels in *Brassica napus* reveals that BnSultr2;2 transcripts were enhanced in leaves of sulfate-deficient plants and under 20–120 μM Cd exposure, but under the same conditions, the BnSultr2;2 expression in roots was severely suppressed (Sun et al. 2005). RT-PCR analysis also demonstrated that BnSultr1;1 was expressed only in roots, and its expression was upregulated by both sulfate deficiency and Cd exposure (Sun et al. 2005). Proteomic analysis in *Arabidopsis* Sultr 4;1 mutant seed revealed overaccumulation of several proteins related to stress response mechanisms (Zuber et al. 2010), of which the most important were the enzymes involved in detoxification processes, a glutathione S-transferase isoform (GST 6), an aldose reductase, a formate dehydrogenase, and a superoxide dismutase. Furthermore, spots corresponding to proteins usually upregulated during stress response and notably oxidative stress, such as glyceraldehyde-3-phosphate dehydrogenase and alcohol dehydrogenase, were increased in abundance in mutant seeds. Significantly, the mutant seeds possessed higher GSSG levels (46 % against 42 % for wild-type seeds), indicating sulfate remobilization from the vacuole to the other cell compartments is important for the seed's defense against abiotic oxidative stress during seed development and storage (Zuber et al. 2010). An upregulation of Sultr4;1 gene expression concomitantly with an oxidative stress response has also been reported in *Arabidopsis* roots and shoots under Cd stress (Herbette et al. 2006). Both constitutively elevated activity of SAT and concentration of GSH are involved in the ability of nickel-hyperaccumulator *Thlaspi goesingense* to tolerate nickel. Further study pointed out that all three isoforms of SAT in *T. goesingense* are insensitive to feedback inhibition by Cys (Na and Salt 2011).

Thiol-Metabolisms and Metalloid Stress

Perhaps, the most breathtaking discovery in recent years regarding roles of thiol-cascade in plant stress response has come from studies on arsenic (As) tolerance of crop plants. As is a ubiquitous toxic metalloid and is highly detrimental to plant growth, development, and nutritional quality of edible parts due to formation of ROS and consequently induction of oxidative stress (Gupta et al. 2008; Srivastava et al. 2009; Talukdar 2012f, h, 2013a, h, k; Zhao et al. 2012; Sharma 2013). Functional genomic approaches are very important in analyzing plant As transport and accumulation. Thus, a new concept termed as "Arsenomics" defined as an approach dealing with transcriptome, proteome, and metabolome alterations during As exposure has emerged (Tripathi et al. 2012b). In rice, As(V) induced genes involved in abiotic stress, detoxification pathways, and secondary metabolic process. Genes involved in secondary cell wall biogenesis, cell cycle, and oligopeptide transport were mainly downregulated (Huang et al. 2012). Several lines of evidence point to the coordinated roles of thiols in As tolerance and detoxification (Chakrabarty et al. 2009; Rai et al. 2011; Talukdar 2011f; Tripathi et al. 2012a, b), and GSH plays central role of this detoxification process (Ahsan et al. 2008; Talukdar 2013b), a generalized scheme of which has been shown in Fig. 3. When plants are supplied As^{V}, typically more than 90 % of As within roots is converted to As^{III} (Finnegan and Chen 2012), and the binding of As^{III} to GSH and/or PCs is the basis for the main detoxification

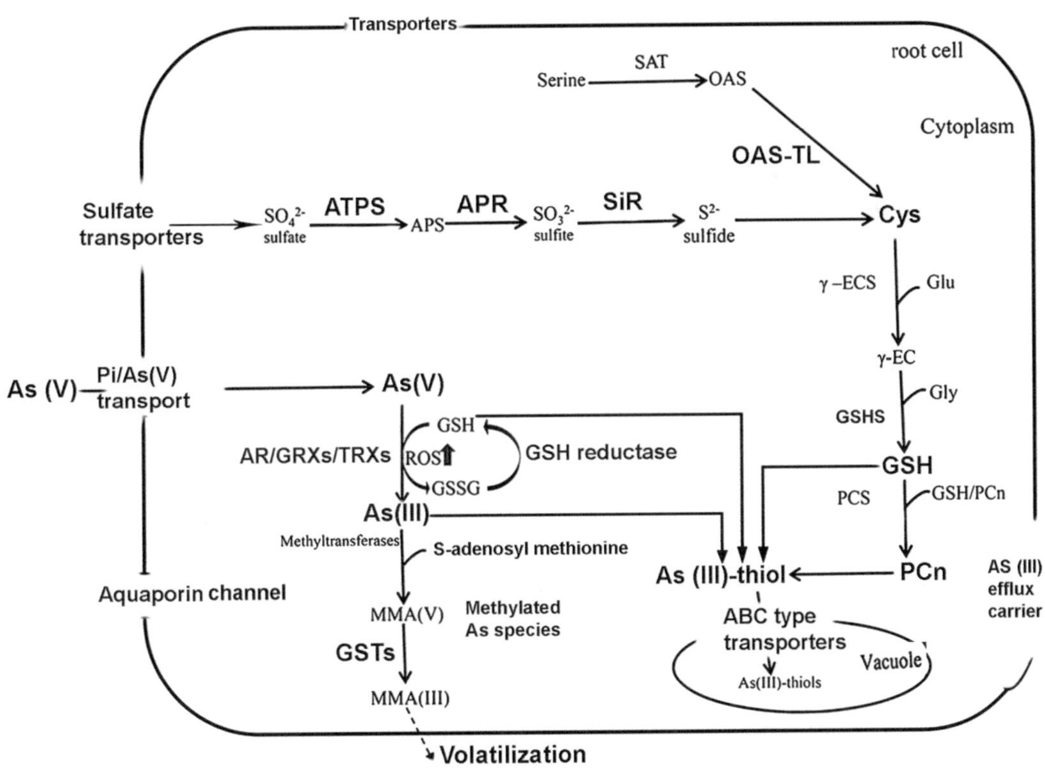

Fig. 3 A generalized scheme for involvement of components of thiol-cascade in arsenic transport and metabolisms in plant root cells

pathway for both As^V and As^{III} (Finnegan and Chen 2012; Zhao et al. 2012). In another study on rice, Tripathi et al. (2012c) observed alteration of various amino acids in rice and its synchronized role with thiolic ligand and potency for As tolerance and detoxification. Thiol biosynthesis-related enzymes were positively correlated to As accumulation-tolerant cultivar while opposite result was noticed in sensitive one (Tripathi et al. 2012b, c). Sung et al. (2009) revealed that ARS5 (arsenate reductase) is a component of the 26S proteasome complex and negatively regulates thiol biosynthesis and As tolerance in *Arabidopsis*. A recent report indicates that As exposure can significantly enhance *Fusarium* wilt infection in grain legumes (Talukdar 2013i) and is a predominant determining factor for high aggressiveness of *Leucaena leucocephala*, an invasive tree in tropical countries (Talukdar 2013d). In both cases, stimulation of GSH-mediated antioxidant defense plays vital roles in maintaining plant growth and seed yield. GSH exclusively requires Cys as one of its building blocks. Therefore, increased Cys biosynthesis to support GSH and PC production would add to the effectiveness of approaches designed to increase nonprotein thiols within plants, a process that would also require inputs from S-metabolism (Finnegan and Chen 2012). A large number of genes involved in thiol synthesis, metabolism, and transport as well as GSH-mediated antioxidant defense have recently been recorded during transcriptomic, proteomic, and metabolic analysis of rice, *Brassica*, lentil, *Pteris*, *Agrostis tenuis*, and aquatic macrophytes challenged with As stress (Norton et al. 2008; Duquesnoy et al. 2009; Srivastava et al. 2009, 2010; Bona et al. 2010; Rai et al. 2011; Tripathi et al. 2012a, b; Talukdar and Talukdar 2013d). The central role played by the GSH and PCs in the detoxification of the metalloid indicates a critical importance for S-metabolism in determining plant survival in As-contaminated soils (Ahsan et al. 2008). Transcriptomic and proteomic analysis revealed that nearly five sulfate transporter genes in rice (Norton et al. 2008; Chakrabarty et al. 2009), four in lentil (Talukdar and Talukdar 2013d), three in *Brassica* (Srivastava et al. 2009), and at least one transporter in *Arabidopsis* (Sung et al. 2009) are upregulated in roots exposed to As. In the Cys synthesizing steps, transcriptional upregulation has been observed for SAT1;1 and SAT1;2 isoforms and at least one OAS-TL each in *Arabidopsis*, rice, *Brassica*, *Hydrilla*, and lentil (Tripathi et al. 2012a, b; Talukdar and Talukdar 2013d). As^V and As^{III} exposure may cause a downregulation of OAS-TL in As-sensitive plants. OAS-TL protein disappeared from maize shoots exposed to As (Requejo and Tena 2006), while OAS-TL activity was repressed in an As-sensitive line of *B. juncea* (Srivastava et al. 2009), compromising Cys biosynthesis and, therefore, As detoxification through GSH and PC. Differential regulations of transcripts were also reported for downstream metabolites like GSH, PCs, and GSH-mediated antioxidant defense comprising different functional isoforms of AsA–GSH cycle enzymes like APX, DHAR, and GR and enzymes outside this cycle like CAT, GPXs, and GSTs in cereals, oilseeds, legumes, aquatic macrophyte *Ceratophyllum demersum*, and medicinal crops under As exposures (Li et al. 2004a, b; Mishra et al. 2008; Talukdar and Talukdar 2013b, c). An arsenate-activated GRX (PvGRX5) isolated from the As-hyperaccumulator fern *Pteris vittata* L. regulates intracellular arsenite and conferred tolerance to transgenic *Arabidopsis* by reducing As accumulation in leaves (Wei et al. 2010). Under AsV stress, the microarray experiment revealed induction of APR3, GST20 (*Tau* class), chloroplast Cu/Zn superoxide dismutase (SOD) (at2g28190), Cu/Zn SOD (at1g08830), and an SOD copper chaperone (at1g12520) in response to treatment (Abercrombie et al. 2008). Like SOD and catalase, GST, GRX and/or peroxidase transcript or protein abundance, or enzymatic activity often increases in response to As exposure (Stoeva et al. 2005; Srivastava et al. 2007; Abercrombie et al. 2008; Ahsan et al. 2008; Norton et al. 2008; Chakrabarty et al. 2009). As an example, in rice, at least 10 GST genes are upregulated in response to As^V exposure, while no more than two GST genes are downregulated (Norton et al. 2008; Chakrabarty et al. 2009). At least one γ-ECS, one PCS, three GST (Lc GSTI,

Lc GSTII, and Lc GSTIII) genes, two GR (Lc GR1 LcGR2), two Cu/Zn SOD (I, II), one each of FeSOD and MnSOD, three APXs (LcAPX1, 2, 3), and two DHAR (LcDHAR1, LcDHAR2) transcripts are differentially regulated in lentil genotypes, differing in As tolerance (Talukdar and Talukdar 2013d). Similarly, in common bean genotypes, differential expression and complementation of SOD isoforms and concomitant effect on GSH-mediated antioxidant defense were observed under As stress (Talukdar and Talukdar 2013b). In As-hyperaccumulator fern *Pteris vittata*, exogenous application of sulfate and GSH enhanced As accumulation (Wei et al. 2010). In *B. juncea* genotypes differing in As tolerance, transcriptional profiling showed an upregulation of sulfate transporters (Sultr2;1 and Sultr4;1), γ-ECS, PCS, and auxin and jasmonate biosynthesis pathway genes, whereas there was a downregulation of ethylene biosynthesis and cytokinin-responsive genes within 6 h of exposure to As^{III} (Srivastava et al. 2009). This suggested that perception of As-induced stress was presumably mediated through an integrated modulation in hormonal functioning that led to both short- and long-term adaptations to combat the stress (Srivastava et al. 2009). It seems likely that in the As-tolerant variety, there was an induction in Cys synthase activity, as well as in the activities of SAT and γ-ECS, the penultimate enzyme in GSH biosynthesis. These increases in enzyme activity were accompanied by increased levels of both Cys and GSH, indicating that increased S-metabolism may be a viable mechanism for increasing As tolerance in plants (Finnegan and Chen 2012). In order to identify the differentially expressed transcripts and the pathways involved in As metabolism and detoxification, *Crambe abyssinica*, the Ethiopian mustard plants, were subjected to arsenate stress, and a PCR-Select Suppression Subtraction Hybridization (SSH) approach was employed (Paulose et al. 2010). A total of 105 differentially expressed subtracted cDNAs were sequenced which were found to represent 38 genes. Transcripts related to thiol-cascade include GSTs which form the largest group in the subtracted cDNA library and enzymes involved in reductive assimilation of S such as APR, APS, and SiR. Most of the transcripts (12 %) of GSTs fall in the *Tau* subfamily (GST-*Tau*), while the remaining sequences (4 %) are similar to the *Phi* subfamily (Paulose et al. 2010). PCs are GSH-derived peptides synthesized in the cytosol where they form PC-metal(loid) complexes that are transported into vacuoles, thus removing these toxic elements from the cytosol. Using transcriptional responses, several vacuolar PCs and other thiol transporters including ABCC-types for long-distance transport of Cd and As have been identified in yeast, Pteris, rice, *Brassica*, and *Arabidopsis* (Indriolo et al. 2010; Song et al. 2010; Mendoza-Cózatl et al. 2011; Huang et al. 2012), and shoot-specific expression of γ-ECS, the first enzyme in GSH biosynthesis, to direct long-distance transport of thiol peptides to roots conferring Cd and As tolerance has been revealed (Li et al. 2004a, b).

Thiol-Metabolisms and Chilling Stress

Role of S-metabolisms and thiol-metabolites in conferring tolerance to chilling stress has been documented (Noctor et al. 2012; Fatma et al. 2013). In maize, a plant with C_4 metabolisms showing Kranz anatomy, chilling stress induces foliar thiol levels and activities of APR, γ-ECS, and GSHS (Kopriva et al. 2009), and total GSH content and the activities of APR and GR are increased in chilling-tolerant maize compared with a sensitive genotype even at standard growth conditions (Kopriva et al. 2009).

Response of Thiol-Metabolisms to S-Status and Selenate Stress

In order to ascertain the effect of sulfate starvation and/or selenate stress on sulfate metabolisms, transcript expression of putative sulfate/selenate transporters from one Se hyperaccumulator, *Astragalus racemosus*, and one closely related nonaccumulator species, *Astragalus drummondii*, was determined by semiquantitative RT-PCR. Transporters belonging to groups 2 and

4 and group 3 of the plant sulfate transporter family were expressed in both root and shoot tissues of *A. racemosus* and *A. drummondii*, whereas the group 1-type transporter genes, which have been shown in other species to be mainly responsible for the initial uptake (Buchner et al. 2004a), were detected only in root tissues. A 16-day time-course experiment on the same taxa revealed (a) modest increases of groups 1 and 4 transcript abundance and no influence on the expression of Sultr2 in root tissues starved for 16 days, (b) no Sultr1 transcript in shoots under sufficient S or under S starvation and a small increase in Sultr4 and no change of Sultr2 abundance in response to S starvation, and (c) that Sultr4 transcripts accumulated more in shoots of nonaccumulating *Astragalus* species compared with the Se-accumulating *Astragalus* species (Cabannes et al. 2011). Global expression profiling of S-starved *Arabidopsis* by DNA macroarray reveals the role of OAS as a general regulator of gene expression in response to S nutrition (Hirai et al. 2003). Koralewska et al. (2008) found differential regulations of gene expressions of sultr 1;1 and sultr 1;2 and APR in *Brassica pekinensis* (Chinese cabbage) by H_2S nutrition and sulfate deprivation. As mentioned earlier, a sulfate transporter induced in sulfate-starved roots plays a central role in *Arabidopsis thaliana* (Takahashi et al. 1997), and several transporters were later identified (Takahashi et al. 2011).

Thiol-Cascade and Biosynthesis of Proteins Involved in Fatty Acids and Lipids Under Oxidative Stress

Besides proteins directly related to the oxidative stress response, an up-accumulation of proteins involved in the biosynthesis of fatty acids and lipids was revealed. Proteomic and transcriptomic analysis of *Arabidopsis* seeds provided molecular evidence for successive processing of seed proteins and its implication in the stress response to S nutrition. In *Arabidopsis* Sultr 4;1 mutant seeds, proteins such as enoyl-[acyl-carrier-protein] reductase, two isoforms of hydroxysteroid dehydrogenase 1 (HSD1), and ketoacyl carrier protein synthase I were up-accumulated (Zuber et al. 2010). An upregulation of a ketoacyl carrier protein synthase was also observed in developing *Arabidopsis* seeds under S-starved conditions. It is well known that under stress conditions, toxic oxygen derivatives are produced that inactivate enzymes and damage important cellular components, such as membranes by lipid peroxidation and fatty-acid de-esterification. It, thus, seems that the upregulation of enzymes involved in fatty acid and lipid biosynthesis may represent a mechanism to repair stress-induced membrane damage. Transgenic *Arabidopsis* lines overexpressing HSD1 have an increased tolerance to salt stress (Zuber et al. 2010).

Thiol-Cascade and Nutritional Fortifications of Crops

Substantial effort has been expended over many years to enrich the plant parts used as food and feed in essential amino acids, most notably lysine in cereal grains and Cys and methionine in legume seeds (Tabe and Droux 2002; Tabe et al. 2010). Two distinct strategies have been used: manipulation of the pathways of amino acid biosynthesis and creation of a storage sink by expression of a protein rich in the relevant amino acid (Tabe et al. 2010). In *Arabidopsis*, mutant over-accumulating OAS has been characterized (Ohkama-Ohtsu et al. 2004) and seed Met content has been enhanced by reducing the activity of HMT2, a Met biosynthetic enzyme (Lee et al. 2008). Remarkably, tissue-specific gene expressions of sulfate transporter families have been studied in relation to nutrition (Buchner et al. 2010). Overexpression of mutated forms of aspartate kinase and cystathionine γ-synthase in tobacco leaves resulted in the high accumulation of Met and threonine (Hacham et al. 2008). Protein quality in legume crops is limited by the suboptimal levels of the essential S-containing amino acids, Met and Cys. Seed development in pea has been positively modulated by increasing the phloem transport of S-methylmethionine and S as well as N metabolisms (Tan et al. 2010). In the case of the S-containing amino acids, many different steps of the pathways of reductive S-assimilation and S amino acid biosynthesis have been manipulated in a range of plant species. Increased efficiency of wool growth and live weight gain was recorded in Merino sheep fed transgenic lupin seed containing sunflower albumin (White et al. 2007). A unique feature of *Phaseolus* and several *Vigna* species is the accumulation of a nonprotein amino acid, *S-methyl-Cys*, to a high concentration in seed, of up to 0.3 % per dry weight, mainly as a γ-Glu dipeptide (Liao et al. 2012). *S*-Methyl-Cys cannot substitute for Met or Cys in the diet (Padovese et al. 2001). Major seed proteins in common bean, the 7S globulin phaseolin and lectin phytohaemagglutinin, are poor in Met and Cys. Proteomic analysis identified several S-rich proteins whose levels are elevated in the absence of phaseolin and major lectins, including the 11S globulin legumin, albumin-2, defensin, albumin-1, and the Bowman–Birk-type proteinase inhibitor (Marsolais et al. 2010). Under these conditions, legumin becomes the dominant storage protein, accounting for at least 17 % of total protein. Integration of proteomic and functional genomic data enabled the identification and isolation of cDNAs encoding these proteins (Yin et al. 2011). These characteristics are reminiscent of the *opaque-2* mutant, which was used to develop quality protein maize (Huang et al. 2009). Till date, most approaches to improve protein quality in grain legumes have involved the transgenic expression of S-rich proteins, sometimes in combination with metabolic engineering of S amino acid pathways. In common bean, the expression of Brazil nut 2S albumin increased the Met concentration by 20 % (Aragao et al. 1999) and transcriptomic profiling identified candidate genes associated with the accumulation of distinct sulfur γ-glutamyl dipeptides in *Phaseolus vulgaris* and *Vigna mungo* seeds (Liao et al. 2013). In lupin and chickpea, expression of sunflower seed albumin (SSA) stimulated S-assimilation. S was shifted from the sulfate to the protein Met pool, elevated by 90 %, while the concentration of Cys was reduced by 10 % (Molvig et al. 1997; Tabe and Droux 2002; Chiaiese et al. 2004). Expression in developing lupin embryos of a SAT from *Arabidopsis thaliana* (AtSAT1 or AtSerat 2;1) was associated with increases of up to 5-fold in the concentrations of OAS, the immediate product of SAT, and up to 26-fold in free Cys, resulting in some of the highest in vivo concentrations of these metabolites yet reported (Tabe et al. 2010). In *Vicia narbonensis*, which accumulates little sulfate in mature seed, co-expression of Brazil nut 2S albumin with a feedback-insensitive, bacterial Asp kinase increased Met and Cys concentrations by 100 % and 20 %, respectively (Demidov et al. 2003). The increased levels of Met and Cys were accompanied by decreases in the concentration of γ-Glu-*S*-ethenyl-Cys (2-fold) and free thiols, particularly γ-GluCys and GSH. About two-third increase in Met and Cys concentration was attributed to an enhanced supply of S to the seed. In soybean, transgenic expression of Brazil nut 2S albumin increased Met concentration by 26 %, while expression of 15 kDa δ-zein increased Met and Cys concentrations by 20 % and 35 %, respectively (Dinkins et al. 2001). Constitutive overexpression of a cytosolic form of OAS-TL in transgenic soybean led to sustained enzymatic activity at the late stages of seed development and resulted in a 70 % increase in total Cys concentration in mature seed (Kim et al. 2012). This was associated with enhanced levels of the endogenous Cys-rich protein, the Bowman–Birk protease inhibitor. Two types of novel protein bodies in transitional cells situated between the vascular tissue and storage parenchyma were obtained from transgenic soybean plant overexpressing an 11 kDa Met-rich delta-zein (Kim and Krishnan 2004). High-level expression of maize γ-zein protein has been obtained (Li et al. 2005), and protease inhibitor activity has been reduced by expression of a mutant Bowman–Birk gene in soybean seed (Livingstone et al. 2007). Furthermore, S-assimilation and Cys biosynthesis has been manipulated towards engineering seed S amino acid content in food and feed (Krishnan 2008; Jez and Krishnan 2009). Although mature seed is the primary target of nutritional biofortification process, studies revealed S-metabolisms and transportations of thiol compounds in developing legume seeds. In soybean, sulfate in pods is transformed into homoglutathione, which is mobilized into developing seed (Anderson and Fitzgerald 2001). While homoglutathione contributes Cys, *S*-methyl-Met is anticipated to be a major form of Met transported to the seed. Recent functional genomic studies have highlighted the occurrence of complete pathways of sulfate assimilation and de novo Cys and methionine biosynthesis in developing seed, both in soybean and common bean (Yi et al. 2010; Yin et al. 2011). Seed-specific transgenic expression of a chloroplastic, feedback-insensitive SAT in lupin increased the concentration of OAS and free Cys (Tabe et al. 2010). But the total concentration of Cys and Met could not be enhanced, even after co-segregation of the SSA transgene. This suggests that sulfate assimilation and Cys biosynthesis are regulated independently from Met biosynthesis, which is part of the aspartic acid-derived amino acid pathway (Galili et al. 2005). A global analysis of transcripts and free amino acids spanning the developmental period of γ-Glu-*S*-methyl-Cys accumulation in seeds of common bean revealed that (a) during seed development, phaseolin and phytohaemagglutinin transcripts are most abundant in cotyledonary stage, while phaseolin as a protein accumulates during maturation; (b) seed desiccation takes place during maturation stage. Hierarchical clustering of expression values for major seed protein transcripts that were differentially expressed between common bean genotypes of SARC1 and SMARC1N-PN1 throughout all four developmental stages were consistent (Liao et al. 2012) with previous proteomic findings (Marsolais et al. 2010); and (c) lower expression values were systematically observed for phaseolin, arcelin, lectin, and most phytohaemagglutinin contigs in SMARC1N-PN1. The microarray data were examined for differential expression of transcripts coding for S-rich proteins (Liao et al. 2012). S-rich proteins previously identified by proteomics as elevated in SMARC1N-PN1 also had increased transcript levels. They include legumin, albumin-2, defensin D1, albumin-1A and albumin-1B, and the Bowman–Birk-type proteinase inhibitor. Transcripts of additional types of S-rich proteins were identified as differentially expressed, including the basic 7S globulin, double-headed trypsin, and Kunitz trypsin protease inhibitors. Soybean basic 7S globulin, known as γ-conglutin in lupin, is a minor Cys-rich globulin with structural similarity to xyloglucan-specific endo-β-1,4-glucanase inhibitor-like protein (Scarafoni et al. 2010; Yoshizawa et al. 2011), which has insulin-binding properties and glucose-lowering nutritional effects (Hanada and Hirano 2004; Magni et al. 2004; Lovati et al. 2012). Transcripts were most elevated for albumin-1A (10-fold), albumin-1E (7-fold), and albumin-1F contigs (11-fold) and the three differentially expressed genes coding for the Bowman–Birk-type proteinase inhibitor 2 and the double-headed trypsin inhibitor/trypsin chymotrypsin inhibitor. Differential expression was observed at distinct developmental stages, also. Albumin-2, albumin-1E, the Bowman–Birk-type proteinase inhibitor 2, and the double-headed trypsin inhibitor contigs had increased transcript levels starting from cotyledonary stage. This was followed by legumin and basic 7S globulin-2 at cotyledonary stage. Defensin D1 was differentially expressed at early maturation stage and albumin-1A and albumin-1B at final maturation stage (Liao et al. 2012). Transcripts of the Kunitz trypsin protease inhibitor were elevated from early maturation stages to final maturation. In parallel with transcripts of Cys-rich proteins, those of three different chaperones involved in the formation of disulfide bridges were also elevated (Liao et al. 2012). Besides, transcripts coding for enzymes associated with several processes of S-metabolism were significantly increased; notable among these are-Sultr 1;2, Sultr 3;3, ATPS1, APSR1, SAT1;1, SAT1;2, and SAT 2;1 (Liao et al. 2012). The transcript profiling results focus on a remarkable coordination in the expression of S-metabolic genes, which includes those participating in sulfate transport and assimilation, de novo Cys, and Met biosynthesis and further confirms the fact that in vegetative tissue, transcription of S uptake and assimilation is regulated by demand, with feedback inhibition by GSH and stimulation by OAS (Liao et al. 2012). In transgenic seeds expressing SSA, upregulation of endogenous

S-rich proteins was associated with reduced levels of GSH in rice (Hagan et al. 2003) and increased levels of OAS in chickpea (Chiaiese et al. 2004). Howarth et al. (2009) identified a unique S deficiency-induced gene *sdi1*, involved in the utilization of stored sulfate pools under S-limiting conditions. This gene has potential as a diagnostic indicator of S nutritional status.

Plant Nuclear Ploidy, Sexual Reproduction, and Thiol-Metabolisms

Unique effect of the genome structure in determining the size of the sulfate transporter gene family has been studied in the Poaceae and Fabaceae. Phylogeny and expression of paralogous and orthologous sulfate transporter genes have been studied in diploid and hexaploid wheat (Buchner et al. 2004b). The partially diploidized tetraploid soybean (*Glycine max*) genome contains in total 28 sulfate transporter genes indicating gene duplications of the individual group isoforms (http://www.phytozome.net/). In contrast, in the diploid *Medicago truncatula* genome (www.medicago-hapmap.org), only 10 sulfate transporter genes have been detected. The "New World" *Astragalus* species, which include *A. crotalaria, A. drummondii, A. racemosus*, and *A. bisulcatus*, are mostly aneuploid, with a chromosome number based on $n = 11$ to 15, in contrast to the "Old World" euploid *Astragalus* ($n = 8$) species, which include *A. glycyphyllos* (Cabannes et al. 2011). Three very closely related group 1 high-affinity sulfate transporter genes (Sultr1a–Sultr1c) were identified in *A. racemosus*. Using the same RT-PCR approaches, a Sultr1b isoform from A. *crotalaria, A. bisulcatus*, and *A. drummondii* and Sultr1a and Sultr1b isoforms from *A. glycyphyllos* were isolated (Cabannes et al. 2011). No Sultr2;1 type was isolated from the *Astragalus* species, but a Sultr 2;2 isoform has been phylogenetically close to *Arabidopsis* Sultr 2;2 (Buchner et al. 2010; Cabannes et al. 2011). However, group 3 Sultr3;4 type was identified in *A. drummondii, A. racemosus*, and *A. bisulcatus*, and a group 4 type was identified from *A. racemosus* (Cabannes et al. 2011).

ROS, reactive nitrogen species (RNS), GSH and other classic buffer molecules or antioxidant proteins, and some thiol/disulfide-containing proteins belonging to the thioredoxin superfamily, like glutaredoxins (GRXs) or thioredoxins (TRXs), form a complex network of redox regulations. These components participate as critical elements not only in the switch between the mitotic to the meiotic cycle but also at further developmental stages of microsporogenesis, regulation of pollen rejection as the result of self-incompatibility, and display precise space-temporal patterns of expression and are present in specific localizations like the stigmatic papillae or the mature pollen (Traverso et al. 2013). Gene expressions of several TRXs such as TRXh1, TRXh4, TRXh5, and GRX in *Arabidopsis*, rice, *Pisum sativum, Crocus sativus* (saffron), and *Nicotiana* have been found to be involved in the development of stigma, style, anther, and pollen–pistil interactions and during growth of pollen tube (reviewed Traverso et al. 2013).

Origin of "Thiolomics": Progress and Future Prospects in Crop Improvement

Continuous increase in global population along with the growing urbanization and impending climate change imposes significant pressure for increasing agricultural crop productivity and nutritional quality. Crop yield is a complex trait and is found to be dependent upon three interdependent factors such as generation of photosynthetic reductant, its assimilation into the carbon product, and then translocation in different plant parts. In recent years, various transgenic-based approaches have been tested to modulate source and sink strength; however, limited success has been achieved in the terms of increased crop yield. As an alternate strategy, the concept of strengthening the plant's built in mechanism using priming mediated physiological tuning, which does not involve any genetic modification, can be useful. Thiourea (TU), a known ROS scavenger and sulfhydryl regulator, governs source-to-sink relationship in plants through sugar dynamics, based upon the fact that most of the steps for generating photoassimilates at source and its translocation towards sink are regulated in a redox state-dependent manner (Srivastava et al. 2010; Pandey et al. 2013). This non-transgenic-based priming method has recently been successfully utilized to enhance crop productivity and nutritional biofortifications and to impart stress tolerance in crops such as mustard, wheat, mung bean, salt grass, potato, and maize (Pandey et al. 2013). High-throughput omics techniques are extensively being exploited in recent times to dissect plant molecular strategies of thiol-cascade in regulating plant growth, development, metabolisms, and stress tolerance. All these events are neatly interwoven with each other, and plant S-metabolisms, thiolic potency, its dynamisms, and regulations in each and every events either directly or indirectly focus the facts that thiol-cascade is in the central of plant growth. The components of this cascade can be effectively manipulated in favor of crop yield and nutritional quality using power of omics technology specifically employed in plant thiol-metabolisms through "thiolomics."

References

Abercrombie JM, Halfhill MD, Ranjan P, Rao MR, Saxton AM, Yuan JS, Stewart CN Jr (2008) Transcriptional responses of *Arabidopsis thaliana* plants to As(V) stress. BMC Plant Biol 8:87

Ahsan N, Lee DG, Alam I, Kim PJ, Lee JJ, Ahn YO, Kwak SS, Lee IJ, Bahk JD, Kang KY, Renaut J, Komatsu S, Lee BH (2008) Comparative proteomic study of arsenic-induced differentially expressed proteins in rice roots reveals glutathione plays a central role during As stress. Proteomics 8:3561–3576

Anderson JW, Fitzgerald MA (2001) Physiological and metabolic origin of sulphur for the synthesis of seed storage proteins. J Plant Physiol 158:447–456

Anjum NA, Gill SS, Umar S, Ahmad I, Duarte AC, Pereira E (2012) Improving growth and productivity of *Oleiferous Brassicas* under changing environment: significance of nitrogen and sulphur nutrition, and underlying mechanisms. Sci World J 2012, Article ID 657808. doi:10.1100/2012/657808

Arabidopsis Genome Initiative (2000) Analysis of the genome sequence of the flowering plant *Arabidopsis thaliana*. Nature 408:796

Aragao FJL, Barros LMG, de Sousa MV, de Sa MFG, Almeida ERP, Gander ES, Rech EL (1999) Expression of a methionine rich storage albumin from the Brazil nut (*Bertholletia excelsa* HBK, Lecythidaceae) in transgenic bean plants (*Phaseolus vulgaris* L., Fabaceae). Genet Mol Biol 22:445–449

Aravind L, Koonin EV (2000) The STAS domain: a link between anion transporters and anti sigma-factor antagonists. Curr Biol 10:53–55

Astolfi S, Zuchi S (2013) Adequate S supply protects barley plants from adverse effects of salinity stress by increasing thiol contents. Acta Physiol Plant 35:175–181

Awazuhara M, Kim H, Goto DB, Matsui A, Hayashi H, Chino M, Kim S-G, Naito S, Fujiwara T (2002) A 235 bp region from a nutritionally regulated soybean specific gene promoter can confer its sulfur and nitrogen response to a constitutive promoter in aerial tissues of *Arabidopsis thaliana*. Plant Sci 163:75–82

Barroso C, Romero LC, Cejudo FJ, Vega JM, Gotor C (1999) Salt specific regulation of the cytosolic O-acetylserine(thiol)lyase gene from *Arabidopsis thaliana* is dependent on abscisic acid. Plant Mol Biol 40:729–736

Birke H, Müller SJ, Rother M, Zimmer AD, Hoernstein SNW, Wesenberg D, Wirtz M, Krauss G-J, Ralf Reski R, Hell R (2012) The relevance of compartmentation for cysteine synthesis in phototrophic organisms. Protoplasma 249(Suppl 2):S147–S155. doi:10.1007/s00709-012-0411-9

Blum R, Meyer KC, Wünschmann J, Lendzian KJ, Grill E (2010) Cytosolic action of phytochelatin synthase. Plant Physiol 153:159–169

Bolchi A, Petrucco S, Tenca PL, Foroni C, Ottonello S (1999) Coordinate modulation of maize sulfate permease and ATP sulfurylase mRNAs in response to variations in sulfur nutritional status: stereospecific downregulation by L-cysteine. Plant Mol Biol 39:527–537

Bona E, Cattaneo C, Cesaro P, Marsano F, Lingua G, Cavaletto M, Berta G (2010) Proteomic analysis of *Pteris vittata* fronds: two arbuscular mycorrhizal fungi differentially modulate protein expression under arsenic contamination. Proteomics 10:3811–3834

Bonner ER, Cahoon RE, Knapke SM, Jez JM (2005) Molecular basis of cysteine biosynthesis in plants: structural and functional analysis of O-acetylserine sulfhydrylase from *Arabidopsis thaliana*. J Biol Chem 280:38803–38813

Borevitz JO, Xia Y, Blount J, Dixon RA, Lamb C (2000) Activation tagging identifies a conserved MYB regulator of phenylpropanoid biosynthesis. Plant Cell 12:2383–2394

Buchner P, Takahashi H, Hawkesford MJ (2004a) Plant sulphate transporters: co-ordination of uptake, intracellular and long-distance transport. J Exp Bot 55:1765–1773

Buchner P, Prosser I, Hawkesford MJ (2004b) Phylogeny and expression of paralogous and orthologous sulphate transporter genes in diploid and hexaploid wheats. Genome 47:526–534

Buchner P, Parmar S, Kriegel A, Carpentier M, Hawkesford MJ (2010) The sulfate transporter family in wheat: tissue-specific gene expression in relation to nutrition. Mol Plant 3:374–389

Bürstenbinder K, Rzewuski G, Wirtz M, Hell R, Sauter M (2007) The role of methionine recycling for ethylene synthesis in *Arabidopsis*. Plant J 49:238–249

Cabannes E, Buchner P, Broadley MR, Hawkesford MJ (2011) A comparison of sulfate and selenium accumulation in relation to the expression of sulfate transporter genes in *Astragalus* species. Plant Physiol 157:2227–2239

Chakrabarty D, Trivedi PK, Misra P, Tiwari M, Shri M, Shukla D, Kumar S, Rai A, Pandey A, Nigam D, Tripathi RD, Tuli R (2009) Comparative transcriptome analysis of arsenate and arsenite stresses in rice seedlings. Chemosphere 74:688–702

Chiaiese P, Ohkama-Ohtsu N, Molvig L, Godfree R, Dove H, Hocart C, Fujiwara T, Higgins TJ, Tabe LM (2004) Sulphur and nitrogen nutrition influence the response of chickpea seeds to an added, transgenic sink for organic sulphur. J Exp Bot 55:1889–1901

Chronis D, Krishnan HB (2004) Sulfur assimilation in soybean (*Glycine max* [L.] Merr.): molecular cloning and characterization of a cytosolic isoform of serine acetyltransferase. Planta 218:417–426

Davidian J-C, Kopriva S (2010) Regulation of sulfate uptake and assimilation—the same or not the same. Mol Plant 3:314–325

Demidov D, Horstmann C, Meixner M, Pickardt T, Saalbach I, Galili G, Muentz K (2003) Additive effects of the feed-back insensitive bacterial aspartate kinase and the Brazil nut 2S albumin on the methionine content of transgenic narbon bean (*Vicia narbonensis* L.). Mol Breed 11:187–201

Dietz K-J, Pfannschmidt T (2011) Novel regulators in photosynthetic redox control of plant metabolism and gene expression. Plant Physiol 155:1477–1485

Dinkins RD, Reddy MSS, Meurer CA, Yan B, Trick H, Thibaud-Nissen F, Finer JJ, Parrott WA, Collins GB (2001) Increased sulfur amino acids in soybean plants overexpressing the maize 15 kDa zein protein. In Vitro Cell Dev Biol Plant 37:742–747

Dixon DP, Edwards R (2009) Selective binding of glutathione conjugates of fatty acid derivatives by plant glutathione transferases. J Biol Chem 284:21249–21256

Domínguez-Solís JR, Lopez-Martin MC, Ager FJ, Ynsa MD, Romero LC, Gotor C (2004) Increased cysteine availability is essential for cadmium tolerance and accumulation in *Arabidopsis thaliana*. Plant Biotechnol J 2:469–476

Droux M (2003) Plant serine acetyltransferase: new insights for regulation of sulphur metabolism in plant cells. Plant Physiol Biochem 41:619–627

Dubousset L, Abdallah M, Desfeux AS, Etienne P, Meuriot F, Hawkesford MJ, Gombert J, Ségura R, Bataillé M-P, Rezé S, Bonnefoy J, Ameline AF, Ourry A, Dily FL, Avice JC (2009) Remobilization of leaf S compounds and senescence in response to restricted sulphate supply during the vegetative stage of oilseed rape are affected by mineral N availability. J Exp Bot 60:3239–3253

Duquesnoy I, Goupil P, Nadaud I, Branlard G, Piquet-Pissaloux A, Ledoigt G (2009) Identification of *Agrostis tenuis* leaf proteins in response to As(V) and As(III) induced stress using a proteomics approach. Plant Sci 176:206–213

Edwards R, Dixon DP (2005) Plant glutathione transferases. Methods Enzymol 401:169–186

Ellis CM, Nagpal P, Young JC, Hagen G, Guilfoyle TJ, Reed JW (2005) Auxin Response Factor1 and Auxin Response Factor2 regulate senescence and floral organ abscission in *Arabidopsis thaliana*. Development 132:4563–4574

Falkenberg B, Witt I, Zanor MI, Steinhauser D, Mueller-Roeber B, Hesse H, Hoefgen R (2008) Transcription factors relevant to auxin signalling coordinate broad-spectrum metabolic shifts including sulphur metabolism. J Exp Bot 59:2831–2846

Fatma M, Iqbal M, Khan R, Masood A, Khan NA (2013) Coordinate changes in assimilatory sulfate reduction are correlated to salt tolerance: involvement of phytohormones. Annu Rev Res Biol 3(3):267–295

Finnegan PM, Chen W (2012) Arsenic toxicity: the effects on plant metabolism. Front Physiol 3:182

Fitzpatrick KL, Tyerman SD, Kaiser BN (2008) Molybdate transport through the plant sulfate transporter SHST1. FEBS Lett 582:1508–1513

Francois JA, Kumaran S, Jez JM (2006) Structural basis for interaction of O-acetylserine sulfhydrylase and serine acetyltransferase in the Arabidopsis cysteine synthase complex. Plant Cell 18:3647–3655

Frendo P, Harrison J, Norman C, Hernández-Jiménez MJ, Van de Sype G, Gilabert A, Puppo A (2005) Glutathione and homoglutathione play a critical role in the nodulation process of *Medicago truncatula*. Mol Plant Microbe Interact 18:254–259

Galant A, Preuss ML, Cameron JC, Jez JM (2011) Plant glutathione biosynthesis: diversity in biochemical regulation and reaction products. Front Plant Sci 2:45. doi:10.3389/fpls.2011. 00045

Galili G, Amir R, Hoefgen R, Hesse H (2005) Improving the levels of essential amino acids and sulfur metabolites in plants. Biol Chem 386:817–831

Gómez LD, Vanacker H, Buchner P, Noctor G, Foyer CH (2004) Intercellular distribution of glutathione synthesis in maize leaves and its response to short-term chilling. Plant Physiol 134:1662–1671

Gupta DK, Tripathi RD, Mishra S, Srivastava S, Dwivedi S, Rai UN, Yang XE, Huanj H, Inouhe M (2008) Arsenic accumulation in root and shoot vis-a-vis its effects on growth and level of phytochelatins in seedlings of *Cicer arietinum* L. J Environ Biol 29:281–286

Haas FH, Heeg C, Queiroz R, Bauer A, Wirtz M, Hell R (2008) Mitochondrial serine acetyltransferase functions as a pacemaker of cysteine synthesis in plant cells. Plant Physiol 148:1055–1067

Hacham Y, Matityahu I, Schuster G, Amir R (2008) Overexpression of mutated forms of aspartate kinase and cystathionine gamma-synthase in tobacco leaves resulted in the high accumulation of methionine and threonine. Plant J 54:260–271

Hagan ND, Upadhyaya N, Tabe LM, Higgins TJV (2003) The redistribution of protein sulfur in transgenic rice expressing a gene for a foreign, sulfur-rich protein. Plant J 34:1–11

Hanada K, Hirano H (2004) Interaction of a 43-kDa receptor-like protein with a 4-kDa hormone-like peptide in soybean. Biochemistry 43:12105–12112

Harada E, Coi Y-E, Tsuchisaka A, Obata H, Sano H (2001) Transgenic tobacco plants expressing a rice cysteine synthase gene are tolerant to toxic levels of cadmium. J Plant Physiol 158:655–661

Hartmann T, Hönicke P, Wirtz M, Hell R, Rennenberg H, Kopriva S (2004) Regulation of sulphate assimilation by glutathione in poplars (*Populus tremula* x P. *alba*) of wild type and overexpressing γ-glutamylcysteine synthetase in the cytosol. J Exp Bot 55:837–845

Hawkesford MJ (2000) Plant responses to sulfur deficiency and the genetic manipulation of sulfate transporters to improve S utilization efficiency. J Exp Bot 5:131–138

Hawkesford MJ (2003) Transporter gene families in plants: the sulphate transporter gene family: redundancy or specialization? Physiol Plant 117:115–163

Hawkesford MJ, Wray JL (2000) Molecular genetics of sulphur assimilation. Adv Bot Res 33:159–223

Heeg C, Kruse C, Jost R, Gutensohn M, Ruppert T, Wirtz M, Hell R (2008) Analysis of the Arabidopsis O-acetylserine(thiol)lyase gene family demonstrates compartment-specific differences in the regulation of cysteine synthesis. Plant Cell 20:168–185

Heiss S, Schafer HJ, Haag-Kerwer A, Rausch T (1999) Cloning sulfur assimilation genes of *Brassica juncea* L.: cadmium differentially affects the expression of a putative low-affinity sulfate transporter and isoforms of ATP sulfurylase and APS reductase. Plant Mol Biol 39:847–857

Hell R, Bergmann L (1990) γ-Glutamylcysteine synthetase in higher plants: catalytic properties and subcellular localization. Planta 180:603–612

Hell R, Wirtz M (2011) Molecular biology, biochemistry and cellular physiology of cysteine metabolism in *Arabidopsis thaliana*. Arabidopsis Book 9:e0154. doi:10.1199/tab.0154

Herbette S, Taconnat L, Hugouvieux V, Piette L, Magniette MLM, Cuine S, Auroy P, Richaud P, Forestier C, Bourguignon J, Renou JP, Vavasseur A, Leonhardt N (2006) Genome-wide transcriptome profiling of the early cadmium response of Arabidopsis roots and shoots. Biochimie 88:1751–1765

Herschbach C, Teuber M, Eiblmeier M, Ehlting B, Ache P, Polle A, Schnitzler J-P, Rennenberg H (2010) Changes in sulphur metabolism of grey poplar (*Populus × canescens*) leaves during salt stress: a metabolic link to photorespiration. Tree Physiol 30:1161–1173

Hesse H, Lipke J, Altmann T, Höfgen R (1999) Molecular cloning and expression analyses of mitochondrial and plastidic isoforms of cysteine synthase (O-acetylserine(thiol)lyase) from *Arabidopsis thaliana*. Amino Acids 16:113–131

Hesse H, Nikiforova V, Gakière B, Hoefgen R (2004) Molecular analysis and control of cysteine biosynthesis: integration of nitrogen and sulfur metabolism. J Exp Bot 55:1283–1292

Hicks LM, Cahoon RE, Bonner ER, Rivard RS, Sheffield J, Jez JM (2007) Thiol-based regulation of redox-active glutamate-cysteine ligase from *Arabidopsis thaliana*. Plant Cell 19:2653–2661

Hirai MY, Fujiwara T, Awazuhara M, Kimura T, Noji M, Saito K (2003) Global expression profiling of sulphur-starved Arabidopsis by DNA macroarray reveals the role of O-acetyl-L-serine as a general regulator of gene expression in response to sulphur nutrition. Plant J 33:651–663

Hirai MY, Yano M, Goodenowe DB, Kanaya S, Kimura T, Awazuhara M, Arita M, Fujiwara T, Saito K (2004) Integration of transcriptomics and metabolomics for understanding of global responses to nutritional stresses in *Arabidopsis thaliana*. Proc Natl Acad Sci USA 101:10205–10210

Hirai MY, Klein M, Fujikawa Y et al (2005) Elucidation of gene-to-gene and metabolite-to-gene networks in Arabidopsis by integration of metabolomics and transcriptomics. J Biol Chem 280:25590–25595

Hirai MY, Sugiyama K, Sawada Y, Tohge T, Obayashi T, Suzuki A, Araki R, Sakurai N, Suzuki H, Aoki K, Goda H, Nishizawa OI, Shibata D, Saito K (2007) Omics-based identification of Arabidopsis Myb transcription factors regulating aliphatic glucosinolate biosynthesis. Proc Natl Acad Sci USA 10:6478–6483

Höfgen R, Hesse H (2008) Sulfur and cysteine metabolism. In: Jez JM (ed) Sulfur: a missing link between soils, crops, and nutrition. ASA-CSSA-SSSA Publishing, Madison, pp 83–104

Hopkins L, Parmar S, Blaszczyk A, Hesse H, Höfgen R, Hawkesford MJ (2005) O-acetylserine and the regulation of expression of genes encoding components for sulfate uptake and assimilation in potato. Plant Physiol 138:433–440

Hossain Z, Komatsu S (2012) Contribution of proteomic studies towards understanding plant heavy metal stress response. Front Plant Sci 3:310

Howarth J, Fourcroy P, Davidian J-C, Smith FW, Hawkesford MJ (2003) Cloning of two contrasting high-affinity sulphate transporters from tomato induced by low sulphate and infection by the vascular pathogen *Verticillium dahliae*. Planta 218:58–64

Howarth JR, Parmar S, Barraclough PB, Hawkesford MJ (2009) A sulphur deficiency-induced gene *sdi1*, involved in the utilization of stored sulphate pools under sulphur-limiting conditions has potential as a diagnostic indicator of sulphur nutritional status. Plant Biotechnol J 7:200–209

Huang S, Frizzi A, Malvar TM (2009) Engineering high lysine corn. In: Krishnan H (ed) Modification of seed composition to promote health and nutrition. American Society of Agronomy, Crop Science Society of America, Soil Science Society of America, Madison, pp 233–248

Huang T-L, Fu S-F, Thuy Nguyen QT, Lin C-Y, Chen Y-C, Huang H-J (2012) Transcriptomic changes and signalling pathways induced by arsenic stress in rice roots. Plant Mol Biol. doi:10.1007/s11103-012-9969-z

Hubberten HM, Klie S, Caldana C, Degenkolbe T, Willmitzer L, Höefgen R (2012) Additional role of *O*-acetylserine as a sulfur status-independent regulator during plant growth. Plant J 70:666–677

Indriolo E, Na G, Ellis D, Salt DE, Banksb JA (2010) A vacuolar arsenite transporter necessary for arsenic tolerance in the arsenic hyperaccumulating fern *Pteris vittata* is missing in flowering plants. Plant Cell 22:2045–2057

Iqbal N, Masood A, Iqbal M, Khan R, Asgher M, Fatma M, Khan NA (2013) Cross-talk between sulfur assimilation and ethylene signaling in plants. Plant Signal Behav 8(1):e22478

Jez JM, Krishnan HB (2009) Sulfur assimilation and cysteine biosynthesis in soybean seeds: towards engineering sulfur amino acid content. In: Krishnan HB (ed) Modification of seed composition to promote health and nutrition. ASA-CSSA-SSSA Publishing, Madison, pp 249–261

Jez JM, Cahoon RE, Chen S (2004) *Arabidopsis thaliana* glutamate-cysteine ligase: functional properties, kinetic mechanism, and regulation of activity. J Biol Chem 279:33463–33479

Jing HC, Hebeler R, Oeljeklaus S, Sitek B, Stühler K, Meyer HE, Sturre MJ, Hille J, Warscheid B, Dijkwel PP (2008) Early leaf senescence is associated with an altered cellular redox balance in *Arabidopsis* cpr5/old1 mutants. Plant Biol 1:85–98

Jones-Rhoades MW, Bartel DP (2004) Computational identification of plant microRNAs and their targets, including a stress-induced miRNA. Mol Cell 14:787–799

Jost R, Berkowitz O, Wirtz M, Hopkins L, Hawkesford MJ, Hell R (2000) Genomic and functional characterization of the oas gene family encoding O-acetylserine (thiol) lyases, enzymes catalyzing the final step in cysteine biosynthesis in *Arabidopsis thaliana*. Gene 253:237–247

Kataoka T, Watanabe-Takahashi A, Hayashi N, Ohnishi M, Mimura T, Buchner P, Hawkesford MJ, Yamaya T, Takahashi H (2004) Vacuolar sulfate transporters are essential determinants controlling internal distribution of sulfate in Arabidopsis. Plant Cell 16:2693–2704

Kawashima CG, Berkowitz O, Hell R, Noji M, Saito K (2005) Characterization and expression analysis of a serine acetyltransferase gene family involved in a key step of the sulfur assimilation pathway in Arabidopsis. Plant Physiol 137:220–230

Kawashima CG, Yoshimoto N, Maruyama-Nakashita A, Tsuchiya YN, Saito K, Takahashi H, Dalmay T (2009) Sulfur starvation induces the expression of microRNA-395 and one of its target genes but in different cell types. Plant J 57:313–321

Khan TA, Mazid M (2011) Nutritional significance of sulphur in pulse cropping system. Biol Med 3(2):114–133

Khan MS, Haas FH, Allboje Samami A, Moghaddas Gholami A, Bauer A, Fellenberg K, Reichelt M, Hansch R, Mendel RR, Meyer AJ, Wirtz M, Hell R (2010) Sulfite reductase defines a newly discovered bottleneck for assimilatory sulfate reduction and is essential for growth and development in *Arabidopsis thaliana*. Plant Cell 22:1216–1231

Kim WS, Krishnan HB (2004) Expression of an 11 kDa methionine rich delta-zein in transgenic soybean results in the formation of two types of novel protein bodies in transitional cells situated between the vascular tissue and storage parenchyma cells. Plant Biotechnol J 2:199–210

Kim WS, Chronis D, Juergens M, Schroeder AC, Hyun SW, Jez JM, Krishnan HB (2012) Transgenic soybean plants overexpressing *O*-acetylserine sulfhydrylase accumulate enhanced levels of cysteine and Bowman–Birk protease inhibitor in seeds. Planta 235:13–23

Klapheck S, Fliegner W, Zimmer I (1994) Hydroxymethyl-phytochelatins [(gamma-glutamylcysteine)n-serine] are metal-induced peptides of the Poaceae. Plant Physiol 104:1325–1332

Kopriva S (2006) Regulation of sulfate assimilation in Arabidopsis and beyond. Ann Bot 97:479–495

Kopriva S, Buchert T, Fritz G, Suter M, Weber M, Benda R, Schaller J, Feller U, Schurmann P, Schunemann V, Trautwein AX, Kroneck PM, Brunold C (2001) Plant adenosine 5′-phosphosulfate reductase is a novel iron-sulfur protein. J Biol Chem 276:42881–42886

Kopriva S, Mugford SG, Matthewman C, Koprivova A (2009) Plant sulfate assimilation genes: redundancy versus specialization. Plant Cell Rep 28:1769–1780

Kopriva S, Mugford SG, Baraniecka P, Lee BR, Matthewman CA, Koprivova A (2012) Control of sulfur partitioning between primary and secondary metabolism in Arabidopsis. Front Plant Sci 3:1–9

Koprivova A, Suter M, op den Camp R, Brunold C, Kopriva S (2000) Regulation of sulfate assimilation by nitrogen in Arabidopsis. Plant Physiol 122:737–746

Koprivova A, North KA, Kopriva S (2008) Complex signaling network in regulation of adenosine 5′-phosphosulfate reductase by salt stress in Arabidopsis roots. Plant Physiol 146:1408–1420

Koralewska A, Stuiver CEE, Posthumus FS, Kopriva S, Hawkesford MJ, De Kok LJ (2008) Regulation of sulfate uptake, expression of the sulfate transporters sultr1;1 and sultr1;2, and APS reductase in Chinese cabbage (*Brassica pekinensis*) as affected by atmospheric H_2S nutrition and sulfate deprivation. Funct Plant Biol 35:318–327

Krishnan HB (2008) Improving the sulfur-containing amino acids of soybeans to enhance its nutritional value in animal feed. In: Jez JM (ed) Sulfur: a missing link between soils, crops, and nutrition. ASA-CSSA-SSSA Publishing, Madison, pp 235–249

Krueger S, Niehl A, Lopez Martin MC, Steinhauser D, Donath A, Hildebrandt T, Romero LC, Höfgen R, Gotor C, Hesse H (2009) Analysis of cytosolic and plastidic serine acetyltransferase mutants and subcellular metabolite distributions suggests interplay of the cellular compartments for cysteine biosynthesis in Arabidopsis. Plant Cell Environ 32:349–367

Krusell L, Krause K, Ott T, Desbrosses G, Krämer U, Sato S, Nakamura Y, Tabata S, James EK, Sandal N, Stougaard J, Kawaguchi M, Miyamoto A, Suganuma N, Udvardi MK (2005) The sulfate transporter SST1 is crucial for symbiotic nitrogen fixation in *Lotus japonicus* root nodules. Plant Cell 17:1625–1636

Kumar S, Asif MH, Chakrabarty D, Tripathi RD, Trivedi P (2011a) Differential expression and alternative splicing of rice sulphate transporter family members regulate sulphur status during plant growth, development and stress conditions. Funct Integr Genomics 11:259–273

Kumar S, Bejiga G, Ahmed S, Nakkoul H, Sarker A (2011b) Genetic improvement of grass pea for low neurotoxin (β-ODAP) content. Food Chem Toxicol 49:589–600

Kumaran S, Yi H, Krishnan HB, Jez JM (2009) Assembly of the cysteine synthase complex and the regulatory role of protein protein interactions. J Biol Chem 284:10268–10275

Kutz A, Müller A, Hennig P, Kaiser WM, Piotrowski M, Weiler EW (2002) A role for nitrilase 3 in the regulation of root morphology in sulphur-starving *Arabidopsis thaliana*. Plant J 30:95–106

Lappartient AG, Vidmar JJ, Leustek T, Glass AD, Touraine B (1999) Inter-organ signaling in plants: regulation of ATP sulfurylase and sulfate transporter genes expression in roots mediated by phloem translocated compound. Plant J 18:89–95

Lee S, Leustek T (1998) APS kinase from *Arabidopsis thaliana*: genomic organization, expression, and kinetic analysis of the recombinant enzyme. Biochem Biophys Res Commun 247:171–175

Lee MS, Huang TF, Toro-Ramos T, Fraga M, Last RL, Jander G (2008) Reduced activity of *Arabidopsis thaliana* HMT2, a methionine biosynthetic enzyme, increases seed methionine content. Plant J 54:310–320

Leustek T, Martin MN, Bick JA, Davies JP (2000) Pathways and regulation of sulfur metabolism revealed through molecular and genetic studies. Annu Rev Plant Physiol Plant Mol Biol 51:141–165

Li H, Johnson P, Stepanova A, Alonso JM, Ecker JR (2004a) Convergence of signaling pathways in the control of differential cell growth in Arabidopsis. Dev Cell 7:1–20

Li Y, Dhankher OP, Carreira L, Lee D, Chen A, Schroeder J, Balish R, Meagher R (2004b) Overexpression of phytochelatin synthase in *Arabidopsis* leads to enhanced arsenic tolerance and cadmium sensitivity. Plant Cell Physiol 45:1787–1797

Li Z, Meyer S, Essig JS, Liu Y, Schapaugh MA, Muthukrishnan S, Hainline BE, Trick HN (2005) High-level expression of maize γ-zein protein in transgenic soybean (*Glycine max*). Mol Breed 16:11–20

Liao D, Pajak A, Karcz AR, Chapman BP, Sharpe AG, Austin RS, Datla R, Dhaubhadel S, Marsolais F (2012) Transcripts of sulphur metabolic genes are co-ordinately regulated in developing seeds of common bean lacking phaseolin and major lectins. J Exp Bot 63:6283–6295

Liao D, Cram D, Sharpe AG, Marsolais F (2013) Transcriptome profiling identifies candidate genes associated with the accumulation of distinct sulfur γ-glutamyl dipeptides in *Phaseolus vulgaris* and *Vigna mungo* seeds. Front Plant Sci 4:60

Lillo C, Lea US, Ruoff P (2008) Nutrient depletion as a key factor for manipulating gene expression and product formation in different branches of the flavonoid pathway. Plant Cell Environ 31:587–601

Liu F, Yoo BC, Lee JY, Pan W, Harmon AC (2006) Calcium regulated phosphorylation of soybean serine acetyltransferase in response to oxidative stress. J Biol Chem 281:27405–27415

Livingstone D, Beilinson V, Kalyaeva M, Schmidt MA, Herman EM, Nielsen NC (2007) Reduction of protease inhibitor activity by expression of a mutant Bowman-Birk gene in soybean seed. Plant Mol Biol 64:387–408

Lopez-Berenguera C, Carvajala M, Garcea-Viguerab C, Alcaraz CF (2007) Nitrogen, phosphorus, and sulfur nutrition in Broccoli plants grown under salinity. J Plant Nutr 30:1855–1870

Lovati MR, Manzoni C, Castiglioni S, Parolari A, Magni C, Duranti M (2012) Lupin seed γ-conglutin lowers blood glucose in hyperglycaemic rats and increases glucose consumption of HepG2 cells. Br J Nutr 107:67–73

Lyubetsky VA, Seliverstov AV, Zverkov OA (2013) Transcription regulation of plastid genes involved in sulfate transport in Viridiplantae. BioMed Res Int 2013, Article ID 413450, 6 pages. http://dx.doi.org/10.1155/2013/413450

Magni C, Sessa F, Accardo E, Vanoni M, Morazzoni P, Scarafoni A, Duranti M (2004) Conglutin γ, a lupin seed protein, binds insulin *in vitro* and reduces plasma glucose levels of hyperglycemic rats. J Nutr Biochem 15:646–650

Marsolais F, Pajak A, Yin F, Taylor M, Gabriel M, Merino DM, Ma V, Kameka A, Vijayan P, Pham H, Huang S, Rivoal J, Bett K, Hernández-Sebastià C, Liu Q, Bertrand A, Chapman R (2010) Proteomic analysis of common bean seed with storage protein deficiency reveals up-regulation of sulfur-rich proteins and starch and raffinose metabolic enzymes, and down-regulation of the secretory pathway. J Proteomics 73:1587–1600

Maruyama-Nakashita A, Takahashi H (2005) Transcriptional regulation of Sultr1 and Sultr1;2 in Arabidopsis roots. In: Saito K, De Kok LJ, Stuhlen I, Hawkesford MJ, Schnug E, Sirko A, Rennenberg H (eds) Sulfur transport and assimilation in plants in the postgenomic era. Backhuys Publishers, Leiden, pp 43–441

Maruyama-Nakashita A, Inoue E, Watanabe-Takahashi A, Yamaya T, Takahashi H (2003) Transcriptome profiling of sulfur-responsive genes in Arabidopsis reveals global effects of sulfur nutrition on multiple metabolic pathways. Plant Physiol 132:597–605

Maruyama-Nakashita A, Nakamura Y, Watanabe-Takahashi A, Yamaya T, Takahashi H (2004) Induction of SULTR1;1 sulfate transporter in Arabidopsis roots involves protein phosphorylation/dephosphorylation circuit for transcriptional regulation. Plant Cell Physiol 45:340–345

Maruyama-Nakashita A, Nakamura Y, Tohge T, Saito K, Takahashi H (2006) Arabidopsis SLIM1 is a central transcriptional regulator of plant sulfur response and metabolism. Plant Cell 18:3235–3251

Masood A, Iqbal N, Khan NA (2012) Role of ethylene in alleviation of cadmium-induced photosynthetic capacity inhibition by sulphur in mustard. Plant Cell Environ 35:524–533

Matamoros MA, Clemente MR, Sato S, Asamizu E, Tabata S, Ramos J, Moran JF, Stiller J, Gresshoff PM, Becana M (2003) Molecular analysis of the pathway for the synthesis of thiol tripeptides in the model legume *Lotus japonicus*. Mol Plant Microbe Interact 16:1039–1046

Mendoza-Cózatl DG, Jobe TO, Hauser F, Schroeder JI (2011) Long-distance transport, vacuolar sequestration and transcriptional responses induced by cadmium and arsenic. Curr Opin Plant Biol 14:554–562

Meyer AJ, Rausch T (2008) Biosynthesis, compartmentation and cellular functions of glutathione in plant cells. In: Hell R, Dahl C, Knaff DB, Leustek T (eds) Sulfur metabolism in phototrophic organisms, vol 27, Advances in photosynthesis and respiration. Springer, Dordrecht, pp 161–184

Mhamdi A, Hager J, Chaouch S, Queval G, Han Y, Taconnat L, Saindrenan P, Gouia H, Issakidis-Bourguet E, Renou J-P, Noctor G (2010) Arabidopsis GLUTATHIONE REDUCTASE1 plays a crucial role in leaf responses to intracellular hydrogen peroxide and in ensuring appropriate gene expression through both salicylic acid and jasmonic acid signaling pathways. Plant Physiol 153:1144–1160

Mishra S, Srivastava S, Tripathi RD, Trivedi PK (2008) Thiol metabolism and antioxidant system complement each other during arsenate detoxification in *Ceratophyllum demersum* L. Aquat Toxicol 86:205–215

Molvig L, Tabe LM, Eggum BO, Moore AE, Craig S, Spencer D, Higgins TJ (1997) Enhanced methionine levels and increased nutritive value of seeds of transgenic lupins (*Lupinus angustifolius* L.) expressing a sunflower seed albumin gene. Proc Natl Acad Sci USA 94:8393–8398

Moran JF, Iturbe-Ormaetxe I, Matamoros MA, Rubio MC, Clemente MR, Brewin NJ, Becana M (2000) Glutathione and homoglutathione synthetases of legume nodules. Cloning, expression, and subcellular localization. Plant Physiol 124:1381–1392

Mou Z, Fan W, Dong X (2003) Inducers of plant systemic acquired resistance regulate NPR1 function through redox changes. Cell 113:935–944

Mugford SG, Yoshimoto N, Reichelt M, Wirtz M, Hill L, Mugford ST, Nakazato Y, Noji M, Takahashi H, Kramell R, Gigolashvili T, Flügge UI, Wasternack C, Gershenzon J, Hell R, Saito K, Kopriva S (2009) Disruption of adenosine-5′-phosphosulfate kinase in Arabidopsis reduces levels of sulfated secondary metabolites. Plant Cell 21:910–927

Mullineaux PM, Rausch T (2005) Glutathione, photosynthesis and the redox regulation of stress-responsive gene expression. Photosynth Res 86:459–474

Na G, Salt DE (2011) Differential regulation of serine acetyltransferase is involved in nickel hyperaccumulation in *Thlaspi goesingense*. J Biol Chem 286:40423–40432

Nikiforova V, Freitag J, Kempa S, Adamik M, Hesse H, Höfgen R (2003) Transcriptome analysis of sulfur depletion in *Arabidopsis thaliana*: interlacing of biosynthetic pathways provides response specificity. Plant J 33:633–650

Nikiforova VJ, Daub CO, Hesse H, Willmitzer L, Hoefgen R (2005a) Integrative gene-metabolite network with implemented causality deciphers informational fluxes of sulfur stress response. J Exp Bot 56:1887–1896

Nikiforova VJ, Kopka J, Tolstikov V, Fiehn O, Hopkins L, Hawkesford MJ, Hesse H, Hoefgen R (2005b) Systems rebalancing of metabolism in response to sulfur deprivation, as revealed by metabolome analysis of Arabidopsis plants. Plant Physiol 138:304–318

Nikiforova VJ, Bielecka M, Gakière B, Krueger S, Rinder J, Kempa S, Morcuende R, Scheible W-R, Hesse H, Hoefgen R (2006) Effect of sulfur availability on the integrity of amino acid biosynthesis in plants. Amino Acids 30:173–183

Nocito F, Lancilli C, Crema B, Fourcroy P, Davidian J-C, Sacchi GA (2006) Heavy metal stress and sulfate uptake in maize roots. Plant Physiol 141:1138–1148

Noctor G, Mhamdi A, Chaouch S, Han Y, Neukermans J, Marquez-Garcia B, Queval G, Foyer CH (2012) Glutathione in plants: an integrated overview. Plant Cell Environ 35:454–484

Noji M, Saito M, Nakamura M, Aono M, Saji H, Saito K (2001) Cysteine synthase overexpression in tobacco confers tolerance to sulfur-containing environmental pollutants. Plant Physiol 126:973–980

Norton GJ, Lou-Hing DE, Meharg AA, Price AH (2008) Rice-arsenate interactions in hydroponics: whole genome transcriptional analysis. J Exp Bot 59:2267–2276

Novero AU, Taylor Paul WJ, Ford R (2008) Isolation and characterization of o-acetylserine (thiol) lyase, an enzyme of the cysteine biosynthetic pathway of vetch (*Vicia sativa* L.). Aust J Crop Sci 2:96–104

Ohkama N, Takei K, Sakakibara H, Hayashi H, Yoneyama T, Fujiwara T (2002) Regulation of sulfur-responsive gene expression by exogenously applied cytokinins in *Arabidopsis thaliana*. Plant Cell Physiol 43:1493–1501

Ohkama-Ohtsu N, Kasajima I, Fujiwara T, Naito S (2004) Isolation and characterization of an Arabidopsis mutant that overaccumulates *O*-acetyl-L-Ser. Plant Physiol 136:3209–3222

Okushima Y, Mitina I, Quach HL, Theologis A (2005a) AUXIN RESPONSE FACTOR 2 (ARF2): a pleiotropic developmental regulator. Plant J 43:29–46

Okushima Y, Overvoorde PJ, Arima K et al (2005b) Functional genomic analysis of the AUXIN RESPONSE FACTOR gene family members in *Arabidopsis thaliana*: unique and overlapping functions of ARF7 and ARF19. Plant Cell 17:444–463

Padovese R, Kina SM, Barros RMC, Borelli P, Marquez UML (2001) Biological importance of gamma-glutamyl-S-methylcysteine of kidney bean (*Phaseolus vulgaris* L.). Food Chem 73:291–297

Pandey M, Srivastava AK, D'Souza SF, Penna S (2013) Thiourea, a ROS scavenger, regulates source-to-sink relationship to enhance crop yield and oil content in *Brassica juncea* (L.). PLoS One 8(9):e73921. doi:10.1371/journal.pone.0073921

Parmar S, Buchner P, Hawkesford MJ (2007) Leaf developmental stage affects sulfate depletion and specific sulfate transporter expression during sulfur deprivation in *Brassica napus* L. Plant Biol 9:647–653

Paulose B, Kandasamy S, Dhankher OP (2010) Expression profiling of *Crambe abyssinica* under arsenate stress identifies genes and gene networks involved in arsenic metabolism and detoxification. BMC Plant Biol 10:108

Phartiyal P, Kim WS, Cahoon RE, Jez JM, Krishnan HB (2006) Soybean ATP sulfurylase, a homodimeric enzyme involved in sulfur assimilation, is abundantly expressed in roots and induced by cold treatment. Arch Biochem Biophys 450:20–29

Phartiyal P, Kim WS, Cahoon RE, Jez JM, Krishnan HB (2008) The role of 5′-adenylylsulfate reductase in the sulfur assimilation pathway of soybean: molecular cloning, gene expression, and kinetic characterization. Phytochemistry 69:356–364

Queval G, Thominet D, Vanacker H, Miginiac-Maslow M, Gakière B, Noctor G (2009) H2O2-activated up-regulation of glutathione in Arabidopsis involves induction of genes encoding enzymes involved in cysteine synthesis in the chloroplast. Mol Plant 2:344–356

Rae AL, Smith FW (2002) Localization of expression of a high affinity sulfate transporter in barley roots. Planta 215:565–568

Rai A, Tripathi P, Dwivedi S, Dubey S, Shri M, Kumar S, Tripathi PK, Dave R, Kumar A, Singh R, Adhikari B, Bag M, Tripathi RD, Trivedi PK, Chakrabarty D, Tuli R (2011) Arsenic tolerances in rice (*Oryza sativa*) have a predominant role in transcriptional regulation of a set of genes including sulphur assimilation pathway and anti- oxidant system. Chemosphere 82:986–995

Requejo R, Tena M (2006) Maize response to acute arsenic toxicity as revealed by proteome analysis of plant shoots. Proteomics 6:S156–S162

Rogg LE, Lasswell J, Bartel B (2001) A gain-of-function mutation in IAA28 suppresses lateral root development. Plant Cell 13:465–480

Rother M, Krauss GJ, Grass G, Wesenberg D (2006) Sulphate assimilation under Cd stress in *Physcomitrella patens*-combined transcript, enzyme and metabolite profiling. Plant Cell Environ 29:1801–1811

Rouached H, Secco D, Arpat AB (2009) Getting the most sulfate from soil: regulation of sulfate uptake transporters in Arabidopsis. J Plant Physiol 166:893–902

Rouhier N, Lemaire SD, Jacquot JP (2008) The role of glutathione in photosynthetic organisms: emerging functions for glutaredoxins and glutathionylation. Annu Rev Plant Biol 59:143–166

Ruiz JM, Blumwald E (2002) Salinity-induced glutathione synthesis in *Brassica napus*. Planta 214:965–969

Saito K (2000) Regulation of sulfate transport and synthesis of sulfur containing amino acids. Curr Opin Plant Biol 3:188–195

Scarafoni A, Ronchi A, Duranti M (2010) γ-Conglutin, the *Lupinus albus* XEGIP-like protein, whose expression is elicited by chitosan, lacks of the typical inhibitory activity against GH12 endo-glucanases. Phytochemistry 71:142–148

Schiavon M, Pittarello M, Pilon-Smits EAH, Wirtz M, Hell R, Malagoli M (2012) Selenate and molybdate alter sulfate transport and assimilation in *Brassica juncea* L. Czern.: implications for phytoremediation. Environ Exp Bot 75:41–51

Schippers JH, Nunes-Nesi A, Apetrei R, Hille J, Fernie AR, Dijkwel PP (2008) The *Arabidopsis* onset of leaf death5 mutation of quinolinate synthase affects nicotinamide adenine dinucleotide biosynthesis and causes early ageing. Plant Cell 20:2909–2925

Sharma I (2013) Arsenic-induced oxidative stress and antioxidant defense system of *Pisum sativum* and *Pennisetum typhoides*: a comparative study. Res J Biotechnol 8:48–56

Shibagaki N, Grossman AR (2006) The role of the STAS domain in the function and biogenesis of a sulfate transporter as probed by random mutagenesis. J Biol Chem 281:22964–22973

Shibagaki N, Rose A, McDermott JP, Fujiwara T, Hayashi H, Yoneyama T, Davies JP (2002) Selenate-resistant mutants of *Arabidopsis thaliana* identify Sultr1;2, a sulfate transporter required for efficient transport of sulfate into roots. Plant J 29:475–486

Shirzadian-Khorramabad R, Jing H-C, Everts GE, Schippers JHM, Hille J, Dijkwel PP (2010) A mutation in the cytosolic O-acetylserine (thiol) lyase induces a genome-dependent early leaf death phenotype in *Arabidopsis*. BMC Plant Biol 10:80

Smith FW, Hawkesford MJ, Ealing PM, Clarkson DT, van den Berg PJ, Belcher AR, Warrilow AGS (1997) Regulation of expression of a cDNA from barley roots encoding a high affinity sulphate transporter. Plant J 12:875–884

Song WY, Park J, Mendoza-Cozatl DG, Suter-Grotemeyer M, Shim D, Hortensteiner S, Geisler M, Weder B, Rea PA, Rentsch D, Schroeder JI, Lee Y, Martinoia E

(2010) Arsenic tolerance in Arabidopsis is mediated by two ABCC-type phytochelatin transporters. Proc Natl Acad Sci USA 107:21187–21192
Srivastava S, Mishra S, Tripathi RD, Dwivedi S, Trivedi PK, Tandon PK (2007) Phytochelatins and antioxidant systems respond differentially during arsenite and arsenate stress in *Hydrilla verticillata* (L.f.) Royle. Environ Sci Technol 41:2930–2936
Srivastava S, Srivastava AK, Suprasanna P, D'Souza SF (2009) Comparative biochemical and transcriptional profiling of two contrasting varieties of *Brassica juncea* L. in response to arsenic exposure reveals mechanisms of stress perception and tolerance. J Exp Bot 60:3419–3431
Srivastava AK, Ramaswamy NK, Suprasanna P, D'Souza SF (2010) Genome-wide analysis of thiourea-modulated salinity stress-responsive transcripts in seeds of *Brassica juncea*: identification of signalling and effector components of stress tolerance. Ann Bot 106:663–674
Stoeva N, Berova M, Zlatev Z (2005) Effect of arsenic on some physiological parameters in bean plants. Biol Plant 49:293–296
Sun X, Sun XM, Yang ZM, Li SQ, Wang J, Wang SH (2005) Expression of *Brassica napus* γ-glutamylcysteine synthetase, low- and high-affinity sulfate transporters in response to excess cadmium. J Integr Plant Biol 47:243–250
Sung DY, Kim TH, Komives EA, Mendoza-Cózatl DG, Schroeder JI (2009) ARS5 is a component of the 26S proteasome complex, and negatively regulates thiol biosynthesis and arsenic tolerance in Arabidopsis. Plant J 59:802–812
Suter M, von Ballmoos P, Kopriva S, den Camp RO, Schaller J, Kuhlemeier C, Schurmann P, Brunold C (2000) Adenosine 5′-phosphosulfate sulfotransferase and adenosine 5′-phosphosulfate reductase are identical enzymes. J Biol Chem 275:930–936
Tabe LM, Droux M (2002) Limits to sulfur accumulation in transgenic lupin seeds expressing a foreign sulfur-rich protein. Plant Physiol 128:1137–1148
Tabe LM, Venables I, Grootemaat A, Lewis D (2003) Sulfur transport and assimilation in developing embryos of chickpea (*Cicer arietinum*). In: Davidian J-C, Grill D, de Kok LJ, Stulen I, Hawkesford MJ, Schnug E, Rennenberg H (eds) Sulfur transport and assimilation in plants. Backhuys Publishers, Leiden, pp 335–337
Tabe L, Wirtz M, Molvig L, Droux M, Hell R (2010) Overexpression of serine acetyltransferase produced large increases in *O*-acetylserine and free cysteine in developing seeds of a grain legume. J Exp Bot 61:721–733
Takahashi H, Yamazaki M, Sasakura N, Watanabe A, Leustek T, Engler JA, Engler G, Van Montagu M, Saito K (1997) Regulation of sulfur assimilation in higher plants: a sulfate transporter induced in sulfate-starved roots plays a central role in *Arabidopsis thaliana*. Proc Natl Acad Sci USA 94:11102–11107
Takahashi H, Watanabe-Takahashi A, Smith FW, Blake-Klaff M, Hawkesford MJ, Saito K (2000) The roles of three functional sulphate transporters involved in uptake and translocation of sulphate in *Arabidopsis thaliana*. Plant J 23:171–182
Takahashi H, Watanabe-Takahashi A, Yamaya T (2003) T-DNA insertion mutagenesis of sulfate transporters in Arabidopsis. In: Davidian J-C, Grill D, de Kok LJ, Stulen I, Hawkesford MJ, Schnug E, Rennenberg H (eds) Sulfur transport and assimilation in plants. Backhuys Publishers, Leiden, pp 339–340
Takahashi H, Kopriva S, Giordano M, Saito K, Hell R (2011) Sulfur assimilation in photosynthetic organisms: molecular functions and regulations of transporters and assimilatory enzymes. Annu Rev Plant Biol 62:157–184
Talukdar D (2008) Cytogenetic characterization of seven different primary tetrasomics in grass pea (*Lathyrus sativus* L.). Caryologia 61:402–410
Talukdar D (2009a) Dwarf mutations in grass pea (*Lathyrus sativus* L.): origin, morphology, inheritance and linkage studies. J Genet 88(2):165–175
Talukdar D (2009b) Recent progress on genetic analysis of novel mutants and aneuploid research in grass pea (*Lathyrus sativus* L.). Afr J Agric Res 4:1549–1559
Talukdar D (2010a) Reciprocal translocations in grass pea (*Lathyrus sativus* L.). Pattern of transmission, detection of multiple interchanges and their independence. J Hered 101:169–176
Talukdar D (2010b) Cytogenetic characterization of induced autotetraploids in grass pea (*Lathyrus sativus* L.). Caryologia 63:62–72
Talukdar D (2010c) Fluorescent-banded karyotype analysis and identification of chromosomes in three improved Indian varieties of grass pea (*Lathyrus sativus* L.). Chromosome Sci 13:3–10
Talukdar D (2011a) Cytogenetic analysis of a novel yellow flower mutant carrying a reciprocal translocation in grass pea (*Lathyrus sativus* L.). J Biol Res-Thessaloniki 15:123–134
Talukdar D (2011b) Isolation and characterization of NaCl-tolerant mutations in two important legumes, *Clitoria ternatea* L. and *Lathyrus sativus* L.: induced mutagenesis and selection by salt stress. J Med Plants Res 5(16):3619–3628
Talukdar D (2011c) Morpho-physiological responses of grass pea (*Lathyrus sativus* L.) genotypes to salt stress at germination and seedling stages. Legume Res 34(4):232–241
Talukdar D (2011d) The aneuploid switch: extra-chromosomal effect on antioxidant defense through trisomic shift in *Lathyrus sativus* L. Indian J Fundam Appl Life Sci 1(4):263–273
Talukdar D (2011e) Flower and pod production, abortion, leaf injury, yield and seed neurotoxin levels in stable dwarf mutant lines of grass pea (*Lathyrus sativus* L.) differing in salt stress responses. Int J Curr Res 2(1):46–54
Talukdar D (2011f) Effect of arsenic-induced toxicity on morphological traits of *Trigonella foenum-graecum* L.

and *Lathyrus sativus* L. during germination and early seedling growth. Curr Res J Biol Sci 3(2):116–123
Talukdar D (2012a) Ascorbate deficient semi-dwarf *asfL1* mutant of *Lathyrus sativus* exhibits alterations in antioxidant defense. Biol Plant 56(4):675–682
Talukdar D (2012b) Flavonoid-deficient mutants in grass pea (*Lathyrus sativus* L.): genetic control, linkage relationships, and mapping with aconitase and S nitrosoglutathione reductase isozyme loci. Sci World J 2012, Article ID 345983, 11 pages. doi:10.1100/2012/345983
Talukdar D (2012c) A glutathione-overproducing mutant in grass pea (*Lathyrus sativus* L.): alterations in glutathione content, modifications in antioxidant defense response to cadmium stress and genetic analysis using primary trisomic. Int J Recent Sci Res 3(4):234–243
Talukdar D (2012d) An induced glutathione-deficient mutant in grass pea (*Lathyrus sativus* L.): modifications in plant morphology, alteration in antioxidant activities and increased sensitivity to cadmium. Biorem Biodiv Bioavail 6:75–86
Talukdar D (2012e) Exogenous calcium alleviates the impact of cadmium-induced oxidative stress in *Lens culinaris* Medic. Seedlings through modulation of antioxidant enzyme activities. J Crop Sci Biotechnol 15(4):325–334
Talukdar D (2012f) Total flavonoids, phenolics, tannins and antioxidant activity in seeds of lentil and grass pea. Int J Phytomedicine 4(4):537–542
Talukdar D (2012g) Modulation of plant growth and leaf biochemical parameters in grass pea (*Lathyrus sativus* L) and fenugreek (*Trigonella foenum-graecum* L.) exposed to NaCl treatments. Indian J Fundam Appl Life Sci 2(3):20–28
Talukdar D (2012h) Changes in neurotoxin, β-N-OXALYL- L α, β-diaminopropionic acid (β-ODAP), level in grass pea (*Lathyrus sativus* L.) genotypes under arsenic treatments. J Appl Biosci 38(2):148–153
Talukdar D (2013a) Bioaccumulation and transport of arsenic in different genotypes of lentil (*Lens culinaris* Medik.). Int J Pharma Bio Sci 4(1):B694–B701
Talukdar D (2013b) Arsenic-induced oxidative stress in the common bean legume, *Phaseolus vulgaris* L. seedlings and its amelioration by exogenous nitric oxide. Physiol Mol Biol Plants 19(1):69–79
Talukdar D (2013c) Studies on antioxidant enzymes in *Canna indica* plant under copper stress. J Environ Biol 34:93–98
Talukdar D (2013d) Bioaccumulation, growth and antioxidant defense responses of *Leucaena* species differing in arsenic tolerance. Int J Bot Res 3(1):1–18
Talukdar D (2013e) Allelopathic effects of *Lantana camara* L. on *Lathyrus sativus* L.: oxidative imbalance and cytogenetic consequences. Allelopathy J 31(1):71–90
Talukdar D (2013f) Selenium priming selectively ameliorates weed– induced phytotoxicity by modulating antioxidant defense components in lentil (*Lens culinaris* Medik.) and grass pea (*Lathyrus sativus* L.). Annu Rev Res Biol 3(3):195–212
Talukdar D (2013g) Cytogenetics of a reciprocal translocation integrating distichous pedicel and tendril-less leaf mutations in *Lathyrus sativus* L. Caryologia 66(1):21–30
Talukdar D (2013h) Arsenic-induced changes in growth and antioxidant metabolism of fenugreek. Russ J Plant Physiol 60(5):652–660
Talukdar D (2013i) Arsenic exposure modifies *Fusarium* wilt tolerance in grass pea (*Lathyrus sativus* L.) genotypes through modulation of antioxidant defense response. J Plant Sci Mol Breed 2(4), 12 pages. doi:http://dx.doi.org/10.7243/2050-2389-2-4
Talukdar D (2013j) Balanced hydrogen peroxide metabolism is central in controlling NaCl-induced oxidative stress in medicinal legume, fenugreek (*Trigonella foenum-graecum* L.). Biochem Mol Biol 1(2):34–43
Talukdar D (2013k) Plant growth and leaf antioxidant metabolism of four elite grass pea (*Lathyrus sativus*) genotypes, differing in arsenic tolerance. Agric Res 2(4):330–339
Talukdar D (2013l) Comparative morpho-physiological and biochemical responses of lentil and grass pea genotypes under water stress. J Nat Sci Biol Med 4(2):396–402
Talukdar D (2013m) Growth responses and leaf antioxidant metabolism of grass pea (*Lathyrus sativus* L.) genotypes under salinity stress. ISRN Agron 2013, Article ID 284830, 15 pages. doi:http://dx.doi.org/10.1155/2013/284830
Talukdar D, Biswas AK (2005) Induced seed coat colour mutations and their inheritance in grass pea (*Lathyrus sativus* L.). Indian J Genet 65:135–136
Talukdar D, Biswas AK (2007a) Seven different primary trisomics in grass pea (*Lathyrus sativus* L.). I Cytogenetic characterization. Cytologia 72(4):385–396
Talukdar D, Biswas AK (2007b) Inheritance of flower and stipule characters in different induced mutant lines of grass pea (*Lathyrus sativus* L.). Indian J Genet Plant Breed 67(4):396–400
Talukdar D, Talukdar T (2013a) Catalase-deficient mutants in lentil (*Lens culinaris* Medik.): perturbations in morpho-physiology, antioxidant redox and cytogenetic parameters. Int J Agric Sci Res 3(2):197–212
Talukdar D, Talukdar T (2013b) Superoxide-dismutase deficient mutants in common beans (*Phaseolus vulgaris* L.): genetic control, differential expressions of isozymes, and sensitivity to arsenic. BioMed Res Int, ePub 8/8/13, 2013, Article ID 782450, 11 pages. doi:http://dx.doi.org/10.1155/2013/782450
Talukdar T, Talukdar D (2013c) Response of antioxidative enzymes to arsenic-induced phytotoxicity in leaves of a medicinal daisy, *Wedelia chinensis* Merrill. J Nat Sci Biol Med 4(2):383–388
Talukdar D, Talukdar T (2013d) Coordinated response of sulfate transport, cysteine biosynthesis and glutathione-mediated antioxidant defense in lentil

(*Lens culinaris* Medik.) genotypes exposed to arsenic. Protoplasma. doi:10.1007/s00709-013-0586-8

Talukdar T, Talukdar D (2014) Leaf photosynthesis and antioxidant defense in male and hermaphrodite tree of a critically endangered legume, *Gymnocladus assamicus* Kanjilal ex P.C. Kanjilal. Plant Gene Trait 5(1):1–10

Talukdar D, Biswas SC, Biswas AK (2002) An induced flower colour mutant in grass pea (*Lathyrus sativus* L.). Indian J Genet 62:162

Tan Q, Zhang L, Grant J, Cooper P, Tegeder M (2010) Increased phloem transport of *S*-methylmethionine positively affects sulfur and nitrogen metabolism and seed development in pea plants. Plant Physiol 154:1886–1896

Taylor M, Chapman R, Beyaert R, Hernández-Sebastià C, Marsolais F (2008) Seed storage protein deficiency improves sulfur amino acid content in common bean (*Phaseolus vulgaris* L.): redirection of sulfur from gamma-glutamyl-*S*-methyl-cysteine. J Agric Food Chem 56:5647–5654

Tomatsu H, Takano J, Takahashi H, Watanabe-Takahashi A, Shibagaki N, Fujiwara T (2007) An *Arabidopsis thaliana* high-affinity molybdate transporter required for efficient uptake of molybdate from soil. Proc Natl Acad Sci USA 104:18807–18812

Traverso JA, Pulido A, Rodríguez-García MI, Alché JD (2013) Thiol-based redox regulation in sexual plant reproduction: new insights and perspectives. Front Plant Sci 4:465

Tripathi P, Mishra A, Dwivedi S, Chakrabarty D, Trivedi PK, Singh RP, Tripathi RD (2012a) Differential response of oxidative stress and thiol metabolism in contrasting rice genotypes for arsenic tolerance. Ecotoxicol Environ Saf 79:189–198

Tripathi P, Tripathi RD, Singh RP, Dwivedi S, Chakrabarty D, Prabodh K, Trivedi PK, Adhikari B (2012b) Arsenite tolerance in rice (*Oryza sativa* L.) involves coordinated role of metabolic pathways of thiols and amino acids. Environ Sci Pollut Res. doi:10.1007/s11356-012-1205-5

Tripathi RD, Tripathi P, Dwivedi S, Dubey S, Chatterjee S, Chakrabarty D, Trivedi PK (2012c) Arsenomics: omics of arsenic metabolism in plants. Front Physiol 3:275

van de Mortel JE, Schat H, Moerland PD, Loren V, van Themaat E, van der Ent S, Blankestijn H, Ghandilyan A, Tsiatsiani S, Aarts MGM (2008) Expression differences for genes involved in lignin, glutathione and sulphate metabolism in response to cadmium in *Arabidopsis thaliana* and the related Zn/Cd-hyperaccumulator *Thlaspi caerulescens*. Plant Cell Environ 31:301–324

Van Hoewyk D, Takahashi H, Inoue E, Hess A, Tamaoki M, Pilon-Smits EA (2008) Transcriptome analyses give insights into selenium-stress responses and selenium tolerance mechanisms in Arabidopsis. Physiol Plant 132:236–253

Vauclare P, Kopriva S, Fell D, Suter M, Sticher L, von Ballmoos P, Krahenbuhl U, den Camp RO, Brunold C (2002) Flux control of sulphate assimilation in *Arabidopsis thaliana*: adenosine 5′-phosphosulphate reductase is more susceptible than ATP sulphurylase to negative control by thiols. Plant J 31:729–740

Vaz Patto MC, Skiba B, Pang ECK, Ochatt SJ, Lambein F, Rubiales D (2006) Lathyrus improvement for resistance against biotic and abiotic stresses: from classical breeding to marker assisted selection. Euphytica 147:133–147

Vernoux T, Wilson RC, Seeley KA, Reichheld JP, Muroy S, Brown S, Maughan SC, Cobbett CS, Van Montsgu M, Inzé D, May MJ, Sung ZR (2000) The *ROOT MERISTEMLESS/CADMIUM SENSITIVE2* gene defines a glutathione-dependent pathway involved in initiation and maintenance of cell division during post-embryonic root development. Plant Cell 12:97–110

Vidmar JJ, Tagmount A, Cathala N, Touraine B, Davidian J-C (2000) Cloning and characterization of a root specific high-affinity sulfate transporter from *Arabidopsis thaliana*. FEBS Lett 475:65–69

Wachter A, Wolf S, Steininger H, Bogs J, Rausch T (2005) Differential targeting of GSH1 and GSH2 is achieved by multiple transcription initiation: implications for the compartmentation of glutathione biosynthesis in the Brassicaceae. Plant J 41:15–30

Watanabe M, Kusano M, Oikawa A, Fukushima A, Noji M, Saito K (2008a) Physiological roles of the beta-substituted alanine synthase gene family in Arabidopsis. Plant Physiol 146:310–320

Watanabe M, Mochida K, Kato T, Tabata S, Yoshimoto N, Noji M, Saito K (2008b) Comparative genomics and reverse genetics analysis reveal indispensable functions of the serine acetyltransferase gene family in Arabidopsis. Plant Cell 20:2484–2496

Watanabe M, Hubberten H-M, Saito K, Hoefgen R (2010) General regulatory patterns of plant mineral nutrient depletion as revealed by serat quadruple mutants disturbed in cysteine synthesis. Mol Plant 3:438–466

Wawrzyńska A, Kurzyk A, Mierzwińska M, Płochocka D, Wieczorek G, Sirko A (2013) Direct targeting of Arabidopsis cysteine synthase complexes with synthetic polypeptides to selectively deregulate cysteine synthesis. Plant Sci 207:148–157

Wei S, Ma LQ, Saha U, Mathews S, Sundaram S, Rathinasabapathi B, Zhou Q (2010) Sulfate and glutathione enhanced arsenic accumulation by arsenic hyperaccumulator *Pteris vittata* L. Environ Pollut 158:1530–1535

White PJ, Bowen HC, Marshall B, Broadley MR (2007) Extraordinarily high leaf selenium to sulfur ratios define 'Se-accumulator' plants. Ann Bot (Lond) 100:111–118

Wickett NJ, Forrest LL, Budke JM, Shaw B, Goffinet B (2011) Frequent pseudogenization and loss of the plastid-encoded sulfate-transport gene *cys*A throughout the evolution of liverworts. Am J Bot 98:1263–1275

Wirtz M, Droux M, Hell R (2004) O-acetylserine (thiol) lyase: an enigmatic enzyme of plant cysteine biosyn-

thesis revisited in *Arabidopsis thaliana*. J Exp Bot 55:1785–1798
Wirtz M, Birke H, Heeg C, Mueller C, Hosp F, Throm C, Koenig S, Feldman-Salit A, Rippe K, Petersen G, Wade RC, Rybin V, Scheffzek K, Hell R (2010) Structure and function of the hetero-oligomeric cysteine synthase complex in plants. J Biol Chem 285:32810–32817
Wirtz M, Beard KF, Lee CP, Boltz A, Schwarzlander M, Fuchs C, Meyer AJ, Heeg C, Sweetlove LJ, Ratcliffe RG, Hell R (2012) Mitochondrial cysteine synthase complex regulates O-acetylserine biosynthesis in plants. J Biol Chem 287:27941–27947
Wu Y, Zhao Q, Gao L, Yu X-M, Fang P, Oliver DJ, Xiang C-B (2010) Isolation and characterization of low-sulphur-tolerant mutants of Arabidopsis. J Exp Bot 61:3407–3422
Xiang C, Oliver DJ (1998) Glutathione metabolic genes coordinately respond to heavy metals and jasmonic acid in Arabidopsis. Plant Cell 10:1539–1550
Yamaguchi Y, Nakamura T, Harada E, Koizumi N, Sano H (1999) Differential accumulation of transcripts encoding sulfur assimilation enzymes upon sulfur and/or nitrogen deprivation in *Arabidopsis thaliana*. Biosci Biotechnol Biochem 63:762–766
Yi H, Galant A, Ravilious GE, Preuss ML, Jez JM (2010) Sensing sulfur conditions: simple to complex biochemical regulatory mechanisms in plant thiol metabolism. Mol Plant 3:269–279
Yin F, Pajak A, Chapman R, Sharpe A, Huang S, Marsolais F (2011) Analysis of common bean expressed sequence tags identifies sulfur metabolic pathways active in seed and sulfur-rich proteins highly expressed in the absence of phaseolin and major lectins. BMC Genomics 12:268
Yoshida S, Tamaoki M, Ioki M, Ogawa D, Sato Y, Aono M, Kubo A, Saji S, Saji H, Satoh S, Nakajima N (2009) Ethylene and salicylic acid control glutathione biosynthesis in ozone-exposed *Arabidopsis thaliana*. Physiol Plant 136:284–298
Yoshimoto N, Takahashi H, Smith FW, Yamaya T, Saito K (2002) Two distinct high-affinity sulfate transporters with different inducibilities mediate uptake of sulfate in Arabidopsis root. Plant J 29:465–473
Yoshimoto N, Inoue E, Saito K, Yamaya T, Takahashi H (2003) Phloem-localizing sulfate transporter, Sultr1;3, mediates redistribution of sulfur from source to sink organs in Arabidopsis. Plant Physiol 131:1511–1517
Yoshimoto N, Inoue E, Watanabe-Takahashi A, Saito K, Takahashi H (2007) Posttranscriptional regulation of high-affinity sulfate transporters in Arabidopsis by sulfur nutrition. Plant Physiol 145:378–388
Yoshizawa T, Shimizu T, Yamabe M, Taichi M, Nishiuchi Y, Shichijo N, Unzai S, Hirano H, Sato M, Hashimoto H (2011) Crystal structure of basic 7S globulin, a xyloglucan-specific endo-β-1,4-glucanase inhibitor protein-like protein from soybean lacking inhibitory activity against endo-β-glucanase. FEBS J 278:1944–1954
Zagorchev L, Seal CE, Kranner I, Odjakova M (2013) A central role for thiols in plant tolerance to abiotic stress. Int J Mol Sci 14:7405–7432
Zhang C, Meng Q, Zhang M, Huang F, Gai J, Yu D (2008) Characterization of O-acetylserine(thiol)lyase-encoding genes reveals their distinct but cooperative expression in cysteine synthesis of soybean [*Glycine max* (L.) Merr.]. Plant Mol Biol Rep 26:277–291
Zhao FJ, Stroud JL, Khan MA, McGrath SP (2012) Arsenic translocation in rice investigated using radioactive73As tracer. Plant Soil 350:413–442
Zuber H, Davidian JC, Aubert G, Aimé D, Belghazi M, Lugan R, Heintz D, Wirtz M, Hell R, Thompson R, Gallardo K (2010) The seed composition of arabidopsis mutants for the group 3 sulfate transporters indicates a role in sulfate translocation within developing seeds. Plant Physiol 154:913–926

Chloroplast Omics

L.A. de Luna-Valdez, P. León-Mejía,
S. Encarnación-Guevara, and A.A. Guevara-García

Contents

Abstract

The chloroplast is the most remarkable organelle of plant cells; it is the site of a myriad of different chemical reactions; among chloroplast's many functionalities is photosynthesis, perhaps the most fundamental biological process on the biosphere. The chloroplast has been subject of a plethora of research efforts that try to understand the molecular mechanisms that regulate its biochemical capabilities, development, and evolutionary origin. Omic technologies have provided researchers with tools to study different aspects of biology from a global perspective, and, not surprisingly, chloroplast research has taken advantage of them. This chapter explores how chloroplasts organize their genomes and regulate their transcriptomes, proteomes, and metabolomes, trying to focus on classical knowledge and reviewing new datasets obtained through large-scale research projects that shed light on chloroplast functionality.

Keywords

Chloroplast • Chlorogenomes • Chlorotranscriptomes • Chloroproteomes • Chlorometabolomes • Chloroplast's system biology

L.A. de Luna-Valdez, Ph.D. • P. León-Mejía, Ph.D.
A.A. Guevara-García, Ph.D. (✉)
Instituto de Biotecnología, Universidad Nacional Autónoma de México, Apartado Postal 510-3, 62250 Cuernavaca, Morelos, Mexico
e-mail: aguevara@ibt.unam.mx

S. Encarnación-Guevara, Ph.D.
Centro de Ciencias Genómicas, Universidad Nacional Autónoma de México, Cuernavaca 62271, Morelos, Mexico

D. Barh et al. (eds.), *PlantOmics: The Omics of Plant Science*,
DOI 10.1007/978-81-322-2172-2_18, © Springer India 2015

Introduction

Chloroplasts are semiautonomous organelles that came in the eukaryotic life scene around 1.6 billion years ago, through a process of primary endosymbiosis between cyanobacteria and ancient heterotrophic mitochondria-bearing eukaryotes (Yoon et al. 2004). From that moment on, chloroplasts have integrated themselves in almost every aspect of the biology of plants; chloroplasts perform functions that impact a wide range of processes like ecological traits (volatile emissions), specific molecular events (chloroplast-to-nucleus signaling), and biomass production.

As several other well-studied endosymbionts, at the morphological level, chloroplasts are surrounded by two lipid bilayers that enclose a fluid compartment called stroma (Keeling 2010). The stroma contains a most distinctive net of internal membranes known as thylakoids, whose lumen is also fluid. The high degree of compartmentalization existing in chloroplasts is essential for the proper production of the plethora of metabolites and chemical processes that find source in the membranes and compartments of the chloroplasts. For instance take the carbon fixation process, commonly known as photosynthesis; this process is a series of energy-driven chemical reactions that mediate the assimilation of inorganic carbon atoms into sugar molecules. Photosynthesis is composed of two conceptually independent sets of reactions, the light-dependent and light-independent reactions; light-dependent reactions take place in the thylakoid membrane and involve the conversion of light energy to chemical energy through a series of electron donors and acceptors that are associated to the thylakoid membrane. Starting from antenna and reaction center pigments (such as chlorophylls and carotenes), the electrons travel through proteins like plastoquinones and cytochromes to finally reduce NADPH (nicotinamide adenine dinucleotide phosphate). Along the way from the antenna to the NADPH, electron carriers mediate the pumping of H^+ from the chloroplast stroma to the lumen of the thylakoids, this way forming an electrochemical gradient that is used by an ATP (adenosine triphosphate) synthase complex oriented towards the plastid stroma to generate ATP in that compartment. During light-independent reactions, enzymes in the stroma use the ATP and NADPH produced during the light-dependent reactions to generate sugars, from inorganic CO_2 molecules (Blankenship 2002); it is important to note that compartmentalized changes in pH are key to regulate enzymes important for the light-independent reactions to take place, such as ribulose-bisphosphate carboxylase-oxygenase whose activation is facilitated by the generation of a pH gradient through the thylakoid membrane (Campbell and Ogren 1990; Chen et al. 2010). In this simple example, it results evident how the existence of several compartments in the chloroplasts is essential to the performance of one given function.

Since photosynthesis is the source of all the organic carbon found in every single metabolite, chloroplast function results essential to almost every process in the cell, and synthesis pathways of major metabolites and the photosynthetic activity of the plant tissues are fine tuned (Paul and Pellny 2003). Even though dependent on carbon fixed during photosynthesis, other important chemical reactions take place inside chloroplasts such as lipids, isoprenoids, amino acids, proteins, and complex carbohydrate synthesis (Kannangara et al. 1971; Lichtenthaler 1999; Kirk and Leech 1972; Siddell and Ellis 1975), making the chloroplast a source of energy, structural components, and a plethora of signals that regulate several aspects of plant development.

Given the importance of chloroplasts for life, several efforts have been made worldwide in order to grasp understanding of the intricate regulatory networks underlying its origin, function and development. Here we present a review of the current knowledge on how chloroplasts organize their genomes and regulate their transcriptomes, proteomes, metabolomes, and even some new and not so well-studied fields, such as lipidomics.

Chloroplast Genomes

In 1920 Hans Winkler adapted the term genome to refer to the haploid chromosome set of a given organism. More than 50 years later, in 1976, Walter Fiers reported the complete sequence of the bacteriophage MS2 RNA genome (Fiers et al. 1976); in 1977 Frederick Sanger reported the first sequence of a DNA (deoxyribonucleic acid) genome (Phage Φ X-174; Sanger et al. 1977). In the following years, the genomic sequences of representative organisms from the three different domains of life were reported, being *Haemophilus influenzae*, *Saccharomyces cerevisiae*, and *Methanococcus jannaschii* the firsts on their respective domains (Fleischmann et al. 1995; Goffeau et al. 1996; Bult et al. 1996). Furthermore, in 1986 the first two chloroplast genomes came into view when two different Japanese research teams reported full sequences for the chloroplast genomes of *Marchantia polymorpha* and *Nicotiana tabacum* (Ohyama et al. 1986; Shinozaki et al. 1986). Nowadays, the next generation of DNA sequencing technologies has advanced in precision, time consumption and cost, to the point that genomic information for 10,904 organisms is now publicly available; among them only 197 genomes correspond to plants (http://www.ncbi.nlm.nih.gov/genome/browse/).

It is now common knowledge that eukaryotic organelles like mitochondria and plastids originated by means of a process called endosymbiosis, in which prokaryotic cells were engulfed by pro-eukaryotic cells and then underwent a process of functional specialization. Organelles generated this way have a very unique trait compared to other cell organelles: they have their own DNA genome, which is a reminiscent of their past prokaryotic genomes. The chloroplast genome (cpDNA) varies little in topology and content among plants, it is generally accepted that cpDNA molecules are made of quadripartite circular molecules of around 145 kbps in size, this molecules are organized in two single-copy segments of different length (long single copy, LSC; short single copy, SSC) which are separated by two inverted repeat segments (IR) (Fig. 1) (Palmer 1991). Even if it is widely accepted that cpDNA molecules are circular molecules, over the years evidence has been accumulated showing that the circle does not define the most common conformation of cpDNA molecules. As a matter of fact, early evidence showed that circular cpDNA molecules represent only the 37 % of the total cpDNA molecules extracted from pea chloroplasts (Kolodner and Tewari 1972). For decades the view of circular cpDNA molecules dominated, and the theory suggested that most of the cpDNA molecules suffer random breakage during the extraction process and then concatenated, this way explaining the oligomeric linear forms of cpDNA molecules found in all the experiments reported. In contrast, in 2004, Oldenburg and Bendich reported high-resolution digital images of ethidium-stained plastid DNA extracted from maize meristematic tissue, showing the cpDNA as complex branched linear molecules whose size was several times larger than the genome size, it was also demonstrated that the ends of such molecules are specific rather than random, as expected from molecules randomly broken during the extraction process (Oldenburg and Bendich 2004). Furthermore, it is now proposed that cpDNA replication does not proceed by the canonical cairn structure and rolling circle mechanism; instead, a recombination-

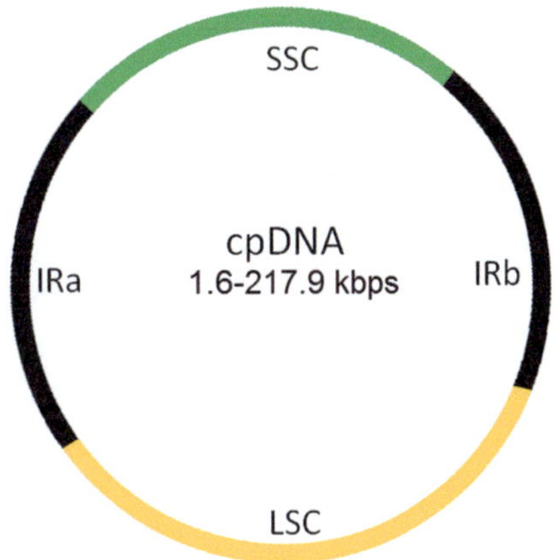

Fig. 1 **Map of a standard cpDNA molecule**. Graphical representation of a standard circular cpDNA molecule. Annotated LSC (large single copy), SSC (short single copy), and IR (inverted repeat) elements

dependent mechanism is thought to mediate replication, since this mechanism better explains the formation of the reported head-to-tail-branched concatemers (Bendich 2004). Evidence of these linear complexes of cpDNA has been reported for several species such as *A. thaliana*, maize, pea, and watermelon (Bendich 1991, 2004; Rowan et al. 2004). Even though there is plenty of evidence to show the linear nature of cpDNA molecules, to this date, chloroplast genomes are still reported as circular molecules, whether the idea of a circular cpDNA molecule is very well rooted in the scientific community or it is used as a way to homogenize the way chloroplast genomes are reported, future literature must adapt to accommodate the findings about linear cpDNA molecules.

Despite the controversial information regarding the shape of cpDNA molecules, it is very well known that the standard cpDNA molecule has around 128 different genes organized in several clusters; these cpDNA-encoded genes include protein (84), tRNA (37), and rRNA (8) coding sequences (Saski et al. 2005; Wu et al. 2009). Table 1 compiles publicly available information on genome size and gene content for chloroplastic genomes of several economically important plants. To this date, 555 entries exist that report sequences for cpDNA of different species (32 Chlorophyta and 523 Streptophyta) in the Viridiplantae clade (http://www.ncbi.nlm.nih.gov/genomes/ORGANELLES/organelles.html). From the Streptophyta information available at NCBI, *Conopholis americana (American cancer-root)* and *Pelargonium hortorum (Garden geranium)* represent the smallest and the largest cpDNA sequences, respectively. The cpDNA of *Conopholis americana* has only 45.63 kbps, while the cpDNA of *Pelargonium hortorum* is 217.9 kbps in length (Fig. 2). In accordance to the great difference existing in genome length, differences exist in the number of genes coded by the cpDNA of each species, the cpDNA of *Pelargonium hortorum* contains 181 genes (131 proteins, 40 tRNAs, 10 rRNAs), while *Conopholis americana* cpDNA codes only for 44 genes (21 proteins, 18 tRNAs, 4 rRNAs); functional characterization of the proteins coded by each gene reveals that *Pelargonium hortorum* codes for proteins with a wide set of functionalities, including photosynthesis, transcription, translation, and energy metabolism (Fig. 2). On the other hand, *Conopholis americana* cpDNA lacks most of the plastid-encoded photosynthetic and energy metabolism genes, presenting genes coding for the plastid translation machinery; this information correlates with the parasitic lifestyle of this plant species (Fig. 2a). The information regarding the specific genes encoded by all the cpDNA sequences available to this date is compiled in the cpBase: The Chloroplast Genome Database (http://chloroplast.ocean.washington.edu/). Chloroplast genome information has seeded the development of one particular field of plant biology, the chloroplast phylogenomics; this approach tries to uncover the phylogenetic relationships existing between different plant species by analyzing their cpDNA sequences. Chloroplast phylogenomics, aided by the available high-throughput next-generation technologies of DNA sequencing, has been applied successfully to resolve the controverted hypothesis dealing with the proper placement of the genus ginkgo in the phylogenetic tree of land plants, strongly supporting the previously proposed monophyly between ginkgo and cycad groups (Wu et al. 2013). Moreover, cpDNA sequences generated by state-of-the-art sequencing methods have been used to explore the angiosperm phylogenetic tree; it was found that the most basal lineage of angiosperms is represented by aquatic and herbaceous plants, whose surviving relatives are represented by plants of the genuses *Trithuria* and *Amborella* and the family Nymphaeaceae, instead of the previous belief that pointed the plants in the genus *Amborella* as the only surviving organisms related to the root of the angiosperm phylogenetic tree (Goremykin et al. 2013).

Developing photosynthetic tissues contain around one thousand cpDNA molecules per plastid; though the functionality of such a high copy number during development is not clear, two major hypotheses try to explain this phenomenon. First, high copy number may compensate random sorting of cpDNA molecules during plastid division; second, the increased gene dosage

Table 1 Gene content of chloroplastic genomes of several crop species

Species	Accession	Size (Kbps)	Genes		
			Protein	tRNA	rRNA
Adiantum capillus-veneris	NC_004766	150.568	87	35	8
Agrostis stolonifera	NC_008591	136.584	85	40	8
Anthriscus cerefolium	NC_015113	154.719	85	37	8
Brassica napus	NC_016734	152.86	87	37	8
Capsella bursa-pastoris	NC_009270	154.49	85	37	8
Capsicum annuum	NC_018552	156.781	86	38	8
Carica papaya	NC_010323	1.601	84	37	8
Castanea mollissima	NC_014674	160.799	83	37	8
Cedrus deodara	NC_014575	119.299	75	35	4
Cicer arietinum	NC_011163	125.319	75	29	4
Citrus sinensis	NC_008334	160.129	87	45	8
Coffea arabica	NC_008535	155.189	85	45	8
Coix lacryma-jobi	NC_013273	140.745	104	40	8
Colocasia esculenta	NC_016753	162.424	86	37	8
Cucumis melo subsp. melo	NC_015983	156.017	88	37	8
Cucumis sativus	NC_007144	155.293	85	37	8
Elaeis guineensis	NC_017602	156.973	86	38	8
Festuca arundinacea	NC_011713	136.048	80	38	8
Glycine max	NC_007942	152.218	83	37	8
Gossypium hirsutum	NC_007944	160.301	83	37	8
Helianthus annuus	NC_007977	151.104	85	43	8
Hevea brasiliensis	NC_015308	161.191	84	37	8
Hordeum vulgare subsp. vulgare	NC_008590	136.462	83	48	8
Lactuca sativa	NC_007578	152.765	84	37	7
Lathyrus sativus	NC_014063	121.02	74	30	4
Liriodendron tulipifera	NC_008326	159.886	84	37	8
Nicotiana tabacum	NC_001879	155.943	98	37	8
Olea europaea	NC_013707	155.872	85	37	8
Oryza sativa Japonica	NC_001320	134.525	108	38	8
Phaseolus vulgaris	NC_009259	150.285	83	36	8
Ricinus communis	NC_016736	163.161	86	37	8
Sesamum indicum	NC_016433	153.324	87	37	8
Solanum tuberosum	NC_008096	155.296	84	45	8
Sorghum bicolor	NC_008602	140.754	84	48	8
Spinacia oleracea	NC_002202	150.725	96	37	8
Theobroma cacao	NC_014676	160.619	81	37	8
Triticum aestivum	NC_002762	134.545	83	42	8
Vigna unguiculata	NC_018051	152.415	84	38	8
Vitis vinifera	NC_007957	160.928	84	45	8
Zea mays	NC_001666	140.384	111	38	8

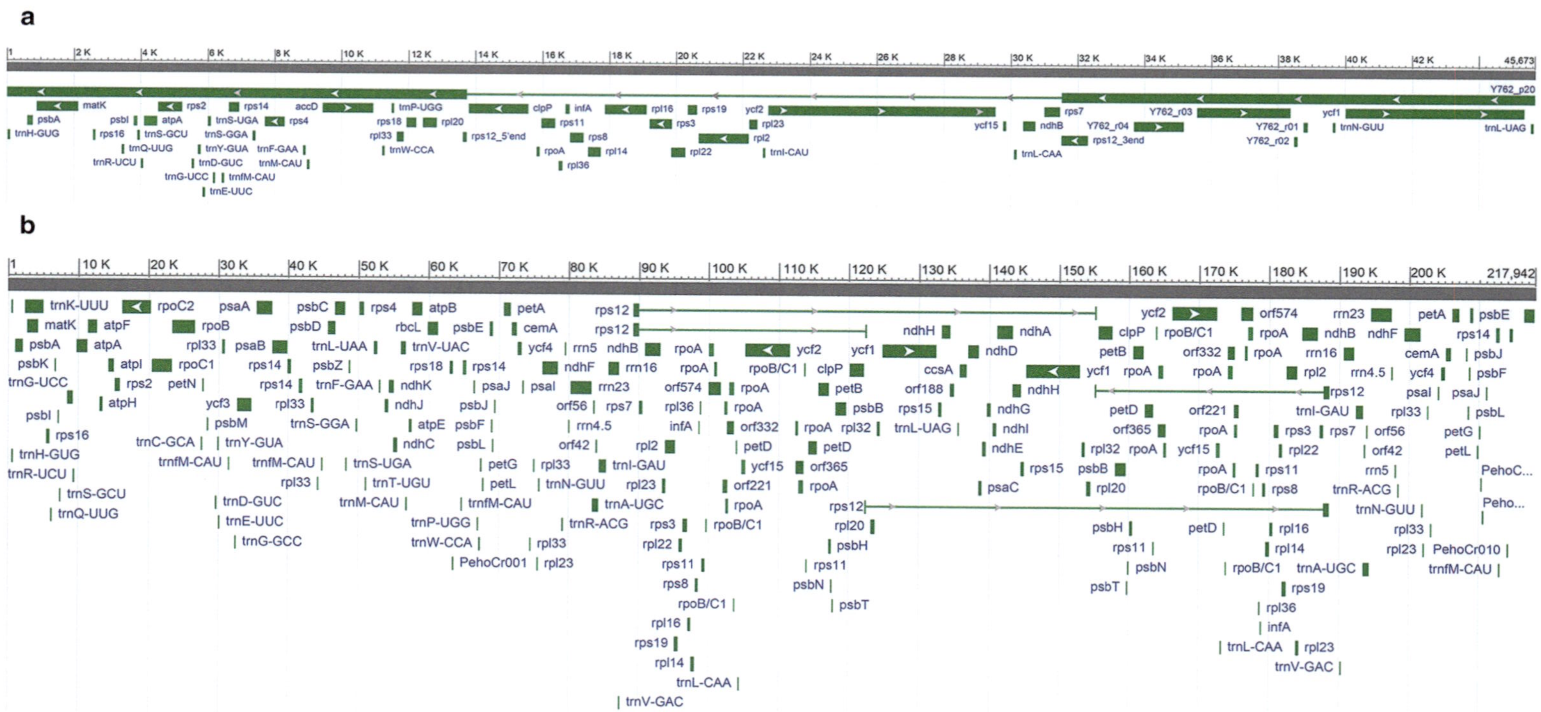

Fig. 2 **cpDNA maps for two plant species**. Graphical representation of the cpDNA maps for *Conopholis americana* (**a**) and *Pelargonium hortorum* (**b**). The genes coded by each genome are represented by *green bars*

may be necessary in early developmental stages to maintain the high demand of photosynthetic proteins necessary for the establishment of photoautotrophic metabolism (Bendich 1987). Recently, analysis of the maize albino mutant *w2* (defective in a chloroplast-specific DNA polymerase) has proven the functional importance of gene dosage in maize chloroplasts; it is shown that protein subunits of photosynthetic enzymes are affected to levels not matching to the decrease in their respective mRNA levels, suggesting the lowered availability of rRNAs (and hence plastid ribosomes) and tRNAs affects the rate of protein synthesis and directly relating the defective plastid function to gene dosage (Udy et al. 2012).

Besides the knowledge available related to cpDNA-encoded genes, there is also data showing that there is plenty of noncoding regions in cpDNA molecules. Around 42.9 % of the total length of the *Nicotiana* LSC and SSC regions is noncoding DNA (10.6 % introns and 32.3 % intergenic regions) (Shaw et al. 2007), and the precise knowledge of the main function of these noncoding sequences and the coding sequences they might regulate is pivotal for the development of transplastomic plant lines, a field that has been successful in overcoming specific crop-related problems.

In 1988 Boynton et al. reported a method based on the bombardment of *Chlamydomonas reinhardtii* cells with cpDNA-coated tungsten microprojectiles, which made possible the transformation of the chloroplastic genome of *C. reinhardtii*. Only a few years later in 1990 and 1993, respectively, the concept was applied and improved for tobacco cpDNA by Svab et al., successfully generating the first tobacco transplastomic line resistant to the antibiotic spectinomycin (Svab and Maliga 1993).

Transplastomic plant lines have several advantages over the classical transgenic plants:

- Since chloroplasts are maternally inherited, transgenic plastids are not disseminated into the environment by pollen, allowing the propagation of transgenic crops without risking naturally occurring varieties (Svab et al. 1990).
- High levels of protein accumulation when transgenes are stably integrated in cpDNA, caused by the elevated copy number of cpDNA molecules (De Cosa et al. 2001).
- Lack of position effect. Integration of transgenes into cpDNA is mediated by a site-specific recombination mechanism, this way preventing random undesired effects produced by stochastic transgene insertions (Svab et al. 1990).
- No transgene silencing has been reported so far.
- Multigene cloning in a single transformation event (Quesada-Vargas et al. 2005).

Recently, the information available on the functionality of the noncoding regions of cpDNA sequences of several plant species and the polycistronic nature of the chloroplastic RNAs, has led to the optimization of chloroplast genetic engineering tools. For instance, expression cassettes have been optimized to direct the insertion of entire operons in cpDNA molecules via recombination with different species-specific segments of cpDNAs or highly conserved IR segments, under the regulation of the promoters, 5′ and 3′ regulatory regions of different chloroplast-encoded genes, which allow high expression levels of the genes of interest in transplastomic plants. RNA-processing sites, translation signals, and amino acid sequences that affect protein turnover are also added to the expression cassettes to enhance translation and protein accumulation (Verma and Daniell 2007). Figure 3 shows a diagram depicting the basic elements necessary for the construction of cpDNA transformation cassettes, detailed information about the different cpDNA transformation vectors, and the regulatory elements contained within them which will be discussed in further chapters of this book.

Several years ago in 1998, the idea of a universal chloroplast transformation vector was proposed by Danielle et al., using the sequences of the IR-encoded genes *trnA* and *trnI* as flanking sequences for the recombination process and the conserved noncoding spacer region between those genes as insertion site. This system was designed using tobacco cpDNA as model, and it

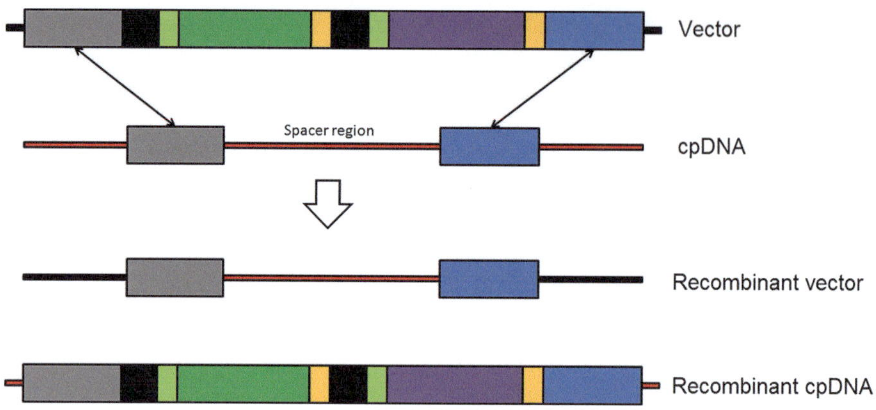

Fig. 3 cpDNA transformation cassette. Representation of the different cis-elements included in chloroplast DNA transformation cassettes. Flanking sequences (*blue* and *gray boxes*) for recombination reaction with the cpDNA; promoter sequences (*black boxes*); 5′- and 3′-UTR regions (*green* and *yellow boxes*) to ensure proper RNA processing; gene of interest (*purple box*) and spacer region (*white box*)

worked for transformation of potato and tomato chloroplasts, but at a lower transformation efficiency than the achieved for tobacco plastid transformation (Sidorov et al. 1999; Ruf et al. 2001). Later studies comparing chloroplast genomes of different groups of phylogenetically related plants, found only a few or no identical spacer regions to be used in the design of vectors to allow the cpDNA transformation of several different yet related species, making clear that the lack of enough complete cpDNA sequences still poses a great problem for cpDNA genetic engineering (Saski et al. 2007).

At the present time, several transplastomic lines are available for different species displaying economically important agronomical traits, such as insect resistance (McBride et al. 1995; Kota et al. 1999; De Cosa et al. 2001; Hou et al. 2003; Dufourmantel et al. 2005; Chakrabarti et al. 2006), herbicide resistance (Danielle et al. 1998), pathogen resistance (Jin et al. 2012), drought tolerance (Lee et al. 2003), salt tolerance (Kumar et al. 2004), vitamin production (Yabuta et al. 2012), bioplastics production (Bohmert-Tatarev et al. 2011), and carotenoid production (Apel and Bock 2009). Furthermore, several efforts have been made to improve crop yield through the engineering of RuBisCO (ribulose-1,5-bisphosphate carboxylase/oxygenase) enzyme and are extensively reviewed by Parry et al. (2012) and Hanson et al. (2012).

Along with the data provided by transcriptomics and proteomics, genomics has been useful for the development of genetically modified organisms, a technological keystone that has boosted research in virtually all fields of biology and is now impacting on economically important subjects.

Chloroplast Transcriptomes

The transcriptome is best defined as the entire set of RNA transcripts found in a cell, tissue, or organ at one specific developmental stage or physiological condition. Hence, the field arising to understand the transcriptome is called transcriptomics; the main goals of transcriptomics consist on (1) the identification of all the RNA species that exist on a given transcriptome, be it coding, noncoding, and small RNA molecules; (2) the quantitation of the changes in transcript abundance levels under different conditions or developmental stages; and (3) the determination of transcript structure, including 5′ and 3′ ends, splicing patterns, and other posttranscriptional modifications such as editing.

Chloroplast transcription is mediated by two different kinds of RNA polymerases, a single-subunit viral-like nuclear-encoded RNA polymerase (NEP) and a multi-subunit bacterial-like plastid-encoded RNA polymerase (PEP) (Allison et al. 1996). These enzymes have different target promoters, and based on the RNA polymerase in charge of its transcription, chloroplast genes can be classified into three groups: genes only transcribed by PEP are class I, genes transcribed by both PEP and NEP are called class II, and those genes transcribed only by NEP are class III (Hajdukiewicz et al. 1997). Even though the PEP core subunits are plastid-encoded, several other noncore subunits of PEP are nuclear encoded, and evidence suggests that PEP might interact with up to 50 accessory subunits. Furthermore, specificity of PEP binding to its target promoters depends on the interaction with several different nucleus-encoded σ factors, these factors recognize bacterial-like promoters harboring −10 and −35 boxes (Gruissem and Zurawski 1985). In contrast to what is seen in bacteria, where essential and nonessential σ factors exist, there is no evidence that proves the existence of essential σ factors in higher plants. For instance, *A. thaliana* nuclear genome contains six genes that code for functional σ factors, but the analysis of several insertional mutants and antisense lines for factors *AtSig1-5* only showed weak defects on plant development (pale-green pigmentation) (Schweer et al. 2010); however, *AtSig6* mutant lines display a strong pigment-accumulation phenotype during cotyledon stage, but this phenotype is however not present in later developmental stages, suggesting that expression of this factor is essential only during early developmental stages (Ishizaki et al. 2005; Loschelder et al. 2006; Schweer et al. 2006). Nevertheless, these observations strongly suggest an overall nonessential role for σ factors in plastid transcription and indicate redundancy between the σ factors studied. Furthermore, it has been recently proposed that phosphorylation of several amino acid residues on σ factors is important for regulation of their function, positive or negative effects on promoter specificity and transcription depending on the amino acids phosphorylated (Link 2003; Baginsky and Link 2005). Also, it has been reported that PEP- and NEP-dependent gene expression is mainly regulated in a development-dependent fashion, being NEP target genes expressed during earlier stages of chloroplast development, followed by PEP expression of target genes in later developmental stages (Courtois et al. 2007; Swiatecka-Hagenbruch et al. 2008). According to the current knowledge, most housekeeping genes are class II, while the genes coding for photosystem I and II proteins are all class I, and finally only a few housekeeping genes are class III (Hajdukiewicz et al. 1997; Swiatecka-Hagenbruch et al. 2007).

As it was previously stated, chloroplast genes are organized as operons and then transcribed as polycistronic RNAs; around 60 operons have been identified in tobacco cpDNA (Sugita and Sugiura 1996). To this day, the most studied chloroplastic operon is the *psbB* operon of *Chlamydomonas* and tobacco, which contains five different genes that code for important photosynthetic proteins. The genes *petB* and *petD* contained in this operon have one group II intron each, with the particularity that these two introns require a set of six proteins (APO1, APO2, CAF1, CAF2, CRS2, and CFM3) to undergo proper splicing (Watkins et al. 2011; Barkan 2011; Asakura et al. 2008). Furthermore, specific intercistronic stabilization PPR (pentatricopeptide repeat)-like proteins (HCF107, Mbb1, and HCF152) are needed in order to protect the UTR (untranslated region)-contained sequences that enhance RNA stabilization and later translation (Vaistij et al. 2000; Felder et al. 2001; Hammani et al. 2012; Zhelyazkova et al. 2012a). Figure 4 shows a diagram depicting the basic processes involved in RNA maturation of *psbB* operon genes. Taking all this data in consideration, it is clear that several different mechanisms are involved in the metabolism of native RNAs, providing a complex regulatory network for fine-tuning mature RNA stability and/or translation.

In addition, most chloroplast genomes of angiosperms are reported to have around 20 group II introns and around 30–40 editing sites (Schmitz-Linneweber and Barkan 2007). RNA editing is a transcript maturation step that

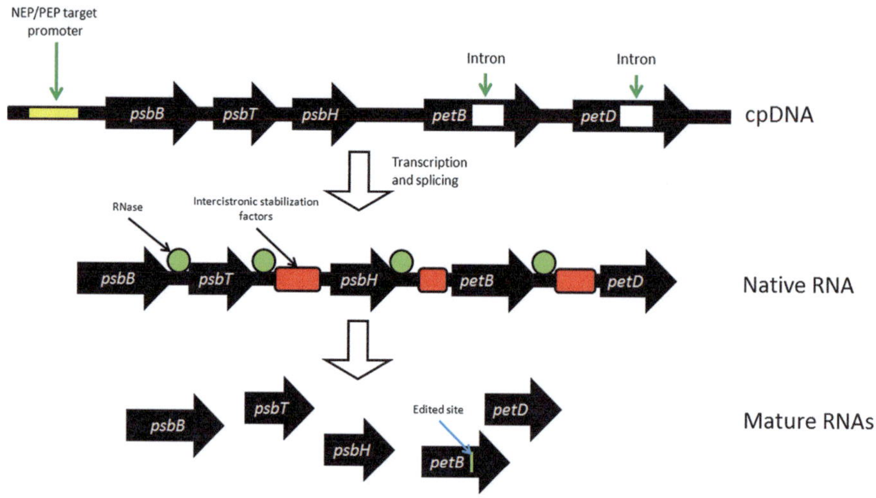

Fig. 4 Structure and posttranscriptional processing of the *psbB* operon. Organization of the *psbB*, *psbT*, *psbH*, *petB*, and *petD* genes (linked *black arrows*) in the *psbB* operon, under the regulation of the same promoter region (highlighted in *yellow*). *petB* and *petD* group II introns (*white spaces*) are posttranscriptionally spliced, and several different RNA-binding proteins (*green circles* and *red boxes*) stabilize and further process the polycistronic native RNA into individual mRNAs (*single black arrows*) that may undergo editing (highlighted in *green*)

involves the deamination of specific cytidine, generating C-to-U changes in mature mRNAs; the C-to-U changes usually lead to change in the amino acid coded by the modified codon (Chateigner-Boutin and Small 2010). It is known that the degree of site editing changes when plants are exposed to different stressing conditions; however, recent studies using novel technologies for transcriptome sequencing (RNA-seq) have found that, under normal growth conditions, most of the chloroplast editing sites are edited with efficiencies over the 74 %. However, as much as 45 % of the total mRNA count was found unedited for the translation start codon of the *ndhD* gene (Ruwe et al. 2013). Despite the importance of editing for protein synthesis, very little is known about the molecular mechanism underlying the process. Very recently, pentatricopeptide repeat proteins (PPR) have been implicated in the recognition of the cytosine bases to be edited, since it was demonstrated that PPR proteins bind to RNA segments of 10–20 nucleotides long that are upstream to the editing site (Chateigner-Boutin and Small 2010; Okuda et al. 2006). To this date, controversy exists about the biological role of transcript editing; however, some observations point that it might be a mechanism to cope with mutations, since replacement of tobacco chloroplast gene *psbF* (not edited in tobacco) with the spinach orthologue generated plants with slow growth and low chlorophyll accumulation (Bock et al. 1994).

Providing yet another RNA metabolism regulation module are chloroplastic noncoding RNAs (ncRNAs). Noncoding RNAs are widely distributed among eukaryotes and prokaryotes; they are involved in several different steps of RNA metabolism, from transcription regulation by chromatin modification to gene silencing by directed mRNA degradation. In plastids, it was thought that only a few ncRNAs were present, but this idea is now being challenged (thanks to new technologies such as RNA-seq) by recent findings that point out the accumulation of several ncRNAs derived from cpDNA in *A. thaliana*, rice, barley, and Chinese cabbage (Hotto et al. 2011; Wang et al. 2011; Chen et al. 2006; Zhelyazkova et al. 2012b). Figure 5 depicts the sites that give rise to ncRNAs in the plastid genomes of *A. thaliana* and barley. It is important to note that the precise functions and biogenesis mechanisms of the newly discovered ncRNAs are still largely unknown, though there are experiments showing that chloroplastic ncRNAs may

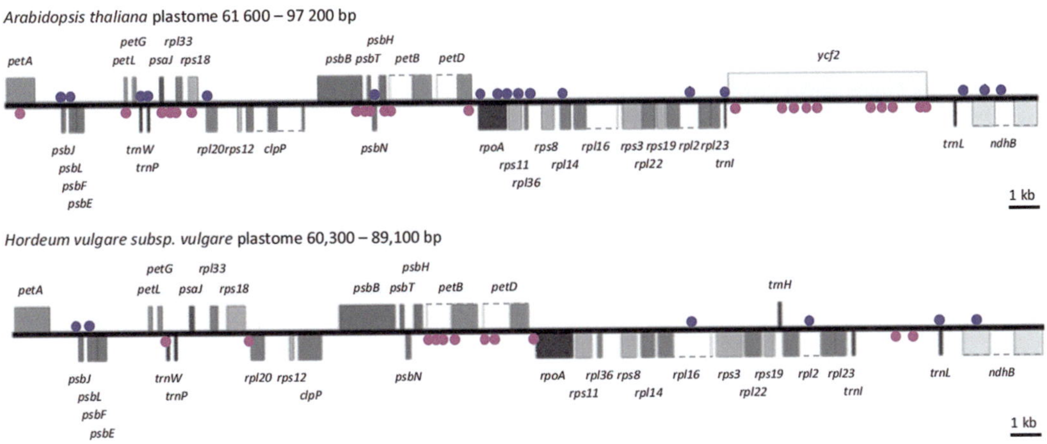

Fig. 5 Noncoding RNA coding sites. Sites reported as sources of ncRNAs in cpDNAs of *Arabidopsis* and barley. *Purple circles* represent ncRNAs encoded on the (+) strand and *magenta circles* represent ncRNAs encoded on the (–) strand (Adapted from Hotto et al. 2012)

be generated by the protection against exonuclease activity exerted by the PPR-like proteins involved in polycistronic processing (Ruwe and Schmitz-Linneweber 2012).

There are several available methods for the analysis of transcriptomes of plants under different stress conditions, such as DNA microarrays, SAGE (serial analysis of gene expression), and RNA-seq. Very recently, the chloroplast transcriptome of *A. thaliana* has been explored during seed development, stratification, germination, and early seedling development. It was found that abundance of most plastid mRNAs increase during maturation and decrease in dry seeds, while the abundance of mRNAs coding for the plastid gene expression machinery do not follow this pattern (Allorent et al. 2013). Another study shows that several mRNAs (including PEP subunits) are present in dry seed plastids and those mRNAs increase their abundance after stratification, at the same time antisense RNAs start accumulating during stratification and become easily detectable upon cold release. Furthermore, it was found that the expression of several housekeeping genes (PEP subunits, ribosomal proteins, and *matK*) starts during stratification, while the expression of genes related to PSI, PSII, and electron transport proteins start only upon cold release and light incidence (Demarsy et al. 2012). It was also found that little changes in mRNA expression occur once the chloroplasts have achieved photosynthetic competence, regardless of the developmental stage of the plant (Demarsy et al. 2012).

Several studies have given insight to the genes relevant to stress processes like tolerance to salt stress, drought, herbicides, and pathogens; the identified genes are perfect candidates for application in crop genetic engineering. For instance, the plant responses to pathogen infection have been explored through transcriptomics approaches in several different crop species such as canola, peanut, soybean, barley, tomato, rice, potato, grape, and wheat (Zhao et al. 2007; Luo et al. 2005; Moy et al. 2004; Zierold et al. 2005; Gibly et al. 2004; Zhou et al. 2010; Restrepo et al. 2005; Bruggmann et al. 2005; Figueiredo et al. 2008). Some studies in *A. thaliana* showed that infection with *Pseudomonas syringe* produce expression changes of several metabolic genes related with carbon metabolism, such expression changes seem to be coupled to the deviation of energy resources from biomass production to pathogen eradication (Scheideler et al. 2002); this data may be used in the near future to generate crop lines capable of efficient pathogen fighting at lower energy cost. Furthermore, in 2003 a study was conducted to investigate the responses of rice plants to cold, salt, and drought stress; it was found that 73 genes were induced by plant exposure to stress (36 cold-induced genes, 62 drought-induced genes, 57 salt-induced genes). Among the stress-induced genes are several transcriptional factors, carbohydrate, and amino acid metabolism proteins, all enzymes whose expression may be engineered in order to improve rice yield under the mentioned stress conditions (Rabbani et al. 2003). Table 2 shows a list of the 73 stress upregulated genes found in the cited study.

Currently, a plethora of examples of transcriptomics-based studies dealing with common crop problems are available, and extensive reviews on the topic are published on a regular basis. However, the knowledge of chloroplast transcriptomes (being relatively simple and mainly focused on photosynthesis) largely lead to the emergence of a very interesting field directly related to the improvement of crops: photosynthesis/RuBisCO engineering.

Despite the existence of different alternative metabolic pathways to assimilate CO_2 to generate biomass, the carboxylating activity of RuBisCO sits at the core of the photosynthetic activity of all plant species. In spite of its great importance for plant survival, RuBisCO has some traits that make it an extremely inefficient enzyme. For instance, RuBisCO is able to catalyze a side reaction with oxygen (known as photorespiration) that leads to the formation of 2-phosphoglycolate, a metabolite that has to go through a long series of chemical reactions comprised in different

Table 2 Cold stress upregulated genes in rice (Rabbani et al. 2003)

Functional category	Number of genes	Description
Transcription factor	6	bZip DNA-binding protein, C_2H_2-type zinc finger DNA-binding protein, C_3HC_4-type RING finger protein, Myb-type DNA-binding protein, NAC-type DNA-binding protein
Receptor-like protein kinase	1	Receptor-like protein kinase
Protein phosphatase	1	Protein phosphatase 2C
Compatible solutes	6	LEA protein, dehydrin, lectin
Detoxification	3	Catalase, *O*-methyltransferase, aldehyde dehydrogenase
Photosynthesis	1	Chlorophyll *a*-/*b*-binding protein
Membrane protein	1	Chloroplast membrane protein
Carbohydrate metabolism	7	Glycoside hydrolase, glycosyl transferase, phosphoglycerate kinase, pyruvate dehydrogenase kinase 1, trehalose-6-phosphate phosphatase, UDP-Glc-4-epimease, carboxyphosphoenolpyruvate mutase
Electron transport system	1	Thioredoxin
Amino acid metabolism	2	4-Hydroxyphenylpyruvate dioxygenase, *S*-adenosylmethionine decarboxylase
Fatty acid metabolism	3	Choline kinase, lipase, and lipoxygenase
Nucleotide metabolism	1	Adenylate kinase
Hormone biosynthesis	1	Zeaxanthin epoxidase
F-box protein	1	F-box protein
Protease inhibitor	1	Protease inhibitor
Protease	1	Papain Cys protease
Dehydrogenase	3	3-Hydroxyacyl-CoA dehydrogenase, dihydroorotate dehydrogenase, glutamate dehydrogenase
Iron homeostasis	2	Ferritin metallothionein-like type 2
Cytoskeleton	2	Actin, actin-depolymerizing factor
Transporter	1	Sugar transporter
Unknown protein	28	Unknown protein

plant cell compartments (chloroplast, mitochondrion, peroxisome, and cytosol) to get its carbon atoms back into the Calvin cycle, at the cost of $NADH_2$, ATP, fixed ammonia, and CO_2 (Foyer et al. 2009). Besides photorespiration, RuBisCO is a slow enzyme that requires posttranslational modifications and conformational remodeling to stay active; hence, a great amount of RuBisCO enzyme is necessary to maintain adequate photosynthesis rates and to support plant growth; actually, around 50 % of the total soluble protein extracted from leaves is RuBisCO (Parry et al. 2012). Taken together, these observations and the pivotal role of RuBisCO for plant yield pose this enzyme or its modifying enzymes as straightforward candidates for engineering in order to achieve the creation of crop-specific RuBisCO enzymes efficient enough to greatly improve plant yield in specific environmental conditions. Several efforts have been made in this matter and will be discussed here.

As it was previously stated, RuBisCO makes up to 50 % of the total amount of leaf protein and contains around 25 % of the total leaf nitrogen; being nitrogen an essential and expensive plant nutrient, in 1994 some experiments were conducted with tobacco antisense lines with around 20 % less RuBisCO content than wild-type plants, reducing nitrogen demand in around 11 % without negatively affecting CO_2 fixation (Stitt and Schulze 1994). Besides lowering plant nitrogen demand by modulating RuBisCO levels, efforts have been made for the improvement of some of its enzymatic traits such as CO_2 affinity, CO_2/O_2 specificity, and reaction speed by amino acid substitutions in the large subunit of tobacco RuBisCO (Zhu et al. 2010). Furthermore, there are several reports that link heat stress to low rates of CO_2 assimilation and decreased levels of RuBisCO activity caused by heat-dependent RuBisCO activase inactivation (Figs. 6 and 7). RuBisCO activase has optimal activity at temperatures below 40 °C, but this trait varies among plant species depending on the average temperatures of their natural environments (Figs. 6 and 7) and correlates with the optimal temperatures for photosynthesis (Carmo-Silva and Salvucci 2011; Carmo-Silva et al. 2012). Given this data, RuBisCO activase provides opportunities for genetic engineering, in the sense that creating more thermostable variants of this enzyme would help photosynthesis to be less sensitive to heat stress; despite the apparent advantages that RuBisCO activase-engineered crops would pres-

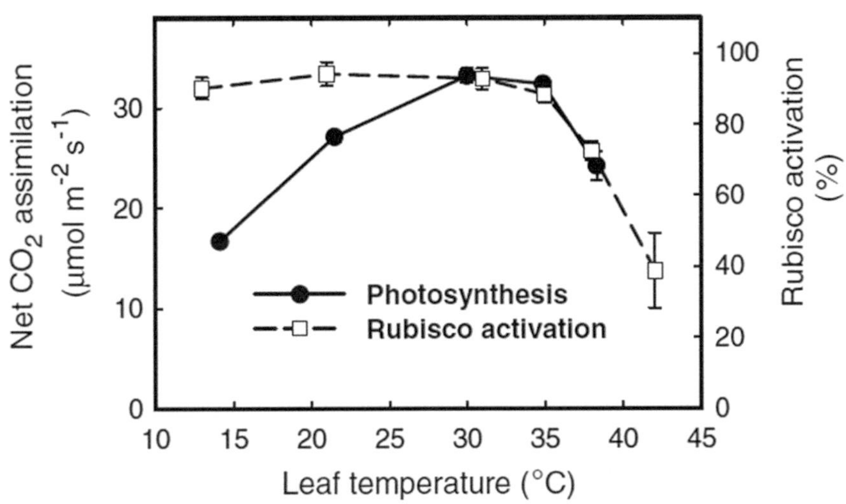

Fig. 6 Effect of heat stress on RuBisCO activation and photosynthesis. The effect of heat stress on RuBisCO activation (*squares*) and photosynthesis (*filled circles*) was evaluated by measuring the net CO_2 assimilation rate and RuBisCO in vitro activity of plants under non-photorespiratory conditions (2 % O_2) (Adapted from Carmo-Silva et al. 2012)

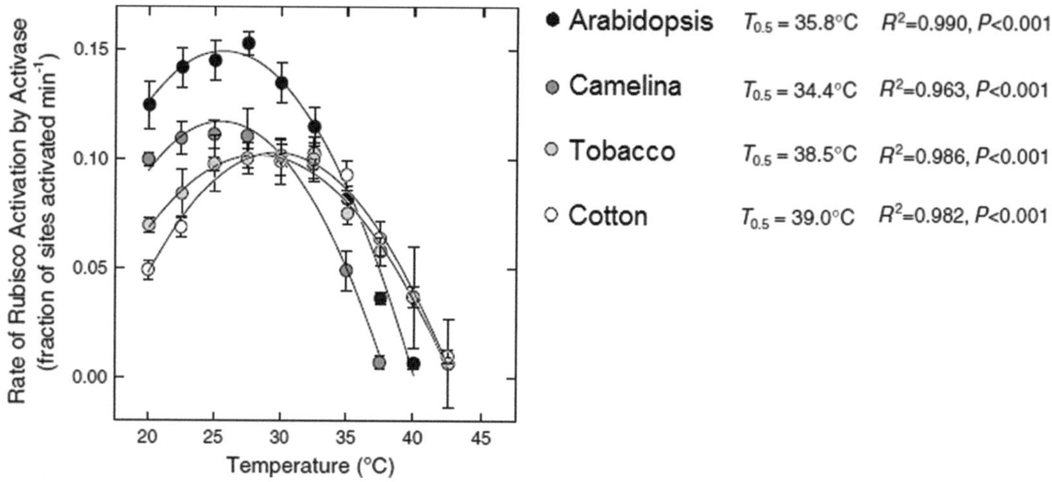

Fig. 7 Effect of heat stress on RuBisCO activation by RuBisCO activase. The effect of heat stress on RuBisCO activase was monitored at different temperatures in different plant species, Arabidopsis (*black filled circles*), Camelina (*dark-gray filled circles*), tobacco (*light-gray filled circles*), and cotton (*void circles*) (Adapted from Carmo-Silva and Salvucci 2011)

ent in an over-warming planet, such plant lines are still in the very early stages of development (Yamori et al. 2012).

Transcriptomics has proved to be a very useful tool for the identification of candidate genes to engineer in order to improve the performance of crops under diverse stressing conditions; RuBisCO itself provides a great example of the gap existing between transcript accumulation, protein expression, and protein activity; by extrapolating this problem to other genes of interest, it becomes clear that the lack of knowledge poses challenges that cannot be overcome by genomics or transcriptomics alone.

Chloroplast Proteomes

The proteome refers to the entire set of proteins expressed by a given group of cells under specific conditions or developmental stage; the term was coined in 1994 by Marc Wilkins. On the other hand, the term proteomics was coined in 1997 by P. James by blending together the terms protein and genome; this term refers to the large-scale study of proteins. Nowadays, proteomics has proved to be a very valuable tool for the progress of basic and applied sciences. Thanks to the development of different mass-spectrometry systems, researchers have gotten to explore qualitatively and quantitatively the global protein expression profiles at a level of detail never reached before. Research in proteomics of higher plant chloroplasts has achieved considerable success, and the chloroplast proteomes of different plant species are known, including specific subsets like envelope, stroma, and thylakoid proteomes (Pineda et al. 2010; Friso et al. 2004); thanks to particular protein extraction and sample fractionation protocols that enable enrichment of the samples with proteins of specific physico-chemical traits and low abundance, such as membrane integral proteins whose hydrophobicity was for a long time a challenge when trying to analyze thylakoid membrane proteomes. The methods to enrich samples with membrane proteins often include steps of fractionation with organic solvents, 1-D or 2D SDS-PAGE (sodium dodecyl sulfate-polyacrylamide gel electrophoresis) or reverse-phase HPLC (high-performance liquid chromatography) (Friso et al. 2004).

Chloroplast proteomes have been reported for several plant species under different stressing conditions like *A. thaliana*, pea, barley, zucchini, sugar beet, rice and wheat (Andaluz et al. 2006; Cui et al. 2005; Zhou et al. 2006; Curto et al.

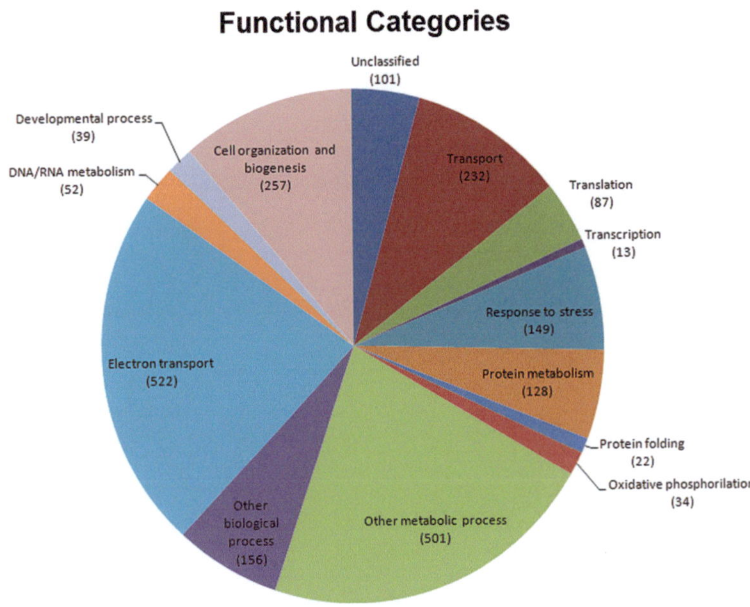

Fig. 8 Probable functions of wheat plastid proteins. Plastid protein extracts from wheat plants were analyzed and functions predicted for 767 different proteins (Data extracted from Kamal et al.2012)

2006; Ciambella et al. 2005; Aro et al. 2005); some examples will be discussed below.

In 2012 Kamal et al. reported a set of 767 unique proteins present in extracts from purified Korean winter wheat chloroplasts and sorted the identified proteins in 14 functional categories (Fig. 8), the majority of proteins were found to be involved in electron transport, cell organization and biogenesis, and metabolic processes. Furthermore, in 2007 it was reported a proteomic analysis dealing with rice chloroplast biogenesis (Kleffmann et al. 2007); the study used shotgun and 2D-PAGE-based proteomics to ensemble a rice etioplast proteome containing 477 etioplast-specific proteins, it was also found that the transition from proplastid to chloroplast induced by light is marked by a shift in protein expression that impacts the metabolic capabilities of the plastid, making it change from heterotrophic to autotrophic metabolism. In the dark, plastid protein expression was found to be focused on carbohydrate and amino acid metabolism, but 2 h after illumination proteins with functions related to carbohydrate metabolism, photosynthesis, and plastid gene expression increased their abundance, while proteins with functions like amino acid and lipid metabolism lowered their abundance, and proteins involved in nucleotide metabolism, redox regulation, and tetrapyrrole synthesis remained unchanged. Kleffmann et al. also found that proteins involved in the plastid translation machinery accumulated during this transition, the elongation factors P and Tu accumulate along with the proteins that make up the ClpP system for protein turnover. It is believed that the accumulation of translation and protein degradation proteins in the developing plastids responds to the need of replacing damaged proteins by the high levels of photooxidative stress that chloroplast proteins are subjected to. Furthermore, proteins involved in the stabilization of mRNAs increase their abundance along this transition stage, while proteins related to mRNA turn over are less abundant in mature plastids of rice. As it was discussed in the previous section, the knowledge of the genomic elements and protein effectors in RNA stability is crucial for the development of chloroplast genetic engineering methodologies, and the data available on this matter may provide opportunities for improving steps in chloroplast biogenesis that might result in enhancements of plant yield.

The analysis of chloroplast subproteomes has resulted in the better understanding of previously known functions of specific organelle compartments. In the past, it was uncertain at what extent

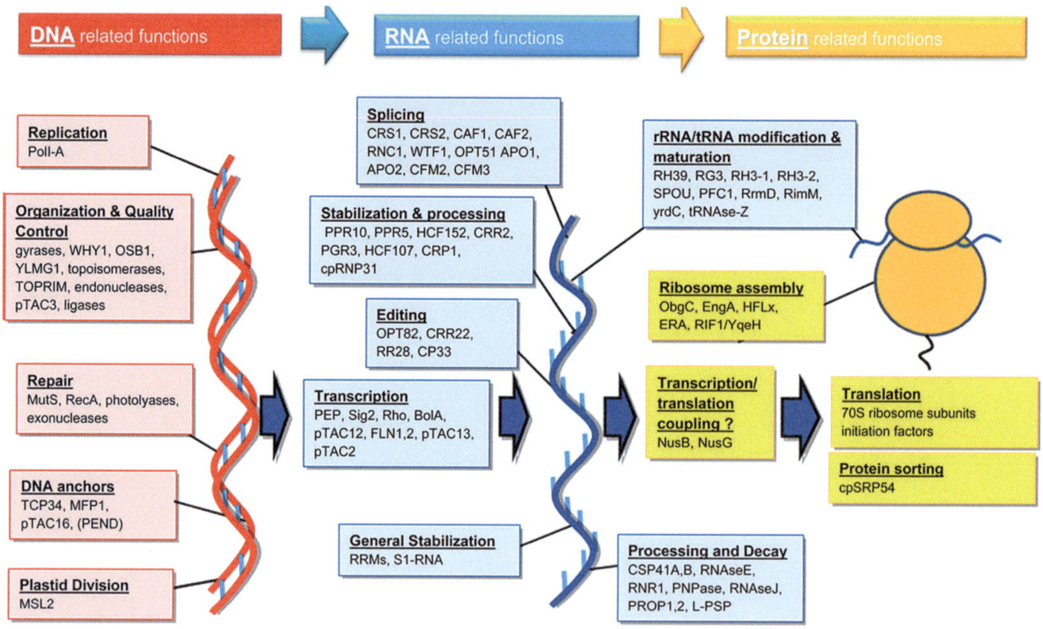

Fig. 9 **Model of the nucleoid function**. Proteins related to transcription, translation, RNA processing, and DNA maintenance were found to be enriched in chloroplast nucleoids, extending the previous model of nucleoid function. DNA-related processes (shown in *red*; replication, organization, repair, and plastid division), RNA-related processes (shown in *blue*; transcription, processing, stabilization, splicing, editing), protein-related processes (shown in *yellow*; translation, ribosome assembly, protein sorting) (Taken from Majeran et al. 2012)

the plastid nucleoids were the site of posttranscriptional processes of RNA maturation and ribosome assembly. However, in 2012, Majeran et al. reported a proteomic analysis of isolated maize plastid nucleoids; in the study, it was possible to identify a group of 127 nucleoid-specific proteins. Most of the identified proteins mapped to processes that are classified as DNA related (DNA anchoring, organization, repair, and replication) and RNA related (transcription, editing, splicing, stabilization, and maturation), making clear that plastid nucleoids are the site of chloroplast gene expression and RNA metabolism; besides DNA- and RNA-related processes, nucleoids were also found to be enriched in ribosomal and translation-related proteins, these proteins represented about 20 % of the total protein mass of nucleoids. These findings lead to the proposal of a novel model in which ribosome assembly and protein translation takes place in the nucleoid, in a process that may be couple to transcription (Fig. 9; Majeran et al. 2012).

Also, the effects of stress on chloroplast proteomes have been explored. For instance, in 2006 the thylakoid proteome of sugar beet was explored under iron deficiency conditions, showing that the abundance of proteins participating in photosynthetic electron transport is largely affected by iron starvation, while carbon metabolism-related proteins (RuBisCO small and large subunits, RuBisCO activase, carbonic anhydrase, phosphoglycerate kinase, aldolase, phosphoribulokinase, transketolase, and ribulose-1,7-bisphosphatase) displayed greater abundance under the stress conditions (Andaluz et al. 2006). Furthermore, the proteome of rice leaves under cold stress was reported to contain 60 unique proteins, whose functions are related to protein synthesis and folding, cell wall synthesis, protein degradation, energy production, and signal transduction. Interestingly more than 40 % of the detected proteins were identified as chloroplast proteins by bioinformatic tools of transit peptide prediction, immediately posing plastids as the most affected organelles by cold stress and possibly as mediators of plant responses to cold stress (Cui et al. 2005).

Other aspects of plastid biology have been analyzed using proteomic approaches. For instance, chloroplast envelopes of pea and maize were isolated and their proteomes were analyzed and compared. The result was a comparison of the plastid proteomes of C_3 and C_4 plants; it was found that metabolite transporter proteins (specifically the triosephosphate/phosphate translocator and the phosphoenolpyruvate/phosphate translocator) are enriched in C_4 chloroplast proteomes probably due the higher metabolic fluxes that occur during the C_4 type of photosynthesis (Bräutigam et al. 2008). This data provides opportunities for the improvement of projects dealing with the engineering of carbon assimilation in C_4 plants.

Yet another interesting aspect of proteomics is the analysis of the phosphoproteome, a particular subproteome that consists of all the proteins that are phosphorylated under certain conditions. To this date, very little research has focused on the analysis of phosphorylated proteins from isolated plant organelles, due to low protein abundance and the low stability that phosphoproteins display under the several fractionation steps needed for organelle isolation and protein extraction; hence, rapid protocols are used for the extraction of phosphorylated proteins from whole tissues or organs, ensuring a high phosphoproteome coverage (Espina et al. 2008). Given the technical limitations that phosphoproteomic approaches face, the data of phosphoprotein abundance profiles for different organelles must be extracted from genome-scale phosphoproteomic approaches; in order to assign phosphoproteins to specific cell locations, name it the chloroplast, the whole phosphoproteome must be compared to information coming from several different sources such as organellar subproteomes and targeting experiments with GFP (green fluorescent protein)-fused proteins. Using this approach, in 2009, Reiland et al. reconstructed a phosphoproteome map for *A. thaliana* chloroplasts from shoots and rosette leaves; it was found that the most frequently phosphorylated amino acids in plastids are serine and threonine (accounting for around 80 % and 20 % of the total phosphorylated amino acids, respectively) and no evidence of tyrosine phosphorylation was found among the 174 analyzed plastid-localized phosphoproteins. Furthermore, it was found that most of the phosphoproteins have functions related to photosynthesis (26), metabolism (30), and gene expression (16); several chloroplast phosphoproteins were found in this study that support older observations and hypothesis. For instance, it was suggested that phosphorylation of the RNA-binding proteins RNP29 and RNP33 is key for RNA stabilization and both proteins were detected in this study (Reiland et al. 2009). Taken together, it is clear that phosphoproteomics is a valuable tool for the confirmation and discovery of phosphoproteins for the many functions performed by the chloroplasts; however, no further efforts have been made recently to explore in detail the phosphoproteome of plastids of other plant species.

The analysis of proteomes, particularly plastid proteomes and the recent systematic efforts to make all proteomics information publicly available through the creation of new and extensive databases, provides opportunities to explore how different processes and stress responses are regulated at the protein level, hence providing the protein candidates necessary for the generation of novel engineered research tools or crops best fit for specific environmental conditions.

Chloroplast Metabolomes

Extending the concept from transcriptomics and proteomics, metabolomics is a field that seeks the profiling of all the small-molecule metabolites present in a biological sample under a fixed condition or developmental stage, that way giving a comprehensive set of data that represents a detailed overview of the precise metabolic and physiological state of the sample analyzed. Several analytical methods are often applied in metabolomics-based studies; sample fractionation is achieved through gas chromatography, HPLC, and capillary electrophoresis (Schauer et al. 2005; Gika et al. 2007; Soga et al. 2003), whereas metabolite detection is usually performed using mass-spectrometry technologies or nuclear magnetic resonance. However, metabolomics faces a great challenge when it comes to the analysis of its molecules of interest; in contrast to genomics, transcriptomics and proteomics, metabolomics has to deal with a very heterogeneous set of molecules with different properties, while the other omics analyze only polymers consisting of a very limited number of monomers with a predictable range of physicochemical properties. For this reason, several different sample fractionation and metabolite extraction methods must be used when trying to profile the entire metabolome of a tissue or organ. As a result, there are no reported standard methods for the analysis of whole metabolomes from plant samples, and extensive identification databases similar to those used for proteomics are nonexistent for metabolites, and the existing efforts are mainly focused on animal cell models. Other common problems related to metabolite determination are the high turnover rates displayed by some metabolites and the elevated rate of translocation that occurs between different cell compartments (Stitt et al. 1983; Weber and Fischer 2007).

Despite the many disadvantages associated with metabolomic approaches, this technologies have been applied several times with different purposes in plant biology, like the safety assessment of GMOs (genetically modified organisms) (Baker et al. 2006; Kogel et al. 2010; Kusano et al. 2011), the discovery of stress-related compounds (Leiss et al. 2009; Lawo et al. 2011; Aliferis and Jabaji 2012), and gene function discovery (Bino et al. 2004; Yonekura-Sakakibara et al. 2008). Also, several efforts have focused on the study of metabolite content of plastids using nonaqueous fractionation (NAF) methods and classical biochemical assays for the determination of metabolites (Gerhardt and Heldt 1984; Riens et al. 1991). However, recent studies exist that employ state-of-the-art high-throughput analytical procedures (such as mass spectrometry) for the profiling of the metabolite content of specific cell compartments isolated by NAF; those studies have reached a point in which it is possible to describe the metabolome of some plant cell compartments, among the studied cell compartments is of course the chloroplast (Geigenberger et al. 2011). In 2011, it was possible to reconstruct a metabolite map of the compartmentalized metabolome of *A. thaliana* leaves under regular growth conditions; from that study, it is clear that three subcellular compartments can be very well defined through nonaqueous fractionation coupled to mass spectrometry, the chloroplast, the vacuole, and the cytosol. In this study, it was found that 344 analytes are plastid-specific, while 334 metabolites are shared with the cytosol and 24 with the vacuole; the analytes were sorted into different categories including primary metabolism (7 chloroplastic, 12 cytosolic, 9 vacuolar specific metabolites), secondary metabolism (10 chloroplastic, 161 cytosolic, 158 vacuolar), lipophilic compounds (326 chloroplastic, 288 cytosolic, 0 vacuolar), and other functions (1 chloroplastic, 0 cytosolic, 0 vacuolar) (Krueger et al. 2011). A computational analysis of this data revealed that the chloroplast is enriched in amino acids and galactolipids, and some amino acids are shared with the cytosol while little overlap exists between the metabolites present at the plastid and those at the vacuole. On the other hand, a plethora of metabolites was mapped to the cytosol, stating this compartment as a gateway for metabolite flux (Klie et al. 2011). In spite of the great amount of data generated by large-scale metabolic profiling studies, only few biologically relevant hypotheses can be generated without information about the specific identities of the detected metabolites, and as this information is still out of reach for the average research project, the impact of chloroplast metabolome determination on the understanding of plant biology will be evaluated in the future.

However, one growing field within metabolomics is the identification of alterations in lipid metabolism under different conditions, a research area commonly known as lipidomics. Being the lipidome part of the metabolome, it suffers from many of the limitations usually associated to metabolomic studies, such as the lack of standard protocols for sample fractionation and metabolite isolation and identification; as a result, no specific large-scale lipidomics experiments have been driven to characterize the lipid accumulation profile of intact chloroplasts. However, the study described above sheds light on the matter, suggesting that the lipidome of chloroplasts under normal growth conditions is composed mainly of galactolipids and only few species of phospholipids (Klie et al. 2011), an observation that is consistent with data generated using classical biochemical strategies for lipid content determination, which points out that the outer and inner membranes of rye, pea, and spinach chloroplasts are composed by a high proportion of monogalactosyldiacylglycerols (47.9 % inner membrane; 20.1 % outer membrane) and digalactosyldiacylglycerols (30 % for both membranes) with no sterols or cerebrosides present (Uemura and Steponkus 1997; Block et al. 1983; Cline et al. 1981).

Plant metabolomics is a field yet in the very early stages of technical and analytical development; hence, full chloroplastic metabolite profiles are not available yet for any plant species, but given the great correlations that can be made between metabolomics, proteomics, transcriptomics, and genomics, it is clear that plastid metabolomics will develop faster in the years to come.

Chloroplast System Biology

Systems biology is a multidisciplinary field of science that features a holistic perspective of biological phenomena instead of the classical reductionist way of addressing questions in biology. Therefore, systems biology makes use of mathematical models to analyze large datasets generated through different approaches such as genomics, transcriptomics, proteomics, and metabolomics to generate hypotheses about dynamic biological systems. The origins of this field of biological research can be tracked down in history to 1952 when a numerical simulation was published by Alan Hodgkin and Andrew Huxley, they constructed a mathematical model that explained the action potential propagation along the axon of a neuronal cell (Hodgkin and Huxley 1952). Hodgkin and Huxley model reconstructed a cellular phenomenon from the known interaction between two molecular components of the cell, a potassium and sodium channel. Nowadays and thanks to the boost that omic technologies have received over the past decade and the increasing capabilities of data processing that computing equipment has attained, systems biology is a field of research on its own, and several institutions dedicated to systems biology have been created in the world.

To this date, much effort has been put in the understanding of anterograde and retrograde signaling pathways between chloroplasts and nucleus; this efforts have been merged together to generate hypotheses about the identity of the key molecular mediators of these signaling events, which is perhaps one of the most enigmatic questions in plant biology. In order to better understand anterograde signaling, different microarray and comparative proteomics experiments have been conducted. For instance, in 2005, a large and comprehensive microarray experiment was designed to evaluate the expression profiles of 3,292 nuclear genes that code for chloroplastic proteins under 101 experimental conditions or genetic backgrounds, the analysis of the data clustered 1,590 genes in 23 independent regulons; furthermore, analysis of individual regulons made clear that photosynthesis and ribosome assembly are highly linked, a possibility never explored before (Biehl et al. 2005).

Microarray-based experiments have also been successfully performed in order to shed light on the events regulating retrograde signaling during the early stages of chloroplast development. In 2007, Koussevitzky et al. reported an approach to compare wild-type plants and two *gun* (*genomes uncoupled*) mutants (*gun1* and *gun5*) at the transcriptional level; it was found that 330 genes are regulated during plastid development and further genomics analysis of the dataset led to the identification of a shared motif in the promoter regions of the studied genes that represent the core sequence of the ABA (abscisic acid) response elements, strongly suggesting that transcription factors associated with ABA signaling might be involved in the regulation of retrograde signaling (Koussevitzky et al. 2007).

Yet another example of systems biology applied to chloroplast research is the generation of computational interactomes using proteins predicted to be translocated to the chloroplast. Yu et al. assembled a chloroplastic proteome composed of 7,592 proteins, by comparing datasets from 9 different resources, further computational analysis determined a set of 1,808 proteins as the core proteome of *A. thaliana chloroplasts*, including the 88 proteins encoded by the chloroplast genome. The rest of the proteins were regarded as putative chloroplastic proteins. By comparison with genome-wide *A. thaliana* interactomes and data derived from gene chips, Yu et al. described 22,925 interaction pairs involving 2,214 proteins; 1,043 of these proteins were from the chloroplast core proteome and the remaining 1,171 from the putative chloroplastic protein group. Further analysis revealed the existence of a large network consisting of 3,109 protein interactions between 309 proteins and 84 minor independent networks consisting of more than two interacting partners. The validation of the interactome map generated consisted in the reconstruction from core protein networks of the entire protein complexes that make up the photosystems I and II, along with their respective light harvesting complexes, plus the ATP synthase and cytochrome b_6f protein complexes, this way reassuring the reliability of the assembled interactome. In a further validation effort, 12 interactions were randomly chosen to be experimentally assessed by yeast two-hybrid assays, and the physical interactions were confirmed (Yu et al. 2008).

Even though systems biology is a not-so-young field of research and many integrative efforts have been made to try to explore chloroplast function and development with a global perspective, most of these state-of-the-art studies have been conducted using *A. thaliana* as model organism; hence, comprehensive information coming from different plant models such as maize or rice is needed for taking the development of analytical tools one step further and assemble a core proteome or interactome of the flowering plants, instead of species-specific maps. Even further, with the addition of information from more plant species, it would be possible to generate maps that depict the core transcriptomes, proteomes, interactomes, and metabolomes of the entire Plantae kingdom, which is of course the ulterior goal of systems biology.

Concluding Remarks

Plastids are the site of thousands of different chemical reactions that provide essential metabolites for plant growth and development. Given the pivotal role of chloroplasts in plant biology, the knowledge of the molecular mechanisms regulating their biogenesis, development, and functioning lays at the core of every crop improvement initiative or any other project that pretends to shed light on fundamental processes for plant life, since the most important molecular regulators (or their ancillary molecules) are usually prime targets for engineering and the information obtained about their regulation can be often extrapolated to several different species.

Nowadays, computational biology is usually very well blended with the tools to analyze omic-generated data, to the point that researchers no longer understand omics without the computational tools to analyze the raw data; take, for instance, proteomics, without image-analysis software (such as Melanie or PD-Quest; http://www.genebio.com/products/melanie/, http://www.bio-rad.com) to process and compare gel images, 2D-PAGE gels would be nothing different to a canvas full of stains, or without the search engines available online (like Mascot; http://www.matrixscience.com/), the raw mass-spectrometry data would yield very little information about the protein content of a given sample and no data about the identity of the detected proteins would be available. In spite of the plethora of analytical tools existing and given the overwhelming amount of data that is being generated on a daily basis through genomics, transcriptomics, proteomics, metabolomics, or any other novel omic approach, one of the most important challenges of modern science is to develop tools that allow researchers to integrate all the information available from different sources, to generate new biologically relevant hypothesis that impact on the global understanding of life and the hundreds of processes that make the development and physiology of an organism by itself, and when interacting with other individuals of the same or different species.

Just as virtually all fields of biology, the study of chloroplasts has been boosted by the emergence of global analysis systems to explore genomes, transcriptomes, proteomes, metabolomes, etc., being transcriptomics and proteomics the main source of engineering targets, while genomics provides the tools for the application of the engineered proteins or transcripts. To date, the greatest problem for the application of "omic-derived" knowledge to crop engineering is the lack of data generated from enough species to create universal improvement systems instead of the species-specific systems available today.

References

Aliferis K, Jabaji S (2012) FT-ICR/MS and CG-EI/MS Metabolomics networking unravels global potato sprout's responses to *Rhizoctonia solani* infection. PLoS One 7:e42576

Allison L, Simon L, Maliga P (1996) Deletion of *rpoB* reveals a second distinct transcription system in plastids of higher plants. EMBO J 15:2802–2809

Allorent G, Courtois F, Chevalier F, Lerbs-Mache S (2013) Plastid gene expression during chloroplast differentiation and dedifferentiation into non-photosynthetic plastids during seed formation. Plant Mol Biol 82:59–70

Andaluz S, López-Millán A, De Las Rivas J, Aro E, Abadía J, Abadía A (2006) Proteomic profiles of thylakoid membranes and changes in response to iron deficiency. Photosynth Res 89:141–155

Apel W, Bock R (2009) Enhancement of carotenoid biosynthesis in transplastomic tomatoes by induced lycopene-to-provitamin A conversion. Plant Physiol 151:59–66

Aro E, Soursa M, Rokka A, Allahverdiyeva Y, Paakkarinen V, Saleem A, Battchikova N, Rintamaki E (2005) Dynamics of photosystem II a proteomic approach to thylakoid protein complexes. J Exp Bot 56:347–356

Asakura Y, Bayraktar O, Barkan A (2008) Two CRM protein subfamilies cooperate in the splicing of group IIB introns in chloroplasts. RNA 14:2319–2332

Baginsky S, Link G (2005) Redox regulation of chloroplast gene expression. In: Demmig-Adams B, Adams W, Mattoo A (eds) Photoprotection, photoinhibition, gene regulation, and environment. Springer, Dordrecht, pp 269–287

Baker J, Hawkins N, Ward J, Lovegrove A, Napier J, Shewry P, Beale M (2006) A metabolomic study of substantial equivalence of field-grown genetically modified wheat. Plant Biotechnol J 4:381–392

Barkan A (2011) Expression of plastid genes: organelle-specific elaborations on a prokaryotic scaffold. Plant Physiol 155:1520–1532

Bendich A (1991) Moving pictures of DNA released upon lysis from bacteria, chloroplasts, and mitochondria. Protoplasma 160:121–130

Bendich A (1987) Why do chloroplasts and mitochondria contain so many copies of their genome? Bioessays 6:279–282

Bendich A (2004) Circular chloroplast chromosomes: the grand illusion. Plant Cell Online 16:1661–1666

Biehl A, Richly E, Noutsos C, Salamini F, Leister D (2005) Analysis of 101 nuclear transcriptomes reveals 23 distinct regulons and their relationship to metabolism, chromosomal gene distribution and co-ordination of nuclear and plastid gene expression. Gene 344:33–41

Bino R, Hall R, Fiehn O, Kopka J, Saito K, Draper J, Nikolau B, Mendes P, Roessner-Tunaliu U, Beale M, Trethewey R, Lange B, Wurtele E, Sumner L (2004) Potential of metabolomics as a functional genomics tool. Trends Plant Sci 9:418–425

Blankenship R (2002) Molecular mechanisms of photosynthesis. Blackwell Science, Oxford

Block M, Dorne A, Joyard J, Douce R (1983) Preparation and characterization of membrane fractions enriched in outer and inner envelope membranes from spinach chloroplasts. J Biol Chem 258:13281–13286

Bock R, Kossel H, Maliga P (1994) Introduction of a heterologous editing site into the tobacco plastid genome: the lack of RNA editing leads to a mutant phenotype. EMBO J 13:4623–4628

Bohmert-Tatarev K, McAvoy S, Daughtry S, Peoples O, Snell K (2011) High levels of bioplastic are produced in fertile transplastomic tobacco plants engineered with a synthetic operon for the production of polyhydroxybutyrate. Plant Physiol 155:1690–1708

Boynton J, Gillham N, Harris E, Hosler J, Johnson A, Jones A, Randolph-Anderson B, Robertson D, Klein T, Shark K et al (1988) Chloroplast transformation in Chlamydomonas with high velocity microprojectiles. Science 240:1534–1538

Bräutigam A, Hoffmann-Benning S, Weber A (2008) Comparative proteomics of chloroplast envelopes from C_3 and C_4 plants reveals specific adaptations of the plastid envelope to C_4 photosynthesis and candidate proteins required for maintaining C_4 metabolite fluxes. Plant Physiol 148:568–579

Bruggmann R, Abderhalden O, Reymond P, Dudler R (2005) Analysis of epidermis- and mesophyll-specific transcript accumulation in powdery mildew-inoculated wheat leave. Plant Mol Biol 58:247–267

Bult C, White O, Olsen G, Zhou L, Fleischmann R, Sutton G, Blake J, FitzGerald L, Clayton R, Gocayne J, Kerlavage A, Dougherty B, Tomb J, Adams M, Reich C, Overbeek R, Kirkness E, Weinstock K, Merrick J, Glodek A, Scott J, Geoghagen N, Weidman J, Fuhrmann J, Nguyen D, Utterback T, Kelley J, Peterson J, Sadow P, Hanna M, Cotton M, Roberts K, Hurst M, Kaine B, Borodovsky M, Klenk H, Fraser C, Smith H, Woese C, Venter J (1996) Complete genome sequence of the methanogenic archaeon, Methanococcus jannaschii. Science 273:1058–1073

Campbell W, Ogren W (1990) Electron transport through photosystem I stimulates light activation of ribulose bisphosphate carboxylase/oxygenase (Rubisco) by Rubisco activase. Plant Physiol 94:479–484

Carmo-Silva A, Salvucci M (2011) The activity of Rubisco's molecular chaperone, Rubisco activase, in leaf extracts. Photosynth Res 108:143–155

Carmo-Silva A, Gore M, Andrade-Sanchez P, French A, Hunsaker D, Salvucci M (2012) Decreased CO_2 availability and inactivation of Rubisco limit photosynthesis in cotton plants under heat and drought stress in the field. Environ Exp Bot 83:1–11

Chakrabarti S, Lutz K, Lertwiriyawong B, Svab Z, Maliga P (2006) Expression of the cry9Aa2 B.t. gene in tobacco chloroplasts confers resistance to potato tuber moth. Transgenic Res 15:481–488

Chateigner-Boutin A, Small I (2010) Plant RNA editing. RNA Biol 7:213–219

Chen Z, Zhang J, Kong J, Li S, Fu Y, Li S, Zhang H, Li Y, Zhu Y (2006) Diversity of endogenous small non-coding RNAs in *Oryza sativa*. Genetica 128:21–31

Chen Y, Chen J, Wang P, Mi H, Chen G, Xu D (2010) Reversible association of ribulose-1, 5-bisphosphate carboxylase/oxygenase activase with the thylakoid membrane depends upon the ATP level and pH in rice without heat stress. J Exp Bot 61:2939–2950

Ciambella C, Roepstorff P, Aro E, Zolla L (2005) A proteomics approach for investigation of photosynthetic apparatus in plants. Proteomics 5:746–757

Cline K, Andrews J, Mersey B, Newcomb E, Keegstra K (1981) Separation and characterization of inner and outer envelope membranes of pea chloroplasts. Proc Natl Acad Sci USA 78:3595–3599

Courtois F, Meredino L, Demarsy E, Mache R, Lerbs-Mache S (2007) Phage-type RNA polymerase RPOTmp transcribes the *rrn* operon from the PC promoter at early developmental stages in *Arabidopsis*. Plant Physiol 145:712–721

Cui S, Huang F, Wang J, Ma X, Cheng Y, Liu J (2005) A proteomic analysis of cold stress responses in rice seedlings. Proteomics 5:3162–3172

Curto M, Camafeita E, López J, Maldonado A, Rubiales D, Jorrín J (2006) A proteomic approach to study pea (*Pisum sativum*) responses to powdery mildew (*Erysiphe pisi*). Proteomics 6:163–174

Danielle H, Datta R, Varma S, Gray S, Lee S (1998) Containment of herbicide resistance through genetic engineering of the chloroplast genome. Nat Biotechnol 16:345–348

De Cosa B, Moar W, Lee S, Miller M, Daniell H (2001) Overexpression of the Bt cry2Aa2 operon in chloroplasts leads to formation of insecticidal crystals. Nat Biotechnol 19:71–74

Demarsy E, Buhr F, Lambert E, Lerbs-Mache S (2012) Characterization of the plastid-specific germination and seedling establishment transcriptional programme. J Exp Bot 63:925–939

Dufourmantel N, Tissot G, Goutorbe F, Garcxon F, Muhr C, Jansens S, Pelissier B, Peltier G, Dubald M (2005) Generation and analysis of soybean plastid transformants expressing Bacillus thuringiensis Cry1Ab protoxin. Plant Mol Biol 58:659–668

Espina V, Edmiston K, Heiby M, Pierobon M, Sciro M, Merritt B, Banks S, Deng J, VanMeter A, Geho D, Pastore L, Sennesh J, Petricoin E, Liotta L (2008) A portrait of tissue phosphoprotein stability in the clinical tissue procurement process. Mol Cell Proteomics 7:1998–2018

Felder S, Meierhoff K, Sane A, Meurer J, Driemel C, Plucken H, Klaff P, Stein B, Bechtold N, Westhoff P (2001) The nucleus encoded HCF107 gene of Arabidopsis provides a link between intercistronic RNA processing and the accumulation of translation-competent psbH transcripts in chloroplasts. Plant Cell 13:2127–2141

Fiers W, Contreras R, Duerinck F, Haegeman G, Iserentant D, Merregaert J, Min Jou W, Molemans F, Raeymaekers A, Van den Berghe A, Volckaert G, Ysebaert M (1976) Complete nucleotide sequence of bacteriophage MS2 RNA: primary and secondary structure of the replicase gene. Nature 260:500–507

Figueiredo A, Fortes A, Ferreira S, Sebastiana M, Choi Y, Sousa L, Acioli-Santos B, Pessoa F, Verpoorte R, Pais M (2008) Transcriptional and metabolic profiling of grape (*Vitis vinifera* L.) leaves unravel possible innate resistance against pathogenic fungi. J Exp Bot 59:3371–3381

Fleischmann R, Adams M, White O, Clayton R, Kirkness E, Kerlavage A, Bult C, Tomb J, Dougherty B, Merrick J (1995) Whole-genome random sequencing and assembly of Haemophilus influenzae Rd. Science 269:496–512

Foyer C, Bloom A, Queval G, Noctor G (2009) Photorespiratory metabolism: genes, mutants, energetics, and redox signaling. Annu Rev Plant Biol 60:455–484

Friso G, Giacomelli L, Ytterberg A, Peltier J, Rudella A, Sun Q, Wijk K (2004) In-depth analysis of the thylakoid membrane proteome of *Arabidopsis thaliana* chloroplasts: new proteins, new functions, and a plastid proteome database. Plant Cell Online 16:478–499

Geigenberger P, Tiessen A, Meurer J (2011) Use of non-aqueous fractionation and metabolomics to study chloroplast function in *Arabidopsis*. Methods Mol Biol 775:135–160

Gerhardt R, Heldt H (1984) Measurement of subcellular metabolite levels in leaves by fractionation of freeze-stopped material in nonaqueous media. Plant Physiol 75:542–547

Gibly A, Bonshtien A, Balaji V, Debbie P, Martin G, Sessa G (2004) Identification and expression profiling of tomato genes differentially regulated during a resistance response to *Xanthomonas campestris pv. vesicatoria*. Mol Plant Microbe Interact 17:1212–1222

Gika H, Theodoridis G, Wingate J, Wilson I (2007) Within-day reproducibility of an LC-MS-based method for metabonomic analysis: application to human urine. J Proteome Res 6:3291–3303

Goffeau A, Barrell B, Bussey H, Davis R, Dujon B, Feldmann H, Galibert F, Hoheisel J, Jacq C, Johnston M, Louis E, Mewes H, Murakami Y, Philippsen P, Tettelin H, Oliver S (1996) Life with 6000 genes. Science 274:546, 563–567

Goremykin V, Nikiforova S, Biggs P, Zhong B, Delange P, Martin W, Woetzel S, Atherton R, McLenachan P, Lockhart P (2013) The evolutionary root of flowering plants. Syst Biol 62:50–61

Gruissem W, Zurawski G (1985) Identification and mutational analysis of the promoter for a spinach chloroplast transfer RNA gene. EMBO J 4:1637–1644

Hajdukiewicz P, Allison L, Maliga P (1997) The two RNA polymerases encoded by the nuclear and the plastid compartments transcribe distinct groups of genes in tobacco plastids. EMBO J 16:4041–4048

Hammani K, Cook W, Barkan A (2012) RNA binding and RNA remodeling activities of the half-a-tetratricopeptide (HAT) protein HCF107 underlie its effects on gene expression. Proc Natl Acad Sci USA 109:5651–5656

Hanson M, Gray B, Ahner B (2012) Chloroplast transformation for engineering of photosynthesis. J Exp Bot 64:731. doi:10.1093/jxb/ers325

Hodgkin A, Huxley A (1952) A quantitative description of membrane current and its application to conduction and excitation in nerve. J Physiol 117:500–544

Hotto A, Schmitz R, Fei Z, Ecker J, Stern D (2011) Unexpected diversity of chloroplast noncoding RNAs as revealed by deep sequencing of the *Arabidopsis* transcriptome. G3 1:559–570

Hotto A, Germain A, Stern D (2012) Plastid non-coding RNAs: emerging candidates for gene regulation. Trends Plant Sci 17:737–744

Hou B, Zhou Y, Wan L, Zhang Z, Shen G, Chen Z, Hu Z (2003) Chloroplast transformation in oilseed rape. Transgenic Res 12:111–114

Ishizaki Y, Tsunoyama Y, Hatano K, Ando K, Kato K, Shinmyo A, Kobori M, Takeba G, Nakahira Y, Shiina T (2005) A nuclear-encoded sigma factor Arabidopsis SIG6, recognizes sigma-70 type chloroplast promoters and regulates early chloroplast development in cotyledons. Plant J 42:133–144

Jin S, Zhang X, Daniell H (2012) Pinellia ternata agglutinin expression in chloroplasts confers broad spectrum resistance against aphid, whitefly, Lepidopteran insects, bacterial and viral pathogens. Plant Biotechnol J 10:313–327

Kamal A, Cho K, Komatsu S, Uozumi N, Choi J, Woo S (2012) Towards an understanding of wheat chloroplasts: a methodical investigation of thylakoid proteome. Mol Biol Rep 39:5069–5083

Kannangara C, Henningsen K, Stumpf P, Appelqvist L, von Wettstein D (1971) Lipid biosynthesis by isolated barley chloroplasts in relation to plastid development. Plant Physiol 48:526–531

Keeling P (2010) The endosymbiotic origin, diversification and fate of plastids. Philos Trans R Soc Biol Sci 365:729–748

Kirk P, Leech R (1972) Amino acid biosynthesis by isolated chloroplasts during photosynthesis. Plant Physiol 50:228–234

Kleffmann T, von Zychlinski A, Russenberger D, Hirsch-Hoffmann M, Gehrig P, Gruissem W, Baginsky S (2007) Proteome dynamics during plastid differentiation in Rice. Plant Physiol 143:912–923

Klie S, Krueger S, Krall L, Giavalisco P, Flügge U, Willmitzer L, Steinhauser D (2011) Analysis of the compartmentalized metabolome – a validation of the non-aqueous fractionation technique. Front Plant Sci 2:55

Kogel K, Voll L, Schäfer P, Jansen C, Wu Y, Langen G, Imani J, Hofmann J, Schimiedl A, Sonnewald S, Wettstein D, Cook R, Soonewald U (2010) Transcriptome and metabolome profiling of field-grown transgenic barley lack induced differences but show cultivar-specific variances. Proc Natl Acad Sci USA 107:6198–6203

Kolodner R, Tewari K (1972) Molecular size and conformation of chloroplast deoxyribonucleic acid from pea leaves. J Biol Chem 247:6355–6364

Kota M, Daniell H, Varma S, Garczynski S, Gould F, Moar W (1999) Overexpression of the Bacillus thuringiensis (Bt) Cry2Aa2 protein in chloroplasts confers resistance to plants against susceptible and Bt-resistant insects. Proc Natl Acad Sci USA 96:1840–1845

Koussevitzky S, Nott A, Mockler T, Hong F, Sachetto-Martins G, Surpin M, Lim J, Mittler R, Chory J (2007) Signals from chloroplasts converge to regulate nuclear gene expression. Science 316:715–719

Krueger S, Giavalisco P, Krall L, Steinhauser M, Bussis D, Usadel B, Flugge U, Fernie A, Willmitzer L, Steinhauser D (2011) A topological map of the compartmentalized *Arabidopsis thaliana* leaf metabolome. PLoS One 6:e17806

Kumar S, Dhingra A, Daniell H (2004) Plastid-expressed betaine aldehyde dehydrogenase gene in carrot cultured cells, roots, and leaves confers enhanced salt tolerance. Plant Physiol 136:2843–2854

Kusano M, Redestig H, Hirai T, Oikawa A, Matsuda F, Fukushima A, Arita M, Watanabe S, Yano M, Hiwasa-Tanase K, Ezura H, Saito K (2011) Covering chemical diversity of genetically-modified tomatoes using metabolomics for objective substantial equivalence assessment. PLoS One 6:e16989

Lawo N, Weingart G, Schuhmacher R, Forneck A (2011) The volatile metabolome of grapevine roots: first insights into the metabolic response upon *Phylloxera* attack. Plant Physiol Biochem 49:1059–1063

Lee S, Kwon H, Kwon S, Park S, Jeong M, Han S, Byun M, Daniell H (2003) Accumulation of trehalose within transgenic chloroplasts confers drought tolerance. Mol Breed 11:1–13

Leiss K, Maltese F, Choi Y, Verpoorte R, Klinkhamer P (2009) Identification of chlorogenic acid as a resistance factor for thrips in *Chrysanthemum*. Plant Physiol 150:1567–1575

Lichtenthaler H (1999) The 1-deoxy-d-xylulose-5-phosphate pathway of isoprenoid biosynthesis in plants. Plant Mol Biol 50:47–65

Link G (2003) Redox regulation of chloroplast transcription. Antioxid Redox Signal 5:79–88

Loschelder H, Schweer J, Link B, Link G (2006) Dual temporal role of plastid sigma factor 6 in Arabidopsis development. Plant Physiol 142:642–650

Luo M, Dang P, Bausher M, Holbrook C, Lee R, Lynch R, Guo B (2005) Identification of transcripts involved in resistance responses to leaf spot disease caused by *Cercosporidium personatum* in peanut (*Arachis hypogea*). Phytopathology 95:381–387

Majeran W, Friso G, Asakura Y, Qu X, Huang M, Ponnala L, Watkins K, Barkan A, van Wijk K (2012) Nucleoid-enriched proteomes in developing plastids and chloroplasts from maize leaves: a new conceptual framework for nucleoid functions. Plant Physiol 158:156–189

McBride K, Svab Z, Schaaf D, Hogan P, Stalker D, Maliga P (1995) Amplification of a chimeric Bacillus gene in chloroplasts leads to an extraordinary level of an insecticidal protein in tobacco. Biotechnology (N Y) 13:362–365

Moy P, Qutob D, Chapman B, Atkinson I, Gijzen M (2004) Patterns of gene expression upon infection of soybean plants by Phytophthora sojae. Mol Plant Microbe Interact 17:1051–1062

Ohyama K, Fukuzawa H, Kohchi T, Shirai H, Sano T, Sano S, Umesono K, Shiki Y, Takeuchi M, Chang Z, Aota S, Inokuchi H, Ozeki H (1986) Chloroplast gene organization deduced from complete sequence of liverwort *Marchantia polymorpha* chloroplast DNA. Nature 322:572–574

Okuda K, Nakamura T, Sugita M, Shimizu T, Shikanai T (2006) A pentatricopeptide repeat protein is a site recognition factor in chloroplast RNA editing. J Biol Chem 281:37661–37667

Oldenburg D, Bendich A (2004) Most chloroplast DNA of maize seedlings in linear molecules with defined ends and branched forms. J Mol Biol 335:953–970

Palmer J (1991) Plastid chromosomes: structure and evolution. In: Hermann RG (ed) The molecular biology of plastids. Cell culture and somatic cell genetics of plants vil7A. Springer, Vienna, pp 5–53

Parry M, Andralojc P, Scales J, Salvucci M, Carmo-Silva E, Alonso H, Whitney S (2012) Rubisco activity and regulation as targets for crop improvement. J Exp Bot 64:717. doi:10.1093/jxb/ers336

Paul M, Pellny T (2003) Carbon metabolite feedback regulation of leaf photosynthesis and development. J Exp Bot 54:539–547

Pineda M, Sajnani C, Barón M (2010) Changes induced by the *Pepper mild mottle tobamovirus* on the chloroplast proteome of *Nicotiana benthamiana*. Photosynth Res 103:31–45

Quesada-Vargas T, Ruiz O, Daniell H (2005) Characterization of heterologous multigene operons in transgenic chloroplasts. Transcription, processing and translation. Plant Physiol 138:1746–1762

Rabbani M, Muruyama K, Abe H, Khan M, Katsura K, Ito Y, Yoshiwara K, Seki M, Shinozaki K, Yamaguchi-Shinozaki K (2003) Monitoring expression profiles of rice genes under cold, drought, and high-salinity stress and Abscisic acid application using cDNA microarray and RNA gel-blot analyses. Plant Physiol 133:1755–1767

Reiland S, Messerli G, Baerenfaller K, Gerrits B, Endler A, Grossmann J, Gruissem W, Baginsky S (2009) Large-scale Arabidopsis phosphoproteome profiling reveals novel chloroplast kinase substrates and phosphorylation networks. Plant Physiol 150:889–903

Restrepo S, Myers K, Pozo O, Martin G, Hart A, Buell C, Fry W, Smart C (2005) Gene profiling of a compatible interaction between Phytophthora infestans and *Solanum tuberosum* suggests a role for carbonic anhydrase. Mol Plant Microbe Interact 18:913–922

Riens B, Lohaus G, Heineke D, Heldt H (1991) Amino acid and sucrose content determined in the cytosolic, chloroplastic, and vacuolar compartments and in the phloem sap of spinach leaves. Plant Physiol 97:227–233

Rowan B, Oldenburg D, Bendich A (2004). Curr Genet 46(3):176–181

Ruf S, Hermann M, Berger I, Carrer H, Bock R (2001) Stable genetic transformation of tomato plastids and expression of a foreign protein in fruit. Nat Biotechnol 19:870–875

Ruwe H, Schmitz-Linneweber C (2012) Short non-coding RNA fragments accumulating in chloroplasts: footprints of RNA binding proteins? Nucleic Acids Res 40:3106–3116

Ruwe H, Castandet B, Schmitz-Linneweber C, Stern D (2013) Arabidopsis chloroplast quantitative editotype. FEBS Lett 587:1429–1433

Sanger F, Air G, Barrell B, Brown N, Coulson A, Fiddes C, Hutchison C, Slocombe P, Smith M (1977) Nucleotide sequence of bacteriophage phi X174 DNA. Nature 265:687–695

Saski C, Lee S, Daniell H, Wood T, Tomkins J, Kim H, Jansen R (2005) Complete chloroplast genome sequence of Gycine max and comparative analyses with other legume genomes. Plant Mol Biol 59:309–322

Saski C, Lee S, Fjellheim S, Guda C, Jansen R, Luo H, Tomkins J, Rognli O, Daniell H, Clarke J (2007) Complete chloroplast genome sequences of *Hordeum vulgare*, *Sorghum bicolor* and *Agrostis stolonifera*, and comparative analyses with other grass genomes. Theor Appl Genet 115:571–590

Schauer N, Steinhauser D, Strelkov S, Schomburg D, Allison G, Moritz T, Lundgren K, Roessner-Tunali U, Forbes M, Willmitzer L, Fernie A, Kopka J (2005) GC-MS libraries for the rapid identification of metabolites in complex biological samples. FEBS Lett 579:1332–1337

Scheideler M, Schlaich N, Fellenberg K, Beissbarth T, Hauser N, Vingron M, Slusarenko A, Hoheisel J (2002) Monitoring the switch from housekeeping to pathogen defence metabolism in *Arabidopsis thaliana* using cDNA Arrays. J Biol Chem 277:10555–10561

Schmitz-Linneweber C, Barkan A (2007) RNA splicing and RNA editing in chloroplasts. In: Bock R (ed) Cell and molecular biology of plastids. Springer, Berlin/Heidelberg, pp 213–248

Schweer J, Loschelder H, Link G (2006) A promoter switch that can rescue a plant sigma factor mutant. FEBS Lett 580:6617–6622

Schweer J, Türkeri H, Kolpack A, Link G (2010) Role and regulation of plastid sigma factors and their functional interactors during chloroplast transcription – recent lessons from *Arabidopsis thaliana*. Eur J Cell Biol 89:940–946

Shaw J, Lickey E, Schilling E, Small R (2007) Comparison of whole chloroplast genome sequences to choose noncoding regions for phylogenetic studies in angiosperms: the tortoise and the hare III. Am J Bot 94:275–288

Shinozaki K, Ohme M, Tanaka M, Wakasugi T, Hayashida N, Matsubayashi T, Zaita N, Chunwongse J, Obokata J, Yamaguchi-Shinozaki K, Ohto C, Torazawa K, Meng B, Sugita M, Deno H, Kamogashira T, Yamada K, Kusuda J, Takaiwa F, Kato A, Tohdoh N, Shimada H, Sugiura M (1986) The complete nucleotide sequence of the tobacco chloroplast genome: its gene organization and expression. EMBO J 5:2043–2049

Siddell S, Ellis R (1975) Characteristics and products of protein synthesis in vitro in etioplasts and developing chloroplasts from pea leaves. Biochem J 146:675–685

Sidorov V, Kasten D, Pang S, Hajdukiewicz P, Staub J, Nehra N (1999) Stable chloroplast transformation in potato: use of green fluorescent protein as a plastid marker. Plant J 19:209–216

Soga T, Ohashi Y, Ueno Y (2003) Quantitative metabolome analysis using capillary electrophoresis mass spectrometry. J Proteome Res 2:488–494

Stitt M, Schulze E (1994) Does Rubisco control the rate of photosynthesis and plant growth? An exercise in molecular ecophysiology. Plant Cell Environ 17:465–487

Stitt M, Wirtz W, Heldt H (1983) Regulation of sucrose synthesis by cytoplasmic fructose bisphosphatase and sucrose phosphate synthase during photosynthesis in varying light and carbon dioxide. Plant Physiol 72:767–774

Sugita M, Sugiura M (1996) Regulation of gene expression in chloroplasts of higher plants. Plant Mol Biol 32:315–326

Svab Z, Maliga P (1993) High-frequency plastid transformation in tobacco by selection for a chimeric *aadA* gene. Proc Natl Acad Sci USA 90:913–917

Svab Z, Hajdukiewicz P, Maliga P (1990) Stable transformation of plastids in higher plants. Proc Natl Acad Sci USA 87:8526–8530

Swiatecka-Hagenbruch M, Liere K, Börner T (2007) High diversity of plastidial promoters in *Arabidopsis thaliana*. Mol Genet Genomics 277:725–734

Swiatecka-Hagenbruch M, Emanuel C, Hedtke B, Liere K, Börner T (2008) Impaired function of the phage-type RNA polymerase RpoTp in transcription of chloroplasts genes is compensated by a second phage-type RNA polymerase. Nucleic Acids Res 36:785–792

Udy D, Belcher S, Williams-Carrier R, Gualberto J, Barkan A (2012) Effects of reduced chloroplast gene copy number on chloroplast gene expression in maize. Plant Physiol 160:1420–1431

Uemura M, Steponkus P (1997) Effect of cold acclimation on the lipid composition of the inner and outer membrane of the chloroplast envelope isolated from rye leaves. Plant Physiol 114:1493–1500

Vaistij F, Goldschmidt-Clermont M, Wostrikoff K, Rochaix J (2000) Stability determinants in the chloroplast psbB/T/H mRNAs of *Chlamydomonas reinhardtii*. Plant J 21:469–482

Verma D, Daniell H (2007) Chloroplast vector systems for biotechnology applications. Plant Physiol 145:1129–1143

Wang L, Yu X, Wang H, Lu Y, Ruiter M, Prins M, He Y (2011) A novel class of heat-responsive small RNAs derive from the chloroplast genome of Chinese cabbage (*Brassica rapa*). BMC Genomics 12:289–303

Watkins K, Rojas M, Friso G, van Wijk K, Meurer J, Barkan A (2011) APO1 promotes the splicing of chloroplast group II introns and harbors a plant-specific zinc-dependent RNA binding domain. Plant Cell 23:1082–1092

Weber A, Fischer K (2007) Making the connections – the crucial role of metabolite transporters at the interface between chloroplast and cytosol. FEBS Lett 581:2215–2222

Wu F, Kan P, Lee B, Daniell H, Lee W, Lin C, Lin S, Lin S (2009) Complete nucleotide sequence of Dendrocalamus latiflorus and Bambusa oldhamii chloroplast genomes. Tree Physiol 29:847–856

Wu C, Chaw S, Huang Y (2013) Chloroplast phylogenomics indicates that Ginkgo biloba is sister to cycads. Genome Biol Evol 5:243–254

Yabuta Y, Tanaka H, Yoshimura S, Suzuki A, Tamoi M, Maruta T, Shigeoka S (2012) Improvement of vitamin E quality and quantity in tobacco and lettuce by chloroplast genetic engineering. Transgenic Res. doi:10.1007/s11248-012-9656-5

Yamori W, Masumoto C, Fukayama H, Makino A (2012) Rubisco activase is a key regulator of non-steady-state photosynthesis at any leaf temperature and, to a lesser extent, of steady-state photosynthesis at high temperature. Plant J 71:870–880

Yonekura-Sakakibara K, Tohge T, Matsuda F, Nakabayashi R, Takayama H, Niida R, Watanabe-Takahashi A, Inoue E, Saito K (2008) Comprehensive flavonol profiling and transcriptome coexpression analysis leading to decoding gene-metabolite correlation in *Arabidopsis*. Plant Cell 20:2160–2176

Yoon H, Hackett J, Ciniglia C, Pinto G, Bhattacharya D (2004) A molecular timeline for the origin of photosynthetic eukaryotes. Mol Biol Evol 21:809–818

Yu Q, Li G, Wang G, Sun J, Wang P, Wang C, Yang Z (2008) Construction of a chloroplast protein interaction network and functional mining of photosynthetic proteins in *Arabidopsis thaliana*. Cell Res 18:1007–1019

Zhao J, Wang J, An L, Doerge R, Chen Z, Grau C, Meng J, Osborn T (2007) Analysis of gene expression profiles in response to Sclerotinia sclerotiorum in Brassica napus. Planta 227:13–24

Zhelyazkova P, Hammani K, Rojas M, Voelker R, Vargas-Suarez M, Borner T, Barkan A (2012a) Protein-mediated protection as the predominant mechanism for defining processed mRNA termini in land plant chloroplasts. Nucleic Acids Res 40:3092–3105

Zhelyazkova P, Sharma C, Förstner K, Liere K, Vogel J, Börner T (2012b) The primary transcriptome of barley chloroplasts: numerous noncoding RNAs and the dominating role of the Plastid-Encoded RNA polymerase. Plant Cell 24:123–136

Zierold U, Scholz U, Schweizer P (2005) Transcriptome analysis of mlo-mediated resistance in the epidermis of barley. Mol Plant Pathol 6:139–151

Zhou W, Eudes F, Laroche A (2006) Identification of differentially regulated proteins in response to a compatible interaction between the pathogen *Fusarium graminearum* and its host, *Triticum aestivum*. Proteomics 6:4599–4609

Zhou Y, Xu M, Zhao M, Xie X, Zhu L, Fu B, Li Z (2010) Genome-wide gene responses in a transgenic rice line carrying the maize resistance gene Rxo1 to the rice bacterial streak pathogen, Xanthomonas oryzae pv. oryzicola. BMC Genomics 11:78–88

Zhu G, Kurek I, Liu L (2010) Engineering photosynthetic enzymes involved in CO2-assimilation by gene shuffling. In: Rebeiz CA, Benning C, Bohnert HJ, Daniell H, Hoober JK, Lichtenthaler HK, Portis AR Jr, Tripathy BC (eds) The chloroplast: basics and application, Advances in photosynthesis and respiration. Springer, Dordrecht, pp 307–322

Transplastomics: A Convergence of Genomics and Biotechnology

Muhammad Sarwar Khan

Contents

Abstract

Transplastomics are developed predominantly for biotechnological applications since heterologous proteins can be expressed to high levels with *bona fide* structures and because of maternal inheritance of tailored traits in most of cultivated plants as rare gene leakage through pollens is experimentally witnessed. Further, advances in plastome sequencing and research have been exponential in the post-genomic era; hence, expressing multiple genes to develop biologically functional pharmaceuticals under strong promoters and translation control elements in operons is made possible. This chapter summarizes the developments from plastid genomics to gene expression and briefly describes how transplastome facilitates expression of vaccines, therapeutics, and plantibodies, in addition to tailoring agronomic traits in plants.

Keywords

Transplastomics • Plastome • Gene expression • Genomics • Biotechnology • Transplastomic biotechnology • Agronomic traits • Health traits

The Chloroplast Genome

Chloroplasts develop either from proplastids in meristematic tissues or from other differentiated plastids, such as chromoplasts, amyloplasts, and

M.S. Khan, Ph.D. (✉)
Centre for Agricultural Biochemistry and Biotechnology (CABB), University of Agriculture, Faisalabad, Pakistan
e-mail: sarwarkhan_40@hotmail.com

D. Barh et al. (eds.), *PlantOmics: The Omics of Plant Science*,
DOI 10.1007/978-81-322-2172-2_19, © Springer India 2015

leucoplasts, on exposure to light. In a fully developed leaf cell, there may be as many as 100 chloroplasts, each with about 100 copies of the plastid genome, giving in total ~10,000 copies of the plastid genome per cell. Although, the plastid genome is very small with respect to the nuclear genome, it makes about 10–20 % of the total cellular DNA content because only two copies of nuclear genome exist in a diploid plant cell while it contains thousands of copies of plastid genome, thus extraordinarily increasing the ploidy level of the plastid genome: 50,000 copies of plastid DNA in a wheat leaf and 10,000 copies of plastid DNA in a single pea leaf (Bendich 1987). The chloroplast genome is a unique and double-stranded circular molecule of DNA which varies in size from 120 to 220 kb depending on the plant species (Palmer 1991; Sugiura et al. 1986; Sugiura 1992). The observed difference in size between different plastid DNA molecules is mainly due to the length of the inverted repeats. Two copies of a large inverted repeat (Palmer and Thompson 1982) divide the genome into four segments: the repeats, a small single-copy region, and a large single-copy region.

The complete nucleotide sequences of plastid DNA have been reported for a number of organisms, disclosing an enormous amount of functional and evolutionary information. The gene order in the tobacco chloroplast genome is perhaps the representative of land plants (Sugiura 1995). The chloroplast genomes encode about 30 transfer RNAs; 23S, 16S, 5S, and 4.5S ribosomal RNAs; about 21 ribosomal proteins; and 4 subunits of RNA polymerase. These RNAs together with ribosomal proteins enable the plastid to synthesize its own proteins. The chloroplast genome also encodes 30 proteins required for proper assembly of thylakoid complexes (Sugiura 1992) and the large subunit of Rubisco which catalyzes the first reaction in the pathway of carbon fixation. In addition, a set of eleven genes referred to as chlororespiratory genes, which resemble the genes for the respiratory chain NADH dehydrogenases in mitochondria, is found in the chloroplast genome. The genes *clpP* and *accD* encoding a putative subunit of an ATP-dependent protease and a subunit of acetyl-CoA carboxylase, respectively, have been identified by homology to *E. coli* proteins (Gray et al. 1990; Li and Coronan 1992). Genes required for light-independent conversion of protochlorophyllide to chlorophyllide have been identified in the chloroplast genome of black pine (Wakasugi et al. 1994; Khan 2013). In addition, all chloroplast genomes contain open reading frames (ORFs), some of which are conserved between species (Shimada and Sugiura 1991). The conserved ORFs have been called *ycfs*; *y* stands for hypothetical, *c* for chloroplast, and *f* for open reading frame. In recent years, most of these *ycfs* have been functionally analyzed through reverse genetics approaches.

The Chloroplast Genomics

Research on chloroplast genomics can be carried out at complete genome sequencing and functional genomics levels. The sequencing of the chloroplast genome of tobacco was commenced in the 1970s and completed in 1986. Based on complete nucleotide sequence information of tobacco plastome, initially 82 different genes from the tobacco chloroplast genome were identified. However, during the past 15 years, 31 additional genes have been identified, with only two genes per year (Wakasugi et al. 2001). This illustrates clearly that the identification of novel genes is an extremely strenuous task, requiring creative experimental approaches.

Complete sequences of two chloroplast genomes of liverworts *Marchantia polymorpha* and the angiosperm plant *Nicotiana tabacum* were the first targets for chloroplast functional genomics (Ohyama et al. 1986; Shinozaki et al. 1986). Striking similarities of chloroplast genomes with bacterial genomes (Schwarz and Kossel 1980) illustrate the prokaryotic origin of chloroplasts. Hence, based on sequence homology of chloroplast genes with bacterial genes, many plastid open reading frames were assigned with tentative functions. Nevertheless, the functions of those potential plastid genes that lack significant homology with known prokaryotic genes remained indescribable. For detailed functional analyses of plastid genes, reverse genetics analysis is considered to be the most powerful tool in chloroplast functional genomics.

Since chloroplasts have acquired active homologous recombination and copy correction mechanisms (Cerutti et al. 1992), hence designing gene deletion or insertion or mutation cassettes and targeting the chloroplast genome exploiting chloroplast transformation approaches, the remaining *ycfs*/genes have been identified (Kuroda and Maliga 2002; Khan et al. 2007). Hence, the transformation technologies for chloroplasts (Boynton et al. 1988; Svab et al. 1990; Svab and Maliga 1993) have paved the way for addressing functional aspects of plastid genes and open reading frames by reverse genetics. Two model plants are being extensively exploited for chloroplast transformation technologies for reverse genetics studies, the unicellular green alga *Chlamydomonas reinhardtii* (Boynton et al. 1988) and *Nicotiana tabacum* (Svab et al. 1990; Svab and Maliga 1993). Reverse genetics technique works on two principles: (1) introduction of point mutations by site-directed mutagenesis of plastid genes and (2) inactivation of plastid-encoded genes by insertional or deletional mutagenesis ("gene knockout"). Using standard reverse genetics approaches based on homologous recombination, a number of genes and open reading frames have been targeted for their functions, including PSI, PSII, ndh, rpo, Ori, accD genes, and ycfs. However, some other approaches have also been employed that utilize copy correction and Cre/Lox systems to analyze plastid genes of unknown functions (Khan et al. 2007; Kuroda and Maliga 2003).

During functional genomics era, the mechanism of chloroplast gene expression has been the most interesting target to manage, but recently it is revealed that chloroplast gene expression is much more complex than previously thought, because there are multiple classes of promoters and RNA polymerases, multiple RNA processing steps (RNA cleavage/trimming, *cis/trans* splicing, RNA editing, and RNA stability), and multiple mechanisms for translational initiation (reviewed in Sugiura et al. 1998). In vitro systems supporting accurate transcription (Kapoor and Sugiura 1999), translation (Hirose and Sugiura 1996, 1997), and RNA editing (Hirose and Sugiura 2001) are now available from tobacco chloroplasts. These systems together with tobacco chloroplast transformation techniques (Svab and Maliga 1993; Khan and Maliga 1999) will provide powerful tools to elucidate further gene expression processes and nuclear factors responsible for chloroplast genome expression.

The Chloroplast Gene Expression

The chloroplast gene expression has many similarities to the gene expression in prokaryotes (Igloi and Kössel 1992; Gruissem and Tonkyn 1993). Two separate RNA polymerases, the plastid-encoded plastid RNA polymerase (PEP) and the nuclear-encoded plastid RNA polymerase (NEP), are responsible for transcribing the plastid genes. The photosynthesis genes are transcribed by an RNA polymerase containing plastid-encoded subunits homologous to α, β, and β' subunits of *E. coli* RNA polymerase (Shinozaki et al. 1986; Sijben-Muller et al. 1986; Ruf and Kossel 1988; Purton and Gray 1989). PEP promoters are similar to eubacterial δ^{70}-type promoters: the core is comprised of two hexameric sequences corresponding to the eubacterial −35 (TTGACA) and −10 (TATAAT) promoter elements. The hexamer are spaced 17–19 nucleotides apart and transcription initiation 5–7 nucleotides downstream of the −10 box sequence (as recorded in Gruissem and Tonkyn 1993).

The activity of RNA polymerase was characterized in its soluble and DNA-bound forms (Igloi and Kössel 1992; Gruissem and Tonkyn 1993), and the expression of the *rpo* genes was confirmed by detection of corresponding subunits in highly purified enzyme preparations from maize chloroplasts (Hu and Bogorad 1990; Hu et al. 1991) and by Western blotting of extracts from spinach chloroplasts (Briat et al. 1987). Evidence for the expression of the *rpo* genes in the form of the corresponding RNA (Hudson et al. 1988; Ruf and Kossel 1988) and of specific proteins in soluble chloroplast extracts (Ruf and Kossel 1988; Purton and Gray 1989) confirmed that the core subunits of a chloroplast RNA polymerase are encoded in the chloroplast genome. Further, the δ [70]-like factors required for promoter recognition (Tiller et al. 1991) are encoded in the nucleus (Tanaka et al. 1997).

The other polymerase, the NEP is related to the mitochondria and phage-type T3/T7 RNA polymerases (Lerbs-Mache 1993). PEP is derived from the RNA polymerase of the ancestral bacterium. It is assumed that the phage-type plastid RNA polymerase evolved by duplication of the nuclear gene encoding the mitochondrial enzyme and retargeting of the gene product to plastid (Hedtke et al. 1997). Several plastid promoters have been shown to direct the transcription of genes in prokaryotic cells (Thompson and Mosig 1988). Transcript levels from these promoters were decreased by cycloheximide, a cytoplasmic protein synthesis inhibitor, providing further evidence for a non-consensus-type plastid promoter (Kapoor et al. 1997). More recently, it has been shown that many plastid genes and operons have at least one promoter each for *E. coli*-like RNA polymerase and nuclear-encoded plastid RNA polymerase (Hajdukiewics et al. 1997).

Transcription of plastid genes by one or both RNA polymerases reflects their function. PEP transcribes Photosystem I and II genes; therefore, it plays an important role in chloroplast gene expression. In the absence of the PEP, non-photosynthetic proplastids are still maintained indicating that essential housekeeping genes are transcribed by the NEP. Indeed, most non-photosynthetic genes have promoters for both RNA polymerases. Only a few genes are known to be transcribed exclusively from an NEP promoter, *accD*, encoding a subunit of the acetyl-CoA carboxylase in dicots (Hajdukiewics et al. 1997). It is assumed that the phage-type plastid RNA polymerase evolved from the mitochondria enzyme (Hedtke et al. 1997) and the transcription of PEP genes by the NEP was probably a critical step in the nucleus indirectly taking control of the transcription of plastid genes, thereby fully integrating plastids in multicellular plants (Shiina et al. 2005).

Reproducible chloroplast transformation approaches have facilitated the study of chloroplast gene expression elements including promoters and UTRs (Untranslated Regions) analyzed by fusing with reporter genes (Monde et al. 2000; Khan and Maliga 1999). Heterologous translation-enhancing sequences, for example, bacteriophage T7 gene *10* leader sequence known to promote high-level protein accumulation in bacteria (Studier et al. 1990), have been fused with reporter genes, and Western blot analyses revealed the accumulation of ~16 to 18 % of total soluble proteins (Kuroda and Maliga 2001; Khan and Maliga 1999). Further studies have confirmed this where translational fusion of the 14 N-terminal amino acids of the chloroplast *rbcL* and *atpB* genes to a reporter sequence resulted in different levels of reporter protein accumulation, but they are not attributable to differences in transcript abundance. Furthermore, silent mutations in the fused N-terminal coding sequences were found to decrease reporter protein accumulation without influencing RNA level (Kuroda and Maliga 2001).

The Transplastomics in Outline

Chloroplast transformation henceforth will be referred to as a plastid transformation since chloroplasts are developed from other plastid types, as described earlier. The chloroplast transformation is carried out either by targeting plastids in dividing cells or mature cells of fully expanded leaves. Plastid transformation vectors are derivatives of *E. coli* plasmids with cloned plastid DNA sequences that flank both sides of a selectable marker gene and gene/s of interest with cloning sites. The flanking sequences serve as targeting regions that catalyze the integration of the marker gene and the gene of interest into the plastid genome at a predetermined site by two events of homologous recombination. Since the backbone of the vector that carries *E. coli* DNA sequences does not carry any plastid replication origin, hence it is subsequently lost (Maliga 2004).

The genetic transformation of plastid genome requires (1) a method to deliver the DNA of the plastid transformation vector through the cell wall, the plasma membrane, and the double membranes of the organelle, (2) a plastid-specific selectable marker gene to promote sorting of transformed and wild-type genome copies, and (3) a highly efficient tissue culture system. To date, the particle bombardment, a physical method of gene delivery, is widely used to engineer plastid genomes in a number of plant species (Daniell et al. 2002; Bock and Khan 2004; Khan et al. 2011). In the method, DNA is coated onto the surface of the inert metal particles, which are subsequently placed onto the surface of the macrocarriers. The macrocarriers are placed along with a stopping screen in the macrocarrier assembly. A rupture disk of appropriate pressure, normally of 1,100 psi, is used to develop the pressure that propels the particles down toward the particles carrying a macrocarrier. The particles hit the surface of the target leaf with a velocity optimized for high transformation efficiencies. The bombarded leaves are then placed in the dark. Leaves are cut into small pieces and placed on antibiotic-containing regeneration medium after 48 h of bombardments. Antibiotic-resistant shoots start appearing within 4–6 weeks of bombardment. Tiny shoots are excised from the bleached leaf sections and shifted to jars for proliferation; leaves from the proliferated shoots are either for genetic analysis or for further round of selection and regeneration to purify the transformed genome from wild-type genomes. During the process, sorting at genome, plastid, and cell levels is carried out, and the purified homoplasmic shoots for transgene are identified for further analyses as per designed experiments.

Genetic markers are either lethal or nonlethal markers used for primary selection of transformation events. These genetic markers provide resistance against spectinomycin, streptomycin, and kanamycin which inhibit protein synthesis on prokaryotic-type plastid ribosomes. These drugs inhibit greening, faster proliferation, and shoot formation in tobacco culture, which are the selection parameters to identify transplastomic clones on a selective medium. Spectinomycin selection is nonlethal that is widely used in selecting the transformed cells. However, kanamycin selection is a lethal selection that has also been used for the selection of transplastomic lines (Carrer et al. 1993). The *aadA* that encodes aminoglycoside 3′′-adenylyltransferase and confers resistance to spectinomycin or streptomycin (Goldschmidt-Clermont 1991; Svab et al. 1990), the *neo* gene that encodes neomycin phosphotransferase II (Carrer et al. 1993) and confers resistance to kanamycin, and the *aphA-6* gene that encodes aminoglycoside phosphotransferase and confers resistance to kanamycin and amikamycin (Bateman et al. 2000) are commonly used selectable marker genes.

In addition to selectable markers, reporter genes also contribute to the development of the technology by serving as tools for visual monitoring of transgene expression in transformed cells, tissues, and organisms. A number of genes have been used to study gene expression in plants, e.g., the genes encoding β-glucuronidase, GUS (*uid*A) and β-galactosidase (*lac*Z), chloramphenicol acetyltransferase *(cat)* and neomycin phosphotransferase (*nptII*), nopaline synthase (*nos*), octopine synthase (*ocs)*, and luciferase (*luc*), as reporter genes. Of these, *uidA* has been expressed transiently (Seki et al. 1995) and stably in tobacco chloroplasts (Staub and Maliga 1994). However, histochemical detection of GUS in chloroplasts requires prolonged incubation because the chloroplast envelope membranes act as a selective barrier to substrate penetration into the chloroplasts. The use of nontoxic marker to identify transgenic cells after transformation is an effective procedure for discerning transformed cell/organs and removing untransformed or non-expressing cells, tissues, or organs. Several chloroplast promoters have been shown to direct the transcription initiation of reporter genes in prokaryotic cells (Thompson and Mosig 1988). The green fluorescent protein *(gfp)* of the jellyfish, *Aequorea victoria*, has been introduced as a reporter gene in plants (Khan and Maliga 1999). The *gfp* provides an easily scored genetic marker in plants and major uses in monitoring gene expression, protein localization, and screening of transformation events at high resolution. It allows the direct imaging of the fluorescent gene product in living cells without the need for prolonged and lethal histochemical staining procedures (Chalfie et al. 1994). The chromophore forms autocatalytically in the presence of oxygen and fluoresces green (508 nm) on absorption of blue or UV light of 395 nm. This protein has successfully been expressed in *E. coli* and chloroplasts of tobacco, potato, and rice (Khan and Maliga 1999) using chloroplast-specific expression signals. In a previous study, it was observed that a bacterial promoter was able to express reporter gene *gfp* successfully (Khan 2001). The development of a gene encoding bifunctional proteins can minimize the use of different set of promoters and terminators and may result in plastid DNA fragment deletion through homologous recombination due to homology with plastid DNA and physical separation of genes. Such a gene will facilitate both the selection and visual screening of recipient cells; therefore, a bifunctional protein was engineered through translational fusion of aa*d*A and *gfp* genes called FLARE-S. This bifunctional protein facilitates plastid transformation to rice in addition to tobacco, where plastid transformation is not associated with a readily identifiable phenotype (Khan and Maliga 1999).

The Transplastomic Biotechnology

After the plastid transformation was achieved in *Chlamydomonas reinhardtii* (Boynton et al. 1988) in 1988, a stable chloroplast transformation was achieved in tobacco using *aadA*, a gene of bacterial origin, which encodes aminoglycoside 3″-adenylyltransferase and confers resistance to spectinomycin and streptomycin (Svab and Maliga 1993) and a visual reporter gene encoding green fluorescent protein from jellyfish (Khan 1997), which facilitated the extension of plastid transformation to nongreen plastids with transient (Hibbered et al. 1998) and stable expression (Khan and Maliga 1999) of the gene. During the period, chloroplast transformation was also carried out to study the function of plastid genes and to express genes coding for industrially valuable enzymes, biomaterials, biopharmaceutical proteins, antibodies, antibiotics, vaccine antigens, and genes that confer important agronomic traits (Bock 2001; Daniell et al. 2002; Khan 2012). Now plastid transformation has been established in number of plant species, as detailed in Table 1. Salient examples of biotechnological applications of the technology are described in the following sections of the chapter.

Transplastomics Conferring Resistance Traits

A number of crystal toxin proteins of *Bacillus thuringiensis* have been expressed and commercialized through nuclear genomes of plants in the recent years, owing to their advantages over traditional chemical insecticides. However, low levels of expression of toxin proteins have shown concerns of developing resistance in feeding insects against those toxins. Such evolving levels of resistance development in insects may be addressed by expressing those proteins from chloroplast genomes, known to express transproteins to high levels due to polyploidy of the genome (Daniell et al. 2002). The chloroplast genome was exploited for its biotechnological applications for the first time when a native *Bacillus* gene was expressed to an unprecedented level from tobacco chloroplasts (McBride et al. 1995), and the plants were extremely toxic to larvae of *Heliothis virescens*, *Helicoverpa zea*, and *Spodoptera exigua* as demonstrated by the insect feeding experiments; there was no need of codon optimization of toxin genes since they are of prokaryotic origin (Kota et al. 1999; De Cosa et al. 2001). In these experiments, Daniell and colleagues reported transgene expression to the level of 45 % of the total soluble protein (TSP) when the transgene, *cry2Aa2*, was expressed in an operon along with two open reading frames, as was natively expressed in the *Bacillus*, which potentially encodes a chaperonin protein, capable of correct folding of the toxin protein that ultimately leads toward crystallization of the toxin protein inside the chloroplasts (De Cosa et al. 2001). Hence, exceedingly difficult to control insects (10-day old cotton bollworm, armyworm) were killed 100 % when fed on transgenic tobacco leaves, extensively reviewed elsewhere (Daniell et al. 2002). Encouraged from such experiments on model plant tobacco,

Table 1 Targeting sequences from plastid genomes and the transplastomic plants

Crop plants	Targeted tissues	Targeting sites	References
Tobacco	Leaves	*rbcL-accD*	Svab and Maliga (1993)
Arabidopsis	Leaves	*trnV-rps12/7*	Sikdar et al. (1998)
Rice	Embryogenic cells	*trnV-rps12/7*	Khan and Maliga (1999)
Potato	Leaves	*rbcL-accD*	Sidorov et al. (1999)
		trnV-3,rps12	
Tomatoes	Leaves	*trnfM-trnG*	Ruf et al. (2001)
Lesquerella	Leaves	*trnV-rps12/7*	Skarjinskaia et al. (2003)
Oilseed rape	Cotyledon petioles	*rps7-ndhB*	Hou et al. (2003)
Cotton	Embryogenic calli	*trnI-trnA*	Kumar et al. (2004b)
Petunia	Leaves	*rbcL-accD*	Zubkot et al. (2004)
Carrot	Embryogenic cells	*trnI-trnA*	Kumar et al. (2004a)
Soybean	Embryogenic tissues	*trnV-rps12/7*	Dufourmantel et al. (2004)
Lettuce	Protoplasts	*trnI-trnA*	Lelivelt et al. (2005)
Cauliflower	Protoplasts	*rbcL-accD*	Nugenta et al. (2006)
Poplar	Calli	*trnI-trnA*	Okumura et al. (2006)
Cabbage	Leaves	*rrn16S-rrn23S*	Liu et al. (2007)
Sugar beet	Leaf petioles	*rrn16-rps12*	De Marchis et al. (2009)
Eggplant	Green stem segments	*trnV-3′rps12*	Singh et al. (2010)
Alfalfa	Leaves and calli	*trnI-trnA*	Wei et al. (2011)
Sugarcane	Embryogenic calli	*trnI-trnA*	Mustafa (2011)
Wheat	Immature embryos and immature inflorescences	*atpB-rbcL*	Cui et al. (2011)

Dufourmantel et al. (2005) expressed *cry1Ab* gene from soybean chloroplasts under the control of complete promoter of 16S ribosomal RNA gene with few mutations, fused with a ribosome binding site (RBS) from the bacteriophage t7 gene 10 leader (Ye et al. 2001) and the 3′ untranslated region of *rbcL* gene from tobacco. Transgenic soybean plants showed strong insecticidal activity in velvet bean caterpillars.

When it comes to developing transgenic plants to fight against pathogens, the progress made to date is promising for antifungal activity (Cary et al. 2000); nevertheless, bacteria manage to adapt to plant defense mechanisms. Therefore, it is needed to express such molecules from plant genomes that could bind to the bacterial surface and cause their lysis. One of the molecules' AMP (antimicrobial peptide) is an amphipathic alpha-helix that can bind to negatively charged phospholipids from outer membranes of bacteria and fungi. In a concentration-dependent fashion, these molecules aggregate to form pores in the membrane and cause microbial lysis. Considering the concentration-dependent action of this peptide, it was expressed in tobacco chloroplasts reportedly to levels of 21.5–43 % of the total soluble protein (De Gray et al. 2001) and retained biological activity against *Pseudomonas syringae* and other pathogens.

Weeds are a serious threat to crop production, competing for food and light. One of the accepted strategies to manage weeds is the expression of herbicide-resistant genes from plant genomes since crops are susceptible to broad-spectrum herbicides. Glyphosate is a widely used broad-spectrum herbicide that acts as a potent inhibitor of the plant aromatic amino acid biosynthetic pathway by competitively inhibiting the key enzyme 5-enolpyruvylshikimate-3-phosphate synthase (EPSPS) and does not distinguish crops from weeds, thereby restricting its use. EPSPS is a plastid-targeting nuclear-encoded gene in plants; hence, the feasibility of expressing the gene from plastid genome was explored (Daniell et al. 1998). In the attempt, a wild-type EPSPS gene from petunia was expressed in tobacco, and the transgenic plants were reported to be resistant to glyphosate 10-fold higher than the lethal concentration. In another study, *Agrobacterium* EPSPS gene (C4) was expressed from tobacco chloroplasts, and the plants exhibited tolerance to field dose of glyphosate (Ye et al. 2001). Pal Maliga and colleagues explored the possibility of expressing a bacterial gene, *bar*, encoding the herbicide inactivating phosphinothricin acetyltransferase (PAT) enzyme from tobacco chloroplasts to confer tolerance to glufosinate (Lutz et al. 2001). The transgenic plants expressing the enzyme to the level of 7 % of the total soluble protein conferred field level tolerance to phosphinothricin (PPT), the active ingredient in the Liberty (Lutz et al. 2001); nevertheless, attempts to develop herbicide resistance genes as selectable markers for chloroplast transformation have failed. Reasons are not known why direct selection of chloroplast transformation events on PPT-containing medium has failed since direct selection for PPT-resistant nuclear transformants in maize (Spencer et al. 1990), wheat (Vasil et al. 1992), rice (Cao et al. 1992), barley (Wan and Lemaux 1994), and sugarcane (Khan et al. 2011) has been reported.

Transplastomics Conferring Medicinal Traits

The human serum albumin was the first recombinant plant-derived protein expressed from transgenic tobacco and potato plants in 1990 (Sijmons et al. 1990). Since the transgenic crops were allowed to cultivate for open field trials in 1992, the U S Department of Agriculture approved planting of recombinant protein containing transgenic crops in every state. In the proof of concept, several therapeutic proteins, including growth hormones, cytokines, antibodies, recombinant enzymes, and human and veterinary vaccines, have been expressed from plants (Twyman et al. 2003).

Resistance conferring proteins have been expressed successfully to exceptionally high levels from chloroplast genome of tobacco, representing an unprecedented opportunity to manufacture affordable modern medicines and make these available cost-effectively at the world level. In a proof of concept, fragment C of the tetanus toxin (TetC), which is already known to be a good antigen against *Clostridium tetani*, the causal agent of tetanus, was expressed from the tobacco chloroplast genome to levels of 10–25 % of the total soluble protein (Tregoning et al. 2003), using varied gene contents. Similarly, promising progress has been made with developing a chloroplast-based vaccine for anthrax. Anthrax is a zoonotic disease transmitted from animals to human and is caused by *Bacillus anthracis*, a gram-positive spore-forming organism. The virulent strain of *Bacillus anthracis* carries plasmids: pX01 and pX02. Of these plasmids, pX01 harbors *pagA*, *lef*, and *cya* genes that encode protective antigen (PA), lethal factor (LF), and edema factor (EF), respectively. None of these proteins are toxic when administered individually to cells or animals. However, PA in combination with EF, known as edema toxin, causes edema. Similarly, PA in combination with LF forms LT, the lethal toxin, extensively reviewed elsewhere (Collier and Young 2003). The term "protective antigen" is derived because of this protein's ability to elicit a protective immune response against anthrax. Considering the importance of the subject, Daniell and colleagues expressed PA in transgenic tobacco by inserting the *pagA* gene into the chloroplast genome. Chloroplast integration of the *pagA* gene was confirmed by PCR as well as by doing Southern blotting. Mature leaves grown under continuous illumination contained PA up to 14.2 % of the total soluble protein. The efficacy of the plant-derived PA was compared with that of PA derived from *B. anthracis* in both in vitro and in vivo studies (Koya et al. 2005), and the chloroplast-derived PA was found equally effective to PA derived from *B. anthracis*. Posttranslational modifications are vital for several proteins to be immunogenic; one of the examples is the outer surface lipoprotein A (OspA) from the pathogenic bacterium *Borrelia burgdorferi*, which has been used as a vaccine against Lyme disease. Chloroplast-based OspA protein together with an adjuvant was subcutaneously injected in mice that induced protective antibodies at levels that should be sufficient to protect the animals from *B. burgdorferi* (Glenz et al. 2006). Cholera toxin B (CTB) acts as a strong mucosal adjuvant, and its nontoxic B subunit when fused with antigens for mucosal immunization has immunostimulatory effects (Freytag and Clements 2005). It is therefore encouraging that CTB can be expressed to high levels in tobacco chloroplasts both alone (Daniell et al. 2001; Takahashi et al. 2009) and as a fusion protein (Glenz et al. 2006).

Several chloroplast-derived biopharmaceutical proteins have been reported. A protein-based polymer, GVGVP, that has medical uses such as wound coverings, artificial pericardia, and programmed drug delivery, was stably expressed in chloroplasts of tobacco (Guda et al. 2000). Similarly, human somatotropin (hST), a secretory protein, was expressed from chloroplasts in a soluble and biologically active form (Staub et al. 2000). The key use of hST is in the cure of hypopituitary dwarfism in children; additional indications are treatment of Turner syndrome, chronic renal failure, and human immunodeficiency virus wasting syndrome. Another important therapeutic protein that comprises approximately 60 % of the protein in blood serum is HSA, prescribed in multigram quantities to restore blood volume in trauma and other clinical conditions. Early attempts at expressing HSA have achieved inadequately low levels of HSA (Human Somatotropin, 0.2 % of tsp) in nuclear transgenic plants (Farran et al. 2002). Nevertheless, in chloroplast transgenic plants, the expressed protein was harvested to the levels of 11.2 % of the total soluble protein (Fernandez-San Millan et al. 2003). Similarly, attempts have been made to express interferon alpha 5 and 2 genes from chloroplasts, but the expression levels were found suboptimal (Nurjis and Khan 2011; Khan and Nurjis 2012), suggesting to reengineer the 5′ ends of the genes.

Oral delivery of pharmaceuticals expressed from chloroplasts required extension of the technology edible plants. Recently, chloroplast transformation has been developed in potato (Sidorov et al. 1999), tomato (Ruf et al. 2001), and carrot (Kumar et al. 2004a; Usman and Khan, unpublished). A green fluorescent protein was expressed from plastids of all three plants; however, betaine aldehyde dehydrogenase (BADH) was also expressed from carrot plastids (Kumar et al. 2004a). Thus, transgenes could be expressed from fruit chromoplasts, carrot roots, and chromoplast-containing plant tissues when routine plastid transformation protocols become available.

Perspective

Complete chloroplast genome sequencing made it possible to compare the chloroplast genome sequences with that of bacteria. Based on sequence homology of chloroplast ycfs/genes with bacterial genes, plastid genes except of divergent sequences were assigned functions. Nevertheless, these genes were analyzed using reverse genetics approaches when the chloroplast transformation was developed. Additionally, chloroplast transformation made it possible to express foreign genes for biotechnological applications in the sectors of agriculture, industry, environment, and health. Contrary to dicotyledonous plants, the plastid transformation in monocots is at its early development because of a number of reasons, explained in details elsewhere (Khan 2012).

References

Bateman A, Birney E, Durbin R, Eddy SR, Howe KL, Sonhammer ELL (2000) The Pfam protein families database. Nucleic Acids Res 28:263–266

Bendich AJ (1987) Why do chloroplasts and mitochondria contain so many copies of their genome? BioEssays 6:279–282

Bock R (2001) Transgenic chloroplasts in basic research and plant biotechnology. J Mol Biol 312:425–438

Bock R, Khan MS (2004) Taming plastids for a green future. Trends Biotechnol 22:311–318

Boynton JE, Gillham NW, Harris EH, Hosler JP, Johnson AM, Jones AR, Randolph-Anderson BL, Robertson D, Klein TM, Shark KB, Sanford JC (1988)

Chloroplast transformation in Chlamydomonas with high velocity microprojectiles. Science 240:1534–1538
Briat J-F, Bisanz-Seyer C, Laulhere J-P, Lerbs S, Lescure A-M, Mache R (1987) The RNA polymerase from chloroplasts and its use for in vitro transcription of plastid genes. Plant Physiol 25:273
Cao J, Duan XL, McElroy D, Wu R (1992) Regeneration of herbicide resistant transgenic rice plants following microprojectile-mediated transformation of suspension culture cells. Plant Cell Rep 11:586–591
Carrer H, Hockenberry TN, Svab Z, Maliga P (1993) Kanamycin resistance as a selectable marker for plastid transformation in tobacco. Mol Gen Genet 241:49–56
Cary JW, Rajasekaran K, Jayens JM, Cleveland TE (2000) Transgenic expression of a gene coding a synthetic antimicrobial peptide results in inhibition of fungal growth in vitro and in planta. Plant Sci 154:171–181
Cerutti H, Osman M, Grandoni P, Jagendorf AT (1992) A homolog of *Escherichia coliRecA* protein in plastids of higher plants. Proc Natl Acad Sci U S A 89:8068–8072
Chalfie M, Tu Y, Euskirchen G, Ward WW, Prasher DC (1994) Green fluorescent protein as a marker for gene expression. Science 263:802–805
Collier RJ, Young JA (2003) Anthrax toxin. Annu Rev Cell Dev Biol 19:45–70
Cui C, Song F, Tan Y, Zhou X, Zhao W, Ma F, Liu Y, Hussain J, Wang Y, Yang G, He G (2011) Stable chloroplast transformation of immature scutella and inflorescences in wheat (Triticum aestivum L.). Acta Biochim Biophys Sin 43:284–291
Daniell H, Datta R, Verma S, Gray S, Lee SB (1998) Containment of herbicide resistance through genetic engineering of the chloroplast genome. Nat Biotechnol 16:345–348
Daniell H, Lee S-B, Panchal T, Wiebe PO (2001) Expression of cholera toxin B subunit oligomers in transgenic tobacco chloroplasts. J Mol Biol 311:1001–1009
Daniell H, Khan MS, Allison L (2002) Milestones in chloroplast genetic engineering: an environmentally friendly era in biotechnology. Trends Plant Sci 7:84–91
De Cosa B, Moar W, Lee SB, Miller M, Daniell H (2001) Hyper-expression of the Bt Cry2Aa2 operon in chloroplasts leads to formation of insecticidal crystals. Nat Biotechnol 19:71–74
De Gray G, Rajasekaran K, Smith F, Sanford J, Daniell H (2001) Expression of an antimicrobial peptide via the chloroplast genome to control phytopathogenic bacteria and fungi. Plant Physiol 127:852–862
De Marchis F, Wang Y, Stevanato P, Arcioni S, Bellucci M (2009) Genetic transformation of the sugar beet plastome. Transgenic Res 18:17–30
Dufourmantel N, Pelissier B, Garcon F, Peltier G, Ferullo JM, Tissot G (2004) Generation of fertile transplastomic soybean. Plant Mol Biol 55:479–489
Dufourmantel N, Tissot G, Goutorbe F, Garcon F, Muhr C, Jansens S, Pelissier B, Peltier G, Dubald M (2005) Generation and analysis of soybean plastid transformants expressing *Bacillus thuringiensis* Cry1Ab protoxin. Plant Mol Biol 58:659–668
Farran I, Sanchez-Serrano JJ, Medina JF, Prieto J, Mingo-Castel AM (2002) Targeted expression of human serum albumin to potato tubers. Transgenic Res 11:337–346
Fernandez-San Millan A, Mingo-Castel A, Miller M, Daniell H (2003) A chloroplast transgenic approach to hyper-express and purify human serum albumin, a protein highly susceptible to proteolytic degradation. Plant Biotechnol J 1:71–79
Freytag LC, Clements JD (2005) Mucosal adjuvants. Vaccine 23:1804–1813
Glenz K, Bouchon B, Stehle T, Wallich R, Simon MM, Warzecha H (2006) Production of a recombinant bacterial lipoprotein in higher plant chloroplasts. Nat Biotechnol 24:76–77
Goldschmidt-Clermont M (1991) Transgenic expression of aminoglycoside adenine transferase in the chloroplast: a selectable marker for site-directed transformation of Chlamydomonas. Nucleic Acids Res 19:4083–4089
Gray JC, Hird SM, Dyer TA (1990) Nucleotide sequence of a wheat chloroplast gene encoding the proteolytic subunit of an ATP-dependent protease. Plant Mol Biol 15:947–954
Gruissem W, Tonkyn J (1993) Control mechanisms of plastid gene expression. Crit Rev Plant Sci 12:19–55
Guda C, Lee SB, Daniell H (2000) Stable expression of a biodegradable protein-based polymer in tobacco chloroplasts. Plant Cell Rep 19:257–262
Hajdukiewics P, Allison LA, Maliga P (1997) The two plastid RNA polymerases encoded by the nuclear and plastid compartments transcribe distinct groups of genes in tobacco plastids. EMBO J 16:4041–4048
Hedtke B, Borner T, Weihe A (1997) Mitochondrial and chloroplast phage-type RNA polymerases in Arabidopsis. Science 277:809–811
Hibbered JM, Linley PJ, Khan MS, Gray JC (1998) Transient expression of green fluorescent protein in various plastid types following microprojectile bombardment. Plant J 16:627–632
Hirose T, Sugiura M (1996) Cis-acting elements and trans-acting factors for accurate translation of chloroplast *psbA* mRNAs: development of an in vitro translation system from tobacco chloroplasts. EMBO J 15:1687–1695
Hirose T, Sugiura M (1997) Both RNA editing and RNA cleavage are required for translation of tobacco chloroplast ndhD mRNA: a possible regulatory mechanism for expression of a chloroplast operon consisting of functionally unrelated genes. EMBO J 16:6804–6811
Hirose T, Sugiura M (2001) Involvement of a site-specific *trans*-acting factor and a common RNA-binding protein in the editing of chloroplast mRNAs: develop-

ment of a chloroplast *in vitro* RNA editing system. EMBO J 20:1144–1152

Hou BK, Zhou YH, Wan LH, Zhang ZL, Shen GF, Chen ZH, Hu ZM (2003) Chloroplast transformation in oilseed rape. Transgenic Res 12:111–114

Hu J, Bogorad L (1990) Maize chloroplast RNA polymerase: the 180-, 120-, and 38-kilodalton polypeptides are encoded in chloroplast genes. Proc Natl Acad Sci U S A 87:1531–1535

Hu J, Troxler RF, Bogorad L (1991) Maize chloroplast RNA polymerase: the 78-kilodalton polypeptide is encoded by the plastid rpoCl gene. Nucleic Acids Res 19:3431–3434

Hudson GS, Holton TA, Whitfeld PR, Bottomley W (1988) Spinach chloroplast rpoB/C genes encode three subunits of the chloroplast RNA polymerase. J Mol Biol 200:639–654

Igloi GL, Kössel H (1992) The transcriptional apparatus of chloroplast. Crit Rev Plant Sci 10:525–558

Kapoor S, Sugiura M (1999) Identification of two essential sequence elements in the non-consensus type II PatpB-290 plastid promoter by using plastid transcription extracts from cultured tobacco BY-2 cells. Plant Cell 11:1799–1810

Kapoor S, Suzuki JY, Sugiura M (1997) Identification and functional significance of a new class of non-consensus-type plastid promoters. Plant J 11:327–337

Khan MS (1997) Tobacco chloroplast transformation using microprojectile bombardment. PhD dissertation, University of Cambrige, UK

Khan MS (2001) Utilizing heterologous promoters to express green fluorescent protein from jellyfish in tobacco chloroplasts. Pak J Bot 33:43–52

Khan MS (2012) Plastid genome engineering in plants: present status and future trends. Mol Plant Breed 3:91–102

Khan MS (2013) Towards engineering dark-operative chlorophyll synthesis pathway in transgenic plastids. In: Barh D et al (eds) OMICS: applications in biomedical, agricultural and environmental sciences. Taylor & Francis, New York, pp 423–436

Khan MS, Maliga P (1999) Fluorescent antibiotic resistance marker for tracking plastid transformation in higher plants. Nat Biotechnol 17:910–915

Khan MS, Nurjis F (2012) Synthesis and expression of recombinant interferon alpha-5 gene in tobacco chloroplasts, a non-edible plant. Mol Biol Rep 39:4391–4400

Khan MS, Hameed W, Nozoe M, Shiina T (2007) Disruption of the *psbA* gene by the copy correction mechanism reveals that the expression of plastid-encoded genes is regulated by photosynthesis activity. J Plant Res 120:421–430

Khan MS, Ali S, Iqbal J (2011) Developmental and photosynthetic regulation of Bacillus thuringiensis δ-endotoxin reveals that engineered sugarcane conferring resistance to ` dead heart` contains no toxins in cane juice. Mol Biol Rep 38:2359–2369

Kota M, Daniell H, Varma S, Garczynski SF, Gould F, Moar WJ (1999) Overexpression of the *Bacillus thuringiensis* (Bt) Cry2Aa2 protein in chloroplasts confers resistance to plants against susceptible and Bt-resistant insects. Proc Natl Acad Sci U S A 96:1840–1845

Koya V, Moayeri M, Leppla SH, Daniell H (2005) Plant-based vaccine: mice immunized with chloroplast-derived anthrax protective antigen survive anthrax lethal toxin challenge. Infect Immun 73:8266–8274

Kumar S, Dhingra A, Daniell H (2004a) Plastid-expressed betaine aldehyde dehydrogenase gene in carrot cultured cells, roots, and leaves confers enhanced salt tolerance. Plant Physiol 136:2843–2854

Kumar S, Dhingra A, Daniell H (2004b) Stable transformation of the cotton plastid genome and maternal inheritance of transgenes. Plant Mol Biol 56:203–216

Kuroda H, Maliga P (2001) Sequences downstream of the translation initiation codon are important determinants of the translation efficiency in chloroplasts. Plant Physiol 125:430–436

Kuroda H, Maliga P (2002) Overexpression of the clpP 5-untranslated region in a chimeric context causes a mutant phenotype. Suggesting competition for a clpP-specific RNA maturation factor in tobacco chloroplasts. Plant Physiol 129:1600–1606

Kuroda H, Maliga P (2003) The plastid clpP1 protease gene is essential for plant development. Nature 425:86–89

Lelivelt C, McCabe M, Newell C, DeSnoo C, Dun K, Birch-Machin I, Gray J, Mills K, Nugent J (2005) Stable plastid transformation in lettuce (Lactuca sativa L.). Plant Mol Biol 58:763–774

Lerbs-Mache S (1993) The 110-kDa polypeptide of spinach plastid DNA-dependent RNA polymerase: Single-subunit enzyme or catalytic core of multimeric enzyme complexes? Proc Natl Acad Sci U S A 90:5509–5513

Li S-J, Coronan JE (1992) A putative zing finger protein encoded by a conserved chloroplast gene is very likely a subunit of a biotin-dependent carboxylase. Plant Mol Biol 20:759–761

Liu CW, Lin CC, Chen J, Tseng MJ (2007) Stable chloroplast transformation in cabbage (Brassica oleracea L. var. capitata L.) by particle bombardment. Plant Cell Rep 26:1733–1744

Lutz KA, Knapp J, Maliga P (2001) Expression of bar in the plastid genome confers herbicide resistance. Plant Physiol 125:1585–1590

Maliga P (2004) Plastid transformation in higher plants. Annu Rev Plant Biol 55:289–313

McBride KE, Svab Z, Schaaf DJ, Hoga PS, Stalker DM, Maliga P (1995) Amplification of a chimeric Bacillus gene in chloroplasts leads to an extraordinary level of an insecticidal protein in tobacco. Biotechnology 13:362–365

Monde RM, Schuster G, Stern DB (2000) Processing and degradation of chloroplast mRNA. Biochimie 82:573–582

Mustafa G (2011) Development of plastid transformation in sugarcane. PhD dissertation, Quaid-i-Azam University, Islamabad

Nugenta GD, Coyne S, Nguyen TT, Kavanaghb TA, Dix PJ (2006) Nuclear and plastid transformation of Brassica oleracea var. botrytis(cauliflower) using PEG-mediated uptake of DNA into protoplasts. Plant Sci 170:135–142

Nurjis F, Khan MS (2011) Expression of recombinant interferon alpha-2a in tobacco chloroplasts using microprojectile bombardment. Afr J Biotechnol 10:17016–17022

Ohyama K, Fukuzawa H, Kohchi T, Shirai H, Sano T, Sano S, Umesono K, Shiki Y, Takeuchi M, Chang Z, Aota S, Inokuchi H, Ozeki H (1986) Chloroplast gene organization deduced from complete nucleotide sequence of liverwort Marchantia polymorpha chloroplast DNA. Nature 322:572–574

Okumura S, Sawada M, Park Y, Hayashi T, Shimamura M, Takase H, Tomizawa KI (2006) Transformation of poplar (Populus alba) plastids and expression of foreign proteins in tree chloroplasts. Transgenic Res 15:637–646

Palmer JD (1991) Plastid chromosomes: structure and evolution. In: Bogorad L, Vasil IK (eds) The molecular biology of plastids, vol 7A, Cell culture and somatic cell genetics of plants. Academic, New York, pp 5–53

Palmer JD, Thompson WF (1982) Chloroplast DNA rearrangements are more frequent when a large inverted repeat sequence is lost. Cell 29:537–550

Purton N, Gray JC (1989) The plastid rpoA gene encoding a protein homologous to the bacterial RNA polymerase alpha subunit is expressed in pea chloroplasts. Mol Gen Genet 217:77–84

Ruf M, Kossel H (1988) Structure and expression of the gene coding for the alpha-subunit of DNA-dependent RNA polymerase from the chloroplast gene of *Zea mays*. Nucleic Acids Res 16:5741–5755

Ruf S, Hermann M, Berger IJ, Carrer H, Bock R (2001) Stable genetic transformation of tomato plastids and expression of a foreign protein in fruit. Nat Biotechnol 19:870–875

Schwarz Z, Kossel H (1980) The primary structure of 16S rDNA from Zea mays chloroplast is homologous to E. coli 16S rRNA. Nature 283:739–742

Seki M, Shigemoto N, Sugita M, Sugiura M, Koop H-U, Irifune K, Morikawa H (1995) Transient expression of β-glucuronidase in plastids of various plant cells and tissues delivered by a pneumatic particle gun. J Plant Res 108:235–240

Shiina T, Tsunoyama Y, Nakahira Y, Khan MS (2005) RNA polymerases: promoters and transcription regulators in higher plant chloroplasts. Int Rev Cytol Cell Biol 244:1–68

Shimada H, Sugiura M (1991) Fine structural features of the chloroplast genome: comparison of the sequenced chloroplast genomes. Nucleic Acids Res 19:983–995

Shinozaki K, Ohme M, Tanaka M, Wakasugi T, Hayashida T, Zaita N, Chunwongse J, Obokata J, Yamaguchi-Shinozaki K, Ohto C, Torazawa K, Meng BY, Sugita M, Deno H, Kamogashira T, Yamada K, Kusuda J, Takaiwa F, Kato A, Tohdoh N, Shimada H, Sugiura M (1986) The complete nucleotide sequence of the tobacco chloroplast genome; its gene organization and expression. EMBO J 5:2043–2049

Sidorov VA, Kasten D, Pang SZ, Hajdukiewicz PTJ, Staub JM, Nehra NS (1999) Stable chloroplast transformation in potato: Use of green fluorescent protein as a plastid marker. Plant J 19:209–216

Sijben-Muller G, Hallick RB, Alt J, Westhof P, Herrmann RG (1986) Spinach plastid genes coding for initiation factor IF-1, ribosomal protein S-11 and RNA polymerase a-subunit. Nucleic Acids Res 14:1029–1045

Sijmons PC, Dekker BM, Schrammeijer B, Verwoerd TC, Elzen VD, Peter JM, Hoekema A (1990) Production of correctly processed human serum albumin in transgenic plants. Bio/Technology 8:217–221

Sikdar SR, Serino G, Chaudhuri S, Maliga P (1998) Plastid transformation in Arabidopsis thaliana. Plant Cell Rep 18:20–24

Singh AK, Verma SS, Bansal KC (2010) Plastid transformation in eggplant (Solanum melongena L.). Transgenic Res 19:113–119

Skarjinskaia M, Svab Z, Maliga P (2003) Plastid transformation in Lesquerella fendleri, an oilseed brassicaceae. Transgenic Res 12:115–122

Spencer TM, Gordon-Kamm WJ, Daines RJ, Start WG, Lemaux PG (1990) Bialaphos selection of stable transformants from maize cell culture. Theor Appl Genet 79:625–631

Staub JM, Maliga P (1994) Translation of psbA mRNA is regulated by light via the 5'-untranslated region in tobacco plastids. Plant J 6:547–553

Staub JM, Garcia B, Graves J, Hajdukiewicz PTJ, Hunter P, Nehra N, Paradkar V, Schlittler M, Carroll JA, Spatola L et al (2000) High-yield production of a human therapeutic protein in tobacco chloroplasts. Nat Biotechnol 18:333–338

Studier FW, Rosenberg AH, Dunn JJ, Dubendorff JW (1990) Use of T7 RNA polymerase to direct expression of cloned genes. Methods Enzymol 185:60–89

Sugiura M (1992) The chloroplast genome. Plant Mol Biol 19:149–168

Sugiura M (1995) The chloroplast genome. Essays Biochem 30:49–57

Sugiura M, Shinozaki K, Zaita N, Kusuda M, Kumano M (1986) Clone bank of the tobacco (*Nicotiana tabacum*) chloroplast genome as a set of overlapping restriction endonuclease fragments: mapping of eleven ribosomal protein genes. Plant Sci 44:211–216

Sugiura M, Kusumegi T, Sugishita H, Murakami K, Ideue T, Hirose T (1998) Translational control of photosynthetic genes in tobacco plastids. In: Garab G (ed) Photosynthesis: mechanisms and effects, vol IV. Kluwer Academic, Dordrecht, pp 2943–2946

Svab Z, Maliga P (1993) High-frequency plastid transformation in tobacco by selection for a chimeric aadA gene. Proc Natl Acad Sci U S A 90:913–917

Svab Z, Hajduckiewicz P, Maliga P (1990) Stable transformation of plastids in higher plants. Proc Natl Acad Sci U S A 87:8526–8530

Takahashi I, Nochi T, Yuki Y, Kiyono H (2009) New horizon of mucosal immunity and vaccines. Curr Opin Immunol 21:352–358

Tanaka K, Tozawa Y, Mochizuki N, Shinozaki K, Nagatani A, Wakasa K, Takahashi H (1997) Characterization of three cDNA species encoding plastid RNA polymerase sigma factor heterogeneity in higher plant plastids. FEBS Lett 413:309–313

Thompson RJ, Mosig G (1988) Integration host factor (IHF) represses a Chlamydomonas chloroplast promoter in E. coli. Nucleic Acids Res 16:3313–3326

Tiller K, Eisermann A, Link G (1991) The chloroplast transcription apparatus from mustard. Evidence for three different transcription factors which resemble bacterial sigma factors. Eur J Biochem 198:93–99

Tregoning JS, Nixon P, Kuroda H, Svab Z, Clare S, Bowe F, Fairweather N, Ytterberg J, van Wijk KJ, Dougan G, Maliga P (2003) Expression of tetanus toxin fragment C in tobacco chloroplasts. Nucleic Acids Res 31:1174–1179

Twyman RM, Stoger E, Schillberg S, Christou P, Fischer R (2003) Molecular farming in plants: host systems and expression technology. Trends Biotechnol 21:570–578

Vasil V, Castillo A, Fromm M, Vasil IK (1992) Herbicide resistant fertile transgenic wheat plants obtained by microprojectile bombardment of regenerable embryogenic callus. Biotechnology 10:667–674

Wakasugi T, Tsudzuki J, Ito S, Nakashima K, Tsudzuki T, Sugiura M (1994) Loss of all ndh genes as revealed by sequencing the entire chloroplast genome of black pine, Pinus thunbergii. Proc Natl Acad Sci U S A 91:9794–9798

Wakasugi T, Tsudzuki T, Sugiura M (2001) The genomics of land plant chloroplasts: gene content and alteration of genomic information by RNA editing. Photosynth Res 70:107–118

Wan Y, Lemaux PG (1994) Generation of large numbers of independently transformed fertile barley plants. Plant Physiol 104:37–48

Wei Z, Liu Y, Lin C, Wang Y, Cai Q, Dong Y, Xing S (2011) Transformation of alfalfa chloroplasts and expression of green fluorescent protein in a forage crop. Biotechnol Lett 33:2487–2494

Ye G-N, Hajdukiewics PTJ, Broyles D, Rodriguez D, Xu CE, Nehra NS, Staub JM (2001) Plastid-expressed 5-enolpyruvylshikimate-3-phosphate synthase genes provide high level glyphosate tolerance in tobacco. Plant J 25:261–270

Zubkot MK, Zubko EI, Zuilen KV, Meyer P, Day A (2004) Stable transformation of petunia plastids. Transgenic Res 13:523–530

Plant Mitochondrial Omics: State-of-the-Art Knowledge

Mustafa Malik Ghulam, Sumaira Kousar, and Harsh Vardhan

Contents

Abstract

The semi-autonomous nature of mitochondria provides avenues to understand the evolution of cellular functions, genome and regulatory circuit connecting gene expression and functions. The mitochondria compromised its genome size for functional compatibility of cell, and this transition has been started approximately three billion years ago, when bacteria with unique energy-producing capabilities took residence in a proto-eukaryotic cell. In the course of evolution of this symbiotic relationship, the bacterium transferred many of its genes to the host nucleus, creating the modern nuclear DNA (nDNA) genome. Since their discovery in 1840 (called *bioblast*), great progress has been made in understand-

M.M. Ghulam, Ph.D. (✉)
Agriculture Biotechnology, National Institute for Biotechnology and Genetic Engineering (NIBGE), Faisalabad, Pakistan
e-mail: mgmustafa2005@hotmail.com

S. Kousar, Ph.D.
Industrial Biotechnology, National Institute for Biotechnology and Genetic Engineering (NIBGE), Faisalabad, Pakistan

H. Vardhan, Ph.D.
Faculty of Medecine, Department of Immunology/ Service of pneumology, University of Sherbrooke, Sherbrooke, QC, Canada

D. Barh et al. (eds.), *PlantOmics: The Omics of Plant Science*,
DOI 10.1007/978-81-322-2172-2_20, © Springer India 2015

ing the central role of mitochondria in the regulation of energy metabolism and nonetheless as a key determinant of cellular fate. The mitochondrial genomic sequence database is available for majority of species used as scientific model to address current scientific questions. Herein mitochondrial omics give an edge to understand regulation at transcriptional, post-transcriptional (splicing and RNA editing), translational and post-translational level, energy complex composition, biogenesis and cell death. Where necessary, the readers are referred to related detailed studies.

Keywords

Plant mitochondrial omics • OXPHOS • Mitochondrial transcriptome • Mitochondrial proteome • Mitoribosomes • Retrograde signalling • Apoptosis • RNA splicing • RNA editing

Introduction

Richard Altman was the first scientist to discover the occurrence of mitochondria and called them *bioblast.* A half century later, Altman's bioblasts were named as mitochondria by Carl Benda in 1898 wherein Greek word *mitos* stands for thread and *chondros* means granule. Granulated thread is the appearance of the mitochondria during spermatogenesis. Mitochondria can be created only by the division of the pre-existing mitochondria. Cell does not have the capacity to generate and/or biosynthesize mitochondria de novo. Mitochondria have the ability to change shape and size and their association to the cytoskeleton enables them to travel long distances in the cell. Mitochondria are membrane-bound eukaryotic organelles that produce ATP (adenosine triphosphate) in the process of oxidative phosphorylation and tricarboxylic acid cycle. However in plants, they are also responsible for other tasks like signal transduction and biosynthesis of nucleotides, amino acids, vitamins and lipids (Rebeille et al. 1997; Bartoli et al. 2000; Gueguen et al. 2000). In addition, they are also involved in regulation of programmed cell death and response to increased oxidative stress produced as a result of high salt, cold and drought conditions (Balk et al. 1999; Moller 2001; Sweetlove et al. 2002) (Balk and Leaver 2001). Mitochondrion of land plants is not only almost 100 times larger than the animal mitochondrion but also has obtained complexity due to ongoing DNA rearrangements. The mitochondrial genome is composed of about 13 electron transport and oxidative phosphorylation components, some tRNAs and ribosomal RNAs and proteins, e.g. in *Arabidopsis* 57 (Unseld et al. 1997), in yeast 43 (Talla et al. 2005) while in humans 37 genes (Anderson et al. 1981). In the following paragraphs, we have summarized the state-of-the-art information on transcriptome and proteome and modulation of gene expression in plant mitochondria. Similarly, mitochondrial translational machinery (mitoribosomes) and mitochondrial complexes/supercomplexes and their assembly and regulation are discussed as well. In addition, the central role of mitochondria in determination of cellular fate is also briefly discussed.

Transcriptional Regulation and Transcriptome

In contrast to plant plastids (where two types of RNA polymerase, i.e. multi-subunit plastid-encoded RNA polymerase (PEP) and single-subunit nuclear-encoded RNA polymerase, transcribe the plastome DNA), only one type of RNA polymerase, i.e. nuclear-encoded RNA polymerase (Mercer et al.), is responsible for transcription of plant chondrion. NEP, a single-protein 110-kDa enzyme (Lerbs-Mache 1993; Hess and Borner 1999), is encoded by nuclear *RpoT* (RNA polymerase of the T-phage type) genes and does not require nuclear-encoded sigma-like factors for promoter specificity and identification (Fig. 1). C-terminal sequences of NEP contain catalytically relevant part; form structural domain fingers, thumb and palm and are highly conserved (Hess and Borner 1999; McAllister 1993). In addition to the nuclear-a

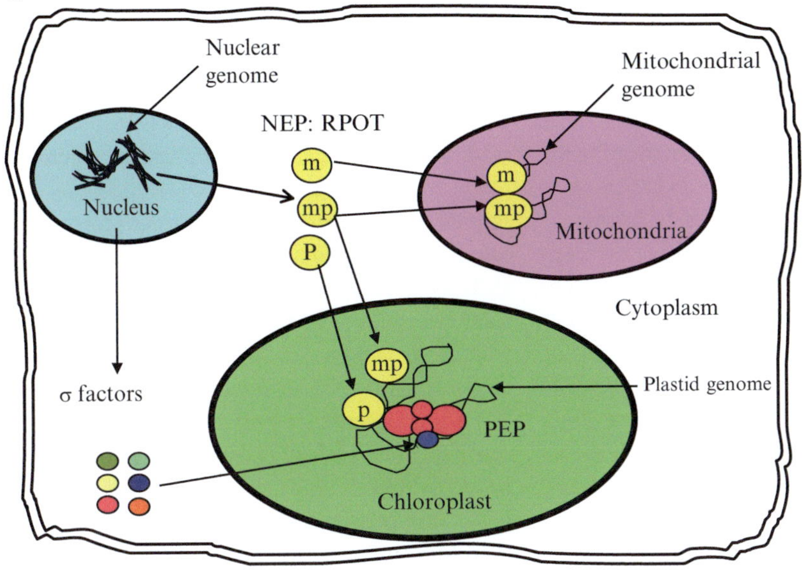

b

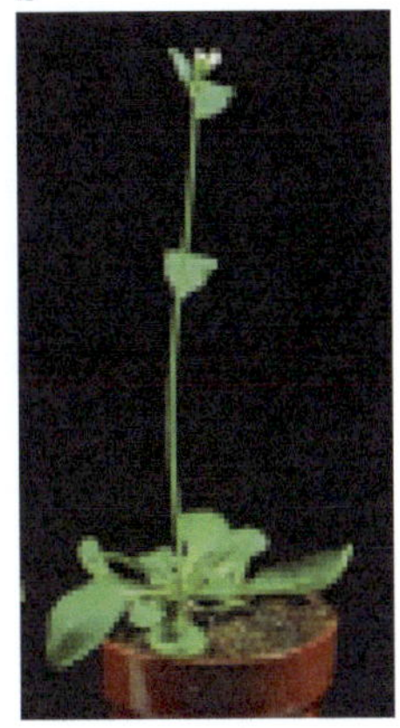

Fig. 1 (**a**) Nuclear- and plastid-encoded RNA polymerases and sigma factors required for PEP promoter specificity in *Arabidopsis thaliana*. *PEP* plastid-encoded RNA polymerase, *NEP* nuclear-encoded RNA polymerase, *m* mitochondria, *mp* mitochondria and plastid, *p* plastid. (**b**) Plant of *Arabidopsis thaliana* (Personal communication Dr. Livia Merendino and Dr. Malik Ghulam Mustafa)

encoded phage-type RNA polymerase (RNAP), another phage RNAP encoded by linear or circular extrachromosomal mtDNA molecules is also found in mitochondria of some higher plants and fungi, but its transcriptional activity has not yet been established (reviewed by Handa 2008). Eudicots nuclear genome hosts three *RpoT* genes which encode for three different NEPs, i.e. *RpoT m* (*m*itochondria), *RpoT mp* (*m*itochondria and *p*lastid) and *RpoT p* (plastid). However, *Nicotiana tabacum* nuclear genome bears six *RpoT* genes (Hedtke et al. 2002; Hedtke et al. 1997, 2000; Kobayashi et al. 2002). Some of the angiosperms have only *RpoTm* and *RpoTp* and hence only one RNA polymerase for chondrion. Dual organelle targeting of *RpoT mp* is either regulated by in-frame shift of translation initiation codon (Richter et al. 2002) or by 5′ UTR in a developmental stage-specific or tissue-specific manner (Kabeya and Sato 2005; Courtois et al. 2007; Christensen et al. 2005).

NEP recognizes consensus sequences like CRTA- (CRTAAGAGA), DDTA- or YYTA-motifs or loose motifs like ATTA and RGTA core where R=A; G, D=A, G; and T, Y=C, T. These sequences are localized in close proximity of translation initiation sites (TIS) (1–5 nucleotides upstream of TIS). In addition, an A/T-rich region is found further upstream of promoter core motifs which positively regulate transcription. Motif sequences and cis-elements together may render gene specificity to the phage RNA polymerase as transcription in mitochondria is gene specific rather than being promoter specific (Kuhn et al. 2009). In contrast to vertebrate mitochondria where transcription starts only from three promoters, i.e. LSP, HSP2 and HSP, transcription in plant mitochondria starts from higher number of promoters which result in higher mono-cistronic/polycistronic transcript ratio (Dombrowski et al. 1999). Mitochondrial genes have multiple promoters whose role and/or need are yet unclear. They may be important to ensure transcription after mitochondrial genome rearrangements (Liere et al. 2011; Kuhn et al. 2005). Preferential transcription from any of these multiple promoters of a single gene under biotic and abiotic stress conditions, at different developmental stages or in tissue-specific manner, has yet not been well established. RpoTmp (early NEP) is believed to transcribe mitochondrial and plastid genes during early seedling development, while RpoTm and RpoTp (late NEP) take over transcription at later developmental stages. RpoTmp is involved in the formation of nad2, nad6 and cox1 and respiratory chain complexes I and IV (Kuhn et al. 2009).

Phage T7 RNA polymerase is capable to recognize, melt and initiate transcription without transcription factors, i.e. no transcription initiation or elongation complex is required. It recognizes the −3 to −11 sequence with C-terminal domain while −12 to −17 by N-terminal domain (Sousa 1996; Cheetham et al. 1999). Although in *Arabidopsis* RpoTm and RpoTp were shown to recognize, melt and initiate transcription and catalyze general transcription elongation from many promoters only on supercoiled DNA in vitro (Kuhn et al. 2007), still it is speculated that plant RpoTm and RpoTmp may still require transcription factors in vivo for promoter recognition, TIS and transcription elongation (Liere et al. 2011). Although homologues of certain transcription factors like mTERF (higher plants), MCT (maize), 32- and 43-kDa proteins in pea and 69-kDa protein in wheat (Liere et al. 2011) have been identified but localization and functionality of such factors in plant mitochondria are poorly characterized.

The mechanism by which plant mitochondrial and nuclear transcription is coordinated to maintain steady-state level of transcripts to synthesize stoichiometric amounts of subunits of different complexes under normal and/or stress conditions is not yet well characterized. Presently available data indicate that transcription in plant mitochondria is loosely regulated and has minimal role in tissue-specific and developmental stage-specific regulation of steady-state level of RNA (Kuhn et al. 2009; Giege et al. 2000; Kuhn et al. 2005). However, post-transcriptional regulation is proposed to play an important role to maintain different tissue-specific steady-state levels of mtRNA (Gagliardi and Leaver 1999; Giege et al. 2005).

A large number of mitochondrial genomes have been sequenced across the living organisms: for a comprehensive list and completely sequenced data of mitochondrial genomes, consult http://megasun.bch.umontreal.ca/ogmp/projects/other/mt_list.html. Similarly, a large number of plant mitochondrial genomes have also been sequenced but the comprehensive transcriptome data is available only for a few plant mitochondria. Forner et al. (2005, 2007) have systematically mapped the 5′ and 3′ termini of the mitochondrial transcripts of *Arabidopsis thaliana* by using circular RT-PCR technique. C-RT-PCR is a very powerful technique for precise measurement of the 5′ and 3′ ends of the mature transcripts and can distinguish transcripts with a difference of even one nucleotide (Fig. 2a–c). They found stable 5′ and 3′ ends for the open reading frames. Most of the genes were found to have multiple 5′ ends with one major form not necessarily originating from transcription initiation. Most of the identified 5′ ends originated as a result of processing of the primary transcripts. In addition, majority of the mature abundant transcripts were produced as a result of cleavage of 5′ ends of the primary transcripts. In contrast to 5′ ends, most of the genes had single 3′ end. All the 5′ and 3′ ends have been summarized by Binder et al. (2011) and Forner et al. (2007). The 3′ UTRs range from 0 to 500 nucleotides. Even some transcripts had major 3′ ends in open reading frames (lack stop codon like in ccmC: –46 from stop codon). Even such transcripts were found to be translationally competent (Raczynska et al. 2006).

Plant mitochondria contain cryptic (originated from intergenic regions due to loose transcriptional control) and defective (produced as a result of inefficient post-transcriptional control) transcripts which are reviewed by Holec et al. (2008). Fujii et al. (2011) have very recently characterized the continuously expressed ORFs present in the intergenic region of mitochondrial genome of rice. Similarly, Islam et al. (2013) have presented the mitochondrial genome and transcriptome sequence analyses of perennial ryegrass (*Lolium perenne* L.), but still it lacks the detailed transcript mapping and promoter identification. In contrast to plant mitochondrial transcriptome data, human mitochondrial transcriptome has been analyzed more comprehensively across different cell lines and tissues (Mercer et al. 2011). The data can be accessed at http://mitochondria.matticklab.com.

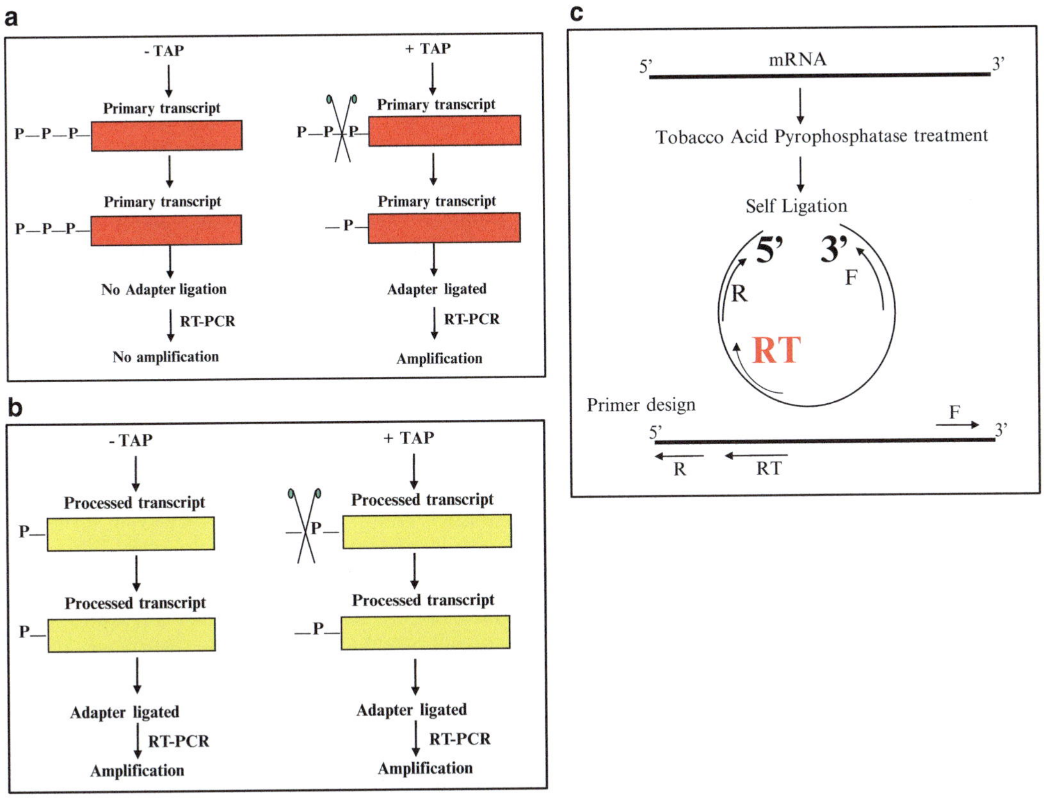

Fig. 2 The principle and method to distinguish the primary and processed transcripts via tobacco acid pyrophosphatase (TAP). (**a**) TAP treatment of primary transcripts. (**b**) TAP treatment of processed transcripts. *P* phosphate, *RT-PCR* reverse transcriptase polymerase chain reaction. (**c**) Illustration of the principal and primer design of circular RT-PCR (Personal communication Dr. Livia Merendino and Dr. Malik Ghulam Mustafa)

Post-transcriptional Regulation

Although not fully established yet, it is probable that post-transcriptional modifications of the plant mitochondrial transcripts (mRNA, tRNA and rRNA) regulate gene expression and, most of which are involved (directly or indirectly) in oxidative phosphorylation, render them translationally and/or functionally competent. Although most of the genes of mitochondrial bacterial endosymbiont have either been lost or transferred to the nucleus, still transcripts of the small number of genes left in the plant mitochondrion undergo splicing (removal of introns), maturation of 5′ and 3′ ends and even editing too which indicates that the mito-omics (mitochondrial genetic system) are regulated in a complex manner and are not as simple as thought at the time of early mitochondrion sequencing.

Complete genome sequence comparison of 15 angiosperm species (Kubo and Newton 2008; Kubo and Mikami 2007) showed not only presence of similar gene content but also high conservation of gene and intron sequences across different angiosperm species. But still the sequenced genomes varied greatly in size. This is because of species-specific intergenic regions' repeated mitochondrial DNA segments (having plastid or nuclear origin or even sequence of unknown origin), collectively tagged as 'junk sequences', but we would name them as 'uncharacterized sequences' of mitochondrial genome unless proven otherwise. As previously named 'junk sequences'—introns—of human genome proved to be involved not only in gene regulation but also in many diseases.

Plant mitochondrial RNAs (rRNAs, tRNAs and mRNAs) are co-transcriptionally or post-transcriptionally processed (to produce mature 3′ and 5′ extremities), spliced (both *cis* and *trans*) and edited prior to becoming competent for translation (mRNAs), structure (rRNAs) and function (tRNA).

RNA Splicing

At the time of their discovery in 1977 (Chow et al. 1977; Berget et al. 1977), introns were defined as 'the non-coding sequences that interrupt protein coding sequences in viral and nuclear genes'. However, by now we know that introns are not necessarily the non-coding sequences and are not only found in coding sequences but also can be present in non-coding regions, 3′ UTRs and 5′ UTRs. In addition, they are found in almost all types of genome (although the prevalence and number vary greatly), i.e. nuclear, plastid, mitochondrial, prokaryotic, eukaryotic and viral genomes.

Majority of the introns found in mitochondrion of land plants are the group II introns. A small number of group I introns (homing ribozymes) are also found in mitochondrion of non-vascular plants. Both the group I and group II introns

1. Contain higher-order conserved structures which are required for accurate and correct splicing.
2. Act as ribozyme, i.e. they can self-splice in the absence of any other enzyme or protein.
3. Some of them act as retro-elements, i.e. an intron RNA that integrates into their host genome at new positions or in another genome with the help of intron-encoded reverse transcriptase or DNA endonuclease (Lambowitz and Zimmerly 2011).
4. Group I introns encode homing DNA endonuclease while group II introns encode reverse transcriptase which facilitates their reintegration into the genome.

The complexity of gene expression and splicing is further enhanced by the need to 'trans-splice' the scattered coding sequences due to rearrangements of segments of mitochondrial DNA.

DNA rearrangements of plant mitochondrion break apart coding sequences of certain genes. But these fractured genes are expressed correctly and remain functional due to the presence of intron sequences which can form conserved intron structures and splice in *trans*. A number of group I and group II introns have been discovered and characterized in nature but only a few of them undergo trans-splicing (Glanz and Kuck 2009). For example, in plant mitochondrion only nine trans-splicing events have been described so far. Seven of them are group II trans-splicing introns in plant mitochondria in nad1, nad2, nad5 and cox2 (*Allium cepa*) genes. Trans-splicing in these genes is not conserved throughout the plant kingdom, i.e. they may *trans-splice* in some and *cis-splice* in other plant mitochondria. For example, nad1 (nad1i394), nad2 (nad2i542), nad5 (nad5i1455, nad5i1477) are cis-splicing in *Isoetes* while *trans-splicing* in seed plants. Similarly, nad1i728 is *cis-splicing* in some flowering plants while *trans-splicing* in others (Qiu and Palmer 2004). The eighth and ninth trans-splicing introns are located at the same position in cox genes in *Isoetes engelmannii* and *Marchantia*, respectively (Grewe et al. 2009). Although Castandet et al. (2010) and Farre et al. (2012) have suggested the requirement of RNA editing for correct splicing reaction in plant mitochondria, the exact mechanism of regulation and/or modulation of the expression of separately located segments of a single gene has not yet been characterized. Similarly, it is not clear how the steady-state level of mRNA of trans-spliced genes is maintained in cases where gene segments are transcribed from different promoters. Some segments become part of the pre-existing transcriptional unit while others recruit new promoters from the non-coding or spacer regions.

Although, none of the land plant mitochondrial intron has been shown to be autocatalytic neither *in vivo* nor *in vitro* probably because of lack of cellular environment in *in vitro* systems and due to technical difficulties in *in vivo* studies, the mechanism of splicing of group I and group II introns has been extensively studied in *Tetrahymena* nuclear ribosomal RNA introns and yeast mitochondrial introns, respectively. Group I and group II introns differ from each other in their structure and biochemical pathways, but still both are ribozymes and both catalyze two trans-esterification reactions in the presence of magnesium ions. They also differ in ribozyme core length as group II introns is about ~500 nt while that of group I introns is ~250 nt (Vicens and Cech 2006; Toor et al. 2006). Group I and group II ribozymes have gained the attention of the scientific community because of their ability to specifically disrupt the gene of interest and/or repair RNA (Long et al. 2003; Lambowitz and Zimmerly 2004; Toro et al. 2007). Ribozyme of group I intron consists of substrate domain (P1–P2), scaffolding domain (P4, P5, P6) and catalytic domain (P3, P7, P8, P9).

Splicing of the group I introns is initiated by the attack of external guanosine on the 5′ splice site which is followed by another trans-esterification reaction in which 3′ OH of upstream exon attacks 3′ splice site resulting in the ligation of two exons and removal of linear intron.

Of the six helical domains (I–VI) of group II introns, domain V (conserved 34 nt) serves as a catalytic core along with sequences of domain I. An adenosine of domain VI near the 3′ end of the intron initiates the splicing of group II introns through the attack of its 2′ OH on the 5′ splice site forming a 2–5′ phospho-di-ester bond. This is followed by the attack of 3′ OH of the upstream exon on the 3′ splice site resulting in the ligation of the two exons and release of intron as lariat. Although the above model is widely accepted among the scientific community, Li-Pook-Than and Bonen (2006) proposed a hydrolytic and a hypothetical model of splicing mechanism of group II introns. Alternative splicing which results in diversity of transcripts and proteins both in animals and plants has not yet been fully established for group I and II introns.

Though some of the components of 'spliceosome' have been described and/or hypothesized which include RNA maturases, reverse transcriptases both arising from the ORFs located within introns, intron-encoded proteins and nuclear-encoded PPR (pentatricopeptide repeat) element proteins, there is no concrete evidence of the spliceosome of plant mitochondrial transcriptome, i.e. it has not yet been isolated and/or fully characterized.

Similarly, temporal and spatial regulation of RNA processing, splicing and editing and their interrelationship has not yet been fully established.

It has recently been shown that introns play an important role in the regulation of expression of ribosomal protein genes of the host gene and paralogs of its host genes in yeast. They can have intergenic and intragenic upregulatory and down-regulatory effects on the gene expression (Parenteau et al. 2011). Do the introns of plant mitochondria play any role in the regulation of the mRNA level of the host gene or neighbouring genes needs to be elaborated/analyzed.

RNA Editing

RNA editing is a post-transcriptionally and/or transcriptionally coupled process in which the sequence of neo-synthesized RNA is altered as compared to the sequence of its template DNA either through conversion, deletion or insertion of ribonucleotides. Although different RNA editing types occur in different types of transcriptomes, i.e. addition/deletion of adenine, insertion of cytidine, C-to-U and A-to-I editing and insertion/deletion of guanosines, in plant mitochondrial transcriptome conversion of cytidine to uracil (through deamination of cytidine) is the most prevalent form of RNA editing (Jacques et al. 1994; Benne et al. 1986; Mahendran et al. 1994; Wagner et al. 1989; Hoch et al. 1991). Uracil to cytidine is another rare type of RNA editing that occurs in plant mitochondria (Shikanai 2006). Editing sites and their number (several hundred) is less conserved in plant mitochondria (Zeng et al. 2007). For example, Chaw et al. (2008) and Grewe et al. (2009) reported almost 1,000 editing sites in the mitochondrial transcriptomes of *Cycas taitungensis* and *Isoetes engelmannii*. While *Arabidopsis thaliana* has 456 (Giege and Brennicke 1999), *Oryza sativa* has 491 (Notsu et al. 2002) and *Brassica napus* has 427 editing sites in mitochondrial transcriptome (Handa 2003). The readers are referred to PREPACT (http://www.prepact.de) for prediction, analysis, annotation and graphical display of RNA editing sites (Lenz et al. 2010). RNA editing can result in the creation of a translation start codon (Hoch et al. 1991; Chapdelaine and Bonen 1991), formation/removal of stop codon (Kugita et al. 2003; Hiesel et al. 1989; Hiesel et al. 1994) and alteration in amino acid sequence to create conserved and functional protein, partial editing sites (only a subset of transcripts is edited) (Brennicke et al. 1999) or even silent editing sites—no change in amino acid but may result in change of translatability due to codon usage differences (most cases of editing at third nucleotide of the codon are silent) (Kempken et al. 1991).

Two components of the mitochondrial editosome, i.e. transcript-specific editing factor (TSEF) and PPR (pentatricopeptide repeat protein), have been identified in plant, but the editing enzyme (if any) or enzymatic mechanism of editing has neither been identified nor characterized. The widely accepted mechanisms for RNA editing are de-amination and transamination (Takenaka et al. 2008). Among the nuclear-encoded TSEFs identified so far, MEF1 (editing sites—ES—*rps4*-956, *nad7*-963 and *nad2*-1,160) (Zehrmann et al. 2009), MEF11 (*cox3*-422, *nad4-124* and *ccb203*-344) and MEF9 (*nad7*-200) (Verbitskiy et al. 2010) are characterized. Similarly, PPR proteins OGR1 having DYW domain were found to be responsible for editing at nad4-C401, nad4-C416, nad4-C433, nad2-C1457, ccmC-458, cox2-167 and cox3-572 (Kim et al. 2009). Characterization of MEF1, MEF9, MEF11 and OGR1 in editing was carried out through nuclear mutants and direct evidence of binding of these factors and/or mechanism of gene specificity are still lacking. Bentolila et al. (2013), Yagi et al. (2013) and Tseng et al. (2013) have further elaborated the role of REME2 (required for efficiency of mitochondrial editing 2) and PPRs (pentatricopeptide proteins) in RNA editing in plant mitochondria, respectively.

Along with the regulation of mitochondrial gene expression at transcript level (discussed above), it is hypothesized that the regulation of mitochondrial gene expression also takes place at translational and post-translational level (complex assembly) (Giege et al. 2005). A very recent study of the *rps10* mutant plants of *Arabidopsis thaliana* further supports the hypothesis where authors have concluded that 'The ultimate coordination of expression of the nuclear and mitochondrial genomes occurs at the complex assembly level' (Kwasniak et al. 2013).

Mitochondrial Proteomics

Diversity of mitochondrial functions leads the scientist to speculate and suggest that the mitochondrial proteome may contain thousands of proteins. In order to decipher the coordinate functioning of different complexes and/or proteins needed to perform multiple tasks of mitochondria, detailed identification and functional attribution to each protein of mitochondrial proteome is of prime importance.

Although mitochondria have their own genome, only a few proteins are encoded by it. Majority of proteins are encoded by nucleus and then transported into mitochondria by complex protein import machinery. It has been shown that proteins are imported into mitochondria through multi-subunit protein complexes called as translocases with combined action of peptidases and molecular chaperons. Once the proteins are translated into the cytosol, they are recognized by the protein import complex and transported into mitochondria. This specific recognition of proteins depends on the presence of N-terminal targeting sequence termed as presequence. The import of proteins into mitochondria occurs by four different types of pathways which include (1) general import pathway based on N-terminal targeting signal sequences; (2) the carrier import pathway which depends on import for proteins of mitochondrial carrier family into the inner membrane; (3) the sorting and assembly pathway which describes the import of β-barrel proteins into the outer membrane through sorting and assembly machinery of outer membranes, also known as topogenesis of β-barrel proteins and (4) the mitochondrial intermembrane space import and assembly pathway which describes the import of proteins into the intermembranes of mitochondria. Due to lack of experimental data of plant mitochondrial proteome, many bioinformatics tools have been used to identify the proteins on the basis of their subcellular location, whereas extensive experimental studies of mitochondrial proteome of the human heart, yeast and mouse have been carried out (Mootha et al. 2003; Sickmann et al. 2003; Taylor et al. 2003).

Initially different experimental approaches like two-dimensional electrophoresis showed that plant mitochondria have 500–1,500 proteins (Kruft et al. 2001; Bardel et al. 2002; Millar et al. 2001), but these data were not reliable as in this approach a single protein can appear as multiple protein spots. To overcome this limitation, later on Heazlewood and his colleagues did the experimental analysis of *Arabidopsis thaliana* mitochondrial proteome by using liquid chromatography in tandem with mass spectrometry (LC-MS/MS) which is a very sensitive method (Heazlewood et al. 2003a). Direct sample analysis of *Arabidopsis* mitochondrial protein lysates resulted in identification of 416 proteins out of which 95 % have been already identified through two-dimensional gel electrophoresis (Kruft et al. 2001; Millar et al. 2001; Werhahn and Braun 2002). They have also confirmed the previously identified protein results which were determined during purification of particular subsets and protein complex of mitochondrial proteins (Sweetlove et al. 2002; Heazlewood et al. 2003a, c; Herald et al. 2003). Out of 416, majority of the proteins (409) are nuclear encoded and only 7 are encoded by mitochondria. Moreover, they have also assigned functions to these proteins by using BLAST on the basis of amino acid sequence similarity with already known proteins. They have shown that some of the proteins are part of the tricarboxylic acid cycle (TCA) and electron transport chain, involved in transcription and translational processes and signalling pathways. Some of the proteins have no amino acid sequence similarity to any protein of known function; hence, they were called as proteins of unknown function.

Later on, another group, Huang and his collaborators, studied the rice mitochondrial proteomics through two-dimensional gel electrophoresis and LC-MS/MS of trypsin-digested samples. By using both gel-based and non-gel-based experimental analysis, they have identified the 322 nonredundant rice mitochondrial proteins (Huang et al. 2009). They have shown that 22 % of the proteins are involved in energy production and 28 % in the tricarboxylic acid cycle and 17 % of the proteins were defined as unknown proteins. The remaining few proteins were found to be involved in signalling process, stress defence, biosynthesis of vitamins and electron transport chain as in *Arabidopsis*. They have shown that *Arabidopsis* and rice have conserved mitochondrial proteins when they compared both proteome results. Almost 80 % of the proteins identified in rice mitochondria have orthologues in *Arabidopsis*. The proteins which are involved in energy and metabolism in *Arabidopsis* have homologues in rice. Similarly, the proteins with unknown function have also conserved amino acid sequences in both *Arabidopsis* and rice which suggests that they have conserved functions (Huang et al. 2009).

Many bioinformatics tools like MitoProt (Claros and Vincens 1996), TargetP (Emanuelsson et al. 2000) and Predator (Small et al. 2004) can be used to determine the protein subcellular location. Recently, a computational model, naive Bayesian network, was used for predicting the *Arabidopsis* mitochondrial proteins. This model is based on the genomic data obtained from eight bioinformatics tools, protein domain properties, multiple orthologous mappings and co-expression patterns obtained from microarray profiles (Cui et al. 2011). They have identified 2,311 mitochondrial proteins by using this bioinformatics strategy and developed CoreMitoP database (http://www.megabionet.org/atpid/webfile/). This database also contains information about the biological functions of these mitochondrial proteins which are similar as predicted in earlier studies.

An important experimental study of *Arabidopsis* mitochondrial proteome was carried out by MS analysis of 2-D gel separable protein spots (Taylor et al. 2011). As a result of this gel-based technique, they have identified 264 mitochondrial proteins, and alongside by using LC-MS/MS, 220 proteins were identified.

It has been shown that mitochondrial proteins change in abundance under a variety of stress conditions both in *Arabidopsis* and rice. Some of the proteins are affected by multiple stress conditions like HSP70 decrease under salt stress and increase under Cd stress and heat in *Arabidopsis*. Similarly, there are some other proteins like GDC P protein which increase under salt, heat and cold conditions. These results are obtained only from preliminary studies but can be helpful for further investigation of the environmental response of the plant mitochondria.

These experimental studies and bioinformatics tools helped scientists to develop databases of mitochondrial proteome which are publically available (http://www.megabionet.org/atpid/webfile/), but still it needs more extensive functional proteomic studies to fully understand the molecular mechanism of plant mitochondria. There are many constraints for better understanding of mitochondrial proteomes like mitochondrial sample purity from different species and fractionation and digestion of mitochondria for their deeper analysis by using different methods like LC-MS/MS. As it is clear from the discussion in the above paragraph that the mitochondrial proteomic studies are at the nascent stage (identification of proteins), a lot of efforts are required in the field for the structural characterization and functional importance.

Translational Machinery: Mitoribosomes

Plant mitochondria possess their own translation machinery composed of ribosomes called mitoribosomes. The constituents of mitoribosomes are proteins (encoded by nuclear and mitochondrial genomes) and ribosomal RNA (rRNA—26S/18S—transcribed from mitochondrion). In animals, all the mitoribosomal proteins were transferred to nucleus during gene loss and transfer (all ribosomal proteins are encoded by nuclear genome), whereas plant mitochondrion has retained a few ribosomal protein genes and even the gene order of their ancestral bacteria (Lang et al. 1999). Mitochondrion of bryophytes contains 16 while flowering plants contain 14 mitoribosomal protein genes including S1–S4, S7, S10–S14, S16, L2, L5 and L16 (Takemura et al. 1992; Adams and Palmer 2003), while the remaining proteins are encoded by the nucleus and are transported to the mitochondria for mitoribosomal assembly. Heterogenic proteins and rRNAs are arranged in two subunits, i.e. large subunit (LSU) and small subunit (SSU). Plant mitoribosome is composed of 80 proteins of which 24 make up the small subunit (SSU) and 42 proteins make up the large subunit. It has recently been shown that silencing of only one nuclear-encoded mitochondrial ribosomal protein gene results in the alteration of the mitochondrial translation (Kwasniak et al. 2013). The results of Kwasniak et al. (2013) indicate that mitochondrial gene expression is regulated at translation and post-translational level through mitoribosomal proteins. The structural details of the mammalian mitochondria have been reviewed by Agrawal and Sharma (2012). Similarly, a 3A° detailed structure of yeast ribosomes has recently been revealed (Ben-Shem et al. 2011), but such detailed structural characterization of plant mitoribosome and its constituents (proteins and rRNA) and modus operandi are still lacking. Although strong and/or definitive experimental evidence is still lacking, a new term, 'specialized ribosomes', has become a hot issue in the field (Xue and Barna 2012). The question of 'specialized mitoribosomes' is still open and need to be discussed and analyzed.

Mitochondrial Energy Metabolomics

The eukaryotic cell came into existence as a result of an endosymbiotic association of unique energy-producing bacteria and a proto-eukaryotic cell, wherein host cell provided physical space and materials to the bacterium, which supplied energy to the host. During the evolution of this symbiotic relationship, the bacterium transferred many of its genes (about 1,500) to the host nucleus, creating the modern nuclear DNA (nDNA) genome (Wallace 2005a). The entire mtDNA is almost exclusively transcribed (it contains very small non-coding areas) and is arranged in nucleoids (Holt et al. 2007).

The mitochondrial genome controls many essential functions to meet energy requirement of cells. The most important among these functions are (1) OXPHOS (oxidative phosphorylation), (2) the production of most of the cell's ROS and (3) the regulation of apoptosis by activating the mitochondrial permeability transmission pore (Wallace 2005b).

OXPHOS System

This is a complex mitochondrial respiratory mechanism to harness electron energy from carbohydrates and fats to phosphorylate ADP to make ATP. The light-dependent metabolic pathway distinguishes plant OXPHOS system from most animals because it includes additional so-called 'alternative' electron transport component, which participates in photorespiration. Therefore, activities of plant mitochondria have to be coordinated with the ones of chloroplasts.

Throughout the evolution, mitochondria and chloroplasts retain the majority of their genome to act semi-autonomously inside eukaryotic hosts. The semi-autonomous status of these organelles resulted in coordination of spatially separated genomes for its biogenesis and its functional activity. About 30 mitochondrial proteins are encoded by the mitochondrial genome, several of which represent components of the OXPHOS system (Unseld et al. 1997). The remaining approximately 2,000 mitochondrial proteins, including most components of the OXPHOS system, are nuclear encoded and have to be post-translationally imported into the organelle (Mackenzie and McIntosh 1999; Adams and Palmer 2003).

Oxidative Phosphorylation (OXPHOS) Apparatus

The OXPHOS apparatus, commonly known as ETC, is comprised of four large protein complexes (I, II, III, IV) that interact with each other via the small lipid ubiquinone (UQ) and the small protein cytochrome c. The electron flow from NADH to oxygen is coupled to proton translocation out of the matrix, which drives phosphorylation of ADP to form ATP by the F-ATP synthase (complex V).

Complex I (CI)

Complex I (NADH ubiquinone oxidoreductase) is the main entrance point for electrons into the respiratory chain under most of the conditions. It transfers two electrons from NADH to ubiquinone, and coupled to this process, four protons are translocated across the inner mitochondrial membrane.

Direct comparisons of CI subunit composition across taxa have revealed divergences between plant CI versus mammalian CI, with eight nuclear-encoded plant CI subunits being plant specific and several others being common between plants and non-mammalian eukaryotes but absent in mammals (Meyer et al. 2008; Klodmann and Braun 2011; Klodmann et al. 2010; Cardol 2011).

The 1,000-kDa complex I consists more than 49 distinct protein subunits, one FMN cofactor and eight Fe–S clusters. The proteins that constitute complex I is encoded by two genomes: nine have a mitochondrial genetic origin in higher plants, while the remaining subunits are encoded in the nucleus. Recently, composition of complex I has been investigated intensively in humans (Ugalde et al. 2004; Vogel et al. 2007; Remacle et al. 2008; Peters et al. 2013) in the vascular plants *Arabidopsis thaliana* and *Oryza sativa* (Heazlewood et al. 2003b, c; Meyer et al. 2008) and in the green algae *Chlamydomonas reinhardtii* (Cardol et al. 2004, 2005). In *Arabidopsis*,

42 distinct complex I subunits are known and some of them interrupted with class II introns. These class II introns are removed post-transcriptionally by mitochondrial CAF-like splicing factor 1 (mCSF1; At4 g31010) (Meyer et al. 2008; Zmudjak et al. 2013). The complex I subunits of *Arabidopsis thaliana* are highly conserved across the organisms; 40 out of 49 subunits have homology with bovine complex I, including 14 orthologues to the prokaryotic enzyme that constitute the 'core' subunits of complex I (Cardol et al. 2005; Peters et al. 2013). The function of most of the extra subunits in mitochondria is yet unclear, but it is speculated that they might be important for enzyme stabilization or protection of the complex towards reactive oxygen species (ROS) (Vogel et al. 2007). For the detailed list of characterized and putative constituents of complex I, refer to Table 1.

Complex II (CII)

Succinate–ubiquinone oxidoreductase, complex II; the smallest among OXPHOS complex, has a central role in mitochondrial metabolism because it is involved in both the tricarboxylic (TCA) acid cycle and the mitochondrial electron transport chain. Complex II mediates the oxidation of succinate to fumarate and the reduction of ubiquinone to ubiquinol. The activity of this complex is not coupled to the generation of an electrochemical gradient of protons across the membrane to drive the synthesis of ATP (Horsefield et al. 2004).

Across the organisms, complex II consists of four core subunits: a flavoprotein (SDH1), an iron–sulphur subunit (SDH2) and two membrane anchor subunits (SDH3 and SDH4). Moreover, four proteins of unknown function are observed which co-migrate with the complex (Eubel et al. 2003; Millar et al. 2004). In *Arabidopsis*, all SDH subunits are encoded in the nuclear genome. Knockout mutants of the SDH1 gene are embryo lethal (Leon et al. 2007), but knockdowns of SDH1 and SDH2 lead to phenotypes associated with altered stomatal aperture, altered mitochondrial ROS production and altered nitrogen use efficiency (Fuentes et al. 2011; Gleason et al. 2011). Several proteins assisting CII assembly have been described in yeast and mammalian cells, but only SDHAF1 and SDHAF2 are considered to be real assembly factors that directly and specifically aid CII assembly (Hao et al. 2009; Ghezzi et al. 2009; Rutter et al. 2010). In *Arabidopsis*, knockdown of the SDHAF2 homologue lowers SDH assembly and markedly reduces root growth. For the detailed list of characterized and putative constituents of complex II, refer to Table 2.

Complex III (CIII)

Cytochrome 'bc' complex or ubiquinol–cytochrome 'c' oxidoreductase (complex III) oxidizes ubiquinol produced by the action of complex I and II. The electrons from ubiquinol are transferred to cytochrome *c* and four protons are pumped for each pair of electrons passing through this multi-subunit complex (Tzagoloff 1995; Berry et al. 2000).

Complex III contains 10 subunits including the dual functional core proteins that act both in CIII function and as the matrix processing peptidase, removing pre-sequences from imported matrix proteins. Cytochrome b, the only one subunit of this complex, is encoded by the plant mitochondrial genome (Unseld et al. 1997), and the remaining nine are encoded by the nuclear genome. In BN-PAGE separations from *Arabidopsis* mitochondria, all of these subunits have been identified and linked back to a set of mostly single-copy genes (Werhahn and Braun 2002; Meyer et al. 2008). Yeast provides an ideal model system to study CIII, due to its ability to survive by fermentation in the absence of the complex, making gene knockout and mutagenesis possible (Smith et al. 2012). CIII assembly follows a modular assembly model including early core subcomplex, late core subcomplex and a dimeric CIII state (Smith et al. 2012). There have been 13 assembly factors involved in aiding the different stages of CIII assembly in yeast. Two of these, BCS1L and TTC19, were also found to have functional homologues in mammalian CIII

Table 1 Complex I

Complex I Genes	Common name	MW (dalton)	Function	Origin	Remarks
At5g11770	N/A	24,043.7	Mitochondrial electron transport, NADH to ubiquinone, oxidation–reduction process, photorespiration, proteasome core complex assembly, protein glycosylation, response to misfolded protein, ubiquitin-dependent protein catabolic process	Nuclear	Mitochondria only
At1g16700	N/A	25,377.3	Oxidation–reduction process	Nuclear	Mitochondria only
At5g37510	CI76, EMB1467, embryo defective 1467	81,524.7	ATP synthesis coupled electron transport, cellular respiration, embryo development ending in seed dormancy, gravitropism, oxidation–reduction process, photorespiration, proteasome core complex assembly, response to misfolded protein, response to oxidative stress, ubiquitin-dependent protein catabolic process	Nuclear	Dual targeted
At5g47890	N/A	10,850.6	Biological process, photorespiration	Nuclear	Dual targeted
AK059007	Not present in TAIR				Mitochondria only
At5g52840	N/A	19,178.8	Pentose-phosphate shunt, photorespiration, proteasome core complex assembly, respiratory electron transport chain, response to misfolded protein, response to salt stress, ubiquitin-dependent protein catabolic process	Nuclear	Dual targeted
At3g03070	N/A	12,234	Phosphatidylinositol biosynthetic process	Nuclear	Mitochondria only
At2g42210	ATOEP16-3, OEP16-3	18,577.1	Photorespiration, proteasome core complex assembly, protein transport, response to misfolded protein, ubiquitin-dependent protein catabolic process	Nuclear	Dual targeted
At3g57785	N/A	12,662.3	Biological process	Nuclear	Dual targeted
At3g62790	N/A	9,893.0	Biological process	Nuclear	Dual targeted
At1g04630	Maternal effect embryo arrest 4, MEE4	16,125.5	Embryo development ending in seed dormancy, photorespiration	Nuclear	Dual targeted
At2g02050	N/A	11,739.7	Glucose catabolic process, mitochondrial electron transport, NADH to ubiquinone, photorespiration, proteasome core complex assembly, response to misfolded protein, ubiquitin-dependent protein catabolic process	Nuclear	Dual targeted
At5g67590	FRO1, frostbite1	17,134.3	Golgi organization, cold acclimation, electron transport chain, glycolysis, hyperosmotic response, photorespiration, proteasome core complex assembly, response to cadmium ion, response to misfolded protein, response to osmotic stress, response to salt stress, response to temperature stimulus, ubiquitin-dependent protein catabolic process, water transport	Nuclear	Mitochondria only

(continued)

Table 1 (continued)

Complex I Genes	Common name	MW (dalton)	Function	Origin	Remarks
At5g18800	N/A	11,967.8	Mitochondrial electron transport, NADH to ubiquinone, photorespiration	Nuclear	Dual targeted
At4g34700	ATCIB22, B22 subunit of eukaryotic mitochondrial complex I, CIB22	13,616.2	Carbohydrate metabolic process, photorespiration, proteasome core complex assembly, response to misfolded protein, ubiquitin-dependent protein catabolic process	Nuclear	Dual targeted
At1g49140	N/A	12,530.3	Photorespiration	Nuclear	Mitochondria only
At4g16450	N/A	11,345	Photorespiration, proteasome core complex assembly, response to misfolded protein, ubiquitin-dependent protein catabolic process	Nuclear	Mitochondria only
At4g20150	N/A	9,207.7	Aerobic respiration, biological process, glycolysis, photorespiration, core complex assembly, response to cadmium ion, response to misfolded protein, ubiquitin-dependent protein catabolic process	Nuclear	Multi-targeted
At3g48680	ATCAL2, gamma CAL2, gamma carbonic anhydrase-like 2	27,955.9	Photorespiration, proteasome core complex assembly, response to abscisic acid stimulus, response to misfolded protein, response to salt stress, starch biosynthetic process, ubiquitin-dependent protein catabolic process, vegetative to reproductive phase transition of meristem	Nuclear	Multi-targeted
At5g66510	Gamma CA3, gamma carbonic anhydrase 3	29,043.4	Pentose-phosphate shunt, photorespiration, proteasome core complex assembly, response to cadmium ion, response to misfolded protein, response to salt stress, second-messenger-mediated signalling, starch biosynthetic process, ubiquitin-dependent protein catabolic process	Nuclear	Dual targeted
At1g47260	APFI, gamma CA2, gamma carbonic anhydrase 2	30,065.1	Anther dehiscence, photorespiration, protein homotrimerization, regulation of reactive oxygen species metabolic process, response to salt stress, starch biosynthetic process	Nuclear	Multi-targeted
At1g67350	N/A	11,791.1	Photorespiration, proteasome core complex assembly, response to misfolded protein, ubiquitin-dependent protein catabolic process	Nuclear	Dual targeted
At2g27730	N/A	11,947.5	Photorespiration, proteasome core complex assembly, response to misfolded protein, ubiquitin-dependent protein catabolic process	Nuclear	Multi-targeted
At1g16700	N/A	25,377.3	Oxidation–reduction process	Nuclear	Mitochondria only

At1g79010	N/A	25,502.5	Oxidation–reduction process	Nuclear	Mitochondria only
AtMg00070	NADH dehydrogenase subunit 9	22,687.4	Cellular respiration, oxidation–reduction process, photorespiration	Mitochondrial	Mitochondria only
AtMg00510	NAD7, NADH dehydrogenase subunit 7	44,577.4	Cellular respiration, oxidation–reduction process, response to cadmium ion	Mitochondrial	Mitochondria only
At5g08530	51-kDa subunit of complex I, CI51	53,449.0	Golgi vesicle transport, N-terminal protein myristoylation, cell growth, cell morphogenesis, mitochondrial electron transport, NADH to ubiquinone, oxidation–reduction process, photorespiration, proteasome core complex assembly, response to misfolded protein, ubiquitin-dependent protein catabolic process	Nuclear	Mitochondria only
AtMg00516	NAD1, NAD1C, NADH dehydrogenase 1, NADH dehydrogenase 1C	35,675.2	Cellular respiration, oxidation–reduction process	Mitochondrial	Mitochondria only
AtMg00285	NAD2, NAD2.1, NAD2A, NADH dehydrogenase 2, NADH dehydrogenase 2.1, NADH dehydrogenase 2A	54,879.2	ATP synthesis coupled electron transport, cellular respiration, oxidation–reduction process	Mitochondrial	Dual targeted
AtMg00990	NAD3, NADH dehydrogenase 3	13,659.8	Cellular respiration, oxidation–reduction process	Mitochondrial	Mitochondria only
AtMg00580	NAD4, NADH dehydrogenase subunit 4	55,231.9	ATP synthesis coupled electron transport, cellular respiration, oxidation–reduction process, photosynthesis	Mitochondrial	Mitochondria only
AtMg00650	NAD4L, NADH dehydrogenase subunit 4 L	10,917.9	ATP synthesis coupled electron transport, cellular respiration, oxidation–reduction process, photosynthesis	Mitochondrial	Mitochondria only
AtMg00513	NAD5, NAD5.1, NAD5A, NADH dehydrogenase 5.1, NADH dehydrogenase 5A, NADH dehydrogenase subunit 5	73,907.2	ATP synthesis coupled electron transport, cellular respiration, oxidation–reduction process	Mitochondrial	Mitochondria only

(continued)

Table 1 (continued)

Complex I Genes	Common name	MW (dalton)	Function	Origin	Remarks
AtMg00270	NAD6, NADH dehydrogenase 6	23,493.8	Cellular respiration, oxidation–reduction process	Mitochondrial	Mitochondria only
At3g08610	N/A	7,337.6	Biological process, photorespiration, proteasome core complex assembly, response to misfolded protein, ubiquitin-dependent protein catabolic process	Nuclear	Mitochondria only
At2g47690	N/A	13,964.6	Photorespiration	Nuclear	Dual targeted
At2g31490	N/A	8,291.5	Photorespiration	Nuclear	Mitochondria only
At2g33220	N/A	16,121.5	Photorespiration	Nuclear	Mitochondria only
At3g03100	N/A	18,316.6	Photorespiration, proteasome core complex assembly, response to misfolded protein, ubiquitin-dependent protein catabolic process	Nuclear	Mitochondria only
At3g18410	N/A	12,438.2	Photorespiration	Nuclear	Mitochondria only
At2g20360	N/A	43,935.4	Cellular metabolic process, photorespiration, proteasome core complex assembly, response to misfolded protein, response to salt stress, ubiquitin-dependent protein catabolic process	Nuclear	Mitochondria only
At1g79010	N/A	25,502.5	Oxidation–reduction process	Nuclear	Mitochondria only
At4g02580	N/A	28,388.4	Mitochondrial electron transport, NADH to ubiquinone, oxidation–reduction process, photorespiration, proteasome core complex assembly, response to misfolded protein, response to oxidative stress, ubiquitin-dependent protein catabolic process	Nuclear	Mitochondria only
AtMg00070	NAD9, NADH dehydrogenase subunit 9	22,687.4	Cellular respiration, oxidation–reduction process, photorespiration	Mitochondrial	Dual targeted
AtMg00510	NAD7, NADH dehydrogenase subunit 7	44,577.4	Cellular respiration, oxidation–reduction process, response to cadmium ion	Mitochondrial	Mitochondria only
At5g08530	51-kDa subunit of complex I, CI51	53,449.0	Golgi vesicle transport, N-terminal protein myristoylation, cell growth, cell morphogenesis, mitochondrial electron transport, NADH to ubiquinone, oxidation–reduction process, photorespiration, proteasome core complex assembly, response to misfolded protein, ubiquitin-dependent protein catabolic process	Nuclear	Mitochondria only

At3g18410	N/A	12,438.2	Photorespiration	Nuclear	Mitochondria only
At2g20360	N/A	43,935.4	Cellular metabolic process, photorespiration, proteasome core complex assembly, response to misfolded protein, response to salt stress, ubiquitin-dependent protein catabolic process	Nuclear	Mitochondria only

http://www.arabidopsis.org/index.jsp.in

Note: Only experimentally proved subunits and intermediate or assembly factor were included in the tables

N/A not applicable or information unavailable; dual targeted, targeted to mitochondria and chloroplast; multi-targeted, mitochondria, chloroplast and other organelles (list of constituents of each complex was modified from bioinformatics analysis of Cardol et al. 2005. The tables have been updated and/or modified where necessary to recent information extracted from TAIR database)

Table 2 Complex II

Complex II Genes	Common name	MW (dalton)	Function	Origin	Remarks
At2g18450	SDH1-2, succinate dehydrogenase 1-2	69,362.7	Electron transport chain, mitochondrial electron transport, succinate to ubiquinone, oxidation–reduction process, tricarboxylic acid cycle	Nuclear	Mitochondria only
At5g40650	SDH2-2, succinate dehydrogenase 2-2	31,140.6	Mitochondrial electron transport, succinate to ubiquinone, oxidation–reduction process, tricarboxylic acid cycle	Nuclear	Mitochondria only
At5g09600	SDH3-1, succinate dehydrogenase 3-1	23,453.8	Mitochondrial electron transport, succinate to ubiquinone	Nuclear	Dual targeted
At2g46505	SDH4, succinate dehydrogenase subunit 4	16,841.3	Aerobic respiration, glycolysis, response to cadmium ion, tricarboxylic acid cycle	Nuclear	Mitochondria only
At1g08480	SDH6, succinate dehydrogenase 6	15,812.6	Biological process, photorespiration	Nuclear	Multi-targeted

http://www.arabidopsis.org/index.jsp.in
Note: Only experimentally proved subunits and intermediate or assembly factor were included in the tables
N/A, not applicable or information unavailable; dual targeted, targeted to mitochondria and chloroplast; multi-targeted, mitochondria, chloroplast and other organelles (list of constituents of each complex was modified from bioinformatics analysis of Cardol et al. 2005. The tables have been updated and/or modified where necessary to recent information extracted from TAIR database)

assembly (Diaz et al. 2011; Smith et al. 2012). Little is known about CIII assembly or functional assembly factors in plants. For the detailed list of characterized and putative constituents of complex III, refer to Table 3.

Complex IV (CIV)

Cytochrome *c* oxidase (COX) or complex IV is the terminal enzyme of the mitochondrial electron transport chain. It comes from heme–copper *aa*3-type terminal oxidase superfamily and catalyzes the sequential transfer of electrons from reduced cytochrome 'c' to dioxygen, consequently generating water. This electron transfer reaction is coupled to electrogenic proton pumping across the inner mitochondrial membrane (Capaldi 1990; Barrientos 2002; Herrmann and Funes 2005).

The CIV consists of 7 or 8 subunits in plants (Peiffer et al. 1990), but the recent list has grown up to 14 proteins in *Arabidopsis* (Millar et al. 2004). However, only eight proteins are ubiquitously present in other organisms with certain amount of homology; further the rest of the six proteins that may represent plant-specific CIV subunits were identified. The CIV subunits assemble sequentially that start with incorporation of subunit I and followed through by several discrete assembly factors in human (Barrientos et al. 2009). The number of assembly factor in yeast goes up to 40 that helps different stages of CIV assembly, but only a few homologues for these factors have been defined in humans (Barrientos et al. 2009; Diaz et al. 2011). A study reveals the capability of plant homologue of yeast assembly factor COX19 to complement the yeast cox19 null mutant and might play a role in the biogenesis of plant cytochrome 'c' oxidase to replace damaged forms of the enzyme (Attallah et al. 2007). However, it seems evident that our knowledge about the assembly of CIV in plants is still incomplete. For the detailed list of characterized and putative constituents of complex IV, refer to Table 4.

Table 3 Complex III

Complex III Genes	Common name	MW (dalton)	Function	Origin	Remarks
At3g02090	MPPBETA	59,611	Glucose catabolic process, pentose-phosphate shunt, photorespiration, proteasome core complex assembly, proteolysis, response to cadmium ion, response to misfolded protein, response to salt stress, ubiquitin-dependent protein catabolic process	Nuclear	Multi-targeted
At3g16480	Mitochondrial processing peptidase alpha subunit, MPPALPHA	54,052.9	Photorespiration, proteasome core complex assembly, proteolysis, response to misfolded protein, ubiquitin-dependent protein catabolic process	Nuclear	Multi-targeted
AtMg00220	Apocytochrome B, COB	44,146.4	Aerobic respiration, respiratory electron transport chain	mitochondrial	Mitochondria only
At3g27240	N/A	33,649.9	Aerobic respiration, glycolysis, pentose-phosphate shunt, photorespiration, proteasome core complex assembly, response to cadmium ion, response to misfolded protein, response to salt stress, ubiquitin-dependent protein catabolic process	Nuclear	Multi-targeted
At4g32470	N/A	14,526.8	Aerobic respiration, glycolysis, mitochondrial electron transport, ubiquinol to cytochrome c, photorespiration, proteasome core complex assembly, response to cadmium ion, response to misfolded protein, ubiquitin-dependent protein catabolic process	Nuclear	Multi-targeted
At3g10860	N/A	8,505.0	Photorespiration	Nuclear	Mitochondria only
At5g13440	N/A	29,901.6	Oxidation–reduction process	Nuclear	Mitochondria only
At2g01090	N/A	7,372.3	Mitochondrial electron transport, ubiquinol to cytochrome c	Nuclear	Mitochondria only
At3g52730	N/A	8,448.6	Mitochondrial electron transport, ubiquinol to cytochrome c, photorespiration	Nuclear	Mitochondria only
At2g40765	N/A	5,975.9	Biological process, photorespiration, response to endoplasmic reticulum stress, systemic acquired resistance	Nuclear	Mitochondria only
At5g66760	SDH1-1, succinate dehydrogenase 1-1	69,655.8	Electron transport chain, mitochondrial electron transport, succinate to ubiquinone, oxidation–reduction process, tricarboxylic acid cycle	Nuclear	Dual targeted
At1g51980	N/A	54,401.4	Aerobic respiration, glucose catabolic process, glycolysis, pentose-phosphate shunt, proteolysis, response to cadmium ion, response to salt stress	Nuclear	Multi-targeted

(continued)

Table 3 (continued)

Complex III Genes	Common name	MW (dalton)	Function	Origin	Remarks
At5g40810	N/A	33,690.0	Photorespiration, proteasome core complex assembly, response to misfolded protein, ubiquitin-dependent protein catabolic process	Nuclear	Dual targeted
At5g13430	N/A	29,607.3	Oxidation–reduction process	Nuclear	Mitochondria only
At5g25450	N/A	14,595.8	Mitochondrial electron transport, ubiquinol to cytochrome c	Nuclear	Mitochondria only
At1g15120	N/A	19,018.8	Aerobic respiration, glycolysis, mitochondrial electron transport, ubiquinol to cytochrome c, photorespiration, proteasome core complex assembly, response to cadmium ion, response to misfolded protein, ubiquitin-dependent protein catabolic process	Nuclear	Mitochondria only
At5g05370	N/A	8,417.9	Ubiquinol–cytochrome c reductase activity	Nuclear	Mitochondria only

http://www.arabidopsis.org/index.jsp.in

Note: Only experimentally proved subunits and intermediate or assembly factor were included in the tables

N/A not applicable or information unavailable; dual targeted, targeted to mitochondria and chloroplast; multi-targeted, mitochondria, chloroplast and other organelles (list of constituents of each complex was modified from bioinformatics analysis of Cardol et al. 2005. The tables have been updated and/or modified where necessary to recent information extracted from TAIR database)

Table 4 Complex IV

Complex IV Genes	Common name	MW (dalton)	Function	Origin	Remarks
AtMg01360	COX1, cytochrome oxidase 1	57,995.9	Aerobic respiration, electron transport chain, oxidation–reduction process, respiratory electron transport chain, transport	Mitochondrial	Dual targeted
AtMg00160	COX2, cytochrome oxidase 2	29,375.6	Electron transport chain, respiratory electron transport chain	Mitochondrial	Dual targeted
AtMg00730	COX3, cytochrome c oxidase subunit 3	29,742.4	Respiratory electron transport chain	Mitochondrial	Mitochondria only
At4g37830	N/A	11,234.6	Cytochrome c oxidase activity	Nuclear	Mitochondria only
At3g17910	EMB3121, embryo defective 3121, SURF1, surfeit1	39,921.1	Respiratory chain complex IV assembly	Nuclear	Mitochondria only
At3g08950	HCC1, homologue of the copper chaperone SCO1	37,194.8	Cellular copper ion homeostasis, copper ion transport, embryo development, methylglyoxal catabolic process to D-lactate, mitochondrial respiratory chain complex IV assembly, protein targeting to mitochondrion, respiratory chain complex IV assembly	Nuclear	Mitochondria only

http://www.arabidopsis.org/index.jsp.in
Note: Only experimentally proved subunits and intermediate or assembly factor were included in the tables
N/A not applicable or information unavailable; dual targeted, targeted to mitochondria and chloroplast; multi-targeted, mitochondria, chloroplast and other organelles (list of constituents of each complex was modified from bioinformatics analysis of Cardol et al. 2005. The tables have been updated and/or modified where necessary to recent information extracted from TAIR database)

Complex V (CV)

The ATP synthase is also considered as complex V of OXPHOS system in addition to earlier described complexes. The membrane-bound ATP synthase is an F_0F type H^+-ATP synthase that catalyzes the steps in oxidative phosphorylation through which ATP is produced. It consists of ATP generating hydrophilic F component which opens up into the matrix and a hydrophobic F component, which channels protons through the membrane while also anchoring the whole complex to the mitochondrial inner membrane (Senior and Tsai 1990; Hamasur and Glaser 1992; Velours and Arselin 2000; Heazlewood et al. 2003a, b). This complex is highly conserved in structural core subunit of enzyme in both prokaryotic and eukaryotic organisms (Millar et al. 2011). The biogenesis of complex V also shows intricate coordination of nucleus and mitochondria, wherein most of F-ATP synthase subunits are encoded in the nucleus (β, γ, δ and ε), while most of the F subunits are encoded in the plant mitochondrial genome and translated in the mitochondrial matrix (a, b, c and A6L) (Jansch et al. 1996; Heazlewood et al. 2003c; Sabar et al. 2003, 2005). F1α subunit is encoded in the mitochondrial genome and highly conserved across most plant species. The changes in mitochondrial-encoded subunits of the F_0F-ATP synthase are frequently associated with cytoplasmic male sterility (CMS) in plants, most likely due to the high ATP demand of floral tissues (Xu et al. 2008). The study investigating knockouts of ATP synthase core subunits has shown its importance in plants, which are lethal in nature. Further it was confirmed in dexamethasone inducible knockdown that the tissue-specific phenotypes slow the rates of mitochondrial ADP: ATP cycling across a range of developmental stages (Robison et al. 2009). Induction of the knockdown during germination in the light leads to seedling lethality, stunting of dark-grown (etiolated) seedlings, downward curling or wavy-edged leaf margins of light-grown plants and ball-shaped unexpanded flowers (Robison et al. 2009), highlighting the high energetic demand of key growth stages. For the detailed list of characterized and putative constituents of complex V, refer to Table 5. For illustration of complete OXPHOS system and potential arrangement of complexes, refer to Fig. 3.

Regulation of the Biogenesis of the OXPHOS Complexes and Supercomplexes

The enzymes in respiratory chain are spatially distributed along inner and outer mitochondrial membrane in a series of protein complexes and supercomplexes. The biogenesis of these complexes requires highest degree of coordinated gene expression between nucleus and mitochondria. Recent study suggests supercomplexes as highly dynamic in nature, which forms respirosome according to growth microenvironment including carbon source. This plasticity model explains that respirosome is neither explained by fluid mosaic nor by static model alone, both are required for most efficient combination (Lapuente-Brun et al. 2013). In plants, genetic information is distributed between three compartments: nucleus, chloroplasts and mitochondria. It is well established that chloroplast and mitochondria are of endosymbiotic origin and on the contrary contain only relatively few protein coding genes, which range between 3 (*Plasmodium* mitochondrion) and 209 (*Porphyra* chloroplast) (Leister 2012). On the other side, their contemporary prokaryotic relatives possess several thousand genes; this shows an ancient, extensive transfer of their original genetic material to the nucleus. In the process mitochondria and chloroplast give up most genes to nucleus which are expressed in nucleus, synthesized in the cytoplasm and post-translationally imported into the organelles. This spatial distribution of genes minimizes functional redundancy of gene expression and alternatively comes up with new integration mechanism for coordinated gene expression and signalling crosstalk. The crosstalk between organelles includes both anterograde (nucleus-to-organelle) and retrograde (organelle-to-nucleus) controls.

Table 5 Complex V

Complex V Genes	Common name	MW (dalton)	Function	Origin	Remarks
AtMg00410	ATP6, ATP6-1, ATPase subunit 6, ATPase subunit 6-1	42,351.9	ATP synthesis coupled proton transport, cellular respiration, mitochondrial ATP synthesis coupled proton transport	Mitochondrial	Mitochondrion, proton-transporting ATP synthase complex, coupling factor F(o)
AtMg01170	ATP6-2	39,738.9	ATP synthesis coupled proton transport	Mitochondrial	Mitochondrion, proton-transporting ATP synthase complex, coupling factor F(o)
AtMg00640	ORF25	21,570.1	ATP synthesis coupled proton transport	Mitochondrial	
AtMg01080	ATP9, mitochondrial F0-ATPase subunit 9	8,841.2	ATP hydrolysis coupled proton transport, ATP synthesis coupled proton transport	Mitochondrial	
At2g07671	N/A	8,841.2	ATP hydrolysis coupled proton transport, ATP synthesis coupled proton transport, proton transport	Nuclear	
At3g52300	'ATP synthase D chain, mitochondrial', ATPQ	19,585.9	ATP synthesis coupled proton transport, photorespiration, proteasome core complex assembly, response to misfolded protein, response to salt stress, ubiquitin-dependent protein catabolic process	Nuclear	Multi-targeted
At5g15320	N/A	5,602.7	Biological process	Nuclear	
At4g30010	N/A	10,438.9	Aerobic respiration, biological process, glycolysis, photorespiration, proteasome core complex assembly, response to cadmium ion, response to misfolded protein, ubiquitin-dependent protein catabolic process	Nuclear	Dual targeted
At2g19680	N/A	13,846.8	ATP synthesis coupled proton transport, proton transport	Nuclear	
AtMg00480	ATP synthase 8, ATP8, ORFB	18,211.0	ATP synthesis coupled proton transport	Mitochondrial	
At5g13450	ATP5, delta subunit of MT ATP synthase	26,321.3	ATP synthesis coupled proton transport, iron ion homeostasis, photorespiration, photosynthesis, proteasome core complex assembly, response to iron ion, response to misfolded protein, response to reactive oxygen species, ubiquitin-dependent protein catabolic process	Nuclear	Multi-targeted
At5g04750	N/A	10,703.1	ATPase inhibitor activity	Nuclear	Dual targeted

(continued)

Table 5 (continued)

Complex V Genes	Common name	MW (dalton)	Function	Origin	Remarks
AtMg01190	ATP synthase subunit 1, ATP1	54,970.6	ATP hydrolysis coupled proton transport, ATP metabolic process, ATP synthesis coupled proton transport, proton transport, response to oxidative stress	Mitochondrial	Dual targeted
At2g07698	N/A	85,932.5	ATP hydrolysis coupled proton transport, ATP metabolic process, ATP synthesis coupled proton transport, proton transport	Nuclear	Multi-targeted
At5g08670	N/A	59,630.3	ATP biosynthetic process, ATP catabolic process, ATP hydrolysis coupled proton transport, ATP metabolic process, ATP synthesis coupled proton transport, proton transport, response to oxidative stress	Nuclear	Multi-targeted
At5g08680	N/A	59,858.5	ATP biosynthetic process, ATP catabolic process, ATP hydrolysis coupled proton transport, ATP metabolic process, ATP synthesis coupled proton transport, proton transport, response to cadmium ion	Nuclear	
At5g08690	N/A	59,712.5	ATP biosynthetic process, ATP catabolic process, ATP hydrolysis coupled proton transport, ATP metabolic process, ATP synthesis coupled proton transport, proton transport	Nuclear	Dual targeted
At2g33040	ATP3, gamma subunit of MT ATP synthase	35,447.6	ATP synthesis coupled proton transport, methylglyoxal catabolic process to D-lactate, photorespiration, proteasome core complex assembly, proton transport, response to misfolded protein, ubiquitin-dependent protein catabolic process	Nuclear	Multi-targeted

At5g47030	N/A	21,547.5	ATP synthesis coupled proton transport, Golgi organization, aerobic respiration, glycolysis, hyperosmotic response, pentose-phosphate shunt, photorespiration, proteasome core complex assembly, response to cadmium ion, response to misfolded protein, response to salt stress, response to temperature stimulus, ubiquitin-dependent protein catabolic process, water transport	Nuclear	
At1g51650	N/A	7,832.0	ATP biosynthetic process, ATP synthesis coupled proton transport, photorespiration	Nuclear	
At2g21870	N/A	27,596.6	Photorespiration, pollen development, proteasome core complex assembly, response to misfolded protein, ubiquitin-dependent protein catabolic process	Nuclear	Multi-targeted
At2g34050	N/A	28,377.9	Protein complex assembly	Nuclear	Mitochondria only
At5g40660	N/A	36,309.2	Proton-transporting ATP synthase complex assembly	Nuclear	Mitochondria only
At3g22370	Alternative oxidase 1A, AOX1A, ATAOX1A, ATHSR3, HSR3, hypersensitivity related 3	39,979.6	Cellular respiration, mitochondria–nucleus signalling pathway, oxidation–reduction process, respiratory gaseous exchange, response to cold	Nuclear	Mitochondria only
At3g22360	Alternative oxidase 1B, AOX1B	37,432.0	Oxidation–reduction process, respiratory gaseous exchange	Nuclear	Mitochondria only
At3g27620	Alternative oxidase 1C, AOX1C	37,816.3	Oxidation–reduction process, respiratory gaseous exchange	Nuclear	Mitochondria only
At1g32350	Alternative oxidase 1D, AOX1D	36,202.1	Abscisic acid-mediated signalling pathway, intracellular signal transduction, oxidation–reduction process, respiratory gaseous exchange, response to ethylene stimulus	Nuclear	Mitochondria only
At5g64210	Alternative oxidase 2, AOX2	40,086.4	Alternative respiration, oxidation–reduction process, respiratory gaseous exchange	Nuclear	Mitochondria only

(continued)

Table 5 (continued)

Complex V Genes	Common name	MW (dalton)	Function	Origin	Remarks
At4g28220	NAD(P)H dehydrogenase B1, NDB1	63,313.6	Oxidation–reduction process	Nuclear	Dual targeted
At4g05020	NAD(P)H dehydrogenase B2, NDB2	69,242.4	Oxidation–reduction process	Nuclear	Mitochondria only
At4g21490	NAD(P)H dehydrogenase B3, NDB3	65,164.0	Oxidation–reduction process	Nuclear	Mitochondria only
At2g20800	NAD(P)H dehydrogenase B4, NDB4	65,371.2	Oxidation–reduction process	Nuclear	Dual targeted
At5g08740	NAD(P)H dehydrogenase C1, NDC1	57,017.7	Cellular response to light stimulus, hydrogen peroxide catabolic process	Nuclear	Multi-targeted
At1g07180	Alternative NAD(P)H dehydrogenase 1, *Arabidopsis thaliana* internal non-phosphorylating NAD(P)H dehydrogenase, ATNDI1, NDA1	56,627.6	Cellular response to light stimulus, myoinositol hexakisphosphate biosynthetic process	Nuclear	Dual targeted
At2g29990	Alternative NAD(P)H dehydrogenase 2, NDA2	56,502.5	Oxidation–reduction process		Dual targeted

http://www.arabidopsis.org/index.jsp.in

Note: Only experimentally proved subunits and intermediate or assembly factor were included in the tables

N/A not applicable or information unavailable; dual targeted, targeted to mitochondria and chloroplast; multi-targeted, mitochondria, chloroplast and other organelles (list of constituents of each complex was modified from bioinformatics analysis of Cardol et al. 2005. The tables have been updated and/or modified where necessary to recent information extracted from TAIR database)

	Complex I NADH Dehydrogenase	Complex II Succinate Dehydrogenase	Complex III Cytochrome bc$_1$ complex	Complex IV Cytochrome Oxidase	Complex V ATP Synthase
Subunits and homology	49 Subunits 14 bacterial 27 eukaryotic 8 Plant	8 Subunits 4 eukaryotic 4 plant specific	10 subunits 10 eukaryotic	14 subunits 8 eukaryotic 6 plant specific	11 subunits 10 eukaryotic 1 plant specific
Intermediate	200, 400, 450 And 650 KD 85-550KD Sub-complexes	unknown	Unknown	Unknown	Matrix F1 Membrane F$_0$
Assembly factor	L-galactono-1, 4-lactone dehydrogenase (GLDH)	SDHAF2	Unknown	COX19	ATP10 ATP11 ATP12

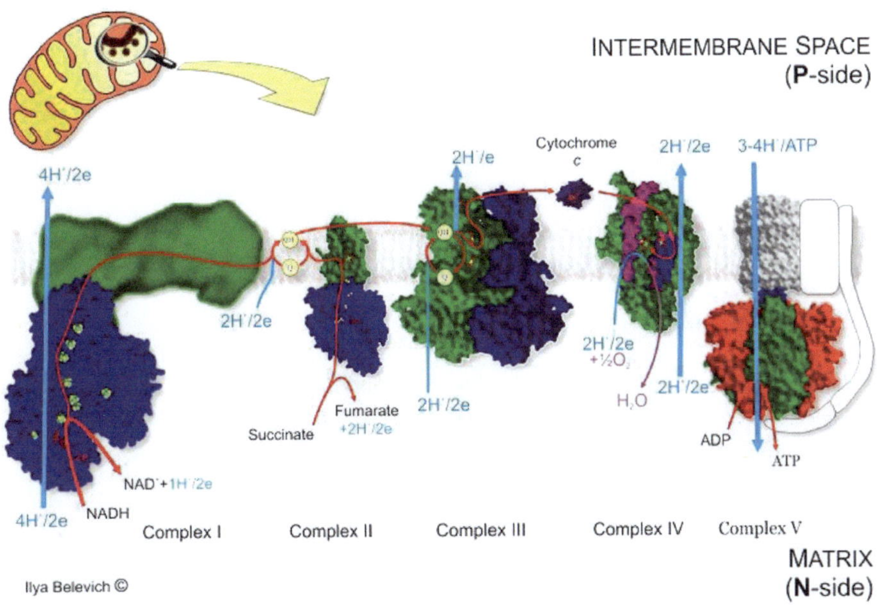

Fig. 3 Tabulated illustration of summary of subunit composition and assembly of OXPHOS system in mitochondria (Modified from Jacoby et al. 2012)

Crosstalk Between Organelles: Anterograde and Retrograde

The anterograde regulatory mechanism coordinates gene expression in chloroplasts and mitochondria and is responsive to endogenous and environmental signals perceived by the nucleus. Retrograde signalling regulates the expression of nuclear organelle genes in response to the metabolic and developmental state of the organelle. Besides the crosstalk between chloroplasts/mitochondria and the nucleus, chloroplast–mitochondrion communication has been established during plant evolution to coordinate the activities of the two organelles which exhibit a high level of metabolic interdependence (Fig. 4) (Woodson and Chory 2008).

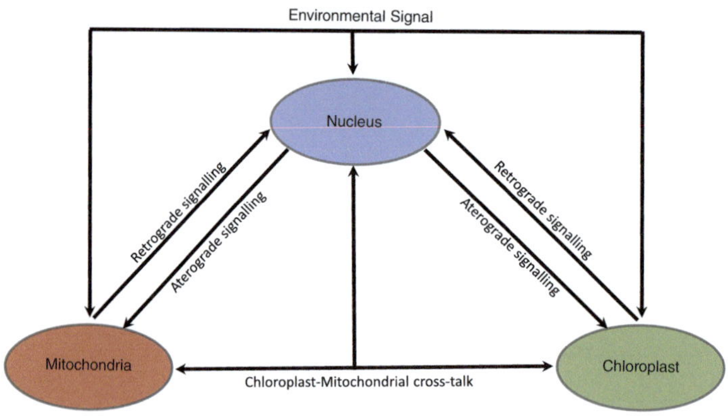

Fig. 4 An overview of crosstalk between organelles in regulation of coordinated gene expression. The simplified diagram illustrates communication between nucleus, mitochondria and chloroplast in response to environmental cues (growth signal, stress, light, reactive oxygen species and many more); details could be found in the main text (Redrawn and modified from Woodson and Chory 2008)

Anterograde Mechanism

The mitochondrial and chloroplast proteome is largely encoded at nucleus, under the control of anterograde regulatory mechanism. This mechanism controls organelle gene expression at multiple levels:

1. Modulation of transcript level of nuclear chloroplast genes (Kleffmann et al. 2004).
2. Post-translational import of proteins into chloroplasts and mitochondria mediated by Tic/Toc and Tim/Tom complexes, respectively. These translocons exist in different isoforms with substrate specificity, potentially participating in the accumulation of tissue-specific plastid and mitochondrial proteomes (Lister et al. 2004; Soll and Schleiff 2004).
3. Transcription, transcript editing, maturation and processing as well as the translation of organelle-encoded proteins and their autoregulation. This is, in large part, mediated by nuclear-encoded factors, allowing nuclear control of the expression of organelle genes (Leon et al. 1998; Choquet and Wollman 2002).
4. Post-translational events, such as the assembly of the multi-protein complexes of the thylakoid membrane or of the inner envelope of mitochondria. These processes require nuclear-encoded assembly factors (Grivell et al. 1999; Vothknecht and Westhoff 2001).
5. Organelle development, for instance organelle division, which is tightly controlled by nuclear-encoded proteins (Osteryoung and Nunnari 2003).

Studies to date support the conclusion that the regulation of plastome and chondrome gene expression occurs majorly at the post-transcriptional level (Leon et al. 1998; Rochaix 2001).

Retrograde Regulation

The retrograde regulation encompasses nuclear gene regulation by organelle-specific signals, like plastid-derived signals regulating nuclear photosynthetic Lhcb genes in photo-oxidized plants, and the redox state of the plastoquinone pool due to variation in excitation state of photosystems controls transcriptional response of selected nuclear photosynthetic gene promoter (Pfannschmidt et al. 2001). The redox state of the stromal electron acceptors of PSI, in fact, seems to be crucial for the regulation of Lhcb genes (Pursiheimo et al. 2001).

In addition to redox signalling, the chlorophyll biosynthetic pathway has also been associated with the control of nuclear gene expression. The intermediates of chlorophyll biosynthetic pathway have been involved in the modulation of transcript accumulation of several nuclear chloroplast genes (Johanningmeier and Howell 1984; Kropat et al. 1997). The *Arabidopsis* Lhcb mutants are resistant to norflurazon-induced photo-oxidative damage, which affects genome uncoupled 1–5 (GUN1–GUN5) proteins involved in tetrapyrrole metabolism. The products of GUN2/HY1 and GUN3/HY2 contribute to heme degradation in the 'Fe branch' of tetrapyrrole biosynthesis, GUN5 encodes the CHLH subunit of the Mg-chelatase (Mochizuki et al. 2001), and GUN4 binds product and substrate of Mg-chelatase and activates Mg-chelatase (Larkin et al. 2003; Larkin et al. 2003).

The gun mutant analysis has supported the plastid retrograde signalling wherein Mg porphyrins and/or enzymes of this biosynthetic pathway act as sensors. Earlier, GUN5 was suggested as a sensor (Mochizuki et al. 2001), but the idea was recently revised in favour of the tetrapyrrole intermediate Mg-protoporphyrin IX, a proposed signalling molecule between chloroplast and nucleus (Strand et al. 2003). In this retrograde signalling model, regulation of Mg-protoporphyrin synthesis and transport is controlled by GUN4 (Larkin et al. 2003; Wilde et al. 2004), while an additional chloroplast protein, LAF6/ABC1, is thought to be involved in the transport and distribution of protoporphyrin IX (Moller 2001).

Furthermore, sugar and reactive oxygen species are very important classes of molecules in retrograde signalling, and the regulatory effect of sugars on nuclear expression of photosynthetic genes and on plant metabolism is well known (Rolland et al. 2002). A low sugar status enhances photosynthesis and induces reserve mobilization and export, whereas an abundant supply of sugars promotes growth and carbohydrate storage. The effect of singlet oxygen generated during dark-to-light shift confined to plastids and activating several stress response pathways was confirmed in the *Arabidopsis* flu mutant (op den Camp et al. 2003). The biological activity of this reactive oxygen species is highly specific and seems to depend on the chemical identity of the molecule and/or the site at which it is generated (Op den Camp et al. 2003). Also photorespiratory H_2O_2 has been suggested to have a direct impact on the transcription of nuclear genes (Vandenabeele et al. 2004).

Crosstalk Between Chloroplast and Mitochondria

The chloroplast and mitochondria are metabolically interdependent, wherein photosynthesis provides substrates for mitochondrial respiration but depends itself on a range of compounds synthesized by mitochondria, including ATP in the dark. Moreover, mitochondrial respiration protects photosynthesis against photo-inhibition by dissipating redox equivalents exported from the chloroplasts (Raghavendra and Padmasree 2003).

The crosstalk between chloroplast and mitochondrion has been actively explored by genetic studies. The exchange of tRNA between two organelles was evident in *C. reinhardtii* mutant for suppressor gene in the chloroplast that acts on a mitochondrial mutation (Bennoun and Delosme 1999). The absence of chloroplast activity in the barley albostrians mutant fully differentiated leaf tissue leads to a compensatory increase in mitochondrial gene copy number and an elevated level of mitochondrial transcripts (Hedtke et al. 1999). Vice versa, mutations affecting specific mitochondrial proteins glycine decarboxylase (gdc) have been analyzed for their effects on chloroplast properties in potato (Heineke et al. 2001) and barley (Igamberdiev et al. 2001). Both mutants show a lowered glycine decarboxylase activity and are impaired in photorespiration, leading to an over-reduction and over-energization of the chloroplast. In the tobacco CMSII mutant, which lacks the major mitochondrial NADH dehydrogenase (complex I), the rate of photosynthesis is decreased, notably during dark–light transitions or when carbon fixation and photorespiration are simultaneously active (Sabar et al. 2000; Dutilleul et al. 2003a). In the same mutant, modulation of nuclear gene expression maintains whole cell redox balance (Dutilleul et al. 2003b). The analysis of the pro rs1 mutant of *Arabidopsis*, defective in an organelle-targeted tRNA synthetase, shows that nuclear photosynthetic genes are specifically downregulated, indicating that mitochondrion–chloroplast crosstalk might involve reprogramming of the expression of nuclear chloroplast genes.

Cell Death: Cellular Defence and Homeostasis

The seminal work of Kerr and co-worker, summarized by—Kerr (2002), opens up multiple facet of cell death pathway, which describes morphological changes associated with regulated (apoptosis) and unregulated (necrosis) cell death in nematode *Caenorhabditis elegans*. Since the initial identification of apoptosis, several types of programmed cell death (PCD) have been identified. Programmed cell death (PCD) is a central process that fine-tunes cellular homeostasis during normal growth and development, its defence mechanism to counter pathogen attack and its response to stress conditions. The PCD basically transits between apoptosis, autophagy and necrosis alike in plant and animal cells, and there is a certain amount of overlap between these different cell death processes. The mitochondria play important role in orchestrating death signals that lead to the initiation of cell death and subsequently release molecules that drive the destruction of the cell. The cytochrome 'c' is released from the mitochondria intermembrane space and is an important event in the early stage of the death process in both plant and animal cells. The release of cytochrome 'c' is controlled in part by selective opening of the permeability transition pore in the mitochondria, and the magnitude of the opening of the permeability transition pore may be a determining factor in the type of programmed cell death a cell undergoes. Similarly, reactive oxygen species (ROS) production by mitochondria (possibly as a result of cytochrome *c* release) has been identified as a driver of cell death in plant cells, and again the type of death programme that is activated may be due to the levels of ROS produced in the dying cell as apoptosis, autophagy or necrosis can be initiated by oxidative stress (Fig. 5).

The extensive research in animal systems has demonstrated the involvement of different forms of cell death in the maintenance of cellular homeostasis and has been linked to foetal developmental events such as the formation of independent digits (Thompson 1995; Zuzarte-Luis and Hurle 2002). The cell death programme also governs many important diseases, which has been linked to either an increase (e.g. Parkinson's disease or Alzheimer's disease) or a decrease (e.g. some types of cancer, systemic lupus erythematosus, and herpes) in rates of apoptosis (Thompson 1995). The autophagic cell death is less studied than apoptosis and seems to occur less often during development. However, it is known to contribute to the development of insect salivary glands and has been linked to heart failure in humans (Knaapen et al. 2001; Berry and Baehrecke 2007). Necrosis-like neuronal cell death has been reported in dying neurons of *C. elegans* (Hall et al. 1997) and also occurs in mammalian cells when caspase activation is blocked or apoptosis is aborted following death programme activation (Leist and Jaattela 2001).

Like the animal counterpart, the plant also shows involvement of PCD in normal growth and development. The xylem vessel formation is the consequence of PCD, wherein tracheary element (TE) cells undergo PCD after secondary wall synthesis, leaving a hollow cell corpse (Fukuda 2000). During embryogenesis in angiosperm, two sister cells arise from the asymmetric division of the zygote and go on to form either the embryo or vacuolated suspensor cells that signal suspensor cell to undergo PCD (McCabe et al. 1997a) (McCabe et al. 1997a). In sunflower, cytoplasmic male sterile (CMS)-associated death shows common features as in mammalian counterpart, like cellular condensation, nuclear DNA cleavage and cytochrome '*c*' release (Balk and Leaver 2001). This indicated a role for the mitochondrion in regulated CMS-associated PCD. The 'self-incompatibility', wherein plants deter inbreeding, is maintained by the interaction of two multi-allelic gene products PrsS (stigma S locus determinant) and PrpS (pollen S locus

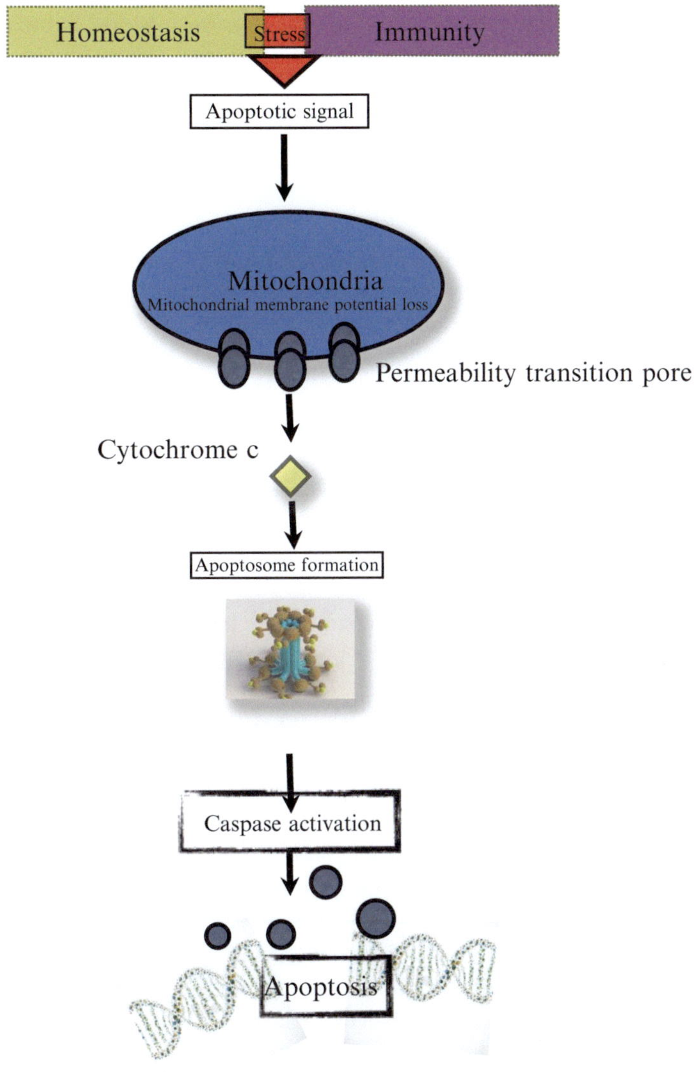

Fig. 5 Central mode of apoptosis. The diagram depicts the basic mode of an apoptosis programme that starts with the apoptotic signal either for cellular homeostasis or clearance of pathogenic stress. The programme starts with the rearrangement of some membrane protein, change in membrane potential and finally release of cytochrome 'c' in cytoplasm. The cytochrome c helps in the formation of apoptosome complex; subsequently it activates effector proteins (caspases) that damages of DNA, RNA and protein leading to cell death

determinant). Upon interaction of the pollen PrpS extracellular loop with the PrsS in the stigma of self-incompatible plants, a Ca^{2+}-dependent signalling network is triggered, which results in growth inhibition and PCD in the pollen tube (Thomas and Franklin-Tong 2004; Wheeler 2009). As well as plant development, PCD plays a significant role in plant stress responses. Apoptotic-like PCD morphology (cellular condensation, which in plant cells is observed as a retraction of the protoplast away from the cell wall) has been shown to occur in cells subjected to various abiotic stress-inducing treatments, such as treatment of carrot or *Arabidopsis thaliana* cells with H_2O_2 (Kawai-Yamada et al. 2004), tunicamycin or brefeldin A treatment of sycamore cells (Crosti et al. 2001; Malerba et al. 2004) and treatment of *A. thaliana* cell cultures with ceramides (Townley et al. 2005). In addition, heat shock is a potent inducer of apoptotic-like PCD morphology in carrot, *A. thaliana* and tobacco (McCabe et al. 1997a; McCabe and Leaver 2000; Vacca et al. 2004). Indeed, McCabe et al. (1997b) showed that the level of apoptotic-like PCD morphology observed in heat-shocked carrot cells shown correlation with the temperature. The studies have shown that at lower temperatures (25–35 °C) cells remained alive, which turns on programmed cell death with gradual increase in temperature and maximum (100 %) at 55 °C. While all cells died with temperature treatments above 55 °C (65–85 °C), the incidence of this corpse morphology declined rapidly and was completely absent after temperature treatments above 75 °C, indicating that these cells died *via* necrosis. This suggested that cell shrinkage is an active process resulting from PCD rather than an uncontrolled collapsing of the cell.

PCD is also a crucial component of responses associated with the biotic stresses caused by pathogen attack. Host cell death occurs in response to many, but not all, plant–pathogen interactions. This cell death can lead to plant resistance or susceptibility to the invading pathogen, largely depending on the food preference of the pathogen (Greenberg and Yao 2004). The hypersensitive response (HR) to avirulent pathogens often terminates in the rapid death of infected or challenged cells, which can result in arrest of pathogen growth (Heath 2000). Elicitors inducing the HR, such as cryptogein, induce an apoptotic-like PCD morphology in challenged soybean and tobacco cells in the HR lesion zone (Levine et al. 1996). Many virulent pathogens also induce programmed cell death. Yao and co-workers, studying the effects of many different pathogens (fungi, bacteria and viruses) in oat, found that apoptotic-like PCD morphologic features occurred in infected cells and sometimes in neighbouring cells at various time points depending on the type of infectious agent (Yao et al. 2002).

Autophagy

In recent time, the autophagy is regarded as a pro-survival process in plants, especially during nutrient stress (Cacas and Diamond 2009). In *S. cerevisiae*, deletions in the orthologues ATG genes have demonstrated that the absence of the autophagy pathway leaves plants more sensitive to low nutrient conditions (Doelling et al. 2002; Hanaoka et al. 2002; Surpin et al. 2003; Yoshimoto et al. 2009; Xiong et al. 2007a). In absence of autophagy plants experience an accelerated senescence and appear more sensitive to chronic oxidative stress (Doelling et al. 2002; Xiong et al. 2007a, b). Further studies in animals have suggested that autophagy is required to remove oxidized proteins and damaged mitochondria (Zhang et al. 2007). If this holds true in plants, it would suggest that the reason autophagy-deficient plants undergo premature senescence is because of an inability to clean up toxic cellular components, such as damaged mitochondria.

The role autophagy in plant defence responses against invading pathogens is a matter of ongoing debate; however, existing evidence suggests its involvement (Liu et al. 2005; Patel and Dinesh-Kumar 2008; Hofius et al. 2009; Yoshimoto et al. 2009). For example, Patel and Dinesh-Kumar (2008) reported that autophagy-deficient (antisense At-*Atg6*) *A. thaliana* plants underwent an uncontrolled hypersensitive response (HR) after infection with avirulent *P. syringae* pv. *tomato* (*Pto*) DC3000 (*AvrRpm1*), where the lesion that normally remains localized to the point of infection instead spread throughout the leaf and killed it. Similar results were previously observed in tobacco plants deficient in Nt-atg6/Beclin1 or Nt-atg7 (after VIGS) following infection with tobacco mosaic virus (Liu et al. 2005) These findings suggest that autophagy is a regulatory mechanism just before commitment stage for PCD that occurs during HR. However, the autophagy can also be a positive regulator of PCD during the HR as shown by infection of autophagy-deficient *A. thaliana* knockout plants with avirulent *Pto*DC3000 (*AvrRpm1*) did not lead to an uncontrolled HR lesion spread (Hofius et al. 2009). Yoshimoto and co-workers suggest that the differences between these studies may relate to differences in pathogen-treated plant or leaf age (Yoshimoto et al. 2009). Whatever the case may be, it seems that in plants, as in animals, autophagy can act in a protective capacity but can also function as a form of PCD.

Autophagy is considered as pro-survival mechanism against acute stress and oxidative stress is one of them (Xiong et al. 2007b). However, it has also been demonstrated that the direct application of H_2O_2 can induce autophagic cell death in plants. As mitochondria are one of the main locations in plant cells for the production of ROS, it is possible that plant mitochondria could directly initiate autophagic cell death. In the future it will be interesting to test the effects of factors that inhibit MPT in plants (such as CsA) on the induction of autophagic cell death. This will help to determine what role, if any, the mitochondrion plays in plant autophagic cell death.

Applications of Plant Mito-Omics

The recent development in mito-omics (genomics, transcriptomics, proteomics and metabolomics) has given impetus to understand vital cellular mechanisms. Numerous studies have tried to establish endosymbiont origin of mitochondria in the past, but now, in light of recent development, mitochondria became so integrated with cell functionality that the thin line also has disappeared. The mitochondrial uniqueness in various mechanisms and essentiality for cellular survival keep them in the centre of cellular metabolism alike in plant and animal world. On the other hand, minor perturbation in mitochondrial functionality leads to major functional loss in terms of growth and development. Along with the modulation of the mito-omics, maintenance of coordinated expression of plant mito-omics, nuclear omics and plastomics in the changing environmental conditions of plant growth and development would in itself be a challenge in future existence.

Plant mitochondria have small genome comparable to that of plastome, are maternally inherited and are found in large numbers per cell—these characters make them comparable to plastids/chloroplast to act as bio-factories and/or bioreactors (for detailed and comprehensive study of plastid transformation, refer to the 'Transplastomics' by Khan M.S. et al. in the same book) for the production of pharmaceuticals and recombinant proteins, but because of ongoing DNA recombination, inconsistent number of mitochondria in intra-tissue and inter-tissue cells, difficulties of transfection and maintenance of the transfected DNA in whole cells have been main hurdles in making plant mitochondrial transformation a success story.

The mitochondrial complexes are the central hub for oxygen utilization in cell, and during the process, oxygen free radicals are generated, which is the leading cause of cellular damage. Alongside, mitochondria have strong antioxidant machinery to get rid of oxygen free radicals and help to prolong the life of cell as a unit and organism as whole system. In future, mitochondria may prove useful to improve the efficiency of cellular energy production and enhance life span. It has been clearly shown that the mitochondria are instrumental in redox and ROS signalling under abiotic and biotic stresses in plant, but the mechanism to modulate the mito-omics to enhance stress tolerance and/or resistance still has to be established, and concerted efforts of the scientific communities across the world are required for the rapid advancement in the field.

References

Adams KL, Palmer JD (2003) Evolution of mitochondrial gene content: gene loss and transfer to the nucleus. Mol Phylogenet Evol 29(3):380–395

Agrawal RK, Sharma MR (2012) Structural aspects of mitochondrial translational apparatus. Curr Opin Struct Biol 22(6):797–803

Anderson S, Bankier AT, Barrell BG, de Bruijn MH, Coulson AR, Drouin J, Eperon IC, Nierlich DP, Roe BA, Sanger F, Schreier PH, Smith AJ, Staden R, Young IG (1981) Sequence and organization of the human mitochondrial genome. Nature 290(5806):457–465

Attallah CV, Welchen E, Pujol C, Bonnard G, Gonzalez DH (2007) Characterization of Arabidopsis thaliana genes encoding functional homologues of the yeast metal chaperone Cox19p, involved in cytochrome c oxidase biogenesis. Plant Mol Biol 65(3):343–355. doi:10.1007/s11103-007-9224-1

Balk J, Leaver CJ (2001) The PET1-CMS mitochondrial mutation in sunflower is associated with premature programmed cell death and cytochrome c release. Plant Cell 13(8):1803–1818

Balk J, Leaver CJ, McCabe PF (1999) Translocation of cytochrome c from the mitochondria to the cytosol

occurs during heat-induced programmed cell death in cucumber plants. FEBS Lett 463(1–2):151–154
Bardel J, Louwagie M, Jaquinod M, Jourdain A, Luche S, Rabilloud T, Macherel D, Garin J, Bourguignon J (2002) A survey of the plant mitochondrial proteome in relation to development. Proteomics 2(7):880–898
Barrientos A (2002) In vivo and in organello assessment of OXPHOS activities. Methods 26(4):307–316
Barrientos A, Fontanesi F, Diaz F (2009) Evaluation of the mitochondrial respiratory chain and oxidative phosphorylation system using polarography and spectrophotometric enzyme assays. Curr Protoc Hum Genet Jonathan L Haines [et al] Chapter 19:Unit19 13. doi:10.1002/0471142905.hg1903s63
Bartoli CG, Pastori GM, Foyer CH (2000) Ascorbate biosynthesis in mitochondria is linked to the electron transport chain between complexes III and IV. Plant Physiol 123(1):335–344
Benne R, Van den Burg J, Brakenhoff JP, Sloof P, Van Boom JH, Tromp MC (1986) Major transcript of the frameshifted coxII gene from trypanosome mitochondria contains four nucleotides that are not encoded in the DNA. Cell 46(6):819–826
Bennoun P, Delosme M (1999) Chloroplast suppressors that act on a mitochondrial mutation in Chlamydomonas reinhardtii. Mol Gen Gene MGG 262(1):85–89
Ben-Shem A, Garreau de Loubresse N, Melnikov S, Jenner L, Yusupova G, Yusupov M (2011) The structure of the eukaryotic ribosome at 3.0 A resolution. Science 334(6062):1524–1529
Bentolila S, Babina AM, Germain A, Hanson MR (2013) Quantitative trait locus mapping identifies REME2, a PPR-DYW protein required for editing of specific C targets in Arabidopsis mitochondria. RNA Biol 10:9
Berget SM, Moore C, Sharp PA (1977) Spliced segments at the 5' terminus of adenovirus 2 late mRNA. Proc Natl Acad Sci U S A 74(8):3171–3175
Berry DL, Baehrecke EH (2007) Growth arrest and autophagy are required for salivary gland cell degradation in Drosophila. Cell 131(6):1137–1148. doi:10.1016/j.cell.2007.10.048
Berry EA, Guergova-Kuras M, Huang LS, Crofts AR (2000) Structure and function of cytochrome bc complexes. Annu Rev Biochem 69:1005–1075
Binder S, Hölzle A, Jonietz C (2011) RNA processing and RNA stability in plant mitochondria. In: Kempken F (ed) Plant mitochondria, vol 1, Advances in plant biology. Springer, New York, pp 107–130. doi:10.1007/978-0-387-89781-3_5
Brennicke A, Marchfelder A, Binder S (1999) RNA editing. FEMS Microbiol Rev 23(3):297–316
Cacas JL, Diamond M (2009) Is the autophagy machinery an executioner of programmed cell death in plants? Trends Plant Sci 14(6):299–300. doi:10.1016/j.tplants.2009.02.008, author reply 300–291
Capaldi RA (1990) Structure and assembly of cytochrome c oxidase. Arch Biochem Biophys 280(2):252–262
Cardol P (2011) Mitochondrial NADH: ubiquinone oxidoreductase (complex I) in eukaryotes: a highly conserved subunit composition highlighted by mining of protein databases. Biochim Biophys Acta 1807(11):1390–1397. doi:10.1016/j.bbabio.2011.06.015
Cardol P, Vanrobaeys F, Devreese B, Van Beeumen J, Matagne RF, Remacle C (2004) Higher plant-like subunit composition of mitochondrial complex I from Chlamydomonas reinhardtii: 31 conserved components among eukaryotes. Biochim Biophys Acta 4(3):212–224
Cardol P, Gonzalez-Halphen D, Reyes-Prieto A, Baurain D, Matagne RF, Remacle C (2005) The mitochondrial oxidative phosphorylation proteome of Chlamydomonas reinhardtii deduced from the Genome Sequencing Project. Plant Physiol 137(2):447–459. doi:10.1104/pp. 104.054148
Castandet B, Choury D, Begu D, Jordana X, Araya A (2010) Intron RNA editing is essential for splicing in plant mitochondria. Nucleic Acids Res 38(20):7112–7121
Chapdelaine Y, Bonen L (1991) The wheat mitochondrial gene for subunit I of the NADH dehydrogenase complex: a trans-splicing model for this gene-in-pieces. Cell 65(3):465–472
Chaw SM, Shih AC, Wang D, Wu YW, Liu SM, Chou TY (2008) The mitochondrial genome of the gymnosperm Cycas taitungensis contains a novel family of short interspersed elements, Bpu sequences, and abundant RNA editing sites. Mol Biol Evol 25(3):603–615
Cheetham GM, Jeruzalmi D, Steitz TA (1999) Structural basis for initiation of transcription from an RNA polymerase-promoter complex. Nature 399(6731):80–83
Choquet Y, Wollman FA (2002) Translational regulations as specific traits of chloroplast gene expression. FEBS Lett 529(1):39–42
Chow LT, Gelinas RE, Broker TR, Roberts RJ (1977) An amazing sequence arrangement at the 5' ends of adenovirus 2 messenger RNA. Cell 12(1):1–8
Christensen AC, Lyznik A, Mohammed S, Elowsky CG, Elo A, Yule R, Mackenzie SA (2005) Dual-domain, dual-targeting organellar protein presequences in Arabidopsis can use non-AUG start codons. Plant Cell 17(10):2805–2816
Claros MG, Vincens P (1996) Computational method to predict mitochondrially imported proteins and their targeting sequences. Eur J Biochem 241(3):779–786
Courtois F, Merendino L, Demarsy E, Mache R, Lerbs-Mache S (2007) Phage-type RNA polymerase RPOTmp transcribes the rrn operon from the PC promoter at early developmental stages in Arabidopsis. Plant Physiol 145(3):712–721
Crosti P, Malerba M, Bianchetti R (2001) Tunicamycin and Brefeldin A induce in plant cells a programmed cell death showing apoptotic features. Protoplasma 216(1–2):31–38
Cui J, Liu J, Li Y, Shi T (2011) Integrative identification of Arabidopsis mitochondrial proteome and its function exploitation through protein interaction network. PLoS One 6(1):0016022

Diaz S, Renault T, Villalba A, Carballal MJ (2011) Disseminated neoplasia in cockles Cerastoderma edule: ultrastructural characterisation and effects on haemolymph cell parameters. Dis Aquat Org 96(2):157–167. doi:10.3354/dao02384

Doelling JH, Walker JM, Friedman EM, Thompson AR, Vierstra RD (2002) The APG8/12-activating enzyme APG7 is required for proper nutrient recycling and senescence in Arabidopsis thaliana. J Biol Chem 277(36):33105–33114. doi:10.1074/jbc.M204630200

Dombrowski S, Hoffmann M, Guha C, Binder S (1999) Continuous primary sequence requirements in the 18-nucleotide promoter of dicot plant mitochondria. J Biol Chem 274(15):10094–10099

Dutilleul C, Driscoll S, Cornic G, De Paepe R, Foyer CH, Noctor G (2003a) Functional mitochondrial complex I is required by tobacco leaves for optimal photosynthetic performance in photorespiratory conditions and during transients. Plant Physiol 131(1):264–275. doi:10.1104/pp. 011155

Dutilleul C, Garmier M, Noctor G, Mathieu C, Chetrit P, Foyer CH, de Paepe R (2003b) Leaf mitochondria modulate whole cell redox homeostasis, set antioxidant capacity, and determine stress resistance through altered signaling and diurnal regulation. Plant Cell 15(5):1212–1226

Emanuelsson O, Nielsen H, Brunak S, von Heijne G (2000) Predicting subcellular localization of proteins based on their N-terminal amino acid sequence. J Mol Biol 300(4):1005–1016

Eubel H, Jansch L, Braun HP (2003) New insights into the respiratory chain of plant mitochondria. Supercomplexes and a unique composition of complex II. Plant Physiol 133(1):274–286

Farre JC, Aknin C, Araya A, Castandet B (2012) RNA editing in mitochondrial trans-introns is required for splicing. PLoS One 7(12):20

Forner J, Weber B, Wietholter C, Meyer RC, Binder S (2005) Distant sequences determine 5' end formation of cox3 transcripts in Arabidopsis thaliana ecotype C24. Nucleic Acids Res 33(15):4673–4682

Forner J, Weber B, Thuss S, Wildum S, Binder S (2007) Mapping of mitochondrial mRNA termini in Arabidopsis thaliana: t-elements contribute to 5' and 3' end formation. Nucleic Acids Res 35(11):3676–3692

Fuentes D, Meneses M, Nunes-Nesi A, Araujo WL, Tapia R, Gomez I, Holuigue L, Gutierrez RA, Fernie AR, Jordana X (2011) A deficiency in the flavoprotein of Arabidopsis mitochondrial complex II results in elevated photosynthesis and better growth in nitrogen-limiting conditions. Plant Physiol 157(3):1114–1127. doi:10.1104/pp. 111.183939

Fujii S, Toda T, Kikuchi S, Suzuki R, Yokoyama K, Tsuchida H, Yano K, Toriyama K (2011) Transcriptome map of plant mitochondria reveals islands of unexpected transcribed regions. BMC Genomics 12(279):1471–2164

Fukuda H (2000) Programmed cell death of tracheary elements as a paradigm in plants. Plant Mol Biol 44(3):245–253

Gagliardi D, Leaver CJ (1999) Polyadenylation accelerates the degradation of the mitochondrial mRNA associated with cytoplasmic male sterility in sunflower. EMBO J 18(13):3757–3766

Ghezzi D, Goffrini P, Uziel G, Horvath R, Klopstock T, Lochmuller H, D'Adamo P, Gasparini P, Strom TM, Prokisch H, Invernizzi F, Ferrero I, Zeviani M (2009) SDHAF1, encoding a LYR complex-II specific assembly factor, is mutated in SDH-defective infantile leukoencephalopathy. Nat Genet 41(6):654–656

Giege P, Brennicke A (1999) RNA editing in Arabidopsis mitochondria effects 441 C to U changes in ORFs. Proc Natl Acad Sci U S A 96(26):15324–15329

Giege P, Hoffmann M, Binder S, Brennicke A (2000) RNA degradation buffers asymmetries of transcription in Arabidopsis mitochondria. EMBO Rep 1(2):164–170

Giege P, Sweetlove LJ, Cognat V, Leaver CJ (2005) Coordination of nuclear and mitochondrial genome expression during mitochondrial biogenesis in Arabidopsis. Plant Cell 17(5):1497–1512

Glanz S, Kuck U (2009) Trans-splicing of organelle introns–a detour to continuous RNAs. BioEssays: News Rev Mol, Cell Dev Biol 31(9):921–934

Gleason C, Huang S, Thatcher LF, Foley RC, Anderson CR, Carroll AJ, Millar AH, Singh KB (2011) Mitochondrial complex II has a key role in mitochondrial-derived reactive oxygen species influence on plant stress gene regulation and defense. Proc Natl Acad Sci U S A 108(26):10768–10773

Greenberg JT, Yao N (2004) The role and regulation of programmed cell death in plant-pathogen interactions. Cell Microbiol 6(3):201–211

Grewe F, Viehoever P, Weisshaar B, Knoop V (2009) A trans-splicing group I intron and tRNA-hyperediting in the mitochondrial genome of the lycophyte Isoetes engelmannii. Nucleic Acids Res 37(15):5093–5104

Grivell LA, Artal-Sanz M, Hakkaart G, de Jong L, Nijtmans LG, van Oosterum K, Siep M, van der Spek H (1999) Mitochondrial assembly in yeast. FEBS Lett 452(1–2):57–60

Gueguen V, Macherel D, Jaquinod M, Douce R, Bourguignon J (2000) Fatty acid and lipoic acid biosynthesis in higher plant mitochondria. J Biol Chem 275(7):5016–5025

Hall DH, Gu G, Garcia-Anoveros J, Gong L, Chalfie M, Driscoll M (1997) Neuropathology of degenerative cell death in Caenorhabditis elegans. J Neurosci 17(3):1033–1045

Hamasur B, Glaser E (1992) Plant mitochondrial F_0F1 ATP synthase. Identification of the individual subunits and properties of the purified spinach leaf mitochondrial ATP synthase. Eur J Biochem 205(1):409–416

Hanaoka H, Noda T, Shirano Y, Kato T, Hayashi H, Shibata D, Tabata S, Ohsumi Y (2002) Leaf senescence and starvation-induced chlorosis are accelerated by the disruption of an Arabidopsis autophagy gene. Plant Physiol 129(3):1181–1193. doi:10.1104/pp. 011024

Handa H (2003) The complete nucleotide sequence and RNA editing content of the mitochondrial genome of rapeseed (Brassica napus L.): comparative analysis of the mitochondrial genomes of rapeseed and Arabidopsis thaliana. Nucleic Acids Res 31(20):5907–5916

Handa H (2008) Linear plasmids in plant mitochondria: peaceful coexistences or malicious invasions? Mitochondrion 8(1):15–25

Hao HX, Khalimonchuk O, Schraders M, Dephoure N, Bayley JP, Kunst H, Devilee P, Cremers CW, Schiffman JD, Bentz BG, Gygi SP, Winge DR, Kremer H, Rutter J (2009) SDH5, a gene required for flavination of succinate dehydrogenase, is mutated in paraganglioma. Science 325(5944):1139–1142

Heath MC (2000) Nonhost resistance and nonspecific plant defenses. Curr Opin Plant Biol 3(4):315–319

Heazlewood JL, Howell KA, Millar AH (2003a) Mitochondrial complex I from Arabidopsis and rice: orthologs of mammalian and fungal components coupled with plant-specific subunits. Biochim Biophys Acta 1604(3):159–169

Heazlewood JL, Howell KA, Whelan J, Millar AH (2003b) Towards an analysis of the rice mitochondrial proteome. Plant Physiol 132(1):230–242. doi:10.1104/pp. 102.018986

Heazlewood JL, Millar AH, Day DA, Whelan J (2003c) What makes a mitochondrion? Genome Biol 4(6):218. doi:10.1186/gb-2003-4-6-218

Hedtke B, Borner T, Weihe A (1997) Mitochondrial and chloroplast phage-type RNA polymerases in Arabidopsis. Science 277(5327):809–811

Hedtke B, Wagner I, Borner T, Hess WR (1999) Inter-organellar crosstalk in higher plants: impaired chloroplast development affects mitochondrial gene and transcript levels. Plant J 19(6):635–643

Hedtke B, Borner T, Weihe A (2000) One RNA polymerase serving two genomes. EMBO Rep 1(5):435–440

Hedtke B, Legen J, Weihe A, Herrmann RG, Borner T (2002) Six active phage-type RNA polymerase genes in Nicotiana tabacum. Plant J 30(6):625–637

Heineke D, Bykova N, Gardestrom P, Bauwe H (2001) Metabolic response of potato plants to an antisense reduction of the P-protein of glycine decarboxylase. Planta 212(5–6):880–887

Herald VL, Heazlewood JL, Day DA, Millar AH (2003) Proteomic identification of divalent metal cation binding proteins in plant mitochondria. FEBS Lett 537(1–3):96–100

Herrmann JM, Funes S (2005) Biogenesis of cytochrome oxidase-sophisticated assembly lines in the mitochondrial inner membrane. Gene 354:43–52

Hess WR, Borner T (1999) Organellar RNA polymerases of higher plants. Int Rev Cytol 190:1–59

Hiesel R, Wissinger B, Schuster W, Brennicke A (1989) RNA editing in plant mitochondria. Science 246(4937):1632–1634

Hiesel R, Combettes B, Brennicke A (1994) Evidence for RNA editing in mitochondria of all major groups of land plants except the Bryophyta. Proc Natl Acad Sci U S A 91(2):629–633

Hoch B, Maier RM, Appel K, Igloi GL, Kossel H (1991) Editing of a chloroplast mRNA by creation of an initiation codon. Nature 353(6340):178–180

Hofius D, Schultz-Larsen T, Joensen J, Tsitsigiannis DI, Petersen NH, Mattsson O, Jorgensen LB, Jones JD, Mundy J, Petersen M (2009) Autophagic components contribute to hypersensitive cell death in Arabidopsis. Cell 137(4):773–783. doi:10.1016/j.cell.2009.02.036

Holec S, Lange H, Canaday J, Gagliardi D (2008) Coping with cryptic and defective transcripts in plant mitochondria. Biochim Biophys Acta 9:566–573

Holt IJ, He J, Mao CC, Boyd-Kirkup JD, Martinsson P, Sembongi H, Reyes A, Spelbrink JN (2007) Mammalian mitochondrial nucleoids: organizing an independently minded genome. Mitochondrion 7 (5):311–321

Horsefield R, Iwata S, Byrne B (2004) Complex II from a structural perspective. Curr Protein Pept Sci 5(2):107–118

Huang S, Taylor NL, Narsai R, Eubel H, Whelan J, Millar AH (2009) Experimental analysis of the rice mitochondrial proteome, its biogenesis, and heterogeneity. Plant Physiol 149(2):719–734

Igamberdiev AU, Bykova NV, Lea PJ, Gardestrom P (2001) The role of photorespiration in redox and energy balance of photosynthetic plant cells: a study with a barley mutant deficient in glycine decarboxylase. Physiol Plant 111(4):427–438

Islam MS, Studer B, Byrne SL, Farrell JD, Panitz F, Bendixen C, Moller IM, Asp T (2013) The genome and transcriptome of perennial ryegrass mitochondria. BMC Genomics 14(1):202

Jacoby RP, Li L, Huang S, Pong Lee C, Millar AH, Taylor NL (2012) Mitochondrial composition, function and stress response in plants. J Integr Plant Biol 54(11):887–906

Jacques JP, Hausmann S, Kolakofsky D (1994) Paramyxovirus mRNA editing leads to G deletions as well as insertions. EMBO J 13(22):5496–5503

Jansch L, Kruft V, Schmitz UK, Braun HP (1996) New insights into the composition, molecular mass and stoichiometry of the protein complexes of plant mitochondria. Plant J 9(3):357–368

Johanningmeier U, Howell SH (1984) Regulation of light-harvesting chlorophyll-binding protein mRNA accumulation in Chlamydomonas reinhardi. Possible involvement of chlorophyll synthesis precursors. J Biol Chem 259(21):13541–13549

Kabeya Y, Sato N (2005) Unique translation initiation at the second AUG codon determines mitochondrial localization of the phage-type RNA polymerases in the moss Physcomitrella patens. Plant Physiol 138(1):369–382

Kawai-Yamada M, Ohori Y, Uchimiya H (2004) Dissection of Arabidopsis Bax inhibitor-1 suppressing Bax-, hydrogen peroxide-, and salicylic acid-induced cell death. Plant Cell 16(1):21–32. doi:10.1105/tpc.014613

Kempken F, Mullen JA, Pring DR, Tang HV (1991) RNA editing of sorghum mitochondrial atp6 transcripts changes 15 amino acids and generates a carboxy-terminus identical to yeast. Curr Genet 20(5):417–422

Kerr JF (2002) History of the events leading to the formulation of the apoptosis concept. Toxicology 181–182:471–474

Kim SR, Yang JI, Moon S, Ryu CH, An K, Kim KM, Yim J, An G (2009) Rice OGR1 encodes a pentatricopeptide repeat-DYW protein and is essential for RNA editing in mitochondria. Plant J 59(5):738–749

Kleffmann T, Russenberger D, von Zychlinski A, Christopher W, Sjolander K, Gruissem W, Baginsky S (2004) The Arabidopsis thaliana chloroplast proteome reveals pathway abundance and novel protein functions. Curr Biol 14(5):354–362

Klodmann J, Braun HP (2011) Proteomic approach to characterize mitochondrial complex I from plants. Phytochemistry 72(10):1071–1080. doi:10.1016/j.phytochem.2010.11.012

Klodmann J, Sunderhaus S, Nimtz M, Jansch L, Braun HP (2010) Internal architecture of mitochondrial complex I from Arabidopsis thaliana. Plant Cell 22(3):797–810

Knaapen MW, Davies MJ, De Bie M, Haven AJ, Martinet W, Kockx MM (2001) Apoptotic versus autophagic cell death in heart failure. Cardiovasc Res 51(2):304–312

Kobayashi Y, Dokiya Y, Kumazawa Y, Sugita M (2002) Non-AUG translation initiation of mRNA encoding plastid-targeted phage-type RNA polymerase in Nicotiana sylvestris. Biochem Biophys Res Commun 299(1):57–61

Kropat J, Oster U, Rudiger W, Beck CF (1997) Chlorophyll precursors are signals of chloroplast origin involved in light induction of nuclear heat-shock genes. Proc Natl Acad Sci U S A 94(25):14168–14172

Kruft V, Eubel H, Jansch L, Werhahn W, Braun HP (2001) Proteomic approach to identify novel mitochondrial proteins in Arabidopsis. Plant Physiol 127(4):1694–1710

Kubo T, Mikami T (2007) Organization and variation of angiosperm mitochondrial genome. Physiol Plant 129(1):6–13. doi:10.1111/j.1399-3054.2006.00768.x

Kubo T, Newton KJ (2008) Angiosperm mitochondrial genomes and mutations. Mitochondrion 8(1):5–14

Kugita M, Yamamoto Y, Fujikawa T, Matsumoto T, Yoshinaga K (2003) RNA editing in hornwort chloroplasts makes more than half the genes functional. Nucleic Acids Res 31(9):2417–2423

Kuhn K, Weihe A, Borner T (2005) Multiple promoters are a common feature of mitochondrial genes in Arabidopsis. Nucleic Acids Res 33(1):337–346

Kuhn K, Bohne AV, Liere K, Weihe A, Borner T (2007) Arabidopsis phage-type RNA polymerases: accurate in vitro transcription of organellar genes. Plant Cell 19(3):959–971

Kuhn K, Richter U, Meyer EH, Delannoy E, de Longevialle AF, O'Toole N, Borner T, Millar AH, Small ID, Whelan J (2009) Phage-type RNA polymerase RPOTmp performs gene-specific transcription in mitochondria of Arabidopsis thaliana. Plant Cell 21(9):2762–2779

Kwasniak M, Majewski P, Skibior R, Adamowicz A, Czarna M, Sliwinska E, Janska H (2013) Silencing of the nuclear RPS10 gene encoding mitochondrial ribosomal protein alters translation in Arabidopsis mitochondria. Plant Cell 25(5):1855–1867

Lenz H, Rudinger M, Volkmar U, Fischer S, Herres S, Grewe F, Knoop V (2010) Introducing the plant RNA editing prediction and analysis computer tool PREPACT and an update on RNA editing site nomenclature. Curr Genet 56(2):189–201

Lambowitz AM, Zimmerly S (2004) Mobile group II introns. Annu Rev Genet 38:1–35

Lambowitz AM, Zimmerly S (2011) Group II introns: mobile ribozymes that invade DNA. Cold Spring Harb Perspect Biol 3(8):a003616

Lang BF, Gray MW, Burger G (1999) Mitochondrial genome evolution and the origin of eukaryotes. Annu Rev Genet 33:351–397

Lapuente-Brun E, Moreno-Loshuertos R, Acin-Perez R, Latorre-Pellicer A, Colas C, Balsa E, Perales-Clemente E, Quiros PM, Calvo E, Rodriguez-Hernandez MA, Navas P, Cruz R, Carracedo A, Lopez-Otin C, Perez-Martos A, Fernandez-Silva P, Fernandez-Vizarra E, Enriquez JA (2013) Supercomplex assembly determines electron flux in the mitochondrial electron transport chain. Science 340(6140):1567–1570

Larkin RM, Alonso JM, Ecker JR, Chory J (2003) GUN4, a regulator of chlorophyll synthesis and intracellular signaling. Science 299(5608):902–906. doi:10.1126/science.1079978

Leist M, Jaattela M (2001) Four deaths and a funeral: from caspases to alternative mechanisms. Nat Rev Mol Cell Biol 2(8):589–598. doi:10.1038/35085008

Leister D (2012) Retrograde signaling in plants: from simple to complex scenarios. Front Plant Sci 3(135):19

Leon P, Arroyo A, Mackenzie S (1998) Nuclear control of plastid and mitochondrial development in higher plants. Annu Rev Plant Physiol Plant Mol Biol 49:453–480. doi:10.1146/annurev.arplant.49.1.453

Leon G, Holuigue L, Jordana X (2007) Mitochondrial complex II Is essential for gametophyte development in Arabidopsis. Plant Physiol 143(4):1534–1546. doi:10.1104/pp. 106.095158

Lerbs-Mache S (1993) The 110-kDa polypeptide of spinach plastid DNA-dependent RNA polymerase: single-subunit enzyme or catalytic core of multimeric enzyme complexes? Proc Natl Acad Sci U S A 90(12):5509–5513

Levine B, Goldman JE, Jiang HH, Griffin DE, Hardwick JM (1996) Bcl-2 protects mice against fatal alphavirus encephalitis. Proc Natl Acad Sci U S A 93(10):4810–4815

Liere K, Weihe A, Borner T (2011) The transcription machineries of plant mitochondria and chloroplasts: composition, function, and regulation. J Plant Physiol 168(12):1345–1360

Li-Pook-Than J, Bonen L (2006) Multiple physical forms of excised group II intron RNAs in wheat mitochondria. Nucleic Acids Res 34(9):2782–2790

Lister R, Chew O, Lee MN, Heazlewood JL, Clifton R, Parker KL, Millar AH, Whelan J (2004) A transcriptomic and proteomic characterization of the Arabidopsis mitochondrial protein import apparatus and its response to mitochondrial dysfunction. Plant Physiol 134(2):777–789. doi:10.1104/pp. 103.033910

Liu Y, Schiff M, Czymmek K, Talloczy Z, Levine B, Dinesh-Kumar SP (2005) Autophagy regulates programmed cell death during the plant innate immune response. Cell 121(4):567–577. doi:10.1016/j.cell.2005.03.007

Long M, Betran E, Thornton K, Wang W (2003) The origin of new genes: glimpses from the young and old. Nat Rev Genet 4(11):865–875

Mackenzie S, McIntosh L (1999) Higher plant mitochondria. Plant Cell 11(4):571–586

Mahendran R, Spottswood MS, Ghate A, Ling ML, Jeng K, Miller DL (1994) Editing of the mitochondrial small subunit rRNA in Physarum polycephalum. EMBO J 13(1):232–240

Malerba M, Cerana R, Crosti P (2004) Comparison between the effects of fusicoccin, Tunicamycin, and Brefeldin A on programmed cell death of cultured sycamore (Acer pseudoplatanus L.) cells. Protoplasma 224(1–2):61–70

McAllister WT (1993) Structure and function of the bacteriophage T7 RNA polymerase (or, the virtues of simplicity). Cell Mol Biol Res 39(4):385–391

McCabe PF, Leaver CJ (2000) Programmed cell death in cell cultures. Plant Mol Biol 44(3):359–368

McCabe MS, Power JB, de Laat AM, Davey MR (1997a) Detection of single-copy genes in DNA from transgenic plants by nonradioactive Southern blot analysis. Mol Biotechnol 7(1):79–84. doi:10.1007/BF02821545

McCabe PF, Valentine TA, Forsberg LS, Pennell RI (1997b) Soluble signals from cells identified at the cell wall establish a developmental pathway in carrot. Plant Cell 9(12):2225–2241

Mercer TR, Neph S, Dinger ME, Crawford J, Smith MA, Shearwood AM, Haugen E, Bracken CP, Rackham O, Stamatoyannopoulos JA, Filipovska A, Mattick JS (2011) The human mitochondrial transcriptome. Cell 146(4):645–658

Meyer EH, Taylor NL, Millar AH (2008) Resolving and identifying protein components of plant mitochondrial respiratory complexes using three dimensions of gel electrophoresis. J Proteome Res 7(2):786–794. doi:10.1021/pr700595p

Millar AH, Sweetlove LJ, Giege P, Leaver CJ (2001) Analysis of the Arabidopsis mitochondrial proteome. Plant Physiol 127(4):1711–1727

Millar AH, Eubel H, Jansch L, Kruft V, Heazlewood JL, Braun HP (2004) Mitochondrial cytochrome c oxidase and succinate dehydrogenase complexes contain plant specific subunits. Plant Mol Biol 56(1):77–90. doi:10.1007/s11103-004-2316-2

Millar AH, Whelan J, Soole KL, Day DA (2011) Organization and regulation of mitochondrial respiration in plants. Annu Rev Plant Biol 62:79–104

Mochizuki N, Brusslan JA, Larkin R, Nagatani A, Chory J (2001) Arabidopsis genomes uncoupled 5 (GUN5) mutant reveals the involvement of Mg-chelatase H subunit in plastid-to-nucleus signal transduction. Proc Natl Acad Sci U S A 98(4):2053–2058. doi:10.1073/pnas.98.4.2053

Moller IM (2001) PLANT MITOCHONDRIA AND OXIDATIVE STRESS: electron transport, NADPH turnover, and metabolism of reactive oxygen species. Annu Rev Plant Physiol Plant Mol Biol 52:561–591

Mootha VK, Bunkenborg J, Olsen JV, Hjerrild M, Wisniewski JR, Stahl E, Bolouri MS, Ray HN, Sihag S, Kamal M, Patterson N, Lander ES, Mann M (2003) Integrated analysis of protein composition, tissue diversity, and gene regulation in mouse mitochondria. Cell 115(5):629–640

Notsu Y, Masood S, Nishikawa T, Kubo N, Akiduki G, Nakazono M, Hirai A, Kadowaki K (2002) The complete sequence of the rice (Oryza sativa L.) mitochondrial genome: frequent DNA sequence acquisition and loss during the evolution of flowering plants. Mol Genet Genomics 268(4):434–445

Op den Camp RG, Przybyla D, Ochsenbein C, Laloi C, Kim C, Danon A, Wagner D, Hideg E, Gobel C, Feussner I, Nater M, Apel K (2003) Rapid induction of distinct stress responses after the release of singlet oxygen in Arabidopsis. Plant Cell 15(10):2320–2332

Osteryoung KW, Nunnari J (2003) The division of endosymbiotic organelles. Science 302(5651):1698–1704. doi:10.1126/science.1082192

Parenteau J, Durand M, Morin G, Gagnon J, Lucier JF, Wellinger RJ, Chabot B, Elela SA (2011) Introns within ribosomal protein genes regulate the production and function of yeast ribosomes. Cell 147(2):320–331

Patel S, Dinesh-Kumar SP (2008) Arabidopsis ATG6 is required to limit the pathogen-associated cell death response. Autophagy 4(1):20–27

Peiffer WE, Ingle RT, Ferguson-Miller S (1990) Structurally unique plant cytochrome c oxidase isolated from wheat germ, a rich source of plant mitochondrial enzymes. Biochemistry 29(37):8696–8701

Peters K, Belt K, Braun HP (2013) 3D gel map of Arabidopsis complex I. Front Plant Sci 4(153)

Pfannschmidt T, Schutze K, Brost M, Oelmuller R (2001) A novel mechanism of nuclear photosynthesis gene regulation by redox signals from the chloroplast during photosystem stoichiometry adjustment. J Biol Chem 276(39):36125–36130. doi:10.1074/jbc.M105701200

Pursiheimo S, Mulo P, Rintamaki E, Aro EM (2001) Coregulation of light-harvesting complex II phosphorylation and lhcb mRNA accumulation in winter rye. Plant J 26(3):317–327

Qiu YL, Palmer JD (2004) Many independent origins of trans splicing of a plant mitochondrial group II intron. J Mol Evol 59(1):80–89

Raczynska KD, Le Ret M, Rurek M, Bonnard G, Augustyniak H, Gualberto JM (2006) Plant mitochondrial genes can be expressed from mRNAs lacking stop codons. FEBS Lett 580(24):5641–5646

Raghavendra AS, Padmasree K (2003) Beneficial interactions of mitochondrial metabolism with photosynthetic carbon assimilation. Trends Plant Sci 8(11):546–553. doi:10.1016/j.tplants.2003.09.015

Rebeille F, Macherel D, Mouillon JM, Garin J, Douce R (1997) Folate biosynthesis in higher plants: purification and molecular cloning of a bifunctional 6-hydroxymethyl-7,8-dihydropterin pyrophosphokinase/7,8-dihydropteroate synthase localized in mitochondria. EMBO J 16(5):947–957

Remacle C, Barbieri MR, Cardol P, Hamel PP (2008) Eukaryotic complex I: functional diversity and experimental systems to unravel the assembly process. Mol Genet Genomics 280(2):93–110. doi:10.1007/s00438-008-0350-5

Richter U, Kiessling J, Hedtke B, Decker E, Reski R, Borner T, Weihe A (2002) Two RpoT genes of Physcomitrella patens encode phage-type RNA polymerases with dual targeting to mitochondria and plastids. Gene 290(1–2):95–105

Robison MM, Ling X, Smid MP, Zarei A, Wolyn DJ (2009) Antisense expression of mitochondrial ATP synthase subunits OSCP (ATP5) and gamma (ATP3) alters leaf morphology, metabolism and gene expression in Arabidopsis. Plant Cell Physiol 50(10):1840–1850. doi:10.1093/pcp/pcp125

Rochaix JD (2001) Assembly, function, and dynamics of the photosynthetic machinery in Chlamydomonas reinhardtii. Plant Physiol 127(4):1394–1398

Rolland F, Moore B, Sheen J (2002) Sugar sensing and signaling in plants. Plant Cell 14(205):S185–S205

Rutter J, Winge DR, Schiffman JD (2010) Succinate dehydrogenase—assembly, regulation and role in human disease. Mitochondrion 10(4):393–401

Sabar M, De Paepe R, de Kouchkovsky Y (2000) Complex I impairment, respiratory compensations, and photosynthetic decrease in nuclear and mitochondrial male sterile mutants of Nicotiana sylvestris. Plant Physiol 124(3):1239–1250

Sabar M, Gagliardi D, Balk J, Leaver CJ (2003) ORFB is a subunit of F1F(O)-ATP synthase: insight into the basis of cytoplasmic male sterility in sunflower. EMBO Rep 4(4):381–386

Sabar M, Balk J, Leaver CJ (2005) Histochemical staining and quantification of plant mitochondrial respiratory chain complexes using blue-native polyacrylamide gel electrophoresis. Plant J 44(5):893–901. doi:10.1111/j.1365-313X.2005.02577.x

Senior DJ, Tsai CS (1990) Esterase activity of high-Km aldehyde dehydrogenase from rat liver mitochondria. Biochem Cell Biol 68(4):758–763

Shikanai T (2006) RNA editing in plant organelles: machinery, physiological function and evolution. Cell Mol Life Sci 63(6):698–708

Sickmann A, Reinders J, Wagner Y, Joppich C, Zahedi R, Meyer HE, Schonfisch B, Perschil I, Chacinska A, Guiard B, Rehling P, Pfanner N, Meisinger C (2003) The proteome of Saccharomyces cerevisiae mitochondria. Proc Natl Acad Sci U S A 100(23):13207–13212

Small I, Peeters N, Legeai F, Lurin C (2004) Predotar: a tool for rapidly screening proteomes for N-terminal targeting sequences. Proteomics 4(6):1581–1590

Smith PM, Fox JL, Winge DR (2012) Biogenesis of the cytochrome bc(1) complex and role of assembly factors. Biochim Biophys Acta 2:276–286

Soll J, Schleiff E (2004) Protein import into chloroplasts. Nat Rev Mol Cell Biol 5(3):198–208

Sousa R (1996) Structural and mechanistic relationships between nucleic acid polymerases. Trends Biochem Sci 21(5):186–190

Strand A, Asami T, Alonso J, Ecker JR, Chory J (2003) Chloroplast to nucleus communication triggered by accumulation of Mg-protoporphyrinIX. Nature 421(6918):79–83. doi:10.1038/nature01204

Surpin M, Zheng H, Morita MT, Saito C, Avila E, Blakeslee JJ, Bandyopadhyay A, Kovaleva V, Carter D, Murphy A, Tasaka M, Raikhel N (2003) The VTI family of SNARE proteins is necessary for plant viability and mediates different protein transport pathways. Plant Cell 15(12):2885–2899. doi:10.1105/tpc.016121

Sweetlove LJ, Heazlewood JL, Herald V, Holtzapffel R, Day DA, Leaver CJ, Millar AH (2002) The impact of oxidative stress on Arabidopsis mitochondria. Plant J 32(6):891–904

Takemura M, Oda K, Yamato K, Ohta E, Nakamura Y, Nozato N, Akashi K, Ohyama K (1992) Gene clusters for ribosomal proteins in the mitochondrial genome of a liverwort, Marchantia polymorpha. Nucleic Acids Res 20(12):3199–3205

Takenaka M, Verbitskiy D, van der Merwe JA, Zehrmann A, Brennicke A (2008) The process of RNA editing in plant mitochondria. Mitochondrion 8(1):35–46

Talla E, Anthouard V, Bouchier C, Frangeul L, Dujon B (2005) The complete mitochondrial genome of the yeast Kluyveromyces thermotolerans. FEBS Lett 579(1):30–40

Taylor SW, Fahy E, Zhang B, Glenn GM, Warnock DE, Wiley S, Murphy AN, Gaucher SP, Capaldi RA, Gibson BW, Ghosh SS (2003) Characterization of the human heart mitochondrial proteome. Nat Biotechnol 21(3):281–286

Taylor NL, Heazlewood JL, Millar AH (2011) The Arabidopsis thaliana 2-D gel mitochondrial proteome: refining the value of reference maps for assessing protein abundance, contaminants and post-translational modifications. Proteomics 11(9):1720–1733

Thomas SG, Franklin-Tong VE (2004) Self-incompatibility triggers programmed cell death in Papaver pollen. Nature 429(6989):305–309. doi:10.1038/nature02540

Thompson CB (1995) Apoptosis in the pathogenesis and treatment of disease. Science 267(5203):1456–1462

Toor N, Robart AR, Christianson J, Zimmerly S (2006) Self-splicing of a group IIC intron: 5' exon recognition and alternative 5' splicing events implicate the stem-loop motif of a transcriptional terminator. Nucleic Acids Res 34(22):6461–6471

Toro N, Jimenez-Zurdo JI, Garcia-Rodriguez FM (2007) Bacterial group II introns: not just splicing. FEMS Microbiol Rev 31(3):342–358

Townley HE, McDonald K, Jenkins GI, Knight MR, Leaver CJ (2005) Ceramides induce programmed cell death in Arabidopsis cells in a calcium-dependent manner. Biol Chem 386(2):161–166. doi:10.1515/BC.2005.020

Tseng CC, Lee CJ, Chung YT, Sung TY, Hsieh MH (2013) Differential regulation of Arabidopsis plastid gene expression and RNA editing in non-photosynthetic tissues. Plant Mol Biol 82(4–5):375–392

Tzagoloff A (1995) Ubiquinol-cytochrome-c oxidoreductase from Saccharomyces cerevisiae. Methods Enzymol 260:51–63

Ugalde C, Vogel R, Huijbens R, Van Den Heuvel B, Smeitink J, Nijtmans L (2004) Human mitochondrial complex I assembles through the combination of evolutionary conserved modules: a framework to interpret complex I deficiencies. Hum Mol Genet 13(20):2461–2472. doi:10.1093/hmg/ddh262

Unseld M, Marienfeld JR, Brandt P, Brennicke A (1997) The mitochondrial genome of Arabidopsis thaliana contains 57 genes in 366,924 nucleotides. Nat Genet 15(1):57–61. doi:10.1038/ng0197-57

Vacca RA, de Pinto MC, Valenti D, Passarella S, Marra E, De Gara L (2004) Production of reactive oxygen species, alteration of cytosolic ascorbate peroxidase, and impairment of mitochondrial metabolism are early events in heat shock-induced programmed cell death in tobacco Bright-Yellow 2 cells. Plant Physiol 134(3):1100–1112. doi:10.1104/pp. 103.035956

Vandenabeele S, Vanderauwera S, Vuylsteke M, Rombauts S, Langebartels C, Seidlitz HK, Zabeau M, Van Montagu M, Inze D, Van Breusegem F (2004) Catalase deficiency drastically affects gene expression induced by high light in Arabidopsis thaliana. Plant J 39(1):45–58

Velours J, Arselin G (2000) The Saccharomyces cerevisiae ATP synthase. J Bioenerg Biomembr 32(4):383–390

Verbitskiy D, Zehrmann A, van der Merwe JA, Brennicke A, Takenaka M (2010) The PPR protein encoded by the LOVASTATIN INSENSITIVE 1 gene is involved in RNA editing at three sites in mitochondria of Arabidopsis thaliana. Plant J 61(3):446–455

Vicens Q, Cech TR (2006) Atomic level architecture of group I introns revealed. Trends Biochem Sci 31(1):41–51

Vogel RO, Smeitink JA, Nijtmans LG (2007) Human mitochondrial complex I assembly: a dynamic and versatile process. Biochim Biophys Acta 10(27):9

Vothknecht UC, Westhoff P (2001) Biogenesis and origin of thylakoid membranes. Biochim Biophys Acta 12:1–2

Wagner RW, Smith JE, Cooperman BS, Nishikura K (1989) A double-stranded RNA unwinding activity introduces structural alterations by means of adenosine to inosine conversions in mammalian cells and Xenopus eggs. Proc Natl Acad Sci U S A 86(8):2647–2651

Wallace DC (2005a) The mitochondrial genome in human adaptive radiation and disease: on the road to therapeutics and performance enhancement. Gene 354:169–180

Wallace DC (2005b) A mitochondrial paradigm of metabolic and degenerative diseases, aging, and cancer: a dawn for evolutionary medicine. Annu Rev Genet 39:359–407

Werhahn W, Braun HP (2002) Biochemical dissection of the mitochondrial proteome from Arabidopsis thaliana by three-dimensional gel electrophoresis. Electrophoresis 23(4):640–646

Wheeler DS (2009) Death to sepsis: targeting apoptosis pathways in sepsis. Crit Care 13(6):1010. doi:10.1186/cc8162

Wilde A, Mikolajczyk S, Alawady A, Lokstein H, Grimm B (2004) The gun4 gene is essential for cyanobacterial porphyrin metabolism. FEBS Lett 571(1–3):119–123. doi:10.1016/j.febslet.2004.06.063

Woodson JD, Chory J (2008) Coordination of gene expression between organellar and nuclear genomes. Nat Rev Genet 9(5):383–395

Xiong Y, Contento AL, Bassham DC (2007a) Disruption of autophagy results in constitutive oxidative stress in Arabidopsis. Autophagy 3(3):257–258

Xiong Y, Contento AL, Nguyen PQ, Bassham DC (2007b) Degradation of oxidized proteins by autophagy during oxidative stress in Arabidopsis. Plant Physiol 143(1):291–299. doi:10.1104/pp. 106.092106

Xu XW, Shi C, He ZQ, Ma CM, Chen WH, Shen YP, Guo Q, Shen CJ, Xu J (2008) Effects of phytoestrogen on mitochondrial structure and function of hippocampal CA1 region of ovariectomized rats. Cell Mol Neurobiol 28(6):875–886. doi:10.1007/s10571-008-9265-2

Xue S, Barna M (2012) Specialized ribosomes: a new frontier in gene regulation and organismal biology. Nat Rev Mol Cell Biol 13(6):355–369

Yagi Y, Tachikawa M, Noguchi H, Satoh S, Obokata J, Nakamura T (2013) Pentatricopeptide repeat proteins involved in plant organellar RNA editing. RNA Biol 10:9

Yao N, Imai S, Tada Y, Nakayashiki H, Tosa Y, Park P, Mayama S (2002) Apoptotic cell death is a common response to pathogen attack in oats. Mol Plant-Microbe Interact 15(10):1000–1007. doi:10.1094/MPMI.2002.15.10.1000

Yoshimoto K, Jikumaru Y, Kamiya Y, Kusano M, Consonni C, Panstruga R, Ohsumi Y, Shirasu K (2009) Autophagy negatively regulates cell death by control-

ling NPR1-dependent salicylic acid signaling during senescence and the innate immune response in Arabidopsis. Plant Cell 21(9):2914–2927. doi:10.1105/tpc.109.068635
Zehrmann A, Verbitskiy D, van der Merwe JA, Brennicke A, Takenaka M (2009) A DYW domain-containing pentatricopeptide repeat protein is required for RNA editing at multiple sites in mitochondria of Arabidopsis thaliana. Plant Cell 21(2):558–567
Zeng WH, Liao SC, Chang CC (2007) Identification of RNA editing sites in chloroplast transcripts of Phalaenopsis aphrodite and comparative analysis with those of other seed plants. Plant Cell Physiol 48(2):362–368
Zhang Z, Liu JJ, Yao FL, He H, Yang J, Xie HF, He T (2007) Morphological changes of non-apoptotic programmed cell death of polymorphonuclear neutrophils induced by ONO-AE-248. Xi bao yu fen zi mian yi xue za zhi Chin J Cell Mol Immunol 23(5):413–415
Zmudjak M, Colas des Francs-Small C, Keren I, Shaya F, Belausov E, Small I, Ostersetzer-Biran O (2013) mCSF1, a nucleus-encoded CRM protein required for the processing of many mitochondrial introns, is involved in the biogenesis of respiratory complexes I and IV in Arabidopsis. New Phytol 199(2):379–394. doi:10.1111/nph.12282
Zuzarte-Luis V, Hurle JM (2002) Programmed cell death in the developing limb. Int J Dev Biol 46(7):871–876

Micromorphomics: A Morphological Dissection to Unveil Environmental Stress

Tulika Talukdar

Contents

T. Talukdar, Ph.D. (✉)
Department of Botany, Krishnagar Government College, Nadia, Krishnagar 741101, West Bengal, India

Department of Botany, APC Roy Govt. College, Siliguri, Darjeeling, West Bengal, India
e-mail: talukdartulip12@gmail.com

Abstract

Earth environment with all its good and bad effects is maintained and disturbed by the interplay between plants and animals. Plants could be silent indicators of healthy or stressful environment in and around themselves through varied morphological manifestations. Environmental stress may be abiotic or biotic.

D. Barh et al. (eds.), *PlantOmics: The Omics of Plant Science*,
DOI 10.1007/978-81-322-2172-2_21, © Springer India 2015

Morphological manifestations of plant's response to any kind of stress may be organ specific or integrative. Among different angiospermic plant organs, leaf (source organ) and fruit (sink organ) micromorphological responses to stress are found to be more dramatic and critical. Along with these, other plant organs like stem, root, flower, etc. also manifest micromorphological changes in response to various environmental stresses. They are certainly being the direct outcome of internal mechanistic alterations of plants to combat stressful environment. Understanding of this organ architecture is the first and foremost step towards unveiling the underlying controlling factors.

Keywords

Environment • Indicator • Micromorphology • Plant defence • Plant organ • Stress

Introduction

Micro is a prefix that comes from the Greek word "μικρός (*mikrós*)" meaning small, denoting a factor of one millionth. More precisely any finer level structure or morphology visible under microscopy is micromorphology. Micromorphomics is the micromorphology-based analysis applied to plants, animals and humans. Its application to illuminating the black box (stress) is currently an emerging branch. Consideration of ecological trends in the evolution of morpho-anatomical traits clearly indicates precise coordination between structure and function in plants. Such coordination becomes strong enough under stressful environment, when structural attributes modified for optimising their function to guarantee growth, survival and reproduction of plants.

The term "stress" can be defined as any disturbance that adversely affects the growth of a plant as a whole. Various abiotic and biotic constraints such as water and nutrient deficiency, adverse climatic conditions, heavy metal contamination, mechanical injury, plant diseases, insect damage, etc. are major stressors that act as growth-limiting factors. In response to stress, plants can adapt to, avoid or may overcome the stress by means of various physiological and biochemical mechanisms. Tolerance involves endurance of the stress such that plants can function normally under both internal and external stress and depends on the development of specialised physiological mechanisms. On the other hand, avoidance involves establishment and maintaining of internal normal conditions under external stressful conditions and more often utilises the morphological devices to shield plants from the effects of extreme conditions. Plant architecture thus can play an important role in stress. Thus, developing plant varieties with appropriate architecture will help to cope with the rapidly increasing stress in environment. So understanding the response of plants to stress would be desirable in the light of global and regional changes not only to forecast population dynamics in natural ecosystems, but also to adjust management practices in agriculture. To analyse the plant's response to stress, measurements from all levels of plant organisation like molecules, organelles, cells, tissues, organs, whole plants and populations often are necessary.

Every plant organ is ideally designed to fulfil metabolic and physiological processes in specific environmental conditions. Even the origin and evolutionary history of plants reflects plenty of examples of how a specific trait and its structural attributes arose to fulfil a specific need. Any undesirable disturbance in environment would create catastrophic changes primarily in specific plant organ adversely affecting its physiological function and subsequently at the whole plant level.

Scientific advancement and its outcome ultimately focus on the effective way to achieve stable environment with less disturbances. So to detect stress as early as possible to minimise its effects on plants and thereafter on mankind, modified micro-features of different plant organs like leaf, fruit and root could be a first-hand indicator (Talukdar 2013b, c). These modified features or symptoms are the direct outcome of interaction between stress factor and plant defence system. More often, degree of interaction determines the intensity of symptoms produced. Therefore, these features validated by microscopic analyses are useful to diagnose and rapid assessment of stress factor. Detection of intrinsic mechanism/s controlling stress-induced architectural changes is very crucial and still is enigmatic. Most of the stress-induced morphological changes are the outcome of reorientation of growth rather than cessation of growth resulting from altered metabolism. Thus identification of underlying physiological and intrinsic metabolic reaction based on the structural signature of stress factors is also possible (André et al. 2006; Martin et al. 2006; Talukdar 2012; Vollenweider et al. 2003, 2006). This mechanistic detail plays a pivotal role to select or create new stress-resistant/tolerant varieties of crops to obtain better productivity (Jaleel et al. 2009a, b; Martinez et al. 2007; Nam et al. 2001). Under heavy metal stress uptake, accumulation of metals at higher concentrations can be cytotoxic in some plant species causing structural and ultrastructural changes affecting the growth and physiological status of the plants (Barcelo et al. 1988; Han et al. 2004; Vazquez et al. 1992; Zhao et al. 2006). So architectural parameters which can be treated as visualisation of internal cytosolic story can other way be used to understand and optimise the overall process of phytoremediation and to screen the effective plant types for phytoremediation.

Leaf Architecture Modifications to Stress

Leaf is the major source organ of higher plants with very few exceptions. Its fascinating array of mesophyll cells and stomata leaf provides thermo-nutrient stability of plants facilitating transpiration and photosynthesis. The following are the major leaf architectural modifications reported in angiosperms in response to different abiotic and biotic stresses.

Leaf Colour Modification

Changes of leaf colour are generally resulted from physiological changes caused by water deficiency or change in nutrient status or chemical changes within plant cell due to pathogen (Jackson 1986). The changes in colour are ultimately associated with the alteration of pigment content and are frequent in dry habitats. They help in reducing solar irradiation and consequently decrease leaf heating and transpiration rates as well as avoid damage to photosystems (Arena et al. 2008; Aronne and De Micco 2001; Jaleel et al. 2009a, b).

Leaf Curling and Rolling

Leaf curling is generally associated with hypertrophy and hyperplasia of mesophyll cells (Evans et al. 1977; Sant'Anna-Santos et al. 2006). Extensive leaf rolling, common in grasses, is associated with drought and salt stress. This feature is affected by the turgidity of bulliform cells. Under high salinity, enlarged and well-developed bulliform cells were observed in *Deschampsia antarctica*, *Leptochloa fusca* and many salt range ecotypes (Gielwanowska et al. 2005; Ola et al. 2012). It plays an important role to avoid water loss and can be regarded as an adaptive defensive strategy against drought, salt and heat stress. Bulliform cells, also known as motor cells, respond rapidly to water or heat stresses by losing their turgor and become flaccid resulting in leaf blade rolling, thus optimising water and temperature condition. Therefore, in some physiological reports leaf rolling was defined as water conservation movement (Srivastava 2001), previously known as nastic movement.

Leaf Necrosis

Tip burn and marginal necrosis of leaf were frequently reported as fluoride injury to vegetation around highly industrialised areas (Fornasiero 2001). Necrosis along veins and leaf chlorosis were noted due to overexposure to heavy metals like zinc (André et al. 2006). Structural injury on leaf induced by heavy metals was decreased at higher shoot position and more pronounced at lower shoot and thus indicates their mode of translocation through vascular tissue.

Stomatal Features

Stomata number of leaf is reduced in high salinity (Çavuşoğlu et al. 2007, 2008; Ola et al. 2012). Decrease in stomata number and density were also noted in severe drought (De Micco and Aronne 2012; Xu and Zhou 2008). Plants of arid or semiarid regions show sunken stomata. However, the occurrence of stomata plugged with cuticular structures has been shown as an adaptation to excess water in plants growing in rainforests and cloud forests. These plugs help in maintaining photosynthetic activity by preventing the formation of a continuous water film (Field et al. 1998). Interestingly, increased stomatal density coupled with decreased stomatal size is considered as a good adaptive feature of plants in response to heavy metal toxicity and polluted environment (Azmat et al. 2009; Melo et al. 2007; Noman et al. 2012). The larger number of small stomata ensures sufficient flow of CO_2 for photosynthesis, keeping transpiration to a minimum.

Stomatal Clustering

Very recently "stomatal clustering" is recognised as a new micromorphological marker for environmental adaptation in terrestrial plants (Gan et al. 2010). Stomatal clustering is an abnormal stomatal distribution/arrangement that is formed by two or more stomata in the leaf epidermis. In normal distribution at least one intervening epidermal cell must be present in between two neighbouring stomata. This convention, also known as "one-cell-spacing rule", optimises the balance between water loss and carbon assimilation by minimising the overlaps of stomatal gaseous diffusion shells (GDS) (Larkin et al. 1997). In contiguous clustering two or more stomata are placed in direct contact without any intervening epidermal cell between neighbouring guard cells. This type was reported in members of Papilionaceae, Sonneratiaceae, Annonaceae and Amaryllidaceae. In many other angiosperm families like Boraginaceae, Cruciferae, Moraceae, Rubiaceae and Theaceae, more common noncontiguous stomatal cluster was reported where most of the stomata are arranged in groups at whole leaf level without any direct contact between guard cells. Few scientists argued that stomatal clusters have positive effect on plants. In water stress condition larger clusters were found in comparison to well-watered soil condition in few *Begonia* species (Hoover 1986). Interestingly, there is a positive correlation between cluster size and multiple epidermis under drought in *Begonia* species and considered as typical drought adaptation trait (Tang et al. 2002). Stomatal clustering plays an important role in water conservation under severe drought. Overlapping of gaseous diffusion shells that resulted in stomatal cluster effectively reduces the total leaf area of evaporation, keeping water loss to a minimum. In few cases higher carbon assimilation rate in leaf was reported than in leaf with normally distributed stomata (Schlüter et al. 2003).

Leaf Trichome

Increase in the number of trichomes per unit area favours plant survival in contaminated environment (Azmat et al. 2009). In addition, increased abaxial and adaxial epidermal thickness contributes significantly in tackling hazardous effects of heavy metal contamination (Gomes et al. 2011; Noman et al. 2012). A twofold increase in trichome length on leaf surface of road-side plants like *Asparagus racemosus*, *Azadirachta indica*, *Bougainvillea spectabilis*, *Cassia fistula*, *Ficus religiosa* and *Nerium indicum* exposed to auto-exhaust pollution was noted (Pal et al. 2002). Most ecological studies suggest that plants growing under stress tend to possess leaves that have more hairs than similar or related plants from normal condition. Leaf pubescence reduces the light absorption during conditions of high temperature and drought, resulting in reduced heat loads which in turn lower leaf temperatures and transpiration rates. Dense hair also protects the plant from predation by insects and larger herbivores. Increases in number of stomata with trichomes on the surface of leaves were observed in many leguminous crops like *Phaseolus mungo* and *Lens culinaris* under Pb toxicity (Azmat et al. 2009). There is also evidence that in *Arabidopsis* trichomes play additional or alternative roles in response to abiotic stress to detoxify heavy metals (Ager et al. 2003). On the other hand, a decrease in trichome length and density upon cadmium exposure was reported in *Cajanus cajan* (Khudsar et al. 2001).

Change in Cuticle Layer

Cuticle is composed of cutin and wax, provides first contact zone between plant surface and environment, maintains structural integrity of plant tissues, protecting plants from foreign invasion (like microbes, other pathogen, insect, etc.), from harmful radiation (Kerstiens 1996; Riederer and Müller 2005) and from dry environment by minimising non-stomatal water loss (Kerstiens 2006; Riederer and Schreiber 2001; Von et al. 2007). Increase in wax amount at enhanced temperature and at low relative humidity has been noticed in rose and other plants (Dixon et al. 1997; Jenks et al. 2001). Increase in cuticle thickness is also a common feature in plants (e. g. peanut, cabbage) facing severe drought and salinity (Shepherd and Griffiths 2006). An increase in cuticular wax is also a very striking feature of cadmium-exposed leaves as noted in barley seedlings (Hollenbach et al. 1997).

Changes in Leaf Epidermal Cells

Rupturing of leaf epidermal cells in high salinity and low pH stress was noted in *Genipa americana* L. of Rubiaceae (Sant'Anna-Santos et al. 2006).

Leaf Thickness

Increased leaf thickness resulting from the thickening of the mesophyll and the cuticle correlated with loss of moisture and is the result of water stress (Bussotti et al. 1995). On the contrary, decrease in lamina thickness and mesophyll thickness with increasing salinity was reported in many plants. Reduced mesophyll area in turn reduces the capacity for re-translocation of mineral nutrients and assimilates in leaf. In few cases palisade mesophyll cell length decreased slightly under salinity as in olive plants (Karimi et al. 2009). Under severe stress degeneration of phloem, necrosis of lower epidermis manifested by cell collapse, cell wall thickening and accumulation of secondary compounds were prevalent. Palisade and spongy cells in mesophyll were adversely affected and showed different degeneration symptoms including cell wall thickening, folding and partial collapse along with accumulation of secondary compounds. In Indian mustard (*Brassica juncea*), high concentration of metal ions (>50 mM) induces shrinkage of epidermal, palisade and spongy parenchyma cells due to accumulation of metal ions (Maruthi Sridhar et al. 2005). These changes were indicative of accelerated cell senescence processes (Vollenweider et al. 2006). However, in response to drought, leaf thickness increased due to increase in parenchyma tissue engaged in water storage (Aronne and De Micco 2001). Higher photosynthetic tissue with enlarged or increased palisade and spongy cells capable of producing higher vital metabolites was reported in few herbaceous plants like *Hibiscus* and *Rosa* in response to stress (Nawaz et al. 2011; Noman et al. 2012). Such contrasting behaviour of photosynthetic tissue in different plants in response to stress leads to speculation that decrease in mesophyll thickness is the injurious effect of stressors hampering growth in susceptible plants, while increased palisade and spongy cell area are considered as healthy adaptability signs to stress in resistant/tolerant plants (Gostin 2009).

Changes in Leaf Temperature

Increased leaf temperature was noted in many plants due to restricted transpiration from water scarcity under drought or due to a vascular disease.

Leaf Vein

Decrease in size and thickness of midrib and large veins along with decrease in number of median and small veins with increasing salinity were noted in kallar grass (*Leptochloa fusca* L. Kunth) (Ola et al. 2012). Gradual decrease in vascular bundle area in the leaf of cogon grass (*Imperata cylindrica*) is reported with increasing salinity level (Hameed et al. 2009). In leaf different types of vein play different physiological functions. The large veins transport water and the small veins mainly load and transport nutrients. Salt stress thus interferes with distribution of nutrients and water, limiting the growth rate of affected plant (Hu et al. 2005).

Leaf Sclerification

Sclerification of leaf increased with increasing salinity to minimise water loss and to confer rigidity. This was reported for many plants like *Festuca* (Ola et al. 2012), *Spartina alterniflora* of Poaceae (Walsh 1990), *Kandelia candel* (Hwang and Chen 1995), cotton (Reinhardt and Rost 1995), *Puccinellia tenuiflora* (Yujing et al. 2000) and *Prosopis strombulifera* of Leguminosae (Reinoso et al. 2004).

Accumulation of Phenol

Accumulation of phenolic compounds in glandular leaf hairs, epidermal cells and parenchymatous cells was noted in plants facing drought, salinity and temperature stress as in *Cistus ladanifer* L. (De Micco and Aronne 2007). Formation of black spot due to high accumulation of phenolics and lignin is also considered as one of the most frequent reactions of plants to industrial pollution like air and water pollution and reported in *Picea abies* and in few members of Fabaceae (Gostin 2009; Wild and Schmitt 1995). Phenolic compounds act as filter against excess radiation during scorching heat. They also protect plants from permanent damage due to grazing and pathogen attacks. In severe cases, higher accumulation of dark phenolic compounds was generally followed by cytoplasm degradation and vacuolar content release that led to cell death (Sant'Anna-Santos et al. 2006; Zobel and Nighswander 1991).

Crystal Formation

Formation of crystals was noted in the intercellular spaces and on the surface of the cell wall of mesophyll tissue in *Vicia faba* L. under high-metal toxicity (Probst et al. 2009. Needle-like crystal deposits in leaves were detected as the unique anatomical feature under arsenic treatment in common bean *Phaseolus vulgaris* (Talukdar 2013a). Crystal formation usually occurs through substitution of mineral ions (e.g. Ca^{2+}) by metals and subsequent transportation and precipitation of displaced minerals as crystals in apoplast (Sarret et al. 2001). These crystals are supposed to be an avoidance mechanism in the ultimate phase of plant response to highly exceeded metals (Probst et al. 2009).

Stem Architecture Modifications to Stress

The stem is a part of the plant that holds up other structures such as the leaves and flowers. This is important as the leaves need to be held up to the sun to get its light for photosynthesis and the flowers need to be held up to be available for pollination. Stems also carry water and minerals up from the roots to the leaves to help with photosynthesis and take food back down to be stored and distributed to the plant as needed. The major stem manifestations noted in different plants under stress are as follows.

Changes in Stem Diameter

Reduction in stem diameter due to reduced cell division and expansion, sometimes accompanied by reduced DNA content in high salt stress, was reported from a wide range of plants (Wignarajahk et al. 1975).

Variation in Stem Epidermal Features

Increase in thickness of cuticle and epidermal cell wall in response to drought and heat stress is another common feature in plants. Increased cell wall thickness was also noted in shoots of *V. faba* L. exposed to Cd or Cu (Liu et al. 2004a) and in marine macroalga exposed to Cu (Andrade et al. 2004). This phenomenon seems to be associated to increased activity of peroxidase which catalyses lignin synthesis (Arduini et al. 1995; Liu et al. 2004a, b) and is induced in higher plants exposed to toxic metals (Prasad 1996).

Changes in Vascular Bundle

The vascular bundle area including xylem (protoxylem and metaxylem) and phloem is reduced with high salinity. This is probably compensated by increase in number of vascular bundle in stem along with increasing salinity (Ola et al. 2012). The reduction of xylem vessel diameter under saline and water-stressed conditions was early observed in cotton and tomato plants (Strogonov 1962) and in wild barley (Huang and Redmann 1995). Narrow vessel only allows slow water flow rate which is valuable because low conduction and low transpiration are needed during the period of drought (Carlquist 1975). Vessel grouping is another phenomenon which recently has been proved to have a positive role in conferring drought resistance in seven *Acer* species (Lens et al. 2011).

Stem Sclerification

Percentage of sclerenchyma (sclerification) increased with increasing salinity, and drought. This was modulated for minimising water loss and for reinforcing mechanical strength that prevent wilting. Consequently, this may lead to may lead to irreversible collapse and damage of cells.

Stem Aerenchyma

Stem air space or aerenchyma is a feature of waterlogged plant. With increasing salinity or drought, decrease in air spaces was noted by filling up with tightly packed parenchyma cells. This increases area of storage tissue with increasing vacuolar volume for storing toxic ions (Akhtar et al. 1998; Ola et al. 2012).

Deposition of Starch

Presence of starch granules in parenchyma cells of stem under high salinity was frequently noted. This might be an effect of carbohydrate accumulation in relation to inhibition of carbohydrate translocation by pollutants and has an important role in the osmotic adjustment in the salt-tolerant plants (Ashraf and Tufail 1995; Murakeozy et al. 2003; Rennenberg et al. 1996).

Crystal Formation

Formation of different types of crystal structures, such as druses, prismatic and crystal sand, were noted in stem cortical cells of *Phaseolus vulgaris* under arsenic toxicity (Talukdar 2013a). Stem crystals were also observed in *Phaseolus mungo* under metal toxicity (Fig. 1). In Indian mustard

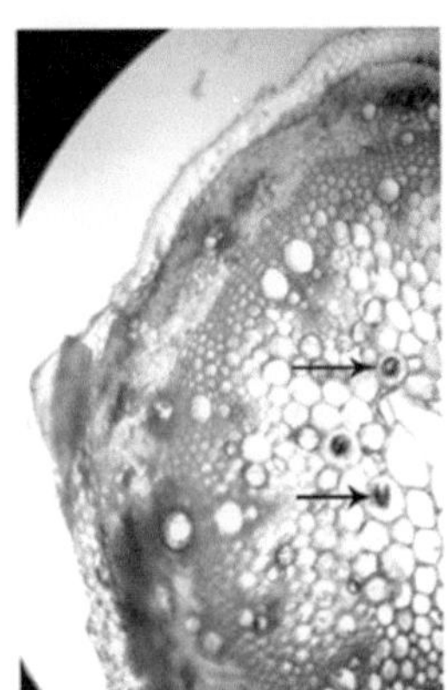

Fig. 1 Light micrograph of transverse section of stem in *Phaseolus mungo* under metal stress showing prismatic crystals (*arrow*). Bar = 50 μm

depositions of electron-dense particles along the walls of vascular bundles of stem were noted under high zinc and cadmium concentration (Maruthi Sridhar et al. 2005). Heavy metal-induced structural changes in the plant parts were also reported in mung bean (Singh et al. 2007), pea (Rodríguez-Serrano et al. 2009) and in radish (Vitória et al. 2006).

Root Architecture Modifications to Stress

Roots are considered a sensor organ that detects changes of water availability in soil and influences the resistance to different stress like drought, salinity and heat at the whole plant level. In offering adaptation, plants showing different root architecture and various combinations of root morphological and anatomical responses have different adaptive strategies. Intensity and occurrence of various traits in different species and even in different cultivars within same species are quite variable (Peña-Valdivia et al. 2010; Shao et al. 2008). The following are the major root architectural modifications reported in angiosperms in response to different abiotic and biotic stresses.

Ramification of Root

Production of ramified root system in response to drought was reported in many crop plants like rice, wheat, soybean, maize, sunflower and others (Jaleel et al. 2009b; Sacks et al. 1997; Tahir et al. 2002). During ramification lateral roots are formed due to auxin accumulation near the root pericycle indicating reorientation of growth (Pasternak et al. 2005). Increased root area is also observed in *Arabidopsis thaliana* (Cruciferae) when the plant is exposed to abiotic stress (Olmos et al. 2006). Such prolific root system increases the water uptake and maintains requisite osmotic pressure through higher proline levels under drought (Djibril et al. 2005; Jaleel et al. 2009b).

Changes in Root Length and Diameter

Elongation of root and decrease in root diameter in response to drought in many wild species were reported. Small diameter root maximises absorptive surfaces, thus increasing rates of water and nutrient uptake (Eissenstat 1992; Reader et al. 1993).

Wall Thickening

To control water loss, thickening of outer cell wall of rhizodermis and the presence of suberised layers of cell at the periphery of root are important root adaptations in drought. In the root, increase in cell wall thickness was also noted under high-metal toxicity in *Vicia faba* L. (Probst et al. 2009). By this ultrastructural modification, plants could limit the metal absorption. In strongly damaged cells, high amounts of electron-dense dark particles of metals were observed within the cell walls and cells as in *Vicia faba* L. and in *Sesbania* root cells (Sahi and Sharma 2005). This accumulation indicates probable entry of metals into the root cells after weakening the primary level of plant defence.

Changes in Root Cortex

The number of cortical layers is reduced under drought conditions for shortening the way between the soil and the stele favouring quick radial water transport (Fahn 1964). Formation of cortical lacunae by rupturing of cortical cells to regulate the inverse flux of water from stele to soil (North and Nobel 1995; Peña-Valdivia et al. 2010; Robards et al. 1979) was noted in *Agave deserti, Opuntia ficus-indica* and in many other plants during drought. Formation of cortical lacunae in turn weakens the root mechanical strength, which is counterbalanced by the formation of additional sclerenchyma, increased vascular system and sometimes by lignifications of pith (Mostajeran and Rahimi-Eichi 2008).

Wall Thickenings

Thickening of endodermal cell wall and formation of additional suberised layer around the stele were noted to prevent desiccation of tissues inside the stele (North and Nobel 1992).

Root Hair Density

Loss of root hairs was noted in model legume *Medicago sativa* under cadmium and mercury toxicity (Villasante et al. 2005). Even complete absence of root hairs were observed in common beans under arsenic toxicity (Talukdar 2013a).

Wall Depositions in Root

Electron-dense depositions were observed all along the cell walls of roots in acknowledged metal accumulator *Brassica juncea* under metal stress (Maruthi Sridhar et al. 2005).

Flower Architecture Modifications to Stress

A flower is a modified stem tip with compressed internodes, bearing structures that are highly modified leaves. The biological function of the flower is the formation of fruits and seeds, hence ensuring maximal reproductive success. The reproductive phase in flowering plants is often highly sensitive to environmental stress than the vegetative phase, adversely affecting the reproductive success of the species leading to severe decrease in crop yield. Sexual reproduction in flowering plants is dependent on the correct delivery of pollen grains to conspecific stigmata. Female reproductive function, that is, the ability of a flowering plant to set viable seed, depends on both stigma receptivity and ovule viability (Dumas et al. 1984). The angiosperm stigma is an efficient structure with both morphological and physiological adaptations that enable pollen capture, hydration and germination. The stigma surface plays a vital part in regulating compatibility relationships within species. The stigma is the first pollen landing site where the recognition events lead to the acceptance of compatible pollen or the rejection of incompatible pollen (Cynthla et al. 1990) and is mostly made up of elongate papilla cells, each of which has a single, finger-like projection that is receptive to pollen binding (Heslop-Harrison 1981). Pollen-stigma adhesion is highly species specific and environment sensitive. Sometimes, environmental sensitivity of reproductive organs varied considerably in different developmental stages of flower. For example, in *Petunia*, the stigmas and styles of young buds were more sensitive to high-temperature stress than those of fully opened flowers. In contrast, the ovary of fully opened flowers appeared to be more sensitive to high-temperature stress than those of young buds and was evidenced by increased SOD (superoxide dismutase) activity.

Understanding how different plants cope with stress during their reproductive phase is critical to managing the future of agricultural productivity. Interestingly, stress tolerance in vegetative and reproductive tissues is not always correlated (Salem et al. 2007). For instance, while the vegetative surfaces of most plants are heavily protected by cuticle from dehydration and pathogen attack, the stigma cannot be protected by a thick cuticle or wax since it must capture and hydrate pollen and allow eventual penetration of the pollen tube. Very few literatures and insights are presently available to comment precisely about the micro-manifestation of stressors on the reproductive organs and their putative role in combating stress. In the given extent of this literature, the most prevalent architectural modifications have been focused.

Changes in Floral Area

Head diameter is decreased in sunflower under drought (Jaleel et al. 2009b). Size of flower in bud condition and in fully opened condition decreased as compared to control under temperature stress in *Petunia* (Wang 2006).

Blossom Drop

Blossom drop is the loss of flowers and known to occur in tomatoes, peppers, snap beans and many other fruiting vegetables under various abiotic and biotic stresses. In tomatoes, blossom drop is usually preceded by the yellowing of the pedicle (Ozores-Hampton and McAvoy 2010). Interestingly, Peet et al. (1998) documented that heat stress increases flower number in tomato under experimental condition.

Stigma Yield

Fresh weight of flower stigma (stigma yield) initially increases with salinity, but in extreme saline condition (>70 mMol), stigma yield decreases rapidly in saffron (*Crocus sativus*) (Torbaghan et al. 2011).

Changes in Stigma Position

High temperature affects the stigma position of flower affecting pollen capturing capacity (Kinet and Peet 1997).

Pollen Viability

Anther-specific development indicates that the male reproductive process is more sensitive to certain environmental stresses than the female reproductive development process (Sakata and Higashitani 2008). For instance, exposure to high temperature resulted in male sterility in barley because of non-viable pollen and failure of anther dehiscence, whereas gynoecium function remained unaffected (Peet et al. 1998). The most common and frequently reported reproductive anomaly under stress is variable percentage of pollen sterility leading to decreased fruit set and is observed in many plants. In tomato temperature stress adversely reduces pollen viability and fruit setting (Pressman et al. 2002). Causes and manifestations of pollen sterility are highly variable among different plant types. In rice high temperature decreases the swelling ability of the pollen grains, resulting in poor theca dehiscence, because this swelling of pollen grains in the locules is the driving force for anther dehiscence (Matsui et al. 1999, 2000; Shah et al. 2011). In contrast, heat-tolerant rice cultivars showed well-developed cavities in anthers and thick locule walls which enable easy rupture of the septa in response to swelling of pollen, resulting in better anther dehiscence and pollen shed. In most of the heat-sensitive cultivars, endothecium layer of anther becomes disorganised, leading to failure of anther dehiscence (Sakata and Higashitani 2008). Sometimes normal round-shaped rice pollen grains possess abnormal tapetum in high temperature, causing pollen adhesion to the stigma. Often premature or delayed degradation of tapetum cells resulted under elevated stress (Sakata and Higashitani 2008). Tapetum degradation was also noted in many plants like sunflower, maize, petunia, wheat and barley (Balk and Leaver 2001; Conley and Hanson 1994). Thus subsequent pollen germination was negatively affected leading to sterility (Endo et al. 2009). In cowpea, male sterility occurs with high temperatures during floral development, due to degeneration of tapetum and lack of endothecial development (Ahmed et al. 1992). The size of the basal dehiscence could be a useful morphological marker of pollen viability. Anthers with large basal dehiscence are more stress tolerant compared to anthers with small basal dehiscence because anthers with large basal pore facilitate pollen release. Moreover, it can be assumed that rice cultivars with large anthers are tolerant to temperature stress because they have a large number of pollen grains per anther, which compensates for the reduction in the number of pollen grains that germinate under high temperature (Matsui and Omasa 2002; Shah et al. 2011). In barley abnormal short anther completely lacking pollen grains were developed under high-temperature stress (Sakata and Higashitani 2008). Abnormal tapetal enlargement was observed in anthers under chill stress in sugar beet and rice (Matsuhira et al. 2007; Nishiyama 1976). Reduced pollen tube growth and seed set were noted under heat and cold stress in *Arabidopsis* and in chickpea (Clarke and Siddique 2004; Zinn et al. 2010). In chickpea high temperature also reduced pollen production per flower, % pollen germination and stigma receptivity, which ultimately lead to reduced pod and seed setting (Devasirvatham et al. 2012). Data indicate that pollen grains were more sensitive to high temperature than the stigma in chickpea. All these structural attributes may play important roles in conferring resistance to heat and other abiotic and biotic stresses. There is high legitimate demand of devoting more attention to develop heat-tolerant cultivar of rice and other cereals using all these structural attributes in the present perspective, since global circulation models predict that increasing greenhouse gases will elevate the average global temperature between 1.1 and 6.4 °C during the twenty-first century (Lobell and Field 2007).

Stamen Modification

Carpeloid stamen and petaloid stamens are male reproductive abnormalities where male reproductive organ (stamen) converted to female reproductive organ (carpel) or to petal. Such changes were reported in wheat, cotton and *Brassica* where early steps of flower formation are impaired (Carlsson et al. 2008; Sakata and Higashitani 2008; Zubko et al. 2001).

Fusion of Anther Lobe

The anthers of healthy hermaphrodite mango flower were bilobed with large number of turgid pollen grains, whereas malformed flowers showed fused lobed anthers with scanty deformed pollen grains. Furthermore, the stigma of healthy flowers exhibited a broad landing pad as compared to malformed stigma which showed hooked and pointed tips with poor stigmatic receptivity. All these impaired morphology of male and female reproductive organs are mainly responsible for restricting the pollen germination and pollen tube growth leading to failure of sexual reproduction (Rani et al. 2013). Such impairment is probably due to augmented level of endogenous ethylene in response to stress.

Fruit Architecture Modifications to Stress

Fruits and seeds are the characteristic reproductive body of plants that carries the miniature undeveloped plants (embryo) and are responsible for the overwhelming evolutionary success of the flowering plants. Fruit is a significant part of the human diet, supplying fibre, minerals, vitamins and other chemopreventive agents such as antioxidants. Hence, in addition to research programmes directed to understanding and improving the qualities of fruit, significant efforts have also been made to protect the fruit in hazardous environment sustaining human health. The ultimate reflection of any type of environmental stress injury is on the crop productivity of plants resulted from reduced fruit and seed yield in terms of quantity and quality. The understanding of associated structural abnormalities is therefore very crucial for manipulating such traits through crop management, breeding or biotechnology in the perspective of stress tolerance (Giovannoni 2001). Data indicates that ethylene plays a major role in mediating stress responses of fruits.

Seed Yield

Achene yield of sunflower (Reddy et al. 2004) and seed yield of common bean, green gram and maize (Webber et al. 2006; Monneveux et al. 2006) are very much affected under drought.

Size of Fruits

Size and weight of the fruit are very badly affected under stress. For instance, water stress decreases fruit size and fruit weight in apple, stone fruit, peach and kiwifruit, resulting from reduced growth rate (Miller et al. 1998; Naor et al. 2008).

Fruit Colour Modification

Colour modification of fruits is another frequently observed phenomenon of plants in response to stress. Water stress can modify fruit colour. For example, severe water stress decreases red skin pigmentation in peach (Lopez et al. 2011). On the contrary, enhancement of red colour is noted in apple with a lowered water status. This may have been due to the advanced accumulation of sugars measured in this fruit, as sucrose plays a major role in anthocyanin development (Kilili et al. 1996; Lopez et al. 2011).

Softening of Fruits

Softening of fruits is an easily detectable change under stress that resulted from increased activity of cell wall-degrading enzymes (Kader 2003).

Black Deposition in Fruits

Considerable blackening of trichome tip and cross wall, pappus tip and margin, and sometimes deposition of electron-dense opaque particles within endospermic cells were observed in fruits of *Wedelia chinensis* (Asteraceae) under arsenic stress (Talukdar 2013b) (Figs. 2 and 3). In exceeded metal concentration due to over deposition of dark materials, the epicarpic cell wall and phytomelanin layer became more thickened. Thickening of the outermost layer of fruit wall might be an avoidance mechanism to limit the entry of toxic metal into ultimate sink organ.

Fruit Wall Features

The exocarp of grape (*Vitis vinifera*) berries developed well-ordered indentations, and mesocarp was collapsed under abiotic stress (Bondada and Keller 2012).

Fig. 2 Light micrograph of fruit trichome of *Wedelia chinensis* Merrill under arsenic stress showing black deposition (*arrow*). Bar = 100 μm

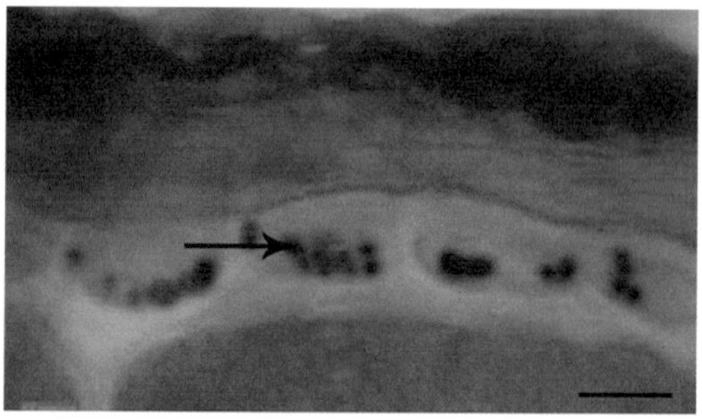

Fig. 3 Light micrograph of endospermic cells in fruit of *Wedelia chinensis* Merrill under arsenic stress showing black deposition (*arrow*). Bar = 25 μm

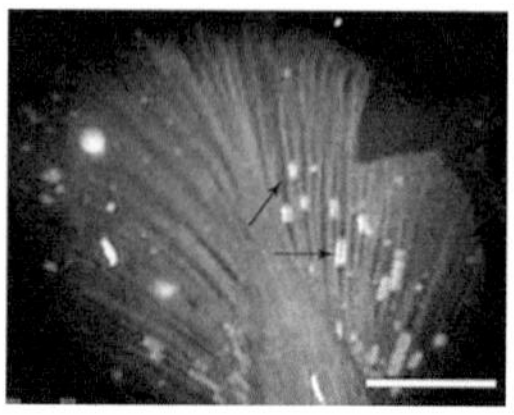

Fig. 4 Light micrograph of fruit pappus of *Wedelia chinensis* Merrill under arsenic stress showing crystals (*arrow*). Bar = 100 μm

Formation of Crystal

Crystallisation in mesocarpic parenchymatous cells and pappus cells of fruit in *Wedelia chinensis* was noted under heavy metal stress (Talukdar 2013b) (Fig. 4).

How Symptoms Symptomise the Black Box?

Nature and distribution of symptoms in different plant organs provides basic information to diagnose the class of stress factor. Among abiotic and biotic stresses, the first one affects the whole plant and also other plants growing nearby (Vollenweider et al. 2003). So multi-organ deformities will indicate abiotic stress origin. In any specific plant organ, evenly distributed symptom gradient indicates abiotic stress origin compared to biotic stress with scattered symptoms without any gradient. For instance, under drought and heavy metal stress, symptom development generally proceeds acropetally. Abiotic and biotic stress factors may be further classified based on detailed analysis of symptoms.

A Case Study

Detection of stress factor based on leaf symptoms:

1. Evenly distributed symptom gradient – abiotic stress factor
2. Scattered symptoms without any gradient – biotic stress factor

Abiotic

1. Symptoms in sun-exposed leaf, intercostals – abiotic airborne stress factor
2. Symptoms in shaded leaf, connected to veins – abiotic soilborne stress factor
3. Leaf rolling – drought
4. Necrotic spot along veins – heavy metals
5. Diffuse mottling – ozone
6. Intercostal stippling – ozone, heavy metals
7. Edge and tip necrosis – fluoro or sulphur emissions, salt, drought/heat, heavy metals, frost, nutrient deficiency
8. Even discolouration – natural autumn senescence, drought, frost, ozone, nutrient deficiency

Biotic

1. Patchy discolouration on leaf, no biotic remnants – bacteria
2. Patchy discolouration on leaf with leaf curling, no biotic remnants – virus
3. Patchy discolouration on leaf with biotic remnants like hyphae, fruit bodies

Fungi

4. Feeding injury – insects
5. Feeding injury on leaf along vein, stippling – mites

Interestingly, a particular stress factor may not always induce the same kind of modification to all exposed plant types. In other words, different plant species respond differently to a specific stress reflecting their underlying differences in adaptive strategies. Moreover, in natural habitat multi-stress factors may often be present in a particular geographic area and subsequently induce different combinations of a variety of symptoms in affected plants. So sample site history also plays a crucial role to restrict the stress factor. Thus with precise diagnostic skill and advent of technology, more insight into the modern-evolved constrained environment is indeed urgently needed to achieve a food-secured world.

References

Ager FJ, Ynsa MD, Dominguez-Solis JR, Lopez-Martin MC, Gotor C, Romero LC (2003) Nuclear microprobe analysis of Arabidopsis thaliana leaves. Nucl Inst Methods Phys Res B 210:401–406

Ahmed FE, Hall AE, DeMason DA (1992) Heat injury during floral development in cowpea (*Vigna unguiculata*). Am J Bot 79:784–791

Akhtar J, Gorham J, Qureshi RH (1998) Does tolerance of wheat to salinity and hypoxia correlate with root dehydrogenase activities or aerenchyma formation? Plant Soil 201:275–284

Andrade LR, Farina M, Amado Filho GM (2004) Effects of copper on *Enteromorpha flexuosa* (Chlorophyta) in vitro. Ecotoxicol Environ Saf 58:117–125

André O, Vollenweider P, Günthardt-George MS (2006) Foliage response to heavy metal contamination in Sycamore Maple (*Acer pseudoplatanus* L.). For Snow Landsc Res 80(3):275–288

Arduini I, Godbold DL, Onnis A (1995) Influence of copper on root growth and morphology of *Pinus pinea* L. and *Pinus pinaster* Ait. seedlings. Tree Physiol 15:411–415

Arena C, Vitale L, De Santo VA (2008) Paraheliotropism in Robinia pseudoacacia L.: an efficient strategy to optimise photosynthetic performance under natural environmental conditions. Plant Biol 10:194–201

Aronne G, De Micco V (2001) Seasonal dimorphism in the Mediterranean Cistus incanus L. subsp. incanus. Ann Bot 87(6):789–794

Ashraf M, Tufail M (1995) Variation in salinity tolerance in sunflower (*Heliunthus annuus* L.). J Agron Crop Sci 174:351–362

Azmat R, Haider S, Nasreen H, Aziz F, Riaz M (2009) A viable alternative mechanism in adapting the plants to heavy metal environment. Pak J Bot 416:2729–2738

Balk J, Leaver CJ (2001) The PETI-CMS mitochondrial mutation in sunflower is associated with premature programmed cell death and cytochrome *c* release. Plant Cell 13:1803–1818

Barcelo J, Vazquez MD, Poschenrieder C (1988) Cadmium induced structural and ultrastructural changes in the vascular system of bush bean stems. Bot Acta 101:254–261

Bondada BR, Keller M (2012) Not all shrivels are created equal – morpho-anatomical AND compositional characteristics differ AMONG different shrivel types that develop during ripening OF grape (*Vitis VINIFERA* L.) berries. Am J Plant Sci 3(7):879–898

Bussotti F, Bottaci A, Bartolesi A, Grossoni P, Tani C (1995) Morpho-anatomical alterations in leaves collected from beech trees (*Fagus sylvatica* L.) in conditions of natural water stress. Environ Exp Bot 35(2):201–213

Carlquist S (1975) Ecological strategies of xylem evolution. University of California Press, Berkeley

Carlsson J, Leino M, Sohlberg J, Sundström JF, Glimelius K (2008) Mitochondrial regulation of flower development. Mitochondrion 8:74–86

Çavuşoğlu K, Kiliç S, Kabar K (2007) Some morphological and anatomical observations during alleviation of salinity (NaCl) stress on seed germination and seedling growth of barley by polyamines. Acta Physiol Plant 29:551–557

Çavuşoğlu K, Kiliç S, Kabar K (2008) Effects of some plant growth regulators on leaf anatomy of radish seedlings grown under saline conditions. J Appl Biol Sci 2:47–50

Clarke HJ, Siddique KHM (2004) Response of chickpea genotypes to low temperature stress during reproductive development. Field Crop Res 90(2–3):323–334

Conley CA, Hanson MR (1994) Tissue specific protein expression in plant mitochondria. Plant Cell 6:85–91

Cynthla JMK, Bong YY, Seabrook JEA (1990) Stigma of *Solanum tuberosum* cv shepody: morphology, ultrastructure, and secretion. Am J Bot 77:1111–1124

De Micco V, Aronne G (2007) Anatomical features, monomer lignin composition and accumulation of phenolics in one-year-old branches of the Mediterranean *Cistus ladanifer* L. Bot J Linn Soc 155:361–371

De Micco V, Aronne G (2012) Morpho-anatomical traits for plant adaptation to drought. In: Aroca R (ed) Plant responses to drought stress. Springer, Berlin/Heidelberg

Devasirvatham V, Gaur PM, Mallikarjuna N, Tokachichu RN, Trethowan RM, Tan DKY (2012) Effect of high temperature on the reproductive development of chickpea genotypes under controlled environments. Funct Plant Biol 39(12):1009–1018

Dixon M, Le Thiec D, Garrec JP (1997) An investigation into the effects of ozone and drought, applied singly and in combination on the quantity and quality of the epicuticular wax of Norway spruce. Plant Physiol Biochem 35(6):447–454

Djibril S, Mohamed OK, Diaga D, Diégane D, Abaye BF, Maurice S, Alain B (2005) Growth and development of date palm (*Phoenix dactylifera* L.) seedlings under drought and salinity stresses. Afr J Biotechnol 4:968–972

Dumas C, Knox RB, Gaude T (1984) Pollen-pistil recognition: new concepts from electron microscopy and cytochemistry. Int Rev Cytol 90:239–272

Eissenstat DM (1992) Costs and benefits of constructing roots of small diameter. J Plant Nutr 15:763–782

Endo M, Tsuchiya T, Hamada K, Kawamura S, Yano K, Ohshima M, Higashitani A, Watanabe M, Kawagishi-Kobayashi M (2009) High temperatures cause male sterility in rice plants with transcriptional alterations during pollen development. Plant Cell Physiol 50:1911–1922

Evans LS, Gmur NF, Kelsch JJ (1977) Leaf surface and histological perturbations of leaves of *Phaseolus vulgaris* and *Helianthus annuus* after exposure to simulated acid rain. Am J Bot 64:903–913

Fahn A (1964) Some anatomical adaptations in desert plants. Phytomorphology 14:93–102

Field TS, Zwieniecki MA, Donoghue MJ, Holbrook NM (1998) Stomatal plugs of Drimys winteri (Winteraceae) protect leaves from mist but not drought. Proc Natl Acad Sci U S A 95:14256–14259

Fornasiero RB (2001) Phytotoxic effects of fluorides. Plant Sci 161(5):979–985

Gan Y, Zhou L, Shen ZJ, Shen ZX, Zhang YQ, Wang GX (2010) Stomatal clustering, a new marker for environmental perception and adaptation in terrestrial plants. Bot Stud 51:325–336

Gielwanowska I, Szczuka E, Bednara J, Górecki R (2005) Anatomical features and ultrastructure of *Deschampsia antarctica* (Poaceae) leaves from different growing habitats. Ann Bot 96(6):1109–1119

Giovannoni J (2001) Molecular biology of fruit maturation and ripening. Annu Rev Plant Physiol Plant Mol Biol 52:725–749

Gomes MP, Marques TCLLDM, Nogueira MDG et al (2011) Ecophysiological and anatomical changes due to uptake and accumulation of heavy metal in Brachiaria decumbens. Sci Agric 68(5):566–573

Gostin IN (2009) Air pollution effects on the leaf structure of some *Fabaceae* species. Not Bot Horti Agrobot Cluj-Napoca 37(2):57–63

Hameed M, Ashraf M, Nargis N (2009) Anatomical adaptations to salinity in cogon grass [*Imperata cylindrica* (L.) Raeuschel] from the Salt Range, Pakistan. Plant Soil 322:229–238

Han FX, Maruthi Sridhar BB, Monts DL, Su Y (2004) Phytoavailability and toxicity of trivalent and hexava-

lent chromium to *Brassica juncea* L. Czern. New Phytol 162:489–499
Heslop-Harrison Y (1981) Stigma characteristics and angiosperm I, taxonomy. Nor J Bot 1:401–420
Hollenbach B, Schreiber L, Hartung W, Dietz K-J (1997) Cadmium leads to stimulated expression of the lipid transfer protein genes in barley: implications for the involvement of lipid transfer proteins in wax assembly. Planta 203:9–19
Hoover WS (1986) Stomata and stomatal clusters in *Begonia*: ecological response in two Mexican species. Biotropica 18:16–21
Hu Y, Fromm J, Schmidhalter U (2005) Effect of salinity on tissue architecture in expanding wheat leaves. Planta 220:838–848
Huang J, Redmann RE (1995) Responses of growth, morphology, and anatomy to salinity and calcium supply in cultivated and wild barley. Can J Bot 73:1859–1866
Hwang YH, Chen SC (1995) Anatomical responses in *Kandelia candel* (L.) Druce seedlings growing in the presence of different concentrations of NaCl. Bot Bull Acad 36:181–188
Jackson RD (1986) Remote sensing of biotic and abiotic plant stress. Annu Rev Phytopathol 24:265–287
Jaleel CA, Gopi R, Azooz MM, Panneerselvam R (2009a) Leaf anatomical modifications in *Catharanthus roseus* as affected by plant growth promoters and retardants. Glob J Mol Sci 4(1):01–05
Jaleel CA, Manivannan P, Wahid A, Farooq M, Jasim Al-Juburi H, Somasundaram R, Panneerselvam R (2009b) Drought stress in plants: a review on morphological characteristics and pigments composition. Int J Agric Biol 11:100–105
Jenks MA, Andersen L, Teusink RS, Williams MH (2001) Leaf cuticular waxes of potted rose cultivars as affected by plant development, drought and paclobutrazol treatments. Physiol Plant 112:62–70
Kader AA (2003) Physiology of CA treated produce. Acta Hort 600:349–354
Karimi E, Abdolzadeh A, Sadeghipour HR (2009) Increasing salt tolerance in Olive. Olea europaea L. plants by supplemental potassium nutrition involves changes in ion accumulation and anatomical attributes. Int J Plant Prod 3(4):49–60
Kerstiens G (1996) Cuticular water permeability and its physiological significance. J Exp Bot 47:1813–1832
Kerstiens G (2006) Water transport in plant cuticles: an update. J Exp Bot 57:2493–2499
Khudsar T, Uzzafar M, Iqbal M (2001) Cadmium induced changes in leaf epidermis, photosynthetic rate and pigment concentrations in *Cajanus cajan*. Biol Plant 44(1):59–64
Kilili AW, Behboudian MH, Mills TM (1996) Postharvest performance of 'Braeburn' apples in relation to withholding of irrigation at different stages of the growing season. J Hortic Sci 71(5):693–701
Kinet JM, Peet MM (1997) Tomato. In: Wien HC (ed) The physiology of vegetable crops. Commonwealth Agricultural Bureau (CAB) International, Wallingford
Larkin JC, Marks MD, Nadeau J, Sack F (1997) Epidermal cell fate and patterning in leaves. Plant Cell 9:1109–1111
Lens F, Sperry JS, Christman MA, Choat B, Rabaey D, Jansen S (2011) Testing hypotheses that link wood anatomy to cavitation resistance and hydraulic conductivity in the genus Acer. New Phytol 190:709–723
Liu HY, Liao BH, Lu SQ (2004a) Toxicity of surfactant, acid rain and Cd2+ combined pollution to the nucleus of *Vicia faba* root tip cells. Chin J Appl Ecol 15(3):493–496
Liu HY, Liao BH, Zhou PH, Yu PZ (2004b) Toxicity of linear alkylbenzene sulfonate and alkylethoxylate to aquatic plants. Bull Environ Contam Toxicol 72(4):866–872
Lobell DB, Field CB (2007) Global scale climate-crop yield relationships and the impacts of recent warming. Environ Res Lett 2:004000
Lopez G, Larrigaudière C, Girona J, Behboudian MH, Marsal J (2011) Fruit thinning in 'conference' pear grown under deficit irrigation: implications for fruit quality at harvest and after cold storage. Sci Hortic 129(1):64–70
Martin D, Vollenweider P, Gunthardt-Goerg MS (2006) Bioindication of heavy metal contamination in vegetable gardens. For Snow Landsc Res 80(2):169–180
Martinez JP, Silva H, Ledent JF, Pinto M (2007) Effect of drought stress on the osmotic adjustment, cell wall elasticity and cell volume of six cultivars of common beans (*Phaseolus vulgaris* L.). Eur J Agron 26:30–38
Maruthi Sridhar BB, Diehl SV, Han FX, Monts DL, Su Y (2005) Anatomical changes due to uptake and accumulation of Zn and Cd in Indian mustard (*Brassica juncea*). Environ Exp Bot 54:131–141
Matsuhira H, Shinada H, Yui-Kurino R, Hamato N, Umeda M, Mikami T, Kubo T (2007) An anther-specific lipid transfer protein gene in sugar beet: its expression is strongly reduced in male-sterile plants with Owen cytoplasm. Physiol Plant 129:407–414
Matsui T, Omasa K (2002) Rice (Oryza sativa L.) cultivars tolerant to high temperature at flowering: anther characteristics. Ann Bot 89:683–687
Matsui T, Omasa K, Horie T (1999) Mechanism of anther dehiscence in rice (Oryza sativa L.). Ann Bot 84:501–506
Matsui T, Omasa K, Horie T (2000) High temperature at flowering inhibits swelling of pollen grains, a driving force for thecae dehiscence in rice (Oryza sativa L.). Plant Prod Sci 3:430–434
Melo HC, Castro EM, Soares AM, Melo LA, Alves JD (2007) Anatomical and physiological alterations in *Setaria anceps* stapf ex Massey and *Paspalum paniculatum* under water deficit conditions. Hoehnea 34:145–153 (in Portuguese with abstract in English)
Miller SA, Smith GS, Boldingh HL, Johansson A (1998) Effects of water stress on fruit quality attributes of kiwifruit. Ann Bot 81:73–81
Monneveux P, Sánchez C, Beck D, Edmeades GO (2006) Drought tolerance improvement in tropical maize

source populations: evidence of progress. Crop Sci 46:180–191

Mostajeran A, Rahimi-Eichi V (2008) Drought stress effects on root anatomical characteristics of rice cultivars (Oryza sativa L.). Pak J Biol Sci 11:2173–2183

Murakeozy EP, Nagy Z, Duhaze C, Bouchereau A, Tuba Z (2003) Seasonal changes in the levels of compatible osmolytes in three halophytic species of inland saline vegetation in Hungary. J Plant Physiol 160:395–401

Nam NH, Chauhan YS, Johansen C (2001) Effect of timing of drought stress on growth and grain yield of extra-short-duration pigeonpea lines. J Agric Sci 136:179–189

Naor A, Naschitz S, Peres M, Gal Y (2008) Responses of apple fruit size to tree water status and crop load. Tree Physiol 28:1255–1261

Nawaz T, Hameed M, Ashraf M, Al- Qurainy F, Ahmad MSA, Younis A, Hayat M (2011) Ecological significance of diversity in leaf tissue architecture of some species/cultivars of the genus *Rosa* L. Pak J Bot 432:873–883

Nishiyama I (1976) Male sterility caused by cooling treatment at the young microspore stage in rice plants. XII. Classification of tapetal hypertrophy on the basis of ultrastructure. Proc Crop Sci Soc Jpn 45:254–262

Noman A, Hameed M, Ali Q, Aqeel M (2012) Foliar tissue architectural diversity among three species of genus *Hibiscus* for better adaptability under industrial environment. Int J Environ Sci 2(4):2212–2222

North GB, Nobel PS (1992) Drought-induced changes in hydraulic conductivity and structure in roots of Ferocactus acanthodes and Opuntia ficus-indica. New Phytol 120:9–19

North GB, Nobel PS (1995) Hydraulic conductivity of concentric root tissues of Agave deserti Engelm. Under wet and drying conditions. New Phytol 130:47–57

Ola H, Elbar A, Reham FE, Eisa SS, Habib SA (2012) Morpho-anatomical changes in salt stressed Kallar grass (*Leptochloa fusca* L. Kunth). Res J Agric Biol Sci 8(2):158–166

Olmos E, Kiddle G, Pellny TK, Kumar S, Foyer CH (2006) Modulation of plant morphology, root architecture, and cell structure by low vitamin C in Arabidopsis thaliana. J Exp Bot 57(8):1645–1655

Ozores-Hampton M, McAvoy G (2010) What causes blossom drop in tomatoes? Tomato Mag 14(4):4–5

Pal A, Kulshreshtha K, Ahmed KJ, Behl HM (2002) Do leaf surface characters play a role in plant resistance to auto-exhaust pollution? Flora 197(1):47–55

Pasternak T, Rudas V, Potters G, Jansen MAK (2005) Morphogenic effects of abiotic stress: reorientation of growth in *Arabidopsis thaliana* seedlings. Environ Exp Bot 53(3):299–314

Peet MM, Sato S, Gardner RG (1998) Comparing heat stress effects on male-fertile and male-sterile tomatoes. Plant Cell Environ 21:225–231

Peña-Valdivia CB, Sánchez-Urdaneta AB, Meza Rangel J, Juárez Muñoz J, García-Nava R, Celis Velázquez R (2010) Anatomical root variations in response to water deficit: wild and domesticated common bean (Phaseolus vulgaris L.). Biol Res 43:417–427

Prasad TK (1996) Mechanism of chilling induced oxidative stress injury and tolerance in developing maize seedlings: changes in antioxidant system, oxidation of proteins and lipids, and protease activities. Plant J 10:1017–1026

Pressman E, Peet MM, Pharr DM (2002) The effect of heat stress on tomato pollen characteristics is associated with changes in carbohydrate concentration in the developing anthers. Ann Bot 90:631–636

Probst A, Liu H, Fanjul M, Liao B, Hollande E (2009) Response of *Vicia faba* L. to metal toxicity on mine tailing substrate: Geochemical and morphological changes in leaf and root. Environ Exp Bot 66:297–308

Rani V, Ansari MW, Shukla A et al (2013) Fused lobed anther and hooked stigma affect pollination, fertilization and fruit set in mango: A scanning electron microscopy study. Plant Signal Behav 8(3)

Reader RJ, Jalili A, Grime JP, Spencer RE, Matthews NN (1993) A comparative-study of plasticity in seedling rooting depth in drying soil. J Ecol 81:543–550

Reddy AR, Chaitanya KV, Vivekanandan M (2004) Drought induced responses of photosynthesis and antioxidant metabolism in higher plants. J Plant Physiol 161:1189–1202

Reinhardt DH, Rost TL (1995) Developmental changes of cotton root primary tissues induced by salinity. Int J Plant Sci 156:505–513

Reinoso H, Sosa L, Ramírez L (2004) Salt-induced changes in the vegetative anatomy of *Prosopis strombulifera* (*Leguminosae*). Can J Bot 82:618–628

Rennenberg H, Herschbach C, Polle A (1996) Consequences of air pollution on shoot-root interactions. J Plant Physiol 148:269–301

Riederer M, Müller C (eds) (2005) Biology of the plant cuticle. Blackwell, Oxford

Riederer M, Schreiber L (2001) Effects of environmental factors on the water permeability of plant cuticles. J Exp Bot 52:2023–2033

Robards AWV, Clarkson DT, Sanderson J (1979) Structure and permeability of the epidermal/hypodermal layers of the sand sedge (Carex arenaria L.). Protoplasma 101:331–347

Rodríguez-Serrano M, Romero-Puertas MC, Pazmiño DM, Testillano PS, Risueño MC, del Río LA, Sandalio LM (2009) Cellular responses of pea plants to cadmium toxicity: cross talk between reactive oxygen species, nitric oxide, and calcium. Plant Physiol 150:229–243

Sacks MM, Silk WK, Burman P (1997) Effect of water stress on cortical cell division rates within the apical meristem of primary roots of maize. Plant Physiol 114:519–527

Sahi SV, Sharma NC (2005) Phytoremediation of lead. In: Shtangeeva I (ed) Trace and ultratrace elements in plants and soils, series advances in ecological researches. Witpress, Southampton/Boston

Sakata T, Higashitani A (2008) Male sterility accompanied with abnormal anther development in plants – genes and environmental stresses with special reference to high temperature injury. Int J Plant Dev Biol 2:42–51

Salem MA, Kakani VG, Koti S, Reddy KR (2007) Pollen-based screening of soybean genotypes for high temperatures. Crop Sci 47:219–231

Sant'Anna-Santos BF, da Silva LC, Azevedo AA, Aguiar R (2006) Effects of simulated acid rain on leaf anatomy and micromorphology of *Genipa americana* L. (Rubiaceae). Braz Arch Biol Technol 49(2):313–321

Sarret G, Vangronsveld J, Manceau A, Musso M, Haen J, Menthonnex JJ, Hazemann JL (2001) Accumulation form of Zn and Pb in *Phaseolus vulgaris* in the presence and absence of EDTA. Environ Sci Technol 35:2854–2859

Schlüter U, Muschak M, Berger D, Altmann T (2003) Photosynthetic performance of an Arabidopsis mutant with elevated stomatal density (*sdd1-1*) under different light regimes. J Exp Bot 54:867–874

Shah F, Huang J, Cui K, Nei L, Shah T, Chen C, Wang K (2011) Impact of high-temperature stress on rice plant and its traits related to tolerance. J Agric Sci 149:545–556

Shao HB, Chu LY, Jaleel CA, Zhao CX (2008) Water-deficit stress-induced anatomical changes in higher plants. C R Biol 33:215–225

Shepherd T, Griffiths DW (2006) The effects of stress on plant cuticular waxes. New Phytol 171:469–499

Singh HP, Batish DR, Kohli RK, Arora K (2007) Arsenic-induced root growth inhibition in mung bean (*Phaseolus aureus* Roxb.) is due to oxidative stress resulting from enhanced lipid peroxidation. Plant Growth Regul 53:65–73

Srivastava LM (2001) Plant growth and development. Academic, San Diego/London

Strogonov BP (1962) Physiological basis of salt tolerance of plants. Israel program for Scientific Translations, Jerusalem (translated from Russian)

Tahir MHN, Imran M, Hussain MK (2002) Evaluation of sunflower (*Helianthus annuus* L.) inbred lines for drought tolerance. Int J Agric Biol 3:398–400

Talukdar D (2012) Modulation of plant growth and leaf biochemical parameters in grass pea (*Lathyrus sativus* L) and fenugreek (*Trigonella foenum-graecum* L.) exposed to NaCl treatments. Indian J Fundam Appl Life Sci 2(3):20–28

Talukdar D (2013a) Arsenic-induced oxidative stress in the common bean legume, Phaseolus vulgaris L. seedlings and its amelioration by exogenous nitric oxide. Physiol Mol Biol Plants 19(1):69–79

Talukdar T (2013b) Fruit microcharacters as potential biomarkers of arsenic toxicity in a medicinal herb, *Wedelia Chinensis* merrill of compositae. Int J Agric Sci Res 3(1):143–150

Talukdar T (2013c) Cypselas diversity of the tribe Cardueae (Asteraceae)-an overview. Lap Lambert Academic Publishing, Saarbrücken

Tang M, Hu YX, Lin JX, Jin XB (2002) Developmental mechanism and distribution pattern of stomatal clusters in *Begonia peltatifolia*. Acta Bot Sin 44:384–390

Torbaghan ME, Torbaghan ME, Ahmadi M (2011) The effect of salt stress on flower yield and growth parameters of saffron (*Crocus sativus* L.) in greenhouse condition. Int Res J Agric Sci Soil Sci 1(10):421–427

Vazquez MD, Poschenrieder CH, Barcelo J (1992) Ultrastructural effects and localization of low cadmium concentrations in bean roots. New Phytol 120:215–226

Villasante CO, Rellán-Álvarez R, Campo FFD, Carpena-Ruiz RO, Hernández LE (2005) Cellular damage induced by cadmium and mercury in *Medicago sativa*. J Exp Bot 56(418):2239–2251

Vitória AP, Da Cunha M, Azevedo RA (2006) Ultrastructural changes of radish leaf exposed to cadmium. Environ Exp Bot 58:47–52

Vollenweider P, Ottiger M, Günthardt-George MS (2003) Validation of leaf ozone symptoms in natural vegetation using microscopical methods. Environ Pollut 124:101–118

Vollenweider P, Cosio C, Günthardt-George MS, Keller C (2006) Localization and effects of cadmium in leaves of a cadmium-tolerant willow (*Salix viminalis* L.). Part II Microlocalization and cellular effects of cadmium. Environ Exp Bot 58:25–40

Von V, Zabka V, Wuppertal A (2007) The plasticity of barley (*Hordeum vulgare*) leaf wax characteristics and their effects on early events in the powdery mildew fungus (*Blumeria graminis* f.sp. *hordei*): interactive adaptations at the physiological and the molecular level. Dissertation, University of Würzburg

Walsh GE (1990) Anatomy of the seed and seedling of *Spartina alterniflora* Lois. (Poaceae). Aquat Bot 38:177–193

Wang YY (2006) Occurrence and characterisation of superoxide dismutases in the female reproductive structures of *Petunia*. Dissertation, University of Canterbury

Webber M, Barnett J, Finlayson B, Wang M (2006) Pricing China's irrigation water. Working Paper, School of Anthropology, Geography and Environmental Studies, The University of Melbourne, Victoria, Australia

Wignarajahk D, Jenning H, Handley JF (1975) The effect of salinity on growth of *Phaseolus vulgaris* L. 1. Anatomical changes in the first trifoliate leaf. Ann Bot 39:1029–1038

Wild A, Schmitt V (1995) Diagnosis of damage to Norway spruce (*Picea abies*) through biochemical criteria. Physiol Plant 93:375–382

Xu Z, Zhou G (2008) Responses of leaf stomatal density to water status and its relationship with photosynthesis in a grass. J Exp Bot 59:3317–3325

Yujing Z, Yong Z, Zizhi H (2000) Studies on microscopic structure of *Puccinellia tenuiflora* stem under salinity stress. Grassl China 5:6–9

Zhao XZ, Yang YS, Shen ZX (2006) Stomatal clustering in *Cinnamomum camphora*. S Afr J Bot 72:565–569

Zinn KE, Tunc-Ozdemir M, Harper JF (2010) Temperature stress and plant sexual reproduction: uncovering the weakest links. J Exp Bot 61(7):1959–1968

Zobel A, Nighswander JE (1991) Accumulation of phenolic compounds in the necrotic areas of Austrian and Red Pine needles after spraying with sulphuric acid: a possible bioindicator of air pollution. New Phytol 117:565–574

Zubko MK, Zubko EI, Ruban AV, Adler K, Mock HP, Misera S, Gleba YY, Grimm B (2001) Extensive developmental and metabolic alterations in cybrids *Nicotiana tabacum* (+*Hyoscyamus niger*) are caused by complex nucleo-cytoplasmic incompatibility. Plant J 25:627–639

Microbiomics: An Approach to Community Microbiology

Pankaj Sharma, Vijaya Brahma, Anamika Sharma, R.K. Dubey, G.S. Sidhu, and P.K. Malhotra

Contents

P. Sharma, Ph.D.
Department of Plant Breeding and Genetics, Punjab Agricultural University, Ludhiana 141004, Punjab, India

V. Brahma, Ph.D.
Toronto General Research Institute, University Health Network, Toronto, ON, Canada

A. Sharma, Ph.D.
Department of Dental Microbiology, Theerthankar Mahaveer University, Moradabad, Uttar Pradesh, India

R.K. Dubey, Ph.D. (✉)
Department of Floriculture and Landscaping, Punjab Agricultural University, Ludhiana 141004, Punjab, India
e-mail: rkdubey.flori@pau.edu

G.S. Sidhu, Ph.D. • P.K. Malhotra, Ph.D.
Schools of Agricultural Biotechnology, Punjab Agricultural University, Ludhiana 141004, Punjab, India

Abstract

Many biologists have known that there are many groups of microorganisms that, although they cannot be cultured, do exist and influence the life of mammals, plants, and other small multicellular organisms. Nowadays, many new sequencing techniques have been developed that, with very high precision, can identify the presence of these microorganisms and how they affect others. The microbiota of humans and plants can be studied, and their effects on health and growth are becoming known using the transcriptome analysis of interaction. Life on earth has evolved over billions of years, and its most initial form—the microorganism—has evolved in gradually changing environmental conditions. They are present everywhere: from high temperatures to freezing conditions, in water and in air, on

D. Barh et al. (eds.), *PlantOmics: The Omics of Plant Science*,
DOI 10.1007/978-81-322-2172-2_22, © Springer India 2015

surfaces and in inner cavities. Despite centuries of research, the world of microorganisms is not fully known, as only a fraction (0.001) of it is able to be cultured, thus leaving a world little known to us. Initially, polymerase chain reaction (PCR)-based techniques, and later 16S ribosomal RNA (rRNA) sequences, allowed a view into this hidden microbial diversity and provided knowledge about uncultured microbes. Microbiomic analyses made it possible to know about and culture members of previously unknown groups of certain microbiomes. Similarly, nucleic acid probes with fluorescent labels can identify even single cells in situ, and sequence-based analyses using random sequencing, as well as cloning of complete microbiomes, has facilitated the reconstruction of genomes. Transcriptomes and proteomes of microbiomes have provided microbiologists with opportunities to switch from diversity studies to that of functional microbiomics. Soil, which is essential for plants, is an ocean of microorganisms that affect plants largely through roots and on aerial surrounds. The study of microbiomics helps in the identification of new groups involved in plant diseases from the rhizosphere microbiome. The application of new groups of microorganisms and their interactions are enormous in the fields of food, human health, and plant health.

Keywords
Microbiomics • Metagenomics • 16S rRNA sequences • Rhizosphere microbiomics • Microscopy

Introduction

The invention of the microscope is associated with the origin and development of the science of microbiology. The first view into the microbial world was via bacterial cells in 1663, the era of Antonie van Leeuwenhoek, who is thought to have designed and developed microscopes. For over 200 years, ever improving microscopy techniques have enabled microbiologists to study the diversity of the microbial world. The findings and discoveries of the botanist Ferdinand Cohn are undoubtedly important, as he classified bacteria and described the life cycle of *Bacillus subtilis* (Geison 1981). Another invention to revolutionize the field of microbiology was the concept of pure culture—developed by Franz Unger in the 1850s—which became the standard bacteriological technique for microbiology (Drews 1999; Mazumdar 1995). From this point on, the microbial world was divided based on the ability to culture the microbes. To date, much is known about microbes, particularly those that have been cultured, as the ability to carry out the necessary quality of work has attracted the attention of scientists. A large number of microbes from a microbial community have evolved with plants, animals, and humans. The human gastrointestinal tract is home to approximately ten times more microbial cells than exists in the rest of the human body. This large number of microbial cells, with more than 1,000 species, constitutes the human gut microbiota, and the total number of genes present in this microbial community is 100 times greater than the genes present in the human genome. This has led microbiologists to understand that humans and other macro-organisms are not individuals, as commonly thought, but instead can be thought of as an ecosystem with thousands of species living together.

This is a very unusual idea, and some scientists have taken the lead in the last decade with unique sampling techniques required for this concept, which differ from traditional sampling methods. With the invention of new sequencing techniques and very high computational powers, some results are convincing the world of this new idea of the co-evolution of microbes and macroorganisms as one ecosystem. Microbiomics has evolved as the study of all the genomes of the microorganisms present in an ecosystem. It is quite complicated to differentiate the terms from an ecological to a microbiological and molecular biology perspective; the definitions themselves are the subject of hot debate. 'Microbiome' has been defined as being all the genomes (or genes) and the interactions of microorganisms present in a particular ecosystem, while 'microbiota' means all the microorganisms present in an ecosystem. With the understanding of the concept of metagenomics, many new genus and species have been identified using new-generation sequencing and computational techniques; this has given rise to some commercially important strains used to treat human gut problems. Plant surfaces are surrounded by diverse microorganisms that affect them in different ways: some can improve plant health, while some have the potential to reduce attacks from and incidences of diseases and insect pests. Soil can be described as an 'ocean' of microbes that largely affect plant roots. Development of suppressiveness in soil to plant pathogens can be improved by understanding the microflora present in the rhizosphere. Water ecosystems are severely impacted by agricultural pesticides and industrial effluents; microorganisms have developed the capacity to degrade these pollutants, and this capacity has been improved by transferring the trait into more efficient native microbes, especially bacteria.

There is great potential in the area of microbiomics in terms of human health and potentially in agricultural and environmental sciences. The area is growing rapidly from its infancy and expanding quickly.

Non-culturable Microorganisms

Culturing methods are very efficient for providing information about definite groups of cultured microorganisms, providing detailed information about specific behaviors and characteristics of microbes in culture. The uncultured world of microorganisms somehow did not gather as much attention, possibly because of the ease in working with, and the level of information coming from, cultured microorganisms. Microbial physiology and genetics was developed with the tamed model microorganisms, and also opened doors to new dimensions for many aspects of biological processes. This was the phenomenon up to the middle of the 1980s. Some believed in the concept of unculturable microorganisms (Staley and Konopka 1985). Estimates for total counts of microbes in a population via microscope and dilution plate techniques differed, and in some cases this difference appeared very large. Grimes et al. (1986) observed differences of as much as four to six times between the plate count method and the estimation of viable cells using acridine orange dye staining. Similarly, Torsvik and Ovreas (2002) found only up to 1.0 % of bacteria were culturable from soils using the standard culture media under normal conditions. The work of Brock and colleagues on uncultured microorganisms was some significant evidence in the line of uncultured members of the microbial community from Yellowstone hot springs. Many bacteria did not grow on culture media, and cultured bacteria did not reflect their physiological activities in their ecological niche. The Yellowstone hot water spring is characterized by particularly hot water, and many of the bacteria present had adapted to the high temperature (which was higher than the melting point of agar used in media for solidification); hence, such bacteria were not cultured on media. These conditions led to the application of some innovative techniques: Brock immersed the glass slides in the spring for different lengths of time and observed them under a microscope. He also used fluorescent antibodies against the known cultured bacteria, which he thought might suppress the uncultured bacteria (Bohlool and Brock 1974; Bott and Brock 1969; Brock 1967). This technique was very successful in estimating the population sizes and growth rates of all community members. Brock and colleagues found that strains of *Sulfolobus* can grow, even at lower temperatures (Mosser et al. 1974). They also identified candidates responsible for photosynthesis in the hot spring; covering the spring led it to lose its pink color, which reflected that *Synechococcus*—a bacterium that is usually pink in culture—was actually responsible for the photosynthesis (Brock and Brock 1968). Further evidence for the uncultured world came from the work of Whang and Hatori (1988), who studied oligotrophs and found that longer culture incubation can improve the growth of certain microorganisms in culture.

Food, when infected with some pathogenic microorganisms can become lethal. The food industry faces this very challenge while trying to maintain quality. Some human pathogenic organisms living on food cannot be cultured until they face some stress, such as hot or cold temperatures or moisture stress (Dahl and Pestka 1985). The idea of living and infectious microorganisms that were nonetheless not able to be cultured comes from the work of Colwell and colleagues, who showed that strains of *Vibrio cholera* were viable and able to cause disease but were unable to be cultured until they had been passed from the mammalian intestine (mouse or human) (Colwell et al. 1990; Colwell and Grimes 2000).

Many such results and advances in technology make it easier for scientists to understand unculturable microbes. Along the same vein was the discovery of the diversity of soil bacteria using DNA–DNA re-association. This work showed that the diversity of bacteria in soil was 100 times greater and more complex than was thought with standard culturing techniques (Torsvik et al. 1990). Further important work proved that gastric ulcers and cancers could be caused by *Helicobacter pylori*. This bacterium was known to exist in the gastric mucosa of dogs in 1893 and confirmed in humans in 1906. In 1938, the correlation between peptic ulcers or cancers and the presence of bacteria was observed, but it was only in the mid-1980s, when *H. pylori* was cultured, that its role with disease was established (Doenges 1938; Marshall et al. 1985a, b). The culturing of *H. pylori* occurred accidently when plates were incubated for 5 days instead of 3 (Buckley and O'Morain, 1998). All the facts and findings of many workers have led to the idea that there are many hidden natures to uncultured microorganisms, and their association with some physiological effects and diseases are still unknown. The culturability of a population of microorganisms can enhance our understanding, but it cannot represent the vast diversity and their effects on human and related activities.

Identifing the Taxa

In Situ Hybridization (Fluorescent Probes)

Identification of a group of microorganisms or any particular species from a population at some time becomes very important. Using fluorescent probes for specific sequences of DNA from a group or species helps in identification and quantification of the group or species from other members of a population. This is a culture-independent technique and needs some prior information on specific genomic sequences of organisms. Fluorescent in situ hybridization (FISH) requires small subunit (SSU) ribosomal RNA (rRNA) oligonucleotide probes; this method combines both SSU rRNA probe hybridization with fluorescent microscopy, which helps in quantification of microbes and their specific importance in the population. Use of FISH is increased in the area of gastrointestinal tract studies, where a large microbial community is not culturable and their relative proportion is not understood. The bacterial genera *Bacteroides*, *Bifidobacterium*, *Streptococcus*, *Lactobacillus*, *Collinsella*, *Eubacterium*, *Fusobacterium*, *Clostridium*, *Veillonella*, *Fibrobacter*, and *Ruminococcus* have been quantified with the help of this technique (Schwiertz et al. 2000; Harmsen et al. 2002; Harmsen and Welling 2002). This technique is combined with automated computerized systems and has been used mainly to learn the proportion and presence of major bacteria in the human gastrointestinal tract (Amann et al. 1990; Jansen et al. 1999).

This technique is very powerful in addressing some ecological issues; it can identify subpopulations, give community structure, there is no need to grow the organisms, and it can give a cell population count (Vaughan et al. 2000). The addition of cytometry to FISH has now improved the automation of cell counting in samples of the gastrointestinal tract; this will sort the microbes, particularly bacteria, and will be further used for diversity studies (Wallner et al. 1997 and Zoetendal et al. 2002).

16S rRNA

Many of the sequenced genes and other portions of genome vary among species. Horizontal transfer, multiple copies of a genomic fragment or genes, or ambiguous rDNA markers make the concept of species confusing with overlapping information (Achtman and Wagner 2008). In practice, a sequence similarity cut-off of 95.97 or

99 % is often used, and the resulting groups are assumed to be identical and are referred to as an operational taxonomic unit (OTU) or phylotype. In microbiome studies, these OTUs are of equal importance to species, as information about species is usually not available (OTUs as a similar genome). Assigning the sequence to a particular OTU is referred to as 'binning', and this is carried out using unsupervised clustering of similar sequences (Schloss et al. 2009; Schloss 2010). The data are trained to assign the sequences to taxonomic bins using any method of mutation rate incorporation, establishing evolutionary relationship, or supervised methods (Wang et al. 2007).

Biological analyses can be performed using computational tools, as the binning process treats individual OTUs as a separate class and each 16S sequence is assigned as a bin to a taxonomic category, thus a microbiome can be presented as a histogram of assigned bins (Hamady and Knight 2009). These histograms can be converted to binary codes on the basis of the presence or absence of a particular bin and studied in other communities and related microbiomes. Bins or OTUs will not appear outside specific microbiomes, thus this approach is typically most useful for low-complexity microbiomes or where OTUs are well designated with their specificity. Boinformaticians can use the histogram of 16S sequence data to analyse the microbiome to compare two related communities; this information can be used for principle component analysis to identify the source of significant variance and correlation with community characteristics (Lozupone and Knight 2005; Johnson and Wichern 2007; Gianoulis et al. 2009). DeLong and colleagues identified a 40 kb fragment in their clones of prokaryotic libraries from sea water. That clone was from an archaebacteria that had not been previously cultured. Similar studies were also conducted using soil, but incurred difficulties related to DNA integrity during the extraction process and purification from soil samples, although this can produce results similar to those with sea water samples (Berry et al. 2003; Courtois et al. 2003).

Information produced by 16S RNA signatures has intensified efforts to culture uncultured bacteria, and identify new bacterial genera and species. This has been facilitated using fluorescent probes in the microbiome. Some members of SAR11 clade from sea water were cultured only after getting information about their presence using the 16S rRNA sequences; these are termed genus *Pelagibactor*. This genus is found in abundance in sea water surfaces, as they form one-third of all prokaryotic cells from sea water (Rappe et al. 2002; Cho and Giovannoni, 2004). SAR11 members (the Pelagibactors) correspond to the *Acitobacteria* phylum found in soil, which comprise 20–30 % of the microbiome based on 16S rRNA sequence analyses, and only three members have been cultured from soil microbiota (Sessitsch et al. 2001; Smit et al. 2001; Janssen et al. 2002; Sait et al. 2002). Information gained from culture-independent methods shows the presence of more genus and species in soil and water microbiota. More intensive study is needed to culture more microorganisms, especially to culture more members of *Acidobacteria* (Lorenz et al. 2002).

Besides identifying new genera and species of microorganisms, 16S rRNA sequence analyses also help identify new genes into the known and unknown microbes. Sequence of a gene that was associated with the Proteobacteria was a bacteriorhodopsin-like gene. This gene product was a photoreceptor that was earlier thought to be associated only with *Archaea* (Stein et al. 1996; Beja et al. 2001). Sequence data from phylogenetic marker data provide information about the physiological behavior of less complex microbiota. Contrary to this is the method of sequence random clones, which provide deep information when carried out on a large scale. Phylogenetic markers used as initial identifiers and for taxonomic information from DNA fragments with some important genes are lesser in number; as more genes are sequenced and cloned they will provide more information and thus will increase the available markers (Tyson et al. 2004; Venter et al. 2004a, b).

Community Sequencing

Community diversity analyses were possible with modern high-throughput techniques, mostly carried out for uncultured microbes.

Metagenomics is considered the study of the combination of all the high-throughput studies with microbiota. Among them, the most useful is whole metagenome shotgun (WMS) sequencing of community microbiomes (Chen and Pachter 2005). The most exciting information from WMS sequencing has come from studies of the human gut microbiome and its role in obesity, energy utilization, and physiological disorders (Turnbaugh and Gordon 2008). Similar to metagenomes of the human gut, WMS sequencing, human stool metatranscriptomics, and medium-throughput human gut metaproteomics has provided important information on the relationship between the human body and its microbiota (Poretsky et al. 2009; Shi et al. 2009; Li et al. 2011; Booijink et al. 2010; Giannoukos et al. 2012). This technique is good for any heterogeneous population, and the low frequency in population and biasness of the 16S rRNA-sequencing methods has no influence (Sogin et al. 2006). Leavue and Tech (2011) studied the leaf and fruit surfaces of grapes using 16S rRNA gene sequencing. Findings indicate that the V5-36 V6 regions of the bacterial 16S rRNA gene do not always offer sufficient resolution to assign sequences at the species level. These studies have opened new horizons for understanding the physiological and functional association of certain types of genes found in abundance and reconstruction of genomes of organisms that were not able to be cultured.

Interaction Molecules/Proteome

A vast variety of new genes are present in the microbiome and interacting with the other proteins produced by members of the community. 'Metatranscriptomics'—a field very similar to metagenomics—uses sequences of reverse-transcribed RNAs; extending this up to the quantification of protein level is 'metaproteomics', while 'metabolomics' deals with the investigation of small metabolites (Turnbaugh and Gordon 2008; Verberkmoes et al. 2009; Li et al. 2011). These fields are in the very early stages of development and use technologies that are almost similar to those used by their cultured counterparts. Comparable computational approaches are to be used for the analyses of sequence data of a community and its interaction under certain environmental conditions; translational metaomics studies are greatly influenced by their host environment. The most studied environment is the human gut; the community structure changes with time and with the food habits of the individual (Nicholson et al. 2005; Dethlefsen and Relman 2010). The microbiota are thus very changeable, as they change metagenomically within hours and metatranscriptionally within minutes to responses against antibiotics and meals (Booijink et al. 2010; Dethlefsen and Relman 2010). The phenotypes that are easily affected by the response of the metagenome have great potential in pharmaceuticals, where the effectiveness of drugs may be affected by the microbiome in the individual's gut. The prebiotics and probiotics that lead to the growth of beneficiary microbes have a range of nutrients to be studied, depending on the climate and infectious disease in the human population of the particular locality (Jia et al. 2008; Guarner and Malagelada 2003). The interaction effects of microbes on human health were well documented before the development of microbiomics; Nobel prize winner Ilya Mechnikov named the yogurt-producing bacterium *Lactobacillus bulgaricus* based on the fact that yogurt-consuming Bulgarians had longer life spans than non-consumers. Recent findings now support the effect of probiotics on human health (Garrett et al. 2010; Martin et al. 2008). Antibiotics affect the microbiota of the colon, and some disease-causing bacteria, such as *Clostridium difficile*, are able to maintain the population from a very low initial spore count. A patient with *C. difficile* infection underwent fecal transplant from her husband and finally recovered the very next day. Metagenomic study revealed that a new microbial community was restored, with its taxonomic variable members. Thus, the human microbiota needs to be studied with reference to its interaction to different factors such as medicines, food, local hygiene, and climate. It requires continued experimentation and strong bioinformatic support to reveal the biological questions arising from interaction of these microbial communities; that information may be helpful in the clinical treatment of individual patients based on specific requirements (Khoruts et al. 2010; Borody 2000).

Culture-Independent Diversity

Diversity Unknown

Microbiota are of interest for research due to their effect on the immunity of the host organism, in turn influencing the health and frequency of disease in a population. Vice versa, the health of an organism also changes the frequency of microbial communities. For example, different places in the body vary in the type and frequency of the associated microbes found. On the surface of human skin, the very few genera present are dominated by Propionibacterium; the human nasal cavity has relatively more genera, with a higher proportion of Corynebacterium. Streptococcus is dominant among the several hundred genera present in the oral cavity. The gut hosts over 500 genera, with microbial density as high as 10^{11} microbial cells per gram tissue. Despite that decades have passed since the discoveries about the importance of gut microflora, these communities are still not well understood. Microbiome present on human skin play an important role in resistance against infection caused by Staphylococcus spp. Vaccination programs against pneumococcus changed after knowledge was gained about the effect of nasal microbial communities on infectious bacterium. Another example is the extreme dysbiosis that occurs in cystic fibrosis and leads to pathogenic infection (Grice et al. 2009; Morgan and Huttenhowr 2012). Microbial communities associated with the human body have a great influence on the health of individuals. The diversity of microbiota and its influence on human physiology are areas of research to understand the meaning of their association with humankind.

Culture Independent

In addition to providing a universal culture-independent means to assess diversity, 16S rRNA sequences also aid in culturing efforts. Microorganisms need specific conditions to grow, which they get readily in their natural environment; in culture-level conditions they might be recalcitrant for several reasons. Some microbes need symbiotic partners, and absence of such partners inhibits their growth. Specific nutrient requirements or surfaces needed to grow may be limiting in some cases. Inhibitory compounds and a combination of temperature, pressure, or atmospheric gas can also be responsible for the unculturable nature of some microbes. Slow growth rates and negative feedback of some metabolites in culture, or rapid dispersion from colonies, are reasons for non-competitiveness in the culture (Simu and Hagstrom 2004). Getting the testing conditions right for a microorganism is critical as well as very laborious and challenging for the scientist. This can be achieved only with the availability of correct quantitative assays.

Human Microbiomics

The human body is known to be dominated with billions of microbial cells, with great variability in number and taxonomical aspects. Some are very well studied, while, for many others, we continue to have no knowledge of their existence and physiological behavior. The human gastrointestinal tract is studied largely as compared with other cavities and surfaces with respect to microbiota. Experiments have proven that some issues related to human health are influenced by microbiota. Obesity is treated as a disease influenced by genetics and personal habits; however, Ley (2010) demonstrated a large change between the microbiota of obese wild-type and non-obese mice and confirmed that, in gnotobiotic mice, the phenotype was transferable with the transfer of microbiota. Reduction in *Bacteroidetes* and a simultaneous increase in *Firmicutes* was found to be a major change between two distinct individuals (Turnbaugh et al. 2006). Shifts of this functional correlation with obesity were observed with functional metagenomics, and relate to the capacity of microbiota to utilize energy and control fat storage. The underlying molecular mechanism is yet to be proved; experimentation can provide information about whether this association was mere correlation or could be used in clinical treatment (Turnbaugh et al. 2009).

Like obesity, a community of great importance once the functional mechanism was understood was the formation of biofilm in the mouth cavity that leads to formation of dental cavities or periodontitis. The microbiota of oral tissues and oral saliva was studied and it was found that the interaction of tooth enamel with microbiota varied, as this interaction changed with the cleaning of the teeth. The complexity of the microbiota was re-established from nothing; first in colonization were *Streptococci*, which consisted of some adhesions and receptors that help in sticking to the bare surface of the tooth, subsequently supporting the binding of a variety of microbes. These bacteria were supported by *Vellonella* and *Actinomyces* sp., which led to the development of a local environment favorable to *Fusobacterium* and *Porphyromonas*. This entire process was mediated by metabolic dependencies of bacterial taxa on each other, surface recognition, and cell-to-cell physical interactions (and intercellular signaling), thus providing a suitable niche for bacteria to evolve a complex microbiota within hours. Intensive computational and experimental work is needed to understand the community function in this kind of system evolvability (Nasidze et al. 2009; Zijnge et al. 2010; Hehemann et al. 2010).

Second Human Genome

The human genome sequencing project provided great insight into molecular-level understanding of the functions of and interactions between genes. Assuming the average size of a bacterial genome is approximately 3 Mb, the human intestine hosts a genome equal to its own total genome. Another human genome is functionally more active due to the high gene density, thus interacting effects are more intense and clinically important. Estimation of this microbial genomic content in the human population should consider that the microbiota of the individual person and the communities inside change greatly with the human population or individuals.

Plant Microbiomics

Crop Microbiomics

Phylloplane Microbiomics

The upper half of plants are exposed to all taxa of fungi, bacteria, and yeast that readily colonize the aerial parts of plants. Knowledge of the microbiota of plant surfaces like phylloplane (leaf surfaces) and the fruit surface has been based on culturing methods and, thus, very limited information was available. Techniques such as denaturing gradient gel electrophoresis and sequencing of 16S rRNA genes improved the understanding of these plant surfaces. Now it is realized that these microbial communities are more complex, and far less is depicted through experimentations about the effect of factors such as environment, type of plant, and microbiota in the aerial surface (Leveau 2009; Delmotte et al. 2009; Redford et al. 2010; Hunter et al. 2011). Microbial compositions of the phylloplane provide insight into protecting the plants from various diseases and insects and help in improving plant health. Leveau and Tech (2011) studied the leaf and fruit surfaces of grapes using 16S rRNA gene sequencing. They used eight leaves and two berries as samples and extracted the microbial DNA from their surfaces. A maximum of prokaryotic sequences (53.3 %) were assigned to the group Protobacteria, followed by *Firmicutes* (15.1 %), Bacteriodetes (10.1 %), and Actinobacteria (8.0 %). Among them, approximately half were from 297 genera that were already known and dominated by the following ten genera: *Sphingomonas*, *Hymenobacter*, *Bacillus*, *Pseudomonas*, *Skermanella*, *Leuconostoc*, *Massilia*, *Methylobacterium*, *Cellvibrio*, and *Curtobacterium*. They also found that 67 % of random sequences from *Pseudomonas* I originated from grape leaves and two were identical to the 16S rRNA gene from *P. caricapapayae*; ten sequences were identical to the 16S rRNA genes from *P. rhizosphaerae* and *P. abietaniphila*, and one was identical to the 16S rRNA genes from the Pseudomonas species *fluorescens*, *corrugata*, *chlororaphis*, *lini*, *congelans*, *tremae*, *kilonensis*, *frederiksbergensis*, *thivervalensis*, *migulae*, *jessenii*, *cedrella*, *veronii*, and *mandelii*. This vast diversity present on the surface has a full scope for their effect on plant health and survival.

Plant Health

Aerial plant surfaces are full of microflora and they do take part in utilization of different substrates. During development of the plant, these organisms may provide some of the nutrients; after the plant has matured, they can go with the produce and start interacting among them for utilization of substrates and can change the quality of produce. Some grape fruit processes are affected by fruit and leaf surface microorganisms. Vinification may encounter interference by some spoilage bacteria such as *Acinetobacter*, *Lactobacillus*, *Oenococcus*, and *Pediococcus* (Bartowski 2009) and may create the aroma of berries (Verginer et al. 2010) or affect wine pH (Bartowski 2009). The wine industry chose to have some natural microflora for ethanol production (Varela et al. 2009).

Phylloplane microbiota interact with one or more members of the ecosystem; these activities (beneficial or deleterious) directly affect the health of the plant. Understanding the composition of the microbial community may provide insight to enable better care of crops and result in more profits for growers.

Disease Resistance

Competition and biocontrol activities are important mechanisms which take place in reducing diseases on the plant. Many fungal and bacterial genera present on the leaf surfaces are known to have biocontrol activities. A foliar pathogen of vineyards with a history of causing losses is *Erysiphe necator*. This pathogen causes powdery mildew of grapes, a very important commercial crop worldwide (Pearson and Goheen 2008). Dimakopoulou et al. (2008) showed that other microbes from the ecological community were responsible for protection against foliar pathogens including powdery mildew pathogen.

The importance of trunk bark as a potential source of inoculums for leaves and fruits of grapes should also be considered. They are physically very close to each other and may even come into direct contact. *Uncinula nector*, a pathogenic fungus of the grape berry, was isolated from the bark of grape vines (Behar et al. 2008; Cortesi et al. 1997; Grove 2004). *Botrytis cinerea, Fusarium laterium, Penicillium* spp., *Phomopsis viticola* (Munkvold and Marois 1993; Barata et al. 2012), and the yeast-like fungus *Aureobasidium pullulans* are known for their biocontrol activities on grape berry pathogens. Based on 16S rDNA sequence analysis, the 224 isolates were assigned to a specific genus, with a classification threshold above 98 %. The frequency of genera and species vary from type of sample and time of sampling. However, there were similarities in the diversity and abundance of isolates from each ecosystem. A total of 24 different genera were identified and the strains belonged to six different bacterial classes (alpha-, beta-, and gamma-Proteobacteria, Actinobacteria, Clostridia, and Bacilli). The most common and abundant genera were Pseudomonads, which may play a role in biocontrol activity (Martins et al. 2013).

Root Rhizosphere Microbiomics

The microbial community in the vicinity of plant roots is important for the functioning of the plant; it can improve the uptake of nutrients by the plant and provide protection against pathogen and insect attacks. Microbiological studies in the soil environment are hampered by the fact that the largest proportion of soil bacteria cannot yet be cultured (Doornbos et al. 2012). Moreover, evolution in the science of metagenomics with modern-day sequencing technologies depicts an improved picture of the microbial community and its activities in the rhizosphere (Hirsch and Mauchline 2012). Microbial activities in complex environments such as the rhizosphere are yet to be exposed, as very little information is available for rhizosphere microbiota of different crop species. Functional microbiomics may reveal some of the concepts of interactions in the rhizospheric environment (Jansson et al. 2012). The new knowledge of rhizospheric microbiome activities will provide tools for manipulating the environment to achieve the genetic potential of crops by improving crop health (Bakker et al. 2013).

Plant Health

Soil microorganisms are chemotactically attracted to plant root exudates, after which they proliferate in this carbon-rich environment (Lugtenberg and Kamilova 2009). Carbon limitation could be demonstrated in bulk soil but not in the rhizosphere using *Pseudomonas fluorescens* strains carrying carbon-limitation reporter systems (Koch et al. 2001). Different plant species have their own configuration of exudates and nutrient uptake from soil; different plant species are expected to have different rhizospheric microbiota. Indeed, plant-specific microbial communities could be isolated from roots in studies comparing, for example, wheat, ryegrass, bentgrass, and clover (Grayston et al. 1998), or wheat and canola (Germida et al. 1998). Further, plant species-specific rhizosphere populations could be isolated within a specific bacterial group such as fluorescent *Pseudomonas* spp., (Lemanceau et al. 1995). More recent studies, in which the rhizosphere microbiomes were characterized based on direct extraction of total community DNA, also provide strong evidence for plant species-specific microbiomes (Kirk et al. 2005; Inceoglu et al. 2013). The roots of wheat, maize, rape, and barrel clover were shown to carry different bacterial communities as a consequence of assimilation of root exudates (Haichar et al. 2008). Bacterial community structures in field-grown potato rhizospheres were affected by the growth stage of the plant (Inceoglu et al. 2013). Further, at the genotype level within a plant species, specificity of the rhizosphere microbiome has been described (Weinert et al. 2011). Micallef et al. (2009) used *A. thaliana* and showed that the rhizosphere of this model plant mediates a significant change in the bacterial community relative to the bulk soil. The rhizosphere effect was illustrated by comparing rhizosphere bacterial communities of tobacco and *A. thaliana* grown on a potting and a clay soil. Total bacterial counts on 1/10 strength tryptic soy agar (TSA) and counts of fluorescent pseudomonads on King's medium B agar (KB) in bulk soil and in the rhizospheres of *A. thaliana* Col-0 and tobacco were recorded. The rhizosphere effect was exemplified by the observation that numbers in the rhizosphere were about 10- to 100-fold higher than the numbers in bulk soil for both plant species. In *Pseudomonas*, specific denaturing gradient gel electrophoresis (DGGE) profiles were shown and compared in a redundancy analysis. For both tobacco and *A. thaliana*, rhizosphere *Pseudomonas* communities differed from those in the bulk soil, and the communities differed between the plant species. The *A. thaliana* root microbiome has been described in detail using pyrosequencing of 16S rRNA gene amplicons (Bulgarelli et al. 2012; Lundberg et al. 2012). Whereas differences between bacterial communities in bulk soil and the rhizosphere were observed in these studies, their focus was on the endophytic compartment. Inside the root, the microbiome clearly differed from the bulk soil and was enriched in Actinobacteria and Proteobacteria (Bulgarelli et al. 2012).

Disease Resistance

The type of plant species also affects the rhizospheric microbiota; roots select or attract a specific set of microbes and decide their frequencies. Under the influence of some disease-causing pathogens, plants act selectively to attract specific microorganisms, particularly bacterial species, into its rhizosphere. This is most clearly observed in so-called disease-suppressive soils, in which disease will not develop despite the presence of a virulent pathogen and a susceptible plant.

Suppressiveness for plant diseases in soil is thus due to the presence of specific microflora in the rhizosphere; only under heavy inoculum load and favorable conditions may disease appear (Mazzola 2002). A well studied example was take all decline (TAD), which developed in continuous wheat cultures after a severe outbreak of the take-all disease caused by *Gaeumannomyces graminis* var. *tritici*. This suppressiveness can be transferred to other disease-conducive soils for TAD disease in wheat, and the application of heat treatment to suppressive soil will cause it to lose its ability to reduce the disease. Under continuous wheat cropping, a specific group of fluorescent pseudomonads that produce 2,4-diacetylphloroglucinol (DAPG) was enriched in the rhizosphere, and these bacteria appeared to be responsible for TAD (Raaijmakers and Weller 1998). Additional bacterial taxa that may be involved in TAD had more recently been identified using 16S rRNA-based techniques (Weller et al. 2002; Schreiner et al. 2010). Mavrodi et al. (2012) studied the plant-specific preferences for bacteria under disease conditions and their effect on reducing the disease. They observed that, under irrigation, the wheat rhizosphere recruited DAPG-producing pseudomonads, whereas, under dry conditions, phenazine-producing pseudomonads were recruited. This observation was correlated with the two different pathogens of wheat, one favoring the irrigated conditions (*G. graminis* var. *tritici*) and the other a pathogen that perpetuates and is more infectious under dry conditions (*Rhizoctonia solani*). *G. graminis* var. *tritici* was found to be more sensitive to DAPG produced by pseudomonads under irrigated conditions, whereas *R. solani* was sensitive to the phenazine, a chemical produced by pseudomonads under dry conditions. Thus, under suppressiveness, plants select the microbiota favoring plant growth by inhibiting pathogen growth. For other soil-borne pathogens, specific compounds that suppress pathogen growth were also found in the soil and root zone of the plant. In *Fusarium* wilt of crop plants, the suppressive soil zone was rich in active phenazines produced by pseudomonads; the efficacy was increased with the synergistic effect of non-pathogenic *Fusarium* spp. competing for nutrients in the suppressive zones (Mazurier et al. 2009). The presence of microbial consortia in suppressive soils was reported from some of the disease-suppressive soils. In the potato, a unique frequency of members of rhizospheric microbiota was recorded in potato common scab suppressive soils (Rosenzweig et al. 2012). In tobacco, black root rot disease is caused by the fungus *Thielaviopsis basicola*; the consortia of bacterial genera of *Pseudomonas*, *Azospirillum*, *Gluconacetobacter*, *Burkholderia*, *Comamonas*, and Sphingomonadaceae was observed in high proportions in the root rot-suppressive soils as compared with disease-conducive soils (Kyselkova et al. 2009). Plants bring a consortia of antagonists into the rhizosphere under disease-conducive conditions as compared with disease suppressiveness. Mendes et al. (2011) identified some such groups in the rhizosphere of sugarbeet against the pathogen *R. solani*. They used PhyloChip analysis in *R. solani*-suppressive and -conducive soils, and were able to identify more than approximately 33,000 bacterial taxonomic groups out of 60,000 tested. Finally, they sorted 17 taxa, including members of β-proteobacteria, γ-proteobacteria, and the firmicutes firmly related to soil suppressiveness. Not only do plants have preferential selection of microbiota in the rhizosphere, there is also a change in expression of the genes of members of rhizospheric microbiota. Changes in expression of genes for interaction with plants in *Pseudomanas* and *Bacillus amyloliquefaciens* in response to root exudates was studied using microarray transcriptome profiling in cultured plants (Mark et al. 2005; Fan et al. 2012). Similarly, Zysko et al. (2012) observed the transcriptional response of pseudomonas on phosphate availability in the rhizosphere of *Lolium.* Barret et al. (2009) showed changes in the gene expression of the Pf29Arp strain of *P. fluorescens* present in *G. graminis* var. *tritici*-infected roots. HCN is produced in the rhizosphere of strawberry, which gives biocontrol activity against the pathogenic fungus *Verticillium dahlia* (DeCoste et al. 2010); the HCN biosynthesis increased with the increased activities of genes of pseudomonas sp LBUM300. Jousset et al. (2011) evidenced that plants produce systemic signaling to improve/increase biocontrol activities against pathogen infection. In the split root system of barley-*pythium ultimum* interaction, one side was infected with fungal pathogen, which enhanced expression of *phlA* gene of *P. fluorescens* CHA0 for production of DAPG on another side of the split root. Expression of the *phlA* gene was induced by phenolic acids produced in response to attack by *Pythium.*

Further, it is more evident that root exudates change the expression of the rhizospheric microbiome. Many new studies are now focusing on the impact of plant variety and rhizospheric variation on interaction and diseases. Wheat take all disease is greatly influenced by the antifungal compound DAPG produced by fluorescent pseudomonas; the production of DAPG depends on plant genotype and pseudomonas strains (Okubara and Bonsall 2008; Kwak et al. 2012).

Insect Resistance/Effect of Insects

Insects are another member of the plant ecosystem; like pathogens, they also influence the microflora related to plants. Microflora are affected by the chemicals secreted by plants in response to insect attack. This can increase the particular group of microbes or pathogen in a particular phylloplane or rhizosphere. Lee et al. (2012) observed a reduction in the bacterial disease caused by *Xanthomonas axonopodis* pv. *vesicatoria* on pepper plants as a result of leaf feeding by aphids; simultaneously, aphid feeding reduced another devastating bacterial pathogen of solanaceous crops, the *Ralstonia solanacearum.* This might have happened due to *Bacillus subtilis*. The population density of this beneficial bacterium increased after aphid foliar feeding. Like the aphid, white fly, a vector of viruses in solanaceous crops, also changes the microbiota of the pepper plant and improves pathogen resistance (Yang et al. 2011). The compounds produced due to whitefly and aphid feeding that result in improved resistance are a subject of current research. There may be some rhizospheric bacteria like pseudomonads that can produce some insecticidal toxins (Pechy-Tarr et al. 2013). Carvalhais et al. (2013) used 16 S rRNA pyrosequencing and reported the use of jasmonate in *A. thaliana*, which led to changes in rhizospheric microbiota to some taxa associated with plant disease reduction.

Forest Microbiomics

Unlike field crops and cropping systems, forests contain a vast variety of vegetation that includes different groups of plants; their interaction with the surrounding environment and among themselves makes specific niches of microbes around forest plants. Soil in the forest is an important component that maintains the flow of energy in the ecosystem and keeps the forest productive with its biodiversity. Forest soils are formed with woody plant parts, litter, and humus; this nutrient-rich environment attracts different microbial taxa. The microbes and their interaction with plants affect the health of plants, thus impacting on carbon utilization as well as on carbon and other nutrient cycles. Soil is a dynamic system, with complex microbiota that takes part in the functioning of the ecosystem and affects the plant in terms of the way it behaves differently to stress conditions. Among all the microbes, fungi play a major role: white and brown wood rot fungi decompose wood; leaf litter fungi decompose leaves and grow as saprophytes. At the center of their activities is the interaction as a community, the role of each member in the microbiota and their effect are usually not known. DNA barcoding using high-throughput techniques merely highlight the diversity. The functional diversity of fungal microbiota is more important than dividing them on the basis of their individual properties such as mycorrhiza, decomposer, or saprophyte. The power of new sequencing technologies to assess structural shifts in the soil microbiota at deep coverage and high phylogenetic resolution provide novel information regarding the resistance and resilience of the forest soil microbiome to compaction (Allison and Martiny 2008; Dominati et al. 2010).

Soil compaction is caused by many disturbances that lead to changes in the soil microbial community. Measuring the microbial community structure and associated functions will help in monitoring the alterations of the soil system. It can evaluate the capacity of the soil system to recover and can hint at future irreversible changes in the soil ecosystem (Van-Camp et al. 2004).

The complete structure of soil microbial diversity was studied in compacted soil (Hartmann et al. 2014) using pyrosequencing. A total of 244 fungal and bacterial groups were studied and found to be affected by soil compaction. The bacterial groups acidobacteria, bacteroidetes, chloroflexi, delta, and beta proteobacteria and firmicutes were more evident in compact soil, thus appearing to prefer to grow in a low-oxygen environment.

The verrucomicrobial and the proteobacterial were affected by soil compaction, and their numbers reduced significantly. Fungal diversity was also affected by soil compaction, as members of the Ascomycotina group increased in number under compact soil conditions and that of Basidiomycotina was reduced; this negatively impacted the growth of ectomycorrhiza and supported the growth of saprophytes (Hartmann et al. 2014).

Environmental Microbiomics

The impact of pollution is inevitable unless it is controlled to enable the sustained development of human society. In order to achieve this pollution-free environment, pollutants and wastes from diverse sources need to be degraded, and biological systems with high catabolic capacity will be important for this purpose. Among all biological processes, those of microorganisms have an astonishing impact on pollutant degradation. New sequencing, proteomics, bioinformatics, and genomics techniques, along with microscopic advancement, provide a huge amount of information to the microbiologist. This flow of information provides in silico methodologies to study the metabolic pathways and regulatory frameworks in microbial communities. The degradation pathway can thus be studied and modified to use organisms to their maximum potential. Functional metagenomics will provide an understanding of regulatory pathways in the context of carbon sources used under particular environments, which will accelerate understanding and development of techniques for bioremediation.

Pesticide Degradation

Strong selective pressures for evolution and growth of pesticide-degrading bacteria lead to their high concentration in polluted waters. This pressure for survival forces the development of new ways of enzymatic degradation and utilization of pesticides as sole carbon sources by bacteria (van der Meer 2008). Mobile genetic elements like plasmids and transposons are called mobilomes, and carry genes for the degradation of pesticides. These mobilomes help bacteria to evolve under high pesticide loads in water; this happens via the exchange of genetic material between similar or different groups of bacteria (Heuer and Smalla 2012). Plasmid IncP-1 and transposone IS1071 carry broad-spectrum xenobiotic catabolic genes; the prevalence of these mobilomes increases under a high pesticide load and enhances the capacity of the ecosystem to catabolize the pesticides and other haloaromatic compounds. The degradation capacity of the microbial community is used to reduce the pesticide load in the ecosystem.

Microbial communities are affected by chemical and other pollutants in water; studies of these effects are very unreliable and uncertain and present a challenge to the environmental microbiologist. Microbial interactions with pollutants in sediments are even more difficult to study. Much less information is available on sediment pesticide–microbe interaction. It was observed that agrochemicals, especially pesticides, present in water body sediments changed the terminal restriction fragment length polymorphism

(T-RFLP) patterns of microbiota existing in contaminated sediments. The change in concentrations of pesticides changed the frequency of the microbial community in polluted water. If the activities of microbiota and biomass of microbial communities are an indicator for the health of water ecosystems, these were largely not affected by pesticide pollutants (Widenfalk et al. 2008).

Water Pollutants

Varieties of pollutants other than pesticides exist in water bodies; no doubt cleaning of these pollutants is important for environmental risks. Microbial degradation is always the method of choice due to its ease of application (Koukkou and Vandera 2011). The astonishing capabilities of microbes in degradation or transformation of chemical pollutants, including hydrocarbons, pharmaceutical substances, heterocyclic compounds, polychlorinated biphenyls, etc., provide a natural tool to manipulate the microbial environment to achieve clean sustainable water bodies. In recent years, a confluence of methodologies in the fields of genomics, metagenomics, and bioinformatics has provided scientists with insights into important pathways for biological degradation of chemical pollutants and provided information to understand the ability of microorganisms to adapt in vast environmental conditions.

Advances in sequencing technologies and development in microbiomics has provided much information to aid in the understanding of the molecular and genetic basis of biodegradation of chemical pollutants. Aromatic compounds are among the most recalcitrant of these pollutants, and lessons can be learned from the recent genomic studies of *Burkholderia xenovorans* LB400 and *Rhodococcus* sp. strain RHA1, the bacterial genomes of which are completely sequenced to date. This has helped in understanding microbial metabolisms, especially those of bacterials for catabolic and non-catabolic adaptations in polluted environments. It also explains the evolution of large bacterial genomes in *Burkholderia xenovorans*.

In *Pseudomonas*, it was demonstrated that lower numbers of aromatic pathways were involved in catabolism of the xenobiotic as well as natural aromatic compounds. During evolution, similar pathways were present in different taxa of bacteria. Recent studies with comparative functional genomics of bacteria established that pathways like Box and Paa were present in more phylogenetic groups. Also, the remarkable capacity of some of the large genome bacteria to catabolise aromatic pollutants was found to be ancient in origin (McLeod and Eltis 2008). Some of the organic pollutants are recalcitrant, and microbial degradation of these chemicals was of great importance. Hydrocarbons, especially halogenated compounds, were thought to degrade only through aerobic processes; this theory changed after isolation of some bacteria capable of anaerobically degrading hydrocarbon and dehalogenating halogen compounds (Cupples et al. 2005). Over the last decade, complete genomes of bacteria capable of anaerobic degradation of organic pollutants have been sequenced. The genome of bacteria *Aromatoleum aromaticum* was sequenced and, the first time, it was a complete sequence of ~4.7 Mb for a bacteria capable of anaerobic degradation of an organic pollutant. About 25 gene clusters were found in the genome of *A. aromaticum* catabolic network responsible for aromatic compound degradation. Niche adaption for the bacteria responsible for anaerobic catabolism of aromatic compounds was only possible with the presence of specific degrading pathways. Complete genome sequencing of *Dehalococcoides ethenogenes* (1.4 Mb) and *Desulfitobacterium hafniense* (5.7 Mb) also provided insights into their ability to dehalogenate halogenated aromatic compounds; this included the dehalogenation of brominated herbicide bromoxynil and iodinated compounds, either used for electron acceptor or in other ways during halorespiration (Heider and Rabus 2008).

Microbial Resource Collection Centers

Role in Maintaining the Microbial Resources

It is evident that new-generation sequencing techniques are opening ways for the identification of many previously unknown and unculturable bacterial genera and species; thus, new culture methods and modifications in existing protocols can be carried out for the isolation and characterization of novel microbes from diverse environments. The preservation without changes in the morphological and physiological characteristics of these newly isolated microbes is obligatory for future reference. Therefore, it is important that, after growing novel microbes, researchers devise appropriate preservation protocol(s) suitable for the microorganisms. Microbial resource centers also play an important role in the development of protocols related to long-term preservation; checking the viability and authenticity of preserved cultures; providing training in areas related to microbial handling, biosafety, and biosecurity; and offering reference strains to the scientific community for quality control and molecular biology research (Heylen et al. 2012). Nowadays, it is becoming very important to study microbial taxonomy, preservation, genomic studies, and microbial ecology to preserve the valuable gene pool and maintain a diversity of microorganisms for future use. Worldwide, countries understand the importance of microbial diversity and are establishing fully equipped microbial resource centers with experts from diverse fields of study for future research and application (Prakash et al. 2013).

The European Consortium of Microbial Resource Centres (EMbaRC) is one example. These centers are self-sustaining, with a large biodiversity of microorganisms. They also provide expert services for different research and have reference strains in their large collection of microorganisms. EMbaRC aims to make methods of strain identification uniform across the consortium centers and provide a service for validation of reference strains. EMbaRC maintains the quality of all consortium centers in Europe, thus fulfilling the research needs of individual centers. In India, similar services are provided by the National Bureau of Agriculturally Important Microorganisms and the Indian Institute of Microbial Technology; there is also the Indian type culture collection center at the Indian Agriculture Research Institute.

Conclusion and Future Direction

Community microbiology has developed from culture-dependent methods to culture-free identification of microbes in a community; this has been possible due to the development of techniques in sequencing and improved knowledge around molecular biology, particularly interactions among different proteins in an environment. Using 16S rRNA techniques, many new genera and species of microbes have been identified, and the relative frequency of many of those that are known have been fixed in some microbial niches. This provides an opportunity to understand the relative role of different members of the microbial community in micro-ecology. Microbiomics provides an opportunity to understand the behavior of microbes in the microbial community and their impact on the community. Human health is an important issue that seems to be largely influenced by the associated microbes present in and on different surfaces and cavities of the human body. Information about this microbial diversity has the potential for treating problems such as obesity and issues related to the gastrointestinal tract. In future, prebiotics and probiotics can be further developed based on specific problems. It can be seen as the clinical problems of a human community living in a particular locality and associated microflora in that community. Family- or individual-based treatments may also be developed based on their associated microflora. Plant health is greatly influenced by the microflora in the rhizosphere or phylloplane. Microbiomics opens a door to study the known and unknown microbial diversity around plant surfaces that either enhances plant growth by producing growth regulators or by inhibiting diseases and insect pests. Studies about treatment with chemicals produced by a specific group of microorganisms may be developed to improve certain microbes around plant surfaces for improved plant growth. Environmental pollutants are effectively combated with certain bacterial communities that have developed the capacity to utilize these pollutants as a sole carbon source. Agrochemicals such as herbicides, pesticides, and fertilizers can be degraded in polluted water bodies by those bacteria with an enhanced capacity to catalyze the chemical pollutants. With the help of new sequencing technologies, scientists can identify the specific microbes or genes present in a community for degrading agrochemicals and other chemical pollutants. New strains may also be developed with enhanced and multiple capacities for catabolism of pollutants.

Microbiomic studies are providing information to different microbial niches and comparing related ecosystems. These studies are improving clinical treatments for some problems such as obesity and gastric ulcers, which have otherwise been treated in different ways. In agriculture, it has been shown that the disease suppressiveness of soils is influenced by the presence of certain microbes; in polluted water bodies, microbiomics is becoming important in the identification of new genes from the microbiomes of water and sediments. This emerging field of microbiomics will greatly influence human life in the near future, as it has the potential to explore new sources of genes or establish the role of some known microbial genes in community.

References

Achtman M, Wagner M (2008) Microbial diversity and the genetic nature of microbial species. Nat Rev Microbiol 6:431–440. doi:10.1038/nrmicro1872

Allison SD, Martiny JBH (2008) Resistance, resilience, and redundancy in microbial communities. Proc Natl Acad Sci U S A 105:11512–11519

Amann RI, Krumholz L, Stahl DA (1990) Fluorescent-oligonucleotide probing of whole cells for determinative, phylogenetic, and environmental studies in microbiology. J Bacteriol 172:762–770

Bakker PAHM, Berendsen RL, Doornbos RF, Wintermans PCA, Pieterse CM (2013) The rhizosphere revisited: root microbiomics. Front Plant Sci 165:1–7

Barata A, Malfeito-Ferreira M, Loureiro V (2012) The microbial ecology of wine grape berries. Int J Food Microbiol 153:243–259

Barret M, Frey-Klett P, Guillerm-Erckelboudt AV, Boutin M, Guernec G, Sarniguet A (2009) Effect of wheat roots infected with the pathogenic fungus *Gaeumannomyces graminis* var. *tritici* on gene expression of the biocontrol bacterium *Pseudomonas fluorescens* Pf29Arp. Mol Plant Microbe Interact 22:1611–1623

Bartowski EJ (2009) Bacterial spoilage of wine and approaches to minimize it. Lett Appl Microbiol 48:149–156

Behar A, Jurkevitch E, Yuval B (2008) Bringing back the fruit into fruit fly – bacteria interactions. Mol Ecol 17:1375–1386

Beja O, Spudich EN, Spudich JL, Leclerc M, DeLong EF (2001) Proteorhodopsin phototrophy in the ocean. Nature 411:786–789

Berry AE, Chiocchini C, Selby T, Sosio M, Wellington EM (2003) Isolation of high molecular weight DNA from soil for cloning into BAC vectors. FEMS Microbiol Lett 223:15–20

Bohlool BB, Brock TD (1974) Immunofluorescence approach to the study of the ecology of Thermoplasma acidophilum in coal refuse material. Appl Microbiol 28:11–16

Booijink CC, Boekhorst J, Zoetendal EG, Smidt H, Kleerebezem M et al (2010) Metatranscriptome analysis of the human fecal microbiota reveals subject-specific expression profiles, with genes encoding proteins involved in carbohydrate metabolism being dominantly expressed. Appl Environ Microbiol 76:5533–5540

Borody TJ (2000) "Flora Power" – fecal bacteria cure chronic C. difficile diarrhea. Am J Gastroenterol 95:3028–3029. doi:10.1111/j.1572-0241.2000.03277.x

Bott TL, Brock TD (1969) Bacterial growth rates above 90 degrees C in Yellowstone hot springs. Science 164:1411–1412

Brock TD (1967) Life at high temperatures. Science 158:1012–1019

Brock TD, Brock ML (1968) Measurement of steady-state growth rates of a thermophilic alga directly in nature. J Bacteriol 95:811–815

Buckley MJM, O'Morain CA (1998) *Helicobacter* biology discovery. Br Med Bull 54:7–16

Bulgarelli D, Rott M, Schlaeppi K, Ver Loren van Themaat E, Ahmadinejad N, Assenza F et al (2012) Revealing structure and assembly cues for *Arabidopsis* root-inhabiting bacterial microbiota. Nature 488:91–95. doi:10.1038/nature11336

Carvalhais LC, Dennis PG, Badri DV, Tyson GW, Vivanco JM, Schenk PM (2013) Activation of the jasmonic acid plant defence pathway alters the composition of rhizosphere bacterial communities. PLoS One 8:e56457. doi:10.1371/journal.pone.0056457

Chen K, Pachter L (2005) Bioinformatics for whole-genome shotgun sequencing of microbial communities. PLoS Comput Biol 1:106–112. doi:10.1371/journal.pcbi.0010024

Cho JC, Giovannoni SJ (2004) Cultivation and growth characteristics of a diverse group of oligotrophic marine gammaproteobacteria. Appl Environ Microbiol 70:432–440

Colwell RR, Grimes DJ (eds) (2000) Nonculturable microorganisms in the environment. ASM Press, Washington, DC

Colwell RR, Tamplin ML, Brayton PR, Gauzens AL, Tall BD, Harrington D, Levine MM, Hall S, Huq A, Sack DA (1990) Environmental aspects of *V. cholerae* in transmission of cholera. In: Sack RB, Zinnaka Y (eds) Advances in research on cholera and related diarrhoeas, 7th edn. KTK Scientific Publications, Tokyo, pp 327–343

Cortesi P, Bisiach M, Ricciolini M, Gadoury DM (1997) Cleistothecia of Uncinula necator – an additional source of inoculum in Italian vineyards. Plant Dis 81:922–926

Courtois S, Cappellano CM, Ball M, Francou FX, Normand P, Helynck G, Martinez A, Kolvek SJ, Hopke J, Osburne MS, August PR, Nalin R, Guerineau M, Jeannin P, Simonet P, Pernodet JL (2003) Recombinant environmental libraries provide access to microbial diversity for drug discovery from natural products. Appl Environ Microbiol 69:49–55

Cupples AM, Sanford RA, Sims GK (2005) Dehalogenation of bromoxynil (3,5-dibromo-4-hydroxybenzonitrile) and ioxynil (3,5-diiodino-4-hydroxybenzonitrile) by desulfitobacterium chlororespirans. Appl Environ Microbiol 71(7):3741–3746

Dahl Sawyer CA, Pestka JJ (1985) Foodservice systems: presence of injured bacteria in foods during food product flow. Annu Rev Microbiol 39:51–67

DeCoste NJ, Gadkar VJ, Filion M (2010) Verticillium dahliae alters *Pseudomonas* spp. populations and HCN gene expression in the rhizosphere of strawberry. Can J Microbiol 56:906–915. doi:10.1139/W10-080

Delmotte N, Knief C, Chaffron S, Innerebner G, Roschitzki B, Schlapbach R, von Mering C, Vorholt JA (2009) Community proteogenomics reveals insights into the physiology of phyllosphere bacteria

Dethlefsen L, Relman DA (2010) Microbes and health sackler colloquium: incomplete recovery and individualized responses of the human distal gut microbiota to repeated antibiotic perturbation. Proc Natl Acad Sci U S A. doi:10.1073/pnas.1000087107

Dimakopoulou M, Tjamos SE, Antoniou PP, Pietri A, Battilani P, Avramidis N, Markakis EA, Tjamos EC (2008) Phyllosphere grapevine yeast *Aureobasidium pullulans* reduces *Aspergillus carbonarius* (sour rot) incidence in wine-producing vineyards in Greece. Biol Control 46:158–165

Doenges JL (1938) Spirochaetes in the gastric glands of *Macacus rhesus* and humans without definite history of related disease. Proc Soc Exp Biol Med 38:536–538

Dominati E, Patterson M, Mackay A (2010) A framework for classifying and quantifying the natural capital and ecosystem services of soils. Ecol Econ 69:1858–1868

Doornbos RF, VanLoon LC, Bakker PAHM (2012) Impact of root exudates and plant defense signaling on bacterial communities in the rhizosphere. Agron Sustain Dev 32:227–243. doi:10.1007/s13593-011-0028-y

Drews G (1999) Ferdinand Cohn: a promoter of modern microbiology. Nova Acta Leopold 80(130):13–14

Fan B, Cravalhais LC, Becker A, Fedoseyenko D, VonWiren N, Borriss R (2012) Transcriptomic profiling of *Bacillus amyloliquefaciens* FZB42 in response to maize root exudates. BMC Microbiol 12:116. doi:10.1186/1471-2180-12-116

Garrett WS, Gordon JI, Glimcher LH (2010) Homeostasis and inflammation in the intestine. Cell 140:859–870

Geison GL (1981) Cohn, Ferdinand Julius. In: Gillispie CC (ed) Dictionary of scientific biography, vol 3. Scribner, New York

Germida JJ, Siciliano SD, DeFreitas JR, Seib AM (1998) Diversity of root-associated bacteria associated with field grown canola (*Brassica napus* L.) and wheat(*Triticum aestivum* L.). FEMS Microbiol Ecol 26:43–50. doi:10.1111/j.1574-6941.1998.tb01560.x

Giannoukos G, Ciulla DM, Huang K, Haas BJ, Izard J et al (2012) Efficient and robust RNA-seq process for cultured bacteria and complex community transcriptomes. Genome Biol 13:R23. doi:10.1186/gb-2012-13-3-r23

Gianoulis TA, Raes J, Patel PV, Bjornson R, Korbel JO et al (2009) Quantifying environmental adaptation of metabolic pathways in metagenomics. Proc Natl Acad Sci U S A 106:1374–1379. doi:10.1073/pnas.0808022106

Grayston SJ, Wang S, Campbell CD, Edwards AC (1998) Selective in fluence of plant species on microbial diversity in the rhizosphere. Soil Biol Biochem 30:369–378. doi:10.1016/S0038-0717(97)00124-7

Grimes DJ, Atwell RW, Brayton PR, Palmer LM, Rollins DM, Roszak DB, Singleton FL, Tamplin ML, Colwell RR (1986) The fate of enteric pathogenic bacteria in estuarine and marine environments. Microbiol Sci 3:324–329

Grove GG (2004) Perennation of Uncinula necator in vineyards of eastern Washington. Plant Dis 88:242–247

Guarner F, Malagelada JR (2003) Gut flora in health and disease. Lancet 361:512–519

Haichar FZ, Marol C, Berge O, Rangel-Castro JI, Prosser JI, Balesdent J et al (2008) Plant host habitat and root exudates shape soil bacterial community structure. ISME J 2:1221–1230. doi:10.1038/ismej.2008.80

Hamady M, Knight R (2009) Microbial community profiling for human microbiome projects: tools, tech-

niques, and challenges. Genome Res 19:1141–1152. doi:10.1101/gr.085464.108
Harmsen HJM, Welling GW (2002) Fluorescence in situ hybridization as a tool in intestinal bacteriology. In: Tannock GW (ed) Probiotics and prebiotics: where are we going? Caister Academic Press, Norfolk, pp 41–58
Harmsen HJM, Raangs GC, He T, Degener JE, Welling GW (2002) Extensive set of 16S rRNA-based probes for detection of bacteria in human feces. Appl Environ Microbiol 68:2982–2990
Hartmann M, Niklaus PA, Zimmermann S, Schmutz S, Kremer J, Abarenkov K, Lüscher P, Widmer F, Frey B (2014) Resistance and resilience of the forest soil microbiome to logging-associated compaction. ISME J 8(1):226–244
Hehemann JH, Correc G, Barbeyron T, Helbert W, Czjzek M et al (2010) Transfer of carbohydrate-active enzymes from marine bacteria to Japanese gut microbiota. Nature 464:908–912. doi:10.1038/nature08937
Heider J, Rabus R (2008) Genomic insights in the anaerobic biodegradation of organic pollutants. In: Diaz E (ed) Microbial degradation, Genomics and molecular biology. Caister Academic Press, Norfolk, pp 25–54. ISBN 978-1-904455-17-2
Heuer H, Smalla K (2012) Plasmids foster diversification and adaptation of bacterial populations in soil. FEMS Microbiol Rev 36:1083–1104
Heylen K, Hoefman S, Vekeman B, Peiren J, De Vos P (2012) Safeguarding bacterial resources promotes biotechnological innovation. Appl Microbiol Biotechnol 94:565–574
Hirsch PR, Mauchline TH (2012) Who's who in the plant root microbiome? Nat Biotechnol 30:961–962. doi:10.1038/nbt.2387
Hunter PJ, Hand P, Pink D, Whipps JM, Bending GD (2011) Both leaf properties and microbe-microbe interactions influence within-species variation in bacterial population diversity and structure in the lettuce (*Lactuca* species) phyllosphere. Appl Environ Microbiol 76:8117–8125
Inceoglu O, VanOverbeek LS, Salles JF, Van Elsas JD (2013) The normal operating range of bacterial communities in soil used for potato cropping. Appl Environ Microbiol 79:1160–1170. doi:10.1128/AEM.02811-12
Jansen GJ, Wildeboer-Veloo AC, Tonk RH, Franks AH, Welling GW (1999) Development and validation of an automated, microscopy-based method for enumeration of groups of intestinal bacteria. J Microbiol Methods 37:215–221
Janssen PH, Yates PS, Grinton BE, Taylor PM, Sait M (2002) Improved culturability of soil bacteria and isolation in pure culture of novel members of the divisions acidobacteria, actinobacteria, proteobacteria, and verrucomicrobia. Appl Environ Microbiol 68:2391–2396
Jansson JK, Neufeld JD, Moran MA, Gilbert JA (2012) Omics for understanding microbial functional dynamics. Environ Microbiol 14:1–3. doi:10.1111/j.1462-2920.2011.02518
Jia W, Li H, Zhao L, Nicholson JK (2008) Gut microbiota: a potential new territory for drug targeting. Nat Rev Drug Discov 7:123–129. doi:10.1038/nrd2505
Johnson RA, Wichern DW (2007) Applied multivariate statistical analysis, 6th edn. Prentice Hall, Englewood Cliffs
Jousset A, Rochat L, Lanoue A, Bonkowski M, Keel C, Scheu S (2011) Plants respond to pathogen infection by enhancing the antifungal gene expression of root-associated bacteria. Mol Plant Microbe Interact 24:352–358. doi:10.1094/MPMI-09-10-0208
Khoruts A, Dicksved J, Jansson JK, Sadowsky MJ (2010) Changes in the composition of the human fecal microbiome after bacteriotherapy for recurrent Clostridium difficile-associated diarrhea. J Clin Gastroenterol 44:354–360. doi:10.1097/MCG.0b013e3181c87e02
Kirk JL, Klironomos JN, Lee H, Trevors JT (2005) The effects of perennial ryegrass and alfalfa on microbial abundance and diversity in petroleum contaminated soil. Environ Pollut 133:455–465. doi:10.1016/j.envpol.2004.06.002
Koch B, Worm J, Jensen LE, Hojberg O, Nybroe O (2001) Carbon limitation induces σs-dependent gene expression in *Pseudomonas fluorescens* in soil. Appl Environ Microbiol 67:3363–3370. doi:10.1128/AEM.67.8.3363-3370.2001
Koukkou A-I, Vandera E (2011) Hydrocarbon-degrading soil bacteria: current research. In: Koukkou A-I (ed) Microbial bioremediation of non-metals: current research. Caister Academic Press, Norfolk, pp 93–117. ISBN 978-1-904455-83-7
Kwak YS, Bonsall RF, Okubara PA, Paulitz TC, Thomashow LS, Weller DM (2012) Factors impacting the activity of 2,4-diacetylphloroglucinol-producing *Pseudomonas fluorescens* against take-all of wheat. Soil Biol Biochem 54:48–56. doi:10.1016/j.soilbio.2012.05.012
Kyselkova M, Kopecky J, Frapolli M, Defago G, Sagova-Mareckova M, Grundmann GL et al (2009) Comparison of rhizobacterial community composition in soil suppressive or conducive to tobacco black root rot disease. ISME J 3:1127–1138. doi:10.1038/ismej.2009.61
Lee B, Lee S, Ryu MR (2012) Foliar aphid feeding recruits rhizosphere bacteria and primes plant immunity against pathogenic and nonpathogenic bacteria in pepper. Ann Bot 110:281–290. doi:10.1093/aob/mcs055
Lemanceau P, Corberand T, Gardan L, Latour X, Laguerre G, Boeuf-gras J-M et al (1995) Effect of two plant species, flax (Linum usitatissimum L.) and tomato (Lycopersicon esculentum Mill.), on the diversity of soil borne populations of fluorescent pseudomonads. Appl Environ Microbiol 61:1004–1012
Leveau JHJ (2009) Life on leaves. Nature 461:741
Leveau JHJ, Tech JJ (2011) Grapevine microbiomics: Bacterial diversity on grape leaves and berries revealed by high-throughput sequence analysis of 16S rRNA amplicons. Acta Horticult 905:31–42

Ley RE (2010) Obesity and the human microbiome. Curr Opin Gastroenterol 26:5–11. doi:10.1097/MOG.0b013e328333d751

Li X, LeBlanc J, Truong A, Vuthoori R, Chen SS et al (2011) A metaproteomic approach to study human-microbial ecosystems at the mucosal luminal interface. PLoS One 6:e26542. doi:10.1371/journal.pone.0026542

Lorenz P, Liebeton K, Niehaus F, Eck J (2002) Screening for novel enzymes for biocatalytic processes: accessing the metagenome as a resource of novel functional sequence space. Curr Opin Biotechnol 13:572–577

Lozupone C, Knight R (2005) UniFrac: a new phylogenetic method for comparing microbial communities. Appl Environ Microbiol 71:8228–8235. doi:10.1128/AEM. 71.12.8228-8235.2005

Lugtenberg B, Kamilova F (2009) Plant-growth-promoting rhizobacteria. Annu Rev Microbiol 63:541–556. doi:10.1146/annurev.micro.62.081307

Lundberg DS, Lebeis SL, Paredes SH, Yourstone S, Gehring J, Malfatti S, Tremblay J, Engelbrektson A, Victor Kunin V, del Rio TG, Edgar RC, Eickhorst T, Ley RE, Hugenholtz P, Tringe SG, Dangl JL (2012) Defining the core *Arabidopsis thaliana* root microbiome. Nature 488(7409):86–90

Mark GL, Dow JM, Kiely PD, Higgings H, Haynes J, Baysse C et al (2005) Transcriptome profiling of bacterial responses to root exudates identifies genes involved in microbe plant interactions. Proc Natl Acad Sci U S A 102:17454–17459. doi:10.1073/pnas.0506407102

Marshall BJ, McGechie DB, Rogers PA, Glancy RJ (1985a) Pyloric campylobacter infection and gastroduodenal disease. Med J Aust 142:439–444

Marshall BJ, Armstrong JA, McGechie DB, Glancy RJ (1985b) Attempt to fulfil Koch's postulates for pyloric *Campylobacter*. Med J Aust 142:436–439

Martin FP, Wang Y, Sprenger N, Yap IK, Lundstedt T et al (2008) Probiotic modulation of symbiotic gut microbial-host metabolic interactions in a humanized microbiome mouse model. Mol Syst Biol 4:157. doi:10.1038/msb4100190

Martins G, Lauga B, Miot-Sertier C, Mercier A, Lonvaud A, Soulas M-L, Soulas G, de Masneuf-Pomaré I (2013) Characterization of epiphytic bacterial communities from grape, leaves, bark and soil of grapevine plants grown, and their relation. PLoS One 8(8):1–9

Mavrodi OV, Mavrodi DV, Parejko JA, Thomashow LS, Weller DM (2012) Irrigation differentially impacts populations of indigenous antibiotic producing Pseudomonas spp. in the rhizosphere of wheat. Appl Environ Microbiol 78:3214–3220. doi:10.1128/AEM.07968-11

Mazumdar PMH (1995) Species and specificity: an interpretation of the history of immunology. Cambridge University Press, New York

Mazurier S, Corberand T, Lemanceau P, Raaijmakers JM (2009) Phenazine antibiotics produced by fluorescent pseudomonads contribute to natural soil suppressiveness to *Fusarium* wilt. ISME J 3:977–991. doi:10.1038/ismej.2009.33

Mazzola M (2002) Mechanisms of natural soil suppressiveness to soil borne diseases. Antonie Van Leeuwenhoek 81:557–564. doi:10.1023/A:1020557523557

McLeod MP, Eltis LD (2008) Genomic insights into the aerobic pathways for degradation of organic pollutants. In: Diaz E (ed) Microbial degradation, Genomics and molecular biology. Caister Academic Press, Norfolk, pp 1–24. ISBN 978-1-904455-17-2

Mendes R, Kruijt M, DeBruijn I, Dekkers E, VanderVoort M, Schneider JHM et al (2011) Deciphering the rhizosphere microbiome for disease suppressive bacteria. Science 332:1097–1100. doi:10.1126/science.1203980

Micallef SA, Shiaris MP, Colon-Carmona A (2009) Influence of *Arabidopsis thaliana* accessions on rhizobacterial communities and natural variation in root exudates. J Exp Bot 60:1729–1742. doi:10.1093/jxb/erp053

Morgan C, Huttenhowr C (2012) Chapter 12: human microbiome analysis. PLoS Comput Biol. doi:10.1371/journal.pcbi.1002808

Mosser JL, Bohlool BB, Brock TD (1974) Growth rates of Sulfolobus acidocaldarius in nature. J Bacteriol 118:1075–1081

Munkvold GP, Marois JJ (1993) Efficacy of natural epiphytes and colonizers of grapevine pruning wounds for biological control of Eutypa dieback. Phytopathology 83:624–629

Nasidze I, Li J, Quinque D, Tang K, Stoneking M (2009) Global diversity in the human salivary microbiome. Genome Res 19:636–643. doi:10.1101/gr.084616.108

Nicholson JK, Holmes E, Wilson ID (2005) Gut microorganisms, mammalian metabolism and personalized health care. Nat Rev Microbiol 3:431–438. doi:10.1038/nrmicro1152

Okubara PA, Bonsall RF (2008) Accumulation of *Pseudomonas*-derived 2,4-diacetylphloroglucinol on wheat seedling roots is influenced by hostcultivar. Biol Control 46:322–331. doi:10.1016/j.biocontrol.2008.03.013

Pearson RC, Goheen AC (eds) (2008) Compendium of grape diseases. APS Press, St. Paul

Pechy-Tarr M, Borel N, Kupfer-Schmied P, Turner V, Binggeli O, Radovanovic D et al (2013) Control and host-dependent activation of insect toxin expression in a root-associated biocontrol pseudomonad. Environ Microbiol 15:736–750. doi:10.1111/1462-2920.12050

Poretsky RS, Hewson I, Sun S, Allen AE, Zehr JP et al (2009) Comparative day/night metatranscriptomic analysis of microbial communities in the North Pacific subtropical gyre. Environ Microbiol 11:1358–1375. doi:10.1111/j.1462-2920.2008.01863.x

Prakash O, Shouche Y, Jangid K, Kostka JE (2013) Microbial cultivation and the role of microbial resource centers in the omics era. Appl Microbiol Biotechnol 97:51–62

Raaijmakers JM, Weller DM (1998) Natural plant protection by 2,4-diacetylphloroglucinol-producing *Pseudomonas* spp. in take-all decline soils. Mol Plant Microbe Interact 11:144–152. doi:10.1094/MPMI.1998.11.2.144

Rappe MS, Connon SA, Vergin KL, Giovannoni SJ (2002) Cultivation of the ubiquitous SAR11 marine bacterioplankton clade. Nature 418:630–633

Redford AJ, Bowers RM, Knight R, Linhart Y, Fierer N (2010) The ecology of the phyllosphere: geographic and phylogenetic variability in the distribution of bacteria on tree leaves. Environ Microbiol 12:2885–2893

Rosenzweig N, Tiedje JM, Quensen JF III, Meng Q, Hao JJ (2012) Microbial communities associated with potato common scab-suppressive soil determined by pyrosequencing analyses. Plant Dis 96:718–725. doi:10.1094/PDIS-07-11-0571

Sait M, Hugenholtz P, Janssen PH (2002) Cultivation of globally distributed soil bacteria from phylogenetic lineages previously only detected in cultivation-independent surveys. Environ Microbiol 4:654–666

Schloss PD (2010) The effects of alignment quality, distance calculation method, sequence filtering, and region on the analysis of 16S rRNA gene-based studies. PLoS Comput Biol 6:e1000844. doi:10.1371/journal.pcbi.1000844

Schloss PD, Westcott SL, Ryabin T, Hall JR, Hartmann M et al (2009) Introducing mothur: open-source, platform-independent, community-supported software for describing and comparing microbial communities. Appl Environ Microbiol 75:7537–7541. doi:10.1128/AEM.01541-09

Schreiner K, Hagn A, Kyselkova M, Moenne-Loccoz Y, Welzl G, Munch JC et al (2010) Comparison of barley succession and take-all disease as environmental factors shaping the rhizobacterial community during take-all decline. Appl Environ Microbiol 76:4703–4712. doi:10.1128/AEM.00481-10

Schwiertz A, Le Blay G, Blaut M (2000) Quantification of different *Eubacterium* spp. in human fecal samples with species-specific 16S rRNA targeted oligonucleotide probes. Appl Environ Microbiol 66:375–382

Sessitsch A, Weilharter A, Gerzabek MH, Kirchmann H, Kandeler E (2001) Microbial population structures in soil particle size fractions of a long-term fertilizer field experiment. Appl Environ Microbiol 67:4215–4224

Shi Y, Tyson GW, DeLong EF (2009) Metatranscriptomics reveals unique microbial small RNAs in the ocean's water column. Nature 459:266–269. doi:10.1038/nature08055

Simu K, Hagstrom A (2004) Oligotrophic bacterioplankton with a novel single-cell life strategy. Appl Environ Microbiol 70:2445–2451

Smit E, Leeflang P, Gommans S, van den Broek J, van Mil S, Wernars K (2001) Diversity and seasonal fluctuations of the dominant members of the bacterial soil community in a wheat field as determined by cultivation and molecular methods. Appl Environ Microbiol 67:2284–2291

Sogin ML, Morrison HG, Huber JA et al (2006) Microbial diversity in the deep sea and the underexplored "rare biosphere". Proc Natl Acad Sci USA 103(32):12115–12120

Staley JT, Konopka A (1985) Measurement of in situ activities of nonphotosynthetic microorganisms in aquatic and terrestrial habitats. Annu Rev Microbiol 39:321–346

Stein JL, Marsh TL, Wu KY, Shizuya H, DeLong EF (1996) Characterization of uncultivated prokaryotes: isolation and analysis of a 40-kilobase-pair genome fragment front a planktonic marine archaeon. J Bacteriol 178:591–599

Torsvik V, Ovreas L (2002) Microbial diversity and function in soil: from genes to ecosystems. Curr Opin Microbiol 5:240–245

Torsvik V, Goksoyr J, Daae FL (1990) High diversity in DNA of soil bacteria. Appl Environ Microbiol 56:782–787

Turnbaugh PJ, Gordon JI (2008) An invitation to the marriage of metagenomics and metabolomics. Cell 134:708–713. doi:10.1016/j.cell.2008.08.025

Turnbaugh PJ, Ley RE, Mahowald MA, Magrini V, Mardis ER et al (2006) An obesity-associated gut microbiome with increased capacity for energy harvest. Nature 444:1027–1031. doi:10.1038/nature05414

Turnbaugh PJ, Hamady M, Yatsunenko T, Cantarel BL, Duncan A et al (2009) A core gut microbiome in obese and lean twins. Nature 457:480–484. doi:10.1038/nature07540

Tyson GW, Chapman J, Hugenholtz P, Allen EE, Ram RJ, Richardson PM, Solovyev VV, Rubin EM, Rokhsar DS, Banfield JF (2004) Community structure and metabolism through reconstruction of microbial genomes from the environment. Nature 428:37–43

van der Meer JR (2008) A genomic view on the evolution of catabolic pathways and bacterial adaptation to xenobiotic compounds. In: Diaz E (ed) Genomics and molecular biology. Caister Academic Press, Madrid, pp 219–269

Van-Camp L, Bujarrabal B, Gentile AR, Jones RJA, Montanarella L, Olazabal C et al (2004) Reports of the technical working groups established under the thematic strategy for soil protection. Office for Official Publication of the European Communities, Luxembourg, p 872

Varela C, Siebert T, Cozzolino D, Rose L, McLean H, Henschke PA (2009) Discovering a chemical basis for differentiating wines made by fermentation with 'wild' indigenous and inoculated yeasts: role of yeast volatile compounds. Aust J Grape Wine Res 15:238–248

Vaughan EE, Schut F, Heilig GHJ, Zoetendal EG, de Vos WM, Akkermans ADL (2000) A molecular view of the intestinal ecosystem. Curr Issues Intest Microbiol 1:1–12

Verberkmoes NC, Russell AL, Shah M, Godzik A, Rosenquist M et al (2009) Shotgun metaproteomics of the human distal gut microbiota. ISME J 3:179–189

Venter JC, Remington K, Heidelberg JF, Halpern AL, Rusch D et al (2004a) Environmental genome shotgun sequencing of the Sargasso Sea. Science 304:66–74. doi:10.1126/science.1093857

Venter JC, Remington K, Heidelberg JF, Halpern AL, Rusch D, Eisen JA, Wu D, Paulsen I, Nelson KE, Nelson W, Fouts DE, Levy S, Knap AH, Lomas MW, Nealson K, White O, Peterson J, Hoffman J, Parsons R, Baden-Tillson H, Pfannkoch C, Rogers YH, Smith HO (2004b) Environmental genome shotgun sequencing of the Sargasso Sea. Science 304:66–74

Verginer M, Leitner E, Berg G (2010) Production of volatile metabolites by grape associated microorganisms. J Agric Food Chem 58(14):8344–8350

Wallner G, Fuchs B, Spring S, Beisker W, Amann RI (1997) Flow sorting of micro organisms for molecular analysis. Appl Environ Microbiol 63:4223–4231

Wang Q, Garrity GM, Tiedje JM, Cole JR (2007) Naive Bayesian classifier for rapid assignment of rRNA sequences into the new bacterial taxonomy. Appl Environ Microbiol 73:5261–5267. doi:10.1128/AEM.00062-07

Weinert N, Piceno Y, Ding GC, Meincke R, Heuer H, Berg G et al (2011) PhyloChip hybridization uncovered an enormous bacterial diversity in the rhizosphere of different potato cultivars: many common and few cultivar dependent taxa. FEMS Microbiol Ecol 75:497–506. doi:10.1111/j.1574-6941.2010.01025.x

Weller DM, Raaijmakers JM, McSpadden Gardener BB, Thomashow LS (2002) Microbial populations responsible for specific soil suppressiveness to plant pathogens. Annu Rev Phytopathol 40:309–348. doi:10.1146/annurev.phyto.40.030402.110010

Whang K, Hattori T (1988) Oligotrophic bacteria from rendzina forest soil. Antonie Van Leeuwenhoek 54:19–36

Widenfalk A, Bertilsson S, Sundh I, Goedkoop W (2008) Effects of pesticides on community composition and activity of sediment microbes–responses at various levels of microbial community organization. Environ Pollut 152(3):576–584

Yang JW, Yi HS, Kim H, Lee B, Lee S, Ghim SY, Ryu CM (2011) Whitefly infestation of pepper plants elicits defence responses against bacterial pathogens in leaves and roots and changes the below-ground microflora. J Ecol 99:46–56

Zijnge V, van Leeuwen MB, Degener JE, Abbas F, Thurnheer T et al (2010) Oral biofilm architecture on natural teeth. PLoS One 5:e9321. doi:10.1371/journal.pone.0009321

Zoetendal EG, Ben-Amor K, Harmsen HJM, Schut F, Akkermans ADL, de Vos WM (2002) Quantification of uncultured *Ruminococcus obeum*-like bacteria in human fecal samples by fluorescent in situ hybridization and flow cytometry using 16S rRNA-targeted probes. Appl Environ Microbiol 68:4225–4232

Zysko A, Sanguin H, Hayes A, Wardleworth L, Zeef LAH, Sim A et al (2012) Transcriptional response of *Pseudomonas aeruginosa* to a phosphate-deficient *Lolium perenne* rhizosphere. Plant Soil 359:25–44. doi:10.1007/s11104-011-1060-z

Cryobionomics: Evaluating the Concept in Plant Cryopreservation

Marcos E. Martinez-Montero and Keith Harding

Contents

M.E. Martinez-Montero, Ph.D. (✉)
Bioplantas Center, Plant Breeding Laboratory, University of Ciego de Ávila, Car. a Moron km 9, CP 69450, Ciego de Avila, Cuba
e-mail: marcosem@bioplantas.cu; cubaplantas@gmail.com

K. Harding, Ph.D.
Damar Research Scientists, Damar, Drum Road, Cuparmuir, Fife KY15 5RJ, Scotland, UK
e-mail: k.harding-damar@tiscali.co.uk

Abstract

Cryopreservation at ultra-low temperatures is used for the long-term conservation of nonorthodox seeds and the germplasm of vegetatively propagated species. Advances in biomolecular or 'omics' technologies are creating a new knowledge base that provides insights into how to solve some of the more difficult cryobiological and conservation challenges. Before routinely implementing cryostorage, it is important to verify that it does not have any genotypic and/or phenotypic destabilising effects and that plants produced from cryopreserved germplasm are true-to-type. The evolving concept of 'cryobionomics' considers two practical aspects: (1) the linkage between cryoinjury and stability in vitro and

D. Barh et al. (eds.), *PlantOmics: The Omics of Plant Science*,
DOI 10.1007/978-81-322-2172-2_23, © Springer India 2015

(2) the behaviour and functionality of plants recovered from cryopreserved germplasm once they are reintroduced into natural environments. Cryobionomics is a working hypothesis that explores the emerging research evidence that connects causal factors related to cryoinjury and loss of viability to the risks of genetic instability. This paper presents the principles of cryopreservation and reviews contemporary omics research literature within the conceptual framework of cryobionomics. The aim is to explore the connections between stability and cryogenic/non-cryogenic stress factors with a view to aiding protocol improvement, optimisation and validation for plant genetic resources conservation.

Keywords
Cryoinjury • Vitrification • Viability • Plant germplasm • Genetic stability • Cell signalling

Introduction

Cryopreservation is the long-term storage of viable biological resources (Day et al. 2008) at ultra-low temperatures (−196 °C) usually in liquid nitrogen (LN). It is the method of choice for ensuring the long-term storage of nonorthodox seeds, the germplasm of vegetatively propagated species and biotechnologically important plant cell lines (Benson 2008a; Engelmann 2004, 2011; Tsai and Hubscher 2004). At cryogenic temperatures, it is generally understood that all cellular divisions and metabolic processes are stopped and that plant germplasm can be maintained indefinitely without genetic alteration. Phytosanitary tested in vitro cultures may be stored in relatively small volumes in cryobanks that preserve genetic stability with limited routine maintenance (Kaczmarczyk et al. 2012; Kaviani 2011). For cryopreservation to be successful, it is vital to prevent the formation of intracellular ice, and hence vitrification is the preferred method for cryopreserving plant germplasm in LN (Sakai 2000). Ideally, cryopreservation protocols should be easy to use and readily available (Kami 2012; Kaviani 2011); they include rapid and controlled rate cooling, colligative cryoprotection and vitrification (chemical cryoprotectants, droplet freezing, droplet vitrification, encapsulation-dehydration and desiccation). These techniques have been successfully applied to some 200 amenable species (Dulloo et al. 2010) using diverse cell and tissue types: cell cultures and suspensions, calluses, apices and somatic and zygotic embryos (Benson et al. 2002; Reed 2008a; Normah et al. 2012). The cryogenic process typically comprises multiple stages including non-cryogenic manipulations, ranging from tissue culture, pregrowth, acclimation, cryoprotection (colligative or vitrification), cooling, thawing or rewarming and recovery (Reed 2008a, b).

In the case of vegetatively propagated species, cryopreservation is widely applied to many economically important plants; protocols are established for root and tuber crops, fruit trees, ornamentals, forestry species and plantation crops from temperate and tropical origin (Engelmann 2000; Kaczmarczyk et al. 2012). Problematic germplasm is either nonorthodox or produces dehydration-sensitive, recalcitrant seeds for which storage problems can be circumvented in some cases by cryopreserving embryos and/or embryonic axes (Engelmann 1991, 1997). Recalcitrance is a dynamic concept which has evolved with advances in seed biology and improvements in cryostorage (Engelmann 2011) that effectively expedite problematic germplasm conservation and utilisation (Normah et al. 2012; Harding et al. 2013).

Contemporary plant cryopreservation research is now underpinned by sophisticated biomolecular and omics technologies to advance the understanding of storage stability (Benson et al. 2013). Cryobionomics was first presented at the Society for Low Temperature Biology (Harding 2002) and in a subsequent review concerning the genetic integrity of cryopreserved plant cells (Harding 2004). The concept was proposed to embrace the diverse research that describes genetic instability in plants derived from cryopreservation and was further developed within the EU projects COBRA and CRYMCEPT to gain insights into the complex relationship between cryoinjury and stability (Harding 2010; Harding et al. 2005). Cryobionomics was later built into the working group activities of the European project CRYOPLANET-COST Action 871 (Grapin et al. 2011; Lynch et al. 2011a; Harding 2010), a South African networking initiative (Berjak et al. 2011a), and has been presented at various scientific venues (Harding et al. 2008a, 2009). The implications of cryobionomics have been considered in a range of applications, from tropical plant germplasm conservation (Benson 2008a; Janardhan 2007; Harding 2010; Berjak et al. 2011a; Harding and Benson 2012) to algal culture collections (Harding et al. 2008b) and crop genebanks (Benson et al. 2011a, b, c). Cryobionomics is an evolving hypothesis that deals with two practical aspects of cryopreservation: (1) the linkage between cryoinjury and genetic stability and (2) the behaviour and functionality of plants following their reintroduction in natural habitats and environments (Harding 2004).

This chapter comprises two sections: the first describes the general principles of cryopreservation and outlines the key stages for protocol development, mainly for vegetatively propagated species and using sugar cane as a case study (Martinez-Montero et al. 2012). The second considers the developing framework of cryobionomics and how plant-related omics technologies may be applied to assist storage protocol improvement. The full spectrum of cryostorage operations is considered, from germplasm donor to the regeneration of cryopreserved plantlets, and assessing their genetic stability and confirmation of trueness-to-type (Perazzo et al. 2000).

Conservation and Cryopreservation

Some of the world's most important food crops are clonally propagated including (Benson et al. 2011b) *Dioscorea* spp. (yam), *Solanum* spp. (potato), *Musa* spp. (banana), *Manihot* spp. (cassava), *Colocasia esculenta* (taro), *Saccharum* sp. hybrids (sugarcane) and *Ipomoea batatas* (sweet potato). Traditionally, in situ genetic conservation is achieved in clonal field collections, but these are labour-intensive and can expose genetic resources to risks, particularly pests, pathogens and/or natural disasters (González-Benito et al. 2004; Engelmann 2011). Species-level conservation can involve several hundreds to thousands of accessions (Panis and Lambardi 2005) demanding high running costs and large areas of land for their sustainable cultivation. Cryopreservation is usually applied to problematic species that produce dehydration-sensitive recalcitrant seeds or genetic resources that cannot be preserved by other means. Cryostorage is also used for clonally propagated crop cultivars for which it is desirable to conserve a specific genetic combination that would be eroded by outbreeding (Halmagyi et al. 2004; Kaczmarczyk et al. 2012; Panis et al. 2005).

Cryopreservation of plant tissues involves storage in LN and/or in the vapour phase of LN (at ~ −135 °C) under conditions that maintain tissue viability following thawing or rewarming (Day et al. 2008; Hamilton et al. 2009; Kaczmarczyk et al. 2012). Cryostorage requires minimal space and negates the need for continual, serial subculturing, thereby reducing the risks of somaclonal variation occurring (González-Benito et al. 2004; Panis and Lambardi 2005). However, cryopreservation has special risk management issues, and it is critical that facilities are managed appropriately to ensure that cryobanks are a realistic backup for the long-term conservation of plant genetic resources. Supply of LN needs to be maintained and levels adequately monitored to assure safe cryoconservation of germplasm *in perpetuity*. However, there are crucial considerations regarding the thermal stability of stored materials (Benson 2008a, b; Benson et al. 2011c, 2013) especially as some theoretical evidence suggests that biochemical and molecular processes may not be completely arrested at ultra-low temperatures (Walters et al. 2004). Although cryopreservation has evident strategic advantages, the resulting freezing/thawing and cooling/rewarming injuries related to membrane structure, cellular functions and loss of totipotency can cause unacceptable reductions in viability and, most significantly, regrowth of whole plants, which still remains a major limiting factor as does storage recalcitrance (Benson 2008a).

Several reviews catalogue the species that have been successfully cryopreserved (Cyr 2000; Engelmann 1997; Engelmann and Takagi 2000; Sakai et al. 2002). There are general guidelines for cryopreservation (Benson et al. 2011c, 2013; Panis 2009; Reed et al. 2004a), but there is no one 'universal protocol' that can be used for all groups of plants, as each species has physiological and biochemical uniqueness that requires protocols to be optimised for individual characteristics (Nadarajan et al. 2007). Moreover, different protocols on balance offer different advantages and disadvantages that need to be fully taken into account before they are routinely implemented (Benson 2008a; Benson et al. 2011a, b, c; Reed 2008a). During the development and/or improvement of a protocol, both cryogenic and non-cryogenic factors need to be carefully balanced to support acceptable levels of viability and recovery that satisfy fitness-for-purpose criteria.

Cryopreservation: General Principles

The conservation of plant species requires a pragmatic approach that empirically defines those critical factors that contribute to successful cryopreservation, described as follows.

Plant Factors, Donor Plants and Physiological Status

As a general rule, germplasm is chosen from either in vivo- or in vitro-grown juvenile plants (Engelmann 1991); meristematic cells are more likely to withstand freezing because they are relatively small in size and contain fewer vacuoles compared to mature tissues, and their cytoplasm is dense indicating a lower water content. Pathogen-tested in vitro cultures are preferred sources of shoot apices and meristems because of their manageable size and reduced levels of contamination as compared to field-grown plants. In vitro cultures are maintained by aseptic subculture, making them more amenable to cryogenic manipulations and post-storage recovery; organised in vitro cultures are relatively less prone to genetic instability compared to callus cultures (Scowcroft 1984).

According to Kami et al. (2010), the size of the tissue is a critical factor; the dissection of meristematic shoot tips is an exacting manual process requiring the trimming of shoot tissue from ~3 × 3 mm to ≥1 × 1 mm. Viability and regrowth of larger shoot tips after exposure to LN appear to decline in comparison to smaller tissues, which may also be influenced by in vitro growth conditions (Keller et al. 2006). Conversely, the lower (<1 mm) limits of dissection also affect viable recovery; generally shoot tissues are dissected to size of 0.5–2 mm. For most cryogenic protocols, the state of the cells and tissues should be optimised to ensure tolerance to dehydration and recovery of vigorous growth after cryopreservation (Dereuddre et al. 1988; Withers 1979). Water status critically affects the ability of germplasm to be stored in LN (Stanwood 1985), and as cells are sensitive to freezing temperatures, moisture content is determined and optimised to support the survival, viability and regrowth of cryopreserved germplasm (Benson et al. 2013).

Preconditioning

Naturally cold hardy plants and some temperate species can withstand exposure to declining temperatures, and their tolerance is exploited to enhance the ability of germplasm to survive LN. This is usually achieved by exposing donor plants and/or their tissue cultures to low (>0 °C) temperatures (Benson 2008a). Cold acclimation is usually applied as a temperature diurnal cycle with a shorter day length (Gale et al. 2013; Reed and Uchendu 2008); the treatment induces an intrinsic tolerance to low temperature and desiccation by triggering genes responsible for cold adaptation (Fowler and Thomashow 2002; Takagi 2000). For temperate herbaceous plants which are minimally cold hardy, low-temperature treatments (0–5 °C) are applied either to donor plants or their dissected meristems; these regimes are effective in supporting post-storage recovery when protocols are optimised (Keller et al. 2008b). Tropical species are temperature sensitive, and one of the most challenging issues is devising preconditioning treatments that stimulate physiological responses to sufficiently enhance tolerance to dehydration and cryogenic procedures.

Preculture

In the case of cold sensitive, tropical plant species, cold hardening is replaced by preculturing the mother plants or excised tissues on sucrose-enriched medium to induce 'dehydration tolerance' and produce acceptable levels of recovery and growth after cryopreservation (Dumet et al. 1993; Engelmann 1991). This can involve culturing of excised embryos and embryonic axes on medium containing various sugars (glucose, fructose, sucrose) or sugar alcohols (mannitol, sorbitol) followed by exposure to cryoprotectants such as dimethylsulphoxide (DMSO).

Osmoprotection

This comprises broad-ranging treatments that generally involve tissues being optimally conditioned on a medium supplemented with sugars or other osmotically active substances. Medium composition, duration and temperature of exposure to osmotic are key determinants, and they are critical factors for successful cryopreservation (Engelmann 1997; Walters et al. 2002).

Cryoprotection and Cryoprotectants

Avoidances of physical and chemical injury during cryopreservation are core principles of cryoprotection, and they are largely influenced by the stages that precede the final immersion of germplasm in LN. A comprehensive account of cryoprotection can be found elsewhere (Fuller 2004; Benson 2008a, b). There are two main types of cryoprotectant (Ciani et al. 2012): (1) those that penetrate the cell and are considered to have colligative properties (DMSO, glycerol, ethylene glycol, propylene glycol) and (2) non-penetrating cryoprotectants which have osmotic activity (sucrose, dextrans, amino acids). However, a more precise term has been introduced to describe cryoprotectants as 'colligative acting'; this is because both cell-penetrating and cell-non-penetrating additives can modulate solute concentration during cryopreservation (Benson et al. 2013). In the case of some types of plant germplasm, a combination of both penetrating and non-penetrating cryoprotectants may be applied (Panis et al. 2005; Panis and Lambardi 2005).

Cryopreservation Techniques

A range of protocols have been widely applied to cryopreserve plant germplasm (Kaviani 2011), and some are based on a singular approach, whereas others use a combination of techniques as described below.

Controlled Rate Cooling

Slow cooling is applied in conjunction with chemical cryoprotection and involves stepwise cooling to a terminal transfer temperature before immersion of cryovials in LN (Reed and Uchendu 2008). By controlling the decreasing temperature at a comparatively slow rate, ice crystals are formed in the extracellular solution and water is removed from the intracellular compartments; this leads to cellular dehydration, and intracellular ice formation is circumvented when cells are exposed to LN (Meryman and Williams 1985; Benson et al. 2013; Morris and Acton 2013). Rapid thawing of samples is usually required to achieve viable recovery. Controlled rate cooling has been successful for moderately hydrated callus and cell cultures (Schrijnemakers and Van Iren 1995; Lynch 2000). The technique can have complex applications where several cooling steps are required. The computerised cooling apparatus is an initial expense, and the programmed run can take several hours to complete for slow rates of cooling. Compared to other techniques, controlled cooling is relatively costly to operate routinely (Ashmore 1997), but it has the advantage that many cryovials can be simultaneously bulk handled during a cooling run, making the operation more time efficient compared to most vitrification methods that require operators to manually handle cryovial on a one-by-one basis. Moreover, from a quality management-SOP perspective, controlled rate programmable cooling enables the programmed run to be systematically monitored, creating an electronic record with critical point thermal data and ice nucleation temperatures (Morris and Acton 2013).

Encapsulation/Dehydration

This is a vitrification-based technique in which shoot meristems or somatic embryos are encapsulated in calcium alginate beads (Fabre and Dereuddre 1990). The protective encapsulation process enables dehydration and desiccation to proceed which would otherwise be highly damaging or lethal to nonencapsulated samples (González-Arnao and Engelmann 2006). The basic protocol comprises encapsulation, preculture of alginate-coated samples in liquid medium with high levels (0.5–0.75 M) of sucrose, evaporative air or silica gel desiccation, placing beads into cryovials, rapid cooling of cryovials in LN and rewarming of the alginate encapsulated tissue. Recovery is usually performed by placing the beads onto standard culture medium without having to extract the shoots or embryos from their alginate coating (Engelmann et al. 2008; Sherlock et al. 2005).

Desiccation

This involves the direct (rapid cooling) immersion of tissues in LN. Desiccation is usually performed in the sterile stream of air produced from a laminar airflow cabinet/workbench, but more precise and reproducible desiccation is achieved by dying material over activated silica gel or using a flow of sterile compressed air (Engelmann 2000; Sherlock et al. 2005). Germplasm is usually recovered by the rewarming of samples under ambient conditions.

Vitrification

The process of vitrification involves the exposure of plant germplasm to highly concentrated cryoprotectant solutions for relatively short durations. To induce dehydration tolerance, tissues are cultured on medium with high levels of sucrose (0.3 M) or sorbitol (1.4 M) and subsequently transferred to a glycerol-sucrose solution (2 M glycerol + 0.4 M sucrose), referred to as the loading solution (Sakai 2000). The most widely used plant vitrification solution was developed by Sakai et al. (1990) and named PVS2, which consists of 30 % (w/v) glycerol, 15 % (w/v) ethylene glycol and 15 % (w/v) DMSO in liquid medium to a final concentration of 0.4 M sucrose (Benson and Harding 2012). Dissected, pretreated shoot tips are placed into cryovials which are plunged into LN. After rewarming, samples are usually placed in unloading solution (1.2 M sucrose) for a short time and then cultured on standard recovery medium.

DMSO Droplet Freezing

The droplet-freezing technique is based on the cryopreservation protocol established for cassava shoot tips by Kartha et al. (1982) and was later applied to conserve potato shoot tips (Schäfer-Menuhr et al. 1996). The technique consists of cryoprotecting dissected shoot tips derived from in vitro plantlets with a 10 % (v/v) DMSO solution for 1–3 h and then rapidly freezing the shoot tips in micro-droplets (2.5 μL) of DMSO solution on aluminium foils placed in cryovials in which the droplet foils are immersed directly into liquid LN. Rapid thawing of samples is necessary to recover viable shoots that are capable of regrowth.

Droplet Vitrification

The technique combines the droplet-freezing method with the vitrification procedure (Sakai et al. 1990), in which shoot meristems are cooled in a droplet of PVS2 cryoprotectant solution. The droplets enclosing the meristems are formed on aluminium foils that are then manually transferred into cryovials filled with LN, after which they are directly immersed into LN (Panis et al. 2005). Droplet methods have the disadvantage that tissues are directly exposed to LN (Benson et al.

2013). Rapid rewarming of cryovials in a heated water bath is necessary to recover viable shoots that are capable of regrowth.

Encapsulation-Vitrification

This is a combination of the encapsulation/dehydration and vitrification procedures as described above, in which samples are encapsulated in alginate beads and then subjected to vitrification by PVS2, after which the beads are placed in cryovials which are immersed in LN. The beads are rewarmed and the vitrification solution removed before transferring to culture medium (Martinez-Montero et al. 2012).

Thawing and Rewarming

The avoidance of ice recrystallisation is essential during rewarming as ice nucleation can occur when samples are slowly rewarmed above the Tg and homogeneous ice nucleation point (≥−40 °C); therefore, rewarming should usually be performed by rapidly transferring cryovials to +40 °C water bath for several minutes until the visible signs of ice disappear. To avoid the destabilisation of the noncrystalline vitrified glassy state, rewarming of samples is usually carried out rapidly to limit devitrification and promote post-storage recovery (Mazur 2004). This practice may cause stress fractures in the glass if rewarming is performed too rapidly (Benson 2008a). To reduce this possibility, a two-phased approach may be required: (1) first a short phase (e.g. 1–2 s at ambient temperatures) to allow glass relaxation without stress fracturing and (2) second rapid warming (at +45 °C) to ensure the speedy transition from glass to liquid without an intervening passage through an ice phase.

Sugarcane, Conservation and Cryopreservation

Commercial sugarcane, belonging to the genus *Saccharum* (*Poaceae*), is an important industrial crop in Latin America, accounting for nearly 70 % of sugar produced worldwide (Lakshmanan et al. 2005). A substantial effort has been directed towards developing sugarcane as a biofactory for high-value products (Wang et al. 2005, 2013; Chen and Dixon 2007). This displays many features of a natural biofactory: rapid, vigorous growth, an efficient carbon fixation pathway and large biomass production. Sugarcane has a well-developed stem-organ storage system with a large pool of hexose sugar and is cultivated throughout the world (Altpeter and Oraby 2010; Wang et al. 2013) particularly as it has the potential to decrease the dependency on the use of fossil fuels (Rein 2007).

In vitro culture of sugarcane has a crucial role in the conservation, breeding and utilisation of genetic variability of the crop, and biotechnological techniques also include cryopreservation, in vitro selection, genetic engineering and the commercial mass production of disease-free sugarcane (Lakshmanan et al. 2005; Altpeter and Oraby 2010). Tissue culture systems have a vital role to facilitate the international exchange of germplasm as the size of the samples is drastically reduced and genetic resources can be transported under sterile conditions.

Sugarcane is an excellent example for demonstrating where a range of cryopreservation protocols have been developed for various types of germplasm: apices of in vitro plantlets using the encapsulation/dehydration technique (González-Arnao et al. 1993; Paulet et al. 1993), cell suspensions (Finkle and Ulrich 1979; Gnanapragasam and Vasil 1990) and embryogenic callus using classical freezing protocols (Eksomtramage et al. 1992; Gnanapragasam and Vasil 1992; Jian et al. 1987), simplified cryopreservation protocols (Martinez-Montero et al. 1998) and vitrification-based techniques for somatic embryos (Martinez-Montero et al. 2008, 2012).

The Genesis of Cryobionomics

The origin for creating cryobionomics as a new paradigm was largely derived from the synthesis of the various issues regarding genetic stability assessments, as a start to finding an acceptable basis for reaching international consensus for the release and reintroduction of plants derived from cryopreserved germplasm into the environment and for their use in biotechnological applications and/or in plant breeding (Harding 2004). In the last few decades, the plant sciences have seen an immense advancement in analytical technology, providing a multiplicity of procedures that have diverse applications in biotechnology (Harding et al. 2013), as molecular markers and genomics, and many are increasingly being used for the characterisation of plant germplasm (Ayad et al. 1997; de Vicente 2004; de Vicente and Andersson 2006; de Vicente and Fulton 2003; Harding and Benson 2012; Karp et al. 1997; Spooner et al. 2005). Despite these technological advancements, limited progress has been made in establishing precise scientific criteria that are internationally agreed and/or acceptable for assessing genetic stability in plants recovered from cryopreserved germplasm.

A wide range of reports indicate that the choice of technique or approach used in assessments of genetic stability is largely determined by the type or nature of the conservation strategy, the needs and experience of the end user, local availability of expertise and resources appropriate for a given species and in vitro/ex situ conservation objectives. The monitoring of genetic stability has been advised as a technical guideline for the management of field and in vitro germplasm collections (Reed et al. 2004a). However, it remains that there is no single technique or approach that can be described as the method of choice and/or guidance to adequately judge genetic stability or acceptable levels of genetic change. The importance of stability has been recognised in several guidelines and standards (Genebank Standards 1994; IPGRI/CIAT 1994; FAO 2012, 2013) for in vitro conservation best practices (Benson et al. 2011c) and in biorepositories and biobanks (Benson et al. 2013).

It is a fundamental requirement of genebank management to monitor and determine sample quality, viability and genetic integrity of materials during in vitro storage (Benson et al. 2011a, b, c). The use of robust standard operating procedures (SOPs) and establishment of best practices are integral parts of a quality management system which can assist the operations of genebank management to achieve its conservation objectives (Benson and Harding 2012; Morris and Acton 2013). In the context of in vitro conservation, the use of cryopreservation for the long-term storage of plant germplasm is ultimately the process that enables the restoration and sustainable utilisation of a species (Berjak et al. 2011a; Harding and Benson 2012). Cryobionomics was coined to signify the reintroduction of species into the environment following cryostorage; accordingly, this term combines and integrates studies of genetic stability with cryopreservation as follows:

- Nomic – a science or field of knowledge or the discipline of the study
- Cryo – reference to the subject ‘cryobiology and cryopreservation’
- Bionomics – a branch of biology dealing with organisms’ habitats and modes of life in their natural environments (Oxford dict.)
- Cryobionomics – the biological science dealing with cryopreserved organisms’ behaviour and habitats following their reintroduction into their natural environment

Cryobionomics may be defined as an interdisciplinary subject that requires phenotypic, histological, cytological, biochemical and molecular biological knowledge of the organism to assess possible cellular/biochemical damage (cryoinjury), impairment of metabolism and loss of reproductive functions. It also proposes to examine the temporal shifts in gene expression that cause disruption of normal regulatory mechanisms, growth and developmental sequences (Harding 2004).

Importantly, the concept provides a working hypothesis to explore the relationship between cryoinjury, viability and genetic stability as it draws attention to the potential impacts of cryoinjury on the genome, transcriptome, proteome and metabolome as shown in Fig. 1. Thus, cryobionomics provides a conceptual framework to investigate the linkages between cryogenic and non-cryogenic stress factors (Berjak et al. 2011a). This approach aims to understand the basis of success or failure following cryostorage by using molecular and physiological approaches to identify robust post-storage performance indicators

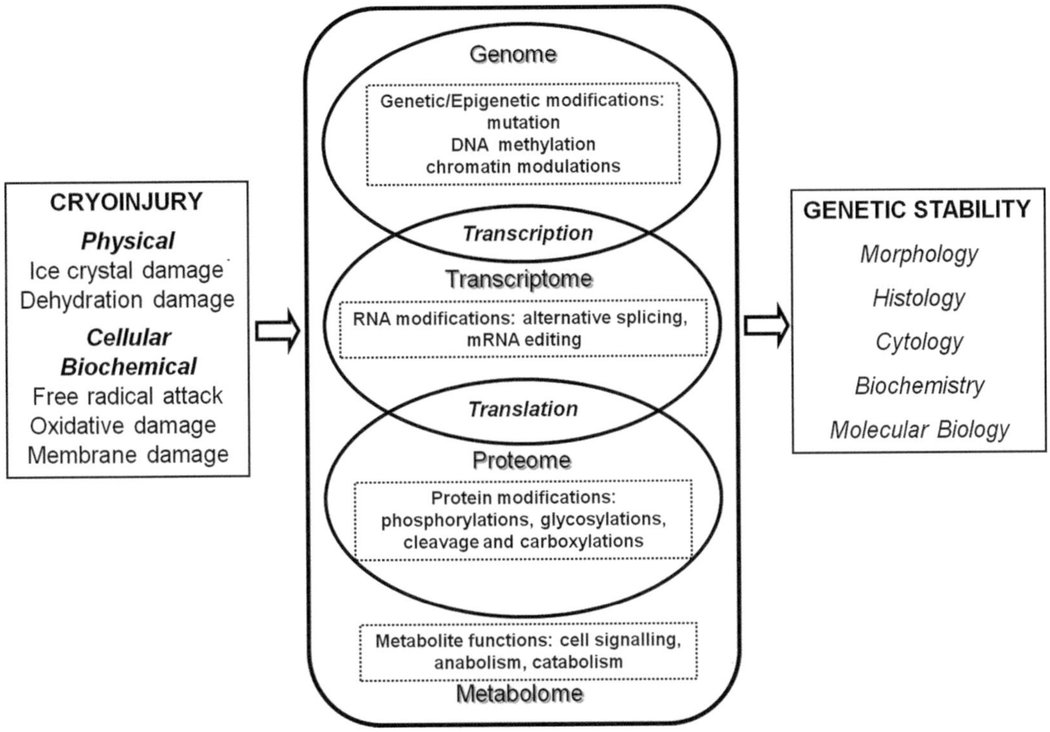

Fig. 1 A schematic illustrating the linkages between the various impacts of cryoinjury on a plant cell indicating the potential changes to the genome, transcriptome, proteome, metabolome and genetic stability (Adapted from Harding et al. 2009)

and assist evidence-based protocol optimisation and improvement (Benson et al. 2013; Harding and Benson 2012; Harding et al. 2009, 2013; Skyba et al. 2010). For example, epigenetic and/or genetic alterations to the genome may affect the functionality of germplasm by disrupting normal patterns of growth and reproduction causing undesirable instability (Johnston et al. 2009; Kaity et al. 2008; Peredo et al. 2008). However, another perspective of cryobionomics is that molecular changes, particularly at the epigenetic level, may be indicative of a positive adaptive response to the stresses incurred during cryopreservation and which may be advantageous to post-storage survival (Johnston et al. 2007b, 2009). Genetic/epigenetic events may be manifested as detectable changes in morphology, cytology, histology and biochemical characteristics and molecular marker profiles that affect the true-to-type nature of plants recovered from cryostorage (Benson et al. 2011a). The extent of these changes may influence the quality of conserved germplasm and the performance of plants recovered from cryobanked materials (Benson et al. 2011c) and reintroduced into natural environments and, similarly, their commercial exploitation and utilisation by biotechnologists and plant breeders (Harding et al. 2009).

Is There a Connection Between Cryoinjury and Viability?

The efficacy of any cryopreservation protocol is largely determined by balancing the plant's intrinsic tolerance and sensitivity to stress with the ability to withstand the severity of cryogenic treatments (Benson 2008a). There are a wide range of techniques (see section "Cryopreservation: General Principles") that are employed to achieve this objective (Reed 2008a), and failure to endure exposure to the extreme conditions imposed during cryopreservation is likely to incur a considerable degree of cellular damage, which is referred to as cryoinjury (Benson 1999; Benson et al. 2013; Fuller et al. 2004; Katkov 2012a, b). Numerous stress-related factors are associated with cryoinjury, and frequently these are a result of suboptimal protocols which can include any combination of the following damaging events:

- Intracellular ice formation during controlled rate cooling and/or LN exposure (Benson et al. 2007; Reed and Uchendu 2008)
- Osmotic injury during cryodehydration and/or thawing/rewarming (Benson et al. 2005, 2013)
- Devitrification and/or ice recrystallisation during rewarming/thawing (Benson 2008a, b)
- Toxicity of cryoprotective agents/mixtures (Sakai et al. 2008; Keller et al. 2008b)
- Oxidative damage by ROS activity (Berjak et al. 2011a, b; Johnston et al. 2007b, 2010)
- Secondary lipid peroxidation products (Johnston et al. 2007a; Martinez-Montero et al. 2012)

It is an operational imperative to optimise cryoprotective conditions, as suboptimal regimes (Normah and Makeen 2008) drastically reduce the viability of recovering plant germplasm (Berjak et al. 2011a), and effective cryoprotection that adequately reduces cryoinjury requires careful protocol optimisation (Nadarajan et al. 2007), often on a species and genotype basis (Benson 2008a; Fuller et al. 2004). The principles of cryoprotection are central to maintaining cell integrity and survival following exposure to LN, and in practice, there are two effective strategies to achieve this objective: (1) colligative cryoprotection coupled with controlled rate cooling and (2) vitrification-based protocols as described above and elsewhere (Benson 2008a, b; Benson et al. 2005, 2011a, b, c, 2013).

Briefly, the 2-factor hypothesis described by Mazur (2004) requires that for a given protocol, it is vital to optimise the cooling rate with the permeability of the cryoprotective additive to obviate deleterious colligative cell damage (lethal cell volume changes, toxic concentration of solutes) and intracellular ice formation. The critical events during cryoprotection by controlled rate cooling are schematically shown in Fig. 2. Additives typically include DMSO, glycerol or methanol in the case of protists (Harding et al. 2010). Ultra-rapid cooling rates are effective

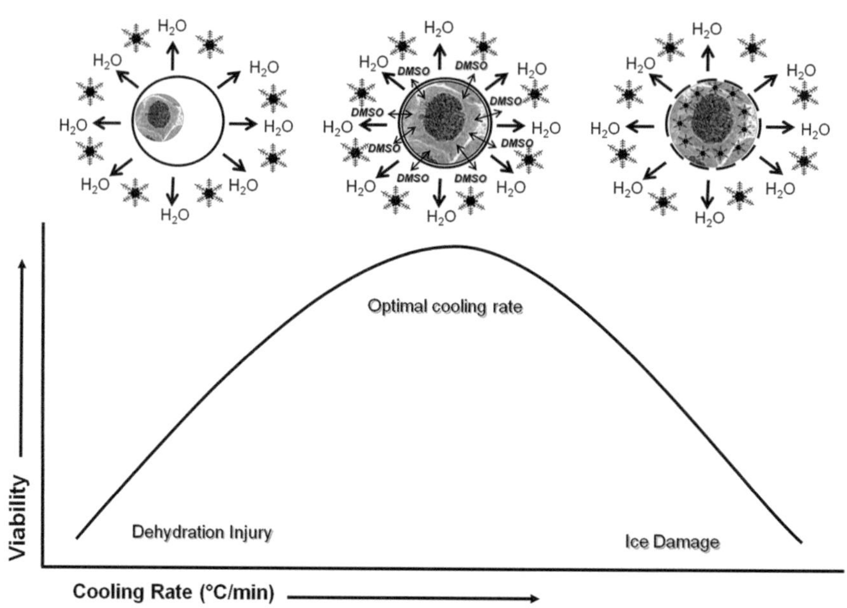

Fig. 2 The 2-factor hypothesis (Mazur 2004) and the effects of cryoinjury on cell viability incurred through either colligative dehydration injury or intracellular ice damage and the intervention of colligative-acting DMSO

because they prevent water molecules from forming ice crystals of a size that would be injurious. In contrast slow, controlled rates of cooling initially cause the formation of extracellular ice, thus creating a differential water (vapour pressure) gradient across the cell membrane. The osmolality of extracellular solutes increase causing a freezing point depression and osmotic pressure differential. To reach equilibrium, intracellular water moves to the outside of the cell, thereby increasing internal osmolality (Elliott et al. 2008) and reducing the water available to form ice; this results in a cryoprotective effect and cryoprotectants expedite the process. A common, colligative-acting, cell-penetrating cryoprotectant that is applied to plant germplasm is DMSO (Benson et al. 2013), which is applied singularly or in combination with glycerol and other additives (Sakai et al. 2008). Extracellular ice formation may be harmful to the integrity of some complex, fine cellular structures, but intracellular ice is lethal to multicellular plant tissues (Fig. 2). Continued slow cooling in extreme cases will increase the removal of water, thereby concentrating the intracellular solutes, causing cell shrinkage by colligative damage or injury to cellular architecture (Fig. 2). Ideally, the residual water remaining may be so negligible that ice crystals do not form, or the viscosity of the cellular solution is so high that the cryoprotected cells most likely vitrify on exposure to LN (Benson et al. 2013). In which case, cells may become preserved in a partial glassy (vitrified) state, although the extracellular matrix may be frozen (Fig. 2). Cryoprotectant permeability and toxicity are species dependent; cryoinjury is incurred through either intracellular ice or colligative damage which results from the failure to optimise nucleation, cooling rates and cryoprotection regimes (Morris and Acton 2013).

Vitrification-based cryoprotection or ice-free cryopreservation can be achieved by increasing cell viscosity to the critical point at which ice formation is inhibited both inside and outside the cell, where the available residual water becomes vitrified on exposure to LN. Vitrification protocols can be complex and the vitrified state is metastable and vulnerable to devitrification during cooling and rewarming. The vitrified state is characterised by the Tg, glass transition temperature, which is the temperature at which a liquid forms a glass (Zámečník et al. 2012). Optimised rewarming is essential to avoid glass relaxation and devitrification; the risks of which are damaging fractures and cracks from glass tension that break fragile structures, ice nucleation and recrystallisation can also occur. Glass stability is highly dependent on moisture content (MC) and the calibration of the drying (dehydration and/or desiccation) time for which MC values do vary but typically fall around 0.4 g water g^{-1} dry weight to achieve optimal survival (Benson et al. 2005).

Cryopreservation is a multistage process, and each step can be potentially injurious, and the cumulative effects of these stresses may well lead to loss of viability and ability to regrow as a whole plant (Berjak et al. 2011a; Elliott et al. 2008). To achieve the complete and/or partial vitrified state, plant tissue must be exposed to extreme biophysical and chemical stresses through the application of high concentrations of additives that can become toxic (Benson et al. 2005, 2013). Intolerance to drying by dehydration and/or desiccation can be lethal or cause significant loss of viability. In the case of suboptimal protocols, osmotic stress, cell shrinkage and damage to cell membranes are inevitable and can lead to the production of secondary lipid peroxidation products (Johnston et al. 2007a; Skyba et al. 2010; Martinez-Montero et al. 2012; Kaczmarczyk et al. 2012) which cause cell death and necrosis in recovering plant germplasm (Kaczmarczyk et al. 2008b). Undoubtedly, there are numerous physiological adaptations that affect the way plants are able to survive the different physical and chemical cryoprotective treatments, which may well implicate various cellular protective responses.

Validity, Variability and Viability of Cryopreservation Recovery Responses

The motivation to conserve the vast diversity of plant species (RBG 2010) arises from understanding the value and nature of threats to plants in their natural environment (Berjak et al. 2011a; Harding et al. 2013; Kaczmarczyk et al. 2011a; Paton 2009). This necessitates the pragmatic use of in vitro and cryostorage techniques (Reed 2008a; Normah et al. 2012) for species that are not easily conserved using other approaches, particularly clonally propagated plants or those that produce nonorthodox seeds, of which tropical species are particularly storage recalcitrant (Harding and Benson 2012). The validity of any given cryopreservation protocol is the proof of its reproducibility to successfully return the maximal number of true-to-type plants from germplasm that has been stored at ultra-low temperatures (Reed et al. 2001, 2004a, b; Keller et al. 2008a). This requirement is essential but highly dependent upon the optimisation of a multiplicity of storage parameters (see section "Cryopreservation: General Principles", Benson and Harding 2012; Benson et al. 2002). The task of protocol optimisation can be an immense undertaking and can also require exacting technical skills (e.g. meristem excision), robust experimental rigour and exhaustive empirical work during the trial and error process that eliminates the negative variables to achieve maximal viability and plant regrowth (Benson 2008a; Kaczmarczyk et al. 2008a, 2011a; Kim and Lee 2012; Martinez-Montero et al. 2012; Nadarajan et al. 2007).

It is generally recognised that the application of cryoprotectants rarely permits 100 % survival of in vitro plant germplasm after freezing and thawing (Kaczmarczyk et al. 2008b; Sakai et al. 2008). The overall cryopreservation process can thus be described as obeying the law of diminishing returns, in that fewer shoot tissues survive cryoprotective and/or cryogenic treatments compared to the original number exposed to these conditions (Johnston et al. 2009; Kaczmarczyk et al. 2008a; Mix-Wagner et al. 2003; Panis et al. 1996, 2005). Notably, there is a considerable level of variability amongst species and/or genotypes where survival can vary greatly (Reed et al. 2004b). Typical 'diminishing returns' in regrowth responses are shown in Fig. 3 for *Ribes* genotypes with differential tolerances to cryostorage (Johnston et al. 2009). The fate and viability of surviving shoot tips (Reed and Uchendu 2008) may also be subject to further decline as they fail

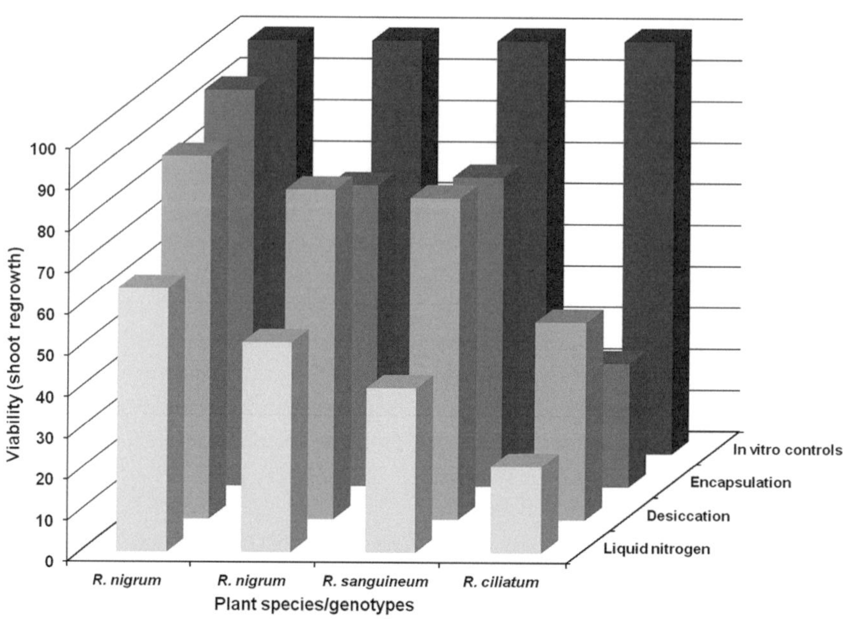

Fig. 3 Viability of *Ribes* species and genotypes (*R. nigrum*, *R. nigrum* cv Ben More, *R. sanguineum* cv Ben Tron and *R. ciliatum* cv King Edward VI) defined as percentage (%) of shoot regrowth after 6 weeks following exposure to different stages of the encapsulation/dehydration protocol (Adapted from Johnston et al. 2009)

to thrive and grow beyond their existing physical structure (Kaczmarczyk et al. 2008b; Keller 2005). These survival profiles have been described as cryopreservation-induced delayed onset cell death in both animal (Baust 2007) and plant systems (Harding et al. 2008a, 2009). Survivors that do continue to regrow might not develop fully as growth is restricted to unorganised callus that produces a few adventitious roots and sporadic green leaves or incidental shoots (Chang and Reed 1999; Kaczmarczyk et al. 2008b). Ideally, when a protocol is successful, it is desirable that the greater number of the surviving shoot tips completely regenerate into fully organised and differentiated plantlets, having a primary shoot formation connecting to a significant rootstock that displays vigorous growth and development (Mix-Wagner et al. 2003). It is a common observation that regrowth and regeneration are often much lower compared to the initial values of surviving cryopreserved shoot tips (Lynch et al. 2012; Kaczmarczyk et al. 2008a; Mix-Wagner et al. 2003; Panis et al. 1996, 2005), and this has implications for cryobank cost/benefits and efficiency (Keller et al. 2013).

The variability in responses is often unpredictable, which highlights the complexity of the temporal (phased) nature of post-storage recovery. Often only a single point in time is used to evaluate viability and recovery, and this has limited comparative value as it is not possible to identify at which stage of recovery the survivors succumb to stress and begin to decline. Evidence-based optimisation of post-storage recovery requires continual monitoring up to 6–8 weeks or longer (species dependent) following cryostorage to fully characterise the factors and interventions that affect sustainable recovery (Benson and Harding 2012; Johnston et al. 2009). Assessment of post-storage viability and recovery usually comprises three stages: (1) initial survival of the meristem, (2) regrowth or regeneration of a viable shoot, and (3) full plant regeneration. There are well-established criteria (greening, swelling, regrowth of roots or one or more shoots) described by Keller (2005) for evaluating the survival of shoot meristems following cryopreservation.

Connecting In Vitro Culture, Cryoinjury, Viability and Genetic Stability

Since the recognition that plant tissue culture plays a key role in supporting in vitro conservation, the subject of genetic stability has been an important concern and is defined as somaclonal variation (SCV) by Scowcroft (1984). Cryopreservation has been used successfully for the long-term, ex situ conservation of germplasm from a diverse range of plant species (Benson 1999; Benson et al. 2002; Martinez-Montero et al. 2012; Harding et al. 2013), but it is important to verify that a storage protocol does not have any destabilising effects and that the plants recovered and regenerated from cryopreserved germplasm are true-to-type (Castillo et al. 2010; Miguel and Marum 2011; Perazzo et al. 2000) and satisfy the fitness-for-purpose performance indicators of end users. The concern regarding genetic stability of in vitro cultures has lead to decades of study resulting in a wide range of reported stability investigations, including different levels of analyses: plant morphology, phenotypes, biochemical characteristics, cytology and histology, genetics and molecular biology (Jaligot et al. 2002; Harding 2004; Harding et al. 2009; Benson et al. 2011a). Although there are numerous reports of SCV for specific cryopreserved species, the exact mechanism and elucidation of the actual causal nature of genetic instability and its relative high frequency in vitro have been mainly speculative and elusive. Overall, the causal factors have evaded precise determination due to the unpredictability, randomness and vagaries of SCV manifestations and symptoms (Harding et al. 2008a; Oh et al. 2007; Miguel and Marum 2011; Rodriguez-Enriquez et al. 2011). These collective observations are perhaps not too surprising considering the multistages involved in cryopreservation. The accumulation of DNA polymorphisms may not only be induced by cryopreservation *per se* but are the result of the whole culture-cryoprotection-regeneration process (Harding 2004). New candidates, for instance, microRNAs, siRNA and the role of the relationship between SCV and DNA methylation, are beginning to emerge as potential markers for SCV. Although it can be rationalised that some level of reversible post-storage variability may be attributed to storage stress adaptations (Harding et al. 2009; Johnston et al. 2009) rather than to genetic instability *per se*. Technological advancements are leading to a better understanding of the mechanisms that regulate patterns of gene expression (Rodriguez-Enriquez et al. 2011; Mosher and Melnyk 2010; Miguel and Marum 2011), and these will have useful applications in the study of cryostorage stability. For example, changes in cellular/metabolism homeostasis following the stressful exposure of plant tissues to cryoprotective treatments may also trigger gene activation and/or deregulation events. The complexity of signalling responses of tissues during cryoprotection may well be the initial factors that predetermine the events that ultimately fashion the outcome of cryoconserved germplasm stability.

Cell Signalling and Gene Expression

The scientific literature is replete with cellular, biochemical and molecular evidence that characterise many in vitro cryopreservation transitory stages (Skyba et al. 2010; Kaczmarczyk et al. 2012). At the onset of the preparatory stages of cryopreservation, the initial processing of donor material for in vitro cultures is likely to invoke the activation of many signalling pathways and the concomitant changes associated with gene expression and metabolism (Harding et al. 2008a). Transference of field- or glasshouse-grown material into tissue culture is often accompanied by disruptive physiological conversions (Kevers et al. 2004) and genomic (Peredo et al. 2008) and epigenetic alterations (Johnston et al. 2005, 2009) that are associated with in vitro change (Cassells and Curry 2001; Joyce and Cassells 2002; Joyce et al. 2003).

Sensing cold (Benson 2008a) and its perception during acclimation by exposing in vitro tissue cultures to low temperatures before cryopreservation trigger crosstalk between signalling networks (Joyce et al. 2003; Chinnusamy et al. 2004; Knight and Knight 2001) and the activation of various gene-regulated mechanisms (Renaut et al. 2006). Cryopreservation-induced abiotic stress in germplasm tissues is likely to elicit a rise in cytosolic free calcium levels; the Ca^{2+}-mediated response involves protein phosphatases and kinases which are mediated through mitogen-activated protein kinase (MAPK) cascades (Knight and Knight 2001; Zhang and Klessig 2001; Pitzschke et al. 2009; Rodriguez et al. 2010).

The signal transduction process that occurs during cold acclimation and the secondary messenger networks are beginning to unfold a complex series of pathways revealed by mutant, functional genomics, proteomic and metabolomic studies (Benson 2008a). The perception of stress by cells and the associated cascades elicited by signalling molecules activates the expression of transcription factors that affect hundreds of stress response genes (Volk 2010). Tissue cultures exposed to conditions of osmotic stress that are caused by dehydration prior to cryopreservation may also elicit many of the phytohormone, abscisic acid (ABA)-mediated responses (Carpentier et al. 2007; Volk 2010). Cold acclimation, stimulated by the incorporation of ABA in culture medium (Reed and Uchendu 2008), could potentially activate inducible gene pathways (Seki et al. 2002) and a stress-responsive MAPK cascade to induce the biosynthesis of ethylene (Kim et al. 2003) with the concomitant changes in reactive oxygen species (ROS) production and membrane fluidity (Zhu et al. 2006). The ROS gene network includes over 150 genes (Suzuki and Mittler 2006), where the ROS homeostasis is often a delicate balance between normal cellular signalling and, in the extreme, the destruction of aerobic life (Harding et al. 2008a, 2009).

Oxidative stress occurs when the redox homeostasis is disrupted and the steady state between prooxidants and antioxidants becomes unbalanced (Johnston et al. 2006; Skyba et al. 2010; Kaczmarczyk et al. 2012) consequently eliciting protective responses mediated by many gene regulatory mechanisms. A common plant abiotic stress response to wounding and osmotic shock is the accelerated generation and accumulation of ROS such as hydrogen peroxide (H_2O_2), the superoxide anion and hydroxyl radicals (Benson and Roubelakis-Angelakis 1994; Kovtun et al. 2000; Martinez-Montero et al. 2012). As part of a transduction network, H_2O_2 is an active signalling molecule, although its accumulation can lead to oxidative stress and the activation of MAPKs (Kovtun et al. 2000). Oxidative stress may be induced as a result of the mechanical damage incurred during shoot-tip meristem dissection (Keller 2005; Volk and Caspersen 2007) and the recovery of wounded, dissected meristems (Johnston et al. 2007b, 2010) especially if they have been exposed to suboptimal osmotic treatments and cryopreservation protocols (Normah and Makeen 2008). Free radical cell damage to plasma membranes can also generate ROS, toxic lipid peroxides and their aldehydic products in cryopreserved tissues (Johnston et al. 2007a; Lynch et al. 2011b; Skyba et al. 2010). A knowledge of the plant's intrinsic ability to regulate ROS and quench free radical activity by phenolics, terpenes, carotenoids and pigment interactions (Johnston et al. 2006; Carpentier et al. 2005) and scavenge free radicals by antioxidant mechanisms (Johnston et al. 2007b) can help to improve post-storage viability (Martinez-Montero et al. 2012). Future research may aid understanding the poorly characterised process of recovery following cryopreservation (Häggman et al. 2008; Volk 2010), especially as Volk et al. (2011) have identified hundreds of genes that are up-regulated and down-regulated following the recovery of shoot tips from LN.

Viability and Genetic Stability

There are numerous, unresolved issues regarding what constitutes in vitro storage recalcitrance and the many variable factors that can negatively impact on the viability of cryopreserved germplasm. Viable regrowth to functional whole plantlets after cryostorage is typically the decisive response (Reed 2008b) as this is required for reintroductions, rehabilitation and long-term conservation of genetic resources (Berjak et al. 2011b). As a quality control measure and risk management practice (Benson and Harding 2012), the assessment of genetic integrity, trueness-to-type and fitness-for-purpose is recommended for plants recovered from cryogenic storage (Harding et al. 2009; Benson et al. 2011a, c). This is because there is always the risk of SCV occurring as a consequence of in vitro conservation, and by taking the appropriate precautionary measures, the frequency of SCV can be reduced to a minimum (Scowcroft 1984). For example, it is a usual requirement for in vitro genetic resources conservation to avoid the dedifferentiated callus phase where possible (Reed 2008b). This risk may be by minimised by moderating the application of plant growth regulators (PGRs) to reduce dedifferentiation and by using differentiated, organised tissues that comprise the original meristem (buds, shoot tips, meristems, embryos) as they are genetically programmed to develop into true-to-type plants (Benson et al. 2011a). Although in practice, this can be difficult as PGRs are sometimes required to stimulate shoot regeneration and regrowth after cryopreservation (Harding and Benson 1994; Faltus et al. 2007a, b). The damaging effects of cryoprotection and LN exposure on tissue cultures can retard the patterns of organised shoot regeneration in favour of callus formation which can often arise from wounded tissues as a result of cryoinjury (Chang and Reed 1999; Kaczmarczyk et al. 2008b; Volk and Caspersen 2007). By diligently monitoring regrowth during the recovery time course (section "Validity, Variability and Viability of Cryopreservation Recovery Responses"), this problem can be eliminated, in part, through the early detection of callus and cutting the dedifferentiated tissues from the original meristem, reducing the risks of a shoot and/or plantlet arising via adventitious regeneration (Keller et al. 2008b; Martinez-Montero et al. 2012).

An examination of the post-cryopreservation recovery responses reported in the published literature invites the question – is there a relationship that links suboptimal protocols with cryoinjury, recovery, genetic stability? Undoubtedly, it is widely understood and accepted that suboptimal conditions and protocols affect the recovery of cryopreserved plant material (Normah and Makeen 2008; Reed 2008b), but given the intricate complexity of the signalling networks and patterns of gene expression (section "Connecting In Vitro Culture, Cryoinjury, Viability and Genetic Stability"), it is not intuitively obvious to reason that the observed low levels of survival and regrowth may predispose germplasm to genetic instability. There is a spectrum of responses that fall into the high (>80 %) and low (<20 %) ends of shoot and plant regrowth for many species (Dussert et al. 2003; Sakai et al. 2008; Kaczmarczyk et al. 2011a, b) including *Actinidia*, *Fragaria* and *Ribes* spp. (Reed 2008c); apple (Towill and Ellis 2008; Zámečník et al. 2012); *Eucalyptus*, *Populus* spp. and silver birch (Häggman et al. 2008); garlic (Zanke et al. 2011; Zámečník et al. 2012); hop (Zámečník et al. 2012); and potato (Keller et al. 2008b; Mix-Wagner et al. 2003; Zámečník et al. 2012). The precautionary measures described above, coupled with optimisation of the protocol and recovery medium that give high levels of regrowth, provide reasonable assurance of stability, but it may not be true for those species/genotypes that exhibit low levels of survival and/or regrowth, signifying that the cryobionomics hypothesis has yet to be fully tested.

Issues of Genetic Stability

The question of genetic stability may be addressed using numerous analytical and investigative approaches and techniques (Kaczmarczyk et al. 2011b), but by its elusive nature, the detection of SCV is a challenge and subject to the issues and limitations of the methods employed for its detection (Harding 2004; Harding et al. 2005). Oxidative stress and the production of ROS (Harding et al. 2008a; Skyba et al. 2010; Berjak et al. 2011b; Kaczmarczyk et al. 2012) exacerbated by the presence of H_2O_2 during cryostorage do raise concerns related to genetic instability, as evidenced by the detection of 8-hydroxy-2′-deoxyguanosine, a marker for oxidative damage in DNA and detected in germplasm exposed to cryogenic treatments (Johnston et al. 2010). Despite these constraints and concerns, there is sufficient cumulative evidence in the reported literature to indicate an overall positive outcome regarding stability following cryopreservation, based on biochemical, cytological, molecular and morphological criteria (Harding et al. 2009; Reed 2008b; Johnston et al. 2010; Lambardi et al. 2008; Häggman et al. 2008).

Nonetheless, scrutinising the accumulating evidence that suggests a more rigorous form of genomic analysis would provide better assurance of stability. Numerous plant species have different levels of ploidy, for instance, sugarcane is a hybrid of different species with a complex octoploid genome and chromosome diploid number ranging from $2n=70–140$ (Asano et al. 2004). Sampling of the plant genome by the contemporary molecular biological marker techniques (Kaczmarczyk et al. 2011b), typically by RFLPs, RAPDs, SSRs and AFLPs, appears to be restricted to a small fraction of the genome. Some estimates indicate this fraction as 0.001 % for potato (Harding 2004), 0.00055 % in chrysanthemum (Martín and González-Benito 2005), 0.001–0.003 % for St. John's wort (Skyba et al. 2010) and ~0.03 % of the genome for papaya (Kaity et al. 2008). These calculations assume that each marker detects DNA variation along its entire fragment length (Harding 2004), but any given variant fragment may only result from a single nucleotide polymorphism (SNP) at the primer binding; therefore, these values overestimate the portion of genome analysed (Kaity et al. 2008). Thus, given the diminutive size of this fraction and the limited sampling of the plant genome (Harding 2004; Kaczmarczyk et al. 2010, 2012), it is quite probable that molecular analyses will not detect any DNA polymorphisms, thereby indicating stability as an experimental false negative. Although this is a difficult issue to resolve (Häggman et al. 2008), advances in plant genomics (Lescot et al. 2008; PGSC 2011) and the rapid progress in next-generation sequencing (NGS) technology (Shendure and Ji 2008; Hawkins et al. 2010; Metzker 2010; Nordborg and Weigel 2008) have contributed to an expanding database of over 6,868 projects (July 2013) with completely sequenced genomes (www.genomesonline.org). These technologies are rapidly evolving, and high-throughput DNA sequencing platforms are now routinely available and in so doing reducing the cost of DNA sequencing. NGS will accelerate progress for the comparative analysis of plant genomes, transcriptomes and proteomes, and on a not too far distant horizon, they will provide definite evidence regarding the fidelity of the primary DNA sequence of plants recovered from cryopreservation.

Cryopreservation and Omics Technologies

Plant cryopreservation is a complex, multistage process which often involves numerous, empirically derived manual and technical manipulations of in vitro materials that typically include the selection and quality of donor material, preconditioning, preculture, cryoprotection, cryopreservation, thawing/rewarming, recovery and regrowth to achieve a successful outcome as shown in Fig. 4 (Reed 2008a; Kaczmarczyk et al. 2012). Standardisation of this 'overall process' to achieve a high level of recovery is an immense undertaking that requires years of concerted investigation using amenable 'model systems' to understand the complexity of the species- and genotype-specific responses to cryopreservation (Johnston et al. 2007b, 2009, 2010). Significant advancements have been made in the last decade (Reed 2008a; Normah et al. 2012), but there remain instances where cryopreservation is constrained by lack of process where its use is highly desirable for the preservation of valued material (Häggman et al. 2008). As developments in cryopreservation progress towards recalcitrant 'non-model' species, an increasing number of difficulties have been encountered with low levels of post-storage survival (Normah et al. 2012). Impediments to progress are complex scientific and technical issues that may not be solely resolved using empirical experimentation, which in itself can impose a limit to progress which is difficult to go beyond.

Contemporary plant cryopreservation research is now supported by advanced biomolecular or 'omics' technologies that create a new knowledge base which will aid progress to solve some of the more difficult cryobiological challenges (Basu 2008; Carpentier et al. 2007, 2008a, b;

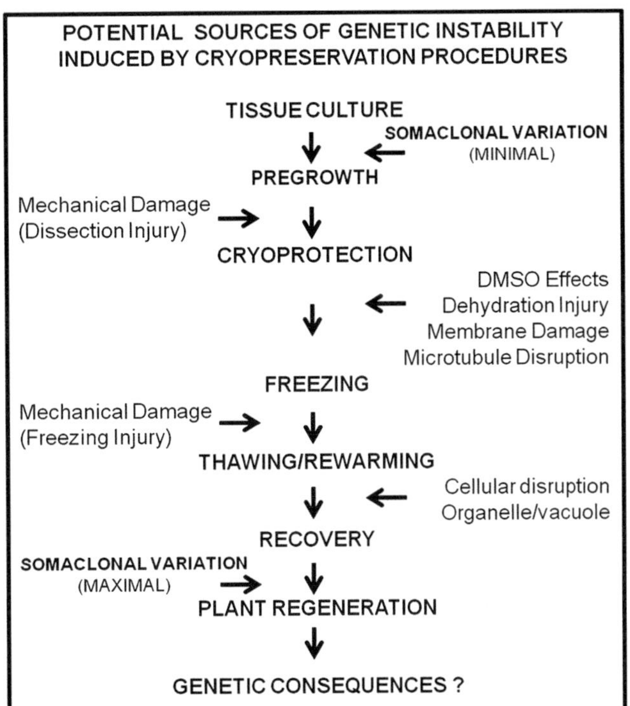

Fig. 4 The multistage process of cryopreservation signifying the possible events that leads to cellular damage and genetic instability (Harding 2004)

Volk 2010). To obtain a more complete picture of the fundamental biological processes that comprise 'cryopreservation overall' systems, biologists are now able to run the same sample through the full range of omics technologies (Morrison et al. 2006). High-throughput omics technologies such as genomics, transcriptomics, proteomics, metabolomics and other omics platforms can readily be applied to the same sample, enabling a detailed investigation into biomolecular changes that accompany the cryopreservation of plant materials. The general ontological terms of omics describe the use of genomics, functional genomics and bioinformatics approaches as generating a wealth of DNA sequence data and information that reveal the complexity of global changes in gene expression which is supported by profiling using powerful transcriptomics techniques. The study of global changes in proteins by proteomics is an essential bridge between the transcriptome and the metabolome, with metabolomics providing a global profile of a wide array of metabolites and cell signalling processes. Gene ontology and bioinformatics are important tools that unify omics platforms across biology (Ashburner et al. 2000; Morrison et al. 2006), and other contributors to this book provide more technical details of respective omics approaches.

Omics Application and Utilities

Benson (2008a) has defined the key areas of the omics era and fundamental molecular genetics that have provided considerable insights to advance contemporary understanding of natural cold-tolerance mechanisms and the critical factors associated with storage recalcitrance and cryobanking of temperate, subtropical and tropical plant species. Cryopreservation exposes plant material to severe stresses (Kaczmarczyk et al. 2012) including cellular perturbation in water flux resulting from osmotic dehydration (Carpentier et al. 2007), desiccation (Benson et al. 2005; Normah and Makeen 2008) and temperature oscillations during freezing/cooling and thawing/rewarming (Benson 2008b; Reed 2008b; Reed and Uchendu 2008), each of which can potentially change gene expression and metabolism (Volk 2010). The expanding areas of genomics (Volk 2010), transcriptomics (Carpentier et al. 2008a), proteomics (Carpentier et al. 2007; González-Arnao et al. 2011) and metabolomics (Ogawa et al. 2008) are increasingly providing evidence that will contribute significantly to understanding plant responses to cold and cryopreservation treatments (Basu 2008; Volk 2010).

Whilst the application of single omics analysis can generate a wealth of unique data, it is the integrative complementation of data sets from different omics platforms that can provide invaluable information of fundamental biological processes (Renaut et al. 2006; Volk 2010; Carpentier et al. 2008a). Undoubtedly, RNA and protein-based measurements are complementary where both transcriptomics and proteomics studies support functional genomics of banana, as a non-model crop species (Carpentier et al. 2008a, b). Proteomic studies use mass spectrometric techniques to obtain peptide maps that may be matched against theoretical spectra derived from primary sequence databases (Carpentier et al. 2005, 2007). Masses of intact tryptic peptides and/or derivatised fragment ions are then correlated with corresponding masses calculated by in silico processing of genomic, expressed sequence tags (ESTs) or protein sequences from database entries (Samyn et al. 2007; Carpentier et al. 2007). The technique is dependent partly on genome sequencing projects which provide data on the specific amino acid sequence of proteins coded by genes of a reference species (Carpentier et al. 2008a, b). This powerful comparative omics approach allows the identification of protein isoforms from plant taxa that have poorly characterised or non-sequenced genomes (Carpentier et al. 2005, 2007). The identification of proteins by sequence similarity searches depends on the content and availability of sufficient sequence information of homologous proteins in the reference database (Carpentier et al. 2008b). The greater the number of recognised peptides from a digested protein, the higher the success of the identification. When more peptides are analysed and matched, there is a greater chance for proteins of less similarity to database sequences to be identified with a limit set at ~50 % identity. Samyn et al. (2007) reported that the 40 identified proteins derived from banana; the majority of cross-species hits were made to proteins from plants with completely sequenced genomes such as rice and *Arabidopsis*. Advances in protein identification are supported by a genomic sequence comparison of distantly related *Musa* species with orthologous regions in the rice genome that provide insights into the *Musa* genome and monocot evolution (Lescot et al. 2008). The abundance of database sequences of closely related organisms enables similarity comparisons of more homologous genes in silico to make cross-species identifications (Carpentier et al. 2007, 2008b). However, the likelihood of protein identification decreases when a plant is distantly related to the reference organism with a sequenced genome. The *pro et contra* of comparative sequence-based, unambiguous identification between transcriptomics and proteomics are subjects of debate (Carpentier et al. 2008a).

Omics and Biomarkers

The central dogma, describes one gene, gives rise to a single protein, for example, in viruses, archaea and prokaryotes; evolution appears to have shaped eukaryotic plants with more proteins, as isoforms which are separated during the 2-dimensional gel electrophoresis in proteomics (Carpentier et al. 2005, 2007). For eukaryotes, it has been estimated that the number of proteins is at least one order of magnitude greater than those of their corresponding genes (Service 2001). These entities exist by virtue of the mechanisms of differential gene loci, multiple alleles, different subunit interactions, alternative splicing or mRNA editing and various post-translational modifications (Samyn et al. 2007). Most common modifications are phosphorylations, glycosylations, cleavage and carboxylations which are involved in a variety of metabolic processes (Renaut et al. 2006). For example, Samyn et al. (2007) report that the separation of a protein spot from *Musa* was identified as two isoforms of a pectinesterase-3 precursor. The derivatised peptides showed the observed mass difference between the two peptides corresponded to the mass difference between the amino acids Phe and Leu/Ile within an identical peptide sequence. The triplets coding for Phe (UUU, UUC) and Leu (UUG, UUA, CUU, CUC, CUG and CUA) are very closely related, suggesting that both isoforms are probably the result of a single nucleotide mutation event in the primary DNA sequence that signify allelic variance.

The comparative aspect of omics and biomarker research is well illustrated by the characterisation of the preconditioning phase, cold acclimation as an essential lead up to cryopreservation (Reed and Uchendu 2008). Exposing in vitro plantlets to low, nonfreezing temperatures causes genetically programmed changes to the physiology and biochemistry of tissues that are vital to survival at low temperature (section "Cell Signalling and Gene Expression"). Cold stress responses in plants are known to be associated with different families of proteins which include the dehydrins (Close 1996, 1997); the LEAs, late embryogenesis abundant proteins (NDong et al. 2002); the AFPs, antifreeze proteins (Griffith and Yaish 2004); the HSP, heat shock protein (Sung et al. 2003); and the CORs, cold-regulated proteins (Jaglo-Ottosen et al. 1998). It is recognised that genomic and transcript profile studies (Basu 2008) provide a wealth of information about the process of cold acclimation (Fowler and Thomashow 2002), but the abundance of mRNA transcripts is not always representative of cognate protein levels (Renaut et al. 2006). Expression studies from a wide range of organisms reveal that there is a poor correlation between mRNA transcript level and protein abundance (Carpentier et al. 2008a).

As a prelude to cryopreservation, the genome is a relatively fixed entity, but the proteome and transcriptome are dynamic; therefore, the picture of a proteome is at a single point in time relative to a specific environmental and/or experimental challenge (Renaut et al. 2006; Carpentier et al. 2008a). By definition, the proteomic characteristics of this type of biological sample will have a temporal component which is subject to time-related change (Morrison et al. 2006) which has particular relevance for cryopreservation. Therefore, acquisition and collection of recalcitrant plant donor material from remote field locations and transportation to the laboratory prior to cryopreservation may cause germplasm characteristics to change (Mansor 2012). Sample collection is a recognised source of variability in transcriptomics and proteomics (Carpentier et al. 2008a). The compilation of transport and process metadata through the use of the 'SPREC tool' was designed as a Standard PREanalytical Code to evaluate these effects on collection samples from remote locations of algal and tropical plant biodiversity where samples are collected for omics and systematic studies and for the purposes of cryostorage (Benson et al. 2011d; Harding and Benson 2012; Harding et al. 2013). There are numerous applications for biomarkers to assist the evaluation of collected samples for conservation research and facilitate the development of protocols to improve cryostorage outcomes (Harding and Benson 2012; Harding et al. 2013).

Omic Insights

Optimal provisions for collecting samples are essential as the quality of plant material (Benson et al. 2011c; Keller et al. 2006) is of paramount importance for cryopreservation and in the preparation and extraction of protein from recalcitrant plant tissues for 2-dimensional gel electrophoresis and proteomic analysis (Carpentier et al. 2005, 2008a). Prior to LN exposure, successful cryopreservation also requires the effective reduction of water in tissues by dehydration which can be achieved by the supplementation of media with sugars, in particular sucrose as a preconditioning treatment (Panis 2008; Engelmann et al. 2008; Keller et al. 2008b; Reed and Uchendu 2008), during preculture (Panis et al. 1996; Petijová et al. 2012), and as an osmotic cryoprotectant (Johnston et al. 2007b). Sucrose-simulated acclimation is an alternative to cold acclimation for the cryoprotective conditioning phase of shoot-tip meristems (Harding et al. 2009). An analysis of the effects of sucrose-mediated osmotic stress on the banana meristem proteome revealed that proteins were significantly up- or down-regulated by the high sucrose treatment (Carpentier et al. 2007). In this study, maintenance of an osmoprotective level of intracellular sucrose and the enhanced expression of some energy-conserving glycolysis genes and the conservation of the cell wall integrity were essential for meristems to acclimate and to survive dehydration. A comparison of dehydration-sensitive and dehydration-tolerant banana genotypes showed unique protein isoform differences associated with energy metabolism for phosphoglycerate kinase, phosphoglucomutase, UDP-glucose pyrophosphorylase and the OSR40 and abscisic stress ripening-like proteins related to stress adaptation. A further 2-dimensional DIGE proteomic study of different osmoprotectants during the acclimation of banana meristems produced hundreds of proteins; 28 were correlated to osmotic stress and 59 proteins were exclusively correlated to the sucrose treatment (Carpentier et al. 2009).

Other proteomic analyses using 2-dimensional gel electrophoresis of *Vanilla planifolia*-dissected apices revealed 206 proteins responded to PVS3 cryoprotective treatments (González-Arnao et al. 2011). Of these, quantitative changes were observed in 15 proteins, with 13 showing increased levels of expression and 2 decreased; qualitative changes were detected in 2 other proteins in the PVS-treated samples. A preliminary analysis of 2-dimensional DIGE proteomics gels showed differences in protein patterns in 15 up- or down-regulated proteins derived from potato plants grown on sorbitol precultured media indicating an alteration of the primary metabolism associated with abiotic stress underlying potato cryopreservation (Criel et al. 2005, 2008). Although cold acclimation is a requirement for some species (Reed and Uchendu 2008), different gene responses may operate in nonhardy species to enable them to survive the extreme stresses imposed by cryopreservation. This aspect was investigated in a transcriptomic study by Volk et al. (2011) that identified 244 up-regulated and

167 down-regulated genes in shoot tips recovered from cryoprotective PVS3 and LN exposure. Several groups of genes were found to be an abiotic stimulus responsive to cold, water deprivation and oxidative stress. In some gene sets, high levels of expression for transcription factors, LEA proteins and ABA proteins were associated with the recovery response to extreme temperature stress following exposure to LN.

Summary

Cryobionomics provides a conceptual framework to examine the significance of cell signalling mechanisms on cellular functions, the impacts of cryoinjury and cryogenic/non-cryogenic stress factors on germplasm viability and the implications for genetic stability following cryostorage. The opportunities for omics research to explore the overall process of cryopreservation are limitless for resolving the outstanding issues of germplasm recalcitrance (Carpentier et al. 2005; Vertommen et al. 2007) and to extend the current status of long-term preservation for economically significant species that fall out of the 'model system' experimental framework (Carpentier et al. 2008a, b; Häggman et al. 2008; Volk 2010). Progress to date has been largely driven by empirical investigative enquiry, whereas current omics platforms are beginning to capture unique insights into the biomolecular events that underpin the numerous stages of cryopreservation. Further interrogative omics research is likely to unfold the 'hidden' mechanisms that invoke tissue culture-specific responses. Armed with the investigative tools of omics and knowing the future challenges of cryopreservation invites researchers to explore this multistage process from beginning to end in hitherto poorly characterised species (Häggman et al. 2008). The advancements of the 'omics toolbox' provide immense scope and multiple prospects for future gene discovery (Tsai and Hubscher 2004) that may be directed towards intelligently developing the rationale for evidence-based, generic decisions that will improve the existing status of cryopreservation and ensure the adequate future proofing of plant cryobanks.

Acknowledgements The authors wish to thank Erica Benson for the critical reading of the manuscript.

References

Altpeter F, Oraby H (2010) Sugarcane. In: Kempken F, Jung C (eds) Genetic modification of plants, biotechnology in agriculture and forestry 64. Springer, Berlin

Asano T, Tsudzuki T, Takahashi S, Shimada H, Kadowaki K (2004) Complete nucleotide sequence of the sugarcane (*Saccharum officinarum*) chloroplast genome: a comparative analysis of four monocot chloroplast genomes. DNA Res 11:93–99

Ashburner M, Ball CA, Blake JA et al (2000) Gene ontology consortium. Gene ontology: tool for the unification of biology. Nat Genet 25:25–29

Ashmore SE (1997) Status report on the development and application of *in vitro* techniques for the conservation and use of plant genetic resources. IPGRI, Rome

Ayad WG, Hodgkin T, Jaradat A et al (1997) Molecular genetic techniques for plant genetic resources. Report of an IPGRI workshop, 9–11 October 1995, IPGRI, Rome

Basu C (2008) Gene amplification from cryopreserved *Arabidopsis thaliana* shoot tips. Curr Issues Mol Biol 10:55–60

Baust JG (2007) Concepts in biopreservation. In: Baust JG, Baust JM (eds) Advances in biopreservation. CRC Press, Boca Raton

Benson EE (1999) Plant conservation biotechnology. Taylor and Francis Ltd, London

Benson EE (2008a) Cryopreservation of phytodiversity: a critical appraisal of theory and practice. Crit Rev Plant Sci 27:141–219

Benson EE (2008b) Cryopreservation theory. In: Reed BM (ed) Plant cryopreservation: a practical guide. Springer, New York

Benson EE, Harding K (2012) Cryopreservation of shoots and meristems: an overview of contemporary methodologies. In: Loyola-Vargas VM, Ocho-Alejo N (eds) Plant cell culture protocols, 3rd edn. Humana Press, London

Benson EE, Roubelakis-Angelakis KA (1994) Oxidative stress in recalcitrant grapevine tissue cultures. Free Radic Biol Med 16:355–364

Benson EE, Harding K, Dumet DJ (2002) Cryopreservation of plant cells, tissues and organs. In: Spier RE (ed) Encyclopaedia for plant cell technology. Wiley Press, London

Benson EE, Johnston J, Muthusamy J, Harding K (2005) Physical and engineering perspectives of *in vitro* plant cryopreservation. In: Gupta SD, Ibaraki Y (eds) Plant tissue culture engineering. Springer, Dordrecht

Benson EE, Harding K, Johnston J (2007) Cryopreservation of shoot-tips and meristems. In: Day JG, Stacey G (eds) Cryopreservation and freeze drying protocols, 2nd edn, Vol 368 Methods in molecular biology. Human Press, Totowa

Benson EE, Betsou F, Amaral R, Santos LMA, Harding K (2011a) Standard PREanalytical codes: a new paradigm for environmental biobanking sectors explored in algal culture collections. Biopreserv Biobank 9:399–410

Benson EE, Harding K, Debouck D, Dumet D, Escobar R, Mafla G, Panis B, Panta A, Tay D, Van Den Houwe I, Roux N (2011b) Part I: global public goods phase 2: project landscape and general status of clonal crop *in vitro* conservation status. Systemwide Genetic Resources Programme, Rome. ISBN 978-92-9043-905-9

Benson EE, Harding K, Debouck D, Dumet D, Escobar R, Mafla G, Panis B, Panta A, Tay D, Van Den Houwe I, Roux N (2011c) Part II: status of *in vitro* conservation technologies for: Andean root and tuber crops, cassava, *Musa*, potato, sweet potato and yam. Systemwide Genetic Resources Programme, Rome. ISBN 978-92-9043-906-6

Benson EE, Harding K, Debouck D, Dumet D, Escobar R, Mafla G, Panis B, Panta A, Tay D, Van Den Houwe I, Roux N (2011d) Part III: multi-crop guidelines for developing *in vitro* conservation best practices for clonal crops. Systemwide Genetic Resources Programme, Rome. ISBN 978-92-9243-833-5

Benson EE, Betsou F, Fuller BJ, Harding K, Kofanova O (2013) Translating cryobiology principles into trans-disciplinary storage guidelines for biorepositories and biobanks: a concept paper. CryoLetters 34:277–312

Berjak P, Bartels P, Benson EE, Harding K, Mycock D, Pammenter NW, Sershen, Wesley-Smith J (2011a) Cryo-conservation of South African plant genetic diversity. In Vitro Cell Dev Biol Plant 47:65–81

Berjak P, Sershen, Varghese B, Pammenter NW (2011b) Cathodic amelioration of the adverse effects of oxidative stress accompanying procedures necessary for cryopreservation of embryonic axes of recalcitrant-seeded species. Seed Sci Res 21:187–203

Carpentier SC, Witters E, Laukens K, Deckers K, Swennen R, Panis B (2005) Preparation of protein extracts from recalcitrant plant tissues: an evaluation of different method for two-dimensional electrophoresis analysis. Proteomics 5:2497–2507

Carpentier SC, Witters E, Laukens K, Van Onckelen H, Swennen R, Panis B (2007) Banana (*Musa* spp.) as a model to study the meristem proteome: acclimation to osmotic stress. Proteomics 7:92–105

Carpentier SC, Coemans B, Podevin N, Laukens K, Witters E, Matsumura H, Terauchi R, Swennen R, Panis B (2008a) Functional genomics in a non-model crop: transcriptomics or proteomics? Physiol Plant 133:117–130

Carpentier SC, Panis B, Vertommen A, Swennen R, Sergeant K, Renaut J, Laukens K, Witters E, Samyn B, Devreese B (2008b) Proteome analysis of non-model plants: a challenging but powerful approach. Mass Spectrom Rev 27:354–377

Carpentier SC, Vertommen A, Swennen R, Panis B (2009) Will proteomics contribute to a better understanding of cryopreservation survival? In: Panis B (ed) Proceedings of the 1st international symposium on cryopreservation in horticultural species, 5–8 April 2009, Leuven

Cassells AC, Curry RF (2001) Oxidative stress and physiological, epigenetic and genetic variability in plant tissue culture: implications for micropropagators and genetic engineers. Plant Cell Tissue Organ Cult 64:145–157

Castillo NRF, Bassil NV, Wada S, Reed BM (2010) Genetic stability of cryopreserved shoot tips of *Rubus* germplasm. In Vitro Cell Dev Biol Plant 46:246–256

Chang Y, Reed BM (1999) Extended cold acclimation and recovery medium alteration improve regrowth of *Rubus* shoot tips following cryopreservation. CryoLetters 20:371–376

Chen F, Dixon R (2007) Lignin modification improves fermentable sugar yields for biofuel production. Nat Biotechnol 25:759–761

Chinnusamy V, Schumaker K, Zhu JK (2004) Molecular genetic perspectives on cross-talk and specificity in abiotic stress signalling in plants. J Exp Bot 55:225–236

Ciani F, Cocchia N, Esposito L, Avallone L (2012) Fertility cryopreservation. In: Wu B (ed) Advances in embryo transfer. InTech, Rijeka. doi:10.5772/38511

Close TJ (1996) Dehydrins: emergence of a biochemical role of a family of plant dehydration proteins. Physiol Plant 97:795–803

Close TJ (1997) Dehydrins: a commonality in the response of plants to dehydration and low temperature. Physiol Plant 100:291–296

Criel B, Panta A, Carpentier S, Renaut J, Swennen R, Panis B, Hausman JF (2005) Cryopreservation and abiotic stress tolerance in potato: a proteomic approach. Commun Agric Appl Biol Sci 70:83–86, Ghent University

Criel B, Panis B, Oufir M, Swennen R, Renaut J, Hausman J-F (2008) Protein and carbohydrate analyses of abiotic stress underlying cryopreservation in potato. In: Laamanen J, Uosukainen M, Häggman H, Nukari A, Rantala S (eds) Cryopreservation of crop species in Europe, proceedings of CRYOPLANET COST Action 871, 20–23 February 2008, Oulu, MTT Agrifood Research Working Papers 153

Cyr DR (2000) Cryopreservation: roles in clonal propagation and germplasm conservation of conifers. In: Engelmann F, Takagi H (eds) Cryopreservation of tropical plant germplasm: current research progress and applications. JIRCAS/IPGRI, Tsukuba/Rome

Day JG, Harding K, Nadarajan J, Benson EE (2008) Cryopreservation conservation of bioresources and ultra low temperatures. In: Walker JM, Rapley R (eds) Molecular biomethods handbook, 2nd edn. Humana Press, Totowa

de Vicente MC (2004) The evolving role of genebanks in the fast-developing field of molecular genetics, Issues in genetic resources No. 11. IPGRI, Rome

de Vicente MC, Andersson MS (2006) DNA banks – providing novel options for genebanks? Topical reviews in agricultural biodiversity. IPGRI, Rome de Vicente MC, Fulton T (2003) Using molecular marker technology in studies on plant genetic diversity. IPGRI, Rome

Dereuddre J, Fabre J, Bassaglia C (1988) Resistance to freezing in liquid nitrogen of carnation (*Dianthus caryophyllus* L. var. Eolo) apical and axillary shoot tips excised from different aged *in vitro* plantlets. Plant Cell Rep 7:170–173

Dulloo ME, Hunter D, Borelli T (2010) *Ex situ* and *in situ* conservation of agricultural biodiversity: major advances and research needs. Not Bot Horti Agrobot Cluj-Napoca 38:123–135

Dumet D, Engelmann F, Chabrillange N, Duval Y (1993) Cryopreservation of oil palm (*Elaeis guineensis* Jacq.) somatic embryo involving a desiccation step. Plant Cell Rep 12:352–355

Dussert S, Engelmann F, Noirot M (2003) Development of probabilistic tools to assist in the establishment and management of cryopreserved plant germplasm collections. CryoLetters 24:149–160

Eksomtramage T, Paulet F, Guiderdoni E, Glaszmann JC, Engelmann F (1992) Development of a cryopreservation process for embryogenic calluses of a commercial hybrid of sugarcane *(Saccharum* sp.) and application to different varieties. CryoLetters 13:239–352

Elliott GD, Chakraborty N, Biswas D (2008) Anhydrous preservation of mammalian cells: cumulative osmotic stress analysis. Biopreserv Biobank 6:253–260

Engelmann F (1991) *In vitro* conservation of tropical plant germplasm – a review. Euphytica 57:227–243

Engelmann F (1997) In vitro conservation methods. In: Callow JA, Ford-Lloyd BV, Newbury HJ (eds) Biotechnology and plant genetic resources. CABI, Oxon

Engelmann F (2000) Importance of cryopreservation for the conservation of plant genetic resources. In: Engelmann F, Takagi H (eds) Cryopreservation of tropical plant germplasm: current research progress and application. JIRCAS/IPGRI, Tsukuba/Rome

Engelmann F (2004) Plant cryopreservation: progress and prospects. In Vitro Cell Dev Biol Plant 40:427–433

Engelmann F (2011) Use of biotechnologies for the conservation of plant biodiversity. In Vitro Cell Dev Biol Plant 47:5–16

Engelmann F, Takagi H (2000) Cryopreservation of tropical plant germplasm – current research progress and applications. JIRCAS/IPGRI, Tsukuba/Rome

Engelmann F, Gonzalez Arnao M-T, Wu Y, Escobar R (2008) Development of encapsulation dehydration. In: Reed BM (ed) Plant cryopreservation: a practical guide. Springer, New York

Fabre J, Dereuddre J (1990) Encapsulation-dehydration: a new approach to cryopreservation of *Solanum* shoots tips. CryoLetters 11:413–426

Faltus M, Bilavčík A, Zámečník J (2007a) Study of phytohormone composition of growth medium for hop plant recovery improvement after cryopreservation. In: Lambardi M, Benelli C (eds) Proceedings of CRYOPLANET, 1st meeting of working group 2, technology, application and validation of plant cryopreservation, 10–12 May, Florence

Faltus M, Bilavčík A, Zámečník J, Svoboda P (2007b) Effect of phytohormone composition of nutrient medium on *in vitro* plant regeneration in hop clones with different sensitivities to indole-3-butyric acid. Adv Hortic Sci 21:219–224

FAO (2012) CGRFA The draft genebank standards for the conservation of non-orthodox seeds and clonally-propagated plants. Rome. http://www.fao.org/agriculture/crops/core-themes/theme/seeds-pgr/conservation/gbs/en/. Accessed 11 July 2013

FAO (2013) CGRFA draft genebank standards for plant genetic resources for food and agriculture (draft genebank standards) which comprise standards for the conservation of orthodox seeds, non-orthodox seeds and vegetatively propagated plants. Rome. http://www.fao.org/agriculture/crops/core-themes/theme/seeds-pgr/en/. Accessed 11 July 2013

Finkle BJ, Ulrich JM (1979) Effect of cryoprotectants in combination on the survival of frozen sugarcane cells. Plant Physiol 63:598–604

Fowler S, Thomashow MF (2002) Arabidopsis transcriptome profiling indicates that multiple regulatory pathways are activated during cold acclimation in addition to the CBF cold response pathway. Plant Cell 14:1675–1690

Fuller BJ (2004) Cryoprotectants: the essential antifreezes to protect life in the frozen state. CryoLetters 25:375–388

Fuller BJ, Lane N, Benson EE (2004) Life in the frozen state. CRC Press, London

Gale S, Benson EE, Harding K (2013) A life cycle model to enable research of cryostorage recalcitrance in temperate woody species: the case of Sitka spruce (Picea sitchensis). CryoLetters 34:30–39

Genebank Standards (1994) Food and Agriculture Organization of the United Nations, Rome. International Plant Genetic Resources Institute, Rome

Gnanapragasam S, Vasil IK (1990) Plant regeneration from a cryopreserved embryogenic cell suspension of a commercial sugarcane hybrid (*Saccharum* sp.). Plant Cell Rep 9:419–423

Gnanapragasam S, Vasil IK (1992) Cryopreservation of immature embryos, embryogenic callus and cell suspension cultures of gramineous species. Plant Sci 83:205–215

González-Arnao MT, Engelmann F (2006) Cryopreservation of plant germplasm using the encapsulation–dehydration technique: review and case study on sugarcane. CryoLetters 27:155–168

González-Arnao MT, Engelmann F, Huet C, Urra C (1993) Cryopreservation of encapsulated apices of sugarcane: effect of freezing procedure and histology. CryoLetters 14:303–308

González-Arnao MT, Durán-Sánchez B, Jiménez-Francisco B, Lázaro-Vallejo CE, Valdés-Rodríguez SE, Guerrero A (2011) Cryopreservation and proteomic analysis of vanilla (*V. planifolia* A.) apices treated with osmoprotectants. Acta Hortic 908:67–72

González-Benito ME, Clavero-Ramírez I, López-Aranda JM (2004) The use of cryopreservation for germplasm

conservation of vegetatively propagated crops. Span J Agric Res 2:341–351
Grapin A, Keller ERJ, Lynch PT, Panis B, Bahillo AR, Engelmann F (2011) Cryopreservation of crop species in Europe. In: Proceedings of the final meeting, COST Action 871 – CRYOPLANET, 8–11 February 2011, Agrocampus Ouest INHP, Angers, COST Action Office, Brussels, ISBN 978-92-898-0051-8. Publications Office of the European Union, Luxembourg. http://www.biw.kuleuven.be/dtp/tro/cost871/Home.htm, http://w3.cost.esf.org/index.php?id=181&action_number=871. Accessed 11 July 2013
Griffith M, Yaish MWF (2004) Antifreeze proteins in overwintering plants: a tale of two activities. Trends Plant Sci 9:399–405
Häggman H, Rusanen M, Jokipii S (2008) Cryopreservation of *in vitro* tissues of deciduous forest trees. In: Reed BM (ed) Plant cryopreservation: a practical guide. Springer, New York
Halmagyi A, Fischer-Kluver G, Mix-Wagner G, Schumacher HM (2004) Cryopreservation of *Chrysanthemum morifolium* (*Dendranthema grandiflora* Ramat.) using different approaches. Plant Cell Rep 22:371–375
Hamilton KN, Ashmore SE, Pritchard HW (2009) Thermal analysis and cryopreservation of seeds of Australian wild *Citrus* species (Rutaceae): *Citrus australasica*, *C. inodora* and *C. garrawayi*. CryoLetters 30:268–279
Harding K (2002) Genetic integrity of cryopreserved plant cells. In: Society low temperature biology symposium, Chromosomes, genes and cryobiology, Medical Society, London
Harding K (2004) Genetic integrity of cryopreserved plant cells: a review. CryoLetters 25:3–22
Harding K (2010) Plant and algal cryopreservation: issues in genetic integrity, concepts in cryobionomics and current applications in cryobiology. In: Proceedings of the Asia pacific conference on plant tissues cultures and agrobiotechnology (APaCPa) 2007, Kuala Lumpur, Malaysia. Asia Pac J Mol Biol Biotechnol 18:151–154
Harding K, Benson EE (1994) A study of growth, flowering and tuberisation in plants derived from cryopreserved potato shoot-tips. CryoLetters 15:59–66
Harding K, Benson EE (2012) Biomarkers from molecules to ecosystems and biobanks to genebanks. In: Normah NM, Chin HF, Reed BM (eds) Conservation of tropical plant species. Springer, London
Harding K, Johnston J, Benson EE (2005) Plant and algal cell cryopreservation: issues in genetic integrity, concepts in cryobionomics and current European applications. In: Benett IJ, Bunn E, Clarke H, McComb JA (eds) Contributing to a sustainable future, Proceedings of the Australian branch of the IAPTC and B, Perth, Western Australia
Harding K, Johnston JW, Benson EE (2008a) Concepts in cryobionomics: a case study of *Ribes* genotype responses to cryopreservation in relation to thermal analysis oxidative stress nucleic acid methylation and transcriptional activity. In: Laamanen J, Uosukainen M, Häggman H, Nukari A, Rantala S (eds) cryopreservation of crop species in Europe, proceedings of CRYOPLANET COST Action 871, 20–23 February 2008, Oulu, MTT Agrifood Research Working Papers 153
Harding K, Müller J, Friedl T, Day JG (2008b) Cryopreservation, genetic stability and the concept of cryobionomics for algal culture collections. Algal culture collections, ACC2008, 8–11 June 2008, Scottish Association of Marine Science, Dunstaffnage Marine Laboratory, Oban. http://www.ccap.ac.uk/oralabstracts.htm. Accessed 11 July 2013
Harding K, Johnston JW, Benson EE (2009) Exploring the physiological basis of cryopreservation success and failure in clonally propagated *in vitro* crop plant germplasm. Agric Food Sci 18:3–16
Harding K, Müller J, Day JG, Lorenz M, Friedl T (2010) Encapsulation-dehydration and colligative cryoprotective strategies and the use of amplified fragment length polymorphism (AFLP) markers to verify the identify genetic stability of cryopreserved *Euglena gracilis*. CryoLetters 31:460–472
Harding K, Benson EE, da Costa Nunes E, Pilatti FK, Lemos J, Viana AM (2013) Can biospecimen science expedite the *ex situ* conservation of plants in megadiverse countries? A focus on the flora of Brazil. Crit Rev Plant Sci 32:411–444. doi:10.1080/07352689.2013.800421
Hawkins RD, Hon GC, Ren B (2010) Next-generation genomics: an integrative approach. Nat Rev Genet 11:476–486
IPGRI/CIAT (1994) Establishment and operation of a pilot *in vitro* active genebank. Report of a CIAT-IBPGR collaborative project using cassava (*Manihot esculenta* Crantz) as a model. A joint publication of IPGRI and CIAT, Cali
Jaglo-Ottosen KR, Gilmour SJ, Zarka DG, Schabenberger O, Thomashow MF (1998) Arabidopsis CBF1 overexpression induces COR genes and enhances freezing tolerance. Science 280:104–106
Jaligot E, Rival A, Beulé T, Dussert S, Verdeil JL (2002) Somaclonal variation in oil palm (*Elaeis guineensis* Jacq.): the DNA methylation hypothesis. Plant Cell Rep 19:684–690
Janardhan BS (2007) Promising achievements and new challenges in agriculture biotechnology. Report for the 4th Asia pacific conference on plant tissue culture and agribiotechnology for better food, health and quality living, 17–21 June 2007, Kuala Lumpur, Curr Sci 93:1052–1054
Jian LC, Sun DL, Sun LH (1987) Sugarcane callus cryopreservation. In: Li PH (ed) Plant cold hardiness. Alan R. Liss Inc., New York
Johnston JW, Harding K, Bremner DH, Souch G, Green J, Lynch PT, Grout B, Benson EE (2005) HPLC analysis of plant DNA methylation: a study of critical methodological factors. Plant Physiol Biochem 43:844–853

Johnston JW, Dussert S, Gale S, Nadarajan J, Harding K, Benson EE (2006) Optimisation of the azinobis-3-ethyl-benzothiazoline-6-sulfonic acid radical scavenging assay for physiological studies of total antioxidant activity in woody plant germplasm. Plant Physiol Biochem 44:193–201

Johnston JW, Harding K, Benson EE (2007a) Antioxidant status and genotypic tolerance of *Ribes in vitro* cultures to cryopreservation. Plant Sci 172:524–534

Johnston JW, Horne S, Harding K, Benson EE (2007b) Evaluation of the 1-methyl-2-phenylindole colorimetric assay for aldehydic lipid peroxidation products in plants: malondialdehyde and 4-hydroxynonenal. Plant Physiol Biochem 45:108–112

Johnston JW, Benson EE, Harding K (2009) Cryopreservation of *in vitro Ribes* shoots induces temporal changes in DNA methylation. Plant Physiol Biochem 47:123–131

Johnston JW, Pimbley I, Harding K, Benson EE (2010) Detection of 8-hydroxy-2′-deoxyguanosine a marker of DNA damage in germplasm and DNA exposed to cryogenic treatments. CryoLetters 31:1–13

Joyce SM, Cassells AC (2002) Variation in potato microplant morphology *in vitro* and DNA methylation. Plant Cell Tissue Organ Cult 70:125–137

Joyce SM, Cassells AC, Jain SM (2003) Stress and aberrant phenotypes in *in vitro* culture. Plant Cell Tissue Organ Cult 74:103–121

Kaczmarczyk A, Rutten T, Melzer M, Keller ERJ (2008a) Ultrastructural changes associated with cryopreservation of potato (*Solanum tuberosum* L.). CryoLetters 29:145–166

Kaczmarczyk A, Shvachko N, Lupysheva Y, Hajirezaei M-R, Keller ERJ (2008b) Influence of alternating temperature preculture on cryopreservation results for potato shoot tips. Plant Cell Rep 27:1551–1558

Kaczmarczyk A, Houben A, Keller ERJ, Mette MF (2010) Influence of cryopreservation on the cytosine methylation state of potato genomic DNA. CryoLetters 31:380–391

Kaczmarczyk A, Rokka V-M, Keller ERJ (2011a) Potato shoot tip cryopreservation. A review. Potato Res 54:45–79

Kaczmarczyk A, Turner SR, Bunn E, Mancera RL, Dixon KW (2011b) Cryopreservation of threatened native Australian species – what have we learned and where to from here? In Vitro Cell Dev Biol Plant 47:17–25

Kaczmarczyk A, Funnekotter B, Menon A, Phang PY, Al-Hanbali A, Bunn E, Mancera RL (2012) Current issues in plant cryopreservation. In: Katkov II (ed) Current frontiers in cryobiology. InTech, Rijeka. doi:10.5772/32860. ISBN 978-953-51-0191-8

Kaity A, Ashmore SE, Drew RA et al (2008) Assessment of genetic and epigenetic changes following cryopreservation in papaya. Plant Cell Rep 27:1529–1539

Kami D (2012) Cryopreservation of plant genetic resources. In: Katkov II (ed) Current frontiers in cryobiology. InTech, Rijeka. doi:10.5772/34414. ISBN 978-953-51-0191-8

Kami D, Kido S, Otokita K, Suzuki T, Sugiyama K, Suzuki M (2010) Cryopreservation of shoot apices of *Cardamine yezoensis in vitro*-cultures by vitrification Method. Cryobiol Cryotechnol 56:119–126

Karp A, Kresovich S, Bhat KV et al (1997) Molecular tools in plant genetic resources conservation: a guide to the technologies, IPGRI technical bulletin No. 2. IPGRI, Rome

Kartha KK, Leung NL, Mroginski LA (1982) *In vitro* growth and plant regeneration from cryopreserved meristems of cassava (*Manihot esculenta* Crantz.). Zeitschrift Pflanzenphysiol Bd 107S:133–140

Katkov II (2012a) Current frontiers in cryopreservation. InTech, Rijeka. doi:10.5772/32047, ISBN 978-953-51-0302-8

Katkov II (2012b) Current frontiers in cryobiology. InTech, Rijeka. doi:10.5772/1962, ISBN 978-953-51-0191-8

Kaviani B (2011) Conservation of plant genetic resources by cryopreservation. Aust J Crop Sci 5:778–800

Keller ERJ (2005) Improvement of cryopreservation results in garlic using low temperature preculture and high-quality *in vitro* plantlets. CryoLetters 26:357–366

Keller ERJ, Senula A, Leunufna S, Grübe M (2006) Slow growth storage and cryopreservation – tools to facilitate germplasm maintenance of vegetatively propagated crops in living plant collections. Int J Refrig 29:411–417

Keller ERJ, Kaczmarczyk A, Senula A (2008a) Cryopreservation for plant genebanks – a matter between high expectations and cautious reservation. CryoLetters 29:53–62

Keller ERJ, Senula A, Kaczmarczyk A (2008b) Cryopreservation of herbaceous dicots. In: Reed BM (ed) Plant cryopreservation: a practical guide. Springer, New York

Keller ERJ, Zanke CD, Senula A, Breuing B, Hardeweg B, Winkelmann T (2013) Comparing costs for different conservation strategies of garlic (*Allium sativum* L.) germplasm in genebanks. Genet Resour Crop Evol 60:913–926

Kevers CT, Franck RJ, Strasser J, Dommes J, Gaspar T (2004) Hyperhydricity of micropropagated shoots: a typically stress-induced change of physiological state. Plant Cell Tissue Organ Cult 77:181–191

Kim HH, Lee SC (2012) Personalisation of droplet-vitrification protocols for plant cells: a systematic approach to optimising chemical and osmotic effects. CryoLetters 33:271–799

Kim CY, Liu Y, Thorne ET, Yang H, Fukushige H, Gassmann W, Hildebrand D, Sharp RE, Zhang S (2003) Activation of a stress-responsive mitogen-activated protein kinase cascade induces the biosynthesis of ethylene in plants. Plant Cell 15:2707–2718

Knight H, Knight MR (2001) Abiotic stress signalling pathways: specificity and cross-talk. Trends Plant Sci 6:262–267

Kovtun Y, Chiu WL, Tena G, Sheen J (2000) Functional analysis of oxidative stress-activated mitogen-activated protein kinase cascade in plants. Proc Natl Acad Sci U S A 97:2940–2945

Lakshmanan P, Geijskes RJ, Aitken KS, Grof CPL, Bonnett GD, Smith GR (2005) Sugarcane biotechnology: the challenges and opportunities. In Vitro Cell Dev Biol Plant 41:345–363

Lambardi M, Ozudogru EA, Benelli C (2008) Cryopreservation of embryogenic cultures. In: Reed BM (ed) Plant cryopreservation: a practical guide. Springer, New York

Lescot M, Piffanelli P, Ciampi AY et al (2008) Insights into the *Musa* genome: syntenic relationships to rice and between *Musa* species. BMC Genomics 9(58):1–20. doi:10.1186/1471-2164-9-58

Lynch PT (2000) Applications of cryopreservation to the long-term storage of dedifferentiated plant cultures. In: Razdan MK, Cocking EC (eds) Conservation of plant genetic resources *in vitro*, Vol 2 Applications and limitations. Science Publishers Inc., Enfield

Lynch PT, Pritchard HW, Nadarajan J, Benson EE, Harding K, Wetten AC (2011a) Country report: United Kingdom, COST ACTION 871. In: Grapin A, Keller ERJ, Lynch PT, Panis B, Bahillo AR, Engelmann F (eds) Final report conference proceedings. Cryopreservation of crop species in Europe, COST Action 871 – CRYOPLANET, COST Action Office, Brussels, ISBN 978-92-898-0051-8. Publications Office of the European Union, Luxembourg

Lynch PT, Siddika A, Johnston JW, Mehra A, Benelli C, Lambardi M, Benson EE (2011b) Effects of osmotic pretreatments on oxidative stress and antioxidant profiles of cryopreserved olive somatic embryos. Plant Sci 181:47–56

Lynch PT, Souch GR, Harding K (2012) Effects of post-harvest storage of *Allium sativum* bulbs on the cryopreservation of stem-discs by encapsulation/dehydration. J Hortic Sci Biotechnol 87:588–592

Mansor M (2012) Diversity and conservation of tropical forestry species in southeast Asia. In: Noor NM, Chin HF, Reed BM (eds) Conservation of tropical plant species. Springer, London

Martín C, González-Benito ME (2005) Survival and genetic stability of *Dendranthema grandiflora* Tzvelev shoot apices after cryopreservation by vitrification and encapsulation-dehydration. Cryobiology 51:281–289

Martinez-Montero ME, Gonzalez-Arnao MT, Borroto-Nordelo C, Puentes-Diaz C, Engelmann F (1998) Cryopreservation of sugarcane embryogenic callus using a simplified freezing process. CryoLetters 19:171–176

Martinez-Montero ME, Martinez J, Engelmann F (2008) Cryopreservation of sugarcane somatic embryos. CryoLetters 29:229–242

Martinez-Montero ME, Gonzalez Arnao MT, Engelmann F (2012) Cryopreservation of tropical plant germplasm with vegetative propagation – review of sugarcane (*Saccharum* spp.) and pineapple (*Ananas comosus* (L.) Merrill) cases. In: Katkov II (ed) Current frontiers in cryopreservation. InTech, Rijeka. doi:10.5772/32047. ISBN 978-953-51-0302-8

Mazur P (2004) Principles of cryobiology. In: Fuller BJ, Lane N, Benson EE (eds) Life in the frozen state. CRC Press, London

Meryman HT, Williams RJ (1985) Basic principles of freezing injury to plant cells: natural tolerance and approaches to cryopreservation. In: Kartha KK (ed) Cryopreservation of plant cells and organs. CRC Press, Boca Raton

Metzker ML (2010) Sequencing technologies – the next generation. Nat Rev Genet 11:31–46

Miguel C, Marum L (2011) An epigenetic view of plant cells cultured *in vitro*: somaclonal variation and beyond. J Exp Bot 62:3713–3725. doi:10.1093/jxb/err155

Mix-Wagner G, Schumacher HM, Cross RJ (2003) Recovery of potato apices after several years of storage in liquid nitrogen. CryoLetters 24:33–41

Morris JG, Acton E (2013) Controlled ice nucleation in cryopreservation – a review. Cryobiology 66:85–92

Morrison N, Cochrane G, Faruque N, Tatusova T, Tateno Y, Hancock D, Field D (2006) Concept of sample in OMICS technology. OMICS J Integr Biol 10:127–137

Mosher RA, Melnyk CW (2010) siRNAs and DNA methylation: seedy epigenetics. Trends Plant Sci 15:204–210

Nadarajan J, Staines HJ, Benson EE, Marzalina M, Krishnapillay B, Harding K (2007) Optimization of cryopreservation for *Sterculia cordata* zygotic embryos using vitrification techniques. J Trop For Sci 19:79–85

NDong C, Danyluk J, Wilson KE, Pocock T, Huner NPA, Sarhan F (2002) Cold-regulated cereal chloroplast late embryogenesis abundant-like proteins. Molecular characterization and functional analyses. Plant Physiol 129:1368–1381

Nordborg M, Weigel D (2008) Next-generation genetics in plants. Nature 456:720–723

Normah MN, Makeen AM (2008) Cryopreservation of excised embryos and embryonic axes. In: Reed BM (ed) Plant cryopreservation: a practical guide. Springer, New York

Normah MN, Chin HF, Reed BM (2012) Conservation of tropical plant species. Springer, London

Ogawa Y, Suzuki H, Sakurai N, Aoki K, Saito K, Shibata D (2008) Cryopreservation and metabolic profiling analysis of Arabidopsis T87 suspension cultured cells. CryoLetters 29:427–436

Oh TJ, Cullis MA, Kunert K, Engelborghs I, Swennen R, Cullis CA (2007) Genomic changes associated with somaclonal variation in banana (Musa spp). Physiol Plant 129:766–774

Panis B (2008) Cryopreservation of monocots. In: Reed BM (ed) Plant cryopreservation: a practical guide. Springer, New York

Panis B (2009) Cryopreservation of *Musa* germplasm. In: Engelmann F, Benson E (eds) Technical guidelines no. 9, 2nd edn. Bioversity International, Montpellier

Panis B, Lambardi M (2005) Status of cryopreservation technologies in plants (crops and forest trees). In: Ruane J, Sonnino A (eds) The role of biotechnology in exploring and protecting agricultural genetic resources. FAO, Rome

Panis B, Totté N, Van Nimmen K, Withers LA, Swennen R (1996) Cryopreservation of banana (*Musa* spp.) meristem cultures after preculture on sucrose. Plant Sci 121:95–106

Panis B, Piette B, Swennen R (2005) Droplet vitrification of apical meristems: a cryopreservation protocol applicable to all *Musaceae*. Plant Sci 168:45–55

Paton A (2009) Biodiversity informatics and the plant conservation baseline. Trends Plant Sci 14:629–637

Paulet F, Engelmann F, Glaszmann J (1993) Cryopreservation of apices of *in vitro* plantlets of sugar cane (*Saccharum* sp. hybrids) using encapsulation/dehydration. Plant Cell Rep 12:525–529

Perazzo G, Panta A, Rodriguez F, Gomez R, Toledo J, Huamán Z, Ghislain M, Golmirzaie AM, Roca W (2000) Clonal true-to-type verification of potato accessions retrieved from *in vitro* conservation and cryopreservation. CIP programme report 1999–2000. CIP, Lima. http://cipotato.org/publications/program_reports/99_00/20clonverif.pdf/view. Accessed 11 July 2013

Peredo EL, Arroyo-Garcia R, Reed BM, Revilla MA (2008) Genetic and epigenetic stability of cryopreserved and cold-stored hops (*Humulus lupulus* L.). Cryobiology 57:234–241

Petijová L, Skyba M, Cellárová E (2012) Genotype-dependent response of St. John's wort (*Hypericum perforatum* L.) shoot tips to cryogenic treatment: effect of pre-culture conditions on post-thaw recovery. Plant Omics 5:291–297

PGSC – The Potato Genome Sequencing Consortium (2011) Genome sequence and analysis of the tuber crop potato. Nature 475:189–195

Pitzschke A, Schikora A, Hirt H (2009) MAPK cascade signalling networks in plant defence. Curr Opin Plant Biol 12:1–6

RBG – Royal Botanic Gardens, Kew (2010) Plants under pressure a global assessment. The first report of the IUCN sampled red list index for plants, plants people possibilities, Royal Botanic Gardens, Kew, UK. http://www.kew.org/science/plants-at-risk/plants-worldwide.htm. Accessed 11 July 2013

Reed BM (2008a) Plant cryopreservation: a practical guide. Springer, New York

Reed BM (2008b) Cryopreservation-practical considerations. In: Reed BM (ed) Plant cryopreservation: a practical guide. Springer, New York

Reed BM (2008c) Cryopreservation of temperate berry crops. In: Reed BM (ed) Plant cryopreservation: a practical guide. Springer, New York

Reed BM, Uchendu E (2008) Controlled rate cooling. In: Reed BM (ed) Plant cryopreservation: a practical guide. Springer, New York

Reed BM, Dumet DJ, Denoma JM, Benson EE (2001) Validation of cryopreservation protocols for plant germplasm conservation: a pilot study using *Ribes* L. Biodivers Conserv 10:939–949

Reed BM, Engelmann F, Dulloo ME, Engels JMM (2004a) Technical guidelines for the management of field and *in vitro* germplasm collections, IPGRI handbooks for genebanks, No. 7. IPGRI, Rome

Reed BM, Kovalchuk I, Kushnarenko S, Meier-Dinkel A, Schoenweiss K, Pluta S, Straczynska K, Benson EE (2004b) Evaluation of critical points in technology transfer of cryopreservation protocols to international plant conservation laboratories. CryoLetters 25:341–352

Rein PW (2007) Prospects for the conversion of a sugar mill into a biorefinery. Proc Int Soc Sugar Cane Technol 26:44–60

Renaut J, Hausman J-F, Wisniewski ME (2006) Proteomics and low-temperature studies: bridging the gap between gene expression and metabolism. Physiol Plant 126:97–109

Rodriguez MCS, Petersen M, Mundy J (2010) Mitogen-activated protein kinase signaling in plants. Ann Rev Plant Biol 61:621–649

Rodriguez-Enriquez J, Dickinson HG, Grant-Downton RT (2011) MicroRNA misregulation: an overlooked factor generating somaclonal variation? Trends Plant Sci 16:242–248

Sakai A (2000) Development of cryopreservation techniques. In: Engelmann F, Takagi H (eds) Cryopreservation of tropical plant germplasm: current research progress and application. JIRCAS/IPGRI, Tsukuba/Rome

Sakai A, Kobayashi S, Oiyama I (1990) Cryopreservation of nucellar cells of navel orange (*Citrus sinensis* Osb. var. *brasiliensis* Tanaka) by vitrification. Plant Cell Rep 9:30–33

Sakai A, Matsumoto T, Hirai D, Charoensub R (2002) Survival of tropical apices cooled to −196°C by vitrification. In: Li PH, Palva ET (eds) Plant cold hardiness, gene regulation and genetic engineering. Kluwer Academic/Plenum Publishers, New York

Sakai A, Hirai D, Niino T (2008) Development of PVS-based vitrification and encapsulation-vitrification protocols. In: Reed BM (ed) Plant cryopreservation: a practical guide. Springer, New York

Samyn B, Sergeant K, Carpentier S, Debyser G, Panis B, Swennen R, Van Beeumen J (2007) Functional proteome analysis of the banana plant (*Musa* spp.) using *de novo* sequence analysis of derivatized peptides. J Proteome Res 6:70–80

Schäfer-Menuhr A, Müller E, Mix-Wagner G (1996) The use of cryopreservation as routine method for the preservation of old potato varieties. Landbauf. Volkenrode 46:65–75

Schrijnemakers EWM, Van Iren F (1995) A two step or equilibrium freezing procedure for the cryopreservation of plant cell suspensions. In: Day JG, McLellan MR (eds) Cryopreservation and freeze-drying protocols, Methods in molecular biology, vol 38. Humana Press, Totowa

Scowcroft WR (1984) Genetic variability in tissue culture: impact on germplasm conservation and utilisation. Report (AGPG:IBPGR/84/152). International Board for Plant Genetic Resources, Rome

Seki M, Narusaka M, Ishida J et al (2002) Monitoring the expression profiles of 7000 Arabidopsis genes under drought, cold, and high-salinity stresses using a full-length cDNA microarray. Plant J 31:279–292

Service RF (2001) Proteomics. High-speed biology search for gold in proteins. Science 294:2074–2077

Shendure J, Ji H (2008) Next-generation DNA sequencing. Nat Biotech 26:1135–1145

Sherlock G, Block W, Benson EE (2005) Thermal analysis of the plant encapsulation/dehydration protocol using silica gel as the desiccant. CryoLetters 26:45–54

Skyba M, Urbanová M, Kapchina-Toteva V, Košuth J, Harding K, Čellárová E (2010) Physiological, biochemical and molecular characteristics of cryopreserved *Hypericum perforatum* L. shoots tips. CryoLetters 31:249–260

Spooner D, van Treuren R, de Vicente MC (2005) Molecular markers for genebank management. IPGRI technical bulletin, No. 10. IPGRI, Rome

Stanwood PC (1985) Cryopreservation of seed germplasm for genetic conservation. In: Kartha KK (ed) Cryopreservation of plant cells and organs. CRC Press, Boca Raton

Sung DY, Kaplan F, Lee K-J, Guy CL (2003) Acquired tolerance to temperature extremes. Trends Plant Sci 8:179–187

Suzuki N, Mittler R (2006) Reactive oxygen species and temperature stresses: a delicate balance between signaling and destruction. Physiol Plant 126:45–51

Takagi H (2000) Recent developments in cryopreservation of shoot apices of tropical species. In: Engelmann F, Takagi H (eds) Cryopreservation of tropical plant germplasm- current research progress and application. JIRCAS/IPGRI, Tsukuba/Rome

Towill LE, Ellis DD (2008) Cryopreservation of dormant buds. In: Reed BM (ed) Plant cryopreservation: a practical guide. Springer, New York

Tsai CJ, Hubscher SL (2004) Cryopreservation of *Populus* functional genomics. New Phytol 164:73–81

Vertommen A, Carpentier SC, Remmerie N, Witters E, Swennen R, Panis B (2007) Towards the identification of protein complexes in banana *Musa* spp. meristems. Commun Agric Appl Biol Sci 72:1–4. Ghent University

Volk GM (2010) Application of functional genomics and proteomics to plant cryopreservation. Curr Genomics 11:24–29

Volk GM, Caspersen AM (2007) Plasmolysis and recovery of different cell types in cryoprotected shoot tips of *Mentha X piperita*. Protoplasma 231:215–226

Volk GM, Henk AD, Chhandak B (2011) Gene expression in response to cryoprotectant and liquid nitrogen exposure in Arabidopsis shoot tips. Acta Hortic 908:55–66

Walters C, Touchell DH, Power P, Wesley-Smith J, Antolin MF (2002) A cryopreservation protocol for embryos of the endangered species *Zizania texana*. CryoLetters 23:291–298

Walters C, Wheeler L, Stanwood PC (2004) Longevity of cryogenically stored seeds. Cryobiology 48:229–244

Wang ML, Goldstein C, Su W, Moore PH, Albert HH (2005) Production of biologically active GM-CSF in sugarcane: a secure biofactory. Transgenic Res 14:167–178

Wang J, Nayak S, Koch K, Ming R (2013) Carbon partitioning in sugarcane (*Saccharum* species). Front Plant Sci Plant Biotechnol 4:201. doi:10.3389/fpls.2013.00201

Withers LA (1979) Freeze preservation of somatic, embryos and clonal plantlets of carrot (*Daucus carota*). Plant Physiol 63:460–467

Zámečník J, Faltus M, Bilavčík A, Kotková R (2012) Comparison of cryopreservation methods of vegetatively propagated crops based on thermal analysis. In: Katkov II (ed) Current frontiers in cryopreservation. InTech, Rijeka. doi:10.5772/34454. ISBN 978-953-51-0302-8

Zanke C, Zamecnik J, Kotlińska T, Olas M, Keller ERJ (2011) Cryopreservation of garlic for the establishment of a European core collection. Acta Hortic 908:431–438

Zhang SQ, Klessig DF (2001) MAPK cascades in plant defence signaling. Trends Plant Sci 6:520–527

Zhu GY, Guens JMC, Dussert S, Swennen R, Panis B (2006) Change in sugar, sterol and fatty acid composition in banana meristems caused by sucrose-induced acclimation and its effects on cryopreservation. Physiol Plant 128:80–94

Nanobiotechnology in Agricultural Development

Saleha Resham, Maria Khalid, and Alvina Gul Kazi

Contents

S. Resham, Ph.D. • M. Khalid, Ph.D.
A.G. Kazi, Ph.D. (✉)
Atta-ur-Rahman School of Applied Biosciences (ASAB), National University of Sciences and Technology (NUST), Islamabad, Pakistan
e-mail: alvina_gul@yahoo.com

Abstract

Nanobiotechnology is the field of science that has recently emerged by conjugation of biotechnology and nanoscience. An extensive range of applications of the field of nanoscience (nanoparticles) have been established in several fields of biosciences and biomedicine with wide applications in industry. Since the potential of this newly emerged field of research and medicine is beyond the scope of this chapter, we will be focusing on their applications in agriculture solely. Since this is a hybrid technique, so it employs all the biotechnological tools for its applications. Their key applications include use in treating plant diseases through site-specific targeting of diseased organs, transforming plants through gold/tungsten nanoparticles coated with engineered plasmid, targeted delivery and controlled release of bioactive substances, etc. Their use in crop protection is just in its infancy. Recently, the concept of using nanoparticles in plant treatment has been established and their applications in the parasitic control in plants are practised successfully. The chapter will focus on the development and use of 'nanodevices' for formulating agriculturally important chemicals (fertilizers) with more useful properties and their direct delivery as well as their applications in various agricultural sectors. Still the limitations are there which hinder their use on large scale (commercially).

D. Barh et al. (eds.), *PlantOmics: The Omics of Plant Science*,
DOI 10.1007/978-81-322-2172-2_24,

Keywords
Nanobiotechnology • Nanodevices • Gold nanoparticle (GNP) • Mesoporous silicon nanoparticles (MSNs)

Introduction

Nanobiotechnology is a novel field which is a conjugation of molecular biotechnology, nanotechnology and material research which are cross fields of physics, biology, chemistry, computer science and engineering. It is a technology of twenty-first century and has a wide range of applications.

The word nano has been derived from Greek language meaning 'dwarf'. Technically, nano means 10^{-9}. Nanotechnology is defined as making and utilization of systems and devices at nanometric scale (0.1–100 nm) which are associated with physical and chemical properties. There are numerous ways in which nanotechnology has been defined. The Royal Society defines nanotechnology as 'the design, characterization, production and application of structures, devices and systems by controlling shape and size at nanometer scale' as shown in Fig. 1 (RSRAE 2004). Nanoscale can be best exemplified by comparing the wavelength of visible light, the size of bacteria (1,000–10,000 nm), the size of viruses (75–100 nm), the size of proteins (5–50 nm), the size of DNA (2 nm) and the size of atoms (0.1 nm).

Nanobiotechnology is defined as 'utilization of biological molecules for making technical devices at nanoscale for exploitation of biological makeup'. Biological interests in this science are due to its 'nano' size that can provide new and useful tools for nanobiologists. The term 'nanobiotechnology' was first used by a biophysicist Lynn W. Jelinski at Cornell University USA.

Aspects of Biology in Nanotechnology

In 1968, an Indian American chemist named Har Gobind Khorana synthesized DNA nucleotides and received Nobel Prize for making synthetic DNA. In 1976, a Californian team constructed a full gene by automated processing of synthesized DNA. Nowadays, these syntheses are staple in medicinal and agriculture biotechnology.

Nanotechnology can exploit biology in extraordinary details that can never be accessed through other means. Biology provides an easy way for making nanostructures such as lipids that can form liquid crystals. In nanobiotechnology, DNA structures are used in making nanostructures because of the quality of DNA ladder that brings atoms together. It is expected that it can serve the computers by mounting into the biochip

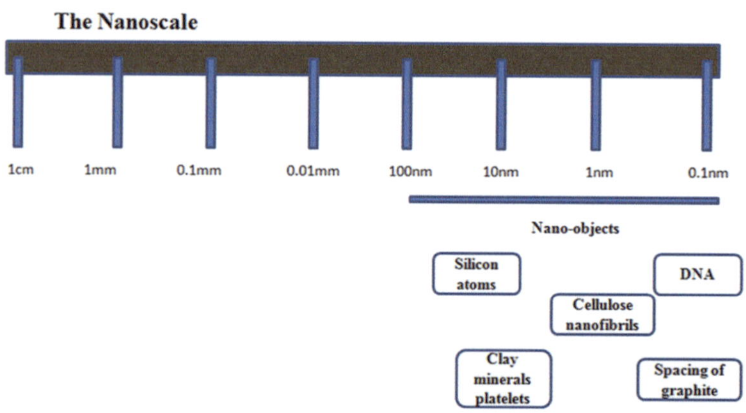

Fig. 1 The nanoscale; biological aspect

having an extraordinary information storage capacity (Alivisatos et al. 1996).

Nanotechnology is a vast field and has a wide range of applications in medicine and agriculture by providing basic tools and innovative devices for gathering genomic information of mammalian, microorganisms and plants. It has come out as an interdisciplinary science that has rapidly found its own place in scientific methodologies containing therapeutic, diagnostic, imaging, tissue engineering and drug delivery (Suh and Tanaka 2011).

There is a tremendous room of research interest in this emerging field. Vigorous research is being carried out in developing a reliable process for the synthesis of nanomaterials over a range of sizes and chemical composition. A wide range of methods is employed in the synthesis of nanoparticles for various applications; classified as chemical and biological methods. Numerous variations of the basic essential methods make and manipulate these nanoparticles to accomplish needs of a specific research objective (Bhattacharyya and Srivastava 2003; Mandal et al. 2002).

Nanoagriculture

After successful application of nanobiotechnology in medicines, it can also play a significant role in improving agriculture in future. Total world inhabitants are around seven billion, and 50 % of the population is residing in Asia. Owing to a large number of environmental factors such as storm, flood and many others, developing countries are mostly affected by shortage of food supply (Joseph and Morrison 2006). Yield of different crops is decreasing day by day due to a number of biotic factors affecting wheat, rice and cotton up to 25 %, 5–10 % and 50 %. The worldwide damage of crops by pests is estimated as 14 %. Weed is also a major problem in reducing yield. There is a need of management through latest nanotechnology techniques (Pimentel 2009) as shown in Fig. 2. Besides biotechnology, nanotechnology has also played an important role in gene transfer in order to protect crops from insects and pests (Torney 2009).

Nanomaterials and Agriculture

Composition of Nanomaterials

Nanomaterials play a significant role in revolutionizing new devices and materials. Some of the nanodevices and their role in agriculture are given below.

Carbon Nanotubes

Carbon nanotubes have potential applications like high-strength composites, energy storage

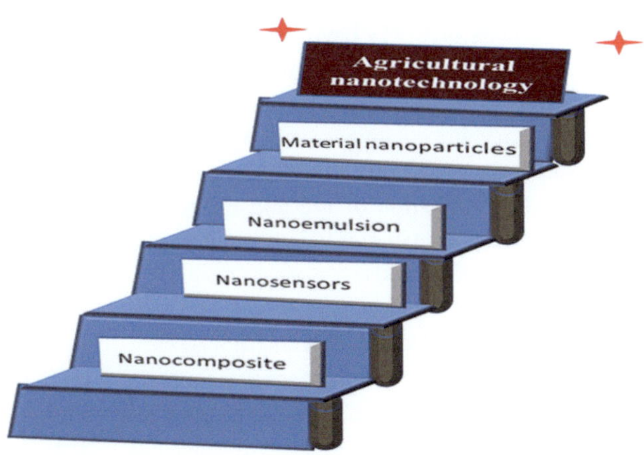

Fig. 2 Nanomaterials used in agricultural nanotechnology

and conductive energy conversion devices; sensors; radiation sources and field emission displays; nanometre-sized semiconductor devices; hydrogen storage media; interconnects; and probes. Polydispersity in nanotube type, nanotube cost and limitations in assembly and processing methods are significant barriers for some applications of single-walled nanotubes (Baughman et al. 2002).

Carbon nanotubes were discovered in 1991 (Ijima 1991); it contributed the highest share due to their unique thermal, electrical and chemical characteristics (Dresselhaus et al. 2004). However, less attention has been paid for their role in plant cell development and physiology rather than human study. Thus, more investigation is needed for CNT-plant interaction to understand its complexity. Nanoagriculture can play an important role in increasing biomass yield of crop plants (Srinivasan and Saraswathi 2010).

The role of CNTs at cellular level in tobacco has been studied by Liu et al., in 2009. Carbon nanotubes work efficiently in penetrating the chemicals as a smart delivery method to the cell especially to cell wall and cell membrane. The cup-stacked cellulase impregnated with carbon nanotubes is passed intracellularly through nanoholes induced by cellulose (Serag et al. 2012). Multiwall carbon nanotubes have been found effective. Its effect is studied on tomato seeds by penetrating the seed coat. The germination and plant growth are found to be affected (Khodakovskaya et al. 2009; Mondal et al. 2011; Tripathi et al. 2011; Villagarcia et al. 2012). Harmful effects of CNTs have been reported in plants (Lin and Xing 2007; Begum et al. 2012). The reduction of fresh root weight has been reported in rice and cucumber seedlings by using multiwall carbon nanotubes (Begum et al. 2012). The rate of germination of rye and maize gets reduced, and increase in root length is reported in maize (Lin and Xing 2007).

Mesoporous Silica Nanoparticles (MSNs)

It has been demonstrated recently that the MSNs can be internalized efficiently by plant and animal cells. Recent progress in morphological control as well as surface functionalization of mesoporous silica nanoparticles (MSNs) has improved biocompatibility of these materials with pore volumes and high surface areas. The structure, morphology, size and surface properties of MSNs have been found to be easily tunable for the purposes of controlled drug release and delivery, drug loading and multifunctionalization.

Mesoporous silica nanoparticles played an important role in enhancing the genetic engineering. These particles coated with chemicals can efficiently deliver gene into plant cells (Park et al. 2008). The ability of surface-functionalized mesoporous nanoparticles to penetrate into the cell wall can also amend the gene expression by delivering DNA and other chemicals in a controlled manner (Torney et al. 2007). Honeycomb MSNs play an important role in transferring DNA into an isolated plant cell and leaves. MSNs with genes and chemical inducer are capped with gold nanoparticle. It results in triggering of gene expression.

Silicon Nanotubes

Mesoporous silica-based materials show a number of attractive features for biomedical applications (Patil et al. 2011) such as:

- Stable mesoporous structures
- Large surface areas
- Tunable pore sizes and volumes
- Good biocompatibility

These MSNs have an average diameter less than 80 nm and exhibit well-ordered porous structure with aqueous stability.

The functionalization of MSNs with *organic moieties* or other nanostructures carries controlled discharge as well as molecular detection potential to these mesoporous substances for gene/drug delivery along with sensing applications, correspondingly. Herein, the review of recent research advancement on the set-up of functional MSN materials amid various mechanisms of controlled discharge to attain zero release in the absence of stimuli and facilitate the utilization of nonselective molecules as screens for the creation of selective sensor systems is provided (Douroumis 2011; Srilatha 2011).

Silver Nanoparticles

Silver is extensively recognized as a catalyst of pro-oxidation of methanol to formaldehyde as well as ethylene to ethylene oxide (Nagy and Mestl 1999). Colloidal silver is of particular significance because of special assets such as good conductivity, chemical stability and catalytic and antibacterial activity. An applied method for silver nanoparticles is chemical reductivity. The reduction of silver ions leads to oligomeric clusters which then form colloidal silver nanoparticles (Wiley et al. 2005). Reduction processes are based on two steps: strong reducing agent is used to produce small nanoparticles which are then enlarged by weak reducing agent; otherwise, small nanoparticles get dispersed and larger nanoparticles are difficult to control (Lee and Meisel 1982).

Green synthesis of AgNPs has three main steps (Sharma et al. 2009):

1. Selection of solvent medium
2. Selection of environmentally benign reducing agent
3. Selection of non-toxic substances for AgNP stability

Silver nanoparticle is effective as an antimicrobial agent. These nanoparticles are about 25 nm in size but have a large surface area which is a characteristic of nanoparticle. It is, when contacted with microbes like fungus and bacteria can affect the growth and cellular metabolism. It suppresses basal metabolism of electron transport system, respiration and transport of substrate in microbial cell membrane. It also inhibits the growth of harmful bacteria, which cause itchiness and infection. It also causes inhibitory effect on powdery mildew in cucumber and pumpkins (Kabir 2011). Different research groups have tested silver nanoparticles for the control of plant pathogens. It has been studied that several nanoparticles like silver, copper and gold are noxious to microorganisms (Slawson et al. 1992). It has a superior antimicrobial efficiency against viruses, bacteria and other eukaryotic microorganisms (Gong et al. 2007). It shows biocidal effects on as many as 16 species of bacteria including *E. coli* (Spadaro et al. 1974). Silver ions as an antibacterial component are used in the formulation of dental resin composites (Herrera et al. 2001 and Yoshida et al. 1999), in ion exchange fibres and in coating of medical devices (Schierholz et al. 1998; Bosetti et al. 2002; Hillyer and Albrecht 2001; Aymonier et al. 2002). Ionized silver is highly reactive and metallic silver is inert. It hinders replication of bacteria by binding to RNA or DNA. It denatures the cell wall of bacteria and leads to death (Lansdown 2002; Castellano et al. 2007).

Gold Nanoparticles

Gold nanoparticles are synthesized by a wide range of methods on the basis of plummeting chloroauric acid in the presence of stabilizing agents. The usual mean is the 'citrate synthesis' method which comprises of reduction of chloroauric acid using *trisodium citrate* resulting in the formation of GNPs (Connor et al. 2005; Tiwari et al. 2011). Gold nanoparticles are the colloidal suspension of gold nanoparticles that range in nanometre size (Tiwari et al. 2011). There are various methods to functionalize gold nanoparticles for a wide variety of uses. Commonly used functionalization approaches rely on the employment of any one or a combination of the groups such as oligo- or polyethylene glycol (PEG), bovine serum albumin (BSA), oligo- or polypeptides, oligonucleotides, antisense or sense RNA molecules, antibodies, cell surface receptors and other similar molecules (Tiwari et al. 2011).

It has been shown that the particle farming of alfalfa plants can be done by mechanically separating the gold nanoparticles stored in tissues passing through the roots when these are grown in soil rich in gold particles.

Structure of Nanoparticles

Nanocomposites

These are the combination of one nanoparticle with other or with a larger bulk material in order to improve barrier, mechanical, thermal and flame retardant properties. They can also be merged into more complex structures having special characteristics. Clay polymer nanocomposite has been used for pretreatment of effluents and removal of pollutants (Rytwo 2012).

Nanosensors

Nanobiosensors were developed in twenty-first century by combining computers, biology, electronics and nanoscience. It is the creation of biosensors of extraordinary sensing capability and shows unrivalled spatial resolution. A nanobiosensor is a nanosensor that contains a biosensor that is immobilized and selects targeted analyte molecules. It is formed at nanoscale in order to analyse data at atomic level. It provides tools for bio-analytical applications (You et al. 2009; Velasco 2009; Fan et al. 2008; Khanna 2008). It plays a role in bioprocessing and food control, biodefence and agriculture.

It consists of three components: detector, transducer and sensitized elements (Shana and Rogers 1994). Nanobiosensors are of the following types:

- *Mechanical nanobiosensor*
  An excited ground has been provided by applying mechanical forces between molecules (Cheng et al. 2006).
- *Optical nanobiosensor*
  Optics is arranged in such a way that a beam of light can pass through a closed path to make an optical nanobiosensor. When analyte binds to the resonator, change is recorded. Resonator is divided into ring resonator and linear resonator (Vo-Dinh 2005).
- *Nanowire biosensors*
  It is a hybrid of two molecules perceptive to outside signals. ssDNA serves as a detector and carbon nanotube serves as a transmitter. By using a chemical and biological molecular legend, surface properties of nanowires can easily be modified (Cui et al. 2001).
- *Ion channel switch biosensor technologies*
  It relies on synthetic personality assembling membrane work as a biological switch to notice signals and brings precise and quantitative test consequences to decrease the time for emergency diagnosis (Cornell 2002).
- *Electronic nanobiosensor*
  It is used for the detection of target DNA binding by electronic means using separated wires on a microchip, which includes multiple sensors (Jain 2005).
- *Viral nanobiosensors*
  These are essential nanoparticles; herpes simplex virus (HSV) and adenovirus activate the assembly of magnetic nanobeads as a nanosensor for clinically relevant viruses (Perez et al. 2003).
- *PEBBEL nanobiosensors*
  It comprises of sensor molecules ensnared in a chemically inert matrix by microemulsion polymerization process and provides spherical sensors in a large range of 20–200 nm. These include sensors for detecting optical change (Clark 1999a, b), pH change and detection of fluorescence (Sumner et al. 2002).
- *Nanoshell biosensor*
  It is used for the detection of analytes surrounded by multifaceted biological media devoid of any sample preparation (Hirsch et al. 2002). It also improves chemical sensing by as much as 10 billion times (Jain 2005).
- *Detector of soil quality and disease analyses*
  Soil diseases in animals and environmentalin animals and environmental factors that are controlled by viruses, bacteria and fungi can be detected by nanobiosensors by taking measurement of relative oxygen consumption of good and bad microbes. It also predicts the breakout of soil disease. It can also predict whether or not soil disease breaks out.
- *Detection of contamination and other molecules*
  It is used for the detection of contamination, pests, nutrient content and stresses such as temperature, drought and pressure. This will help farmers to apply inputs only on the required time. Research has shown that organophosphorus pesticides such as paraoxon and dichlorvos can be checked by liposome-based biosensors. A method has been developed to detect *E. coli*, by using bismuth nanofil on the basis of flow injection analysis (FIA) principle (Zhang et al. 2007). Construction of biochip sensor system was done by utilizing 150 nm Ti nanowell device and two Ti contact pads on $LiNbO_3$ substrate (Seo et al. 2008). By using a photosystem II, nanobiosensors have been developed, which

bind to several herbicides from photosynthetic organism-monitored chemicals to many others and are capable to reveal specific herbicides (Giardi and Piletska 2006).

- *DNA and protein detection*
  Carbon nanotubes as discussed earlier act as a biosensor to detect DNA and protein (Cao et al. 2008). For effective discrimination of DNA sequences, MWNTs/ZnO/CHIT composite film-modified GCE has been used for immobilization of ssDNA probes (Zhang et al. 2008a, b). To detect deep DNA damage, bionanocomposite layer of multiwalled nanotube is deposited on SPCE in chitosan (Galandova et al. 2008). Nano-SiO_2 thiophenol/p-amino film (PATP) is used to detect PAT gene sequences by a label-free EIS method (Ma et al. 2008). For the detection of electronic DNA methylation, transistor-based biosensors are developed to avoid complicated bisulphite treatment and PCR amplification (Maki et al. 2008). The biosensors based on protein nanoparticles also play a role in protein-ligand interaction properties by identifying the interaction of special protein molecules. These protein- and DNA-based biosensors play an important role in detecting plant pathogens, biomarkers and abnormalities in plants linked to mineral deficiencies and in discriminating one plant species from another.

Applied Nanoagriculture

Within a vast field of agriculture, there are various applications (as shown in Fig. 3) of nanotechnology in producing plant protection products. It not only covers the seed treatment but also improves the best utilization of water, improves fertilizer utilization and lowered dose of pesticide applications. The following are the applications of nanotechnology in agriculture.

Nanoagriculture and Plant Disease Control

Besides drought and other abiotic problems, overdose and unequal implication of agrochemicals such as fungicides, plant hormones, pesticides and herbicides are also a reason of decreasing yield of agricultural crops (Joseph and Morrison 2006). Nanobiotechnology plays an important role in improving crop plants and food industries by treating plant diseases. Overuse of pesticides and control of various disease-causing organisms in plants can be ruled out by successfully applying nanoparticles such as silver nanoparticles or the combination of two or more nanoparticles. Such nanotechnological interventions could have more efficient results in sustainable food production by minimizing the chances of disease and pest incidence in plants (Nair and Kumar 2013). There are various stud-

Fig. 3 Applications of Nanoagriculture

ies reporting the use of nanoparticles for managing various plant diseases more precisely utilizing the antimicrobial property of nanoparticles (Ocsoy et al. 2013). Advancement in nanomaterial synthesis, for example, metallic, polymeric and carbon based, has just come under researchers' notice for their application in managing bacterial diseases of plants. In a study conducted in 2013 by Ocsoy and colleagues, the antiseptic activity of Ag@dsDNA@GO complex in the direction of *X. perforans*, a model plant pathogenic bacterium, is improved by the synergistic outcome flanked by AgNPs and GO, as revealed by transmission electron microscope (TEM), scanning electron microscope (SEM), cell viability assay and cell membrane staining. Novel antimicrobials for managing pathogenic bacteria have an effect on agricultural crops, animals and humans and could be formed in conjunction with nanoparticles. Recently, a wide study of the photocatalytic activity of TiO_2/Ag, TiO_2 NPs and TiO_2/Zn on *X. perforans* has been carried out signifying the potential by means of nanoparticles in managing bacterial spot in tomato (Ocsoy et al. 2013). The application of agricultural pesticides, nutrients, fertilizers and antibiotics is through spray application to soil or plants and through feed or injection in case of animals. Pesticides or medicines are delivered either as 'preventative' treatment or make available once the disease-causing organism has multiplied and symptoms are obvious in the plant (Srilatha 2011). Nanostructure catalysts will be available in the future, which can increase the effectiveness but decrease the dose of commercial insecticides and pesticides (Joseph and Morrison 2006).

Previous studies showed that metal nanoparticles can be utilized in making insect repellent, insecticides and pesticides (Gajbhiye et al. 2009; Barik et al. 2008; Owolade et al. 2008; Goswami et al. 2010). For controlled delivery of pesticides and prolonged release, porous hollow silica nanoparticles (PHSNs) are being used with pesticide validamycin for efficient delivery and controlled release (Liu et al. 2006). Nanoemulsion is also useful for pesticide implications and effective against insect pests in agriculture (Wang et al. 2007). Similarly, solid lipid nanoparticles are essential for nanopesticides (Lu et al. 2006). Nanoencapsulation is the latest technology used to protect host plant against insects and pests. The basic principle of this term is the slow and efficient release as well as proper absorption of chemicals such as insecticides to a particular host plant instead of a larger particle for insect pest control (Scrinis and Lyons 2007).

Another type of unique nanoparticle is nanosilica which is formed from silica and has useful applications in drugs; it can also be used as nanopesticides. Pests and insects employ a variety of cuticular lipids as water barrier and avert death from desiccation. In case of nanosilica when applied onto leaves and stems, after contacting with insects, get engrossed into cuticular lipids and cause physical demise of insects (Barik et al. 2008). Modified surface-charged nanosilica of approximately 3–5 nm size can be effectively applied to control agricultural insects and pests. Polyethylene glycol nanoparticles are utilized with garlic essential oil coating against *Tribolium castaneum*. A result of slow and persistent release of these components showed 80 % of control of insect (Yang et al. 2009).

Bioassay has been conducted for silver nanoparticles; *titanium oxide, zinc oxide* and *aluminium oxide* play a role to control insects. Liquid and solid formulations of these nanoparticles have been applied on rice and held in reserve in a plastic box amid adults of *S. oryzae* and was kept for 7 days. On day one, most hydrophilic SNP was affected. On day 2, mortality rate of 90 % was obtained when treated with SNP and ANP, and 86 % and 95 % mortality rate was obtained with ANP and SNP on day 7. Another bioassay for silk worm grasserie disease was carried out. A significant decrease of virus occurred when the plant was treated with ethanolic suspension of hydrophobic *aluminosilicate* nanoparticles (Goswami et al. 2010). In near future, nanotechnology will modernize the pest management. It is predicted that nanotechnology will accelerate the 'green revolution' (Bhattacharyya et al. 2010). *S. oryzae* and *Rhyzopertha dominica* are the two insect pests found on stored food supplies. In order to control these insect pests, insecticidal activity of nanostructured alumina was studied by Teodoro et al. (2010). After continuous exposure of 3 days of alumina-treated water wheat, significant mortality was observed. This nanostructured alumina is considered to be a cheap alternative of any other commercially available insecticides. Further studies possibly will also inflate the nanoparticle-based pest management.

Polymer-based copper and silver nanoparticles are reported as an antifungal against plant pathogenic fungus (Cioffi et al. 2004). Silica silver nanoparticles have been efficiently used against plant pathogenic fungus like *Rhizoctonia solani*, *Colletotrichum gloeosporioides*, *Botrytis cinerea*, *Pythium ultimum* and *Magnaporthe grisea*. These nanoparticles are also effective against powdery mildew (Park et al. 2006). Silver nanoparticles have been found as an antifungal against *Raffaelea spp.* which is highly pathogenic against oak trees (Kim et al. 2009). Silver nanoparticles not only affect the fungus hyphae but also the conidial germination. Efficient antimicrobial activity is shown by copper nanoparticles with soda lime glass powder against gram-negative and gram-positive fungi and bacteria (Esteban-Tejeda et al. 2009). Silver nanoparticles are effective against *M. grisea* and *B. sorokiniana* (Jo et al. 2009), *Fusarium* and *Phoma* (Gajbhiye et al. 2009).

Nanopesticides, nanofungicides and nanoherbicides are being utilized in agriculture. Some nanoparticles within 100–250 nm range size can easily be dissolved in water and some are oil based used in making uniform suspensions of herbicides and pesticides of 200–400 nm size and have various applications in harvested crops (Rickman et al. 1999; Goswami et al. 2010). In addition to these, nanotechnology also plays an important role in food packaging and crop improvement by transferring gene (Bhattacharyya et al. 2011).

Precision Farming and Crop Improvement via Nanotechnology

In order to get maximum yield, inputs such as fertilizers, herbicides and insecticides are not only important but the environmental factors also play an important role to get higher yield. Precise farming can be achieved by checking environmental conditions by using computers, remote sensing devices and global satellite positioning. These can be done in order to check maximum efficiency of crops by finding out location of problems as well as reducing agricultural wastes thus minimizing agricultural pollution. As a result of using sensor nanotechnology, accurate decisions will help farmer to take better decisions (Cioffi et al. 2004).

Nanotechnology plays an important role in enhancing biotechnology; instead of using conventional method for gene transfer; nanoparticles, nanocapsule and nanofibres are used as vectors, which transfer a number of genes and set off expression of gene (Miller and Senjen 2008; Nair et al. 2010). Chitosan is a nanocomplex nonviral vector having low transfection efficiencies (Zhao et al. 2011). It has been modified into octapeptide having a condense DNA with high transfection efficiencies and lower toxicity compared with non-modified chitosan. The integration of DNA through *Jatropha curcas* cells has been studied through this type of nanoparticles (Wang et al. 2011) (Table 1).

Nanofertilizers and Nanoagrochemicals

Nanoparticles also find their applications for use in conjugation with the fertilizers. Studies show that there is an increase in nutrient use efficiency, reduction in soil toxicity, minimization in the potential negative effects associated with overdosage and reduction in the frequency of the nanofertilizer application. Hence, nanotechnology has a high potential for achieving sustainable agriculture, especially in developing countries (Naderi and Danesh-Shahraki 2013). Nanofertilizers mainly delay the release of the nutrients and extend the fertilizer effect period. Nanofertilizers can be facilitated in three ways:

(a) Nutrient can be encapsulated inside nanoporous materials.
(b) It can be coated with thin polymer film.
(c) It can be delivered as particle or emulsion of nanoscale dimension (Rai and Ingle 2012).

Table 1 Nanoparticles and their applications in agriculture

Applications	Nanoparticles	References
Effects against plant pathogenic fungi	WA-CV-WA13B, WA-ATWB13R silver nanoparticles	Kim SW, Jung JH, Lamsal K, Kim YS, Min JS, Lee YS (2012) Antifungal effects of silver nanoparticles (AgNPs) against various plant pathogenic fungi. Mycobiology 40(1):53–58
Nitrogen capture, storage and slow release of fertilizer	Nanoporous zeolites	Leggo PJ (2000) An investigation of plant growth in an organo–zeolitic substrate and its ecological significance. Plant Soil 219:135–146
(a) Capability for the detection of progesterone concentration in cow milk (b) Detecting ovulation	Immunosensors	Carralero V, González-Cortés A, Yañez-Sedeño P, Pingarrón JM (2007a) Development of progesterona immunosensor based on a colloidal gold-graphite-teflon composite electrode. Electroanalysis 19:853–858
Used for controlling pathogens	Nano-hexaconazole	Gopal M, Roy SC, Roy I, Pradhan S, Srivastava C, Gogoi R, Kumar R, Goswami A (2011b) Nanoencapsulated hexaconazole: a novel fungicide and process for making the same. Indian Patent Application No. 205
Used for controlling mites	Nano-sulphur	Gopal M, Chaudary SR, Ghose M, Dasgupta R, Devakumar C, Subrahmanyam B, Shrivastava C, Gogoi R, Kumar R, Goswami A (2011a) Samfungin: a novel fungicide for making the same. Indian Patent Application No. 1599
Used as insecticides, drugs and dyes	2,4-dinitophenol (2,4-DNP), 2,5-dinitrophenol (2,5-DNP), 2,6-trinitrophenol (2,6-DNP), and 2,4,6-TNP (2,4,6-TNP)	Shimazu M, Mulchandani A, Chen W (2001) Simultaneous degradation of organophosphorus pesticides and p-nitrophenol by a genetically engineers Moraxella sp. with surface expressed organophosphorus hydrolase. Biotechnol Bioeng 76(4):318–324
Electrospinning of cotton fibres	Cellulose electrospinner at nanoscale	Electrospinning Nanofibres Can Turn Waste Into New Products. AZoNano – The A to Z of Nanotechnology. 10 September 2003. New York State College of Human Ecology at Cornell. 25 March 2005 http://www.azonano.com/details.asp?ArticleID
Gene is transferred by bombardment of DNA-absorbed gold particles	Gold nanoparticles	Christou P, McCabe DE, Swain WF (1988) Stable transformation of soybean callus by DNA coated gold particles. Plant Physiol 87(3):671–674
Efficient delivery of DNA and chemicals through silica nanoparticles internalizes in plant cells	Silica nanoparticles	Torney F, Trewyn BG, Lin VSY, Wang K (2007) Mesoporous silica nanoparticles deliver DNA and chemicals into plants. Nat Nanotechnol 2:295–300

Nanostructure formulation through mechanisms such as targeted delivery or slow, controlled and conditional release mechanisms could discharge their active ingredients in response to environmental triggers and biological demands in the most precise manner. Studies show that the use of nanofertilizers has provided us with some great worthy advantages: augments nutrient use efficiency, minimizes the potential negative effects associated with overdosage, reduces soil toxicity and most importantly decreases the frequency of the application.

Nanotechnology can improve crop management through careful usage of agrochemicals. Only a very small amount is required for treatment. Currently, a very large amount of chemicals are used which are ultimately lost into the environment. This method of managing chemicals has negative effect on the environment like soil degradation, water pollution and side effects on other species. The nanotechnology will ensure the availability of effectual product in a prescribed way and conditions. It has an ability to reach the active side and cause changes on plant metabolism. Formulation of beta-cypermethrin nanoemulsion is more effective than commercial microemulsion of similar compound (Wang et al. 2007). Development of silica nanoparticles (SNP), which is surface functionalized, caused mortality of 90 % in *Sitophilus oryzae by* evaluating it with conventional silica (Debnath et al. 2011). Using nano-sized calcium carbonate results in sustainable release of validamycin after 2 weeks (Qian et al. 2011). The use of imidacloprid encapsulated with a coating of sodium alginate and chitosan layer by layer increases the speed rate in soil applications (Guan et al. 2010). Agrochemicals tagged with nanoparticles help in smart delivery of chemicals to the targeted site of plant organs and tissues (Faheem et al. 2013).

Nanotechnology Improving Animal Production

Livestock put in 40 % of the total worth of agriculture output internationally. It secures the billion of people from scarcity. Urbanization along with rising income and increasing population growth highly demands meat and other products. The annual growth rate of animal production is 0. 9 % in underdeveloped countries and 2–7 % in developed countries. By using nanotechnology, a lot of problems such as diseases in animals and environmental factors can be solved. Nanotechnology has been classified into five categories which include veterinary medicines, pathogen detection, waste removal and remediation and weed improvement, animal breeding and genetics and can identify preservation and supply chain tracking (Kuzma 2010).

Triazophos within nanoemulsion stayed protected under neutral and acidic conditions but released easily under alkaline conditions. Some nanoparticles have the ability to reach the sap (Nair et al. 2010), while some nanoparticles can move through several sites of a pumpkin. These substances made in nanoform can affect the metabolism of cells. Magnetic nanoparticles cause inhibition of chlorophyll biosynthesis. Nanosensors are incorporated into livestock which may be helpful in animal tracking and detecting other substances such as hormones and drugs during marketing of animals (Nguyen et al. 2012).

Water and Soil Resources Conservation

Nanoparticles are synthesized with catalytic oxidation and reduction objectives in order to accelerate active substances. It results in minimum time utilization as pesticides, plant growth regulators and fertilizers. These nanoparticles can also be used as soil decontaminants (Knauer and Bucheli 2009). Shen et al. (2007) worked on magnetic Fe3aO4-C18 which is a synthetic composite nanoparticle. It is more effective than conventional C18 materials which are a part of organophosphorus pesticides. In crop field, organophosphorus and carbamate degradation can be enhanced by TiO_2 nanoparticles. Some of the nanomaterials which are successfully used are multiwalled carbon nanotube coated electrodes (Sundari and Manisankar 2011), nanoparticle carbon electrodes with magnetic composite and nano-TiO_2 (Kumaravel and Chandrasekaran 2011). Nano-iron and nano-carbons are used for water and soil purification and remediation (Karn et al. 2009).

Agricultural Waste Recycling by Nanotechnology

The agricultural waste is that part of agricultural product, which cannot be utilized for anything. Nanotechnology can help in making this waste useful. For example, some of the cotton has to be discarded as a waste or used for low-quality products. By using a technique known as electrospinning, 100 nm fibres can be used as fertilizer or pesticide absorbant, which allows targeted application at a desired time and location (Susan 2003). Nanotechnology increases the performance of enzymes to convert cellulose into ethanol. Nanoengineering allows cellulose from waste plant part to convert into ethanol. Nanosilica is produced by burning rice husk and is then utilized in making glass or concrete.

Plant Growth and Germination via Nanotechnology

It has been studied that nanomaterials also play an important role in plant growth and germination. An experiment showed that spinach was treated with TiO_2 and non-TiO_2. Nano-TiO_2 showed dry weight of more than 73 %, an increase of 45 % of chlorophyll a formation and three times higher photosynthetic rate when compared with non-TiO_2-treated seeds over 30 days' germination period. This is because of the increased rate of seed resistance and uptake of water and oxygen to increase the rate of germination (Zheng et al. 2005).

Nanotechnology Application in Food Sector

Nanotechnology has played an important role in food production, processing and packaging. Food production through applied nanoagriculture such as applications of nanoagrochemicals has been described earlier. Nanotechnology has played an important role in food processing. Food nanostructures are used for the improvement of tastes and textures. If natural food materials are nanostructured, tasty food with less fat can be produced. The examples include nanostructured mayonnaise and low fat but creamy ice cream and yogurt (Khan 2012).

Similarly, nanostructured food additives, for example, vitamins, flavors, colors, preservatives and antioxidants improve the taste of food and reduce fat, sugar and salt preservatives, (Chaudhry et al. 2008). Bioactivation and nanoencapsulation are useful nanomaterials for the development of nutraceuticals to deliver drugs to specific sites (Kaufmann 2005). There is efficient bioavailability due to nanocrystals, e.g. fatty acids, omega 3, phytosterols, antimicrobial components and carotenoids are effectively absorbed (Graciela and Qin 2012). In Germany, Canola cooking oil is used with nano-phytosterols to reduce cholesterol absorption. Nano-green tea is produced for active availability of selenium to human body (Chaudhry et al. 2010; Coles and Fewer 2013).

Nanotechnology has a wide range of applications in food packaging. If nanomaterials are incorporated into plastic polymers, it leads to the development of novel food packaging with improvement in flexibility, stability and durability. Mostly, silver nanoparticles are used as microbicide for increasing food freshness and preventing contamination (Duncan 2011). Nanoparticles play an important role in the regulation of gases and moisture passages to extend the shelf life and maintain the quality and freshness of packed food materials (Sozer and Kokini 2009). It has been estimated, that by 2015 nanotechnology-based packaging will create 19 % of the share in global consumer goods industry (Nanoposts Report 2008). Nanosensors are used in detection of pathogens and ripening of food products (Kuswandi et al. 2011).

Conclusion

The aim of this review is to explore the potential of the field of nanotechnology with respect to agriculture. Our main focus is on the currently applied uses of nanoparticles for agriculture. Various compositions and structures of nanomaterials can be applied to overcome various problems related to agriculture like insect pest management by utilization of traditional methods, development of improved crop varieties and diluting adverse effects of chemical pesticides. Nanotechnology has a significant influence on the economy and the environment by improving fertilizers and energy. Hence, it has a high potential for achieving sustainable agriculture, especially in developing countries.

References

Alivisatos AP, Johnsson KP, Peng X, Wilson TE, Loweth CJ, Bruchez MP, Schultz PG (1996) Organization of 'nanocrystals molecules' using DNA. Nature 382:609–611

Aymonier C, Schlotterbeck U, Antonietti L, Zacharias P, Thomann R, Tiller JC, Mecking S (2002) Hybrids of silver nanoparticles with amphiphilic hyperbranched macromolecules exhibiting antimicrobial properties. Chem Commun 2002:3018–3019

Barik TK, Sahu B, Swain V (2008) Nano-silica from medicine to pest control. Parasitol Res 103:253–258

Baughman RH, Zakhidov AA, de Heer WA (2002) Carbon nanotubes – the route toward applications. Science 297(5582):787–792

Begum P, Ikhtiari R, Fugetsu B, Matsuoka M, Akasaka T, Watari F (2012) Phytotoxicity of multi-walled carbon nanotubes assessed by selected plant species in the seedling stage. Appl Surf Sci 262:120–124

Bhattacharyya S, Srivastava A (2003) Synthesis of gold nanoparticles stabilized by metal chelator and the controlled formation of close packed aggregates by them. J Chem Sci 115(5–6):613–619

Bhattacharyya A, Bhaumik A, Usha Rani P, Mandal S, Epidi TT (2010) Nano-particles: a recent approach to insect pest control. Afr J Biotechnol 9(24):3489–3493
Bhattacharyya A, Datta PS, Chaudhuri P, Barik BR (2011) Nanotechnology: a new frontier for food security in socio economic development. In: Proceeding of disaster, risk and vulnerability conference 2011 held at School of Environmental Sciences, Mahatma Gandhi University, India in association with the Applied Geoinformatics for Society and Environment, Germany, 12–14 March 2011
Bosetti M, Masse A, Tobin E, Cannas M (2002) Efficacy of silver coated medical devices. J Appl Biomater 23(3):887–892
Cao C, Kim JH, Yoon D, Hwang ES, Kim YJ, Baik S (2008) Optical detection of DNA hybridization using absorption spectra of single-walled carbon nanotubes. Mater Chem Phys 112(3):738–741
Castellano JJ, Shafii SM, Ko F, Donate G, Wright TE, Mannari RJ, Payne WG, Smith DJ, Robson MG (2007) Comparative evaluation of silver-containing antimicrobial dressings and drugs. Int Wound J 4(2):114–122
Chaudhry Q, Scotter M, Blackburn J, Ross B, Boxall A, Castle L, Aitken R, Watkens R (2008) Applications and implications of nanotechnologies for the food sector. Food Addit Contam 25(3):241–258
Chaudhry Q, Castle L, Watkins R (2010) Nanotechnologies in food, RSC nanosciences and nanotechnology no. 14. Springer, Cambridge
Cheng MM, Cuda G, Bunimovich YL, Gaspari M, Heath JR, Hill HD, Mirkin CA, Nijdam AJ, Terracciano R, Thundat T, Ferrari M (2006) Nanotechnologies for biomolecular detection and medical diagnostics. Curr Opin Chem Biol 19(1):10–11
Cioffi N, Torsi L, Ditaranto N, Sabbatini L, Zambonin PG, Tantillo G, Ghibelli L, D'Alessio M, Bleve-Zacheo T, Traversa E (2004) Antifungal activity of polymer-based copper nano-composite coatings. Appl Phys Lett 85:2417–2419
Clark HA (1999a) Optical nanosensors for chemical analysis inside single living cells, 2: sensors for pH and calcium and the intracellular application of PEEBLE sensors. J Anal Chem 71(21):4837–4843
Clark HA (1999b) Optical nanosensors for chemical analysis inside single living cells, 1: fabrication, characterization, and methods for intracellular delivery of PEBBLE sensors. Anal Chem 71(21):4831–4836
Coles D, Fewer FJ (2013) Nanotechnology applied to European food production – a review of ethical and regulatory issues. Trends Food Sci Technol 34:32–43
Connor EE, Mwamuka J, Gole A, Murphy CJ, Wyatt MD (2005) Gold nanoparticles are taken up by human cells but do not cause acute cytotoxicity. Small 1(3):325–327
Cornell BA (2002) Optical biosensors: present and future. In: Lighler F, Taitt CR (eds) Membrane based biosensors. Elsevier, Amsterdam
Cui Y, Wei Q, Park H, Lieber CM (2001) Nanowire nanosensors for highly sensitive and selective detection of biological and chemical species. Science 293(12):89–92
Debnath N, Das S, Seth D, Chandra R, Bhattacharya SC, Goswami A (2011) Entomologic effect of silica nanoparticles against *Sitophilus oryzae* (L.). J Pest Sci 81(1):99–105
Douroumis D (2011) Mesoporous silica nanoparticles as drug delivery system. J Nanomed Nanotechnol 2:102e. doi:10.4172/2157-7439.1000102e
Dresselhaus MS, Dresselhaus G, Jorio A (2004) Unusual properties and structure of carbon nano tubes. Annu Rev Mater Res 34:247–278
Duncan TJ (2011) Applications of nanotechnology in food packaging and food safety: barrier materials, antimicrobials and sensors. J Colloid Interface Sci 363(1):1–24
Esteban-Tejeda L, Malpartida F, Esteban-Cubillo A, Pecharromán C, Moya JS (2009) Antibacterial and antifungal activity of a soda-lime glass containing copper nanoparticles. Nanotechnology 20(8):085103. doi:10.1088/0957-4484/20/8/085103
Faheem A, Nishat A, Shalendra K, Sarvajeet SG, Ritu G, Narendra T, Bon HK (2013) Nanobiotechnology: scope and potential for crop improvement. In: Tuteja N, Gill SS (eds) Crop improvement under adverse conditions. Springer, New York. doi:10.1007/978-1-4614-4633-0_11
Fan X, White IM, Shopova SI, Zhu H, Suter JD, Sun Y (2008) Sensitive optical biosensors for unlabeled targets: a review. Anal Chim Acta 620(1–2):8–26
Gajbhiye M, Kesharwani J, Ingle A, Gade A, Rai M (2009) Fungus mediated synthesis of silvernanoparticles and its activity against pathogenic fungi in combination of fluconazole. Nanomedicine 5(4):282–286
Galandova J, Ziyatdinova G, Labuda J (2008) Disposable electrochemical biosensor with multiwalled carbon nanotubes-chitosan composite layer for the detection of deep DNA damage. Anal Sci 24(6):711–716
Giardi MT, Piletska EV (2006) Biotechnological applications of photosynthetic proteins: biochips, biosensors and biodevices. Springer, New York
Gong P, Li H, He X, Wang K, Hu J, Zhang S, Yang X (2007) Preparation and antibacterial activity of Fe3O4@Ag nanoparticles. Nanotechnology 18:604–611
Goswami A, Roy I, Sengupta S, Debnath N (2010) Novel applications of solid and liquid formulations of nanoparticles against insect pests and pathogens. Thin Solid Films 519:1252–1257
Graciela WP, Qin W (2012) Nanotechnology research methods for food and bioproducts. Wiley-Blackwell, Oxford
Guan HA, Chi DF, Yu J, Li H (2010) Dynamics of residues from a novel nano-imidacloprid formulation in soybean fields. Crop Prot 29(9):942–946
Herrera M, Carrion P, Baca P, Liebana J, Castillo A (2001) In vitro antibacterial activity of glass ionomer cements. Microbios 104(409):141–148
Hillyer JF, Albrecht RM (2001) Gastrointestinal persorption and tissue distribution of differently sized colloidal gold nanoparticles. J Pharm Sci 90:1927–1936

Hirsch LR, Jackson JB, Lee A, Halas NJ, West JL (2002) A whole blood immunoassay using gold nano-shells. Anal Chem 75(23):77–81

Ijima S (1991) Helical micro-tubules of graphitic carbon. Nature 354:56–58

Jain KK (2005) Nanotechnology in clinical laboratory diagnostics. Clin Chim Acta 358(1–2):37–54

Jo Y-K, Kim BH, Jung G (2009) Antifungal activity of silver ions and nanoparticles on phytopathogenic fungi. Plant Dis 93(10):1037–1043

Joseph T, Morrison M (2006) Nanotechnology in agriculture and food. www.nanoforum.org

Kabir L (2011) Inhibition effect of silver nanoparticles against powdery mildew for cucumber and pumpkin. Mycobiology 39(1):26–32

Karn B, Kuiken T, Otto M (2009) Nanotechnology and in situ remediation: a review of the benefits and potential risks. Environ Health Perspect 117(12):1813–1831

Kaufmann SHF (2005) Novel vaccination strategies. Wiley-VCH, Weinham

Khan AS (2012) Nanotechnology: ethical and social implications. CRC Press, Boca Raton

Khanna VK (2008) New-generation nano-engineered biosensors, enabling nanotechnologies and nanomaterials. Sens Rev 28(1):39–45

Khodakovskaya M, Dervishi E, Mahmood M, Xu Y, Li Z, Watanabe F, Biris AS (2009) Carbon nanotubes are able to penetrate plant seed coat and dramatically affect seed germination and plant growth. ACS Nano 3(10):3221–3227 (Article retracted, ACS Nano, 6, 7541 (2012))

Kim SW, Kim KS, Lamsal K, Kim YJ, Kim SB, Jung M, Sim SJ, Kim HS, Chang SJ, Kim JK, Lee YS (2009) An in vitro study of the antifungal effect of silver nanoparticles on oak wilt pathogen *Raffaelea sp.* J Microbiol Biotechnol 19:760–764

Knauer K, Bucheli T (2009) Nano-materials-the need for research in agriculture. Agrarforschung 16(10):390–395

Kumaravel A, Chandrasekaran M (2011) A biocompatible nano TiO(2)/nafion composite modified glassy carbon electrode for the detection of fenitrothion. J Electroanal Chem 650(2):163–170

Kuswandi B, Wicaksono Y, Jayus J, Abdullah A, Heng L, Ahmad M (2011) Smart packaging: sensors for monitoring of food quality and safety. Sens Instrum Food Qual Saf 5(3):137–146

Kuzma J (2010) Nanotechnology in animal production – upstream assessment of applications. Livest Prod Sci 130:14–24

Lansdown ABG (2002) Silver I: its antibacterial properties and mechanism of action. J Wound Care 11:125–138

Lee PC, Meisel D (1982) Adsorption and surface enhanced Raman of dyes on silver and gold sols. J Phys Chem 86:3391

Lin D, Xing B (2007) Phytotoxicity of nanoparticles: inhibition of seed germination and root growth. Environ Pollut 150:243–250

Liu F, Wen LX, Li ZZ, Yu W, Sun HY, Chen JF (2006) Porous hollow silica nanoparticles as controlled delivery system for water-soluble pesticide. Mater Res Bull 41:2268–2275

Liu Q, Chen B, Wang Q, Shi X, Xiao Z, Lin J, Fang X (2009) Carbon nanotubes as molecular transporters for walled plant cells. Nano Lett 9:1007–1010

Lu C, Toepel K, Irish R, Fenske RA, Barr DB, Braro R (2006) Organic diets significantly lower children's dietary exposure to organophosphorus pesticides. Environ Health Perspect 114(2):260–263

Ma Y, Jiao K, Yang T, Sun D (2008) Sensitive PAT gene sequence detection by nano-SiO2/paminothio-phenol self-assembled films DNA electrochemical biosensor based on impedance measurement. Sens Actuator B 131(2):565–571

Ma Y, Kuang L, He X, Bai W, Ding Y, Zhang Z, Zhao Y, Chai Z (2010) Effects of rare earth oxide nanoparticles on root elongation of plants. Chemosphere 78:273–279

Maki WC, Mishra NN, Cameron EG, Filanoski B, Rastogi SK, Maki GK (2008) Nanowire transistor based ultrasensitive DNA methylation detection. Biosens Bioelectron 23(6):780–787

Mandal S, Selvakannan P, Phadtare S, Pasricha R, Sastry M (2002) Synthesis of stable gold hydrosol by the reduction of chloroaurate ions by amino acid, aspartic acid. J Chem Sci 114(5):513–520

Miller G, Senjen R (2008) Out of the laboratory and on to our plates. Nanotechnology in food & agriculture. Friends of the Earth, Australia/Europe/USA. In: Friends of the Earth Europe website, 31 May 2011. Available from http://www.foeeurope.org/activities/nanotechnology/Documents/Nano_food_report.pdf

Mondal A, Basu R, Das S, Nandy P (2011) Beneficial role of carbon nanotubes on mustard plant growth: an agricultural prospect. J Nanoparticle Res 13:4519–4528

Naderi MR, Danesh-Shahraki A (2013) Nanofertilizers and their roles in sustainable agriculture. Int J Agric Crop Sci 5(19):2229–2232

Nagy A, Mestl G (1999) High temperature partial oxidation reactions over silver catalysts. Appl Catal A Gen 188(1):337–353

Nair R, Kumar DS (2013) Plant diseases control and remedy through nanobiotechnology. In: Tutja N, Gill SS (eds) Crop improvement under adverse conditions. Springer, New York

Nair R, Varguese SH, Nair BG, Maekawa T, Yoshida Y, Kumar DS (2010) Nanoparticulate material delivery to plants. Plant Sci 179:154–163

Nanoposts Report (2008) Nanotechnology and consumer goods market and applications to 2015, 2008.

Nguyen DN, Ngo TT, Nguyen QL (2012) Highly sensitive fluorescence resonance energy transfer (FRET)-based nanosensor for rapid detection of clenbuterol. Adv Nat Sci Nanosci Nanotechnol 3(3). doi:10.1088/2043-6262/3/3/035011

Ocsoy I, Paret ML, Ocsoy MA, Kunwar S, Chen T, You M, Tan W (2013) Nanotechnology in plant disease management: DNA directed silver nanoparticles on graphene oxide as an antibacterial against *Xanthomonas perforans*. ACS Nano 7(10):8972–8980

Owolade OF, Ogunleti DO, Adenekan MO (2008) Titanium dioxide affects disease development and yield of edible cowpea. Elect J Environ Agric Food Chem 7(50):2942–2947

Park HJ, Kim SH, Kim HJ, Choi SH (2006) A new composition of nanosized silica-silver for control of various plant diseases. Plant Pathol J 22:295–302

Park IY, Kim IY, Yoo MK, Choy YJ, Cho MH, Cho CS (2008) Mannosylated polyethylenimine coupled mesoporous silica nanoparticles for receptor mediated gene delivery. Int J Pharm 359:280–287

Patil A, Chirmade UN, Slipper I, Lamprou DA, Urquhart A, Douroumis D (2011) Encapsulation of water insoluble drugs in mesoporous silica nanoparticles using supercritical carbon dioxide. J Nanomed Nanotechnol 2:111. doi:10.4172/2157-7439.1000111

Perez JM, Simeone FJ, Saeki Y, Josephson L, Weissleder R (2003) Viral-induced self-assembly of magnetic nanoparticles allows the detection of viral particles in biological media. J Am Chem Soc 125(34):10192–10193

Pimentel D (2009) Pesticide and pest control. In: Peshin P, Dhawan AK (eds) Integrated pest management: innovation-development process. Springer, Dordrecht

Qian K, Shi TY, Tang T, Zhang SL, Liu XL, Cao YS (2011) Preparation and characterization of nano-sized calcium carbonate as controlled release pesticide carrier for validamycin against *Rhizoctonia solani*. Microchim Acta 173(1–2):51–57

Rai M, Ingle A (2012) Role of nanotechnology in agriculture with special reference to management of insect pests. Appl Microbiol Biotechnol 94(2):287–293

Rickman D, Luvall JC, Shaw J, Mask P, Kissel D, Sullivan D (1999) Precision agriculture: changing the face of farming. Geotimes feature article. www.ghcc.msfc.nasa.gove/precisionag/. Accessed 19 Nov 2011

RSRAE The Royal Academy of Engineering (2004) Nanoscience and nanotechnologies: opportunities and uncertainties. RS policy document, The Royal Society and Royal Academy of Engineering, London 19/04 (July 2004)

Rytwo G (2012) The use of clay polymer nanocomposites in waste water pretreatment. Sci World J (7), 498503. doi:10.1100/2012/498503

Schierholz JM, Lucas LJ, Rump A, Pulverer G (1998) Efficacy of silver coated medical devices. J Hosp Infect 40:257–262

Scrinis G, Lyons K (2007) The emerging nano-corporate paradigm nanotechnology and the transformation of nature, food and agri-food systems. Int J Social Agric Food 15(2):22–44

Seo S, Dobozi-King M, Young RF, Kish LB, Cheng M (2008) Patterning a nanowell sensor biochip for specific and rapid detection of bacteria. Microelectron Eng 85(7):1484–1489

Serag MF, Kaji N, Tokeshi M, Baba Y (2012) Introducing carbon nanotubes into living walled plant cells through cellulase induced nanoholes. RSC Adv 2:398–400

Shana A, Rogers KR (1994) Biosensors. Meas Sci Technol 5(5):461–472

Sharma VK, Yngard RA, Lin Y (2009) Silver nanoparticles: green synthesis and their antimicrobial activities. Adv Colloid Interface Sci 145(1–2):83–96

Shen HY, Zhu Y, Wen XE, Zhuang YM (2007) Preparation of Fe3O4-C18 nanomagnetic composite materials and their cleanup properties for organophosphorus pesticides. Anal Bioanal Chem 387(6):2227–2237

Slawson RM, Van Dyke MI, Lee H, Trevors JT (1992) Germanium and silver resistance, accumulation, and toxicity in microorganisms. Plasmid 27(1):72–79

Sozer N, Kokini JL (2009) Nanotechnology and its applications in the food sector. Trends Biotechnol 27(2):82–89

Spadaro JA, Berger TJ, Barranco SD, Chapin SE, Becker RO (1974) Antibacterial effects of silver electrodes with weak direct current. Microb Agent Chemother 6:637–642

Srilatha B (2011) Nanotechnology in agriculture. J Nanomed Nanotechnol 2:123

Srinivasan C, Saraswathi R (2010) Nanoagriculture – carbon nanotubes enhance tomato seed germination and plant growth. Curr Sci 99:274–275

Suh KS, Tanaka T (2011) Nanomedicine in cancer. Transl Med 1:103e. doi:10.4172/2161-1025.1000103e

Sumner JP, Aylott JW, Monson E, Kopelman R (2002) A fluorescent PEBBLE nanosensor for intracellular free zinc. Analyst 127:11–16

Sundari PA, Manisankar P (2011) Development of nano poly (3-methylthiophene)/multiwalled carbon nanotubes sensor for the efficient detection of some pesticides. J Braz Chem Soc 22(4):746–755

Susan SL (2003) Waste fiber can be recycled into valuable products using new technique of electrospinning, Cornell researchers report. Cornell Chronicle. http://www.news.cornell.edu/releases/Sept03/electrospinning.ACS.ssl.html

Teodoro S, Micaela B, David KW (2010) Novel use of nano-structured alumina as an insecticide. Pest Manag Sci 66(6):577–579

Tiwari PM, Vig K, Dennis VA, Singh SR (2011) Functionalized gold nanoparticles and their biomedical applications. Nanomaterials 1(1):31–63

Torney F (2009) Nanoparticle mediated plant transformation. Emerging technologies in plant science research. Interdepartmental Plant Physiology Major Fall Seminar Series Physics, p 696

Torney F, Trewyn BG, Lin VS, Wang K (2007) Mesoporous silica nanoparticles deliver DNA and chemicals into plants. Nat Nanotechnol 2:295–300

Tripathi S, Sonkar SK, Sarkar S (2011) Growth stimulation of gram (*Cicer arietinum*) plant by water soluble carbon nanotubes. Nanoscale 3:1176

Velasco MN (2009) Optical biosensors for probing at the cellular level: a review of recent progress and future prospects. Semin Cell Dev Biol 20(1):27–33

Villagarcia H, Dervishi E, Silva K, Biris AS, Khodakovskaya MV (2012) Surface chemistry of carbon nanotubes impacts the growth and expression of water channel protein in tomato plants. Small 8:2328–2334

Vo-Dinh T (2005) Optical nanosensors for detecting proteins and biomarkers in individual living cells. Methods Mol Biol 300:383–402

Wang LJ, Li XF, Zhang GY, Dong JF, Eastoe J (2007) Oil-in-water nanoemulsions for pesticide formulations. J Colloid Interface Sci 314(1):230–235

Wang Q, Chen J, Zhang H, Lu M, Qiu D, Wen Y, Kong Q (2011) Synthesis of water soluble quantum dots for monitoring carrier-DNA nanoparticles in plant cells. J Nanosci Nanotechnol 11(3):1533–4880

Wiley B, Sun Y, Mayers B, Xi Y (2005) Controlled synthesis of metal nanostructures. Chem Eur J 11:454–463

Yang FL, Li XG, Zhu F, Lei CL (2009) Structural characterization of nanoparticles loaded with garlic essential oil and their insecticidal activity against *Tribolium castaneum* (Herbst) (Coleoptera: Tenebrionidae). J Agric Food Chem 57(21):10156–10162

Yoshida K, Tanagawa M, Matsumoto S, Yamada T, Atsuta M (1999) Antibacterial activity of resin composites with silver containing materials. Eur J Oral Sci 107(4):290–296

You C, Bhagawati M, Brecht A, Piehler A (2009) Affinity capturing for targeting proteins into micro and nanostructures. Anal Bioanal Chem 393(6–7):1563–1570

Zhang W, Tang H, Geng P, Wang Q, Jin L, Wu Z (2007) Amperometric method for rapid detection of *Escherichia coli* by flow injection analysis using a bismuth nano-film modified glassy carbon electrode. Electrochem Commun 9(4):833–838

Zhang W, Yang T, Huang D, Jiao K, Li G (2008a) Synergistic effects of nano-ZnO/multi-walled carbon nano- tubes/chitosan nanocomposite membrane for the sensitive detection of sequence specific of PAT gene and PCR amplification of NOS gene. J Membr Sci 325:245–251

Zhang W, Yang T, Huang DM, Jiao K (2008b) Electrochemical sensing of DNA immobilization and hybridization based on carbon nanotubes/nano zinc oxide/chitosan composite film. Chin Chem Lett 19:589–591

Zhao X, Zhaoyang L, Wenguang L, Wingmoon L, Peng S, Richard YTK, Keith DKL, William WL (2011) Octaarginine-modified chitosan as a nonviral gene delivery vector: properties and in vitro transfection efficiency. J Nanoparticle Res 13(2):1572–1896

Zheng L, Hong F, Lu S, Liu C (2005) Effect of nano-TiO2 on strength of naturally aged seeds and growth of spinach. Biol Trace Elem Res 104:83–91

Plant Pharmacogenomics: From Drug Discovery to Personalized Ethnomedicine

Mustafeez Mujtaba Babar, Najam us Sahar Sadaf Zaidi, and Alvina Gul Kazi

Contents

M.M. Babar, Ph.D. • N. us Sahar Sadaf Zaidi, Ph.D.
A.G. Kazi, Ph.D. (✉)
Atta-ur-Rahman School of Applied Biosciences (ASAB), National University of Sciences and Technology (NUST), Islamabad, Pakistan
e-mail: alvina_gul@yahoo.com

Abstract

Plants have been an integral part of ethnomedicine and are being exploited for their therapeutic benefits for centuries. They have not only contributed to the drug discovery and development process but also supported the conventional therapeutic strategies by providing effective curative alternatives. The genetic makeup of the botanical source tends to affect the qualitative and quantitative properties of a phytochemical. These plant products, like other medicinal agents, possess beneficial and adverse drug reactions, which vary widely among individuals. This variation is mainly attributable to the difference in the genetic makeup of an individual. To investigate these differences, the study of plant pharmacogenomics becomes inevitable. Plant pharmacogenomics deals with the study of the genetic processes involved in the differential pharmacological responses to a particular phyto-

D. Barh et al. (eds.), *PlantOmics: The Omics of Plant Science*,
DOI 10.1007/978-81-322-2172-2_25, © Springer India 2015

chemical and vice versa. Interindividual variation in drug transport, metabolism, and interaction at the receptor site are some of the representations of these genetic variants. Understanding of the interrelationship of the genetic diversity of individuals and the administered plant product can help develop ideal therapeutic strategy for the treatment of a disease. Recent advances in the fields of genomics, transcriptomics, and proteomics have made the functional understanding of the genome efficient and accurate. The scope of pharmacogenomics spreads from simple monogenic traits to complex pathways involving hundreds of alleles, influencing both pharmacokinetic and pharmacodynamics parameters, thus helping in the effective translation of molecular data into clinical findings. Once the correlation between the genetic makeup of an individual and the expected therapeutic response to a particular plant product is established, targeted, personalized pharmacotherapy can be developed. This review systemically analyzes the recent developments in plant pharmacogenomics and its contributions in the field of molecular and pharmaceutical sciences.

Keywords
Ethnomedicine • Drug development • Personalized medicine • Plant pharmacogenomics • Plant pharmacology

Abbreviations

ACE	Angiotensin-converting enzyme
ADR	Adverse drug reactions
APO	Apoenzyme
BRCA	Breast cancer (susceptibility gene)
BSA	Body surface area
CAM	Complementary and alternate medicine
DDD	Drug design and discovery
HER 2	Human epidermal growth factor receptor-2
HIF	Hypoxia-inducible factor
HTS	High-throughput screening
IBD	Identity by descent
LOH	Loss of heterozygosity
PCR	Polymerase chain reaction
QSAR	Quantitative structure-activity relationship
RFLP	Restriction fragment length polymorphism
ROS	Reactive oxygen species
SNP	Single nucleotide polymorphism
TCTP	Translationally controlled tumor protein

Introduction

Genetic variations form the basis of morphological and physiological diversity in human beings. This may, in part, be exhibited as differential response to exogenous agents and external stimuli. Employment of therapeutic remedies for the cure of diseased conditions has been an integral part of the human social system. The chief source of acquiring health benefits has been the plant-based therapy. Many medicinal systems rely on plants for the manufacture of drugs or for the extraction of chemical constituents used in the preparation of medicines. These phytochemicals can be used for the treatment of a wide variety of ailments affecting different organ systems of the body including the cardiovascular system, nervous system, respiratory system, and reproductive system. Many individuals, however, respond to the administration of these drug molecules in a variety of ways. The degree of pharmacological response, therapeutic effectiveness, and the appearance of adverse effects of the drugs are the main representatives of the differential therapeutic responses. These are mainly attributable to the genetic polymorphism in the drug-processing systems of the body and are covered under the scope of pharmacogenomics.

Pharmacogenomics may, hence, be referred to as the study of molecular pharmacology in relation to human genomics. Various cellular components are involved in the processing of natural and synthetic drugs. In the genetic polymorphism of a drug's molecular targets, the proteins which are involved in the transport of the drug molecule and the enzymatic systems involved in the biotransformation process are the main players in presenting variation in the pharmacological responses. In addition to the genetic factors, the properties of the active moiety of a drug and other components of the dosage form also affect the therapeutic presentations. Plant pharmacogenomics, as a branch, deals with the differential responses observed among individuals to phytotherapy. Moreover, being extracted from living sources, the genetic nature of the plant varieties tends to affect the chemoprofile of the plant products. Additionally, the human genetic data pertaining to the plant-based therapies, though limited, has proved that the drug disposition and metabolism processes can serve as potential regulators for altering the pharmacological profile of the drug. Pharmacogenomics can, hence, be employed in the drug development by interacting at myriad levels in the process. Target identification, validation, testing, and evaluation are some of the avenues which can employ the tools of pharmacogenomics. Additionally, the availability of genetic tests and predetermination of the therapeutic effects of a drug have helped in bringing pharmacogenomics from the bench to the bedside. Personalized medicines can now be developed owing to the contribution of the knowledge of this field.

This chapter summarizes the importance of plant-based products in various medicinal systems. The genetic basis of the differential response to the drug therapies and the fundamental concepts of pharmacogenomics are covered. Pharmaceutical considerations of the plant-based therapies in association with the genetic makeup of individuals and the employment of genetic data to study the varying responses of the individuals are brought under discussion. Finally, the pharmacogenomics tools employed in the drug and personalized medicine development process, with particular reference to ethnopharmacology, are reviewed.

Pharmacological Diversity in Plant Kingdom

Plant-based products have formed an integral part of man's medicinal system. They provide rich sources of drugs that can be employed to cure many diseased conditions, from local infections to generalized systemic ailments. With the industrial development and increase in population, the newer systems of medicine now rely on synthetic preparation of drugs as they yield greater output. However, many of the contemporary systems of medicines of the world, like Ayurvedic, Chinese, African, homeopathy, and Unani systems are still dependent upon plants for the extraction of principal ingredients. Plants contain a diversity of secondary metabolites. These phytochemicals have proven antimicrobial, anti-inflammatory, anticancer, analgesic, anxiolytic, and a wide array of other bioactivities. It is believed that more than a tenth of the discovered plant species are used in drugs (McChesney et al. 2007). Moreover, around 50 % of the prescribed drugs are either produced from plants or are derivatives of plant products (Kinghorn and Balandrin 1993).

Phytotherapy is not limited to the third-world countries. Today, many practitioners of complementary and alternate medicine (CAM) are contributing to the health system of industrialized countries as well. Around one-third of the population is dependent upon these systems in the United States of America and the United Kingdom (Ernst 1996; Wootton and Sparber 2001). Plants form the largest part of the drugs of natural origin. Greater therapeutic benefits, safety, availability, and cost-effectiveness make plant-based therapies the preferred choice of many individuals. Apart from their use in plant-based therapy, the allopathic system of medicine owes its origin to natural products. Ethnomedicine has served as the basis for the discovery of many of the lead compounds now used in the modern medicinal systems (Fig. 1).

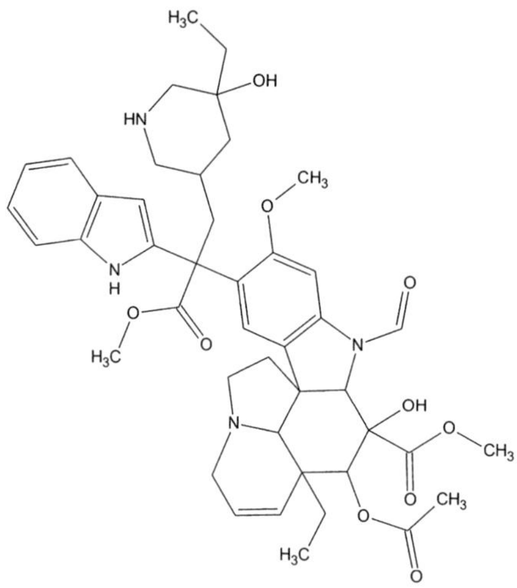

Vincristine
(***anticancer agent***)

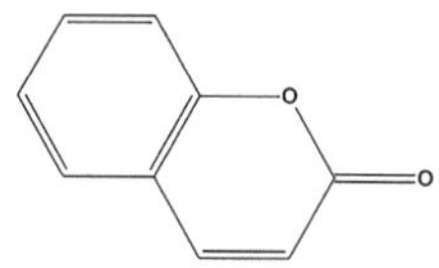

Coumarin
(***anticoagulant agent***)

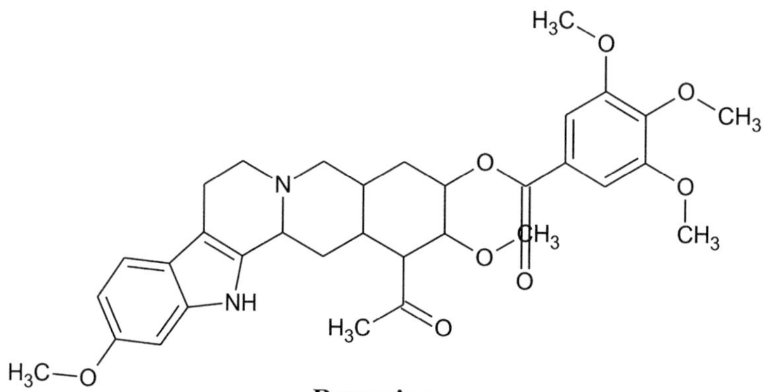

Reserpine
(***antihypertensive and antipsychotic***)

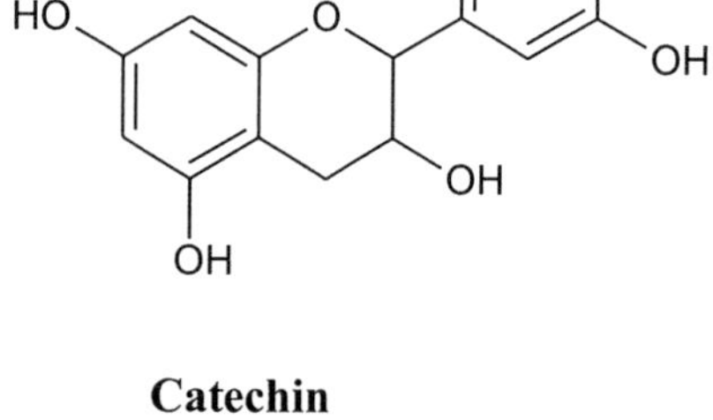

Catechin
(***antioxidant***)

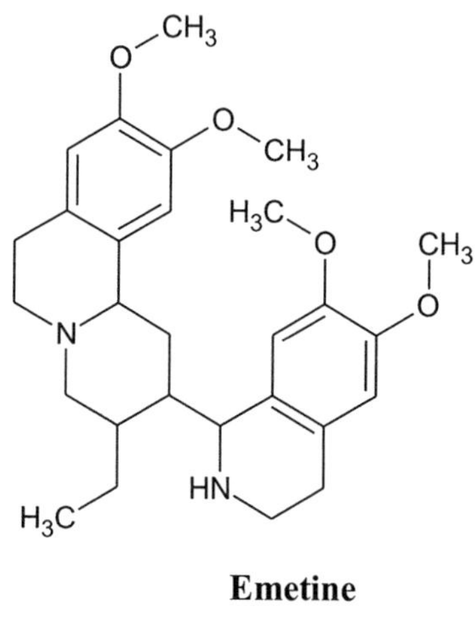

Emetine
(***antiprotozoal***)

Fig. 1 Chemical structures of some therapeutically effective phytochemicals

The development in the purification, analytical, and bioactivity procedures has led to the characterization of many phytochemicals. The focus has, hence, shifted from the use of crude drug forms to the isolation and application of specific chemical compounds. Many products, either active or prodrugs, contain a variety of chemical groups. Flavonoids, carotenoids, glucosinolates, catechins, phytoestrogens, and terpenoids are some of the phytochemical groups identified. These functional moieties have been associated with the treatment of specific diseased conditions. Lignans, for instance, have proved their efficacy as antineoplastic, anti-inflammatory, and cardiovascular protectant agents (Qu et al. 2005; Pusztai et al. 2010; Peterson et al. 2010; Pan et al. 2009). Sharing these therapeutic properties with lignans, coumarins additionally provide analgesic, antiviral, and antiasthmatic effects (Daniel 2006; Keri et al. 2010; Gliszczyńska and Brodelius 2012; Deng et al. 2006). Another medically important class of plant-derived compounds is alkaloid. Alkaloidal salts have gained acceptance by nearly all the medicinal systems in practice today. They are responsible for providing antiarrhythmic, antihypertensive, diuretic, and antimicrobial effects (Agrawal et al. 2010; Moyer et al. 2010; Mojab et al. 2010; Saleem et al. 2010).

Phytochemicals not only provide diversity in chemical groups they incorporate, but they target a wide variety of cell types and follow a number of mechanisms to present their therapeutic benefits. They tend to be effective in both prokaryotic and eukaryotic cells. A number of chemicals, including essential oils, phenolics, flavonoids, and alkaloids, target prokaryotic cells like bacteria and cause bacterial death. They target various metabolic enzymes and structural units of bacteria to show their antimicrobial efficacy (Radulovic et al. 2013; Simões et al. 2009). Similarly, plant-derived antifungals have proved to be very effective against yeasts and molds. Within the complex animal body, these phytochemicals target various macromolecules to show their therapeutic effects. Human cardiovascular, respiratory, nervous, integumentary, and digestive systems, histologically, vary widely from one another. However, the plant products exploit many different types of cellular systems to manifest their therapeutic benefits. Prevention of oxidation by neutralization of reactive oxygen species (ROS) helps in providing anticancer, anti-inflammatory, and cardioprotectant effects (Hernández et al. 2009). A number of plant products help in releasing the endothelial-derived relaxing factor (EDRF) that acts as a vasodilator to help in the hypertensive conditions (Schmitt and Dirsch 2009). Additionally, the immune potentiation effect of many plant products by stimulation of various transcription factors has also been proved (Bremner and Heinrich 2010; Licciardi and Underwood 2010). An important consideration, however, is that the genotype of the plant species also affects the nature, type, and yield of a phytochemical. *Withania somnifera*, for example, is employed in antitumor and antitubercular remedies. Different varieties of the plant contain different grades of alkaloids (withanolides) depending upon the genetic, environmental, and demographic factors of a plant (Joshi et al. 2004) (Table 1).

Phytotherapy essentially follows a similar pharmacological profile as those of newer drug forms. Following the extraction of crude drug, a variety of formulations may be designed. Commonly employed dosage forms include decoctions, pills, balms, elixirs, and syrups. After the administration of the drug into the body, it is absorbed through a biological membrane. Gastrointestinal tract and skin are the main sites of absorption of a drug. This phase is followed by binding of the drug with the plasma proteins resulting in the distribution of the drug to various target tissues of the body. Concurrently, the drug is being exposed to various enzymes in the body to be metabolically converted into alternate forms. This conversion produces a more hydrophilic drug, yielding either an active drug or aiding the elimination of the drug from the body by the process of excretion in later steps. Once the drug has reached the target cell, it interacts with the cellular receptors and enters the cell. Within the cell, the phytochemical interacts with the molecular machinery to exhibit its therapeutic effects. However, a number of environmental and

Table 1 Pharmacogenomics aspects of some phytochemicals

Phytochemicals	Botanical sources	Main therapeutic applications	Gene targeted
Glucosinolates, isothiocyanates	Cruciferous plants	Anticancer	IDO (Banerjee et al. 2007), NRF-2 (Cho et al. 2006), CYP2E1 (Lampe 1999)
Polyphenols	Multiple families	Anti-inflammatory, anticancer, antioxidant	NF-KB (Schauss 2012; Menaa et al. 2012), apoE (Milenkovic et al. 2012), COX (Diebolt et al. 2001), SIRT (Chung et al. 2010)
Alkaloids	*Cinchona*, henbane, strychnine	Antiarrhythmic, anticholinergic, analgesic, antihypertensive, neuronal	CGRP (Negro et al. 2012), CYP3A (Klein and Zanger 2013), PIG3, p53 (Lin et al. 2011; Tsai et al. 2008), NRF-2 (Cho et al. 2008), MYD88 (Tu et al. 2012)
Allicin	Alliaceae plants	Antibacterial, antiviral	MTb 85B (Hasan et al. 2007), WAF1 (Aggarwal et al. 2004), INK4 (Ha and Yuan 2004)
Ginsenosides (saponins and glycosides)	Araliaceae plants	Vasodilation, anticoagulation, cell proliferation	CAMs (Cho et al. 2013; Wahid et al. 2010), CDK (Lee et al. 2009), Rg-1 (Lü et al. 2004)

Abbreviations: *apoE* apolipoprotein E, *CAM* cell adhesion molecules, *CDK* cyclin-dependent kinase, *CGRP* calcitonin gene-related peptide, *COX* cyclooxygenase, *CYP* cytochrome, *IDO* indoleamine 2,3 dioxygenase, *INK* tumor suppressor protein, *MTb* metallothionein, *MYD* myeloid differentiation, *NF-KB* nuclear factor kappa-light-chain-enhancer of activated B cells, *NRF-2* nuclear factor (erythroid-derived 2)-like 2, *p53* (tumor) protein 53, *PIG3* p53 inducible gene, *Rg* regeneration gene, *SIRT* sirtuin (silent mating-type information regulation 2 homologue) 1, *WAF* cyclin-dependent kinase inhibitor 1

inherent factors hamper the provision of pharmacological benefits in a predefined way. Keeping the external factors constant, same drug therapy yields varying therapeutic responses and adverse drug reactions (ADRs) in susceptible individuals. Genetic diversity is the major intrinsic factor that results in altered pharmacological effects to a drug. This interindividual genetic variation, hence, makes the study of human genome, role of drug-processing proteins, mutations in their genes, and epigenetic factors, in association with the therapeutic outcomes observed, all the more important.

Genetic Polymorphism

Phytochemicals are the chemical entities obtained from plants. The genetic background of the botanical source has profound effect on the overall chemical profile of the plant. Both qualitative and quantitative phytomedicinal traits are affected as a result of these genetic differences. In addition, the concept of plant pharmacogenomics extends further to understanding the variation in response to an administered phytochemical agent, once the drug has been administered into the human body.

The advent of modern molecular biology techniques led to the discovery of a number of genetic factors that are, now, considered responsible for presenting variable therapeutic response. Drugs tend to exploit various mechanisms in the human body to show their therapeutic benefits. These targets may be located on the cellular surfaces, in the form of receptors, or within the cell, as members of signaling cascade or other proteins. After the interaction of a phytochemical with its target, a cascade of events takes place to present the final pharmacological effect. However, the overall therapeutic response is not only dependent upon the effect of the drug molecule on the body, but certain other factors are also involved. These players are responsible for changing the overall pharmacological profile of a drug. In addition, they also contribute to the appearance of interindividual variation in therapeutic responses. Of these, genetic polymorphism in drug-processing systems is the most important contributing factor. Genes encoding cellular receptors, hormones, plasma proteins, drug transporters, and metabolism possess a high degree of variation, resulting in the difference of pharmacological action of a drug molecule. A difference in only 1 % of the gene sequences of these factors has been related with the appearance of altered response to a pharmacotherapy (Meyer 2000). Underlying these differences are a number of racial, geographic, ethnic, and demographic features which have been discussed in the subsequent sections.

Genetic polymorphism tends to affect both the pharmacokinetic and pharmacodynamics properties of a drug. Pharmacokinetics involves all the steps carried out by the body to make the drug reach its final target site and elimination of toxic components from the body. These include the absorption through the gastrointestinal tract, distribution via plasma proteins, metabolism of the drug, and the final excretion from the body. Pharmacodynamics, on the other hand, covers the interaction of drug molecule with the receptor site and the pharmacological action of the drug at the cellular and subcellular levels. Genetic polymorphism yielding changes in structural or functional properties of any of the pharmacokinetic or pharmacodynamics parameters results in the appearance of unexpected therapeutic responses. Among these processes, altered drug targets, drug transport mechanisms, and metabolic processes have been mainly associated with differential response to plant-based drug therapies.

Drug Targets

Targeting specific molecular mechanisms helps in achieving better outcomes with minimal adverse effects. The massive content of the human genome along with the genetic polymorphism results in the need to identify and develop specific targets. Recent developments in the fields of genomics and proteomics have helped in identifying the genes and proteins responsible for many diseases. These targets serve as biomarkers that can be used for identifying a particular disease or a particular variant of diseased gene. Proteins, as surface receptors, enzymes, ion channels, or secreted proteins, are involved in drug binding and, hence, in the initiation of pharmacological cycle of the drug molecule. Mutations at specific sites in these genes are responsible for presenting altered therapeutic responses to a phytotherapy.

Among the plant-derived anticancer agents, differential response has been documented in breast cancer patients owing to angiotensin-converting enzyme (ACE) genotype variations (Yuan et al. 2005). Ephedrine used as a bronchodilator for the treatment of asthma is an alkaloid derived from *Ephedra equisetina*. The drug acts on the β2 adrenergic receptors. There are nine variants of the receptor responsible for differential binding and, hence, therapeutic effects of the drug (Sears and Lötvall 2005). Polyphenols are a group of phytochemicals comprising coumarins, flavonoids, tannins, and stilbenes. These groups utilize dopamine, serotonin, and estrogen receptors to exhibit their pharmacological action in neurological and cardiovascular conditions. Mutations in the genetic makeup of these receptors yield altered targets for drug action. Consequently, there is a difference in therapeutic response to these drugs (Fig. 2).

Drug Transport

On reaching the desired target, a particular intracellular drug concentration needs to be attained for optimum therapeutic effects. However, different membrane transport systems may prevent the drug from achieving the effective levels within the cell. Moreover, in genetically predisposed individuals, efflux mechanisms can cause the drug draining from the cell. This results in the development of resistance or tolerance in susceptible individuals presenting, consequently, as therapeutic failure.

Among the many drugs that are made therapeutically ineffective by the efflux mechanism are the anticancer drugs of natural origin, vincristine and vinblastine. Resistance tends to develop against these drugs by the involvement of a cell membrane protein. The P-glycoprotein 1 or multidrug resistance protein 1 (MDR1) is a transporter responsible for the exchange of various species across the cell membrane. Mutations in the gene result in altered permeability of this compound.

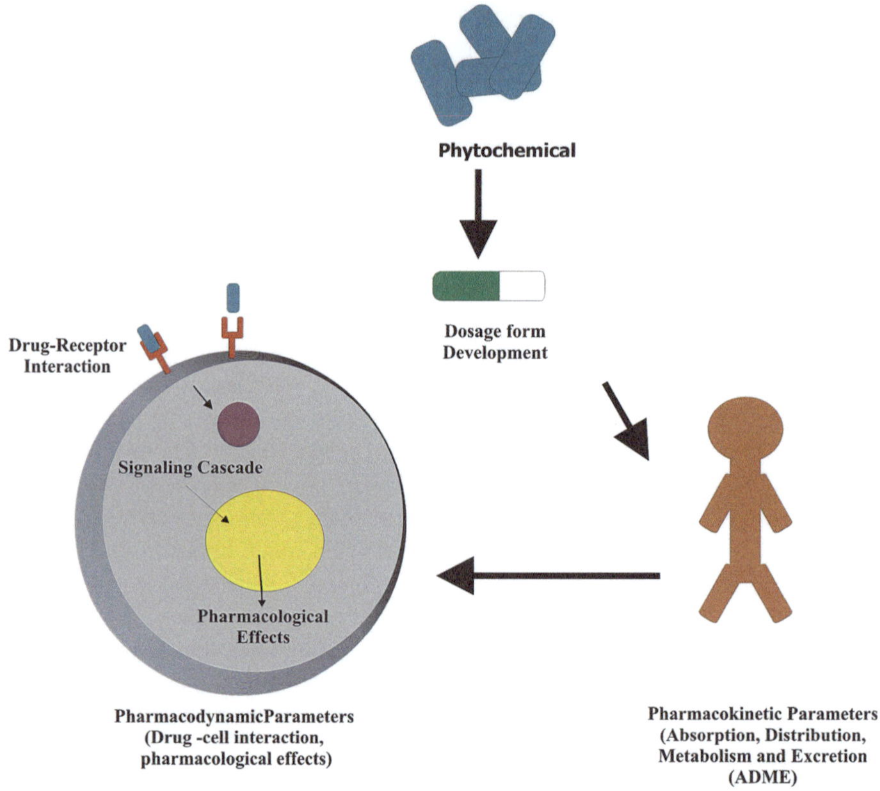

Fig. 2 Pharmacological aspects of therapeutic action of phytochemicals

There are at least 13 different characterized variants of the gene among different ethnic groups (Drysdale et al. 2000). Similarly, MDR1 causes the decreased drug accumulation of digoxin in the target cells (Gottesman et al. 1996). Digoxin is a phytochemical obtained chiefly from *Digitalis* spp. and therapeutically employed in various heart conditions. Decreased drug accumulation leads to therapeutic failure of the drug. A number of other phytochemicals, including alkaloids, terpenoids, polyphenols, and quinones, have been observed to interact with this transport system as well (Yazaki 2006). Other membrane proteins involved in the drug transportation and resistance processes include multidrug resistance-associated protein 2, drug exporter-2, and drug resistance ATPase-1/ATPase-2. They have been characterized to play a role in the egress of drugs of natural and synthetic origin from the cell (Schinkel and Jonker 2012; Van Bambeke et al. 2000; Saier 2002) (Fig. 3).

Drug Metabolism

Among drug disposition processes, an important contributor to the altered pharmacological response in view of genetic polymorphism is the drug metabolism. Metabolism refers to the enzymatic and chemical conversion of a drug molecule, so as to change the chemical nature of the drug making it liable to be excreted from the body easily. Occurring chiefly in the liver cells, drug metabolism comprises three steps: chemical modification, conjugation, and supplementary alteration. The first and the last phases bring about a chemical change in the chemical structure of the drug molecule, while conjugation involves the addition of an additional chemical moiety to the drug. The nature and rate of these biochemical steps determine the duration for which drug molecule will remain in the body and, hence, the extent of the pharmacological action of the drug.

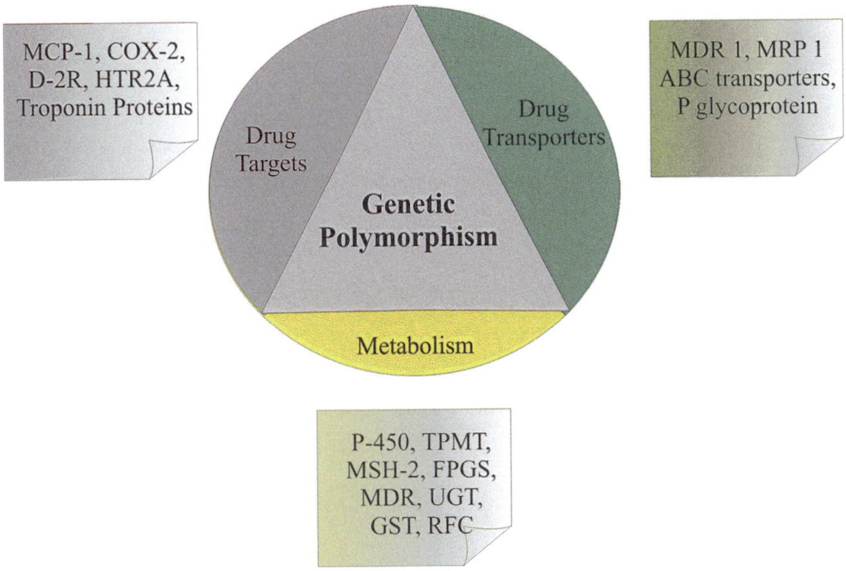

Fig. 3 Genetic polymorphism in drug targets, transporters, and metabolizing systems (Abbreviations: *ABC* ATP-binding cassette transporters, *COX2* cyclooxygenase-2, *D-2R* dopamine receptor 2, *FPGS* folylpolyglutamate synthase, *GST* glutathione S-transferase, *HTR 2A* serotonin receptor, *MCP* membrane cofactor protein, *MDR/MRP* multidrug resistance gene/protein, *MSH* melanocyte-stimulating hormone, *CP450* cytochrome P-450, *RFC* reduced folate carrier, *TPMT* thiopurine methyltransferase, *UGT* UDP glucuronosyltransferases)

One of the most important enzyme machineries involved in drug metabolism is the hepatic cytochrome P-450 (CYP-450) system. It oxidizes the drugs to yield a more polar species which results in the rapid excretion of the drug from the body. Many drugs exploit this metabolic process for their ultimate elimination from the body. CYP-450, however, shows a high rate of genetic variability among individuals. The molecular mechanisms affecting the enzymatic activity include the alteration in promoter sequences, presence of termination codons, amino acid substitutions, and formation of unstable proteins. These genetic alterations collectively are responsible for making an individual a slow, normal, or fast metabolizer and, hence, altering the drug's resident time in the body. Various isoforms of this enzyme system including CYP2D6, CY1A2, CYP2C9, and variants of CYP3A play a major role in phase I metabolic reactions. Additionally, enzymatic systems encoding glutathione S-transferases, N- acetyl transferases, and uridine 5′-triphosphate glucuronosyltransferases play a major part in the conjugation steps. Therefore, underlying genetic polymorphisms in these enzymatic systems can be associated with altered drug metabolizing process (Table 2).

Phytochemicals, like other xenobiotics, undergo similar mechanisms of metabolism. Secoisolariciresinol is a lignin extracted from flaxseed and used in traditional medicine for the treatment of atherosclerosis and diabetes. The drug tends to exhibit its effect only after being enzymatically converted into the active form. Owing to the genetic polymorphism in drug metabolizing enzymes, significant interindividual therapeutic differences have been reported among patients using flaxseed (Clavel et al. 2006; Kuijsten et al. 2005). Similarly, remarkable differences have been noted between patients using psoralen therapy for the treatment of skin conditions (Shephard et al. 1999). Psoralen, a furocoumarin, is obtained from the seeds of *Psoralea* spp., figs, and parsley. The product is employed as a therapeutic agent in psoriasis, eczema, and vitiligo (Fig. 4).

Plant-derived products are, hence, subjected to similar drug-processing steps in the body as

Table 2 Genetic polymorphism in drug targets, transporter, and biotransformation process leads to interindividual variations in drug responses

	Drug targets	Drug transporters	Metabolic enzymes
Candidates	MCP-1, COX-2, B-adrenoceptor, 5HT, HER 2	MDR-1, MRP 1, ABC transporters	CYP-450, pseudocholinesterase, acetyl transferase, COMT, ADH
Genetic modifications	Activating mutation, gene duplication, gene deletion, exon skipping, unstable protein, amino acid substitutions		
Phenotypic presentation	Up-/downregulation, over-/underexpression, polymorphism		Fast/slow/poor metabolizers
Drugs	Vinblastine, paclitaxel, doxorubicin	Digoxin, glycosides, trastuzumab	Anticancer, antimicrobials, ethanol, and other drugs metabolized in the liver

Abbreviations: *5HT* serotonin, *ABC* ATP-binding cassette, *ADH* alcohol dehydrogenase, *COMT* catecholamine O-methyltransferase, *COX* cyclooxygenase, *CYP* cytochrome, *HER 2* human epidermal growth factor receptor-2, *MCP* monocyte chemoattractant protein, *MDR/MRP* multidrug resistance gene/protein

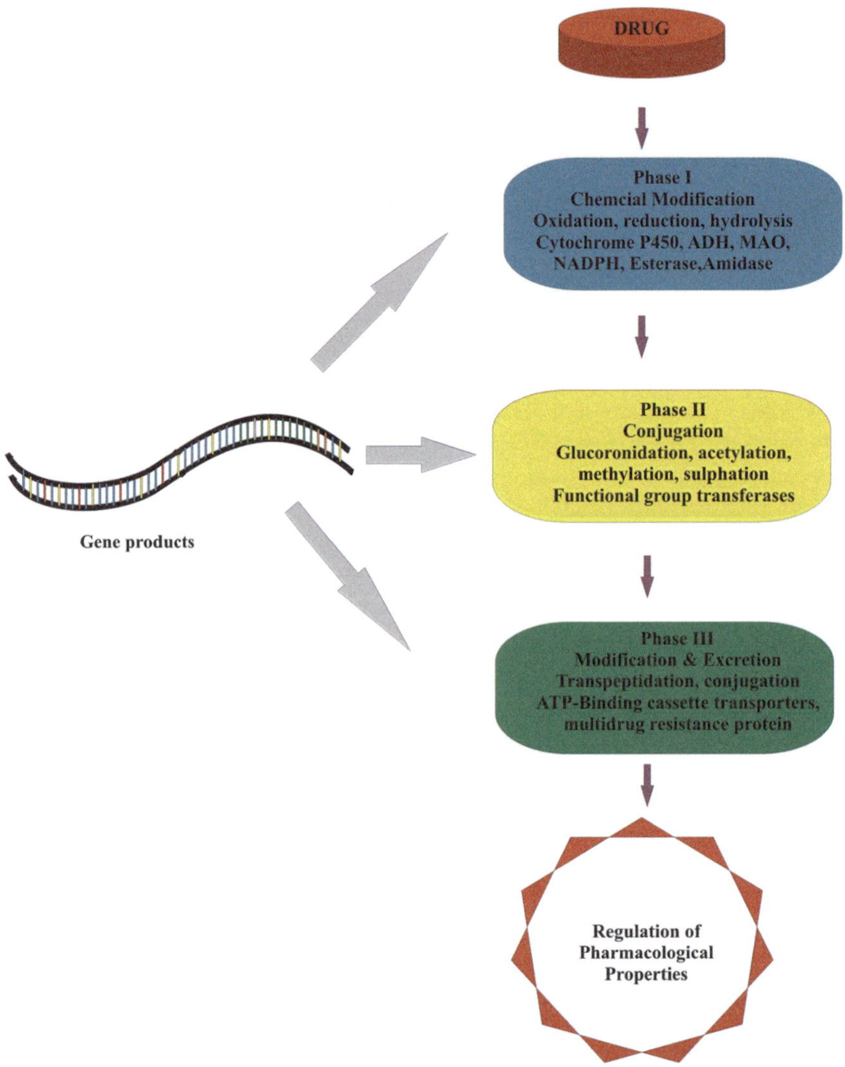

Fig. 4 Genetic control of drug metabolism processes

the synthetic drugs undergo. Genetic variations in these active players result in the appearance of marked alterations in the drug bioavailability and therapeutic effects. Study of these factors with reference to the use of phytochemicals would help enhance the characterization and therapeutic outcomes of a plant-based drug therapy.

Phytochemicals, Pharmacogenetics, and Pharmacogenomics

Human beings respond differently to plant products administered for a particular ailment. This diversity in therapeutic response of drugs is studied under the scope of pharmacogenetics and pharmacogenomics. Pharmacogenetics deals with the evaluation of inherited traits in conjunction with the observed pharmacological responses. Pharmacogenomics, on the other hand, involves the study of genetic variation, the roles of genes, and their ultimate impact on the pharmacological properties of a drug. Both the fields, together, contribute to understanding the genetic factors responsible for exhibiting interindividual difference in therapeutic responses. Polymorphism in genes encoding metabolism, drug-receptor interaction, transportation, and elimination are the major causes of altered pharmacokinetic and pharmacodynamics profiles of drugs.

The advent of molecular biology and genetic engineering techniques aided the molecular dissection of various diseased conditions. The discovery of new genes and associated genetic variations helped in establishing them as the etiological agents for various diseases. Developments in transcriptomics, proteomics, and metabolomics have led to the establishment of our understanding that most diseases are polygenic traits, i.e., they are controlled by more than one gene, and not monogenic. These genes can, hence, serve as potential targets for therapeutic strategies. Variations in these genes can lead to altered therapeutic responses. Apart from the genetic manipulation by drug molecules, mutations in genes involved in drug-processing systems can present different therapeutic outcomes to a phytochemical.

Genetic mutations can cause variation in the morphological or physiological characteristics of an individual. With respect to drug response, phenotypic changes in metabolic rates, drug targets, and drug transporters may significantly alter the therapeutic outcomes of a phytotherapy. Attributable mainly to altered allelic forms, variations in drug response do not require mutation in large genetic sequences. Single nucleotide polymorphisms (SNPs), i.e., the change of only one nucleotide in the genetic material can have profound effects on drug processing. Drug targets, transporters, and metabolizing enzymes are more viable to the genetic variation resulting in the appearance of different therapeutic profiles in genetically diverse individuals. Carotenoids, for instance, are a group of phytochemicals possessing effective antioxidant properties. They are metabolized by an enzyme, beta-carotene 15,15′-monooxygenase 1, the activity of which ultimately affects the bioavailability and plasma levels of the drug. SNPs in this gene have been associated with differential response to the administered carotenes (Wang et al. 2013). Similarly, organic anion-transporting polypeptide (OATP) is involved in the elimination of phytochemicals and other xenobiotics (Lee and Kim 2004). Genetic variation in OATP system results in altered drug disposition properties and, hence, therapeutic efficacy (Riedmaier et al. 2012). Some other phytochemicals and their associated single nucleotide polymorphisms and therapeutic outcomes have been discussed in Table 3. Exon skipping of drug metabolizing enzymes is also related to diverse pharmacological effects. The process involves inability to read the actual coding sequence of a gene resulting in the failure of translation into the correct protein. Dihydropyrimidine dehydrogenase (DPYD), for instance, is an enzyme involved in drug metabolism and subjected to exon skipping phenomenon in genetically predisposed individuals (Ciccolini et al. 2009; Meinsma et al. 1995). Flavonoids and essential oils are metabolized by this enzymatic system, and genetic variation in DPYD may be associated with altered bioactivities of plant

Table 3 Phytochemicals and associated single nucleotide polymorphism in human genes and associated pharmacogenomics effects

Drug	Pharmacological class	Botanical source	Disease	Gene	SNPs	Clinical outcomes	Refs
Caffeine	Cardiac stimulant	*Coffea* spp.	Coronary artery disease	CYP1A2	Rs762551 -163C>A	Slow metabolism leads to risk of development of nonfatal metabolism	Cornelis et al. (2006)
Camptothecin and topotecan	Anticancer	*Camptotheca acuminata*	Various cancers	ABCG2	Q141K	Altered pharmacokinetic properties, increased bioavailability	Zamboni et al. (2006) and Sparreboom et al. (2005)
Capsaicin	Analgesic, anticancer	*Capsicum* spp.	Gallbladder cancer (GBC)	(i) Apolipoprotein B, (ii) CETP	(i) Rs693 (C/T), (ii) rs708272 (C/T)	Decreased risk of GBC in wild type	Báez et al. (2010)
Cocaine	Psychoactivation	*Erythroxylon coca, Cannabis sativa*	Mental disorders	DRD2	Rs2283265/rs1076560	Increased metabolism and addiction	Moyer et al. (2010)
Codeine	Analgesic	*Papaver somniferum*	Generalized pain	CYP2D6	*17 (1023T), *29 (1659A, 3183A), and *41 (2988A)	Altered metabolic capacity, more dose required	Shord et al. (2009)
Coumarin	Anticoagulant	*Dipteryx odorata, Galium odoratum, Cinnamomum aromaticum*	Clotting disorders	(i) CYP2C9, (ii) VKORC1	(i) Exon 3 (CGT>TGT), exon 7 (ATT>CTT) (ii) promoter polymorphism −1639G>A	Increased drug bioavailability, lesser dose required	Gage (2006)
Digoxin	Cardiotonic	*Digitalis lanata*	Heart conditions (atrial fibrillation/flutter)	MDR1	Nucleotide: C3435T (no change in amino acids)	Increase plasma levels resulting in lower dose requirement	Verstuyft et al. (2003)
Docetaxel	Anticancer	*Taxus brevifolia (semisynthetic)*	Nasopharyngeal carcinoma	(i) SLCO1B3, (ii) ABCB1	(i) Rs11045585 (IVS12-5676A>G), (ii) rs1045642 (3435C>T)	Increase plasma level and bioavailability	Chew et al. (2011)
Etoposide	Anticancer	*Podophyllum peltatum (derivative)*	Various malignancies	ABCG2	Q126stop, F208S, S248P, E334stop, S441N, F489L	Cellular tolerance, increased dose requirement	Sharom (2008)
Green tea	Antioxidants	*Camellia sinensis*	Breast cancer	COMT	rs4680 Val158Met	Altered risk of breast cancer	Shrubsole et al. (2009)
Isothiocyanates (from glucosinolates)	Anticancer	*Brassica, Capparis, Carica, Drypetes*	Colorectal cancer, Breast cancer	Cyclin D1 (CCND1)	A870G	Decreased risk of cancer	Probst-Hensch et al. (2006)

Methamphetamine	Psychoactive	*Acacia berlandieri*	Psychiatric disorders	COMT (3′-UTR)	A/G for rs4680/rs165599	Altered addiction potential	Jugurnauth et al. (2011)
Morphine	Analgesic	*Papaver somniferum*	Neuro-/cancer pain	*COMT*	Rs4680 (Val158Met)	Requires lower morphine doses for treating cancer pain	Rakvåg et al. (2008)
Nicotine	Psychoactive	*Nicotiana tobaccum*	Addiction	CYP2A6	L160H, G479V, R128Q, I471T, R485L, S224P, V365M	Decreased dependence	Nakajima et al. (2004) and Fukami et al. (2004)
Paclitaxel	Anticancer	*Taxus brevifolia*	Gastric cancer	TP53 codon 72	Arg/Pro and Pro/Pro	Increased drug dose required	Kim et al. (2009)
Phytoestrogens	Anticancer	*Pueraria mirifica, Panax ginseng*	Breast cancer	BCRP	Q126stop, Q141K	Increase the cellular responsiveness to anticancer drugs	Sugimoto et al. (2005)
Phytosterols	Lipid lowering agent	Nuts, vegetables, cereals, and fruits	Hypercholesterolemia	CYP7A1 (7α-hydroxylase)	rs3808607 -204A>C	Enhanced cholesterol lowering capability	De Castro-Oros et al. (2011)
Polycyclic aromatic hydrocarbons (PAHs)	Cancer-causing agents	Cigarette smoke	Breast cancer	TP53	IVS6+62 A>G, IVS3 16 bp insertion/deletion	Decreased risk of breast cancer	Gaudet et al. (2008)
Tobacco carcinogens	Cancer-causing agents	*Nicotiana* spp.	Cancer	(i) CYP2A13 (ii) CHEK2	(i) R257C (ii) L157T	Protection against toxicity/cancer	Zhang et al. (2002) and Brennan et al. (2007)
Vincristine	Anticancer	*Catharanthus roseus*	Lymphoma	FCGR3A	rs396991/nucleotide: T559G	Higher response rate	Kim et al. (2006)
Warfarin	Anticoagulant	*Galium odoratum (derivative)*	Coagulation disorders	GGCX	rs11676382 (C/G variant)	Increased plasma levels, reduction in warfarin dose	King et al. (2010)

Abbreviations: *ABCG2* ATP-binding cassette subfamily G member 2, *ABDB1* ATP-binding cassette subfamily B, *APO* apolipoprotein, *BCRP* breast cancer resistance protein, *CETP* cholesteryl ester transfer protein, *CHEK2* checkpoint kinase 2, *COMT* catechol-O-methyltransferase, *Cyp* cytochrome, *DRD2* dopamine receptor D2, *FCGR3A* Fc fragment of IgG, low affinity IIIa, *GGCX* gamma-glutamyl carboxylase, *MDR1* multidrug resistance gene, *SLCO1B3* solute carrier organic anion transporter family member 1B3, *SNP* single nucleotide polymorphism, *TP53* tumor protein 53, *VKORC1* vitamin K epoxide reductase complex subunit 1 products (Carnesecchi et al. 2004; Manosroi et al. 2006). Deletions, substitutions, or addition of nucleotides into the genetic region is another important factor. Once transcription into RNA has been done, the presence of SNPs in promoter sequence and appearance of early stop codons lead to genetic errors resulting in the formation of altered or dysfunctional proteins. Early termination codons in the genes encoding various members of cytochrome P-450 enzyme system have been associated with altered biotransformation properties of plant-derived drugs including nicotine and other related compounds (Oscarson 2001; Ingelman-Sundberg 2002). While translating, amino acid substitutions or formation of unstable molecules yield altered proteins. These variations, hence, cause the development of different protein products, both structural and functional. Repression of expression patterns of certain drug metabolizing enzymes like glutathione S-transferases has been associated with decreased release of isothiocyanates (ICTs) from the body, resulting in increase in toxicity and appearance of ADRs in susceptible individuals (Seow et al. 2005). ICTs are the phytochemicals found in a variety of plant species and are employed as anticancer agents (Gamet-Payrastre et al. 2000). In addition, a group of ADRs referred to as idiosyncratic reactions are often associated with certain genetic basis, but no evidence has yet been found (Knowles et al. 2000). Consequently, whether kinetic or dynamic in nature, these variations lead to an appearance of considerable interindividual difference in response to a therapy (Severino and Zompo 2004).

In addition to the genetic factors, environmental factors also affect an individual's response to a drug. Certain individuals share some common living conditions and tend to react differently to a prescribed therapy (Kalow 2006). Activation of rapid drug metabolism in response to alcohol and cigarette smoking has been documented in smokers. Additionally, exposure to certain mutagenic agents, including chemotherapeutic agents, tends to exhibit changes in the metabolic processes, resulting in the altered drug response to other drugs (Evans and Relling 1999). The ethnicity of an individual also causes variation in therapeutic responses, for instance, Caucasian population tends to show different drug metabolism patterns than Arabs (Baillie and Rettie 2011; Scott et al. 2011; Phan et al. 2009). Familial exposure to a certain agent, like high-intensity UV radiations or higher altitudes, makes the individuals prone to certain diseases. The drug responses are also altered in these patients. Additionally, the presence of certain nutrients in daily diets causes changes in in vivo drug-processing rates. Grape fruit, fruit juices, vitamins, dietary supplements, and fatty foods significantly contribute to the altered pharmacological effects of a drug (Hanley et al. 2011; Dresser and Bailey 2003; Custodio et al. 2008). Moreover, the anatomical and physiological processes of individuals, including gender, lean body weight, body surface area (BSA), and age may cause differential therapeutic responses to the same phytochemical (Soldin et al. 2011; Hanley et al. 2010; Sawyer and Ratain 2001; Shi and Klotz 2011). Among the pathological determinants, diseases affecting the absorption, distribution, metabolism, and excretion tend to affect the overall rate at which a drug is processed and therapeutic effects are observed. Peptic ulcers, hypoproteinemia, cardiovascular disorders, and hepatic and renal diseases are the important conditions in this case. Coadministration of certain drugs, including phytochemicals, also tends to affect the efficacy of other drugs (Wu et al. 2012; Efferth and Koch 2011). Though not inherited, these characteristics manipulate the genetic characteristics of an individual and may be considered responsible for altering the pharmacological profile of an administered plant-based drug component.

Once the genetic predisposition of an individual to a particular drug has been established, the occurrence of a disease in an individual and the degree to which the physiological or morphological characteristics will be altered can be predicted. Moreover, pharmacogenetics can help evaluate the susceptibility and sensitivity of a patient to a particular drug therapy. Additionally, dose adjustment, dosage regimen design, and monitoring of therapeutic responses can be done. Hence, pharmacogenetics and pharmacogenomics involve the study of underlying genetic factors that affect the therapeutic responses to a plant-derived drug molecule and the application of tools for improving the outcomes.

Pharmacogenetic Markers in Phytochemical Bioavailability

The validation of pharmacological properties of a plant-based drug therapy emphasizes the need that the drugs may effectively be delivered to the target site. In order to ensure the achievement of therapeutic benefits, drug molecules need to be effectively made bioavailable at the target site. The physicochemical properties, potency, and strength of a phytochemical significantly affect the drug absorption and binding capability. A drug's partition coefficient and solubility also affect the overall bioavailability and, hence, the therapeutic response. These properties make the drug act differentially in various cellular and tissue environments. A number of physiological and anatomical factors affect the drug processing, utilizing the physicochemical properties of a drug. They do not directly interact with the genetic factors but play their role in pharmacological outcomes by varying the drug disposition. A highly lipophilic phytochemical, for instance, carotenoids, may show different drug distribution patterns in water and fat components of the body. Hence, an increased availability and accumulation of the drug may take place at the preferred sites of distribution, causing the appearance of adverse effects and/or therapeutic failures. The physicochemical properties of the drug in relation with the genomic background of an individual gave rise to the birth of chemical genomics (Salemme 2003). The approach employs the use of in silico analysis for determining the target sites and designing of ideal drugs against these sites. Virtual screening and interactions are observed to determine the drug candidate. The identified agent is then evaluated for its pharmacokinetic profile, before being finally developed for use in susceptible individuals.

Apart from a relatively neutral nature of the drug, a number of parameters need to be met for ensuring the proper drug disposition characteristics and ideal therapeutic responses. Efficacy and safety are the underlying principles that an ideal drug molecule should be possessing. Targeting specific molecular mechanisms in a cell helps in providing the therapeutic benefits expected of a drug therapy. Alongside, adverse drug reactions are prevented. Pharmacogenomics helps in establishing the causative agents for a particular disease and determining the variants that may respond differently to a particular drug therapy. Providing individuals with these drugs can help in increasing the effectiveness and decreasing the adverse effects of the drug. Moreover, the provision of an individualized therapy to a patient can help prevent the misuse of drugs in unwanted cases. One basic advantage offered by pharmacogenomics technology is the identification of biological markers to interpret the genetic predisposition of an individual to a designed therapy. The presence or absence of these biomarkers can facilitate the categorization of individuals as responders or nonresponders of a pharmacotherapy (Vernon et al. 2006). Polyphenols, a large group of phytochemicals, are effective protective agents against various cancers. Cyclooxygenase-2 and prostate-specific antigen are two of the significant markers used to evaluate the polyphenol-based phytotherapy (Thomasset et al. 2007). Therefore, individuals carrying a particular variation of a biomarker may be prescribed a particular therapy. Nonresponders may not be treated with a similar regimen paving the way for targeted drug therapy.

Application of pharmacogenomics for the achievement of ideal pharmaceutical properties stresses the need to identify and characterize the genetic basis of altered therapeutic response. The implications of kinetic and dynamic processes of a drug have already been discussed in earlier sections. Additionally, metabolic profiling of the drug substance needs to be performed. It involves the study of different metabolic products produced as a result of biotransformation processes in the body (van Duynhoven et al. 2012). The metabolites formed in different individuals may not only vary in their chemical composition, but significant differences in the pharmacological actions are also noted. These may be exhibited in the form of genetic, developmental, physiological, or pathological modifications (Clarke and Haselden 2008). Hence, once the metabolic profiling of a phytochemical has been done, only then can the exact response to a drug be determined and a therapeutic regiment be designed. Moreover, execution of standardized genetic tests is an integral part of any pharmacogenomics-based drug designing process. Genetic tests help in establishing certain biomarkers for a disease and the association of that marker with the therapeutic strategy employed. The expression of human epidermal growth factor receptor-2 (HER 2/neu), for instance, has been used for guiding the development of a particular regimen of anticancer agents in breast cancer patients (Phillips et al. 2004). Vinca alkaloids, vincristine and vinblastine, and taxanes, paclitaxel and docetaxel, are some of the plant-derived anticancer agents dependent upon HER 2 biomarker studies (Pritchard et al. 2008; de Hoon et al. 2012). On successful determination of the factors responsible for providing differential therapeutic response to a drug, clinical and genomic data is integrated (Fig. 5).

Implementation of pharmacogenomics technology would involve the incorporation of a number of ethical issues related to the therapy (Buchanan et al. 2002; Mordini 2004). The conventional drug therapies do not depend upon the attainment of genetic data from every individual. Pharmacogenomics, however, requires genetic tests to be performed before a suitable candidate drug can be prescribed. This approach, therefore, requires the need of informed consent of the patient before beginning the drug therapy. Safeguard should be provided to the privacy and confidentiality of the results obtained. Moreover, as the concept employs that a particular individual would respond to only a select few drugs, the availability of the drug and access to the diagnostic-cum-therapeutic facilities need to be ensured to every individual on a particular drug therapy. Certain genetic variations are observed only in particular races. In case the pharmacogenomics-based drug products are introduced in the market, it should be made necessary to develop products for these diverse groups of patients, with unique ethnic backgrounds (Lee 2005). Clinicians similarly need to be vigilant in assessing the genetic tests and evaluating the need of a pharmacogenomics-based product.

There are, however, certain limitations these products may suffer while getting regulatory approval (Rothstein and Epps 2001). The necessity of genetic tests and data may further increase the time of drug development. Moreover, the cost of development and therapy may increase (Shah 2003). The rate of approvals may become even slower due to the need to provide additional data. Conversely, pharmacogenomics would help increase the overall value of a therapy as only the individuals who would respond to the therapy would be prescribed the particular drug. Moreover, this concept may facilitate the development of specific drug therapies from a variety of sources, both natural and synthetic, and help in the achievement of ideal therapeutic targets.

Pharmacogenomics of Plant Products

Plant-based pharmaceutical products have made an integral part of the currently practiced medicinal systems. The practitioners of ethnomedicine

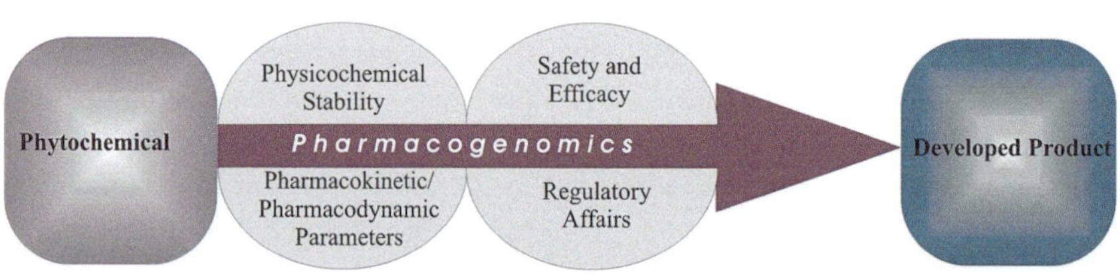

Fig. 5 Pharmaceutical considerations in product development

prescribe specific therapies to individuals based upon their age, gender, and physical characteristics. However, advancements in the field of genomics have provided the genetic basis of the therapeutic outcomes of these drugs. The genetic predisposition of an individual can help in the prediction of therapeutic outcomes even before the initiation of a particular therapeutic regimen. Research is underway to relate the ethnopharmacological aspects with the genomic data to provide efficient models for disease prognosis.

Genetic Manipulation by Ethnomedicine

Drugs alter the genetic expression of their targets and present variable pharmacological effects in different individuals. Phytochemicals tend to follow similar pharmacokinetic, pharmacodynamics, and pharmaceutical profiles as other synthetically prepared formulations. Genetic variation in drug targets, transporters, and metabolism process cause significant changes in the pharmacological response to any drug. The CYP-450 enzyme system, for example, is involved in the metabolism of a variety of plant products including alkaloids, terpenes, and essential oils. A number of different mutations have been identified in this enzyme system resulting in altered therapeutic responses (Mann 2006; Cascorbi 2003). Administration of a crude drug affects the overall genotypic and phenotypic characteristics of an individual by the regulation of certain genes. The change in these properties is responsible for exhibiting the pharmacological properties of a drug. Two of the commonly employed phytochemicals in both traditional and modern medicine are the vinca alkaloids, vincristine and vinblastine. They target the polymerization of microtubules by specifically targeting the β-tubulin subunit to specifically demonstrate their anticancer effects. Regulation of hypoxia-inducible factor-1 (HIF-1α) through various genetic routes is responsible for the induction of the therapeutic effects (Nagle and Zhou 2006). Manipulation of the responsible genes is, hence, the underlying basis of the mechanism of action of these drugs. Variation in the interindividual therapeutic responses is related to the genetic polymorphism in these genes. The study of pharmacogenomics in association with the phytochemicals, hence, becomes imperative.

Methods

The advent of modern analytical and molecular biology procedures have aided in the identification of many phytochemicals and their application in the field of clinical sciences to achieve better therapeutic outcomes. Time-consuming, repetitive extract-and-test procedures have now been replaced by the high-throughput screening (HTS). The process involves the use of chemical, pharmacological, and computational methods for the discovery of active phytochemicals from a complex mixture of substances. Following the extraction, the product is subjected to both chemical and pharmacological tests. These assays evaluate the physicochemical and biological properties of the substance. Molecular tests follow these assays to identify the cellular and molecular mechanisms that would be involved in showing the respective pharmacological actions. Availability of cell and tissue culture techniques facilitates the in vitro study of mechanisms involved in the pharmacological response. Additionally, gene cloning and expression methods help yield the information related to the molecular machinery involved.

To further investigate the molecular mechanisms and binding patterns of the drug with the target protein, computational biology and bioinformatics tools may be employed. The putative target biomolecules, involved in a particular disease, are designed and developed in silico. They are, then, characterized based upon their involvement as a causative agent for an ailment. Translationally controlled tumor protein (TCTP), for instance, is a conserved, eukaryotic protein that is associated in a number of cellular mechanisms including cell stabilization, microtubule formation, and apoptosis (Efferth 2006). It has been established as a major player in causing cancer (Koziol and Gurdon 2012). On establishing the genetic cause, the target-based drug design procedures can be employed to isolate and characterize these active agents by bioactivity-based screening processes. Polyphenols, a group of phytochemicals, have been observed to prevent cancer by causing manipulations in the TCTP protein (Stagos et al. 2012). Among other tools, microarray-based techniques can help in identifying the particular targets in the genome that might be up- or downregulated in response to a particular phytotherapy. Hence, pharmacogenomics can provide a number of routes for establishing the ligand-target interaction studies and in providing genetic basis for selecting an appropriate therapeutic entity for an individual.

Pharmacology and Plant Products

Pharmacology provides the evidence-based study of molecular mechanisms involved in pharmacotherapy and their observation in terms of clinical outcomes. Owing to the high rate of use of plant-based products, the determination of their pharmacological properties becomes inevitable. Phytochemicals target a variety of molecular and cellular mechanisms to exhibit their therapeutic effects. Table 1 shows the variety of mechanisms of actions a phytochemical utilizes to show its pharmacological response. Another important consideration while studying phytochemicals is the dosage form developed. Generally the use of decoctions and other extracts of plants is employed for therapeutic purposes. These extracts contain, in addition to the active pharmaceutical ingredient or the drug substance, other phytochemicals that might potentiate or inhibit the effect of the main drug product. Reports have suggested that isolated independent chemicals in comparison to extracts of substances show different expression profiles and regulate different molecular mechanisms (Yang et al. 2004). Along with demonstrating variable therapeutic response, these plant products have been observed to target different tissue types too, showing differential expression in various tissues. These interindividual differences are affected by the genomic makeup of an individual. Plant chemicals, in particular, as the sources of many pharmaceutical products need to be evaluated for their effects on the genomic profiles of patients. Some of the commonly employed plant products including *Ginkgo biloba*, *Catharanthus roseus*, *Rauwolfia serpentina*, and *Papaver somniferum*, among many others, have been involved in the genetic manipulation of many physiological systems (Liu et al. 2007; Smith and Luo 2004; Comín-Anduix et al. 2001; Pathak 2011; Ferrari 2010; Dasgupta 2010). Genetic polymorphisms, hence, affect the overall clinical presentations in response to a therapy. Once the pharmacology of various plant products in association with genomic profiles of various individuals has been established, pharmacogenomics can be employed to achieve better scientific and therapeutic outcomes.

Box 1: Some Commonly Used Plants and Their Associated Pharmacogenomics Aspects

Nicotiana tobaccum: It is the principal source of nicotine, a CNS stimulant. Nicotine is processed by a number of hepatic enzymes of the cytochrome P-450 A2 subclass. They convert the nicotine into cotinine using these enzymatic players (Murphy et al. 2013). Allelic variations in CYPA2 members lead to altered drug metabolism and, hence, bioavailability in the body (Swan et al. 2005).

Ginkgo biloba: Ginseng is an important source of a pharmacologically active group known as ginsenosides. It acts as an effective CVS, CNS, immunomodulatory, and antineoplastic agent. They act on a number of membrane proteins to show their pharmacological properties. TNF-α, NF-κB, and IL-6 are the major targets of ginsenosides for their therapeutic benefits (Zhou et al. 2006). Polymorphism in these targets can yield to altered pharmacological profile of the drug.

(continued)

Box 1: (continued)

Centella asiatica: It is an Indian plant known by the vernacular name of gotu kola. The extract contains four active ingredients including asiaticosides, asiatic acid, asiaticoside 6, and SM2, which can prevent the death of the beta-amyloid cells (Rao et al. 2012). Hence, they can be used as neuroprotective agent. However, genetic modification in these target cells may lead to altered therapeutic responses to these phytochemicals.

Rauwolfia serpentina: Reserpine, the principal alkaloid, obtained from the plant has been associated with hypotensive and antipsychotic effects. Catecholamines and serotonin (5-HT) are the major targets of the phytochemical (Konno et al. 2010). Polymorphism in the drug-processing enzymes can lead to altered effects of the drug in genetically predisposed individuals.

Pharmacogenomics and Drug Discovery and Development

Pharmacogenomics helps in the discovery and development of new drug molecules. They facilitate the identification of potential targets by establishing the biomarkers in a genome. Bioinformatics tools may be employed to propose the mechanisms involved in the drug-protein interactions, in silico. Rapid molecular methods can then be utilized to verify the target in wet lab. Pharmacological responses can be evaluated based upon the principles of differential binding capacities and expression profiles in different genetic variants.

Systems Biology and Pharmacogenomics

The conventional drug design and discovery (DDD) process involved the targeting of structural and functional proteins. Pharmacogenomics-based development aims at acquiring specific targets at the genetic level, the products of which can potentially behave as the regulators of therapeutic response. Hence, various targets may be acquired from the human genome and interventional strategies may be designed against them, aiding the drug identification and validation process.

Pharmacogenomics has helped in improving both forward and backward drug design process. Forward drug design focuses on the development of drug based on the known structure of the target molecule. The structure of the target molecule can be determined by various crystallography techniques (Murray and Blundell 2010). Once known, the spatial arrangement of the molecule can be established, and the design of a drug candidate can be proposed. The chemical structure of the candidate, in association with the structure activity relationship, can be optimized (Schneider and Fechner 2005). Conversely, the reverse drug design process involves designing the chemical structure of the ligand molecule, based upon the structure homology of the established drug molecules (An et al. 2005). This is followed by the confirmation of the biological properties of the candidate drug. These approaches are now commonly referred to as the target- and ligand-based drug deigning processes, respectively. Based upon the differences in the genetic makeup, reverse pharmacology-based designing can help in the development of novel and specific drug molecules against different variants of the candidate molecule. Plant-derived antimalarial and CNS drugs have been developed using similar approaches (Willcox et al. 2011; Guerrini et al. 2010).

Genetic information may be utilized to study the drug affinity for a receptor site. Genetic algorithms facilitate the generation of homologies to identify the putative sites for drug targeting (Kitano 2007). The spatial arrangement of the protein molecules can then be determined to evaluate the drug binding capacity of the proposed drug molecule (Vuignier et al. 2010). Various fitness modules are employed for the selection of the most appropriate target. In silico techniques, hence, provide an effective alternative to the expensive, time-consuming, and laborious methods based upon the trial-and-error-based selection of both the drug targets and the ligands. Once the ideal interaction profiles are obtained, molecular and biochemical techniques may be employed to mimic the proposed plan in wet lab.

Genetic Associations and Drug Processing

Based on the determination of actual genetic makeup of an individual, specific molecules can be developed for use in various diseased conditions. These entities exploit the physiological and biochemical pathways for targeting an ailment. Microarray technology has been considered to be one of the most important contributors to the employment of genomic data to clinical sciences. Microarrays can provide highly effective means to determine the differential replication and expression patterns among individuals (Dubey and Kumar 2013). These arrays consist of genetic material embedded on a substrate. Complementarity in the genetic sequences can help in the identification of a particular variant of the gene. Similarly, protein–drug interactions can be identified in vitro. In case a drug molecule binds to a target protein on these artificial plates, a characteristic signal is noted representing the affinity of the potential drug molecule with the target protein (Bouzyk et al. 2012). Using this technology, phytochemicals can be evaluated at a faster pace (Ovesná et al. 2008). The use of DNA microarrays is particularly helpful in studying gene expression patterns among various individuals. Concentration-based signaling while reading out the microarrays helps in evaluation of particular gene expression levels obtained from different tissues or organisms. They may, additionally, be employed for the detection of physiological and pathological variants of a particular disease (Sepulveda 2012). An isolated drug from a natural or synthetic source can be administered to an individual, and the changes in the genomic patterns of an individual can be noted (Naoghare and Myong Song 2010) (Table 4).

Combinatorial chemistry employs the use of molecular and biological data to manipulate and optimize the chemistry of the candidate drug molecules (Moulin et al. 2012; Yuen 2013). Chemical derivatization is an important methodology available to drug developers for finding alternate therapeutic agents for a particular condition. Drugs isolated from the natural origin after being subjected to the structure elucidation process can be forwarded for developing drugs with improved pharmacokinetic and pharmacodynamics properties. As a result, changes in partition coefficient, drug disposition parameters, and pharmacological effects may be noted. Derivatives of penicillin and morphine are some of the established applications of this process (Brown 1987; Rice 1985). Earlier chemical derivatization relied totally upon the laborious

Table 4 Genetics and informatics tools employed in pharmacogenomics

Method	Tools employed
Genotyping techniques	Polymerase chain reaction, restriction fragment length polymorphism, random amplified polymorphic detection, allele-specific oligonucleotides, hybridization studies in microarrays, denaturing high-performance liquid chromatography, pyrosequencing
Functional analysis of genetic variation	Transfection assays, allele-specific variation, in vivo analysis
Informatics	Genome sequencing, alignment, 3D structure modeling, docking studies, genome comparison, reconstruction of metabolic/regulatory pathway construction/fingerprinting, evolutionary biology
Data management (Innocenti 2005)	Clinical outcome, pharmacokinetic, pharmacodynamics, molecular and cellular functions data
Miscellaneous techniques	Nanotechnology, subunit/conjugate/combo vaccine, liposomes, biosensor technology

synthesis and testing procedures. Bioinformatics has, today, provided an alternate to this methodology. Identifying and isolating a particular gene responsible for a disease followed by the determination of the chemical structure of the candidate drug and employment of molecular docking techniques can help in the determination of ideal binding sites and interactions of the drug with the target protein (Taylor 2012). The drug development process can now be specified for the development of drugs for the individuals with certain genetic, ethnic, and racial background (Seenivasagam et al. 2013).

Identifying Drug Targets Through Pharmacogenomics

The modern face of drug discovery and development process employs the identification of certain genes that may be responsible for a particular disease. The currently licensed approaches generally target enzymes, hormones, cell receptors, or genetic material of a cell. Target identification is one of the major advantages offered by pharmacogenomics. Structure elucidation of the phytochemical followed by in silico protein modeling techniques help in enhancing the overall output of the plant-based DDD process (Barreiro et al. 2012).

The underlying principles that govern the pharmacogenomics DDD approach are that there may be interindividual differences in the therapeutic responses to the same drug. This forms the basis of in silico rational drug designing process. The phenomenon involves the identification of a pharmacophore or a pharmacologically active moiety based upon either the sites with which a drug would react on the target or through structural mimicry of therapeutic agents (Sacan et al. 2012). Concurrently, chemical databases are searched and compounds with similar functional groups and spatial arrangements are identified. The isolated compound is then subjected to virtual high-throughput screening process to study the drug receptor interactions (Cooper 2012; Zoete et al. 2009). Based upon their physical and chemical profiles, the compounds are scored and ranked. The pharmacokinetic properties, absorption, distribution, metabolism, and excretion are then predicted (Rowland et al. 2011). For the purpose, quantitative structure–activity relationship (QSAR) approach is followed which is based upon the fact that changes in the pharmacological properties of the drugs would be observed as the functional groups are modified or their positions are changed within the same compound (Kramer and Lewis 2012). QSAR has been employed for a number of plant-based products to analyze, design, and derive appropriate alternative chemical entities. Phytochemicals including flavonoids, phenolics, and stilbenes have been subjected to QSAR studies to analyze their structure-activity relationships (De-Eknamkul et al. 2011; Roy and Popelier 2008; Tripathi and Saxena 2008). These tools can, therefore, help in optimizing the geometry of the candidate drugs and selecting the confirmation of the drug that would best fit the target protein. In case the structure of the target protein is known by various experimental procedures, like crystallography or spectroscopy, the drug development process is simplified and easily validated. However, if the structure is not known, then pharmacophore mapping or perception approach is followed, which follows the use of the candidate drug for investigating the particular site at which the drug would act (Yang 2010; Deora et al. 2012). The identified molecule can then be optimized according to the pharmacological needs.

Pharmacogenomics in Preclinical Evaluation Studies

The difference between the modeled and the actual drug–protein interactions necessitates the physical development of the drug candidate and the isolation of the target protein responsible for a particular diseased condition. Molecular cloning and expression techniques can help identify the genetic basis of the disease and, after expression, help in the isolation of the target in vitro. Analytical techniques including x-ray crystallography and nuclear magnetic resonance (NMR) can help in the characterization of the proteins

(Wyss et al. 2012; Murray and Rees 2009). Once confirmed, in vitro drug binding assays are performed. These assays, classified on the basis of the physical nature of the substrate into solid or liquid phase, encompass a wide variety of analytical procedures. Sandwich hybridization ligand binding assay, for instance, is employed mainly for the nucleic acid therapeutics and makes use of the complementarity of the genetic codes for studying the binding of the target and the ligand (Wang and Findlay 2009). Plasma protein binding assays involve the incubation of the developed biopharmaceutical with the plasma protein followed by its centrifugation and filtration (Walter et al. 2002). The drug-protein complexes are, therefore, separated based upon the differences in their molecular weights and passage through the filter membrane. Other target validation assays, including nucleic acid aptamers, antibody detection assays, proximity ligation assays, and immune PCR ligand binding assays, rely on the hybridization of the nucleic acids or the proteins based upon the sequence complementarity (Blank and Blind 2005; Kellar and Iannone 2002; Zhang et al. 2009; Finan and Zhao 2007). These sensitive techniques can evaluate the differences in the binding capacities even in case there is a difference of less than 10 nucleotides in the genetic sequences. Hence, these analytical and binding assays can facilitate the development of targeted therapeutics against many diseases. Once in vitro confirmation has been made regarding the drug–protein bindings, animal testing has to be performed. Diseased models of various animals can be developed using the genetic manipulation techniques like mutation breeding, nuclear transfer, mutagenesis, transgenesis, and cisgenesis (Strachan and Read 1999). Drug molecules can then be tested in vivo, and pharmacokinetic and pharmacodynamics parameters can be established, both qualitatively and quantitatively. Functional assays and competition binding assays can be of particular relevance to evaluate the drug response in living systems (Glubb and Innocenti 2013; Croston 2002) (Fig. 6).

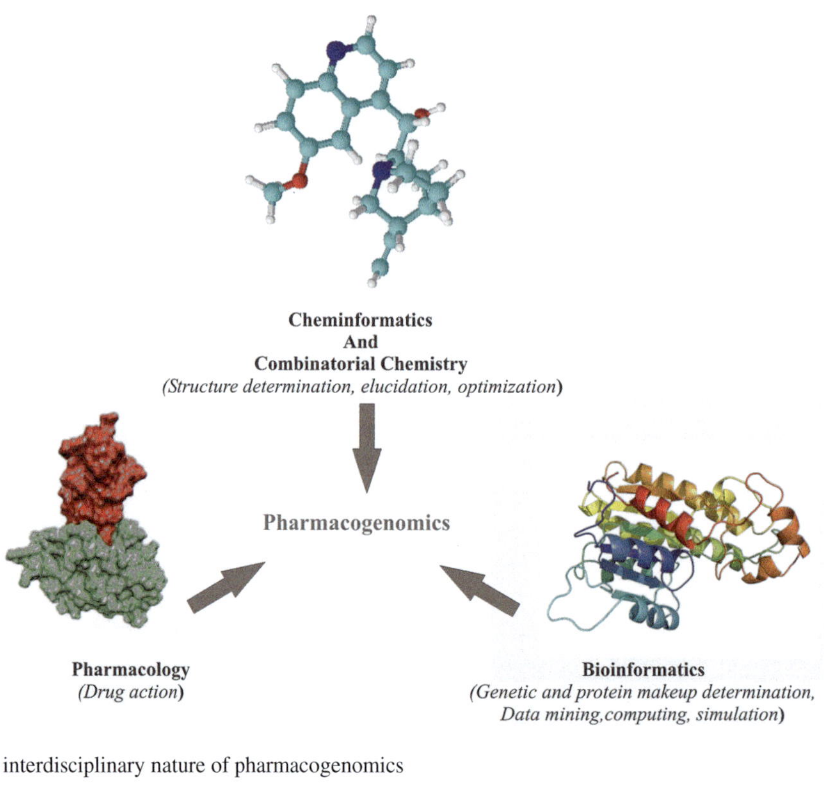

Fig. 6 The interdisciplinary nature of pharmacogenomics

Pharmacogenomics in Clinical Studies

The last phase of the drug development process involves the testing of the candidate drug in human subjects. Various phases of the clinical trials are performed as per the rules of the regulatory authorities (Liou et al. 2012). Phase I of the clinical trials involves the study of the safety of the drug. Once the safety has been established, the drug is transferred to Phase II, which involves the study of efficacy in addition to the safety profile. Phase III involves the establishment of the appropriate dose and the margin of safety for the candidate drug. The number of individuals enrolled in each phase increases as the drug is forwarded from one stage to the next. The drug is then marketed and evaluated for the therapeutic and cost-effectiveness. Based upon the fact that many diseased conditions have polygenic basis, pharmacogenomics studies can help in the evaluation of the differential response to a drug. Genetic polymorphism in the targeted gene can help in the classification of individuals into responders and nonresponders to a particular drug therapy. The drug developed can, hence, be associated and specified for use in individuals or subgroups belonging to a specific genetic pool. Adverse drug reactions and therapeutic failures can be prevented in individuals who would not respond to the therapy.

Pharmacogenomics can help in designing and developing drugs. Moreover, the tests employed for the characterization of drug responses can help in evaluating the efficacy and safety of the drugs making the drug discovery process faster and more effective. However, the goal of developing patient-specific medicines lies in the understanding of the genetic basis of differential response to a drug, investigation of these differences, and the development of such methodologies that ensure that the right drug has been provided to the right individual.

Pharmacogenomics for Personalized Medicine

Personalized medicine or the concept of providing the most appropriate medicine in an accurate formulation and dose to an individual, based upon his/her genetic makeup, offers a solution to the problems of therapeutic failures and adverse drug reactions. Genetic polymorphism has been associated with a number of disease presentations in susceptible individuals. The pharmacogenomics approach can benefit the patients by detecting the candidate genes responsible, identifying the correct drug and dose for the individual, and monitoring of the pharmacological responses as a result of the drug (Tremblay and Hamet 2012). Personalized medicine, hence, incorporates the knowledge of a patient's genetic and ethnic variables in the drug selection process by a health care provider.

Pharmacological Profiles and Genetic Diversity

Health care practitioners have been very well aware of the differences in the therapeutic responses in various individuals. The discovery of the genes as the basis of disease leads to the understanding that pharmacological responses vary among different individuals based upon the genetic variation among these individuals. These genes encode for various target proteins including the cell surface receptors, signal mediators, hormones, and cellular proteins. Genetic polymorphism in these targets confers differences in the interindividual therapeutic responses. Nucleoside analogues employed in cancer, for instance, show a significant difference in individuals possessing different allelic variants of the target molecules (Yee et al. 2013; Elnaggar et al. 2012). Histaminergic receptors, H3, are involved in the release of neurotransmitters in the central and peripheral nervous systems (Gbahou et al. 2012). The underlying genetic variation can cause altered response to drugs, if these receptors are targeted. Therefore, the genetic differences are to be kept under consideration while making a drug selection decision (Hancock et al. 2003). Apart from these pharmacodynamics aspects, the ADME profile of the drug molecule may also vary among individuals resulting in changes in drug disposition and elimination processes.

Molecular Markers in Diseases

Molecular assays may be performed to classify the individuals into different genetic groups. Modern biotechnology and molecular biology techniques have led to the discovery of specific sequences, on the genome, which code for a particular disease. Minute differences in these regions can render a human being susceptible to an ailment. Using the molecular techniques like restriction fragment length polymorphism (RFLP), loss of heterozygosity (LOH), identity by descent (IBD), and high-throughput sequencing can be employed for detecting very minute differences in the genetic makeup (Shi et al. 1999; Heil et al. 2012; Wheeler et al. 2012; Ji et al. 2012; Hardiman 2012). Mutations observed in these genes can be employed as the basis of detection of genetic variant and susceptibility evaluation of an individual to a particular drug therapy. Similarly, microarray technique can be employed for the detection of the level of gene expression in various cells, tissues, or individuals (Sepulveda 2012).

Accurate diagnostic tests form the basis of personalized medicine as they could prevent the nonresponders from taking a particular drug. Companion diagnostics with certain anticancer drugs, like trastuzumab, to identify the subpopulation of individuals bearing a specific kind of genetic variant of the receptor is now very common (Cho and Roukos 2013). Similarly, mutations in BRCA 1 and BRCA 2 are associated with an increased incidence of breast cancer (Liu et al. 2012). Establishing these molecular markers can prevent the adverse drug reactions and promote effective use of therapeutic agents.

Pharmacogenomics in Diagnostic, Prophylactic, and Therapeutic Products

On identifying a molecular marker, the next step is their employment as targets of pharmacological interventions. They can be used as diagnostic agents to evaluate an individual's susceptibility to a disease. Genotyping can be performed for the determination of a particular genetic variation. Moreover, tailor-made vaccines can be synthesized to generate immunity against various pathogenic organisms. The difference in the genetic makeup can serve as an efficient tool for disease and risk profiling. Moreover, evaluation can be performed about a patient's response to a particular drug therapy, resulting in the decrease in the use of unwanted drugs, decrease in the adverse drug reactions, and increased cost-effectiveness.

Significant developments have been made in the employment of pharmacogenomics to practice. Nearly 10 % of the registered drugs now contain information related to the gene-based differential therapeutic responses (Frueh et al. 2008). Similarly, testing for molecular markers has now become common as compared to a decade ago (Bouzyk et al. 2012; Stanek et al. 2012). Currently, the main progress has been achieved in the cancer chemotherapies. Oncologists now rely highly upon the testing of individuals for the presence of certain genetic traits before making any decision (McLeod 2013). Prescription decisions are made once the desired information has been obtained. The researchers are focusing on the development of similar strategies to investigate other diseases. Therapeutic responses in Alzheimer's disease, for instance, are generally related to the genetic variation in apolipoprotein E (APOE) (Aslibekyan et al. 2013; Cacabelos 2012). Differential responses are observed in individuals with different social and ethnic backgrounds, which may contribute to the treatment choices. Hence, pharmacogenomics approach can help in developing highly sensitive diagnostic tests, identifying drug-responder populations, and developing the basis for rationalizing the prescription process.

Disease Association in Plant Pharmacogenomics

Genetic polymorphism contributes to the difference in the pharmacological responses among individuals. Traditional healers, though not knowing the genetic basis, prescribe alternate drugs, dosage forms, and dose regimen to patients with distinctive physical characteristics and individual needs. Hence, the concept of personalized medicine owes its basis to the alternate medicine practitioners (Patwardhan and Mashelkar 2009). The Human Genome Project provided the scientific basis to strengthen this idea.

Differences in therapeutic responses are attributable to the proteins regulating various pharmacokinetic and pharmacodynamics players of the drug-processing system. Artemisinin-based compounds are potent antimalarial and anticancer agents (Chaturvedi et al. 2010). They are thought to act by interacting with a number of different genes including those responsible for antiproliferative effects, apoptotic signals, oxidative stress, proto-oncogenes, and tumor suppressor genes (Tin et al. 2012; Sertel et al. 2013; Percário et al. 2012; Firestone and Sundar 2009). Variation in any of these genes can cause a differential response to the artemisinin-based antitumor properties. Similarly, varying degrees of pharmacological responses are observed to herbs employed in Chinese, Korean, Ayurvedic, and African medicine, in individuals bearing genetic differences in the drug-processing enzymes. Plant-based remedies are, hence, subjected to similar pharmacological activities as the newer synthetic drugs are. However, lack of appropriate scientific data to establish correlation with the information from the health care practitioners hinders the incorporation of ethnomedicine in the mainstream of clinical practice.

Future Perspective and Conclusion

Apart from providing hope to treat the rare and neglected cases, pharmacogenomics does raise a number of concerns. Ethical issues are worth mentioning among these constraints. A patient's genetic information, including disease susceptibility and genetic background, has to be extracted and used in the decision-making process, which renders the additional need to safeguard this data. Additionally, as new and specific drugs would be targeted to a particular subpopulation of patients, the industry's economic concerns and profit margins would need to be addressed. Regulatory authorities would have to devise newer protocols for drug analysis and testing. Pharmacogenomics focuses on the transfer from the current protein-based therapeutic targets to the gene-based strategies. Hence, health care professionals need to be trained to make decisions accordingly. Additionally, all the genetic contributors of a disease must be known, before targeting a particular system.

Nevertheless, pharmacogenomics holds the promise of development and provision of rational therapeutic strategies. It helps in isolating new druggable targets in individuals of different ethnical, racial, and genetic backgrounds. The scarcity of health resources in the third world countries and dependence of individuals on the complementary and alternate systems of medicine urge the need to exploit the pharmacogenomics basis of phytochemicals. The main elements of the biomedical and clinical sciences, including practitioners, researchers, public health, and funding agencies, need to focus on the development of schemes to employ the genetic data to clinical practice.

References

Aggarwal BB, Takada Y, Oommen OV (2004) From chemoprevention to chemotherapy: common targets and common goals. Expert Opin Investig Drugs 13(10):1327–1338

Agrawal M, Nandini D, Sharma V, Chauhan N (2010) Herbal remedies for treatment of hypertension. Int J Pharma Sci Res 1:1–21

An G, Jeong D-H, Jung K-H, Lee S (2005) Reverse genetic approaches for functional genomics of rice. Plant Mol Biol 59(1):111–123

Aslibekyan S, Straka RJ, Irvin MR, Claas SA, Arnett DK (2013) Pharmacogenomics of high-density lipoprotein-cholesterol-raising therapies. Expert Rev Cardiovasc Ther 11(3):355–364

Báez S, Tsuchiya Y, Calvo A, Pruyas M, Nakamura K, Kiyohara C, Oyama M, Yamamoto M (2010) Genetic variants involved in gallstone formation and capsaicin metabolism, and the risk of gallbladder cancer in Chilean women. World J Gastroenterol 16(3):372

Baillie TA, Rettie AE (2011) Role of biotransformation in drug-induced toxicity: Influence of intra-and inter-species differences in drug metabolism. Drug Metab Pharmacokinet 26(1):15–29

Banerjee T, Duhadaway J, Gaspari P, Sutanto-Ward E, Munn D, Mellor A, Malachowski W, Prendergast G, Muller A (2007) A key in vivo antitumor mechanism of action of natural product-based brassinins is inhibition of indoleamine 2, 3-dioxygenase. Oncogene 27(20):2851–2857

Barreiro EJ, Fraga CA, Lima LM (2012) Natural products as lead compounds in medicinal chemistry. In: Cechinel-Filho V (ed) Plant bioactives and drug discovery: principles, practice, and perspectives, vol 17. Wiley, Hoboken, pp 81–126

Blank M, Blind M (2005) Aptamers as tools for target validation. Curr Opin Chem Biol 9(4):336–342

Bouzyk MM, Tang W, Leyland-Jones B (2012) Pharmacogenomics to link genetic background with therapeutic efficacy and safety. In: Fu H (ed) Chemical genomics, vol 1. Cambridge University Press, Cambridge, pp 293–300

Bremner P, Heinrich M (2010) Natural products as targeted modulators of the nuclear factor-KB pathway. J Pharm Pharmacol 54(4):453–472

Brennan P, McKay J, Moore L, Zaridze D, Mukeria A, Szeszenia-Dabrowska N, Lissowska J, Rudnai P, Fabianova E, Mates D, Bencko V, Foretova L, Janout V, Chow WH, Rothman N, Chabrier A, Gaborieau V, Odefrey F, Southey M, Hashibe M, Hall J, Boffetta P, Peto J, Peto R, Hung RJ (2007) Uncommon CHEK2 mis-sense variant and reduced risk of tobacco-related cancers: case control study. Hum Mol Genet 16(15):1794–1801. doi:10.1093/hmg/ddm127

Brown A (1987) Discovery and development of new b-lactam antibiotics. Pure Appl Chem 59:475–484

Buchanan AE, Califano A, Kahn JP, McPherson E, Robertson JA, Brody BA (2002) Pharmacogenetics: ethical issues and policy options. Kennedy Inst Ethics J 12(1):1–15

Cacabelos R (2012) Pharmacogenomics of central nervous system (CNS) drugs. Drug Dev Res 73(8):461–476

Carnesecchi S, Bras-Gonçalves R, Bradaia A, Zeisel M, Gossé F, Poupon M-F, Raul F (2004) Geraniol, a component of plant essential oils, modulates DNA synthesis and potentiates 5-fluorouracil efficacy on human colon tumor xenografts. Cancer Lett 215(1):53–59

Cascorbi I (2003) Pharmacogenetics of cytochrome p4502D6: genetic background and clinical implication. Eur J Clin Investig 33(s2):17–22

Chaturvedi D, Goswami A, Saikia PP, Barua NC, Rao PG (2010) Artemisinin and its derivatives: a novel class of anti-malarial and anti-cancer agents. Chem Soc Rev 39(2):435–454

Chew S-C, Singh O, Chen X, Ramasamy RD, Kulkarni T, Lee EJ, Tan E-H, Lim W-T, Chowbay B (2011) The effects of CYP3A4, CYP3A5, ABCB1, ABCC2, ABCG2 and SLCO1B3 single nucleotide polymorphisms on the pharmacokinetics and pharmacodynamics of docetaxel in nasopharyngeal carcinoma patients. Cancer Chemother Pharmacol 67(6):1471–1478

Cho WC, Roukos DH (2013) Trastuzumab emtansine for advanced HER2-positive breast cancer and beyond: genome landscape-based targets. Expert Rev Anticancer Ther 13(1):5–8

Cho H-Y, Reddy SP, Kleeberger SR (2006) Nrf2 defends the lung from oxidative stress. Antioxid Redox Signal 8(1–2):76–87

Cho J-M, Manandhar S, Lee H-R, Park H-M, Kwak M-K (2008) Role of the Nrf2-antioxidant system in cytotoxicity mediated by anticancer cisplatin: implication to cancer cell resistance. Cancer Lett 260(1):96–108

Cho Y-S, Kim CH, Ha T-S, Lee SJ, Ahn HY (2013) Ginsenoside Rg2 inhibits lipopolysaccharide-induced adhesion molecule expression in human umbilical vein endothelial cell. Korean J Physiol Pharmacol 17(2):133–137

Chung S, Yao H, Caito S, Hwang J-w, Arunachalam G, Rahman I (2010) Regulation of SIRT1 in cellular functions: role of polyphenols. Arch Biochem Biophys 501(1):79–90

Ciccolini J, Mercier C, Milano G (2009) Dihydropyrimidine Dehydrogenase (DPYD) gene polymorphism: portrait of a serial killer. In: Innocenti F (ed) Genomics and pharmacogenomics in anticancer drug development and clinical response, vol 1. Humana Press, Totowa, pp 249–265

Clarke CJ, Haselden JN (2008) Metabolic profiling as a tool for understanding mechanisms of toxicity. Toxicol Pathol 36(1):140–147

Clavel T, Doré J, Blaut M (2006) Bioavailability of lignans in human subjects. Nutr Res Rev 19(2):187–196

Comín-Anduix B, Agell N, Bachs O, Ovádi J, Cascante M (2001) A new bis-indole, KARs, induces selective M arrest with specific spindle aberration in neuroblastoma cell line SH-SY5Y. Mol Pharmacol 60(6):1235–1242

Cooper MA (2012) The changing face of screening and drug discovery. Expert Rev Proteomics 9(2):123–124

Cornelis MC, El-Sohemy A, Kabagambe EK, Campos H (2006) Coffee, CYP1A2 genotype, and risk of myocardial infarction. JAMA 295(10):1135–1141. doi:10.1001/jama.295.10.1135

Croston GE (2002) Functional cell-based uHTS in chemical genomic drug discovery. Trends Biotechnol 20(3):110–115

Custodio JM, Wu C-Y, Benet LZ (2008) Predicting drug disposition, absorption/elimination/transporter interplay and the role of food on drug absorption. Adv Drug Deliv Rev 60(6):717–733

Daniel M (2006) Medicinal plants: chemistry and properties. Science Pub Incorporated, Enfield

Dasgupta A (2010) Pharmacology of commonly abused drugs. In: Dasgupta A (ed) Beating drug tests and defending positive results. Springer, New York, NY, USA, pp 11–28

De-Eknamkul W, Umehara K, Monthakantirat O, Toth R, Frecer V, Knapic L, Braiuca P, Noguchi H, Miertus S (2011) QSAR study of natural estrogen-like isoflavonoids and diphenolics from Thai medicinal plants. J Mol Graph Model 29(6):784–794

De Castro-Oros I, Pampin S, Cofan M, Mozas P, Pinto X, Salas-Salvado J, Rodriguez-Rey JC, Ros E, Civeira F, Pocovi M (2011) Promoter variant -204A > C of the cholesterol 7alpha-hydroxylase gene: association with response to plant sterols in humans and increased transcriptional activity in transfected HepG2 cells. Clin Nutr 30(2):239–246. doi:10.1016/j.clnu.2010.07.020

de Hoon JP, Veeck J, Vriens BE, Calon TG, van Engeland M, Tjan-Heijnen VC (2012) Taxane resistance in breast cancer: A closed HER2 circuit? Biochim Biophys Acta - Rev Cancer 1825(2):197–206

Deng S, Li W, Han L (2006) Investigation of the mechanism of total coumarins against asthma allergic inflammation [J]. Chin J New Drugs 9:688–690

Deora GS, Joshi P, Rathore V, Kumar KL, Ohlyan R, Kandale A (2012) Pharmacophore modeling and 3D QSAR analysis of isothiazolidinedione derivatives as PTP1B inhibitors. Med Chem Res 22(7):3478–3484

Diebolt M, Bucher B, Andriantsitohaina R (2001) Wine polyphenols decrease blood pressure, improve NO vasodilatation, and induce gene expression. Hypertension 38(2):159–165

Dresser G, Bailey D (2003) The effects of fruit juices on drug disposition: a new model for drug interactions. Eur J Clin Investig 33(s2):10–16

Drysdale CM, McGraw DW, Stack CB, Stephens JC, Judson RS, Nandabalan K, Arnold K, Ruano G, Liggett SB (2000) Complex promoter and coding region beta 2-adrenergic receptor haplotypes alter receptor expression and predict in vivo responsiveness. Proc Natl Acad Sci U S A 97(19):10483–10488

Dubey P, Kumar D (2013) Microarray technology: basic concept, protocols, and applications. In: Arora DK, Das S, Sukumar M (eds) Analyzing microbes. Springer, Heidelberg, pp 261–279

Efferth T (2006) Molecular pharmacology and pharmacogenomics of artemisinin and its derivatives in cancer cells. Curr Drug Targets 7(4):407–421

Efferth T, Koch E (2011) Complex interactions between phytochemicals. The multi-target therapeutic concept of phytotherapy. Curr Drug Targets 12(1):122–132

Elnaggar M, Giovannetti E, Peters GJ (2012) Molecular targets of gemcitabine action: rationale for development of novel drugs and drug combinations. Curr Pharm Des 18(19):2811–2829

Ernst E (1996) Regulating complementary medicine. Only 0.08% of funding for research in NHS goes to complementary medicine. BMJ 313(7061):882

Evans WE, Relling MV (1999) Pharmacogenomics: translating functional genomics into rational therapeutics. Science 286(5439):487–491

Ferrari S (2010) Biological elicitors of plant secondary metabolites: mode of action and use in the production of nutraceutics. In: Giardi MT, Rea G, Berra B (eds) Bio-farms for nutraceuticals. Springer, Austin, Texas, USA, pp 152–166

Finan JE, Zhao RY (2007) From molecular diagnostics to personalized testing. Pharmacogenomics 8(1):85–99

Firestone GL, Sundar SN (2009) Anticancer activities of artemisinin and its bioactive derivatives. Expert Rev Mol Med 11:e32

Frueh FW, Amur S, Mummaneni P, Epstein RS, Aubert RE, DeLuca TM, Verbrugge RR, Burckart GJ, Lesko LJ (2008) Pharmacogenomic biomarker information in drug labels approved by the United States food and drug administration: prevalence of related drug use. Pharmacotherapy 28(8):992–998

Fukami T, Nakajima M, Yoshida R, Tsuchiya Y, Fujiki Y, Katoh M, McLeod HL, Yokoi T (2004) A novel polymorphism of human CYP2A6 gene CYP2A6*17 has an amino acid substitution (V365M) that decreases enzymatic activity in vitro and in vivo. Clin Pharmacol Ther 76(6):519–527

Gage BF (2006) Pharmacogenetics-based coumarin therapy. Hematol Am Soc Hematol Educ Program 1:467–473.

Gamet-Payrastre L, Li P, Lumeau S, Cassar G, Dupont M-A, Chevolleau S, Gasc N, Tulliez J, Tercé F (2000) Sulforaphane, a naturally occurring isothiocyanate, induces cell cycle arrest and apoptosis in HT29 human colon cancer cells. Cancer Res 60(5):1426–1433

Gaudet MM, Gammon MD, Bensen JT, Sagiv SK, Shantakumar S, Teitelbaum SL, Eng SM, Neugut AI, Santella RM (2008) Genetic variation of TP53, polycyclic aromatic hydrocarbon-related exposures, and breast cancer risk among women on Long Island, New York. Breast Cancer Res Treat 108(1):93–99. doi:10.1007/s10549-007-9573-0

Gbahou F, Rouleau A, Arrang JM (2012) The histamine autoreceptor is a short isoform of the H3 receptor. Br J Pharmacol 166(6):1860–1871

Gliszczyńska A, Brodelius PE (2012) Sesquiterpene coumarins. Phytochem Rev 11(1):77–96

Glubb DM, Innocenti F (2013) Architecture of pharmacogenomic associations: structures with functional foundations or castles made of sand? Pharmacogenomics 14(1):1–4

Gottesman MM, Pastan I, Ambudkar SV (1996) P-glycoprotein and multidrug resistance. Curr Opin Genet Dev 6(5):610–617

Guerrini R, Salvadori S, Rizzi A, Regoli D, Calo G (2010) Neurobiology, pharmacology, and medicinal chemis-

try of neuropeptide S and its receptor. Med Res Rev 30(5):751–777
Ha M, Yuan Y (2004) Allicin induced cell cycle arrest in human gastric cancer cell lines. Zhonghua zhong liu za zhi [Chinese journal of oncology] 26(10):585
Hancock AA, Esbenshade TA, Krueger KM, Yao BB (2003) Genetic and pharmacological aspects of histamine H3 receptor heterogeneity. Life Sci 73(24):3043
Hanley MJ, Abernethy DR, Greenblatt DJ (2010) Effect of obesity on the pharmacokinetics of drugs in humans. Clin Pharmacokinet 49(2):71–87
Hanley MJ, Cancalon P, Widmer WW, Greenblatt DJ (2011) The effect of grapefruit juice on drug disposition. Expert Opin Drug Metab Toxicol 7(3):267–286
Hardiman G (2012) Application of ultra-high throughput sequencing and microarray technologies in pharmacogenomics testing. In: Dasgupta A (ed) Therapeutic drug monitoring, vol 1. Elsevier, London, pp 144–159
Hasan N, Siddiqui MU, Toossi Z, Khan S, Iqbal J, Islam N (2007) Allicin-induced suppression of Mycobacterium tuberculosis 85B mRNA in human monocytes. Biochem Biophys Res Commun 355(2):471–476
Heil SG, van der Ende ME, Schenk PW, van der Heiden I, Lindemans J, Burger D, van Schaik RH (2012) Associations between ABCB1, CYP2A6, CYP2B6, CYP2D6, and CYP3A5 alleles in relation to efavirenz and nevirapine pharmacokinetics in HIV-infected individuals. Ther Drug Monit 34(2):153–159
Hernández I, Alegre L, Van Breusegem F, Munné-Bosch S (2009) How relevant are flavonoids as antioxidants in plants? Trends Plant Sci 14(3):125–132
Ingelman-Sundberg M (2002) Polymorphism of cytochrome P450 and xenobiotic toxicity. Toxicology 181:447–452
Innocenti F (2005) Pharmacogenomics: methods and protocols, vol 311. Humana Press, USA
Ji Y, Biernacka J, Hebbring S, Chai Y, Jenkins G, Batzler A, Snyder K, Drews M, Desta Z, Flockhart D (2012) Pharmacogenomics of selective serotonin reuptake inhibitor treatment for major depressive disorder: genome-wide associations and functional genomics. Pharmacogenomics J 13(5):456–463
Joshi K, Chavan P, Warude D, Patwardhan B (2004) Molecular markers in herbal drug technology. Curr Sci-Bangalore 87:159–165
Jugurnauth SK, Chen CK, Barnes MR, Li T, Lin SK, Liu HC, Collier DA, Breen G (2011) A COMT gene haplotype associated with methamphetamine abuse. Pharmacogenet Genomics 21(11):731–740. doi:10.1097/FPC.0b013e32834a53f9
Kalow W (2006) Pharmacogenetics and pharmacogenomics: origin, status, and the hope for personalized medicine. Pharmacogenomics J 6(3):162–165
Kellar KL, Iannone MA (2002) Multiplexed microsphere-based flow cytometric assays. Exp Hematol 30(11): 1227–1237
Keri RS, Hosamani KM, Shingalapur RV, Hugar MH (2010) Analgesic, anti-pyretic and DNA cleavage studies of novel pyrimidine derivatives of coumarin moiety. Eur J Med Chem 45(6):2597–2605
Kim DH, Du Jung H, Kim JG, Lee J-J, Yang D-H, Park YH, Do YR, Shin HJ, Kim MK, Hyun MS (2006) FCGR3A gene polymorphisms may correlate with response to frontline R-CHOP therapy for diffuse large B-cell lymphoma. Blood 108(8):2720–2725
Kim JG, Sohn SK, Chae YS, Song HS, Kwon K-Y, Do YR, Kim MK, Lee KH, Hyun MS, Lee WS (2009) TP53 codon 72 polymorphism associated with prognosis in patients with advanced gastric cancer treated with paclitaxel and cisplatin. Cancer Chemother Pharmacol 64(2):355–360
King CR, Deych E, Milligan P, Eby C, Lenzini P, Grice G, Porche-Sorbet RM, Ridker PM, Gage BF (2010) Gamma-glutamyl carboxylase and its influence on warfarin dose. Thromb Haemost 104(4):750–754. doi:10.1160/TH09-11-0763
Kinghorn ADE, Balandrin MFE (1993) Human medicinal agents from plants: symposium: 203rd National meeting: papers. American Chemical Society, Washington, DC.
Kitano H (2007) A robustness-based approach to systems-oriented drug design. Nat Rev Drug Discov 5(3):202–210
Klein K, Zanger UM (2013) Pharmacogenomics of cytochrome P450 3A4: recent progress toward the "missing heritability" problem. Front Genet 4:12
Knowles SR, Uetrecht J, Shear NH (2000) Idiosyncratic drug reactions: the reactive metabolite syndromes. Lancet 356(9241):1587–1591
Konno A, Terada N, Okamoto Y (2010) Effect of long-term use of reserpine, a sympathetic neuron blocker, on muscarinic cholinergic receptors in the guinea pig nasal mucosa. ORL J Otorhinolaryngol Relat Spec 50(5):334–339
Koziol MJ, Gurdon JB (2012) TCTP in development and cancer. Biochem Res Int 1–9
Kramer C, Lewis R (2012) QSARs, data and error in the modern age of drug discovery. Curr Top Med Chem 12(17):1896–1902
Kuijsten A, Arts IC, Vree TB, Hollman PC (2005) Pharmacokinetics of enterolignans in healthy men and women consuming a single dose of secoisolariciresinol diglucoside. J Nutr 135(4):795–801
Lampe JW (1999) Health effects of vegetables and fruit: assessing mechanisms of action in human experimental studies. Am J Clin Nutr 70(3):475s–490s
Lee SS-J (2005) Racializing drug design: implications of pharmacogenomics for health disparities. Am J Public Health 95(12):2133–2138
Lee W, Kim RB (2004) Transporters and renal drug elimination. Annu Rev Pharmacol Toxicol 44:137–166
Lee W, Choi J, Kim H, Park J, Lee S, Surh Y (2009) Heat-processed neoginseng, KG-135, down-regulates G1 Cyclin-dependent kinase through the proteasome-mediated pathway in HeLa cells. Oncol Rep 21(2):467
Licciardi PV, Underwood JR (2010) Plant-derived medicines: a novel class of immunological adjuvants. Int Immunopharmacol 11:390–398

Lin ML, Lu YC, Su HL, Lin HT, Lee CC, Kang SE, Lai TC, Chung JG, Chen SS (2011) Destabilization of CARP mRNAs by aloé-emodin contributes to caspasé-8-mediated p53-independent apoptosis of human carcinoma cells. J Cell Biochem 112(4): 1176–1191

Liou S-Y, Stringer F, Hirayama M (2012) The impact of pharmacogenomics research on drug development. Drug Metab Pharmacokinet 27(1):2–8

Liu X-P, Goldring C, Copple IM, Wang H-Y, Wei W, Kitteringham NR, Park BK (2007) Extract of Ginkgo biloba induces phase 2 genes through Keap1-Nrf2-ARE signaling pathway. Life Sci 80(17):1586

Liu G, Yang D, Sun Y, Shmulevich I, Xue F, Sood AK, Zhang W (2012) Differing clinical impact of BRCA1 and BRCA2 mutations in serous ovarian cancer. Pharmacogenomics 13(13):1523–1535

Lü J-P, Ma Z-C, Yang J, Huang J, Wang S-R, Wang S-Q (2004) Ginsenoside Rg1-induced alterations in gene expression in TNF-α stimulated endothelial cells. Chin Med J 117(6):871–876

Mann HJ (2006) Drug-associated disease: cytochrome P450 interactions. Crit Care Clin 22(2):329

Manosroi J, Dhumtanom P, Manosroi A (2006) Anti-proliferative activity of essential oil extracted from Thai medicinal plants on KB and P388 cell lines. Cancer Lett 235(1):114–120

McChesney JD, Venkataraman SK, Henri JT (2007) Plant natural products: back to the future or into extinction? Phytochemistry 68(14):2015–2022. doi:10.1016/j.phytochem.2007.04.032

McLeod HL (2013) Cancer pharmacogenomics: early promise, but concerted effort needed. Science 339(6127):1563–1566

Meinsma R, Fernandez-Salguero P, Van Kuilenburg A, Van Gennip A, Gonzalez F (1995) Human polymorphism in drug metabolism: mutation in the dihydropyrimidine dehydrogenase gene results in exon skipping and thymine uracilurea. DNA Cell Biol 14(1):1–6

Menaa F, Badole S, Menaa B, Menaa A, Bodhankar S (2012) Polyphenols, promising therapeutics for inflammatory diseases? In: Watson RR, Preedy VR (eds) Bioactive food as dietary interventions for arthritis and related inflammatory diseases, Bioactive food in chronic disease states. Academic Press, San Diego, USA, pp 421–430

Meyer UA (2000) Pharmacogenetics and adverse drug reactions. Lancet (London) 356(9242):1667–1671

Milenkovic D, Deval C, Gouranton E, Landrier J-F, Scalbert A, Morand C, Mazur A (2012) Modulation of miRNA expression by dietary polyphenols in apoE deficient mice: a new mechanism of the action of polyphenols. PLoS One 7(1):e29837

Mojab F, Kamalinejad M, Ghaderi N, Vahidipour HR (2010) Phytochemical screening of some species of Iranian plants. Iran J Pharm Res 2(2):77–82

Mordini E (2004) Ethical considerations on pharmacogenomics. Pharmacol Res 49(4):375–379

Moulin E, Cormos G, Giuseppone N (2012) Dynamic combinatorial chemistry as a tool for the design of functional materials and devices. Chem Soc Rev 41(3):1031–1049

Moyer RA, Wang D, Papp AC, Smith RM, Duque L, Mash DC, Sadee W (2010) Intronic polymorphisms affecting alternative splicing of human dopamine D2 receptor are associated with cocaine abuse. Neuropsychopharmacology 36(4):753–762

Murphy SE, Wickham KM, Lindgren BR, Spector LG, Joseph A (2013) Cotinine and trans 3′-hydroxycotinine in dried blood spots as biomarkers of tobacco exposure and nicotine metabolism. J Expo Sci Environ Epidemiol 23:513–518

Murray CW, Blundell TL (2010) Structural biology in fragment-based drug design. Curr Opin Struct Biol 20(4):497–507

Murray CW, Rees DC (2009) The rise of fragment-based drug discovery. Nat Chem 1(3):187–192

Nagle DG, Zhou Y-D (2006) Natural product-based inhibitors of hypoxia-inducible factor-1 (HIF-1). Curr Drug Targets 7(3):355

Nakajima M, Yoshida R, Fukami T, McLeod HL, Yokoi T (2004) Novel human CYP2A6 alleles confound gene deletion analysis. FEBS Lett 569(1–3):75–81

Naoghare PK, Myong Song J (2010) Chip-based high throughput screening of herbal medicines. Comb Chem High Throughput Screen 13(10):923–931

Negro A, Lionetto L, Simmaco M, Martelletti P (2012) CGRP receptor antagonists: an expanding drug class for acute migraine? Expert Opin Investig Drugs 21(6):807–818

Oscarson M (2001) Genetic polymorphisms in the cytochrome P450 2A6 (CYP2A6) gene: implications for interindividual differences in nicotine metabolism. Drug Metab Dispos 29(2):91–95

Ovesná J, Slabý O, Toussaint O, Kodíček M, Maršík P, Pouchová V, Vaněk T (2008) High throughput 'omics' approaches to assess the effects of phytochemicals in human health studies. Br J Nutr 99(E-S1):ES127–ES134

Pan J-Y, Chen S-L, Yang M-H, Wu J, Sinkkonen J, Zou K (2009) An update on lignans: natural products and synthesis. Nat Prod Rep 26(10):1251–1292

Pathak N (2011) Reverse pharmacology of Ayurvedic drugs includes mechanisms of molecular actions. J Ayurveda Integr Med 2(2):49

Patwardhan B, Mashelkar RA (2009) Traditional medicine-inspired approaches to drug discovery: can Ayurveda show the way forward? Drug Discov Today 14(15):804–811

Percário S, Moreira DR, Gomes BA, Ferreira ME, Gonçalves ACM, Laurindo PS, Vilhena TC, Dolabela MF, Green MD (2012) Oxidative stress in malaria. Int J Mol Sci 13(12):16346–16372

Peterson J, Dwyer J, Adlercreutz H, Scalbert A, Jacques P, McCullough ML (2010) Dietary lignans: physiology and potential for cardiovascular disease risk reduction. Nutr Rev 68(10):571–603

Phan VH, Moore MM, McLachlan AJ, Piquette-Miller M, Xu H, Clarke SJ (2009) Ethnic differences in drug

metabolism and toxicity from chemotherapy. Expert Opin Drug Metab Toxicol 5(3):243–257
Phillips KA, Veenstra DL, Ramsey SD, Van Bebber SL, Sakowski J (2004) Genetic testing and pharmacogenomics: issues for determining the impact to healthcare delivery and costs-page. Genet Test 10:426–432
Pritchard KI, Messersmith H, Elavathil L, Trudeau M, O'Malley F, Dhesy-Thind B (2008) HER-2 and topoisomerase II as predictors of response to chemotherapy. J Clin Oncol 26(5):736–744
Probst-Hensch NM, Sun C-L, Van Den Berg D, Ceschi M, Koh W-P, Mimi CY (2006) The effect of the cyclin D1 (CCND1) A870G polymorphism on colorectal cancer risk is modified by glutathione-S-transferase polymorphisms and isothiocyanate intake in the Singapore Chinese Health Study. Carcinogenesis 27(12):2475–2482
Pusztai R, Abrantes M, Sherly J, Duarte N, Molnar J, Ferreira M-JU (2010) Antitumor-promoting activity of lignans: inhibition of human cytomegalovirus IE gene expression. Anticancer Res 30(2):451–454
Qu H, Madl RL, Takemoto DJ, Baybutt RC, Wang W (2005) Lignans are involved in the antitumor activity of wheat bran in colon cancer SW480 cells. J Nutr 135(3):598–602
Radulovic N, Blagojevic P, Stojanovic-Radic Z, Stojanovic N (2013) Antimicrobial plant metabolites: structural diversity and mechanism of action. Curr Med Chem 20(7):932–952
Rakvåg TT, Ross JR, Sato H, Skorpen F, Kaasa S, Klepstad P (2008) Genetic variation in the catechol-O-methyltransferase (COMT) gene and morphine requirements in cancer patients with pain. Mol Pain 4(1):64
Rao RV, Descamps O, John V, Bredesen DE (2012) Ayurvedic medicinal plants for Alzheimer's disease: a review. Alzheimers Res Ther 4(3):1–9
Rice K (1985) The development of a practical total synthesis of natural and unnatural codeine, morphine and thebaine. In: Phillipson JD, Roberts MF, Zenk MH (eds) The chemistry and biology of isoquinoline alkaloids. Springer, Heidelberg, pp 191–203
Riedmaier AE, Nies AT, Schaeffeler E, Schwab M (2012) Organic anion transporters and their implications in pharmacotherapy. Pharmacol Rev 64(3):421–449
Rothstein MA, Epps PG (2001) Ethical and legal implications of pharmacogenomics. Nat Rev Genet 2(3):228–231
Rowland M, Peck C, Tucker G (2011) Physiologically-based pharmacokinetics in drug development and regulatory science. Annu Rev Pharmacol Toxicol 51:45–73
Roy K, Popelier PL (2008) Exploring predictive QSAR models for hepatocyte toxicity of phenols using QTMS descriptors. Bioorg Med Chem Lett 18(8):2604–2609
Sacan A, Ekins S, Kortagere S (2012) Applications and limitations of in silico models in drug discovery. In: Larson RS (ed) Bioinformatics and drug discovery. Humana Press, New York, USA, pp 87–124
Saier MH (2002) Families of transporters and their classification. In: Quick MW (ed) Transmembrane transporters, Single volume, 1st edn. Wiley, Hoboken, pp 1–17
Saleem M, Nazir M, Ali MS, Hussain H, Lee YS, Riaz N, Jabbar A (2010) Antimicrobial natural products: an update on future antibiotic drug candidates. Nat Prod Rep 27(2):238–254
Salemme FR (2003) Chemical genomics as an emerging paradigm for postgenomic drug discovery. Pharmacogenomics 4(3):257–267
Sawyer M, Ratain MJ (2001) Body surface area as a determinant of pharmacokinetics and drug dosing. Investig New Drugs 19(2):171–177
Schauss A (2012) Polyphenols and inflammation. In: Watson RR, Preedy VR (eds) Bioactive food as dietary interventions for arthritis and related inflammatory diseases. Bioactive food in chronic disease states. Academic Press, San Diego, USA, pp 379–392
Schinkel AH, Jonker JW (2012) Mammalian drug efflux transporters of the ATP binding cassette (ABC) family: an overview. Adv Drug Deliv Rev 55:3–29
Schmitt CA, Dirsch VM (2009) Modulation of endothelial nitric oxide by plant-derived products. Nitric Oxide 21(2):77–91
Schneider G, Fechner U (2005) Computer-based de novo design of drug-like molecules. Nat Rev Drug Discov 4(8):649–663
Scott S, Martis S, Peter I, Kasai Y, Kornreich R, Desnick R (2011) Identification of CYP2C19* 4B: pharmacogenetic implications for drug metabolism including clopidogrel responsiveness. Pharmacogenomics J 12(4):297–305
Sears MR, Lötvall J (2005) Past, present and future – β2-adrenoceptor agonists in asthma management. Respir Med 99(2):152–170
Seenivasagam R, Hemavathi K, Sivakumar G, Niranjan V (2013) Discovering novel carriers for oral insulin tablets: a pharmacoinformatics approach. Int J Bioinforma Res Appl 9(2):184–206
Seow A, Vainio H, Yu MC (2005) Effect of glutathione-S-transferase polymorphisms on the cancer preventive potential of isothiocyanates: an epidemiological perspective. Mutat Res/Fundam Mol Mech Mutagen 592(1):58–67
Sepulveda JL (2012) Microarray technology and other methods in pharmacogenomics testing. In: Langman LJ, Dasgupta A (eds) Pharmacogenomics in clinical therapeutics, vol 1. Wiley, Chichester, pp 185–200
Sertel S, Plinkert PK, Efferth T (2013) Activity of artemisinin-type compounds against cancer cells. In: Wagner H, Ulrich-Merzenich G (eds) Evidence and rational based research on Chinese drugs. Springer, Vienna, pp 333–362
Severino G, Zompo MD (2004) Adverse drug reactions: role of pharmacogenomics. Pharmacol Res 49(4):363–373
Shah J (2003) Economic and regulatory considerations in pharmacogenomics for drug licensing and healthcare. Nat Biotechnol 21(7):747–753

Sharom FJ (2008) ABC multidrug transporters: structure, function and role in chemoresistance. Pharmacogenomics 9(1):105–127

Shephard SE, Zogg M, Burg G, Panizzon RG (1999) Measurement of 5-methoxypsoralen and 8-methoxypsoralen in saliva of PUVA patients as a noninvasive, clinically relevant alternative to monitoring in blood. Arch Dermatol Res 291(9):491–499

Shi S, Klotz U (2011) Age-related changes in pharmacokinetics. Curr Drug Metab 12(7):601–610

Shi MM, Bleavins MR, de la Iglesia FA (1999) Technologies for detecting genetic polymorphisms in pharmacogenomics. Mol Diagn 4(4):343–351

Shord SS, Cavallari LH, Gao W, Jeong H-Y, Deyo K, Patel SR, Camp JR, Labott SM, Molokie RE (2009) The pharmacokinetics of codeine and its metabolites in Blacks with sickle cell disease. Eur J Clin Pharmacol 65(7):651–658

Shrubsole MJ, Lu W, Chen Z, Shu XO, Zheng Y, Dai Q, Cai Q, Gu K, Ruan ZX, Gao YT, Zheng W (2009) Drinking green tea modestly reduces breast cancer risk. J Nutr 139(2):310–316. doi:10.3945/jn.108.098699

Simões M, Bennett RN, Rosa EA (2009) Understanding antimicrobial activities of phytochemicals against multidrug resistant bacteria and biofilms. Nat Prod Rep 26(6):746–757

Smith J, Luo Y (2004) Studies on molecular mechanisms of Ginkgo biloba extract. Appl Microbiol Biotechnol 64(4):465–472

Soldin OP, Chung SH, Mattison DR (2011) Sex differences in drug disposition. J Biomed Biotechnol 2011:187103 (14 pages)

Sparreboom A, Loos WJ, Burger H, Sissung TM, Verweij J, Figg WD, Nooter K, Gelderblom H (2005) Effect of ABCG2 genotype on the oral bioavailability of topotecan. Cancer Biol Ther 4(6):650–658

Stagos D, Amougias G, Matakos A, Spyrou A, Tsatsakis A, Kouretas D (2012) Chemoprevention of liver cancer by plant polyphenols. Food Chem Toxicol 50:2155–2170

Stanek E, Sanders C, Taber KJ, Khalid M, Patel A, Verbrugge R, Agatep B, Aubert R, Epstein R, Frueh F (2012) Adoption of pharmacogenomic testing by US physicians: results of a nationwide survey. Clin Pharmacol Ther 91(3):450–458

Strachan T, Read AP (2010) Genetic manipulation of animals for modeling disease and investigating gene function. In: Strachan T, Read AP (eds) Human molecular genetics, vol 4. Garland Science, New York, NY, USA, pp 639–676

Sugimoto Y, Tsukahara S, Ishikawa E, Mitsuhashi J (2005) Breast cancer resistance protein: molecular target for anticancer drug resistance and pharmacokinetics/pharmacodynamics. Cancer Sci 96(8):457–465

Swan GE, Benowitz NL, Lessov CN, Jacob P III, Tyndale RF, Wilhelmsen K (2005) Nicotine metabolism: the impact of CYP2A6 on estimates of additive genetic influence. Pharmacogenet Genomics 15(2):115–125

Taylor P (2012) Pharmacogenomics: mapping monogenic mutations to direct therapy. J Clin Invest 122(7):2356–2358

Thomasset SC, Berry DP, Garcea G, Marczylo T, Steward WP, Gescher AJ (2007) Dietary polyphenolic phytochemicals – promising cancer chemopreventive agents in humans? A review of their clinical properties. Int J Cancer 120(3):451–458

Tin AS, Sundar SN, Tran KQ, Park AH, Poindexter KM, Firestone GL (2012) Antiproliferative effects of artemisinin on human breast cancer cells requires the downregulated expression of the E2F1 transcription factor and loss of E2F1-target cell cycle genes. Anti-Cancer Drugs 23(4):370–379

Tremblay J, Hamet P (2012) Role of genomics on the path to personalized medicine. Metabolism 62:S2–S5

Tripathi T, Saxena AK (2008) 2D-QSAR studies on new stilbene derivatives of resveratrol as a new selective aryl hydrocarbon receptor. Med Chem Res 17(2–7):212–218

Tsai Y-S, Lee K-W, Huang J-L, Liu Y-S, Juo S-HH, Kuo W-R, Chang J-G, Lin C-S, Jong Y-J (2008) Arecoline, a major alkaloid of areca nut, inhibits p53, represses DNA repair, and triggers DNA damage response in human epithelial cells. Toxicology 249(2):230–237

Tu C-t, Han B, Yao Q-y, Zhang Y-a, Liu H-c, Zhang S-c (2012) Curcumin attenuates Concanavalin A-induced liver injury in mice by inhibition of toll-like receptor (TLR) 2, TLR4 and TLR9 expression. Int Immunopharmacol 12(1):151–157

Van Bambeke F, Balzi E, Tulkens PM (2000) Antibiotic efflux pumps. Biochem Pharmacol 60(4):457–470

van Duynhoven J, van Velzen E, Westerhuis J, Foltz M, Jacobs D, Smilde A (2012) Nutrikinetics: concept, technologies, applications, perspectives. Trends Food Sci Technol 26(1):4–13

Vernon JA, Johnson SJ, Hughen WK, Trujillo A (2006) Economic and developmental considerations for pharmacogenomic technology. Pharmacoeconomics 24(4):335–343

Verstuyft C, Schwab M, Schaeffeler E, Kerb R, Brinkmann U, Jaillon P, Funck-Brentano C, Becquemont L (2003) Digoxin pharmacokinetics and MDR1 genetic polymorphisms. Eur J Clin Pharmacol 58(12):809–812

Vuignier K, Schappler J, Veuthey J-L, Carrupt P-A, Martel S (2010) Drug–protein binding: a critical review of analytical tools. Anal Bioanal Chem 398(1):53–66

Wahid F, Khan T, Subhan F, Khan M, Kim Y (2010) Ginseng pharmacology: multiple molecular targets and recent clinical trials. Drugs Future 35(5):399–407

Walter G, Büssow K, Lueking A, Glökler J (2002) High-throughput protein arrays: prospects for molecular diagnostics. Trends Mol Med 8(6):250–253

Wang HF, Findlay JW (2009) Alternative and emerging methodologies in ligand-binding assays. In: Khan MN, Findlay JW (eds) Ligandbinding assays: development, validation, and implementation in the drug development arena. Wiley, Hoboken, New Jersey, USA, pp 343–380

Wang TT, Edwards AJ, Clevidence BA (2013) Strong and weak plasma response to dietary carotenoids identified

by cluster analysis and linked to beta-carotene 15, 15′-monooxygenase 1 single nucleotide polymorphisms. J Nutr Biochem 24(8):1538–1546
Wheeler HE, Maitland ML, Dolan ME, Cox NJ, Ratain MJ (2012) Cancer pharmacogenomics: strategies and challenges. Nat Rev Genet 14(1):23–34
Willcox ML, Graz B, Falquet J, Diakite C, Giani S, Diallo D (2011) A "reverse pharmacology" approach for developing an anti-malarial phytomedicine. Malar J 10(Suppl 1):S8
Wootton JC, Sparber A (2001) Surveys of complementary and alternative medicine: part I. General trends and demographic groups. J Altern Complement Med 7(2):195–208. doi:10.1089/107555301750164307
Wu J-J, Ai C-Z, Liu Y, Zhang Y-Y, Jiang M, Fan X-R, Lv A-P, Yang L (2012) Interactions between phytochemicals from traditional Chinese medicines and human cytochrome P450 enzymes. Curr Drug Metab 13(5):599–614
Wyss DF, Wang Y-S, Eaton HL, Strickland C, Voigt JH, Zhu Z, Stamford AW (2012) Combining NMR and X-ray crystallography in fragment-based drug discovery: discovery of highly potent and selective BACE-1 inhibitors. In: Davies TG, Hyvönen M (eds) Fragment-based drug discovery and X-Ray crystallography. Springer, Berlin/Heidelberg, pp 83–114
Yang S-Y (2010) Pharmacophore modeling and applications in drug discovery: challenges and recent advances. Drug Discov Today 15(11):444–450
Yang N-S, Shyur L-F, Chen C-H, Wang S-Y, Tzeng C-M (2004) Medicinal herb extract and a single-compound drug confer similar complex pharmacogenomic activities in mcf-7 cells. J Biomed Sci 11(3):418–422
Yazaki K (2006) ABC transporters involved in the transport of plant secondary metabolites. FEBS Lett 580(4):1183–1191
Yee SW, Kroetz DL, Giacomini KM (2013) The pharmacogenomics of membrane transporters project. In: Ishikawa T, Kim RB, König J (eds) Pharmacogenomics of human drug transporters: clinical impacts. Wiley, Hoboken, New Jersey, USA, pp 73–108
Yuan J-M, Koh W-P, Sun C-L, Lee H-P, Mimi CY (2005) Green tea intake, ACE gene polymorphism and breast cancer risk among Chinese women in Singapore. Carcinogenesis 26(8):1389–1394
Yuen J (2013) Combinatorial chemistry in ethnopharmacology. Mod Chem Appl 1:e104
Zamboni WC, Ramanathan RK, McLeod HL, Mani S, Potter DM, Strychor S, Maruca LJ, King CR, Jung LL, Parise RA, Egorin MJ, Davis TA, Marsh S (2006) Disposition of 9-nitrocamptothecin and its 9-aminocamptothecin metabolite in relation to ABC transporter genotypes. Investig New Drugs 24(5):393–401. doi:10.1007/s10637-006-6335-5
Zhang X, Su T, Zhang Q-Y, Gu J, Caggana M, Li H, Ding X (2002) Genetic polymorphisms of the human CYP2A13 gene: identification of single-nucleotide polymorphisms and functional characterization of an Arg257Cys variant. J Pharmacol Exp Ther 302(2): 416–423
Zhang Y, Wang Y, Wang H, Jiang J-H, Shen G-L, Yu R-Q, Li J (2009) Electrochemical DNA biosensor based on the proximity-dependent surface hybridization assay. Anal Chem 81(5):1982–1987
Zhou Y-H, Yu J-P, Liu Y-F, Teng X-J, Ming M, Lv P, An P, Liu S-Q, Yu H-G (2006) Effects of Ginkgo biloba extract on inflammatory mediators (SOD, MDA, TNF-a, NF-k Bp65, IL-6) in TNBS-induced colitis in rats. Mediat Inflamm 2006(5):92642 (9 pages)
Zoete V, Grosdidier A, Michielin O (2009) Docking, virtual high throughput screening and in silico fragment-based drug design. J Cell Mol Med 13(2): 238–248

Machine Learning Techniques in Plant Biology

Khwaja Osama, Bhartendu Nath Mishra, and Pallavi Somvanshi

Contents

Abstract

There is an increasing amount of various genome-sequencing projects and advancement in generation of plant ESTs has resulted in generation of large quantities of data from different fields of plant biology in the public domain. Therefore, a need arises in the analysis of the available data and integrating them with several information of plant biology like crop improvement, nutrigenomics, biochemical engineering, etc. The biological data are mostly complex and vague, analysis of these data is difficult, and interpretation of interaction in different elements cannot be done by simple mathematical functions. Complex computing approaches like artificial intelligence are being applied to understand and interpret these data. The definition of intelligence is debatable for a long period of time; however, intelligence can be vaguely defined as the ability to learn from previous experiences and to adapt accordingly in relatively

K. Osama, Ph.D. • B.N. Mishra, Ph.D.
Department of Biotechnology, Institute of Engineering and Technology, G.B. Technical University, Sitapur Road, Lucknow 226021, India

P. Somvanshi, Ph.D. (✉)
Department of Biotechnology, Teri University, 10 Institutional Area, Vasant Kunj, New Delhi 110070, India
e-mail: psomvanshi@gmail.com

D. Barh et al. (eds.), *PlantOmics: The Omics of Plant Science*,
DOI 10.1007/978-81-322-2172-2_26, © Springer India 2015

new situations. Artificial intelligence uses machine learning algorithm in which the system generates some adaptive learning approaches in order to achieve some goal of environment. Several machine learning approaches have been applied in plant biology till date. In this chapter we will discuss few machine learning approaches and their applications in plant biology.

Keywords

Artificial neural network • Support vector machine • Artificial intelligence • Hidden Markov model • Genetic algorithm

Introduction

One of the most distinctive qualities of human beings that distinguishes them from other animals is their desire to understand and control the process of nature. This basic nature of humans gives birth to the field of science. Over the years of learning and understanding, humans have developed a majestic structure of knowledge that helps us to understand and predict to varying extent various natural phenomena. Though, there is a limit in our abilities to predict and hence several complex resources have been developed by us to understand and control many aspects of life. Over the course of time and our repeated interactions with great uncontrollable natural forces, we have learned the extent in which we can control some aspects of life and extent to which many aspects are uncontrollable. The aim of creating artificial intelligence and artificial life can be traced back to the very beginnings of the Computer Age. Although the study of intelligence is a more than 2000-year-old discipline, artificial intelligence is one of the newest disciplines formally started in 1956. Artificial intelligence is a vast field comprising large areas like logical reasoning, computation, and probability. Different scientists have defined artificial intelligence in different ways; however, these definitions can be broadly categorized in four classes (Russell et al. 1995) as:

1. Systems that think rationally
2. Systems that act rationally
3. Systems that think like human beings
4. Systems that act like human beings

The machine learning field is a part of the broad field of artificial intelligence and a direct successor of statistical model fitting with useful information from pile of data. The only staggering difference between statistical approach and machine learning is that the former regards description of data to be handled in mathematical terms of probability measure and not in terms of deterministic function such as cluster assignments, prediction functions, etc. The tasks to be solved are practically equivalent. In this area, learning methods are also known as estimation methods. Many researchers have long time ago found the basic philosophy and idea of machine learning to be very closely related to nonparametric forms of estimation. Estimation, unlike the statistical approach, does not require learning frameworks or other related things to statistical approach as the former does not have to do with probabilistic model of any data. As an alternative, it only assumes interest in prediction of new instances, which is a far less ambitious work and requires lesser examples to help the point in order to achieve a required performance.

Human beings learn through their life experience and human brain can solve very complex problems based on that learning. On the contrary, machines follow a set of rules or algorithm to solve a problem. The difference in the workings of a human brain and machine can be explained by a simple example of a 5-year-old child who can easily differentiate between a cow, goat, and sheep but will not be able to solve a mathematical equation. However, a computer can solve the equation quite easily, while differentiating between animals would be much difficult (Fig. 1). In the past few years, various studies have shown that machine learning and statistical approach are unlike approaches but converge at some point of time. It is often possible to express the methods of machine learning in probabilistic framework, and vice versa performance of these methods in view of theoretical study is immensely based on the similar assumption or postulation and has been followed as probability theory. It is inspired by the biological brain; the word "fitting" is exchanged with "learning." The "learning"

Fig. 1 Illustration of difference in the workings of the human brain and machines

process in machine learning can be classified into three categories.

(a) *Supervised Learning:* In this type of learning, a sample of input–output pair is provided to the machine for learning (training data set). Each input set in the training data set is associated with the output set. The task of the machine is to find a deterministic function that maps each input with its associated target values in order to minimize the error in future prediction. While solving a given problem using supervised learning, some of the steps are to be considered (Fig. 2). First, we need to determine the number of variables involved in defining the problem; second, we need to select a training data set that describes the problem completely. In the third step, the training data is presented to the system in a form understandable to it. The "machine" or the model is then trained with the data. The learning takes place by adjusting weights of connections according to training error calculated by comparing model output and the actual output in training data set. The trained model is then validated for its robustness and accuracy with the validation data set which contains data that were not present in the training data set. Based upon the nature of the target values' type of deterministic function, changes and different types of learning can be performed like classification learning (the aim is to find whether the two elements in output space are the same or not); preference learning (where the aim is to find whether two elements in output space are equal or not and if not, which one is preferred over the other), an example of such learning are search results of queries on web search engines; and function learning (where the aim is to optimize a function for a given process).

(b) *Unsupervised Learning:* Unsupervised learning is a machine learning technique in which the data set used for training the system does not contain target vectors. Instead of which training data set contains input vectors and a cost function which is to be minimized during the process of learning. The aim of machine in unsupervised learning is to develop representations of input data that can be used for solving problems like decision making, predicting future inputs, etc. Unsupervised learning is mostly used in the field of estimation problem like dimensionality reduction, clustering, statistical modeling, etc. One simple example of unsupervised learning is clustering where we try to cluster different types of data based upon the input data. The inputs in training data set are used by machine to learn pattern, and any new data which lies beyond the limit of those patterns is considered noise.

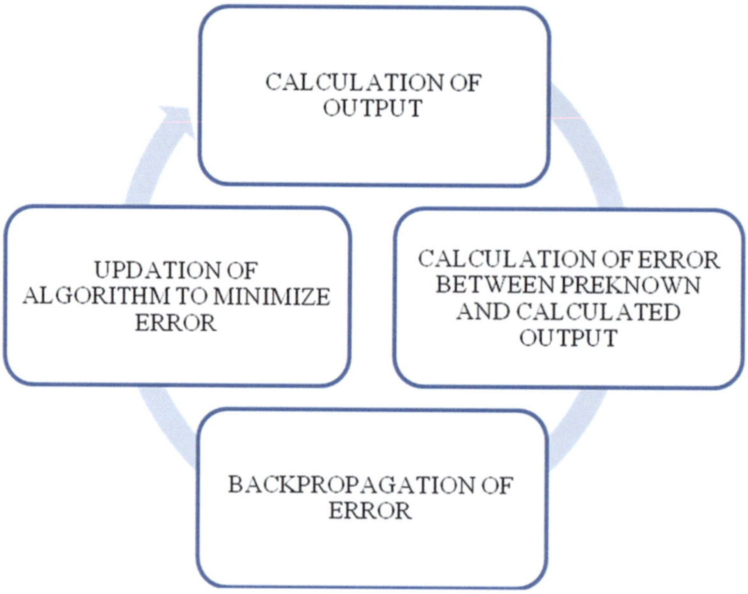

Fig. 2 Steps taken in supervised learning

(c) *Reinforced Learning:* In reinforced learning, the data is not provided to the machine. Instead, the machine interacts with the environment by taking some actions and is rewarded or punished by the reactions from the environment. The machine uses these rewards, corresponding to its actions to learn to act in a way to maximize its future rewards (and minimize punishments). The reinforced learning is concerned about how the machine learns to live with the environment for long term while maximizing its rewards. It defines a function that needs to be maximized during the learning process, and then it finds a strategy to get maximum reward. The reinforced learning uses several algorithms to find this strategy like native brute-force algorithm, value function approaches, direct policy estimation, etc. Reinforced learning is being successfully used in robotics, games, telecommunications, etc.

Machine learning is called black box model as compared to mathematical modeling which is considered to be white box model (Fig. 3). Machine learning is considered black box because in these models are based upon the information or data available from the process but very little theory is known, while in mathematical modeling, models are based upon theoretical knowledge of the process. It will classify or optimize test set according to training set, but it will not provide information of which variables are involved (Mozer and Smolensky 1989; Andrews et al. 1995; Tickle et al. 1998; Alexander and Mozer 1999).

Ideally, machine learning approaches are best suited where abundant amount of data are available, but very less is known about the process. Thus, machine learning is used in fields which are rich in information. These fields are rich in data but the theoretical knowledge is not sufficient for building a model. Thus, these fields solve the problems by principle of induction in inference. However, building a model with the available data and no theoretical knowledge is a difficult task because often these data are incomplete and noisy (Baldi and Brunak 2001). Biological science is essentially an information-rich field and since biological processes are too complex, so very little has been discovered about them; thus, biological science in general or computational biology in particular is an ideal field for application of machine learning.

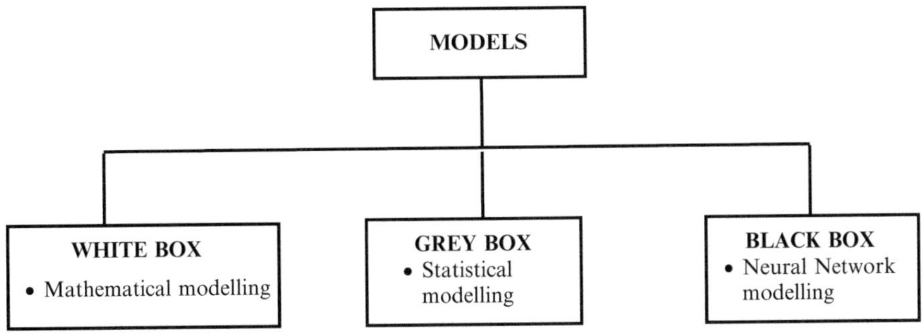

Fig. 3 Classification of models

Problems in Modeling Biological Process

Biological processes vary with time, are nonlinear in nature, and are complex due to composition of many different and interacting elements governed by nondeterministic rules and influenced by external factors (Coruzzi et al. 2009; Gallego et al. 2011). Commonly, most of biological interactions cannot be elucidated by simple stepwise algorithm or a precise formula, particularly when the data set is complex, noisy, vague, uncompleted, or formed by different kinds of data (Prasad and Gupta 2006; Gallego et al. 2011). Many times, behavior of a biological system over a time period is difficult to understand and interpret; additionally, genetic and environmental factors also show biological responses (Karim et al. 1997).

The modeling of these systems is challenging and of extreme importance for scientists and engineers for purposes such as prediction and simulation. Typically, researchers look to create models with two main goals in mind. First, the model should accurately map the input variables to the output variables as observed in real-life situations. Secondly, the model should be a fitting representation of the system's underlying physical characteristics (Resop 2006).

Deterministic mathematical models also known as white box models have been traditionally developed from either physical principles or by statistical regression (Salas et al. 2000). Physical models consist of systems of ordinary or partial differential equations. These models try to represent the underlying physical relationship between variables involved. The benefit of physical models is that they are based on a deep and thorough understanding of the system. However, limitations of these models include the difficulty of setting up and solving complex differential equations analytically, as well as determining equation coefficients and initial and boundary conditions (Coppola et al. 2005). Usually, these equations must be solved using numerical methods, such as the finite element method.

Statistical models on the other hand are designed by finding the equation that best fits a set of experimental data. These models are useful and are generally simple to solve. Statistical regression equations limit the user by requiring a large amount of sample data to estimate parameters of equation and to find the data trend. Also, there are difficulties that arise when manually determining the optimal structure of the statistical equation (Hill et al. 1994).

Artificial intelligence technologies have same or even better potential than traditional statistics in modeling nonlinear relationships in biological data and also have superior prediction powers (Gago et al. 2010a). Recent studies have demonstrated that AI technologies show the same or even better performance than traditional statistics for modeling complex nonlinear relationships hidden in the data and offer superior prediction powers (Landín et al. 2009; Gago et al. 2010a).

Algorithms of Machine Learning

In this section we will briefly discuss some of the highly used algorithms of machine learning.

Dynamic Programming

Dynamic programming is used for problems which can be divided into smaller subproblems and the solution of the bigger problem is found by combining the solutions of smaller problems. Dynamic programming is used almost everywhere in sequence analysis. Sequence alignment analysis algorithms like Needleman–Wunsch and Smith–Waterman are some examples of dynamic programming. Dynamic programming is very well known and the origin of many predictable algorithms for series analysis. Reinforcement or fortification learning algorithms are another very important class of algorithms which can be analyzed as simplification of ideas for dynamic programming.

Gradient Descent

Gradient descent also known as steepest descent is one of the most important breakthroughs in machine learning. It is used for building a best model that minimizes the error. It is simple and easy to use and guarantees to find a minimum of a function if present. Gradient descent is often used with back propagation of information like in back-propagation neural networks. In complex functions where a number of local minima are present, gradient descent mostly ends up finding the local minima rather than global. Therefore, gradient descent algorithm is generally run in multiple iterations with different starting points and learning rates.

Gradient descent is slow and often less effective at small step size, and while convergence speed could be increased with increasing step size, larger step size results in large error. Speed of convergence and efficiency of gradient descent are improved using various variations like conjugate gradient descent (in which the weights are adjusted in directions conjugate to the gradient in order to get fast convergence) and gradient descent with adaptive (in which the learning rate is adjusted during the training in order to produce an optimum convergence rate and error) line search algorithms (Stanimirovic and Miladinovic 2010).

Expectation–Maximization Algorithms

In computational biology, the data available for training probabilistic are often incomplete. Expectation–maximization algorithm is used for parameter estimation in such models. It is a generalization of maximum likelihood estimation in incomplete data case, but expectation–maximization addresses more difficult problems than maximum likelihood.

In the expectation–maximization algorithm, hidden or missing variables are estimated using known or present parameters (the E step), and then these completions are used to reestimate hidden parameters. This step is called the M step because this can be thought as maximization of expected log-likelihood of data. In the complete data case, there is only one global optimum, but in the incomplete data case, there are multiple local optima. Expectation–maximization algorithm cuts the problem into simple subproblems which have single global optima (Chuong and Serafim 2008).

Many models in computational biology have hidden variables. These hidden variables are due to missing or non-recordable or corrupted data. Expectation–maximization algorithms are used in many applications like hidden Markov models, neural networks, etc. and sequence analysis like gene expression clustering (D'haeseleer 2005), motif finding (Lawrence and Reilly 1990), haplotype inference problem (Excoffier and Slatkin 1995), etc.

Markov Chain Monte Carlo Methods

Markov Chain Monte Carlo methods belong to an important class of stochastic methods, which are strongly related to statistical physics and are highly sought methods for machine learning and Bayesian inference. These methods use probability distribution to solve problems. Markov Chain Monte Carlo (MCMC) method aims to find the solution of two basic problems: first, to use a probability distribution to generate random inputs in a defined domain and second, to estimate the expectations of a deterministic function over the distribution. This takes several steps, and after number of steps, the quality of the given sample improves as function for a number of steps to be taken at some point of time. If the distribution of inputs is not uniform, the approximation will not be accurate; also, size of input affects the accuracy of approximation; if a number of input samples are low, approximation will be poor. Typically, it is really not hard to create a Markov Chain with the preferred properties. In addition, the more difficult thing is to determine the steps, which are needed to converge to static distribution within a suitable error. A good form of chain will also have rapid mixing in which the immobile distribution is reached very quickly starting from any arbitrary position. The most common application of algorithm used in Markov Chain Monte Carlo methods is numerically calculating multidimensional integrals.

Simulated Annealing

Optimization of problem is a difficult task often practically impossible. As the problem gets large, area required to search for optimum also becomes large and a huge number of possible solutions were searched to find the optimum one. These are a large number of solutions still for modern computing. Often, while finding the global optima for a given problem, optimization algorithm gets stuck in local optima. Simulated annealing is a random search method for global optimization problem. This method is inspired by the annealing process of metals. Annealing involves heating and cooling of metals to change their physical properties. When the metal is heated, molecules in metal have high energy and they vibrate highly, but when it cools slowly, the vibration of metal molecules also slows down and metals' new structure gets fixed. In simulated annealing, the search of optimum is started at high energy and then it is lowered slowly. At high energy, the algorithm will accept solutions with greater frequency, accepting more solutions worse than the current solution. This provides algorithm the ability to jump local minimums. As the algorithm progresses, energy is slowly lowered, reducing the algorithm's frequency of finding solutions worse than the current one. Thus, the algorithm focuses on a search space to find global minima. Simulated annealing may become more efficient than other algorithms such as exhaustive enumeration, if the aim is merely to find an acceptably good solution in a fixed amount of time, or rather the best of all the possible solutions.

Evolutionary and Genetic Algorithms

Evolutionary algorithms are computer programs, which can solve complex and complicated mathematical and statistical problems using Darwin's theory of evolution. Several fixed-length vectors also known as individuals compete with each other to search for an optimal area. These creatures evolve with time to find the optimal solution. Evolutionary algorithms have been started with an initial population of individuals of finite size. Each individual is then associated with a fitness score. A fitness function is used to calculate the fitness score of each individual. The individuals with high fitness score represent the healthier solutions of the problems than that with individuals having low fitness score. After this initial phase, the main cycle of evolution begins. Each individual in the initial population generates one offspring using mutations. These offspring are then given a fitness value. Now, this first-generation children form a population which is considered as present population, and this cycle is repeated many times. These individuals evolve from generation to generation and compete with each other in the same generation to be fittest scorer. The individual with the fittest score is considered to be the fittest individual and is selected to solve the problem.

Artificial Neural Networks

Artificial neural networks are complex mathematical models, which mimic biological neural networks. An artificial neural network like any biological neural network is built by connecting several neurons. An artificial neuron (Fig. 4) is a simple mathematical model that does three functions: multiplication, addition, and activation (Krenker et al. 2011). The neuron first multiplies every input given by its corresponding weights. The network to memorize a given process uses weights corresponding to inputs and inter-node connections; these weights determine the conductivity of inputs through the network. The weighted inputs given to an artificial neuron are then added and passed through an activation function, also called a transfer to the next neuron. This function can be sigmoidal, linear, hyperbolic, tangent, or radial basis, and the type of activation can be selected according to specific problem. The most common neuronal nonlinear activation function used in biological systems is sigmoid in nature.

A single neuron is a simple mathematical function which is not useful in solving big problems. When these neurons are connected to each other to form an artificial neural network, the real potential of these models is visible. Unlike biological neural networks, artificial neurons are connected in a defined architecture. This arrangement of artificial neurons is called topology of network. Several standard network topologies have been defined by scientists in the past for different problems. For a particular problem, appropriate topology has to be decided and we need to fine-tune the topology itself and its parameters. The fine-tuning of network topology includes teaching the network in solving a given problem. Artificial neural networks like biological neural network can learn their behavior on the basis of inputs that they get from their environment (Kos et al. 2011).

This teaching or fine-tuning of neural network is called training of artificial neural network. Many different training mechanisms have been used in neural networks. These affect the accuracy of models and also influence speed at which the networks converge. The training of artificial neural networks can be classified as supervised learning and unsupervised learning. In supervised method, the desired output or target values are provided by an external source, and then the network output is matched with target values for optimizing the network weights and correcting network functioning. Artificial neural network uses a delta rule for training (Widrow and Hoff 1960). To train a network with a given set of training samples containing input data set x^p and target data set d^p, network calculates output y^p for every input values and subtract it from target values to calculate error $(d^p - y^p)$:

$$y^p = \sum_j w_j x_j + \theta \tag{1}$$

where θ is the bias of the network.

Delta rule uses a cost function based on these errors to modify weights. The final error is then calculated using a cost or error function which can be mean square, root mean square, least mean

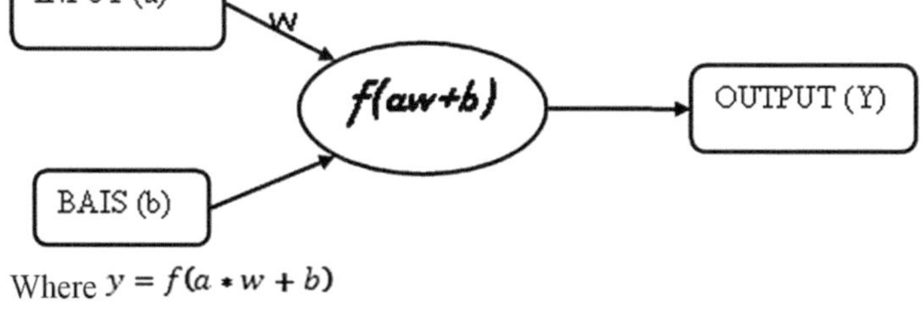

Fig. 4 Working principle of an artificial neuron

square, etc. The total error using least mean square error can be defined by

$$E = \sum_{p} E^{p} = \frac{1}{2}\sum_{p}\left(d^{p} - y^{p}\right)^{2} \quad (2)$$

P represents the range of input data set, and E^p represents the error on the whole range of input data set. Weights are then adjusted to reduce error by gradient descent method and delta rule finds the value of new weights. The weights are changed proportionally to the negative of derivative of error measured for the current iteration with respect to each weight:

$$\Delta_{p} w_{j} = -g\frac{\partial E^{p}}{\partial w_{j}} \quad (3)$$

where γ is a constant of proportionality and $\Delta_p w_j$ represents the change in target for pattern p. The derivative is

$$\frac{\partial E^{p}}{\partial w_{j}} = \frac{\partial E^{p}}{\partial y^{p}}\frac{\partial y^{p}}{\partial w_{j}} \quad (4)$$

For a linear activation function,

$$\frac{\partial y^{p}}{\partial w_{j}} = x_{j} \quad (5)$$

and

$$\frac{\partial E^{p}}{\partial y^{p}} = -\left(d^{p} - y^{p}\right) \quad (6)$$

Thus,

$$\Delta_{p} W_{j} = \gamma \partial^{p} x_{j} \quad (7)$$

where x_j is the input vector and $\partial^p = (d^p - y^p)$ is the difference between the network output and the actual output or target for pattern p.

The delta rule modifies weights according to a proportionally negative derivative of error, i.e., if on increasing weight error decreases, then delta rule keeps on increasing weight till error reaches minimum or starts increasing, and if on increasing weight error decreases, delta rule decreases weights till error reaches minimum or starts decreasing.

In supervised learning technique, input variables and a cost function are provided to network, but no output variables are provided. In this technique, network parameters are set on the basis of input data and cost function. In pattern classification using unsupervised learning, a self-organizing network identifies the silent features of input data set; however, unlike supervised learning, In supervised learning there is no defined set of categories into which the patterns can be classified (Prasad and Gupta 2006).

A trained network is then validated for accuracy and robustness by simulating for a validation data set. Validation data set contains some input variables present in training data set and some new input variables. If network accurately simulates validation data set, then it is considered a trained or learned network.

Sometimes after learning if the network performance for training data is best but for test data set its performance is poor, it is called overlearning or over-fitting of network. Network size plays an important role in overlearning of a network; a large network over-fits small problem. So, an optimum size of network (number of nodes) has to be decided for a given problem. After training of network, it can be used for solving problems. Artificial neural networks are used for problems like function approximation, regression analysis, time series prediction, classification, pattern recognition, decision making, data processing, filtering, clustering, etc.

Structure of Artificial Neural Network

Artificial neural networks are inspired from biological nervous system and consist of a network of artificial neurons. An artificial neuron is a simple mathematical model which is not capable of solving complex real-life problems. The ability of artificial neural network in solving complex problems is due to arrangement of these neurons in the form of a network. The information in a neural network is processed through its building blocks in a nonlinear and parallel manner.

Arrangement of neurons in a network is called its topology or architecture of artificial neural network. An artificial neuron is called a node in a network; these nodes are arranged in the form of layers. Typically, the most commonly used artificial neural network is a three-layered network. The first layer is an input layer, the second layer is a hidden layer, and the third layer is an output layer. Generally, the three-layered neural network is shown as typical example because it is capable of solving practically all types of problems. Three-layered networks are capable of solving all problems but larger networks can solve these problems more efficiently.

The connections between nodes and number of nodes per layer are defined by the approach, which is adopted to solve or interpret a given problem. Internodal connections direct direction of information flow through a neural network. On the basis of direction, the flow of information neural networks can be classified as: feed-forward neural network and cascade forward for unidirectional flow of information and recurrent or feedback for bidirectional flow of information. In feed-forward neural network, information flows from one layer to another in one direction, i.e., from input to hidden to output layer (Fig. 5). In cascade forward neural network, the flow of information is unidirectional except that there is an extra weight connection from input to each layer (Fig. 6). However, in networks having bidirectional flow of information (recurrent neural network), information flows both in forward and backward directions (Fig. 7). In a complete recurrent neural network, each node is connected to every node including itself. Recurrent neural networks are bulky and complex compared to feed-forward neural networks due to massive parallel processing of information. These networks require large computational space and are not easy to understand, unstable, and noise sensitive (Mandic and Chambers 2001). However, recurrent neural networks are closer to biological neural networks. Due to recurrent connections in these networks, output of a network can be used as input of the same network. This property provides the ability of prediction of future outcomes to recurrent neural networks. Several other networks are being designed using fuzzy logic and other techniques; some of these networks are self-organizing neural network, regression neural networks, fuzzy neural networks, etc. (Hayashi et al. 1993; Yao 1999; Yang 2006).

In 1980s back-propagation algorithm developed and helped in increasing popularity of artificial neural networks as function optimizers. Back-propagation algorithm is used to train networks using experimental data. Learning and updating of weights became easy with back-propagation algorithm. In this algorithm, network errors are back propagated to the network in order to train it. Back propagation can also be considered as a generalization of delta rule for nonlinear activation functions and multilayer networks (Kruschke and Movellan 1991).

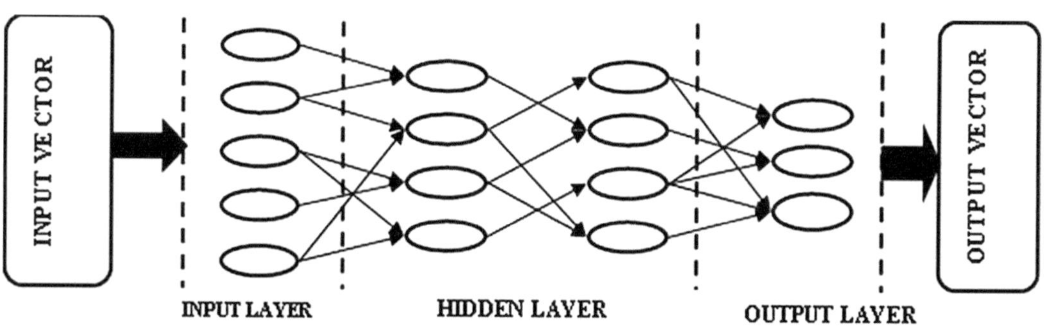

Fig. 5 Architecture of feed-forward neural network

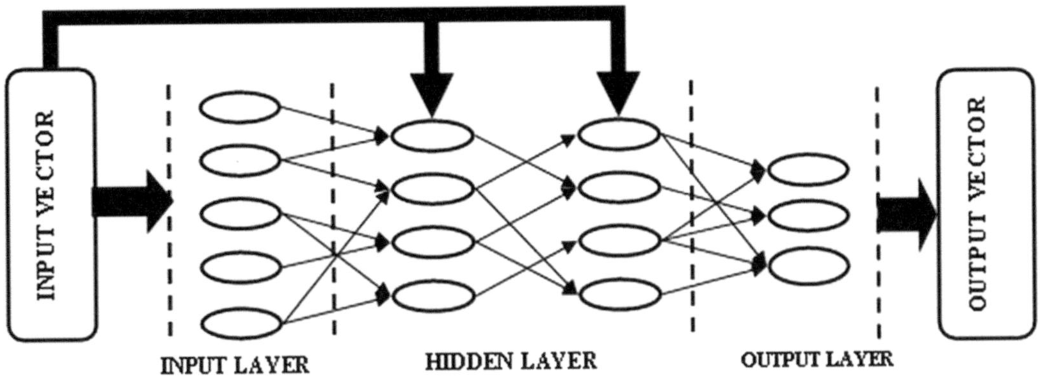

Fig. 6 Architecture of cascade forward neural network

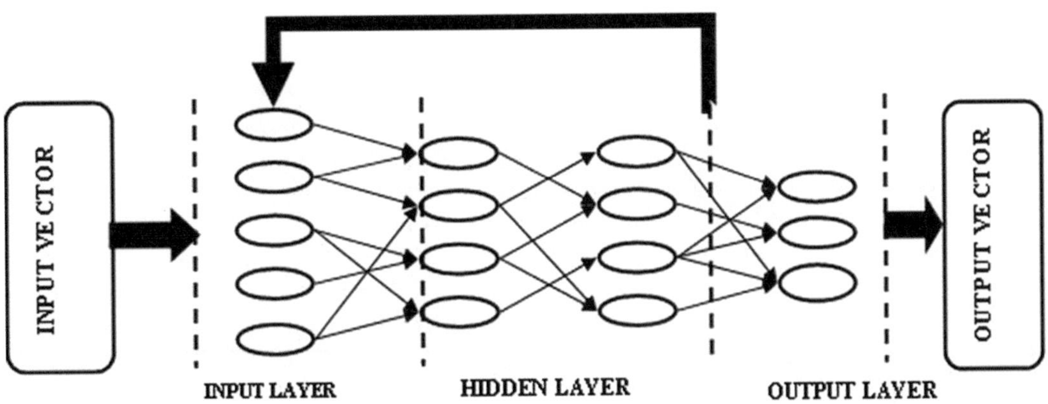

Fig. 7 Architecture of recurrent neural network

Key Steps in Applying Artificial Neural Networks

Data Preprocessing

Transformation and normalization are two widely used preprocessing methods. Transformation involves manipulating raw input data to reduce its dimensionality, while normalization is a transformation performed on input data to distribute the data evenly and scale it (mostly in range of −1 to 1 or 0 to 1) into an acceptable range for the network.

Network Selection

It involves selection of network model, number and size of hidden layers, initial weight matrix, etc.

Training Selection

It needs to start with network topology and initial weight, and train the network on your training data set. When the network reaches the satisfactory minimum error, it saves the weights.

Testing and Interpretation of Results

The trained network is applied on test data set to find the error. If it is not satisfactory, the training process and/or network architecture needs to be modified.

Applications of Artificial Neural Networks in Plant Biology

Artificial neural networks have been researched and used for applications in many different fields. Many of these areas are using artificial neural networks to solve problems, previously thought to be impossible or very difficult with traditional methods. Some of the reasons for the great interest shown in neural networks are their property of being a universal function approximator, i.e., their applicability on different types of problems (White 1992), and their flexibility, robustness, and no need of prior knowledge. A neural network can be applied to practically any problem unlike statistical methods where a mathematical relation between input and output variables is necessary and responses have to be determined for each problem (Zealand et al. 1999). Artificial neural networks prove to be a better choice than regression models for noisy data (Denton 1995). Thus, neural networks are powerful tools for modeling complex processes. They can model highly nonlinear complex systems like biological processes and other real-life problems (Gevrey et al. 2003) (Table 1).

Less applications of artificial neural network to plant biology are present in literature as compared to other fields, viz., pharmaceutical science (Colbourn 2003; Shao et al. 2006; York et al. 2009), ecology (Hilbert and Ostendorf 2001; Adriaenssens et al. 2004), agriculture (Huang 2009), etc. Initial applications of artificial neural networks in plant science include optimization studies. Earlier studies used neural networks with image analysis to identify live or dead plant cells (Fukuda et al. 1991), analyze developmental stages of somatic embryos (Uozumi et al. 1993), etc. Some researchers (Honda et al. 1997) used hybrid fuzzy neural model to predict the length of shoots regenerated from rice callus to be transferred from the growth medium to sugar-free medium for acclimatization. Radius, length, width, roundness, area, and perimeter of the digital images of somatic embryo taken from CCD camera were used as input data for the network. The results of neural network analysis were compared to that of multiple regression analysis. Neural network predicted shoot lengths with 95 % accuracy with an average error of 1.3 mm. In another work, fuzzy neural network was used for modeling the production of Ginjo sake by fermentation in bioreactor (Hanai et al. 1997).

The use of neural network technology in plant and agriculture biology has increased in the last 20 years (Huang 2009). Neural networks have been used for predicting crop yield and modeling pest control treatments according to environmental conditions. Optimization of pesticide concentration and periods of treatment improves the economy of production and minimizes toxic residual levels of agricultural products (Jiménez et al. 2008). Other authors developed neural network models to predict crop yield for crops like corn (Kaul et al. 2005), sugar beet (Kehagias et al. 1998), soybean (Kaul et al. 2005) and winter wheat for different cultivation conditions and distribution of crops in different climate distributions. These works provide important information about the effect of climate change on vegetation of different areas and thus help in conservation of these areas (Hilbert and Ostendorf 2001).

Neural networks have been used in combination with a bioelectric recognition assay to detect plant viruses (Frossyniotis et al. 2008). The authors used biosensors to detect electric signals from plant cells suspended in a gel matrix. The responses of plant cells on interaction with viruses were recorded and used for training of a neural network, and a classification model of cultured cell was prepared. In another study, a multilayer neural network was used with genetic algorithm for the detection of plant virus (Glezakos et al. 2010).

Neural networks have been used in tissue culture on determining the effect of different parameters like carbon source, pH, etc. on plant growth (Prasad and Gupta 2006). A growth model was

Table 1 Some applications of artificial neural network in plant biology

S. No.	Application	Network architecture	Plant species	Database	References
1.	Growth modeling of alfalfa shoots	Feed-forward neural network with Kalman filter	*Medicago sativa*	Dry weight, leaf number, and root initiation stage	Tani et al. (1992)
2.	Classification of embryo types from non-embryos and predicting embryo-derived plantlet formation	Feed-forward	*Apium graveolens*	Area, length to width ratio, circularity, and distance dispersion of plant cell cultures	Uozumi et al. (1993)
3.	Biomass estimation of cell cultures	Feed-forward neural network with gradient descent training method	*Daucus carota*	Sucrose, glucose, and fructose level of medium	Albiol et al. (1995)
4.	Simulation of temperature distribution in culture vessel	Fuzzy neural network with back-propagation algorithm		Spatial temperature distribution of culture vessel	Suroso et al. (1996)
5.	Identification of live and dead plant cells	Three-layered neural network		Size and color of cells	Fukuda et al. (1991)
6.	Clustering of regenerated plantlets into groups	Adaptive resonance theory		Mean brightness values, maximum pixel count, and gray level of maximum pixel count in RBG regions	Zhang et al. (1999)
7.	Predicting shoot length of regenerated callus	Three-layered fuzzy neural network with Kalman filter	*Oryza sativa*	Radius, length, width, roundness, area, and perimeter of the somatic embryo images	Honda et al. (1997)
8.	Detection of plant viruses	Feed-forward back-propagation network with BFGS quasi-Newton optimization algorithm		Electric response from plant cells	Frossyniotis et al. (2008)
9.	Model in vitro rhizogenesis and subsequent acclimatization	Feed-forward neural network	*Vitis vinifera*		Gago et al. (2010c)
10.	Classification of sweet potato embryos	Back-propagation neural network	*Ipomoea batatas*	Embryo area, length, and symmetry; polar coordinates of an embryo's perimeter with respect to its centroid	Molto and Harrell (1993)
11.	Classification of somatic embryos into normal and abnormal	Feed-forward neural network	*Daucus carota*	Morphological features	Ruan et al. (1997)

(continued)

Table 1 (continued)

S. No.	Application	Network architecture	Plant species	Database	References
12.	Prediction of culture condition for optimal productivity	Feed-forward back-propagation neural network	*Glycyrrhiza glabra*	Inoculum size, fresh wt, density, culture temperature, pH, and time of inoculation	Mehrotra et al. (2008)
13.	Growth modeling of hairy roots in nutrient mist reactor	Feed-forward and cascade forward and recurrent neural networks	*Artemisia annua*	Mist On/Off cycle time, initial packing density, media volume, initial sucrose concentration in media, and culture time	Osama et al. (2013)
14.	Sorting of regenerated plants on the basis of their photometrical behavior	Adaptive resonance theory	*Gladiolus*	Photometrical behavior of their leaves in red, blue, and green color regimes	Mahendra et al. (2004)
15.	Prediction of culture conditions for maximum biomass growth	Generalized regression network with radial basis function	*Glycyrrhiza glabra*	Inoculum density, medium pH, sucrose conc., media volume	Prakash et al. (2010)

developed for the study in effect for CO_2 and sucrose content on in vitro shoots of alfalfa (Tani et al. 1992). Gago et al. (2010c) developed a neural network model for in vitro rhizogenesis and acclimatization of micropropogated *Vitis vinifera* L. plants. Effect of cultivars, IBA concentration, and exposure time of plant to IBA on acclimated plant were studied. It was found that they have significant effect on root numbers, number of nodes, and height of the acclimatized plantlets, with exposure time being a more prominent factor. Study was done in different cultivars and the model did good predictions for all cultivars.

Albiol et al. (1995) compared a deterministic mathematical model with a neural network model and found that neural network modeling was cost effective and time efficient and required smaller data set. Neural networks have been used in classification and pattern recognition in plant tissue culture (Prasad and Gupta 2006). Zhang et al. (1999) used neural networks with image analysis for selection of embryos in embryonic tissue culture of Douglas fir. In another study, machine vision was used for distinguishing between white and bright-yellow callus formed in sugarcane callus culture. The technique was successfully used to obtain regeneration frequency of callus culture (Honda et al. 1999). ANNs have been used for sorting regenerated plantlets according to trichometric features of leaves. The sorting of plantlets was done using adaptive resonance theory (ART) having unsupervised architecture (Mahendra et al. 2004). This approach provided a means of selecting plants suitable for *ex vitro* transfer and helps in quality control of commercial micropropagation.

Hairy roots are a good source of plant secondary metabolites. These roots show genetic and biosynthetic stability and require no exogenous growth hormone. However, culture of these hairy roots for high secondary metabolite associated with better biomass production requires optimization of several physical and chemical parameters that affect the growth and productivity of these roots. A feed-forward back-propagation neural network model was developed (Mehrotra et al. 2008) for prediction of in vitro culture conditions for hairy root growth. The model used inoculum size, fresh weight, density, culture temperature, pH, and time of inoculation as input parameters and final fresh weight of roots as final parameters. The trained neural network model was able to predict the final biomass for a particular culture condition efficiently. Later, Prakash et al. (2010) developed a regression and feed-forward neural network model for prediction of optimal culture conditions for prediction of hairy root maximum biomass yield. It was found that both networks predicted culture conditions efficiently; however, regression neural network was more accurate.

Scale-up of hairy roots in bioreactors is a very difficult task. During their growth hairy roots form clumps which causes heat and mass transfer restrictions. In order to solve these problems, agitators and aerators have to be used in bioreactors, but these equipment cause shear stress on roots, which results in injury and callus formation. Neural networks have also been used for modeling of bioreactors for hairy root growth. Neural network model (Osama et al. 2013) was developed for modeling of hairy root growth in a nutrient mist reactor. The significant features for culture parameters, viz., inoculum size, mist ON time, mist OFF time, initial packing density, media volume, initial sucrose concentration in media, and time of culture, were considered as input of the network. The final biomass of hairy roots on dry weight basis was taken as network output. Three network architectures, viz., feed-forward, cascade forward, and recurrent, were tested and all these networks were found efficient with recurrent neural network being the most accurate.

The knowledge derived through ANNs can be easily increased by training the model by adding to database new inputs (salt concentration, type of medium, other plant hormone, etc.) and/or outputs (plantlet weight, chlorophyll and carotene content, stomata analysis, etc.).

Support Vector Machine

A support vector machine is an abstract machine that uses supervised learning to solve classification problems. These are relatively new, general formulation for learning machines. These techniques learn to assign labels to objects. To control the generalization ability of a learning machine, one has to control two different factors: error rate on the training data and capacity of learning machine as measured by its Vapnik–Chervonenkis dimension, which is a non-negative integer that measures the expressive power for the family of classification functions realized by the learning machine (Haykin 2003). A special property of SVMs is that they simultaneously minimize the empirical classification error and maximize the geometric margin. Hence, they are also known as maximum margin classifiers. A classification task usually involves with training and testing data, which consists of some data instances. Each instance in the training set contains one target value called class labels and several "attributes" called features. The goal of SVM is to produce the model that predicts the target value of data instances in the testing set, which are given only the attributes.

The training data provided for training contains n ($i = 1, 2, 3\ldots n$) input vectors denoted by x_i, with each of vectors paired with corresponding labels and are denoted by y_i. The labels in the training data lie in two classes, and for classification of these data in different classes, support vector machine uses an oriented hyperplane. This hyperplane separates the two classes of data points on either side of it. The data points on one side of the hyperplane are labeled positive and on the other side negative. The directed hyperplane is defined by the maximally distant hyperplane from the data points on both of its sides. Thus, the points closest to the hyperplane on each side are the most influential for defining its position, and therefore, these points are called support vectors. The perpendicular distance from support vectors and hyperplane is called margin.

A nonlinear transformation function $\Phi(.)$ is defined to map the input space to a higher dimensional feature space. The oriented separating hyperplane is given as $w.\ \Phi(x) + b = 0$ where $w.\ \Phi(x)$ is the product of data points and weights that determine their orientation and b is the bias or offset of hyperplane from the origin. If for a point x_i, $w.\ \Phi(x_i) + b = 1$, it lies on one side of the hyperplane, and if $w.\ \Phi(x_i) + b = -1$, the point lies on the other side. The support vector machine uses the optimization theory to minimize an objective function which is half the distance

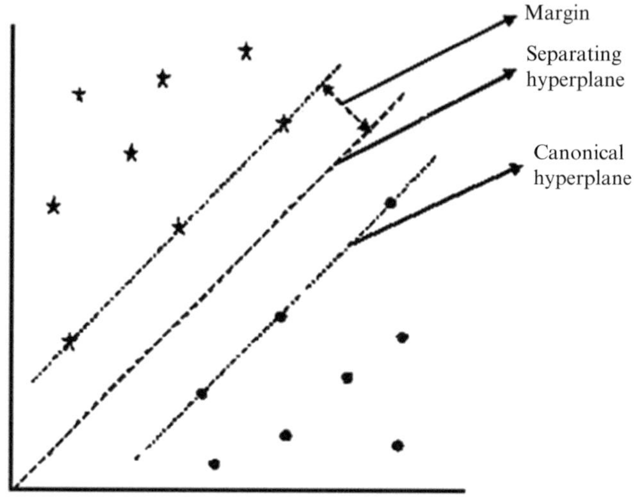

Fig. 8 Classification using support vector machine

between two canonical hyperplanes (hyperplanes passing through support vectors) (Fig. 8).

As mentioned above, most of the traditional neural network models seek to minimize the training error by implementing the empirical risk minimization principle, whereas the support vector machines implement the structural risk minimization principle which attempts to minimize an upper bound on the generalization error by striking a right balance between training error and capacity of machine. Support vector machine also provides guaranteed global optimal solution (Haykin 2003).

Applications of Support Vector Machine in Plant Biology

Biological applications of support vector machines involve classifying objects such as protein and DNA sequences and microarray expression profiles. Performance of support vector machine is mostly similar to or better than traditional machine learning approaches (Hua and Sun 2001). At present there is no algorithm for finding the optimum network architecture, i.e., ideal number of hidden layers, best activation function, etc. This is usually done by trial and error method and is time consuming and often less effective (Shigidi and Garcia 2003). Generally, neural network structures are developed according to past experiences; this requires considerable skills, and efficiency of these networks depends largely on their training (Gonzalez 2000). Another common approach is simply selecting an arbitrarily large number of neurons as models (Xiang et al. 2005). However, a large number of nodes may lead to poor generalization and large computational requirement (Archer and Wang 1993). Support vector machines can thus be an alternative for more accurate classifications.

Protein interactions play a very significant role in any of the biological operation. Prediction of these interactions is a point of key focus for researchers. However, very little has been achieved in this area due to expensive and time-consuming experimental approaches. A support vector machine-based model was developed (Lin et al. 2009) to predict potential *Arabidopsis* (*A. thaliana*) protein interactions based on a variety of indirect evidences. The potential interactions were predicted based on 14 features derived from four types of indirect evidence (coexpression, domain interaction, colocalization and shared annotations). The confidence of predicted interaction was estimated to be 46.87 % and these interactions were expected to cover 29.02 % of the entire interactome. The model successfully recognized 28.91 % of new interactions, similar to its expected sensitivity (29.02 %).

Knowledge of locations in protein expression is important for better understanding of defined cellular processes at organellar and cellular levels. A complete map of a plant proteome is clearly a major goal for plant research community in terms of determining the function and regulation of each encoded protein. An integrative support vector machine-based localization predictor called AtSubP was developed (Kaundal et al. 2010) which was based on the combinatorial presence of diverse protein features, viz., amino acid composition, sequence-order effects, terminal information, position-specific scoring matrix, and similarity search-based, position-specific, iterated Basic Local Alignment Search Tool information. The model predicted seven subcellular compartments through fivefold cross-validation test and achieved an overall sensitivity of 91 % with high-confidence precision and Matthew's correlation coefficient values of 90.9 % and 0.89, respectively.

Eichner et al. (2011) used a support vector-based model for identification of alternate splicing in *A. thaliana* by detecting intron retention and exon skip from tiling arrays. The model used existing EST and cDNA sequences for supervised training. The method developed in this work expands the scarce repertoire of analysis tools for identification of alternative mRNA splicing from whole-genome tiling arrays.

Other authors used image analysis with support vector machine for detection and tracking of plant cell division by in vivo imaging. Cell division in plants takes place mostly in meristems which contain stem cells that give rise to all cell types by regular cell division. However, the control mechanism of cell division is not understood properly and is the center of interest of developmental biologists doing in vivo research in plant cell division (Marcuzzo et al. 2008b). However, for automated machine, vision images have to be partitioned into multiple segments. The goal of segmentation is to simplify and/or change the representation of an image into something that is more meaningful and easier to analyze. Some authors developed support vector machine-based model for classification of plant root cells of *A. thaliana* (Marcuzzo et al. 2008a, 2009). The images were segmented using watershed algorithm and result was improved by merging adjacent regions. The selection of individual cells was obtained using a support vector machine (SVM) classifier, based on a cell descriptor constructed from the shape and edge strength of the cells' contour.

Recently, support vector machines have been used in agricultural research. Support vector machine-based models have been used for classification of crop types, seed pollen grains, etc. Crop type classification is an important application of remote sensing technology, and since the advent of remote sensing technologies, several studies on crop type classification have been published. A support vector machine-based model was developed (Karimi et al. 2006) for detection of weed and nitrogen stress in corn. The results of support vector machine-based classification were compared to that of neural network, and SVM-based classifier was found more accurate. These models use imaging spectroscopy for studying hyperspectral images. Extraction of end-members from these remote-sensed images is a difficult task. A number of algorithms based on notion of spectral mixture modeling have been proposed to accomplish the complex task of finding appropriate end-members for spectral unmixing in hyperspectral data (Martinez et al. 2006). A support vector machine-based end-member extraction (SVM-BEE) model was developed (Filippi and Archibald 2009) for hyperspectral agricultural mapping. This model accurately and rapidly yields a computed representation of hyperspectral data that can accommodate multiple distributions. The efficacies of SVM-BEE, N-FINDR and SMACC algorithms in extracting end-members from real, predominantly agricultural scene were compared. SVM-BEE estimated vegetation and other end-members for all classes in the image, which N-FINDR and SMACC failed to do. Shi et al. (2012) used SVM for classifying agricultural data from public agricultural data set and concluded that SVM outperformed other popular algorithms, like naive Bayes and artificial neural network, in terms of the F_1 measure. Different machine learning algorithms and traditional maximum likelihood algorithm were compared for classification of crops (Nitze et al. 2012). Researchers compared support vector machine, artificial neural network, random forest and maximum likelihood algorithms for classification of ten different crop types. Support vector machine was found to exhibit better results than the other algorithms.

Hidden Markov Model

Andrey Markov gave a concept of a mathematical system that undergoes change from one state to another, between finite numbers of possible states. In this process, next state depends upon the current state and not on sequence of events that preceded it. This phenomenon is called Markov property. Markov model is a probabilistic graphical model considering Markov property. The simplest Markov model is a Markov chain which is like any random process with Markov property. In Markov chain model, the system is assumed to be autonomous and the transition states are fully observable. However, in a hidden Markov model (HMM), the states of transition are not fully observable but output depending upon those states is observable. These models are named because of their two properties. First, it assumes that states Y_t are not observable or hidden from the observer, and second, it assumes that the states of transition and output follow the Markov property, i.e., state Y_t at time t does not depend upon the previous states and also the output A_t does not depend upon the state Y_t (Fig. 9).

These models are used to analyze different types of time series problems in different areas like speech recognition (Juang and Rabiner 1991), ion channel recording (Venkataramanan and Sigworth 2002), optical character recognition (Agazzi and Kuo 1993), computational biology (Krogh et al. 1994), etc.

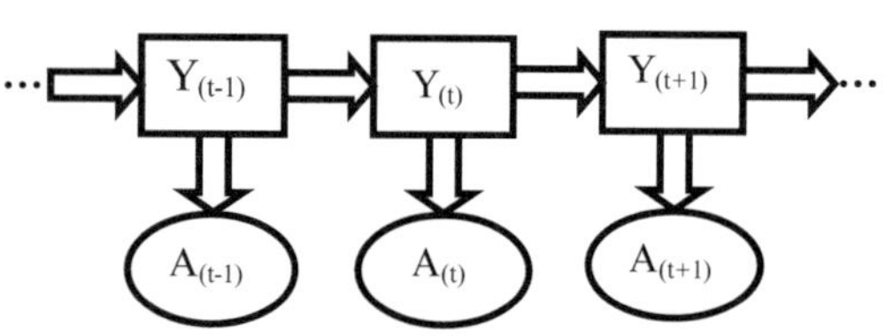

Fig. 9 Simple hidden Markov model

Hidden Markov model is a tool for estimating probability distributions of a sequence of observations over a time series, finite time series t. The observations can be represented as discrete alphabets, integers, real-valued numbers, etc. In biological applications, the HMM observations are generally discrete alphabets (the 20-letter amino acid, 4-letter nucleotide, 64-letter codon triplet, etc.). A HMM used for solving real-life problems contains many transition states and hidden variables connected to each other. Generally, in biological problems, a unidirectional HMM is used which is also called left–right model. In this type of model, the direction of transition of state is from left to right, and it prevents any transition to a state if transition from that state to another state has taken place (machine learning bioinformatics). There are three basic questions one can ask immediately for a HMM: the likelihood question (how likely is the output for HMM), decoding question (probable sequence of transition states followed by the model to reach the output), and the learning question (how to revise values of transitions and emissions from the given information given that they are not known with complete certainty).

Applications of Hidden Markov Model in Plant Biology

HMM is widely used in speech recognition, natural language modeling, and on-line handwriting recognition. HMM is widely being used in a variety of biological problems like gene finding, secondary structure prediction, gene annotation, etc.

With development in automated whole-genome sequencing, complete genome sequences are becoming more and more abundant. The first and most important task after getting a new genome is to find a protein coding sequence. One of the most successful gene finders was GeneMark (Borodovsky and McIninch 1993; Borodovsky et al. 1994, 1995), which in its first version was based on frame-dependent nonhomogeneous

Markov models. The accuracy of gene finder depends on various factors, the most important being training. A HMM-based gene finder named SNAP was developed and is easily adaptable to a number of organisms (Korf 2004). This gene finder was used to evaluate genomes of *A. thaliana* and *Oryza sativa*. Earlier, three HMM-based gene finders—Exonomy, Unveil and GlimmerM—were presented (Majoros et al. 2003). These gene finders were trained with *Oryza sativa* and *A. thaliana* along with other organisms. Later, a HMM profile was developed (Feng and Xue 2006) to search the proteome of *O. sativa* L. ssp. japonica for serine carboxypeptidase (SCP) and serine carboxypeptidase-like (SCPL) protein. A total of 71 SCPs and SCPLs were found in rice.

Alternative splicing contributes to genome complexity and proteome diversity. Thus, study of alternative splicing sites in a genome is a topic of great interest. Experimental research in this field is both costly and time consuming. A hidden Markov model was used for genome-wide detection and analysis of alternative splicing for nucleotide binding site/leucine-rich repeat sequences in rice (Gu and Guo 2007). HMM-based searches were performed for nucleotide binding site and leucine-rich repeat (NBS-LRR) domain. Out of 875 NBS-LRR sequences obtained from The Institute for Genomic Research (TIGR), 119 (13.6 %) sequences had alternative splicing. Conversely, 71 intron retention events, 20 exon skipping events, 16 alternative termination events, 25 alternative initiation events, 12 alternative 5′ splicing events and 16 alternative 3′ splicing events were identified.

In another study, HMM was used in a neural network-based combinatorial model for prediction of optimal culture condition for maximum biomass yields in *Rauwolfia serpentina* hairy root cultures. Neural network approaches can be evaluated through spatial variations; there is no proper resolution for temporal variations. Nonlinear biological responses are affected by both spatial as well as temporal differences. Therefore, a stochastic approach where time-based differences are taken as random variables to evaluate the whole bioprocess should be considered (Mehrotra et al. 2013). In this study, five HMMs were derived for five test culture conditions and connected to the input layer of three-layered feed-forward neural network. The results of combinatorial ANN-HMM model simulation was compared with ANN model and it was observed that only 2.99 % deviation from experimental result was recorded from combinatorial model against 44 % recorded from ANN model.

Genetic Algorithms

Genetic algorithms were invented in the late 1960s by John Holland and colleagues and students at the University of Michigan in between 1960s and 1970s. Furthermore, John Holland was the first researcher who not only challenged but succeeded to put computational evolution and development on a firm hypothetical footing. Until this, theoretical foundation, based on the ideas of schemas, was the basis of almost all succeeding theoretical works on genetic algorithms. In the past few years, there have been extensive interactions among various researchers studying different computational methods, and existing boundaries between genetic algorithms, evolutionary programming, evolution strategies,, and related approaches have broken down to some extent. At present, the interpretations of genetic algorithm have changed to a very far meaning from John Holland's original conception. Genetic algorithms are an adaptive search heuristic algorithm based on the evolutionary ideas of genetic and natural selection. The basic idea of genetic algorithms is designed to simulate various processes in natural system, which is essential for evolution, specifically those processes which follow the doctrines laid down by Charles Darwin of fittest survival. They represent a sharp exploitation of random search space to solve any particular problem. This search heuristic is consistently used to generate useful solutions to search problems and optimization. GAs are the higher class of evolutionary algorithms, which is used to generate the solutions to optimization problems using the techniques evolved by natural evolution such as mutation, inheritance, crossover, and selection. Genetic algorithms not only provide alternate method for problem solving but also outperforms consistently other traditional methods. Many problems of real world involve optimal parameters which might become cumbersome for traditional methods but ideally suited to genetic algorithms. Because of its outstanding performance in optimization, genetic algorithms have been wrongly regarded as a function optimizer. There are many ways of viewing the genetic algorithms and perhaps the idea that most users come to make use of genetic algorithms for a problem solver is a restrictive view.

In contrast with evolutionary programming and evolution strategies, John Holland's original aim was to establish and design algorithms for solving specific problems, but to study the phenomenon of adaptation occurring in nature and to develop some ways to import the mechanism of natural adaptation into computer systems to run simulation and study the effect. John Holland's 1975 book *Adaptation in Natural and Artificial Systems* presented the genetic algorithm as a generalization of natural evolution and also gave theoretical structure for adaptation under the genetic algorithm. Genetic algorithm is a popular method for moving one inhabitant of chromosomes or individual of population (which is a computer program or algorithm capable of providing a potential solution of the problem) to a new inhabitant using a kind of natural selection with the influx of genetics and stimulated operators of crossover, inversion, and mutation. The individuals or the computer programs used in genetic algorithms are basically a set of rules arranged in the form of trees. These tree-like structures also called parse tree can be mutated and recombined with new variants. Thus, the machine evolves to find a solution of a complex problem using simple equations. These equations give different outputs from different inputs characteristic for different classes (Kell et al. 2001). The individuals or rules (algorithm/programs) from the population which are chosen by selection operator are allowed to reproduce, and a fitter individual is more likely to produce further offspring than the less fit one. For any defined problem to be solved, the simplest of genetic algorithm works as follows:

(a) Start with a randomly generated population of computer programs/algorithms.
(b) Evaluate the fitness f(x) of each individual in the population.
(c) Select individuals from the present population, the probability of selection being an increasing function of fitness. Selection of individuals is done with replacement, which means any particular chromosome can be selected more than once for it to become a parent one.
(d) Modify the individuals by mutation, recombination, or crossover.
    1. With the crossover rate or its probability, cross over the pair at an assorted selected pair to form two offspring. If no crossover takes place, form two offspring that are exact copies of respective parents. The thing, which is here to be noted, is that crossover rate is simply the probability of two-parent crossing over in a single point. There is some multi-point crossover of the genetic algorithm in which the crossover rate is the number of valid points for crossover to take place.
    2. Mutate the resulted offspring at locus with probability P_m, mutation rate or probability, and set the individual in new population.
(e) After this, replace the present population with new population.
(f) Go back to step b.

Each resultant of this process is called "generation"; genetic algorithm (GA) is typically iterated for anything in between 50 and 500 or could be even more generations. The entire set of required generation is called as "run." At the end of run, there is likelihood to get one or more fit chromosomes from the population. Randomness plays a vital role in run; that's why different runs with different number seeds are more likely to produce different behaviors of generations. Researchers on genetic algorithm often scale down observation based on various simulations averaged over many different runs on the same problem.

Applications of Genetic Algorithms in Plant Biology

Major application of genetic algorithms in fermentation technology and plant science has been for optimization of bioprocess. In a study a mobile robot was developed for harvesting fruit automatically (Noguchi and Terao 1997). Genetic algorithms were used to optimize methods for finding small paths in a given space. Neural networks have been for long used for optimization of bioprocess in fermentation technology. In plant biology, hybrid models of neural networks and genetic algorithms have been used for optimization of conditions for storage of fruits (Morimoto et al. 1997; Morimoto and Hashimoto 2000) and for detection of plant virus using biosensors to observe virus reactions (Glezakos et al. 2010). More complex in vitro culture processes such as shoot proliferation, root formation (rhizogenesis), and plantlet acclimatization have been modeled by ANNs and successfully optimized by genetic algorithms in woody fruit plants, such as kiwifruit (Gago et al. 2010a) and grapevine (Gago et al. 2010b).

Future Prospects

Biological systems are complex and nondeterministic and depend upon genetic and environmental factors. Due to the advent of modern technologies, we can generate a large amount of biological data. These large data have to be analyzed and interpreted to understand important relations between different factors. Machine learning techniques provide a very good scope to analyze these large biological data, interpret the obtained information, and give deep insight into the biological processes. These technologies can be used to develop models that can explain the relationship between different factors and biological responses, which can further be used to predict future responses in specific situations.

Machine learning techniques are good for determining nonlinear relationships followed in biological systems; thus, these techniques make better models than statistical techniques. Techniques like neural network require less prior data and are more accurate compared to statistical techniques (Gago et al. 2010b). Combinations of these techniques could be used for developing more accurate models which can predict outcome of tissue culture experiments, optimize and control bioprocess operations at large-scale, predict crop yields according to climate changes, etc.. These techniques are easy to understand and a plant biologist can very easily use these techniques by having a very good understanding of mathematical and statistical modeling.

References

Adriaenssens V, De Baets B, Goethals P, De Pauw N (2004) Fuzzy rule-based models for decision support in ecosystem management. Sci Total Environ 319(1):1–12

Agazzi OE, Kuo SS (1993) Hidden Markov model based optical character recognition in the presence of deterministic transformations. Pattern Recogn 26(12):1813–1826

Albiol J, Campmajó C, Casas C, Poch M (1995) Biomass estimation in plant cell cultures: a neural network approach. Biotechnol Prog 11:88–92

Alexander J, Mozer M (1999) Template-based procedures for neural network interpretation. Neural Netw 12:479–498

Andrews R, Diederich J, Tickle AB (1995) Survey and critique of techniques for extracting rules from trained artificial neural networks. Knowl-Based Syst 8:373–389

Archer NP, Wang S (1993) Application of the back propagation neural network algorithm with monotonicity constraints for two-group classification problems. Decis Sci 24:60–67

Baldi P, Brunak S (2001) Bioinformatics: the machine learning approach. MIT Press, Cambridge, MA

Borodovsky M, McIninch J (1993) GENMARK: parallel gene recognition for both DNA strands. Comput Chem 17:123–133

Borodovsky M, Rudd KE, Koonin EV (1994) Intrinsic and extrinsic approaches for detecting genes in a bacterial genome. Nucleic Acids Res 22(2):4756–4767

Borodovsky M, Mclninch JD, Koonin EV, Rudd KE, Médigue C, Danchin A (1995) Detection of new genes in a bacterial genome using Markov models for three gene classes. Nucleic Acids Res 23:3554–3562

Chuong BD, Serafim B (2008) What is the expectation maximization algorithm? Nat Biotechnol 26(8):897–899

Colbourn E (2003) Neural computing: enable intelligent formulations. Pharm Technol Suppl 16–20

Coppola EA, Rana AJ, Poulton MM, Szidarovszky F, Uhl VW (2005) A neural network model for predicting aquifer water level elevations. Ground Water 43(2):231–241

Coruzzi GM, Burga AR, Katari MS, Gutiérrez RA (2009) Systems biology: principles and applications in plant research. Annu Plant Rev 35:3–40

Denton JW (1995) How good are neural networks for casual forecasting? J Bus Forecast Methods Syst 14(2):17–20

D'haeseleer P (2005) How does gene expression clustering work? Nat Biotechnol 23:1499–1502

Eichner J, Zeller G, Laubinger S, Ratsch G (2011) Support vector machines-based identification of alternative splicing in Arabidopsis thaliana from whole-genome tiling arrays. BMC Bioinform 12(1):5

Excoffier L, Slatkin M (1995) Maximum-likelihood estimation of molecular haplotype frequencies in a diploid population. Mol Biol Evol 12:921–927

Feng Y, Xue Q (2006) The serine carboxypeptidase like gene family of rice (Oryza sativa L. ssp. japonica). Funct Integr Genomics 6(1):14–24

Filippi AM, Archibald R (2009) Support vector machine-based endmember extraction. IEEE Trans Geosci Remote Sens 47(3):771–791

Frossyniotis D, Anthopoulos Y, Kintzios S, Moschopoulou G, Yialouris C (2008) Artificial neural network selection for the detection of plant viruses. World J Agric Sci 4(1):114–120

Fukuda T, Shiotani S, Aral F, Takeuchi N, Sasaki K, Kimoshita T (1991) Cell recognition by image processing. 1st report. Recognition of dead or alive plant cells by neural network. Trans Jpn Soc Mech Eng Ser 57:77–84

Gago J, Landín M, Gallego P (2010a) Strengths of artificial neural networks in modeling complex plant processes. Plant Signal Behav 5(6):743–774

Gago J, Landín M, Gallego PP (2010b) Artificial neural networks modeling the *in vitro* rhizogenesis and acclimatization of *Vitis vinifera* L. J Plant Physiol 167:1226–1231

Gago J, Martínez-Núñez L, Landín M, Gallego P (2010c) Artificial neural networks as an alternative to the traditional statistical methodology in plant research. J Plant Physiol 167:23–27

Gallego PP, Gago J, Landín M (2011) Artificial neural networks technology to model and predict plant biology process. In: Suzuki K (ed) Artificial neural networks-methodological advances and biomedical applications. Intech, Rijeka, Carotia, pp 197–216

Gevrey M, Dimopoulos I, Lek S (2003) Review and comparison of methods to study the contribution of variables in artificial neural networks. Ecol Model 160(3):249–264

Glezakos TJ, Moschopoulou G, Tsiligiridis TA, Kintzios S, Yialouris CP (2010) Plant virus identification based on neural networks with evolutionary preprocessing. Comput Electron Agric 70:263–275

Gonzalez S (2000) Neural networks for macroeconomic forecasting: a complementary approach to linear regression models, Working paper. Department of Finance, Canada, pp 2000–2007

Gu L, Guo R (2007) Genome-wide detection and analysis of alternative splicing for nucleotide binding site-leucine-rich repeats sequences in rice. J Genet Genomics 34(3):247–257

Hanai T, Katayama A, Honda H, Kobayaski T (1997) Automatic fuzzy modelling for Ginjo sake brewing process using fuzzy neural network. J Chem Eng Jpn 30:94–100

Hayashi Y, Buckley JJ, Czogala E (1993) Fuzzy neural network with fuzzy signals and weights. Int J Intell Syst 8:527–537

Haykin S (2003) Neural networks: a comprehensive foundation, fourth Indian reprint. Pearson Education, Singapore

Hilbert DW, Ostendorf B (2001) The utility of artificial neural networks for modelling the distribution of vegetation in past, present and future climates. Ecol Model 146:311–327

Hill T, Marquez L, O'Connor M, Remus W (1994) Artificial neural network models for forecasting and decision making. Int J Forecast 10:5–15

Honda H, Takikawa N, Noguchi H, Hanai T, Kobayashi T (1997) Image analysis associated with a fuzzy neural network and estimation of shoot length of regenerated rice callus. J Ferment Bioeng 84:342–347

Honda H, Ito T, Yamada J, Hanai T, Matsuoka M, Kobayashi T (1999) Selection of embryogenic sugarcane callus by image analysis. J Biosci Bioeng 87(5):700–702

Hua S, Sun Z (2001) A novel method of protein secondary structure prediction with high segment overlap measure: support vector machine approach. J Mol Biol 308(2):397–408

Huang Y (2009) Advances in artificial neural networks – methodological development and application. Algorithms 2:973–1007

Jiménez D, Pérez-Uribe A, Satizábal H, Barreto M, Van Damme P, Marco T (2008) A survey of artificial neural network-based modeling in agroecology. In: Prasad B (ed) Soft computing applications in industry, STUDFUZZ. Springer, Berlin/Heidelberg, pp 247–269

Juang B-H, Rabiner LR (1991) Hidden Markov models for speech recognition. Technometrics 33(3):251–272

Karim MN, Yoshida T, Rivera SL, Saucedo VM, Eikens B, Oh G-S (1997) Global and local neural network models in biotechnology: application to different cultivation processes. J Ferment Bioeng 83(1):1–11

Karimi Y, Prasher S, Patel R, Kim S (2006) Application of support vector machine technology for weed and nitrogen stress detection in corn. Comput Electron Agric 51(1):99–109

Kaul M, Hill RL, Walthall C (2005) Artificial neural networks for corn and soybean yield prediction. Agric Syst 85:1–18

Kaundal R, Saini R, Zhao PX (2010) Combining machine learning and homology-based approaches to accurately predict subcellular localization in Arabidopsis. Plant Physiol 154(1):36–54

Kehagias A, Panagiotou H, Maslaris N, Petridis V, Petrou L (1998) Predictive modular neural networks methods for prediction of sugar beet crop yield. In: IFAC conference on control applications and ergonomics in agriculture, Athens, pp 1–5

Kell DB, Darby RM, Draper J (2001) Genomic computing. Explanatory analysis of plant expression profiling data using machine learning. Plant Physiol 126:943–951

Korf I (2004) Gene finding in novel genomes. BMC Bioinform 5:59

Krenker A, Bešter J, Kos A (2011) Introduction to the artificial neural networks. In: Suzuki K (ed) Artificial neural networks-methodological advances and biomedical applications, Intech, Rijeka, Carotia, pp 3–18

Krogh A, Brown M, Mian IS, Sjolander K, Haussler D (1994) Hidden Markov models in computational biology: applications to protein modeling. J Mol Biol 235(5):1501–1531

Kruschke JK, Movellan JR (1991) Benefits of gain: speeded learning and minimal hidden layers in back-propagation networks. IEEE Trans Syst Man Cybern 21(1):273–280

Landín M, Rowe R, York P (2009) Advantages of neuro-fuzzy logic against conventional experimental design and statistical analysis in studying and developing direct compression formulations. Eur J Pharm Sci 38:325–331

Lawrence CE, Reilly AA (1990) An expectation maximization (EM) algorithm for the identification and characterization of common sites in unaligned biopolymer sequences. Proteins 7:41–51

Lin M, Hu B, Chen L, Sun P, Fan Y, Wu P, Chen X (2009) Computational identification of potential molecular interactions in Arabidopsis. Plant Physiol 151(1):34–46

Mahendra V, Prasad VSS, Dutta Gupta S (2004) Trichromatic sorting of in vitro regenerated plants of gladiolus using adaptive resonance theory. Curr Sci 87:348–353

Majoros WH, Pertea M, Antonescu C, Salzberg SL (2003) GlimmerM, exonomy and unveil: three ab initio eukaryotic genefinders. Nucleic Acids Res 31(13):3601–3604

Mandic DP, Chambers J (2001) Recurrent neural networks for prediction: learning algorithms, architectures and stability. Wiley, Chichester/New York

Marcuzzo M, Quelhas P, Campilho A, Mendonça AM (2008a) Automatic cell segmentation from confocal microscopy images of the Arabidopsis root. ISBI 2008 5th IEEE international symposium on Biomedical Imaging: From Nano to Macro. 712–771

Marcuzzo M, Quelhas P, Campilho A, Mendonça AM, Campilho A (2008b) A hybrid approach for Arabidopsis root cell image segmentation. In: Campilha A, Kamel M (eds) Image analysis and recognition. Springer Berlin Heidelberg, pp 739–749

Marcuzzo M, Quelhas P, Campilho A, Maria Mendonça A, Campilho A (2009) Automated Arabidopsis plant root cell segmentation based on SVM classification and region merging. Comput Biol Med 39:785–793

Martinez PJ, Pérez RM, Plaza A, Aguilar PL, Cantero MC, Plaza J (2006) Endmember extraction algorithms from hyperspectral images. Ann Geophys 49(1):93–101

Mehrotra S, Prakash O, Mishra BN, Dwevedi B (2008) Efficiency of neural networks for prediction of in vitro culture conditions and inoculum properties for optimum productivity. Plant Cell Tissue Organ Cult 95:29–35

Mehrotra S, Prakash O, Khan F, Kukreja A (2013) Efficiency of neural network-based combinatorial model predicting optimal culture conditions for maximum biomass yields in hairy root cultures. Plant Cell Rep 32:309–317

Molto E, Harrell RC (1993) Neural network classification of sweet potato embryos. Optics Agric Forest, Proc SPIE 1836:239–249

Morimoto T, Hashimoto Y (2000) An intelligent control for greenhouse automation, oriented by the concepts of SPA and SFA – an application to a post-harvest process. Comput Electron Agric 29:3–20

Morimoto T, De Baerdemaeker J, Hashimoto Y (1997) An intelligent approach for optimal control of fruit-storage process using neural networks and genetic algorithms. Comput Electron Agric 18:205–224

Mozer MC, Smolensky P (1989) Using relevance to reduce network size automatically. Connect Sci 1:3–16

Nitze I, Schulthess U, Asche H (2012) Comparison of machine learning algorithms random forest, artificial neural network and support vector machine to maximum likelihood for supervised crop type classification. Proceedings of the 4th GEOBIA Janeiro, Brazil, 35–40

Noguchi N, Terao H (1997) Path planning of an agricultural mobile robot by neural network and genetic algorithm. Comput Electron Agric 18:187–204

Osama K, Somvanshi P, Pandey AK, Mishra BN (2013) Modelling of nutrient mist reactor for hairy root growth using artificial neural network. Eur J Sci Res 97(4):516–526

Prakash O, Mehrotra S, Krishna A, Mishra BN (2010) A neural network approach for the prediction of in vitro culture parameters for maximum biomass yields in hairy root cultures. J Theor Biol 265:579–585

Prasad V, Gupta SD (2006) Applications and potentials of artificial neural networks in plant tissue culture. In: Gupta SD, Ibaraki Y (eds) Plant tissue culture engineering. Springer Netherlands, pp 47–67

Resop JP (2006) A comparison of artificial neural networks and statistical regression with biological

resources applications. University of Maryland, College Park, USA
Russell SJ, Norvig P, Canny JF, Malik JM, Edwards DD (1995) Artificial intelligence: a modern approach. Prentice Hall, Englewood Cliffs
Ruan R, Xu J, Zhang C, Chi C-M, Hu W-S (1997) Classification of plant somatic embryos by using neural network classifiers. Biotechnol Prog 13:741–746
Salas J, Markus M, Tokar A (2000) Streamflow forecasting based on artificial neural networks. In: Govindaraju RS, Roa AR (eds) Artificial neural networks in hydrology. Springer Netherlands, pp 23–51
Shao Q, Rowe RC, York P (2006) Comparison of neurofuzzy logic and neural networks in modelling experimental data of an immediate release tablet formulation. Eur J Pharm Sci 28(5):394–404
Shi L, Duan Q, Ma X, Weng M (2012) The research of support vector machine in agricultural data classification. In: Li D, Chen Y (eds) Computer and computing technologies in agriculture V. Springer Berlin Heidelberg, pp 265–269
Shigidi A, Garcia LA (2003) Parameter estimation in groundwater hydrology using artificial neural networks. J Comput Civ Eng 17:281–289
Stanimirovic PS, Miladinovic MB (2010) Accelerated gradient descent methods with line search. Numer Algorithms 54(4):503–520
Suroso MH, Tani A, Hoami N, Takigawa H, Nishiura Y (1996) Inverse technique for analysis of convective heat transfer over the surface of plant culture vessel. Trans ASAE 39:2277–2282
Tani A, Murase H, Kiyota M, Honami N (1992) Growth simulation of alfalfa cuttings in vitro by Kalman filter neural network. Acta Hort 319:671–676
Tickle AB, Andrews R, Golea M, Diederich J (1998) The truth will come to light: directions and challenges in extracting the knowledge embedded within trained artificial neural networks. IEEE Trans Neural Netw 9:1057–1068
Uozumi N, Yoshino T, Shiotani S, Suehara K-I, Arai F, Fukuda T, Kobayashi T (1993) Application of image analysis with neural network for plant somatic embryo culture. J Ferment Bioeng 76:505–509
Venkataramanan L, Sigworth F (2002) Applying hidden Markov models to the analysis of single ion channel activity. Biophys J 82:1930
White H (1992) Artificial neural networks: approximation and learning theory. Blackwell Publishers, Inc., Oxford/Cambridge
Widrow B, Hoff M (1960) Adaptive switching circuits. In: 1960 IRE WESCON convention record, vol 4. IRE, New York, pp 96–104
Xiang C, Ding SQ, Lee TH (2005) Geometrical interpretation and architecture selection of MLP. IEEE Trans Neural Netw 16:84–96
Yang ZR (2006) A novel radial basis function neural network for discriminant analysis. IEEE Trans Neural Netw 17:604–612
Yao X (1999) Evolving artificial neural networks. Proc IEEE 87:1423–1447
Zealand CM, Burn DH, Simonovic SP (1999) Short term streamflow forecasting using artificial neural networks. J Hydrol 214(1):32–48
Zhang C, Timmis R, Hu W-S (1999) A neural network based pattern recognition system for somatic embryos of Douglas fir. Plant Cell Tissue Organ Cult 56:25–35

Applications of Bioinformatics in Plant and Agriculture

M.A. Iquebal, Sarika Jaiswal, C.S. Mukhopadhyay, Chiranjib Sarkar, Anil Rai, and Dinesh Kumar

Contents

M.A. Iquebal, Ph.D. • S. Jaiswal, Ph.D. (✉)
C. Sarkar, Ph.D. • A. Rai, Ph.D. • D. Kumar, Ph.D.
Centre for Agricultural Bioinformatics, Indian Agricultural Statistics Research Institute, New Delhi, India
e-mail: aijaiswal@gmail.com

C.S. Mukhopadhyay, Ph.D.
School of Animal Biotechnology, Guru Angad Dev Veterinary and Animal Sciences University, Ludhiana, Punjab 141004, India

Abstract

The high-throughput technologies generating large-scale biological data, as well as the development of related computational tools, have united global efforts and brought revolutionary changes to the research of biology during the last decade. Today, biologists work in association with scientists from a broad spectrum of disciplines to unravel how complex biological systems work. Bioinformatics is a multidisciplinary field that makes use of computers to store and analyse molecular biology information with integration of statistical algorithms. The genome sequencing of a number of organisms has led to the discovery of many fascinating things. Today, the world feels the need of this discipline to save resources and time. This chapter emphasises on a number of applications of bioinformatics in agriculture in view of functional genomics, data mining techniques, genome-wide association studies, high-performance computing facilities in agriculture and various bioinformatics tools/databases important for breeders, biotechnologists and pathologists. Agricultural genomics leads to the global understanding of plant/animal and pathogen biology, and its

D. Barh et al. (eds.), *PlantOmics: The Omics of Plant Science*,
DOI 10.1007/978-81-322-2172-2_27, © Springer India 2015

application would be beneficial for agriculture.

Keywords
Agriculture • Bioinformatics • HPC • GWAS • Machine learning

The great challenge in biological research today is how to turn data into knowledge. I have met people who think data is Knowledge but these people are then striving for a means of turning knowledge into understanding. (Sydney Brenner (2002) The Scientist. 16(6):12)

Introduction

Demand for agricultural products is increasing with the rapid increase of population to provide food and daily needs. The world has the production potential to cope with population demand. The projections of feeding a world population of 9.1 billion people in 2050 would require raising overall food production by around 70 % by 2050. Production in the developing countries needs to almost double, which implies significant increase in the production of many key commodities. For example, annual cereal production would have to grow by almost one billion tonnes and meat production by over 200 million tonnes to a total of 470 million tonnes in 2050 (Agriculture Development Economics Section, Rome, 2009). However, the developing countries are more dependent on agricultural imports. To improve the conditions, food security in many unprivileged areas requires substantial increases in local production. As evident, the area used for crop cultivation is limited, and there is hardly any scope of increasing cultivated area for agriculture to make up the demand. The only way out is the use of modern technologies in agricultural research and their application in the field level to increase production of agricultural commodities. The advent of the genomic era with massive amount of biological information with the aid of advances in the fields of molecular biology and genomics will certainly pave the way to such problems. Figure 1 shows the growth of biological data from its first release in 1987 to its 93rd

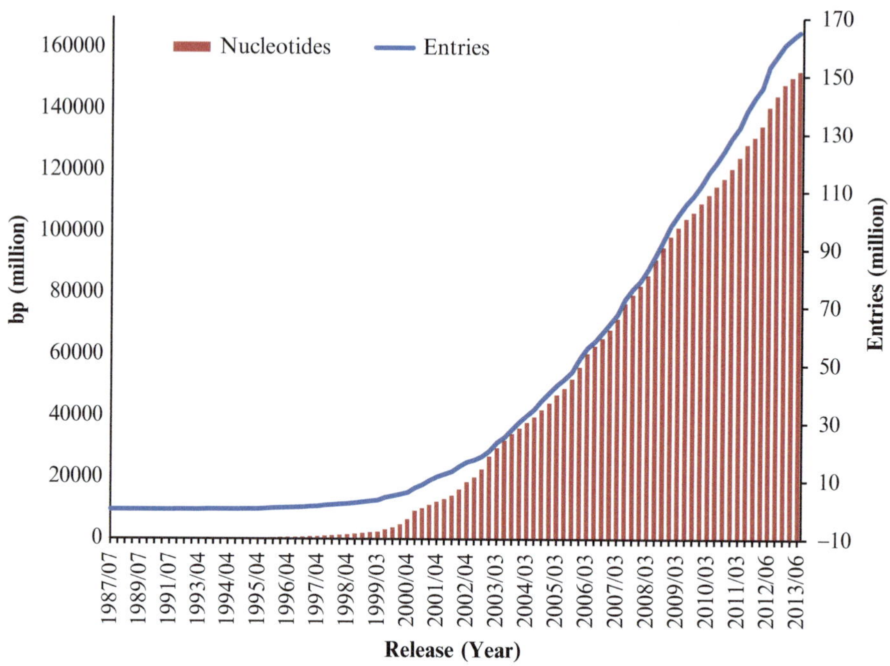

Fig. 1 Growth of biological data over the years (Source: http://www.ddbj.nig.ac.jp/breakdown_stats/dbgrowth-e.html#dbgrowth-graph)

release in 2013 (http://www.ddbj.nig.ac.jp/breakdown_stats/dbgrowth-e.html#dbgrowth-graph). It is obvious from Fig. 2 that the bases and sequences have increased 2.2 million fold and 2.5 million fold, respectively, from the year 1982 to 2013 (http://www.ncbi.nlm.nih.gov/genbank/statistics). With the science of informatics in biology, i.e. *bioinformatics*, handling of the information along with extraction of biological meaning from it can be met. This new knowledge will have profound impacts on various fields, viz. agriculture, human health, environment, energy and biotechnology. The increasing data volume requires computational methodologies to manage the data deluge. Hence, the prime requirement and challenge in this genomics era is to store and handle the staggering volume of information through computer databases giving birth to the discipline of *bioinformatics*. In other words, *bioinformatics* leads to quantitative analysis of information related to biological data with the aid of computers.

The history of the discipline of *bioinformatics* is interesting and dates back chronologically from the 1960s, when even the term *bioinformatics* was not even coined. The first breakthrough in this area can be traced under the project undertaken by Margaret Dayhoff in 1965, for the development of the first protein sequence database called *Atlas of Protein Sequence and Structure*. Later on, in the 1970s, the Protein Data Bank was established by Brookhaven National Laboratory which archived three-dimensional protein structures starting with hardly less than a dozen protein structures, compared to more than 30,000 structures as of today (Dayhoff et al. 1972). The next mega contribution was by Needleman and Wunsch in 1970 (Needleman and Wunsch 1970) with the sequence alignment algorithm, which was the fundamental step in paving the way to routine sequence comparisons and database searching practice by modern biologists. Further, the first protein structure prediction algorithm which later pioneered a series of developments in protein structure prediction was

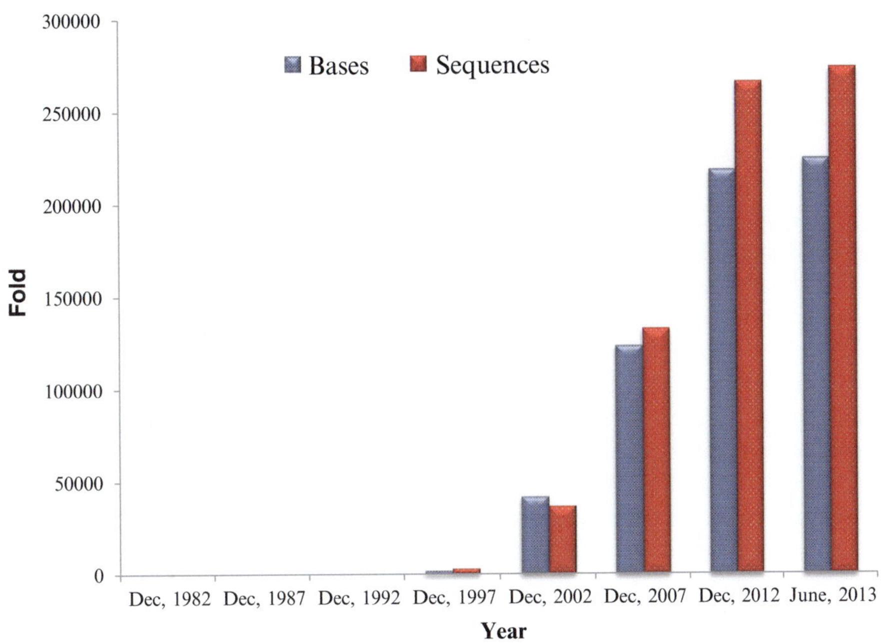

Fig. 2 Fold increase in biological data over the years (Source: http://www.ncbi.nlm.nih.gov/genbank/statistics)

proposed by Chou and Fasman in 1974 (Chou and Fasman 1974). The era of the 1980s saw the establishment of GenBank and the development of fast database searching algorithms, viz. FASTA by William Pearson and BLAST by Stephen Altschul et al. 1990 (Altschul et al. 1990). Then came the real breakthrough in this field with the human genome project (HGP) in the late 1980s. With the growing adaptability of widespread Internet in the 1990s, exchange and dissemination of biological data was made instant and easily accessible. Besides these, there are many more major boosts in the field of *bioinformatics*. The basic reason behind this subject gaining prominence is the advancement of genome studies that produced unprecedented amount of biological data. This data explosion generated a sudden demand for efficient computational tools to manage and analyse the data. The criteria which are looked upon by researchers to easily access these data stored in databases are

(a) Easy access to the information
(b) Extracting of only those information which are required to answer a specific biological question

Hence, the development of computational tools dependent on knowledge generated from a wide range of disciplines, viz. mathematics, statistics, computer science, information technology and molecular biology, gave birth to *bioinformatics*.

A biological database is a well-organised catalogue which stores and organises data so that information retrieval is made easy with various search criteria. Each record (or *entry*) contains a number of fields that hold the actual data items. To retrieve a particular record from the database, the user can specify the desirable information (called *value*) to be found in a particular field and expect the computer to retrieve the whole data record. This process is called *query*. The knowledge discovery is very much required in biological databases which refer to the identification of connections between pieces of information that were not known when information was first entered. To quote the example, databases containing information of the raw sequence can perform extra computational tasks to identify sequence homology or conserved motifs which facilitate the discovery of new biological insights from raw data. The extraction of valid information from wealthy data from various genome sequencing projects is the major challenge before the scientific community. The traditional approach of molecular biology research was restricted to the experimental laboratory bench, but the exponential growth of molecular data in this genomic era has realised the need to introduce computational and statistical approaches into the research process. Various new tools and databases in molecular biology help to carry out research at genomic, proteomic, transcriptomic and metabolomic levels.

It is possible to study the genomic and molecular level of agricultural crops with the help of modern technologies. The yields are reduced due to attack of insect pests, diseases caused by pathogens and other environmental factors in field situations. Different technologies can be used to make the crop resistant to diseases and pests and biotic and abiotic factors. All these technologies are dependent on the study of genomic, proteomic and metabolic pathways of different agricultural crops as well as the insect pests and pathogens and will not only help to know the cellular organisation of plant and insect pathogens but also the mode of attack of insect pathogens and the plant variety resistant to the disease/pest.

Tools for Bioinformatics and Applications

The pre-genome era has significantly contributed for the various biological databases. With the clear view of the structure of DNA by Crick and Watson, the field of molecular biology exploded and the increasing amount of information needed to be carefully managed and organised. The advent of sequencing methods in the late 1970s (Sanger and Coulson 1975; Maxam and Gilbert 1977) and initiation of the genome sequencing programs in the late 1980s led to massive flow of biological data which needed data management efforts.

Biological databases are broadly classified into sequence and structure databases. Sequence databases are applicable to both nucleic acid sequences and protein sequences, while structure database is applicable to only proteins. The era of molecular databases actually started in the 1960s with the online version of the Atlas of Protein Sequence and Structure by Margaret Dayhoff known popularly as the Protein Identification Resource (http://pir.georgetown.edu/). Numerous molecular biology databases are reported under the major categories like nucleotide sequence databases, RNA sequence databases, protein sequence databases, structure databases, genomics databases (non-vertebrate, human and other vertebrates), metabolic and signalling pathways, animal genes and diseases, microarray data and other gene expression databases, proteomics resources, organelle databases, immunological databases and so on. Table 1 describes the development of various biological databases in chronological order. Many biological databases bloomed and flourished, being very informative to researchers, and by the late 1990s, all of them operated primarily autonomously.

There are many reports of whole genome sequencing of agriculturally important crops and animals like rice, tomato, pigeon pea, chicken, bovine, etc. (Hillier et al. 2004; International Rice Genome Sequencing Project 2005; Sonstegard and Connor 2004; Barbazuk et al. 2005; Gill et al. 2004; Varshney et al. 2011; The Tomato Genome Consortium 2012) for their economic importance as well as being biomedical models (Prather et al. 2003; Anthony et al. 2003). After genome sequencing, one aims at identification and demarcation of the functional elements in the genome, i.e., *structural annotation* and linking these genomic elements to biological function known as *functional annotation*. For many of the genomes, their assemblies have lacunae leading to poor gene model predictions. Livestock genomes are known to have low build numbers compared with model organisms like humans and mice. Moreover, with lesser genomic data as compared to model organisms, the genome structural annotation is less likely to be successful for agricultural species (Eyras et al. 2005). Table 2 represents the whole genome sequencing projects in plants.

Gene Ontology (GO) is part of the Open Biomedical Ontologies (OBO) and a model for numerous other biological ontology projects that aim similarly to achieve structured, standardised vocabularies for describing various biological systems. It is a structured network consisting of defined terms and relationships between them that describe the molecular function, biological process and cellular component of gene products (http://www.geneontology.org/GO.doc.html). GO is the de facto standard for functional annotation and most popularly used for modelling microarray and other functional genomics data, but its use is still restricted in agricultural species (Lewis 2005) due to quite poor GO annotation. Further annotations for grasses and microbes are provided by Gramene (Ware et al. 2002) and TIGR (Yuan et al. 2003), respectively, but the European Bioinformatics Institute Gene Ontology Annotation (EBI-GOA) project provides most GO annotations for other agriculturally important species as well. Also, most GO annotations that do exist for agricultural proteins are reported as *inferred from electronic annotation* (IEA). IEA is basically applied to broad GO terms and results in very general superficial GO functional information. To override the limitations, *AgBase* (McCarthy et al. 2006) was developed to enable genome-wide structural and functional annotation and modelling of microarray and other functional genomics data in agricultural species. It integrates structural and functional annotations and provides tools in a simplified pipeline, allowing researchers to rapidly and effectively model and derive biological significance from microarray and other functional genomics data sets. AgBase is a functional genomics resource for agricultural species including Gene Ontology annotations, aided with various tools for analysis of large scale datasets. Both experimentally derived structural annotation and functional data in a unified resource are provided here. It is unique since the structural data provided is experimentally derived, the structural and functional data is provided from a unified resource, and tools for analysis of this data are freely available via *AgBase*.

Table 1 Biological databases

Sites	URL	Description
Protein Identification Resource (PIR)	http://iubio.bio.indiana.edu/soft/help/old/pir-server.help	It provides very rapid access to the database information and, with the addition of several sequence manipulation routines, serves as extremely powerful research tools
PDB Current Holdings	http://www.rcsb.org/pdb/holdings.html	The number of deposited structures includes structures that have not yet been released (i.e. structures on hold), structures that have been obsolete since they were released and theoretical models. The number of released structures shown does not include obsolete structures
Jackson Laboratory	http://www.jax.org/about/milestones.html	It is the first mammalian genetics database
GenBank	http://www.ncbi.nlm.nih.gov/Genbank/index.html	It is a vast repository and a public database of nucleic acid sequences, literature and genome-specific resources. Besides, it provides several biocomputational tools for sequence analysis and FTPs for sequence retrieval
NCBI-dbVar	http://www.ncbi.nlm.nih.gov/dbvar/	It is a database maintained by NCBI for genomic structural variations
NCBI Human Genome Browser	http://www.ncbi.nlm.nih.gov/mapview/map_search.cgi?chr=hum_chr.inf&query	The NCBI Map Viewer provides graphical displays of features on the human reference genome sequence assembly maintained by the GRC and the alternate HuRef genome assembly, as well as cytogenetic, genetic, physical and radiation hybrid maps. Map features seen along the sequence include genes, transcripts, contigs (the 'Contig' map), the BAC tiling path (the 'Component' map), STSs, FISH-mapped clones, ESTs and transcripts from several different organisms, Gnomon predicted gene models, etc.
SWISS-PROT	http://www.ebi.ac.uk/swissprot	It is a manually annotated reviewed section of the UniProtKB. It is a high-quality annotated and nonredundant protein sequence database, which brings together experimental results, computed features and scientific conclusions
Caenorhabditis elegans (AceDB)	http://www.acedb.org/	This database provides tools to give flexibility for the manipulation, display and annotation of the genomic data
The Institute for Genomic Research (TIGR)	http://www.tigr.org/about/history.shtml	The main focus is on improved molecular understanding of human health and disease
FlyBase	http://www.flybase.org	1993, primary repository of genetic and molecular data for the insect family Drosophilidae. These data types include mutant phenotypes, molecular characterisation of mutant alleles and other deviations, cytological maps and wild-type expression patterns
Ted Codd	http://www.nap.edu/readingroom/books/far/ch6.html	It marks the rise of relational databases
National Centre for Biotechnology Information (NCBI)	http://www.ncbi.nlm.nih.gov/	It is a database of primary genomic data repository and provides tools for genomic data analysis
NIH Human Microbiome Project	http://www.hmpdacc.org/resources/data_browser.php/	The Human Microbiome Project (HMP) is the initiative of the US National Institutes of Health with the goal of identifying and characterising the microorganisms which are found in association with both healthy and diseased humans
PubMed	http://www.ncbi.nlm.nih.gov/sites/entrez?db=PubMed	PubMed comprises more than 22 million citations for biomedical literature from MEDLINE, life science journals and online books. Citations may include links to full-text content from PubMed Central and publisher websites

(continued)

Table 1 (continued)

Sites	URL	Description
NCBI GenBank Taxonomy Database	http://www.ncbi.nlm.nih.gov/Taxonomy/taxonomyhome.html/	It provides taxonomical information of an organism
European Bioinformatics Information (EBI)	http://www.ebi.ac.uk	The EBI, a part of EMBL, is an academic research institute located on the Wellcome Trust Genome Campus in Cambridge (UK). It serves as a public repository of molecular data. It also provides free online bioinformatics software and tools
European Molecular Biology Laboratory (EMBL)	http://www.ebi.ac.uk/embl/	The EMBL Nucleotide Sequence Database (also known as EMBL-Bank) constitutes Europe's primary nucleotide sequence resource. Main sources for DNA and RNA sequences are direct submissions from individual researchers, genome sequencing projects and patent applications
DNA Databank of Japan (DDBJ)	http://www.ddbj.nig.ac.jp/	It is one of the three summit databanks that construct DDBJ/EMBL/GenBank International Nucleotide Sequence database through close collaboration of EBI in Europe and NCBI in the USA
Protein Data Bank (PDB)	http://www.rcsb.org/pdb/home/home.do	The PDB archive contains information about experimentally determined structures of proteins, nucleic acids and complex assemblies. As a member of the wwPDB, the RCSB PDB curates and annotates PDB data according to agreed-upon standards
Pfam	http://pfam.sanger.ac.uk/	The Pfam database is a large collection of protein families, each represented by multiple sequence alignments and hidden Markov models
PRINTS	http://www.bioinf.man.ac.uk/dbbrowser/PRINTS/index.php	PRINTS is a compendium of protein fingerprints. A fingerprint is a group of conserved motifs used to characterise a protein family; its diagnostic power is refined by iterative scanning of a SWISS-PROT/TrEMBL composite
BLOCKS	http://bioinformatics.weizmann.ac.il/blocks/	BLOCKS are multiply aligned ungapped segments corresponding to the most highly conserved regions of proteins. Block Searcher, Get Blocks and Block Maker are aids to detection and verification of protein sequence homology
Protein Information Resource (PIR)	http://pir.georgetown.edu/	It is the integrated protein informatics resource for genomics, proteomics and system biology research
PROSITE	http://www.expasy.ch/prosite/	PROSITE consists of documentation entries describing protein domains, families and functional sites as well as associated patterns and profiles to identify them. PROSITE is complemented by ProRule, a collection of rules based on profiles and patterns, which increases the discriminatory power of profiles and patterns by providing additional information about functionally and/or structurally critical amino acids
ProDom	http://prodom.prabi.fr/prodom/current/html/home.php	ProDom is a comprehensive set of protein domain families automatically generated from the UniProt Knowledge Database
InterPro	http://www.ebi.ac.uk/interpro/	InterPro provides functional analysis of proteins by classifying them into families and predicting domains and important sites
Multiple EM for Motif Elicitation (MEME)	http://meme.nbcr.net/meme/	MEME is a tool for discovering motifs in a group of related DNA or protein sequences

(continued)

Table 1 (continued)

Sites	URL	Description
Motif Alignment and Search Tool	http://tools.genouest.org/tools/meme/doc/examples/mastexample_output_files/mast.html#version	It is the program developed by Bailey and Gribskov in 1997 to find the most probable order and spacing of the patterns
SMART	http://smart.embl-heidelberg.de/	SMART is a classification scheme for identifying and analysing protein domains. The database is maintained by the EMBL
TrEMBL	http://www.expasy.ch/sprot/	It is a curated protein sequence database which strives to provide a high level of annotation (such as the description of the function of a protein, its domain structure, post-translational modifications, variants, etc.), a minimal level of redundancy and a high level of integration with other databases
Research Collaboratory for Structural Bioinformatics (RCSB)	http://home.rcsb.org/	The Research Collaboratory for Structural Bioinformatics (RCSB) undertakes research works directed towards understanding the function of biological systems through the study of the 3-D structure of biological macromolecules
RNAdb	http://research.imb.uq.edu.au/rnadb/	This database is a comprehensive mammalian noncoding RNA database (RNAdb) containing sequences and annotations for tens of thousands of noncoding RNAs. These include a wide range of microRNAs, small nucleolar RNAs and larger mRNA-like ncRNAs
Comparative RNA database	http://www.rna.ccbb.utexas.edu/	The Comparative RNA Web (CRW) Site disseminates information about RNA structure and evolution that has been determined using comparative sequence analysis
Genomic tRNA database	http://gtrnadb.ucsc.edu/	This genomic tRNA database contains tRNA gene predictions made by the program tRNAscan-SE on complete or nearly complete genomes
European rRNA database	http://www.psb.ugent.be/rRNA	This database compiles all complete or nearly complete SSU (small subunit) and LSU (large subunit) ribosomal RNA sequences, in aligned format. The alignment takes into account the secondary structure information derived by comparative sequence analysis
miRNA database	http://www.mirbase.org/	The miRBase database is a searchable database of published miRNA sequences and annotation. Each entry in the miRBase sequence database represents a predicted hairpin portion of a miRNA transcript with information on the location and sequence of the mature miRNA sequence (termed miR)
Genomes OnLine Database (GOLD)	http://www.genomesonline.org/cgi-bin/GOLD/index.cgi	GOLD is a worldwide web resource for comprehensive access to information regarding genome and metagenome sequencing projects and their associated metadata, around the world
Animal Genome Size Database	http://www.genomesize.com/	Animal Genome Size Database, Release 2.0, is a comprehensive catalogue of animal genome size data. Haploid DNA contents are currently available for 4,972 species (3,231 vertebrates and 1,741 non-vertebrates) based on 6,518 records from 669 published sources

(continued)

Table 1 (continued)

Sites	URL	Description
ArkDB	http://www.thearkdb.org/arkdb/	The ArkDB database system aims to provide a comprehensive public repository for genome mapping data from farmed and other animal species (viz. cat, deer, chicken, cow, duck, horse, pig, quail, etc.). It thus targets to provide a route into genomic and other sequences from the initial viewpoint of linkage mapping, RH mapping, physical mapping or QTL mapping data
Database of Genomic Variants archive (DGVa)	http://www.ebi.ac.uk/dgva/	The Database of Genomic Variants archive (DGVa) is a repository that provides archiving, accessioning and distribution of publicly available genomic structural variants, in all species
Kyoto Encyclopedia of Genes and Genomes (KEGG)	http://www.genome.jp/kegg/	It is a database resource for understanding high-level functions and utilities of the biological system, such as the cell, the organism and the ecosystem, from molecular-level information, especially large-scale molecular data sets generated by genome sequencing and other high-throughput experimental technologies
BovMap Link	http://locus.jouy.inra.fr/cgi-bin/bovmap/intro.pl	The site contains links to BovMap database and other genomic resources
EcoCyc	http://ecocyc.org/	It is a scientific database for the bacterium *Escherichia coli* K-12 MG1655. The EcoCyc project performs literature-based curation of the entire genome and of transcriptional regulation, transporters and metabolic pathways
Comprehensive Microbial Resource (CMR)	http://cmr.jcvi.org/tigr-scripts/CMR/CmrHomePage.cgi	It is a free website used to display information on all of the publicly available, complete prokaryotic genomes
Mouse Genome Informatics	http://www.informatics.jax.org	A very useful site that contains information, data and tools for mouse genome and analytical aspects (viz. gene, SNP, orthology, phenotype and disease model, expression, tumour, function, pathways, etc.)
NAGRP Pig	http://www.animalgenome.org/pig/maps/index.html	This is the website of NAGRP Pig Genome Coordination program (USDA, USA)
Vector Database	http://www.addgene.org/vector-database/	Vector Database is a digital collection of vector backbones assembled from publications and commercially available sources. This is a free resource for the scientific community that is compiled by Addgene. Only the plasmids deposited at Addgene are available for purchase through this website
Saccharomyces Genome Database (SGD)	http://www.yeastgenome.org/	SGD provides comprehensive integrated biological information for the budding yeast *Saccharomyces cerevisiae* along with search and analysis tools to explore these data, enabling the discovery of functional relationships between sequence and gene products in fungi and higher organisms
Intronless Gene Database	http://www.bioinfo-cbs.org/igd/description1.php	A highly curated database of eukaryotic intronless genes
Database resource of livestock	http://www.animalgenome.org/community/other.html	This website links to Animal Genomics Research Web Sites and Database Resources for almost all domestic animals and pets
Integrated Taxonomic Information System	http://www.itis.gov/	To find authoritative taxonomic information on plants, animals, fungi and microbes of North America and the world

(continued)

Table 1 (continued)

Sites	URL	Description
BuffSatDb	http://cabindb.iasri.res.in/buffsatdb/	By in silico microsatellite mining of whole genome, it is the first whole genome STR relational database of water buffalo with 910,529 microsatellite markers, developed using PHP and MySQL database. Being the first buffalo STR database in the world, this would not only pave the way in resolving current assembly problem but shall be of immense use for global community in QTL/gene mapping critically required to increase knowledge in the endeavour to increase buffalo productivity, especially for third-world countries where the rural economy is significantly dependent on buffalo productivity
PIPEMicroDB	http://cabindb.iasri.res.in/pigeonpea/	PIgeonPEa Microsatellite DataBase (PIPEMicroDB) is an automated primer designing tool for pigeon pea genome, based on chromosome-wise as well as location-wise search of primers, housing 123387 STRs

Numerous other bioinformatics software and tools are being used to maintain, analyse and retrieve the astronomical amount of molecular data. Following are some of the very useful software and tools applied in the research related to agricultural genomics:

Electrophoresis Image Analysis Software

A number of software is available to quantify, analyse and format the one- or two-dimensional electrophoresis gel images. DNA or RNA from different sources (*viz.* genomic DNA, PCR products, RE digested DNA, vectors, etc.) can be subjected to analysis. This software is used to provide mass and molecular weight analysis. It comparatively studies different bands, processes the bands, removes smiling effects, straightens lanes, alters lane settings, fits the bands within lanes to Gaussian curve and is used for data compilation The software used are.

1. *GelQuant software*: for one-dimensional gel analysis (http://www.amplcom.au/gq_frames.htm)
2. *MCID Elite*: image analysis software for autoradiography, densitometry, fluorescence, etc. (http://www.imagingresearch.com/products/MCID.asp)

Primer Designing

Primer designing is the first stage to step into the field of molecular genetics and PCR-based genome analysis. While designing primers, the specificity and efficiency of the primer(s) need attention. The *specificity* takes care of mispriming which may occur when primers are poorly designed. This results in 'non-specific amplification'. The specificity of a primer is determined by the length and sequence of the oligo, vis-à-vis by the sequence pattern (repetitive or single copy, part of a multi-gene family or not, etc.) of the template used in the PCR reaction. The *efficiency* of a primer pair is determined by the fold increase of amplicon in each cycle. An efficient primer pair implies almost a twofold increase in PCR product for each cycle of the PCR. The specificity and efficiency of a primer depend on several factors which must be taken into account while designing primers. The features of a 'good' primer constitute the rules adumbrated below:

(a) *Optimal primer length*: General PCR primers range between 18 and 24 bases. However, for multiplexing purpose the length may be as long as 30–35 bp, while primers used for random priming (viz. RAPD) are kept short, viz. 8- (octamers) to 12-mer, to promote random priming.

Table 2 Whole genome sequencing projects in plants

Crops	URL	Feature
Arabidopsis thaliana	http://www.arabidopsis.org/	It includes the complete genome sequence along with gene structure, gene product information, metabolism, gene expression, DNA and seed stocks, genome maps, genetic and physical markers
Populus trichocarpa	http://genome.jgi-psf.org/Poptr1_1/Poptr1_1.home.html	With a genome of size 500 million letters of genetic code, *Populus trichocarpa* was sequenced eight over to attain the highest quality standards
Arabidopsis lyrata	http://genome.jgi-psf.org/Araly1/Araly1.home.html	*A. lyrata* and *A. thaliana* are so close that ancestral states of polymorphisms in *A. thaliana* can be directly inferred from comparisons with *A. lyrata*, which will lead to better understanding of mutation and selection in plants
Capsella rubella	http://www.jgi.doe.gov/sequencing/why/3066.html	Sequencing *Arabidopsis lyrata* and *Capsella rubella*, close relatives of *Arabidopsis thaliana*
Brassica database	http://brassicadb.org/brad/	It is a web-based database of genetic data at the whole genome scale for important *Brassica* crops
Solanum lycopersicum	http://solgenomics.net/	It provides genome and sequence with detailed information of genome map, markers, phenotypes and breeder tool box
Solanum tuberosum	http://www.potatogenome.net/index.php/Main_Page	The Potato Genome Sequencing Consortium (PGSC) is an international group for sequencing of complete potato genome
Medicago truncatula	http://www.medicago.org/genome/	384 inbred lines spanning the range of *Medicago* diversity are being resequenced using Illumina next-generation technology
Lotus japonicus	http://www.kazusa.or.jp/lotus/	*Lotus japonicus* genome browser
Mimulus guttatus	http://www.jgi.doe.gov/sequencing/why/3062.html	*Mimulus guttatus*, a leading model system for studying ecological and evolutionary genetics in nature, is sequenced
Glycine max	http://www.phytozome.net/soybean.php	Large-scale shotgun sequencing of soybean
Manihot esculenta	http://www.phytozome.org/cassava.php	The goals of this project is to generate a draft sequence of the *Cassava* genome
Vitis vinifera	http://www.genoscope.cns.fr/externe/GenomeBrowser/Vitis/	Grape genome browser
Aquilegia formosa	http://www.jgi.doe.gov/sequencing/why/51280.html	Sequencing of *Aquilegia formosa* helps in ecological and evolutionary studies
Eucalyptus grandis	http://bioinformatics.psb.ugent.be/genomes/view/Eucalyptus-grandis	Whole genome shotgun sequencing and assembly will be carried out by JGI
Carica papaya	http://asgpb.mhpcc.hawaii.edu/papaya/	It is the Hawaii Papaya Genome Project, developing the resources necessary to map and clone papaya genes in an effort to improve the economic value and efficiency of agricultural cultivation
Ricinus communis	http://castorbean.jcvi.org/	Castor genome has been sequenced and assembled at 4X draft of the ~400 Mbp using a whole genome shotgun strategy

(continued)

Table 2 (continued)

Crops	URL	Feature
Oryza sativa japonica	http://rgp.dna.affrc.go.jp/E/IRGSP/index.html	International Rice Genome Sequencing Project (IRGSP) aims to obtain a high-quality, map-based sequence of the rice genome using the cultivar 'Nipponbare' of *Oryza sativa* ssp. *japonica*
Oryza sativa indica	http://rice.genomics.org.cn/rice/index2.jsp	The Beijing Genomics Institute (BGI) has sequenced rice genome. The comprehensive data of *Oryza sativa* L. spp. *indica* are also carefully curated
Zea mays	http://www.maizegdb.org/	MaizeGDB is a community-oriented, long-term, federally funded informatics service to researchers focused on the crop plant and model organism *Zea mays*
Sorghum bicolour	http://genome.jgi-psf.org/Sorbi1/Sorbi1.home.html	It contains transcript and protein sequences from the fully sequenced plant genome of *Sorghum bicolour*
Brachypodium distachyon	http://www.brachypodium.org/	First International *Brachypodium* Conference aims to bring together the global *Brachypodium* community to provide reports and discuss recent achievements in *Brachypodium*
Triticum aestivum	http://www.wheatgenome.org/	International Wheat Genome Sequencing Consortium (IWGSC) dedicated to sequencing wheat genome for advance in structural and functional genome analysis
Hordeum vulgare	http://www.public.iastate.edu/~imagefpc/IBSC%20Webpage/IBSC%20Template-home.html	International Barley Sequencing Consortium (IBSC) aims to physically map and sequence the barley gene space
Physcomitrella patens	http://genome.jgi-psf.org/Phypa1_1/Phypa1_1.home.html	Having the full *Physcomitrella* genome sequence is a great asset for reconstructing the evolution of plant genomes and for informing the community in the functional genomics of plants
Selaginella moellendorffii	http://genome.jgi-psf.org/Selmo1/Selmo1.home.html	The sequence of the *Selaginella* genome by JGI provides scientists an important reference genome necessary for deciphering the evolution of biochemical
Cyanidioschyzon merolae	http://merolae.biol.s.u-tokyo.ac.jp/	It is *Cyanidioschyzon merolae* Genome Project

(b) *Melting temperature* (T_m): It is the temperature at which 50 % of the DNA duplex dissociates become single stranded. In the case of primers, it can be defined as the temperature at which 50 % of the primer and its complementary nucleotides of the template are hybridised. A primer should anneal to the template before the template strands renature. Ideally it ranges between 52 and 62 °C; however, the G/C content of the primer is critical in determining the melting temperature.

(c) *Primer pair T_m mismatch*: Permissible T_m difference between the primers is less than 5 °C, preferably within 2 °C.

(d) *Non-specific amplification/cross-homology*: Primer pairs may amplify non-specifically (not intended to amplify) either due to chance factor or due to similarity of the sequence of some genes belonging to the same gene family or due to the presence of repeat sequence in the designed primers. The designed primers should be checked through Primer-BLAST or by Nucleotide BLAST (http://www.ncbi.nlm.nih.gov/) against nonredundant sequence database of NCBI (discussed later).

(e) *Primer G/C content*: In general, the optimal G/C content is between 45 and 55 %, with an acceptable range of 40–60 %. The G/C content ultimately determines the annealing temperature.

(f) *G/C clamp*: The 3′-terminus of the primer is very important, since the DNA amplification occurs in 5′- to 3′-direction. The G/C bond is more stable than A/T due to three hydrogen bonds in G/C (instead of two hydrogen bonds in A/T).

(g) *Max 3′-end stability*: It is rendered by the maximum ΔG of the 5 bases from the 3′-end of primers. Higher 3′-end stability improves priming efficiency; however, too high stability could negatively affect specificity because of 3′-terminal partial hybridisation induced non-specific extension. Hence, ΔG value less than –9 should be avoided.

(h) *Analysing Gibbs free energy (delta G)*: It measures the amount of work that can be extracted from a process operating at a constant pressure. Thus, it indicates the spontaneity of the reaction. In case of oligos, the ΔG represents the stability of secondary structure. It is the amount of energy required to break the secondary structure. More negative value (–6 kcal/mol versus –3 kcal/mol) for ΔG indicates stable, undesirable hairpins (for the first one).

(i) *Secondary structures in primers*: Secondary structures are the various combinations of the primers formed among themselves (self and heterologous) vis-à-vis loop-like structure production by the same primer. ΔG is the energy required to break the secondary structure, and larger negative values indicate a higher propensity for false priming as the 3′-end can initiate polymerisation even if the remainder of the primer does not bind well.
   - Hairpins: The acceptable ΔG for 3′-end hairpin should be higher than –2 kcal/mole (viz. –1.5, –1.0 kcal/mole will be better), while the ΔG should be more than –3 kcal/mole for internal hairpin.
   - Self-dimer (homodimer): The acceptable ΔG is more than –5 kcal/mole for 3′-end self-dimer and more than –6 kcal/mole for internal self-dimers. ΔG is a measure of the spontaneity of formation of a dimer between the internal regions of two same-sense primers. Therefore, ΔG is the energy required to break the secondary structure. Larger negative values (i.e. away from zero towards the negative side) indicate a higher inclination for identical primers to hybridise to each other rather than to the template (http://www.sigmaaldrich.com/).
   - Cross-dimer (heterodimer): The acceptable ΔG is more than –5 kcal/mole for 3′-end cross-dimer and –6 kcal/mole for internal cross-dimer.

(j) *Stretches of nucleotides:* Primers with long ‘polyG’ or ‘polyC’ stretches can promote mispriming. Runs of same bases, such as polynucleotides, increase the probability of primer–dimer and hairpin loop formation. PolyA and polyT opens the primer–template complex and makes it looser. The maximum acceptable number of runs of di- or mononucleotide is 4.

Table 3 shows the list of freely available online programs/applications and software related to primer designing.

Promoter Prediction

Promoter analysis helps to envisage certain fundamental questions regarding gene expression like determining the actual transcription start site (TSS), whether the upstream sequence overlaps another promoter, alternative splicing of a given gene, effect of mutation in promoter sequence on expression profile and the trait of interest, etc. Besides, promoter prediction has got direct bearing on expression studies and other relevant experiments like methylation study, sodium bisulphite sequencing, etc. Table 4 lists the various bioinformatics tools for this purpose.

Restriction Enzymes Analysis

The restriction enzymes have been widely used in genetic engineering, cloning experiments, polymorphism detection and for SNP detection. A number of software are in vogue to search for Types I and II REs as listed in Table 5.

Table 3 Tools and software for primer designing

Site name	URL	Description
FastPCR	http://primerdigital.com/fastpcr.html	It is an integrated tool for PCR primers or probe design, in silico PCR, oligonucleotide assembly and analyses, alignment and repeat searching. This program can be downloaded and run in personal computers
AutoPrime	http://www.autoprime.de/AutoPrimeWeb	It is a very useful software for designing Reverse Transcription Real-Time PCR (Q-RT-PCR) primers that are specific to the exon–intron boundaries
Primer3	http://frodo.wi.mit.edu/	It is a freely available online software for designing primers and probe from a DNA sequence. It is popular due to availability of several parameters to design primers with high specificity and accuracy
The PCR Suite	http://pcrsuite.cse.ucsc.edu/	It is an online primer designing software, hosted by UCSC, that allows users to design primers specific to various types of templates, viz. overlapping amplicons on a template, primers around SNP (in a GenBank), primers flanking exons and cDNA
Uniprime2	http://habanero.ucd.ie/uniprime2/	It is a website for universal primer designing
Primo Pro 3.2	http://www.changbioscience.com/primo/primo.html	It is an online primer designing software with a notable feature to reduce background noise by exercising check on mispriming on nontarget DNA sequence. It also introduces a batch mode option for high-throughput PCR primer design
MethPrimer	http://www.urogene.org/methprimer/index1.html	It is very useful site for designing primers for methylation PCR (denatured, single-stranded DNA (ssDNA) is modified with sodium bisulphite followed by PCR amplification using two pairs of primers, with one pair specific for methylated DNA, the other unmethylated DNA)
Oligo Analyzer Version 3.1	http://eu.idtdna.com/analyzer/applications/oligoanalyzer/	This online tool is provided by IDT for analysing the properties of the oligos as well as for predicting the likelihood of self- and heterodimer formation by oligos
IDT Antisense Design	http://www.idtdna.com/Scitools/Applications/AntiSense/Antisense.aspx	It is a tool to synthesise antisense oligos for a specific target sequence of interest
Primer-BLAST	http://www.ncbi.nlm.nih.gov/tools/primerblast/index.cgi?LINK_LOC=BlastHome	It is extensively used for designing primer and checking the specificity of a given primer
Oligo Properties Calculator	http://www.promega.com/biomath/calc11.htm	This tool calculates base count, thermodynamic properties (ΔS & ΔH), T_m, GC% values of a given oligo
Oligonucleotide Properties Calculator	http://www.basic.northwestern.edu/biotools/oligocalc.html	This tool displays the reverse complementary sequence, physical properties (length, molecular weight, GC%), T_m, thermodynamic constants and hairpin and self-dimer production by a given primer/sequence
UNAFold	http://www.idtdna.com/Scitools/Applications/unafold/	The likelihood of secondary structure formation by the single-stranded target is checked by UNAFold software of IDT

Table 4 Tools for promoter prediction and analysis

Site name	URL	Description
Transcriptional Regulatory Element Database	http://rulai.cshl.edu/cgi-bin/TRED/tred.cgi?process=home	This database is a good resource to obtain training data sets for genome-wide *cis*-regulatory element prediction, gene functional studies and exploring gene regulatory networks
Promoter Analysis Pipeline (PAP)	http://bioinformatics.wustl.edu/webTools/PromoterAnalysis.do	It is used for analysing set of co-expressed genes vis-à-vis predicting the transcriptional regulatory mechanisms
Promoter 2.0 Prediction Server	http://www.cbs.dtu.dk/services/Promoter/	It predicts transcription start sites of vertebrate PolII promoters in DNA sequences. It has been developed as an evolution of simulated transcription factors that interact with sequences in promoter regions and is built on principles that are common to neural networks and genetic algorithms
Promoters and Terminators	http://molbiol-tools.ca/Promoters.html	This site maintains links for different software and tools (viz. PromScan, SCOPE, Promoser, ARNold, WebGeSTer) for scanning and predicting promoters and transcription terminators in eukaryotes and prokaryotes

Table 5 Tools for restriction enzymes analysis

Site name	URL	Description
RestrictionMapper	http://www.restrictionmapper.org/	This is an online, freely available tool for mapping restriction endonuclease sites on a DNA sequence
Webcutter 2.0	http://rna.lundberg.gu.se/cutter2/	This is an online RE site detection software for linear and circular DNA
Mapper	http://arbl.cvmbs.colostate.edu/molkit/mapper/index.html	It is a Java platform-based online software to map the RE sites on a target sequence
NEBcutter	http://tools.neb.com/NEBcutter2/index	This software is an RE site mapper, hosted by New England Biolabs
Web Map	http://pga.mgh.harvard.edu/web_apps/web_map/start	This software maps RE sites for a given sequence (circular or linear). It also produces the reverse complementary of input sequence, which can be assigned for mapping of RE sites

Phylogenetic Studies

Phylogenetic analysis is of immense importance in evolutionary studies as well as determining closeness of molecular sequences. A large number of packages are available for this purpose as enlisted in Table 6.

Statistical Genetics

Data generated in molecular genetics are subjected to statistical analysis to draw logical inference on the population parameters. Table 7 shows several such software packages used for different types of data and various purposes.

Genome Annotation and Gene Prediction

It is a very important tool in bioinformatics that enables us to understand several features and functions of a biological entity, viz. function(s) of the protein, post-translational modification(s), domains and sites, secondary and quaternary structures, similarities to other proteins, disease(s) associated with deficiencies in the pro-

Table 6 Tools for phylogenetic analysis

Tool	Link	Description
Phylogeny Inference Package (v. 3.5)	http://cmgm.stanford.edu/phylip/	This is for inferring phylogenies and carrying out certain related tasks. At present it contains 31 programs, which carry out different algorithms on different kinds of data
Molecular Evolutionary Genetic Analysis (v. 5.1 beta)	http://www.megasoftware.net/	A package for analysing sequence data for pairwise and multiple sequence alignment, phylogenetic tree (including neighbour-joining, maximum parsimony, UPGMA, maximum likelihood and minimum evolution-based) construction and estimation of evolutionary parameters
APE	http://www.ird.fr/	R-Project package for analysis of phylogenetics and evolution
Armadillo Workflow Platform	http://adn.bioinfo.uqam.ca/armadillo/	Workflow platform dedicated to phylogenetic and general bioinformatics analysis using distance, maximum likelihood, maximum parsimony and Bayesian methods
BAli-Phy	http://www.biomath.ucla.edu/msuchard/bali-phy/	Simultaneous Bayesian inference of alignment and phylogeny
BATWING	http://www.mas.ncl.ac.uk/~nijw/	Bayesian Analysis of Trees With Internal Node Generation using Bayesian inference, demographic history and population splits
BayesPhylogenies	http://www.evolution.rdg.ac.uk/BayesPhy.html	Bayesian inference of trees using Markov Chain Monte Carlo methods
BayesTraits	http://www.evolution.rdg.ac.uk/BayesTraits.html	Analyses of trait evolution among groups of species for which a phylogeny or sample of phylogenies is available
Bayesian Evolutionary Analysis Sampling Trees (BEAST)	http://beast.bio.ed.ac.uk/Main_Page	Bayesian inference, relaxed molecular clock and demographic history
Bosque	http://bosque.udec.cl/	Integrated graphical software to perform phylogenetic analyses, from the importing of sequences to the plotting and graphical edition of trees and alignments using distance and maximum likelihood methods
BUCKy	http://www.stat.wisc.edu/~ane/bucky/index.html/	Bayesian concordance using modified greedy consensus of unrooted quartets
ClustalW	http://www.ch.embnet.org/software/ClustalW.html	Progressive multiple sequence alignment based on distance matrix/nearest neighbour
fastDNAml	ftp://ftp.bio.indiana.edu/molbio/evolve/fastdnaml/fastDNAml.html	Optimised maximum likelihood (nucleotides only) on maximum likelihood approach
Geneious		Geneious provides genome and proteome research tools based on neighbour-joining, UPGMA, MrBayes plugin, PHYML plugin
HyPhy	http://www.hyphy.org/	Hypothesis testing using phylogenies by maximum likelihood, neighbour-joining, clustering techniques and distance matrices
IQPNNI	http://www.cibiv.at/software/iqpnni/	Iterative ML tree search with stopping rule by maximum likelihood and neighbour-joining
jModelTest 2	https://code.google.com/p/jmodeltest2/	A high-performance computing program to carry out statistical selection of best-fit models of nucleotide substitution
LisBeth	http://lis-upmc.snv.jussieu.fr/lis/?q=en/resources/software/lisbeth	Three-item analysis for phylogenetics and biogeography

(continued)

Table 6 (continued)

Tool	Link	Description
Molecular Evolutionary Genetics Analysis (MEGA)	http://www.megasoftware.net/	This tool is based on distance, parsimony and maximum composite likelihood methods
Mesquite	http://www.mesquiteproject.org/mesquite/mesquite.html	Mesquite is a software for evolutionary biology, designed to help biologists analyse comparative data about organisms. Its emphasis is on phylogenetic analysis, but some of its modules concern population genetics, while others do non-phylogenetic multivariate analysis
MetaPIGA2	http://www.metapiga.org/	Maximum likelihood phylogeny inference multicore program for DNA and protein sequences and morphological data. Analyses can be performed using an extensive and user-friendly graphical interface or by using batch files. It also implements tree visualisation tools, ancestral sequences and automated selection of best substitution model and parameters
ModelGenerator	http://bioinf.nuim.ie/modelgenerator/	Model selection (protein or nucleotide) based on maximum likelihood approach
MrBayes	http://mrbayes.sourceforge.net/index.php	This is for posterior probability estimation with the concept of Bayesian inference
Network	http://www.fluxus-engineering.com/sharenet.htm	Free phylogenetic network software
Phylogenetic Analysis by Maximum Likelihood (PAML)	http://abacus.gene.ucl.ac.uk/software/paml.html	This tool is based on maximum likelihood and Bayesian inference
PartitionFinder	http://robertlanfear.com/partitionfinder/	Combined selection of models of molecular evolution and partitioning schemes for DNA and protein alignments
Phylogenetic Analysis Using Parsimony (PAUP)	http://paup.csit.fsu.edu/	This is the phylogenetic analysis based on maximum parsimony, distance matrix and maximum likelihood
PHYLogeny Inference Package (PHYLIP)	http://evolution.genetics.washington.edu/phylip.html	This is the phylogenetic analysis based on maximum parsimony, distance matrix and maximum likelihood
PhyloQuart	http://www.lirmm.fr/~vberry/PHYLOQUART/phyloquart.php	Quartet implementation (uses sequences or distances) using quartet method
ProtTest 3	https://bitbucket.org/diegodl/prottest3/wiki/Home	A high-performance computing program for selecting the model of protein evolution that best fits a given set of aligned sequences
PyCogent	http://pycogent.org/	Software library for genomic biology
QuickTree		Tree construction optimised for efficiency
RAxML-HPC	http://www.exelixis-lab.org/	Randomized Axelerated Maximum Likelihood for High Performance Computing (nucleotides and amino acids)
TreeGen	http://www.cbrg.ethz.ch/services/TreeGen	Tree construction given precomputed distance data
TreeAlign	http://mobyle.pasteur.fr/cgi-bin/portal.py?#forms	Efficient hybrid method based on distance matrix and approximate parsimony
Treefinder	http://www.treefinder.de/	Fast ML tree reconstruction, bootstrap analysis, model selection, hypothesis testing, tree calibration, tree manipulation and visualisation, computation of sitewise rates, sequence simulation, many models of evolution (DNA, protein, rRNA, mixed protein, user-definable), GUI and scripting language
TREE-PUZZLE	http://www.tree-puzzle.de/	This tool is based on maximum likelihood and other statistical analyses

Table 7 Tools for statistical genetics

Site name	URL	Description
QTL mapping software	http://www.rqtl.org/	A QTL mapping software (in experimental crosses) in R environment
Haploview	http://www.broadinstitute.org/scientific-community/science/programs/medical-and-populationgenetics/haploview/haploview	This is designed to simplify and expedite the process of haplotype analysis by providing a common interface to several tasks relating to such analyses, like haplotype population frequency estimation, single SNP and haplotype association tests, permutation testing for association significance, implementation of Paul de Bakker's Tagger tag SNP selection algorithm, etc.
Python for Population Genomics (PyPop)	http://www.pypop.org/	This aids in large-scale population genetic analyses like (1) conformity to Hardy–Weinberg expectations, (2) tests for balancing or directional selection and (3) estimates of haplotype frequencies (and their distributions) and measures and tests of significance for linkage disequilibrium (LD)
Haplotype analysis (HapStat)	http://www.bios.unc.edu/~dlin/hapstat/	It is a user-friendly software interface for the statistical analysis of haplotype–disease association which allows users to estimate or test haplotype effects and haplotype–environment interactions by maximising the (observed-data) likelihood that properly accounts for phase uncertainty and study design
PopGene	http://www.ualberta.ca/~fyeh/popgene_download.html	A user-friendly computer freeware for the analysis of genetic variation among and within populations using codominant and dominant markers, developed by Francis Yeh, Rongcai Yang and Timothy Boyle
Micro-Checker	http://www.microchecker.hull.ac.uk/	It is a Windows application that checks for microsatellite null alleles and scoring errors. It also provides null allele estimates and adjusts allele and genotypes frequencies
Microsatellite Analyzer (MSA) 4.05	http://i122server.vuwien.ac.at/MSA/MSA_download.html	MSA is a universal, platform-independent, data analysis tool designed to handle large microsatellite data sets. It calculates the standard suit of descriptive statistics and provides input files for other software packages
Bayesian QTL mapping software for inbred lines	http://www.rni.helsinki.fi/~mjs/	This contains link to Bayesian QTL mapping software
Quantitative Trait Loci Mapping Software	http://www.stat.wisc.edu/~yandell/statgen/software/biosci/qtl.html	This provides discussion and provision of links to some important QTL mapping software
NTSYSpc, Numerical Taxonomy System, Version 2.2	http://www.exetersoftware.com/cat/ntsyspc/ntsyspc.html	It is commercially available and is used to discover pattern and structure in multivariate data. Thus it is helpful for clustering of data and dendogram construction
GenAlEX 6.5 software	http://biology.anu.edu.au/GenAlEx/Welcome.html	This offers a wide range of population genetic analysis options for the full spectrum of genetic markers within the Microsoft Excel environment on both PC and Macintosh computers

tein, sequence conflicts, variants and several other relevant information. Literally 'annotation' means to combine comments, notations, references and citations in a given format that collectively designates the features about a gene or a protein (Manikanandakumar 2009). Genome annotation is thus a pragmatic approach of ascribing pertinent biological information to sequences. The purpose of genome annotation is to predict the novel genes and characterise these sequences bio-computationally.

Nucleotide sequences belonging to coding region can be predicted either by *homology-based approach* or *ab initio approaches.* The homology-based approach has got some limitations due to requirement of wider coverage of transcriptomic and proteomic data vis-à-vis generation of false-positive results. The alternative approach, especially for the organisms devoid of reference genome (like buffalo, at present), is ab initio gene finding. Complex probabilistic models are used for gene finding in both prokaryotes and eukaryotes. Some of the popular software used for gene finding are GLIMMER (http://cbcb.umd.edu/software/glimmer/) and GeneMark (http://opal.biology.gatech.edu/GeneMark/) in prokaryotes and GENSCAN and geneid (http://genome.crg.es/software/geneid/) in eukaryotes. A few recent approaches like mSplicer, CONTRAST or mGene also use machine learning techniques like support vector machines for successful gene prediction. A list of gene prediction software is tabulated in Table 8.

Biological Data Mining Techniques

Data mining is the science of extracting useful information from large data sets or databases using computational techniques from statistics, machine learning and pattern recognition. The complexity of biological data varies from simple strings (nucleotides/amino acid sequences) to complex graphs (biochemical networks/pathways); from 1-dimensional (sequence data) to 3-dimensional (protein and RNA structures). Considering the amount and complexity of the data, it is becoming tough for an expert to compute and compare the entries within the current databases. Classification techniques are increasingly being used to address problems in computational biology and bioinformatics. Novel computational techniques to analyse high-throughput data in the form of sequences, gene and protein expressions, pathways and images are becoming vital for understanding diseases and future drug discovery. Machine learning techniques such as Markov models, support vector machines, neural networks and graphical models have been successful in analysing life science data because of their capabilities in handling randomness and uncertainty of data noise and in generalisation. Classification techniques in bioinformatics are an indispensable resource for computer scientists, engineers, biologists, mathematicians, researchers, clinicians and physicians.

Machine Learning Algorithms

Machine learning is supposed to be the descendant of statistical model fitting dealing with computers for optimization of performance using example data or past experience. This falls under the category of supervised learning. Machine learning utilizes statistical theory while building computational models since it has to make inference from a sample. Training set comprises of finite observations labelled with class knowledge. The observations are represented in terms of feature and each observation is a set of measurements, one for each such feature. This constitutes a feature vector. Each observation can be viewed in multidimensional space spanned by the features known as feature space. In the field of bioinformatics, machine learning is gaining much applicability due to ability of learning algorithms to construct classifiers to explain the complexity of biological data.

In two-class supervised classification, there is a *feature vector* $X \in Rn$ whose components are called *predictor variables* and a *label* or *class* variable $C \in \{0, 1\}$. Hence, the task is to induce classifiers from *training data*, which consists of a set of N independent observations

Table 8 Gene prediction tools

Site name	URL	Description
Promoter 2.0 Prediction Server	http://www.cbs.dtu.dk/services/Promoter/	Promoter 2.0 predicts transcription start sites of vertebrate PolII promoters in DNA sequences. It has been developed as an evolution of simulated transcription factors that interact with sequences in promoter regions. It builds on principles that are common to neural networks and genetic algorithms
Splice predictors	http://deepc2.psi.iastate.edu/cgibin/sp.cgi	Tool to identify potential splice sites in (plant) pre-mRNA by sequence inspection using Bayesian statistical models
FGENESH	http://linux1.softberry.com/berry.phtml?topic=fgenesh&group=help&subgroup=gfind	This tool is for predicting multiple genes in genomic DNA sequences
GeneMark	http://exon.gatech.edu/	It is a family of gene prediction programs developed at Georgia Institute of Technology, Atlanta, Georgia, USA
HMMgene 1.1 web server	http://www.cbs.dtu.dk/services/HMMgene/hmmgene1_1.php	HMMgene is a program for prediction of genes in anonymous DNA. The program predicts whole genes, so the predicted exons always splice correctly. It can predict several whole or partial genes in one sequence, so it can be used on whole cosmids or even longer sequences
GAS	http://users.ox.ac.uk/~ayoung/gas.html	GAS is a UNIX- or DOS-based downloadable, command-line oriented integrated computer program designed to automate and accelerate the acquisition and analysis of genomic data
GENSCAN	http://genes.mit.edu/GENSCAN.html	GENSCAN is a freely available software used for identification of complete gene structures in genomic DNA and can be used for predicting the locations and exon–intron structures of genes in genomic sequences from a variety of organisms
Gene Locator and Interpolated Markov ModelER (GLIMMER)	http://www.cbcb.umd.edu/software/glimmer/	It is a system for finding genes in microbial DNA, especially the genomes of bacteria, archaea and viruses. This tool uses interpolated Markov models (IMMs) to identify the coding regions and distinguish them from noncoding DNA
Genome Bioinformatics Research Lab	http://genome.imim.es/geneid.html	The site harbours the geneid program which is used to predict genes, exons, splice sites and other signals along a DNA sequence. This site is also hyperlinked with gene prediction on whole genome which is a precomputed whole genome prediction data set
geneid	http://genome.crg.es/software/geneid/	It predicts genes in anonymous genomic sequences designed with a hierarchical structure
mGene	http://www.mgene.org/	It is a computational tool for the genome-wide prediction of protein-coding genes from eukaryotic DNA sequences
Ensembl Genome Browser	http://asia.ensembl.org/index.html?redirect=mirror;source=www.ensembl.org	The Ensembl project produces genome databases for vertebrates and other eukaryotic species and makes this information freely available online
Ensemble	http://www.ensemble.org/	Ensemble, a genome browser like NCBI and UCSC, is a centralised repository of the Whole Genome Sequence of human and other species (vertebrate and model organisms)

(continued)

Table 8 (continued)

Site name	URL	Description
UCSC Human Genome Browser	http://genome.cse.ucsc.edu/cgibin/hgGateway/	It is an interactive genome browser dedicated to human genome sequence
UCSC Genome Browser	http://genome.ucsc.edu/	This is an online genome browser hosted by the University of California, Santa Cruz (UCSC). It is an interactive website offering access to genome sequence data from a variety of vertebrate and invertebrate species and major model organisms, integrated with a large collection of aligned annotations

Source: http://en.wikipedia.org/wiki/List_of_gene_prediction_software

$DN = \{(\mathbf{x}^{(1)}, c^{(1)}), \ldots, (\mathbf{x}^{(N)}, c^{(N)})\}$ drawn from the joint probability distribution $p(\mathbf{x}, c)$. The classification model will be used to assign labels to new instances according to the value of its predictor variables.

To simplify, supervised classification is a technique based on the principles of machine learning in which parameters of inferring a function is estimated based on training data such that a set of input vector, which consists of realised values of explanatory factors, is being used to get desired output values of dependent factors with desired accuracy. This function is also called as *classifier*. This inferred function is expected to predict correct output value for any valid input vector. This means it requires the learning algorithm to generalise from the training data to unseen situations with desired accuracy. In order to develop reliable inferred function, the following steps need to be followed:

1. Selection of an appropriate training data set as a representative of the real world problem under consideration along with representative sets of output values
2. Selection of input features (factors) capable of predicting the output with desired accuracy but preferably not too large in number
3. Determination of structure of the function and corresponding learning algorithms based on optimised performance through cross-validation techniques on subset of training data set which is also known as validation set
4. Evaluation of the accuracy of the learned function after parameter adjustments on a test data set which is different from the training set

Although large numbers of supervised learning algorithms are available in literature with their advantages and disadvantages, there is no single algorithm which can be used on all types of data sets.

Clustering

Clustering falls under the category of unsupervised learning and is the method of identifying natural, underlying classes from a set of observations. This is used where there is no knowledge of class. Clustering methods may be either iterative or hierarchical. The *k*-means algorithm falls under the iterative approach. The methodology begins with a set of *k* randomly chosen clusters of observations and iteratively calculating the center of each cluster which is also known as centroid. Further, each observation is assigned to the cluster defined by the closest centroid. This procedure repeats until no more observations change clusters. This algorithm requires less memory and hence is fast. But it depends on initial number and configuration of clusters.

The other popular hierarchical clustering method is agglomerative hierarchical clustering. This algorithm starts with the observations as one cluster and consecutively merges the two most similar clusters till all come within a large cluster. The distance between two clusters is calculated either as the average distance between all pairs of observations in the two clusters (average linkage) or longest/shortest distance between two observations in the two clusters (complete/single linkage). This results into a dendrogram. As compared to the *k*-means algorithm, this method is slower and uses more memory.

Bayes Classification

Bayes classification rule is based on probability. It states that an observation is to be assigned to the class with highest probability, provided the probability distribution of feature vectors in each class. This rule often results in optimal error rate for classification, that is, fraction of training observations classified to the wrong class. The true probability distribution is generally unknown and its estimation is required. This problem of estimating distributions from training data makes it unpopular. For estimation of probability distributions, parametric and non-parametric methods exist. Parametric method considers the assumption of distribution structure and calculates its parameters from the data while non-parametric method is based on constructing histograms from the data using k-nearest neighbour density estimation or simulation methods like Monte Carlo simulation or bootstrapping.

Artificial Neural Networks

An artificial neural network (ANNs) is the popular method for nonlinear problems, based on networks of *perceptron*. A perceptron is a simple computational unit that multiplies each input value with a weight and sums up the products. In principle, the output from the perceptron is zero if the sum is less than a particular threshold. A simple ANN model consists of three layers of nodes, viz. input, hidden and output layers, and allows connection of each node in one layer with every other node in the next layer. Training of network is done by back-propagation algorithm for estimating the functional relationship between inputs and outputs using supervised learning and by means of adjusting/estimating the weights (strength of connections between the nodes) associated between the nodes at all iterations in order to minimise the sum of squared errors. The net input into a node is given by

$$\text{Net input}_i = \sum \left(w_{ij} * \text{output}_j \right) + u_i$$

where wij are weights connecting neuron j to neuron i, output$j$ is the output from the unit j and ui is a threshold for neuron i. Each unit takes its net input and applies an activation function to it. For example, suppose the output of the jth unit $g(\sum w_{ij}x_i)$, where $g(.)$ is activation function, where x_i is output of the ith unit connected to unit j. The important activation functions generally used are identity, tanh, logistic, exponential and sine (Hertz et al. 1991). The schema of this process can be represented as in Fig. 3.

The most popular form of ANN architecture is the multilayer perceptron (MLP). Typically, it consists of a set of source nodes that constitute the input layer, one or more hidden layers of computation nodes and an output layer of computation nodes. The input signal propagates through the network in a forward direction on a layer by layer basis. Another quite popular ANN architecture is radial basis function (RBF), which has a very strong mathematical foundation rooted in regularisation theory for solving ill-conditioned problems. An RBF, almost invariably, consists of three layers: a transparent input layer, a hidden layer with sufficiently large number of nodes and an output layer. As its name implies, radially symmetric basis function is used as activation

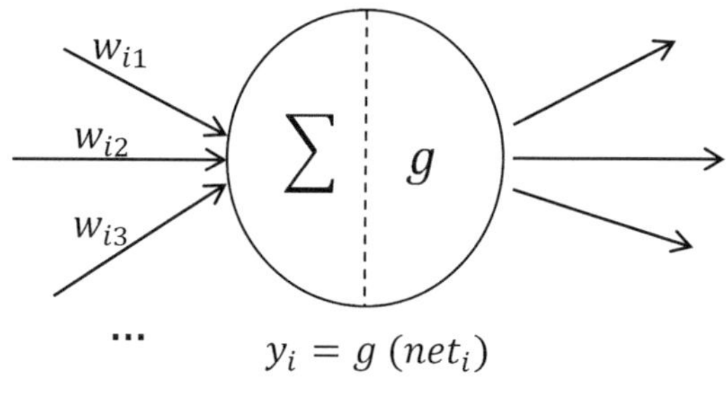

Fig. 3 A perceptron's schematic input/output structure

function of hidden nodes. Learning or training is used to describe the process of finding values of these weights. A learning algorithm adjusts connection weights until the system converges to approximately reproduce the output. The optimal weights may be obtained by using gradient descent algorithm (GDA) and *Broyden–Fletcher–Goldfarb–Shanno* (BFGS) or conjugate gradient descent algorithm (CGDA) with a view of minimising the sum of the squared error function of the network output.

Support Vector Machines

Another popular method for nonlinear problems (nonlinear) in biological data is *support vector machines* (SVMs). The SVMs first map the observations in the feature space into another space using a kernel function. A maximally separating hyperplane is then constructed based on the observations closest to the region that separates the two classes (the *support vectors*). The performance of SVMs greatly relies on the choice of kernel function and to what degree the kernel function is able to map the original classification problem into a linearly separable one.

This technique was introduced by Vapnik in 1995 (Vapnik 1995). This technique is based on finding linear hyperplanes in input space and kernel space for avoiding overfitting. Let training sample data consist of n pairs $(x_1, y_1), (x_2, y_2), \ldots, (xn, yn)$ with $(xi \in Rp)$ and $yi \in \{-1, 1\}$, and then the SVM classifier finds hyperplane (P_0) bisecting the closest points of the data which is linearly separable (Fig. 4). P_0 is defined as

$$\{x : f(x) = xb + b_0 = 0\} \text{and } \; b \; = 1.$$

Classifier creates a parallel hyperplane P_1 defined as

$$(P_1)\{x : f(x) = xb + b_0 = -1\}.$$

On a point in class −1 closest to (P_0) and second hyperplane (P_2) the expression is as follows

$$(P_2)\{x : f(x) = xb + b_0 = 1\} \text{on a point in class closest to} (P_0).$$

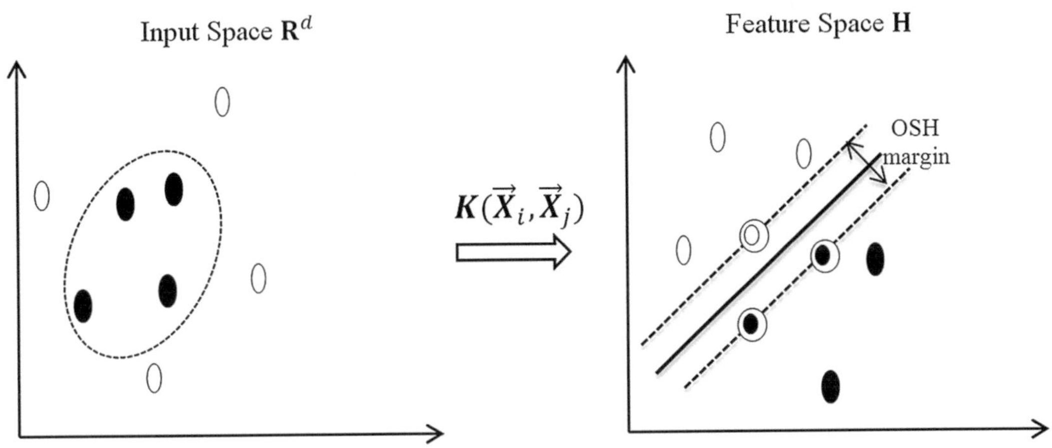

Fig. 4 Schematic diagram representing SVM classifier. Mapping is determined by a kernel function $K(xi, xj)$

The optimum hyperplane for separating the data can be formed by maximising the perpendicular distance between two parallel supporting planes P_1 and P_2, i.e. M. The resulting classifier can be given by

$$\hat{y} = \sin g(xb + b_0).$$

Since the classes are separable, $m=2/\|\boldsymbol{\beta}\|$ where maximisation of M leads to minimisation of $\|\boldsymbol{\beta}\|/2$. Therefore, this problem can be reduced to minimisation of $\phi(\boldsymbol{\beta})=\|\boldsymbol{\beta}\|/2$, subject to $yi(\mathbf{x}i\boldsymbol{\beta}+\beta_0)\geq 1$ for all $\{(\mathbf{x}i, yi); i=1,2,..,n\}$.

In case the data set is not separable, then this technique maps the data into higher-dimensional space where training set is separable via some transformation:

$$K: x \rightarrow \phi(x).$$

A kernel function $K(xi,xj)=<\phi(\mathbf{x}i),\phi(\mathbf{x}j)>$ computes the inner product in some expanded feature space. Linear or Gaussian kernels are widely used. Different mappings construct different SVMs. The mapping $\varphi(.)$ is performed by kernel function $K(xi,xj)$ which defines an inner product in the space H. Some typical choices of kernel function are (Cristianini and Shawe-Taylor 2000):

(a) $K(xi,xj)=xiTxj$ (linear SVM)
(b) $K(xi,xj)=(\gamma xiTxj+r)d$(polynomial SVM of degree d)
(c) $K(xi,xj)=\exp\{-\gamma\|xi-xj\|^2\}$ (radial basis function kernel)
(d) $K(xi,xj)=\tanh(\gamma xiTxj+r)$ sigmoid

where r, d and $\gamma>0$ are the kernel parameters.

SVM can handle large feature spaces; can effectively avoid overfitting by controlling the margin; can automatically identify a small subset made up of informative points, i.e. support vectors; etc. The choice of proper kernel function is an important issue for SVM training because the power of SVM comes from the kernel representation that allows the nonlinear mapping of input space to a higher-dimensional feature space. The use of appropriate decision function can give better classification.

Methods of Error Estimation

Cross-validation is primarily a way of measuring the predictive performance of a statistical model. Every statistician knows that the model fit statistics are not a good guide to how well a model will predict; high R^2 does not necessarily mean a good model. One way to measure the predictive ability of a model is to test it on a set of data not used in estimation. Data miners call this a *test set* and the data used for estimation is the *training set*.

k-Fold Cross-validation

Create a k-fold partition of the data set. For each of the k experiments, use k-1 folds for training and the remaining for testing (as shown below).

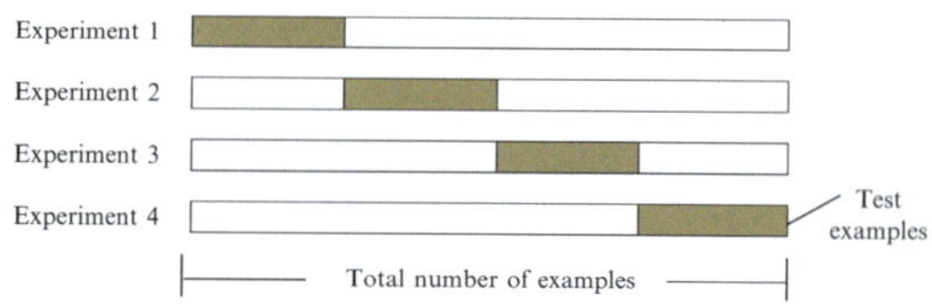

The advantage of k-fold cross-validation is that all the examples in the data set are eventually used for both training and testing. The true error is estimated as the average error rate.

Leave-One-Out Cross-Validation

Leave-one-out is the degenerate case of k-fold cross-validation, where k is chosen as the total number of examples. For a data set with N examples, perform N experiments. For each experiment use N–1 examples for training and the remaining example for testing. As usual, the true error is estimated as the average error rate on test examples.

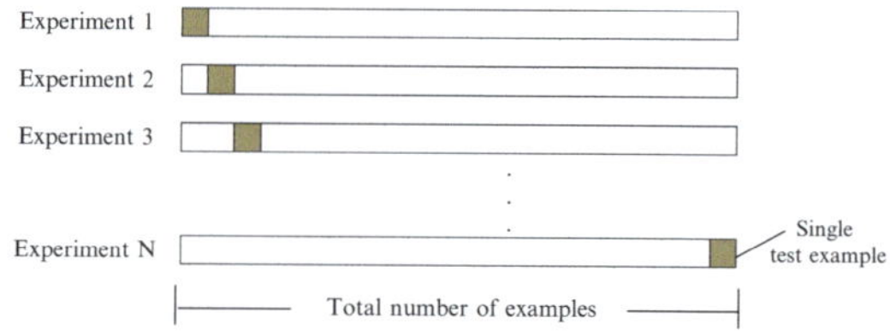

Bootstrapping, Bagging and Boosting

Bootstrapping is a general resampling method that allows statistical inference about a summary statistic (e.g. sample mean) from a data set without knowing the sample distribution. The idea is to randomly draw with replacement a large number of new data sets from the original data set and to calculate the summary statistic from each such bootstrap sample. This provides several values for the summary statistic which may be used to infer, for example, its variance or confidence interval. Bagging and boosting are general methods for improving the classification performance of any supervised method. Bagging (bootstrap aggregation) uses bootstrapping to sample a large number of training sets from the original set of examples. A model is induced from each such bootstrap sample and combined (aggregated) during classification to obtain what is often a better classification performance. Boosting is a similar method in which a weight is associated with each training example. Models are iteratively induced from the training set according to these weights and used to reclassify the examples. The weights are subsequently updated to put more emphasis on incorrectly classified examples. If the applied learning method cannot utilise the weights directly, bootstrap training sets may be constructed according to the weights (i.e. each example is drawn with a probability corresponding to the weight).

Measures of Performance

Several measures are available for the statistical estimation of the accuracy of prediction models. Sensitivity, specificity, precision or positive predictive value (PPV), negative predictive value (NPV), accuracy and Matthew's correlation coefficient (MCC) are the terms which are most commonly associated with a classification test, and these statistically measure the performance of the test. In a binary classification, a given data set is divided into two categories on the basis of whether they have common properties or not by identifying their significance (error calculations are based on the confusion matrix as follows):

	Actual			
Predicted		Positive	Negative	
	Positive	TP	FP	PPV
	Negative	FN	TN	NPV
		Sensitivity	Specificity	

The sensitivity indicates the 'quantity' of predictions, i.e. the proportion of real positives correctly predicted. The specificity indicates the 'quality' of predictions, i.e. the proportion of true negatives correctly predicted. The PPV indicates the proportion of true positives in predicted positives, 'the success rate', while NPV is the proportion of true negatives in predicted negatives. These measures are defined as follows:

$$\text{Sensitivity} = \frac{\text{TP}}{(\text{TP}+\text{FN})} * 100.$$

$$\text{Specificity} = \frac{\text{TN}}{(\text{TN}+\text{FP})} * 100.$$

$$\text{PPV} = \frac{\text{TP}}{(\text{TP}+\text{FP})} * 100.$$

$$\text{NPV} = \frac{\text{TN}}{(\text{TP}+\text{FN})} * 100.$$

$$\text{Accuracy} = \frac{\text{TP}+\text{TN}}{(\text{TP}+\text{FP}+\text{TN}+\text{FN})} * 100.$$

$$\text{MCC} = \frac{(\text{TP} * \text{TN} - \text{FP} * \text{FN})}{\sqrt{(\text{TP}+\text{FP})(\text{TP}+\text{FN})(\text{TN}+\text{FP})(\text{TN}+\text{FN})}} * 100.$$

A good binary classification test always results with high values for the above factors, whereas a poor binary classification test results with low values for these. If sensitivity is high and specificity is low, then there is no need to bother about the positive candidates, but the negative candidates must be re-examined to eliminate false positives (negative candidates mistakenly selected). But if sensitivity is low and specificity is high, then there is no need to bother about the negative candidates, but the positive candidates must be re-examined to eliminate false negatives (positive candidates mistakenly rejected). An average binary classification test always results with average values which are almost similar for all the three factors.

Application of Next-Generation Sequencing in Agriculture and Its Challenges

Food security of the growing population is a great challenge before the world, where agricultural lands are decreasing. Traditional plant and animal breeding leads to increase in yield with precision of selection very limited to certain traits. Precision of selection and rate of genetic gain have been a great challenge rendering limited increase in agricultural productivity. With advancement of genomics, the predictive breeding is preferred as it accelerates the rate of genetic gain with very high precision and the time needed is short and it is economical as well. Researchers and breeders use genetic markers to construct linkage maps, which help to identify useful genes. They are also vital to marker-assisted selection which is now genome-wide association study (GWAS). Such selections are based on >50 thousand SNPs present on each and every sample of the organism to be used in selection.

The human genome project in the 1990s paved the way for many genome sequencing projects, surely also revolutionising the DNA sequencing technologies over the last two decades. New technologies generate billions of relatively short (~100-base) sequence reads at very low cost, increasing sequencing efficiency by approximately 100,000-fold. Now, next-generation sequencing (NGS) machines can sequence the entire human genome in a few days, inspiring a flood of new sequencing projects. But these high-speed technologies are synonymous to many technical challenges for sequence alignment and assembly programs in response to their product. Repetitive DNA sequences (i.e. sequences that are similar or identical to sequences elsewhere in the genome) are abundant in a broad range of species, from bacteria to mammals. High levels of repetitiveness are found across all kingdoms of life. Plant genomes are reported to contain high proportions of repeats (e.g. transposable elements cover >80 % of the maize genome). The short-lived fish *Nothobranchius furzeri* has 21 % of its genome occupied by tandem repeats, and bacterial genomes (e.g. *Orientia tsutsugamushi*) can exhibit repeat content up to 40 %.

The sequence alignment and assembly challenges are tough to handle for short reads and high data volumes from next-generation sequencing projects. Irregularities and ambiguity in sequence alignment and assembly are encountered due to such short repeats when we look into it with computational perspective, leading to biases and errors while interpreting results. Mere ignorance of repeats does not lead to solution and may indicate that important biological phenomena are missed. In response to this influx of new laboratory methods, many novel computational tools have been developed to map NGS reads to genomes and to reconstruct genomes and transcriptomes. Current NGS platforms produce shorter reads (i.e. 50–150 bp) than Sanger sequencing but with vastly greater numbers of reads (almost six billion per run). With the speed and rise in the development of the NGS technologies, its application in various disciplines such as comparative genomics and evolution, forensics, epidemiology and applied medicine for diagnostics and therapeutics has increased dramatically. Table 9 compares the various next-generation sequencing methods. Li et al. (2010) used next-generation sequencing technology alone, to successfully generate and assemble a draft sequence of the giant panda (*Ailuropoda melanoleuca*) genome.

There are challenges in de novo genome assembly and RNA-seq analysis posed by repeats for genome resequencing projects. Genome resequencing allows researchers to study genetic variation by analysing many genomes from the

Table 9 Comparison of next-generation sequencing methods

Method	Single-molecule real-time sequencing (Pacific Bio)	Ion semiconductor (Ion Torrent sequencing)	Pyrosequencing (454)	Sequencing by synthesis (Illumina)	Sequencing by ligation (SOLiD sequencing)	Chain termination (Sanger sequencing)
Read length	2,900 bp average	200 bp	700 bp	50–250 bp	50 + 35 or 50 + 50 bp	400–900 bp
Accuracy	87 % (read length mode), 99 % (accuracy mode)	98 %	99.9 %	98 %	99.9 %	99.9 %
Reads per run	35–75 thousand	Up to five million	One million	Up to three billion	1.2–1.4 billion	N/A
Time per run	30 min to 2 h	2 h	24 h	1–10 days, depending upon sequencer and specified read length	1–2 weeks	20 min to 3 h
Cost per one million bases (in US$)	$2	$1	$10	$0.05–$0.15	$0.13	$2,400
Advantages	Longes*t* read length. Fast. Detects 4mC, 5mC, 6 mA	Less expensive equipment. Fast	Long read size. Fast	Potential for high sequence yield, depending upon sequencer model and desired application	Low cost per base	Long individual reads. Useful for many applications
Disadvantages	Low yield at high accuracy. Equipment can be very expensive	Homopolymer errors	Runs are expensive. Homopolymer errors	Equipment can be very expensive	Slower than other methods	More expensive and impractical for larger sequencing projects

same or from closely related species. The primary requirement is for a high-quality reference genome onto which all of the short NGS reads can be mapped. After sequencing a sample to deep coverage, it is possible to detect SNPs, copy number variants (CNVs) and other types of sequence variation without the need for de novo assembly. The computational task involves aligning millions or billions of reads back to the reference genome using one of several short-read alignment programs (Table 10). Bowtie and the Burrows–Wheeler Aligner (BWA), the two most efficient aligners, achieve throughputs of 10–40 million reads per hour on a single computer processor. In spite of this recent progress, a major challenge remains when trying to decide what to do with multi-reads, i.e. reads that map to multiple locations. Three choices of an algorithm exist for dealing with multi-reads. First is to ignore them, i.e. discarding all multi-reads. The second option is the best match approach, where the alignment with the fewest mismatches is reported. If there are multiple, equally good best match alignments, then an aligner will either choose one at random or report all of them. Thirdly is to report all alignments up to a maximum number, *d*, regardless of the total number of alignments found. A variant on this strategy is to ignore multi-reads that align to >*d* locations. For simplicity, 'ignore' strategy for multi-reads is majorly followed. However, this strategy limits analysis to unique regions in the genome, discarding many multi-gene families as well as all repeats, which might result in biologically important variants being missed.

After following the mapping strategies of the reads, the next step is SNP detection using a program such as GATK, MAQ, SAMtools, SOAPsnp or VarScan (Tables 10 and 11). Further, the alignment probability for each multi-read is computed using a Bayesian genotyping model that decomposes the likelihood of a read mapping to a given locus into its component likelihoods. Computational tools can discover multiple types of variants in NGS data, including deletions, insertions, inversions, translocations and duplications. VariationHunter was one of the first algorithms to incorporate both read-depth and read-pair data for accurate CNV discovery. He et al. (2011) described a new method that was designed to find CNVs even in repeat-rich regions.

De Novo Genome Assembly

NGS read lengths (50–150 bp) are considerably shorter than the capillary-based (Sanger) sequencing methods (800–900 bp). These short read lengths lead to difficult assembly, as mentioned earlier. NGS technology generates higher depth of coverage at far lower cost than Sanger sequencing, and, as a result, current strategies for assembly attempt to use deeper coverage to compensate for shorter reads. However, repetitive sequences create substantial difficulties that coverage depth cannot always overcome. For de novo assembly, repeats that are longer than the read length create gaps in the assembly, which is the major problem caused by repeats coupled with the short length of NGS sequences. Genome assembly algorithms begin with a set of reads and attempt to reconstruct a genome as completely as possible without introducing errors. In addition to creating gaps, repeats can be erroneously collapsed on top of one another and can cause complex, misassembled rearrangements. Many new de novo assemblers have emerged to tackle this problem. These assemblers fall into one of two classes, i.e. *overlap-based assemblers* and *de Bruijn graph assemblers*, both of which create graphs (of different types) from the read data.

Gnerrea et al. (2011) reported the development of an algorithm for genome assembly, ALLPATHS-LG, and its application to massively parallel DNA sequence data from the human and mouse genomes, generated on the Illumina platform. ALLPATHS-LG is the extensive improvement of the previous ALLPATHS program with key innovations on concepts of handling repetitive sequences, error corrections, use of jumping data, efficient memory usage and low-coverage regions. For massively parallel sequencing data, they specified provisional sequencing model for de novo assembly (Table 12).

Table 10 Computational tools for next-generation sequencing genome alignment and assembly (Treangen and Salzberg 2012)

Scope	Program	Repeat-relevant parameters	Website
SV or CNV detection	BreakDancer	Specify the mapping quality threshold for ambiguous reads: -q	http://sourceforge.net/projects/breakdancer
	CNVnator	None available or none required	http://sv.gersteinlab.org/cnvnator/
	He et al. (2011)	Algorithm only, able to estimate CNV counts in repeat-rich regions	None
	PEMer	Maximum alignments per multi-read:--ax_duplicates_per_score	http://sv.gersteinlab.org/pemer
	VariationHunter	None available or none required	http://compbio.cs.sfu.ca/strvar.htm
SNP detection	GATK	None available or none required	http://www.broadinstitute.org/gsa/wiki/index.php/Downloading_the_GATK
	SAMtools	In repetitive regions, avoid calling 'A': -avcf ref.fa aln.bam	http://samtools.sourceforge.net
	SOAPsnp	None required; multi-reads supported by read aligner parameters	http://soap.genomics.org.cn/soapsnp.html
	Sniper	Read mapping policy:--ll,--niq,--est	http://kim.bio.upenn.edu/software/sniper.shtml
	VarScan	None available or none required	http://varscan.sourceforge.net
Short-read alignment	Bowtie	Randomly distribute reads across repeats:--est –M 1 -strata	http://bowtie-bio.sourceforge.net
	BFAST	Reports all locations by default	http://bfast.sourceforge.net
	Burrows–Wheeler Aligner (BWA)	Report one random hit for repetitive reads: -n 1	http://bio-bwa.sourceforge.net
	mrFAST	Reports all locations by default, for best match:--est	http://mrfast.sourceforge.net
	SOAPaligner	Report all locations: -r 2	http://soap.genomics.org.cn/soapaligner.html
De novo assembly	ALLPATHS-LG	None required: incorporated into library insert size recipe	http://www.broadinstitute.org/software/allpaths-lg/blog/?page_id=12
	CABOG	Reassemble misclassified nonunique unitigs: doToggle = 1	http://wgs-assembler.sf.net
	SGA	Resolve small repeats at end of reads: -r 20	http://github.com/jts/sga
	SOAPdenovo	Use reads to solve small repeats: -R	http://soap.genomics.org.cn/soapdenovo.html
	Velvet	Use long reads to resolve repeats: -long, -exp_cov auto	http://www.ebi.ac.uk/~zerbino/velvet

Table 11 Computational tools for next-generation sequencing transcriptome analysis (Treangen and Salzberg 2012)

Scope	Program	Repeat-relevant parameters	Website
Spliced read alignment	GSNAP	–	http://share.gene.com/gmap
	MapSplice	–	http://www.netlab.uky.edu/p/bioinfo/MapSplice
	RUM	–	http://www.cbil.upenn.edu/RUM
	SpliceMap	–	http://www.stanford.edu/group/wonglab/SpliceMap
	TopHat	–	http://tophat.cbcb.umd.edu
Reference-guided transcript assembly	Cufflinks	Improve repeat read mapping estimate:--ulti-read-correct	http://cufflinks.cbcb.umd.edu
	ERANGE	Use multi-read fractions:--ith multifraction	http://woldlab.caltech.edu/rnaseq
	G-Mo.R-Se	None required; multi-reads supported by read aligner parameters	http://www.genoscope.cns.fr/externe/gmorse
	Myrna	None required; multi-reads supported by read aligner parameters	http://bowtie-bio.sourceforge.net/myrna
	Scripture	None required; multi-reads supported by read aligner parameters	http://www.broadinstitute.org/software/scripture
De novo transcript assembly	Multiple-k	None required or none available	http://www.surget-groba.ch/downloads
	Rnnotator	None required or none available	None
	Trinity	Separate transcripts derived from paralogues:--un_butterfly	http://trinityrnaseq.sourceforge.net
	Trans-ABySS	None required or none available through command line	http://www.bcgsc.ca/platform/bioinfo/software/trans-abyss
	Velvet-Oases	Use long reads to resolve repeats: -long, -exp_cov auto	http://www.ebi.ac.uk/~zerbino/oases

The 'Program' column contains the name of program or algorithm. The 'Repeat-relevant parameters' column is a list of parameters that adjust how repeats are treated. The programs have many other parameters, but more careful treatment of repeats would start with modification of these

Table 12 Provisional sequencing model for de novo assembly in ALLPATHS-LG algorithm (Gnerrea et al. 2011)

Libraries, insert types[a]	Fragment size, bp	Read length, bases	Sequence coverage, ×	Required fragment
Fragment	180[b]	≥100	45	Yes
Short jump	3,000	≥100 preferable	45	Yes
Long jump	6,000	≥100 preferable	5	No[c]
Fosmid jump	40,000	≥26	1	No[c]

[a]Inserts are sequenced from both ends, to provide the specified coverage
[b]More generally, the inserts for the fragment libraries should be equal to ~1.8 the sequencing read length
In this way, the reads from the two ends overlap by ~20 % and can be merged to create a single longer read. The current sequencing read length is ~100 bases
[c]Long and Fosmid jumps are a recommended option to create greater continuity

Genome-Wide Association Studies

Genetic markers have always been a cornerstone in the areas identifying the disease-carrier alleles and selection of plants and animals for improved production, reproduction, disease resistance and growth performances. Marker data analysis experienced a humble start from single-locus marker-aided selection process and then traversed towards the much coveted genome-wide association study (GWAS) or whole genome association study (WGAS) during the last three decades. The molecular markers (single or a few taken together) are used for screening the individuals with desirable (economic trait loci) or for culling the carriers of deleterious (disease carriers) alleles in order to select the best individuals to propagate their genome to the next generation. The advent of high-throughput sequencing technologies has revolutionised the marker-assisted selection process. Now, the whole genome of the individual is screened for the loci contributing to the trait of interest. In contrast to methods which specifically test one or a few genomic regions, the GWAS investigates the entire genome. The approach is therefore said to be genome-wide and non-candidate-specific in contrast to the candidate-gene approach. GWAS is a process for inspection and screening of detectable common genetic variants (single-nucleotide polymorphisms) in individuals to identify the variant(s) associated with the trait under study. GWAS compares the DNA profiles of individuals having altered trait (viz. disease, improved production, reproduction or growth parameters) with the control ones (healthy ones or with normal or below average parameters). The DNA specimen from each of the individuals is subjected to microarray analysis for detection of specific SNPs that are more prevalent in any one group. The associated SNPs mark a region on the genome. GWAS identifies SNPs and other variants in DNA which are associated with a trait; however, it cannot specify the causal genes on their own. GWA studies have been able to identify the quantitative trait loci (QTL) affecting many common complex diseases, demonstrating the utility of this approach for dissecting the genetic basis of polygenic traits (McCarthy et al. 2008).

Agricultural production and food security in the developing world face numerous threats, depending upon the regions, which are relatively vulnerable to climate change. GWAS is becoming the dominant paradigm for investigating the genetics of natural phenotypic variation. GWAS has got several basic applications in molecular animal breeding like association between the variations in genotypes and phenotypes to identify the causal genetic mechanism, identification of QTL underlying many common, complex diseases and association of a trait with a region in the genome, in order to map the clinically and/or economically important QTLs.

GWAS has basically been employed for human diseases but still has been successful in mapping causal variants in organisms like *Arabidopsis thaliana*, the model organism for such studies. Genomic analysis of diverse populations is increasingly being used to expose the genetic basis of complex traits, including agroclimatic traits of crop species. Genome-wide single-nucleotide polymorphism (SNP) scans of population genetic parameters in agricultural crops have been used to identify loci under selection (Jiao et al. 2012; Hufford et al. 2012). GWAS has also been used to elucidate the genetic basis of agronomic traits in rice (Huang et al. 2010) and maize (Jiao et al. 2012). Nucleotide diversity scans (Casa et al. 2006; Bouchet et al. 2012) and association studies (de AlencarFigueiredo et al. 2010) have been carried out in sorghum. GWAPP tool enables researchers working with *Arabidopsis thaliana* to perform GWAS on their phenotypes (http://gwas.gmi.oeaw.ac.at/index.html#!homePage).

High-Performance Computing in Agricultural Research

Advancement in computer science enabled the development of high-performance computing (HPC) system for massive data handling and supercomputing functionalities across different domains of science. The biological data in the field of agriculture science across the globe is increasing at an exponential rate with the advent of relatively less expensive data generation technologies. Therefore, application of HPC systems for knowledge extraction and pattern detection from this high-dimensional biological data set is the need of the hour. Apart from this, applications of structural biology, metabolic pathway analysis and ultimately the system biology approaches require advanced computing and visualisation tools for understanding the biological systems.

In general, the HPC system consists of a cluster (farm) of high-end servers connected together with a master–slave configuration. The data processing on this system needs development of parallel computing algorithms and tools for maximum utilisation of its computing power. In order to utilise the full potential of the HPC system, these algorithms need to be optimised in a systematic way. In this system, a meta-job is divided into smaller modular components which are distributed and synchronised across different servers as per the requirement by master node. The intermediate calculations are stored on a common memory space, and after completion of the job, the results are sent to the user.

The analysis of biological data has many computational challenges such as:

1. Multiple comparison, which leads to the non-polynomial problems as it is not possible to search all possible targets through finite/polynomial computational algorithms
2. High dimensionality, as the biological data is being generated with many targets with space data points
3. Small n and large p problem, as the number of factors is large as compared to sample size
4. Noisy high throughput, as biotechnological data has a number of unavoidable noises
5. Integration of multiple heterogeneous data sets

Biological problems involve well-behaved polynomial time algorithms but imbibe massive computational requirements for analysis of the large data sets, for example, the assembly of the human genome in 2001 from numerous short segments of sequence data required approximately 10,000 CPU hours (Venter et al. 2001). Apart from this, the biological problems such as protein structure prediction, evolutionary studies, etc. based on molecular data need very complex and computer-intensive algorithms. These problems of biological data analysis can only be addressed through implementation of appropriate HPC system. These systems provide sufficient capability for evaluating biomolecular hypotheses. Algorithms for protein folding are sophisticated with good computational cost for biomolecular modelling of the physical processes.

Three innovative parallel computing approaches are applied to run these algorithms: first, the massive computational system known as Blue Gene/L developed by IBM; second, specialised hardware, especially for molecular dynamics like MD-GRAPE (Molecular Dynamics GRAvity PipE) or PetaFLOPS developed by IBM's Research Division and Institute of Chemical Research, or RIKEN, in Japan; and thirdly, cycle-scavenging approaches such as the Folding@home project at Stanford University.

In case of phylogenetic analysis, Genome Rearrangement Analysis through Parsimony and other Phylogenetic Algorithms (GRAPPA) is a prime example of the potential of high-performance algorithm development and implementation of HPC systems in computational environment. The billion-fold speed-up with GRAPPA allowed expansion of data sets from 10 taxa to 18 taxa. Due to easy accessibility of the HPC system, the research activities, which were earlier considered impossible due to computational challenges, have now become feasible in biological and biomedical research.

Whole genome sequencing and bulk SNP discovery of different populations to construct HapMap are imperative. Such steps are heavily dependent on bulk genomic data generation and processing which can only be handled using HPC. The HPC system can be effectively used in agriculture for enhancing production and productivity and ensuring nutritional security of the society. The major challenges of agriculture such as biotic and abiotic stresses, water conservation, low productivity and deficient nutritional quality can be easily addressed following *omics* approaches using HPC system. This will lead to the development of superior varieties/products in agriculture to feed the growing population of the world on sustainable basis.

In order to accelerate the agriculture research and development in this field, the Indian Council of Agricultural Research (ICAR), New Delhi, Government of India, initiated a research project entitled 'Establishment of National Agriculture Bioinformatics Grid (NABG) in ICAR' under the National Agricultural Innovation Project (NAIP) at the Centre for Agricultural Bioinformatics (CABin), Indian Agricultural Statistics Research Institute, New Delhi. The HPC systems have high-end configuration of 70 teraFLOPS, 256 nodes with 2 masters Linux cluster, 16 nodes with one master node Windows cluster, 16 nodes with one master GPU cluster, 500 gigaFLOPS SMP-based computing, 200 terabytes NFS server, 100 terabytes high-performance parallel file system-based storage and 200 terabytes archival storage (Fig. 5). Out of these four systems, two have been registered among the top

Fig. 5 High-performance computing facility at IASRI, New Delhi

supercomputers of India (http://topsupercomputers-india.iisc.ernet.in/jsps/june2013/index.html). This will provide a platform for the interdisciplinary research in cross-species genomics in agriculture in the country.

References

Altschul SF, Gish W, Miller W, Myers EW, Lipman DJ (1990) Basic local alignment search tool. J Mol Biol 215(3):403–410

Anthony RV, Scheaffer AN, Wright CD, Regnault TR (2003) Ruminant models of prenatal growth restriction. Reprod Suppl 61:183–194

Barbazuk WB, Bedell JA, Rabinowicz PD (2005) Reduced representation sequencing: a success in maize and a promise for other plant genomes. BioEssays 27:839–848

Baridam BB (2012) More work on K-means clustering algorithm: the dimensionality problem. Int J Comput Appl 44(2):23–30

Bouchet S, Pot D, Deu M et al (2012) Genetic structure, linkage disequilibrium and signature of selection in sorghum: lessons from physically anchored DArT markers. PLoS One 7(3):e33470

Casa AM, Mitchell SE, Jensen JD et al (2006) Evidence for a selective sweep on chromosome 1 of cultivated sorghum. Crop Sci 46(S1):S27–S40

Chou PY, Fasman GD (1974) Prediction of protein conformation. Biochemistry 13(2):222–245. doi:10.1021/bi00699a002

Cristianini N, Shawe-Taylor J (2000) An introduction to support vector machines and other kernel-based learning methods. Cambridge University Press, Cambridge/New York

Dayhoff MO, Eck RV, Park CM (1972) A model of evolutionary change in proteins. In: Dayhoff MO (ed) Atlas of protein sequence and structure, 5th edn. National Biomedical Research Foundation, Washington, DC

de AlencarFigueiredo LF, Sine B, Chantereau J et al (2010) Variability of grain quality in sorghum: association with polymorphism in Sh2, Bt2, SssI, Ae1, Wx and O2. Theor Appl Genet 121(6):1171–1185

Eyras E, Reymond A, Castelo R, Bye JM, Camara F, Flicek P, Huckle EJ, Parra G, Shteynberg DD, Wyss C, Rogers J, Antonarakis SE, Birney E, Guigo R, Brent MR (2005) Gene finding in the chicken genome. BMC Bioinforma 6(1):131

Gill BS, Appels R, Botha-Oberholster AM et al (2004) A workshop report on wheat genome sequencing: international genome research on wheat consortium. Genetics 168(2):1087–1096

Gnerrea S, MacCallum I, Przybyiski D et al (2011) High-quality draft assemblies of mammalian genomes from massively parallel sequence data. Proc Natl Acad Sci USA 108(4):1513–1518

He D, Hormozdiari F, Furlotte N, Eskin E (2011) Efficient algorithms for tandem copy number variation reconstruction in repeat-rich regions. Bioinformatics 27(11):1513–1520

Hertz J, Krogh A, Palmer RG (1991) Introduction to the theory of neural computing. Addison-Wesley Publishing Company, Redwood City

Hillier LW, Miler W, Birney E et al (2004) Sequence and comparative analysis of the chicken genome provide unique perspectives on vertebrate evolution. Nature 432(7018):695–716. doi:10.1038/nature03154

How to Feed the World in 2050 (2009) Office of the Director, Agricultural Development Economics Division, Economic and Social Development Department www.fao.org/wsfs/forum2050/wsfs-background-documents/wsfs-expert-papers/en/

Huang X, Wei X, Sang T et al (2010) Genome-wide association studies of 14 agronomic traits in rice landraces. Nat Genet 42(11):961–967

Hufford MB, Xu X, van Heerwaarden J et al (2012) Comparative population genomics of maize domestication and improvement. Nat Genet 44(7):808–811

International Rice Genome Sequencing Project (2005) The map-based sequence of the rice genome. Nature 436(7052):793–800

Jiao Y, Zhao H, Ren L et al (2012) Genome-wide genetic changes during modern breeding of maize. Nat Genet 44(7):812–815

Lewis SE (2005) Gene ontology: looking backwards and forwards. Genome Biol 6(1):103

Li R, Fan W, Tian G et al (2010) The sequence and de novo assembly of the giant panda genome. Nature 463:311–317

Manikanandakumar K (2009) Dictionary of bioinformatics. MJP Publishers, Chennai

Maxam AM, Gilbert W (1977) A new method for sequencing DNA. Proc Natl Acad Sci U S A 74:560–564

McCarthy FM, Wang N, Magee GB, Nanduri B, Lawrence ML, Camon EB, Barrell DG, Hill DP, Dolan ME, Williams WP, Luthe DS, Bridges SM, Burgess SC (2006) AgBase: a functional genomics resource for agriculture. BMC Genomics 7:229

McCarthy MI, Abecasis GR, Cardon LR et al (2008) Genome-wide association studies for complex traits: consensus, uncertainty and challenges. Nat Genet 9:356–369

Needleman SB, Wunsch CD (1970) A general method applicable to the search for similarities in the amino acid sequence of two proteins. J Mol Biol 48(3):443–453

Prather RS, Hawley RJ, Carter DB, Lai L, Greenstein JL (2003) Transgenic swine for biomedicine and agriculture. Theriogenology 59(1):115–123

Protein Identification Resource http://pir.georgetown.edu/

Sanger F, Coulson AR (1975) A rapid method for determining sequences in DNA by primed synthesis with DNA polymerase. J Mol Biol 94:441–448

Sonstegard TS, Connor EE (2004) Sequencing and mapping of the bovine thyroid hormone response element

(spot14) gene. Germplasm Release. Accession number AY656814
The Tomato Genome Consortium (2012) The tomato genome sequence provides insights into fleshy fruit evolution. Nat Lett 485:635–641. doi:10.1038/nature11119
Treangen TJ, Salzberg SL (2012) Repetitive DNA and next-generation sequencing: computational challenges and solutions. Nature 13:36–46
Vapnik V (1995) The nature of statistical learning theory. Springer, New York
Varshney RK, Chen W, Li Y et al (2011) Draft genome sequence of pigeonpea (*Cajanus cajan*), an orphan legume crop of resource-poor farmers. Nat Biotechnol 30:83–89
Venter J, Adams MD, Myers EW et al (2001) The sequence of the human genome. Science 291(5507):1304–1351. doi:10.1126/science.1058040
Ware DH, Jaiswal P, Ni J et al (2002) Gramene, a tool for grass genomics. Plant Physiol 130(4):1606–1613
Yuan Q, Ouyang S, Liu J, Suh B, Cheung F, Sultana R, Lee D, Quackenbush J, Buell CR (2003) The TIGR rice genome annotation resource: annotating the rice genome and creating resources for plant biologists. Nucleic Acids Res 31(1):229–233

Plant Systems Biology: Insights and Advancements

Tulika Bhardwaj and Pallavi Somvanshi

Contents

Abstract

In order to study the complex biology of plants, systems biology focuses on every aspect of the highly interacting components (adjustment of metabolism in response to stresses, different physiological properties due to mutation and epigenetic effects, etc.) by the development of high-throughput data generation technologies, i.e. 'omics', massive databases, employment of specialised bioinformatics tools and algorithms. Mathematical models are also used to study the structure and dynamics of interacting networks. This approach is essential for understanding how a plant system works which is advantageous for further research.

Keywords

Omics • Systems biology • Interactomics • Metabolomics • Virtual cell • Molecular pharming

T. Bhardwaj, Ph.D. • P. Somvanshi, Ph.D. (✉)
Department of Biotechnology, Teri University, 10 Institutional Area, Vasant Kunj, New Delhi 110070, India
e-mail: psomvanshi@gmail.com

D. Barh et al. (eds.), *PlantOmics: The Omics of Plant Science*,
DOI 10.1007/978-81-322-2172-2_28, © Springer India 2015

Introduction

Plants dwelling at the base of biological food chain are of fundamental significance in providing solutions to some of the most daunting ecological and environmental problems faced by our planet. The reductionist views of molecular biology provide only a partial understanding to the phenotypic knowledge of plants. Systems biology offers a comprehensive view of plant systems by employing a holistic approach integrating the molecular data at various hierarchical levels. In this chapter, we discuss the basics of systems biology including the various 'omics' approaches and their integration, the modelling aspects and tools needed for plant systems research. A particular emphasis is given to the recent analytical advances, updated published examples of plant systems biology studies and the future trends.

Biological Systems and Systems Biology in Plants

The understanding of systems had an enormous impact on what are loosely regarded as human sciences, including economics, sociology, psychology and medicine. A system can be generally defined as a network of interacting elements, receiving certain inputs and producing certain outputs (Hammer et al. 2004). It is a network of mutually dependent and thus interconnected components comprising a unified whole. Every system exhibits emergent behaviour, a unique property possessed only by the whole system and not shared to any great degree by the individual components. The properties of systems are the result of two important characteristics:

- Systems have a hierarchical structure.
- The structure is held together by numerous linkages to construct very complex networks (Trewavas 2006).

Figure 1 illustrates a familiar but oversimplified biological hierarchy. Each level in the hierarchy above that of molecules is an emergent property resulting from the very complex interactions between the constituents of the lower level. Each level in turn contains numerous recognisable subsystems: some simple, some complex, but each presenting emergent properties that can also be arranged in a hierarchy of organisation.

In cells the aggregation of subunits of multienzyme complexes or calcium/calmodulin with dependent enzymes creates the simple emergent property of novel enzyme activity. For example, tubulin or actin polymerisation in the test tube creates the emergent behaviour of isolated microtubules or filaments, while the organised cellular behaviour of cyclins and other numerous regulatory proteins describes the emergent property of the cell cycle (Kohn 1999; Strogatz 2001). This process, like circadian phenomena or growth, results from an integrated, organised collective of complex feedback controls, protein phosphorylation regulation, second messenger distribution, structural interactions, organelle interactions and

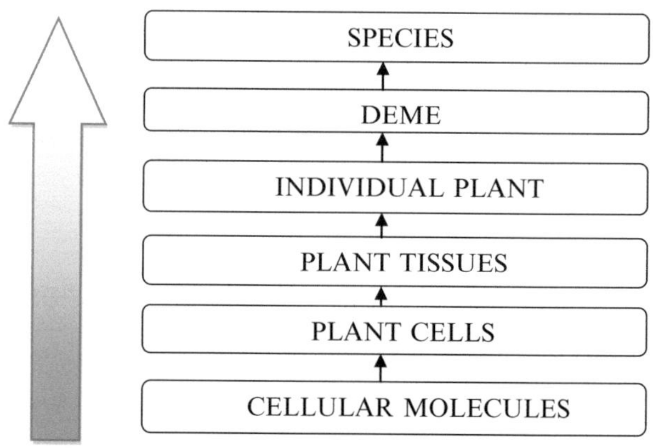

Fig. 1 Biological systems arranged in a hierarchy of increasing organisational complexity (Trewavas 2006)

other as yet uncovered control mechanisms. This level of organisation is much more complex than that of individual enzymes (Davidson et al. 2002).

Weiss define the critical characteristics of biological systems using the recent recognition of their hierarchical structure (Weiss 1973). Two interconnected points were made.

1. All systems express much greater variation at lower levels than at higher levels of organisation.
2. The output of individual pathways is more ordered within a system than would be expected from random operation of those pathways outside the system (Williams 1956).

Biological communities are the living components of ecosystems. A profound understanding of these biological communities is essential to make predictions about the effects of developmental programmes, natural perturbations or directed human-induced changes on the composition and functioning of biological systems (Wardle et al. 2004; Van der Putten 2009; Van der Putten et al. 2001; Kefi et al. 2007; de Ruiter et al. 2005; Keurentjes et al. 2012). The practice of integrating physiological, morphological, molecular, biochemical and genetic information has long been applied to biological research and in diverse fields such as plant breeding and ecology (Trewavas 2006; Yuan et al. 2008). The development of modern systems biology was driven by the need to assimilate the large amounts of data generated by genome-scale studies into biologically meaningful interpretations.

Systems biology has generated revolutions in ecology, population biology and evolutionary studies and is slowly moving towards biochemistry, development, genetics and whole-plant biology. It is currently undergoing enormous expansion, but there seems little awareness of either the history of systems biology or the behaviour of systems that makes them exciting to study. Understanding the dynamics of plant communities requires knowledge on how individual plants perceive, interpret and integrate environmental signals, how they organise a coherent response and how they sense the presence of competitors, natural enemies, antagonists or mutualistic symbionts (Wardle et al. 2004; Van der Putten 2009; Van der Putten et al. 2001; Kefi et al. 2007; de Ruiter et al. 2005; Keurentjes et al. 2012).

In systems biology approach, plants are studied at different levels of biological organisation in a hierarchical way with special attention to the interdependence between the various levels of biological organisation. These levels include (1) molecular signalling pathways at the subcellular level; (2) networks of physiological processes at the cellular level, (3) plant growth and development at the individual level; (4) genetic variation among individuals within a species at the population level; (5) plant performance, competition and trophic interactions at the community level and (6) trade-off dynamics that ultimately influence properties and processes at the ecosystem level (Simpson et al. 2010; Keurentjes et al. 2012).

Definition of Plant Systems Biology

Plant systems biology is defined as ‘the study of interactions among biological components using models and/or networks to integrate genes, metabolites, proteins, regulatory elements and other biological components’.

It is also defined as ‘the study of structure, dynamics and control of plant components’ (Williams 1956).

Omics and Data Integration

‘Omics’ research approaches have produced enormous amount of data for living systems which are necessary for the development of systems biology to integrate multidimensional biological information into networks and models. It is not synonymous to the postgenomics omics technologies such as microarrays for high-throughput generation of large-scale data. Genome sequencing enables functional or comparative studies of plant genomes. Genome information also leads to the study of the mRNA transcripts and proteins of an organism as a whole, which are referred to as transcriptomics and proteomics, respectively. Research of all or most of the metabolites in an organism is referred to as *metabolomics*. The study of genome-scale interactions among proteins is referred to as *interactomics*. The advances in RNA interference and other mutagenesis technologies have enabled high-throughput phenotype screens for genes; this is referred to as *phenomics* (Yuan et al. 2008).

Apart from the chief omics approaches, some recent approaches include lipidomics (comprehensive study of the lipid entities of the organism) (Welti et al. 2007) and hormonomics (the entire set of endogenous hormones in a plant). The low molecular weight plant hormones include auxin, ABA, cytokinin, gibberellins, ethylene, brassinosteroids, jasmonates, salicylic acid (Davies 2010) and a newly identified one strigolactone, acting as a shoot-branching inhibitor (Gomez-Roldan et al. 2008; Umehara et al. 2008), lectinomics (bioinformatics studies of carbohydrate binding proteins, lectins) and various others. Also, a new concept which has gained much attention in this era is that of phenomics, the high-throughput systemic analysis of phenotypes, which has probably the biggest application in plant biotechnology (Edwards and Batley 2004).

The biggest problems when constructing genome-scale models of biological systems is that the data produced by employing above technologies is noisy. This is an inherent property of the high-throughput techniques used for the acquisition of massive data, interactions between proteins, etc. The main reason behind this is the false-positive interactions in data sets due to different experimental or *in silico* approaches. Data integration is also important as we try to build comprehensive models of a system. The sheer amount of information published daily for any of the model organisms makes it almost impossible to manually store, classify and integrate the data for systems modelling (Coruzzi and Gutierrez 2009) (Fig. 2).

There are generally four methods that are used as a solution for data integration:

1. *Hypertext navigation* allows the user to query only one database, but the results often contain hyperlinks to the equivalent entry in another database. This method is more common for websites that are based on information retrieval.
2. *Data warehouse* retrieves data from multiple resources, translates the formats and puts them in one database. This allows for much faster and more complex queries taking advantage of all the data loaded in one database and translates from one format to another.

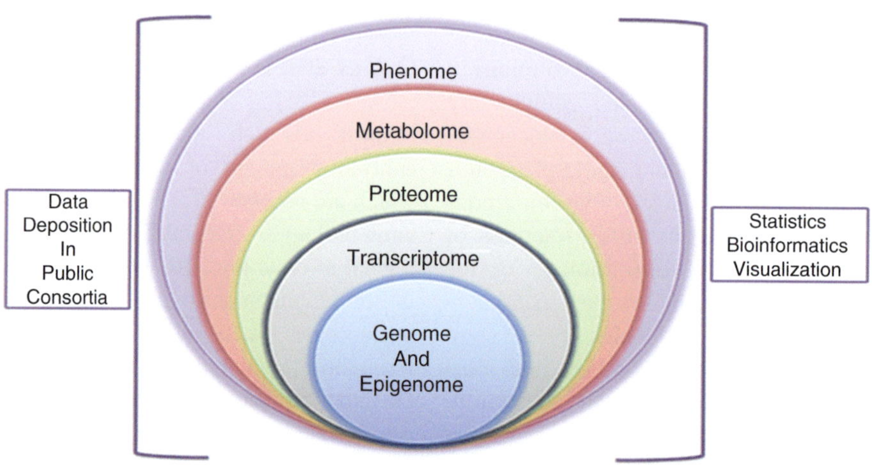

Fig. 2 Integration of heterogeneous multiple 'omics' data (Choi and Pavelka 2011)

The major drawback of this method is that it would be very difficult to keep up with all the new resources becoming available and keeping it all updated. The examples of data warehouses include: Atlas, BioMart, BioWarehouse, Columba, SYSTOMONAS, BioDWH, VINEdb, Booly and GNCPro (Turenne 2011).

3. *Unmediated multiDB queries* allow the databases to remain separate but the query itself extends across all of the databases.
4. *Federated database* is a combination of data warehouse and unmediated multiDB queries. It allows the databases to be separate, but it contains a federated schema, which dynamically translates to queries in the individual schemas (Karp 1996; Coruzzi and Gutierrez 2009).

Protein–Protein Interactions and Interactomics

The regulation and execution of biological processes requires specific interactions of numerous proteins. Tightly regulated protein interaction networks mediate cellular responses to environmental cues and direct the implementation of developmental programmes. The selectivity of protein–protein interactions and their appropriate temporal and spatial regulation determine the developmental potential of the cell and its response to endogenous and exogenous signals. On the molecular level, differential protein–protein interactions are thought to determine the operation of complex regulatory circuits and signal transduction systems (Walter et al. 2004; Terentiev et al. 2009). Several methods have been developed to identify, examine and visualise protein interactions and protein complexes in living cells like yeast two-hybrid system, tandem affinity purification-mass spectrometry, mass spectrometry and the phage display method. Some microscopy- and computer-based methods are also used (Field and Song 1989; Stephens and Banting 2000; Terentiev et al. 2009).

The use of mathematical and computer modelling methods allows investigation of processes and events that are difficult to study even using highly efficient experimental methods (You 2004; Yuan et al. 2008). If mathematical methods are based on description and analysis of intra- and intercellular processes and events using a system of mathematical equations, computer methods mainly serve for creation of algorithms to simulate biological processes, to construct and visualise them. Different variants of molecular dynamics (MD) technique allow dynamic modelling of detailed mechanisms of intracellular biochemical processes and intermolecular interactions (Shaitan et al. 2006; Terentiev et al. 2009). Molecular dynamics methods are based on calculations of trajectories of atoms in molecules via solution of a system of classical equations of motion using Newton's laws.

All molecular networks including gene–protein interaction, metabolic and signalling ones, are used for studying cell functioning. Analysis of these networks reveals in them functional modules and elucidates the role of each network component in cell functioning. Groups of physically interacting proteins that function in the cell in cooperation and coordination, controlling interrelated processes taking place in the organism, form protein interaction networks. Disruption of protein–protein interactions can result in the emergence of various diseases including tumour, neurodegenerative, cardiovascular, autoimmune, etc. Therefore, investigation of interacting partners and analysis of protein networks formed by protein–protein interactions comprise an important instrument in the diagnosis of diseases and in revealing the mechanism of their emergence and development, as well as the efficiency of different therapeutic approaches (Mayer 1999; Houtman et al. 2005; Terentiev et al. 2009).

The whole set of protein–protein interactions of a given organism are referred to as the interactome. Structural organisation of interactomes and total number of interactions in them are among important factors that determine complexity of biological systems. The number of copies of a certain protein per cell can vary from several tens to millions (Ghaemmaghami et al. 2003; Terentiev et al. 2009). The size of the human interactome is approximately tenfold larger than that in *D. melanogaster* and can be three times higher than that of *C. elegans* (Stumpf et al. 2008; Terentiev et al. 2009). Determination of physically interacting protein pairs makes it possible to design interactome maps as graphs consisting of nodes and of links between them that indicate paired interactions. The interactome maps are considered as keys to obtain knowledge on protein functioning (Cusick et al. 2005; Terentiev et al. 2009). Data obtained in vitro are used to construct static interactome maps and their further analysis which makes it possible to describe dynamic protein–protein interactions existing in vivo. Construction of interactome maps is also useful for the determination of the role of individual proteins and their interactions in the emergence and development of diseases and their diagnosis, as well as for identification of possible drug targets and monitoring of treatment efficiency. Biochemical methods like chemical cross-linking, combined fractionation during chromatography and co-immunoprecipitation were used for investigation of protein–protein interactions which were later replaced by highly efficient and high-throughput experimental methods like yeast two-hybrid assay (Y2H), phage display and tandem affinity purification-mass spectrometry (TAPMS) and were elaborated for interactome determination in various organisms (Causier 2004; Puig et al. 2001; Rigaut et al. 1999; Ho et al. 2002; Terentiev et al. 2009). Different microscopy techniques and different mathematical and computer methods also open broad possibilities for dynamic proteomics.

The cascade mechanism of signalling mediated by mitogen-activated protein kinase (MAPK) was analysed by computer-based approaches. On the other hand complicated intra- and intercellular processes were studied by mathematical modelling (Ivanov and Ivanova 2006; Terentiev et al. 2009). It becomes widely used for the description of events and processes taking place in a living cell and serves as a universal language for interpreting experimental data and predicting properties and behaviour of biomacromolecules under various conditions. Application of special computer programs such as Cytoscape can be used to comparatively visualise experimental data and to use them together with information contained in annotated databases on molecular networks (Shannon et al. 2003; Terentiev et al. 2009)

Transcriptomics

Transcriptomes as assessed by either microarrays or next-generation sequencing have produced an unprecedented data flood regarding transcript identity and levels in plant systems (Usadel and Fernie 2013). The transcriptome is the set of all the parts of the genome that are expressed as RNA transcripts in one or several populations of cells in a given time and a given environmental condition. Earlier northern blotting was the only choice for molecular biologists to measure the expression of thousands of genes in one experiment, but then microarray technology was introduced, but its high-cost input precluded this technology from scientific community (Schena et al. 1995). Arabidopsis was chosen to demonstrate microarray technology because of its small genome and rich EST collection in 1995. But the major drawback is the analysis of the data, there are several public databases containing microarray data: Genevestigator (https://www.genevestigator.ethz.ch/), NASCArrays, ArrayExpress (http://www.ebi.ac.uk/arrayexpress), The Gene Expression Omnibus (http://www.ncbi.nlm.nih.gov/geo/) and Stanford Microarray Database (http://genome-www5.stanford.edu) are the most popular. But much information and potential knowledge is untapped by adopting this approach. To overcome these problems, a computational solution where in recent extension of the principal component analysis variants STATIS and dual-STATIS (Lavit et al. 1994; Abdi et al. 2012; Klie et al. 2012) is applied to study the time-resolved response of *Arabidopsis thaliana* to perturbations in the prevailing light and temperature conditions (Urano et al. 2009).

In addition to mRNAs, ribosomal RNAs (rRNAs) and tRNAs (transfer RNAs) are of great interest in studying the expression of the large and heterogeneous population of small RNAs in plants. These small RNAs have important roles for regulation of gene expression as well as other roles. Unfortunately, EST libraries and microarray gene chips were not designed to detect small RNAs such as microRNAs (miRNAs). However, with the advancement of sequencing technologies, we are now able to measure miRNAs and other small RNAs (sRNAs), quantify their expression in different cell types and treatments and begin to understand their functional roles in plants (Lu et al. 2005).

Gene-to-Metabolite Networks

The study to analyse gene profiling and metabolite profiling data and its mining under different conditions to study the interaction among them are gene-to-metabolite networks. Their outputs are mainly visualised after the statistical and computer-based analysis by considering a factor at a single moment such as distance between the gene and metabolite. The study of gene-to-metabolite networks is more complex in plants because of their greater diversity and larger numbers of metabolites produced by plants (Lee et al. 2002). Early research used gene-to-metabolite networks to dissect the dynamic responses during sulphur and nitrogen starvation in *Arabidopsis* (Hirai et al. 2004; Yuan et al. 2008). The gene-to metabolite networks have also been constructed for plant species with limited available genome information, such as Madagascar periwinkle (*Catharanthus roseus*) (Rischer et al. 2006; Yuan et al. 2008). The network analysis in this case led to the discovery of novel candidate genes for terpenoid indole alkaloid biosynthesis.

The work integrated microarray-based gene profiling with liquid chromatography–mass spectrometry (MS) and Fourier transform–ion cyclotron MS-based metabolite profiling using multivariate analysis methods including self-organising map and principal components analysis to derive gene-to-metabolite associations (Nikiforova et al. 2005; Hirai et al. 2004; Yuan et al. 2008). Several unknown desulfoglucosinolate sulfotransferases and candidate transcriptional factors regulating anthocyanin biosynthesis (Langebartels et al. 2005) were identified using gene-to-metabolite profiling studies which were previously unknown (Hirai et al. 2005; Yuan et al. 2008). In order to study for stress responses, plant defence and hormone-induced responses, gene-to-metabolite networks have been characterised by employing systems biology approaches (Goossens et al. 2003; Zulak et al. 2007; Carrari et al. 2006; Yuan et al. 2008). In silico analysis such as meta-analysis of microarray data identifies coregulated genes with higher correlation efficiency, and metabolic pathways were linked to these groups in order to make a relation among different groups. MapMan is a software which enables statistical treatment of multiple microarray datasets to display the significantly changed genes in the corresponding metabolic pathway (Usadel et al. 2005; Yuan et al. 2008). This approach was successfully applied to identify the genes and metabolic pathways involved in the response to nitrogen deficiency and during diurnal cycles (Blasing et al. 2005; Scheible et al. 2004; Yuan et al. 2008).

The association of genes with metabolites enables the discovery of new genes involved in metabolite biosynthesis, transport, regulation and modification, in addition to their regulation (Hirai et al. 2005; Yuan et al. 2008). Improvement in the gene function annotation becomes possible with the discovery of key regulatory components in the biological process by employing systems biology approaches (Tohge et al. 2005; Yuan et al. 2008). The gene-to-metabolite network clarified how biological processes are interrelated, enabling substantial improvements in omics data interpretation, better prediction of outcomes for system perturbations and conceptual reconstruction of interactive biological systems with multiple components, including enzyme activities, gene expression and metabolite levels (Hesse and Hoefgen 2006; Yuan et al. 2008).

But there are some technological and computational limitations always associated to these analyses such as calculation of the dynamics and mainly the structure visualisation. These challenges can be met either employing statistical parameters, for example, correlation analysis or graphs. Technological problems can be solved by using highly improved mass spectroscopy techniques or the next-generation sequencing (NGS) facilities. But the employment of advanced NGS methods requires efficient and high-memory supercomputer and knowledge about algorithms and tools.

Plant Metabolomics

In the postgenomic era, metabolomics is expected to be the newest useful omics science for functional genomics. Metabolomics represents the exhaustive profiling of metabolites contained in the organisms. Proteomics and transcriptomics are both considered to be a flow of media concerning genetic information. Recently, it has been proved that slight changes in the metabolome can be explained by perturbations (environmental changes, physical stress, abiotic stress, nutritional stress, mutation and transgenic events) imposed on plants. Despite being an immature technology, it has also been used as a powerful tool for precise phenotyping, analysis of large mutants and transgenic libraries of model experimental plants such as *Arabidopsis*, etc. Nowadays, venture business companies for plant biotechnology are using plant metabolomics technology to drive the large-scale exhaustive screening of T-DNA tagging transgenic libraries of *Arabidopsis* to determine the functionality of genes. The main goal of these companies is to establish the relationship between the useful features and their corresponding genes (Fukusaki and Kobayashi 2005).

The complete set of small-molecule metabolites (such as metabolic intermediates, hormones and other signalling molecules and secondary metabolites) within a biological system refers as the *metabolome*. One of the main problems when studying the metabolome is the extreme heterogeneous chemical nature of the constituents of the metabolome especially in plants. They are known to have a tremendous enzymatic capacity needed for the production of an estimated ~200,000 different small molecules (Fiehn 2002). Finding the equilibrium between coverage and accuracy of measurement is necessary in the metabolomics field.

For most of the commercial plants such as wheat, barley, maize, soya bean and potato, genomic information is not necessary for metabolomics. This is one of the most important advantages of metabolomics compared with transcriptomics and proteomics (Kuhn et al. 1997). However, metabolomics is a complicated interdisciplinary research field that requires biosciences, analytical chemistry, organic chemistry, chemometrics and informatics knowledge. Its analysis requires the following steps: plant cultivation, sampling, derivatisation and pretreatment, separation and quantification, data conversion and data mining by multivariate analysis (principal component analysis, hierarchical cluster analysis, self-organising map).

Among them several other methods are also available like soft independent modelling of class analogy (SIMCA). It is a method for classification and prediction of unknown samples by means of principal component models that are prepared in each category of training sets. K-nearest neighbour and K-mean cluster analysis are also available for sample classification. In addition to this, a de facto protocol for data mining in metabolomics was also established (Fukusaki and Kobayashi 2005).

Model Plant Systems Biology

Systems approaches enable plant scientists to understand the structural stability of plants, their control and design structure and how these lead to robust and resilient behaviour. These capabilities are the result of a complex biological system in which control operates at many different levels (Alistair et al. 2012). These complexities were resolved by the incorporation of models. Modelling studies of networks at different levels of biological organisation share similar approaches, and integrative studies need to identify the links between the different levels. Information about these links enables predictions to understand and interconnect the consecutive levels (Goymer 2008; Keurentjes et al. 2012), and new experimental data can be used to improve and validate connections and interrelationships. Models can be constructed from different perspectives such as incorporating network theories, transcriptional networks, gene regulatory networks, etc. (Harel and Pnueli 1985). The choice of selection depends on the requirement of the experiment and research. Attempts to capture plant performance over multiple levels of complexity were solved by models in the greed to cope with different regulatory mechanisms. In top–down models statistical modelling is often the most favoured approach, whereas in bottom–up models dynamic or constraint-based modelling might be more appropriate (Keurentjes et al. 2012). Models incorporating the individual level in plant ecophysiological dynamic modelling trace back almost 50 years from now and can be very complex (Keurentjes et al. 2012). Geometric simulation models were used to visualise and predict the effect of parameter settings at a higher level of aggregation in order to analyse the plant growth. Three-dimensional (3D) architecture of the plant was used to study the structural–functional relationship in the plant modelling by representing all the organs of a plant spatially (De Las Rivas et al. 2004; Evers et al. 2010). The most widely used method to create functional–structural plant models is the Lindenmayer system method (Jones et al. 2003) facilitating the communication with the environment and allows the several necessary inclusion including light interception and penetration, photosynthesis, assimilate distribution throughout the plant structure, transport of compounds from and to organs and production and release of volatiles in a 3D environment (van Der Putten et al. 2009; Keurentjes et al. 2012).

Modelling and Simulation in Plant System Dynamics

The systems interest to biological sciences dates back to the days of Wiener (1948) and Forrester (1958). In the context of biology, biochemical systems theory (Voit 2000) and metabolic control theory (Heinrich and Schuster 1996) proposed general mathematical models of biological systems at and around a steady state (equilibrium). Systems biology aims to elucidate how complex behaviours of biological systems emerge from the properties of the components and interactions in the systems. It uses a combination of experimental techniques and computational approaches to gain global insights into complex biological systems. The experimental techniques employed in systems biology tend to have high-throughput capabilities by the employment of techniques, such as protein–protein interactions (Yuan et al. 2008; Wang et al. 2012), transcriptional regulations (protein–DNA interactions) (Wang et al. 2012; Le et al. 2006; Hallikas and Taipale 2006) and genetic interactions (Wang et al. 2012; Tong et al. 2004), which enable the researchers to determine the abundance or activity of numerous components at the same time. Experimental data from high-throughput technologies and small-scale studies provide a rich source for understanding the system-level mechanisms of biological processes (Wang et al. 2012; Papin et al. 2005).

Depending on the available information and the complexity of biological systems, several approaches can be followed to reconstruct the system in a mathematical model:

- *Top–down approaches* start by reconstructing a possible topology of the network under consideration at a low level of complexity and providing a broad overview of the system. For example, statistical analyses and static network models are applied to high-throughput omics data (Wang et al. 2012; Sobie et al. 2011; Kahlem and Birney 2006; Albert 2007).
- *Bottom–up approaches* start with a highly detailed model of a single module of the network and address the question of how variation in the output of this module affects processes described by other modules at a higher level of organisation (Keurentjes et al. 2012). Internal node within the network is broadly considered. This class of methods usually starts with hypotheses of biological mechanisms generated from individual small-scale experiments. Continuous dynamic modelling is the most widely used bottom–up method (Wang et al. 2012; Sobie et al. 2011; Aldridge et al. 2006; Karlebach and Shamir 2008).

Both approaches follow the central paradigm of linearity between cause and consequence, which has proven to be very effective in physical sciences. Discrete dynamic modelling such as

Boolean network models (Wang et al. 2012; Thakar and Albert 2010; Assmann and Albert 2009; Albert and Wang 2009), multi-valued logical models (Aldridge et al. 2009; Morris et al. 2011; Wang et al. 2012) and Petri nets (Chaouiya 2007; Ruths et al. 2008; Wang et al. 2012) does not require kinetic parameters and is able to provide qualitative dynamic descriptions of system behaviours.

Graphical Methods

A network/graph, in systems biology, has two basic parts: the elements of the system are represented as *graph nodes* (also called vertices), and the interactions are represented as *edges*, that is, lines connecting pairs of nodes. Edges may be directed (originating from a source (starting node) to a sink (ending node) and represent unidirectional flow of material or information) or nondirected (representing mutual interactions where the directional flow of information is not known). In biological networks, nodes (or vertices) represent the molecules present inside a cell (e.g. proteins, RNAs and/or metabolites), and links (or edges) between nodes represent their biological relationships (e.g. physical interaction, regulatory connections, metabolic reactions) (Blais and Dynlacht 2005). The nodes of the interaction network represent population of biomolecules, whose abundance varies in time and in response to the internal and environmental perturbations.

In order to create models, variables such as concentration, expression and activity which indicate the state of each node and set of equations indicating how the state change reacts to stimuli are considered. Models may be static or dynamic depending on their behaviour in the system with time. The four common types of networks in plant systems biology include (a) gene-to-metabolite networks, (b) protein–protein interaction networks, (c) transcriptional regulatory networks and (d) gene regulatory networks, wherein the first three types are often static, while the gene regulatory network frequently is dynamic (Yuan et al. 2008) (Fig. 3).

Mathematical Dynamic Methods

The complexity of biological systems is compounded by the fact that they are open and react to time-varying input received from their environment. A reactive system must respond to each stimulus as it occurs, often needing to respond to many stimuli concurrently (Harel 2003; Harel and Pnueli 1985; Sadot et al. 2013). The structure of the system is also typically dynamic, with its components being repeatedly created and destroyed during the system's lifespan, adding yet another level of complexity (Sadot et al. 2013).

To facilitate further analysis and systems-level understanding, this information is often integrated into large-scale models using a variety of mathematical and computational approaches. A general class of models is so-called executable models. Such a model defines how, given certain events, the system transitions from one state to another. State-based formalisms can be used to construct computational models that describe the complex dynamics of reactive systems, including biological systems. Such models are typically highly nonlinear and nondeterministic and can simulate very large systems (Fisher and Henzinger 2007; Sadot et al. 2013).

Dynamic computational models are powerful tools for developing and testing hypotheses about complex biological systems (Gutenkunst et al. 2004; Kitano 2002; Locke et al. (2005); Voit et al. 2006). It has even been suggested that such models will soon replace databases as the primary means for exchanging biological knowledge (Gutenkunst et al. 2004; Aldridge et al. 2006).

The main advantage of building a mathematical model is that molecular mechanisms that are supposed to govern the respective process need to be formulated explicitly (Gutenkunst et al. 2004). A major challenge with such models, however, is that they often possess tens or even hundreds of free parameters whose values can significantly affect model behaviour (Ingram et al. 2006; Mayo et al. 2006; Gutenkunst et al. 2004). While high-throughput methods for discovering interactions are well-developed (Sachs et al. 2005; Gutenkunst et al. 2004), high-throughput meth-

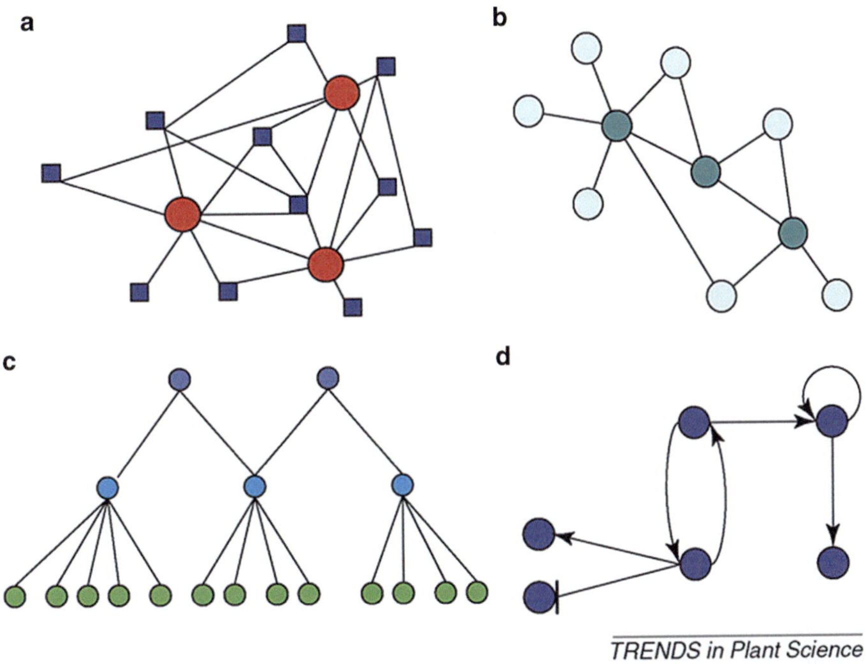

Fig. 3 Types of network in plant system biology. (**a**) Gene- to- metabolite network. (**b**) Protein-protein interaction network. (**c**) Transcriptional regulatory networks. (**d**) Gene regulatory networks (Yuan et al. 2008)

ods for measuring biochemical parameters remain limited (Maerkl and Quake 2007; Gutenkunst et al. 2004). Furthermore, using values measured in vitro and in vivo application may introduce substantial inaccuracies (Teusink et al. 2000; Minton 2001; Gutenkunst et al. 2004). In approaches typically more focused on steady-state distributions of fluxes in metabolic networks, metabolic control analysis has been used to quantify the sensitivity of model behaviour with respect to parameter variation (Fell 1997; Gutenkunst et al. 2004), and flux balance analysis and related techniques have probed the robustness of metabolic networks (Wiback et al. 2004; Famili et al. 2005; Gutenkunst et al. 2004).

Modelling and simulation offer the possibility of integrating information, performing *in silico* experiments and generating predictions and novel hypotheses so as to better understand complex biological systems. However, the quality of the results will highly depend on the predictive capabilities of the model at hand. In this regard, the selection of an adequate modelling framework for the system under consideration and for the questions to be addressed is crucial (Wolkenhauer et al. 2004; Chis et al. 2011) together with the capacity to anchor model sophistication with experimental data (Janes and Lauffenburger 2006; Chis et al. 2011). In this respect, parameter estimation by means of data fitting has become a critical step in the model-building process (Banga and Balsa-Canto 2008; Chis et al. 2011).

In the context of systems biology, dynamical models consisting of ordinary differential equations (ODE) are a frequently used approach that facilitates to analyse the mechanism of action in a systematic manner.

For each observable $y(t,\theta)$, the corresponding experimental data is $y(t)$. For concentration measurements by biochemical assays, it is reasonable to assume that the measurement noise is multiplicative log-normally distributed.

Mathematical Frameworks to Build Dynamical Models

The mathematical frameworks to build dynamic models follow steady-state equation. The state S of the model at time t is just the set of all the variables "$x1, x2, \ldots, xN$" at time t:

$$S(t) = \{x1(t), x2(t) \ldots x7(t) \ldots x(t)\} \quad (1)$$

which can be considered as a point in the (N-dimensional) state space of the system. In principle, the variables are measurable quantities, such as mRNA, protein or hormone concentrations. If the state is changing with time t, then the model is dynamical, and the time-varying components of the state are the variables $x1(t)$, $x2(t), \ldots, xN(t)$. The form of model we shall study is:

$$S(t_2) = f(S(t_2)); p_1, p_2 \ldots p_M \quad (2)$$

A function f is assigned to a state of variable and $p_1, p_2, \ldots, p_M$ are model parameters. The parameters are numerical values that encode information about the system and do not vary in time. A particular choice of parameter values can be thought of as a point in parameter space. The components $[x_i(t)]$ and interactions can be inferred from a wide range of data sources, such as genetic and RNA interference screens, mRNA profiling, protein–protein interaction screens and analysis of transcription factor binding. Each has strengths and limitations, and integration of multiple data sources is important for reliably inferring the interactions in a network (Middleton et al. 2012).

Modelling the Dynamics of Cellular Processes

The majority of cellular processes can be described by networks of biochemical reactions. The dynamics of these processes, i.e. the time evolution of the concentrations of the involved molecular compounds, can often be modelled by systems of ODEs (Wolkenhauer 2008; Raue et al. 2013):

$$\frac{d}{dt}x(t,\theta) = fx(x(t,\theta), u(t), \theta) = N.v(x(t,\theta), u(t), \theta) \quad (3)$$

The variables x correspond to the dynamics of the concentration of n molecular compounds such as hormones, proteins in different phosphorylation states, mRNA or complexes of the former. The right-hand side of Eq. (3) can usually be decomposed into a stoichiometry matrix N and reaction rate equations v of the molecular interactions (Heinrich and Schuster 1996). A time-dependent experimental treatment that alters the dynamical behaviour of the system can be incorporated by the function $u(t)$. The initial state of the system is described by $x(0,\theta) = fx(\theta)$. Often, these initial conditions represent a steady-state solution to Eq. (3) that indicates that the system is in equilibrium (Raue et al. 2013).

Quantitative Measurements for Model Calibration

All desired molecular compounds can be measured directly or individually. In order to compare the model dynamics simulated for candidate parameter values θ, the dynamic variables x are mapped to m observables

$$y(t,\theta) = fy(t, x(t,\theta), x) \quad (4)$$

via a function fy. The observables y are the quantities that can be measured in experiments at time points t. They may depend on additional parameters that are included in θ such as scaling or offset parameters in case of relative data or measurement background (Raue et al. 2013).

Boolean Method

This method is also known as *discrete dynamic modelling* or *method to model complex species in absence of quantitative information.* A Boolean network consists of a set of nodes whose state is binary and is determined by other nodes in the network through Boolean functions. In terms of complexity, these networks lie between static network models and continuous dynamic models (Kahlem and Birney 2006; Wang et al. 2012). They are mainly considered for large biological systems, for example, genetic regulatory networks (Kauffman 1969; Thomas 1973; Wang et al. 2012). Boolean networks have been used as models in reverse engineering of biological networks, e.g. to infer regulatory interactions and signalling pathways from gene expression or proteomics data (Saez-Rodriguez et al. 2009, 2011; Lee and Tzou 2009; Wang et al. 2012).

In the case of Boolean networks, variables represent the state of a gene by the means of Boolean operators either be 0 or 1. The change of the state and the interconnection between two nodes are represented by the equations developed with the means of Boolean functions (AND, OR and NOT). The present state of the operator must be checked and regular update procedures were carried to solve the process (Thomas and D'Ari 1990; Middleton et al. 2012). The majority of the procedures are initiated by considering the initial state of the node as the steady one as it is applicable to both synchronous and asynchronous system.

Simpler synchronous update schemes are often used if one wishes to study only the steady states of the system, whereas asynchronous is better suited to study the dynamics of the system. For an asynchronous update, the choice of which gene is updated at each time step can be made in several ways, which may be defined as follows: (1) by specifying an order in which genes are updated; (2) based on a specific delay for each gene, which is compared with a global clock and reset after a change of state or (3) as a stochastic event. At each time step, one gene is chosen randomly and updated (Li et al. 2006; Middleton et al. 2012). In fact, more general random update functions can be used (Shmulevich et al. 2002; Middleton et al. 2012).

Advantages

Boolean models may:

- Require no quantitative parameters at all, all regulatory functions being specified using logical gates (e.g. AND, NOT, OR)
- Require time delay parameters for asynchronous updates
- Require weights for each input of regulatory function

Examples of Model and Translational Studies

Metabolic Engineering

The science that combines systematic analysis of metabolic pathways with molecular biological techniques to improve cellular properties by designing and implementing rational genetic modifications is known as *metabolic engineering*. It deals with the measurement of metabolic fluxes and elucidation of their control as determinants of metabolic function and cell physiology. Departing away from the traditional reductionist paradigm of cellular metabolism, it takes a holistic view. It is used as a framework for the analysis of genome-wide differential gene expression data along with data on protein content and in vivo metabolic fluxes (Stephanopoulos 1999). The insights of the integrated view of metabolism generated by metabolic engineering will have profound implications in biotechnological applications, as well as in devising rational strategies for target selection for screening candidate drugs or designing gene therapies. In addition to this, metabolic engineering provides examples of applications in the production of primary and secondary metabolites, improving cellular properties and biomedical engineering (Vemuri and Aristidou 2005).

The production of a certain substance from the cell is increased with the application of metabolic engineering by optimising cell's genetic and regulatory processes. These processes are systematic analysis of biological and chemical reactions controlled by enzymes that allow cells to convert raw materials into molecules necessary for the cell's survival. Metabolic engineering specifically allows to mathematically model these networks, calculate a yield of useful products and highlight the parts of the network that constrain the production of these products (Yang et al. 1998). The ultimate goal of metabolic engineering is to be able to employ microorganisms to produce economically valuable substances on an industrial scale in a cost-effective manner. Current examples include producing beer, wine, cheese, recombinant proteins (Srivastava et al. 2013), pharmaceuticals and other biotechnology products.

At the industrial scale, metabolic engineering is becoming more convenient and cost effective. According to the Biotechnology Industry Organization, more than 50 biorefinery facilities are being built across North America, employing metabolic engineering to produce biofuels and chemicals from renewable biomass which can help in reducing greenhouse gas emissions (Keasling 2010). The significant contribution of the metabolic engineering is the emphasis it places on the metabolic fluxes and their control under in vivo conditions. It is initially conceived as the ad hoc pathway manipulation and quickly became the natural outlet for the analytical skills of engineers who saw the opportunity for introducing rigour in this process by utilising the available platform of metabolic control analysis. The combination of analytical methods to quantify fluxes and their control with molecular biological techniques to implement suggested genetic modification is the essence of metabolic engineering.

The flux is a fundamental determinant of cell physiology and the most critical parameter of a metabolic pathway. While a metabolic pathway is defined as the study of feasible and observable biochemical reaction steps connecting a specific set of input and output metabolites in a sequential manner. The pathway flux is then defined as the rate at which the input metabolites are processed to form output metabolites. And the determination of metabolic fluxes in vivo has been termed as metabolic flux analysis (MFA). Metabolic flux analysis reveals the degree of pathway engagement in the overall metabolic process. There are three steps in the process for the systematic investigation of the metabolic flux and their control:

- The first is to develop the means to observe as many pathways as possible and to measure their fluxes.
- The second step is to introduce well-defined perturbations to the bioreaction network and to determine the pathway fluxes after the system relaxes to its new steady state.
- The third step is the flux control determination and analysis of flux perturbation results.

Metabolic engineering offers one of the best ways for meaningfully engaging chemical engineers in biological research, for it allows the direct application of the core subjects of kinetics, transport and their thermodynamics to the analysis of the reactions of metabolic networks (Koffas et al. 1999). Increased attention to the need to provide comprehensive cellular descriptions by integrating the plethora of individual pieces of information anticipated explosion of information from genomics research. In this context, metabolic engineering provides a valuable forum for upgrading the quality of biological information and synthesising it for the purpose of developing useful products and processes (Stephanopoulos 1999).

Specific areas of industrial production where metabolic engineering make significant contributions are the production of presently petroleum-derived thermoplastics by fermentation as well as by expression in whole plants, the production of new materials and the production of new biologically active agents such as polyketides. The production of gums, solvents, proteins, diverse antibiotics, foods, alcohols, biogas, oligopeptides, lipids, oils, pigments, vitamins, amino acids and bacterial cellulases is the partial list of product classes that have been produced biologically and presently are the target of metabolic manipulations mainly in microorganisms. The major driving forces behind manufacturing applications are the continuing increase in the production volume of carbohydrate raw material worldwide, continuing decline in the manufacturing cost of biotechnologically produced products and the power of technologies developed by modern molecular biology (Stephanopoulos 1999).

In industrial context, the practical ultimate goal of the metabolic engineering is the design and creation of optimal biocatalyst, optimising in terms of maximising the yield and productivity of desired product. In addition to this, it will have a significant effect on medical field by focusing on the design of new therapies by identifying specific targets for drug development and by contributing to the design of gene therapies (Stephanopoulos 1999).

Plants Response to Abiotic Stress

The plant molecular responses to abiotic stresses involve interactions and crosstalk with many molecular pathways (Takahashi et al. 2004; Cramer et al. 2011). One of the earliest signals in many abiotic stresses involves ROS and reactive nitrogen species (RNS), which modify enzyme activity and gene regulation (Molassiotis and Fotopoulos 2011; Wilkinson and Davies 2009; Cramer et al. 2011). There are a large number of studies on the oxidative effects of ROS on plant responses to abiotic stress, but only a few studies documenting the nitrosative effects of RNS (Mittler et al. 2011; Cramer et al. 2011). Hormones are also important regulators of plant responses to abiotic stress (Morgan and Drew 1997). Among them the two most important are abscisic acid (ABA), a central regulator of many plant responses to environmental osmotic stresses, and ethylene, involved in many stress responses (Stepanova and Alonso 2009) including drought, ozone, flooding (hypoxia and anoxia), heat, chilling, wounding and UV-B light (Molassiotis and Fotopoulos 2011; Mittler et al. 2011; Morgan and Drew 1997; Cramer et al. 2011).

In the postgenomic era, comprehensive analyses using three systematic approaches or omics have increased our understanding of the complex molecular regulatory networks associated with stress adaptation and tolerance (Grant et al. 2011). The first one is 'transcriptomics' for the analysis of coding and noncoding RNAs and their expression profiles. The second one is 'metabolomics' that is a powerful tool to analyse a large number of metabolites. The third one is 'proteomics' in which protein and protein modification profiles offer an unprecedented understanding of regulatory networks (Cramer et al. 2011). Integration of the different omics analyses facilitates abiotic stress signalling studies allowing for more robust identifications of molecular targets for future biotechnological applications in plants. Transcriptome analysis technologies have revealed the global transcriptomes of plants exposed to abiotic stresses such as dehydration, cold, heat, high salinity, osmotic stress and ABA. These analyses indicate that these stresses increase or decrease transcript abundance from not only previously identified stress-responsive genes but also from thousands of unannotated non-protein-coding regions.

An important application of transcriptomics data is co-expression analysis of target genes using online analytical tools, such as ATTED-II (Usadel et al. 2009; Cramer et al. 2011). This approach is very promising for understanding gene–gene correlations and finding master genes in target conditions. Datasets from 1,486 microarray experiments were used for the construction of regulatory model of *Arabidopsis* using InferGene application (Carrera et al. 2009; Cramer et al. 2011). Ten genes were predicted to be the most central regulatory hubs influencing the largest number of genes. Included in this set were transcription factor genes involved in auxin (KAN3), gibberellins (MYB29), abscisic acid (MYB121), ethylene (ERF1) and stress responses (ANAC036). The top 12 gene subnetworks were computed out of which four of these were related to biotic and abiotic stresses (Sakurai et al. 2007).

The drought response of loblolly pine roots was investigated and results in a number of hubs in the transcriptional network (Lorenz et al. 2011; Cramer et al. 2011). These genes are involved in phospholipid metabolism, ABA biosynthesis and signalling and cytokinin metabolism; they appear to be important in stress mediation.

To define six modules for *Arabidopsis* responses to abiotic stress, weighted co-expression analysis was used. Two hubs in the common response module were an ankyrin-repeat protein and genes involved in Ca signalling. They created a compendium of genomic signatures and linked them to their co-expression analysis (Weston et al. 2008; Cramer et al. 2011). The effects of hydrogen peroxide (H_2O_2) signalling during high light stress using microarray analyses were analysed. They found that H_2O_2 was not only heavily involved in signalling in high light stress along with salinity, water deficit, heat and cold stress (Vanderauwera et al. 2005; Cramer et al. 2011).

A time-series experiment was conducted to study the effects of UVB light on *Arabidopsis* using both metabolomics and transcriptomics analyses. They found that plants responded in two phases with an upregulation of primary metabolites in the first phase and the induction of protective secondary metabolites, especially phenolics, in the second phase (Kusano et al. 2011; Cramer et al. 2011).

The effect of cold temperatures over time were studied using transcriptomics, metabolomics and enzyme activity. The early changes approximately 6 h in enzyme activities were poorly correlated with transcript abundance, but after 78 h these correlations were greatly improved. Much of the long-term changes in metabolism could be ascribed to the CBF regulon (Usadel et al. 2009; Cramer et al. 2011).

Integrated analyses of the transcriptome and the metabolome successfully demonstrate connections between genes and metabolites, elucidating a wide range of signal output from ABA under dehydration (Urano et al. 2009; Cramer et al. 2011; Urano et al. 2009) and the DREB1/CBF transcription factors in response to low temperature (Cook et al. 2004; Maruyama et al. 2009; Cramer et al. 2011). Metabolite profiling reveals that ABA accumulates during dehydration, regulating the accumulation of various amino acids and sugars such as glucose and fructose. In particular, the dehydration-inducible accumulation of BCAAs (branched-chain amino acids), saccharopine, proline and agmatine is correlated with the dehydration-inducible expression of their key biosynthetic genes (BCAT2, LKR/SDH, P5CS1 and ADC2, respectively), which are regulated by endogenous ABA (Urano et al. 2009; Cramer et al. 2011).

The above explained are some of the good examples of systems biology and omics approaches that have been used to identify key genes regulating stress tolerance. Some of the experiments related to the integration of systems biology results in advantage to the crop fields also.

An SNAC1 gene was identified from microarray experiments of stress treatments on rice (Hu et al. 2006; Cramer et al. 2011). SNAC1 is an NAC transcription factor that induces the expression of a number of stress-tolerance genes and improves the drought and salt tolerance of rice in the field. The transgenic plants exhibited increased sensitivity to ABA and reduced water loss. The linkage of key regulatory hubs to phenotypic traits will allow for more rapid progress in the genetic manipulation and production of crop plants. Due to the importance of salinity stress in agriculture, there are many metabolomic studies to assess the metabolic effect of salinity in a variety of crop and related plant species; including tomato (Lenz et al. 2011; Cramer et al. 2011), grapevine (Kaufmann et al. 2011; Cramer et al. 2011), poplar (Usadel et al. 2009; Cramer et al. 2011), sea lavender *(Limonium latifolium)* (Hirai et al. 2004; Cramer et al. 2011) and rice (Hirai et al. 2004; Cramer et al. 2011).

Defence System

Plants respond to herbivory through various morphological, biochemical and molecular mechanisms to counter the effects of herbivore attack. There is a wide range of biochemical mechanisms of defence (direct or indirect) against the herbivores (Abdul et al. 2012). The defensive compounds are either produced constitutively or in response to plant damage that can affect feeding, growth and survival of herbivores (Howe and Jander 2008). In addition, plants also release volatile organic compounds that attract the natural enemies of the herbivores (War et al. 2012). Direct defences are mediated by plant characteristics affecting herbivore's biology, for example, mechanical protection on the surface of the plants (in the form of hairs, trichomes, thorns, spines and thicker leaves) and by the production of toxic chemicals (such as terpenoids, alkaloids, anthocyanins, phenols and quinones) that decrease the development of the herbivores and even death (Hanley et al. 2007). Indirect defences against insects are characterised by the release of a blend of volatiles that specifically attract natural enemies of the herbivores by providing food (e.g.

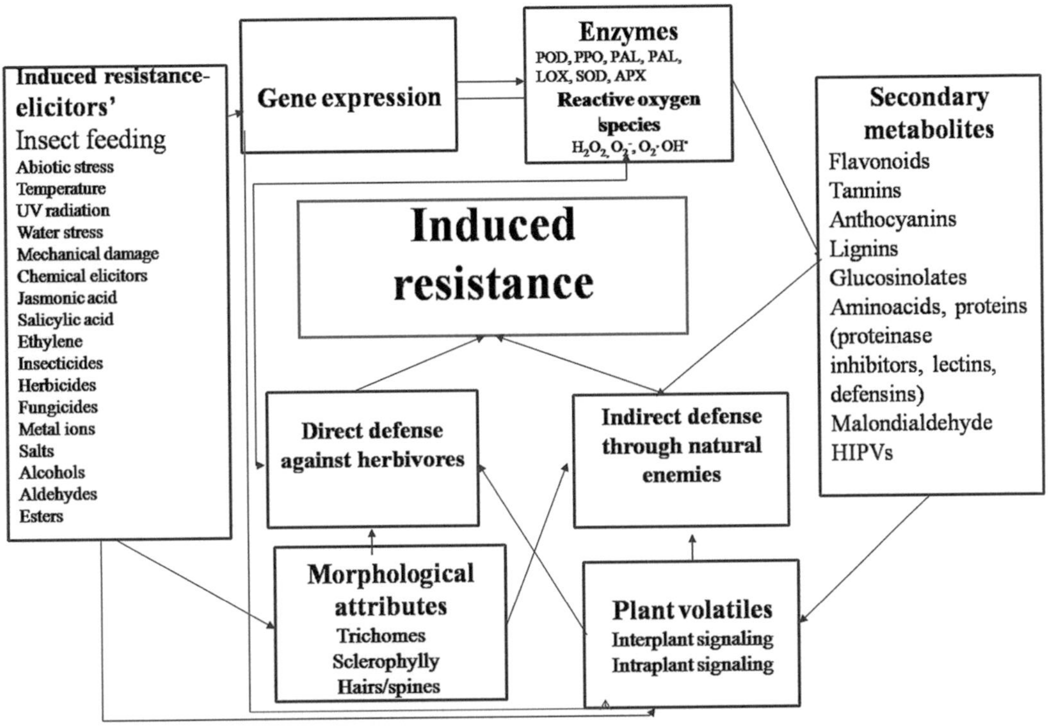

Fig. 4 Mechanism of induced resistance in plants. *POD* peroxidase, *PPO* polyphenol oxidase, *PAL* phenylalanine ammonia lyase, *TAL* tyrosine alanine ammonia lyase, *LOX* lipoxygenase, *SOD* superoxide dismutase, *APX* ascorbate peroxidase, *HIPVs* herbivore induced plant volatiles

extra floral nectar) and housing to optimise the effectiveness of the natural enemies (Arimura et al. 2009) (Fig. 4).

The first physical barrier is provided by the means of some structural traits such as leaf surface wax, thorns or trichomes and cell wall thickness and lignifications, and the subsequent production of the secondary metabolites forms the next barriers by acting as a toxin and reducing the growth, development and digestibility (Agrawal et al. 2009). In *Nicotiana attenuata*, trypsin proteinase inhibitors and nicotine expression contributed synergistically to the defensive response against *Spodoptera exigua* (Steppuhn and Baldwin 2007). Structural defences include morphological and anatomical traits that confer a fitness advantage to the plant by directly deterring the herbivores from feeding (Agrawal et al. 2009) and range from prominent protuberances on a plant to microscopic changes in cell wall thickness as a result of lignification and suberisation (Hanley et al. 2007; He et al. 2011).

Structural traits such as spines and thorns, trichomes, toughened or hardened leaves, incorporation of granular minerals into plant tissues and divaricated branching play a leading role in plant protection against herbivory (Hanley et al. 2007; He et al. 2011; Chamarthi et al. 2010). Sclerophyll referred to as hardened leaves reduces the palatability and digestibility of the tissues and results in the decreasing herbivore damage (Hanley et al. 2007; Handley et al. 2005).

Trichomes play an imperative role in plant defence against many insect pests and involve both toxic and deterrent effects (Chamarthi et al. 2010; Handley et al. 2005); they do not affect the normal growth and development of a plant but reduce the palatability of the plant tissues in which they are produced (Howe and Jander 2008). The defensive (secondary) metabolites can be either constitutive stored as inactive forms such as *phytoanticipins* or induced in response to the insect or microbe attack *phytoalexins* (Takahashi et al. 2004). Insect attack leads to qualitative and quantitative changes in proteins due to alteration of gene expression under stress including which in turn play an important role in signal transduction and oxidative defence (Usha Rani and Jyothsna 2010; Gulsen et al. 2010).

Many plant proteins ingested by insects remain intact in the midgut and move across the gut wall into the haemolymph. The function of a particular protein is affected by any alteration in its amino acid's sequence or content. One of the most abundant defensive classes of proteins in plants are proteinase inhibitors (PIs) with their higher concentration in storage organs such as seeds and tubers, and 1–10 % of their total proteins comprise of PIs inhibiting different types of enzymes. Rapid and transient generation of ROS is a common phenomenon in plants on account of oxidative stress due to biotic and abiotic factors (War et al. 2012).

Modelling Multigenic Traits

Their is a need to use plant systems biology to model multigenic traits. Phenotypes were used to model QTLs under different environmental conditions to derive genotype to environmental interactions, in which one or more portion of the genome is associated with phenotypes. The introduction of several systems biology approaches helps in the prediction of multigenic traits such as modelling of QTLs; using the systems biology approach has helped to predict multigenic traits such as leaf growth and nitrogen accumulation in maize grain. This approach come over the limitation factors related to gene, protein and metabolite information. Comprehensive models using mathematical models and computer-based applications integrating information about QTLs, gene, protein, metabolite and phenotyping information enable more efficient breeding programme to improve traits in crop species such as maize, soya bean and tomato. The so-called expression QTL and metabolite QTL can be used to integrate QTL information with gene expression and metabolite profiling, respectively, and the exploration of the genetic basis for metabolite diversity across different *Arabidopsis* lines (Kamlage et al. 2008). Systems biology is ideally suited to resolve complex interactions that are defined by multigenic traits (Yuan et al. 2008).

Virtual Cell

The 'Virtual Cell' developed at the University of Connecticut Health Centre is a unique computational tool that allows the biologists, physicist and mathematicians to engage in computational cell biology. A biologically oriented graphical user interface is provided for the assembly of models by specifying the molecules, reactions and structures involved. It enables the construction of complex spatial models of biological interest. From these models simulations are then produced by the software, and predictions of these simulations can be directly compared with the experimental results. In case, if the simulation does not match, the model must be an incomplete or faulty description of experimental data and must be modified. And if the simulations are consistent with the experimental results, new simulations with the different conditions can be used to predict the results of new experiments that can further test the limits of the applicability of model.

It is also designed to be useful for experienced modellers such as bioengineers and mathematical biologists. It allows the direct entry of the mathematical equations that describes a model using a declarative language (Virtual Cell Mathematics Description Language). The conversion of mathematics into C++ coding language was done which can further be sent to numeric solvers. Thus this dual interface makes the virtual cell highly effective for promoting and supporting interactions between experimental biologists and experimental modellers in order to bridge the cultural gap between these two separate communities (Loew and Schaff 2011).

The Virtual Cell software is decomposed into three main components:

1. Modelling framework
2. Mathematics framework
3. WWW interface–biologically oriented interface

The biologically oriented user interface allows experimentalists to create models, define cellular geometry, specify simulations and analyse simulation results. There is a Math Editor component that has been integrated within the biological interface. The design of the biological to mathematical mapping allows for separate use of biology and math components and includes mathematical simplifications using pseudo-state approximation and mass conservation relationships. This allows for direct specification of mathematical problems, performing simulations and analysis on those systems. The stand-alone mathematics user interface is also a powerful tool for modelling reaction–diffusion systems (Rastogi et al. 2012). The Virtual Cell modelling and simulation software framework are extensively used for the following:

- Calcium dynamics in neuroblastoma cells
- Published mathematical descriptions using Virtual Cell
- Calcium wave in fertilised eggs
- Nuclear envelope breakdown
- Mitochondrial diffusion
- RNA trafficking

Molecular Pharming

The production of proteins or other metabolites valuable to medicine or industry in plants is defined as molecular pharming. The use of terminologies such as 'pharming', 'biopharming' or 'molecular pharming' has same meaning in biotechnology sector. Both short and long peptide chains can be synthesised either chemically or by living cells. Plants can synthesise a wide variety of proteins that are free of mammalian toxins and pathogens (Abumhadi et al. 2005). It represents a novel source of molecular medicines (plasma proteins), enzymes, growth factors, vaccines and recombinant antibodies. For the production of antibodies directed against dental caries, rheumatoid arthritis, cholera, *E. coli* diarrhoea, malaria, certain cancers, Norwalk virus, HIV, rhinovirus, influenza, hepatitis B virus and herpes simplex virus, transgenic plants have been used (Thomas et al. 2002).

Plants are an alternative expression system to animals for the molecular farming of antibodies (Schillberg et al. 2003). Plant-derived vaccines have been produced against *Vibrio cholerae*, enterotoxigenic *E. coli*, hepatitis B virus, Norwalk virus, rabies virus, human cytomegalovirus, rotavirus and respiratory syncytial virus F (Thomas et al. 2002). Many of these plant-derived antigens were purified and used as injectable vaccines (Miller 2003). But protection has actually been better with the edible vaccine than with the commercially available vaccine (Lamphear et al. 2004). Milk proteins which could be used to improve child health like ß-casein, lactoferrin and lysozyme and protein polymers that could be used in surgery and tissue replacement (Ma et al. 2003) were produced by the application of recombinant DNA technology.

Expression of thioredoxin in foods such as cereal grains would increase the digestibility of proteins and decrease its allergic capability (Thomas et al. 2002). Human collagen can also be produced in transgenic tobacco plants and that the protein is spontaneously processed and assembled into its typical triple-helical conformation. The production of chicken egg white avidin in transgenic corn using an avidin gene whose sequence had been optimised for expression in corn (Hood et al. 1997). The resultant avidin had properties almost identical to those of avidin from chicken egg white (Horn et al. 2004).

A wide range of pharmaceutical drugs, including vaccines for infectious diseases and therapeutic proteins for treatment of such things as cancer and heart disease, are produced by the application of biopharming (Chapple 2000). Lettuce is also being investigated as a production host for edible recombinant vaccines and has been used in one series of clinical trials for a vaccine against HBV (Ma et al. 2003). Tomatoes were used to produce the first plant-derived rabies vaccine (Ma et al. 2003). Structurally authentic and biologically active human growth hormone, serum albumin, a tetanus toxin fragment and the cholera toxin B subunit have been produced at high levels in tobacco chloroplasts (Ma et al. 2003). Recombinant antibody of a single-chain Fv against carcinoembryonic antigen was produced in rice and wheat (Stöger et al. 2000).

But there are certain environmental concerns while using food and nonfood crops. All field crops are subject to ingestion by wildlife. It would even be difficult to guarantee that all wildlife could be kept away from greenhouse crops. The products that accumulate in these plants could be toxic to an animal or could lead to more subtle physiological or behavioural effects. Biodiversity can also be affected as there must be some alterations in the exudates from the root system due to the change in the surrounding soil composition because of the production of genetically modified plants. The practice of molecular farming could also lead to a change in the proportion of crops grown in certain regions.

Plant Secondary Metabolite Production

Plant secondary metabolites are a diverse group of molecules that are involved in the adaptation of plants to their environment but are not part of the primary biochemical pathways of cell growth and reproduction. Protease inhibitors, lectins, alkaloids, nonprotein amino acids, cyanogenic glycosides, saponins and tannins are the major plant secondary metabolites or phytochemicals that occur in plants. They play a significant role in defence against herbivores and pathogens, regulation of symbiosis and control of seed germination. Plant secondary metabolites are unique sources for food additives, flavours, pharmaceuticals and industrially important pharmaceuticals (Ravishankar and Venkataraman 1990; Ravishankar and Rao 2000). Secondary metabolites are involved in regulating the chemical inhibition of competing plant species, an integral part of the interactions of species in plant and animal communities and the adaptation of plants to their environment (Makkar et al. 2007). In higher plants a wide variety of secondary metabolites are synthesised from primary metabolites (e.g. carbohydrates, lipids and amino acids). They are needed in plant defence against herbivores and even environmental stress (Bennett and Wallsgrove 1994) (Fig. 5).

Even abiotic and environmental stresses (pathogen attack, UV irradiation, high light, wounding, nutrient deficiencies, temperature and herbicide treatment) influence growth and secondary metabolite production in higher plants (Dixon and Paiva 1995). Ozone exposure has been shown to increase conifer phenolic concentrations (Rosemann et al. 1991), but low ozone exposure had no effect on monoterpenes and resin acid concentrations (Kainulainen et al. 1998). Deficiencies in nitrogen and phosphate lead to the accumulation of phenyl propanoids and lignifications (Dixon and Paiva 1995). Calcium is a ubiquitous molecule involved in various signal transduction pathways in plants. Calcium has been found to increase in response to stress such as light, salinity, cold and drought (Tuteja and Mahajan 2007). Metabolic activity and plant ontology are highly influenced by varying temperature, for example, high temperatures can induce premature leaf senescence, carotenoids and β-carotene. Brassicaceae were found to be slightly decreased after thermal treatments (Morison and Lawlor 1999).

The antioxidant, antiviral, antibacterial and anticancer effects of plant secondary metabolites influence the plant molecular biologists and plant breeders globally to carry out their research. For example:

- Molecular biologists have made genetic modifications in proanthocyanidin biosynthesis in forage plants with the aim of eliminating bloat.
- Improving the efficiency of conversion of plant protein into animal protein.
- Reduce greenhouse gases.
- Reduce gastrointestinal parasites.
- Rapeseeds (canola) with low levels of glucosinolates and erucic acid and cottonseed with low gossypol have been developed and commercialised by plant breeders.
- Genetically modified rice-expressing insecticidal cowpea trypsin inhibitor has also been produced.

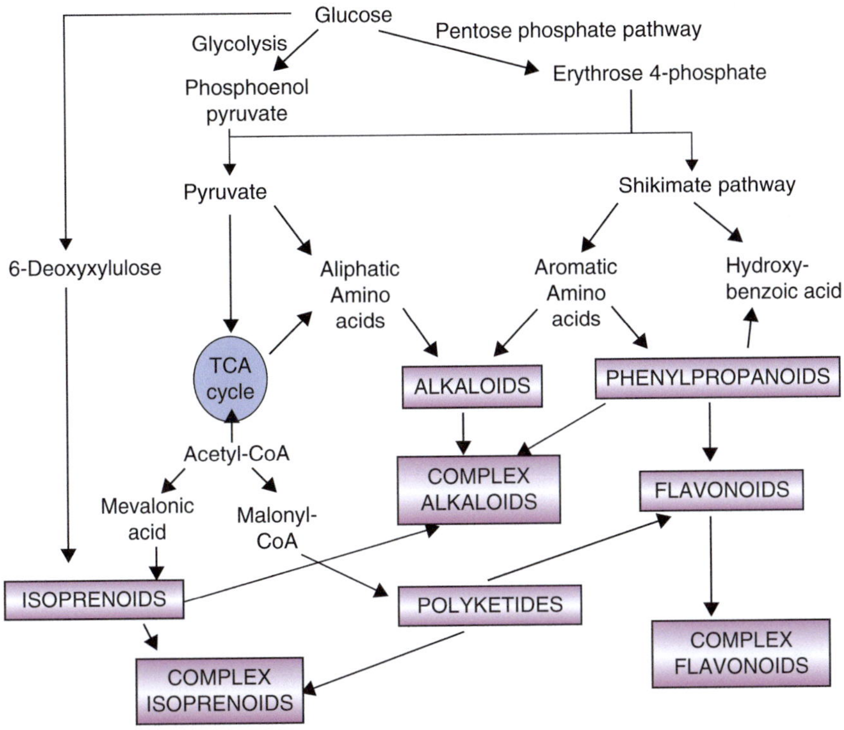

Fig. 5 Primary and secondary metabolites derived from carbon metabolism in plants

The emerging molecular genetic approaches have tremendous potential to unravel the regulatory genes that control plant secondary metabolite biosynthesis and production. But an increased knowledge of the enzymes specific for the pathway is needed for further research in the production of secondary metabolites.

Algorithms for System Analysis

1. *Stochastic optimization algorithms* apply sophisticated heuristics that randomly sample parameter space to evaluate the objective function. Due to the stochastic approach, the methods are less likely to get stuck in local minima. This is particularly advantageous for applications that are characterised by many local optima. Hence, they increase the probability of locating the global optimum despite local optima.
2. *Deterministic optimization algorithms* take steps that successively decrease the value of the objective function beginning from an initial guess for the parameter values (Press et al. 1990). They evaluate derivatives of the objective function. This leads to more rapid convergence to the optimum compared to stochastic algorithms. However, depending on the initially assumed parameter values, a deterministic optimization algorithm may converge to a local rather than global optimum. This limitation of deterministic optimization algorithms can be overcome by performing many independent optimization runs from randomly selected initial parameter guesses. This 'multi-start' approach facilitates a broad coverage of the parameter search space in order to find the global optimum.
3. *Hybrid optimization algorithms* use a combination of both strategies. First, promising candidate sets of parameter values are generated

using a stochastic strategy. The candidate sets are then further improved by a deterministic strategy (Raue et al. 2013).

Tools for System Analysis

In order to estimate the plant performance over multiple levels of complexity, models are needed that can cope with different regulatory mechanisms. In top–down models, statistical modelling is often the most favoured approach, whereas in bottom–up models dynamic or constraint-based modelling might be more appropriate. Techniques such as multivariate statistics and machine learning methods might be used to reconstruct a network topology at the initiation of modelling cycle, and dynamic models might be used to predict the dynamic behaviour of the network in further steps (Puchalka et al. 2008; Williams et al. 2010; Yuan et al. 2008).

Methods can be based on a hierarchical approach, which takes into account the structure of the model and the model's sensitivity to the parameter settings, which might be combined with a Monte Carlo or global sensitivity analysis approach of homology modelling (Vrugt et al. 2009). Once the network topologies have been estimated, (semi-)predictive dynamic models can be built in a bottom–up approach using, for instance, hybrid Petri nets, Boolean networks or differential equations (Mattheij and Molenaar 2002; Neutel et al. 2002; Polynikis et al. 2009; Davidich and Bornholdt 2008). Statistical network models can relatively easily be derived with regression techniques (Ter Braak and Prentice 2004; Bonneau et al. 2006; Keurentjes et al. 2007; Marbach et al. 2010), relevance networks based on association scores, Gaussian graphical models allowing identification of conditional independence or Bayesian networks used to represent probabilistic relationships (Neapolitan 2004; Jordan 1999; Keurentjes et al. 2007).

Bioinformatics leads towards the discovery of tools for network visualisation, modelling environments, pathway construction and visualisation tools, systems biology platforms and repositories of the models (Joyce and Palsson 2006; Turenne 2011). The core systems biology networks include SynBioWave (Staab et al. 2010), Cell Illustrator (Nagasaki et al. 2010), Moksiskaan (Laakso and Hautaniemi 2010), MEMOSys (Pabinger et al. 2011), Babelomics (Al-Shahrour et al. 2006), MetNet (Sucaet et al. 2012), etc. The systems biology model repositories include BioModels Database (Le Novere et al. 2006) or JWS (Olivier and Snoep 2004). Both are public, centralised databases of curated, published, quantitative kinetic models of biochemical and cellular systems.

Pathway databases are used for modelling systems, since they offer a clear-cut way of building network topologies by the annotated reaction systems (Usadel et al. 2009). The various pathway databases for systems analyses include KEGG (Kanehisa et al. 2012), BioCyc (Caspi et al. 2010), Aracyc (Mueller et al. 2003), Pathway Interaction Database (PID) (Schaefer et al. 2009) and BioCarta (Nishimura 2001). Also, several comprehensive modelling environments are available, like Gepasi (Mendes 1997), Virtual Cell (Loew and Schaff 2001), Osprey (Breitkreutz et al. 2003), *Arabidopsis* eFP browser (Winter et al. 2007), COPASI (Hoops et al. 2006), R (http://www.R-project.org), MatLab and InfoBiotics workbench (Blakes et al. 2011), E-Cell (Tomita et al. 1999) and Systems Biology Workbench (Sauro et al. 2003).

Conclusion

Plant as a system concerns each of its molecular constituents (DNA, RNA, proteins, metabolites, ions); the expanding development of high-throughput data generation technologies such as genomics, transcriptomics, proteomics, metabolomics, etc. made it possible to apply a systems biology paradigm in plant science. Systems approaches enable plant scientists to understand the structural stability of plants, their control and design structure. These capabilities are the result of a complex biological system in which control operates at many different levels. This understanding of complex system was made possible with the integration of various model-based methods including computer proficiency and mathematical calculations. Systems biology and systems ecology have in common the concept that networks form the core of the modelling language, and this is clearly reflected in the examples of study of the plant defence mechanisms and their response to abiotic stresses. The dynamic methods of modelling, parameters and algorithms involved results in a unique way to represent a complex system and provide a new direction to the researchers. The inner view of the molecular networks reflects their importance and advantages in medicine and biomedical field. By understanding systems integration biology, we are capable to accelerate crop adaptation for food, feed, biofuels and industrial and pharmaceutical production.

References

Abdi H, Williams LJ, Valentin D, Bennani-Dosse M (2012) STASIS and DISTATIS: optimum multi-table principal component analysis and three way metric multidimensional scaling. Wiley Interdiscip Rev Comput Stat 4:124–167

Abdul RW, Paulraj MG, Ahmad T, Buhroo AA, Hussain B, Ignacimuthu S, Sharma HC (2012) Mechanisms of plant defense against insect herbivores. Plant Signal Behav 7(10):1306–1320

Abumhadi N, Kamenarova K, Gecheff K, Atanassov A (2005) Molecular farming in plants: an approach of agricultural biotechnology. J Cell Mol Biol 4:77–86

Agrawal AA, Fishbein M, Jetter R, Salminen JP, Goldstein JB, Freitag AE et al (2009) Phylogenetic ecology of leaf surface traits in the milkweeds (Asclepias spp.): chemistry, ecophysiology, and insect behavior. New Phytol 183:848–867

Albert R (2007) Network inference, analysis, and modeling in systems biology. Plant Cell 19(11):3327–3338

Albert R, Wang RS (2009) Discrete dynamic modeling of cellular signaling networks. Methods Enzymol 467:281–306

Aldridge BB, Burke JM, Lauffenburger DA, Sorger PK (2006) Physicochemical modelling of cell signalling pathways. Nat Cell Biol 8(11):1195–1203

Aldridge BB, Saez-Rodriguez J, Muhlich JL, Sorger PK, Lauffenburger DA (2009) Fuzzy logic analysis of kinase pathway crosstalk in TNF/EGF/insulin-induced signaling. PLoS Comput Biol 5:e1000340

Alistair M, Middleton EF, Markus RO, Teva V (2012) Modeling regulatory networks to understand plant development: small is beautiful. Plant Cell 24:3876–3891

Al-Shahrour F, Minguez P, Tarraga J et al (2006) BABELOMICS: a systems biology perspective in the functional annotation of genome-scale experiments. Nucleic Acids Res 34:W472–W476

Arimura GI, Matsui K, Takabayashi J (2009) Chemical and molecular ecology of herbivore-induced plant volatiles: proximate factors and their ultimate functions. Plant Cell Physiol 50:911–923

Assmann SM, Albert R (2009) Discrete dynamic modeling with asynchronous update, or how to model complex systems in the absence of quantitative information. Methods Mol Biol 553:207–225

Banga JR, Balsa-Canto E (2008) Parameter estimation and optimal experimental design. Essays Biochem 45:195–210

Bennett RN, Wallsgrove RM (1994) Secondary metabolites in plant defence mechanisms. New Phytol 127:617–633. Biophys J 88: 1616–1625

Blais A, Dynlacht BD (2005) Constructing transcriptional regulatory networks. Genes Dev 19(13):1499–1511

Blakes J, Twycross J, Romero FJ et al (2011) The Infobiotics Workbench: an integrated in silico modelling platform for Systems and Synthetic Biology. Bioinformatics 27(23):3323–3324

Blasing OE et al (2005) Sugars and circadian regulation make major contributions to the global regulation of diurnal gene expression in Arabidopsis. Plant Cell 17(12):3257–3281

Bonneau R et al (2006) The Inferelator: an algorithm for learning parsimonious regulatory networks from systems-biology data sets de novo. Genome Biol 7:R36

Breitkreutz BJ, Stark C, Tyers M (2003) Osprey: a network visualization system. Genome Biol 4(3):R22

Carrari F et al (2006) Integrated analysis of metabolite and transcript levels reveals the metabolic shifts that underlie tomato fruit development and highlight regulatory aspects of metabolic network behavior. Plant Physiol 142:1380–1396

Carrera J, Rodrigo G, Jaramillo A, Elena SF (2009) Reverse-engineering the Arabidopsis thaliana transcriptional network under changing environmental conditions. Genome Biol 10(9):R96

Caspi R, Altman T, Dale JM et al (2010) The MetaCyc database of metabolic pathways and enzymes and the BioCyc collection of pathway/genome databases. Nucleic Acids Res 38(1):473–479

Causier B (2004) Studying the interactome with the yeast two-hybrid system and mass spectrometry. Mass Spectrom Rev 23:350–367

Chamarthi SK, Sharma HC, Sahrawat KL, Narasu LM, Dhillon MK (2010) Physico-chemical mechanisms of resistance to shoot fly, Atherigona soccata in sorghum, Sorghum bicolor. J Appl Entomol 135:446–455

Chaouiya C (2007) Petri net modelling of biological networks. Brief Bioinform 8:210–219

Chapple C (2000) Molecular 'pharming' with plant P450s. Trends Plant Sci 5(7):271–272

Chis O, Banga JR, Canto EV (2011) Structural identifiability of systems biology models: a critical comparison of methods. PLoS ONE 6(11):e27755

Choi H, Pavelka N (2011) When one and one gives more than two: challenges and opportunities of integrative omics. Front Genet 2:105

Cook D, Fowler S, Fiehn O, Thomashow MF (2004) A prominent role for the CBF cold response pathway in configuring the low-temperature metabolome of Arabidopsis. Proc Natl Acad Sci U S A 101(42):15243–15248

Coruzzi G, Gutierrez R (2009) Plant system biology. Annu Plant Rev 35

Cramer GR, Urano K, Delrot S, Pezzotti M, Shinozaki K (2011) Effects of abiotic stress on plants: a systems biology perspective. BMC Plant Biol 11:163

Cusick ME, Klitgord ME, Vidal M, Hill DE (2005) Interactome: gateway into systems biology. Hum Mol Genet 14:R171–R181

de Ruiter PC et al (2005) Ecology. Food web ecology: playing Jenga and beyond. Science 309:68–71

Davidich M, Bornholdt S (2008) The transition from differential equations to Boolean networks: a case study in simplifying a regulatory network model. J Theor Biol 255:269–277

Davidson EH et al (2002) A genomic regulatory network for development. Science 295:1669–1678

Davies PJ (ed) (2010) Plant hormones: biosynthesis, signal transduction, action. Plant Hormones. Springer, pp. 1–15

De Las Rivas J et al (2004) Evolution of oxygenic photosynthesis: genome-wide analysis of the OEC extrinsic proteins. Trends Plant Sci 9:18–25

Dixon RA, Paiva N (1995) Stressed induced phenylpropanoid metabolism. Plant Cell 7:1085–1097

Edwards D, Batley J (2004) Plant bioinformatics: from genome to phenome. Trends Biotechnol 22(5):232–237

Evers JB et al (2010) Simulation of wheat growth and development based on organ-level photosynthesis and assimilate allocation. J Exp Bot 61:2203–2216

Famili I, Mahadevan R, Palsson BØ (2005) k-cone analysis: determining all candidate values for kinetic parameters on a network scale Biophys J 88(3):1616–1625

Fell D (1997) Understanding the control of metabolism. Ashgate Publishing, London, 300 p

Fiehn O (2002) Metabolomics–the link between genotypes and phenotypes. Plant Mol Biol 48(1–2):155–171

Field S, Song OK (1989) A novel genetic system to detect protein–protein interactions. Nature 340:245–246

Fisher J, Henzinger TA (2007) Formal Methods in Systems Biology Executable cell biology. Nat Biotechnol 25:1239–1249

Forrester JW (1958) Industrial dynamics: A major breakthrough for decision makers. Harv Bus Rev 36(4):37–66

Fukusaki E, Kobayashi A (2005) Plant metabolomics: potential for practical operation. J Biosci Bioeng 100:347–354

Ghaemmaghami S, Huh W-K, Bower K, Howson RW, Belle A, Dephoure N, O'Shea EK, Weissman JS (2003) Global analysis of protein expression in yeast. Nature 425:737–741

Gomez-Roldan V, Fermas S, Brewer PB et al (2008) Strigolactone inhibition of shoot branching. Nature 455(7210):189–194

Goossens A et al (2003) A functional genomics approach toward the understanding of secondary metabolism in plant cells. Proc Natl Acad Sci U S A 100:8595–8600

Goymer P (2008) Network biology: why do we need hubs? Nat Rev Genet 9:650–651

Grant RC, Kaoru U, Serge D, Mario P, Kazuo SC et al (2011) Effects of abiotic stress on plants: a systems biology perspective. BMC Plant Biol 2011(11):163

Gulsen O, Eickhoff T, Heng-Moss T, Shearman R, Baxendale F, Sarath G et al (2010) Characterization of peroxidase changes in resistant and susceptible warm-season turfgrasses challenged by Blissus occiduus. Arthropod-Plant Interact 4:45–55

Gutenkunst RN, Waterfall JJ, Casey FP, Brown KS, Myers CR, Sethna JP (2004) Universally sloppy parameter sensitivities in systems biology models. PLoS Comput 3(10):e189

Hallikas O, Taipale J (2006) High-throughput assay for determining specificity and affinity of protein-DNA binding interactions. Nat Protoc 1:215–222

Hammer GL, Sinclair TR, Chapman SC, Oosterom EV (2004) On systems thinking, systems biology, and the in silico plant. Plant Physiol V 134:909–911

Handley R, Ekbom B, Agren J (2005) Variation in trichome density and resistance against a specialist insect herbivore in natural populations of Arabidopsis thaliana. Ecol Entomol 30:284–292

Hanley ME, Lamont BB, Fairbanks MM, Rafferty CM (2007) Plant structural traits and their role in antiherbivore defense. Perspect Plant Ecol Evol Syst 8:157–178

Harel D (2003) A grand challenge for computing: full reactive modeling of a multi-cellular animal. Verification, model checking, and abstract interpretation. Lect Notes Comput Sci 2937:323–324

Harel D, Pnueli A (1985) On the development of reactive systems. Logics Model Concurr Syst 13:477–498

He J, Chen F, Chen S, Lv G, Deng Y, Fang W et al (2011) Chrysanthemum leaf epidermal surface morphology and antioxidant and defense enzyme activity in response to aphid infestation. J Plant Physiol 168:687–693

Heinrich R, Schuster S (1996) The regulation of cellular systems. Chapman & Hall, London

Hesse H, Hoefgen R (2006) On the way to understand biological complexity in plants: S-nutrition as a case study for systems biology. Cell Mol Biol Lett 11:37–56

Hirai MY, Yano M, Goodenowe DB, Kanaya S, Kimura T, Awazuhara M, Arita M, Fujiwara T, Saito K (2004)

Integration of transcriptomics and metabolomics for understanding of global responses to nutritional stresses in Arabidopsis thaliana. Proc Natl Acad Sci U S A 101(27):10205–10210

Hirai MY et al (2005) Elucidation of gene-to-gene and metabolite-togene networks in Arabidopsis by integration of metabolomics and transcriptomics. J Biol Chem 280:25590–25595

Ho Y, Gruhler A, Heilbut A, Bader GD, Moore L, Adams SL, Millar A, Taylor P, Bennett K, Boutilier K, Yang L, Wolting C, Donaldson I, Schandorff S, Shewnarane J, Vo M, Taggart J, Goudreault M, Muskat B, Alfarano C, Dewar D, Lin Z, Michalickova K, Willems AR, Sassi H, Nielsen PA, Rasmussen KJ, Andersen JR, Johansen LE, Hansen LH, Jespersen H, Podtelejnikov A, Nielsen E, Crawford J, Poulsen V, Sorensen BD, Matthiesen J, Hendrickson RC, Gleeson F, Pawson T, Moran MF, Durocher D, Mann M, Hogue CW, Figeys D, Tyers M (2002) Systematic identification of protein complexes in Saccharomyces cerevisiae by mass spectrometry. Nature 415:180–183

Hood EE, Witcher DR, Maddock S, Meyer T, Baszczynski C, Bailey M, Flynn P, Register J, Marshall L, Bond D, Kulisek E, Kusnadi A, Evangelista R, Nikolov Z, Wooge C, Mehigh RJ, Hernan R, Kappel WK, Ritland D, Li CP, Howard JA (1997) Commercial production of avidin from transgenic maize: characterization of transformant, production, processing, extraction and purification. Mol Breed 3:291–306

Hoops S, Sahle S, Gauges R et al (2006) COPASI—a complex pathway simulator. Bioinformatics 22(24):3067–3074

Horn ME, Woodard SL, Howard JA (2004) Plant molecular farming: systems and products. Plant Cell Rep 22:711–720

Houtman JCD, Barda-Saad M, Samelson LE (2005) Examining multiprotein signaling complexes from all angles. FEBS J 272:5426–5435

Howe GA, Jander G (2008) Plant immunity to insect herbivores. Annu Rev Plant Biol 59:41–66

Hu H, Dai M, Yao J, Xiao B, Li X, Zhang Q, Xiong L (2006) Overexpressing a NAM, ATAF, and CUC (NAC) transcription factor enhances drought resistance and salt tolerance in rice. Proc Natl Acad Sci U S A. 103(35):12987–12992

Ingram PJ, Stumpf MPH, Stark J (2006) Network motifs: structure does not structure does not determine function. BMC Genomics 5(7):108

Ivanov VV, Ivanova NV (2006) Mathematical models of the cells and cell-associated objects. In: Chui CK (ed) Mathematics in science and engineering. Elsevier, 206 p

Jacob F (1974) The logic of living systems. Plant Cell 18:2420–2430

Janes K, Lauffenburger D A(2006) A biological approach to computational models of proteomic networks. Curr Opin Chem Biol 10:73–80. http://www.cdpcenter.org/files/pubs/janes_biological.pdf

Jones JW et al (2003) The DSSAT cropping system model. Eur J Agron 18:235–265

Jordan MI (1999) Learning in graphical models. Adaptive Computation and Machine Learning series, MIT Press, Cambridge

Joyce AR, Palsson BO (2006) The model organism as a system: integrating 'omics' data sets. Nat Rev Mol Cell Biol 7:198–210

Kahlem P, Birney E (2006) Dry work in a wet world: computation in systems biology. Mol Syst Biol 2:40

Kainulainen P, Holopainen JK, Holopainen T (1998) The influence of elevated CO_2 and O_3 concentrations on Scots pine needles: changes in starch and secondary metabolites over three exposure years. Oecologia 114:45560

Kamlage B, Poorter H, Stitt M (2008) Multilevel genomic analysis of the response of transcripts, enzyme activities and metabolites in Arabidopsis rosettes to a progressive decrease of temperature in the non-freezing range. Plant Cell Environ 31(4):518–547

Kanehisa M, Goto S, Sato Y et al (2012) KEGG for integration and interpretation of large-scale molecular data sets. Nucleic Acids Res 40(D1):109–114

Karlebach G, Shamir R (2008) Modelling and analysis of gene regulatory networks. Nat Rev Mol Cell Biol 9:770–780

Karp PD (1996) A strategy for database interoperation. J Comput Biol 2(4):573–583

Kauffman SA (1969) Metabolic stability and epigenesis in randomly constructed genetic nets. J Theor Biol 22:437–467

Kaufmann K, Smaczniak C, de Vries S, Angenent GC, Karlova R (2011) Proteomics insights into plant signaling and development. Proteomics 11(4):744–755

Keasling DJ (2010) Advanced biofuel production in microbes. Biotechnol J 5:147–162

Kefi S et al (2007) Spatial vegetation patterns and imminent desertification in Mediterranean arid ecosystems. Nature 449:213–217

Keurentjes JJB et al (2007) Regulatory network construction in Arabidopsis by using genome-wide gene expression quantitative trait loci. Proc Natl Acad Sci U S A 104:1708–1713

Keurentjes JJ, Angenent GC, Dicke M, Vı´tor AP, Santos MD, Olenaar JM, Wim H (2012) Redefining plant systems biology: from cell to ecosystem. Trends Plant Sci 16:4

Kitano H (2002) Computational systems biology. Nature 420:206–210

Klie S, Caldana C, Nikoloski Z (2012) Compromise of multiple time-resolved transcriptomics experiments identifies tightly regulated functions. Front Plant Sci 3:249

Koffas M, Roberge C, Lee K, Stephanopoulos G (1999) Metabolic engineering. Annu Rev Biomed Eng 1:535–557

Kuhn C et al (1997) Macromolecular trafficking indicated by localization and turnover of sucrose transporters in enucleate sieve elements. Science 275:1298–1300

Kusano M, Tohge T, Fukushima A, Kobayashi M, Hayashi N, Otsuki H, Kondou Y, Goto H, Kawashima M, Matsuda F, Niida R, Matsui M, Saito K, Fernie AR (2011) Metabolomics reveals comprehensive reprogramming involving two independent metabolic responses of Arabidopsis to UV-B light. Plant J 67(2):354–369

Kohn KW (1999) Molecular interaction map of the mammalian cell cycle control and DNA repair systems. Mol Biol Cell 10:2703–2734

Laakso M, Hautaniemi S (2010) Integrative platform to translate gene sets to networks. Bioinformatics 26(14):1802–1803

Lamphear BJ, Jilka JM, Kesl L, Welter M, Howard JA, Streatfield SJ (2004) A corn-based delivery system for animal vaccines: an oral transmissible gastroenteritis virus vaccine boosts lactogenic immunity in swine. Vaccine 22(19):2420–2424

Langebartels C, Gruissem W, Inze D, Van BF (2005) Genome-wide analysis of hydrogen peroxide-regulated gene expression in Arabidopsis reveals a high light-induced transcriptional cluster involved in anthocyanin biosynthesis. Plant Physiol 139(2):806–821

Le Novere N, Bornstein B, Broicher A et al (2006) BioModels Database: a free, centralized database of curated, published, quantitative kinetic models of biochemical and cellular systems. Nucleic Acids Res 34(1):689–691

Lee WP, Tzou WS (2009) Computational methods for discovering gene networks from expression data. Brief Bioinform 10:408–423

Lee TI et al (2002) Transcriptional regulatory networks in Saccharomyces cerevisiae. Science 298(5594):799–804

Lenz T, Fischer JJ, Dreger M (2011) Probing small molecule-protein interactions: a new perspective for functional proteomics. J Proteomics 75(1):100–115

Li S, Assmann SM, Albert R (2006) Predicting essential components of signal transduction networks: a dynamic model of guard cell abscisic acid signaling. PLoS Biol 4:e312

Locke JCW, Southern MM, Kozma-Bognr L, Hibberd V, Brown PE et al (2005) Extension of a genetic network model by iterative experimentation and mathematical analysis. Mol Syst Biol 1:0013

Loew LM, Schaff JC (2001) The virtual cell: a software environment for computational cell biology. Trends Biotechnol 10:401–406

Loew LM, Schaff JC (2011) Virtual Cell; a software environment for computational cell biology. Trends Biotechnol 19(10):401–406

Lorenz WW, Alba R, Yu YS, Bordeaux JM, Simoes M, Dean JF (2011) Microarray analysis and scale-free gene networks identify candidate regulators in drought-stressed roots of loblolly pine (P. taeda L.). BMC Genomics 24(12):264

Lavit C, Escoufier Y, Sabatier R, Traissac P (1994) The, ACT (STASIS method). Computational 18:97–119

Lu Y, Hong S, Tjonahen E, Serhan CN (2005) Mediator lipidomics: databases and search algorithms for PUFA-derived Mediators. J Lipid Res 46:790–802

Ma JK-C, Drake PMW, Christou P (2003) The production of recombinant pharmaceutical proteins in plants. Genetics 4:794–805

Maerkl SJ, Quake SR (2007) A systems approach to measuring the binding energy landscapes of transcription factors. Science 315(5809):233–237

Makkar HPS, Francis G, Becker K (2007) Bioactivity of phytochemicals in some lesser-known plants and their effects and potential applications in livestock and aquaculture production systems. Animal 1:1371–1391

Marbach D et al (2010) Revealing strengths and weaknesses of methods for gene network inference. Proc Natl Acad Sci U S A 107:6286–6291

Maruyama K, Takeda M, Kidokoro S, Yamada K, Sakuma Y, Urano K, Fujita M, Yoshiwara K, Matsukura S, Morishita Y, Sasaki R, Suzuki H, Saito K, Shibata D, Shinozaki K, Yamaguchi-Shinozaki K (2009) Metabolic pathways involved in cold acclimation identified by integrated analysis of metabolites and transcripts regulated by DREB1A and DREB2A. Plant Physiol 150(4):1972–1980

Mattheij RMM, Molenaar J (2002) Ordinary differential equations in theory and practice. SIAM classics in Applied Mathematics 43. SIAM, Philadelphia

Mayer BJ (1999) Protein–protein interactions in signaling cascades. Mol Biotechnol 13:201–213

Mayo AE, Setty Y, Shavit S, Zaslaver A, Alon U (2006) Plasticity of the cisregulatory input function of a gene. PLoS Biol 4:e45

Mendes P (1997) Biochemistry by numbers: simulation of biochemical pathways with Gepasi. Trends Biochem Sci 22:361–363

Middleton AM, Úbeda-Tomás S, Griffiths J, Holman T, Hedden P, Thomas SG, Phillips AL, Holdsworth MJ, Bennett MJ, King JR, Owen MR (2012) Mathematical modeling elucidates the role of transcriptional feedback in gibberellin signaling. Proc Natl Acad Sci U S A 109:7571–7576

Miller HI (2003) Will we reap what biopharming sows? Nat Biotechnol 21(5):480–481

Minton AP (2001) The influence of macromolecular crowding and macromolecular confinement on biochemical reactions in physiological media. J Biol Chem 276:10577–10580

Mittler R, Vanderauwera S, Suzuki N, Miller G, Tognetti VB, Vandepoele K, Gollery M, Shulaev V, Van BF (2011) ROS signaling: the new wave? Trends Plant Sci 16(6):300–309

Molassiotis A, Fotopoulos V (2011) Oxidative and nitrosative signaling in plants: two branches in the same tree? Plant Signal Behav 6(2):210–214

Morgan PW, Drew MC (1997) Ethylene and plant responses to stress. Physiology 100:620–630

Morison JIL, Lawlor DW (1999) Interactions between increasing CO_2 concentration and temperature on plant growth. Plant Cell Environ 22:659–682

Morris MK, Saez-Rodriguez J, Clarke DC, Sorger PK, Lauffenburger DA (2011) Training signaling pathway maps to biochemical data with constrained fuzzy logic: quantitative analysis of liver cell responses to inflammatory stimuli. PLoS Comput Biol 7:e1001099

Mueller LA, Zhang P, Rhee SY (2003) AraCyc: a biochemical pathway database for Arabidopsis. Plant Physiol 132(2):453–460

Nagasaki M, Saito A, Jeong E et al (2010) Cell Illustrator 4.0: a computational platform for systems biology. Silico Biol 10(1):5–26

Neapolitan RE (2004) Learning Bayesian networks. Prentice Hall, Upper Saddle River, NJ

Neutel AM et al (2002) Stability in real food webs: weak links in long loops. Science 296:1120–1123

Nikiforova VJ et al (2005) Integrative gene-metabolite network with implemented causality deciphers informational fluxes of sulphur stress response. J Exp Bot 56:1887–1896

Nishimura D (2001) BioCarta. Biotech Softw Internet Rep 2:117–120

Olivier BG, Snoep JL (2004) Web-based kinetic modelling using JWS Online. Bioinformatics 20:2143–2144

Pabinger S, Rader R, Agren R et al (2011) MEMOSys: bioinformatics platform for genome-scale metabolic models. BMC Syst Biol 5(1):20

Papin JA, Hunter T, Palsson BO, Subramaniam S (2005) Reconstruction of cellular signalling networks and analysis of their properties. Nat Rev Mol Cell Biol 6:99–111

Polynikis A et al (2009) Comparing different ODE modelling approaches for gene regulatory networks. J Theor Biol 261:511–530

Press W, Teukolsky S, Flannery B, Vetterling W (1990) Numerical recipes: FORTRAN. Cambridge University Press, Cambridge

Puchalka J et al (2008) Genome-scale reconstruction and analysis of the Pseudomonas putida KT2440 metabolic network facilitates applications in biotechnology. PLoS Comput Biol 4:e1000210

Puig O, Caspary F, Rigaut G, Rutz B, Bouveret E, Bragado-Nilsson E, Wilm M, Seraphin B (2001) The tandem affinity purification (TAP) method: a general procedure of protein complex purification. Methods 24:218–229

Rastogi SC, Rastogi P, Mendiratta N (2012) Bioinformatics methods and applications: genomics proteomics and drug discovery ed. (3). PHI Learning Private Limited, Delhi

Raue A, Schilling M, Bachmann J, Matteson A, Schelke M, Kaschek D, Hug S, Kreutz C, Brian DH, Fabian JT, Ursula K, Jens T (2013) Lessons learned from quantitative dynamical modeling in systems biology. PLoS ONE 8(9):e74335

Ravishankar GA, Rao SR (2000) Biotechnological production of phytopharmaceuticals. J Biochem Mol Biol Biophys 4:73–102

Ravishankar GA, Venkataraman LV (1990) Food applications of plant cell cultures. Curr Sci 57:381–383

Rigaut G, Shevchenko A, Rutz B, Wilm M, Mann M, Seraphin B (1999) A generic protein purification method for protein complex characterization and proteome exploration. Nat Biotechnol 17:1030–1032

Rischer H et al (2006) Gene-to-metabolite networks for terpenoid indole alkaloid biosynthesis in Catharanthus roseus cells. Proc Natl Acad Sci U S A 103:5614–5619

Rosemann D, Heller W, Sandermann H (1991) Biochemical plant responses to ozone. II. Induction of stilbene biosynthesis in Scots pine (Pinus sylvestris L.) seedlings. Plant Physiol 97:1280–1286

Ruths D, Muller M, Tseng JT, Nakhleh L, Ram PT (2008) The signaling petri net-based simulator: a non-parametric strategy for characterizing the dynamics of cell-specific signaling networks. PLoS Comput Biol 298:799–804

Sachs K, Perez O, Pe'er D, Lauffenburger DA, Nolan GP (2005) Causal protein-signaling networks derived from multiparameter single-cell data. Science 308:523–529

Sadot A, Sarbu S, Kesseli J, Amir-Kroll H, Zhang W, Nykter M, Shmulevich I (2013) Information-theoretic analysis of the dynamics of an executable biological model. PLoS One 8(3):e59303

Saez-Rodriguez J, Alexopoulos LG, Epperlein J, Samaga R, Lauffenburger DA, Klamt S, Sorger PK (2009) Discrete logic modelling as a means to link protein signalling networks with functional analysis of mammalian signal transduction. Mol Syst Biol 5:331

Saez-Rodriguez J, Alexopoulos LG, Zhang M, Morris MK, Lauffenburger DA, Sorger PK (2011) Comparing signaling networks between normal and transformed hepatocytes using discrete logical models. Cancer Res 71:5400–5411

Sakurai N, Suzuki H, Aoki K, Goda H, Nishizawa OI, Shibata D, Saito K (2007) Omics-based identification of Arabidopsis Myb transcription factors regulating aliphatic glucosinolate biosynthesis. Proc Natl Acad Sci U S A 104(15):6478–6483

Sauro HM, Hucka M, Finney A et al (2003) Next generation simulation tools: the Systems Biology Workbench and BioSPICE integration. Omics 7(4):355–372

Schaefer CF, Anthony K, Krupa S et al (2009) PID: the pathway interaction database. Nucleic Acids Res 37(1):674–679

Scheible WR et al (2004) Genome-wide reprogramming of primary and secondary metabolism, protein synthesis, cellular growth processes, and the regulatory infrastructure of Arabidopsis in response to nitrogen. Plant Physiol 136:2483–2499

Schillberg S, Fischer R, Emans N (2003) Molecular farming of antibodies in plants. Naturwissenschften 90:145–155

Shaitan KV, Tourleigh Ye V, Golik DN, Tereshkina KV, Levtsova OV, Fedik IV, Shaitan AK, Li A, Kirpichnikov MP (2006) Dynamics and molecular design of bio- and nanostructures. Ros Khim Zh 50:53–65

Shannon P, Markiel A, Ozier O, Baliga NS, Wang JT, Ramage D, Amin N, Schwikowski B, Ideker T (2003) Cytoscape: a software environment for integrated models of biomolecular interaction networks. Genome Res 13:2498–2504

Shmulevich I, Dougherty ER, Kim S, Zhang W (2002) Probabilistic Boolean Networks: a rule-based uncertainty model for gene regulatory networks. Bioinformatics 18:261–274

Simpson SJ et al (2010) Modelling nutritional interactions: from individuals to communities. Trends Ecol Evol 25:53–60

Sobie EA, Lee YS, Jenkins SL, Iyengar R (2011) Systems biology—biomedical modeling. Sci Signal 4(190):2

Srivastava A, Somvanshi P, Mishra BN (2013) Reconstruction and visualization of carbohydrate, N-glycosylation pathways in *Pichia pastoris* CBS7435 using computational and system biology approaches. Syst Synth Biol 7(7):22

Staab PR, Walossek J, Nellessen D, Grünberg R, Arndt KM, Müller KM (2010) SynBioWave–a real-time communication platform for molecular and synthetic biology. Bioinformatics 26(21):2782–2783

Stepanova AN, Alonso JM (2009) Ethylene signaling and response: where different regulatory modules meet. Curr Opin Plant Biol 12(5):548–555

Stephanopoulos G (1999) Metabolic fluxes and metabolic engineering. Metab Eng 1:11

Stephens DJ, Banting G (2000) The use of yeast two-hybrid screens in studies of protein: protein interactions involved in trafficking. Traffic 1:763–768

Steppuhn A, Baldwin IT (2007) Resistance management in a native plant: nicotine prevents herbivores from compensating for plant protease inhibitors. Ecol Lett 10:499–511

Stöger E, Vaquero C, Torres E, Sack M, Nicholson L, Drossard J, Williams S, Keen D, Perrin Y, Christou P, Fischer R (2000) Cereal crops as viable production and storage systems for pharmaceutical scFv antibodies. Plant Mol Biol 42(4):583–590

Stumpf MPH, Thorne T, de Silva E, Stewart R, An HJ, Lapper M, Wiuf C (2008) Estimating the size of the human interactome. Proc Natl Acad Sci U S A 105:6959–6964

Sucaet Y, Wang Y, Li J et al (2012) MetNet Online: a novel integrated resource for plant systems biology. BMC Bioinforma 13(1):267

Schena M, Shalon D, Davis RW, Brown PO (1995) Quantitative monitoring of gene-expression patterns with a complementary-DNA microarray. Science 270:467–470

Strogatz SH (2001) Exploring complex networks. Nature 410:268–276

Takahashi S, Seki M, Ishida J, Satou M, Sakurai T, Narusaka M, Kamiya A, Nakajima M, Enju A, Akiyama K, Yamaguchi-Shinozaki K, Shinozaki K (2004) Monitoring the expression profiles of genes induced by hyperosmotic, high salinity, and oxidative stress and abscisic acid treatment in Arabidopsis cell culture using a full-length cDNA microarray. Plant Mol Biol 56(1):29–55

Ter Braak CJF, Prentice IC (2004) A theory of gradient analysis. Adv Ecol Res 34:235–282

Teusink B, Passarge J, Reijenga CA, Esgalhado E, van der Weijden CC et al (2000) Can yeast glycolysis be understood in terms of in vitro kinetics of the constituent enzymes? Testing biochemistry. Eur J Biochem 267:5313–5329

Thakar J, Albert R (2010) Boolean models of within-host immune interactions. Curr Opin Microbiol 13:377–381

Thomas R (1973) Boolean formalization of genetic control circuits. J Theor Biol 42:563–585

Thomas R, D'Ari R (1990) Biological feedback. CRC Press, Boca Raton

Thomas B, Van Deynze A, Bradford K (2002) Production of therapeutic proteins in plants. Agricultural Biotechnology in California Series. Division of Agriculture and Natural resources, University of California Pub 8078

Tohge T et al (2005) Functional genomics by integrated analysis of metabolome and transcriptome of Arabidopsis plants over-expressing an MYB transcription factor. Plant J 42:218–235

Tomita M, Hashimoto K, Takahashi K et al (1999) E-CELL: software environment for whole-cell simulation. Bioinformatics 15(1):72–84

Tong AH et al (2004) Global mapping of the yeast genetic interaction network. Science 303:808–813

Trewavas A (2006) A brief history of systems biology. 'Every object that biology studies are a system of systems.' Francois Jacob (1974). Plant Cell 18:2420–2430

Turenne N (2011) Role of a web-based software platform for systems biology. J Comput Sci Syst Biol 4:035–041

Tuteja N, Mahajan S (2007) Calcium signaling network in plants: an overview. Plant Signal Behav 2:79–85

Terentiev AA, Moldogazieva NT, Shaitan KV (2009) Dynamic proteomics in modeling of the living cell. Protein–protein interactions. Biochemistry 74(13): 1587–1607

Umehara M, Hanada A, Yoshida S et al (2008) Inhibition of shoot branching by new terpenoid plant hormones. Nature 455(7210):195–200

Urano K, Maruyama K, Ogata Y, Morishita Y, Takeda M, Sakurai N, Suzuki H, Saito K, Shibata D, Kobayashi M, Yamaguchi-Shinozaki K, Shinozaki K (2009) Characterization of the ABA-regulated global responses to dehydration in Arabidopsis by metabolomics. Plant J 57(6):1065–1078

Usadel B et al (2005) Extension of the visualization tool MapMan to allow statistical analysis of arrays, display of corresponding genes, and comparison with known responses. Plant Physiol 138:1195–1204

Usadel B, Fernie AR (2013) The plant transcriptome-from integrating observations to models. Front Plant Sci 4:48. doi:10.3389/fpls.2013.00048

Usadel B, Obayashi T, Mutwil M, Giorgi FM, Bassel GW, Tanimoto M, Chow A, Steinhauser D, Persson S, Provart NJ (2009) Co-expression tools for plant biology: opportunities for hypothesis generation and caveats. Plant Cell Environ 32(12):1633–1651

Usha Rani P, Jyothsna Y (2010) Biochemical and enzymatic changes in rice as a mechanism of defense. Acta Physiol Plant 32:695–701

Van der Putten WH (2009) A multitrophic perspective on functioning and evolution of facilitation in plant communities. J Ecol 97:1131–1138

Van der Putten WH et al (2001) Linking above- and below ground multitrophic interactions of plants, herbivores, pathogens and their antagonists. Trends Ecol Evol 16:547–554

Vanderauwera S, Zimmermann P, Rombauts S, Vandenabeele S, Langebartels C, Gruissem W, Inzé D, Van Breusegem F (2005) Genome-wide analysis of hydrogen peroxide-regulated gene expression in Arabidopsis reveals a high light-induced transcriptional cluster involved in anthocyanin biosynthesis. Plant Physiol 139(2):806–821

Vemuri GM, Aristidou AA (2005) Metabolic engineering in the -omics era: elucidating and modulating regulatory networks. Microbiol Mol Biol Rev 69:197–216

Voit EO (2000) Computational analysis of biochemical systems: a practical guide for biochemists and molecular biologists. Cambridge University Press, Cambridge

Voit E, Neves AR, Santos H (2006) The intricate side of systems biology. Proc Natl Acad Sci U S A 103(25):9452–9457

Vrugt JA et al (2009) Accelerating Markov chain Monte Carlo simulation by differential evolution with self-adaptive randomized subspace sampling. Int J Nonlinear Sci Numer Simul 10:273–290

Wang G et al (2012) Process-driven inference of biological network structure: feasibility, minimality, and multiplicity. PLoS One 7:e40330

Walter M, Chaban C, Schütze K, Batistic O, Weckermann K, Na¨ ke C, Blazevic D, Grefen C, Schumacher K, Oecking C, Harter K, Jo¨ rg K (2004) Visualization of protein interactions in living plant cells using bimolecular fluorescence complementation. Plant J 40:428–438

War AR, Paulraj MG, Ahmad T, Buhroo AA, Hussain B, Ignacimuthu S, Sharma HC (2012) Mechanisms of plant fefense against insect herbivores. Plant Signal Behav 7(10):1306–1320

Wardle DA et al (2004) Ecological linkages between aboveground and belowground biota. Science 304:1629–1633

Weiss P (1973) The science of life. Futura Publishing, New York

Welti R, Shah J, Li W et al (2007) Plant lipidomics: discerning biological function by profiling plant complex lipids using mass spectrometry. Front Biosci 12:2494–2506

Weston DJ, Gunter LE, Rogers A, Wullschleger SD (2008) Connecting genes, coexpression modules, and molecular signatures to environmental stress phenotypes in plants. BMC Syst Biol 2:16

Wiback S, Famili I, Greenberg HJ, Palsson BØ (2004) Monte Carlo sampling can be used to determine the size and shape of the steady-state flux space. J Theor Biol 228:437–447

Wilkinson S, Davies WJ (2009) Drought, ozone, ABA and ethylene: new insights from cell to plant to community. Plant Cell Environ 33(4):510–525

Williams RJ (1956) Biochemical individuality. The key for the genetotrophic concept. Wiley, New York

Williams TC et al (2010) A genome-scale metabolic model accurately predicts fluxes in central carbon metabolism under stress conditions. Plant Physiol 154:311–323

Winter D, Vinegar B, Nahal H et al (2007) An "Electronic Fluorescent Pictograph" browser for exploring and analyzing large-scale biological data sets. PLoS ONE 2(8):e718

Wolkenhauer O (2008) Systems biology. Portland Press, London

Wolkenhauer O, Ullah M, Kolch W, Cho KH (2004) Modelling and simulation of intracellular dynamics: choosing an appropriate framework. IEEE Trans Nanobioscience 3(3):200–207

Wiener N (1948) Cybernetics, or control and communication in the animal and the machine. The Technology Press/Wiley, Cambridge/New York

Yang YT, Bennet GN, San KY (1998) Genetic and metabolic engineering. EJB Electron J Biotechnol 1(3):134–141

You L (2004) Toward computational systems biology. Cell Biochem Biophys 40:167–184

Yuan JS, Galbraith DW, Dai SY et al (2008) Plant systems biology comes of age. Trends Plant Sci 13(4):165–171

Zulak KG et al (2007) Gene transcript and metabolite profiling of elicitor-induced opium poppy cell cultures reveals the coordinate regulation of primary and secondary metabolism. Planta 225:1085–1106

Plantomics and Futuromics

Eric Davies and Debmalya Barh

Contents

E. Davies, Ph.D.
Department of Plant Biology, North Carolina State University, Raleigh, NC, USA
e-mail: edavies.pv@gmail.com

D. Barh, Ph.D. (✉)
Department of Genomics, Institute of Integrative Omics and Applied Biotechnology (IIOAB), Nonakuri, West Bengal, India
e-mail: dr.barh@gmail.com

Omics and Plantomics

There are two frequently used fundamental terms: "ome" and "omics." We feel that the "ome" should refer to the totality of the facts, those bits of information that make up the entirety of the topic (i.e., comprehensive coverage), whereas "omics" will refer to the synthesis of those facts leading to understanding (i.e., comprehension). The concept of omics in biology began with the need to understand the basic units of biology, the genes, and their nucleotide sequences, and huge amounts of money, effort, and intellectual capacity were poured into the human genome project with the idea of sequencing the entire genome and eventually knowing what each gene was doing. Thus omics began as a vast project of sequencing and enumeration, listing, cataloging, assigning, and above all analyzing and interpreting the vast amount of sequencing data. The entire process required the development of methods to sequence and obtain the data and protocols to assess and compile and infer the sequences. Thus biology joined physics, chemistry, mathematics, and computational science as an integrative "big" science. The genome project was followed soon after by another project in enumeration, the sequencing of proteins, and thus proteomics was born. This was followed by transcriptomics, which is the compilation and functional understanding of mRNA sequences. With the realization that the ultimate role of most proteins (enzymes) was to generate the plethora

D. Barh et al. (eds.), *PlantOmics: The Omics of Plant Science*,
DOI 10.1007/978-81-322-2172-2_29, © Springer India 2015

of metabolites, the field of metabolomics was developed, and this has led in turn to phenomics, the physical and biochemical traits of an organism. "Omics" has now become a general suffix to describe and assess the totality of any discipline (primarily in biology), and so there is an almost infinite number of "omes" that could be developed. In line, we have introduced the term "plantomics" that should include all plant-related "omes" and "omics."

Where Does Plantomics Go from Here?

Here we attempt to delineate some of the approaches that might be taken in the study under plantomics. It is sometimes possible to predict the future by tracing the footprints from the past into the present and guessing where they might lead in the future. Several approaches are available for this (or any) scientific field with the extremes being "splitting" and "joining." The splitters (or analyzers) will identify ever more defined topics of limited scope, while the joiners (or synthesizers) will develop ever more broad and inclusive topics showing inter-relationships between previously unconnected topics. We will describe what we envisage for the "splitter" or detailed approach first and then describe a more intermediate level and finish with the "joiner" (big picture, global) approach.

(a) The splitter (detail) approach

There is a continuum between (a) detail, (b) intermediate, and (c) global approaches, and so the divisions are somewhat arbitrary, but we will attempt to place the contents of this volume into what we deem is their appropriate location. The "splitter" approach can be purely plant based and thus fit squarely into the discipline of plantomics and is the approach taken in chapters "Cytogenomics and Mutagenomics in Plant Functional Biology and Breeding" (plant cytogenomics), "Plant Epigenetics and Crop Improvement" (plant epigenetics), "Plant miRNomics: Novel Insights in Gene Expression and Regulation" (plant miRNomics), "Plant Glycomics: Advances and Applications" (plant glycomics), "Plant Lipidomics: Signalling and Analytical Strategies" (plant lipidomics), "Plant Secretomics: Unique Initiatives" (plant secretomics), "Phenomics: Technologies and Applications in Plant and Agriculture" (plant phenomics), "Signalomics: Diversity and Methods of Analysis of Systemic Signals in Plants" (plant signalomics), "Thiolomics: Molecular Mechanisms of Thiol-cascade in Plant Growth and Nutrition" (thiolomics), "Chloroplast Omics: Global Strategies for Study of Plastid Biology" (chloroplastomics), "Transplastomics: A Convergence of Genomics and Biotechnology" (transplastomics), "Plant Mitochondrial Omics: State of the Art Knowledge" (plant mitochondrial omics), "Micromorphomics: A Morphological Dissection to Unveil Environmental Stress" (micromorphomics), "Cryobionomics: Evaluating the Concept in Plant Cryopreservation" (cryobionomics), and "Plant Pharmacogenomics: From Drug Discovery to Personalized Ethnomedicine" (plant pharmacogenomics).

In the future, we envisage the development of even more circumscribed fields of endeavor based on structure or functions/processes unique to plants. These could include those based on plant organs, such as phytomics (phytochemicals), floromics (flowers), phyllomics (leaves), and rhizomics (roots), or more defined regions, pollenomics, pistillomics, endospermomics, and cotyledonomics, or on functions (NCPP-omics), glyoxalatomics, plant mitomics, and meiomics, plant motility, plant communication, plant memory, and plant intelligence for instance.

(b) The intermediate approach

This more broadly based approach can be purely plant derived, or it can include animal or microbial equivalencies, and this is the approach taken in chapters "Omics of Model Plants" (model plants), "Instrumental Techniques and Methods: Their Role in Plant Omics" (techniques), "Next Generation Sequencing and Assembly of Plant Genomes" (genome sequencing), "Functional Genomics: Applications in Plant Science" (functional genomics), "Plant Proteomics: Technologies and Applications" (plant proteomic methodology), "Plant Metabolomics: An Overview of Technology Platforms for Applications in Metabolism" (plant metabolomics methodology), "Plant Cytomics: Novel Methods to View Molecules on the Move" (plant cytomics), "Plant Physiomics: Photo-electro-chemical and Molecular Retrograde Signaling in Plant Acclimatory and Defence Responses" (plant physiomics), "Microbiomics: An Approach to Community Microbiology" (microbiomics), "Nanobiotechnology: Applications in Plant and Agriculture" (nanobiotechnology), "Machine Learning Approaches in Plant Biology" (machine learning), "Applications of Bioinformatics in Plant and Agriculture" (bioinformatics), and "Plant Systems Biology: Insights and Advancements" (plant systems biology).

We envisage somewhat more broadly based (yet still focused) fields incorporating knowledge from plants, animals, and microbes in such topics as biometalomics (e.g., metals in enzyme function, oxygen transport, electron transfer reactions, etc.), bioergomics (energy generation and use in living organisms), biomics (study of world biomes), and the related ecolomics (synthesis of ecological interactions at all levels), along with some of the sections under splitter approach without the limiting plant epithet, such as epigenetics, glycomics, and lipidomics, among others.

(c) The global approach or "futuromics"

In the same way that early "omes" began as cataloging of individual bits of information that got assembled into some form of understanding using the "omics" approach so we anticipate that "futuromics" will be a "meta-omics," the assembling of the understandings derived from different (but related) fields. As per our knowledge, there are three identifiable items in the known universe: energy, matter, and information. The ultimate goal of the futuromics would be to develop the meta-disciplines of ergomics (energy), materialomics (matter), and informatomics (information), and plantomics would play a leading role in these meta-disciplines. We will discuss "ergomics" to a certain detail, while "materialomics" and "informatomics" will be discussed briefly.

Ergomics

We envisage that the not-yet-invented meta-discipline of universal energy relations, or ergomics, would have plantomics as its basis and would combine the information and understanding of apparently disparate disciplines including the sciences of biology, medicine, chemistry, physics, engineering, geology, climatology, as well as the disciplines of economics, commerce, law, government, politics, education, etc. Of the sciences, the biology components would include: subcellular biology (areas such as physiology, biochemistry, biophysics, cytomics, organelle biology, cytoplasmic streaming, nanomicroscopy), whole organism animal biology (growth, energy expenditure in muscle movement, energy efficiency in digestion), and whole plant energetics (growth, differentiation, mass transport). Chemistry would involve endo- and exothermic reactions, distillation of fuels, and greenhouse gas absorption or recovery, while physics would involve electricity, microwave irradiation, atomic energy, fusion, fission, dark energy, gravitation, and cosmic background radiation, among others. Engineering would involve research in and construction of solar, wind, tidal, hydro, and thermal energy plants and systems for absorbing greenhouse gases, while geology would play a major role in the discovery of additional fossil fuel sources. The participation by climatologists is especially necessary because of the tight link between (fossil fuel) energy consumption and global warming/climate change. We foresee members of all these fields becoming involved in the search for alternative, renewable, nontoxic energy sources.

Participation by non-scientists is absolutely necessary for full implementation of viable energy policies. Economics will need not only to perform cost-benefit analyses of all aspects of energy generation and consumption but also to compare various ways to lessen global warming (climate change) and weigh these costs with the inevitable gigantic costs of not mitigating against global warming. The private sector (commerce, industry, banking, law) also needs to be involved to commercialize (and perhaps patent) the new ergomics findings and recommendations and needs to be aware of the damage of short-term, bottom line (quarterly) planning, while government, politicians, and lobbying groups need to support and not oppose scientific findings that focus on the long-term good health of the planet and the life forms inhabiting it. Finally, education, especially science education and critical thinking, is absolutely necessary not only for the young but also for the adults in science, commerce, government, politics, law, etc. Indeed, we contemplate a more perfect world where the adults in vital decision-making positions would not be scientific nincompoops, but would need to have their scientific understanding measured and validated before they could hold such positions.

Specific Roles for Plantomics

Life on Earth is powered almost entirely by the sun, primarily through photosynthesis conducted by higher plants, algae, and cyanobacteria with current photosynthesis providing food, fiber, fabrics, and biofuels and with past photosynthesis providing fossil fuels. Accordingly, the basis of ergomics is the understanding of this process from a biophysical, biochemical, physiological, and ecological perspective. The preliminary goal would be to establish the field of "photosynthomics," where the individual components of the light-harvesting complex were known, as was its supramolecular architecture and the pathways of transmission of electrons and photons, the reduction of high energy electron acceptors, the generation of ATP, and the release of oxygen. A further goal would be in the use of nanotechnology, biomechanics, and related fields to reconstruct this complex, so that it functions in vitro with at least as much efficiency as it does within the chloroplast, and a close to final goal would be to incorporate these complexes into paints for vehicles (buses, trucks, cars, trains, airplanes, ships) and buildings (homes, hospitals, factories, offices, universities) so individuals, companies, governments, etc., would become net energy producers rather than consumers. Of course, neither the initial in vivo biological goal nor the later in vitro nanotechnology goal will happen without careful planning and investment of time, energy, money, and intellectual capacity (i.e., exactly what was needed for the development of genomics). Whether governments, industry, banks, ultra-wealthy individuals, or conglomerates of these provide the resources is immaterial, but these resources are a total requisite.

Materialomics

All materials are either organic (generated by current or previous biological processes) or inorganic (generated by chemical and physical processes), and the only genuinely renewable materials are biologically based. As with ergomics, the fundamentals of the discipline would be plantomics, in particular that aspect dealing with the generation of biomass through the process of photosynthesis from past, present, or future activities. While ergomics would focus on the light reactions of photosynthesis, materialomics would concentrate on the dark reactions of C3, C4, and CAM plants and all the subsequent metabolic pathways up to and including the production of the so-called secondary metabolites. Plants provide the fundamental six "Fs", food, feed fiber, fabrics, fuel, and "farmaceuticals," and there is no easy way to see how the role of plants in food, feed, and farmaceutical production can be circumvented. However, production of fiber, fabrics, and fuel by nonbiological processes is likely to increase especially in the search for ultralight and ultra-strong fabrics and non-fossil fuel sources such as hydrogen. From the biological perspective, the fields of metabolomics, enzymology, and genetic modification will be paramount, and it is too early to tell whether many different plant species will be modified for the manufacture of specialized products or whether certain plants such as *Arabidopsis*, or even tobacco, which has become a major source of plantibodies, will become generalized manufacturers.

Informatomics

The role of plantomics in this discipline is not as apparent as in the previous two. However, all living organisms contain information in their DNA and RNA, and all organisms extract information from the environment to modify their behavior, and all organisms transmit information within themselves and to other organisms. Examples of behavior modifications in plantomics would be the perception of light to determine flowering time, as well as perception of light and gravity to determine orientation. Examples of information transmission within the whole plant include electrical signals (action potentials and variation potentials) and at the subcellular level, cytoskeleton dynamics and nuclear-plastid communication. Indeed, one of the more fascinating aspects of the role of plants in informatomics is the possibility of transmitting GSM (mobile phone) signals to modulate gene expression to plants growing at great distances from the GSM source, as hinted at in chapter "Signalomics: Diversity and Methods of Analysis of Systemic Signals in Plants" (plant signalomics).

MIX
Papier aus verantwortungsvollen Quellen
Paper from responsible sources
FSC® C105338

If you have any concerns about our products,
you can contact us on
ProductSafety@springernature.com

In case Publisher is established outside the EU,
the EU authorized representative is:
Springer Nature Customer Service Center GmbH
Europaplatz 3, 69115 Heidelberg, Germany

Printed by Libri Plureos GmbH
in Hamburg, Germany